1

zürich m94

Proceedings of the International Congress of Mathematicians

August 3–11, 1994
Zürich, Switzerland

Birkhäuser Verlag
Basel · Boston · Berlin

Editor:
S. D. Chatterji
EPFL
Département de Mathématiques
1015 Lausanne
Switzerland

The logo for ICM 94 was designed by
Georg Staehelin, Bachweg 6, 8913 Ottenbach, Switzerland.

A CIP catalogue record for this book is available from the
Library of Congress, Washington D.C., USA

Deutsche Bibliothek Cataloging-in-Publication Data

International Congress of Mathematicians <1994, Zürich>:
Proceedings of the International Congress of Mathematicians
1994 : August 3 - 11, 1994, Zürich, Switzerland / [Ed.: S. D.
Chatterji]. - Basel ; Boston ; Berlin : Birkhäuser.

ISBN 978-3-0348-9897-3 ISBN 978-3-0348-9078-6 (eBook)
DOI 10.1007/978-3-0348-9078-6

NE: S. D. Chatterji [Hrsg.]

Vol. 1 (1995)

Softcover reprint of the hardcover 1st edition 1995
P.O. Box 133
CH-4010 Basel
Switzerland

Printed on acid-free paper produced from chlorine-free pulp. TCF ∞
Layout, typesetting by *mathScreen online*, Allschwil

9 8 7 6 5 4 3 2 1

Table of Contents

Volume I

Volume II

Preface

The Proceedings of the International Congress of Mathematicians 1994, held in Zürich from August 3rd to 11th, 1994, are published in two volumes. Volume I contains an account of the organization of the Congress, the list of ordinary members, the reports on the work of the Fields Medalists and the Nevanlinna Prize Winner, the plenary one-hour addresses, and the invited addresses presented at Section Meetings 1–6. Volume II contains the invited addresses for Section Meetings 7–19. A complete author index is included in both volumes.

The five invited lectures organized by the ICMI (International Commission on Mathematical Instruction) and the five invited lectures organized by the ICHM (International Commission on History of Mathematics) which were a part of the scientific programme have not been included in these Proceedings. Also not included are the short communications presented in poster sessions during the Congress; summaries of those communications which were received in due time were printed in a separate volume *Abstracts of Short Communications*; another volume containing the available abstracts of the plenary addresses, the invited section lectures and the ICMI and ICHM lectures was also prepared. Both volumes of abstracts were given to all ordinary members at the time of their registration at the Congress.

Lausanne, April 1995 — The Editor

Past Congresses

1897	Zürich
1900	Paris
1904	Heidelberg
1908	Roma
1912	Cambridge, UK
1920	Strasbourg
1924	Toronto
1928	Bologna
1932	Zürich
1936	Oslo
1950	Cambridge, USA
1954	Amsterdam
1958	Edinburgh
1962	Stockholm
1966	Moskva
1970	Nice
1974	Vancouver
1978	Helsinki
1982	Warszawa (held in 1983)
1986	Berkeley
1990	Kyoto

Past Fields Medalists and Rolf Nevanlinna Prize Winners

Recipients of Fields Medals

1936
Lars V. Ahlfors
Jesse Douglas

1950
Laurent Schwartz
Atle Selberg

1954
Kunihiko Kodaira
Jean-Pierre Serre

1958
Klaus F. Roth
René Thom

1962
Lars Hörmander
John W. Milnor

1966
Michael F. Atiyah
Paul J. Cohen
Alexander Grothendieck
Stephen Smale

1970
Alan Baker
Heisuke Hironaka
Sergei P. Novikov
John G. Thompson

1974
Enrico Bombieri
David B. Mumford

1978
Pierre R. Deligne
Charles F. Fefferman
Grigorii A. Margulis
Daniel G. Quillen

1982
Alain Connes
William P. Thurston
Shing-Tung Yau

1986
Simon K. Donaldson
Gerd Faltings
Michael H. Freedman

1990
Vladimir G. Drinfeld
Vaughan F. R. Jones
Shigefumi Mori
Edward Witten

Rolf Nevanlinna Prize Winners

1982 Robert E. Tarjan
1986 Leslie G. Valiant
1990 Alexander A. Razborov

The Kongresshaus in Zürich

Organization of the Congress

The 1994 International Congress of Mathematicians (ICM) was held in Zürich, Switzerland, at the invitation of the Swiss Mathematical Society (SMS), representing the Swiss mathematical community, and under the auspices of the International Mathematical Union (IMU) whose official approval was announced at the 1990 ICM held in Kyoto. The SMS invitation was made possible by the support promised at the very outset by the appropriate authorities of the Federal Government, the Government of the Canton of Zürich and the Municipality of the city of Zürich. Financial help given by these and other public and academic bodies as well as the donations of many private corporations and individuals were crucial for the realization of the Congress; a list of the donors is given in these Proceedings. Naturally, the registration fees paid by the participants of the Congress were an essential element in financing the organization.

The members of the Organizing Committee are listed in the following pages; also listed are the members of the Honorary Committee, the Finance Committee, the Scientific Committee and the Administrative Staff.

The scientific program of the Congress was in the hands of the Programme Committee appointed by the IMU. Its members were Louis Nirenberg (Chairman), Simon K. Donaldson, Vladimir Drinfeld, Pierre de la Harpe, Richard Karp, Hanspeter Kraft, Andrew J. Majda, Michel Raynaud and Y. Sinai.

Recipients of the Fields Medals and the Rolf Nevanlinna Prize were selected by the respective committees appointed by the IMU. The Fields Medal Committee consisted of David Mumford (Chairman), Masaki Kashiwara, Barry Mazur, Alexander Schrijver, Dennis Sullivan, Jacques Tits and S. R. S. Varadhan. The Rolf Nevanlinna Prize Committee consisted of Jacques-Louis Lions (Chairman), H. W. Lenstra, R. Tarjan, M. Yamaguti and J. Matiyasevic.

The Organizing Committee was responsible for all the other activities of the Congress. MCI Travel (Zürich) handled accomodation and related arrangements as the official travel agency of the Organizing Committee.

The opening and the closing ceremonies as well as all the one-hour Plenary Addresses were held in the Zürich Kongreßhaus. The forty-five minute section lectures were organized in various parallel sessions in the auditoria of the ETH Zürich and the University of Zürich. There were 16 plenary lectures and 148 section lectures on the program. In addition, there were five lectures organized by the International Commission on Mathematical Instruction and five lectures organized by the International Commission on the History of Mathematics which were scheduled along with the section lectures. Poster sessions arranged at the ETH Zürich permitted the presentation of numerous short communications; summaries of 782 of these, received before a fixed deadline, were printed in a separate abstracts volume and a further 100 additional contributions were actually presented at the poster sessions. There were also several informal seminars as well as a symposium organized on Thursday, August 4, by the Association for Women in Mathematics and the European Women in Mathematics.

In the afternoon of August 3, lectures reporting on the works of the Fields Medalists and the Nevanlinna Prize Winner were presented; an account will be found in these Proceedings.

A total of 2476 participants from 92 countries along with 363 accompanying members attended the Congress; 77 exhibitors were present.

The Organizing Committee was able to give financial support for the participation of the prize winners, the officials of the IMU, 19 of the invited speakers and some 200 participants from Eastern Europe. The IMU, through its special Development Fund, paid the travel expenses of 79 young scholars from developing countries whose living expenses were covered by the Organizing Committee.

All participants were invited to a number of social events. A reception was offered by the city of Zürich in the Kongreßhaus on the evening of the opening day of the Congress, on Wednesday, August 3. A Buffet-Banquet was given in the Irchel campus of the Universiy of Zürich, on the evening of Friday, August 5. Tuesday evening, August 9, a violin recital by Hansheinz Schneeberger, accompanied by Gérard Wyss, in the Tonhalle (Kongreßhaus) and a performance by the pantomime group Mummenschanz together with the folk music group Trio da Besto in the Kongreßsaal (Kongreßhaus) were proposed.

Evening reception at Irchel Campus of the University of Zürich

The Organizing Committee of the Congress

Honorary Committee

Mrs. Ruth Dreifuss, Head of the Federal Department of Home Affairs
Mrs. Hedi Lang, President of the Cantonal Government, Zürich
Dr. Albertine Trutmann, Chief of the University Section, Department of Education, Canton Zürich
Mr. Josef Estermann, Mayor of Zürich
Prof. Jakob Nüesch, President of ETHZ
Prof. Hans H. Schmid, Rector of the University of Zürich
Prof. Paul Walter, President of the Swiss Academy of Sciences
Prof. André Aeschlimann, President of the Research Council of the Swiss National Science Foundation
Prof. Armand Borel, Princeton
Prof. Hans Bühlmann, Zürich
Prof. Beno Eckmann, Zürich
Prof. Hans Künzi, Zürich
Prof. Jürgen Moser, Zürich

Program Committee (appointed by the IMU)

Louis Nirenberg, Courant Institute, New York, USA (*Chairman*)

Simon K. Donaldson, Oxford University, United Kingdom
Vladimir Drinfeld, FTINT, Kharkov, Ukraine
Pierre de la Harpe, Université de Genève, Switzerland
Richard Karp, University of California, Berkeley, USA
Hanspeter Kraft, Universität Basel, Switzerland
Andrew J. Majda, Princeton University, USA
Michel Raynaud, Université de Paris Sud, France
Y. Sinai, Landau Institute for Theoretical Physics, Moscow, Russia

Organizing Committee

Henri Carnal, Universität Bern (*Chairman*)
Christian Blatter, ETHZ (*Secretary*)

Andrew D. Barbour, Universität Zürich
Peter Baumann, Universität Zürich
Erwin Bolthausen, Universität Zürich
S. D. Chatterji, EPF Lausanne

Hans Jarchow, Universität Zürich
Max Albert Knus, ETHZ
François Sigrist, Université de Neuchâtel
Hans Heiner Storrer, Universität Zürich
Jörg Waldvogel, ETHZ

Finance Committee

Hans Bühlmann, former President of ETHZ
Hans Künzi, former Regierungsrat, Canton Zürich
Carl August Zehnder, ETHZ
Henri Carnal, President of ICM 94

Scientific Committee

S. D. Chatterji, EPF Lausanne (*Chairman*)

D. Arlettaz (Uni Lausanne), A. Barbour (Uni Zürich), O. Besson (Uni Neuchâtel), M. Bronstein (ETHZ), M. Brodmann (Uni Zürich), M. Burger (Uni Lausanne), P. Buser (EPFL), D. Coray (Uni Genève), B. Dacorogna (EPFL), A. Derighetti (Uni Lausanne), J.-C. Hausmann (Uni Genève), H. Hofer (ETHZ), J. Hüsler (Uni Bern), G. Jäger (Uni Bern), U. Kirchgraber (ETHZ), T. Liebling (EPFL), P. Littelmann (Uni Basel), J.-J. Loeffel (Uni Lausanne), J.-C. Pont (Uni Genève), H. M. Reimann (Uni Bern), C. Riedtmann (Uni Bern), V. Schroeder (Uni Zürich), M. Struwe (ETHZ), A.-S. Sznitman (ETHZ), A. Valette (Uni Neuchâtel), G. Wüstholz (ETHZ)

Administrative Staff

Bettina Collart
Michael Goßmann

The mathematics assistants at ETHZ and at the University of Zürich
Assistants from other Swiss universities
Staff members from the Mathematics Departments at ETHZ and at the University
Technical personnel of ETHZ and of the University

The Organizing Committee extends its warmest thanks to these and numerous other collaborators, without whose generous help the Congress would never have been possible.

List of Donors

The Organizing Committee is greatly indebted to all those who have supported the Congress, either by monetary contribution, or by donating goods or services of various kinds. It is only on the firm basis of these generously donated contributions that it has been possible to launch this event and carry it through.

Public and academic bodies

Canton Zürich, University Foundation
Swiss Confederation, ETH domain
International Mathematical Union
Swiss National Science Foundation
City of Zürich
INTAS (International Association for the Promotion of Cooperation with
Scientists from the Independent States of the Former Soviet Union, Brussels)
Swiss Academy of Sciences
Swiss Mathematical Society

Private corporations

International Science Foundation ("Soros Foundation")
Kontaktgruppe für Forschungsfragen, Basel
(Ciba-Geigy, F. Hoffmann-La Roche, Lonza, Sandoz)
Springer-Verlag, Heidelberg, Germany
Schweizerische Rückversicherung, Zürich
Schweizerischer Bankverein, Basel
Schweizerische Kreditanstalt, Zürich
Anonymous
Basler Versicherungsgesellschaft, Basel
Winterthur Versicherungen, Winterthur
Rentenanstalt, Zürich
Türler Uhren- und Schmuckgeschäfte, Zürich
Zürcher Handelskammer, Zürich
Wolfram Research Inc., Champaign IL, USA
Mövenpick Unternehmungen, Adliswil (ZH)
Bank Julius Bär, Zürich
Zürcher Kantonalbank, Zürich
IBM Schweiz, Zürich
Schweizerische Industriegesellschaft, Neuhausen
Patria Leben, Basel

Georg Wagner Stiftung, Basel
Hatebur Umformmaschinen, Reinach (BL)
Hans Denzler & Co., Reinach (BL)
Elektrowatt AG, Zürich
Asea Brown Boveri AG, Zürich
"La Suisse" Assurances, Lausanne
Sulzer AG, Winterthur
Gerling Rückversicherungsgruppe, Zug
Alusuisse-Lonza Holding, Zürich
Huber + Suhner AG, Pfäffikon (ZH)

Individuals

Members of the Swiss Mathematical Society who generously responded to the special ICM 94 fund drive

Opening Ceremonies

The opening ceremonies of the Congress were held in the Zürich Kongreßhaus in the morning of Wednesday, August 3, 1994, starting at 9.30. The Brass Quintet of the Zürich Conservatory of Music provided the musical accompaniment, opening with a suite from Banchetto musicale 1617 by Johann Hermann Schein (1586–1630). Professor Jacques-Louis Lions, President of the International Mathematical Union (IMU), opened the Congress with the following speech:

Madame la Conseillère Fédérale,
Mesdames, Messieurs,

J'ai l'honneur de saluer:

Madame Ruth Dreifuss, Conseillère Fédérale,
Madame Hedi Lang, Présidente du Gouvernement Cantonal Zürichois,
Monsieur Dr. Thomas Wagner, représentant du Conseil Municipal de la Ville de Zürich,
Monsieur Dr. Alfred Gilgen, Directeur de l'Instruction Publique du Canton de Zürich,

je les remercie très vivement de leur présence ainsi que les Professeurs

Clive Kuenzle, Prorecteur de l'Université de Zürich,
Ralf Hütter, Vice-Président de l'Ecole Polytechnique Fédérale de Zürich,
Dominique de Werra, Vice-Président de l'Ecole Polytechnique Fédérale de Lausanne,
André Aeschlimann, Président du Conseil de la Recherche au Fonds National; et les Professeurs Chandrasekharan et Moser, anciens Présidents de l'Union Mathématique Internationale.

Je poursuis en Anglais.

Excellencies,
Ladies and Gentlemen,

Already now mathematics, in addition to its intrinsic importance, is one of the keys for the development of other sciences and of industry. Everything indicates that this already fundamental role will increase during the next century. This implies responsibilities for us and for our governments to:

1) continue, and even increase the support to mathematical research of the highest quality,

J.-L. Lions, President of the IMU

The Brass Quintet of the Zürich Conservatory of Music at the opening ceremonies

2) further develop collaborations and exchanges with other disciplines and with industry,
3) help as much as possible the Mathematical Instruction and the Mathematical Research in countries suffering of difficult economic situations, and to
4) explain as clearly as possible what we are doing to a not too specialized public.

These four points are the aims pursued by the Executive Committee of IMU, according to the wishes of the General Assembly in Kobe in 1990, and Lucerne in 1994. These are also the aims of the Commissions of IMU, namely ICMI and CDE and of the WMY 2000, launched by IMU at IMPA, Rio de Janeiro, co-sponsored by UNESCO and the Third World Academy of Sciences.

These four points are also reflected in the present Congress.

The Program Committee has been nominated by the Executive Committee of IMU and by the Swiss Organizing Committee. It had the responsibility of the selection of the speakers.

The Chairman of the Program Committee is Professor Louis Nirenberg and the members are: S. K. Donaldson, V. Drinfeld, P. de la Harpe, R. Karp, H. Kraft, A. Majda, M. Raynaud, M. Sato and Y. Sinai.

Je passe maintenant la parole au Professeur Henri Carnal, Président du Comité d'Organisation du Congrès International des Mathématiciens 1994, qui devient le Président du Congrès par acclamation.

Professor Henri Carnal then took the floor and welcomed the audience as follows:

Madame la Conseillère fédérale,
Frau Regierungspräsidentin,
Herr Stadtrat,
Mr. President,
Ladies and Gentlemen,

In the name of the Organizing Committee of ICM 94, I am very glad to welcome you to Zürich today.

As you know, this is the third time that the International Congress of Mathematicians takes place in this city. In the year 1897, at the very first of these meetings, the plenary speakers were Hurwitz, Felix Klein, Peano and Poincaré, four outstanding scientists, whose names and achievements are still familiar to us, even after one hundred years of tremendous changes in science and society.

This lasting quality of mathematical ideas is certainly one of the most fascinating aspects of our science. Another aspect, which is at least as important, is the universality of mathematical activity, a feature that Hermann Weyl emphasized in a speech in 1932, when the Congress came to Zürich for the second time. Then, at the beginning of the darkest period of our century, the world's scientific community was called upon to develop a new sense of solidarity, arising from a joint search for truth.

H. Carnal, President of the Congress

J. J. Burckhardt at the opening ceremonies

Six decades later, we face a completely different world, balanced (precariously) between order and chaos. When we began our preparations for this event, in the summer of 1989, the borders in Europe seemed to be topologically and even metrically invariant, so that we didn't include them in the list of problems that we might have to cope with. Since then, we have witnessed the birth of many new countries and of many new mathematical societies. We are very glad to note that most of them have found their way to Zürich. We are especially pleased to be able to welcome the representatives of the Bosnian Mathematical Society and to thank the authorities and the people who helped to organize their journey out of Sarajevo.

As well as the political instability of the last few years, we have also seen serious world economic problems, from which Switzerland has not been completely excluded, and it was by no means evident that we would find the financial support we needed for the Congress. We are therefore deeply grateful to all of those who have helped us, and with unexpected generosity. I would like especially to mention the federal authorities, the ETHs of Zürich and Lausanne, the Swiss National Science Foundation, the Cantonal Government of Zürich and its University department, the City of Zürich, many private companies in insurance, banking, chemicals, commerce and industry and, last but not least, the International Science Foundation in Washington. I wish to thank my colleagues in the Finance Committee who not only helped us to raise funds, but who also suggested many ways to make this event more attractive.

We have tried to use the money we received with the greatest possible efficiency, but we have been forced to the conclusion that optimization problems in real life are very different from what they are in theory! We hope nonetheless that the positive aspects outweigh the negative ones, and that you will enjoy the former and forgive us the latter.

Let me end with two pleasant remarks about the Congress of 1932: The first one is that we have among us Prof. J. J. Burckhardt, who was active in the organization 62 years ago. By extrapolation, we may assume that some of today's participants will be able to attend the fourth congress in Zürich in the middle of the next century. The second remark relates to a comment that I found in a historical survey on the International Mathematical Congresses concerning that meeting of 1932: "In a country which at that time didn't allow women to vote, it was distinguished by the inclusion of a woman mathematician — Emmy Noether ... But the number of women who have been invited to speak at international congresses since Noether does not differ much from 0!" I am therefore happy to observe not only that the number of plenary lectures by women will this time be greater than 0, and even greater than 1, but also that the highest federal and cantonal authorities are both represented here by women. This shows that we can always hope for positive changes!

And now I request your help in the election by acclamation of the Honorary President of the Congress. I propose to you the former secretary of the IMU and founder of the Mathematical Research Institute at the ETH, Prof. Beno Eckmann.

Professor Beno Eckmann addressed the audience as follows:

Frau Bundesrätin, sehr geehrte Damen und Herren,

Ich danke Ihnen und dem Kongress-Komitee herzlich für die grosse Ehre, die ich im Namen der Schweizer und im besondern der Zürcher Mathematik annehme. Je vous remercie ainsi que le Comité du Congrès tres sincèrement pour le grand honneur que vous venez de me témoigner, et je souhaite á vous tous la très cordiale bienvenue. Vorrei ringraziare cordialmente il Comitato e tutti i presenti per il grande onore reso a me con questa nomine; e saluto in modo particolare i matematici di lingua italiana.

I will now try to continue in English. I am not at all able to express myself in our fourth national language, Romantsch, which anyway is not likely to be understood by many in this audience.

Ladies and Gentlemen,

I have to confess that I did not participate in the tremendous work of preparing this Congress. So, in any case from that viewpoint, I do not deserve being elected Honorary President. I can accept, however, the honor with not too bad a conscience: indeed, I have been very active in the preparation of two earlier Congresses, namely, 1958 Edinburgh and 1962 Stockholm, when I was Secretary of IMU (the International Mathematical Union), 1956 to 1961. It can be said that this was a very important and interesting period for international collaboration in all aspects of mathematics.

May I recall first of all that just at that time many countries — some of them very large and important —, which did not up until then adhere to the IMU, became members. One can imagine that quite some difficulties of a political, personal, and financial nature had to be overcome; but it was a gratifying challenge. For the Union became a truly worldwide family. Today, clearly the Union must be faced with problems of a quite different nature. Secondly, a decision was taken which today seems most natural, namely, that the Scientific Program of the International Congress be prepared by the IMU, since that task could not be handled any longer by a single country. Stockholm was the first Congress where the new scheme was adopted, after several — very friendly — discussions. Nowadays the functioning of the international collaboration in mathematics can certainly be considered as a model for many other fields.

Something else has, since these times, considerably changed local and global mathematical life. I think, of course, of the computer, as a tool within our science and as a marvellous means of communication. I believe there are very few mathematicians who have not taken advantage of and derived great benefit from the fabulous possibilities of this tool. But we should not forget that the most important tool of a mathematician is the fellow mathematician! And that is why we all are here today: to exchange ideas, views, and results, and to listen to each other.

With regard to the computer I have heard over and over again the saying: *Whether mathematicians like it or not, the computer is here to stay*. I do not agree with that formulation. We like the computer and we use it. But today I find it important to turn that phrase around and say: *Whether the computer likes it or not, mathematics is here to stay*. This means mathematics as the art of creating concepts out of the vague intuition and evidence of the real world and of everyday life; and then to put these concepts to work and experiment with them by all available means — including, of course, the computer; to see relations, conjectures and theorems emerge; and to prove the same by the good old traditional proof, which is at the heart of our science. For mathematics is, and remains, an abstract intellectual enterprise, despite the fact that natural sciences and technology, and much more, bear witness to its practical usefulness. Sometimes it is the same person who speculates and conjectures, provides proofs, and makes applications to our real world; but more often this is done by different people and at different times. Personal collaboration always remains an essential feature.

Most beautiful evidence of all the above is given by the scientific program of our Congress, and by the impressive work of the laureates of the Fields Medal and the Nevanlinna Prize, which are the most prestigious distinctions in mathematics. It will be my duty and immense pleasure to hand over the medals and the Prize to the winners. I should make it clear that their names have not yet been disclosed; they are not even known to myself, in any case not officially. I will learn later this morning whether my guess and the traditional mathematicians' gossip will prove to be correct. Nevertheless let me congratulate them in advance. I share their feelings of pride and accomplishment, and I am looking forward to their continued success — hoping that I will be able for some time to come, to understand what they are doing. I also share the feelings of the many who are disappointed because they did not get the medal; there is simply too much excellent work being done!

Mathematical research indeed seems to live in a golden age. As for the mathematical education of coming generations, however, I must say that I see some danger: there are worldwide trends trying to completely replace rigorous reasoning and proving by computer visualisation and experimentation. This is not the place to elaborate on the theme of the central importance of rigorous proof. Instead let me quote Hermann Weyl (who spent a long and very important period of his scientific life in Zürich): *Mathematics, besides language and music, is one of the pri-*

mary manifestations of the free creative power of the human mind and it is the universal organ for world-understanding through theoretical construction. It has to remain an essential element of the knowledge and abilities we have to teach, of the culture we have to transmit to the next generations.

May I just add: To achieve more we probably dare not hope; to achieve less we certainly must not try.

Thank you.

After a musical interlude ("Changing Moods" by Gordon Jacob (1895–1984)), Minister Ruth Dreifuss, Head of the Federal Department of Home Affairs, gave the following speech:

Ladies and Gentlemen,

A hundred years ago, in 1897, the first International Congress of Mathematicians was held in Zürich. In 1932, the Congress met in Switzerland for the second time. On that occasion, the Fields Medal was introduced as your Nobel Prize equivalent. Today, our country hosts your Congress for the third time. No other country has been honoured in such a way by your scientific community and I am sure, that the "genius loci" will show his gratitude for your fidelity and ensure the success of your work.

I feel personally very honoured to open your Congress. It's a rare opportunity to host the world's leading masters of this art and to come into contact with their scientific debate.

If the subject of your Congress were cancer research or modern history, for a lay person it would be simpler to understand what the discussion is about. In contrast, mathematics at first sight seems to be an abstract tool for its own purpose or an exclusive art.

Two years ago, in Rio de Janeiro, under the sponsorship of the UNESCO, "World Mathematical Year 2000" was launched. On that occasion, the International Mathematical Union defined a vision for mathematics which stresses the relationship between science and society. The Declaration of Rio de Janeiro states that "pure and applied mathematics are one of the main keys to understanding the world and its development". I am sure that society needs these keys.

But since I am not a mathematician myself, I wonder what doors they open, and what society will find behind them. Therefore I would like to learn from you how mathematicians view their role in society.

With the relationship between science and society in mind, I sent three questions to over a dozen of the world's most eminent mathematicians and I am very grateful for all the answers I received. For the first two questions, I referred to the distinction between pure and applied mathematics cited in the Declaration of Rio.

The first question concerns pure mathematics. Pure mathematics seems to function within a realm of complete independence. Its results have their purpose not in their usefulness to society, but in their truth.

The clarity of this truth finds a beauty which elevates pure mathematics to an art form. But, in contrast to a harpist who delights others by her music, I fear the pure mathematician cannot make his art accessible to a wider public. My question then was: How can pure mathematics justify its art to the State which finances it?

For Beno Eckmann, mathematics "sets the standard for every objective thought" and according to Friedrich Hirzebruch "without mathematics there would be no structured logical thinking".

For Raoul Bott, "the treasure the [mathematician] hunts is at the very core of all ... precise inquiry into the world .... As such [his] search must be a central concern of any enlightened state".

I agree and I am convinced of the need of mathematical thinking as a fundamental component of the modern world. Historically mathematics has been a key to open the doors to enlightment. Today, pure mathematics can still be considered as the guardian of the grail of logical thinking.

But as Roland Bulirsch puts it, "mathematics is invisible culture". Further Jürgen Moser says that "mathematics may not be accessible for the enjoyment of a broad audience". If this culture of pure mathematics is invisible and inaccessible how then can one show its practical use and demonstrate its tangible results ?

Armand Borel explains that "mathematics resembles an iceberg: beneath the surface is the realm of pure mathematics, hidden from the public view .... Above the water is the tip, the visible part which we call applied mathematics".

According to Phillip Griffiths, "one of the deep mysteries of life is the way in which the best pure mathematics, pursued for its own sake, inexplicably and unpredictably turns out to be useful".

Jürgen Moser adds that "the difficulty in getting this message across lies in the longer timespan needed to recognize the significance of mathematical discoveries .... Sometimes twenty or more years have to elapse .... Politicians unfortunately often think in much shorter terms."

This is certainly true not only for politicians but for society as a whole. In modern times we insist on increasingly shorter timespans for everything in our life. We ask for immediate return on investment. We want real time information. The life-span of technologies is getting shorter and shorter. Cost efficiency and speed have become the basic criteria to judge any human activity. This is dangerous because it's shortsighted.

In such an environment it is very important to continue to recognize that knowledge is a value in itself. Mathematics or philosophy or any basic research develops only thanks to this principle which is an important part of our civilization. If we start to forget it, we jeopardize the roots of our progress. The future is unpredictable. We cannot judge knowledge on the basis of its immediate usefulness. As an example, the work of Vaughan Jones, who connected three-dimensional knot theory with functional analysis, was awarded the Fields Medal at your last Congress in Kyoto on the basis of its intrinsic merit. Later, his theory was utilized by physicists in statistical mechanics and by biologists to explain the structure of DNA. It is only through the recognition and support of basic research that society can ensure the continued and full development of scientific progress.

Let us turn to applied mathematics. Today applied mathematics has become a basis for all other sciences and has a tremendous impact on life in modern societies. Applied mathematics is hereby both highly relevant and useful to society but it has lost its innocence. However, in contrast to the debate on the responsibility of nuclear physics and of genetechnology, it seems to me that there has been little ethical discussion on the role of mathematics in society. Thus here is my second question: Has mathematics avoided such discussions ?

There are mathematicians who claim moral neutrality for their science. René Thom for example writes me that "mathematics by itself is ethically neutral".

But Sir Michael Atiyah reminded me in his answer that the "atomic bomb was only built after extensive mathematical calculations", and Jürgen Moser adds that "the renowned mathematicians von Neumann and Ulam played an important role" in this project.

Armand Borel asks "should one see the fact that mathematics is at the base of artillery or guided bombs as an ethical problem?" Yes, I think one should.

It is true that "most mathematicians are remote from the decisions of the application" of their work, as Friedrich Hirzebruch puts it. Beno Eckmann goes even further, when he says: "For mathematics itself this [ethical and political] discussion is not relevant .... As a purely intellectual activity, it could not be influenced by such a discussion. Of course, those who apply mathematics have to face [this] discussion".

However, I do not think that making a distinction between abstract theory and practical application can altogether eliminate the ethical problem. We owe much of our progress in society to mathematicians and we have to recognize their merits while at the same time they have to assume their responsibilities.

Raoul Bott has expressed his argument against ethical neutrality, writing to me "that the age of innocence has come to an end for us all".

I am convinced this is true not only for science, but for most human activities. Today, thanks to science, our society has developed a tremendous power to control nature. This power enables us to take our destiny in our hands. But this power forces us to assume the responsibilities bound to it. If the age of innocence has come to an end, we have to recognize that it is the age of responsibility that has replaced it.

Let's turn now to my last question: If, as Minister of Science, I had the possibility to create 10 new professorships in Swiss universities, how many of them should I give to mathematics and why?

Phillip Griffiths is generous with his science and answers: "They should all go to mathematical scientists".

So is Gerd Faltings: nine chairs for mathematics but — as he likes music — he leaves the tenth chair to the harpists.

Sir Michael Atiyah, Friedrich Hirzebruch and Jürgen Moser request four or five chairs for mathematics. That is about the average of all the answers. In fact, in Switzerland today only one chair out of twenty is for mathematics.

Some replies focus exclusively on the needs of natural sciences. This is surprising. When one considers the complexities of the problems that face society, I am convinced that their solution will require a supported and dedicated effort of the social sciences and humanities, in close collaboration with natural sciences.

In view of the growing importance of science I understand why scientists ask for more means, why they want more professorships than they have. Scientists are increasingly expected to find solutions to all of our problems. It is more than legitimate that you ask for the necessary means from society.

Science and research are crucial today. You don't have to convince me of this as minister of science, but together we have to convince the public and the Parliament. We have to convince the taxpayer. This is a difficult task when public budgets are running huge deficits.

One problem is that the growing impact of science in society is not felt when we drive a car or use a phone. Most people are not aware of the scientist whose work is behind everything in our everyday life. Ask for instance any Swiss "Whose portrait is on the ten franc note?" They won't be able to tell you. They have never noticed that it is Leonhard Euler. Probably they don't even know who Euler is.

It is the task of the scientific community to tell the public why science matters. It is your task and it is mine.

I wish you all the best for your Congress. Thank you.

Dr. A. Gilgen, Head of the Department of Education, Canton Zürich, greeted the Congress participants on behalf of the Cantonal Government, recalling the names of celebrated mathematicians who had worked at the University of Zürich or at the ETH Zürich.

Dr. Th. Wagner, Zürich City Councillor, saluted the Congress participants on behalf of the authorities of the city of Zürich, depicting the image of Zürich as an open and international city of Congresses.

After a musical interlude ("Suite for Brass Quintet" by Edvard Grieg (1843–1907)) Professor David Mumford, Chairman of the Fields Medal Committee, announced the names of the recipients of the Fields Medals as follows:

I would like to thank our Swiss hosts very warmly for organizing so flawlessly and for giving us such a beautiful locale for the 1994 ICM. I am here as the chairman of the Fields Medal Committee for this Congress, whose other members are:

Luis Caffarelli
Masaki Kashiwara
Barry Mazur
Alexander Schrijver
Dennis Sullivan
Jacques Tits
S.R.S. Varadhan.

I should add that we consulted many many others in making our decisions.

As the committee compiled lists of names of candidates and their accomplishments, we found ourselves both pleased and awed by the great fecundity of recent mathematics, and by the great number of possible candidates representing a great number of areas of mathematical research. What to me is the most miraculous aspect of our field is that it is growing in so many directions: limbs sprout new growth and new shoots go off in unexpected dimensions. There is growth by deep and subtle proofs of old problems, and by the discovery and exploration of wholly new phenomena with new models.

Our response to this is to try to reward excellence in as many areas as possible. Fields himself realized that at least two medals were needed "because of the multiplicity of the branches of mathematics" and, as you know, this has grown to three or four medals. With at most four, we had to make quite a few very painful choices.

Fields also said in his 1932 memorandum on the medals: *It is understood that in making the awards, while it was in recognition of work already done, it was at the same time intended to be an encouragement for further achievement on the part of the recipients and a stimulus to renewed effort on the part of others ... with a view to encouraging further development along these lines.*

We have followed previous committees in interpreting his intent by restricting ourselves to considering candidates who are at most forty. His words also bring up an issue which is central to the future of our field: in many countries, governments have been attempting in the last few years to channel mathematical research along lines that bureaucrats deem to be productive and useful. Note that Fields' recommendation is instead to let mathematics develop by its internal forces, to let its success encourage further success. I agree with him that, in the long run, this will produce more results for both mathematics and for society.

Finally, we must bear in mind how clearly hindsight shows that past recipients of the Fields' medal were only a selection from a much larger group of mathematicians whose impact on mathematics was at least as great as that of the chosen.

So now, with great pleasure, let me announce the recipients whose work, in the view of the whole committee, embodies the best in mathematics today. In alphabetical order, they are:

Jean Bourgain (IHES, University of Illinois, Institute for Advanced Study),
Pierre-Louis Lions (Université de Paris-Dauphine)
Jean-Christophe Yoccoz (Université de Paris-Sud, Orsay)
Efim Zelmanov (University of Wisconsin, University of Chicago)

The medalists came forward and received their medals and prizes from Professor B. Eckmann.

Prof. Jacques-Louis Lions, Chairman of the Rolf Nevanlinna Prize Committee (the full composition of the Committee is given under Organization of the Congress), announced Avi Wigderson as the recipient of the Rolf Nevanlinna Prize; the prize winner received his prize from Prof. B. Eckmann.

The opening ceremonies concluded with the music of Jean-François Michel (born 1957) "Trois Pastels sur la Belle Epoque".

D. Mumford, President of the Fields Medal Committee

The prize winners with Minister Ruth Dreifuss; from left to right:
J. Bourgain, A. Wigderson, J.-C. Yoccoz, P.-L. Lions, E. Zelmanov

The Honorary President B. Eckmann with P.-L. Lions

Closing Ceremonies

The closing ceremonies were held in the Zürich Kongreßhaus in the afternoon of Thursday, August 11, 1994, starting at 15.00, immediately after the final lecture by Andrew Wiles.

Professor Jacques-Louis Lions, President of the International Mathematical Union, began the closing ceremonies with the following words:

Ladies and Gentlemen,

I am sure to convey the feelings of all participants when thanking and congratulating all those — in particular Prof. Carnal and Prof. Blatter — who worked so hard to make this Congress perfectly organized and scientifically of exceptional interest.

It is now my responsibility to give you the names of colleagues who have been elected to the Executive Committee and in the other committees of the International Mathematical Union at the General Assembly which took place in Luzern just before this Congress.

International Mathematical Union — IMU Executive Committee

President:	David Mumford	(Harvard University, USA)
Vice-presidents:	V. Arnold	(Steklov Institute, Russia)
	A. Dold	(University of Heidelberg, Germany)
Secretary:	J. Palis	(IMPA, Brazil)
Members:	J. Arthur	(University of Toronto, Canada)
	S. Donaldson	(Oxford University, United Kingdom)
	B. Engquist	(KTH Stockholm, Sweden)
	S. Mori	(RIMS, Kyoto University, Japan)
	K. Parthasarathy	(Indian St. Inst., New Delhi, India)

The past president, J.-L. Lions, is an ex-officio member of the Executive Committee.

Commission on Development and Exchange — CDE

Chairman:	Rolando Rebolledo	(University Cat. de Chile)
Secretary:	Pierre Bérard	(Inst. Fourier, France)
Members:	A. A. Ashour	(Cairo University, Egypt)
	C. H. Clemens	(University of Utah, USA)
	Kung Ching Chang	(University of Beijing, China)
	Cesar Camacho	(IMPA, Brazil)
	Jean Mawhin	(University of Louvain, Belgium)
	Mitsuo Morimoto	(Sophia University, Japan)

The ex-officio members are: the past chairman of CDE, M. S. Narasimhan, the president of IMU, and the secretary of IMU.

International Commission on Mathematical Instruction — ICMI Executive Committee

President:	Miguel de Guzman	(University of Complutense, Spain)
Vice-presidents:	Jeremy Kilpatrick	(University of Georgia, USA)
	Anna Sierpinska	(University of Montreal, Canada)
Secretary:	Morgens Niss	(Roskilde University, Denmark)
Members:	Colette Laborde	(Grenoble, France)
	Gilash Leder	(Monash University Melbourne, Australia)
	Carlos Vasco	(University Nac., Colombia)
	Zhang Dian-Zhou	(East China Normal University, Shanghai)

The ex-officio members are: the past president of ICMI, J.-P. Kahane, the president of IMU, and the secretary of IMU.

International Commission on History of Mathematics — ICHM (jointly with the International Union of the History and Philosophy of Science — IUHPS)

Two representatives of IMU were elected:

Karen Hunger Parshall	(USA)
Laura Toti–Rigatelli	(Italy)

Various resolutions were voted at the General Assembly. I would like you to know in particular of four of them (all of them will be published).

Resolution 2
The General Assembly thanks the Turn of the Century Committee for its report. It asks the new Executive Committee to proceed with the planning of World Mathematical Year 2000, and to organize and coordinate activities such as:

a) an invitation to a selected group of outstanding mathematicians to present their views on topics they expect to be central to mathematical activity in the next century.
b) the selection of a number of symposia, some possibly organized together with other scientific bodies, dedicated to mathematics, its applications and to its role in society, and
c) events held under the auspices of ICMI, CDE and ICHM.

The Executive Committee is asked to explore the possibilities provided by communications technology to unite activities occurring around the world.

Resolution 3
The General Assembly recommends that the name of the Chair of the Program Committee be made public on appointment.

Resolution 7
The General Assembly expresses its gratitude to UNESCO and ICSU for the help and support that these organizations have provided for various mathematical activities, in particular those carried out in cooperation with the Commission for Development and Exchange (CDE) and the International Commission on Mathematical Instruction (ICMI) of the IMU.

Resolution 10
The General Assembly expresses its gratitude to the Swiss organizers of the 1994 Congress, for their hospitable reception, and for the excellent arrangements for this meeting of the Assembly.

I have now a last reponsibility, a very pleasant one. I announce that the General Assembly has unanimously approved that the next ICM 98 will take place in Berlin. And I leave the floor to Professor Hirzebruch.

Professor Friedrich Hirzebruch invited the audience to the next International Congress of Mathematicians with these words:

Herr Ehrenpräsident, lieber Beno; meine Herren Präsidenten, meine Damen und Herren!

Just as Professor Lions did, I would like to express thanks to the organizing committee, in particular its Chairman Professor Carnal, for the wonderful work they did from which we all profited so much. It will be difficult to follow their example and to keep up the high standards. The third International Congress of Mathematicians took place in Heidelberg at the beginning of our century in 1904. Since then the Congress was not held in Germany. After the terrible period of World War II there were attempts to invite the Congress to Germany, beginning in the sixties; these attempts failed, always for understandable reasons. Let me say a few words about ICM 1904, to illustrate how times have changed. Who supported the Congress? His Majesty Kaiser Wilhelm gave 5000 M. from his private funds, his ministry "der geistlichen, Unterrichts- und Medizinalangelegenheiten" another 5000 M. His Royal Highness, the Grand Duke of Baden, the German State to which Heidelberg belonged, gave 3000 M., a well known publishing house contributed 2000 M. The registration fee for members was 20 M. I communicated this information to the Federal Ministry of Research and Technology in Bonn and to the Governing Mayor of Berlin and asked them to be as generous as the Kaiser and the Grand Duke, and they were. Some more information on the Heidelberg Congress: In 1904 the number of participating mathematicians was 336, among them 30 from Russia (about 10%, like in the Zürich Congress now), 25 from the Austrian-Hungarian empire, 24 from France, 15 from the United States, 12 from Switzerland and 2 from Japan. Felix Klein from Göttingen and Julius Molk

Closing ceremonies

F. Hirzebruch at the closing ceremonies

from Nancy presented the first volume of the Encyclopedia of Mathematical Sciences (German and French editions). Lectures were given, for example, by Minkowski, Hilbert, Borel, Voronoi, Fricke, Mittag-Leffler, Hadamard, Sommerfeld, Weber, Prandtl, Cappelli, Macaulay, Levi-Cività. Felix Klein lectured in the ICM section, as we call it nowadays, on "Über eine zeitgemäße Umgestaltung des mathematischen Unterrichts an den höheren Schulen", a title which can be used without change at any congress.

For Berlin we expect 10 times as many mathematicians as in Heidelberg from all parts of the world. We plan to provide fellowships for young mathematicians and for mathematicians — young and senior — who come from countries with difficult financial conditions. We hope that the formerly divided city of Berlin will be a symbol for improved worldwide cooperation. The mathematical landscape in the Berlin area is highly developed with three universities in Berlin and one in Potsdam and with two research institutions, the Konrad-Zuse-Center and the Weierstrass Institute for Applied Analysis and Stochastics. Many Berlin mathematicians are eager to make the Congress a success. Much work was already done by the Berlin members of the Provisional Organizing Committee whom I thank very much, in particular Professor Grötschel, Vice-President of the Konrad-Zuse-Center and President of the German Mathematical Society. By his efforts it was possible to preregister for ICM 1998 here in Zürich using the world-wide web. I hope many of you have done so.

It is a great pleasure and honor for me to invite on behalf of the German Mathematical Society the International Congress of Mathematicians to Berlin for the period August 18 to 28, 1998. The united city has many theatres and museums. It has forests and lakes. There is much to do. The average maximum daily temperature in Berlin in August is 23 degrees. Specialists in probability have told me, however, that in August 1998, Berlin may even be hotter than Zürich in 1994. But never mind, I hope to see you all in Berlin in any weather.

Professor Henri Carnal then addressed the audience as follows:

Ladies and Gentlemen,

I know that most of you have heard many, many talks in the last few days. So, let's have a very short closing ceremony!

My colleagues in the Organizing Committee and I are today in the mood of mathematicians who have worked for years on a problem and who have come to the publication of their results. At the beginning, as it often happens in mathematics, we thought that we could just solve our problem by adapting well-known methods to the new boundary conditions. But, after a while, we realized that the boundary conditions were the core of the problem, e.g. the unrealistic exchange rate for the Swiss franc and, even more, the new political situation in Eastern Europe.

From that part of the world, we received over 1000 requests from scholars who wanted to participate and who asked for support. We would have liked to have invited all of them, but it was unfortunately only possible for a minority. So the number of registered participants stayed at 2326, which is below our initial expectation. You should add to this figure not only 363 accompanying members, 77 exhibitors and some 200 day visitors, but also the 150 young mathematicians from Swiss institutes who helped us during the last two weeks. They worked hard, certainly harder than they, and even we, had expected and they would have deserved more than one yellow T-shirt. I want to thank them very warmly here.

And now, as I said before, we feel like people who have just presented a partial solution to a challenging problem: we could have done better, but we are still very proud to have brought a non-trivial contribution to a long sequence of events which started almost one hundred years ago and which will certainly continue into the next century. We would be quite happy to know that you feel the same way. Thank you.

Professor Henri Carnal then declared the Congress closed.

Scientific Program

The Work of the Fields Medalists and the Rolf Nevanlinna Prize Winner

Invited One-Hour Addresses at the Plenary Sessions

---

*) The material presented is largely covered by a recent survey under this title by L. Babai in: First European Congress of Mathematics, Paris 1992 (A. Joseph et al., eds.), Vol. I, Progress in Math. **119**, Birkhäuser Verlag, Basel und Boston 1994, pp. 31–91.

Invited Forty-Five Minute Addresses at the Section Meetings

Section 1. Logic

Section 2. Algebra

Section 3. Number theory

Section 4. Geometry

Section 5. Topology

Section 6. Algebraic geometry

Section 7. Lie groups and representations

Section 8. Real and complex analysis

Section 9. Operator algebras and functional analysis

Section 10. Probability and statistics

Section 11. Partial differential equations

Section 12. Ordinary differential equations and dynamical systems

Section 13. Mathematical physics

Section 14. Combinatorics

Section 15. Mathematical aspects of computer science

Section 16. Numerical analysis and scientific computing

Section 17. Applications of mathematics in the sciences

Section 18. Teaching and popularization of mathematics

Section 19. History of mathematics

List of Participants (Ordinary Members)

Abbas Casim, Switzerland
Abbott Harvey L., Canada
Abdukhalikov Kanat S., Kazakhstan
A'Campo Norbert, Switzerland
A'Campo–Neuen Annette, Switzerland
Abe Shingo J., Japan
Abe Takehisa, Japan
Abib Odinette R., Senegal
Aboltins Aivars J., Latvia
Abramini Silvia, Switzerland
Abreu Paolo, Switzerland
Adachi Norio, Japan
Adams Jeffrey D., USA
Adams Malcolm R., USA
Adelmeyer Moritz, Switzerland
Adem Alejandro, USA
Adler Joel, Switzerland
Aebischer Beat, Switzerland
Agrachev Andrei, Russia
Aguilar Marcelo, Mexico
Aharonov Dov, USA
Ahluwalia Daljit S., USA
Ahmad Nasir M., Switzerland
Ahsan Javed, Saudi Arabia
Aiba Akira, Japan
Aigner Martin, Germany
Akhmetov Marat, Kazakhstan
Akiba Shigeo, Japan
Akkouchi Mohamed, Maroc
Akpan Edet P., Italy
Al Ali M.I.M., Jordan
Alber Hans-Dieter, Germany
Alber Mark, USA
Albertini Claudia, Switzerland
Albrecht Raphael Jakob, Germany
Albuquerque Paul, Switzerland
Alessandrini Lucia, Italy
Alestalo Peka J., Finland
Alexandrov Alexandre D., Russia
Alexandrov V.A., Russia
Alexandrova Nadezhda, Russia
Alexevskii Andrei V., Russia
Alibert Daniel, France
Aliev Bairam, Russia
Aljadeff Eli, Israel
Almatszaz , Iran
Altschuler Daniel, Switzerland
Alvarez-Scherer Ma. de la Paz, Mexico
Amann Herbert, Switzerland
Ambrosetti Antonio, Italy
Amiraliyev Gabil M., Azerbajan
Anastassiou George A., USA
Andersen Henning H., Denmark
Anderson James W., USA
Anderson Michael, USA
Ando Hideo, Japan
Ando Matthew, USA
Ando Shiro, Japan
Anglin William S., Canada
Anichini Guiseppe, Italy
Annaby Mahmoud H., Egypt
Antal Peter, Switzerland
Arai Hitoshi, Japan
Araki Huzihiro, Japan
Archinard Etienne, Switzerland
Arima Satoshi, Japan
Arlettaz Dominique, Switzerland
Arora Manmohan S., Bahrain
Arpaia Pasquale J., USA
Arsham Hossein, USA
Arslanov Marat, Russia
Artemiadis Nicolas K., Greece
Arthur James G., Canada
Asada Teruko, Japan
Asamoto Noriko, Japan
Ash J. Marshall, USA
Ashbaugh Mark, USA
Ashour Samir A., Egypt
Askitas Nikos, Switzerland
Aslaksen Helmer, Singapore
Asmuss Svetlana, Latvia
Asoo Yasuhiro, Japan
Astala Kari, Finland
Atan Kamel A., Malaysia

Aubert Jean-Christophe, Switzerland
Auckly Dave, USA
Avdispahic Muharem, Bosnia
Avellaneda Marco, USA
Aviolat Frédéric, Switzerland
Avkhadiev Farit, Russia
Awartani Marwan M., Israel
Axler Sheldon, USA
Ayoub Ayoub B., USA
Ayupov Shavkat, Uzbekistan
Azarpanah Fariborz, Iran
Azérad Pascal, Switzerland
Azhari Abdelhak, Maroc
Azimi Parviz, Iran
Baak Eric, Switzerland
Baas Nils A., Norway
Babai László, Hungary
Babbitt Donald G., USA
Bachmann Franz, Switzerland
Badiozzaman Abdul Jabbar, Iran
Bahturin Yuri, Russia
Bakaev Nikolai, Russia
Baker Alan, Switzerland
Baker Mark D., France
Baladi Viviane, Switzerland
Ball John M., France
Balmace Jose M. P., Philippines
Balmer Paul, Switzerland
Banaszak Grzegorz M., Poland
Banchoff Thomas F., USA
Bandelou Christoph, Germany
Bangert Victor, Germany
Bär Christian, Germany
Baranikov Sergey, Russia
Baranov Alexander S., Russia
Barbour Andrew D., Switzerland
Barbu Gheorge, Romania
Barel Zeev, USA
Barge Jean, France
Baribaud Claire, Switzerland
Barner Klaus, Germany
Baron Simson, Israel
Barrett David, USA
Barsegian Grigor A., Armenia
Barth Karl F., USA
Bartolozzi Federico, Italy
Bartz Krystyna M., Poland
Basakk Gopal P., India
Basmajian Ara S., USA
Bass Richard, USA
Bassi Franco, Switzerland
Batchelor Philipp, Switzerland
Bates Jean, UK
Bauer Franziska, Switzerland
Bauer Ingrid C., Italy
Baumann Bernd, Germany
Baumann Peter, Switzerland
Baumgärtel Hellmut, Germany
Baumgartner James, USA
Baur Walter, Germany
Bayer-Fluckiger Eva, France
Bayrooti Afshin, USA
Bazaliy Boris V., Ukraine
Beale Thomas, USA
Beck Jonathan, USA
Becker Eberhard, Germany
Becker Jochen, Germany
Becker Ronald I., South Africa
Beem John, USA
Behforooz Hossein, USA
Behr Helmut, Germany
Bekbaev Ural, Uzbekistan
Belinskii E., Israel
Bell Rowen B., USA
Bellaiche André, France
Bellissard Jean, France
Belyi Vladimir I., Ukraine
Bendikov Alexander, Russia
Benkart Georgia, USA
Benmiled Slimane, France
Benson David B., USA
Bercovier Michel, Israel
Berestovskii Valerii, Russia
Berezansky Yurij M., Ukraine
Berg Christian, Denmark
Berger Clemens, France
Berger Roland, France
Berglez Peter, Austria
Bergweiler Walter, Germany
Bernard Jérôme, France
Bernardi Marco L., Italy
Bernardi Marco Paolo, Italy
Bernik Vasily I., Belarus
Berrut Jean-Paul, Switzerland
Bertin Marie José, France
Bertrand Anne, France
Beschler Edwin F., USA
Bessenrodt Christine, Germany
Besson Olivier, Switzerland
Bezard Max J., France
Beznea Lucian, Romania
Bezuglyi Sergey, Ukraine

Bhattacharya Prabir, USA
Bhattacharyya Prodipeswar, India
Bickel Thomas, Switzerland
Bielawski Roger, Poland
Bierstedt Klaus D., Germany
Biler Piotr, Poland
Bingham M.S., UK
Binswanger Klemens, Switzerland
Birman Graciela, Argentina
Biró András, Hungary
Bismut Jean-Michel, France
Biswas Indranil, Italy
Biswas Paritosh, India
Bitterlin Marcel, Switzerland
Bizhanova Galina I., Kazakhstan
Björn Anders, Sweden
Blackburn Norman, UK
Blakemore Carroll F., USA
Blanc André G., France
Blanchard Philippe, Germany
Blanchet Christian, France
Blatter Christian, Switzerland
Blazic Novica, Yugoslavia
Bleiler Steven A., USA
Blondel Vincent D., Sweden
Bloznelis Mindaugas P., Lithuania
Blum Lenore, USA
Böcherer Siegfried, Germany
Bochorishvili Ramaz, Georgia
Böckle Gebhard C., USA
Bodenmann Ruedi, Switzerland
Bödigheimer Carl-Friedrich, Germany
Bodnarescu M.V., Germany
Bogachev Vladimir I., Russia
Bogaevski Ilia, Russia
Bokhee Im, Korea
Bolibruch A.A., Russia
Bolla Marianna, Hungary
Bollow Norbert, Switzerland
Bollow-Mannzen Anna, Germany
Bolotin Sergey V., Russia
Bombieri Enrico, Switzerland
Bonckaert Patrick, Belgium
Bongiorno Benedetto, Italy
Boo Per-Anders V., Sweden
Booß-Bavnbek Bernhelm, Denmark
Borcherds Richard, USA
Borel Armand, USA
Borisenko Alexander, Ukraine
Borisenko Alexsei A., Russia
Borodin Allan B., Israel
Borwein David, Canada
Botana Francisco, Spain
Botta Nicola, Switzerland
Bottazzini Umberto, Italy
Bouche Thierry, France
Bouharb Jalel, France
Boujot Jean Paul, France
Bourgain Jean, France
Bourgeois Claude, Switzerland
Bourguignon Jean-Pierre, France
Boutot Jean-François, France
Bovier Pierre-Alain, Switzerland
Boyd James N., USA
Brannan David, UK
Branner Bodil, Denmark
Braumann Carlos, Portugal
Braumann M.M., Portugal
Braun-Angott Peter, Germany
Bredimas A.P., France
Bremigan Ralph J., USA
Bremner Murray R., Canada
Brešar Matej, Slovenia
Briem Eggert, Denmark
Brinkmann Hans, Germany
Brinzanescu Vasile, Germany
Brion Michel, France
Brodmann Markus, Switzerland
Brodsky Mikhail, USA
Brookfield Gary, Germany
Brown Gary, USA
Brown Gordon E., USA
Brown Leon, USA
Brüning Erwin A., South Africa
Bruno Alexander, Russia
Brunovský Pavol, Slovakia
Brunstad Solmund G., Norway
Brüstle Thomas, Switzerland
Brzezinski Juliusz, Sweden
Bshouty Daoud H., Israel
Buchanan Thomas, Germany
Bucki Andrzej J., USA
Budak Talin, Turkey
Büeler Benno, Switzerland
Bühler Ulrich, Germany
Bühlmann Peter, Switzerland
Bühlmann Hans, Switzerland
Buhmann Martin, Switzerland
Bujalance José A., Spain
Bulitko Valerii K., Ukraine
Bundschuh Peter, Germany
Bunke Ulrich, Germany

Burdick Bruce, USA
Burger Marc, Switzerland
Bürgisser Peter, Switzerland
Burlet Oscar, Switzerland
Burns Daniel M., USA
Burns Robert G., Canada
Buser Peter, Switzerland
Bushnell Colin J., UK
Butler Lynne M., USA
Bylinski Czeslaw, Poland
Caffarelli , Italy
Cakir Isa, Switzerland
Cappell Sylvain, USA
Caraman Petru, Romania
Cardell Leif, Sweden
Carlsson Renate, Germany
Carnal Henri, Switzerland
Carvalho Sonia Pinto de, Brazil
Casacuberta Carles, Spain
Case Bettye Anne, USA
Caselles Vicent, Spain
Castellet Manuel, Spain
Catalano Domenico, Switzerland
Catanese Fabrizio, Italy
Catepillan Ximena P., USA
Cattaneo Uberto G., Switzerland
Caumel Yves, France
Cawagas Raoul, Philippines
Cawley Elise E., USA
Cejalvo Flor V., Philippines
Čelikovský Sergej, Czech Rep.
Čerin Zvonko T., Croatia
Chadam John, Canada
Chahal Jasbir S., USA
Chaleyat-Maurel Mireille, France
Chambers Llevelyn G., UK
Chandrasekharan Komaravolu, Switzerland
Chandrinos Konstantinos, Greece
Chang Der-Chen E., USA
Chang Kun Soo, Korea
Chang Kung-Ching, China P.R.
Chang Mei-Chu, USA
Chatterji Srishti D., Switzerland
Chemin Jean-Yves, France
Chen Gui-Qiang, USA
Chen Larry Lung-Kee, USA
Chen Zhi Qiang, Switzerland
Cheng Qing-Ming, Italy
Cheng Shun-Jen, Taiwan
Chérix Pierre-Alain, Switzerland
Chernyavskaya Nina, Israel
Chevallier Benoît, France
Chillag David, Israel
Chillingworth David, UK
Chin Chou-Hsieng, China P.R.
Cho Sung Je, Korea
Choban Mitrofan M., Moldova
Chopyk Volodymyr I., Ukraine
Chou Ching Sung, Taiwan
Christ Michael, USA
Christian Sunyach, France
Chu Cho-Ho, UK
Chu Wenchang, Italy
Chung Fan R.K., USA
Chyba Monique, Switzerland
Ciarlet Philippe, France
Cibils Claude, Switzerland
Cieliebak Kai, Switzerland
Cĭesĭelski Zbigniew, Poland
Clark David Alan, USA
Clarke Francis W., UK
Claudi Alsina, Spain
Clemence Donicic P., Zimbabwe
Clerici Mario, Switzerland
Clozel Laurent, France
Coates John H., UK
Coelho Flávio U., Brazil
Coffey Joseph, USA
Cohn Paul M., UK
Cohn Richard M., USA
Colbois Bruno, Switzerland
Colella Phillip, USA
Colin de Verdière Yves, France
Collart Bettina, Switzerland
Collart Stéphane, Switzerland
Collins Donald J., UK
Colliot-Thélène Jean-Louis, France
Collot Francis, France
Conder Marston D.E., New Zealand
Conlon Lawrence W., USA
Connett William C., France
Conrad Mark, Germany
Constantin Peter, USA
Constantinescu Corneliu, Switzerland
Contessa Maria, Italy
Conway John, USA
Coray Daniel, Switzerland
Cordero-Vourtsanis Minerva, USA
Cordova Yevenes Antonio, Germany
Corduneanu Constantin C., USA
Corlette Kevin D., USA
Cornea Aurel, Germany

Cotsiolis Athanase, Greece
Courcelle Olivier, France
Coutinho Severino, Brazil
Crespin Daniel, Venezuela
Cristea Mihai C., Romania
Croft Hallard T., UK
Curcic Vladan, Switzerland
Cutkovsky Steven, India
Dacorogna Bernard, Switzerland
Dadarlat Marius, USA
Dafermos Constantine, USA
Daguman Rosario Z., Philippines
Dahmardah Habib-Olah, Iran
Dahmen Wolfgang, Germany
Dale Knut, Norway
Dalmasso Juan Carlos, Argentina
Dan Nicosor, France
Danaee Ali, Iran
Dani Shrikrishna G., India
Dark Rex S., Ireland
Dăscălescu Sorin, Romania
Daubechies Ingrid, USA
Davermann Robert J., USA
Davis Chandler, Canada
Davis James F., USA
Dawson Donald A., Canada
Dayton Barry H., USA
De Guzman Miguel, Spain
De la Harpe Pierre, Switzerland
Debinska-Nagorska Anna, Poland
Debnath Lokenath, USA
Debrunner Hans E., Switzerland
Dechéne Lucy, USA
Deguire Paul, Canada
Delatte David A., USA
Delerins Richard, France
Delessert André, Switzerland
Deligne Pierre R., USA
Demailly Jean-Pierre, France
Denzler Jochen, Germany
Derighetti Antoine, Switzerland
Déruaz Marcel A., Canada
Desolneux-Moulis Nicole, France
Desquith Etienne, Ivory Coast
Deuber Walter, Germany
Dewan Kumkum, India
Di Piazza Luise, Italy
Di Pompeo Vincenzo, Switzerland
Diarra Bertin, France
Diaz Casado Lorenzo J., Brazil
Dimovski Dončo, Macedonia
Ding Jiu, USA
Ding Kequan, USA
Dinges H., Germany
Divis Zita M., USA
Dixon John, Canada
Djafari-Rouhani Behzad, Iran
Djakov Plamen B., Turkey
Dlab Vlastimil, Canada
Dmitriyeva Irene, Ukraine
Dodziuk Jozef, USA
Dôku Isamu, USA
Dold Albrecht, Germany
Dolgachev Igor, USA
Domenig Thomas, Switzerland
Domokos Mátyás, Hungary
Dong Jianping, USA
Donig Jörg, Germany
Donoho David, USA
Doraisamy Logeswary, Malaysia
Dorofeev Sergey, Switzerland
Doty Stephen R., USA
Douady Régine, France
Douchet Jacques, Switzerland
Douglas Robert J., USA
Douglas Ronald G., USA
Dow Alan S., Canada
Doyen Jean, Belgium
Drasin David, USA
Drasny Gábor, Hungary
Dreyfus Tommy P., Israel
Driver Kathy A., South Africa
Drmota Michael, Austria
Droms Carl, USA
Du Cloux Fokko, France
Duggal Krishan, Canada
Dumitrescu Horia, Romania
Dumortier Freddy, Belgium
Durfee Alan H., USA
Dürig Willi, Switzerland
Dvoretzky Aryeh, Israel
Dwilewicz Roman, Canada
Dymacek Wayne, USA
Dzhumadildaev Askar S., Kazakhstan
Dzinotyiweyi Heneri A.M., Zimbabwe
Dzung Wong Boaswan, Switzerland
Ebihara Madoka, Germany
Eckmann Beno, Switzerland
Eckmann Jean-Pierre, Switzerland
Edwards David A., UK
Edwards Robert, USA
Eguchi Kazuo, Japan

Ehrig Gertraud, Germany
Eichelsbacher Peter, Switzerland
Eida Atsuhiko, Japan
Eisele Toni, Switzerland
Ejov Vladimir V., Germany
Ekhaguere Godwin O.S., Nigeria
Ekong Samuel, France
El Zein Fouad, France
Elbaz-Vincent Philippe, France
Elduque Alberto, Spain
Eliahou Shalom, Switzerland
Elizarov Alexander, Russia
Elkies Noam, USA
Elliott George A., Canada
Embrechts Paul, Switzerland
Emmanouil Ioannis, USA
Enemoto Kazuyuki, Japan
Engelbert Hans-Jürgen, Germany
Engeler Erwin, Switzerland
Enriquez Benjamin, France
Epkenhans Martin, Germany
Epstein Henri, France
Epstein Mordechai R., Israel
Erdős Paul, Czech Rep.
Eriksson Folke S., Sweden
Erkama Timo P., Finland
Erlandsson Thomas, Sweden
Erle Dieter, Germany
Eto Kazufumi, Japan
Eudave-Munoz Mario, Mexico
Evtushik Leonid, Russia
Ezeilo James O.C., Nigeria
Faddeev Ludvig D., Russia
Fadkic Huse, Bosnia
Faltings Gerd, Germany
Fan Kwai-Man, Hong Kong
Fan Paul S., USA
Fan Peng, USA
Fan Tian-You, USA
Farahmand Kambiz, UK
Farinelli Simone, Switzerland
Farran Hani, Kuwait
Fathi Albert, France
Fauck Jana, Germany
Faxén Birger J., Sweden
Fazekas Attila, Hungary
Fedenko Anatoliy S., Russia
Fedorov Yuri, Russia
Feehan Paul M.N., USA
Feher Franziska, Germany
Fehlmann René, Switzerland
Feichtinger Hans G., Austria
Feit Walter, USA
Felder Giovanni, Switzerland
Fel'shtyn Alexander, Germany
Feng Keqin, China P.R.
Feng Xuning, China P.R.
Feng Yu Yu, China P.R.
Fenn Roger, UK
Feroe John A., USA
Ferrand Jacqueline, France
Ferretti Roberto, Switzerland
Fialowski Alice, Hungary
Fidler Jiří, Czech Rep.
Fiedler Mirsolav, Czech Rep.
Figueiredo Jr. Ruy T., Brazil
Fillmore Peter Arthur, Canada
Fischer Gerd, Germany
Fischer Wolfgang, Germany
Fisk Steve, USA
Fjelstad Paul, USA
Flach Nicole, Switzerland
Flament Dominique, France
Flato Moshé, France
Flatto Leopold, USA
Fleischer Anja, Switzerland
Flensted-Jensen Morges, Denmark
Flexor Marguerite, France
Flor Peter, Austria
Flores Fabian, China P.R.
Florio José, Mexico
Florit Luis A., Brazil
Flucher Martin, Switzerland
Föllmer Hans, Germany
Fontana Marco, Italy
Fook Leong, Malaysia
Forrer Hans, Switzerland
Forsberg Mikael, Sweden
Fossum Robert M., USA
Franjou Vincent, France
Franks John, USA
Frei Alfons, Switzerland
Frei Günther, Canada
Freire Alexandre, USA
Frenkel Edward, USA
Fric Roman, Slovakia
Friedl Katalin, Hungary
Friedlander John, Canada
Friedlander Leonid, USA
Frigon Marlène, Canada
Fritsch Rudolf, Germany
Fröberg Rolf, Sweden

Fröhlich Jürg, Switzerland
Frölicher Alfred, Switzerland
Fu Joseph, USA
Fuchs Wolfgang H., USA
Fuglede Bent, Denmark
Fujii Kazuyuki, Japan
Fujita Hiroshi, Japan
Fujiwara Koji, USA
Fukaya Kenji, Japan
Fukuda Takashi, Japan
Fumio Maitani, Japan
Funaya Bokuro, Japan
Füredi Zoltán, USA
Furrer Hans-Jörg, Switzerland
Furta Stanislav, Russia
Furukawa Yasukuni, Japan
Furutsu Hirotoshi, Japan
Gabai David, USA
Galkin Valery A., Russia
Gangbo Wilfrid, USA
Gardiner Anthony, UK
Gardner Robert Brown, USA
Gärtner Jürgen, Germany
Gauthier Yvon, Canada
Gavosto Estela A., USA
Gavrilov Lubomir, France
Ge Liming, USA
Gebel Michael, Germany
Geffen Nima, Israel
Gelca Razvan, USA
Geller Daryl, USA
Gelpke Verena, Switzerland
Geluk Jaap, The Netherlands
Gentile Giovanni, Switzerland
Gentz Barbara, Switzerland
Geoffriau François, France
Geoghegan Ross, USA
Gérard Olivier, France
Gerlach Eberhard, USA
Ghahramani Saeed, USA
Ghanaat Patrick, Switzerland
Gheorghe Barbu V., Romania
Gill Jelena, USA
Gillard Roland, France
Gilligan Bruce C., Canada
Gillman Leonard, USA
Gillot Antonio L., Guatemala
Ginosar Yuval, Israel
Ginovian Mamikon, Armenia
Ginzburg Abraham, Israel
Giordano Thierry, Canada
Girlich Hans-Joachim, Germany
Giroux André, Canada
Giusti Marc F., France
Givental Alexander B., France
Glaus Christian, Switzerland
Glavan Vasi, Moldova
Glemser Helmut, Switzerland
Gliklikh Yuri E., Russia
Gloden Raoul, Italy
Glover Henri, USA
Glutsuk Aleksei A., Russia
Gmür Bruno, Switzerland
Gobert Jules, Belgium
Göing Anja, Switzerland
Goldie C.M., UK
Goldreich Oded, Israel
Goldstein Myron, USA
Golodets V.Ya., Ukraine
Goloubeva Valentina A., Russia
Golub Gene H., USA
Gomez Francisco, Spain
Gómez Urgellés Joan, Spain
Gompf Robert E., USA
Goncharov Alexander B., USA
Gonzáles-Jimenez Santos, Spain
Gonzalez Jorge A., Cuba
Gootman Elliot C., USA
Göpfert Alfred, Germany
Gorbachuk Miroslav, Ukraine
Gordeziani David, Georgia
Gordon B.Brent, USA
Gordon Cameron M., USA
Goroff Daniel L., USA
Gorunovich Vladimir V., Ukraine
Goryainov Victor V., Ukraine
Goßmann Michael, Switzerland
Goto Shiro, Japan
Gottlieb Daniel H., USA
Gottwald Siegfried, Germany
Gowers W. Timothy, France
Graf Gian Michele, Switzerland
Grafakos Loukas G., USA
Graham Ronald, USA
Gramsch Bernhard, Germany
Grand Arno, Switzerland
Grandjean Oliver, Switzerland
Grant Douglass L., Canada
Granville Andrew, USA
Gratza Bernhard, Switzerland
Gray Eva Maria, USA
Gray Jeremy, UK

Gray John W., USA Gray Mary W., USA
Green Dominic, UK
Greene Curtis, USA
Greenspoon Arthur, USA
Grieder Ralph, Switzerland
Grieder Stefan, Switzerland
Grigorchuk R.I., Russia
Grigor'yan Alexander, Germany
Grillakis Manoussos, USA
Gritsenko Alekseevich, Russia
Gröchenig Karlheinz U., USA
Groh Gabor, Switzerland
Gromov Mikhael, France
Gropp Harald, Germany
Grosse-Erdmann K.-G., Germany
Grötschel Martin, Germany
Grubbs Mary M., USA
Gruber Peter M., Austria
Grundman Helen G., USA
Grunenfelder Luzius, Canada
Grüter Michael, Germany
Guaraldo Rosalind, USA
Gubeladze Joseph, Georgia
Gubler Walter, Switzerland
Guerry Graziella, France
Guiasu Silviu, Canada
Guichardet Alain, France
Guidon Thomas, Switzerland
Guidotti Patrick, Switzerland
Gunawan Hendra, Indonesia
Gundlach Karl-Bernhard, Germany
Guo Lei, China P.R.
Guo Lijia, USA
Gursky Matthew, USA
Gusein-Zade Sabir M., Russia
Gustavs Bruno, Switzerland
Gutknecht Martin H., Switzerland
Gutlyanskii Vladimir, Ukraine
Haagerup Uffe, Denmark
Haas B., Switzerland
Hada Ichiro, Japan
Haefliger André, Switzerland
Hag Kari Jorum, Norway
Hag Per, Norway
Haida Minoru, Japan
Haines Roger, The Netherlands
Hajarnavis C.R., UK
Hajduk Boguslaw, Poland
Hajnal Peter, Hungary
Hakopian Hakop, Armenia
Halbeisen Lorenz, Switzerland
Hales Thomas C., USA
Haller Jakob, Switzerland
Hampel Frank R., Switzerland
Han Maoan, China P.R.
Han Sun Hyuk, Korea
Hanamura Masaki, USA
Hančl Jaroslav, Czech Rep.
Handa Toru, Japan
Hanin Leonid G., Israel
Hanouzet Bernard, France
Hansen Vagn, Denmark
Harbater David, USA
Hardt Robert, USA
Har'El Zvi, Israel
Harkleroad Leon, USA
Hartz David G., USA
Harzheim Egbert R., Germany
Hasegawa Keizo, Japan
Hasegawa Kenji, Japan
Hashiguchi Norikazu, France
Hassler Urs, Switzerland
Hasumi Morisuke, Japan
Hatori Asako, Japan
Hatori Osamu, Japan
Hattori Akio, Japan
Hattori Toshiaki, Japan
Hauser Raphael, Switzerland
Hausmann Jean-Claude, Switzerland
Haviar Miroslav, Slovakia
Hayashi Hiroshi, Japan
Hayashi Susumu, Japan
Hayes Sandra, USA
Hazewinkel Michiel, The Netherlands
He Wu, Brazil
Hearst William R., USA
Heber Jens O., Germany
Hedberg Lars Inge, Sweden
Heeb Thierry, Switzerland
Hegyvári Norbert, Hungary
Heimberg Gérard, Switzerland
Heinonen Juha M., USA
Heintze Ernst, Germany
Helemskii Alexander, Russia
Helversen-Pasotto Anna, France
Hengartner Walter, Canada
Henning Michael A., South Africa
Henrichs Rolf W., Germany
Hepp Klaus, Switzerland
Herald Christopher, USA
Herbertson Magnus, Sweden
Herendi Tamas, Hungary

Héritier Serge, Switzerland
Herman Michael R., France
Hermann Carl-Friedrich, Germany
Hernandez Carlos M., USA
Hersch Joseph, Switzerland
Heusener Michael, Germany
Heyer Herbert K.W., Germany
Heywood Philip, UK
Hidaka Fumio, Japan
Hidber Christian, Switzerland
Hieber Matthias, Switzerland
Higashiyama Teiko, Japan
Hill Raymond, UK
Hilsum Michel, France
Hinz Andreas M., Germany
Hironaka Yumiko, Japan
Hirsch Michael D., USA
Hirzebruch Friedrich E.P., Germany
Hochmuth Reinhard, Germany
Hofer Helmut, Switzerland
Hogbe-Nlend Henri, France
Hogendijk Jan, The Netherlands
Holá Lŭbica, Slovakia
Holmann Harald R.A., Switzerland
Holme Audun, Norway
Holzapfel Rolf-Peter, Germany
Holzwart Hartmut, Germany
Hopkins Michael, USA
Horiuchi Toshio, Japan
Hörmander Lars V., Sweden
Horn Mary, USA
Hostettler Maria, Switzerland
Houh Chorng, USA
Houillot Royer Josette, France
Houston Johnny, USA
Hruby Jarolav, Czech Rep.
Hsieh Din-Yu, Hong Kong
Hu Yi, USA
Hua Xinhou, Germany
Huang Jing-Song, USA
Hubbard John, USA
Huber Martin, Switzerland
Huber-Dyson Verena, Canada
Hughes Anne C., USA
Hughes Hallet Deborah, USA
Hughes Kenneth, South Africa
Hui King, Taiwan
Hummel Christoph, Switzerland
Hungerbühler Norbert, Switzerland
Hunt John H.V., South Africa
Hurder Steven E., USA
Hurley Donal J., Ireland
Hürlimann Werner, Switzerland
Hurrelbrink Jurgen, USA
Hürzeler Markus, Switzerland
Huse Tomohiro, Japan
Husemöller Dale, France
Hüsler Jürg R., Switzerland
Hussin Véronique, Canada
Hutchinson Alan, UK
Hütter Ralf, Switzerland
Hwang Chii-Ruey, Taiwan
Hwang Jun-Muk, USA
Hwang Tea-Yuan, Taiwan
Iarrobino Anthony, USA
Ibisch Horst, France
Igodt Paul, Belgium
Iitaka Shigeru, Japan
Ikegami Teruo, Japan
Ilinski Kirill, Russia
Illge Reinhard, Germany
Illman Sören A., Finland
Ilolov Mamadsho, Tajikistan
Im Hof Hans-Christoph, Switzerland
Imamoḡlu Özlem, USA
Imhof Jean-Pierre, Switzerland
Inandar S.P., India
Indurain-Eraso Esteban, Spain
Ion Patrick, USA
Ionescu Paul-Cristodor, Romania
Iranmanesh Ali, Iran
Isbell John, USA
Ishigami Yoshiyasu, Japan
Ishihara Shigeru, Japan
Ishii Shihoko, Japan
Ishikawa Saneaki, Japan
Ishikawa Takeshi, Japan
Ishikawa Tsuneo, Japan
Istad Roy Martin, Norway
Itenberg Ilia, France
Ito Ryuichi, Japan
Ito Yoshihiko, Japan
Itoh Tatsuo, Japan
Itoh Yasuhiko, Japan
Iturriaga Renato A., Mexico
Ivanauskas Feliksas, Lithuania
Ivanov A.P., Russia
Ivanov Alexandr O., Russia
Ivanšić Ivan, Croatia
Izumi Masaki, Italy
Jackson Allyn, USA
Jaco William H., USA

Jacob Niels, Germany
Jafari Farhad, USA
Jaffe Arthur, USA
Jäger Gerhard, Switzerland
Jahanshani Mohammad, Iran
Jahren Bjørn, Norway
Jaiani George V., Georgia
Jakubczyk Bronislaw, Poland
James Donald G., USA
Janas Jan, Poland
Janno Jaan, Estonia
Jannsen Uwe, Germany
Jansen Karl-Heinz, Germany
Janssen Gerhard, Germany
Jarchow Hans, Switzerland
Jaroszewska Magdalena, Poland
Jatsko Anriej, Poland
Jauame Mohammed, Iran
Jauslin Hans-Rudolf, France
Jaworowski Jan, USA
Jeanquartier Pierre, Switzerland
Jellouli Habib, France
Jeltsch Rolf, Switzerland
Jensen Arne, Sweden
Jensen Christian U., Denmark
Jerison David, USA
Jerison Meyer, USA
Jerrum Mark, UK
Jeurnink Gerard A.M., The Netherlands
Ježková Jana, Sweden
Jha Basant Kumar, India
Jiang Mei-Yue, China P.R.
Jiang Yunping, USA
Jiayu Li, China P.R.
Jimbo Shuichi, Japan
Jing Naihuan, USA
Johannson Klaus, USA
Johansson Bo I., Sweden
Johnson Roy A., USA
Johnson William B., USA
Johnstone Peter T., UK
Jones Vaughan, USA
Jordan Tibor, Hungary
Jori Alessandro, Switzerland
Joris Henri, Switzerland
Jost Jürgen, Switzerland
Joyet Pierre, Switzerland
Juhnke Friedrich, Germany
Jun Hu, USA
Just Andrzej, Poland
Kaarli Kalle, Estonia
Kadison Lars, Denmark
Kadison Richard V., USA
Kaftal Victor, USA
Kahane Jean Pierre, France
Kahn Donald W., USA
Kahn Jeffry N., USA
Kakie Kunio, Japan
Kakiichi Yoshiaki, Japan
Kalai Gil, Israel
Kälin Markus, Switzerland
Kalkbrener Michael, Switzerland
Källström Rolf, Sweden
Kalyabin Gennadiy A., Russia
Kamada Masaru, Switzerland
Kamalian Armen, Armenia
Kamenski Mikhail, Russia
Kamimura Yutaka, Japan
Kamowitz Herbert, USA
Kanas Stanislawa, Poland
Kaneko Makoto, Japan
Kaneko Masanobu, Germany
Kang Ming-Chang, Taiwan
Kang Seok-Jin, Korea
Kankaanrinta Marja Kaarina, Finland
Kantor Jean-Michel, France
Kapouleas Nicolaos, USA
Kappeler Thomas, USA
Kappos Efthimios, UK
Kapur Aruna, India
Karagulian Grigor, Armenia
Karamzadeh O.A.S., Iran
Karapetyan Arman, Armenia
Karbe Manfred, Germany
Karlovich Yuri, Ukraine
Karno Zbigniew, Poland
Károlyi Gyula, Germany
Karoubi Max, France
Karpenko Nikita, Russia
Kashiwabara Takuji, France
Kashiwada Toyoko, Japan
Kassay Gábor, Romania
Katase Kiyoshi, Japan
Kato Mikio, Japan
Katona Gyula Y., Hungary
Katori Makoto, Japan
Katsov Jefim, USA
Katsuda Atsushi, Japan
Katsurada Hidenori, Japan
Kaul Saroop, Canada
Kaup Burchard, Switzerland
Kaup Ludger, Germany

Kawaguchi Hirokai, Japan
Kawahigashi Yasuyuki, Japan
Kawasaki Takeshi, Japan
Kawohl Bernhard, Germany
Ke Wen-Fong, Taiwan
Kearton Cherry, UK
Kegel Otto, Germany
Keleti Tamás, Hungary
Keller Bernhard M., France
Keller Hans Arwed, Switzerland
Keller Joseph, USA
Keller Philipp, Switzerland
Kellerhals Ruth, Germany
Kersten Ina, Germany
Kerzman Norberto L., USA
Khammash Ahmed A., Saudi Arabia
Kharin Stanislav, Kazakhstan
Kharlampovich Olga, Canada
Khatoonabadi Mahmood, Iran
Khimchiachvili Georgi, Georgia
Khisamiev Nazif, Kazakhstan
Khorunzhy Alexei M., Ukraine
Khosrovshahi Gholamreza B., Iran
Khruslov Eugene Y., Ukraine
Kiechle Hubert, Germany
Kiener Bernhard, Switzerland
Kienle Lothar, Germany
Kiesel Rüdiger, Germany
Kikuchi Isamu, Japan
Kikuchi Kazunori, Japan
Kikuchi Shigeki, Ireland
Kilpatrick Jeremy, USA
Kim Young Ho, Korea
Kimoto Yasushi, Japan
Kimura Noriaki, Japan
Kirchberg Eberhard, Germany
Kirchgraber Urs, Switzerland
Kirichenko Vadim F., Russia
Kirkovits Magdalen Sz., Hungary
Kirwan Frances, UK
Kishimoto Akitaka, Japan
Kishimoto Kazuo, Japan
Kissin Edward V., UK
Kitada Yasuhiko, Japan
Klamka Jerzy, Poland
Klassen Eric P., Germany
Kleisli Heinrich, Switzerland
Klimentov S.B., Russia
Klingenstein Petra, Switzerland
Klotzek Benno, Germany
Klüppelberg Claudia, Switzerland
Knill Oliver, USA
Knopfmacher Arnold, South Africa
Knorrenschild Michael, Germany
Knörrer Horst Hans, Switzerland
Knus Max-Albert, Switzerland
Kobayashi Keiko, Japan
Kobayashi Masanori, France
Kobayashi Toshiyuki, Japan
Koch Helmut, Germany
Kochman Stanley O., Canada
Kodama Tetsuo, Japan
Kogiso Takeyoshi, Japan
Kohatsu-Higd Arturo, Mexico
Kohayakawa Yoshiharu, Brazil
Köhler Franz, Germany
Köhnen Walter, Germany
Kohr Gabriela, Romania
Kolaneci Fejzi, Albania
Kolk Johan A., The Netherlands
Kolodiazhny Volodimir M., Ukraine
Kolomiets Yuriy V., Ukraine
Kolountzakis Mihail N., USA
Kolyada Sergeiĭ F., Ukraine
Komatsu Hikosaburo, Japan
Komornik Vilmos, France
Konderak Jerzy Julian, Italy
Kondrat'eva M.F., Russia
König Steffen, Germany
König Wolfgang, Switzerland
Konnov Valery, Russia
Konstantinova Elena, Russia
Kontsevich Maxim, Germany
Kopanskii Alexander, Moldova
Korenblum Boris, USA
Korobov Valery, Poland
Korolyuk Vladimir, Ukraine
Kosarew Siegmund, France
Kosatchevsjaia Elena, Russia
Koschorke Ulrich, Germany
Koshi Shozo, Japan
Kosmann-Schwarzbach Yvette, France
Kostant Ann S., USA
Kostin Vladimir, Russia
Kostov Vladimir, France
Kostova Elena V., France
Kôta Osamu, Japan
Kotake Takeshi, Japan
Kotschick Dieter, Switzerland
Kotta Ülle, Estonia
Kovačec Alexander, Portugal
Kovács Katalin, Hungary

Kovalev Alexander M., Ukraine
Kovalevsky Alexander, Ukraine
Kowalski Oldřich, Czech Rep.
Krabbe Gregers, USA
Kraev Yegor, Switzerland
Kraft Hanspeter, Switzerland
Kraft Roger L., USA
Kramer Jürg, Switzerland
Krasoń Piotr, Poland
Krause Décio, Brazil
Kravchenko Viktor G., Ukraine
Kravchuk Helena V., Ukraine
Kravvaritis Dimitrios, Greece
Krein Selim G., Russia
Krengel Ulrich, Germany
Kriener Markus, Switzerland
Krikorian Raphaël, France
Krishnaswarni Alladi, USA
Kristensen Leif, Denmark
Kröger Pawel, Germany
Kruglikov Boris S., Russia
Kruzhkov Stanislav, Russia
Krzyż Jan G., Poland
Ktitarev Dmitrij, Russia
Kubilius Kestutis, Lithuania
Kubo Adisato, Japan
Kudaibergenov Kanat Zh., Kazakhstan
Kufner Alois, Czech Rep.
Kuku Aderemi O., Nigeria
Kulenovic M.R.S., Bosnia
Kulikov Victor S., Russia
Kulkarni Sudhir H., Germany
Külshammer Burkhard, Germany
Kumabe Masahiro, Japan
Kumahara Keisaku, Japan
Kumlin Peter, Sweden
Kündgen André, Germany
Künsch Hans, Switzerland
Künzi Hans, Switzerland
Künzi Hans-Peter, Switzerland
Künzi Urs-Martin, Switzerland
Künzle Alfred, Germany
Kuo Tsang-hai, Taiwan
Kuperberg Krystyna, USA
Kurbatov V.G., Russia
Kur'erov Yury N., Russia
Kuribayashi Akikazu, Japan
Kuribayashi Yukio, Japan
Kurshan Robert P., USA
Kurta Vasilii V., Russia
Kurusa Arpad, Hungary
Kusama Tokitake, Japan
Kushpel Alexander, Ukraine
Kusunoki Yukio, Japan
Kutev Nickolay D., Germany
Kutzko Philip C., USA
Kuwata Masato, France
Kwak Minkyu, Korea
Kytmanov Alexander, Russia
Laborde Colette, France
Laburta Maria P., Spain
Labuschagne Louis E., Switzerland
Lacomblez Chantal, France
Lacroix Yves, France
Ladyzhenskaya Olga A., Russia
Laffey Thomas, Ireland
Lahtinen Aatos, Finland
Lakatos Piroska, Hungary
Lakshmibai Venkairamani, USA
Lalli Bikkar Singh, Canada
Lambrinos Panos Th., Greece
Lamprecht Erich, Germany
Lance E. Christopher, UK
Landis Eugene, Russia
Landis Yelena, USA
Lanford Oscar E. III, Switzerland
Lang Urs, Germany
Langevin Rémi N., France
Lanini Flavia, Switzerland
Lannes Jean, France
Lap James T., USA
Lapidot Eitan, Israel
Lapidus Michel L., USA
Lapin Alexander V., Russia
Lara-Aparicio Miguel, Mexico
Lascaux Patrick M., France
Lau Anthony T.M., Canada
Lau Kee-Wai, Hong Kong
Laubenbacher Reinhard, USA
Läuchli Hans, Switzerland
Laumon Gérard, France
Laurinčikas Antanas, Lithuania
Lauter Robert, Germany
Lavoie Jean-Louis, Canada
Lawson H. Blaine, USA
Laywine Charles F., Canada
Lazar Aldo J., Israel
Lê Dũng Tráng, France
Le Gall Jean-François, France
Le Ngu Viet, Vietnam
Leahu Alexe, Moldova
Leble Sergei B., Russia

Leborgne Daniel, France
LeBrun Claude R., USA
Lecko Adam, Poland
Ledgard Ronald C., UK
Ledrappier François, France
Ledzewicz Urszula A., USA
Lehmann Detlef, Switzerland
Lehner Daniel, Switzerland
Lehrer Gustav, Australia
Lehto Olli, Finland
Leighton Tom, USA
Leite Maria-Luiza, Brazil
Leites Dimitry, Sweden
Lejeune-Jalabert Monique, France
Lemaire Lucrick, Belgium
Lempert László, USA
Leng W. S., UK
Lennard Christopher, USA
Leonenko Nikolay, Ukraine
Lermer Karl, Switzerland
Lescot Paul, France
Lester June A., Canada
Leupp Marcel, Switzerland
Leuzinger Enrico, Switzerland
Levashkin Sergey, Russia
Levesque Claude, Canada
Levin Genadi M., Germany
Levin Leonid A., Israel
Levitt Gilbert, France
Levner Eugene, Israel
Lewin Mordechai, Israel
Lewis Ira Wayne, USA
Lewis James D., Canada
Li Chun, Australia
Li Delang, China P.R.
Li Jian-Shu, USA
Li Jun, USA
Li Yanyan, USA
Li Yingchen, USA
Li Yun Feng, China P.R.
Libgober Anatoly S., USA
Lickorish W. B. Raymond, UK
Lieb Elliott, USA
Liebling Thomas M., Switzerland
Liedahl Steven, Israel
Lieman Daniel B., USA
Lifanov Ivan K., Russia
Liflyand Elijah R., Israel
Lima Suely, Brazil
Lindenstrauss Joram, Israel
Ling San, Singapore
Linton Fred E.J., USA
Lions Jacques Louis, France
Lions Pierre L., France
Lippus Jüri, Estonia
Littelmann Peter, Switzerland
Litvinchuk Georgue S., Portugal
Litvine Igor N., Lesotho
Liu Fon-Che, China P.R.
Liu Jinn-Liang, Taiwan
Liu Regina, USA
Liu Ying Ming, China P.R.
Lizama Carlos, Chile
Ljeskovac Milos, Switzerland
Lockhart Deborah F., USA
Loday Jean-Louis, France
Loday-Richaud Michèle, France
Loeffel Jean-Jacques, Switzerland
Logofet Dmitriĭ O., Russia
Loher Thomas, Switzerland
Lomonaco Luciano, Italy
Lomtatidze Alexander, Georgia
Longo Roberto, Italy
Lopez Molina Juan Antonio, Spain
Lorenzini Dino J., USA
Loreti Paola, Italy
Loring Terry A., USA
Lott John, USA
Lotto Ben, USA
Louveau Alain, France
Lovász László, USA
Lozanov-Crvenkovic Zagorka S., Yugoslavia
Luboobi Livingstone, Uganda
Lubotzky Alex, Israel
Luchko Yurii F., Belarus
Luchsinger Christof, Switzerland
Lück Wolfgang, Germany
Ludwig Garry, Canada
Luecke John, USA
Luik Eberhard, Germany
Lumiste Ülo, Estonia
Luna Charita A., Philippines
Luna D., France
Lundquist Michael, USA
Luo Feng, USA
Lupu Gheorghe, Romania
Luse Dzidra, Latvia
Luukkainen Jouni, Finland
Luzzatto Stefano, Italy
Lyashenko Andrej A., Italy
Lyubashenko Vladimir, UK
Lyubeznik Gennady, USA

Lyubich Mikhail, USA
Lyubich Yuri, Israel
Ma Fengshi, South Africa
Ma Li, China P.R.
Ma Wen-Xiu, China P.R.
Ma Wen-Xiu, Hong Kong
Ma Zhi-Ming, China P.R.
Mächler Martin, Switzerland
Macintyre A.J., UK
Madsen Ib H., Denmark
Maennel Hartmut, Germany
Magajna Bojan Peter, Slovenia
Magenes Enrico, Italy
Magid Andy R., USA
Magidor Menachem, Israel
Maier Helmut, Germany
Maiorov Vitaly, Israel
Maire Henri, Switzerland
Maistrenko Yuri L., Ukraine
Majadas Javier, Spain
Majda Andrew J., USA
Major Peter, Hungary
Makienko Peter M., Russia
Malamud Mark M., Ukraine
Malcolmson Peter, USA
Maley Miller F., USA
Malfait Wim, Belgium
Maliaukiene Livija, Lithuania
Maligranda Lech, Sweden
Mall Daniel, Switzerland
Maltsev Arkadii, Austria
Maltsiniotis Georges, France
Malyshev Peter V., Ukraine
Mammana Carmelo, Italy
Mampassi Benjamin, Senegal
Mandai Takeshi, Japan
Mandallaz Daniel M.E., Switzerland
Manders Kenneth L., USA
Manhas Jasbir Singh, India
Manolache Nicolae, Switzerland
Manturov Oleg V., Russia
Manuilov Vladimir M., Russia
Manz Ueli, Switzerland
Marafino John T., USA
Marano Miguel Antonio, Spain
Marcantognini Stefania A., Venezuela
Mardešić Sibe, Croatia
Margalef-Roig Juan, Spain
Margerin Christophe M., France
Marini Mauro, Italy
Márki László, Hungary
Marsden Jerrold E., Canada
Marti Jürg, Switzerland
Marti Kurt, Germany
Martignoni Nicolas, Switzerland
Martikainen A., Germany
Martínez Consuelo, Spain
Martínez-Finkelshtein Andrei, Cuba
Maruyama Fumitsuna, Japan
Maryukov Michaèl N., Russia
Mascioni Vania D., Switzerland
Massa Silvio, Italy
Masser David, Switzerland
Masuda Tekashi, Japan
Masuda Tetsuya, Japan
Masur Howard A., USA
Matano Hiroshi, Japan
Máté László, Hungary
Matevossyan Hovik, Russia
Mathew Tarek, USA
Mathieu Martin, Germany
Matic Gordana, USA
Matiyasevich Yuri V., Russia
Matravers David R., UK
Matsui Kiyoshi, Japan
Matsui Mai, Japan
Matsumoto Kazuko, Japan
Matsumoto Shigenori, Japan
Matsumoto Shiro, Japan
Matsumura Hideyuki, Japan
Matsushita Yasuo, Japan
Matsuto Mie, Japan
Matsuzaki Katsuhiko, Japan
Matteotti Carlo, Switzerland
Mattner Lutz, Germany
Matuszewski Roman, Poland
Matuura Takahide, Japan
Matvejchuk Marjan, Russia
Matzinger Heinrich, Switzerland
Matzinger Heinrich F., Switzerland
Maude Ronald, UK
Mauduit Christian, France
Maumary Serge, Switzerland
Mayer David, Mexico
Mayer Karl-Heinz, Germany
Mayorquin Jesus G., Mexico
Mazur Tomasz, Poland
McBride Adam C., UK
McCrory Clinton, USA
McCulloh Leon R., USA
McDonald Bernard R., USA
McGehee Richard P., USA

McIntyre D.W., New Zealand
McLaughlin David, USA
McLaughlin Joyce, USA
McNeil Alexander, Switzerland
Mdzinarishvili Leonard, Georgia
Mechenov Alexander, Russia
Medvedeva Natalija, Russia
Megginson Robert, USA
Megyesi Gábor, USA
Meier David, Switzerland
Meier Rolf Heinz, Switzerland
Meier-Wunderli Heinrich, Switzerland
Meister Andreas, Switzerland
Mel'nyk Taras A., Ukraine
Melnik Vyacheslav, Ukraine
Melville Duncan J., USA
Mendes Pedro, Brazil
Mercier Armel, Canada
Merino Sandro, Switzerland
Merkl Franz, Switzerland
Mestre Jean-François, France
Meszaros Csaba, Hungary
Meyer Johannes H., South Africa
Meyer Yves F., France
Meyers Leroy F., USA
Michael Ernest, USA
Michaelis Walter, USA
Michelsohn Marie-Louise, USA
Michelucci Marialetizia, Italy
Michor Peter Wolfram, Austria
Mihai Ion, Romania
Mikaelian Hamlet S., Armenia
Mikhailyuk Evgeniy A., Ukraine
Mikhalev Alexander, Russia
Mikhalkin Grigory B., Germany
Mikulska Margaret, USA
Millett Kenneth C., USA
Milman Vitali, Israel
Milnes Paul, Canada
Milnor John W., USA
Min Lequan, China P.R.
Minda David C., USA
Mio Washington, USA
Misaki Norihiro, Japan
Mishura Yuliya S., Ukraine
Mislin Guido, Switzerland
Misra Kailash C., USA
Mitsumdatsu Yoshihiko, France
Mittal M.L., India
Miura Takeo, Japan
Miyadera Isao, Japan
Miyaoka Yoichi, Japan
Miyaura Suga, Japan
Miyazaki Kenichi, Japan
Mizutani Tadayoshi, Japan
Mochizuki Masaya, Japan
Mohammadi Hassanabadi A., Iran
Mohebi Hossein, Iran
Moine Stéphane, Switzerland
Mok Ngaiming, France
Mokhov Oleg I., Russia
Molchanov Vladimir F., Russia
Moll Victor, Switzerland
Monk James D., USA
Montano Roberto F., Guatemala
Montiel Sebastian, Spain
Moon John W., Canada
Mooney John W., UK
Moore Charles N., USA
Moore Gregory, USA
Moore Hal G., USA
Morales Bernardo René, Guatemala
Morales Jorge, USA
Mordasini Francesco, Switzerland
Moreira Carlos G., Brazil
Morel Anne-Thérèse, Switzerland
Morelli Robert, USA
Móricz Ferenc, Hungary
Morimoto Hiroko, Japan
Morimoto Kanji, Japan
Morimoto Mitsuo, Japan
Moritoh Shinya, Japan
Moriyoshi Hitoshi, Japan
Moroz B.Z., UK
Morozov Andrei S., Russia
Morris Sidney A., Australia
Morrison David, USA
Moser Jürgen, Switzerland
Moser-Jauslin Lucy, France
Motoko Kotani, Germany
Mountford Thomas S., USA
Moussavi Ahmad, Iran
Moustafa Magdi Sami, Egypt
Mouton Hendrik D., South Africa
Mrowka Tomasz S., USA
Mukai Juno, Japan
Müller Franz, France
Muller Paul, Austria
Mumford Davis B., USA
Murakami Haruo, Japan
Murakami Hitoshi, Japan
Muro Masakdzu, Japan

Murolo Claudio, France
Murray Margaret, USA
Murthy M. Pavamam, Switzerland
Mustafin Tulendiy, Kazakhstan
Muster Stefan, Switzerland
Myslivets Simona, Russia
Näätänen Marjatta, Finland
Naevdal Geir, Norway
Naffah Nadim, Switzerland
Nagahara Takashi, Japan
Nagai Osamu, Japan
Nagasawa M., Switzerland
Nagayama Haruya, Japan
Naguibeda Tatiana, Switzerland
Nagura Toshinobu, Japan
Naito Hirotada, Japan
Najman Branko, Croatia
Nakai Toru, Germany
Nakamura Kirio, Japan
Nakamura Yoshio, Japan
Nakanishi Shizu, Japan
Nakanishi Yasutaka, Japan
Nakatsuka Harunori, Japan
Nakayama Eiji, Japan
Nanbu Tokumori, Japan
Nanthakumar Ampalavanar, USA
Naomitsu Kitani, Japan
Narasimhan Mudumbai S., Italy
Narayanaswami P.P., Canada
Nasim Cyril, Canada
Nassopoulos Georgios F., Greece
Nasyrov Samyon R., Russia
Natanzon Sergei M., Russia
Natchev Nako A., Bulgaria
Naumkin Pavel I., Russia
Navarro Milagros P., Philippines
Nayatani Shin, Germany
Nazarova Ludmila, Ukraine
Nebres Bienvenido F., Philippines
Nedoma Jiří, Czech Rep.
Neeman Amnon, USA
Nehaniv Chrystopher L., Japan
Nenashev Alexandre, Russia
Nesetril Jaroslav, Czech Rep.
Netsvetaev Nikita Yu, Russia
Neuberger Barbara, USA
Neuberger John W., USA
Neuenschwander Daniel, Switzerland
Neuenschwander Erwin, Switzerland
Neumann Bernhard H., Australia
Neumann-Coto Max, Mexico
Neustadter Siegfried, USA
Newman Charles M., USA
Nezit Pierre, Ivory Coast
Ng Chi-Keung, UK
Nguyen Thanh Hai, Russia
Nguyen Tien Dung, Italy
Nguyen Tri Dinh, Vietnam
Niango Donatien, Ivory Coast
Niçka Dhimitraq, Albania
Niculescu Constantin, Romania
Nikichine Nicolas, France
Nikoltjeva-Hedberg Margarita, Sweden
Ninnemann Olaf, Germany
Nipp Kaspar, Switzerland
Nisan Noam, Israel
Nishida Koji, Japan
Nishimura Kazuko, Japan
Nishiyama Akishige, Japan
Niss Mogens, Denmark
Nohel John A., Switzerland
Noma Atsushi, Japan
Nori Madhav V., USA
Norton Alec, USA
Novák Břetislav, Czech Rep.
Nowak Krzysztof, Poland
Nowakowski Andrzej, Poland
Nummelin Esa, Finland
Octavio Alfredo, Venezuela
Odell Edward, USA
Odlyzko Andrew M., USA
Ōeda Kazuo, Japan
Oertel Frank, Switzerland
Oezekes Hasan, Turkey
Ogata Hayao, Japan
Ogata Yoshiko, Japan
Oh Yong-Geun, USA
Ohba Kiyoshi, Japan
Ohezin Sergei P., Russia
Ohlsson Bonnie, Sweden
Ohmiya Mayumi, Japan
Ohno Shûichi, Japan
Ohsawa Takeo, Japan
Oinarov Ryskul, Kazakhstan
Oka Hirokzu, Japan
Okada Tatsuya, Japan
Okecha George E., Zimbabwe
Okoh Frank, USA
Okoya Samuel S., Nigeria
Okubo Tokuyuko, Japan
Okumura Hirozo, Japan
Okuyama Yukihiko, Japan

Olbrich Martin, Germany
Olevskii Victor A., Israel
Olmos Carlos, Argentina
Omarjev Moubinool, France
Onita Yoshihiro, Japan
Onn Uri, Israel
Ono Kaoru, Japan
Ono Ken, USA
Ooe Takeshi, Germany
Orazov Mered, Turkmenistan
Orlov Vladimir, Russia
O'Shea Donal B., USA
Osher Stanley, USA
Osmatescu Petre, Moldova
Osmolovskii Nikolai P., Russia
Ossé Akimou, Switzerland
Osterwalder Konrad, Switzerland
Otsuka Kayo, Japan
Otsuka Kenichi, Japan
Otsuki Tominosuke, Japan
Ouchi Osamu, Japan
Ould Jidoumou Ahmedou, Mauritania
Ovchinnikov Vladimir, Russia
Pagliara Vito, Switzerland
Pagurova Vera, Russia
Palanques-Mestre August, Spain
Palis Jacob, Brazil
Pan Jie, USA
Pandey Jagdish N., India
Pandian R.D., USA
Panov Evgeniy, Russia
Pantoja José, Chile
Panyushev D.I., Germany
Papadopol Peter V., USA
Papakonstantinou Vassilios, Greece
Papanicolaou Vassilis G., USA
Parameswaran Aryampilly, India
Parameswaran Sankaran, India
Pardoux Etienne, France
Parimala Raman, India
Park Chan-Young, Korea
Parks Harold R., USA
Parmár Vijay, Italy
Parmenter M., Canada
Parmentier Serge V., France
Parshall Karen, USA
Parthasarathy Kalyanapuram, India
Partova Edita, Slovakia
Parusnikov Vladimir, Russia
Păsărescu Maria Angela, Hungary
Pasarescu Ovidiu, Romania
Paseman Gerhard R., USA
Passare Mikael, Sweden
Paul Pedro, Spain
Pavlov Igor V., Russia
Pawalowski Krzysztof M., Poland
Paxia Guiseppe, Italy
Pecher Hartmut, Germany
Pehlivan Serpil, Turkey
Pekonen Osmo, Finland
Pelczar Andrzej M., Poland
Pellegrinelli Andrea Fernando, Switzerland
Peller Vladimir, USA
Perelman Grigori, USA
Peric Veselin, Yugoslavia
Perišić Vesna, Germany
Perkins Ed, Canada
Perotti Alessandro, Italy
Perrin-Riou Bernadette, France
Perthame Benoît, France
Peschke George, Canada
Peterburgsky Irina, USA
Petersen Wesley P., Switzerland
Peterson Christopher S., USA
Petridis Yiannis N., Germany
Petro Petraq, Albania
Pettersson Rolf, Sweden
Petz Dénes, Hungary
Pfister Albrecht, Germany
Pfluger Pia, The Netherlands
Philippin Gérard A., Canada
Phillips Christoper N., USA
Phong Duong H., USA
Piene Ragni, Norway
Pier Jean-Paul, Luxembourg
Pierantoni Margherita, Switzerland
Pietrocola Norma C., Argentina
Pilipchuk Valery N., Ukraine
Pillay Anand, USA
Pinchover Yehuda, Israel
Piot Michel, Switzerland
Pittet Christophe, France
Pittie Harsh V., India
Pituk Michal, Hungary
Pizzo Aldo Bruno, Argentina
Planche Pierre, Switzerland
Plantiko Rüdiger, Switzerland
Platonov Vladimir, Canada
Pohl Bert, Switzerland
Pohozaev Stanislav, Russia
Pointet François, Switzerland
Poláčik Peter, Slovakia

Polianski Anatoli, Russia
Pollak Henry O., USA
Pollington Andrew D., USA
Polombo Albert, France
Pomerance Carl, USA
Pop Marius, Romania
Popa Sorin T., USA
Popescu Florica V., Romania
Popescu Mihai E., Romania
Popov Andrey G., Russia
Pöppe Christoph, Germany
Portenko Nikolai, Ukraine
Porteous Hugh L., UK
Porubský Štefan, Czech Rep.
Potthast John, USA
Poulkou Anthippi, Greece
Pourkazemi Mohammad H., Iran
Povel Tobias, Switzerland
Precup Radu, Romania
Preissler Gabi, Germany
Prikarpats Anatoli, USA
Prokhorov Dimitrii, Russia
Prokhorov Vasili A., Belarus
Prokhorov Yuri G., Russia
Proppe Harald, Canada
Pudlák Pavel, Czech Rep.
Puel Jean-Pierre, France
Puppe Voler, Germany
Pustyl'nikov Lev D., Russia
Puusemp Peeter, Estonia
Qing Jie, USA
Quadrat Jean Pierre, France
Quagebeur Johan, Belgium
Quine Jack R., USA
Quoc Thang Le Tu, Italy
Racke Reinhard, Germany
Rademacher Hans-Bert, Germany
Radjabalipour Mehdi, Iran
Radulescu Florin, USA
Radyno Yakov V., Belarus
Radzievskii Grigori, Ukraine
Raghavan K.- N., Italy
Raitums Uldis, Latvia
Rajabov Nusrat, Tajikistan
Rakhimberdiev Marat, Kazakhstan
Rákosník Jiří, Czech Rep.
Ram Arun, USA
Ramanaiah G., India
Ramanan Sundararaman, India
Ramanathan Harshavardhan, UK
Ramírez-Alfonsin Jorge, France
Ramm Alexander G., USA
Rao Bopeng, France
Raouyane Mohammed, Maroc
Rapoport Michael, Germany
Rappoport Juri M., Russia
Rassias Themistocles M., Greece
Ratner Marina, USA
Raugel Genevieve, France
Raychowdhury Pratip N., USA
Razzhevaikin Valerii N., Russia
Recalde Marco V.C., Ecuador
Recamán-Santos Bernardo, Colombia
Redjel Saïd, Germany
Reed Jon, Norway
Rees S. Mary, UK
Reifenberg Michèle, Switzerland
Reiher Gesine, Germany
Reimann Hans-Martin, Switzerland
Reinert Gesine, Switzerland
Reinfelds Andrejs, Latvia
Rejali Ali, Iran
Ren Jiagang, China P.R.
Rennenkampf J., Switzerland
Rentsen Enhbat, Mongolia
Repovš Dušan, Slovenia
Ressel Paul, Germany
Reutenauer Christophe, Canada
Rhodius Adolf, Germany
Riahi Hasna, Tunesia
Ricardo Perez-Marco, France
Richards Donald, USA
Rickman Seppo U., Finland
Riedl Reinhard, Switzerland
Riedtmann Christine, Switzerland
Rieffel Eleanor G., USA
Rieffel Marc A., USA
Riesel Hans I., Sweden
Ringrose John R., UK
Rips Eliyahu, USA
Risler Emmanuel, France
Ritter Gunter, Germany
Rivera Ortun Maria José, Spain
Rivkind Valeri Jakob, Russia
Rizou Eleni, Greece
Roan Shi-Shyr, Taiwan
Robbiani Marcello, Switzerland
Robbins Neville, USA
Robert Alain M., Switzerland
Robert Raoul, France
Roberts Leslie G., Canada
Roberts Paul C., USA

Robinson Margaret M., USA
Robson J. Chris, UK
Roche Claude, France
Roczen Marko, Germany
Rodin Vladimir, Russia
Roessler Damian, France
Roever Frédéric, Switzerland
Roiter Andrey, Ukraine
Rokhlin Vladimir, USA
Rokyta Mirko, Czech Rep.
Rolfsen Dale, Canada
Römisch Werner, Germany
Rooney Paul G., Canada
Roos Jan-Erik, Sweden
Roseman Joseph J., Israel
Rosenknop John Z., Germany
Rosinger Elemér E., South Africa
Rossmann Wulf, Canada
Rothberger Fritz, Canada
Rothstein Mitchell, USA
Rousseau Christiane, Canada
Roux Andre D., France
Rovenskij Vladimir Y., Russia
Rovinsky Olga, Israel
Roy Rahul, India
Ru Min, USA
Rubin Boris, Israel
Rubinstein Joachim Hyam, Australia
Rubinstein Zalman, Israel
Ruciński Andrzej, Poland
Rudakov Alexei N., Russia
Rudin Mary, USA
Rudin Walter, USA
Rudolph Mynhard, South Africa
Ruedin Laurent, Switzerland
Rueff Marcel, Switzerland
Ruggiero Rafael, Mexico
Ruh Ernst, Switzerland
Ruiz Ceferino, Spain
Ruiz-Soto Galo, Mexico
Rummler Hansklaus, Switzerland
Rumyantsev V.V., Russia
Rung Josef, Germany
Runovskiĭ Konstantin V., Ukraine
Ruoff Dieter, Canada
Rüttimann Gottfried T., Switzerland
Rvachev Vladimir, Ukraine
Ryan Frank B., USA
Ryff John V., USA
Saad Germaine H., USA
Saakian Artur, Armenia
Sad Paulo R., Brazil
Sadosky Cora, USA Sadov Sergey, Russia
Safuanov Ildar S., Russia
Sakai Akira, Japan
Sakai Shoichiro, Japan
Sakata Hiroshi, Japan
Sălăgean Grigore, Romania
Salamon Dietmar A., UK
Salberger Per, Switzerland
Salem Eliane, France
Salinas Oscar M., Argentina
Salinger David L., UK
Samoilenko Anatolij M., Ukraine
Samovol Vladimir, Russia
Sangare Daouda, Ivory Coast
Sanjari-Farsipour Nahid, Iran
Santilli Ruggero Maria, USA
Sanugi Bahrom B., Malaysia
Sanz-Serna J.M., Spain
Saphar Pierre David, Israel
Sarychev Andrej, Portugal
Sasano Kazuhiro, Japan
Sasao Seiya, Japan
Sastry Pramathanath, India
Satake Ichiro, Japan
Sato Ken-Iti, Japan
Sato Yumiko, Japan
Satoh Yuukiti, Japan
Sauerberg James, USA
Sawada Ken, Japan
Sawada Toshio, Japan
Sawae Ryuichi, Japan
Saxena Subhash, USA
Saxl Jan, UK
Sbordone Carlo, Italy
Scarmiglione Pippo, Switzerland
Schaefer Philip W., USA
Schafer Alice T., USA
Schafer Richard D., USA
Schätti Guido, Switzerland
Schechter Samuel, USA
Scheiderer Claus, Germany
Scherback Inna G., Russia
Scherer Beat, Switzerland
Scherer Jérôme, Switzerland
Schiffels Gerhard, Germany
Schimmerling Ernest, USA
Schlatter Andreas, Switzerland
Schleicher Dierk, Germany
Schlichenmaier Martin, Germany
Schlichting Günter H., Germany

Schmalz Gerd, Germany
Schmeelk John, USA
Schmid Rudolf, USA
Schmidt Manfred, Switzerland
Schmidt Stefan Erik, Germany
Schmidt Werner, Germany
Schmitt Alexander, Switzerland
Schmock Uwe, Switzerland
Schmutz Paul, Switzerland
Schnare Paul, USA
Schneebeli Hans Rudolf, Switzerland
Schneider Joel E., USA
Schneider Michael, Germany
Scholz Erhard, Germany
Schoenhage Arnold, Germany
Schott Dieter, Germany
Schreiber Bertram M., USA
Schroeder Viktor, Switzerland
Schulze Bert-Wolfgang, Germany
Schultens Jennifer Carol, USA
Schulz-Baldes Hermann, France
Schulz-Rinne Carsten, Switzerland
Schuppli Reto, Switzerland
Schwaller Thomas E., Germany
Schwartz Alan L., USA
Schwartzman Efim, Israel
Schwarz Gerry, USA
Schwarz Matthias, Switzerland
Schwarz Willi, Germany
Schwarz Wolfgang, Germany
Scott Rich, USA
Seddighi Karim, Iran
Sedykh Vyacheslav, Russia
Seibert Peter, Mexico
Seidel J.J., The Netherlands
Seifullaev Rustam K., Azerbajan
Seiler Wolfgang K., Germany
Sekiguchi Takeshi, Japan
Selfridge John L., USA
Seligman George B., USA
Selinger V.-M., Romania
Semenov Evgeny, Russia
Semmes Stephen W., France
Semmler Klaus-Dieter, Switzerland
Sen Dipak K., Canada
Senderovizh Rivka, Israel
Sendov Blagovest, Bulgaria
Senn Walter M., Switzerland
Sentenac Pierrette, France
Senti Samuel, Switzerland
Seregin Gregory, Russia
Sergiescu Vlad, France
Serre Jean-Pierre, France
Sesay Thamba W., USA
Seshadri Conjevaram S., Italy
Sethi Pannalal, India
Sevennec Bruno, France
Sevenster Arjen, The Netherlands
Seydi Hamet, Senegal
Seymour Paul, USA
Shadwick William, Canada
Shahvarani-Semnani Ahmad, Iran
Shaidenko Künzi Anna V., Switzerland
Shalev Aner, Israel
Shamolin Maxim V., Russia
Shandra Igor Georgevich, Russia
Shaneson Julius L., USA
Shapiro Michael V., Mexico
Sharma Rajendra K., India
Shatah Jalal, USA
Shatz Stephen, USA
Shaw Ronald, UK
Shaw Sen-Yen, Taiwan
Sheinman Oleg K., Russia
Shen Lian, Switzerland
Shepelsky Dmitry G., France
Shevchenko Aleksandr N., Ukraine
Shi Wujie, China P.R.
Shibano Hiroki, Japan
Shigeru Kobayashi, Japan
Shima Kazuhisa, Japan
Shimada Nobuo, Japan
Shimizu Yuji, France
Shimoda Yasuhiro, Japan
Shimomura Katsunori, Japan
Shintani Toshitada, Japan
Shiraiwa Kenichi, Japan
Shiratani Katsumi, Japan
Shishikura M., Japan
Shishkov Andrey E., Ukraine
Shishmarev I. A., Russia
Shitanda Yoshimi, Japan
Shoshitaishvili Alexander, Russia
Shtikonas Artueras, Lithuania
Shu Chi-Wang, USA
Shu Linghsueh, USA
Shubin Mikhail, USA
Shum Kar-Ping, Hong Kong
Siburg Karl Friedrich, Switzerland
Sidoravicius Vladas, Brazil
Siebenmann Laurent C., France
Sierpinska Anna, Canada

Sieveking Malte, Germany
Sigrist François, Switzerland
Silva Ana Maria, Spain
Silvestrov Dmitrii S., Sweden
Silvestrov Sergei D., Sweden
Simbirsky Mikhail N., Ukraine
Simon Károly, UK
Simonovits Miklós, Hungary
Simson Daniel, Poland
Sinaceur Hourya, France
Sinclair Robert, Switzerland
Singh Dinesh, India
Siu Man Keum, Hong Kong
Sjögren Peter, Sweden
Skjelbred Tor, Norway
Sklyar Grigory M., Poland
Skorobogatov Alexei N., Russia
Skrypnik Igor, Ukraine
Skula Ladislav, Czech Rep.
Slade Gordon, Canada
Slapničar Ivan, Croatia
Smirnov Georgi, Italy
Smirnov Mikhail, USA
Smith Karen E., USA
Smith Perry B., USA
Smith Stuart P., USA
Soares Marcio Gomes, Brazil
Sochacki James S., USA
Soergel Wolfgang, Germany
Sogge Christopher D., USA
Soibelman Yan, Germany
Söllner Hildegard, Germany
Solodov Victor V., Russia
Sologuren Santiago, Bolivia
Solynin Alexander, Russia
Sonin Isaac, USA
Sonn Jack, Israel
Sontag Eduardo D., USA
Sós Vera, Hungary
Soto-Andrade Jorge, Chile
Sottile Frank, Canada
Souganidis Panagiotis E., USA
Soulé Christophe, France
Spain Philip G., Israel
Spatzier Ralf, USA
Spencer Joel H., USA
Spera Mauro, Italy
Spies Jakob, Switzerland
Spitkovsky Ilya, USA
Spruck Joel, USA
Srebro Uri, Israel
Sridharan Ramaiyengar, India
Srinivasan Hema, India
SrisatkunaraJah Sivakolundu, Sri Lanka
Srivastava T. N., Canada
Stagnaro Ezio, Italy
Stahl Herbert, Germany
Stakgold Ivar, USA
Stammbach Urs, Switzerland
Stan Ioan, Romania
Stangler Sieghart S., Germany
Stanik Rotraut, Germany
Stankiewicz Jan , Poland
Starkov Alexander, Russia
Starr Edith, USA
State Luminita, Romania
Stefanescu Mirela, Romania
Stefánsson Jón R., Iceland
Steffen Eckhard, Germany
Steger Tim Joshua, USA
Steiner Philippe A.-J., Switzerland
Steiner Richard J., UK
Steiner Walter J., Switzerland
Steinhorn Charles I., USA
Stephenson Kenneth R., USA
Sterk Henri J., The Netherlands
Sternheimer Daniel, France
Stiller Peter F., USA
Stillwell John C., Australia
Stoffer Daniel, Switzerland
Stojanovic Mirjana Novica, Yugoslavia
Stolz Stephen A., USA
Storrer Hans Heiner, Switzerland
Storvick David A., USA
Stotzer Marie-Odile, Switzerland
Strassen Volker, Germany
Stratigos Peter D., USA
Stray Arne, Norway
Strebel Kurt, Switzerland
Strebel Ralph, Switzerland
Stroethoff Karel M., USA
Strooker Jan R., The Netherlands
Struwe Michael, Switzerland
Strygin Vadim V., Russia
Studer Gerold, Switzerland
Su Meiyu, USA
Suda Hiroshi, Japan
Sueyoshi Yutaka, Japan
Suez Tiferet M., Israel
Suger Eduardo J., Guatemala
Suhov Yuri, UK
Sujatha Ramdurai, India

Sukhinin Eugene, Russia
Sukla Indu Lata, India
Suleiman Mohamed B., Malaysia
Sultangazin Umirzak, Kazakhstan
Sun Jialin, China P.R.
Sund Terje, Norway
Sundararaman Duraiswamy, USA
Surchat Daniel, Switzerland
Suresh Venapally, India
Suslin Andrei, Russia
Suter Ruedi, Switzerland
Suter Ueli, Switzerland
Suwa Noriyuki, Japan
Suzuki Michio, USA
Suzuki Nyoyoshi, Japan
Suzuki Shinichi, Japan
Suzuki Yuki, Japan
Svaiter Benar Fux, Brazil
Šverák Vladimír, USA
Swanson Irena, USA
Swishchuk Anatoly, Ukraine
Sychev Michail, Russia
Syta Halina, Ukraine
Szabó Tibor, Hungary
Szabo Zoltan I., Botswana
Szalay Michael, Hungary
Szanto Agnes, USA
Szarek Stanislaw J., USA
Szczepanski Susan, USA
Szigeti Zoltán, Hungary
Sznitman Alain-Sol, Switzerland
Tabachnikov Serge, USA
Taflin Erik, France
Taguchi Yuichiro, USA
Tahvildar-Zadeh Abdolreza S., USA
Taimina Daina, Latvia
Takahashi Reiji, Japan
Takahdshi Sechiko, Japan
Takashima Keizo, Japan
Takebayashi Tadayoshi, Japan
Takei Yoshitsugu, Japan
Takemoto Hideo, Japan
Takenouchi Osamu, Japan
Takeo Fukushi, Japan
Takesaki Masamichi, USA
Takeuchi Jiro, Japan
Takeuchi Kiyoshi, Japan
Takizawa Seiji, Japan
Takizawa Takenobu, Japan
Tali Anne, Estonia
Tall David, UK
Talvacchia Janet, USA
Tamamura Akie, Japan
Tammer Elisabeth-Ch., Germany
Tamrazov Promarz M., Ukraine
Tan Audrey M., UK
Tanaka Hiroshi, Japan
Tanasi Corrado, Italy
Tanemura Hideki, Switzerland
Tang Zi-Zhou, China P.R.
Tangmanee Suwon, Thailand
Tankeev Sergueir, Russia
Tapia Cesar G., Philippines
Tapia Joseph A., France
Targonski György, Switzerland
Targonski Jolán H., Switzerland
Tasaki Hiroyuki, Japan
Taubes Clifford, USA
Tauts Ants, Estonia
Tavgen Oleg I., Belarus
Taylor Laurence R., USA
Taylor Richard, UK
Taylor S. James, USA
Tedeev Anatoly F., Ukraine
Teicher Mina, Israel
Teleman Andrei, Switzerland
Terai Nobuhiro, Japan
Terng Chuu-Lian, USA
Terrier Jean-Marc, Canada
Tersian Stepan A., Bulgaria
Test Peter, Germany
Testerman Donna M., Switzerland
Theis Antoinette, France
Thomi Peter, Switzerland
Tian Youliang, USA
Tiep Pham Huu, Germany
Tietz Horst, Germany
Timoney Richard M., Ireland
Timourian James G., Canada
Tita Nicolae, Romania
Titani Sakoto H., Japan
Titi Edriss S., USA
Tits Jacques L., France
Tkachev Vladimir G., Russia
Tobin Seán, Ireland
Togari Yoshio, Japan
Tognola Diego, Switzerland
Tokizawa Masamichi, Japan
Tokunaga Kiyohisa, Japan
Toledano Laredo Valerio, UK
Tomescu Rodica T., Romania
Tomida Masamichi, Japan

Tominaga Yasuo, Japan
Tomiyama Jun, Japan
Ton-That Tuong, USA
Tonchev Vladimir D., USA
Tondeur Philippe M., USA
Toomire Bruce V., USA
Topsoe Flemming, Denmark
Toró Tiberius, Romania
Torres Fernando E., Italy
Torres Rodolfo H., USA
Totaro Bert, USA
Toure Saliou, Ivory Coast
Tralle Alexei, Poland
Travesa Artur, Spain
Trefethen Lloyd N., Switzerland
Trenčevski Kostadin, Macedonia
Treshchev Dimitry V., Russia
Triulzi Mauro, Switzerland
Trochold Heinz, Germany
Troger Hans, Austria
Troitskaya Saule D., Russia
Troitsky Evgenii V., Russia
Tronel Gérard H., France
Troyanov Marc, Switzerland
Trubowitz Eugene, Switzerland
Tsang Kai-Man, Hong Kong
Tsarkov Yevgeny F., Latvia
Tsatsanis Peter, Canada
Tsikalenko Tatiana V., Ukraine
Tsuboi Shoji, Japan
Tsuboi Takashi, Japan
Tsuchiya Nobuo, Japan
Tukia Pekka, Finland
Tuncali Huseyin M., Canada
Tuomela Jukka, Finland
Turo Jan, Poland
Turquette Atwell R., USA
Turtschi Adrian, Switzerland
Tuzhilin Alexei A., Russia
Tylli Hans-Olav J., Finland
Tzafriri Lior, Israel
Udrişte Constantin, Romania
Uetake Yoichi, Japan
Ugochukwu Uko L., Nigeria
Ukon Michihisa, Japan
Ulecia Maria T., Spain
Ullrich Peter, Germany
Umezu Yumiko, Japan
Upmeier Harald O., Germany
Ures Raul M., Uruguay
Ursu Ioan, Romania
Vaarmann Otu, Estonia
Valdivia Oscar, Spain
Valere Bouche Liane G., France
Valette Alain, Switzerland
Valibouze Annick A.F., France
Van den Bergh Michel, Belgium
Van der Kallen Wilberd, The Netherlands
Van der Waall Robert W., The Netherlands
Van der Walt Andries P., South Africa
Van der Weide Hans, The Netherlands
Van Geel Jan, Belgium
Van Mouche Pierre H.M., The Netherlands
Van Nooyen Ronald R.P., The Netherlands
Vandevoorde Robert, Belgium
Vaněček Antonín, Czech Rep.
Vanini Paolo, Switzerland
Varadarajan Veeravalli S., USA
Varadhan S.R.S., USA
Vasil'ev Alexander, Russia
Vasilevski Nikolai, Mexico
Vasiliev Victor, Russia
Vavilov Nikolai A., Germany
Venema Gerard A., USA
Veretennikov Alexander, Russia
Vershik Anatoly M., Russia
Vershinin Vladimir, Spain
Viana Marcelo, Brazil
Vichmann Frederik, Estonia
Viharos Laszlo, Hungary
Vila-Freyer Ricardo, Mexico
Villani Vinicio, Italy
Villella José Agustin, Argentina
Violette Donald, Canada
Virchenko Nina, Ukraine
Viterbo Claude, France
Vodop'yanov Serguey, Russia
Vogel Wolfgang E., New Zealand
Vogelsang Volker, Germany
Vogt Arthur, Switzerland
Voiculescu Dan, USA
Voisin Claire, France
Volevich Leonid R., Russia
Volodin Andrei I., Russia
Von Rennenkampff Julia, Switzerland
Von Renteln Michael, Germany
Vondraček Zoran, Croatia
Voss Konrad, Switzerland
Vougiouklis Thomas N., Greece
Vourtsanis Yiannis, USA
Vucāns Jānis, Latvia
Vulakh Leonid, USA

Wada Hidekazu, Japan
Waldspurger Jean-Loup, France
Waldvogel Jörg, Switzerland
Walker Peter L., UK
Wallace David A.R., UK
Walschap Gerard, Belgium
Walter Wolfgang, Germany
Walus Yvonne E., South Africa
Walz Michel, Switzerland
Wan Zhe-Xian, Sweden
Wang Hwai-chiuan, Taiwan
Wang Shuzhou, USA
Wang Xiang, Switzerland
Wang Zhi-Qang, USA
Wanner Gerhard, Switzerland
Warchall Henry Alexander, USA
Washiyama Toshiyuki, Japan
Wassermann Arthur G., USA
Wasuwanich Sudhira, Thailand
Watanabe Hisao, Japan
Watatani Yasuo, Japan
Watters John F., UK
Weber Claude, Switzerland
Webster Sidney M., USA
Wefelscheid Heinrich, Germany
Wehrhahn Karl, Australia
Weibel Anton, Switzerland
Weibel Charles A., USA
Weibel Reto, Switzerland
Weikart Rudi, USA
Weill Georges G., USA
Weinberger Shmuel A., USA
Weintraub Steven H., USA
Weisfeld Morris, USA
Weiss Asia, Canada
Weiss Gary, USA
Weit Yitzhak, Israel
Welk Reiner, Germany
Widman Kjell-Ove, Switzerland
Wiedmann Jens, Switzerland
Wigderson Avi, Israel
Wild Marcel, Germany
Wildemann Petra, Switzerland
Wiles Andrew, USA
Wilhelmy Lutz, Switzerland
Wilkens David L., UK
Wilkins David Raynor, Ireland
Williams Mark, USA
Winkler Jörg Herbert, Germany
Witherspoon Sarah J., Canada
Witte David, USA
Wittstock Gerd, Germany
Wodzicki Mariusz, USA
Wolbers Marcel, Switzerland
Wolfart Jürgen, Germany
Wolff Gerhard, Germany
Wood Jay A., USA
Wortmann Sigrid, USA
Wrikat F.A.D., Jordan
Wu Jyh-Yang, Taiwan
Wu Sijue, USA
Wu Yihren, USA
Wulfsohn Aubrey, Canada
Wünsch Volkmar, Germany
Wüstholz Gisbert, Switzerland
Wysocki Krzysztof, Poland
Xi Nanhua, Germany
Xu Ping, China P.R.
Xu Xiaoping, Hong Kong
Yadrenko Mikhail I., Ukraine
Yakubov Eduard, Israel
Yamagishi Kikumichi, Japan
Yamaguti Masaya, Japan
Yamakawa Aiko, Japan
Yamaki Hiroyoshi, Japan
Yamamoto Shūichi, Japan
Yamanoshita Tsuneyo, Japan
Yamashima Shigeho, Japan
Yamashita Hajime, Japan
Yan Fengsheng, Switzerland
Yang Lo, China P.R.
Yasugi Mariko, Japan
Ye Rugang, Switzerland
Ye Xiang Dong, China P.R.
Yegorchenko Irina A., Ukraine
Yesiltepe Hayrettin, The Netherlands
Ylinen Kari E., Finland
Yoccoz Jean-Christophe, France
Yoichi Miyaoka, Japan
Yona Ilan, Israel
Yonezawa Yasushi, Japan
Yoshida Katsuaki, Japan
Yoshida Kenji, Japan
Yoshida Tomoyuki, Japan
You Yuncheng, USA
Young Lai-Sang, USA
Yousif Mohamed F., USA
Yu Jie Tai, USA
Yu Rupert W., France
Yuan Ya-Xiang, China P.R.
Yui Noriko, Canada
Yukich Joseph E., USA

Zafarani Jafar, Iran
Zaharopol Radu, USA
Zaigraev A.Yu., Poland
Zainodin Hahi-Jubok, UK
Zajtz Andrzej, Poland
Zalaletdinov Roustam M., Uzbekistan
Zalesskii Pavel, Belarus
Zarinelli Alain, Switzerland
Zastrow Andreas, UK
Zdravkovskda Smilka, USA
Zehnder Eduard, Switzerland
Zekri Richard, France
Żelazko Wiesław T., Poland
Zelmanov Efim I., USA
Zemánek Jaroslav, Poland
Zerrin Göktürk, Turkey
Zhang De-Qi, Singapore
Zhang Jian James, USA
Zhang Jian, China P.R.
Zhang Jiping, China P.R.
Zhang Liqun, China P.R.
Zhang Shaowei, China P.R.
Zhang Xianke, China P.R.
Zheng Yuxi, USA
Zhensykbaev Alexander, Kazakhstan
Zhong Xiaohui, USA
Zhou Xingwei, China P.R.
Zhou Zuo Ling, China P.R.
Zhu Chengbo, Singapore
Zhu Kehe, USA
Zhukova Alessandra Nina, Russia
Ziemer L., Switzerland
Ziemian Bogdan, Poland
Zilli Giovanni, Italy
Zimmermann Benno, Switzerland
Zorich Anton, Italy
Zorii Natalia V., Ukraine
Zubelli Jorge P., Brazil
Zuikova Alla G., Ukraine
Zumsteg Stefan, Switzerland
Zuo Kang, Germany
Zürcher Matthias, Switzerland
Zvyagin Viktor, Russia
Zwahlen Bruno, Switzerland

Total: 2476 Ordinary Members.

Membership by Country

Albania	3	Ireland	7
Argentina	7	Israel	54
Armenia	8	Italy	48
Australia	7	Ivory Coast	5
Austria	9	Japan	236
Azerbajan	2	Jordan	2
Bahrain	1	Kazakhstan	12
Belarus	6	Korea	8
Belgium	12	Kuwait	1
Bolivia	1	Latvia	8
Bosnia	3	Lesotho	1
Botswana	1	Lithuania	6
Brazil	20	Luxembourg	1
Bulgaria	3	Macedonia	2
Canada	77	Malaysia	5
Chile	3	Maroc	3
China P.R.	33	Mauritania	1
China Taiwan	13	Mexico	16
Colombia	1	Moldova	5
Croatia	6	Mongolia	1
Cuba	2	Netherlands	17
Czech Rep.	17	New Zealand	3
Denmark	15	Nigeria	5
Ecuador	1	Norway	14
Egypt	3	Philippines	8
Estonia	10	Poland	38
Finland	16	Portugal	5
France	165	Romania	28
Georgia	7	Russia	139
Germany	199	Saudi Arabia	2
Greece	11	Senegal	3
Guatemala	4	Singapore	4
Hong Kong	8	Slovakia	6
Hungary	32	Slovenia	3
Iceland	1	South Africa	14
India	29	Spain	29
Indonesia	1	Sri Lanka	1
Iran	23	Sweden	30

Switzerland	335
Tajikistan	2
Thailand	2
Tunesia	1
Turkey	5
Turkmenistan	1
Uganda	1
United Kingdom	64
Ukraine	55
Uruguay	1
USA	463
Uzbekistan	3
Venezuela	3
Vietnam	2
Yugoslavia	4
Zimbabwe	3

Total number of countries represented: 92

The Work of the Fields Medalists and the Rolf Nevanlinna Prize Winner

The work of the Fields Medalists and the Rolf Nevanlinna Prize Winner was presented as follows:

Fields Medalists

The Rolf Nevanlinna Prize Winner

The Work of Jean Bourgain

Luis Caffarelli

Institute for Advanced Study
Princeton, NJ 08540, USA

Introduction

Bourgain's work touches on several central topics of mathematical analysis: the geometry of Banach spaces, convexity in high dimensions, harmonic analysis, ergodic theory, and finally, nonlinear partial differential equations (P.D.E.'s) from mathematical physics. In all of these areas, he made spectacular inroads into questions where progress had been blocked for a long time.

This he did by simultaneously bringing into play different areas of mathematics: number theory, combinatorics, probability, and showing their relevance to the problem in a previously unforeseen fashion.

To give a flavor of his work, I have concentrated on his recent research, of about the last ten years.

The solution of the Λ_p problem

A great part of the work of Bourgain, in the study of the geometry of Banach spaces, concentrated on the question: Given a Banach space of finite dimension n, how large a section can we find that resembles a Hilbert subspace?

Maybe his most relevant paper in this field is his solution of the $\Lambda(p)$ problem: Given a subset Λ of the set of characters of a compact Abelian group, Λ is a p-set $(p > 2)$ if the L^p and L^2 norms are equivalent in the suspace of $L^p(G)$ generated by Λ.

The longstanding question was: Do $\Lambda(p)$ and $\Lambda(q)$ sets coincide?

Bourgain answers this problem in the negative with the following sharp estimates:

Among n given characters there is a subset of optimal size $[n^{2/p}]$ for which

$$(*) \qquad \left\| \sum a_i \varphi_i \right\|_{L^p} \leq C(p) \left(\sum |a_i|^2 \right)^{1/2}.$$

Through a lacunary argument, one can construct a $\Lambda(p)$ set, which is not $\Lambda(q)$ for any $q > p$.

Proceedings of the International Congress
of Mathematicians, Zürich, Switzerland 1994

The converse of Santalo's inequality

Another product of Bourgain's studies is his proof of Santalo's inequality:

Given K the unit ball of a norm on R^n and K^* its dual, Bourgain and Milman prove

$$\text{vol}\,(K)\ \text{vol}\,(K^*) \geq C^n |B|^2$$

for some absolute constant $0 < c < 1$.

This has applications to number theory and computer sciences.

Ergodic theory

In ergodic theory, Bourgain developed a completely new theory, where averages under very sparse (polynomial) families of iterations are studied (and shown to converge).

The basic theorem, from which the general setting follows by a well-known transformation, due to Calderon, is the maximal theorem for $\ell^2(\mathbb{Z})$.

THEOREM. *Let $f \in \ell^2(\mathbb{Z})$, and ℓ be a positive integer; let*

$$Mf(n) = \sup_{N>0} \left|\frac{1}{N}\sum_{k=1}^{N} f(n+k^\ell)\right|.$$

Then

$$\|Mf\|_{\ell^2} \leq C\|f\|_{\ell^2}$$

i.e., the maximal function of the partial averages of the k^ℓ iterations of the 1-translation is bounded in ℓ^2.

Oscillatory integrals

An important family of ideas introduced by Stein in harmonic analysis concerns the study of the restriction of classical operators (maximal functions, Hilbert and Fourier transforms) to curves in space (parabolics, circles) that have special relevance to the study of partial differential equations (singular integrals of parabolic type, spherical averages related to the wave equation, etc.).

In this area I'll mention two fundamental contributions of Bourgain:

The circle maximal function

$$Mf(y) = \sup_r \frac{1}{A(S_r)} \int_{x\in S_r(0)} f(y+x)\,dA$$

was shown by Stein to be a bounded operator in L^p for some range $p(n)$ for $n \geq 3$.

The two-dimensional case ($p \geq 2$) remained open for a long time until Bourgain closed the gap.

As the "solid" maximal function in any dimension can be written as an average of spherical maximal functions in a fixed low dimension, this allows us in particular to prove bounds for the "solid" maximal function independent of dimensions.

The second contribution refers to the restriction of Fourier transforms to spheres, or, related to it, the properties of the characteristic function of the ball as a Fourier multiplier (a natural generalization of taking partial sums of Fourier series).

In two dimensions, these problems were well understood (C. Fefferman).

In higher dimensions some range of continuity is expected around L^2, and a series of results was obtained by Tomas and Stein.

Bourgain considerably sharpened these results, but what is more important than the exact ranges, is the fact that, in doing so, he introduced completely new techniques, where instead of relying on L^2 theory (i.e., decomposing functions and operators in "L^2 pieces") he sharpened the geometric understanding of Besicovitch-type maximal operators and Kakeya sets.

Nonlinear partial differential equations

Bourgain's contributions to nonlinear partial differential equations are very recent, and it is somewhat difficult to decide where to stop in this presentation because results from him and many others in the field (Kenig, Klainerman, Machedon, Ponce, Vega) have been pouring in, in good part thanks to the revitalization of the field brought in by Bourgain's approach.

Let us say that he obtained very sharp results for the well-posedness of the nonlinear Schrödinger equation,

$$iu_t + \triangle u + u|u|^\alpha = 0$$

for non smooth data.

Previous to his work, there was mainly local well-posedness in H^s for large enough s.

By introducing new, suitable space-time functional spaces, Bourgain started a new, more deep and elegant way of treating dispersive equations.

In closing, let me reiterate that some of the outstanding qualities of Bourgain are his power to use whatever it takes — number theory, probabilistic methods, covering techniques, sharp decompositions — to understand the problem at hand, and his versatility, which allowed him to deeply touch so many areas in such a short period of time.

The Work of Pierre-Louis Lions

S. R. S. Varadhan

Courant Institute, New York University
251 Mercer Street, New York, NY 10012, USA

Pierre-Louis Lions has made unique contributions over the last fifteen years to mathematics. His contributions cover a variety of areas, from probability theory to partial differential equations (PDEs). Within the PDE area he has done several beautiful things in nonlinear equations. The choice of his problems has always been motivated by applications. Many of the problems in physics, engineering and economics when formulated in mathematical terms lead to nonlinear PDEs; these problems are often very hard. The nonlinearity makes each equation different. The work of Lions is important because he has developed techniques that, with variations, can be applied to classes of such problems. To say that something is nonlinear does not mean much; in fact it could even be linear. The entire class of nonlinear PDEs is therefore very extensive and one does not expect an all-inclusive theory. On the other hand, one does not want to treat each example differently and have a collection of unrelated techniques. It is thus extremely important to identify large classes that admit a unified treatment.

In dealing with nonlinear PDEs one has to allow for nonclassical or nonsmooth solutions. Unlike the linear case one cannot use the theory of distributions to define the notion of a weak solution. One has to invent the appropriate notion of a generalized solution and hope that this will cover a wide class and be sufficient to yield a complete theory of existence, uniqueness, and stability for the class.

Due to the very limited time that is available, I shall focus on three areas within nonlinear PDE where Lions has made major contributions. The first is the so called "viscosity method". This development is a long story that started with some work in collaboration with Crandall. Over many years, in partial collaboration with others (besides Crandall, Evans and Ishii), Lions has developed the method, which is applicable to the large class of nonlinear PDEs known as fully nonlinear second order degenerate elliptic PDEs. The class contains very many important subclasses that arise in different contexts.

By solving a nonlinear PDE one is trying to solve an equation involving an unknown function and its derivatives. Let u be a function in a region G in some R^n and let $Du, D^2u, \ldots, D^k u$ be its derivatives of order up to k. A nonlinear PDE is an equation of the form

$$F[x, u(x), (Du)(x), (D^2u)(x), \ldots, (D^k(u)(x))] = 0 \quad \text{in} \quad G$$

with some boundary conditions on ∂G. Such a PDE is said to be nonlinear and of order k. The viscosity method applies in cases where $k = 2$ and $F(x, u, p, H)$

Proceedings of the International Congress
of Mathematicians, Zürich, Switzerland 1994

has certain monotonicity properties in the arguments u and H. More precisely, it is nondecreasing in u and nonincreasing in H. Here u is a scalar and H is a symmetric matrix of size $n \times n$ with the natural partial ordering for symmetric matrices.

Some of the many examples of such functions are described below.

Linear elliptic equations:

$$-\sum_{i,j} a^{ij}(x) \frac{\partial^2 u}{\partial x_i \partial x_j}(x) + f(x) = 0$$

where the matrix $a^{ij}(x)$ is uniformly positive definite.

In this case the function F is given by

$$F(x, u, p, H) = -\text{Trace}[a(x)H] + f(x) .$$

First order equations:

$$f(x, u(x), (Du)(x)) = 0$$

These include Hamilton-Jacobi equations where it all started. One added a term of the form $\epsilon\Delta$ to the equation and constructed the solution in the limit as ϵ went to zero. The theory owes its name to its early origins.

If one has a family F_α of such functions one can generate a new one by defining

$$F = \sup_\alpha F_\alpha .$$

If one has a two-parameter family $F_{\alpha\beta}$ of such functions one can generate a new one by defining

$$F = \sup_\alpha \inf_\beta F_{\alpha\beta} .$$

Such examples arise naturally in control theory and game theory and are referred to as Hamilton-Jacobi-Bellman and Isaacs equations.

In order to understand the notion of a generalized solution it is convenient to talk about supersolutions and subsolutions. Suppose u is a subsolution, i.e.

$$F(x, u(x), (Du)(x), (D^2 u(x))) \le 0$$

and we have another function ϕ, which is smooth, such that $u - \phi$ has a maximum at some point $\hat{x}$. Then by calculus $Du(\hat{x}) = D\phi(\hat{x})$ and $D^2(u)(\hat{x}) \le D^2(\phi)(\hat{x})$. From the monotonicity properties of F it follows that

$$F(\hat{x}, u(\hat{x}), (Du)(\hat{x}), (D^2 u(\hat{x}))) \ge F(\hat{x}, u(\hat{x}), (D\phi)(\hat{x}), (D^2\phi(\hat{x}))) .$$

Therefore

$$F(\hat{x}, u(\hat{x}), (D\phi)(\hat{x}), (D^2\phi(\hat{x}))) \le 0.$$

The last inequality makes sense without any smoothness assumption on u. We can try to define a nonsmooth subsolution as a u that satisfies the above for arbitrary smooth ϕ and $\hat{x}$ provided $u - \phi$ has a maximum at $\hat{x}$. The definition of a supersolution is similar, and a solution is one that is simultaneously a super and a subsolution.

Let us consider first a Dirichlet boundary value problem where we want to find a u that solves our equation and has boundary value zero.

The main step is to establish the key comparison theorem (with a long history that began with the work of Crandall and Lions and saw an important contribution from Jensen) that if u is a subsolution and if v is a supersolution in a bounded domain G and if $u \leq v$ on the boundary ∂G then $u \leq v$ in $G \cup \partial G$. This requires some mild regularity conditions on F as well as some nondegeneracy conditions. After all, we have not ruled out $F \equiv 0$. Once such conditions are imposed one can establish the key comparison theorem. From this point on, the theory proceeds in a way similar to the classical Perron's method for solving the Dirichlet problem. Assuming that there is at least one subsolution $\bar{u}$ and at least one supersolution $\bar{v}$ with the the given boundary value, one establishes that

$$W(x) = \sup\{w(x) : \bar{u} \leq w \leq \bar{v}\ ,\ w \text{ is a subsolution}\}$$

is a solution. The comparison theorem is of course enough to guarantee uniqueness. The constructibility of $\bar{u}$ and $\bar{v}$ depends on the circumstances and is relatively easy to establish.

The richness of the theory is in its flexibility. One can prove stability results of various kinds and the validity of various approximation schemes. One can modify the theory to include Neumann boundary conditions. This is tricky because one has to interpret the normal derivative suitably for a function that has no smoothness requirements and the boundary condition can be nonlinear as well. Treating parabolic equations is not any different. One can just consider t as another variable.

I would suggest the survey article by Crandall, Ishii, and Lions that appeared in the Bulletin of the American Mathematical Society in 1992 for those who want to read more about this area.

The second body of work that I want to discuss has to do with the Boltzmann equation and similar equations. During the last six or seven years Pierre-Louis Lions has played the central role in new developments in the theory of the Boltzmann equation and similar transport equations. These are important in kinetic theory and arise in a wide variety of physical applications. We will for simplicity stay within the context of the Boltzmann equation. In R^3 we have a collection of particles moving along and interacting through "collisions" among themselves. As we do not want to keep track of the positions and velocities of the particles individually, we abstract the situation by the density $f(x, v)$ of particles that are at position x with velocity v. Even if there is no interaction, the density $f(x, v)$ will change in time due to uniform motion of the particles. The time-dependent density $f(t, x.v)$ will satisfy the equation

$$\frac{\partial f}{\partial t} + v.\nabla_x f = 0.$$

The collisions will change this equation to

$$\frac{\partial f}{\partial t} + v.\nabla_x f = Q(f, f).$$

Here Q is a quadratic quantity that represents binary collisions and its precise form depends on the nature of the interaction. Generally it looks like

$$Q(f, f) = \int\int_{R^3 \times S^2} dv_* \, d\omega B(v - v_*, \omega)\{f' f'_* - f f_*\}.$$

The notation here is standard: v and v_* are the incoming velocities and v' and v'_* are the outgoing velocities. B is the collision kernel. For given incoming velocities v and v_*, ω on the sphere S^2 parametrizes all the outgoing velocities compatible with the conservation of energy and momenta.

$$v' = v - (v - v_*, \omega)\omega, \qquad v'_* = v + (v - v_*, \omega)\omega$$

and f', f_*, f'_* are $f(t, x, v)$ with v replaced by the correspondingly changed v', v_*, and v'_*.

This problem of course has a long history. Smooth and unique solutions had been obtained for small time or globally for initial data close to equilibrium. Carleman had studied spatially homogeneous solutions. But a general global existence theorem had never been proved. The work of Lions (in collaboration with DiPerna) is a breakthrough for this and many other related transport problems of great physical interest.

Let me spend a few minutes giving you some idea of the method as developed by Lions and others (mostly his collaborators).

Although the nonlinearity looks somewhat benign it causes a serious problem in trying to establish any existence results. The collision term is quadratic and involves both positive and negative terms. To carry out any analysis one must control each piece separately. One gets certain a priori estimates from the conservation of mass and energy. The Boltzmann H-theorem gives an important additional control if one starts with an initial data with finite entropy. If we denote by Q^+ and Q^- the positive and negative terms in the collision term with considerable effort one is able to obtain only local L_1 bounds on $(1 + f)^{-1}Q^{\pm}(f, f)$. The weak solutions are therefore formulated in terms of $\log(1 + f)$. As there are no smoothness estimates in x one has to show that the velocity integrals contained in Q provide the compactification needed to make the weak limit behave properly.

This idea of "velocity averaging", which is central to these methods, is easy to state in a simple context. Suppose we have a function $g(x, v)$ in $R^N \times R^N$ and for some reasonable function $a(v)$ we have a local L_p estimate on $a(v).\nabla_x g(x, v)$. Then for a good test function ψ the velocity integral

$$\int_{R^N} \psi(v) g(x, v) \, dv$$

is in a suitable Sobolev space. Another important step that is needed in dealing with the Vlasov equation is the ability to integrate vector fields with minimal regularity. In nonlinear problems you have to learn to live with the regularity that the problem gives you. The writeup by Lions in the Proceedings of the last ICM (Kyoto 1990) provides a survey with references.

The third and final topic that I would like to touch on is the contribution Lions has made to a class of variational problems. There are many nonlinear PDEs that are Euler equations for variational problems. The first step in solving such equations by the variational method is to show that the extremum is attained. This requires some coercivity or compactness. If the quantity to be minimized has an "energy"-like term involving derivatives, then one has control on local regularity along a minimizing sequence. This usually works if the domain is compact. If the domain is noncompact the situation is far from clear. Take for instance the problem of minimizing

$$\int_{R^N} |(\nabla f)(x)|^2 dx - \int\int V(x-y) f^2(x) f^2(y)\; dx\, dy$$

over functions f with L_2 norm λ (fixed positive number). Here $V(.)$ is a reasonable function decaying at ∞. Because of translation invariance, the minimizing sequence must be centered properly in order to have a chance of converging. The key idea is that, in some complicated but precise sense, if the minimizing sequence cannot be centered, then any member of the sequence can be thought of as two functions with supports very far away from each other. If we denote the infimum by $\sigma(\lambda)$, then along such sequences the infimum will be $\sigma(\lambda_1) + \sigma(\lambda_2)$ with $\lambda_1 + \lambda_2 = \lambda$ $0 < \lambda_1, \lambda_2 < \lambda$ rather than $\sigma(\lambda)$. If independently one can show that $\sigma(\lambda)$ is strictly subadditive, then one can prove the existence of a minimizer. This idea has been developed fully and applied successfully by Lions to many important and interesting problems.

See the papers in Annales de l'Institut Henri Poincaré, Analyse Non Linéaire 1984 by Lions for many examples where this point of view is successfully used.

References

[1] Michael G. Crandall, Hitoshi Ishi, and Pierre-Louis Lions, *Users Guide to viscosity solutions of second order partial differential equations*, Bull. Amer. Math. Soc. (New Series) **27**, no. 1 (1992), 1–67.

[2] Pierre-Louis Lions, *On Kinetic Equations*, Proceedings of the ICM, Kyoto, 1990, vol. II, 1173–1185.

[3] ——, *The concentration-compactness principle in the calculus of variations. The locally compact case*, Parts 1 and 2, Annales de l'IHP, Analyse non-linéaire, vol. I (1984), 109–145 and 224–283.

Présentation de Jean-Christophe Yoccoz

ADRIEN DOUADY

Ecole Normale Supérieure
45, rue d'Ulm, F-75230 Paris Cedex, France

1. Curriculum

Jean-Christophe Yoccoz est un pur produit, et du meilleur cru, du système français. Ancien élève de l'Ecole Normale Supérieure où il fut reçu 1er en 1975, reçu 1er à l'Ecole Polytechnique la même année, 1er ex-aequo à l'Agrégation de Mathématiques en 1977, il soutient sa Thèse de Doctorat d'Etat en 1985 et fut invité à enseigner le Cours Peccot au Collège de France en 1987. Aujourd'hui, âgé de 37 ans, il est Professeur à l'Université de Paris-Sud (Orsay), membre de l'I.U.F. (Institut Universitaire de France) et de l'URA "Topologie et Dynamique" du CNRS à Orsay.

Il effectua son Service National en Coopération à l'IMPA à Rio de Janeiro, et cela devait le marquer profondément. Il visite régulièrement le Brésil, mais aussi le Centre International de Trieste. D'ailleurs sa femme est brésilienne.

Yoccoz est un étudiant de Michel Herman, et c'est ainsi qu'il est devenu peut-être le meilleur spécialiste de la Théorie des Systèmes Dynamiques.

2. La Théorie des Systèmes Dynamiques

Cette théorie cherche à décrire *l'évolution à long terme* d'un système quand on en connait la loi d'évolution élémentaire. Le temps peut y être continu ou discret.

Dans le cas d'un *temps continu*, la loi d'évolution infinitésimale se traduit par une équation différentielle, qui est donnée par un *champ de vecteurs*, et le problème est de comprendre l'évolution à long terme des solutions. On obtient parfois des attracteurs étranges.

Un exemple typique est le problème de la *stabilité du Système Solaire*, qui a amené Poincaré à fonder la théorie au tournant du siècle.

Dans le cas d'un *temps discret*, l'évolution élémentaire est donnée par une application f, qui donne l'état du système au temps $n+1$ en fonction de l'état au temps n. Il s'agit alors d'itérer f un grand nombre de fois.

Quand deux applications f et g décrivent le même phénomène dans des représentations différentes, elles sont *conjuguées* par l'application h qui traduit le changement de représentation. Toute conjugaison peut être interprétée de cette façon. Deux applications conjuguées ont donc les mêmes propriétés dynamiques. Par suite la classification des applications à conjugaison près est un problème central dans la théorie.

Proceedings of the International Congress
of Mathematicians, Zürich, Switzerland 1994

3. Conjugaison C^∞ à la rotation

L'exemple le plus simple est celui où l'espace des états est un cercle, et où l'application à itérer est indéfiniment différentiable ainsi que son inverse, autrement dit un *difféomorphisme* C^∞. Pour une telle application f, Poincaré a défini le *nombre de rotation* $\alpha = \text{Rot}\,(f) \in \mathbf{T} = \mathbf{R}/\mathbf{Z}$. Les question est alors: quand-est-ce que f est C^∞-conjuguée à la rotation $\mathcal{R}_\alpha\colon t \mapsto t+\alpha$?

Si α en rationnel, disons $\alpha = \frac{p}{q}$, ceci exige que l'on ait $f^q = I$, ce qui ne se produit essentiellement jamais. Le cas intéressant est donc celui où α est irrationnel. Il a été étudié par Denjoy — qui a montré que f est toujours topologiqement conjuguée à $\mathcal{R}_\alpha$ —, Birkhoff, Arnold, Herman et beaucoup d'autres, et bien sûr Yoccoz. Ils ont tous insisté sur l'importance des *propriétés arithmétiques* de α.

Pour α rationnel, il se produit des résonnances. Si α est irrationnel, il se produit presque toujours des compensations et on observe certaines régularités. Mais si α, tout en étant irrationnel, se trouve extrêmement proche de rationnels avec des dénominateurs modérément grands, il arrive qu'une résonnance s'amorce et qu'avant qu'elle soit amortie une autre prenne le relais, et on peut obtenir une situation très compliquée. Ce qui importe est donc la distance $\delta_q(\alpha)$ de α à l'ensemble des rationnels à dénominateur borné par q, et la façon dont cette distance tend vers 0 quand q tend vers l'infini.

4. Conditions diophantiennes

On dit que α est diophantien si $\delta_q(\alpha)$ est minoré par une expression de la forme $\frac{c}{q^\beta}$.

Pour un difféomorphisme C^∞ du cercle de nombre de rotation α, Herman a montré que f est nécessairement C^∞-conjuguée à la rotation $\mathcal{R}_\alpha$ si α est diophantien d'exposant 2. Ce résultat constituait une percée importante. En fait Herman avait démontré un théorème plus fort: le même résultat sous une hypothèse plus faible, satisfaite pour presque toute valeur de α.

Dans sa thèse, Yoccoz a amélioré le théorème de Herman: il a donné une démonstration plus simple et obtenu le résultat sous l'hypothèse que α est diophantien sans restriction d'exposant - hypothèse plus faible que celle de Herman.

Des contre-exemples de Herman montrent que ce résultat est optimal.

5. Le cas R-analytique

On peut se poser la même question dans le cadre $\mathbf{R}$-analytique.

Yoccoz a démontré dans sa thèse qu'un difféomorphisme $\mathbf{R}$-analytique du cercle à nombre de rotation α diophantien est nécessairement $\mathbf{R}$-analytiquement conjugué à la rotation $\mathcal{R}_\alpha$. Récemment il a donné une description de l'ensemble exact des nombres de rotation α ayant cette propriété. C'est un ensemble compliqué: alors que les ensembles qu'on définit de cette façon sont en général du type F_σ, celui-ci est seulement $F_{\sigma\delta}$.

Les fonctions C^∞ et $\mathbf{R}$-analytique ont une consistance differente. Quand on travaille dans le cadre $\mathbf{R}$-analytique, la première chose que l'on fait est d'étendre les applications aux valeurs complexes de la variable. Une application $f\colon \mathbf{T} \to \mathbf{T}$ s'étend ainsi à un voisinage annulaire Ω de $\mathbf{T}$ dans le cylindre $\mathbf{C}/\mathbf{Z}$, et si f est $\mathbf{R}$-analytiquement conjugué à $\mathcal{R}_\alpha$ il y a un anneau A dans Ω qui et invariant par f. Pour $z \in A$, la fermeture de l'orbite de A est une courbe $\mathbf{R}$-analytique, correspondant à un cercle parallèle à l'équateur $\mathbf{T}$ de $\mathbf{C}/\mathbf{Z}$. L'épaisseur minimale de A, son module, ce qui se produit au voisinage de sa frontière sont autant de propriétés sur lesquelles le raisonnement géométrique a prise.

6. Réciproque du Théorème de Bruno

La question de linéarisabilité locale des difféomorphismes holomorphes au voisinage d'un point fixe est étroitement liée à la précédente. C'est la suivante:

Une fonction $f\colon z \mapsto a_1 z + a_2 z^2 + a_3 z^3 + \cdots$, est-elle holomorphiquement conjuguée au voisinage de 0 à sa partie linéaire $z \mapsto a_1 z$?

Le résultat est facile si $|a_1| \neq 1$ (Schröder, Böttcher), le cas intéressant est celui où a_1 est de la forme $e^{2i\pi\alpha}$. Il a été étudié par Fatou - qui a traité le cas où α est rationnel, Cremer - qui a donné des exemples de non-linéarisabilité, Siegel - qui a montré que f est linéarisable dès que α est Diophantien (et ce quelle que soit la queue $a_2 z^2 + \cdots$), Bruno - qui a amélioré le théorème de Siegel en démontrant le résultat sous l'hypothèse plus faible $\sum \frac{\operatorname{Log} q_{n+1}}{q_n} < \infty$ (où les $\frac{p_n}{q_n}$ sont les réduites du développement de α en fraction continue), et enfin Yoccoz qui a démontré la réciproque du théorème de Bruno.

Siegel et Bruno travaillaient en force, résolvant le problème formellement et majorant les coefficients de la conjugante. Yoccoz a une approche plus géométrique et plus fine. Il y a une construction que l'on appelle renormalisation, qui associe à une application f d'angle α une application f_1, ayant un angle α_1, dont le développment en fraction continue est le même que celui de α mais décalé avec perte du premier terme. Par une étude quantitative pousséee des propriétés de cette opération et de ses itérées, Yoccoz a obtenu une démonstration très éclairante du théorème de Bruno, et il a pu prouver la réciproque: que pour tout α ne satisfaisant pas à la condition de Bruno on peut chosir la queue de façon que f ait des points périodiques arbitrairement proches de 0, ce qui exclut la linéarisabilité. En fait la queue la plus simple ($f = a_1 z + z^2$) fait l'affaire.

Restait une question: La non-linéarisabilité est-elle toujours due à la présence de petits cycles? Les exemples construits par Cremer et Yoccoz pouvaient le laisser croire. La question a été résolue par la négative par Perez-Marco, un élève de Yoccoz qui a encore affiné sa méthode. Elle fait très curieusement intervenir la condition $\sum \frac{\operatorname{Log}\operatorname{Log} q_{n+1}}{q_n} < \infty$, plus faible que celle de Bruno.

7. Jeu de Cadres

La Géométrie et l'Analyse interviennent dans toute les parties de la Théorie des Systèmes Dynamiques. Mais elles ont une façon particulière d'interagir dans les Systèmes Dynamiques Complexes, grâce aux inégalités de Schwarz, Koebe, Groetzsch etc, inégalités puissantes qu'on peut appliquer sous des hypothèses purement topologiques. C'est une méthode que Yoccoz a énormément développée.

8. *MLC*: La taille des membres

L'essentiel de notre connaissance des propriétés dynamiques de la famille des polynômes quadratiques complexes est concentré dans les propriétés topologiques de son lieu de connexité M, connu sons le nom d'*ensemble de Mandelbrot.* Ses propriétés combinatoires sont maintenant bien comprises, et Thurston en a proposé un modèle synthétique. Mais pour savoir que M est effectivement homéomorphe à son modèle, il manque une information: que M est localement connexe.

L'ensemble M contient des copies de lui-même. Yoccoz a montré que M est localement connexe en tout point "non infiniment renormalisable", c'est à dire qui n'est pas contenu dans l'intersection d'une suite décroissante de copies de M. Pour démontrer la conjecture MLC complète, il reste aujourd'hui à montrer que l'intersection d'une suite décroissante de copies de M est réduite à un point.

Le premier cas est constitué par les points de la cardioïde: L'ensemble M est formé d'une grande cardioïde Γ, remplie, et de *membres* attachés aux points de Γ d'argument interne rationnel (les arguments internes définissent la paramétrisation naturelle de Γ).

Yoccoz a montré que le diamètre d'un membre $M_{p/q}$ attaché au point d'argument interne p/q est majoré par une expression de la forme $\frac{c}{q}$ où c est une constante. Ce résultat n'est surement pas optimal (d'après Hubbard ou pourrait espérer $\frac{\operatorname{Log} q}{q^2}$), mais il suffit à montrer que M est localement connexe aux points de Γ. Par la même méthode, il obtient la connexité locale en tout point qui est sur le bord d'une composante hyperbolique.

9. MLC: les puzzles de Yoccoz

Pour montrer que M est localement connexe aux points c qui ne sont ni infiniment renormalisables ni sur le bord d'une composante hyperbolique, Yoccoz emploie la méthode dite des "*Puzzles de Yoccoz*". Selon de principe que je défends en Dynamique Complexe.

On laboure dans le plan dynamique.

On moissonne dans le plan des paramètres.

Il y a en effet des figures dans le plan dynamique qui se trouvent reproduites plus on moins fidèlement dans le plan des paramètres.

Le point de départ est un article de Branner-Hubbard, qui traite d'une certaine famille de polynômes cubiques (voir l'exposé de Lyubich à ce Congrès). Il leur faut montrer que certains ensembles dans le plan des paramètres, dont on s'attend à ce qu'ils soient réduits à un point, le sont effectivement. D'après l'inégalité de Grötzsch, il suffit d'enfermer un tel ensemble dans une suite d'anneaux emboités dont la somme des modules est infinie. C'est ce qu'ils font, mais d'abord dans le plan dynamique où les anneaux considérés sont des revêtements les uns des autres, de sorte qu'à une constante près les modules sont des inverses d'entiers et la divergence résulte d'une étude combinatoire très poussée.

Dans leur cas, le transfert dans le plan des paramètre est facile, car les anneaux considérés s'y retrouvent reproduits conformément.

La Conjecture MLC est aussi en un sens un énoncé de la forme "les points sont effectivement des points". Dans le plan des paramètres qui contient M, on peut définir des *pièces* limitées par des rayons externes et des arcs d'équipotentielles. Une telle pièce découpe dans M un ensemble connexe, et pour démontrer

MLC en un point c il suffit démontrer que l'intersection des pièces qui sont des voisinages de c est réduite à c.

La situation est analogue à celle de Branner-Hubbard, et la démonstration dans le plan dynamique peut se faire suivant les mêmes lignes. L'essentiel de la difficulté réside dans le passage au plan des paramètres, et Yoccoz réalise là un tour de force d'Analyse. En effet, en dehors de M et de K, il y a une correspondance conforme entre le plan dynamique et le plan des paramètres, mais sur ces ensembles il n'y a plus de correspondance (dans M il y a des petites copies de M qui ne se retrouvent pas dans le plan dynamique), les

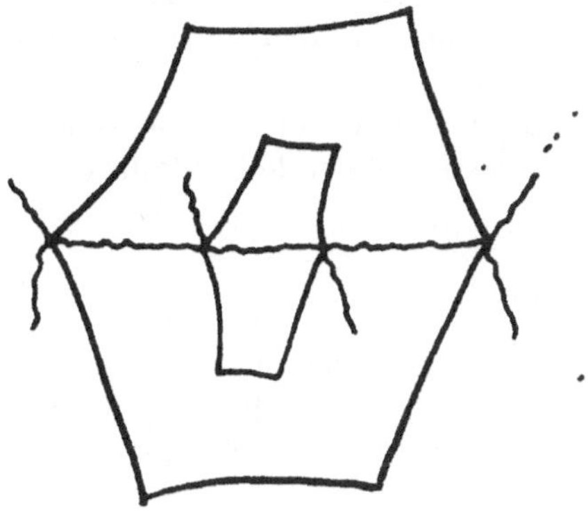

Plan dynamique

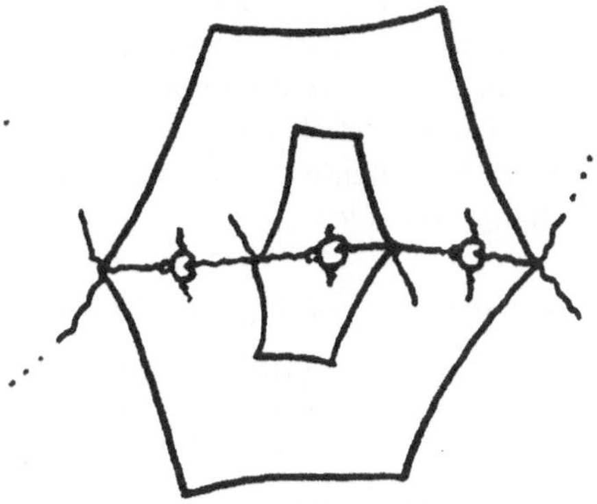

Plan des paramètres

anneaux n'ont pas même module et il faut faire de l'Analyse fine pour montrer que le rapport des modules est borné, et que la divergence est donc préservée.

Yoccoz n'a pas fait taper son manuscrit, mais on peut lire une démonstration dans les rédactions qu'en ont faites Milnor dans un preprint, et Hubbard (Three theorems of Yoccoz) dans le livre dédié à Milnor.

10. Conjugaison C^∞

Je me suis étendu longuement sur la Dynamique Complexe, parce que c'est ce que je comprends le mieux, mais les travaux de Yoccoz en Dynamique Réelle sont tout aussi importants. La plupart sont en collaboration.

Palis et Yoccoz ont obtenu un système complet d'invariants de conjugaison C^∞ pour les difféomorphismes de Morse-Smale.

- des invariants locaux qui décrivent les formes normales aux points attractifs ou répulsifs;
- des invariants globaux qui comparent les coordonnées adaptées à ces formes normales là où les bassins se recouvrent.

Le cas d'une dynamique Nord-Sud sur S^n en facile: le second invariant est le changement de cartes. Mais dans le cas général il y a des points-selle. Palis et Yoccoz montrent que ces points-selle ne produisent pas de nouveaux invariants en raison d'un théorème de singularités inessentielles: Si deux difféomorphismes de Morse-Smale sont C^∞-conjugués sur la réunion des bassins attractifs et répulsifs, la conjugaison s'étend de façon C^∞ à la variété toute entière.

11. Autres travaux avec Palis

Yoccoz a écrit au moins trois autres articles avec Palis. Un sur les centralisateurs des difféomorphismes, où ils démontrent que, sous certaines conditions assez générales, en partant d'un difféomorphisme hyperbolique f on peut obtenir par une perturbation arbitrairement petite un difféomorphisme f, qui ne commute qu'avec lui-même et ses itérés, et tel que tout difféomorphisme f_2 suffisamment voisin de f_1 ait la même propriété.

Un autre article sur les bifurcations homoclines complète un résultat de Newhouse et établit une réciproque à un résultat de Palis-Takens: si la dimension de Hausdorff de l'ensemble hyperbolique créant cette bifurcation est plus grande que un, les applications structurellement stables ne sont pas prévalentes au voisinage, détruisant un vieux rêve de Thom.

A en croire Michel Serres, dans une telle collaboration, il y a toujours un renard qui va à la chasse et un sanglier qui creuse. Combien de fois Palis et Yoccoz ont-ils échangé les rôles?

12. Travaux avec Le Calvez et Raphael Douady

Avec Le Calvez, autre étudiant de Herman, Yoccoz a démontré qu'il n'y a pas d'homéomorphisme minimal de l'anneau $S^1 \times \mathbf{R}$. Autrement dit il n'y a pas d'homéomorphisme de la sphère S^2 préservant les deux poles, et tel que tout autre point ait une orbite dense.

Les méthodes sont celles de la topologie en dimension 2. Le lemme central est que, au voisinage d'un point fixe qui n'est ni attractif ni répulsif, et pour lequel il y a un voisinage ne contenant aucune orbite complète, l'application se comporte du point de vue de l'indice comme $z \mapsto e^{2\pi i\, p/q}\, z(1 - \overline{z}^{rq})$ pour certains entiers r et q.

Je veux aussi citer un article avec Raphael Douady. Pour un difféomorphisme f du cercle conjugué à $\mathcal{R}_\alpha$ par une application h de classe φ^1, la mesme μ_s de densité $(h')^{1-s}$ satisfait $f_*((f')^s\, \mu_s) = \mu_s$. Douady et Yoccoz montrent qu'il existe une mesure unique satisfaisant cette propriété des que f, difféomorphisme de classe φ^2, a un nombre de rotation irrationnel, même si la conjugante est seulement un homéomorphisme.

13. Cocorico

Ce retour aux difféomorphismes du cercle termine notre brève visite guidée des travaux de Yoccoz.

Avec deux médailles Fields pour la France, même s'il s'agit d'une coincidence nous pouvons pavoiser. Mais dans une occasion pareille il convient de se rappeler un proverbe de nos jardiniers

Si la rose est belle, c'est que le fumier est gras.

Il est de la responsabilité de chaque communauté nationale, en Mathématiques, de veiller à ce que la qualité de l'enseignement en mathématiques, en particulier au niveau secondaire, soit préservée. Pour nous Français, au moment où des réductions d'horaires draconniennes menacent, cette tâche sera rude.

On the Work of Efim Zelmanov

WALTER FEIT

Department of Mathematics, Yale University
New Haven, CT 06511, USA

0 Introduction

Efim Zelmanov has received a Fields Medal for the solution of the restricted Burnside problem. This problem in group theory had long been known to be related to the theory of Lie algebras. In fact, to a large extent *it is* the problem in Lie algebras. A precise statement of it can be found in Section 2 below.

In proving the necessary properties of Lie algebras, Zelmanov built on the work of many others, though he went far beyond what had previously been done in this direction. For instance, he greatly simplified Kostrikin's results [K] which settled the case of prime exponent and then extended these methods to handle the prime power case.

However, while the case of exponent 2 is trivial, the case of exponent 2^k for arbitrary k is the most difficult case that needed to be addressed. The results from Lie algebras that work for exponent p^k with p an odd prime are not adequate for exponent 2^k. This indicated that a new approach was necessary here. Zelmanov was the first to realize that in the case of groups of exponent 2^k the theory of Jordan algebras is of great significance. Even though Vaughan-Lee later removed the need for Jordan algebras [V], it seems probable that this proof could not have been discovered without them, as the ideas used arise most naturally from Jordan algebras.

Zelmanov had earlier made fundamental contributions to Jordan algebras and was an expert in this area, thus he was uniquely qualified to attack the restricted Burnside problem.

Below the background from the theory of Jordan algebras and some of Zelmanov's contributions to this theory are first discussed (I am grateful to McCrimmon and Jacobson for much of this material). See [J1] and [J2] for the general theory of Jordan algebras. Then the Burnside problems are described and some of the things that were earlier known about them are listed. Section 4 contains some consequences of the restricted Burnside problem. Finally, some relevant results from Lie and Jordan algebras are mentioned (in a necessarily sketchy manner).

Zelmanov himself has written a set of expository notes on these topics [Z11]. It contains all the appropriate definitions and some of the material used in the proof of the restricted Burnside problem. It also includes material on several related questions, such as the Kurosh-Levitzky problem. Of course, ultimately, the details are the heart of the matter, and for these the reader should consult [Z8] and [Z9], or [V].

Proceedings of the International Congress
of Mathematicians, Zürich, Switzerland 1994

This circle of ideas illustrates the unity of mathematics once again. Although many formal identities are used to settle the restricted Burnside problem, it seems unlikely that they could have been discovered without the conceptual framework provided by the seemingly unrelated and diverse fields of Lie and Jordan algebras.

1 Jordan algebras

Jordan algebras were introduced in the 1930s by the physicist P. Jordan in an attempt to find an algebraic setting for quantum mechanics, essentially different from the standard setting of hermitian matrices. Hermitian matrices or operators are not closed under the associative product xy, but are closed under the symmetric products $xy + yx$, xyx, x^n. An empirical investigation indicated that the basic operation was the Jordan product

$$x \cdot y = \frac{1}{2}(xy + yx),$$

and that all other properties flowed from the commutative law $x \cdot y = y \cdot x$ and the Jordan identity $(x^2 \cdot y) \cdot x = x^2 \cdot (y \cdot x)$. (For example, the Jordan triple product $\{xyz\} = \frac{1}{2}(xyz + zyx)$ can be expressed as $x \cdot (y \cdot z) + (x \cdot y) \cdot z - (x \cdot z) \cdot y$, though the tetrad

$$\{xyzw\} = \frac{1}{2}(xyzw + wzyx)$$

cannot be expressed in terms of the Jordan product.) Jordan took these as axioms for the variety of Jordan algebras. Algebras resulting from the Jordan product in an associative algebra were called special, so the physicists were seeking algebras that were exceptional (= nonspecial). In a fundamental paper [JNW] Jordan, von Neumann, and Wigner classified all finite-dimensional formally-real Jordan algebras. These are direct sums of five types of simple algebras: algebras determined by a quadratic form on a vector space (a special subalgebra of the Clifford algebra of the quadratic form) and four types of algebras of hermitian $(n \times n)$-matrices over the four composition algebras (the reals, complexes, quaternions, and octonions). The algebra of hermitian matrices over the octonions is Jordan only for $n \leq 3$, and is exceptional if $n = 3$, so there was only one exceptional simple algebra in their list (now known as the Albert algebra, of dimension 27). At the end of their paper Jordan, von Neumann, and Wigner expressed the hope that by dropping the assumption of finite dimensionality one might obtain exceptional simple algebras other than Albert algebras.

Algebraists developed a rich structure theory of Jordan algebras over fields of characteristic $\neq 2$. First, the analogue of Wedderburn's theory of finite-dimensional associative algebras was obtained by Albert. Next this was extended by Jacobson to an analogue of the Wedderburn-Artin theory of semisimple rings with minimum condition on left or right ideals. In this, the role of the one-sided ideals was played by inner ideals, defined as subspaces B such that $U_b x$ is in B for all x in the algebra A and all b in B where $U_a = 2L_a^2 - L_{a^2}$ and L_a is the left multiplication by a in the Jordan algebra A. If A is the Jordan subalgebra of an associative algebra then $U_b x = bxb$ in the associative product. Using the definition of semi-simplicity (called nondegeneracy) that A contains no $z \neq 0$ such that $U_z = 0$, Jacobson showed that every nondegenerate Jordan algebra with d.c.c. on inner ideals is the direct sum of simple algebras that are of classical type (analogues of those found in [JNW]: (Type I) Jordan algebras of nondegenerate quadratic forms; (Type II) algebras $H(A, *)$ of hermitian elements in a $*$-simple artinian associative algebra A (($n \times n$)-matrices over a division algebra with involution, or over a direct sum of a division algebra and its opposite under the exchange involution, or matrices over a split quaternion algebra with standard involution); (Type III) 27-dimensional exceptional Albert algebras; (Type IV) Jordan division algebras, defined by the condition that U_a is invertible for every $a \neq 0$.

Up to this point, the structure theory treated only algebras with finiteness conditions because the primary tool was the use of primitive idempotents to introduce coordinates. In 1975 Alfsen, Schultz, and Stömer obtained a Gelfand-Naimark theorem for Jordan C^*-algebras, and once again the basic dimensional structure theorem, but here again it was crucial that the hypotheses guaranteed a rich supply of idempotents.

In three papers [Z1], [Z2], [Z3], Zelmanov revolutionized the structure theory of Jordan algebras. These deal with prime Jordan algebras, where A is called prime if $U_B C = 0$ for ideals B and C in A implies that either B or $C = 0$. In [Z1] Zelmanov proved the remarkable result that a prime Jordan algebra without nil ideals (improved in [Z3] to prime and nondegenerate) is either i-special (a homomorphic image of a special Jordan algebra) or is a form of the 27-dimensional exceptional algebra. This applied in particular to simple algebras. The proof required the introduction of a host of novel concepts and techniques as well as sharpening of earlier methods, e.g. the coordinatization theorem of Jacobson and analogues of results on radicals due to Amitsur.

The paper [Z3] is devoted to the study of i-special Jordan algebras. Zelmanov showed that a prime nondegenerate i-special algebra is special, and he determined their structure as either of hermitian type or of Clifford type. Paper [Z2], which preceded [Z3] obtained these results for Jordan division algebras.

The principal tool in both papers is the study of the free associative algebra $\Phi\langle X\rangle$ on $X = \{x_1, x_2, \ldots\}$. This becomes a Jordan algebra $\Phi\langle X\rangle^+$ by replacing the given associative multiplication ab by $a \cdot b = \frac{1}{2}(ab + ba)$. The subalgebra $SJ\langle X\rangle$ of $\Phi\langle X\rangle^+$ generated by X is called the free special Jordan algebra.

We also have the subalgebra $H\langle X\rangle$ of $\Phi\langle X\rangle^+$ of symmetric elements ($a^* = a$) under the involution in $\Phi\langle X\rangle$ fixing the elements of X. It was shown by Paul Cohn in 1954 that $SJ\langle X\rangle \subsetneqq H\langle X\rangle$ and $H\langle X\rangle$ is the subalgebra of $\Phi\langle X\rangle^+$ generated by X and all the tetrads $x_i x_j x_k x_l$ with $i < j < k < l$. Zelmanov has obtained a completely unanticipated supplement to Cohn's theorem: the existence of elements f in $SJ\langle X\rangle$ such that if $I(f)$ denotes the ideal generated by f then

$$\{I(f), p, q, r\} \in SJ\langle X\rangle$$

for p, q, r in $SJ\langle X\rangle$. This is used to sort out the two types of i-special algebras: Clifford types characterized by the identity $f \equiv 0$ and hermitian types by the nonidentity $f \not\equiv 0$.

One of the consequences of Zelmanov's theory is that the only exceptional simple Jordan algebras, even including infinite-dimensional ones, are the forms of the 27-dimensional Albert algebras. This laid to rest the hope that had been raised by Jordan, von Neumann, and Wigner in [JNW]. Another consequence of Zelmanov's results is that the free Jordan algebra in three or more generators has zero divisors (elements a such that U_a is not injective). This is in sharp contrast to the theorem of Malcev and Neumann that any free associative algebra can be imbedded in a division algebra.

Motivated by applications to analysis and differential geometry, Koecher, Loos, and Myberg extended the structure theory of Jordan algebras to triple systems and Jordan pairs. Zelmanov applied his methods to obtain new results on these.

Lie methods were used in these papers based on the Tits-Koecher construction. The final work in this line of investigation was [Z4] in which Zelmanov applied the theory of Jordan triple systems to study graded Lie algebras with finite gradings in which the homogeneous parts could be infinite dimensional.

To encompass characteristic 2 (which is essential for applications to the restricted Burnside problem) it is necessary to deal with quadratic Jordan algebras [JM]. These were introduced by McCrimmon in [Mc] as the natural extension of Jordan algebras to algebras over any commutative ring. This amounted to replacing the product $a \cdot b = \frac{1}{2}(ab + ba)$ in an associative algebra by the product $U_a b$.

In the joint paper with McCrimmon [ZM], the results of [Z3] were extended to quadratic Jordan algebras.

2 Burnside problems

We begin with some definitions and notation.

A group is *locally finite* if every finite subset generates a finite group. In 1902 Burnside [B1] studied torsion groups and asked when such groups are locally finite. The most general form of the question is the Generalized Burnside Problem (GBP).

(GBP) Is a torsion group necessarily locally finite?

Equivalently

(GBP)′ Is every finitely generated torsion group finite?

A group G has a *finite exponent* e if $x^e = 1$ for all x in G and e is the smallest natural number with this property. Clearly a group with a finite exponent is a torsion group. A more restricted version of GBP, which already occurs in Burnside's work, is the ordinary Burnside Problem (BP).

(BP) Is every group that has a finite exponent locally finite?

There is a universal object $B(r, e)$, (the Burnside group of exponent e on r generators), which is the quotient of the free group on r generators by the subgroup generated by all eth powers. BP is equivalent to

(BP)′ Is $B(r, e)$ finite for all natural numbers e and r?

Burnside proved that groups of exponent 2 (trivial) and exponent 3 are locally finite. In 1905 Burnside [B2] showed that a subgroup of $GL(n, \mathbf{C})$ of finite exponent is finite. Schur in 1911 [Sc] proved that a finitely generated torsion subgroup of $GL(n, \mathbf{C})$ has finite exponent, and hence a torsion subgroup of $GL(n, \mathbf{C})$ is locally finite. This was very important as it showed that answers to BP or GBP would necessarily involve groups not describable in terms of linear transformations over $\mathbf{C}$. Other methods were required. In handling groups of exponent 3 Burnside had used only the multiplication table of a group. However, his methods were totally inadequate to handle, for instance, groups of prime exponent greater than 3.

During the 1930s people began to study finite quotients of $B(r, e)$ and considered the following statement.

(RBP) $B(r, e)$ has only finitely many finite quotients.

This is equivalent to

(RBP)′ $B(r, e)$ has a unique maximal finite quotient $RB(r, e)$.

W. Magnus called the question of the truth or falsity of RBP *the restricted Burnside problem.* If such a unique maximal finite quotient $RB(r, e)$ exists for some e and r, then necessarily every finite group on r generators and exponent e is a homomorphic image of $RB(r, e)$. If $RB(r, e)$ exists for some e and all r we say that RBP is true for e.

3 Results

In 1964 Golod [G] constructed infinite groups for every prime p, which are generated by 2 elements and in which every element has order a power of p, thus giving a negative answer to GBP. A few years later in 1968 Adian and Novikov [AN] showed that $B(2, e)$ is infinite for e odd and $e > 4380$, thus giving a negative answer to BP. The bound has been improved since then as $B(r, e)$ is finite for $e = 2, 3, 4$, or 6, but in no other case with $r > 1$ is it known to be finite.

In a seminal paper Hall and Higman [HH] in 1956 proved a series of results concerning RBP. Let π be a set of primes. Consider the following two statements.

(1) There are only finitely many finite simple π-groups of any given exponent.
(2) The Schreier conjecture is true for π-groups, i.e. for any finite simple π-group G, $\mathrm{Aut}(G)/G$ is solvable.

A special case of one of their results is the following.

THEOREM [HH]. *Suppose that statements* (1) *and* (2) *are true for the set* π. *Then if for every prime* p *in* π, *and natural numbers* m *and* r, $RB(r, p^m)$ *exists; then RBP is true for any exponent* e *that is a* π*-number.*

The classification of the finite simple groups shows that (1) and (2) are true for any set of primes π. Hence the truth of RBP will follow once it is proved that $RB(r, p^m)$ exists for all primes p and all natural numbers m and r.

In 1959 Kostrikin announced that $RB(r, p)$ exists for p a prime and any natural number r. Kostrikin's original argument had some difficulties. He published a corrected and updated version of his proof in his book [K], which contains numerous references to Zelmanov.

In 1989 Zelmanov announced that RBP is true for all exponents p^m with p any prime, and hence for all exponents by the remarks above. The proof appeared in 1990–91 in Russian. The English translation appeared in [Z8] and [Z9].

It should be mentioned that analogous questions have been raised for associative, Lie, and Jordan algebras. Golod's work was actually motivated by the associative algebra question and the counterexamples for groups arose as corollaries. The questions for Lie and Jordan algebras will be discussed below.

4 Some consequences

This section contains some consequences of RBP. The ideas used in the proof, in addition to the actual result, have also been applied widely.

The next three results were proved by Zelmanov [Z10] as direct consequences of RBP.

THEOREM 1. *Every periodic pro-p-group is locally finite.*

COROLLARY 2. *Every infinite compact (Hausdorff) group contains an infinite abelian subgroup.*

THEOREM 3. *Every periodic compact (Hausdorff) group is locally finite.*

Theorem 3 was conjectured by Platonov [Ko].

Shalev showed that RBP implies the following.

THEOREM 4 [Sh]. *A pro-p-group is p-adic analytic if and only if there exists a natural number n such that the wreath product $Z_p \wr Z_{p^n}$ is not a homomorphic image of any subgroup of G.*

The "only if" part of Theorem 4 is elementary, but the converse is equivalent to RBP. Since then, Zelmanov jointly with others, has made several further contributions to the study of pro-p-groups, see e.g. [ZS], [ZW].

5 Lie algebras

Let G be a finite group of exponent p^k, p a prime. Let $G = G_0$ and $G_{i+1} = [G, G_i]$ for all i. Choose s with $G_s \neq \langle 1\rangle$, $G_{s+1} = \langle 1\rangle$. Then

$$G = G_0 > \cdots > G_{s+1} = \langle 1\rangle$$

is the lower central series of G. Define

$$L(G) = \sum_{i=0}^{s} G_i/G_{i+1}$$

as abelian groups. Then $L(G)$ becomes a Lie ring with $[a_i G_i, a_j G_j] = [a_i, a_j]G_{i+j+1}$, and $L(G)$ has the same nilpotency class as G. Furthermore $L(G)/pL(G)$ is a Lie algebra over Z_p.

Let L be a Lie algebra.

L satisfies the *Engel identity* (E_n) if $ad(x)^n = 0$ for all x in L.

An element x in L is *nilpotent* if $ad(x)^n = 0$ for some n.

If G has exponent p then $L(G)$ is a Lie algebra over Z_p that satisfies (E_{p-1}). Kostrikin proved

THEOREM 5 [K]. *If L is a Lie algebra over Z_p that satisfies (E_{p-1}) then L is locally nilpotent.*

Theorem 1 implies the existence of $RB(r,p)$ and so yields RBP for prime exponent. Observe that for prime exponent $e = p$, the case $p = 2$ is trivial, so that it may be assumed that $p > 2$. This is in sharp contrast to prime power exponents $e = p^k$, where $p = 2$ is the most complicated case.

An element a of L is a *sandwich* if $[[L,a],a] = 0$ and $[[[L,a],L],a] = 0$. L is a *sandwich algebra* if it is generated by finitely many sandwiches. This concept was introduced by Kostrikin and is of fundamental importance for the proof of RBP. A first critical result is

THEOREM 6 [ZK]. *Every sandwich Lie algebra is locally nilpotent.*

Theorem 6 is essential for the proof of Theorem 5.

The main result in [Z8] is rather technical but it has the following consequence.

THEOREM 7 [Z8]. *Every Lie ring satisfying an Engel condition is locally nilpotent.*

More importantly, it implies

THEOREM 8. *$RB(r,p^k)$ exists for p an odd prime.*

Once again an essential part of the proof requires Theorem 2. Let L be a Lie algebra over an infinite field of characteristic p that satisfies an Engel condition. The way to apply Theorem 2 is to construct a polynomial $f(x_1,\dots,x_t)$ that is not identically zero, such that every element in $f(L)$ is a sandwich in L. Actually such a polynomial is not constructed but its existence for $p > 2$ follows only after a very complicated series of arguments, which constitute the bulk of the paper [Z8]. This of course settles RBP for odd exponent. (It might be mentioned that the classification of finite simple groups is not required here, only that groups of odd order are solvable.)

6 The case of exponent 2^k

The outline of the proof of RBP for exponent 2^k is similar to that for exponent p^k with $p > 2$ described in the previous section. However, the construction of the function f is vastly more complicated. It is here that quadratic Jordan algebras play an essential role, most especially the results of [ZM]. The details are extremely technical and cannot be presented here. The reader should consult [Z9] for a complete proof.

General References

[AN] S. I. Adian and P. S. Novikov, *On infinite periodic groups, I, II, III*, Math. USSR-Izv. **2** (1968), 209–236, 241–479, 665–685.

[B1] W. Burnside, *On an unsettled question in the theory of discontinuous groups*, Quart. J. Math. **33** (1902), 230–238.

[B2] ——, *On criteria for the finiteness of the order of a group of linear substitutions*, Proc. London Math. Soc. (2) **3** (1905), 435–440.

[G] E. Golod, *On nil algebras and residually finite p-groups*, Math. USSR-Izv. **28** (1964), 273–276.

[HH] P. Hall and G. Higman, *On the p-length of p-soluble groups and reduction theorems for Burnside's problem*, Proc. London Math. Soc. **6** (1956), 1–40.

[J1] N. Jacobson, Structure and representation of Jordan algebras, Amer. Math. Soc., Providence, RI 1968.

[J2] ——, Structure theory of Jordan algebras, Univ. of Arkansas Lecture Notes **5** (1981).

[JM] N. Jacobson and K. McCrimmon, *Quadratic Jordan algebras of quadratic forms with base points*, J. Indian Math. Soc. **35** (1971), 1–45.

[JNW] P. Jordan, J. von Neumann, and F. Wigner, *On an algebraic generalization of the quantum mechanical formalism*, Ann. of Math. (2) **35** (1934), 29–73.

[K] A. I. Kostrikin, Around Burnside, (Russian) MR (89d,20032).

[Ko] Kourovka Notebook, 12th ed., Inst. Mat. SO AN SSSR Novosibirsk, 1992. Problem 3.41.

[Mc] K. McCrimmon, *A general theory of Jordan rings*, PNAS **56** (1966), 1072–1079.

[Sc] I Schur, Über Gruppen periodischer Substitutionen, Collected Works **I**, 442–450, Springer, Berlin, Heidelberg, and New York 1973.

[Sh] A. Shalev, *Characterization of p-adic analytic groups in terms of wreath products*, J. Algebra **145** (1992), 204–208.

[V] M. R. Vaughan-Lee, The restricted Burnside problem, 2nd ed., Oxford University Press, Oxford, 1993.

References to Zelmanov's Work

[Z1] E. Zelmanov, *Primary Jordan algebras*, Algebra and Logic **18** (1979), 103–111.

[Z2] ——, *Jordan division algebras*, Algebra and Logic **18** (1979), 175–190.

[Z3] ——, *On prime Jordan algebras II*, Siberian Math. J. **24** (1983), 73–85.

[Z4] ——, *Lie algebras with a finite grading*, Math. USSR-Sb. **52** (1985), 347–385.

[Z5] ——, *On Engel Lie Algebras*, Soviet Math. Dokl. **35** (1987), 44–47.

[Z6] ——, *Engelian Lie Algebras*, Siberian Math. J. **29** (1989), 777–781.

[Z7] ——, *Weakened Burnside problem*, Siberian Math. J. **30** (1989), 885–891.

[Z8] ——, *Solution of the restricted Burnside problem for groups of odd exponent*, Math. USSR-Izv. **36** (1991), 41–60.

[Z9] ——, *A solution of the restricted Burnside problem for 2-groups*, Math. USSR-Sb. **72** (1992), 543–564.

[Z10] ——, *On periodic compact groups*, Israel J. Math. **77** (1992), 83–95.

[Z11] ——, Nil rings and periodic groups, Korean Math. Soc. Lecture Notes 1992, Korean Math. Soc. Seoul, South Korea.

[ZK] —— and A. I. Kostrikin, *A theorem on sandwich algebras*, Proc. Steklov Inst. Math. **183** issue **4** (1991), 121–126.

[ZM] —— and K. McCrimmon, *The structure of strongly prime quadratic Jordan algebras*, Adv. in Math. **69** (1988), 113–122.

[ZS] —— and Shalev, *Pro-p-groups of finite coclass*, Math. Proc. Cambridge Philos. Soc. **111** (1992), 417–421.

[ZW] —— and J. S. Wilson, *Identities of Lie algebras for pro-p-groups*, J. Pure Appl. Algebra **81** (1992), 103–109.

On some works of Avi Wigderson

YURI MATIYASEVICH

Steklov Institute of Mathematics, Russian Academy of Sciences
Fontanka 27, St. Petersburg 191011, Russia

I am to present the research performed by Professor Avi Wigderson. This is both a pleasant and a difficult errand, and there are two sources of this difficulty.

First, Avi Wigderson has made a lot of wonderful contributions to diverse areas of the mathematical foundations of computer science. I have time to outline only a few of them. The selection is entirely mine. If another member of the Committee were chosen for this presentation, he well might speak about different works of Avi Wigderson. The choice is indeed very large.

The second source of difficulty in presenting Avi Wigderson's works is due to the fact that in many cases they are based on complex well-balanced definitions that are too technical to be reproduced here (the balance is often between a statement to be uninteresting or to be not true). That is why my presentation will be on informal, intuitive level. (More technical details related to Avi Wigderson's research area are presented to this Congress in his paper and also in the papers by L. Babai, by O. Goldreich, and by M. Levin.)

Some of Avi Wigderson's impressive results are connected with the so-called *zero-knowledge interactive proofs*. This was a new and rather paradoxical kind of mathematical evidence introduced by Goldwasser, Micali, and Rackoff [5]. A zero-knowledge interactive proof allows a mathematician A, called *the Prover*, to *convince* another mathematician B, called *the Verifier*, that a certain mathematical statement is true without providing a formal proof, and moreover, without giving any hints as to such a proof.

For example, suppose that mathematician A established that some large integer N is composite by finding its nontrivial factorization $N = PQ$. Now he or she can, in a dialog with another mathematician B, convince the latter that N is indeed composite but without revealing the values of the factors.

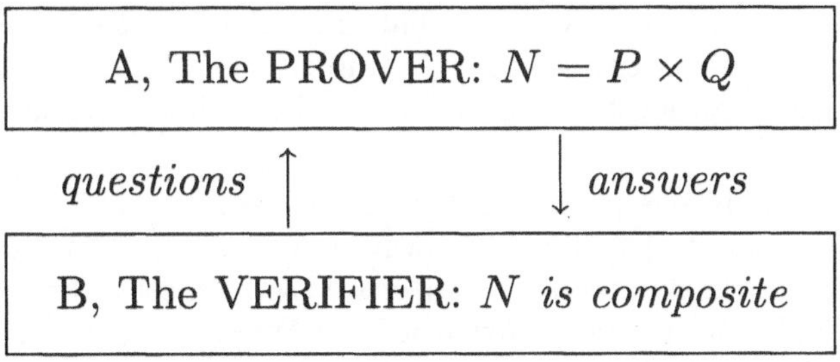

Proceedings of the International Congress
of Mathematicians, Zürich, Switzerland 1994

Of course, in mathematics there cannot be a secret in the strong sense. As soon as two mathematicians have agreed on the axioms and on the rules of logical deduction, it is only a matter of time and space to find a proof for any provable statement. (In our example mathematician B could find a factorization of N himself or herself.) Speaking more technically, zero-knowledge means that the amount of the work required for the Verifier to find an actual proof *after* the dialog with the Prover is on average essentially the same as it was *before* the dialog.

In fact, an interactive proof is not a proof in the strong sense because such a proof can be given, in principle, even to a statement that is not true at all. However, the probability of such a misproof goes to zero exponentially with the number of questions asked and hence the probability of giving an interactive "proof" for a false statement can be easily made arbitrarily small. This is just what gives the convincing power to interactive proofs.

For such a convincing but zero-knowledge proof to become possible, the Prover and the Verifier should agree in advance upon what kinds of questions can be asked. In technical terms such an agreement is called *protocol.*

When the notion of zero-knowledge was first introduced, zero-knowledge protocols were known only for statements of a rather restricted nature. The actual scope of zero-knowledge proofs remained unclear. Avi Wigderson, in collaboration with Goldreich and Micali [2], established the following fundamental fact: *zero-knowledge proofs are possible for so-called class* **NP**. This class is one of the main subjects of study in modern computer science. Roughly speaking, class **NP** consists of statements that become easy to verify when a small amount of additional information is supplied.

(In our example, it is easy to verify that N is composite as soon as its factors, P and Q, are given. This example was chosen as an easy-to-explain specimen of a statement from **NP**. In fact, thanks to the magic of number theory, in this particular case no dialog is required for mathematician B to become convinced of the mere fact that N is composite. However, a zero-knowledge interactive proof can do much more, namely, *convince* mathematician B that mathematician A *knows* nontrivial factors P and Q.)

This result on the existence of zero-knowledge proofs for class **NP** was based on the conjecture of the existence of so called *one-way functions.* Roughly speaking, F is such a function if it is easy to calculate its value $y = F(x)$ for given x but it is difficult to find an x for given y. For example, it is easy to calculate an integer from its prime factorization but it is *believed* to be a difficult task to factor a large integer.

However, nobody has proved so far that factoring is in fact a difficult task. Moreover, the existence of a single one-way function has not been proved so far. Nevertheless, it is a widely accepted conjecture used almost as an axiom and hence its use was completely justified. However, Avi Wigderson, in collaboration with Ostrovski [6], also studied what would follow from the nonexistence of one-way functions. They established that *the assumption of the existence of one-way functions was indeed essential for the existence of nontrivial zero-knowledge interactive proofs.*

This is so for the original type of interactive proofs. Wishing still to avoid the yet unproved conjecture, Avi Wigderson in collaboration with Ben-Or, Goldwasser, and Kilian [4], introduced a new kind of interactive proof. In such *multi-prover interactive proofs* two or more persons convince another one by conversation with the latter. In this scheme the Verifier cannot check whether the answers are correct but he or she can check that the answers from different Provers are consistent. That is why the Verifier should be sure that during the interactive proof there is no exchange of information among the Provers, and this nonmathematical assumption replaces the yet unproved conjecture on the existence of one-way functions.

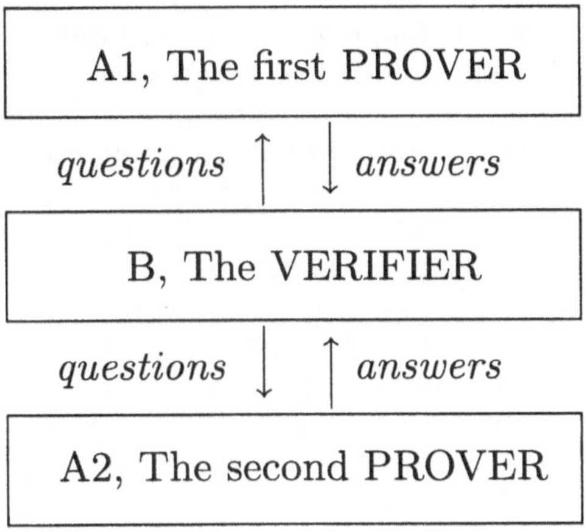

(This difference between single-user and multi-user interactive proofs can be viewed as mathematical justification of the saying *"Two heads are better than one"* which seems to have counterparts in many languages. In a sense, almost all papers of Avi Wigderson contribute to the justification of such sayings, at least with respect to mathematics, because almost all his papers were written in collaboration with other mathematicians.)

Zero-knowledge proofs are a very interesting subject of study by themselves but they have also found numerous applications, in particular, for *fault-tolerant distributed computations.* Such computations involve several communicating agents, some of whose work could be faulty. In this area we found

> A TYPICAL THEOREM. A particular computational task can be achieved by a computational network of k participating agents provided that strictly fewer than ck of them are faulty.

Avi Wigderson, in collaboration with Goldwasser and Ben-Or [3] and with Goldreich and Micali [1], obtained very general results of this kind. These results are striking in two aspects. First, they are uniform; namely, the values of c do not depend on the particular computational task (but may vary for different assumptions about the network). Second, the values of c obtained are the best possible, namely, given exactly ck or more faulty agents, there exist computations that cannot be accomplished.

Interactive proofs use randomness in an essential way. (Clearly, the Prover must be unable to predict the forthcoming questions of the Verifier.) For a long time randomness has been recognized as a specific computational resource. Avi Wigderson contributed to the study of the computational power of randomness and to the construction of pseudo-random generators.

Formal definition of zero-knowledge, of what constitutes a good pseudo-random generator, and definitions of many other important concepts are based on estimates of the amount of computational work required for particular tasks. Such estimates depend heavily on the computational model used, and here Avi Wigderson shows his universal capacities. He contributed to both lower and upper bounds on the complexity of computations on very different computational devices: from powerful parallel computers to very restricted Boolean circuits. While nontrivial upper bounds are of practical interest, the lower bounds in this area are usually much more difficult to prove. In fact, to get an upper bound one needs to find *only one* algorithm, let ingenious, whereas to get a lower bound one has to observe in some way *all* conceivable algorithms for a particular task.

Avi Wigderson's more recent investigations are connected with computational aspects of special dynamical systems motivated by genetic algorithms and kinetic gas models. Here he had already contributed to transferring to nonlinear systems results that previously were known only in the linear case. I do not doubt that soon we shall hear about his new striking results in this area as well.

References

[1] O. Goldreich, S. Micali, and A. Wigderson, *How to play any mental game*, Proceedings of ACM 19th Annual Symposium on Theory of Computing, (1987), 218–229.

[2] O. Goldreich, S. Micali, and A. Wigderson, *Proofs that yield nothing but their validity, or all languages in NP have zero-knowledge proof systems*, J. Assoc. Comput. Mach., 38:1 (1991), 691–729.

[3] S. Goldwasser, M. Ben-Or, and A. Wigderson, *Completeness theorems for non-cryptographic fault-tolerant distributed computing*, Proceedings of ACM 20th Annual Symposium on Theory of Computing, (1988), 1–10.

[4] S. Goldwasser, J. Kilian, M. Ben-Or, and A. Wigderson, *Multi-prover interactive proofs: How to remove intractability assumptions*, Proceedings of ACM 19th Annual Symposium on Theory of Computing, (1988), 113–131.

[5] S. Goldwasser, S. Micali, and C. Rackoff, *Knowledge complexity of interactive proofs*, Proceedings of ACM 17th annual Symposium on Theory of Computing, (1985), 291–304.

[6] R. Ostrovski and A. Wigderson, *Nontrivial zero-knowledge implies one-way functions*, Proc. of the 2nd Israel Symposium on Theory of Computing and Systems, (1993), 1–13.

Invited One-Hour Addresses at the Plenary Sessions

Harmonic Analysis and Nonlinear Partial Differential Equations

JEAN BOURGAIN

Inst. for Advanced Study
Princeton, NJ 08540
USA

I.H.E.S.
F-91440 Bures-sur-Yvette
France

University of Illinois
Urbana, IL 61801
USA

1 Introduction

The aim of this report is to describe some recent research in the area of nonlinear evolution equations. The choice of the topics is largely influenced by the author's own interests and it is in no way a complete survey of this field, which would be nearly impossible to achieve in a single exposition. Some very outstanding achievements in recent years such as, for instance, the work of Christodoulou and Klainerman [C-K] on global solutions of the Einstein equations will not be discussed here.

We will be mainly concerned with nonlinear Hamiltonian equations on bounded domains (Dirichlet or periodic boundary conditions say) and the following issues:

(i) The Cauchy problem; i.e., local and global wellposedness results for individual data

(ii) Behavior of solutions for time $\to \infty$

(iii) Behavior of the flow in phase space.

These issues are rather well understood in the integrable case, because of the presence of a large set of invariants of motion. The integrable Hamiltonian evolution equations form a small and distinguished class, including, for instance, the

1D cubic nonlinear Schrödinger (NLS) equation $iu_t + u_{xx} + u|u|^2 = 0$

Korteweg de Vries (KdV) equation $u_t + \partial_x^3 u + uu_x = 0$

Modified KdV equation $u_t + \partial_x^3 u + u^2 u_x = 0$.

These invariants of motion allow us to control for a given data $u(0) = \phi$ the solution $u(t)$ for all time. In the general Hamiltonian case on the other hand, one only disposes of a few conserved quantities, namely the Hamiltonian itself, sometimes the L^2-norm $\|u(t)\|_{L^2}$. Hence, to establish global existence of solutions even for smooth data, one needs to study the local wellposedness problem for data of low regularity, because the existence time should only depend on the conserved quantities. This procedure leads to estimates on higher smoothness norms that

Proceedings of the International Congress
of Mathematicians, Zürich, Switzerland 1994

are exponential in time, thus $\|u(t)\|_{H^s} < C^{|t|}$, for some $C > 1$ depending on the initial data. A natural problem is whether this is the "true" behavior or only a crude estimate, and the lack of a rigorous mathematical approach here is in sharp contrast to the situation in integrable models. On the other hand, certain information on the global behavior of the flow of the equation in an appropriate phase space may be obtained from methods of statistical mechanics and symplectic geometry. The Gibbs measure construction from statistical mechanics gives a normalization procedure of the formally invariant Liouville measure on an infinite-dimensional phase space and permits us to obtain Poincaré recurrence properties for the flow. Other symplectic invariants, called symplectic capacities, originating from Gromov's pioneering work [Grom], allow us to study "squeezing properties" and energy transitions in the symplectic normalization of phase space. These normalizations of the phase space are however such that the resulting theories deal with low-regularity solutions and consequently, as a first step, require us again to establish the existence of the flow for such data. Our investigations on the Cauchy problem have been mainly pursued for periodic boundary conditions (i.e. the space variable ranges in a d-dimensional torus $\mathbb{T}^d$), which is also the context for the discussion above. In fact, the literature on the initial value problems (IVPs) in the periodic case is far less extensive than that on the line and the theory is less developed. It turned out that the analysis in the periodic situation is significantly different (due for instance to the absence of dispersion) and requires new ideas, some of which also eventually lead to an advance on the corresponding problem for the line. Several results in this direction will be discussed in the next section.

A third important method borrowed from classical mechanics is the KAM method to establish persistency of time periodic or quasi-periodic solutions of small Hamiltonian perturbations of linear or integrable equations. The main contributor in adapting the KAM technology to the PDE setting is Kuksin [Kuk_1]. His work gives satisfactory results for 1D problems with Dirichlet boundary conditions. A different approach, avoiding some of the limitations of the KAM technique, has been elaborated by Craig and Wayne [$\text{C-W}_{1,2}$] and the author [B_1] and permits us to deal with 1D periodic boundary conditions.

2 Initial value problems for KdV type equations

There are numerous investigations of the Cauchy problem for the standard KdV equation on $\mathbb{R}$

$$\begin{cases} u_t + \partial_x^3 u + u u_x = 0 \\ \\ u(0) = \phi(x) \end{cases} \tag{2.1}$$

using either fixpoint techniques or inverse scattering methods. The advantage of the fixpoint approach is its large range of applicability, and it is the only method we consider here. The setup is given by Duhamel's integral formula

$$u(t) = S(t)\phi - \int_0^t S(t-\tau)\, w(\tau)\, d\tau \quad , \quad w = u u_x \tag{2.2}$$

where $S(t)\psi$ solves the linear problem

$$\begin{cases} u_t + \partial_x^3 u = 0 \\ u(0) = \psi \end{cases} \tag{2.3}$$

and is explicitly given by the oscillatory integral ($\widehat{\psi}$ denotes the Fourier transform of ψ)

$$S(t)\psi(x) = \int \widehat{\psi}(\lambda)\, e^{i(\lambda x + \lambda^3 t)}\, d\lambda \tag{2.4}$$

on the line ($\mathbb{R}$-case) and the exponential sum

$$S(t)\psi(x) = \sum_{n\in\mathbb{Z}} \widehat{\psi}(n)\, e^{i(nx+n^3t)} \tag{2.5}$$

in the periodic case ($\mathbb{T}$-case).

Solving (2.2) by a fixpoint argument in the $\mathbb{R}$-case is mainly based on the regularizing properties of the linear group $S(t)$, such as Stricharti's inequality and Kato's smoothing. In the periodic case, no smoothing properties may be expected. We introduced in [B_2] new space-time norms defined in terms of the Fourier transform of u. These norms exploit some arithmetical features, which form a substitute for the smoothing properties of (2.4). In fact, the regularity gains here are due both to the linear part of the equation and the specific structure of the nonlinear part. This method and its application in conjunction with earlier techniques lead to the "best" known results on the IVP for the KdV equation.

THEOREM 2.6. ($\mathbb{T}$) *There is local wellposedness for data $\phi \in H^s(\mathbb{T})$, $s \geq -\frac{1}{2}$; global wellposedness for data $\phi \in L^2(\mathbb{T})$; the solutions resulting from L^2-data are almost periodic in time (see* [B_3]*).*

($\mathbb{R}$) *There is local wellposedness for data $\phi \in H^s(\mathbb{R})$, $s > -\frac{3}{4}$; global wellposedness for data $\phi \in L^2(\mathbb{R})$ (see* [K-P-V_1], [K]*).*

The solution depends real analytically on the data.

REMARKS.

(i) By wellposedness, we mean the construction of a unique solution for a certain class of data (coinciding with the classical solution in the smooth case) and depending continuously on the data.

(ii) The almost periodicity of KdV solutions is a subject with a long history that we will not recall here. Important steps are due to Gardner-Kruskal-Miura [G-K-M], Lax [Lax], Novikov [Nov], and McKean-Trubowitz [M-T]. The statement for L^2-data is a consequence of the work of [M-T] and the existence of regular L^2-flow.

A direct generalization of KdV are equations of the form

$$u_t + \partial_x^3 u + f(u)u_x = 0 \tag{2.7}$$

where f is a smooth function of u. The Hamiltonian is given by

$$H(\phi) = \tfrac{1}{2} \int (\phi_x)^2 - \int F_2(\phi) \tag{2.8}$$

where F_2 is the second primitive of f ($F_2(0) = F_2'(0) = 0$) and equation (2.7) is equivalent to

$$u_t = D_x \frac{\partial H}{\partial u}\ . \tag{2.9}$$

The mean $\int \phi$, the L^2-norm $\int \phi^2$, and $H(\phi)$ are preserved under the flow. In the general case ($f(u) = u, u^2$ are special), these are the only invariants of motion at our disposal. Construction of global solutions based on these a priori bounds requires a local theory for H^1-data. In the $\mathbb{R}$-case, results along these lines appear in the works of Kenig, Ponce, and Vega (see [K-P-V$_2$] for instance). We state the theorem in the periodic case (see [B$_2$]).

THEOREM 2.10. *The* IVP

$$u_t + \partial_x^3 u + f(u)u_x = 0 \quad ; \quad u(0) = \phi$$

is globally well posed for sufficiently smooth data as long as the H^1-norm remains bounded. This is in particular the case for small data.

The proof uses many of the techniques developped for the periodic KdV case. We just want to mention one additional point, which is a certain "renormalization" of the nonlinearity, in the spirit of "Wick-ordering" discussed below. Rewrite (2.7) in the form

$$u_t + \partial_x^3 u + v(t)u = [v(t) - f(u)]\ u_x \quad \text{with} \quad v(t) = \int_{\mathbb{T}} f(u)\, dx \tag{2.11}$$

redefining the linear and nonlinear parts of the equation. Observe that for $f(u) = u$ or u^2, $v(t)$ is time independent. The different setup (2.11) seems necessary for a regularizing interaction between linear and nonlinear terms in a fixpoint argument.

REMARK. There has been recent work by Klainerman and Machedon [K-M$_1$] on nonlinear wave equations in the same spirit as Theorem 2.10. In particular, a local H^1-theory is developed considering appropriate space-time norms, and a careful analysis of the nonlinear term is needed. See also [K-M$_2$] for IVP results related to Yang-Mills equations.

A 2-dimensional generalization of the KdV equation is given by the Kadomstev-Petviashvili equation (KP). The KP-II equation

$$u_t + \partial_x^3 u + uu_x + D_x^{-1} u_{yy} = 0 \tag{2.12}$$

is integrable. However, the conserved integrals

$$\int u \, dx \, dy$$

$$\int u^2 \, dx \, dy$$

$$\int \left[u_x^2 - \tfrac{1}{3}\, u^3 - (D_x^{-1} u_y)^2\right] dx \, dy$$

do not imply immediately a priori bounds on the solution, except for the L^2-norm (because of sign features). Techniques closely related to those used in proving the periodic results for the KdV equation (Theorem 2.6) yield the following.

THEOREM 2.13. [B_4] *The* KP-II *equation is globally well posed for data* $\phi \in H^s(\mathbb{T}^2)$ *or* $\phi \in H^s(\mathbb{R}^2)$, $s \geq 0$.

There is a rich algebraic theory around the KP-II equation and explicit solutions may be expressd in terms of logarithmic derivatives of θ-functions associated to certain (possibly infinite genus) Riemann surfaces. This theory has been developed by Its, Novikov, and Krichever among others (see [Kr]). Very recently, Knörrer and Trubowitz proved the following analogue of the [M-T] result for the KdV equation.

THEOREM 2.14. [K-Tr] *Solutions of the periodic* KP-II *equations for smooth periodic data are almost periodic in time.*

3 Nonlinear Schrödinger equations and invariant Gibbs measures

We consider the nonlinear Schrödinger (NLS) equation

$$iu_t + \Delta u \pm u|u|^{p-2} = 0 \tag{3.1}$$

with periodic boundary conditions. Thus, u is a complex function on $\mathbb{T}^d \times I$ (local) or $\mathbb{T}^d \times \mathbb{R}$ (global). The equation may be rewritten in Hamiltonian format as

$$u_t = i\,\frac{\partial H}{\partial\, \bar{u}} \tag{3.2}$$

where $H(\phi) = \frac{1}{2}\int_{\mathbb{T}^d} |\nabla\phi|^2 \mp \frac{1}{p}\int_{\mathbb{T}^d} |\phi|^p$. Both the Hamiltonian $H(\phi)$ and the L^2-norm $\int |\phi|^2$ are preserved under the flow. The $1D$ case $p = 4$ is special (1D cubic NLS) because it is integrable and there are many invariants of motion. This aspect will however play no role in the present discussion. The possible sign choice $\pm$ in (3.1) corresponds to the focusing (resp. defocusing) case. In the focusing case, the Hamiltonian may be unbounded from below and blowup phenomena may occur (for $p \geq 2 + \frac{4}{d}$). The canonical coordinates are $(\operatorname{Re}\phi, \operatorname{Im}\phi)$ or alternatively $(\operatorname{Re}\widehat{\phi}, \operatorname{Im}\widehat{\phi})$. The formal Gibbs measure on this infinite-dimensional phase is given by

$$d\gamma_\beta = e^{-\beta H(\phi)} \prod_x d\phi(x) = e^{\pm\frac{\beta}{p}\int|\phi|^p} \cdot e^{-\frac{\beta}{2}\int|\nabla\phi|^2} \prod_x d\phi(x). \tag{3.3}$$

($\beta > 0$ is the reciprocal temperature and we may take $\beta = 1$ in this discussion.)

From Liouville's theorem, (3.3) defines an invariant measure for the flow of (3.1). Making this statement precise requires us to clarify the following two issues:

(i) The rigorous construction (normalization) of the measure (3.3)

(ii) The existence problem for the flow of (3.1) on the support of the measure.

The first issue is well understood in the defocusing case. The case $D = 1$ is trivial, the case $D = 2$, p even integer is based on the Wick-ordering procedure (see [G-J]), and the normalization for $D = 3$, $p = 4$ is due to Jaffe [Ja]. In the focusing case, only the case $D = 1$ is understood [L-R-S] and normalization of the measure is possible for $p \leq 6$, restricting ϕ to an appropriate ball in $L^2(\mathbb{T})$.

The construction of a flow is clearly a PDE issue. The author succeeded in this in the $D = 1$ and $D = 2$, $p = 4$ cases ([B$_5$], [B$_6$]). For $D = 2$, $p = 4$ there is a natural PDE counterpart of the Wick-ordering procedure and equation (3.1) has to be suitably modified (this modification seems physically inessential however). We may summarize the results as follows.

THEOREM 3.4. *($D = 1$)* (i) *In the defocusing case, the measure* (3.3) *appears as a weighted Wiener measure, the density being given by the first factor. The same statement is true in the focusing case for $p \leq 6$, provided one restricts the measure to an L^2-ball $[\|\phi\|_2 \leq B]$. The choice of B is arbitrary for $p < 6$ and B has to be sufficiently small if $p = 6$.*

(ii) *Assuming the measure exists, the corresponding* 1D *equation* (3.1) *is globally well posed on a K_σ set A of data, $A \subset \bigcap_{s<\frac{1}{2}} H^s(\mathbb{T})$, carrying the Gibbs measure γ_β. The set A and the Gibbs measure γ_β are invariant under the flow.*

REMARKS.

(i) In dimension 1, the L^2-restriction is acceptable, because L^2 is a conserved quantity and a typical ϕ in the support of the Wiener measure is a function in $H^s(\mathbb{T})$, for all $s < \frac{1}{2}$. Instead of restricting to an L^2-ball, one may alternativelymultiply with a weight function with a suitable exponential decay in $\|\phi\|_2$.

(ii) Let for each $N = 1, 2, \dots$

$$P_N\phi = \phi_N = \sum_{|n|\leq N} \widehat{\phi}(n)\, e^{i\langle n,x\rangle} \tag{3.5}$$

be the restriction operator to the N first Fourier modes. Finite dimensional versions of the PDE model are obtained considering "truncated" equations

$$\begin{cases} iu_t^N + u_{xx}^N \pm P_N\left(u^N|u^N|^{p-2}\right) = 0 \\ u^N(0) = P_N\phi\,. \end{cases} \tag{3.6}$$

It is proved that for typical ϕ, the solutions u^N of (3.6) converge in the space $C_{H^s(\mathbb{T})}[0,T]$ for all time T and $s < \frac{1}{2}$ to a (strong) solution of

$$\begin{cases} iu_t + u_{xx} \pm P\left(u|u|^{p-2}\right) = 0 \\ u(0) = \phi\ . \end{cases} \tag{3.7}$$

THEOREM 3.8. *($D = 2$, $p = 4$) (i) Denote $\widetilde{H}_N$ the Wick-ordered Hamiltonians, obtained replacing*

$$|\phi_N|^4 \quad by \quad |\phi_N|^4 - 4a_N|\phi_N|^2 + 2|a_N|^2 \qquad \left(a_N = \sum_{|n|\le N} \frac{1}{|n|^2+\rho} \sim \log N\right).$$

The corresponding measures $e^{-\beta\widetilde{H}_N(\phi)}\prod d\phi$ converge for $N \to \infty$ to a weighted 2-dimensional Wiener measure whose density belongs to all L^p-spaces. Denote by $\widetilde{\gamma}_\beta$ this "Wick-ordered" Gibbs measure.

(ii) *The measure $\widetilde{\gamma}_\beta$ is invariant under the flow of the "Wick-ordered" equation*

$$iu_t + \Delta u - \left(u|u|^2 - 2u\int |u|^2\right) = 0 \tag{3.9}$$

which is well defined. More precisely, denoting by u^N the solutions of

$$\begin{cases} iu_t^N + \Delta u^N - P_N\left(u^N|u^N|^2 - 2u^N \int |u^N|^2\right) = 0 \\ u^N(0) = P_N\phi \end{cases} \tag{3.10}$$

the sequence

$$u^N(t) - \sum_{|n|\le N} \widehat{\phi}(n)\ e^{i(\langle n,x\rangle + |n|^2 t)} \tag{3.11}$$

converges for typical ϕ in $C_{H^s(\mathbb{T}^2)}[0,T]$ for some $s > 0$, all time T, to

$$u(t) - \sum \widehat{\phi}(n)\ e^{i(\langle n,x\rangle + |n|^2 t)}. \tag{3.12}$$

REMARKS.

(i) We repeat that the novelty of Theorem 3.8 lies in the second statement on the existence of a flow. The first statement is a classical result.

(ii) The second terms in (3.11), (3.12) are the solutions to the linear problem

$$\begin{cases} iu_t + \Delta u = 0 \\ u(0) = \phi\ . \end{cases}$$

Here a typical ϕ is a distribution, not a function. However the difference (3.12) between solutions of the linear and nonlinear equation is an H^s-function for some $s > 0$, which is a rather remarkable fact.

(iii) The failure in $D = 2$ of typical ϕ to be an L^2-function makes the [L-R-S] construction for $D = 1$ inadequate to deal with the $D = 2$ focusing case. Some recent work on this issue is due to Jaffe, but for cubic nonlinearities in the Hamiltonian only. The problem for $D = 2$, $p = 4$ in the focusing case is open and intimately related to blowup phenomena ($p = 4$ is critical in 2D).

The $1D$ cubic NLS equation appears as the limit of the $1D$ Zakharov model (ZE)

$$\begin{cases} iu_t = -u_{xx} + nu \\ n_{tt} - c^2 n_{xx} = c^2 \left(|u|^2\right)_{xx} \end{cases} \tag{3.13}$$

when $c \to \infty$. The physical meaning of u, n, c are resp. the electrostatic envelope field, the ion density fluctuation field, and the ion sound speed. This model is discussed in [L-R-S]. Defining an auxiliary field $V(x,t)$ by

$$\begin{cases} n_t = -c^2\, V_x \\ V_t = -n_x - |u|_x^2 \end{cases} \tag{3.14}$$

we may write (3.13) in a Hamiltonian way, where

$$H = \tfrac{1}{2} \int \left[|u_x|^2 + \tfrac{1}{2}\,(n^2 + c^2\, V^2) + n|u|^2\right]\, dx \tag{3.15}$$

and $(\operatorname{Re} u, \operatorname{Im} u)$, $\left(\widetilde{n}, \widetilde{V}\right)$ with $\widetilde{n} = 2^{-1/2} n$, $\widetilde{V} = 2^{-1/2} \int^x V$ as pairs of conjugate variables. Considering the associated Gibbs measure

$$e^{-\beta H} \cdot \chi_{\left\{\int |u|^2 dx \leq B\right\}} \prod_x d^2 u(x)\; d\widetilde{n}(x)\; d\widetilde{V}(x) \tag{3.16}$$

one gets the 1D cubic NLS Gibbs measure as marginal distribution of the u-field.

THEOREM 3.17. [B$_7$] *The* $1D$ *(ZE) is globally well posed for almost all data* $\left(u_0, \widetilde{u}_0,\, \widetilde{V}_0\right)$ *in the support of the Gibbs measure, which is invariant under the resulting flow.*

REMARKS.

(i) In the study of invariant Gibbs measures, it suffices to establish local wellposedness of the IVP for typical data in the support of the measure. One may then exploit the invariance of the measure as a conservation law and generate a global flow. For instance, for the 1D NLS $iu_t + u_{xx} \pm u|u|^{p-2} = 0$, there is for $p = 4$ a global wellposedness result for L^2-data (L^2 is conserved).

However, for $p > 4$, we only dispose presently of a local result (in the periodic case) for data ϕ satisfying

$$\begin{cases} \phi \in H^s \ , \ s > 0 \qquad (p \leq 6) \\ \phi \in H^s \ , \ s > s_* \ , \ p = 2 + \frac{4}{1-2s_*} \qquad (p > 6) \end{cases} \tag{3.18}$$

and a global flow is established from the invariant measure considerations.

(ii) There have been other investigations in 1D on invariant measures, mostly by more probabilistic arguments. In this respect, we mention the works of McKean-Vaninski and in particular Mckean [McK] on the 1D cubic NLS. These methods are more general but give less information on the flow.

4 Symplectic capacities, squeezing and growth of higher derivatives

The works of Gromov and Ekeland, Hofer, Zehnder, and Viterbo lead to new finite-dimensional symplectic invariants, different from Liouville measure on the phase space. Let us recall the following construction of a symplectic capacity for open domains O in $\mathbb{R}^n \times \mathbb{R}^n$, $dp \wedge dq$. Call a smooth function f m-admissible ($m > 0$) if $f = 1$ on a neighborhood of O and $f = 0$ on a nonempty subdomain of O. Denote V_f the associated Hamiltonian vector field $\left(\frac{\partial f}{\partial p} \ , \ -\frac{\partial f}{\partial q}\right)$. Define the symplectic invariant

$$\begin{aligned} c_{2n}(O) = \ &\inf \{m > 0 \mid V_f \text{ has nontrivial periodic orbit of period } \leq 1 \ , \\ &\text{whenever } f \text{ is } m\text{-admissible for } O\}. \end{aligned} \tag{4.1}$$

Then $c_{2n}(\cdot)$ is monotonic and translation invariant and scales as $c_{2n}(\tau O) = \tau^2 c_{2n}$ (O). The main property is that

$$c_{2n}(B_\rho) = \pi\rho^2 = c_{2n}\left(\Pi_\rho\right) \tag{4.2}$$

where B_ρ is the ball $B_\rho = \left\{|p|^2 + |q|^2 < \rho^2\right\}$ and $\prod_\rho$ a cylinder, say $\prod_\rho = \{p_1^2 + q_1^2 < \rho^2\}$. As a corollary, there is no symplectic squeezing of a ρ-ball in a cylinder of width ρ', $\rho' < \rho$.

Exploiting such an invariant in Hamiltonian PDE requires an infinite-dimensional setting. Notice that although the theory described above is finite-dimensional, a conclusion such as (4.2) is dimension free. An appropriate "finite-dimensional approximation" appears to be possible if the flow S_t of the considered equation is of the form

$$\text{linear operator } + \text{ "smooth compact operator"} \tag{4.3}$$

or, more generally, if the evolution of individual Fourier modes on a finite time interval is approximately the same as in a truncated model $\dot{v} = J\nabla H(v, x, t)$, $v = P_N v$. Here the cutoff N should only depend on the required approximation, the time interval $[0, T]$, and the size of the initial data in phase space. Here and also in (4.3), the phase space has to be defined in a specific way, corresponding to the finite-dimensional normalizations. Hence, the flow properties derived this way relate to a specific "symplectic Hilbert space", for instance

L^2 for nonlinear Schrödinger equations (in any dimension)

$H^{1/2} \times H^{1/2}$ for nonlinear wave equations (in any dimension)

$H^{-1/2}$ for KdV type equations,

and "nonsqueezing" refers to that particular space.

THEOREM 4.4. ([B_8], [Kuk_2]) *There is nonsqueezing of balls in cylinders of smaller width*

(i) *For nonlinear wave equations* $u_{tt} = \nabla u + p(u; t, x)$ *with smooth nonlinearity of arbitrary polynomial growth in* u *in dimension* 1 *and polynomial in* u *of degree* ≤ 4 *(resp.* ≤ 2*) in dimension* 2 *(resp.* 3, 4*).*

(ii) For certain $1D$ *nonlinear Schrödinger equations.*

The interest of the squeezing or nonsqueezing properties lies in its relevance to the energy transition to higher modes, more precisely whether, for instance, part of the energy may leave a given Fourier mode, which would correspond to squeezing in a small cylinder. The nonsqueezing implies also the lack of uniform asymptotic stability of bounded solutions; i.e., $\operatorname{diam} S_t(B_\rho)$ does not tend to 0 for $t \to \infty$ if $\rho > 0$.

The drawback of those results is that they do not relate to properties of the flow in a classical sense, because of the phase space topology. On the other hand, Kuksin showed recently that in fact certain squeezing of balls in cylinders may occur in spaces of higher smoothness, if one considers for instance a nonlinear wave equation $u_{tt} = \rho\Delta u + p(u)$ where ρ is a small parameter (small dispersion). The squeezing phenomena appear in some finite time and are stronger when $\rho \to 0$.

As far as the behavior of individual smooth solutions concerns, some examples are obtained in [B_2] and [B_8] of Hamiltonian PDE (in NLS or KdV form) defined as a smooth perturbation of a linear equation, showing in particular that higher derivatives of solutions $u(t)$ for smooth data $u(0) = \phi$ need not be bounded in time. For instance

PROPOSITION 4.5. *There is a Hamiltonian* NLS *equation with smooth and local nonlinearity such that* $S_t(B^s(\delta))$, $t > 0$, *is not a bounded subset of* H^{s_0}, *for any* $s < \infty$, $\delta > 0$. *Here* $B^s(\delta)$ *denote* $\{\varphi \in H^s \mid \|\varphi\|_s < \delta\}$ *and* s_0 *is numerical.*

Another example, closely related to the discussion in the next section, is the following. Considering a linear Schrödinger equation

$$-iu_t = -u_{xx} + V(x)u \tag{4.6}$$

where $V(x)$ is a real smooth periodic potential and the periodic spectrum $\{\lambda_k\}$ of $-\frac{d^2}{dx^2} + V$ satisfies a "near resonance" property

$$\operatorname{dist}(\lambda_{n_j}, \mathbb{Z}\lambda_{n_0}) \to 0 \quad \text{rapidly for} \quad j \to \infty \tag{4.7}$$

for some subsequence $\{n_j\}$. We construct a Hamiltonian perturbation $\Gamma(u) = \frac{\partial}{\partial \bar{u}} G$ such that the solution $u_{\varepsilon,q}$ of the IVP

$$\begin{cases} -iu_t = -u_{xx} + V(x)u + \varepsilon\, \Gamma(u) \\ \qquad u(0) = q \end{cases} \tag{4.8}$$

satisfies

$$\inf_{q \in O} \sup_t \|u_{\varepsilon,q}(t)\|_{H^{s_0}} \to \infty \quad \text{for} \quad \varepsilon \to 0. \tag{4.9}$$

Here s_0 is again some positive integer and O is some nonempty open subset of $H^{s_0}(\mathbb{T})$.

5 Persistency of periodic and quasi-periodic solutions under perturbation

One of the most exciting recent developments in nonlinear PDE is the use of the classical KAM-type techniques to construct time quasi-periodic solutions of Hamiltonian equations obtained by perturbation of a linear or integrable PDE. This subject is rapidly developing. Results so far are only obtained in 1D and in this brief discussion, we only consider perturbations of linear equations. We work in the real analytic category. Important contributions are due to Kuksin [Kuk$_1$], using the standard KAM scheme and more precisely infinite-dimensional versions of Melnikov's theorem on the persistency of n-dimensional tori in systems with $N > n$ degrees of freedom. His work yields a rather general theory and we mention only some typical examples of applications to 1D nonlinear wave or Schrödinger equations

$$w_{tt} = \left(\frac{\partial^2}{\partial_{x^2}} - V(x;a)\right) w - \varepsilon\, \frac{\partial \varphi}{\partial_w}\,(x, w; a) \tag{5.1}$$

$$-iu_t = -u_{xx} + V(x,a)u + \varepsilon\, \frac{\partial \varphi}{\partial_{|u|^2}}\,(x, |u|^2; a)\, u. \tag{5.2}$$

Here $V(x,a)$ is a real periodic smooth potential, depending on n outer parameters $a = (a_1, \dots, a_n)$. Denote $\{\lambda_j(a)\}$ the Dirichlet spectrum of the Sturm-Liouville operator $-\frac{d^2}{dx^2} + V(x,a)$. Thus, $\lambda_j(a) = \pi^2 j^2 + 0(1)$ and we assume the following nondegeneracy condition

$$\det\{\partial\lambda_j(a)/\partial a_k \mid 1 \le j, k \le n\} \ne 0 \tag{5.3}$$

(this condition is a substitute for the classical "twist" condition). Denoting $\{\varphi_j\}$ the corresponding eigenfunctions, the $2n$-dimensional linear space

$$Z^0 = \text{span }\{\varphi_j, i\varphi_j \mid 1 \le j \le n\} \tag{5.4}$$

is invariant under the the flow of equation (5.2) for $\varepsilon = 0$ and foliated into invariant n-tori

$$T^n(I) = \left\{ \sum_{j=1}^n (x_j^+ + ix_j^-)\varphi_j \mid (x_j^+)^2 + (x_j^-)^2 = 2I_j \;,\; j = 1, \dots, n \right\} \tag{5.5}$$

which are filled with quasi-periodic solutions of (5.2) for $\varepsilon = 0$. A typical result from [Kuk_1] is that under assumption (5.3), for most parameter values of a there is an invariant torus $\sum_{a,I}^{\varepsilon} (\mathbb{T}^n)$ near the unperturbed torus $\sum_{a,I}^{0}$ given by (5.5) and filled with quasi-periodic solutions of (5.2). The frequency vector ω_ε of a perturbed solution will be $c\varepsilon$ close to $\omega = (\lambda_1, \dots, \lambda_n)$ of the unperturbed one.

The methods in [Kuk_1] leave out the case of periodic boundary conditions, because of certain limitations of the KAM method (second Melnikov condition) excluding multiplicities in the normal frequencies. A different approach has been recently used by Craig and Wayne [$\text{C-W}_{1,2}$], based on the Lyapunov-Schmidt decomposition and leading to time periodic solutions of perturbed equations under periodic boundary conditions. This method consists in splitting the problem into a (finite-dimensional) resonant part (Q-equation) and an infinite-dimensional nonresonant part (P-equation). In the PDE-case (contrary to the case of a finite-dimensional phase space), small divisor problems appear when solving the P-equation by a Newton iteration method, also in the time periodic case. Writing u in the form

$$u = \sum_{m,k} \widehat{u}(m,k)\ e^{im\lambda t}\ \varphi_k(x) \tag{5.6}$$

and letting the linearized operator act on the Fourier coefficients $\widehat{u}(m,k)$, one gets operators of the form

$$(m\lambda - \lambda_k) + \varepsilon\, T \tag{5.7}$$

where the first term is diagonal and T is essentially given by Toeplitz operators with exponentially decreasing matrix elements. The main task is then to obtain reasonable bounds on their inverses. The problem is closely related to a line of research around localization in the Anderson model and in particular the works of Fröhlich, Spencer, and Surace with quasi-periodic potentials (see [F-S-W], [Sur]). In this case, the operator T in (5.7) is replaced by $-\Delta$, $\Delta =$ lattice Laplacian, and the first term plays the role of the potential.

The author succeeded very recently in dealing with the quasi-periodic case by the same methods [B_1]; giving thus a new proof of the KAM theorem where one avoids Melnikov's second condition. Also the case of periodic boundary conditions and quasi-periodic solutions for (5.1), (5.2) may be treated this way. Observe that in the quasi-periodic setting, the diagonal part of (5.7) becomes now $\langle m, \lambda \rangle - \lambda_k$ where for instance $\langle m, \lambda \rangle = m_1\lambda_1 + m_2\lambda_2$. The singularities here are more severe and a large part of the difficulty already appears in the classical finite-dimensional case.

The Lyapunov-Schmidt method is significantly more flexible than KAM, and other applications, possibly to the 2D problem, should be expected.

Added in Proof: The author succeeded more recently in developing a theory of quasi-periodic solutions for NLS equations of the form (5.2) in 2D (see [B9]). For the special case of time periodic solutions, the work of [C-W_1] may be extended to any dimension, leading for instance to periodic solutions of the NLW equation $u_{tt} - \Delta u + \rho u + u^3 = 0$, for typical ρ (cf. [B_{10}]).

References

(strictly for the purpose of previous exposition)

[B_1] J. Bourgain, *Construction of quasi-periodic solutions for Hamiltonian perturbations of linear equations and applications to nonlinear PDE*, International Math. Research Notices, N. 11, 1995, 1–23.

[B_2] J. Bourgain, *On the Cauchy problem for periodic KdV type equations*, preprint IHES (Juin 1993), to appear in Journal of Fourier Analysis, Colloque J.P. Kahane.

[B_3] J. Bourgain, *Fourier transform restriction phenomena for certain lattice subsets and applications to nonlinear evolution equations*, Geom. Functional Anal., **3** no. 2, 107–156; **3** no. 3, 209–262 (1993).

[B_4] J. Bourgain, *On the Cauchy problem for the Kadomstev-Petviashvili equation*, Geom. Functional Anal., **3** n^0 4 (1993), 315–341.

[B_5] J. Bourgain, *Periodic nonlinear Schrödinger equations and invariant measures*, preprint IHES (1993), Comm. Math. Phys. **166** (1994), 1–26.

[B_6] J. Bourgain, *Invariant measures for the 2D-defocusing nonlinear Schrödinger equation*, preprint IHES (1994), to appear in Comm. Math. Phys.

[B_7] J. Bourgain, *On the Cauchy and invariant measure problem for the periodic Zakharov system*, Duke Math. J., **76**, no. 1 (1994), 175–202.

[B_8] J. Bourgain, *Aspects of long time behaviour of solutions of nonlinear Hamiltonian evolution equations*, preprint IHES (1994), to appear in Geom. Functional Anal.

[B_9] J. Bourgain, *Quasi-periodic solutions of Hamiltonian perturbations of 2D linear Schrödinger equations*, preprint IHES 1994.

[B_{10}] J. Bourgain, *Construction of periodic solutions of nonlinear wave equations in higher dimensions*, to appear in Geom. Functional Anal.

[C-K] Christodoulou and S. Klainerman, *The global nonlinear stability of the Minkowski space*, Ann. of Math. Stud., to appear.

[C-W_1] W. Craig and C. Wayne, *Newton's method and periodic solutions of nonlinear wave equations*, Comm. Pure Appl. Math., **46** (1993), 1409–1501.

[C-W_2] W. Craig and C. Wayne, *Periodic solutions of nonlinear Schrödinger equations and the Nash-Moser method*, preprint.

[E-H] I. Ekeland and H. Hofer, *Symplectic topology and Hamiltonian dynamics II*, Math. Z., **203** (1990), 553–567.

[F-S-W] J. Fröhlich, T. Spencer, and P. Wittwer, *Localization for a class of one dimensional quasi-periodic Schrödinger operators*, Comm. Math. Phys., **132** no. 1 (1990), 5–25.

[G-K-M] M. Gardner, K. Kruskal, and R. Miura, *Korteweg-de Vries equation and generalizations (II). Existence of conservation laws and constants of motion*, J. Math. Phys., **9:8** (1968), 1204–1209.

[G-J] J. Glimm and A. Jaffe, Quantum Physics, Springer Verlag, Berlin and New York (1987).

[Grom] M. Gromov, *Pseudo holomorphic curves on almost complex manifold*, Invent. Math., **82** (1985), 307–347.

[Ja] A. Jaffe, private notes.

[K] C. Kenig, private communication.

[K-P-V_1] C. Kenig, G. Ponce, and L. Vega, *The Cauchy problem for the Korteweg-de Vries equation in Sobolev spaces of negative indices*, Duke Math. J. **71** (1993), 1–21.

[K-P-V_2] C. Kenig, G. Ponce, and L. Vega, *Wellposedness and scattering results for the generalized Korteweg-de Vries equation via the contraction principle*, Comm. Pure Appl. Math., **46** (1993), 527–560.

[K-M_1] C. Klainerman and M. Machedon, *Space-Time Estimate for Null Forms and the Local Existence Theorem*, Comm. Pure Appl. Math., **46** no. 2 (1993), 1221–1268.

[K-M_2] C. Klainerman and M. Machedon, *Finite energy solutions of the Yang-Mills equations in $\mathbb{R}^{3+1}$*, preprint.

[K-T] H. Knörrer and E. Trubowitz, preprint.

[Kr] I. Krichever, *The periodic problems for two-dimensional integrable systems*, ICM Proc., (1990), 1353–1362.

[Kuk_1] S. Kuksin, Nearly integrable infinite-dimensional Hamiltonian systems, Lecture Notes in Math., **1556**, Springer, Berlin and New York (1993).

[Kuk_2] S. Kuksin, *Infinite-dimensional symplectic capacities and a squeezing theorem for Hamiltonian PDE*, to appear in Comm. Math. Phys.

[Kuk_3] S. Kuksin, *On squeezing and flow of energy for nonlinear wave equations*, IHES preprint, (summer 1994).

[Lax] P. Lax, *Periodic solutions of the KdV equations*, Comm. Pure Appl. Math., **26** (1975), 141–188.

[L-R-S] J. Lebowitz, R. Rose, and E. Speer, *Statistical mechanics of the nonlinear Schrödinger equation*, J. Statist. Phys., **50** (1988), 657–687.

[McK] H. McKean, preprint.

[M-T] H. McKean and E. Trubowitz, *Hill's operator and hyper elliptic function theory in the presence of infinitely many branch points*, Comm. Pure Appl. Math., **29** (1976), 143–226.

[Nov] S. Novikov, *The periodic problem for the Korteweg-de Vries equation*, Funktsional Anal. Prilozhen, **8** (1974), 54–66.

[Sur] S. Surace, *The Schrödinger equation with a quasi-periodic potential*, Trans. Amer. Math. Soc., **320** no. 1 (1990), 321–370.

Sphere Packings, Lattices, Codes, and Greed

JOHN H. CONWAY

Department of Mathematics
Princeton University
Fine Hall – Washington Road
Princeton, NJ 08544-1000, USA

The problem of determining the greatest density to which n-dimensional space can be filled by nonoverlapping unit spheres is solved only for the first three values of n (namely $n = 0, 1, 2$), and so we must impose further conditions if we are to make any progress at the moment.

The lattice-packing problem, when we demand that the vector sum and difference of any two sphere-centers must be another center, was solved by Blichfeldt more than sixty years ago in all dimensions up to 8, but in all those years there has been no advance on the 9-dimensional problem.

About fifty years ago, in an unsuccessful attack on this problem, Chaundy made the unwarranted assumption that an optimal $(n+1)$-dimensional lattice must necessarily contain an optimal n-dimensional one. Although this is now known to fail for some $n < 11$, Sloane and I turned it into a definition of what we called the "laminated lattices", and investigated these in all dimensions up to 48.

The laminated lattices serve as benchmarks for the general sphere-packing problem; thus, I shall define them and briefly summarize our results. By a sphere-packing lattice I mean one in which each point is distant at least 2 from all other points (so that it can be used to pack unit spheres).

DEFINITION. The 0-dimensional lattice is laminated. The $(n+1)$-dimensional laminated lattices are precisely all the $(n+1)$-dimensional sphere-packing lattices of maximal density that contain at least one n-dimensional laminated lattice.

THEOREM. *The unique 24-dimensional laminated lattice is the celebrated lattice discovered in 1969 by John Leech, and for $n < 24$ every n-dimensional laminated lattice is a section of the Leech lattice. The inclusions between these lattices in consecutive dimensions are as shown in Figure 1. There are precisely 23 distinct laminated lattices of dimension 25 (one for each type of "deep hole" in the Leech lattice). In each dimension from 26 to 48 the density of all laminated lattices is known, and at least one such lattice has been found.*

Figure 2 illustrates the first few laminated lattices. In the illustrations for dimensions n up to 3, we have shaded the sphere at the origin, and put spots at the centers of n neighboring spheres for which the corresponding vectors generate

Proceedings of the International Congress
of Mathematicians, Zürich, Switzerland 1994

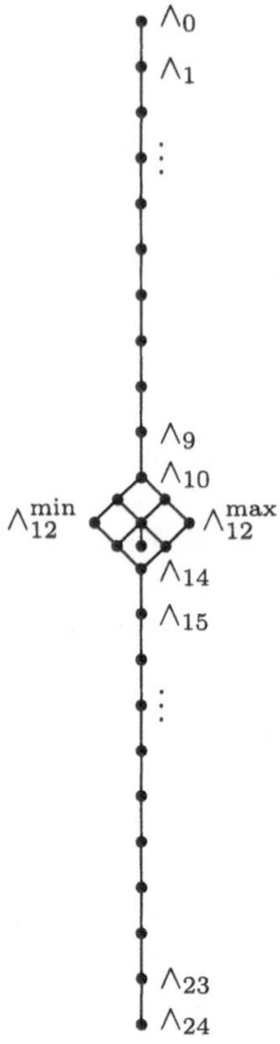

Figure 1.
Laminated lattices
to 24 dimensions.

the lattice. If we join two spots whose spheres touch, and leave them unjoined when the corresponding vectors are orthogonal, then the diagrams that indicate the shapes of the lattices in up to 8 dimensions are very familiar — they are the Coxeter-Dynkin diagrams of certain root lattices. In 9 dimensions we need a new convention — the broken line indicates a pair of vectors at angle arccos(1/4).

These root lattices have very simple definitions. The root lattice A_n consists of all the points specified by $n+1$ integer coordinates with zero sum; for D_n we have n integer coordinates with even sum. We write $(D_n)^{+t}$ for the union of D_n and its coset determined by the vector $(1/2, 1/2, 1/2, \ldots, 1/2, t/2)$, and write just D_n^+ when $t = 1$. Then (for $n < 9$) E_n consists precisely of those vectors of D_8^+ whose last $9-n$ coordinates are equal.

The laminated lattices in dimensions up to 9 are A_0, A_1, A_2, $A_3 = D_3$, D_4, $D_5 = E_5$, E_6, E_7, $E_8 = D_8^+$, and D_9^{+0}. They were all known to Khorkhine and Zolotarev in 1880. Most of the remaining laminated lattices in dimensions up to 24 were found by John Leech in about 1970. The numbers of laminated lattices in dimensions 26–48 are almost certainly very large indeed: Sloane and I gave a probabilistic estimate of at least 75,000 for the number of 26-dimensional laminated lattices of a certain very special type.

Denser sphere-packing lattices than the laminated ones are known in dimensions 11, 12, 13, and 32–48, but most of the others are probably optimal. In 1980

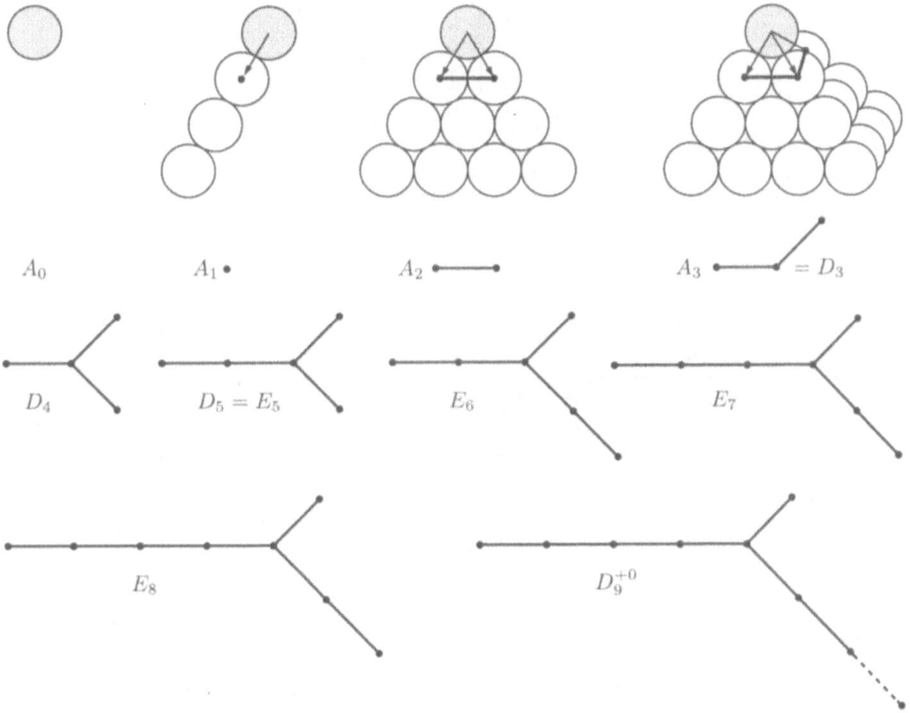

Figure 2. Laminated lattices and Dynkyn diagrams.

M. R. Best discovered a nonlattice packing in 10 dimensions that has a higher density than any 10-dimensional lattice packing currently known.

All of this is recorded in my book with Sloane: *Sphere Packings, Lattices and Groups* (Springer). So what further information on these topics has been discovered in the last four years?

Noam Elkies has improved the records in many dimensions beyond 48 by using the lattice structures of the Mordell-Weil groups of certain algebraic curves. Wu-Yi Hsiang has made a strongly disputed claim to have solved the general 3-dimensional sphere-packing problem. On the basis of a certain "Postulate", Sloane and I have found all the optimal sphere packings in dimensions up to 9. In the rest of the first half of this communication, I shall briefly describe only the latter result.

Our "Postulate n", which requires a slight modification in 9 dimensions, is that the centers of the spheres in an optimal n-dimensional packing ($n > 1$) can be grouped into parallel m-spaces that each contain the centers of an optimal m-dimensional packing, where m is the largest power of 2 that is strictly less than n.

The situation is familiar in the 3-dimensional case. It seems that in all optimal 3-dimensional packings the spheres form 2-dimensional layers in which they are arranged hexagonally as in Figure 3. If the centers of the spheres in one horizontal layer are the points marked 0 in the figure, then those of an adjacent layer must be above either those marked 1 or those marked 2. But there is complete symmetry

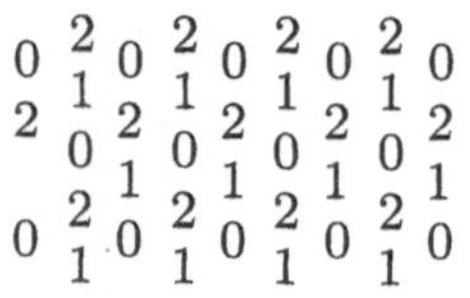

Figure 3. The three positions for layers.

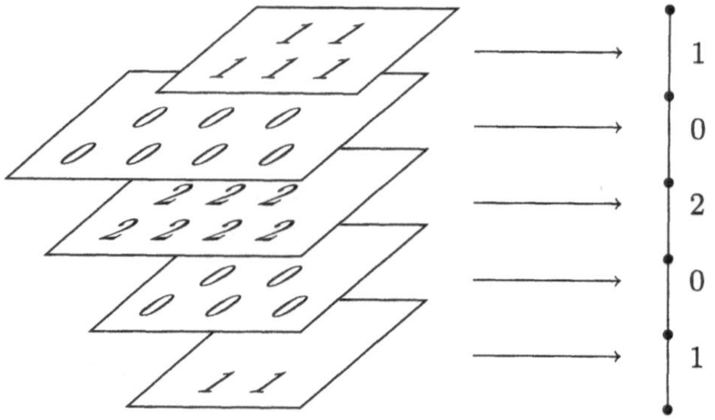

Figure 4. How a packing corresponds to a coloring.

between the three sets of points 0, 1, 2, and so we see inductively that the centers of *any* layer must lie vertically above one of these three sets of points.

Figure 4 shows how we can code this by giving a 3-coloring of the 1-dimensional sphere packing whose centers are obtained by projecting those of the 3-dimensional one onto a vertical line. Here a (1-dimensional) sphere colored n (for $n = 0, 1, 2$) represents all the spheres of a 2-dimensional layer centered above all the points marked n in Figure 3. Just two of these packings are uniform — the root lattice A_3, or face-centered cubic (f.c.c.) packing, which we get from the coloring $\ldots, 0, 1, 2, 0, 1, 2, 0, 1, 2, \ldots$, and the hexagonal close packing (h.c.p.), from the coloring $\ldots, 0, 1, 0, 1, 0, 1, 0, 1, 0, \ldots$.

This method works because of the symmetry between the three sets of points 0, 1, 2. Each of these sets is a lattice whose "deep holes" (the points of space at maximal distance from the lattice) form the union of the other two sets. They are in fact the three cosets of the root lattice A_2 in its dual.

In 4 dimensions, both the horizontal and vertical spaces are 2 dimensional. It follows from our Postulate 4 that in an optimal 4-dimensional packing, the "heights" (the positions in "vertical" space) will form a scaled copy of the optimal 2-dimensional packing A_2, which has a 3-coloring that specifies the placing of the layers above the "horizontal" space.

However, the 3-coloring of A_2 (Figure 3) is unique! So it follows from our Postulate that the optimal 4-dimensional packing is also unique. This is the 4-dimensional root lattice D_4. It has four cosets 0, 1, 2, 3 in its dual, and the set of deep holes in any one of these is formed by the union of the other three.

Our Postulates now imply that all optimal packings in dimensions 5, 6, 7, 8 are specified by 4-colorings of the optimal packings in dimensions 1, 2, 3, 4,

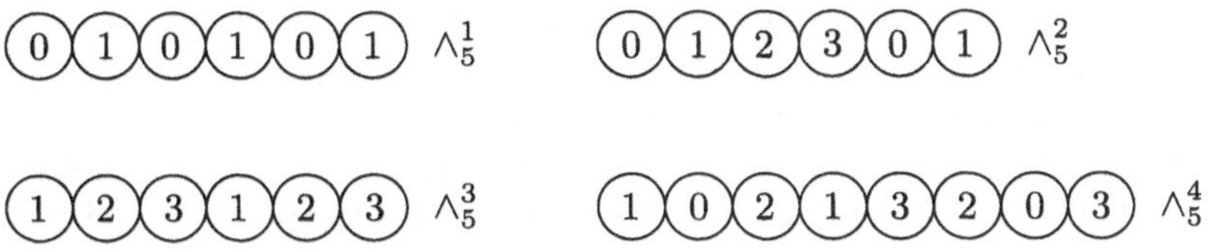

Figure 5. The uniform packings in 5 dimensions.

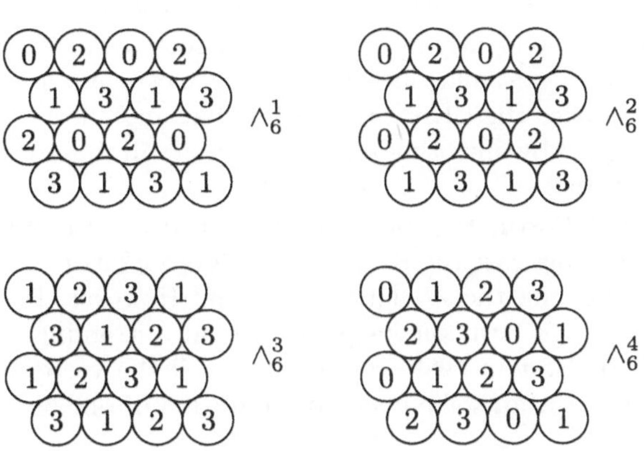

Figure 6. The uniform packings in 6 dimensions.

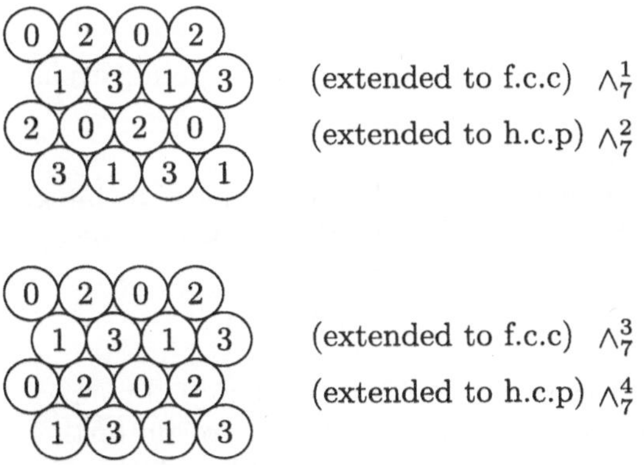

Figure 7. The uniform packings in 7 dimensions.

respectively. In each of the dimensions 5, 6, and 7 there are just four uniform packings, arising from the colorings shown in the respective figures. However, the 4-coloring of the 4-dimensional packing D_4 is unique, and so the Postulate entails that the only optimal 8-dimensional packing is the root lattice E_8.

In 9 dimensions, there are several new features. The deep holes in the E_8 lattice are not the union of its cosets in its dual (it is in fact self-dual), but of 135 particular cosets of E_8 in $(1/2)E_8$. The successive E_8 layers need not be obtained from each other by translation alone, but perhaps by translation combined with rotation. There are in fact precisely 382,185 choices for the position and orientation of each successive layer.

However, the most interesting new fact is that there are some remarkable new packings — the "fluid diamond" packings $D_9(v)$ (consisting of D_9 and its translate by a suitable vector v) — that, among other things, disprove our Postulate 9. That's not all they do — the spheres in these packings form two equinumerous sets (the "gold" and "silver" spheres) that can (by varying the vector parameter v) be moved around independently of each other in such a way that at most instants no silver sphere touches a gold one. There is in fact a motion that fixes all the gold spheres, but moves the silver ones so far that any chosen one can reach the place initially occupied by any other one, although at all times the packing remains (conjecturally) optimal!

It appears that Postulate 9 only just fails, because the fluid diamond packings include as a limiting case the Khorkhine-Zolotarev lattice packing D_9^{+0}, which *is* obtainable by stacking the E_8 lattice packing. However, Postulate 10 is irredeemably false, and the best known packing is an intriguing nonlattice packing discovered by M. R. Best in 1980. It consists of all the vectors whose 10 coordinates can be obtained from some cyclic permutation of one of the words

$$\begin{array}{cccc}(01112), & (21132), & (21310), & (01330), \\ (03110), & (23130), & (23312), & (03332)\end{array}$$

by replacing each digit by a pair of integers according to the scheme

$$\begin{array}{llll}0 \to & \text{even, even} & 2 \to & \text{odd, odd} \\ 1 \to & \text{even, odd} & 3 \to & \text{odd, even.}\end{array}$$

Lexicographic Codes

I now turn to an apparently totally different topic. The integral lexicographic code ("lexicode") of distance d is defined by the following "greedy algorithm". We start with the word

$$\ldots 0\,0\,0\,0\,0$$

(all "words" in this theory are semi-infinite strings of nonnegative integer "digits", almost all zero). Then we proceed inductively to add further words, at each stage choosing the lexicographically earliest word that differs in at least d digits from all preceding ones.

We illustrate by taking $d = 3$.

$$\begin{array}{c}
\ldots 000000 \\
\ldots 000111 \\
\ldots 000222 \\
\ldots 000333 \\
\ldots 000444 \\
\ldots \\
\ldots 000nnn \\
\ldots \\
\ldots 001012 \\
\ldots 001103 \\
\ldots 001230 \\
\ldots 001321 \\
\ldots 001456 \\
\ldots \\
\ldots 002023 \\
\ldots 002132 \\
\ldots \\
\ldots 003031 \\
\ldots \\
\ldots 004048 \\
\ldots \\
\ldots \\
\ldots 010013 \\
\ldots
\end{array}$$

There is a quite remarkable theorem about codes of this type:

THE LEXICODE THEOREM.
Any lexicode, when equipped with natural termwise definitions of addition and scalar multiplication, is a vector space.

Rather than prove this theorem, I want to explore its consequences, so I will take it for granted and rename it the *Lexicode Axiom.*

One consequence is that the termwise sum of any two words from any lexicode is another word in that lexicode: for example it asserts that the sum

$$\begin{array}{r}
\ldots 000111 \\
+\ldots 001012
\end{array}$$

should be in the lexicode we took as our example. But in fact $\dots 001123$ is *not* in that lexicode, because $\dots 001103$ *is*, and because any two distinct words of that code must have distance at least 3.

What is wrong? The answer is that the termwise definitions of addition and scalar multiplication referred to in the *Lexicode Axiom*, although "natural", are not quite the ones you might have expected! What happens is that the underlying addition and multiplication operations in the integers are not the customary ones. How could they be? With the customary definitions of addition and multiplication, the integers do not even form a field.

What are the new operations? The best way to find out is to turn the *Lexicode Axiom* around once again, and rename it the *Lexicode Definition*! Let's see how this works.

THEOREM 0. $0+0=0$.

Proof. Suppose that $0+0=z$. Then we have the addition sum

$$\begin{aligned} &\dots 000000 \\ +&\dots 000000 \\ =&\dots zzzzzz, \end{aligned}$$

and for the latter word to be in the lexicode, it must have almost all its digits zero, so that $z=0$. □

It now follows that the zero of our field is "0", and so we have $0+n=n=n+0$ for all n.

THEOREM 1. *We have* $1+1=0$, $1+2=3$.

Proof. We have the addition sum

$$\begin{aligned} &\dots 000111 \\ +&\dots 001012 \\ =&\dots 0011xy \end{aligned}$$

where $x=1+1$, $y=1+2$. But $\dots 001103$ is in our lexicode, and so must be the answer to this sum, whence $x=0$, $y=3$. □

THEOREM 2. *Our field has characteristic two.*

Proof. By multiplying the equation $1+1=0$ by a suitable constant, we find that $n+n=0$ for any given n. □

THEOREM 3. *We have* $3+2=1$.

Proof. $3+2=(1+2)+2=1+(2+2)=1+0=0$. □

THEOREM 4. *We have* $4+0=4$, $4+1=5$, $4+2=6$, $4+3=7$.

Proof. These assertions follow from the easy addition sum

$$\ldots 0\,0\,0\,4\,4\,4\,4 \;+\; \ldots\; 0\,0\,1\,0\,1\,2\,3 \;=\; \ldots\; 0\,0\,1\,4\,5\,6\,7$$

in the distance 4 lexicode. □

The entire addition table of our field can be established by a precisely similar argument:

THEOREM 5. *If A is any one of the numbers*

$$1,\ 2,\ 4,\ 8,\ 16,\ 32,\ 64,\ 128,\ 256,\ 512,\ 1024,\ \ldots$$

and B is any strictly smaller number, then $A+B$ takes its usual value, while $A+A=0$.

Before proving this, we show how it can be used to work out an arbitrary addition-sum, taking $13+11$ as an example. By repeated use of the theorem, we find

$$13 = 8+4+1, \qquad 11 = 8+2+1,$$

whence (again using the theorem)

$$13+11 = (8+8)+4+2+(1+1) = 4+2 = 6.$$

Proof of Theorem 5. From this part of the addition table

0	1	2	3	4	5	6	7
1	0	3	2	5	4	7	6
2	3	0	1	6	7	4	5
3	2	1	0	7	6	5	4
4	5	6	7	0	1	2	3
5	4	7	6	1	0	3	2
6	7	4	5	2	3	0	1
7	6	5	4	3	2	1	0

we shall show how to continue. The eight words obtained from the above by prefixing $\ldots\ 0\ 0\ 0\ 1$ must all be in the distance 8 lexicode, because the first of them is, and the others are obtained by adding

$$\ldots\ 0\,0\,n\,n\,n\,n\,n\,n\,n\,n$$

for $n = 1, \ldots, 7$.

It then easily follows that the next word in this code is

$$\ldots\ 0\ 0\ 0\ 1\ 8\ 9\ 10\ 11\ 12\ 13\ 14\ 15$$

so that $8+0=8$, $8+1=9$, $8+2=10$, $\ldots$, $8+7=15$, from which we deduce the addition table up to $15+15$. □

THEOREM 6. *We have* $6 = 4.4$.

Proof. In the distance 5 lexicode we find the words

$$w = \ldots\ 0\ 0\ 1\ 0\ 1\ 2\ 3\ 4$$

and

$$4w = \ldots\ 0\ 0\ 4\ 0\ 4\ 8\ 12\ 6.$$ □

There is an analogue of Theorem 5 for multiplication.

THEOREM 7. *If A is any of the numbers*

$$2,\ 4,\ 16,\ 256,\ 65536,\ 4294967296,\ \ldots$$

and B is any smaller number, then $A \cdot B$ takes its usual value, while $A \cdot A$ is the usual value of $3A/2$.

We shall not prove this, but just show how to use it to work out arbitrary multiplications. We have

$$5 \cdot 12 = (4+1)(8+4) = 4 \cdot 8 + 8 + 4 + 1 = 4 \cdot 8 + 13$$

and in this

$$4 \cdot 8 = 4 \cdot 4 \cdot 2 = 6 \cdot 2 = (4+2) \cdot 2 = 8 + 3 = 11$$

so that finally $5 \cdot 12 = 11 + 13 = 6$.

Further Remarks About Our Field

Readers who are familiar with the game of nim will recognize that the addition of our field is "nim-addition", namely addition without carry in the binary notation. So I call the multiplication "nim-multiplication", and the field, the "nim field". It is indeed a field, and a very interesting one. The reader might like to verify that $1/4 = 15$, that the fifth roots of unity are 1, 8, 13, 14, 10, and that we have

$$2^2 = 3,\ 4^4 = 5,\ 16^{16} = 17,\ 256^{256} = 257, \ldots.$$

The definitions extend naturally to infinite ordinal numbers, and we find for example that Ω, the first infinite ordinal, is a cube root of 2, and that the ordinal usually called Ω^Ω is a fifth root of 4, and so on! The ordinal numbers form an algebraically closed field under these operations — the finite ones form the quadratic closure of the field of order two.

Lexicodes, Sphere Packings, and Games

How are lexicodes related to sphere packings? The answer is, they ARE sphere-packings! For example, the set of all integer sequences that differ in at most one place from a given one $\ldots f\ e\ d\ c\ b\ a$ is a solid sphere in a certain space, and the words of the distance 3 lexicode are the centers of a perfect packing of this space by spheres.

How are they related to games? Let us define a two-player game on the set of such sequences by allowing either player to move from any such sequence to any lexicographically earlier one that differs from it in at most two digits. Then the winning strategy in this game is simply to move always to a lexicode word! (If we replace "at most two" by "at most one", we get a game equivalent to nim, and so explain the connection with nim-addition.)

Are laminated lattices related to games? I think so. If two players play a game on the points (x, y) of the first quadrant in which the move is to replace (x, y) by any lexicographically earlier point distant strictly less than 1 from it, then the winning strategy is to move always to a point of the lattice shown in our final figure.

However, the definition of this game is slightly wrong, because in 3 dimensions the winning positions are the centers of the hexagonal close packing rather than the face-centered cubic lattice. I hope to find the correct definition, for which an analogue of the lexicode theorem will force the solution to be a lattice, which should be one of the laminated lattices. and so the Leech lattice in 24 dimensions.

Ingrid Daubechies, a plenary speaker

Wavelets and Other Phase Space Localization Methods

INGRID DAUBECHIES

Princeton University
Princeton, NJ 08544, USA

1. Introduction

Mathematicians have various ways of judging the merits of new theorems and constructions. One very important criterion is esthetic — some developments just "feel" right, fitting, and beautiful. Just as in other venues where beauty or esthetics are discussed, taste plays an important role in this, but I think I am not alone in being especially excited when apparently different fields suddenly meet in a new concept, a new understanding. It is often of the sparks of such encounters that our esthetic enjoyment of mathematics is born.

Another important criterion for according merit to some particular piece of mathematics is the extent to which it can be useful in applications; this is the criterion almost exclusively used by nonmathematicians. Mathematicians themselves do not discount the importance of mathematics for applications (after all, if we were producing only beauty, there wouldn't be as many teaching positions allotted to us), but often beauty is considered the real grail, with applicability second-best. Although we have come some way since Hardy's *A Mathematician's Apology*, we often still believe, maybe subliminally, that the two criteria are exclusive — that mathematics, when really close to applications, cannot be beautiful and is often even "dirty."

I believe that this does not have to be so; a wish for beauty and simplicity, and a desire to bring different fields together, can equally well drive developments in "applicable" mathematics.

When mapping out this presentation, I initially thought that I wanted to speak about wavelets, but I soon realized that other developments, aside from or beyond wavelets, should have their place here as well, and the scope was enlarged to add the "other phase space localization methods." Let me start by explaining what I mean by this.

I shall use the term "phase space" when a special type of description is meant, involving several complementary variables. It is really a term that is appropriated here from physics. Imagine that you want to describe the motion of a planet in the solar system. A simple way to do this is to give, as a function of time, its position in space as well as its momentum. This is a phase space description: the two complementary variables are position and momentum, and you are describing the motion by a curve in phase space. "Phase space localization" is no problem here: both position and momentum can be measured, with arbitrarily high precision. ("Phase space" is also used in a more general sense for other dynamical systems,

Proceedings of the International Congress
of Mathematicians, Zürich, Switzerland 1994

but that is a different story.) The situation is different if we look at a quantum system, say an electron in a solid state crystal, where measuring position and momentum both, simultaneously, with arbitrarily high precision, cannot be done: the uncertainty principle forbids it. Nevertheless, it is still very useful to think in terms of phase space, or momentum and position, when comparing a quantum system with its classical analog, for instance. This poses a problem to the theoretical physicist: How to give a description, localized in phase space, despite the uncertainty principle? The mathematical model for a quantum mechanical particle assigns to a physical state a wave function $\psi(x)$, where x is the position variable; an equivalent description is given by the Fourier transform $\hat{\psi}(p)$, where p is interpreted as the momentum variable. Trying to "localize in phase space" amounts therefore to pinning down, as well as possible, a function's local properties and the local properties of its Fourier transform simultaneously — something analysts have been doing for decades under the name *microlocalization.*

The same problem also crops up in electrical engineering, or in statistics: for instance, when trying to understand signals depending on time, such as a recorded audio signal, it is often useful to gauge its spectrum or frequency content, again modeled naturally by the Fourier transform of the data. But the make-up of such signals, in terms of their different frequency characteristics, seems to change with time. This is immediately clear when you think of a music score which, after all, tells the musician to play different notes (= frequencies) at different times. Once again, the intuitive notion of the mathematical tool needed involves localization in phase space, with the two complementary variables now in the form of time and frequency.

Similarly, the computer scientist or engineer working with images (such as any image on your television screen) finds it helpful to break it up in smaller pieces (localization in space) and to look at the different spatial frequencies present in those pieces: again a phase space localization, now in two dimensions.

Because similar problems occur in different disciplines, it is not surprising that the answers developed, often independently, have some similarity as well. What I want to describe here is how the synthesis of different points of view and different approaches has led in some cases to new developments, making the whole much more than the sum of its parts.

Before embarking on a more detailed discussion, I would like to point out that this presentation will summarize essential contributions by many people besides myself. At the ICM '90 in Kyoto, both R. Coifman and Y. Meyer gave talks related to this one; at this ICM, related talks include those by W. Dahmen, D. Donoho, and V. Rokhlin. For a more complete list of important contributors, I refer the reader to the references and their references. I would like to take this opportunity to thank especially R. Coifman, A. Cohen, A. Grossmann, S. Mallat, and Y. Meyer, from all of whom I learned a lot.

2. Wavelets

Most of this presentation will concern the development of wavelets, in particular of orthonormal wavelet bases, our growing understanding of their mathematical properties, and the ways in which they can be applied.

What are wavelets? To keep things simple, I shall restrict myself mostly to one dimension; with slight modifications, everything here can be generalized to higher dimensions (the few exceptions will be pointed out explicitly). I shall also almost systematically *not* try to give the most general conditions under which my statements hold, preferring to strip down the technicalities so as to lay bare the essential ideas.

A typical example of a family of wavelets $\psi_{j,k}(x)$ is given by

$$\psi_{j,k}(x) = 2^{-j/2}\psi(2^{-j}x - k) = 2^{-j/2}\psi(\frac{x - 2^j k}{2^j}) \quad , \quad j,k \in \mathbb{Z} \; , \tag{1}$$

where ψ is a function with reasonable decay (say, $|\psi(x)| < C(1 + |x|)^{-(1+\epsilon)}$), with some smoothness (as measured by the decay of the Fourier transform $\hat{\psi}$, say $|\hat{\psi}(\xi)| < C(1 + |\xi|)^{-(1+\epsilon)}$), and such that $\int \psi(x) = 0$. For particular choices of ψ, the $\psi_{j,k}$ constitute a basis often orthonormal for $L^2(\mathbb{R})$; I shall mainly restrict myself to this case (although there are many interesting applications that use wavelets that are not linearly independent, which fall outside this framework). The first known example of a function ψ for which the $\psi_{j,k}$ give an orthonormal basis is the Haar wavelet, known since 1910,

$$\psi(x) = \begin{array}{rll} 1 & \text{if} & 0 \le x < 1/2 \\ -1 & \text{if} & 1/2 \le x < 1 \\ 0 & \text{otherwise} & ; \end{array} \tag{2}$$

this does not satisfy the smoothness requirement above. Much smoother constructions were found only in the 1980s: Stromberg (1982), Meyer (1985), Battle (1987), Lemarié (1988), and Daubechies (1988) are some examples. The first construction, by Stromberg, did not attract a lot of attention at the time, although it later turned out to be very useful, not only for the harmonic analyst, but also computationally. Meyer rediscovered that dilations and translations of a single smooth and decaying function, as in (1) above, could give rise to orthonormal bases for $L^2(\mathbb{R})$; in his example both ψ and $\hat{\psi}$ are C^∞ and $\hat{\psi}$ has compact support. The constructions by Battle and Lemarié use $\psi \in C^m$, where m can be arbitrarily large but finite; moreover ψ has exponential decay. (Stromberg's ψ has similar properties.) These first ad hoc constructions became much more transparent with the development by Mallat (1989) and Meyer of multiresolution analysis, a framework that linked wavelets with approximation theory. Interestingly, this construction was triggered by analogies with tools in vision theory, with which Mallat was familiar. Multiresolution analysis was then used in Daubechies (1988) to construct a basis of type (1) where ψ is still in C^m but compactly supported.

3. Multiresolution Analysis

The multiresolution analysis framework views the expansion of f in $L^2(R)$ with respect to an orthonormal wavelet basis,

$$f = \sum_{j,k \in \mathbb{Z}} < f, \psi_{j,k} > \psi_{j,k} \quad , \tag{3}$$

as a decomposition of f into successive layers, each more detailed than the previous one. That is, we write

$$L^2(\mathbb{R}) = \overline{\bigcup_{j \in \mathbb{Z}} V_j} \quad ,$$

where the spaces V_j constitute a nested sequence of approximation spaces,

$$\cdots \subset V_2 \subset V_1 \subset V_0 \subset V_{-1} \subset V_{-2} \subset \cdots$$

$$\bigcap_{j \in \mathbb{Z}} V_j = \{0\} \ .$$

For fixed j, summing the terms in (3) over k gives exactly the layer to be peeled away from $P_{j-1}f := \text{Proj}_{V_{j-1}} f$ to reach the coarser approximation $P_j f := \text{Proj}_{V_j} f$,

$$P_{j-1}f = P_j f \ + \ \sum_k < f, \psi_{j,k} > \psi_{j,k} \quad . \tag{4}$$

For the Haar basis, the corresponding spaces V_j are given by

$$V_j = \{f \in L^2(\mathbb{R}) \ ; f|_{[2^j k, 2^2(k+1)[} \ = \ \text{constant for each } k \in \mathbb{Z}\}.$$

For the constructions of Stromberg, Battle, and Lemarié, the multiresolution hierarchy consists of spaces of spline functions,

$$\begin{aligned} V_j \ = \ & \{f \in L^2(\mathbb{R}) \ ; f \in C^m \text{ and} \\ & f|_{[2^j k, 2^2(k+1)[} \ = \ \text{polynomial of degree } m+1, \text{ for each } k \in \mathbb{Z}\}. \end{aligned} \tag{5}$$

Additional requirements are that the spaces V_j are all scaled versions of each other,

$$f \in V_j \Leftrightarrow f(2^j \cdot) \in V_0$$

(as is obviously the case in the examples above) and that the central space V_0 is invariant under integer translation. This invariance follows automatically from the final requirement, that there exists a function ϕ in V_0, commonly called the *scaling* function, such that the $\phi(. - k) = 2^{-j/2}\phi(2^{-j}x - k), k \in \mathbb{Z}$, constitute an orthonormal basis for V_j. In the Haar basis case, $\phi(x)$ is taken to be $\chi_{[0,1[}(x)$, the characteristic function of $[0, 1[$; in the spline examples, ϕ is a spline function of the appropriate order and with exponential decay. The work of Lemarié (1993) and Auscher (1992) proves that *any* wavelet basis of type (1) is associated with such a multiresolution analysis, provided that ψ has some smoothness and decay. (Note that this result does not completely translate to higher dimensions.)

As a result of the nesting property of the V_j, we have that $\phi \in V_0 \subset V_{-1}$, so that ϕ can be written as a linear combination of the orthonormal basis functions $\phi_{-1,k}$ in V_{-1}:

$$\phi(x) = \sqrt{2} \sum_n h_n \phi(2x - n) \quad . \tag{6}$$

Similarly $\psi \in V_{-1}$, so that

$$\psi(x) = \sqrt{2} \sum_n g_n \phi(2x - n) \quad . \tag{7}$$

For other scales j, (6) and (7) can be rewritten as

$$\phi_{j,k} = \sum_n h_n \phi_{j-1,2+2k} \quad , \quad \psi_{j,k} = \sum_n g_n \phi_{j-1,2+2k} \quad . \tag{8}$$

4. Fast Algorithm for a Decomposition into Wavelets

The layered structure of the underlying multiresolution analysis translates into a fast algorithm for the wavelet decomposition of functions. In numerical applications, the function f to decompose will be given with a finite resolution only: for instance, in the form of samples. That is, we really know only the projection of f onto one of the spaces V_{j_0} in the scale; all information pertaining to finer structure (corresponding to $\psi_{j,k}$ with $j \leq j_0$) cannot be recovered (unless we have a priori information on f). Let us rescale our length unit so that $j_0 = 0$. Then we suppose that we know

$$P_0 f = \sum_k s_{0,k} \phi_{0,k} = \sum_k \langle f, \phi_{0,k} \rangle \phi_{0,k} \quad ,$$

where the $s_{0,k}$ are either given or computed from the data, depending on the application. Because of (4), we have

$$P_0 f = P_1 f + \sum_k \langle f, \psi_{1,k} \rangle \psi_{1,k} \quad ,$$

and with the help of (8), the $d_{1,k} := \langle f, \psi_{1,k} \rangle$ can be computed by

$$d_{1,k} = \langle f, \sum_n g_n \phi_{0,n+2k} \rangle = \sum_n g^*_{n-2k} s_{0,n} \quad .$$

Similarly, we can also compute $s_{1,k} := \langle f, \phi_{1,k} \rangle$ (the coefficients of $P_1 f$ in the orthogonal basis $\{\phi_{1,k}, k \in \mathbb{Z}\}$ of V_1) by using (6) again,

$$s_{1,k} = \sum_n h^*_{n-2k} s_{0,n} \quad .$$

$P_1 f$ can be decomposed further into $P_2 f$ and a combination of the $\psi_{2,k}$; this can be repeated for successively higher values of j, resulting in

$$d_{j,k} := \langle f, \psi_{j,k} \rangle = \sum_n g^*_{n-2k} s_{j-1,n} \tag{9}$$

$$s_{j,k} := \langle f, \phi_{j,k} \rangle = \sum_n h_{n-2k} s_{j-1,n} \quad . \tag{10}$$

These formulas consist of a convolution of the sequence s_{j-1} with $\overline{h} = (h^*_{-n})_{n\in\mathbb{Z}}$ or $\overline{g} = (g^*_{-n})_{n\in\mathbb{Z}}$, followed by retaining only the even entries of the result. Schematically, this is represented by

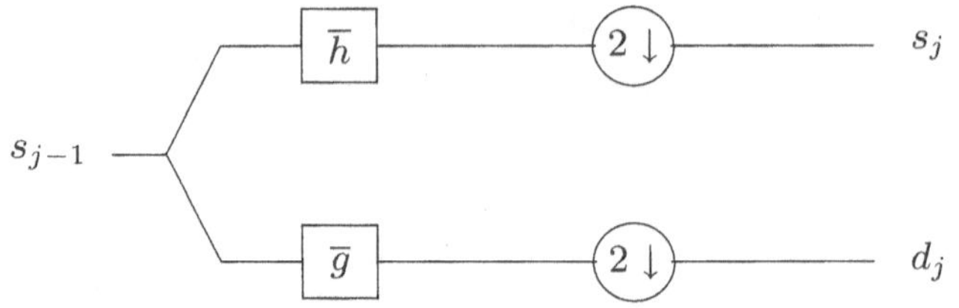

, (11)

where the symbol $\boxed{a}$ stands for convolution with the sequence a, and $2\downarrow$ is "decimation by a factor 2."

The transition from the sequence $(s_{j-1,n})_{n\in\mathbb{Z}}$ to the two sequences $(s_{j,k})_{k\in\mathbb{Z}}$, $(d_{j,k})_{k\in\mathbb{Z}}$ corresponds to a change of basis in V_{j-1}, from $\{\phi_{j-1},n \in \mathbb{Z}\}$ to $\{\phi_{j,k}, \psi_{j,k}; k \in \mathbb{Z}\}$. The inverse operation corresponds to the adjoint unitary operator, and we have

$$s_{j-1,n} = \sum_k [h_{n-2k}s_{j,k} + g_{n-2k}d_{j,k}] \ . \tag{12}$$

Each of the two terms in the right-hand side of (12) can be viewed as the result of first "upsampling by 2," i.e. taking the given sequence as the even entries of a new sequence in which all the odd entries are zero, followed by a convolution. Schematically, this becomes

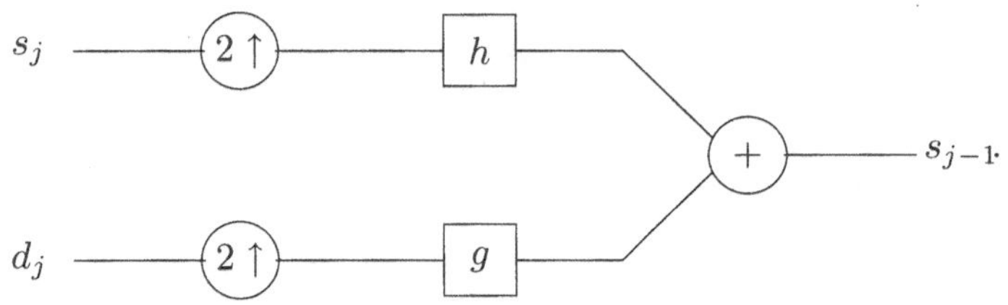

(13)

For the Haar basis one finds $h_0 = \frac{1}{\sqrt{2}} = h_1$, $g_0 = \frac{1}{\sqrt{2}} = -g_1$, with all other $h_n, g_n = 0$. The decomposition steps (9) and (10) then correspond to breaking up the sequence s_{j-1} into pairs, and replacing every pair of numbers by its average (a coarser level approximation, giving s_j) and the difference between the two numbers (the detail d_j). The reconstruction (12) then adds the sum and difference to recover the first number, whereas a subtraction gives the second number in every pair. The resulting algorithm is fast: starting from a sequence s_0 with N entries, we compute sums and differences for $\frac{N}{2}$ pairs to obtain s_1 and d_1. The $\frac{N}{2}$ entries in s_1 give $\frac{N}{4}$ pairs, for each of which we have another sum and difference to compute, and so on. The total number of computations is therefore $2\frac{N}{2} + 2\frac{N}{4} + \cdots \simeq 2N$ (where we have swept edge effect terms under the rug if N is not a power of 2, but they don't matter here: their contribution to the complexity is $O(\log N)$). If we have K nonvanishing h_n, g_n instead of only 2, then the total number of computations is KN, still linear in N. This type of wavelet transform has therefore a lower complexity than the FFT, which uses $O(N \log N)$ computations.

5. What Do Wavelets Buy You?

So we have a fast algorithm for a neat kind of basis, in which all basis functions are shifted and dilated versions of just one template (or a few templates, in some generalizations or in higher dimensions). Why should anyone care? In fact, surprisingly many people do care, and many fields have something to tell us about these wavelet bases.

To harmonic analysts, wavelet bases are a convenient way to carry out a Littlewood-Paley (LP) decomposition. In a traditional LP decomposition of a function f, one writes

$$f = \sum_{j=-\infty}^{\infty} \Delta_j f = f_0 + \sum_{j=0}^{\infty} \Delta_j f \quad ,$$

where the Fourier transform $(\Delta_j f)^\frown(\xi)$ of each $\Delta_j f$ is nonvanishing only for, say, $2^{j-1} \leq |\xi| \leq 2^{j+1}$. One way of obtaining such $\Delta_j f$ is to construct a smooth function w, supported on $\frac{1}{2} \leq |\xi| \leq 2$, such that, for $1 \leq |\xi| \leq 2$, $w(\xi)+w(\xi/2) = 1$, and to define $(\Delta_j f)^\frown(\xi) = \hat{f}(\xi)w(2^{-j}\xi)$. The different $\Delta_j f$ decouple different frequency ranges of f; yet, unlike the Fourier transform itself, they retain some spatial information. This information is sufficient, for instance, to characterize the Hölder spaces C^s: even though it is impossible to characterize (i.e. give an "if and only if" condition) the Hölder exponent of f by the decay of its Fourier transform, nevertheless decay conditions on the $\Delta_j f$, as a function of their frequency range label j, permit such a characterization. More precisely, for any $f \in L^\infty$, we have

$$f \in C^s \Leftrightarrow \sup_{j \in \mathbb{N}} 2^{js} \parallel \Delta_j f \parallel_{L^\infty} < \infty \quad . \tag{14}$$

Similarly, LP decompositions can be used for much more sophisticated estimates (Stein (1993), Frazier *et al.* (1990)). A wavelet decomposition carves up f likewise in dyadic frequency blocks, with $Q_{-j} f := \sum_k < f, \psi_{-j,k} > \psi_{-j,k}$ corresponding to $\Delta_j f$. This means that many achievements of LP decompositions have their mirror image in wavelet terms. For instance, if the wavelet ψ and the scaling function ϕ are in C^r and have sufficiently rapid decay, then we have, for all $s < r$, a characterization of the Hölder spaces similar to (14). Specifically, for $f \in L^\infty$,

$$f \in C^s \Leftrightarrow \sup_{j \in \mathbb{N}} 2^{j(s+\frac{1}{2})} \sup_{k \in \mathbb{Z}} | < f, \psi_{-j,k} > | < \infty \quad . \tag{15}$$

The similarity with (14) is obvious (the extra $\frac{1}{2}$ in the exponent is due to the normalization we chose for the $\psi_{j,k}$ in (1)); more sophisticated estimates using LP-type decompositions translate into wavelet estimates analogously. Wavelets then provide a way to write powerful techniques in harmonic analysis in a language that can also be read as an algorithm. On the other hand, their very convenient orthogonality properties in L^2 also lead to shortcuts in proofs in harmonic analysis (see e.g. Meyer (1990)).

To electrical engineers, wavelet bases are a mathematical framework that links up with a filtering technique developed earlier, called subband filtering. Diagram (11) is in fact the electrical engineering notation for a filter bank with two channels, one *low pass* (the transition $s_{j-1} \to s_j$) and one *high pass* ($s_{j-1} \to d_j$); the downsampling makes this a *critically downsampled* filtering procedure, meaning that after the operation we end up with exactly as many entries as before. Diagram (13) then means that we have in fact a *perfect reconstruction, critically downsampled 2-channel* filter bank. The standard reference in electrical engineering for such filter banks is Smith and Barnwell (1986); similar constructions also appear in Mintzer (1985) and Vetterli (1986). Before these perfect reconstruction filter banks, electrical engineers had constructed similar filter banks that were almost perfect, in the sense that the reconstructed sequence is very close to the original. Such near-perfect filter banks are still designed and used for many applications; giving up perfect reconstruction leads to more degrees of freedom in the design and, if things are done right, to perceptually equally good results.

All this was developed without any input from mathematicians, with the result that electrical engineers sometimes and understandably feel that the present popularity of wavelets gives a lot of "undeserved" credit to mathematicians for reinventing the wheel while engineers were already driving cars. This view would be correct if there were nothing more to wavelets than the algorithm. The realization that the perfect reconstruction banks are linked to a rich underlying mathematical structure, associated with powerful and deep mathematical theorems is a different matter, however. Even for applications of interest to electrical engineering, this link has led to new applications that use the mathematical insights, and that would not have been developed from only the subband filtering concept (examples are Mallat and Hwang (1992), Wornell and Oppenheim (1992)).

To the computer scientist or engineer interested in studying vision, the multiresolution analysis framework, with its different levels of detail, is very reminiscent of multiscale models in vision analysis, such as Witkin (1983), or in a more algorithmic version, the pyramids of Burt and Adelson (1983). (As mentioned above, it was Mallat's background in vision theory that inspired him to re-interpret wavelet bases via the mathematical concept of multiresolution analysis.) Independently of and in parallel with the wavelet development, Adelson had in fact already switched from the (redundant) pyramid schemes to cascaded subband filtering for image analysis (see Adelson *et al.* (1987)).

Approximation theorists also recognized familiar concepts in wavelet theory: the space V_j, with their varying degrees of resolution, are basic standard fare in approximation theory. The example in (5) of spline spaces V_j really stems from approximation theory (de Boor (1978)). Similarly, formulas (10) and (12) are reminiscent of subdivision schemes, a technique developed to generate smooth curves and surfaces (Cavaretta *et al.* (1991)), Dyn *et al.* (1987)). There is a "philosophical" difference between many theorems in approximation theory and, for instance, the way function spaces are characterized via wavelets. Wavelet coefficients $< f, \psi_{j,k} >$ capture the *difference* between the successive approximations $P_{j-1}f$

and $P_j f$, rather than studying f via the $P_j f$. Similarly, subdivision schemes typically contain in their coarse-to-fine-formulas, only the first term in the right hand side of (12). Nevertheless, approximation theorists immediately recognized the kinship of wavelets, and are very active in the field now.

Wavelet techniques have some similarity as well with the multipole algorithms developed by Rokhlin (1985) for fast numerical computations. (See also Rokhlin's presentation in this ICM.) In a multipole expansion, the quantity to be computed (such as, e.g., the total gravitational potential energy of a completely arbitrary distribution of a large number of particles) is taken apart into many contributions living on different scales; for many of them, a coarse scale description suffices. Moreover, the taking apart is done in a hierarchical way. All this results in a fast algorithm. The wavelet algorithms in Beylkin *et al.* (1991) and subsequent work (e.g. Beylkin (1993)) work on the same principle as these fast multipole expansion techniques.

Finally, and as promised, wavelets buy you a time-frequency decomposition: once a function f is decomposed as in (3), it is written as a superposition of building blocks, the wavelets $\psi_{j,k}$, each of which is well localized in frequency (in a frequency band of width proportional to 2^{-j}, i.e. $2^{-j}\alpha \leq |\xi| \leq 2^{-j}\beta$) and in time (around the position $2^j k$, with a resolution proportional to k). Note that this means that high-frequency wavelets have very sharp time resolution, whereas low-frequency wavelets are much more spread out in time but have sharp frequency resolution. A decomposition of this type is well suited to signals f that consist of short-lived high-frequency *transients* superposed on more placid longer-lived low-frequency components. Many signals are of this type. But many more are really more complicated, and require a battery of tools of which wavelets are only one; we shall come back to this later.

6. Back to the Algorithm

For many applications, the powerful mathematical properties of wavelets can be exploited only if the associated algorithm is truly efficient. We saw earlier that the total complexity of a decomposition into orthonormal wavelets is KN, if there are K nonvanishing h_n, g_n in the associated filters. However, many "natural" orthonormal wavelet bases correspond to filters with infinitely many nonvanishing h_n, ruining the complexity estimate. This is the case, for instance, if the V_j are taken to be spline spaces of higher order than 1. In this case, the most natural choice for the function ϕ, the translates $\phi(x-n)$ of which should span all of V_0, would seem to be the B-spline function, obtained by convolving $\chi_{[0,1]}$ with itself $k-1$ times (for splines of order k). For this choice, the $\phi(x-n)$ are not orthogonal, so that ϕ needs to be replaced by an "orthogonalized" version, which is however now supported on all of $\mathbb{R}$ (with exponential decay), leading to infinitely many $h_n \neq 0$.

So how does one get MRA with finitely many nonvanishing h_n? The answer lies in the filtering approach from electrical engineering. If one takes (9), (10), and (12) as the point of departure, rather than as a corollary of the MRA structure, then one finds easily that the h_n should satisfy

$$\sum_n h_n h^*_{n+2k} = \delta k, 0 \quad , \tag{16}$$

while the g_n can be chosen as

$$g_n = (-1)^n h^*_{-n+1} \quad .$$

Smith and Barnwell (1986) had found ways to construct finite sequences h that satisfy (16). Such sequences do not necessarily correspond to an L^2-function ϕ such that (6) holds, however; necessary and sufficient conditions for this correspondence were found by Cohen (1990) and by Lawton (1991). If these conditions are satisfied, then there always exists an MRA associated with h. To obtain high approximation order for the MRA-ladder and smoothness, one needs to impose additional conditions on the sequence h, of the form

$$\sum_n (-1)^n h_n n^l = 0 \qquad l = 0, \ldots, L-1 \quad . \tag{17}$$

Daubechies (1988) constructs such finite sequences h and proves that by this method one can obtain compactly supported ϕ, ψ that are C^k, where k is arbitrarily large (but finite). These functions $\phi(x)$, $\psi(x)$ are not given by an explicit analytic expression, although the Fourier transform of ϕ can be written as an infinite product,

$$\hat{\phi}(\xi) = \hat{\phi}(0) \prod_{j=1}^{\infty} m_0(2^{-j}\xi) \quad , \tag{18}$$

with $m_0(\xi) = 2^{-1/2} \sum_n h_n e^{-in\xi}$.

One can use (6) to make a detailed study of their different, sometimes intriguing, properties. For instance, it turns out that the Hölder exponent of ϕ in a point x in its support depends on the frequency of the digits 1 and 0 in the binary expansion of x, as shown in Daubechies and Lagarias (1992); this means that these ϕ have *multifractal* properties (Daubechies and Lagarias (1994); see also Jaffard (1994)).

Although the wavelet bases constructed in Daubechies (1988) have been used in various applications, they are by no means ideal for all circumstances, and many other constructions have been carried out that improve on them in some respects, while giving up on other properties. For instance, one can give up some of the orthogonality in the constructions above, and construct a Riesz basis rather than an orthonormal basis of wavelets (together with the dual Riesz basis), as in Chui and Wang (1991), Chui and Wang (1992), Auscher (1989), or Cohen *et al.* (1992); this relaxing of orthonormality buys more smoothness and/or symmetry for the wavelets. Another useful construction restricts these wavelet bases to an interval while retaining their powerful mathematical properties (see e.g. Cohen *et al.* (1993), and Andersson *et al.* (1994)). Not all applications require absolutely that the filter h be finite; if m_0, defined as in (18), can be written as the quotient of two trigonometric polynomials, then there still exist fast algorithms to implement convolution with h, and such filters and wavelet bases have been proposed as well (Lemarié and Malgouyres (1989), Evangelista (1992), Herley and Vetterli (1993)) — in fact, the original construction by Stromberg (1982) falls into this class.

7. Higher Dimensions

So far, we have been working in one dimension only. There exist several possible generalizations to higher dimensions. Usually they involve several wavelets $\psi^1, \ldots, \psi^K$, and the wavelet basis is then given by the collection $\psi^n_{j,k}(x)2^{-jd/2}\psi^n(2^{-j}x - k)$, $j \in \mathbb{Z}$, $k \in \mathbb{Z}^d$, $n = 1, \ldots, K$. The easiest construction starts from a one-dimensional multiresolution analysis, with scaling function ϕ and wavelet ψ, and uses these to build one scaling function Φ and $2^d - 1$ wavelets Ψ^k in d-dimensions, by taking products of $\phi(x_k)$ and $\psi(x_m)$. For $d = 2$, for instance, one takes $\Phi(x_1, x_2) = \phi(x_1)\phi(x_2)$, $\Psi^1(x_1, x_2) = \psi(x_1)\phi(x_2)$, $\Psi^2(x_1, x_2) = \phi(x_1)\psi(x_2)$, $\Psi^3(x_1, x_2) = \psi(x_1)\psi(x_2)$. This corresponds to a two-dimensional multiresolution where the spaces $\mathbf{V}_j$ are tensor products $V_j \otimes V_j$, and the $\Psi^n_{j,k}$, $k \in \mathbb{Z}^2$, $n = 1, 2, 3$, then exactly span $\mathbf{W}_j$, the orthogonal complement of $\mathbf{V}_j$ on $\mathbf{V}_{j-1}$. The higher-dimensional Ψ^k and Φ inherit, of course, recursion relations similar to (6) and (7) from their one-dimensional progenitors, so that the algorithms remain basically as simple as in one dimension. There exist other, fancier constructions as well, with "nonseparable" higher-dimensional wavelets, possibly with a dilation matrix A replacing the simple scaling by 2, but the simple tensor product multiresolution analysis above is the most used. One can also introduce special bases of multidimensional wavelets, such as the divergence-free wavelet bases of Battle and Federbush (1993) or Lemarié-Rieusset (1992), useful for decomposing divergence-free vector fields.

In most of what follows, I will stick to the one-dimensional notation, but all statements (unless qualified) will be true for these d-dimensional wavelets as well.

8. Mathematical Properties

A first important property of wavelet bases is that they provide unconditional bases for many classical function spaces. A family of functions $\{g_\alpha; \alpha \in A\}$ is an unconditional basis for a Banach space $B \subset S'$ if it is a Schauder basis and there exists a criterion to decide whether $f \in B$ by using only the *absolute values* $|< f, g_\alpha >|, \alpha \in A$. Equivalently, the g_α constitute an unconditional basis if, whenever $\sum_{\alpha \in A} c_\alpha g_\alpha \in B$, multiplying the coefficients c_α with arbitrarily chosen $\epsilon_\alpha = \pm 1$ always leads to another element of B, i.e. $\sum_{\alpha \in A} \epsilon_\alpha c_\alpha g_\alpha \in B$. It turns out that the orthonormal wavelet bases (or more generally, Riesz bases of wavelets) give such unconditional bases for L^p $(1 < p < \infty)$, the Sobolev spaces W^s, the Besov spaces $B^{p,s}_q$, the Hölder spaces C^s, as well as for the Hardy space H^1 and its dual BMO (see Meyer (1990)). For instance, (15) gives a characterization of $f \in C^s$, using only the $|< f, \psi_{j,k} >|$, if we know a priori that $f \in L^\infty$. This last requirement can be dropped if we also impose that $\sup_{k \in \mathbb{Z}} |< f, \phi_{0,k} >| < \infty$; this then means that $\{\phi_{0,k}; k \in \mathbb{Z}\} \cup \{\psi_{-j,k}; k \in \mathbb{Z}, j \in \mathbb{N}\}$ is an unconditional basis for the (inhomogeneous) Hölder space C^s (provided $\phi \in C^r$ with $r > s$).

Intuitively speaking, wavelet expansions do so well, in such a variety of frameworks, because their smoothness allows them to adjust well to smooth functions (or to smooth portions of functions), their scaling properties allow them to "zoom in" on singularities, and their good spatial concentration allows them to handle decay well.

It follows that wavelet expansions for a function f can converge in many different topologies (depending on which spaces f belongs to). They can converge in even other ways as well. For instance, if we restrict ourselves to an interval, and order the wavelet basis properly (exhausting every scale first before moving on to the next finer scale), then the truncated sums of the correspondingly ordered wavelet expansions will converge in L^1 on the interval. (L^1 does not have unconditional bases, so some ordering is necessary.) When looking at pointwise convergence, one finds easily (provided ϕ and ψ satisfy a minimum of decay and smoothness conditions, as always) that the wavelet expansion of f converges in all points of continuity of f. It is also true that for L^2-functions f, the wavelet expansion of f converges pointwise almost everywhere (more precisely: in every Lebesgue point of f). For compactly supported ϕ, this last point follows from standard harmonic analysis arguments once one realizes that $\sup_j \sum_k |< f, \phi_{j,k} >||\phi_{j,k}(x)|$ is essentially a maximal function for f, bounded above (up to a constant factor) by the standard Hardy-Littlewood maximal function. The result is also true for less constrained ϕ (Auscher (1989), Kelly *et al.* (1994)), and it carries over (as usual) to other L^p-spaces as well.

9. Applications

Among the many successful applications of wavelets, only a few can be presented here. Particularly attractive (at least to me) are those where the mathematical properties of wavelets play an essential role in their effectiveness. A first example was the matrix or operator compression in Beylkin *et al.* (1991). The matrices $A_{i,j}$ they consider are finely sampled versions, $A_{i,j} = K(i\alpha, j\alpha)$ of an integral kernel $K(x,y)$ corresponding to a Calderón-Zygmund operator, i.e. K satisfies bounds of the type $|K(x,y)| \leq C|x-y|^{-1}, |\partial_x K(x,y)| + |\partial_y K(x,y)| \leq C|x-y|^{-2}$ (with often similar bounds for higher order derivatives). For the matrix $A_{i,j}$, this means that the matrix elements vary smoothly with i,j as long as (i,j) stays away from a region around the diagonal; near the diagonal wilder behavior is allowed. Replacing the sequence $A_{i,j}$ by its wavelet coefficients (obtained by "filtering" in both horizontal and vertical directions, with the fast algorithms explained above) results in a new matrix in which the majority of entries are exceedingly small. Thresholding them by ϵ (i.e. the entries smaller than ϵ are replaced by 0) gives a sparse matrix, so that computing (a truncated version of) the action of A on a vector can be done much faster. The beauty is that one can actually control the damage done by thresholding — not a trivial matter, since a large number of small errors can still add up to a sizeable total error. If the tresholding is done a little bit more carefully than by simple truncation (some sum rules need to be respected), then Beylkin, Coifman, and Rokhlin proved that the truncated matrix $A_\epsilon^{\text{trunc}}$ obtained by thresholding and then returning, via the inverse algorithm, to the "real world" from the "wavelet coefficient world," satisfies $\| A - A_\epsilon^{\text{trunc}} \| \leq C\epsilon^\gamma$ in L^2-*operator* norm, with C, γ independent of ϵ and also independent of the size of the matrix. The proof essentially repeats the argument of the "T(1) theorem" by David and Journé (1984).

The orthonormality of wavelet bases, as well as their different scales, are exploited by Elliott and Majda (1994) in an application closer to physics. They use wavelets as a tool to generate random velocity fields that accurately model fractal, self-similar fields, important in turbulent diffusion.

Applications to a very different field (although, strictly speaking, not of wavelet bases but of another type of wavelet representation) can be found in the work by Mallat and Hwang (1992). In one application, they seek to remove noise from very noisy images. This noise has particularly large effects on the fine scale wavelet coefficients, which also contain the information necessary to keep "sharp" edges in the image — discarding these corrupted fine scale coefficients altogether would result in a less noisy image, but it would also look blurred. Mallat and Hwang exploit the characterization of singularities given by the rate at which local wavelet coefficients decay as a function of scale, to sort out the chaff from the grain in the fine scale coefficients, leading to a restored denoised image with sharp edges.

Yet a different set of applications is in the work of Donoho (1993). He also discusses denoising. The starting point is a function f, supposed to belong to a Banach space B (which describes the class of problems of interest in a particular application); f is known only through noisy samples or estimations. Suppose that $(g_\alpha)_{\alpha\in A}$ is an unconditional basis for B. Then the data for f can be translated into noisy estimates for the coefficients of the expansion of f into the g_α. The denoising consists in a thresholded shrinking of these coefficients (all the ones below a threshold are set to zero, the ones above the threshold are multiplied with a nonzero coefficient < 1 depending on their size) and reconstruction. Donoho proves that if the g_α constitute an unconditional basis for B, then the worst-case error for this method cannot be significantly larger than the worst-case error for any other method, however fancy. Because wavelet bases are unconditional bases for many function spaces, they provide therefore a near-optimal method for a large variety of frameworks.

Wavelet bases are also, because of their adaptivity, a good tool to use in nonlinear approximation of e.g. piecewise smooth functions; see e.g. DeVore *et al.* (1992), Donoho (1993). Linear approximation theory discusses how well successive truncations of an expansion approach the desired function. For instance, if $(g_n)_{n\in\mathbb{N}}$ is a basis for B, then linear approximation is concerned with the behavior, as a function of N, of $\text{dist}_B(f, \Sigma_N)$, where Σ_N is the linear subspace $\Sigma_N = \{f = \sum_{n=1}^N c_n g_n; c_n \in \mathbb{C}\}$. In nonlinear approximation, the N-th approximation of f still involves N terms, but they need not correspond to the first N basis functions. That is, one studies $\text{dist}_B(f, S_N)$, where $S_N = \{f = \sum_{n\in I_{f,N}} c_n g_n; c_n \in \mathbb{C}, \#I_{f,N} = N\}$; S_N is no longer a linear subspace of B. An example of how this affects things: if f is a piecewise C^s function with good decay, and possibly discontinuities between the pieces, and if we choose a wavelet basis (with $\phi, \psi \in C^r$ with $r > s$), then $\text{dist}_{L^2}(f, \Sigma_N) \sim CN^{-1/2}$, but $\text{dist}_{L^2}(f, S_N) \sim CN^{-s}$: the nonlinear approximation does not suffer from the presence of the discontinuities. In contrast, if one chooses a Fourier basis, one finds that both $\mathrm{dist}_{L^2}(f, \Sigma_N)$ and $\mathrm{dist}_{L^2}(f, S_N)$ decay like $N^{-1/2}$.

10. Shortcomings of Wavelets

In spite of all their good qualities, wavelets are, of course, not the universal panacea. They are markedly inefficient for coherently oscillating components. Wavelet bases also suffer from being very translationally noninvariant, and no entirely satisfactory solution has been found, so far, to deal with boundary problems in higher dimensions for nonrectangular domains. Other recently developed harmonic analysis tools are much better at dealing with oscillations: wavelet packets and localized trigonometric bases.

11. Wavelet Packets

The algorithm that we sketched above for a decomposition into wavelets consists of concatenating diagram (11) several times, starting a new stage from the preceding "s_j" output. The "d_j"-branches are left untouched. We could also choose to attach another splitting diagram (11) to the "d_j"-branches; this still results in fast algorithms, corresponding to a decomposition into different functions, called wavelet packets. The wavelet bases we saw before are just one (extreme) example of wavelet packet bases. As explained before, the wavelet bases correspond to a Littlewood-Paley decomposition: in the frequency domain, $\widehat{\psi_{j,k}}(\xi)$ is essentially concentrated in and near the region $2^j\pi \leq |\xi| \leq 2^{j+1}\pi$. When the extra splittings are introduced that lead to wavelet packets, they correspond to further splits of these frequency blocks. One can, for instance, choose to keep splitting the branch of the wavelet algorithm diagram that would normally have ended in the "d_j"; if we split j times, at every intermediate step splitting all the subbranches that have been sprouted from the d_j-branch, then we will have subdivided the region $2^j\pi \leq |\xi| \leq 2^{j+1}\pi$ into 2^j subregions. If we do this for all $j \geq 0$, we end up with wavelet packets that all have the same "width," for their Fourier transforms as well as in "physical" space; these are therefore much closer to a standard windowed Fourier type basis than to the dyadic frequency decomposition given by wavelets. By choosing to split fewer times, one can generate a wide variety of wavelet packet bases that are intermediary between the "pure" wavelet bases and these Fourier-type wavelet packet bases.

Among all these bases, one can adaptively choose the one that is most "efficient" for a given function f (meaning, coarsely speaking, that the decomposition into this basis is achieved by a few large coefficients that represent most of the L^2-norm of f, with a small "tail" in the other coefficients) by basing the decision whether or not to split, at every step in the algorithm, on the results obtained for f. Detailed descriptions of these wavelet packet bases, first constructed by Coifman and Meyer, and of their mathematical properties and the associated algorithms can be found in Coifman *et al.* (1992), Coifman and Wickerhauser (1993), Wickerhauser (1994), and references therein. Note that when many splittings are carried out, the carving up of the frequency domain is not really as "clean" as the description above indicates; see Coifman *et al.* (1992).

12. Localized Trigonometric Bases

Wavelet packet bases are already much better at dealing with oscillations than wavelets. Even better are the localized cosine or sine bases constructed by Coifman and Meyer (1991), related to the independently constructed overlapped cosine transforms of Malvar (1990). These are orthogonal bases of the type

$$f_{k,l}(t) = w_k(t)\sin(\Omega_{k,l}t) ,$$

where the functions w_k are window functions, well localized in space (e.g., with compact support) but with possibly varying widths W_k, and the $\Omega_{k,l}$ are a corresponding discrete sequence of frequencies; in first approximation, the $\Omega_{k,l}$ behave like $\pi l/W_k$. More precisely, every $w_k(t)$ is supported on an interval $[a_k - \epsilon_k, a_{k+1} + \epsilon_{k+1}]$, and is $\equiv 1$ on the smaller interval $[a_k + \epsilon_k, a_{k+1} - \epsilon_{k+1}]$; here we assume $\cdots < a_{k-1} < a_k < a_{k+1} < \cdots$, with the ϵ_j chosen so that, for all k, $a_k + \epsilon_k \le a_{k+1} - \epsilon_{k+1}$. In the transition regions $[a_k - \epsilon_k, a_k + \epsilon_k]$, the window functions w_k and w_{k-1} must satisfy the complementarity requirement $w_{k-1}^2(x) + w_k^2(x) = 1$ as well as the symmetry condition $w_{k-1}(a_k - t) = w_k(a_k + t)$ (for $|t| \le \epsilon_k$). The width W_k is then defined as $W_k = a_{k+1} - a_k$, and the $f_{k,l}$ are given by

$$f_{k,l}(t) = (2/W_k)^{1/2} w_k(t) \sin[\frac{\pi}{W_k}(l + \frac{1}{2})(t - a_k)] \quad .$$

It is quite surprising that the functions w_k and the frequencies $\Omega_{k,l}$ can be chosen in such a way that the $f_{k,l}$ are all smooth (even C^∞) and nevertheless provide an orthonormal basis for $L^2(\mathbb{R})$. The construction is ingenious, but it doesn't use any modern techniques — this construction could have been carried out in the eighteenth century, and maybe the biggest surprise is that it wasn't. A remarkable feature of the construction is that neighboring window functions can be "merged," leading to the replacement of the $f_{k,l}$ and $f_{k+1,l'}$ by different functions $\tilde{f}_{k,l''}$; together with the remaining (and untouched!) $f_{n,l}(n < k$ or $n > k + 1)$ these then provide a different orthonormal basis. As in the case of wavelet packets, this choice between two options (to merge or not) can be exploited to construct a whole family of different bases, all "living" within one fast algorithm, so that the "best basis" can be chosen adaptively. See Coifman and Wickerhauser (1993), or Wickerhauser (1994).

13. Libraries of Bases

In practice, functions are usually quite complicated, and even these "best basis" algorithms do not necessarily give the most efficient decomposition. A simple example is a nicely oscillating function with just one superposed spike — the oscillations are best represented with a localized trigonometric basis or a wavelet packet basis, whereas the spike is "asking for" a wavelet representation. To address this, Mallat proposed a "pursuit" algorithm (Mallat and Zhang (1993)), adapted by Coifman and Meyer into an algorithm using libraries of bases. Based on the function to be decomposed, one first selects the best basis from a library, which can contain wavelet packets, various localized trigonometric bases, and other possible bases as well (as long as they are associated with fast algorithms). One monitors the coefficients computed for a decomposition in this basis, ranked by decreasing size. Beyond a certain threshold (which can depend on the total norm of the remaining tail, or the slowing down of the decay rate of the coefficients in this tail), one calls it quits — the selected basis was good for the first components but may not be optimal now. Reconstructing the first components and subtracting from the original leads to a remainder, for which one starts anew: again a best basis is selected, and one sticks to this basis until it becomes less satisfactory, etc .... This process can be repeated several times (see Coifman and Wickerhauser (1993)). This type of approach leads to very flexible and efficient time-frequency, or phase space decompositions.

14. Conclusion

In the last ten years, mathematical tools have emerged that combine insights from harmonic analysis with fast algorithms. They turn out to be very powerful for many applications, especially when used in conjunction with each other, and in combination with many existing tools. Not surprisingly, they can be linked with many other earlier insights in a variety of fields; one way of viewing them is as the synthesis of these varied strands. The result of this synthesis is more than just the sum of its parts, and as these new tools are becoming a familiar part of many a researcher's toolbox, they will turn up in many applications.

References

[1] Adelson, E. H.; Simoncelli, E. P.; and Hingorani, R., 1987, *Orthogonal pyramid transforms for image coding*, Proc. SPIE, Cambridge, **845** (Oct.), 50–58.

[2] Andersson, L.; Hall, N.; Jawerth, B.; and Peters, G., 1994, *Wavelets on closed subsets of the real line*, in Schumaker, Larry L., and Webb, G. (eds.), Topics in the Theory and Applications of Wavelets. Boston: Academic Press.

[3] Auscher, P., 1989, *Ondelettes fractales et applications*, Ph.D. thesis, Université de Paris IX – Dauphine, Paris, France.

[4] Auscher, P., 1992, *Toute base d'ondelettes régulières de $L^2(\mathbb{R})$ est issue d'une analyse multirésolution régulière*, C. R. Acad. Sci. Paris, Série I, **315**, 1227–1230.

[5] Battle, G., 1987, *A block spin construction of ondelettes, Part I: Lemarié functions*, Comm. Math. Phys., **110**, 601–615.

[6] Battle, G., and Federbush, P., 1993, *Divergence-free wavelets*, Michigan Math. J., **40**, 181.

[7] Beylkin, G., 1993, *Wavelets and fast numerical algorithms*, pp. 89–117 of Daubechies, I. (ed.), Different perspectives on wavelets. Proc. Sympos. Appl. Math., no. 47, Amer. Math. Soc., Providence, RI.

[8] Beylkin, G.; Coifman, R.; and Rokhlin, V., 1991, *Fast wavelet transforms and numerical algorithms*, Comm. Pure Appl. Math., **44**, 141–183.

[9] Burt, P., and Adelson, E. H., 1983, *The Laplacian pyramid as a compact image code*, IEEE Trans. Comm., **31**, 482–540.

[10] Cavaretta, A. S.; Dahmen, W.; and Micchelli, C. A., 1991, *Stationary subdivision*, Mem. Amer. Math. Soc., **93**(453), 1–186.

[11] Chui, C. K., and Wang, J. Z., 1991, *A cardinal spline approach to wavelets*, Proc. Amer. Math. Soc., **113**, 785–793.

[12] Chui, C. K., and Wang, J. Z., 1992, *On compactly supported spline wavelets and a duality principle*, Trans. Amer. Math. Soc., **330**, 903–915.

[13] Cohen, A., 1990, *Ondelettes, analyses multirésolutions et traitement numérique du signal*, Ph.D. thesis, Université Paris – Dauphine. Also *Ondelettes et traitement numérique du signal*, (1992), Masson, Paris; English translation to appear with Chapman and Hall, London.

[14] Cohen, A.; Daubechies, I.; and Feauveau, J. C., 1992, *Biorthogonal bases of compactly supported wavelets*, Comm. Pure Appl. Math., **45**, 485–560.

[15] Cohen, A.; Daubechies, I.; and Vial, P., 1993, *Wavelets on the interval and fast wavelet transforms*, Appl. Comput. Harmonic Anal., **1**, 54–81.

[16] Coifman, R., and Meyer, Y., 1991, *Remarques sur l'analyse de Fourier à fenêtre*, C. R. Acad. Sci. Paris Série I, **312**, 259–261.

[17] Coifman, R.; Meyer, Y.; and Wickerhauser, M. V., 1992, *Size properties of wavelet packets*, pp. 453–470 of M. B. Ruskai, et al. (eds.), Wavelets and Their Applications, Boston: Jones and Bartlett.

[18] Coifman, R., and Wickerhauser, M. V., 1993, *Wavelets and adapted waveform analysis: a toolkit for signal processing and numerical analysis*, pp. 119–153 of Daubechies, I. (ed.), Different perspectives on wavelets, AMS Short Course Lecture Notes, Amer. Math. Soc., Providence, RI.

[19] Daubechies, I., 1988, *Orthonormal bases of compactly supported wavelets*, Comm. Pure Appl. Math., **41**, 909–996.

[20] Daubechies, I., and Lagarias, J. C., 1992, *Two-scale difference equations, II. Local regularity, infinite products of matrices and fractals*, SIAM J. Math. Anal., **23**(4), 1031–1079.

[21] Daubechies, I., and Lagarias, J. C., 1994, *On the thermodynamic formalism for multifractal functions*, Rev. Math. Phys., **6**, 1033–1070.

[22] David, G., and Journé, J. L., 1984, *A boundedness criterion for generalized Calderón-Zygmund operators*, Ann. of Math. (2), **120**, 371–397.

[23] de Boor, C., 1978, *A practical guide to splines*, in Appl. Math. Sci. no. 27, New York: Springer.

[24] DeVore, R. A.; Jawerth, B.; and Popov, V., 1992, *Compression of wavelet decompositions*, Amer. J. Math., **114**, 737–785.

[25] Donoho, D., 1993, *Unconditional bases are optimal bases for data compression and for statistical estimation*, Appl. Comput. Harmonic Anal., **1**, 100–115.

[26] Dyn, N.; Gregory, A.; and Levin, D., 1987, *A 4-point interpolatory subdivision scheme for curve design*, Comput. Aided Geom. Design, **4**, 257–268.

[27] Elliott, F., and Majda, A., 1994, *A wavelet Monte Carlo method for turbulent diffusion with many spatial scales*, J. Comp. Phys. **113**, 82–109.

[28] Evangelista, G., 1992, *Wavelet transforms and digital filters*, pp. 396–412 of Meyer, Y. (ed.), Wavelets and Applications, Paris: Masson.

[29] Frazier, M.; Jawerth, B.; and Weiss, G., 1990, *Littlewood-Paley theory and the study of function spaces*, CBMS – Conference held at Auburn University in 1989, published by Amer. Math. Soc., Providence, RI.

[30] Herley, C., and Vetterli, M., 1993, *Wavelets and recursive filter banks*, IEEE Trans. Signal Proc., **41**, 2536–2556.

[31] Jaffard, S., 1994, *Multifractal formalism for functions. I & II*, preprints, CERMA, Ecole Nationale des Ponts et Chaussées, Noisy-le-Grand, France.

[32] Kelly, S. E.; Kon, M. A.; and Raphael, L. A., 1994, *Local convergence for wavelet expansions*, J. Funct. Anal., **126**(1), 102–138.

[33] Lawton, W., 1991, *Necessary and sufficient conditions for constructing orthonormal wavelet bases*, J. Math. Phys. **32**, 57–61.

[34] Lemarié, P. G., 1988, *Une nouvelle base d'ondelettes de* $L^2(\mathbb{R}^n)$, J. Math. Pures Appl. (9), **67**, 227–236.

[35] Lemarié, P. G., 1993, *Sur l'existence des analyses multirésolutions en théorie des ondelettes*, Rev. Mat. Iberoamericana, **8**, 457–474.

[36] Lemarié, P. G., and Malgouyres, G., 1989, Ondelettes sans peine, technical memorandum.

[37] Lemarié-Rieusset, P. G., 1992, *Analyses multi-résolutions non orthogonales, commutation entre projecteurs et dérivation, et ondelettes vecteurs à divergence nulle*, Rev. Mat. Iberoamericana, **8**, 221–237.

[38] Mallat, S., 1989, *Multiresolution approximation and wavelet orthonormal bases of* $L^2(\mathbb{R})$, Trans. Amer. Math. Soc., **315**, 69–88.

[39] Mallat, S., and Hwang, W. L., 1992, *Singularity detection and processing with wavelets*, IEEE Trans. Inform. Theory, **38**, 617–643.

[40] Mallat, S., and Zhang, Z., 1993, *Matching pursuits with time-frequency dictionaries*, IEEE Trans. on Signal Proc., **41**, 3397–3415.

[41] Malvar, H., 1990, *Lapped transforms for efficient transform/subband coding*, IEEE Trans Acoust. Speech Signal Proc., **38**, 969–978.

[42] Meyer, Y., 1985, *Principe d'incertitude, bases hilbertiennes et algèbres d'opérateurs*, Séminaire Bourbaki, 1985–1986, no. 662.

[43] Meyer, Y., 1990, Ondelettes et opérateurs, I: Ondelettes, II: Opérateurs de Calderón-Zygmund, III: Opérateurs multilinéaires. Paris: Hermann. English translation prepared by Cambridge University Press.

[44] Mintzer, F., 1985, *Filters for distortion-free two-band multirate filter banks*, IEEE Trans. Acoust. Speech Signal Proc., **33**, 626–630.

[45] Rokhlin, V., 1985, *Rapid solution of integral equations of classical potential theory*, J. Comput. Phys., **60**, 187.

[46] Smith, M. J., and Barnwell, D. P., 1986, *Exact reconstruction for tree-structured subband coders*, IEEE Trans. Acoust. Speech Signal Proc., **34**, 434–441.

[47] Stein, E., 1993, Harmonic Analysis: Real-variable Methods, Orthogonality, and Oscillatory Integrals, Princeton, NJ: Princeton University Press.

[48] Stromberg, J. O., 1982, *A modified Franklin system and higher order spline systems on* $\mathbb{R}^n$ *as unconditional bases for Hardy spaces*, pp. 475–493 of W. Beckner, et al. (eds.), Conference in Harmonic Analysis in Honor of A. Zygmund, vol. II, Belmont, CA: Wadsworth Math. Series.

[49] Vetterli, M., 1986, *Filter banks allowing perfect reconstruction*, Signal Process, **10**, 219–244.

[50] Wickerhauser, M. V., 1994, Adapted Wavelet Analysis from Theory to Software, Wellesley, MA: AK Peters.

[51] Witkin, A., 1983, Scale space filtering. in Proc. Int. Joint Conf. Artificial Intell.

[52] Wornell, G. W., and Oppenheim, A. V., 1992, *Wavelet-based representations for a class of self-similar signals with application to fractal modulation*, IEEE Trans. Inform. Theory, **38**, 785–800.

The Fractional Quantum Hall Effect, Chern-Simons Theory, and Integral Lattices

R. G. M∪ϕ

Research Group in Mathematical Physics:
J. Fröhlich (speaker and coordinator),
A. H. Chamseddine, F. Gabbiani, T. Kerler, C. King,
P. A. Marchetti, U. M. Studer, E. Thiran

ETH-Zürich, CH-8092 Zürich,
Switzerland

"There's so much fun to be had.... I don't want you to take this stuff too seriously. I think we should just have fun imagining it, and not worry about it — there's no teacher going to ask you questions at the end."(R. P. Feynman)

1 Chern-Simons theory

Chern-Simons theory has come to play an important rôle in three-dimensional topology because of its connections with Ray-Singer analytic torsion [47], the Gauss linking number [25], [14], [57], the Jones polynomial in knot theory [35] and its generalizations [63], [23], and three-manifold invariants [63], [12]. Recently, Chern-Simons forms and actions over noncommutative spaces [7] have been defined [45], [6] and turn out to provide a unifying perspective for topological gauge theories in odd *and* even dimensions [6].

The comparatively trivial abelian pure Chern-Simons theories (which reproduce the Gauss linking number and analytic torsion) have turned out to be fundamental building blocks for a theory of the fractional quantum Hall effect [61], [31], [59], [20], [29], [49]. This effect is one of the more exciting effects in condensed matter physics, discovered and explored between 1980 and the present [58], [54], [9], [44]. It has also been observed that $SU(2)$-Chern-Simons theories come up in problems of condensed matter physics connected with the theory of spin liquids; see e.g. [26].

Thus, it is well justified to start this report with a short review of the definition and some mathematical properties of Chern-Simons theory.

Let M be an oriented, framed three-manifold (the framing of M corresponds to a choice of a trivialization of the tangent bundle of M). Below, we shall consider the example where $M = \mathbb{R}^3$. Let G be a compact Lie group, or let $G = \mathbb{R}^N$. Let E denote the total space of a principal G-bundle with base space M, and let ∇ be a connection on E. Locally, we may describe ∇ in terms of its components, A (the "gauge potential"), in some local trivialization of E. These components are 1-forms on M with values in Lie G (the Lie algebra of G). The Chern-Simons

Proceedings of the International Congress
of Mathematicians, Zürich, Switzerland 1994

3-form on M is defined, locally, by the formula

$$CS^{(3)}(A) \;=\; \mathrm{tr}\,(A \wedge dA \,+\, \frac{2}{3}\, A \wedge A \wedge A), \tag{1.1}$$

where tr$(\cdot)$ is a trace on Lie G that is invariant under the adjoint action of G on Lie G. The Chern-Simons action functional S is defined, formally, by

$$S(A) \;=\; \frac{1}{4\pi} \int_M CS^{(3)}(A). \tag{1.2}$$

Unfortunately, this definition does not make sense in general. To understand the problems with (1.2), we consider the example where $M = S^3$ and $G = SU(N)$. We choose an orthonormal basis $\{T_\alpha\}_{\alpha=1}^{D_N}$, $D_N = N^2 - 1$, in $A_{N-1} = \text{Lie } SU(N)$ and choose tr$(\cdot)$ such that

$$\mathrm{tr}\,(T_\alpha\, T_\beta) \;=\; -\frac{k}{2}\,\delta_{\alpha\beta}, \tag{1.3}$$

$k \in \mathbb{R}$. Because $\pi_3(G) = \mathbb{Z}$, the action $S(A)$ in eq. (1.2), with tr$(\cdot)$ as in (1.3), is defined only modulo $2\pi k\mathbb{Z}$. It follows that $\exp\, i\, S(A)$ is a well-defined, single-valued functional of the connection ∇ if and only if $k \in \mathbb{Z}$. Similar remarks apply to general compact Lie groups.

Assuming now that tr$(\cdot)$ has been chosen such that $\exp\, i\; S(A)$ is a well-defined functional of ∇, quantized Chern-Simons theory is defined as a mathematically precise interpretation of the formal Feynman "functional measure"

$$dP\,(A) \;:=\; Z^{-1}\, \exp\, i\, S(A)\, \mathcal{D}A, \tag{1.4}$$

where $\mathcal{D}A$ is a formal Lebesgue measure on the affine space of connections on E, and the normalization factor Z (the partition function) is chosen such that $\int dP(A) = 1$. One would hope to extract from (1.4) a precise definition of $dP(A)$ as a complex measure on the space $\mathcal{A}$ of orbits of gauge potentials under the action of the group of gauge transformations.

The functional $\exp\, i\, S(A)$ does not require choosing a metric on M, and one might expect, therefore, that $dP(A)$ is independent of a choice of a metric on M. Unfortunately, this is a wrong expectation. The definition of "$\mathcal{D}A$" involves the choice of a metric on M, and, in order to eliminate dependence of $dP(A)$ on that metric, one must add to $S(A)$ a "counterterm", which is given by the Chern-Simons action of the Levi-Civita spin connection [63], [5]. One may then hope to arrive at a definition of $dP(A)$ that depends only on the framing of M and hence yields what is called a topological gauge theory [63], [62].

The kinds of functionals on $\mathcal{A}$ one would like to integrate with the "measure" $dP(A)$ are *Wilson loops*: let $\mathcal{L}$ be a loop in M (i.e., a smooth embedding of S^1 in M), and let R be an irreducible, unitary representation of G. We define

$$W_R[\mathcal{L}] \;:=\; Tr_R R\, \big[P\, \exp\, \zeta \int_{\mathcal{L}} A\big], \tag{1.5}$$

where P indicates path ordering, and ζ is some positive constant ("field strength renormalization" constant) to be determined. For a smooth Lie G-valued 1-form A, the R.S. of (1.5) can be defined via Chen's iterated integrals, i.e., through its *Dyson series.*

As it stands, the expression on the R.S. of eq. (1.4) is nonsense. A conventional strategy used to make sense of (1.4) is to *fix a gauge* and apply the Faddeev-Popov procedure [10] to interpret $\mathcal{D}A$. "Fixing a gauge" consists in choosing connection-dependent, local trivializations of E in such a way that the gauge potentials A satisfy certain constraints. We wish to exemplify gauge fixing in a special case, following [23]: we choose $G = SU(N)$ and $M = \mathbb{R}^3$. Points $x \in M$ are represented by (Cartesian) coordinates (x^+, x^-, t), with x^+, x^-, t in $\mathbb{R}$. We expand the gauge potential A in the basis $\{dx^+, dx^-, dt\}$ of 1-forms:

$$A(x) \;=\; a_+(x)dx^+ + a_-(x)dx^- + a_0(x)dt, \tag{1.6}$$

where $a_i(x) \in A_{N-1}$, $i = +, -, 0$. We choose a basis $\{T_\alpha\}_{\alpha=1}^{D_N}$ in A_{N-1} and a trace $\mathrm{tr}(\cdot)$ on A_{N-1} as specified in (1.3). Then

$$a_i(x) \;=\; \sum_{\alpha=1}^{D_N} a_i^\alpha(x)\, T_\alpha,$$

where $a_i^\alpha(x)$ is a function on M, $\forall i, \alpha$. One easily shows that the condition

$$a_-(x) \;=\; 0 \tag{1.7}$$

fixes a gauge (called "light-cone" or "axial" gauge). In this gauge, the Chern-Simons action S of eq. (1.2) takes the form

$$S(A) \;=\; \frac{1}{4\pi} \int \mathrm{tr}\,(a_+ \partial_- a_0)\; dx^+ \wedge dx^- \wedge dt. \tag{1.8}$$

This action is *quadratic* in A. One may therefore attempt to interpret the measure $dP(A)$ in (1.4) as a "complex Gaussian measure". Well, it actually is a "complex Gaussian", but it isn't a measure. However, all we really need to be able to do is to calculate *moments* of $dP(A)$. Let $\langle(\cdot)\rangle$ denote formal integration $\int dP(A)(\cdot)$ with respect to $dP(A)$. The first moments $\langle a_i^\alpha(x)\rangle$ *vanish* and the second moments $\langle a_i^\alpha(x)\, a_j^\beta(y)\rangle$ can be expressed in terms of the partial derivative of a Green function of the d'Alembertian $\partial_+\partial_-$ with respect to x^+. Together, they determine all higher moments ("Wick's theorem"). It is advantageous to complexify the planes $\{t = \text{const.}\}$, use complex coordinates, $z = x^+ \in \mathbb{C}$, $\bar{z} = x^- \in \mathbb{C}$, and analytically continue the moments of $dP(A)$ in x^+. The physicists call this "Wick rotation". Wick rotation is convenient, but not indispensable, in the following calculations. The Wick-rotated second moments are:

$$\begin{aligned}
\langle a_-^\alpha(x)\; a_j^\beta(y)\rangle &= 0, \quad \text{for all } j, \alpha, \beta,\\
\langle a_+^\alpha(x)\; a_+^\beta(y)\rangle &= 0, \quad \text{for all } \alpha, \beta,\\
\langle a_0^\alpha(x)\; a_0^\beta(y)\rangle &= 0, \quad \text{for all } \alpha, \beta,
\end{aligned}$$

and

$$\langle a_+^\alpha(z,t)\, a_0^\beta(w,s)\rangle \;=\; 2\lambda\, \delta^{\alpha\beta}\, \delta(t-s)\, \frac{1}{z-w}, \tag{1.9}$$

with $\lambda = -1/k$. Expectations $\langle(\cdot)\rangle$ of more complicated functionals of A can be calculated from (1.9) by using Wick's theorem. In particular, we may calculate expectations of "Wilson lines" and Wilson loops from (1.9) (e.g. by expanding them in a Dyson series).

Let $I_1, \ldots, I_m$ be a partition of $\{1, \ldots, n\}$, $m = 1, 2, \ldots$, $n = 1, 2, \ldots$. To every index set I_ℓ we assign a representation R_ℓ of $SU(N)$. Each index $j \in I_\ell$ labels a smooth curve

$$\gamma_j(t) \;=\; \{z_j(t') \in \mathbb{C} \,:\, t_0 \le t' \le t\}$$

in the complex plane that determines a smooth curve $\sigma_j(t)$ in $\mathbb{R}^3$ given by

$$\sigma_j(t) \;=\; \{(\mathrm{Re}\; z_j(t'),\, \mathrm{Im}\; z_j(t'), t') \,:\, z_j(t') \in \gamma_j(t),\, t_0 \le t' \le t\}. \tag{1.10}$$

We define a "Wilson line operator" $w_j(t)$ by setting

$$w_j(t) \;:=\; R_\ell\big[P \exp \zeta \int_{\sigma_j(t)} A\big], \tag{1.11}$$

where $\zeta > 0$ is a field strength renormalization constant. This operator is a holonomy matrix of the connection ∇ with components A and acts on the representation space V_{R_ℓ} of $SU(N)$. It is easy to see that

$$dw_j(t) \;=\; \zeta\, d\alpha_j(t)\, w_j(t), \tag{1.12}$$

where

$$\alpha_j(t) \;:=\; dR_\ell\big[\int_{t_0}^{t} \{a_+(z_j(t'),t')\dot{z}_j(t') + a_0(z_j(t'),t')\}\, dt'\big],$$

with $\dot{z}(t) = dz(t)/dt$, and dR_ℓ the representation of A_{N-1} determined by R_ℓ; $j \in I_\ell$, $\ell = 1, \ldots, m$.

The *basic object* in a mathematically precise definition of $SU(N)$ pure Chern-Simons theory on $\mathbb{R}^3$ is

$$\phi_n(t, t_0) \;:=\; \langle w_1(t) \,\otimes \cdots \otimes\, w_n(t)\rangle\,, \tag{1.13}$$

which is an endomorphism of the vector space

$$\mathcal{V}_n \;:=\; V_{R^{(1)}} \otimes \cdots \otimes V_{R^{(n)}}, \tag{1.14}$$

with $R^{(j)} = R_\ell$, for $j \in I_\ell$, $n = 1, 2, 3, \ldots$. One may attempt to calculate $\phi_n(t, t_0)$ by deriving a differential equation for it. We define

$$\Omega_{ij} \;:=\; \sum_{\alpha=1}^{D_N} \mathbb{I} \otimes \cdots \otimes dR^{(i)}(T_\alpha) \otimes \cdots \otimes dR^{(j)}(T_\alpha) \otimes \cdots \otimes \mathbb{I}, \tag{1.15}$$

for all i, j, with $1 \leq i < j \leq n$. Using (1.12), one shows — see [23] — that

$$\dot{\phi}_n(t, t_0) = \kappa \sum_{1 \leq i < j \leq n} \frac{\dot{z}_i(t) - \dot{z}_j(t)}{z_i(t) - z_j(t)} \Omega_{ij} \phi_n(t, t_0), \tag{1.16}$$

where $\kappa = \zeta^2 \lambda$. Eq. (1.16) is the celebrated *Knizhnik-Zamolodchikov equation*[38]. An alternative method to calculate $\phi_n(t, t_0)$ would be to expand all Wilson line operators $w_j(t)$ in their Dyson series and to calculate the resulting terms by using Wick's theorem and (1.9) [16].

Let M_n denote the subset of $\mathbb{C}^n$ consisting of n-tuples, $\underset{\sim}{z} = (z_1, \ldots, z_n)$, of complex numbers, with $z_i \neq z_j$, for $i \neq j$, and let $\widetilde{M}_n$ be the universal cover of M_n. Let K be the space of $\mathcal{V}_n$-valued functions on $\widetilde{M}_n$. On K we may define a connection 1-form ω by setting

$$\omega = \kappa \sum_{1 \leq i < j \leq n} d \, \log(z_i - z_j) \, \Omega_{ij}. \tag{1.17}$$

This connection is called the Knizhnik-Zamolodchikov connection. It is easy to verify that ω is *flat*, i.e.,

$$d\omega + \omega \wedge \omega = 0.$$

This is a consequence of the infinitesimal pure braid relations

$$[\Omega_{ij}, \Omega_{k\ell}] = 0, \quad [\Omega_{ij}, \Omega_{jk} + \Omega_{ki}] = 0, \tag{1.18}$$

where i, j, k, and ℓ are all distinct. Eq. (1.16) may now be written as

$$d \, \phi_n = \omega \, \phi_n, \tag{1.19}$$

which is the equation for a parallel transporter.

Let $(z_1, \ldots, z_n)$ be a point in M_n, and let π be an arbitrary permutation of $\{1, \ldots, n\}$ leaving the subsets $I_1, \ldots, I_m$ invariant. Let $\sigma_j = \sigma_j(t_1)$ be a curve in $\mathbb{R}^3$, as in (1.10), starting at the point (Re z_j, Im z_j, t_0) and ending at (Re $z_{\pi(j)}$, Im $z_{\pi(j)}, t_1$), for $j = 1, \ldots, n$. The family of all n-tuples $\{\sigma_1, \ldots, \sigma_n\}$ of such curves that do not intersect each other is a union of disjoint homotopy classes of curves labelled by elements b of a subgroup $B_n(I_1, \ldots, I_m)$ of the braid group, B_n, on n strands defined by the property that the cosets of elements of $B_n(I_1, \ldots, I_m)$ modulo the normal subgroup of pure braids are permutations π of $\{1, \ldots, n\}$ leaving $I_1, \ldots, I_m$ invariant. Let $b \in B_n(I_1, \ldots, I_m)$, and let $\{\sigma_1, \ldots, \sigma_n\}$ be n curves in the homotopy class b. Let $\phi_n(b; t_1, t_0)$ be a solution of the Knizhnik-Zamolodchikov eq. (1.16) for the curves $\{\sigma_1, \ldots, \sigma_n\}$, with initial condition $\phi_n(b; t_0, t_0) = \mathbb{1} \big|_{\mathcal{V}_n}$. Then

$$b \mapsto \phi_n(b; t_1, t_0) \tag{1.20}$$

defines a representation ϕ_n of $B_n(I_1, \ldots, I_m)$ on $\mathcal{V}_n$. This is a consequence of the identity

$$\phi_n(b_2 \circ b_1; t_2, t_0) = \phi_n(b_2; t_2, t_1) \, \phi_n(b_1; t_1, t_0)$$

(representation property) and the flatness of the Knizhnik-Zamolodchikov connection ω.

Let

$$g \mapsto R_{(n)}(g) := R^{(1)}(g) \otimes \cdots \otimes R^{(n)}(g), \; g \in SU(N),$$

be the representation of $SU(N)$ on $\mathcal{V}_n$. Because the Knizhnik-Zamolodchikov connection ω is $SU(N)$-invariant, the representation ϕ_n of $B_n(I_1, \ldots, I_m)$ on $\mathcal{V}_n$ commutes with the representation $R_{(n)}$ of $SU(N)$ on $\mathcal{V}_n$. Let $\mathcal{I}_n$ be the subspace of $\mathcal{V}_n$ consisting of $SU(N)$-invariant tensors, i.e., for $\xi \in \mathcal{I}_n$, $R_{(n)}(g)\xi = \xi$, $\forall g \in SU(N)$. The space $\mathcal{I}_n$ inherits the scalar product of $\mathcal{V}_n$. It is an invariant subspace for ϕ_n. It is interesting to ask whether the representation ϕ_n of $B_n(I_1, \ldots, I_m)$ on $\mathcal{V}_n$, or its subrepresentation $\phi_n \big|_{\mathcal{I}_n}$, are *unitary* in the scalar product of $\mathcal{V}_n$. The answer is that they are *not* unitary. However, ϕ_n may contain a *unitary subrepresentation*: suppose that

$$\kappa = \pm \frac{1}{k + c_2}, \quad k = 1, 2, 3, \ldots, \tag{1.21}$$

where c_2 is the eigenvalue of the quadratic Casimir operator in the adjoint representation of the group G, normalized such that $c_2 = N$, for $G = SU(N)$. Let $U_q(\text{Lie } G)$ denote the usual quantum deformation of the universal enveloping algebra of Lie G with deformation parameter $q = \exp i\pi\kappa$ [34]. We assume that the representations R_ℓ, $\ell = 1, \ldots, m$, have positive q-dimensions; see e.g. [21]. One may then define a certain quotient $\mathcal{I}_n^{(q)}$ of $\mathcal{V}_n$ of U_q (Lie G)-invariant tensors, which is expected to be invariant under the representation ϕ_n of $B_n(I_1, \ldots, I_m)$; see e.g. Chapter 6 of [21]. The miracle is that $\phi_n \big|_{\mathcal{I}_n^{(q)}}$ appears to define a *unitary* representation of $B_n(I_1, \ldots, I_m)$ on $\mathcal{I}_n^{(q)}$. For $G = SU(2)$, proofs have been sketched in [52], [39]. More details can be inferred from the explicit formulas in [23], [11] and the general results in Chapters 5 and 6 of [21]. For $G = SU(N)$, $N \geq 3$, a proof may, perhaps, be constructed on the basis of the results in [23], [21], [60], [37], but has apparently not appeared in the literature. The result described above is expected to hold for *arbitrary* compact, simple Lie groups G, but proofs are not available yet. The mathematical setting within which a proof might be constructed is that of braided tensor categories (more precisely "quantum categories" [21]) and of generalized hypergeometric functions [46]; see also the contributions of Felder and Wasserman to these proceedings, and references given there. A mathematically precise definition of quantized pure Chern-Simons theory on $M = \mathbb{R}^3$, with κ as in (1.21), would consist of converting the conjectures just described into theorems. Quantum-mechanical state vectors of this theory would be vectors in the spaces $\mathcal{I}_n^{(q)}$, $n = 0, 1, 2, \ldots$ $(\mathcal{I}_0^{(q)} := \mathbb{C})$, and it would determine unitary representations ϕ_n of the groups $B_n(I_1, \ldots, I_m)$ on $\mathcal{I}_n^{(q)}$, for all $I_1, \ldots, I_m$, and all n. The "physics-inspired" literature on these matters is somewhat confusing, with many incomplete proofs for fairly obvious conjectures.

The analysis sketched above for $G = SU(N)$ becomes very simple when $G = \mathbb{R}^N$, $N = 1, 2, \ldots$ (abelian pure Chern-Simons theory). See Section 3. Chern-Simons theories with $G = \mathbb{R}^N$ are the basic building blocks in the theory of the fractional quantum Hall effect. (It will turn out that G is actually given by $\mathbb{R}^N / \Gamma$, where Γ is an integral Euclidian lattice.)

Chern-Simons theory becomes a more interesting, *dynamical* quantum field theory if the manifold M is a full cylinder (and $k = 1, 2, 3, \dots$). In this situation, it is equivalent to Lie G Kac-Moody algebra at level k and its representation category. See [63], and [43], [24], [16] for more details. In the context of the quantum Hall effect, the Kac-Moody currents acquire physical significance as "edge currents".

But let us return to the representations ϕ_n of the braid groups $B_n(I_1, \dots, I_m)$ on the spaces $\mathcal{I}_n$, for generic values of the parameter κ, and sketch their connection with polynomial invariants of knots and links. We choose $n = 2p$ to be an even integer and assume that

$$R^{(j+p)} = R^{(j)^\vee}, \quad j = 1, \dots, p, \tag{1.22}$$

where $R^\vee$ is the representation of $SU(N)$ conjugate to R. Let π be a permutation of $\{1, \dots, 2p\}$ with $\pi(j+p) = j+p$, $R^{(\pi(j))} = R^{(j)}$ (j and $\pi(j)$ are in the same subset I_ℓ of $\{1, \dots, 2p\}$) for $j = 1, \dots, p$. Let $\{e_\alpha^{(R)}\}$ be an orthonormal basis of the representation space V_R. We define vectors $\xi(\pi) \in \mathcal{I}_{2p}$ by setting

$$\xi(\pi) = \sum_{\alpha_1, \dots, \alpha_p} e_{\alpha_{\pi(1)}}^{(R^{(1)})} \otimes \cdots \otimes e_{\alpha_{\pi(p)}}^{(R^{(p)})} \otimes e_{\alpha_1}^{(R^{(1)})} \otimes \cdots \otimes e_{\alpha_p}^{(R^{(p)})}. \tag{1.23}$$

Let b be an element of the braid group B_{2p} with the property that the coset of b modulo pure braids on $2p$ strands is given by the permutation π. We consider the scalar products

$$\big\langle \xi(\pi), \;\; \phi_{2p}(b; t_1, t_0)\, \xi(id.) \big\rangle. \tag{1.24}$$

These numbers are invariants of framed links. Quotients of these scalar products by analogous scalar products, with $SU(N)$ replaced by $\mathbb{R}$, yield the evaluation of an invariant of oriented links on the oriented link determined by the element $b \in B_{2p}$ and colored by the representations $R^{(1)}, \dots, R^{(p)}$. The special case where $R^{(1)} = \cdots = R^{(p)} = R$ is the N-dimensional, fundamental representation of $SU(N)$ has been analyzed in detail in [23], with generalizations appearing in Section 6.3 of [24].

The scalar products (1.24) can be calculated *perturbatively*, by expanding $\phi_{2p}(b; t_1, t_0)$ in a Taylor series in κ. The Taylor coefficients can be found by either solving the Knizhnik-Zamolodchikov equation for ϕ_{2p} iteratively (see the appendix in [23]) or, equivalently, by expanding the Wilson line operators $w_j(t)$ defined in (1.11) in their Dyson series, plugging the Dyson series into the R.S. of (1.13) and using Wick's theorem and (1.9). These Taylor coefficients are given in terms of multiple integrals along the curves $\sigma_1(t), \dots, \sigma_{2p}(t)$. They are special cases of what has become known under the name of Vassiliev invariants [56]: If, in eq. (1.19), a specific Knizhnik-Zamolodchikov connection ω is replaced by the "universal flat connection" defined by (1.17), with $\{\Omega_{ij}\}$ the "universal solution" of (1.18), one obtains the Vassiliev invariants of links.

It is natural to conjecture that the invariants built from (1.24) depend on the choice of the gauge group G in a nontrivial and interesting way. For a review of recent results concerning this topic see e.g. [2].

Now it is time to shift gears and talk about physics.

2 Quantum Hall effect and integral lattices

Experimentally, the quantum Hall effect is observed in two-dimensional systems of electrons confined to a planar region Ω and subject to a strong, uniform magnetic field $\vec{B}_c$ transversal to Ω, as indicated in *Figure 1.*

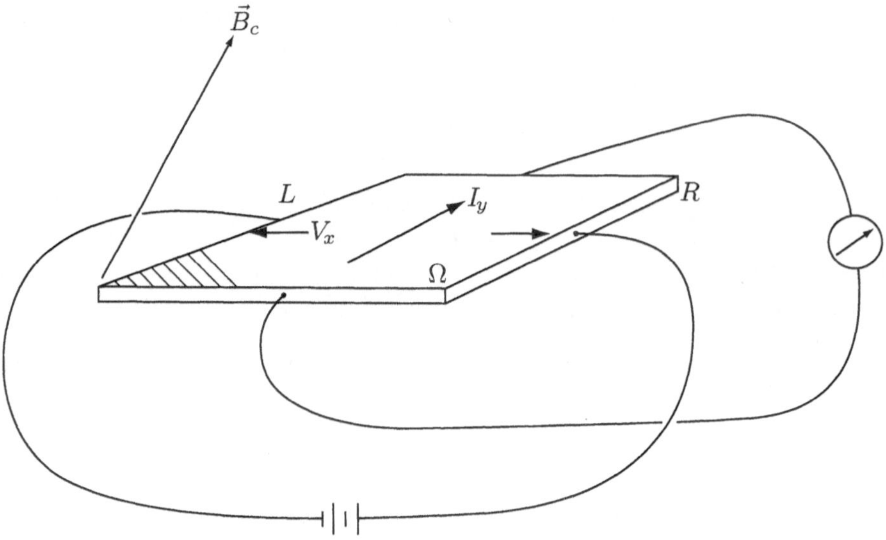

Figure 1

By tuning the y-component I_y of the total electric current to some value and then measuring the voltage drop V_x in the x-direction of the plane of the system, i.e., the difference in the chemical potentials of the electrons at the two edges R and L, one can calculate the *Hall resistance*

$$R_H = -\frac{V_x}{I_y}, \tag{2.1}$$

and finds that, for a fixed density n of electrons and at temperatures close to 0 K (absolute 0), the value of R_H is independent of the current I_y. It depends only on the external magnetic field $\vec{B}_c$. If the electrons are treated classically one finds, by equating the electrostatic to the Lorentz force, that

$$R_H = \frac{B_c}{ecn}, \tag{2.2}$$

where B_c is the z-component of $\vec{B}_c$ perpendicular to the plane of the system, e is the elementary electric charge, and c is the velocity of light.

By also measuring the voltage drop V_y in the y-direction, one can determine the longitudinal resistance, R_L, from the equation

$$R_L = \frac{V_y}{I_y}.$$

Neither classical nor quantum theory makes simple predictions about the behavior of R_L, but $R_L > 0$ means that there are dissipative processes in the system.

Two-dimensional systems of electrons are realized, in the laboratory, as *inversion layers* that form at the interface between an insulator and a semiconductor when an electric field (gate voltage) perpendicular to the interface, the plane of the system, is applied. An example of a material is a sandwich (a "heterojunction") made from GaAs and $Ga_xAl_{1-x}As$. The quantum-mechanical motion of the electrons in the z-direction perpendicular to the interface (identified with the x-y plane) is then constrained by a deep potential well with a minimum on the interface. Quantum theory predicts that electrons of sufficiently low energy, i.e., at low enough temperatures, remain bound to the interface and form a very nearly two-dimensional system.

In a theoretical analysis of the Hall effect it is advantageous to consider the connection between the electric current density $\mathbf{j}(\mathbf{x}) = (j^1(\mathbf{x}), j^2(\mathbf{x}))$ and the electric field $\mathbf{E}(\mathbf{x}) = (E_1(\mathbf{x}), E_2(\mathbf{x}))$ at an arbitrary point $\mathbf{x} = (x^1, x^2) \equiv (x, y)$ of Ω which is given by the *Ohm-Hall law*

$$\mathbf{E}(\mathbf{x}) = \rho\, \mathbf{j}(\mathbf{x}), \quad \rho = \begin{pmatrix} \rho_{xx} & -\rho_H \\ \rho_H & \rho_{yy} \end{pmatrix}, \tag{2.3}$$

where $\rho_{xx} = R_L(\ell_y/\ell_x)$, $\rho_{yy} = R_L(\ell_x/\ell_y)$ are the two longitudinal resistivities, $\rho_H = R_H$ is the Hall resistivity, and ℓ_x, ℓ_y are the widths of the system in the x- and y-directions, respectively. This is a phenomenological law valid on macroscopic distance scales and at low frequencies.

It is convenient to introduce a dimensionless quantity, the so-called *filling factor* ν, by setting

$$\nu = n/(eB_c/hc), \tag{2.4}$$

where $\frac{hc}{e}$ is the quantum of magnetic flux. Then the classical Hall law (2.2) says that R_H^{-1} rises *linearly* in ν, $R_H^{-1} = \frac{e^2}{h}\nu$, the constant of proportionality being given by a constant of nature, $\frac{e^2}{h}$. Because, experimentally, B_c can be varied and n can be varied (by varying the gate voltage), this prediction of classical theory can be put to experimental tests. Experiments at very low temperatures and for rather pure inversion layers yield the following very surprising data shown in Figure 2 [58], [54], [9].

These data tell us the following:

(1) $\sigma_H := \frac{h}{e^2}\, R_H^{-1}$ (the dimensionless Hall conductivity) has *plateaux at certain rational heights.* The plateaux at integer height occur with an astronomical precision of $1{:}10^8$ (defining a new standard for conductivity and yielding perhaps the most precise experimental value for the fine structure constant $\alpha = 2\pi\, e^2/hc \simeq 1/137$). The plateau quantization is insensitive to sample preparation and geometry.

(2) When (ν, σ_H) belongs to a plateau the longitudinal resistance R_L very nearly *vanishes.* This means that, for such values of ν and σ_H, there are no dissipative processes in the system.

The remarkable nature of these facts has been expressed by Laughlin [41] as follows: "The exactness of these results and their apparent insensitivity to the type

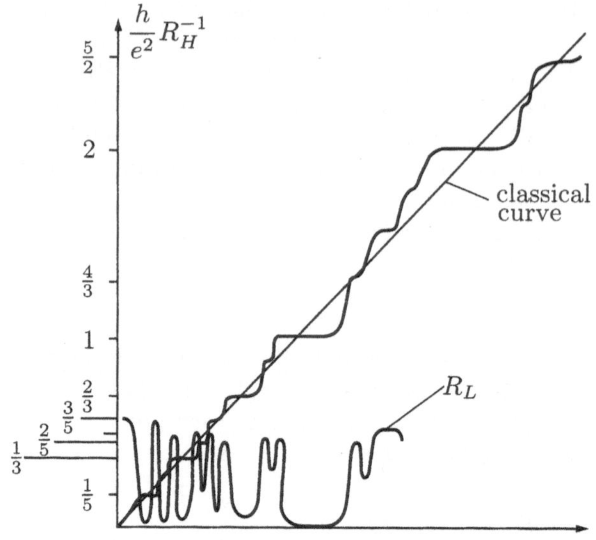

Figure 2

or location of impurities suggest that the effect is due, ultimately, to a fundamental principle."

It is the main purpose of this lecture to uncover some aspects of that principle. We shall be modest and focus our attention on the explanation of why σ_H must be a rational number when R_L vanishes, which rational numbers may occur, and what properties the system has when $R_L = 0$ and σ_H takes an allowed rational value.

As a first step, we formulate the classical electrodynamics of a two-dimensional system of electrons in an external electromagnetic field $(\vec{E}, \vec{B}_{\text{tot.}})$ when $R_L = 0$, and for an arbitrary value of σ_H. Here $\vec{E}$ is an external electric field, and $\vec{B}_{\text{tot.}} = \vec{B}_c + \vec{B}$, where $\vec{B}_c$ is a constant, external magnetic field transversal to the plane of the system, and $\vec{B}$ is a small, nonconstant perturbation of $\vec{B}_c$. As long as we do not describe the dynamics of the spins of the electrons — which are quantum-mechanical degrees of freedom — the laws of electromagnetism in such a system only involve $\mathbf{E} = (E_1, E_2)$, the component of $\vec{E}$ parallel to the plane of the system, and $B_{\text{tot.}} = B_c + B$, the component of $\vec{B}_{\text{tot.}}$ perpendicular to the plane of the system. Because R_L is assumed to vanish, eq. (2.3) can be rewritten as

(i) *Hall's law.*
$j^k(x) = \sigma_H \, \varepsilon^{k\ell} E_\ell(x)$, $x = (\mathbf{x}, t)$, with $k, \ell = 1, 2$, and $\varepsilon = \begin{pmatrix} 0 & 1 \\ -1 & 0 \end{pmatrix}$, in units where $e = h = 1$.

More fundamental are the following two laws:

(ii) *Charge conservation.*
$\frac{\partial}{\partial t} j^0(x) + \boldsymbol{\nabla} \cdot \mathbf{j}(x) = 0$ (continuity equation for the electric charge density j^0 and the electric current density $\mathbf{j}$).

(iii) *Faraday's induction law.*
$\frac{\partial}{\partial t} B(x) + \boldsymbol{\nabla} \wedge \mathbf{E}(x) = 0.$

Combining (i), (ii), and (iii), we find that

$\frac{\partial}{\partial t}\, j^0(x) = \sigma_H\, \frac{\partial}{\partial t}\, B(x)$.

Defining j^0 to be the difference between the total electric charge density and the uniform background density, n, we obtain the following result [20].

(iv) *Charge-flux relation.*

$j^0(x) = \sigma_H\; B(x)$.

The laws (i)–(iv) are generally covariant and metric independent (topological) [20]. Integrating (iv) over all of space Ω, we conclude that

$$q_{\mathrm{el}} \;=\; \sigma_H\; \Phi, \tag{2.5}$$

where $q_{\mathrm{el}} = \int_\Omega d^2\mathbf{x}\; j^0(\mathbf{x},t)$ is the total (excess) electric charge of the system, and $\Phi = \int_\Omega d^2\mathbf{x}\; B(\mathbf{x},t)$ is the total (excess) magnetic flux passing through the system.

These simple, beautiful laws, (i)–(iv), are the starting point of our analysis. They remain valid in a quantum-mechanical treatment of the electrons, see Section 3, that leads to rather remarkable conclusions. Let me anticipate the main results of our analysis and discuss their consequences. To do this, I must recall what *integral Euclidian lattices* are.

Let V be a vector space over the rational number field equipped with a positive-definite inner product $\langle\cdot,\cdot\rangle$. In V we choose a basis $\{\mathbf{e}_i\}_{i=1}^N$, $N = \dim V$, with integral Gram matrix K, where

$$K_{ij} \;=\; K_{ji} \;=\; \langle \mathbf{e}_i, \mathbf{e}_j\rangle \;\in\; \mathbb{Z}, \tag{2.6}$$

for all $i, j = 1, \ldots, N$. The basis $\{\mathbf{e}_i\}_{i=1}^N$ generates an *integral Euclidian lattice* Γ defined by

$$\Gamma \;=\; \Big\{\mathbf{q} \;=\; \sum_{i=1}^{N} q^i\, \mathbf{e}_i \;:\; q^i \;\in \mathbb{Z},\;\; \forall\, i\Big\}\,. \tag{2.7}$$

The lattice Γ^* *dual* to Γ, i.e., the lattice of integer-valued linear forms on Γ, is given by

$$\Gamma^* \;=\; \Big\{\mathbf{n} \;=\; \sum_{i=1}^{N} n_i\, \boldsymbol{\varepsilon}^i \;:\; n_i \in \mathbb{Z}, \forall\, i\Big\}, \tag{2.8}$$

where $\{\varepsilon^i\}_{i=1}^N$ is the basis of V dual to $\{\mathbf{e}_i\}_{i=1}^N$, i.e.,

$$\boldsymbol{\varepsilon}^i \;=\; \sum_{j=1}^{N} (K^{-1})^{ij}\; \mathbf{e}_j\;, \tag{2.9}$$

and

$$(K^{-1})^{ij} \;=\; \langle \boldsymbol{\varepsilon}^i, \boldsymbol{\varepsilon}^j\rangle \;=\; \frac{1}{\Delta}\, \widetilde{K}^{ij}, \tag{2.10}$$

where

$$\Delta \;=\; \det\, K \;=\, \big|\; \Gamma^*/\Gamma\; \big| \tag{2.11}$$

is the discriminant of Γ, and $\widetilde{K}$ is the matrix of cofactors (Kramer's rule).

The matrix K is positive-definite, with rank $(K) = N$, if and only if $\langle\cdot,\cdot\rangle$ is positive-definite. The lattice Γ is called *odd* iff it contains an element $\mathbf{q}$, with $\langle\mathbf{q},\mathbf{q}\rangle \in 2\,\mathbb{Z}+1$. Thus, Γ is odd iff K_{ii} is odd, for at least one $i \in \{1,\ldots,N\}$.

We are now in a position to state our *main contention.* Consider a two-dimensional system of electrons in a uniform, external magnetic field $\vec{B}_c$ at a temperature $T \approx 0$ K, with the property that R_L vanishes. Following Laughlin, we call such a system an *incompressible quantum Hall fluid*, abbreviated as IQHF. We claim that the physics of an IQHF on very large distance scales and at very low frequencies (i.e., in the so-called *scaling limit*) is coded into the data $(\Gamma_e, \mathbf{Q}_e)$ and $(\Gamma_h, \mathbf{Q}_h)$, where

(i) Γ_e and Γ_h are two integral, odd Euclidian lattices, and
(ii) for $x = e, h$, $\mathbf{Q}_x$ is a *primitive, odd* vector in Γ^*_x.

A vector $\mathbf{Q} \in \Gamma^*$ is called *primitive*, or visible, iff g.c.d. $\left(\langle\mathbf{Q},\mathbf{e}_j\rangle\right)_{j=1}^{N} = 1$, and $\mathbf{Q}$ is called *odd* iff

$$\langle\mathbf{Q},\mathbf{q}\rangle \equiv \langle\mathbf{q},\mathbf{q}\rangle \mod 2, \quad \forall\mathbf{q}\in\Gamma\,. \tag{2.12}$$

The dimensionless Hall conductivity σ_H is then given by

$$\sigma_H = \sigma_e - \sigma_h, \tag{2.13}$$

where

$$\sigma_x = \langle\mathbf{Q}_x,\mathbf{Q}_x\rangle, \quad \text{for } x = e,h. \tag{2.14}$$

This proves immediately that σ_H is a *rational number.* We shall denote it by

$$\sigma_H = \frac{n_H}{d_H}\,, \quad \text{with g.c.d. } (n_H, d_H) = 1.$$

At this point, there is the danger that our theory predicts far too many possible rational values of σ_H. However, what our theory really says is that if $R_L = 0$ then σ_H must belong to a certain subset $\mathbb{S}$ of the rational numbers, and that if $R_L = 0$ at some value of σ_H belonging to $\mathbb{S}$ then the properties of the system are encoded in some pair, $(\Gamma_e, \mathbf{Q}_e)$ and $(\Gamma_h, \mathbf{Q}_h)$, of integral Euclidian lattices and primitive vectors in their duals. Typically it happens that there are *many* pairs, $(\Gamma_e, \mathbf{Q}_e)$ and $(\Gamma_h, \mathbf{Q}_h)$, corresponding to a given value of σ_H in $\mathbb{S}$. Whether a pair $(\Gamma_e, \mathbf{Q}_e)$, $(\Gamma_h, \mathbf{Q}_h)$ describes an incompressible quantum Hall fluid that can be realized in a laboratory is a complicated analytical problem of quantum mechanics to which our theory can only give a tentative answer! Thus, it is very likely that many points in $\mathbb{S}$ do *not* correspond to the Hall conductivity σ_H of a *real* IQHF.

The subscripts "e" and "h" refer to the following *physics*: the basic charge carriers in a quantum Hall fluid (QHF) can be mobile *electrons* of electric charge $-e$. If $R_L = 0$ the fluid is then described by a pair $(\Gamma_e, \mathbf{Q}_e)$. They could also be mobile *holes* ("missing electrons") of charge $+e$, in which case the IQHF is described by $(\Gamma_h, \mathbf{Q}_h)$. Finally, an IQHF could be composed of *two* fluids, one consisting of mobile electrons, the other one consisting of mobile holes. It is a natural, *physical* idea that, for small values of the filling factor, these two fluids do not mix. We shall assume this henceforth (but see [22], [27] for a more general analysis also involving (indecomposable) Lorentzian lattices). The IQHF is then described by a pair $(\Gamma_e, \mathbf{Q}_e)$, $(\Gamma_h, \mathbf{Q}_h)$. As the electric charge of an electron is $-e$ and the one of a hole is $+e$, there is a relative minus sign between σ_e and σ_h in eq. (2.13)[1]. As there is a complete symmetry between electrons and holes, it is sufficient to develop the theory of QH fluids composed of electrons, and we set $\sigma_H := \sigma_e$ and drop the subscript "e" henceforth.

A pair $(\Gamma, \mathbf{Q})$, where Γ is an integral, odd Euclidian lattice and $\mathbf{Q}$ is a primitive, odd vector in Γ^* satisfying (2.12), is called a *chiral quantum Hall lattice*(cQHL). Our task is to classify cQHL's and to compare the predictions of the theory with experimental data.

The success of the theory is quite impressive: In Figure 3 we display measured values of σ_H when $R_L \approx 0$ (i.e., for IQHF's) in the range $0 < \sigma_H \leq 1$ that have been reported in the literature [9], [53] (for so-called single-layer, narrow-well IQHF's). We divide the data into separate "windows", Σ_p, $p = 1, 2, 3, \dots$, and each window Σ_p is the union of a left window $\Sigma_p^<$ and a right window $\Sigma_p^>$. Well-established plateau values of σ_H (i.e., values of σ_H corresponding to some IQHF) are indicated as a $\bullet$. Values of σ_H where R_L has a clearly visible local minimum ≈ 0, and σ_H has an inflection point as a function of the filling factor ν are indicated as a $\circ$. Very weak, or controversial data are indicated by $\bullet\!\!.$. Finally, the symbol *p.t.* indicates that, to such a value of σ_H, there correspond several distinct IQHF's, i.e., there are *phase transitions* between two or more different IQHF's with the same σ_H.

The remarkable fact is that these data — in particular the *absence* of data points — are very accurately reproduced by our theory of cQHL, see [28], [22], [27], if a *heuristic principle of stability* of a cQHL is introduced: the stability of a cQHL is intended to be a measure for the stability of the corresponding quantum-mechanical state of an IQHF under small perturbations, such as changes of the filling factor ν, see (2.4), or of the density of "impurities" in the system, etc. In order to formulate our stability principle for cQHL's mathematically, we must introduce some numerical *invariants* of cQHL's. The most primitive invariant of a cQHL $(\Gamma, \mathbf{Q})$ is the *dimension* N of the lattice Γ. Next, let

$$\Gamma = \bigoplus_{j=1}^{k} \Gamma_j \tag{2.15}$$

be the finest decomposition of the lattice Γ into an orthogonal direct sum of sublattices $\Gamma_j, j = 1, \dots, k$, and let

$$\mathbf{Q} = \sum_{j=1}^{k} \mathbf{Q}^{(j)}, \quad \mathbf{Q}^{(j)} \in \Gamma_j^*, \tag{2.16}$$

be the decomposition of $\mathbf{Q}$ associated to (2.15). We say that a cQHL $(\Gamma, \mathbf{Q})$ is *primitive* iff $\mathbf{Q}^{(j)}$ is a (nonvanishing) primitive vector of Γ_j^*, for all $j = 1, \dots, k$. This

[1]Historically, the existence of holes in semiconductors was first discovered in measurements of the sign of R_H!

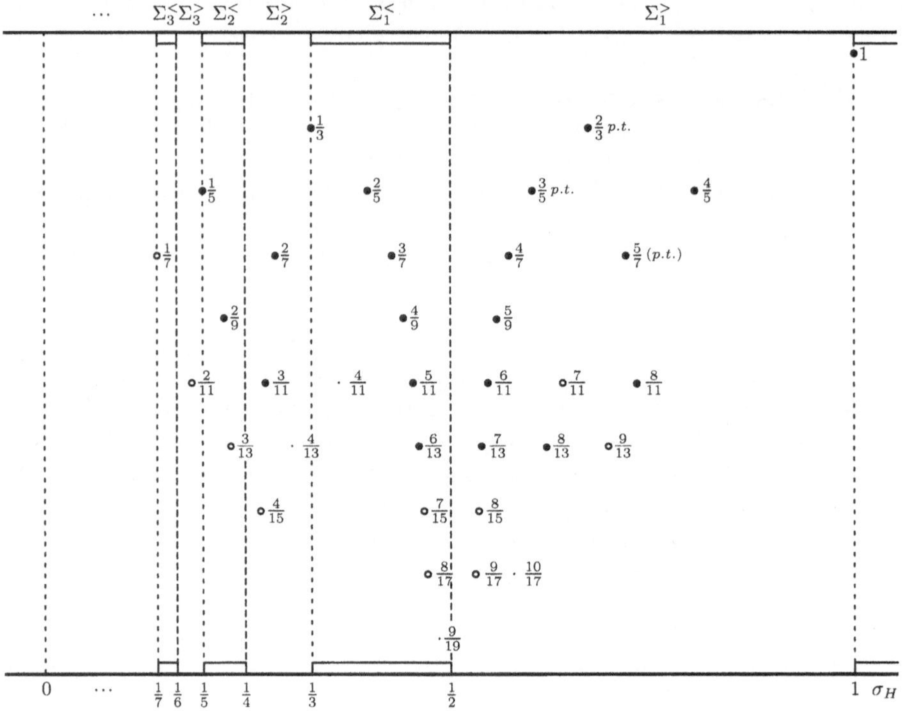

Figure 3. Observed Hall fractions σ_H in the range $0 < \sigma_H \leq 1$ and their experimental status in single-layer quantum Hall systems.

means that the pairs $(\Gamma_j, \mathbf{Q}^{(j)})$ are *indecomposable* cQHL's. Every indecomposable cQHL $(\Gamma_0, \mathbf{Q}_0)$ has a basis $\{\mathbf{q}_1, \dots \mathbf{q}_{N_0}\}$ with the property that $\langle \mathbf{Q}_0, q_\ell \rangle = -1$, for all $\ell = 1, \dots, N_0$. The set of all such bases is denoted by $\mathcal{B}(\Gamma_0, \mathbf{Q}_0)$. We then define an invariant $\ell_{\max.}$ (called "relative-angular-momentum invariant" [28]) by setting

$$\ell_{\max.}(\Gamma_0, \mathbf{Q}_0) \; := \; \min_{\{\mathbf{q}_i\}_{i=1}^{N_0} \in \mathcal{B}(\Gamma_0, \mathbf{Q}_0)} \Big(\max_{1 \leq i \leq N_0} \langle \mathbf{q}_i, \mathbf{q}_i \rangle \Big). \tag{2.17}$$

If $(\Gamma, \mathbf{Q})$ is a *decomposable*, primitive cQHL, i.e.,

$$(\Gamma, \mathbf{Q}) \; = \; \bigoplus_{j=1}^{k} \big(\Gamma_j, \mathbf{Q}^{(j)}\big), \tag{2.18}$$

as in (2.15), (2.16), we define

$$\ell_{\max.}(\Gamma, \mathbf{Q}) \; = \; \max_{1 \leq j \leq k} \; \ell_{\max.}\big(\Gamma_j, \mathbf{Q}^{(j)}\big). \tag{2.19}$$

Our *stability principle* for cQHL's says that an incompressible quantum Hall fluid corresponding to a primitive, chiral quantum Hall lattice $(\Gamma, \mathbf{Q})$ is the *more stable*, the *smaller* the value of the invariant $\ell_{\max.}(\Gamma, \mathbf{Q})$ and the *smaller* its dimension N. Available experimental data suggest that

$$\ell_{\max.}(\Gamma, \mathbf{Q}) \leq 7, (\text{or } 9), \tag{2.20}$$

for an *arbitrary* cQHL $(\Gamma, \mathbf{Q})$ describing a *physically realizable* IQHF. This is confirmed, qualitatively, by heuristic theoretical and numerical arguments [27]. Furthermore, there are fairly convincing, but heuristic theoretical arguments suggesting that, for a *real* IQHF with a nonzero density of impurities of finite strength, the dimension N of the corresponding cQHL is bounded above by a *finite* integer, N_*, depending on the filling factor ν, the density of electrons, and the density and strength of the "impurities", with $N_* \to \infty$, as the density of "impurities" tends to 0.

It is an elementary result in the theory of chiral quantum Hall lattices that the total number of cQHL's, $(\Gamma, \mathbf{Q})$, with $\ell_{\max.}(\Gamma, \mathbf{Q}) \leq \ell_*$ and $N = \dim \Gamma \leq N_*$, for arbitrary finite values of ℓ_*, N_*, is *finite* (though rapidly growing in ℓ_*, N_*).

A simple consequence of the Cauchy-Schwarz inequality tells us that the Hall conductivity σ_H of an IQHF corresponding to a cQHL $(\Gamma, \mathbf{Q})$ obeys the inequality

$$\sigma_H \equiv \sigma_H(\Gamma, \mathbf{Q}) = \langle \mathbf{Q}, \mathbf{Q} \rangle \geq \ell_{\max.}(\Gamma, \mathbf{Q})^{-1}. \tag{2.21}$$

This bound has interesting consequences (confirming a prejudice of Mark Kac [36]): if $\sigma_H \in \Sigma_p$, i.e.,

$$\frac{1}{2p+1} \leq \sigma_H (\Gamma, \mathbf{Q}) < \frac{1}{2p-1},$$

then

$$\ell_{\max.}(\Gamma, \mathbf{Q}) \geq 2p+1. \tag{2.22}$$

Our stability principle for cQHL's then says that the *most stable* IQHF's with $\sigma_H \in \Sigma_p$ are those described by cQHL's $(\Gamma, \mathbf{Q})$ satisfying

$$\ell_{\max.}(\Gamma, \mathbf{Q}) = 2p+1 \quad (N \text{ as small as possible}). \tag{2.23}$$

Combining the universal upper bound (2.20), i.e., $\ell_{\max.}(\Gamma, \mathbf{Q}) \leq 7$, with the bound (2.21), we conclude that there should not exist any physically realizable IQHF's with $\sigma_H < \frac{1}{7}$, and that, for σ_H in the window Σ_3, $\ell_{\max.}(\Gamma, \mathbf{Q})$ *must* take the smallest possible value compatible with (2.21), i.e., $\ell_{\max.}(\Gamma, \mathbf{Q}) = 7$. These conclusions are compatible with the data displayed in Figure 3.

The family of all primitive cQHL's $(\Gamma, \mathbf{Q})$, with $\sigma_H(\Gamma, \mathbf{Q}) \in \Sigma_p$ and $\ell_{\max.}(\Gamma, \mathbf{Q}) = 2p+1$ (the smallest possible value), is henceforth denoted by $\mathcal{H}_p$. In [22] we have proven an easy, yet remarkable theorem that says that there exist *bijections*, called "*shift maps*",

$$\mathcal{S}_p : \mathcal{H}_1 \to \mathcal{H}_{p+1}, \ p = 1, 2, 3, \ldots, \tag{2.24}$$

between the cQHL's in $\mathcal{H}_1$ and those in $\mathcal{H}_{p+1}$, with the properties that

$$\sigma_H\big(\mathcal{S}_p(\Gamma,\mathbf{Q})\big)^{-1} = \sigma_H(\Gamma,\mathbf{Q})^{-1} + 2p,$$

and

$$\ell_{\text{max.}}\big(\mathcal{S}_p(\Gamma,\mathbf{Q})\big) = \ell_{\text{max.}}(\Gamma,\mathbf{Q}) + 2p. \tag{2.25}$$

Furthermore, we have proven a somewhat deeper, but still rather easy *uniqueness theorem*[22]: let

$$\mathcal{H}_p^< := \big\{(\Gamma,\mathbf{Q}) \in \mathcal{H}_p \;:\; \sigma_H(\Gamma,\mathbf{Q}) \in \Sigma_p^<\big\}. \tag{2.26}$$

Then *all* the cQHL's $(\Gamma,\mathbf{Q})$ in $\mathcal{H}_p^<$ are known explicitly: the possible values in $\Sigma_p^<$ of the Hall conductivity σ_H corresponding to IQHF's described by cQHL's in $\mathcal{H}_p^<$ are given by the formula

$$\sigma_H = \frac{N}{2p\,N+1}, \tag{2.27}$$

and to each $N = 1,2,3,\ldots$, with σ_H given by (2.27), there corresponds a *unique* cQHL, $(\Gamma_{N,p},\mathbf{Q})$, of dimension N, and there are no further cQHL's in $\mathcal{H}_p^<$!

Note that it follows from the bound (2.20) on $\ell_{\text{max.}}$ that $\mathcal{H}_p^<$ contains *all* possible cQHL's with $\sigma_H \in \Sigma_p^<$ (as given by (2.27)), for $p = 3$.

The lattices $(\Gamma_{N,p},\mathbf{Q})$ with $\sigma_H(\Gamma_{N,p},\mathbf{Q}) = \langle\mathbf{Q},\mathbf{Q}\rangle = N(2pN+1)^{-1}$ can be described as follows: the lattice $\Gamma_{N,p}$ has a basis $\{\mathbf{q},\mathbf{e}_1,\ldots,\mathbf{e}_{N-1}\}$ with the property that

$$\langle\mathbf{Q},q\rangle = -1, \quad \langle\mathbf{Q},\mathbf{e}_j\rangle = 0, \quad j = 1,\ldots,N-1, \tag{2.28}$$

and with a Gram matrix K given by

$$K = \begin{pmatrix} 2p+1 & -1 & & & \\ -1 & 2 & -1 & & 0 \\ & -1 & 2 & -1 & \\ 0 & & \ddots & & -1 \\ & & & -1 & 2 \end{pmatrix}, \tag{2.29}$$

where $2p+1 = \langle\mathbf{q},\mathbf{q}\rangle$, and $K_{i+1,j+1} = \langle\mathbf{e}_i,\mathbf{e}_j\rangle$ are the matrix elements of the A_{N-1}-Cartan matrix. Thus, the Witt sublattice [8] of $\Gamma_{N,p}$ is the A_{N-1}-root lattice, and it is natural to call the series $(\Gamma_{N,p},\mathbf{Q}) \in \mathcal{H}_p^<$, $N = 1,2,3,\ldots$, of cQHL's the *fundamental A-series* in the window Σ_p. The cQHL's $(\Gamma_{N,p},\mathbf{Q})$ described here are typical examples of a general class of so-called *maximally symmetric* cQHL's [28], [27], which can be classified. The shift map $\mathcal{S}_{p-1}$ acts on the A-series in $\mathcal{H}_1^<$ by replacing $K_{11} = 3$ by $K_{11} = 2p+1$ and leaving the other matrix elements in the Gram matrices unchanged.

If you compare these results with the data in the windows $\Sigma_p^<$ of Figure 3 and recall that an IQHF is the less stable, the larger the values of p and N of the corresponding cQHL, the agreement between theory and experiment is remarkable. Is there a problem with the data point at $\sigma_H = \frac{4}{11} \in \Sigma_1^<$? There are no cQHL's with $\sigma_H = \frac{4}{11}$ and $\ell_{\max.} = 3$, but there actually are at least two distinct, low-dimensional cQHL's, with $\sigma_H = \frac{4}{11}$ and $\ell_{\max.} = 5(!)$, one obtained by applying the shift map $\mathcal{S}_1$ to the lattice $\mathbb{Z} \oplus 3\mathbb{Z}$, hence of dimension 2, and another one of dimension 7 (among, perhaps, further lattices of high dimension). Because, for these lattices, $\ell_{\max.}$ does *not* have the minimal value, 3, allowed in the window Σ_1, an IQHF with $\sigma_H = \frac{4}{11}$ is expected to be quite unstable against perturbations.

To the mathematician, the results just described may look disappointing, because they do not involve interesting lattices. The situation changes when we study the cQHL's belonging to the family $\mathcal{H}_p^> := \mathcal{H}_p \backslash \mathcal{H}_p^<$, corresponding to the range $\Sigma_p^>$ of values of σ_H. Because the shift map $\mathcal{S}_{p-1}$ is a bijection between $\mathcal{H}_1^>$ and $\mathcal{H}_p^>$, $p = 2, 3, 4, \ldots$, the classification of the *most stable* cQHL's with $\sigma_H \in \Sigma_p^>$, that is of all the lattices in $\mathcal{H}_p^>$, reduces to the classification of lattices in $\mathcal{H}_1^>$. But this is not an easy job. Although the number of cQHL's in $\mathcal{H}_1^>$ of dimension $N < N_*$ is *finite*, it grows rapidly in N_*.

In order to make progress, one may attempt to translate physical properties of IQHF's (related e.g. to electron spin, or to the spectrum of quasi-particles in such systems) into mathematical properties of quantum Hall lattices (related to the structure of their Witt sublattices and of the so-called glue group; see [28], [22], [27]). This enables one to introduce subfamilies of quantum Hall lattices, likely to describe physically realizable IQHF's, whose classification is feasible.

A prominent *finite* series of cQHL's in $\mathcal{H}_1^>$ is the one corresponding to the values

$$\sigma_H = \frac{2}{3}, \frac{3}{5}, \frac{4}{7}, \frac{5}{9}, \frac{6}{11}. \tag{2.30}$$

It is called the *E-series*, for the following reasons. Let $\mathcal{O} \oplus \Gamma_W$ denote the Kneser shape [8] of an integral lattice Γ,

$$\mathcal{O} \oplus \Gamma_W \subseteq \Gamma \subseteq \Gamma^* \subseteq \mathcal{O}^* \oplus \Gamma_W^*,$$

where Γ_W is the Witt sublattice generated by vectors of squared lengths 1 and 2. To every σ_H in the E-series (2.30) there corresponds a cQHL $(\Gamma, \mathbf{Q})$ with the property that the $\mathcal{O}$-sublattice in its Kneser shape is a one-dimensional, odd lattice, denoted $\mathcal{O}_k$, $\Gamma_W \equiv \Gamma_{E_k}$ is an E_k-root lattice, with $k = 7, 6, 5, 4, 3$, and $\mathbf{Q} \in \mathcal{O}_k^*$ is orthogonal to Γ_W [28]. Here we define the lattices Γ_{E_k} as the root lattices of the Lie algebras corresponding to the following Dynkin diagrams:

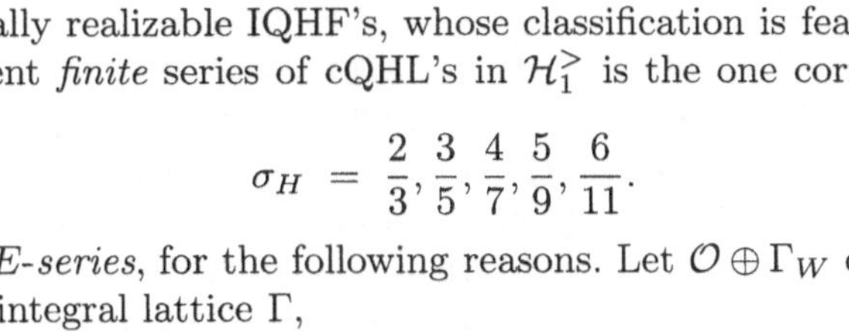

There is also a cQHL with $\sigma_H = \frac{7}{13}$ and $\ell_{\max.} = 3$. It has a two-dimensional $\mathcal{O}$-sublattice, and its Witt sublattice is the A_1-root lattice. This cQHL may be viewed as an irregular endpoint of the E-series. For there is *no* cQHL with $\sigma_H = \frac{8}{15}$ and $\ell_{\max.} = 3$ in dimension $N \leq 4$, or with discriminant $\triangle \leq 15$ and $N \leq 9$.

A lattice Γ is obtained from its Kneser shape, $\mathcal{O} \oplus \Gamma_W$, by gluing, namely by adding cosets of vectors in $\mathcal{O}^* \oplus \Gamma_W^*$, to $\mathcal{O} \oplus \Gamma_W$. The lattices Γ_k obtained from $\mathcal{O}_k \oplus \Gamma_{E_k}$, where Γ_{E_k} is the E_k-root lattice, $k = 7, 6, 5$, are unlikely to correspond to physically realizable IQHF's, as their dimensions (and the number of quasi-particles of the corresponding IQHF's) are large. However, they contain quantum Hall *sublattices*, with the *same* values for σ_H and $\ell_{\max.}$, which *are* realistic. For example, for $k = 7$, $\sigma_H = \frac{2}{3}$, the cQHL obtained from $\mathcal{O}_7 \oplus \Gamma_{E_7}$ by gluing contains a decomposable, two-dimensional QH sublattice, $3\mathbb{Z} \oplus 3\mathbb{Z}$, and an indecomposable, three-dimensional QH sublattice, whose Witt sublattice is the A_1-root lattice which, physically, could describe electron spin [28], or an internal symmetry that we call "isospin" symmetry — as well as less realistic sublattices of dimension 4, 5, 6, and 7. All these sublattices yield cQHL's with $\sigma_H = \frac{2}{3}$, $\ell_{\max.} = 3$. We thus predict that there should be *at least three* rather stable IQHF's with $\sigma_H = \frac{2}{3}$. They differ from each other in the rôle played by electron spin (which can be tuned by tilting the external magnetic field $\vec{B}_c$) or by "isospin". One therefore expects a magnetic-field driven phase transition between different IQHF's with $\sigma_H = \frac{2}{3}$. These predictions of our theory are in remarkable agreement with experimental data.

There is also a *D-series* of cQHL's, leading, e.g., to values of $\sigma_H = \frac{n_H}{d_H}$ with an *even* denominator d_H: $\sigma_H = \frac{1}{2}$ (arbitrary D_n), and $\sigma_H = \frac{4}{12-n}$, corresponding to $\Gamma_W = \Gamma_{D_n}$ with $n \leq 7$. Let $(\Gamma, \mathbf{Q})$ be a primitive cQHL. It has been shown in [28] that the sublattice of Γ orthogonal to $\mathbf{Q}$ cannot contain any self-dual lattice.

Besides the D- and the E-series, there is also an A_{N-1}-series of cQHL's in $\mathcal{H}_1^>$ that could describe single-layer IQHF's if N is an odd integer ≥ 5. They yield the values

$$\sigma_H = \frac{N}{N+4} \tag{2.31}$$

of the Hall conductivity $\left(\frac{5}{9}, \frac{7}{11}, \frac{9}{13}, \ldots\right)$.

Furthermore, we have classified *all* two-dimensional, three-dimensional, and four-dimensional cQHL's in $\mathcal{H}_1^>$; see [27]. (With an efficient computer program one could extend these results to $N = 5, 6$.) They correspond to the values $\frac{1}{2}, \frac{2}{3}, \frac{3}{5}, \frac{5}{7}, \frac{7}{13}$ ($N = 3$), and $\frac{2}{3}, \frac{3}{4}, \frac{3}{5}, \frac{4}{5}, \frac{4}{7}, \frac{5}{7}, \frac{6}{7}, \frac{5}{8}, \frac{5}{9}, \frac{6}{11}, \frac{8}{11}, \frac{10}{11}, \frac{11}{13}, \frac{13}{17}, \frac{14}{19}, \frac{6}{21}$, and $\frac{26}{31}$ ($N = 4$).

Besides the lattices discussed above, there are plenty of *decomposable* cQHL's in $\mathcal{H}_1^>$ obtained as the direct sum of *two* cQHL's of the fundamental A-series of cQHL's in $\mathcal{H}_1^<$. They correspond to the sequence

$$\sigma_H = \frac{4NM + N + M}{(2N+1)(2M+1)}, \quad N, M = 1, 2, 3, \ldots, \tag{2.32}$$

of values of the Hall conductivity. Because there is a very stable single-layer IQHF with $\sigma_H = 1$, described by $(\Gamma = \mathbb{Z}, \mathbf{Q} = 1)$, one does not expect to see plateaux in the Hall conductivity around the points given in (2.32), for values of N and M larger than 2 or 3.

Finally, our theory provides candidates of IQHF's described by *pairs* $(\Gamma_e, \mathbf{Q}_e)$ and $(\Gamma_h, \mathbf{Q}_h)$ of cQHL's corresponding to values of $\sigma_H = \sigma_e - \sigma_h$ in the window $\Sigma_1^>$. These IQHF's would be charge-conjugate to those described by the fundamental A-series in $\mathcal{H}_1^<$. They are obtained by setting $\Gamma_e = \mathbb{Z}$, $\mathbf{Q}_e = 1$, $\Gamma_h = \Gamma_{N,1}$; see (2.27), (2.28). One finds that

$$\sigma_H = \sigma_e - \sigma_h = 1 - \frac{N}{2N+1}, \quad N = 1, 2, 3, \ldots. \tag{2.33}$$

For $N \leq 6$, these values of σ_H coincide with the ones of the E-series. The existence (and uniqueness) of these pairs of cQHL's makes it plausible that $\sigma_H = \frac{6}{11}, \frac{7}{13}, \frac{8}{15}, \frac{9}{17}$ are values of the Hall conductivity of physically realizable IQHF's.

Those values of σ_H that correspond to *several* cQHL's in $\mathcal{H}_1^>$ (e.g. $\frac{2}{3}, \frac{3}{5}, \frac{4}{5}, \frac{5}{7}$, ...) tend to be values where, experimentally, phase transitions are observed.

We emphasize that, logically, our theory predicts the values of σ_H that *cannot* appear in IQHF's — indeed, it predicts plenty of gaps if bounds on $\ell_{\max.}$ and N are imposed. (For example, it tells us that values of $\sigma_H = \frac{n_H}{d_H}$, with d_H very large, require large values of either $\ell_{\max.}$ or N and hence should *not* be observed!) Furthermore, it tells us that *if* an *allowed* value of σ_H *is* observed in an IQHF, the structure of the IQHF can be described by a certain set of cQHL's. That's all our theory does if no *heuristic* principles are added to it.

Next, we propose to sketch how the physics of IQHF's leads us to study the mathematics of chiral quantum Hall lattices.

3 From incompressible quantum Hall fluids to chiral quantum Hall lattices via Chern-Simons theory

The starting point of our analysis is the idea to look for a theoretical description of the physics of an IQHF in the limiting regime of large-distance and long-time (low-frequency) scales. This limiting regime is called the *scaling limit* of the system, and experience shows that the theoretical description of physical systems simplifies in the scaling limit. An IQHF can be characterized by the following physical properties.

(P1) The temperature T of the system is close to 0 K. The longitudinal resistance, R_L, of an IQHF at $T = 0$ K vanishes, and the total electric charge is a good quantum number to label quantum-mechanical state vectors of the system [28], [19]. The charge of the groundstates of the system is normalized to be zero.

(P2) In the scaling limit, the total electric charge and current densities of an IQHF are the sum of $N = 1, 2, 3, \ldots$ separately conserved charge and current densities describing electron and/or hole transport in N separate "channels" distinguished by conserved quantum numbers. In our analysis, N will be treated as a free parameter. (Physically, N turns out to depend on the filling factor ν and other parameters characterizing the system.)

(P3) In units where $e = h = 1$, the electric charge of an electron/hole is $-1/1$. A local excitation of the system composed of electrons and holes and of total electric charge $q_{\text{el.}}$ satisfies *Fermi-Dirac statistics* if $q_{\text{el.}}$ is *odd* and *Bose-Einstein statistics* if $q_{\text{el.}}$ is *even*.

The quantum statistics of any local excitation of the system of electric charge $q_{\text{el.}} \in 2\,\mathbb{Z} + 1$ must be Fermi-Dirac statistics (i.e., the Pauli principle must hold), and if $q_{\text{el.}} \in 2\mathbb{Z}$ it must be Bose-Einstein statistics.

(P4) The quantum-mechanical state vector describing an *arbitrary* physical state of an IQHF is *single valued* in the position of all those excitations that are multi-electrons/-holes.

The properties (P1)–(P4), believed to be true in every IQHF, are physical properties. Part of the art of theoretical physics is to translate physical properties, deduced from experiments, into precise mathematical hypotheses. This cannot be done in the form of theorems and requires intuition. But once this exciting part of the job is completed, one must attempt to use mathematical theorems to derive new predictions on the behavior of a physical system.

The assumption that the longitudinal resistance R_L of an IQHF vanishes is translated into the mathematical assumption that the energy spectrum of the quantum-mechanical Hamiltonian describing the dynamics of the system exhibits what is called a *mobility gap* δ above the groundstate energy which is *strictly positive*, uniformly in the size of the system. This is actually an assumption that one can try to derive from the underlying microscopic Schrödinger quantum mechanics of nonrelativistic electrons. This is a difficult, but *not* hopelessly difficult, problem of analysis; see [15] and references given there.

The quantum-mechanical electric charge and current densities of a physical system are operator-valued distributions

$$j(x) \;=\; \left(j^0(x),\, j^1(x), \ldots, j^d(x)\right), \tag{3.1}$$

where d is the dimension of physical space, and $x = (\vec{x}, t)$ is a space-time point. They satisfy the continuity equation (conservation of electric charge)

$$\frac{\partial}{\partial t}\, j^0(x) + \vec{\nabla} \cdot \vec{j}(x) \;=\; 0. \tag{3.2}$$

Let $J(x) = *j(x)$ be the d-form dual to j. Then (3.2) says that

$$dJ(x) \;=\; 0. \tag{3.3}$$

For a two-dimensional system confined to a disk $\Omega \subseteq \mathbb{R}^2$, the Poincaré lemma tells us that (3.3) implies that

$$J(x) \;=\; db(x), \tag{3.4}$$

where $b(x)$ is a 1-form; b is determined by J up to the gradient of a scalar distribution χ, i.e., b has the properties of an abelian gauge field. By property (P2)

$$J(x) \;=\; \sum_{i=1}^{N} Q_i\, J^i(x), \tag{3.5}$$

where Q_i is the unit of electric charge transported by the current J^i, and J^i satisfies the continuity equation

$$d\, J^i(x) \;=\; 0, \quad \text{for } i = 1, \ldots, N, \tag{3.6}$$

so that, by Poincaré's lemma,

$$J^i(x) \;=\; d\,b^i(x), \quad i \;=\; 1, \ldots, N. \tag{3.7}$$

The *key idea* is to describe the physics of an IQHF in the scaling limit in terms of an *effective field theory* of the gauge fields $\mathbf{b}(x) = \left(b^1(x), \ldots, b^N(x)\right)^T$. Because, by property (P1), an IQHF has a strictly positive mobility gap δ, that effective field theory can only be a *topological field theory.* The presence of a nonzero, external magnetic field transversal to the plane to which the electrons of an IQHF are confined implies that the quantum dynamics of the system violates the symmetries of parity (reflections in lines) and time-reversal. The only topological field theory of the gauge fields $\mathbf{b}(x)$ breaking these symmetries and respecting invariance under the gauge transformations

$$\mathbf{b}(x) \;\mapsto\; \mathbf{b}(x) + d\boldsymbol{\chi}(x) \tag{3.8}$$

is *abelian Chern-Simons theory*, with $G = \mathbb{R}^N$. This has been shown in [29], [26]. (The same conclusion can be reached by starting from the laws (i)–(iv), Section 2, preceding eq. (2.5), of electrodynamics in quantum Hall fluids [20], or by studying gauge anomaly cancellations [59], [26].) The action functional of abelian Chern-Simons theory is given by

$$S_\Lambda(\mathbf{b}) \;=\; \frac{1}{4\pi} \int_\Lambda \mathbf{b}^T \wedge C d\mathbf{b} \;+\; \Gamma_{\partial\Lambda}(\mathbf{b}), \tag{3.9}$$

where $\Lambda = \Omega \times \mathbb{R}$ is the three-dimensional space-time of the system, $C = (C_{ij})_{i,j=1}^N$ is some metric on "field space" $\mathbb{R}^N$, and $\Gamma_{\partial\Lambda}(\mathbf{b})$ is the two-dimensional, anomalous chiral action only depending on the restriction of the gauge fields $\mathbf{b}$ to the boundary $\partial\Lambda$ of Λ; see [50]. Note that, individually, the two terms on the r.h.s. of (3.9) are not invariant under gauge transformations (3.8) not vanishing on $\partial\Lambda$. The boundary action $\Gamma_{\partial\Lambda}(\mathbf{b})$ is chosen such that their sum *is* gauge invariant (and is essentially determined by this requirement [50]). It is quadratic in $\mathbf{b}\,\big|_{\partial\Lambda}$.

Quantum Hall fluids are quantum-mechanical systems, and hence the Chern-Simons theory, with action functional S_Λ given in eq. (3.9), must be quantized. Because S_Λ is quadratic in $\mathbf{b}$, quantization may proceed via Feynman functional integrals. This task is not a big deal; see Section 1, and [25], [63], [23]. It turns out that the only *dynamical* degrees of freedom of the theory are localized on $\partial\Lambda$ and describe chiral $\widehat{u(1)}$-currents [43], [16]. Their dynamics is described by the term $\Gamma_{\partial\Lambda}(\mathbf{b})$, (after having taken into account the equations of motion of Chern-Simons theory). The number of clockwise moving currents is equal to the number of positive (negative) eigenvalues of the metric C; the number of counterclockwise moving currents is equal to the number of negative (positive) eigenvalues of C, (depending on the direction of $\vec{B}_c$). These are the experimentally observed edge currents first predicted by Halperin [32]. We shall focus our attention on the analysis of IQHF's with edge currents of only one chirality. Then C may be chosen to be positive-definite.

As sketched in Section 1, states in the quantum-mechanical Hilbert space of Chern-Simons theory can be viewed as solutions ϕ of the Knizhnik-Zamolodchikov equations [23] in $n = 0, 1, 2, \dots$ variables. For our abelian Chern-Simons theory introduced in (3.9), these equations take the form

$$\frac{d\phi}{dt} = \left\{ \sum_{1 \le i < j \le n} \langle \mathbf{q}_i, \mathbf{q}_j \rangle \frac{\dot{z}_i - \dot{z}_j}{z_i - z_j} + \sum_{i=1}^{n} \langle \mathbf{q}_i, \mathbf{q}_{\partial\Omega} \rangle \, \dot{z}_i \, h'(z_i) \right\} \phi, \tag{3.10}$$

where

$$\mathbf{q}_i = (q_i^1, \dots, q_i^N)^T \in \mathbb{R}^N, \quad i = 1, \dots, n,$$

are n N-tuples of charges, mathematically: characters of $\mathbb{R}^N$, localized at the points $z_1, \dots, z_n$, resp., $\mathbf{q}_{\partial\Omega}$ is an N-tuple of boundary charges,

$$\langle \mathbf{q}, \mathbf{q}' \rangle = \sum_{i,j=1}^{N} q^i C_{ij} q'^j, \tag{3.11}$$

$z_1(t), \dots, z_n(t)$ are n paths in the domain Ω of the complex plane parameterized by a real parameter t, with $\dot{z}_i(t) = \frac{dz_i(t)}{dt}$, and h is a harmonic function on Ω, with $h' = \frac{dh}{dz}$; see [16].

The solutions of eq. (3.10) are functions on the universal covering space $\widetilde{M}_n$ of the space $\Omega^n \backslash \mathcal{D}$, where $\mathcal{D}$ is the diagonal $\{z_i = z_j, \text{ for some } i \ne j\}$. At $t = t_1$, with $z_i \equiv z_i(t_1)$, for $i = 1, \dots, n$, the solution $\phi_{t_1} = \phi(z_1, \mathbf{q}_1, \dots, z_n, \mathbf{q}_n)$ of (3.10) is given by

$$\begin{aligned} \phi(z_1, \mathbf{q}_1, \dots, z_n, \mathbf{q}_n) &= \text{const.} \left[\prod_{1 \le i < j \le n} (z_i - z_j)^{\langle \mathbf{q}_i, \mathbf{q}_j \rangle} \right] \\ &\quad \times \exp\left(\sum_{i=1}^{n} \langle \mathbf{q}_i, \mathbf{q}_{\partial\Omega} \rangle \, h(z_i) \right), \end{aligned} \tag{3.12}$$

with $(z_1, \dots, z_n)$ viewed as a point of $\widetilde{M}_n$, i.e., $(z_1, \dots, z_n)$ stands for $\big(z_1(t_1), \dots, z_n(t_1)\big)$, together with the homotopy class of the path $\big(z_1(t), \dots, z_n(t)\big)_{t \in [t_0, t_1]}$; see Section 1.

To see that the characters q_j^i, $i = 1, \dots, N$, are charges, we consider the charge operators

$$\int_{D_j} J^i = \oint_{\partial D_j} b^i \tag{3.13}$$

of the Chern-Simons theory, where D_j is a disk in Ω containing z_j, but not containing z_k, $k \ne j$. From the results in [23] one easily derives that

$$\Big(\int_{D_j} J^i \Big) \, \phi\,(z_1, \mathbf{q}_1, \dots, z_n, \mathbf{q}_n) = q_j^i \, \phi(z_1, \mathbf{q}_1, \dots, z_n, \mathbf{q}_n), \tag{3.14}$$

i.e., $\phi(z_1, \mathbf{q}_1, \ldots, z_n, \mathbf{q}_n)$ is an eigenvector of the ith charge operator $\int_{D_j} J^i$, with eigenvalue q^i_j, for $i = 1, \ldots, N$, $j = 1, \ldots, n$. By eq. (3.5) the operator detecting the total electric charge in the disk D_j is given by

$$\int_{D_j} J = \sum_{i=1}^{N} Q_i \int_{D_j} J^i \equiv \int_{D_j} \langle \mathbf{Q}, \mathbf{J} \rangle, \tag{3.15}$$

and, by (3.14), ϕ is an eigenvector of $\int\limits_{D_j} J$ with eigenvalue

$$q_{\text{el.}}(D_j, \phi) = \sum_{i=1}^{N} Q_i \, q^i_j = \langle \mathbf{Q}, \mathbf{q}_j \rangle. \tag{3.16}$$

Suppose that $\mathbf{q}_i = \mathbf{q}_j \equiv \mathbf{q}$, for some $i \neq j$. Let us continue the solution ϕ along the path $\big(z_1(t), \ldots, z_n(t)\big)$ from $t = t_1$ to $t = t_2$, assuming that $\dot{z}_k(t) = 0$, for $k \neq i, j$, $t_1 \leq t \leq t_2$, and that $\big(z_i(t), z_j(t)\big)_{t_1 \leq t \leq t_2}$ exchanges z_i and z_j along counterclockwise oriented arcs not including any point z_k, for $k \neq i, j$. Then

$$\phi_{t_2} = \exp\big(i\pi \langle \mathbf{q}, \mathbf{q} \rangle\big) \, \phi_{t_1}, \tag{3.17}$$

i.e., the half-monodromy (called "Aharonov-Bohm phase factor" by the physicists) of the solution ϕ of (3.10) in the pair z_i, z_j is given by

$$\exp\big(i\,\pi \langle \mathbf{q}, \mathbf{q} \rangle\big). \tag{3.18}$$

Similarly, if $\dot{z}_i(t) = 0$, $t_1 \leq t \leq t_2$, $i \neq k$, and $\big(z_k(t)\big)_{t_1 \leq t \leq t_2}$ describes a counterclockwise oriented loop around the point z_ℓ not including any point z_i, $i \neq k, \ell$, then

$$\phi_{t_2} = \exp\big(i2\pi \langle \mathbf{q}_k, \mathbf{q}_\ell \rangle\big) \, \phi_{t_1}, \tag{3.19}$$

i.e., the monodromy of the solution ϕ of (3.10) in the pair z_k, z_ℓ is given by

$$\exp\big(i2\pi \langle \mathbf{q}_k, \mathbf{q}_\ell \rangle\big). \tag{3.20}$$

The groundstate of an incompressible quantum Hall fluid (IQHF) described by the Chern-Simons theory (3.9) is the vector $\phi \equiv \phi_0 \equiv 1$ ($n = 0$ in (3.12)); the charge densities J^i are normalized such that

$$\int_{\Omega} J^i \, \phi_0 = 0.$$

The states $\phi(z_1, \mathbf{q}_1, \ldots, z_n, \mathbf{q}_n)$ given in (3.12) might correspond to *excited states* of the IQHF. To make this idea precise, we must find conditions on the characters, or charge vectors $\mathbf{q}_1, \dots, \mathbf{q}_n$ that guarantee that properties (P1)–(P4) of an IQHF are valid. Thus, suppose that

$$q_{\mathrm{el.}}(\mathbf{q}_j) = \sum_{i=1}^{N} Q_i\, q_j^i = \langle \mathbf{Q}, \mathbf{q}_j \rangle$$

is an *odd* integer. By property (P3), a physical excitation with charges $\mathbf{q}_j$ must then satisfy Fermi-Dirac statistics. Hence the half-monodromy (3.18) must satisfy

$$\exp\left(i\pi \langle \mathbf{q}_j, \mathbf{q}_j \rangle\right) = -1,$$

i.e.,

$$\langle \mathbf{q}_j, \mathbf{q}_j \rangle \in 2\,\mathbb{Z} + 1. \tag{3.21}$$

Similarly, if $\langle \mathbf{Q}, \mathbf{q}_j \rangle$ were *even*, the half-monodromy (3.18) would have to be $+1$, and hence

$$\langle \mathbf{q}_j, \mathbf{q}_j \rangle \in 2\,\mathbb{Z}. \tag{3.22}$$

Summarizing (3.21) and (3.22), we have that

$$\langle \mathbf{Q}, \mathbf{q} \rangle = \langle \mathbf{q}, \mathbf{q} \rangle \quad \mod 2, \tag{3.23}$$

whenever $\langle \mathbf{Q}, \mathbf{q} \rangle \in \mathbb{Z}$.

Next, suppose that $q_{\mathrm{el.}}(\mathbf{q}_j) \in \mathbb{Z}$, for some j (i.e., $\mathbf{q}_j$ corresponds to a multi-electron/-hole excitation of the fluid). By property (P4), the state vector $\phi(z_1, \mathbf{q}_1, \dots, z_j, \mathbf{q}_j, \dots, z_n, \mathbf{q}_n)$ must then be a *single-valued* function of z_j (for fixed z_i, $i \neq j$), provided $\mathbf{q}_1, \dots, \mathbf{q}_n$ are the charge vectors of (finite-energy) physical excitations of the IQHF. Thus, by (3.20),

$$\langle \mathbf{q}_j, \mathbf{q}_i \rangle \in \mathbb{Z}, \quad \text{for all } i \neq j. \tag{3.24}$$

Next, if $\mathbf{q}$ is the charge vector of a localized physical excitation of an IQHF then so is $-\mathbf{q}$, by a principle of charge conjugation. Furthermore, if $\mathbf{q}$ and $\mathbf{q}'$ are the charge vectors of two localized physical excitations of an IQHF then so is $\mathbf{q}+\mathbf{q}'$, because one may let their positions approach each other arbitrarily closely. Thus, the charge vectors of localized physical excitations of an IQHF form an additive group, denoted $\Gamma_{\mathrm{phys.}}$. By (3.23) and (3.24), the charge vectors $\mathbf{q}$ with $q_{\mathrm{el.}}(\mathbf{q}) = \langle \mathbf{Q}, \mathbf{q} \rangle \in \mathbb{Z}$ form an integral sublattice, Γ, in $\Gamma_{\mathrm{phys.}}$. Finally, by eq. (3.24) (which expresses property (P4)),

$$\Gamma_{\mathrm{phys.}} \subseteq \Gamma^*, \tag{3.25}$$

where Γ^* is the lattice dual to Γ. Because

$$q_{\mathrm{el.}}(\mathbf{q}) = \langle \mathbf{Q}, \mathbf{q} \rangle \in \mathbb{Z}, \quad \text{for all } \mathbf{q} \in \Gamma,$$

we conclude that $\mathbf{Q} \in \Gamma^*$. Furthermore, a single electron or hole is a physical excitation of an IQHF. Thus, there exists a vector $\mathbf{q} \in \Gamma$, with

$$\langle \mathbf{Q}, \mathbf{q} \rangle = 1,$$

i.e., $\mathbf{Q}$ is a *primitive* vector of Γ^*.

Suppose that $\Gamma_{\text{phys.}} \supsetneqq \Gamma$. Then there exists some local excitation of the IQHF with a charge vector $\mathbf{q} \in \Gamma_{\text{phys.}}$ such that $\mathbf{q} \bmod \Gamma \neq 0$. The electric charge $q_{\text{el.}}(\mathbf{q}) = \langle \mathbf{Q}, \mathbf{q} \rangle$, of this excitation is then necessarily nonintegral (in units where $e = 1$), and its quantum statistics, as described by the half-monodromy $\exp\big(i\pi \langle \mathbf{q}, \mathbf{q} \rangle\big) \neq \pm 1$, is neither Fermi-Dirac nor Bose-Einstein statistics. It determines abelian, unitary representations of the braid groups B_n, $n = 2, 3, 4, \dots$, and is therefore called abelian braid statistics. Thus, if $\Gamma_{\text{phys.}} \supsetneqq \Gamma$, there are local excitations in an IQHF with *fractional electric charge* and *braid statistics* ("Laughlin vortices").

Our analysis has enabled us to safely land on the notion of *chiral quantum Hall lattices*. It should be emphasized, once more, that the general analysis described here does *not* imply that Γ is a *Euclidian* lattice. The quadratic form $\langle \cdot, \cdot \rangle$ could be indefinite; see [22]. For simplicity, this general situation is not considered here and is presumably not relevant physically.

We are still missing one important point: that the Hall conductivity is given by

$$\sigma_H = \big\langle \mathbf{Q}, \mathbf{Q} \big\rangle. \tag{3.26}$$

To prove eq. (3.26), we study the response of an IQHF to a perturbation given by a small magnetic field $\vec{B}$ in the interior of the region Ω. Let B be the component perpendicular to Ω, and let $A = \sum_{\mu=0}^{2} A_\mu dx^\mu$ be an electromagnetic vector potential on Λ with

$$B = (dA)_{12}. \tag{3.27}$$

Now, recall that Q_i is the unit of electric charge transported by the current J^i. Thus, J^i couples to the electromagnetic vector potential A through a term

$$\frac{1}{2\pi} \int\limits_{\Lambda} J^i \wedge Q_i\, A = -\frac{1}{2\pi} \int\limits_{\Lambda} b^i \wedge Q_i\, dA$$

(up to a boundary term). The action functional of the IQHF in the scaling limit is therefore given by

$$S_\Lambda(\mathbf{b}) = \frac{1}{4\pi} \int\limits_{\Lambda} \mathbf{b}^T \wedge C\, d\mathbf{b} - \frac{1}{2\pi} \int\limits_{\Lambda} \mathbf{b}^T \wedge \mathbf{Q}\, dA,$$

up to a boundary term only depending on $\mathbf{b}\,|_{\partial\Lambda}$ and $A\,|_{\partial\Lambda}$. The equations of motion obtained by variation of S_Λ with respect to $\mathbf{b}$ are found to be

$$db^j(x) = \sum_{i=1}^{N} (C^{-1})^{ji} Q_i\, dA(x), \tag{3.28}$$

for x in the interior of Λ. Thus,

$$\begin{aligned} J_{12}(x) &= Q_j\, J^j_{12}(x) = Q_j (db^j)_{12}(x) \\ &= \Big(\sum_{i,j=1}^{N} Q_j (C^{-1})^{ji} Q_i \Big) (dA)_{12}(x) \\ &= \big\langle \mathbf{Q}, \mathbf{Q} \big\rangle (dA)_{12}(x). \end{aligned}$$

Integrating this equation over Ω, we find, using (3.27), that

$$q_{\mathrm{el.}} = \int_{\Omega} J_{12} = \langle \mathbf{Q}, \mathbf{Q} \rangle \int_{\Omega} B \equiv \langle \mathbf{Q}, \mathbf{Q} \rangle \, \Phi .$$

Comparing this identity with eq. (2.5), we conclude that $\sigma_H = \langle \mathbf{Q}, \mathbf{Q} \rangle$, which proves eq. (3.26). Following [51], [1], [4], one can show that σ_H can also be expressed in terms of a first Chern number of a vector bundle of Chern-Simons groundstates on a two-dimensional torus of magnetic fluxes — this is physically somewhat contrived, though — or as a "generalized index", [20]. These matters will be discussed in more detail elsewhere.

We conclude this report with a list of important invariants of cQHL's $(\Gamma, \mathbf{Q})$ and their physical interpretations. For details and proofs, see [28], [22].

(I) Invariants of Γ

Invariant	**Physical quantity**
dim Γ	number of independently conserved currents ("channels").
$\triangle = \mid \Gamma^*/\Gamma \mid$	number of fractionally charged Laughlin vortices (assuming that $\Gamma_{\mathrm{phys.}} = \Gamma^*$);
$\triangle \langle \mathbf{Q}, \mathbf{Q} \rangle$ mod 8	————————
genus of Γ	monodromies, $\{\exp(i\, 2\pi \langle \mathbf{q}, \mathbf{q}' \rangle) \, : \, \mathbf{q}, \mathbf{q}' \in \Gamma^*\}$ of fractionally charged Laughlin vortices.
Witt sublattice, Γ_W	root lattice of simply laced Lie algebra of nonabelian symmetries of IQHF in scaling limit.

(II) Invariants of $(\Gamma, \mathbf{Q})$

Invariant	**Physical quantity**
$\sigma_H = \langle \mathbf{Q}, \mathbf{Q} \rangle$	Hall conductivity.
orbit of $\mathbf{Q}$ under orthogonal trsfs. of Γ	assignment of electric charges to quasi-particles.
"level" $\ell =$ g.c.d. $(\triangle, \triangle\, \sigma_H)$	————————
$\ell_{\mathrm{max.}}(\Gamma, \mathbf{Q})$ (see (2.17))	relative angular momentum of a pair of electrons.
$q^* = \min_{\substack{\mathbf{q} \in \Gamma^* \\ \langle \mathbf{Q}, \mathbf{q} \rangle \neq 0}} \mid \langle \mathbf{Q}, \mathbf{q} \rangle \mid$	smallest fractional electric charge $\neq 0$.

These invariants and their physical counterparts permit us to elucidate fairly specific physical properties of IQHF's. But this goes beyond the present report.

4 Epilogue: Origins of the problems discussed in this lecture

In 1986, we became interested in two seemingly unrelated topics: three-dimensional gauge theories with a Chern-Simons term in their Lagrangian (or action), and the braid statistics of charged particles described by such theories, on one hand, and the fractional quantum Hall effect, on the other hand. It had already been suggested that these two topics are related to each other [61], [31], but it appeared that nobody understood the relationship in precise terms.

Between the fall of 1986 and 1990, we focused our attention primarily on the problems of understanding Chern-Simons gauge theory, the related two-dimensional conformal field theories, the general theory of braid statistics and of quantized symmetries in two- and three-dimensional quantum field theory, and some mathematical problems in knot theory and the theory of braided tensor categories related to low-dimensional quantum field theory. Our main results on these topics appeared in [23], [21], [24], [17], [30]; see also [13], [42].

In studying Chern-Simons-Higgs theories [25], Fröhlich and Marchetti understood that abelian, pure Chern-Simons theory was, in essence, just a way of reproducing the Gauss linking number. In 1987, during a sabbatical at I.H.É.S., Fröhlich was taught the basics of subfactor and knot theory by Jones. Jones expressed the intriguing idea that, in analogy to the Gauss linking number, more general knot invariants should be calculable from some "field theories" defined on links. Thanks to the presence of Felder and Gawędzki at I.H.É.S., Fröhlich also acquired some rudimentary knowledge in two-dimensional conformal field theory.

These strands of ideas naturally merged and led to some preliminary understanding of braid statistics in low-dimensional quantum field theory and its connection with the theory of knots and links [14]. Seminar notes of Jones and a preprint by Turaev [55] were very helpful in attempting to make those insights more precise. They soon led to the conjecture that, just as abelian pure Chern-Simons theory gives rise to the Gauss invariant of links, nonabelian pure Chern-Simons theory ought to give rise to more interesting link invariants. Apparently, Schwarz independently arrived at the same conjecture, around the same time (1987) [48]. Unfortunately, it appeared to be difficult to identify those invariants. It is well known that, in 1988, Witten independently came up with the same ideas, identified the link invariants emerging from nonabelian Chern-Simons theory, and went on to define new invariants for three-dimensional manifolds [63]. His work provided new motivation for us (Fröhlich and King) to return to the ideas leading to the original conjecture. We found a way of deriving the so-called Knizhnik-Zamolodchikov (KZ-)equations [38] from formal Chern-Simons functional integrals; see Section 1. We showed how to calculate some knot polynomials generalizing the Jones polynomial from solutions of the KZ-equations. The existence of appropriate solutions of the KZ-equations was proven by using convergent power series expansions in $\lambda = \pm(k+c_2)^{-1}$, where k is the level of some Kac-Moody algebra and c_2 is the dual Coxeter number of the underlying Lie algebra [23]. Our results gave substance to Jones' idea of constructing invariants of links from some "field theory" defined on links.

The KZ-equations are the equations for horizontal sections of certain vector bundles equipped with flat connections, called KZ-connections. The construction of KZ-connections is based on solutions of the so-called infinitesimal pure braid relations (a special case of which are the classical Yang-Baxter equations [3]). In fact, *every* solution of the infinitesimal pure braid relations gives rise to a KZ-connection. Horizontal sections of vector bundles can be constructed, locally, with the help of Chen's iterated integrals, more appropriately called Dyson series by the physicists. This method was used in [23].

Later on, the results and methods of [23] – see also Section 6.3 of [24] – were confirmed and put in a more general context of Vassiliev invariants [56] in [40].

In 1990, Morf taught us the basic facts about the (fractional) quantum Hall effect. A paper by Halperin [32] made it clear to us that there is a fundamental relationship between the quantum Hall effect and the theory of Kac-Moody algebras. We found that the quantum Hall effect is actually described by abelian pure Chern-Simons theories [20]. This insight, combined with the theory of the chiral anomaly in two-dimensional gauge theory, provided a completely general explanation of Halperin's findings (in a more general context than the one he had envisaged); see also [26]. Similar results were found, independently and somewhat earlier, by Wen [59] and were later confirmed by many other groups; see e.g. [49].

The work of Fröhlich and King on Chern-Simons theory now turned out to be very useful: it said that physical state vectors of incompressible quantum Hall fluids ($R_L = 0, \sigma_H$ on a plateau) could be constructed in terms of solutions of KZ-equations derived from certain abelian pure Chern-Simons theories. The known monodromy of solutions of the KZ-equations provided an essential clue to understanding the rôle played by the theory of integral quadratic forms on lattices in the theoretical analysis of incompressible quantum Hall fluids. Our analysis led us to the notion of chiral quantum Hall lattices. A partial classification of those chiral quantum Hall lattices that appear in the analysis of incompressible quantum Hall fluids was accomplished in joint work of Fröhlich and Thiran, with contributions by Kerler and Studer. Incidentally, such lattices also appear in algebraic topology (algebraic surfaces in algebraic four-manifolds). Our enterprise has taken quite a lot of time and effort. We are grateful to L. Michel for explaining to us many basic facts concerning integral lattices. Our results have appeared in [29], [26], [28], [22], [27].

Now that the classification of incompressible quantum Hall fluids in terms of chiral quantum Hall lattices has reached a satisfactory stage, it would be time to develop analytical proofs of existence of incompressible quantum Hall fluids. Interesting ideas on this problem have appeared in [64]. The strategy followed there leads to rather beautiful variational problems on spaces of sections of some line bundles — somewhat similar to the vortex problems in Higgs models [33] — which are described in [15].

Another line of research concerns the definition of Chern-Simons actions on noncommutative spaces, in the sense of Connes [7], and the analysis of the corresponding Chern-Simons theories [6]. This leads to a unifying point of view on topological field theory [63], [62]. The interplay between noncommutative geometry and quantum field theory appears to be a promising area for future work [18].

I believe we had "fun imagining it" — even though the job has sometimes been pretty hard.

Acknowledgements. The work summarized in this lecture would not have been possible without the inspiring help provided by V. F. R. Jones and R. Morf. J. Fröhlich also thanks A. Connes, G. Felder, K. Gawędzki, and L. Michel for many stimulating discussions, and M. Berger and D. Ruelle for hospitality at the I.H.É.S. during fruitful periods.

References

[1] J. E. Avron, R. Seiler, and B. Simon, Phys. Rev. Lett. **51**, 51 (1983);
J. E. Avron, R. Seiler, and L. Yaffe, Comm. Math. Phys. **110**, 33 (1987);
J. E. Avron, R. Seiler, and B. Simon, Phys. Rev. Lett. **65**, 2185 (1990).

[2] D. Bar-Natan, *On the Vassiliev knot invariant*, Preprint, 1992;
D. Altschüler and L. Freidel, *On universal Vassiliev invariants*, preprint ETH-TH/94-07, 1994, subm. to Comm. Math. Phys.

[3] A. A. Belavin and V. G. Drinfel'd, Functional Anal. Appl. **16**, 1 (1982).

[4] J. Bellissard, *K-Theory of C^*-Algebras in solid state physics*, in Statistical Mechanics and Field Theory: Mathematical Aspects, T. C. Dorlas, N. M. Hugenholtz, and M. Winnink (eds.), Lecture Notes in Phys. vol. 257, Berlin, Heidelberg, New York: Springer-Verlag, 1986;
J. Bellissard, A. van Elst, and H. Schulz-Baldes, *The non-commutative geometry of the quantum Hall effect*, preprint, 1994;
H. Kunz, Comm. Math. Phys. **112**, 121 (1987).

[5] A. H. Chamseddine and J. Fröhlich, Comm. Math. Phys. **147**, 549 (1992).

[6] A. H. Chamseddine and J. Fröhlich, *The Chern-Simons action in non-commutative geometry*, J. Math. Phys., to appear.

[7] A. Connes, Publ. Math. I.H.E.S. **62**, 41 (1985); Non-Commutative Geometry, Academic Press, to appear.

[8] J. H. Conway and N. J. A. Sloane, Sphere Packings, Lattices and Groups, New York, Berlin, Heidelberg: Springer-Verlag, 1988; Proc. Roy. Soc. London Ser. A **418**, 17 (1988); ibid. **418**, 43 (1988); ibid. **419**, 259 (1988).

[9] R. R. Du, H. L. Störmer, D. C. Tsui, L. N. Pfeiffer, and K. W. West, Phys. Rev. Lett. **70**, 3944 (1993);
W. Kang, H. L. Störmer, L. N. Pfeiffer, K. W. Baldwin, and K. W. West, Phys. Rev. Lett. **1**, 3850 (1993).

[10] L. D. Faddeev and V. N. Popov, Phys. Letters B **25**, 29 (1967).

[11] G. Felder, J. Fröhlich, and G. Keller, Comm. Math. Phys. **124**, 647 (1989).

[12] A. Floer, Comm. Math. Phys. **118**, 215 (1988); Bull. Amer. Math. Soc. **16**, 279 (1987).

[13] K. Fredenhagen, K.-H. Rehren, and B. Schroer, Comm. Math. Phys. **125**, 201 (1989); *Superselection sectors with braid group statistics and exchange algebras, II*, preprint, 1992;
R. Longo, Comm. Math. Phys. **126**, 217 (1989).

[14] J. Fröhlich, *Statistics of fields, the Yang-Baxter equation, and the theory of knots and links*, in Non-perturbative Quantum Field Theory (Cargèse 1987), G. 't Hooft et al. (eds.), New York: Plenum Press, 1988.

[15] J. Fröhlich, *Mathematical aspects of the quantum Hall effect*, in Proc. of the first ECM (Paris 1992), Progress in Math., Basel, Boston: Birkhäuser-Verlag, 1994.

[16] J. Fröhlich, *Chern-Simons theory on manifolds with boundary*, unpubl.

[17] J. Fröhlich and F. Gabbiani, Rev. Math. Phys. **2**, 251 (1990).

[18] J. Fröhlich and K. Gawędzki, *Conformal field theory and geometry of strings*, to appear in CRM (Montréal): Proceedings and Lecture Notes, 1994.
[19] J. Fröhlich, R. Götschmann and P.-A. Marchetti, *Bosonization of Fermi systems in arbitrary dimension in terms of gauge forms*, J. Phys. A **28**, 1169 (1995).
[20] J. Fröhlich and T. Kerler, Nuclear Phys. B **354**, 369 (1991).
[21] J. Fröhlich and T. Kerler, Quantum groups, quantum categories and quantum field theory, Lecture Notes in Math., vol. **1542**, Berlin, Heidelberg, New York: Springer-Verlag, 1993.
[22] J. Fröhlich, T. Kerler, U. M. Studer, and E. Thiran, ETH-preprint, 1994.
[23] J. Fröhlich and C. King, Comm. Math. Phys. **126**, 167 (1989).
[24] J. Fröhlich and C. King, Internat. J. Modern Phys. A **4**, 5321 (1989).
[25] J. Fröhlich and P.-A. Marchetti, Lett. Math. Phys. **16**, 347 (1988); Comm. Math. Phys. **121**, 177 (1989).
[26] J. Fröhlich and U. M. Studer, Rev. Modern Phys. **65**, 733 (1993).
[27] J. Fröhlich, U. M. Studer, and E. Thiran, ETH-preprint, 1994.
[28] J. Fröhlich and E. Thiran, *Integral quadratic forms, Kac-Moody algebras, and fractional quantum Hall effect*, preprint ETH-TH/93-22, to appear in J. Statist. Phys. (1994).
[29] J. Fröhlich and A. Zee, Nuclear Phys. B **364**, 517 (1991).
[30] F. Gabbiani and J. Fröhlich, Comm. Math. Phys. **155**, 569 (1993).
[31] S. M. Girvin, *Summary, omissions and unanswered questions*, Chap. 10, in The Quantum Hall Effect, R. E. Prange and S. M. Girvin (eds.), New York, Berlin, Heidelberg: Springer-Verlag, 1987; (2nd edition 1990).
[32] B. I. Halperin, Phys. Rev. B **25**, 2185 (1982).
[33] A. Jaffe and C. Taubes, Vortices and monopoles, Progress in Physics, vol. 2, Basel, Boston: Birkhäuser Verlag, 1980.
[34] M. Jimbo, Lett. Math. Phys. **10**, 63 (1985); Lett. Math. Phys. **11**, 247 (1986); V. G. Drinfel'd, *Quantum groups*, in: Proc. of ICM Berkeley 1986, A.M. Gleason (ed.), Providence, RI: Amer. Math. Soc. Publ., 1987.
[35] V. F. R. Jones, Bull. Amer. Math. Soc. **12**, 103 (1985); Ann. of Math. (2) **126**, 335 (1987).
[36] M. Kac, *On applying mathematics: Reflections and examples*, Quart. Appl. Math. **30**, 17 (1972).
[37] D. Kazhdan and G. Lusztig, *Tensor structures arising from affine Lie algebras, I&II*, preprints.
[38] V. G. Knizhnik and A. B. Zamolodchikov, Nuclear Phys. B **247**, 83 (1984).
[39] T. Kohno, Invent. Math. **82**, 57 (1985); Ann. Inst. Fourier (Grenoble) **37**, 139 (1987).
[40] M. Kontsevich, *Vassiliev's knot invariants*, Adv. in Sov. Math., to appear.
[41] R. Laughlin, Phys. Rev. B **23**, 5632 (1981).
[42] G. Mack and V. Schomerus, Nuclear Phys. B **370**, 185 (1992); V. Schomerus, *Quantum symmetry in quantum theory*, Ph.D. thesis, DESY-93-18 (1993).
[43] G. Moore and N. Seiberg, Phys. Lett. B **220**, 422 (1989).
[44] R. E. Prange and S. M. Girvin (eds.) The Quantum Hall Effect, (2nd edition), New York, Berlin, Heidelberg: Springer-Verlag, 1990.
[45] D. Quillen, *Chern-Simons form and cyclic cohomology*, in The Interface of Mathematics and Particle Physics, D. Quillen, G. Segal, and S. Tsou (eds.), Oxford: Oxford University Press 1990.
[46] V. Schechtman and A. Varchenko, Invent. Math. **106**, 134 (1991);

B. Feigin, V. Schechtman, and A. Varchenko, Lett. Math. Phys. **20**, 291 (1990); Comm. Math. Phys. **163**, 173 (1994).
[47] A. S. Schwarz, Lett. Math. Phys. **2**, 247 (1978).
[48] A. S. Schwarz, *New topological invariants arising in the theory of quantized fields*, Baku Int. Topological Conference, 1987; quoted in Bull. Amer. Math. Soc. **30**, 197 (1994).
[49] M. Stone (ed.), Quantum Hall Effect, Singapore: World Scientific Publ. Co., 1992.
[50] S. B. Teiman, R. Jackiw, B. Zumino, and E. Witten, Current Algebra and Anomalies, Singapore: World Scientific Publ. Co., 1985.
[51] D. J. Thouless, M. Kohmoto, M. P. Nightingale, and M. den Nijs, Phys. Rev. Lett. **49**, 405 (1982);
Q. Niu and D. J. Thouless, Phys. Rev. B **35**, 2188 (1987);
D. J. Thouless, in ref. [44].
[52] A. Tsuchiya and V. Kanie, Lett. Math. Phys. **13**, 303 (1987); Adv. Stud. Pure Math. **16**, 297 (1988).
[53] D. C. Tsui, Phys. B **164**, 59 (1990);
H. L. Störmer, Physica B **177**, 401 (1992);
T. Sajoto, Y. W. Suen, L. W. Engel, M. B. Santos, and M. Shayegan, Phys. Rev. B **41**, 8449 (1990);
V.J. Goldman and M. Shayegan, Surf. Science **229**, 10 (1990);
R. G. Clark, S. R. Haynes, J. V. Branch, A. M. Suckling, P. A. Wright, P. M. W. Oswald, J. J. Harris, and C. T. Foxon, Surf. Science **229**, 25 (1990);
J. P. Eisenstein, H. L. Störmer, L. N. Pfeiffer, and K. W. West, Phys. Rev. B **41**, 7910 (1990);
L. W. Engel, S. W. Hwang, T. Sajoto, D. C. Tsui, and M. Shayegan, Phys. Rev. B **45**, 3418 (1992);
J. P. Eisenstein, R. L. Willett, H. L. Störmer, L. N. Pfeiffer, and K. W. West, Surf. Sci. **229**, 31 (1990);
Y. W. Suen, L. W. Engel, M. B. Santos, M. Shayegan, and D. C. Tsui, Phys. Rev. Lett. **68**, 1379 (1992);
J. P. Eisenstein, G. S. Boebinger, L. N. Pfeiffer, K. W. West, and Song He, Phys. Rev. Lett. **68**, 1383 (1992).
[54] D. C. Tsui, H. L. Störmer, and A. C. Gossard, Phys. Rev. Lett. **48**, 1559 (1982); Phys. Rev. B **25**, 1405 (1982).
[55] V. Turaev, Invent. Math. **92**, 527 (1988).
[56] V. A. Vassiliev, *Cohomology of knot spaces*, in Theory of Singularities and its Applications, V. I. Arnol'd (ed.), Advances in Soviet Mathematics, Amer. Math. Soc. Publ., 1990.
[57] A. Verjovsky and R.F. Vila Freyer, Comm. Math. Phys. **163**, 73 (1994).
[58] K. von Klitzing, G. Dorda, and M. Pepper, Phys. Rev. Lett. **45**, 494 (1980).
[59] X. G. Wen and Q. Niu, Phys. Rev. B **41**, 9377 (1990); X.G. Wen, Phys. Rev. B **41**, 12, 838 (1990).
[60] H. Wenzl, Invent. Math. **92**, 349 (1988).
[61] F. Wilczek, Phys. Rev. Lett. **48**, 1144 (1982); **49**, 957 (1982); D. Arovas, J. R. Schrieffer, and F. Wilczek, Phys. Rev. Lett. **53**, 722 (1984).
[62] E. Witten, Comm. Math. Phys. **117**, 353 (1988).
[63] E. Witten, Comm. Math. Phys. **121**, 351 (1989).
[64] S. C. Zhang, T. Hansson, and S. Kivelson, Phys. Rev. Lett. **62**, 82 (1989);
N. Read, Phys. Rev. Lett. **62**, 86 (1989);
D. H. Lee and S. C. Zhang, Phys. Rev. Lett. **66**, 122 (1991).

Wave Propagation

JOSEPH B. KELLER

Department of Mathematics, Stanford University
Stanford, CA 94305, USA

1. Introduction

The mathematical theory of wave propagation is the study of partial differential equations, or systems of such equations, with wave-like solutions. An example of such an equation is the wave equation

$$\Delta u(\mathbf{x},t) - \frac{1}{c^2(\mathbf{x})} u_{tt}(\mathbf{x},t) = 0 . \tag{1.1}$$

Here the unknown $u(\mathbf{x},t)$ is a real- or complex-valued function of a point $\mathbf{x}$ in R^n and of the time t. The given positive function $c(\mathbf{x})$ is the speed of propagation of waves at the point $\mathbf{x}$.

Equations like (1.1) are satisfied by the acoustic pressure in a gas or liquid, by each component of an electromagnetic field, by each component of the displacement of a string or membrane from its rest position, and by many other physical quantities. In these examples, the number n of space dimensions is one for a string, two for a membrane, and three for acoustic and electromagnetic fields. In these cases, the physical quantity is real, so it is represented by a real solution, or the real part of a complex-valued solution.

In quantum mechanics, the wave function $u(\mathbf{x},t)$ is complex valued and the number of space dimensions is three times the number of particles in the physical system under consideration, so it can be very large. The differential equation for u is the Schrödinger equation. It differs from (1.1) in having u_{tt} replaced by iu_t, and in containing terms in which u is undifferentiated.

When $c(\mathbf{x})$ is a constant, (1.1) has the plane wave solution

$$u(\mathbf{x},t) = Ae^{i(\mathbf{k}\cdot\mathbf{x}-\omega t)} . \tag{1.2}$$

The constants $A, \mathbf{k}$, and ω are called respectively the amplitude, the wavevector, and the angular frequency of the wave. In order for (1.2) to satisfy (1.1), the angular frequency ω must be related to the wavenumber $k = |\mathbf{k}|$ by the equation $\omega^2 = k^2c^2$. The solution (1.2) is called a plane wave because at each value of t it is constant on the planes $\mathbf{k}\cdot\mathbf{x} =$ constant. These planes are called wavefronts, and they propagate in the direction of $\mathbf{k}$ with the velocity c. The wave is periodic in the direction of $\mathbf{k}$ with the period $\lambda = 2\pi/k$, which is called the wavelength. We note that λ is small when k is large.

Proceedings of the International Congress
of Mathematicians, Zürich, Switzerland 1994

Because (1.1) is linear in u, more general solutions can be formed by superposition of plane wave solutions:

$$u(\mathbf{x},t) = \int \left[A_1(\mathbf{k})e^{i(\mathbf{k}\cdot\mathbf{x}-\omega t)} + A_2(\mathbf{k})e^{i(\mathbf{k}\cdot\mathbf{x}+\omega t)}\right] d\mathbf{k} \, . \tag{1.3}$$

In fact the two amplitude functions $A_1(\mathbf{k})$ and $A_2(\mathbf{k})$ can be adjusted to make u, given by (1.3), and its derivative u_t, take on any specified values at $t = 0$. Then (1.3) is the solution of the initial value problem, also called the Cauchy problem, for (1.1) with $c =$ constant.

More interesting problems occur when (1.1) holds in a subdomain of R^n, with conditions imposed upon u at the boundary of the subdomain. We shall now describe the two classical boundary value problems that arise in this way. Similar problems occur when $c(\mathbf{x})$ is not constant, as we shall see.

2. Boundary Value Problems

First we consider a solution of (1.1), with $c =$ constant, in a bounded domain D in R^n. On the boundary ∂D of D we require that $u = 0$. Furthermore, we require that u be the product of a function $v(\mathbf{x})$ multiplied by a function of t. We find at once that the function of t must be $e^{\pm i\omega t}$, so that u must have the form

$$u(\mathbf{x},t) = v(\mathbf{x})e^{\pm i\omega t} \, . \tag{2.1}$$

Now the problem is to find a constant ω and a function $v(\mathbf{x})$ not identically zero, satisfying the following equations:

$$\Delta v + k^2 v = 0 \, , \quad \mathbf{x} \in D \tag{2.2}$$

$$v = 0 \, , \quad \mathbf{x} \in \partial D \, . \tag{2.3}$$

In the Helmholtz equation (2.2), $k^2 = \omega^2/c^2$. This problem is called an eigenvalue problem. The values of k^2 for which it has solutions are called eigenvalues, and the corresponding solutions $v(\mathbf{x})$ are called eigenfunctions.

In one dimension, each eigenvalue determines a possible frequency of vibration of a vibrating string with fixed endpoints. The corresponding eigenfunction determines the amplitude of vibration at each position $\mathbf{x}$ along the string. Similarly in two dimensions, the solutions determine the frequencies and modes of vibration of a membrane, such as a drumhead, held fixed at ∂D. In three dimensions, the solutions determine the frequencies and modes of vibration of fluid in a soft container with the surface ∂D. At ∂D, the acoustic pressure is forced to vanish.

The second problem we consider is called the scattering problem. In it we seek a solution u of (1.1) with $c =$ constant, in the exterior of a bounded domain D. This solution must consist of the incident plane wave (1.2), which is specified, and a scattered wave of the form (2.1) with the negative sign, which is to be found. Thus, u has the form

$$u(\mathbf{x},t) = Ae^{i(\mathbf{k}\cdot\mathbf{x}-\omega t)} + v(\mathbf{x})e^{-i\omega t} \, . \tag{2.4}$$

On the boundary of D we require that $u = 0$, and at infinity we impose a condition that guarantees that $v(\mathbf{x})$ does indeed represent a wave travelling outward from D.

The problem for v can be formulated as follows: given $\mathbf{k}$ and A, find the scattered wave $v(\mathbf{x})$ satisfying

$$\Delta v + k^2 v = 0 \quad , \quad \mathbf{x} \in R^n - D \tag{2.5}$$

$$v = -Ae^{i\mathbf{k}\cdot\mathbf{x}} \quad , \quad \mathbf{x} \in \partial D \tag{2.6}$$

$$\lim_{r\to\infty} r^{\frac{n-1}{2}} (v_r - ikv) = 0 \,. \tag{2.7}$$

In (2.7), which is called the Sommerfeld radiation condition, r denotes distance from $\mathbf{x}$ to an origin in D and v_r denotes the radial derivative of v.

The eigenvalue problem determines a discrete spectrum of positive real values of k^2 that depend upon the size and shape of D and upon the dimension n. In the scattering problem however, the spectrum consists of all positive real values of k^2.

3. Classical Methods of Solution

Most of the classical methods of solving the eigenvalue problem, the scattering problem, and other boundary value problems, are based upon separation of variables. In this procedure, the point $\mathbf{x}$ is represented in terms of curvilinear coordinates and the differential equation (2.2) is expressed in terms of these coordinates. For example, in $n = 2$ dimensions we write $\mathbf{x} = (\xi, \eta)$ and express Δ in terms of ξ and η. Then we seek a solution that is a product of a function of ξ times a function of η, $v = f(\xi)g(\eta)$.

If there are product solutions, the variables separate in (2.2). This leads to two ordinary differential equations, one for $f(\xi)$ and another for $g(\eta)$:

$$L_1 f(\xi) = \mu f(\xi) \quad , \quad L_2 g(\eta) = -\mu g(\eta) \,. \tag{3.1}$$

Here L_1 and L_2 are second order ordinary differential operators in ξ and η respectively, and μ is a constant called the separation constant. All the special functions of classical mathematics and mathematical physics arise as the solutions of these ordinary differential equations: Bessel, Hankel, Mathieu, Legendre, Hermite, Laguerre, and other functions.

To complete the specification of the solutions of (3.1), we must impose the boundary condition (2.3), the radiation condition (2.7), regularity conditions at singularities of the coordinates, periodicity conditions if coordinates are cyclic, etc. These conditions determine a spectrum of values of μ and corresponding solutions of (3.1).

If the μ spectrum is discrete, say $\mu_1, \mu_2, \ldots$, we denote the solutions corresponding to μ_j by $f_j(\xi)$ and $g_j(\eta), j = 1, 2, \ldots$. Then we can write the solution v as a series of product solutions with undetermined coefficients c_j:

$$v(\xi, \eta) = \sum_{j=1}^{\infty} c_j f_j(\xi) g_j(\eta) \,. \tag{3.2}$$

When the spectrum of μ is continuous, we denote the solutions by $f(\xi, \mu)$ and $g(\eta, \mu)$. Then we represent v as an integral of product solutions with the undetermined coefficient function $c(\mu)$:

$$v(\xi, \eta) = \int c(\mu) f(\xi, \mu) g(\eta, \mu) \, d\mu \, . \tag{3.3}$$

The coefficients c_j or $c(\mu)$ must be determined from the inhomogeneous data of the problem, such as the prescribed value of v on the boundary, a source term in the equation, initial data in the case of the wave equation (1.1), etc. Various discrete and continuous transforms have been developed for this purpose: Fourier, Laplace, Hankel, Mellin, Lebedev, etc.

Once c_j or $c(\mu)$ is determined, (3.2) or (3.3) provides a representation of the solution v as an infinite series or as an infinite integral. It is then necessary to evaluate the series or integral in order to determine the quantitative behavior of $v(x)$. When k is small, so that the wavelength is large compared to the dimensions of D, the series (3.2) converges rapidly and the integral (3.3) can be evaluated readily. However, when k is large, which corresponds to a high frequency, the wavelength is small compared to the dimensions of D. Then the terms in the series (3.2) are oscillatory, as is the integrand in (3.3). In that case, there is a great deal of cancellation in the series and in the integral, and it is difficult to evaluate them.

To overcome this difficulty, methods for the asymptotic evaluation of series, such as the Euler-Maclaurin sum formula and the Poisson summation method, have been adapted. A new method, the Watson transformation, has also been developed. For asymptotic evaluation of the integral, Laplace's method has been used, Kelvin's method of stationary phase was developed, the saddle point method has been used, etc.

In addition to the methods described above, there are other classical methods of solution based upon Green's functions, integral equations, images, etc.

4. Limitations of the Classical Methods

At the end of the nineteenth century and in the beginning of the twentieth century, it was discovered that separation of variables is possible in only very few coordinate systems. This was shown by expressing the Laplacian in (2.2) in general curvilinear coordinates and then deducing conditions on the metric coefficients in order that separation be possible. In the plane, for example, (2.2) is separable only when the coordinate lines are linear or quadratic curves, i.e. conic sections, and a similar conclusion holds in R^3.

This limitation on the coordinate systems also limits the domains D to which separation of variables is applicable. This can be seen by noting that when a product solution $v = f(\xi)g(\eta)$ vanishes at one point ξ_0, η_0 then either $f(\xi_0) = 0$ or $g(\eta_0) = 0$. Thus, v vanishes all along the coordinate line $\xi = \xi_0$ or along the line $\eta = \eta_0$. It is only when ∂D consists of one or more such lines that a product solution can vanish on it. The consequence of this fact is this: the method of separation of variables can be used to solve a boundary value problem only when the boundary consists of portions of coordinate lines (or surfaces) of a coordinate system in which the differential equation is separable.

Within these limits, a large number of scattering problems have been solved. It is from these solutions that our physical insight has developed, and it is from them that we have obtained many quantitative results.

Some of the objects (or domains) for which the scattering problem has been solved are listed below, together with the name of the solver and the approximate date:

circular cylinder	Rayleigh	1881
sphere (electromagnetic waves)	Mie	1908
parabolic cylinder	Epstein	1914
thin strip slit in plane screen }	Morse and Rubinstein	1938
circular disc circular hole in plane screen }	Bouwkamp	1944
circular cone (electromagnetic)	Hansen and Schiff	1945
plane angular sector	Kraus and Levine	1963

Asymptotic evaluation of the solutions for high frequency or short wavelength was done for the following cases:

circular cylinder	Debye	1908
	Nussenzweig	1970
sphere	Watson	1912
parabolic cylinder	Fock	1946
	Rice	1950
	Keller	1956
elliptic cylinder	Levy and Keller	1960

A new modification of the classical method, useful for short wavelengths, was developed by Sommerfeld to solve the problem of diffraction by a half plane, i.e. a thin screen with a straight edge:

half plane	Sommerfeld	1896
wedge	McDonald	1902

During the 1940s, another new modification was developed based upon the Wiener-Hopf method for solving integral equations with convolution kernels. This was done by Schwinger and by Vainstein, and the method has been used by many others. This work is described in the book of Noble (1965) on this method.

5. Geometrical Optics

In addition to the classical methods described above, there is a different classical method called geometrical optics, which is used for describing the propagation of light. It was known to the ancient Greeks, and is presented in the works of Euclid. It is called geometrical optics because it determines the paths, called rays, along which light travels. This method is based upon the following three laws.

Law of Propagation (Euclid). In a homogeneous medium light travels in a straight line.
Law of Reflection (Euclid). A ray that hits a reflecting surface produces a reflected ray. The reflected ray makes the same angle with the normal to the surface as does the incident ray, and the two rays lie on opposite sides of the normal. Alhazen, in the eleventh century, noted that the two rays and the normal are coplanar.
Law of Refraction (Snell 1626, Descartes 1637). A ray that hits the interface between two different media produces a refracted ray on the other side of the interface from the incident ray. The angle β between the refracted ray and the normal is related to the angle α between the incident ray and the normal by Snell's law: $n_1 \sin\alpha = n_2 \sin\beta$. Here n_1 is the refractive index of medium one containing the incident ray and n_2 is that of medium two containing the refracted ray. We now know that $n_i = c_0/c_i$ where c_0 is a reference speed, usually that of light in a vacuum, and c_i is the speed in medium i. The two rays are coplanar with the normal and lie on opposite sides of it. Ptolemy, in the first century, gave an approximation to Snell's law that is valid for small angles, namely $n_1\alpha = n_2\beta$.

These three laws suffice for the calculation of the focal lengths of mirrors and lenses, and for the design of optical instruments. They were used, for example, by Gauss in the early 1800s in his analysis of imaging in axially symmetric optical systems.

There is a different formulation of the laws of optics, based upon the calculus of variations. For propagation it was known to Euclid, because he knew that a straight line is the shortest distance between two points. Thus, the ray from P to Q is the shortest path from P to Q. For reflection, Heron of Alexandria, in the second century, showed that the incident ray from P to a plane mirror, plus the reflected ray from the mirror to Q, is the shortest path from P to Q with one point on the plane mirror.

Finally Fermat (1661) found a variational formulation of the law of refraction. This led to his *principle of least time*, which states that in a medium with the refractive index $n(\mathbf{x})$, the light ray from P to Q minimizes the integral L:

$$L = \int_P^Q n\,[\mathbf{x}(s)]\,ds = c_0 \int_P^Q \frac{ds}{c[\mathbf{x}(s)]} = c_0 \int dt\ . \tag{5.1}$$

The last integral in (5.1) shows that L is c_0 times the time required for light to travel from P to Q along the path $\mathbf{x}(s)$. L is called the optical length of the path. We now know that L is stationary at the light ray, but not necessarily a minimum.

Fermat's Principle yields the laws of propagation and refraction, and it determines the rays in inhomogeneous media. It also yields the law of reflection when it is applied to paths with one point on a given (mirror) surface. Thus, it determines all the rays of geometrical optics.

Hamilton (1833) gave various alternative formulations of the laws of optics. One is the system of six first order ordinary differential equations for the rays in R^3. These are now known as Hamilton's equations, and are most familiar in mechanics. Another is the eiconal equation, a nonlinear first order partial differential equation for the eiconal function $S(\mathbf{x})$:

$$(\nabla S)^2 = n^2(\mathbf{x})\ . \tag{5.2}$$

The surfaces $S(\mathbf{x})$ = constant are orthogonal to a normal congruence of rays. These surfaces are called wavefronts. The difference between the values of S on two wavefronts is equal to the optical length L of any ray of the congruence between them.

Geometrical optics does not suffer from the limitation to special geometries that restricts the method of separation of variables. However, it has its own limitation, namely that it describes only the rays along which light travels, but it gives no quantitative information about the amplitude or phase of the light field.

6. Modern Developments

The modern theory of partial differential equations began to develop in the last half of the nineteenth century, and it has flourished in the twentieth century. It has focused on the questions of existence and uniqueness of solutions of boundary value problems, of initial value problems, of initial and boundary value problems, etc. In addition it has considered whether the solutions depend continuously on the data of the problem. A further concern has been regularity of solutions, i.e. how many continuous derivatives they have and the Hölder continuity of those derivatives. It also deals with the singularities of solutions, the locus of such singularities, the Hausdorff dimension of the set of singularities, etc. For linear partial differential equations, this theory is presented in an elegant and general form in the treatise of Hörmander (1983).

In addition to these mainly qualitative developments, there have been two quantitative developments. One is the use of computers and the related numerical analysis. The other is the development of asymptotic analysis, which is the topic of the remainder of this survey.

An asymptotic expansion of a wavelike solution of the Helmholtz equation (2.2) is an expression of the form

$$v(\mathbf{x},k) \sim e^{ikS(\mathbf{x})} \sum_{j=0}^{\infty} \frac{1}{(ik)^j} A_j(\mathbf{x}) \ . \tag{6.1}$$

This expression is not assumed to be a convergent series, but rather an asymptotic expansion of v, valid as k tends to infinity. This means that as $k \to \infty$,

$$v(\mathbf{x},k) - e^{ikS(\mathbf{x})} \sum_{j=0}^{J} (ik)^{-j} A_j(\mathbf{x}) = o\left(k^{-J}\right) \ . \tag{6.2}$$

In other words, the difference between v and any partial sum vanishes more rapidly than the last term retained, as $k \to \infty$. Therefore such an expansion can be useful in representing v when the wavenumber k is large, or equivalently when the wavelength $2\pi/k$ is small.

Carlini (1815) introduced an expression somewhat like (6.1) to solve Bessel's equation, which is an ordinary differential equation of second order. Jacobi (1817) explained and elaborated upon this work. Then Liouville and Green in the 1830s used expansions of the form (6.1) to solve general linear second order ordinary differential equations. However, despite repeated attempts, no one succeeded in showing that these series converged. Then Poincaré (1886) suggested that they were not convergent but asymptotic, and he introduced the definition (6.2). Steltjes (1886) introduced the same concept to deal with certain formal series that arose in the evaluation of integrals. Finally, in 1893 Korn showed, for certain linear second order ordinary differential equations, that the formal series were indeed asymptotic to solutions.

Shortly after the introduction of quantum mechanics in 1925, this method of solution was rediscovered by Wentzel, by Kramers, and by Brillouin. Therefore physicists call it the WKB method, and sometimes the WKBJ method because Jeffreys had used it in the early 1920s.

Other asymptotic expansions, different from (6.1), have also been used to represent solutions. Often a number of different expansions are used to represent a single solution, each valid in a different region of space. These expansions must match together where these different regions overlap. The use of such combinations of expansions is called the method of matched asymptotic expansions. It is widely used in wave propagation, and it is a basic tool of modern applied mathematics.

7. Asymptotic Expansions and Geometrical Optics

In 1916 Sommerfeld and Runge attempted to show the connection between geometrical optics and the wave equation. They began with the Helmholtz equation in an inhomogeneous medium with the refractive index $n(\mathbf{x})$:

$$\Delta v + k^2 n^2(\mathbf{x}) v = 0 . \tag{7.1}$$

They represented v by the first term of the expansion (6.1), substituted it into (7.1), and equated to zero the coefficient of k^2, which was the highest power of k. This yielded the eiconal equation for the exponent, or phase function, $S(\mathbf{x})$:

$$(\nabla S)^2 = n^2(\mathbf{x}) . \tag{7.2}$$

This is just the equation (5.2) obtained by Hamilton for the eiconal function of geometrical optics, so a connection between the wave equation and geometrical optics was established.

By equating to zero the coefficient of k, they obtained an equation for $A_0(\mathbf{x})$:

$$2\nabla S \cdot \nabla A_0 + A_0 \Delta S = 0 . \tag{7.3}$$

Luneberg (1944) and Friedlander (1947) used the full expansion (6.1) and obtained equations for the other A_n:

$$2\nabla S \cdot \nabla A_n + A_n \Delta S = -\Delta A_{n-1} \quad , \quad n \geq 1 . \tag{7.4}$$

The equation for A_0 is called the (first) transport equation, and the equations for the other A_n are called the higher transport equations.

The eiconal equation can be solved by the method of characteristics, which Hamilton developed for this purpose. The characteristics are just the rays, and the solution for S at the point $\mathbf{x}$ on the ray $\mathbf{x}(\tau)$ is

$$S(\mathbf{x}) = S(\mathbf{x}_0) + \int_{\tau_0}^{\tau} n\left[\mathbf{x}(\tau')\right] d\tau' \; . \tag{7.5}$$

Here $\mathbf{x}_0 = \mathbf{x}(\tau_0)$ is some point on the ray and $S(\mathbf{x}_0)$ is the value of S there. The integral in (7.5) is just the optical length (5.1), which occurs in Fermat's principle.

Each transport equation can be written as a first order ordinary differential equation along a ray, and then it can be solved explicitly. The solution for A_0 can be expressed in the form

$$n(\mathbf{x})A_0^2(\mathbf{x})d\sigma(\mathbf{x}) = n(\mathbf{x}_0)A_0^2(\mathbf{x}_0)d\sigma(\mathbf{x}_0) \; . \tag{7.6}$$

Here $d\sigma(\mathbf{x})$ denotes the normal cross-sectional area at $\mathbf{x}$ of a narrow tube of rays, and $d\sigma(\mathbf{x}_0)$ denotes the corresponding area at $\mathbf{x}_0$. Thus, the relation (7.6) expresses conservation of energy in a tube of rays: the flux of energy $nA^2d\sigma$ at $\mathbf{x}$ is the same as that at $\mathbf{x}_0$. From (7.6) we can solve for $A(\mathbf{x})$:

$$A_0(\mathbf{x}) = A_0(\mathbf{x}_0)\left[\frac{n(\mathbf{x}_0)}{n(\mathbf{x})}\frac{d\sigma(\mathbf{x}_0)}{d\sigma(\mathbf{x})}\right]^{1/2} \; . \tag{7.7}$$

Upon using (7.5) for $S(\mathbf{x})$ and (7.7) for $A_0(\mathbf{x})$ in the first term of (6.1), we obtain

$$v(\mathbf{x},k) \sim e^{ik[S(\mathbf{x}_0)]+\int_{\tau_0}^{\tau} n[\mathbf{x}(\tau')]d\tau'} A_0(\mathbf{x}_0)\left[\frac{n(\mathbf{x}_0)}{n(\mathbf{x})}\frac{d\sigma(\mathbf{x}_0)}{d\sigma(\mathbf{x})}\right]^{1/2} \; . \tag{7.8}$$

This expression for v involves only quantities determined by geometrical optics, in addition to k and the initial value $A_0(\mathbf{x}_0)$. Therefore we call it the geometrical optics field. It is the field associated with a particular ray through $\mathbf{x}$.

As Luneberg (1944) observed, the total field at x is the sum of the fields on all the rays through $\mathbf{x}$. They are the direct rays from the source to $\mathbf{x}$, the reflected rays, the refracted rays, and the multiply reflected and/or multiply refracted rays, if there are any. Therefore the solution $v(\mathbf{x},k)$ is represented by the sum of asymptotic expansions of the form (6.1), with one such expansion associated with each ray through $\mathbf{x}$. We call each of these expansions a wave. Then we can say that v is represented as a sum of direct, reflected, refracted, and multiply reflected and/or refracted waves.

Upon examining asymptotic expansions of various exact solutions of scattering problems, we find that they do indeed contain all these waves. However, they usually contain additional waves, called diffracted waves, which are not associated with the rays of geometrical optics. We shall explain how the preceding theory can be extended to include those waves. This requires an extension of geometrical optics, which we shall present first.

8. The Geometrical Theory of Diffraction

Geometrical optics does not specify what happens when a ray hits an edge or vertex on a boundary or interface. This is so because there is no unique normal to the boundary or interface at such places, so the laws of reflection and refraction do not apply. Therefore we have introduced the following two new laws to describe what happens in such cases, and a third law to describe what happens when a ray is tangent to a boundary or interface (Keller 1953).

Law of Edge Diffraction. A ray that hits the edge of a boundary or interface produces a one parameter family of rays that we call edge diffracted rays. The angle β between an edge diffracted ray and the tangent to the edge at the point of diffraction is equal to the angle α between the incident ray and the edge, if they lie in the same medium. If they lie in different media then $n_1 \cos\alpha = n_2 \cos\beta$. The edge diffracted rays and the incident ray lie on opposite sides of the plane normal to the edge.

Law of Vertex Diffraction. A ray that hits a vertex of a boundary or interface produces a two parameter family of vertex diffracted rays. They leave the vertex in all directions.

Law of Surface Diffraction. A ray that hits a boundary or interface tangentially produces a surface diffracted ray that is a geodesic on the surface in the metric $n(x)ds$. It sheds rays along its tangent at each point. At each point on an interface it also sheds rays into the second medium in accordance with Snell's law.

In addition to these diffracted rays, we have introduced *complex rays*, which are complex-valued solutions of the ray equations. They can be defined when $n(x)$ is analytic or piecewise analytic.

The principle of "least" time can be extended to yield all these new rays. To extend it we introduce various classes of curves C_j, $j = 1, 2, \ldots$. Then we define the rays from P to Q in C_j to be the curves that make the optical length L from P to Q stationary among all curves in C_j. If C_1 is the class of curves with no points on boundaries or interfaces, it yields the direct rays from P to Q. If C_2 consists of curves with one point on a boundary or interface it yields the reflected and refracted rays from P to Q. The class C_3 of curves with one point on an edge yields the edge diffracted rays and the law of edge diffraction. The class C_4 of curves with a point at a vertex yields the vertex diffracted rays. The class of curves C_5 with an arc on a boundary or interface yields surface diffracted rays. Multiply reflected, refracted, and diffracted rays can be obtained in a similar way.

The study and use of these new rays, the diffracted rays and the complex rays, is called the geometrical theory of diffraction. There are normal congruences of such rays, and the wavefronts orthogonal to them are called diffracted wavefronts. There is an eiconal or phase function $S(x)$ associated with each family of diffracted wavefronts, and it satisfies the eiconal equation (7.2). The initial value problem for the wavefronts associated with edge diffracted rays involves specification of the value of $S(x)$ on the edge. Similarly, for the phase function associated with vertex and surface diffracted rays, the initial values must be specified at the vertex or on the surface.

We have now presented three equivalent ways of determining the diffracted rays and the corresponding diffracted wavefronts. The first way is to use the laws governing the rays. The second way is to use the extended form of Fermat's principle. The third way is to solve the eiconal equation, including the multiple branches of the solution that branch at the edges, vertices, and surfaces. From normal congruences of rays we can construct the wavefronts and the solution of the eiconal equation, and conversely from the wavefronts we can determine the rays. All three of these methods yield the usual rays and wavefronts of geometrical optics as well as the diffracted rays and diffracted wavefronts. Thus, all three formulations of the geometrical theory of diffraction are extensions of the corresponding formulations of geometrical optics.

9. Asymptotic Expansions of Diffracted Waves

Next we shall indicate how to use this geometrical theory to obtain asymptotic solutions of boundary value problems for (7.1). Just as in the case of geometrical optics, we construct asymptotic expansions of the form (6.1) by the method described in Section 7. We construct one such expansion for each wave in the problem, i.e. for each normal congruence of rays or equivalently for each branch of the solution of the eiconal equation, or for each family of wavefronts. As before, we represent the solution $v(x,k)$ as a sum of asymptotic expansions $v_p(x,k)$ of the form (6.1), with one expansion associated with each ray through x. Thus, we write

$$v(x,k) \sim \sum_p v_p(x,k) \sim \sum_p e^{ikS_p(x)} \sum_{j=0}^{\infty} \frac{1}{(ik)^j} A_{jp}(x) \, . \tag{9.1}$$

Now the sum includes expansions corresponding to all the diffracted rays or waves through x, as well as to all the usual rays of geometrical optics.

The initial values $S_1(x_0)$ and $A_{j1}(x_0)$ on incident rays are specified, if the rays come from infinity. If the rays come from a source, the initial values are determined by the solution of a local problem containing the source. The initial values on a reflected or transmitted ray are obtained from the values of $S(x)$ and $A_j(x)$ on the incident ray that produces it, by means of a reflection or transmission coefficient. This coefficient is also determined by the solution of a local problem. Similarly, the initial values on a diffracted ray are determined from the values of $S(x)$ and $A_j(x)$ on the corresponding incident ray by means of a diffraction coefficient.

There are edge diffraction coefficients, vertex diffraction coefficients, and surface diffraction coefficients. Each of them is determined by the solution of a suitable problem, called a canonical problem, which incorporates the local geometry of the boundaries or interfaces near the point of diffraction, and the corresponding local properties of $n(x)$. These solutions often can be found by the classical methods of separation of variables, etc., described in the previous sections. The use of local solutions is an instance of the method of matched asymptotic expansions, which was mentioned earlier. It describes the solution near the point of diffraction, where the ray expansion (9.1) becomes singular because of the vanishing of the cross-sectional area $d\sigma(x)$ in the denominator of the expression (7.7) for $A_0(x)$. By matching the local solution to the ray expansion, the diffraction coefficients can be determined.

This procedure shows that any edge diffraction coefficient D is proportional to $k^{-1/2}$, whereas any vertex diffraction coefficient C is proportional to k^{-1}. These same conclusions also follow from dimensional analysis. They show that the field on both edge diffracted and vertex diffracted rays vanishes as $k \to \infty$. That on a vertex diffracted ray vanishes more rapidly; in fact it vanishes as fast as the field on a ray doubly diffracted by edges. Furthermore, the value of v on a surface diffracted ray decays exponentially with distance along the surface, with a decay rate proportional to $k^{1/3}$. The field on a complex ray has a complex phase $S(x)$, so it decays at a rate $kImS(x)$ proportional to k. For simplicity, the dependence upon k described in this paragraph is not shown in (9.1).

These results show that all diffracted fields, i.e. the values of v associated with diffracted waves, vanish as $k \to \infty$. The values of v on the ordinary geometrical optics rays do not vanish as $k \to \infty$, so they are stronger than the diffracted fields for large values of k. However, there are no geometrical optics rays in shadows, so the diffracted fields are the only fields present there.

The method described in Sections 6–9 can be used to solve both scattering and eigenvalue problems. It can also be used to solve other partial differential equations. For scattering problems, the method is relatively straightforward, although there are of course, many complications. But for eigenvalue problems some further considerations are needed, as we shall now explain.

10. Eigenvalue Problems

In an eigenvalue problem in a domain D, there are no incident rays, and the value of k is unknown, so it is not clear how to start using the preceding method. Therefore, we must determine all the rays or waves in the solution simultaneously. Thus, in the domain D we seek an asymptotic expansion of the form (9.1) consisting of a sequence of waves $p = 1, \dots, N$. The pth wave is defined in a domain D_p, which may be all of D or a subdomain of D. On each part of the boundary of D_p, either the pth wave is produced by some other wave or it produces some other wave. Therefore, its phase $S_p(x)$ must equal that of the other wave on the corresponding part of the boundary. Within D_p, $S_p(x)$ must satisfy the eiconal equation (7.2).

It is convenient to introduce an N-sheeted space Σ consisting of the N domains D_p. The domains D_p and $D_{p'}$ are joined together at those parts of their boundaries where one of these waves produces the other. On this space Σ, the function $\nabla S(x)$ is single valued, with the value $\nabla S_p(x)$ on the sheet D_p. When we integrate $\nabla S(x)$ to obtain $S(x)$, we obtain a multiple-valued function. The corresponding amplitude function $A_0(x)$, given by (7.7), may also be multiple valued. But the solution $v(x,k)$ must be single valued. This gives rise to the condition

$$k\Delta S \sim i\Delta \log A_0 = 2\pi n + o(1)\,. \tag{10.1}$$

Here $\Delta S(x)$ denotes the difference between two values of $S(x)$ at a point x on the multi-sheeted space Σ, and similarly for $\Delta A_0(x)$, while $o(1)$ denotes terms which tend to zero as $k \to \infty$, and n is an integer.

We can write $\Delta S(x)$ as a line integral along a closed curve C on Σ in the form

$$\Delta S(x) = \oint_C \nabla S \cdot d\ell\,. \tag{10.2}$$

Every C on Σ is a linear combination with integer coefficients, of basis curves $C_1, C_2, \dots, C_B$ of the fundamental group Σ. Here B, the first Betti number of Σ, is the order of the fundamental group. Therefore (10.1) will hold for every C if the following B conditions hold with integers n_j:

$$k \oint_{C_j} \nabla S \cdot d\ell - i(\Delta \log A_0)_{C_j} = 2\pi n_j \quad , \quad j = 1, 2, \ldots, B \,. \tag{10.3}$$

The function $\log A_0(x)$ changes its value by $-i\pi/2$ as a wave passes through a caustic surface, i.e. an envelope of rays, as can be shown by a local analysis. Therefore, we denote by ν_j the number of times the curve C_j crosses a caustic. Then $-i(\Delta \log A_0)_{C_j} = -\frac{\pi}{2}\nu_j$ and (10.3) becomes

$$k \oint_{C_j} \nabla S \cdot d\ell = 2\pi \left(n_j + \frac{\nu_j}{4}\right) \quad , \quad j = 1, 2, \ldots, B \,. \tag{10.4}$$

This result applies when v satisfies the boundary condition $\partial_n v = 0$ on ∂D. For other boundary conditions, $(\Delta \log A_0)_{C_j}$ contains an additional term associated with each time C_j crosses ∂D.

To use (10.4) we must find N solutions S_j of the eiconal equation that join together pairwise along portions of the boundary of D, or along caustic curves or caustic surfaces inside D. These solutions must depend upon $B-1$ parameters. Then k and these $B-1$ parameters must be determined to satisfy (10.4) for each choice of the nonnegative integers n_j, $j = 1, \ldots, B$. Finding the N solutions S_j is equivalent to finding N families of wavefronts, each of which reflects into another family at ∂D, or that join pairwise at caustics. Alternatively we could find N normal congruences of rays which reflect into one another at ∂D or which join one another at caustics.

In quantum mechanics, conditions like (10.4) are called quantum conditions. Bohr (1913) presented the first such condition for the motion of an electron in a hydrogen atom, with $B = 1$ and $\nu_1 = 0$. Sommerfeld (1916) and Wilson (1916) presented B such conditions for separable systems with B degrees of freedom, and Einstein (1917) gave these conditions for nonseparable systems, all with $\nu_j = 0$. After the Schrödinger equation was formulated, the Sommerfeld-Wilson conditions for separable systems were rederived from it by separation of variables. In some cases in which the separated equations had turning points, this derivation yielded corrected Sommerfeld-Wilson conditions with $\nu_j \neq 0$. The present author's derivation, described above, which does not require separability, led to the widespread use of this method by chemists and physicists. They call it the EBK method, for Einstein, Brillouin (who used the single-valuedness argument), and the author.

The conditions (10.4) were later derived independently by Maslov (1965). Then Arnold (1970) showed that the ν_j are invariants (i.e. independent of the curves C_j). The integer ν_j is sometimes called the Keller-Maslov index.

11. Conclusion

Many authors have contributed to the further development of the asymptotic methods described above. Some of them are R. K. Luneberg, F. G. Friedlander,

M. Kline, I. Kay, R. M. Lewis, R. Buchal, B. Levy, B. Seckler, N. Bleistein, J. Cohen, R. Handelsman, B. Matkowsky, D. Ahluwalia, F. Karal, L. Felsen, D. Ludwig, L. Hörmander, C. Morawetz, C. Bloom, W. Miranker, N. Kazarinoff, P.D. Lax, R.M. Phillips, W. Straus, J. Ralston, A. Majda, M. Taylor, F. Ursell, Y. Kravtsov, L. Babich, P. Ufimtsev, R. Melrose, M. Zworski, J. Rauch, V. Lazutkin, V. Maslov, and M. Berry. They have proved the validity of these methods in many cases, extended them to other equations and systems of equations, applied them to special problems, developed computer programs to determine the various kinds of diffracted rays and to calculate the corresponding phases and amplitudes, etc. In addition, some of these methods have been extended to weakly nonlinear waves.

Some of the early work on these methods is described in the papers listed below, which also contain references to many other works.

References

[1] J. B. Keller, *A geometrical theory of diffraction, calculus of variations and its applications*, Proc. Sympos. Appl. Math., **8** (1958), 27–52; Math. Revs., **20** (1959), 103.

[2] J. B. Keller, *Corrected Bohr-Sommerfeld quantum conditions for nonseparable systems*, Annals Physics, **4** (1958), 180–188; Math. Revs., **20** (1959), 934.

[3] J. B. Keller, *Rays, waves and asymptotics*, Bull. Amer. Math. Soc. **84** (1978), 727–750.

[4] J. B. Keller, *One hundred years of diffraction theory*, IEEE Trans. Antennas and Propagation, **AP-33** (1985), 200–214.

[5] J. B. Keller, *Semiclassical mechanics*, SIAM Rev., **27** (1985), 485–504.

[6] J. B. Keller and S. I. Rubinow, *Asymptotic solution of eigenvalue problems*, Annals Physics, **9** (1960), 24–75.

Homological Algebra of Mirror Symmetry

MAXIM KONTSEVICH

Max Planck Institut für Mathematik
Gottfried-Claren-Straße 26
D-53225 Bonn, Germany

and

Mathematics Department
University of California
Berkeley, CA 94720, USA

Mirror symmetry (MS) was discovered several years ago in string theory as a duality between families of 3-dimensional Calabi-Yau manifolds (more precisely, complex algebraic manifolds possessing holomorphic volume elements without zeros). The name comes from the symmetry among Hodge numbers. For dual Calabi-Yau manifolds V, W of dimension n (not necessarily equal to 3) one has

$$\dim\ \mathrm{H}^p(V,\Omega^q) = \dim\ \mathrm{H}^{n-p}(W,\Omega^q)\ .$$

Physicists conjectured that conformal field theories associated with mirror varieties are equivalent. Mathematically, MS is considered now as a relation between numbers of rational curves on such a manifold and Taylor coefficients of periods of Hodge structures considered as functions on the moduli space of complex structures on a mirror manifold. Recently it has been realized that one can make predictions for numbers of curves of positive genera and also on Calabi-Yau manifolds of arbitrary dimensions.

We will not describe here the complicated history of the subject and will not mention many beautiful constructions, examples, and conjectures motivated by MS. On the contrary, we want to give an outlook of the story in general terms and propose a conceptual framework for a possible explanation of the mirror phenomenon. We will restrict ourselves to a half of MS considering it as a relation between symplectic structures on one side and complex structures on another side. Actually, we will deal only with a half of this half, ignoring the holomorphic anomaly effects (see [BCOV]) in the symplectic part (A-model) and the polarization of Hodge structures in the complex part (B-model). For an introduction to mirror symmetry we recommend [M] and [Y].

At the moment there are only a few completely solid statements, essentially because there was no universal definition of the "number of curves" for a long time.

Comparison of symplectic and complex geometry

We start with a recollection of well-known facts concerning symplectic and complex manifolds. Numbers followed by S indicate facts on symplectic manifolds; numbers followed by C indicate facts on complex manifolds.

Proceedings of the International Congress
of Mathematicians, Zürich, Switzerland 1994

Let V be a compact smooth $2n$-dimensional manifold.

1.S. A symplectic structure on V is given by a reduction of the structure group $GL(2n, \mathbf{R})$ of the tangent bundle T_V to the subgroup $Sp\,(2n, \mathbf{R})$ satisfying certain integrability conditions (the associated 2-form ω is closed or, equivalently, the associated Poisson bracket on smooth functions satisfies the Jacobi identity).

1.C. A complex structure on V is given by a reduction of the structure group $GL(2n, \mathbf{R})$ of the tangent bundle T_V to the subgroup $GL(n, \mathbf{C})$ satisfying certain integrability conditions (the Newlander-Nirenberg theorem).

Notice that both groups $Sp\,(2n, \mathbf{R})$ and $GL(n, \mathbf{C})$ are homotopy equivalent to $U(n)$. Thus, they have the same algebra of characteristic classes generated by Chern classes $c_i \in \mathrm{H}^{2i}(BU(n), \mathbf{Z})$, $1 \leq i \leq n$.

Basic examples of compact symplectic or complex manifolds are complex projective algebraic manifolds endowed with the pullback of the Fubini-Studi-Kähler form on the projective space.

2.S. First-order deformations of symplectic structures on V are in one-to-one correspondence with $\mathrm{H}^2(V, \mathbf{R})$. The deformation theory is unobstructed and the local moduli space of symplectic structures on V can be identified with a domain in the affine space $\mathrm{H}^2(V, \mathbf{R})$ via map $\omega \mapsto [\omega] \in \mathrm{H}^2(V, \mathbf{R})$ (Moser).

2.C. First-order deformations of complex structures on V near a fixed one are in one-to-one correspondence with $\mathrm{H}^1(V, T_V^{\mathrm{hol}})$, where T_V^{hol} denotes the sheaf of holomorphic vector fields on V (Kodaira-Spencer theory). If $c_1(V) = 0$ and V admits a Kähler structure then the deformation theory of V is unobstructed and the local moduli space can be identified with a domain in the affine space $\mathrm{H}^1(V, T_V^{\mathrm{hol}})$ (the Bogomolov-Tian-Todorov theorem).

The following two facts concern only complex manifolds.

3.C. For a complex manifold V admitting a Kähler structure there is a pure Hodge structure on the singular cohomology groups:

$$\mathrm{H}^k(V, \mathbf{Z}) \otimes \mathbf{C} \simeq \bigoplus_{p+q=k} \mathrm{H}^q(V, \Omega^p) \ .$$

4.C. With a complex algebraic manifold V one can associate the abelian category $\mathrm{Coh}(V)$ of coherent sheaves on V and the triangulated category $\mathcal{D}^b(\mathrm{Coh}(V))$ (the bounded derived category).

Our aim in this talk is to propose candidates for 3.S and 4.S in the context of symplectic geometry. The mirror symmetry should be a correspondence (partially defined and multiple valued) between symplectic and complex manifolds (both with $c_1 = 0$) identifying structures 2–4.

To get a feeling of what is going on it is instructive to look at a simplest case of the mirror symmetry, which is already highly nontrivial.

2-dimensional tori (after Dijkgraaf)

Let Σ be a complex elliptic curve and $p_1, \ldots, p_{2g-2}$ be pairwise distinct points of Σ, where $g \geq 2$ is an integer. We consider holomorphic maps $\phi : C \longrightarrow \Sigma$ from compact connected smooth complex curves C to Σ, which have only one double ramification point over each point $p_i \in \Sigma$ and no other ramification points. By the

Hurwitz formula the genus of C is equal to g. The set $X_g(d)$ of equivalence classes of such maps of degree $d \geq 1$ is finite, and for each $\phi : C \longrightarrow \Sigma$ its automorphism group

$$\mathrm{Aut}\ (\phi) := \{f : C \to C \mid \phi \circ f = \phi\}$$

is finite. For $g \geq 2$ we introduce the generating series in one variable q as follows:

$$F_g(q) := \sum_{d \geq 1} \left(\sum_{[\phi] \in X_g(d)} \frac{1}{\mid \mathrm{Aut}\ (\phi) \mid} \right) q^d \in \mathbf{Q}[[q]] \ .$$

The following statement is now rigorously established because of the efforts of several people (Dijkgraaf, Douglas, Zagier, Kaneko):

$$F_g \in \mathbf{Q}[E_2, E_4, E_6] \ ,$$

where E_k are the classical Eisenstein series,

$$E_k = 1 - \frac{2k}{B_k} \sum_{n \geq 1} \left(\sum_{a|n} a^{k-1} \right) q^n \ .$$

E_k is a modular form of weight k for even $k \geq$ and E_2 is *not* a modular form. Here $B_2 = 1/6$, $B_4 = -1/30$, $B_6 = 1/42$, $\ldots$ are Bernoulli numbers. If one associates with E_k, $k = 2, 4, 6$, the degree k, then F_g has degree $6g - 6$.

One can regard Σ as a symplectic 2-dimensional manifold $(S^1 \times S^1, \omega)$ with the area $\int_\Sigma \omega$ equal to $-\log(q)$, $0 < q < 1$, and interpret weights q^d of ramified coverings as

$$\exp\left(-\text{area of } C \text{ with respect to the pullback of } \omega\right) \ .$$

Mirror symmetry in this example is the claim that the generating function for certain invariants of symplectic structures on $S^1 \times S^1$ is a "nice" function on the moduli space of complex structures on $S^1 \times S^1$. The 2-dimensional torus is a self-dual manifold for MS.

Notice that the standard local coordinate $q = \exp(2\pi i \tau)$, $\mathrm{Im}\, \tau > 0$, on the moduli space of elliptic curves,

$$\text{elliptic curve} = \mathbf{C} / (\mathbf{Z} + \mathbf{Z}\tau) \ ,$$

can be written as

$$q = \exp\left(2\pi i \frac{\int_{\gamma_1} \Omega}{\int_{\gamma_0} \Omega} \right) ,$$

where γ_0, γ_1 are two generators of $\mathrm{H}_1(\text{elliptic curve}, \mathbf{Z})$ and Ω is a nonzero holomorphic 1-form.

Quintic 3-folds (after [COGP])

Here we describe the first famous prediction of physicists. Let V be a nonsingular hypersurface in complex projective space $\mathbf{P}^4$ given by an equation $Q(x_1, \ldots, x_5) = 0$ of degree 5 in 5 homogeneous variables $(x_1 : \cdots : x_5)$. This complex manifold carries a top degree holomorphic differential form which is nondegenerate at all points (a holomorphic volume element):

$$\Omega = \frac{1}{dQ} \sum_{i=1}^{5} (-1)^i x_i dx_1 \wedge \cdots \wedge \widehat{dx_i} \wedge \cdots \wedge dx_5 \ .$$

Clemens conjectured that smooth rational curves on a generic quintic 3-fold are isolated. Recently it was checked up to degree 9. It is natural to count the number N_d of rational curves on V of fixed degree d. In fact, there are singular rational curves on V of degree 5, and one has to take them into account as well. At the end of the next section we will propose an algebro-geometric formula for the "physical" number of curves on V, both smooth and singular, without assuming the validity of the Clemens conjecture.

The mirror symmetry prediction is the following. First of all, we define the virtual number of curves of degree d as

$$N_d^{\text{virt}} := \sum_{k|d} \frac{1}{k^3} N_{d/k} \in \mathbf{Q} \ .$$

The reason for this formula is that in string theory one counts not just curves in V but maps from rational curves to V. Any map $\mathbf{P}^1 \to V$ of positive degree is the composition of a rational map $\mathbf{P}^1 \to \mathbf{P}^1$ and of an embedding $\mathbf{P}^1 \hookrightarrow V$. It was argued first by Aspinwall and Morrison [AM] that the factor associated with multiple coverings of degree k should be equal to $1/k^3$.

The complete generating function for rational curves is

$$F(t) := \frac{5}{6} t^3 + \sum_{d \geq 1} N_d^{\text{virt}} \exp{(dt)} \ .$$

The first summand here represents the contribution of maps of degree 0 (i.e., maps to a point of V).

On the mirror side we consider functions

$$\psi_0(z) = \sum_{n=0}^{\infty} \frac{(5n)!}{(n!)^5} z^n$$

$$\psi_1(z) = \log z \cdot \psi_0(z) + 5 \sum_{n=1}^{\infty} \frac{(5n)!}{(n!)^5} \left(\sum_{k=n+1}^{5n} \frac{1}{k} \right) z^n$$

$$\psi_2(z) = \frac{1}{2} (\log z)^2 \cdot \psi_0(z) + \ldots$$

$$\psi_3(z) = \frac{1}{6} (\log z)^3 \cdot \psi_0(z) + \ldots$$

which are solutions of the linear differential equation

$$\left(\left(z\frac{d}{dz}\right)^4 - 5z(5z\frac{d}{dz}+1)(5z\frac{d}{dz}+2)(5z\frac{d}{dz}+3)(5z\frac{d}{dz}+4)\right)\psi(z) = 0 \ .$$

More precisely,

$$\sum_{i=0}^{3}\psi_i(z)\epsilon^i + O(\epsilon^4) = \sum_{n=0}^{\infty}\frac{(1+5\epsilon)\,(2+5\epsilon)\,\ldots\,(5n+5\epsilon)}{((1+\epsilon)\,(2+\epsilon)\,\ldots\,(n+\epsilon))^5}\, z^{n+\epsilon} \ .$$

Functions $\psi_i(z)$ are periods $\int_{\gamma_i}\omega$ of the Calabi-Yau manifold $W = W(z)$, which is a resolution of singularities of the following singular variety:

$$\{(x_1 : x_2 : x_3 : x_4 : x_5)|\ x_1^5 + x_2^5 + x_3^5 + x_4^5 + x_5^5 = z^{-1/5}x_1x_2x_3x_4x_5\}/(\mathbf{Z}/5\mathbf{Z})^3 \ .$$

Here the group $(\mathbf{Z}/5\mathbf{Z})^3$ is the group of diagonal matrices preserving W

$$\{\ \mathrm{diag}(\xi_1,\ldots,\xi_5)\,|\ \xi_i^5 = 1,\ \prod_{i=1}^{5}\xi_i = 1\,\}/\{\,\xi\,\mathrm{Id}\ |\ \xi^5 = 1\,\}$$

and γ_i are certain singular homology classes with complex coefficients. The family of varieties $W(z)$ depending on 1 parameter is mirror dual to a universal family of smooth quintic 3-folds depending on 101 parameters.

The prediction of physicists is that

$$F\left(\frac{\psi_1}{\psi_0}\right) = \frac{5}{2}\cdot\frac{\psi_1\psi_2 - \psi_0\psi_3}{\psi_0^2} \ .$$

One of the miracles in this formula is that

$$\exp\left(\frac{\psi_1}{\psi_0}\right) \in \mathbf{Z}[[z]] \ .$$

Also, numbers N_d computed via the mirror prediction are positive integers.

It is interesting that the contribution of *individual* nonparametrized rational curves on 3-dimensional Calabi-Yau manifolds is connected with variations of mixed Hodge structures in a fashion analogous to the mirror symmetry predictions. Namely, according to the Aspinwall and Morrison formula [AM] we have the following generating function:

$$F(t) = \sum_{d\geq 1}\frac{\exp\,(dt)}{d^3} \ .$$

We introduce functions ψ_*:

$$\psi_0(z) = 1\,,\ \ \psi_1(z) = \log z\,,\ \ \psi_2(z) = \frac{1}{2}\,(\log z)^2\,,\ \ \psi_3(z) = \mathrm{Li}_3(z) := \sum_{d\geq 1}\frac{z^d}{d^3}\,,$$

which are solutions of the linear differential equation

$$\frac{d}{dz}\left(\frac{1-z}{z}\left(z\frac{d}{dz}\right)^3\psi(z)\right) = 0 \ .$$

Functions F and ψ_* are related by the evident formula

$$F\left(\frac{\psi_1}{\psi_0}\right) = \frac{\psi_3}{\psi_0} \ .$$

Gromov-Witten invariants

We describe here a not yet completely constructed theory that has a potentially wider domain of applications than mirror symmetry. It is based on pioneering ideas of Gromov [G] on the role of $\overline{\partial}$-equations in symplectic geometry, and certain physical intuitions proposed by Witten [W1], [W2]. There are many evidences that the following picture from [KM] is correct.

Let (V, ω) be a closed symplectic manifold, $\beta \in \mathrm{H}_2(V, \mathbf{Z})$ be a homology class, and $g, n \geq 0$ be integers satisfying the inequality $2 - 2g - n < 0$. Gromov-Witten classes

$$I_{g,n;\beta} \in \mathrm{H}_D\left(\overline{\mathcal{M}}_{g,n}(\mathbf{C}) \times V^n; \mathbf{Q}\right)$$

are homology classes with rational coefficients of degree

$$D = D(g, n, \beta) = (\dim V - 6)(1 - g) + 2n + 2\int_\beta c_1(T_V) \ .$$

Here $\overline{\mathcal{M}}_{g,n}$ denotes the Deligne-Mumford compactification of the moduli stack of smooth connected algebraic curves of genus g with n marked points. Recall that an algebraic curve C with marked points $p_1, \ldots, p_n$ is called *stable* if

(1) all singular points of C are ordinary double points,
(2) marked points p_i are pairwise distinct and smooth, $p_i \in C^{\text{smooth}}$,
(3) the group of automorphisms of $(C : p_1, \ldots, p_n)$ is finite, or, equivalently, the Euler characteristic of each connected component of $C^{\text{smooth}} \setminus \{p_1, \ldots, p_n\}$ is negative.

The arithmetic genus of stable curve C is defined by the formula

$$g_a(C) := \dim \mathrm{H}^1(C, \mathcal{O}) = -\chi(C^{\text{smooth}})/2 + 1 \ .$$

The stack $\overline{\mathcal{M}}_{g,n}$ is the moduli stack of stable marked curves of arithmetic genus g with n marked points. The associated coarse moduli space $\overline{\mathcal{M}}_{g,n}(\mathbf{C})$ is a compact complex orbifold.

One expects that $I_{g,n;\beta}$ is invariant under continuous deformations of the symplectic structure on V.

Analogously, we expect that the Gromov-Witten invariants can be defined for nonsingular projective algebraic varieties over arbitrary fields and they take values in the Chow groups with rational coefficients instead of the singular homology groups.

Intuitively, the geometrical meaning of Gromov-Witten classes in the symplectic case can be described as follows. Let us choose an almost-complex structure on V compatible in the evident way with the symplectic form ω. Notice that the space of almost-complex structures compatible with the fixed ω is contractible. Denote by $X_{g,n}(V, \beta)$ the space of equivalence classes of $(C; x_1, \ldots, x_n; \phi)$, where C is a smooth complex curve of genus g with pairwise distinct marked points x_i, and $\phi : C \longrightarrow V$ is a pseudo-holomorphic map (i.e., a solution of the Cauchy-Riemann equation $\overline{\partial}\phi = 0$) such that the image of the fundamental class of C

is equal to β. There is a natural map from $X_{g,n}(V,\beta)$ to $\mathcal{M}_{g,n}(\mathbf{C}) \times V^n$ associating with $(C; x_*; \phi)$ the equivalence class of $(C; x_*)$ and the sequence of points $(\phi(x_1), \ldots, \phi(x_n))$. One can show easily that the dimension of the space $X_{g,n}(V,\beta)$ at each point is bigger than or equal to $D(g,n,\beta)$. Also, under appropriate genericity assumptions, $X_{g,n}(V,\beta)$ is *smooth* of dimension $D(g,n,\beta)$. We want to define $I_{g,n;\beta}$ as the image of the fundamental class of a compactification $\overline{X_{g,n}(V,\beta)}$. The problem here is to find a correct compactification and to define the "fundamental class" if there are components of dimensions bigger than expected. Also, one has to prove that classes $I_{g,n;\beta}$ do not depend on the choice of an almost-complex structure.

There are now two approaches to the rigorous construction of Gromov-Witten classes. The first one is due to Ruan and Tian [RT] and it suffices for the genus zero case. This construction works only for so-called semi-positive manifolds (including Fano and Calabi-Yau manifolds), but it gives classes with integral coefficients. The idea of this construction is to perturb generically $\overline{\partial}$-equations and check that there are no strata of dimension larger than $D(0,n,\beta)$ in Gromov's compactification of the space of pseudo-holomorphic curves. In fact, Ruan and Tian define not GW-classes but the number of maps from a fixed complex curve to V satisfying general incidence conditions (counted with signs). Using algebraic results on the structure of $\mathrm{H}^*(\overline{\mathcal{M}}_{0,n})$ it is possible to reconstruct a whole genus-zero part of Gromov-Witten classes (see [KM]). Another construction [K2] is based on a new compactification of the moduli space of maps and should work, presumably, for all genera, for all symplectic manifolds, and also for all nonsingular projective varieties over arbitrary fields. At least, one can produce now purely algebro-geometric definitions of genus-zero Gromov-Witten invariants in the case of complete intersections in projective spaces. Its advantage is that it will not use any general position argument, and its weak point is the lack of control on integrality of arising classes.

As an example we give a definition of "numbers of rational curves" on a quintic 3-fold. Denote by $\overline{\mathcal{M}}_{0,0}(\mathbf{P}^4, d)$ the moduli stack of equivalence classes of maps $\phi : C \longrightarrow \mathbf{P}^4$, where C is a connected curve of arithmetic genus zero with only ordinary double points as singular points (i.e., C is a tree of rational curves) such that each irreducible component of C mapping to a point has at least 3 singular points. The parameter d, $d \geq 1$, denotes the degree of the image of the fundamental class $[C]$ in $\mathrm{H}_2(\mathbf{P}^4, \mathbf{Z}) \simeq \mathbf{Z}$. It is proven in [K2] that $\overline{\mathcal{M}}_{0,0}(\mathbf{P}^4, d)$ is a smooth proper algebraic stack of finite type. The set of its complex points is a compact complex orbifold of dimension $5d+1$.

We define a vector bundle $\mathcal{E}_d$ of rank $5d+1$ over $\overline{\mathcal{M}}_{0,0}(\mathbf{P}^4, d)$. The fiber of $\mathcal{E}_d$ at $\phi : C \longrightarrow \mathbf{P}^4$ is equal to $\mathrm{H}^0(C, \phi^*\mathcal{O}(5))$. Notice that if a quintic 3-fold V is given by an equation Q of degree 5 in 5 variables, $Q \in \Gamma(\mathbf{P}^4, \mathcal{O}(5))$, then there is an associated section $\widetilde{Q}$ of $\mathcal{E}_d$ whose zeros are exactly maps into V. In general, there are connected components of the set of zeros of $\widetilde{Q}$ of positive dimensions arising from multiple covering maps to rational curves in V. Nevertheless, we define the "virtual" number of curves by the formula

$$N_d^{\mathrm{virt}} := \int_{\overline{\mathcal{M}}_{0,0}(\mathbf{P}^4,d)} c_{5d+1}(\mathcal{E}_d) \ .$$

Here the integral is understood in the orbifold sense. Thus, the numbers $N_d^{\text{virt}} \in \mathbf{Q}$ in general are not integers. We are sure that our formula will give the same numbers as physicists predict. This formula was checked up to degree 4. Also we obtained in [K2] a closed expression for the generating function for the numbers N_d^{virt}. Hence, the mirror prediction in the quintic case is reduced to an explicit identity.

There is an extension of the definition above to the case of complete intersections in toric varieties and for counting of higher genus curves in flag varieties.

Axioms

The system of axioms formulated in [KM] is a formalization of what physicists call 2-dimensional topological field theory coupled with gravity (see [W2]). We reproduce here only one of the axioms from [KM], which is the basic one. Other axioms encode more evident properties of Gromov-Witten classes, like the invariance under permutation of indices, etc.

It will be convenient to associate with the class $I_{g,n;\beta}$ a linear map

$$J_{g,n;\beta} : (\mathrm{H}^*(V,\mathbf{Q}))^{\otimes n} \longrightarrow \mathrm{H}^*(\overline{\mathcal{M}}_{g,n},\mathbf{Q})$$

using the Künneth formula and the Poincaré duality. A *splitting* axiom describes the restriction of Gromov-Witten classes to boundary divisors of $\overline{\mathcal{M}}_{g,n}$. Namely, for $g_1, g_2 \geq 0$ and $n_1, n_2 \geq 0$ such that

$$g_1 + g_2 = g, \ \ n_1 + n_2 = n + 2, \ \ 2 - 2g_i - n_i < 0 \text{ for } i = 1, 2$$

there exists a natural inclusion $i_{g_*,n_*} : \overline{\mathcal{M}}_{g_1,n_1} \times \overline{\mathcal{M}}_{g_2,n_2} \hookrightarrow \overline{\mathcal{M}}_{g,n}$

$$(C_1; x_1, \ldots, x_{n_1}) \times (C_2; y_1, \ldots, y_{n_2}) \longmapsto (C_1 \bigcup_{x_1 = y_1} C_2; x_2, \ldots, x_{n_1}, y_2, \ldots, y_{n_2}) .$$

The following diagram should be commutative:

$$\begin{array}{ccc}
\mathrm{H}^*(V)^{\otimes n} & \xrightarrow{\ \simeq\ } & \mathrm{H}^*(V)^{\otimes(n_1-1)} \otimes \mathrm{H}^*(V)^{\otimes(n_2-1)} \\
{\scriptstyle J_{g,n;\beta}}\big\downarrow & & \big\downarrow{\scriptstyle \otimes\Delta} \\
\mathrm{H}^*(\overline{\mathcal{M}}_{g,n}) & & \mathrm{H}^*(V)^{\otimes n_1} \otimes \mathrm{H}^*(V)^{\otimes n_2} \\
{\scriptstyle (i_{g_*,n_*})^*}\big\downarrow & & \big\downarrow{\scriptstyle \sum_{\beta_1+\beta_2=\beta} J_{g_1,n_1;\beta_1} \otimes J_{g_2,n_2;\beta_2}} \\
\mathrm{H}^*(\overline{\mathcal{M}}_{g_1,n_1} \times \overline{\mathcal{M}}_{g_2,n_2}) & \underset{\text{Künneth}}{\xrightarrow{\ \simeq\ }} & \mathrm{H}^*(\overline{\mathcal{M}}_{g_1,n_1}) \otimes \mathrm{H}^*(\overline{\mathcal{M}}_{g_2,n_2})
\end{array}$$

Here all cohomologies are taken with coefficients in $\mathbf{Q}$ and Δ denotes the Poincaré dual to the fundamental class of the diagonal $V \subset V \times V$. The geometric meaning of this axiom is clear: a map ϕ of the glued curve C from the image of i_{g_*,n_*} is the same as two maps ϕ_1, ϕ_2 from C_1 and C_2 with $\phi_1(x_1) = \phi_2(y_1)$.

The splitting axiom in the case $g_1 = g_2 = 0$ was checked by Ruan and Tian for semi-positive manifolds and by me for complete intersections using the stable map approach.

Associativity equation

For a compact symplectic manifold (V,ω) denote by $\mathcal{H} := \oplus_k \mathrm{H}^k(V,\mathbf{C})$ the total cohomology space of V considered as a $\mathbf{Z}$-graded vector space (super vector space) and also as a complex supermanifold. This means that the underlying topological space of $\mathcal{H}$ is $\mathrm{H}^{\mathrm{even}}(V,\mathbf{C})$ and functions $\mathcal{O}(\mathcal{H})$ on $\mathcal{H}$ are holomorphic functions on $\mathrm{H}^{\mathrm{even}}(V,\mathbf{C})$ with values in the exterior algebra generated by $\left(\mathrm{H}^{\mathrm{odd}}(V,\mathbf{C})\right)^*$.

Using Gromov-Witten classes for genus zero we define the following function (pre-potential) on $\mathcal{H}$:

$$\Phi(\gamma) := \sum_{\beta\in \mathrm{H}_2(V,\mathbf{Z})} e^{-\int_\beta \omega} \sum_{n\geq 3} \frac{1}{n!} \int_{I_{0,n;\beta}} 1_{\overline{\mathcal{M}}_{0,n}} \otimes \gamma \otimes \cdots \otimes \gamma .$$

Here γ denotes an even element of $\mathcal{H}\otimes\Lambda$, where Λ is an arbitrary auxiliary supercommutative algebra (as usual in the theory of supermanifolds). The element $1_{\overline{\mathcal{M}}_{0,n}}$ is the identity in the cohomology ring of $\overline{\mathcal{M}}_{0,n}$.

CONJECTURE. *The series Φ is absolutely convergent in a neighborhood $\mathcal{U}$ of 0 in $\mathcal{H}$, if the cohomology class $[\omega]\in H^2(V,\mathbf{R})$ is sufficiently positive.*

Without assuming the validity of this conjecture we can work not over the field $\mathbf{C}$ but over the field of fractions of the semigroup ring $\mathbf{Q}[B]$, where B is the semigroup generated by classes β such that $\int_\beta \omega' \geq 0$ for all symplectic froms ω' close to ω. Other homology classes are excluded because they cannot be represented by pseudo-holomorphic curves.

The function Φ in its definition domain $\mathcal{U}$ satisfies a system of nonlinear differential equations of the third order (due to R. Dijkgraaf, E. Verlinde, H. Verlinde, and E. Witten, see [W2]). Let us choose a basis x_i of the space $\mathcal{H}$ and denote by x^i the corresponding coordinate system on $\mathcal{H}$. Denote by (g_{ij}) the matrix of the Poincaré pairing, $g_{ij} := \int_V x_i \wedge x_j$, and by (g^{ij}) the inverse matrix. For all i,j,k,l, we have (modulo appropriate sign corrections for odd-degree classes):

$$\sum_{m,m'} \frac{\partial^3\Phi}{\partial x^i \partial x^j \partial x^m} g^{mm'} \frac{\partial^3\Phi}{\partial x^k \partial x^l \partial x^{m'}} = \sum_{m,m'} \frac{\partial^3\Phi}{\partial x^i \partial x^k \partial x^m} g^{mm'} \frac{\partial^3\Phi}{\partial x^j \partial x^l \partial x^{m'}} .$$

This equation can be reformulated as the condition of associativity of the algebra given by the structure constants $A_{ij}^k := \sum_{k'} g^{kk'} \partial_{ijk'}\Phi$. In invariant terms it means that Φ defines a supercommutative associative multiplication on the tangent bundle to $\mathcal{H}$ (the quantum cohomology ring).

The associativity equation follows from the splitting axiom and from a certain linear relation among components of the compactification divisor of $\overline{\mathcal{M}}_{0,n}$. Denote by D_S for $S\subset\{1,\dots,n\}$, $2\leq \#S\leq n-2$, the divisor in $\overline{\mathcal{M}}_{0,n}$, which is the closure of the moduli of stable curves $(C;p_1,\dots,p_n)$ consisting of two irreducible components C_1, C_2 such that $p_i\in C_1$ for $i\in S$ and $p_i\in C_2$ for $i\notin S$.

LEMMA. *We have the following identity in $H^2(\overline{\mathcal{M}}_{0,n},\mathbf{Z})$*

$$\sum_{\substack{S:1,2\in S\\ 3,4\notin S}} [D_S] = \sum_{\substack{S:1,3\in S\\ 2,4\notin S}} [D_S] .$$

Both sides in the equality above are pullbacks under the forgetful map $\overline{\mathcal{M}}_{0,n} \to \overline{\mathcal{M}}_{0,4}$ of points $D_{\{1,2\}}$, $D_{\{1,3\}} \in \overline{\mathcal{M}}_{0,4} \simeq \mathbf{P}^1$. It is clear that any two points on $\mathbf{P}^1$ are rationally equivalent as divisors.

Conversely, one can show using the splitting axiom that one can reconstruct the whole system of genus-zero GW-classes starting from Φ. The equation of the associativity is a necessary and sufficient condition for the existence of such reconstruction (see [KM]).

The associativity equation was studied by Dubrovin [D]. He discovered that it is a completely integrable system in many cases (but not for CY manifolds). For example, for $V \simeq \mathbf{P}^2$ the associativity equation is equivalent to the Painlevé VI equation. It means that via a simple recursion formula we can compute the number of rational curves of degree d in the projective plane passing through generic $3d-1$ points.

Notice that by dimensional reasons, the associativity equation is an empty condition for 3-dimensional Calabi-Yau manifolds, because the virtual dimension of the space of rational curves is zero, curves do not intersect each other, and the degeneration argument is inapplicable.

Let us introduce a connection on the tangent bundle $T_{\mathcal{U}}$ by the formula $\nabla = \nabla_{0|\mathcal{U}} + A$, where ∇_0 is the standard connection of the affine space $\mathcal{H}$. The associativity equation implies the flatness of ∇.

Variations of Hodge structures

Suppose that $c_1(V) = 0$, and V carries at least one integrable complex structure compatible with ω such that $\mathrm{H}^{2,0}(V) = 0$. For any such complex structure we have a Hodge decomposition $\oplus \mathrm{H}^k(V, \mathbf{C}) = \oplus \mathrm{H}^{p,q}$. We expect that all cycles $I_{g,n;\beta}$ are Hodge cycles of (complex) dimension equal to $(n + \dim_{\mathbf{C}} V - 3)$. It follows that the restriction of ∇ to the convergence domain of the series Φ in the second cohomology group:

$$\mathcal{U}^{\mathrm{cl}} := \mathcal{U} \cap \mathrm{H}^2(V, \mathbf{C}) \subset \mathrm{H}^2(V, \mathbf{C}) = \mathrm{H}^{1,1}$$

maps $\mathrm{H}^{p,q} \otimes \mathcal{O}(\mathcal{U}^{\mathrm{cl}})$ to $\mathrm{H}^{p+1,q+1} \otimes \Omega^1(\mathcal{U}^{\mathrm{cl}})$. We call $\mathcal{U}^{\mathrm{cl}}$ the classical moduli space because it is locally isomorphic to a complexification of the moduli space of symplectic structures on V.

We introduce filtrations $\oplus_{p \leq p_0} \mathrm{H}^{p,q}$ on trivial bundles over $\mathcal{U}^{\mathrm{cl}}$ with fibers equal to $\oplus_{p-q \text{ is fixed}} \mathrm{H}^{p,q}$. Hence, we have flat connections and filtrations on holomorphic vector bundles over a complex manifold satisfying the Griffiths transversality conditions. We call such data a *complex* variation of pure Hodge structures. One can prove by using formal arguments with Hodge-Tate groups that the equivalence classes of such complex variations of pure Hodge structures do not change under deformations of the complex structure on V.

For general symplectic manifolds V with $c_1(V) = 0$ we can consider just the two trivial vector bundles $\mathrm{H}^{\mathrm{even}}$ and $\mathrm{H}^{\mathrm{odd}}$ on $\mathcal{U}^{\mathrm{cl}} := \mathcal{U} \cap \mathrm{H}^2(V, \mathbf{C}) \subset \mathcal{H}$ endowed with the flat connection induced from ∇ and the filtration by subbundles $\oplus_{k \leq k_0} \mathrm{H}^k$.

Algebro-geometric complex variations of pure Hodge structures are defined as subquotients of variations of pure Hodge structures on cohomology groups of complex projective algebraic manifolds depending algebraically on parameters.

MIRROR CONJECTURE. *Complex variations of pure Hodge structures constructed using Gromov-Witten invariants of symplectic manifold V as above are locally equivalent to algebro-geometric variations.*

In almost all known examples such variations of Hodge structures should be locally equivalent to variations of Hodge structures on total cohomology bundles of a mirror family of complex manifolds with $c_1 = 0$. Exceptions come mostly from CY-manifolds V such that $\dim \mathrm{H}^1(V, T_V) = 0$, i.e. rigid manifolds. In this case dual manifolds with rotated Hodge diamond could not exist, because $\dim \mathrm{H}^1(W, \Omega^1_W) \neq 0$ for Kähler manifolds. Physicists proposed as candidates certain substructures of Hodge structures on cohomology groups of Fano varieties (=algebraic manifolds with an ample anti-canonical bundle). Also, calculations of numbers of curves on projective spaces suggest that in general there exists some relation between the pre-potential of *non* Calabi-Yau manifolds and algebro-geometric variations of Hodge structures.

In the case of a quintic V in $\mathbf{P}^4$ the function Φ is the sum of two terms: the contribution of maps to points of V and the contribution of rational curves in V (and their multiple covers). We introduce coordinates t^i, $i = 0, 1, 2, 3$, in 1-dimensional spaces $\mathrm{H}^{i,i}(V)$ and odd coordinates ξ^j, η^j, $j = 1, \ldots, 102$, in $\mathrm{H}^3(V, \mathbf{C})$. In these coordinates we have (modulo adding a polynomial of degree 2)

$$\Phi(t^i, \xi^j, \eta^j) = \frac{5}{6} \sum_{i+j+k=3} t^i t^j t^k + t^0 \sum_j \xi^j \eta^j + \sum_{d \geq 1} N_d^{\mathrm{virt}} \exp(dt^1) .$$

One can deduce an example from [COGP] from this formula.

The flat coordinates x^i on the moduli space of complex structures on dual manifolds are equal to the ratios of periods $\left(\int_{\gamma_i} \Omega\right) / \left(\int_{\gamma_0} \Omega\right)$, where Ω is a holomorphic volume element on the mirror manifold W and γ_i are elements of $H_*(W, \mathbf{C})$ locally constant with respect to the Gauss-Manin connection on the homology bundle.

There exists a generalization of the mirror correspondence to higher genera. First of all, the dimension formula for degrees of Gromov-Witten classes shows that one can expect a nonnegative dimension for the space of genus g curves for Calabi-Yau varieties V only in the following cases:

(1) $g = 0$ and an arbitrary dimension $n := \dim V$ (this is what we have described right now),
(2) $g = 1$ and arbitrary n,
(3) $g \geq 2$ and $n \leq 3$.

The Harvard group of physicists in the remarkable paper [BCOV] proposed a procedure ("quantum Kodaira-Spencer theory") giving numbers of curves for cases $g = 1$ or $n = 3$. It relates GW-invariants with certain structures on the moduli of dual varieties, which are more complicated than just variations of Hodge structures and are not understood mathematically yet. The example of Dijkgraaf (elliptic curves) is a 1-dimensional version of this theory.

In the rest of this paper we give an outline of a program relating mirror symmetry to general structures of homological algebra. The central ingredient here is a fundamental construction of Fukaya based on ideas of Donaldson, Floer, and Segal.

Extended moduli spaces

When we restrict the flat bundle $T_{\mathcal{U}}$ to the subspace $\mathcal{U}^{cl} = \mathrm{H}^2(V, \mathbf{C})$, much information will be lost. It seems very reasonable to extend the moduli space of symplectic structures to the whole domain $\mathcal{U}$ in $\mathcal{H}$ in which the potential Φ is defined. Hence, the tangent space to the extended moduli space at classical points $\mathcal{U}^{cl}$ should be equal to $\mathcal{H} = \oplus \mathrm{H}^k$.

Now we want to construct an extended moduli space $\mathcal{M}$ for a complex Calabi-Yau W containing the ordinary moduli space of complex structures on W. The natural candidate for the tangent bundle to $\mathcal{M}$ at classical points $\mathcal{M}^{cl} :=$ moduli (W) should be equal to the direct sum $\oplus \mathrm{H}^p(W, \wedge^q T_W)$. The problem of constructing $\mathcal{M}$ was already dicussed by E. Witten (see [W3]).

We anticipate that $\oplus \mathrm{H}^p(W, \wedge^q T_W)$ can be interpreted as the total Hochschild cohomology of the sheaf $\mathcal{O}_W$ of algebras of holomorphic functions on W.

For an algebra A/k over a field its Hochschild cohomology $\mathrm{HH}^*(A) = \mathrm{H}^*(A, A)$ is defined as $\mathrm{Ext}^*_{A-\mathrm{mod}-A}(A, A)$. The second Hochschild cohomology $\mathrm{HH}^2(A)$ classifies infinitesimal deformations of A. Notice that each A-bimodule M defines a functor from the category of A-modules into itself:

$$M \otimes_A : \ A - \mathrm{mod} \longrightarrow A - \mathrm{mod}\,, \quad N \mapsto M \otimes_A N$$

and A corresponds to the identity functor $\mathrm{Id}_{A-\mathrm{mod}}$.

Analogously, we define the Hochschild cohomology of the structure sheaf $\mathcal{O}_W$ of a scheme W over k (or of an analytic space) as the global Ext-functor

$$\mathrm{HH}^*(\mathcal{O}_W) := \mathrm{Ext}^*_{W \times W}\left(\delta_*(\mathcal{O}_W), \delta_*(\mathcal{O}_W)\right)\,,$$

where $\delta : W \hookrightarrow W \times W$ is the diagonal embedding. Another definition of the Hochschild cohomology for algebraic varieties (in fact, equivalent to ours) was proposed by Gerstenhaber and Schack [GS]. The following fact proven in hidden form in [GS] seems to be new in algebraic geometry:

THEOREM. *For smooth (and not necessarily compact) variety W over a field of characteristic zero there is a canonical isomorphism*

$$HH^n(\mathcal{O}_W) \simeq \bigoplus_{k+l=n} H^k(W, \bigwedge^l T_W)\,.$$

For smooth W the second Hochschild cohomology $\mathrm{HH}^2(\mathcal{O}_W)$ splits into the direct sum of ordinary first-order deformations $\mathrm{H}^1(W, T_W)$, noncommutative deformations $\mathrm{H}^0(W, \bigwedge^2 T_W)$ of the sheaf $\mathcal{O}_W$ of associative algebras (global Poisson brackets on W), and a slightly more mysterious piece $\mathrm{H}^2(W, \mathcal{O}_W)$. This third part can be interpreted as locally trivial first-order deformations of the sheaf of abelian categories of $\mathcal{O}_W$-modules.

In the next section we will propose an interpretation of the total Hochschild cohomology as the tangent space to "extended moduli space" $\mathcal{M}$ containing the classical moduli space $\mathcal{M}^{\mathrm{cl}}$ as a part.

A_∞-algebras and categories

A_∞-algebras were introduced by Stasheff in 1963 (see [S]). Let $A = \oplus A^k$ be a **Z**-graded vector space. The structure of the A_∞-algebra on A is an infinite sequence of linear maps $m_k : A^{\otimes k} \to A$, $k \geq 1$, $\deg m_k = 2 - k$, satisfying the (higher) associativity conditions:

(1) $m_1^2 = 0$ (we can consider m_1 as a differential and (A, m_1) as a complex),
(2) $m_1(m_2(a \otimes b)) = m_2(m_1(a) \otimes b) \pm m_2(a \otimes m_1(b))$, ($m_2$ is a morphism of complexes),
(3) $m_3(m_1(a)\otimes b\otimes c) \pm m_3(a\otimes m_1(b)\otimes c) \pm m_3(a\otimes b\otimes m_1(c)) \pm m_1(m_3(a\otimes b\otimes c)) = m_2(m_2(a\otimes b)\otimes c) - m_2(a\otimes m_2(b\otimes c))$, ($m_2$ is associative up to homotopy),
(4) and so on ... .

In one sentence one can define the A_∞-algebra structure on A as a co-derivation in the graded sense d, $d^2 = 0$ of degree 1 on the co-free co-associative co-algebra without a co-unit co-generated by the **Z**-graded vector space $A[1]$, $A[1]^k := A^{k+1}$.

A morphism of A_∞-algebras (from A to B) is an infinite collection of linear maps $A^{\otimes k} \to B$, $k \geq 1$, satisfying some equations analogous to the defining equations for individual A_∞-algebras. In terms of co-free co-algebras it is the same as a differential graded homomorphism. A homotopy equivalence of A_∞-algebras is a morphism whose linear part induces an isomorphism of cohomology groups with respect to the differential m_1.

In general, A_∞-algebras are closely related to differential graded algebras. Namely, a dg-algebra is the same as an A_∞-algebra with $m_3 = m_4 = \cdots = 0$. Conversely, for an A_∞-algebra A one can construct using the bar-construction a differential graded algebra B homotopy equivalent to A.

An additive category over a field k is a category C with finite direct sums such that all sets of morphisms $\mathrm{Hom}_C(X,Y)$ are endowed with structure of vector spaces over k and where the composition of morphisms is a bilinear map. In a sense, one can approximate additive categories by algebras of endomorphisms of their objects. Analogously, one can define a differential graded category as an additive category with the structure of complexes on $\mathrm{Hom}_C(X,Y)$ such that the composition is a morphism of complexes.

An A_∞-category C is a collection of objects and **Z**-graded spaces of morphisms $\mathrm{Hom}_C(X,Y)$ for each two objects endowed with higher compositions of morphisms satisfying relations parallel to the defining relations of A_∞-algebras. We require the existence of identity morphisms $\mathrm{Id}_X \in \mathrm{Hom}_C(X,X)$, which obey the usual properties of identity for composition m_2 and vanish under substitution in other (higher) compositions. We can also require the existence of finite direct sums in C in an obvious sense. Notice that C is not a category in general, because the composition of morphisms is not associative. Nevertheless, one can construct an additive category $H(C)$ from C with the same class of objects by defining new **Z**-graded spaces of morphisms as

$$\mathrm{Hom}_{H(C)}(X,Y) := \frac{\mathrm{Ker}\ (m_1 : \mathrm{Hom}^0_C(X,Y) \to \mathrm{Hom}^1_C(X,Y))}{\mathrm{Im}\ (m_1 : \mathrm{Hom}^{-1}_C(X,Y) \to \mathrm{Hom}^0_C(X,Y))} .$$

There exists a generalization of Hochschild cohomology to the case of A_∞-algebras. The meaning of $\mathrm{HH}^*(A)$ for $* > 0$ is the space of equivalence classes of first-order deformations of A_∞-structure on A over $\mathbf{Z}$-graded bases. We hope that there exists an appropriate version of Hochschild cohomology for some good class of A_∞-categories as well.

One can show under some mild assumptions that if the formal $\mathbf{Z}$-graded moduli space $\mathcal{M}$ of A_∞-categories is smooth then there exists the canonical structure of an associative commutative algebra on the tangent bundle to $\mathcal{M}$. In the case of an A_∞-category consisting just of one object X with morphisms $\mathrm{Hom}_C(X, X)$ consisting of an associative algebra A in degreee 0, the product in Hochschild cohomology (i.e., in the tangent space to $\mathcal{M}$)

$$\mathrm{HH}^*(A) := \mathrm{Ext}^*_{A-\mathrm{mod}-A}(A, A)$$

coincides with the usual Yoneda composition of Ext-groups.

Triangulated categories

One of the fundamental tools in homological algebra is the triangulated category $\mathcal{D}(C)$ associated with an abelian category C satisfying certain conditions (Verdier, see [V]). A triangulated category is an additive category endowed with a shift functor and a class of so-called exact triangles, obeying a complicated list of axioms. For example, for C equal to the category of A-modules, where A is an associative algebra, the category $\mathcal{D}(C)$ is equivalent to the category whose objects are complexes of free A-modules and whose morphisms are equal to homotopy classes of differential graded morphisms of degree 0:

$$\mathrm{Hom}_{D(C)}(X, Y) := \mathrm{H}^0(\bigoplus_k \prod_j \mathrm{Hom}_C(X^j, Y^{j+k})) \ .$$

The bounded derived category $\mathcal{D}^b(C)$ is the full subcategory of $\mathcal{D}(C)$ consisting of complexes of A-modules with nonvanishing cohomology groups only in finitely many degrees.

The shift functors at the level of objects just shifts the degree of complexes: $X \longrightarrow X[n]$, $X[1]^k = X^{k+n}$, and $(X[n])[m] = X[n+m]$, $X[0] = X$.

We will not describe Verdier's axiomatics of exact triangles here because it does not look completely satisfactory, although it was generally adopted and widely used. A certain improvement of axioms was proposed by Bondal and Kapranov in [BK]. The main ingredient in their definition is the notion of a twisted complex in a differential graded category.

We can extend the construction of [BK] to the case of A_∞-categories. We assume that an A_∞-category C is endowed with shift functors such that

$$\mathrm{Hom}_C(X[i], Y[j]) = \mathrm{Hom}_C(X, Y)\,[j-i] \ .$$

By definition, a (one-sided) *twisted complex* is a family $(X^{(i)})_{i\in\mathbf{Z}}$ of objects of an A_∞-category C such that $X^{(i)} = 0$ for almost all i together with a collection of morphisms $d_{ij} \in \mathrm{Hom}_C(X^{(i)}, X^{(j)})^{j-i}$ for $i < j$ obeying a generalization of the Maurer-Cartan equation:

$$\text{for fixed } i,\, j \qquad \sum_{\substack{k;i_0,\dots,i_k\\ i_0=i,i_k=j}} m_k(f_{i_0,i_1},\dots,f_{i_{k-1},i_k}) = 0 \ .$$

We define the **Z**-graded space of morphisms between twisted complexes X and Y as

$$\bigoplus_{k,j} \mathrm{Hom}_C(X^{(j)}, Y^{(j+k)})[-k] \ .$$

Using higher compositions in C one can define the structure of an A_∞-category on {twisted complexes of C}. Any higher composition of morphisms of twisted complexes is defined as the sum over all possible products that one can imagine.

One can check without difficulties that the derived category

$$\mathcal{D}^b(C) := H(\text{twisted complexes of } C)$$

satisfies the Verdier axioms for triangulated category.

Fukaya's A_∞-category

In this section we describe a remarkable contsruction of Fukaya [F] with a few minor modifications.

Let V be a closed symplectic manifold with $c_1(T_V) = 0$.

Denote by LV the space of pairs (x, L), where x is a point of V and L is a Lagrangian subspace in T_xV. The space LV is fibered over V with fibers equal to Lagrangian Grassmanians. Thus, the fundamental group of the fibers is isomorphic to **Z**.

The condition on V posed above guarantees that there exists a **Z**-covering $\widetilde{LV}$ of LV inducing a universal cover of each fiber. Let us fix $\widetilde{LV}$.

Objects of Fukaya's category $F(V)$ are Lagrangian submanifolds $\mathcal{L} \subset V$ endowed with a continuous lift of the evident map $\mathcal{L} \longrightarrow LV$ to a map $\mathcal{L} \longrightarrow \widetilde{LV}$. In fact, it is only a first approximation to right objects (see remarks after the definition). For subvarieties $\mathcal{L}_1, \mathcal{L}_2$ intersecting each other transversally at a point $x \in V$ and endowed with lifts to $\widetilde{LV}$, we can define the Maslov index $\mu_x(\mathcal{L}_1, \mathcal{L}_2) \in$ **Z**. Notice that

$$\mu_x(\mathcal{L}_1, \mathcal{L}_2) + \mu_x(\mathcal{L}_2, \mathcal{L}_1) = n := \frac{1}{2}\dim(V) \ .$$

Fukaya defines the space of morphisms $\mathrm{Mor}_F(\mathcal{L}_1, \mathcal{L}_2)$ only if $\mathcal{L}_1, \mathcal{L}_2$ intersect transversally:

$$\mathrm{Hom}_F(\mathcal{L}_1, \mathcal{L}_2) := \mathbf{C}^{\mathcal{L}_1 \cap \mathcal{L}_2}$$

with **Z**-grading coming from the Maslov index.

The differential in $\mathrm{Hom}_F(\mathcal{L}_1, \mathcal{L}_2)$ is a version of Floer's differential. Its matrix coefficient associated with two intersection points $p_1, p_2 \in \mathcal{L}_1 \cap \mathcal{L}_2$ is defined as

$$\sum_{\phi: D^2 \longrightarrow V} \pm \exp(-\text{area of } D^2) \, ,$$

where $\phi : D^2 \longrightarrow V$ is a pseudo-holomorphic map from the standard disc $D^2 := \{z \mid |z| \leq 1\} \in \mathbf{C}$ to V such that $\phi(-1) = p_1$, $\phi(+1) = p_2$, and

$$\phi(\exp(i\alpha)) \in \mathcal{L}_1 \text{ for } 0 < \alpha < \pi, \quad \phi(\exp(i\alpha)) \in \mathcal{L}_2 \text{ for } \pi < \alpha < 2\pi \ .$$

More precisely, we consider *equivalence classes* of maps ϕ modulo the action of the group of holomorphic automorphisms of D^2 stabilizing points 1 and -1:

$$\mathbf{R}_+^\times \subset PSL(2, \mathbf{R}) = \text{ Aut } (D^2) \ .$$

The area of D^2 with respect to the pullback of ω depends only on the homotopy type of $\phi \in \pi_2(V, \mathcal{L}_1 \cup \mathcal{L}_2)$. One expects that for sufficiently large ω the infinite series is absolutely convergent.

The sign $\pm$ in the definition of the Floer differential comes from a natural orientation of the space of pseudo-holomorphic maps. One expects that there will be finitely many such maps for a generic almost-complex structure on V if $\mu_{p_2} - \mu_{p_1} = 1$. Presumably, one can develop a general technique of stable maps for surfaces with boundaries or extend Ruan-Tian's methods.

Analogously, one can define higher order compositions using zero-dimensional components of spaces of equivalence classes modulo $PSL(2, \mathbf{R})$-action of maps ϕ from the standard disc D^2 to V with the boundary $\phi(\partial D^2)$ sitting in a union of Lagrangian subvarieties. More precisely, if $\mathcal{L}_1, \ldots, \mathcal{L}_{k+1}$ are Lagrangian submanifolds intersecting each other transversally and $p_j \in \mathcal{L}_j \cap \mathcal{L}_{j+1}$, $j = 1, \ldots, k$, are chosen intersection points, then we define the composition of corresponding base elements in spaces of morphisms as

$$m_k(p_1, \ldots, p_k) := \sum_{\substack{\phi: D^2 \longrightarrow V,\, q \in \mathcal{L}_1 \cap \mathcal{L}_{k+1} \\ 0 = \alpha_0 < \alpha_1 < \cdots < \alpha_k < \alpha_{k+1} = 2\pi}} \pm \exp\left(- \int_{D^2} \phi^* \omega \right) q \in \text{Hom}_F(\mathcal{L}_1, \mathcal{L}_{k+1}),$$

where $\phi(\exp(i\alpha)) \in \mathcal{L}_j$ for $\alpha_{j-1} < \alpha < \alpha_j$ and $\phi(\exp(i\alpha_j)) = p_j$ for $j = 1, \ldots, k+1$; $p_{k+1} := q$. Again, we expect that there exist only finitely many equivalence classes modulo the action of Aff$(\mathbf{R}) = \text{Stab}_{1 \in D^2} \subset PSL(2, \mathbf{R})$ of such maps in each homotopy class if $\mu_q = \mu_{p_1} + \cdots + \mu_{p_k} + 2 - k$ and the infinite series in the definition of m_k converges absolutely.

Fukaya claims that the identities of the A_∞-category follow from considerations analogous to the proof of the associativity equations in the case of rational curves. He also claims that it is possible to extract an actual A_∞-category with compositions of all morphisms using an appropriate notion of a "generic" Lagrangian manifold. In particular, it is possible to restore the identity morphisms. The main idea is that two Lagrangian submanifolds obtained one from another by a Hamiltonian flow are equivalent with respect to the Floer cohomology.

There is an extension of Fukaya's category. We can consider pairs consisting of a Lagrangian submanifold $\mathcal{L}$ and a unitary local system $\mathcal{E}$ on $\mathcal{L}$ as objects of a new A_∞-category. Morphism spaces will be defined as

$$\text{Hom}_F((\mathcal{L}_1, \mathcal{E}_1), (\mathcal{L}_2, \mathcal{E}_2)) := \bigoplus_{p \in \mathcal{L}_1 \cap \mathcal{L}_2} \text{Hom}\,(\mathcal{E}_{1|p}, \mathcal{E}_{2|p}) \ .$$

In the definition of higher composition we add a new factor to each term equal to the trace of the composition of holonomy maps along the boundary of D^2. Unitarity in this definition is an obligatory condition, otherwise the series defining higher compositions will be inevitably divergent.

It seems that there are further possible extensions of Fukaya's A_∞-category. One can consider as objects Lagrangian foliations, families of Lagrangian submanifolds parametrized by closed oriented manifolds, etc.

Homological mirror conjecture

We propose here a conjecture in slightly vague form, which should imply the "numerical" Mirror conjecture. Let (V, ω) be a $2n$-dimensional symplectic manifold with $c_1(V) = 0$ and W be a dual n-dimensional complex algebraic manifold.

The derived category constructed from the Fukaya category $F(V)$ (or a suitably enlarged one) is equivalent to the derived category of coherent sheaves on a complex algebraic variety W.

More precisely, we expect that there is an embedding of $\mathcal{D}^b(F(V))$ as a full triangulated subcategory into $\mathcal{D}^b(\mathrm{Coh}(W))$. We have the following evidence for that.

(1) By the general philosophy, A_∞-deformations of first order of $F(V)$ should correspond to Ext-groups in a category of functors $F(V) \to F(V)$. The natural candidate for such a category is $F(V \times V)$, where the symplectic structure on $V \times V$ is $(\omega, -\omega)$. The diagonal $V_{\mathrm{diag}} \subset V \times V$ is a Lagrangian submanifold and it corresponds to the identity functor. By a version of Floer's theorem (see [F]) there is a canonical isomorphism between the Floer cohomology $\mathrm{H}^*(\mathrm{Hom}_{F(V\times V)}(V_{\mathrm{diag}}, V_{\mathrm{diag}}))$ and the ordinary topological cohomology $\mathrm{H}^*(V, \mathbf{C})$. The Yoneda product on the Floer cohomology considered as Ext-groups arises from holomorphic maps from D^2 with 3 marked points on ∂D^2 to $V \times V$ with a boundary on V_{diag}. Such maps are the same as holomorphic maps to V from the 2-dimensional sphere $S^2 \simeq \mathbf{C}P^2$ with 3 marked points. Thus, it seems very reasonable to expect that we will get exactly the quantum cohomology product on $\mathrm{H}^*(V)$. We expect that the equivalence of derived categories will imply numerical predictions.

(2) Lagrangian varieties (and local systems on it) form a natural class of local boundary conditions for the A-model in topological open string theory. Also, holomorphic vector bundles form local boundary conditions for the B-model (Witten [W4]). Physicists believe that the whole string theories on dual varieties are equivalent. Thus, we want to say that topological open string theory is more or less the same as a triangulated category.

(3) Both categories $\mathcal{D}^b(F(V))$ and $\mathcal{D}^b(\mathrm{Coh}(W))$ possess a duality: a functorial isomorphism $(\mathrm{Hom}(X, Y))^* \simeq \mathrm{Hom}(Y, X[n])$. On the algebro-geometric side it is Serre duality. For Fukaya's category the definition of compositions is cyclically symmetric. The duality follows from this symmetry and from the idendity $\mu_x(\mathcal{L}_1, \mathcal{L}_2) + \mu_x(\mathcal{L}_2, \mathcal{L}_1) = n$. We developed some time ago a theory of A_∞-algebras with duality in [K1] and proposed a combinatorial

construction of cohomology classes of the moduli spaces of smooth curves $\mathcal{M}_{g,n}$ based on such algebras. This construction has a generalization to an A_∞-category with a duality. Thus, we expect that the Gromov-Witten invariants could be defined in a general purely algebraic situation. We still do not know what is missed in algebraic structures and how to define classes with values in $H^*(\overline{\mathcal{M}}_{g,n}, \mathbf{C})$.

A mirror complex manifold W is usually not unique. For example, one cannot distinguish the B-models on dual Abelian varieties A, A'. It is compatible with our picture because the derived categories of coherent sheaves on A and A' are equivalent via the Fourier-Mukai transform. Also, the B-models on birationally equivalent Calabi-Yau manifolds W, W' are believed to be isomorphic. In all known examples the Hodge structures on total cohomology depend only on a birational type. Thus, we expect that the derived categories of coherent sheaves on W and on W' are equivalent.

Our conjecture, if it is true, will unveil the mystery of mirror symmetry. The numerical predictions mean that two elements of an uncountable set (formal power series with integral coefficients) coincide. Our homological conjecture is equivalent to the coincidence in a *countable* set (connected components of the "moduli space of triangulated categories", whatever it means).

In the last section we show what our program looks like in the simplest case of mirror symmetry.

2-dimensional tori: A return

Let Σ be the standard flat 2-dimensional torus $S^1 \times S^1$ endowed with a symplectic form ω proportional to the standard volume element. Let $\mathcal{L}_1, \mathcal{L}_2, \mathcal{L}_3$ be three simple closed geodesics from pairwise different homology classes and

$$p_1 \in \mathcal{L}_1 \cap \mathcal{L}_2, \ \ p_2 \in \mathcal{L}_2 \cap \mathcal{L}_3, \ \ p_3 = q \in \mathcal{L}_1 \cap \mathcal{L}_3$$

be three intersection points. We will compute now the tensor coefficient of the composition m_2 corresponding to the base vectors p_1, p_2, p_3. Each map ϕ from D^2 to Σ can be lifted to a map $\tilde{\phi}$ from D^2 to $\mathbf{R}^2$ = the universal covering space of Σ. The preimages of circles $\mathcal{L}_i$ on $\mathbf{R}^2$ form three families of parallel straight lines. Thus, the images of lifted maps $\tilde{\phi}$ are triangles with sides on these lines. It is easy to see that the equivalence classes of triangles modulo the action of $\mathbf{Z}^2 = \pi_1(\Sigma)$ are labeled by terms of an arithmetic progression (the lengths of sides of triangles sitting on the pullback of $\mathcal{L}_1$). The areas of triangles are proportional to the squares of elements of this progression. The tensor element of composition m_2 can be written naturally as

$$\sum_{n \in \mathbf{Z}} \exp\left(-(an+b)^2\right)$$

for some real parameters $a \neq 0$ and b, which is a value of the classical θ-function. The associativity equation is equivalent to the standard bilinear identity for θ-functions. It is well known that θ-functions form natural bases of spaces of global sections of line bundles over elliptic curves.

It seems very plausible that the triangulated category constructed from the Fukaya category $F(\Sigma)$ enlarged using unitary local systems is equivalent to the bounded derived category of coherent sheaves on the elliptic curve with the real modular parameter $\tau := \exp(-\text{ area of } (\Sigma))$.

Note added in proof:
After preparing the text of this lecture I realized that there is a serious flaw in Fukaya's preprint. It seems that he forgot about a certain stratum in the Gromov compactification, which usually produces obstructions to the vanishing of the square of the Floer differential (compare with [O]). Algebraically, it means that we have elements m_0 of degree 2 in morphism spaces and the axioms of A_∞-categories are modified slightly. This defect is very unpleasant, but it does not appear in our simple example of the torus or in the case of the diagonal in the square of a symplectic manifold.

References

[AM] P. S. Aspinwall and D. R. Morrison, *Topological field theory and rational curves*, Comm. Math. Phys. **151** (1993), 245–262.

[BCOV] M. Bershadsky, S. Cecotti, H. Ooguri, and C. Vafa, *Kodaira-Spencer theory of gravity and exact results for quantum string amplitudes*, Comm. Math. Phys. **165** (1994), 311–427.

[BK] A. I. Bondal and M. M. Kapranov, *Enhanced triangulated categories*, Math. USSR-Sb. **70** (1991), 93–107.

[COGP] P. Candelas, X. de la Ossa, P. S. Green, and L. Parkes, *A pair of Calabi-Yau manifolds as an exactly soluble superconformal theory*, in [Y], pp. 31–95.

[D] B. Dubrovin, *Integrable systems in topological field theory*, Nuclear Phys. B **379** (1992), 627–685.

[F] K. Fukaya, *Morse homotopy, A^∞-category and Floer homologies*, MSRI preprint No. 020-94 (1993).

[GS] M. Gerstenhaber and S. D. Schack, *Algebraic cohomology and deformation theory*, NATO ASI Ser. C (1988), 11–264.

[G] M. Gromov, *Pseudo holomorphic curves in symplectic manifolds*, Invent. Math. **82** (1985), 307–347.

[K1] M. Kontsevich, *Formal (non)-commutative symplectic geometry*, The Gelfand Mathematical Seminars, 1990–1992 (L. Corwin, I. Gelfand, and J. Lepowsky, eds.), Birkhäuser, Basel and Boston, 1993, pp. 173–187.

[K2] M. Kontsevich, *Enumeration of rational curves via torus actions*, MPI preprint and hep-th/9405035 (1994).

[KM] M. Kontsevich and Yu. Manin, *Gromov-Witten classes, quantum cohomology and enumerative geometry*, Comm. Math. Phys. **164** (1994), 525–562.

[M] D. R. Morrison, *Mirror symmetry and rational curves on quintic threefolds: A guide for mathematicians*, J. Amer. Math. Soc. **6** (1993), 223–247.

[O] Y.-G. Oh, *Floer Cohomology of Lagrangian intersections and pseudo-holomorphic discs I,II*, Comm. Pure Appl. Math. **46:7** (1993), 949–1012.

[RT] Y. Ruan and G. Tian, *Mathematical theory of quantum cohomology*, Math. Res. Lett. **1** (1994), 269–278.

[S] J. Stasheff, *On the homotopy associativity of H-spaces I, II*, Trans. Amer. Math. Soc. **108** (1963), 275–312.

[V] J. L. Verdier, *Catégories dérivées. Quelques resultats (état 0)*, Séminaire de Géométrie Algébraique du Bois-Marie (SGA $4\frac{1}{2}$), Lecture Notes in Math., vol. 569, Springer-Verlag, Berlin and New York, 1977, pp. 262–311.

[W1] E. Witten, *Topological sigma models*, Comm. Math. Phys. **118** (1988), 411–449.

[W2] E. Witten, *Two-dimensional gravity and intersection theory on moduli space*, Surveys Differential Geom. **1** (1991), 243–310.

[W3] E. Witten, *Mirror manifolds and topological field theory*, in [Y], pp. 120–159.

[W4] E. Witten, *Chern-Simons gauge theory as a string theory*, preprint IASSNS-HEP-92/45 and hep-th/9207094.

[Y] S. T. Yau, ed., Essays on Mirror Manifolds, International Press Co., Hong Kong, 1992.

On Some Recent Methods for Nonlinear Partial Differential Equations

PIERRE-LOUIS LIONS

CEREMADE, Université Paris-Dauphine,
Place de Lattre de Tassigny, F-75775 Paris Cedex 16, France

Dedicated to the memory of Ron DiPerna (1947–1989)

1 Introduction

We wish to present here some aspects of a few general methods that have been introduced recently in order to solve nonlinear partial differential equations and related problems in nonlinear analysis.

As is well known, nonlinear partial differential equations have become a rather vast subject with a long history of deep and fruitful connections with many other areas of mathematics and various sciences like physics, mechanics, chemistry, engineering sciences, etc. And we shall not pretend to make any attempt at surveying all recent activities in that field. Also, we shall concentrate on rather theoretical issues leaving completely aside more applied issues such as mathematical modelling, numerical questions that go hand in hand in a fundamental way with the theories. For a discussion of the interaction between nonlinear analysis and modern applied mathematics, we refer the reader to the report by Majda [56] in the preceding Congress.

We shall mainly discuss here recent methods that have been developed recently for the analysis of the major mathematical models of gas dynamics (and compressible fluid mechanics), namely the Boltzmann equation and compressible Euler and Navier-Stokes equations (essentially in the so-called "isentropic regime"). These methods include velocity averaging, regularization by collisions that we shall apply to the solution of the Boltzmann equation (Section 2 below), and compactness via commutators and in particular compensated compactness, which we illustrate on isentropic compressible Euler and Navier-Stokes equations.

This selection of topics (equations and methods) is by no means an exhaustive treatment of all the exciting progresses that have taken place recently in nonlinear partial differential equations: many more important problems have been investigated — see for instance the various reports in this Congress related to Nonlinear Partial Differential Equations — and other methods and theories have been developed. We briefly mention a few in Section 4. And even for the methods that we describe here, much more could be said in particular about applications to other relevant problems.

Proceedings of the International Congress
of Mathematicians, Zürich, Switzerland 1994

We only hope that our selection will serve as a good illustration of recent activities. It will also emphasize some current trends that go far beyond the material discussed here. The first one is the analysis of the qualitative behavior of solutions (regularity, compactness, classification of possible behaviors, etc.). The second one, related to the preceding one, concerns the structure of specific nonlinearities and its interplay with the behavior (or possible behaviors) of solutions. Finally, this requires theories and methods that are connected with many branches of mathematics and analysis in particular.

2 Boltzmann equation

2.1 Existence and compactness results

The Boltzmann equation is given by

$$\frac{\partial f}{\partial t} + v \cdot \nabla_x f \;=\; Q(f,f) \qquad (x,v) \in \mathbb{R}^{2N}\ ,\ t \geq 0 \tag{1}$$

where the unknown f is a nonnegative function on $\mathbb{R}^{2N} \times [0,\infty)$, $N \geq 2$, ∇_x denotes the gradient with respect to x, and we denote by $x \cdot y$ or (x,y) the scalar product in $\mathbb{R}^N$. The nonlinear operator Q can be written as

$$Q(f,f) \;=\; Q^+(f,f) - Q^-(f,f) \tag{2}$$

$$Q^+(f,f) \;=\; \int_{\mathbb{R}^N} dv_* \int_{S^{N-1}} d\omega\, B(v-v_*,\omega) f' f'_* \tag{3}$$

$$Q^-(f,f) \;=\; \int_{\mathbb{R}^N} dv_* \int_{S^{N-1}} d\omega\, B(v-v_*,\omega) f f_* \;=\; f L(f)\ , \quad L(f) = f \underset{v}{*} A \tag{4}$$

where $f_* = f(x,v_*,t)$, $f' = f(x,v',t)$, $f'_* = f(x,v'_*,t)$, $A(z) = \int_{S^{N-1}} B(z,\omega)\, d\omega$, and $B = B(z,\omega)$ is a given nonnegative function of $|z|$ and $|(z,\omega)|$, is called the scattering cross-section or the collision kernel, which depends on the physical interactions of the gas particles (or molecules) and

$$v' = v - (v-v_*,\omega)\omega \quad , \quad v'_* = v_* + (v-v_*,\omega)\omega\ . \tag{5}$$

A typical example (the so-called hard spheres case) of B is given by: $B = |(z,\omega)|$. We always assume that $A \in L^1_{\text{loc}}(\mathbb{R}^N)$ and$(1+|z|^2)^{-1} \cdot \int_{|\xi|<R} A(z-\xi)\, d\xi \to 0$ as $|z| \to \infty$, for all $R \in (0,\infty)$.

Of course, we wish to solve (1) given an initial condition that is the values of f at $t = 0$

$$f|_{t=0} \;=\; f_0 \qquad \text{in} \ \ \mathbb{R}^{2N}\ . \tag{6}$$

The initial value problem (1),(6) is a deceivingly simple-looking first-order partial differential equation with nonlinear (quadratic) nonlocal terms. It is a relevant model for the study of a rarefied gas and is currently used for flights in the upper layers of the atmosphere (Mach 20–24, altitude of 70–120 km). The statistical description of a gas in terms of the evolution of the density f of molecules was originally obtained by Boltzmann [6] (see also Maxwell [57], [58]). There is a long history of important mathematical contributions to the study of (1) by Hilbert [31], Carleman [8], [9] etc. Further details on the derivation of (1) and references to earlier mathematical contributions can be found in Grad [28], Cercignani [10], and DiPerna and Lions [18].

The major mathematical difficulty of (1), (6) is the lack of a priori estimates on solutions: only bounds on f in L^1 (with weights) and on $f \log f$ in L^1 are known! Nevertheless, the following result, taken from [18],[20], holds:

THEOREM 2.1. *Let $f_0 \geq 0$ satisfy:*

$$\int_{\mathbb{R}^{2N}} f_0(1 + |x|^2 + |v|^2 + |\log f_0|)\, dx\, dv < \infty.$$

Then there exists a global weak solution of (1),(6) $f \in C([0,\infty); L^1(\mathbb{R}^{2N}_{x,v}))$ satisfying

$$\sup_{t\in[0,\infty)} \int_{\mathbb{R}^{2N}} f(t)(1 + |x - vt|^2 + |v|^2 + |\log f(t)|)\, dx\, dv < \infty$$

and the following entropy inequality for all $t \geq 0$

$$\int_{\mathbb{R}^{2N}} f(t) \log f(t)\, dx\, dv + \frac{1}{4}\int_0^t ds \int_{\mathbb{R}^N} dx\, D[f] \leq \int_{\mathbb{R}^{2N}} f_0 \log f_0\, dx\, dv \tag{7}$$

where $D[f] = \iint_{\mathbb{R}^{2N}} dv\, dv_* \int_{S^{N-1}} B\, d\omega (f'f'_* - ff_*) \log \frac{f'f'_*}{ff_*}$.

REMARKS 2.1. (i) We do not want to give here the precise definition of global weak solutions as it is a bit too technical. Let us mention that the notion introduced in [18], [20] is modified in Lions [48] (additional properties are imposed on f in [48]).

(ii) Further regularity properties of solutions are an outstanding open problem. It is only known that the regularity of solutions is not "created by the evolution" and has to come from the initial condition f_0. It is tempting to think, in view of the results shown in [48] (see sections 2.2, 2.3 below), that, at least in the model case when $B = \varphi\left(|z|, \frac{|(z,\omega)|}{|z|}\right)$ with $\varphi \in C_0^\infty((0,\infty) \times (0,1))$, f is smooth if f_0 is smooth. Related to the regularity issue is the uniqueness question: uniqueness of weak solutions is not known (it is shown in [48] that any weak solution is equal to a solution with improved bounds assuming that the latter exists!).

(iii) The assumption made upon B corresponds to the so-called angular cut-off.

(iv) Boundary conditions for Boltzmann's equation can be treated: see Hamdache [29] for an analogue of the above result in that case. Realistic boundary conditions require some new a priori estimates and are treated in Lions [46].

(v) Other kinetic models of physical and mathematical interest can be studied by the methods of proof of Theorem 2.1: see for instance DiPerna and Lions [19], Arkeryd and Cercignani [2], Esteban and Perthame [22], and Lions [48].

The strategy of proof for Theorem 2.1 is a classical one, which is almost always the one followed for the proofs of *global existence results*: one *approximates* the problem by a sequence of simpler problems having the same structure (and the same a priori bounds) for which one shows easily the existence of global solutions, and then one tries to *pass to the limit*. This strategy is also useful for the mathematical analysis of numerical methods because one can view numerical solutions as approximated solutions or solutions of approximated problems. This is why the main mathematical problem behind the proof of Theorem 2.2 is the analysis of the behavior of *sequences of solutions* (we could as well consider approximated solutions...) and in particular of passages to the limit in the equation. This is a delicate question because the available a priori bounds only yield *weak convergences* that are not enough to *pass to the limit in nonlinear terms*. This theme will be recurrent in this report (as it was already in Majda's report [56]).

We thus consider a sequence of (weak or even smooth) nonnegative solutions f^n of (1) corresponding to initial conditions (6) with f_0 replaced by f_0^n and we assume

$$\sup_{n\geq 1}\int_{\mathbb{R}^{2N}} f_0^n\big(1+|x|^2+|v|^2+|\log f_0^n|\big)\,dx\,dv \;<\; \infty \tag{8}$$

$$\sup_{n\geq 1}\;\sup_{t\geq 0}\int_{\mathbb{R}^{2N}} f^n(t)\big(1+|x-vt|^2+|v|^2+|\log f^n|\big)\,dx\,dv \;<\; \infty \tag{9}$$

$$\sup_{n\geq 1}\int_0^\infty dt\int_{\mathbb{R}^N} dv\; D[f^n] \;<\; \infty\,. \tag{10}$$

Without loss of generality — extracting subsequences if necessary — we may assume that f_0^n, f^n converge weakly in $L^1(\mathbb{R}^{2N})$, $L^1(\mathbb{R}^{2N}\times(0,T))(\forall\, T\in(0,\infty))$ respectively to f_0, f.

THEOREM 2.2. *We have for all* $\psi\in C_0^\infty(\mathbb{R}_v^N)$, $T, R\in(0,\infty)$

$$\int_{\mathbb{R}^N} f^n\psi\,dv \;\underset{n}{\rightarrow}\; \int_{\mathbb{R}^N} f\psi\,dv \qquad \text{in}\quad L^1(\mathbb{R}_x^N\times(0,T))\,, \tag{11}$$

$$\left.\begin{aligned}&\int_{\mathbb{R}^N} Q^+(f^n,f^n)\psi\,dv \underset{n}{\rightarrow} \int_{\mathbb{R}^N} Q^+(f,f)\psi\,dv\,,\\ &\int_{\mathbb{R}^N} Q^-(f^n,f^n)\psi\,dv \underset{n}{\rightarrow} \int_{\mathbb{R}^N} Q^-(f,f)\psi\,dv\\ &\qquad\qquad \text{in measure for } |x|<R\,,\ t\in(0,T),\end{aligned}\right\} \tag{12}$$

and f is a global weak solution of (1), (6).

THEOREM 2.3. (1) *We have for all* $R, T\in(0,\infty)$

$$Q^+(f^n,f^n) \underset{n}{\rightarrow} Q^+(f,f) \quad \text{in measure for } |x|<R,\ |v|<R,\ t\in(0,T)\,. \tag{13}$$

(2) *If f_0^n converges in $L^1(\mathbb{R}^{2N})$ to f_0, then f^n converges to f in $C([0,T];L^1(\mathbb{R}_{x,v}^{2N}))$ for all $T\in(0,\infty)$.*

REMARKS 2.2. (i) Theorem 2.2 is shown in [18] — a simplification of the proof of the passage to the limit (using (13)) is given in [48]. Theorem 2.3 is taken from [48].

(ii) The heart of the matter in Theorem 2.2 is (11), which is a consequence of the velocity averaging phenomenon detailed in Section 2.2 below. The proof of Theorem 2.3 relies upon the results of Section 2.3 below.

(iii) It is shown in Lions [48] that the conclusion in (2) of Theorem 2.3 implies that f_0^n converges to f_0 in $L^1(\mathbb{R}^{2N})$: in other words, no compactification and in particular no regularization is taking place for $t > 0$. As indicated in [47] (see also the recent result of Desvillettes [16]) this fact might be related to the angular cut-off assumption because grazing collisions seem to generate some compactification ("nonlinear hypoelliptic features" in the model studied in [47]).

2.2 Velocity averaging

A typical example of the so-called velocity averaging results is the following

THEOREM 2.4. *Let $m \geq 0$, let $\theta \in [0,1)$, and let $f, g \in L^p(\mathbb{R}_x^N \times \mathbb{R}_v^N \times \mathbb{R}_t)$ with $1 < p \leq 2$. We assume*

$$\frac{\partial f}{\partial t} + v \cdot \nabla_x f = (-\Delta_{x,t}+1)^{\theta/2}(-\Delta_v+1)^{m/2} g \qquad in \quad \mathcal{D}'(\mathbb{R}^{2N+1}) \,. \tag{14}$$

Then, for all $\psi \in C_0^\infty(\mathbb{R}^N)$, $\int_{\mathbb{R}^N} f(x,v,t)\psi(v)\,dv$ belongs to the (Besov) space $B_2^{s,p}(\mathbb{R}^N \times \mathbb{R})$ — and thus to $H^{s',p}(\mathbb{R}^N)$ for all $0 < s' < s$ — where $s = (1-\theta)\,\frac{p-1}{p}\,(1+m)^{-1}$.

REMARKS 2.3. (i) If $m = 0$, $\int_{\mathbb{R}^N} f\psi\,dv \in H^{s,p}$ with $s = \frac{p-1}{p}$. The above exponent s is optimal in general (this is shown in a work to appear by the author). Similar results are available if $2 < p < \infty$ or in more general settings: we refer the reader to DiPerna, Lions, and Meyer [21].

(ii) Such velocity averages are known in statistical physics (or mechanics) as macroscopic quantities. The above result shows that transport equations induce some improved partial regularity on velocity averages (by some kind of dispersive effect).

(iii) The first results in this direction were obtained in Golse, Perthame, and Sentis [27], Golse, Lions, Perthame, and Sentis [26] (where the case $m = 0$ is considered). The case $m \geq 0$, $p = 2$, was treated in DiPerna and Lions [19] while the general case is due to DiPerna, Lions, and Meyer [21] — a slight improvement of the Besov space can be found in Bézard [5]. Two related strategies of proof are proposed in [21] that both rely on Fourier analysis, one uses some harmonic analysis, namely product Hardy spaces and interpolation theory, while the second one uses classical multipliers theory and careful Littlewood-Paley dyadic decompositions. However, the main idea is rather elementary and described below in extremely rough terms.

As indicated in the preceding remark, we give a caricatural (but accurate!) explanation of the phenomena illustrated by Theorem 2.4. If we Fourier transform

(14) in (x,t), we see that we gain decay (=regularity) in (ξ,τ) — dual variables of (x,t) — provided $|\tau + v\cdot\xi| \geq \delta |v\cdot\xi|$ for some $\delta > 0$. On the other hand, the set of v on which we do not gain that regularity, namely $\{v \in \text{Supp}(\psi) \;/\; |\tau + v\cdot\xi| < \delta|(\tau,\xi)|\}$, has a measure of order δ, and hence contributes little to the integral $\left(\int_{\mathbb{R}^n} \hat{f}(\xi,v,\tau)\,\psi(v)\,dv\right)$. Balancing the two contributions, we obtain some (fractional) regularity.

Of course, such improved regularity yields local compactness (in (x,t)) of the velocity averages and leads (after some work) to (11).

2.3 Gain terms and Radon transforms

We set for $f, g \in C_0^\infty(\mathbb{R}_v^N)$

$$Q^+(f,g) \;=\; \int_{\mathbb{R}^N} dv_* \int_{S^{N-1}} d\omega\, B(v-v_*,\omega)\, f' g'_* \tag{15}$$

and we assume (to simplify the presentation) that B satisfies:
$B(z,\omega) = \varphi\left(|z|\,,\, \frac{|(z,\omega)|}{|z|}\right)$ (this is always the case in the context of (1)) and $\varphi \in C_0^\infty((0,\infty)\times(0,1))$. We denote by $\mathcal{M}(\mathbb{R}^N)$ the space of bounded measures on $\mathbb{R}^N$.

THEOREM 2.5. *The operator Q^+ from $\mathcal{M}(\mathbb{R}^N)\times H^s(\mathbb{R}^N)$ and $H^s(\mathbb{R}^N)\times\mathcal{M}(\mathbb{R}^N)$ into $H^{s+\frac{N-1}{2}}(\mathbb{R}^N)$ is bounded for all $s\in\mathbb{R}$.*

REMARK 2.4. This result is taken from Lions [48] using generalized Radon transforms; a variant of this proof making direct connection with the classical Radon transform has been recently given by Wennberg [72] (this proof, contrarily to the one in [48], does not extend to more general situations such as collision models for mixtures or relativistic models — this case is treated in Andréasson [1]).

The above gain of regularity ($\frac{N-1}{2}$ derivatives) can be shown by writing Q^+ or its adjoint as a "linear combination" of translates of some Radon-like transforms given by

$$R\psi(v) \;=\; \int_{S^{N-1}} B(v,\omega)\,\psi((v,\omega)\omega)\,d\omega \quad , \quad \forall\, \psi\in C_0^\infty(\mathbb{R}^N) \tag{16}$$

or

$$R\psi(v) \;=\; \int_{S^{N-1}} B(v,\omega)\,\psi(v-(v,\omega)\omega)\,d\omega \quad , \quad \forall\, \psi\in C_0^\infty(\mathbb{R}^N)\,. \tag{17}$$

In both cases, one integrates φ over the set $\{(v,\omega)\omega \mid \omega\in S^{N-1}\} = \{v-(v,\omega)\omega \mid \omega\in S^{N-1}\}$, which is the sphere centered at $\frac{v}{2}$ and of radius $\frac{|v|}{2}$. These operators are rather special Fourier integral operators often called generalized Radon transforms (see for instance Phong and Stein [62], Stein [66]). The crucial fact is that the set over which ψ is integrated "moves" with v — except that all these spheres go through 0, but this does not create difficulties because B vanishes if $(v,\omega)\omega = 0$ or if $v-(v,\omega)\omega = 0$. This is the main reason why one can prove that R is bounded from $H^s(\mathbb{R}^N)$ into $H^{s+\frac{N-1}{2}}(\mathbb{R}^N)$ for all $s\in\mathbb{R}$ ($\frac{N-1}{2}$ comes from the stationary phase principle...).

3 Compressible Euler and Navier-Stokes equations

The compressible Euler and Navier-Stokes equations are the basic models for the evolution of a compressible gas. In the case of aeronautical applications, the main difference between the domains of validity of the Boltzmann equation and the Euler-Navier-Stokes systems is the altitude of the aircraft. This indicates that there should be a transition from the Boltzmann model to those mentioned here. Mathematically, this corresponds to replacing B by $\frac{1}{\varepsilon} B$ in (1) and letting ε go to 0 (at least formally): as is well known, one recovers, taking velocity averages of the limit f (i.e. $\rho = \int_{\mathbb{R}^N} f\,dv$, $\rho u = \int_{\mathbb{R}^N} f v\,dv$, $\rho E = \int_{\mathbb{R}^N} f|v|^2\,dv$), the compressible Euler equations (with $\gamma = \frac{N+2}{N}$) — see Cercignani [10] for more details. This heuristic limit (and related limits) remains completely open from a mathematical viewpoint: partial results can be found in Nishida [61], and Ukaï and Asano [71], and recent progress based upon the material described in Section 2 above is due to Bardos, Golse, and Levermore [4]. Related problems are described in Varadhan's report in this Congress.

The compressible Euler and Navier-Stokes equations take the following form:

$$\frac{\partial p}{\partial t} + \operatorname{div}(\rho u) = 0 \qquad x \in \mathbb{R}^N\,,\ t \geq 0 \tag{18}$$

$$\frac{\partial}{\partial t}(\rho u) + \operatorname{div}(\rho u \otimes u) - \lambda\Delta u - (\lambda+\mu)\nabla \operatorname{div} u + \nabla p = 0 \qquad x \in \mathbb{R}^N\,,\ t \geq 0 \tag{19}$$

and an equation for the pressure p (or equivalently for the total energy or the temperature) that we do not wish to write for reasons explained below. The unknowns ρ, u correspond respectively to the density of the gas ($\rho \geq 0$) and its velocity u (where $u(x,t) \in \mathbb{R}^N$). The constants λ, μ are the viscosity coefficients of the fluid: if $\lambda = \mu = 0$, the above system is called the compressible Euler equations, whereas if $\lambda > 0$, $2\lambda + \mu > 0$, it is called the compressible Navier-Stokes equations. Despite the long history of these problems, the global existence of solutions "in the large" is still open for the full (i.e. with the temperature equation) systems except in the case of compressible Navier-Stokes equations when $N = 1$: in that case, general existence and uniqueness results can be found in Kazhikov and Shelukhin [37], Kazhikov [36], Serre [64], [63], and Hoff [34]. This is why we shall restrict ourselves here to the so-called "isentropic" (or barotropic) case where one postulates that p is a function of ρ only, and in order to fix ideas we take

$$p = a\rho^\gamma \quad , \quad a > 0 \quad , \quad \gamma > 1\,. \tag{20}$$

This condition is a severe restriction from the mechanical viewpoint (in the Navier-Stokes case, it essentially means considering the adiabatic case and neglecting the viscous heating). Mathematically, it leads to an interesting model problem that is supposed to capture some of the difficulties of the exact systems.

Of course we complement (18)–(19) with initial conditions

$$\rho|_{t=0} = \rho_0 \quad , \quad \rho u|_{t=0} = m_0 \qquad \text{in} \quad \mathbb{R}^N \tag{21}$$

where $\rho_0 \geq 0$, m_0 are given functions on $\mathbb{R}^N$.

We study the case of compressible isentropic Euler equations in Section 3.1. The analogous problem for Navier-Stokes equations is considered in Section 3.3.

3.1 1D isentropic gas dynamics

We thus consider the following system

$$\frac{\partial \rho}{\partial t}+\frac{\partial(\rho u)}{\partial x}=0 \quad , \quad \frac{\partial(\rho u)}{\partial t}+\frac{\partial}{\partial x}(\rho u^2+a\rho^\gamma)=0 \qquad x\in\mathbb{R}\,,\ t>0 \tag{22}$$

where $\rho \geq 0$, and $a>0$, $\gamma>1$ are given constants. Without loss of generality (by a simple scaling) we can take $a=\frac{(\gamma-1)^2}{4\gamma}$ (to simplify some of the constants below).

As is well known for such systems of nonlinear hyperbolic (first-order) equations, singularities develop in finite time: even if $\rho_0=\overline{\rho}+\rho_1>0$ on $\mathbb{R}$ with $\overline{\rho}\in\mathbb{R}$, $\overline{\rho}>0$, $\rho_1, u_0 \in C_0^\infty(\mathbb{R})$, then u_x and ρ_x become infinite in finite time (see Lax [38], [39], [41], and Majda [54], [55] for more details). In addition, bounded solutions of (22), (21) are not unique and additional requirements known as (Lax) entropy conditions on the solutions are needed (Lax [41],[40], see also the report by Dafermos in this Congress).

In the case of (22), these requirements take the following form (see DiPerna [17], Chen [11], and Lions, Perthame, and Tadmor [53]):

$$\frac{\partial}{\partial t}\left[\varphi(\rho,\rho u)\right]+\frac{\partial}{\partial x}\left[\psi(\rho,\rho u)\right]\ \leq\ 0 \qquad \text{in}\ \ \mathcal{D}'(\mathbb{R}\times(0,\infty)) \tag{23}$$

and φ, ψ are given by

$$\begin{cases} \varphi\ =\ \displaystyle\int_{\mathbb{R}} dv\,\omega(v)(\rho^{\gamma-1}-(v-u)^2)_+^\lambda\,, \\ \psi\ =\ \displaystyle\int_{\mathbb{R}} dv\,[\theta v+(1-\theta)u]\,\omega(v)(\rho^{\gamma-1}-(v-u)^2)_+^\lambda \end{cases} \tag{24}$$

where ω is an arbitrary convex function on $\mathbb{R}$ such that ω'' is bounded on $\mathbb{R}$, $\lambda=\frac{3-\gamma}{2(\gamma-1)}$, $\theta=\frac{\gamma-1}{2}$.

THEOREM 3.1. *Let $\rho_0, m_0 \in L^\infty(\mathbb{R})$ satisfy: $\rho_0\geq 0$, $|m_0|\leq C\rho_0$ a.e. in $\mathbb{R}$ for some $C\geq 0$. Then there exists $(\rho,u)\in L^\infty(\mathbb{R}\times(0,\infty))$ $(\rho\geq 0)$ solution of* (21)-(22) *satisfying* (23).

As explained in Section 2.2, the proof of the existence results depends very much upon the stability and compactness results shown below (in fact one approximates (22) by the vanishing viscosity method; i.e., adding $-\varepsilon\,\frac{\partial^2\rho}{\partial x^2}$, $-\varepsilon\,\frac{\partial^2(\rho u)}{\partial x^2}$ in the equations respectively satisfied by $\rho, \rho u$ where $\varepsilon>0$, and one lets ε go to 0). We thus consider a sequence (ρ^n, u^n) of solutions of (22) satisfying (23) and we assume that (ρ^n, u^n) is bounded uniformly in n in $L^\infty(\mathbb{R}\times(0,\infty))$ $(\rho^n\geq 0$ a.e.). Without loss of generality, we may assume that (ρ^n,u^n) converges weakly in $L^\infty(\mathbb{R}\times(0,\infty))$-weak$*$ to some $(\rho,u)\in L^\infty(\mathbb{R}\times(0,\infty))$ $(\rho\geq 0$ a.e.). The main mathematical difficulty is the lack of any a priori estimate (except for $\gamma=3$, the so-called monoatomic case) that would ensure the pointwise compactness needed to pass to the limit in $\rho^n(u^n)^2$ or $(\rho^n)^\gamma$.

THEOREM 3.2. *$\rho^n, \rho^n u^n$ converge in measure on $(-R,R)\times(0,T)$ (for all $0<R,T<\infty$) to $\rho, \rho u$ respectively. And (ρ,u) is a solution of* (22) *satisfying* (23).

REMARKS 3.1. (i) This result shows that the hyperbolic system (22) has compactifying properties because initially at $t = 0$ we did not require that ρ^n or $\rho^n u^n$ converge in measure.

(ii) Theorem 3.2 is essentially due to DiPerna [17] if $\gamma = \frac{2k+3}{2k+1}$ $(k \geq 1)$, Chen [11] if $1 < \gamma \leq \frac{5}{3}$. It is shown in Lions, Perthame, and Tadmor [53] if $\gamma \geq 3$ and in Lions, Perthame, and Souganidis [51] if $1 < \gamma < 3$. The existence result (Theorem 3.1) for $1 < \gamma < \infty$ is taken from [51].

(iii) The proofs in [53],[51] use two main tools: the method introduced by Tartar [69] (and developed by DiPerna [17]) which combines the compensated-compactness theory of Tartar [68], [69], Murat [60] and the entropy inequalities (23), and the kinetic formulation of (22) introduced in [52], [53] where one adds a new "velocity" variable, and writes the unknowns $(\rho, \rho u)$ in terms of macroscopic quantities (velocity averages) associated with a density $f(x, v, t)$ that has a fixed "profile" in v (a "pseudo-maxwellian"). This formulation connects the Boltzmann theory as described in Section 2 and the study of compressible hydrodynamic (or gas dynamics) macroscopic models. More details on this new approach are to be found in Perthame's report in this Congress. In the next section, we present some aspects of the compensated-compactness theory.

3.2 Compensated compactness and Hardy spaces

One important point in the compensated-compactness theory developed by Tartar [68], [69] and F. Murat [60] is the systematic detection of nonlinear quantities that enjoy "weak compactness" properties. A typical example known as the div-curl example — it is precisely the one used in the proof of Theorem 3.2 — is given by the following result taken from [60].

THEOREM 3.3. *Let* (E^n, B^n) *converge weakly to* (E, B) *in* $L^p(\mathbb{R}^N)^N \times L^q(\mathbb{R}^N)^N$ *with* $1 < p < \infty$, $\frac{1}{q} + \frac{1}{p} = 1$, $N \geq 2$. *We assume that* $\operatorname{curl} E^n$, $\operatorname{div} B^n$ *are relatively compact in* $W^{-1,p}(\mathbb{R}^N)$, $W^{-1,q}(\mathbb{R}^N)$ *respectively. Then,* $E^n \cdot B^n$ *converges weakly (in the sense of measures or in distributions sense) to* $E \cdot B$.

REMARK 3.2. Let us sketch a proof. We write: $E^n = \nabla \pi^n + \tilde{E}^n$ where $\operatorname{div} \tilde{E}^n = 0$, $\tilde{E}^n$ is compact in $L^p(\mathbb{R}^N)$ (Hodge-De Rham decompositions), $\pi^n \in L^p_{\text{loc}}(\mathbb{R}^N)$, $\nabla \pi^n \in L^p(\mathbb{R}^N)$. Then, we only have to pass to the limit in $B^n \cdot \nabla \pi^n = \operatorname{div}(\pi^n B^n) - \pi^n \operatorname{div} B^n$. The first term passes to the limit because π^n is compact in $L^p_{\text{loc}}(\mathbb{R}^N)$ (Rellich-Kondrakov theorem) while the second term also does because $\operatorname{div} B^n$ is relatively compact in $W^{-1,q}(\mathbb{R}^N)$ and $\nabla \pi^n$ is bounded in $L^p(\mathbb{R}^N)$.

As shown in Coifman, Lions, Meyer, and Semmes [12], the above nonlinear phenomenon is intimately connected with some general results in harmonic analysis associated with the (multi-dimensional) Hardy spaces denoted here by $H_p(\mathbb{R}^N)$ $(0 < p \leq 1)$: see Stein and Weiss [67], Fefferman and Stein [23], and Coifman and Weiss [14] for more details on Hardy spaces.

In particular, the following result holds.

THEOREM 3.4. *Let* $E \in L^p(\mathbb{R}^N)$ *satisfy* $\operatorname{curl} E = 0$ *in* $\mathcal{D}'(\mathbb{R}^N)$, *let* $B \in L^q(\mathbb{R}^N)$ *satisfy* $\operatorname{div} B = 0$ *in* $\mathcal{D}'(\mathbb{R}^N)$ *with* $1 < p, q < \infty$, $\frac{1}{r} = \frac{1}{p} + \frac{1}{q} < 1 + \frac{1}{N}$. *Then* $E \cdot B \in H_r(\mathbb{R}^N)$.

REMARKS 3.3. (i) This result is taken from [12] (and was inspired by a surprising observation due to Müller [59]).

(ii) The relations between the weak compactness result (Theorem 3.3) and the regularity result (Theorem 3.4) are made clear in [12] and follow from some general considerations on dilation and translation invariant multilinear forms that enjoy a crucial cancellation property ($\int_{\mathbb{R}^N} E \cdot B \, dx = 0$ in Theorem 3.4 above).

(iii) Theorem 3.4 is one of the tools used in the proof by Hélein [30] of the regularity of two-dimensional harmonic maps.

(iv) It is shown in [12] that any element of $H_1(\mathbb{R}^N)$ can be decomposed in a series $\sum_{n \geq 1} \lambda_n E_n \cdot B_n$ where $\|E_n\|_{L^2} = \|B_n\|_{L^2} = 1$, $\operatorname{div} B_n = \operatorname{curl} E_n = 0$, $\sum_{n \geq 1} |\lambda_n| < \infty$.

If we denote by R_k the Riesz transform ($= \partial_k(-\Delta)^{-1/2}$), then, under the conditions of Theorem 3.4, there exists $\hat{\pi} \in L^p(\mathbb{R}^N)$ such that $E = R\hat{\pi}$. And $E \cdot B = B \cdot R\hat{\pi} = B \cdot R\hat{\pi} + (R \cdot B)\hat{\pi}$ because $R \cdot B = (-\Delta)^{-1/2} \operatorname{div} B = 0$. Then we can recover the case $r = 1$ in Theorem 3.4 using the H_1-BMO duality and the result on commutators due to Coifman, Rochberg, and Weiss [13]: indeed, we then obtain $f(R_k g) + (R_k f)g \in H_1(\mathbb{R}^N)$ for each $k \geq 1$, $f \in L^p(\mathbb{R}^N)$, $g \in L^q(\mathbb{R}^N)$, $1 < p < \infty$, $\frac{1}{p} + \frac{1}{q} = 1$.

3.3 Isentropic Navier-Stokes equations

We now consider the system

$$\begin{cases} \dfrac{\partial \rho}{\partial t} + \operatorname{div}(\rho u) = 0 \,, \\ \dfrac{\partial \rho u}{\partial t} + \operatorname{div}(\rho u \otimes u) - \lambda \Delta u - (\lambda + \mu) \nabla \operatorname{div} u + a \nabla \rho^\gamma = 0 \,, \\ \qquad x \in \mathbb{R}^N \,,\ t > 0 \,, \end{cases} \tag{25}$$

where $a > 0$, $1 < \gamma < \infty$, $\lambda > 0$, $2\lambda + \mu > 0$, $\rho(x,t) \geq 0$ on $\mathbb{R}^N \times (0, \infty)$, with the initial conditions (21) that are required to satisfy

$$\begin{cases} \rho_0 \in L^1(\mathbb{R}^N) \cap L^\gamma(\mathbb{R}^N) \,, \quad \rho_0 \geq 0 \,, \\ m_0 = \sqrt{\rho_0} v_0 \quad \text{a.e. with} \quad v_0 \in L^2(\mathbb{R}^N) \,. \end{cases} \tag{26}$$

THEOREM 3.5. *We assume (26) and* $\gamma > \frac{3}{2}$ *if* $N = 2$, $\gamma \geq \frac{9}{5}$ *if* $N = 3$, $\gamma > \frac{N}{2}$ *if* $N \geq 4$. *Then there exists a solution* $(\rho, u) \in L^\infty(0, \infty; L^\gamma(\mathbb{R}^N)) \cap L^2(0, T; H^1(B_R))$ $(\forall\ R, T \in (0, \infty))$ *of (25),(21) satisfying in addition:* $\rho \in C([0, \infty); L^p(\mathbb{R}^N))$ *if*

$1 \leq p < \gamma$, $\rho |u|^2 \in L^\infty(0,\infty; L^1(\Omega))$, $\rho \in L^q(\mathbb{R}^N \times (0,T))$ *for* $1 \leq q \leq \gamma + \frac{2\gamma}{N} - 1$ *if* $N \geq 2$.

$$\left\{ \begin{aligned} \int_\Omega \frac{1}{2}\rho(t)|u(t)|^2 + \frac{a}{\gamma-1}\rho(t)^\gamma \, dx + \int_0^t ds \int_\Omega \lambda |\nabla u|^2 + (\lambda+\mu)(\operatorname{div} u)^2 \, dx \\ \leq \int_\Omega \frac{1}{2}\frac{|m_0|^2}{\rho_0} + \frac{a}{\gamma-1}\rho_0^\gamma \, dx \end{aligned} \right. \tag{27}$$

for almost all $t \geq 0$.

REMARKS 3.4. (i) This result is taken from Lions [45] (see also [50]). If $N = 1$, more general results are available and we refer to Serre [63], Hoff [32],[33].

(ii) If $N \geq 2$, the uniqueness and further regularity of solutions are completely open as is the case of a general $\gamma > 1$. The case $\gamma = 1$ is also an interesting mathematical problem (see [49]).

(iii) Of course, the equations contained in (25) hold in the sense of distributions.

(iv) The preceding result is rather similar to the results obtained by Leray [42], [43], [44] on the global existence of weak solutions of three-dimensional incompressible Navier-Stokes equations satisfying an energy inequality like (27). Despite many important contributions (like the partial regularity results obtained by Caffarelli, Kohn, and Nirenberg [7]), the uniqueness and regularity of solutions are still open questions.

As explained in the previous sections, the above existence result is based upon a convergence result for sequences of solutions ρ^n, u^n satisfying uniformly in n the properties mentioned in the above result. Hence, without loss of generality, we may assume that (ρ^n, u^n) converge weakly to (ρ, u) in $L^\gamma(\mathbb{R}^N \times (0,T)) \times L^2(0,T; H^1(B_R))$ $(\forall\ R, T \in (0,\infty))$. Then it is shown in [45], [49] that if ρ_0^n $(= \rho^n|_{t=0})$ converges in $L^1(\mathbb{R}^N)$, then ρ^n converges in $C([0,T]; L^p(\mathbb{R}^N)) \cap L^q(\mathbb{R}^N \times (0,T))$ for all $T \in (0,\infty)$, $1 \leq p < \gamma$, $1 \leq q < \gamma + \frac{2\gamma}{N} - 1$. And (ρ, u) is a solution of (25) with the properties listed in the preceding result. It is also shown in [45], [49] (see also Serre [65]) that the analogue of Theorem 3.2 for the system (25) *does not hold*: in other words, the compactification that took place for the hyperbolic system (22) *is lost* when we add viscous terms while we could expect (from a linear-linearized inspection) that the introduction of viscous terms regularizes the problem! These delicate and surprising phenomena depend in a subtle way on the nonlinearities of the systems we consider. Let us also mention that the proof of the above convergence result is rather delicate and uses in particular the structure of the convective derivatives $(\frac{\partial}{\partial t} + u \cdot \nabla_x)$ that lead with the analysis detailed in [45],[49] to terms like

$$\rho^n\, R_i R_j(\rho^n u_i^n u_j^n) - \rho^n\, u_i^n\, R_i R_j(\rho^n u_j^n),$$

which are shown to converge weakly to $\rho\, R_i R_j(\rho u_i u_j) - \rho u_i\, R_i, R_j(\rho u_j)$ under the sole weak convergence stated above on ρ^n, u^n. This weak continuity follows from regularizing properties of the commutators $[u_i^n, R_i R_j]$. It is worth noting that the incompressible limit of such compressible models yields $\rho^n = \text{cst}$ (say 1), $\operatorname{div} u^n = 0$, in which case the above term reduces to $R_i R_j(u_i^n u_j^n)$ and $R_j(u_i^n u_j^n) = (-\Delta)^{-1/2}\{u^n \cdot \nabla u^n\}$ because $\operatorname{div} u^n = 0$. Obviously, $\operatorname{curl}(\nabla u^n) = 0$, $\operatorname{div} u^n = 0$, and $u^n \cdot \nabla u^n$ is precisely a nonlinear expression for which the compensated-compactness theory applies (see Section 3.2).

4 Perspectives, trends, problems and methods

Let us immediately emphasize that this brief section will select topics in a biased way that reflects the author's tastes.

First of all, we have mentioned above some of the progress made recently and many remaining open questions in gas dynamics and fluid mechanics. There is much more to say and in particular we have not touched here the incompressible models (Euler and Navier-Stokes equations) — see the reports by Beale, Chemin, Constantin, and Avellaneda in this Congress and Majda [56]. Even if many fundamental questions are left open, progress is being made (step by step).

We should also make clear that the topics covered here do not reflect fully the scope of nonlinear partial differential equations and in particular those arising from applications, the variety of mathematical problems and methods developed recently, and their relationships with other fields of mathematics. Let us briefly mention a few more examples of themes covering several related areas that all have important scientific and technological implications: (i) propagation of fronts and interfaces, geometric equations, viscosity solutions, image processing (see the reports by Spruck, Souganidis, and Osher in this Congress), (ii) quantum chemistry, N-body problems, density-dependent and meanfield models, binding, thermodynamic limits, (iii) twinning and defects in solids and crystals, phase transitions, Young measures (see for instance Ball and James [3], James and Kinderlehrer [35], and the report by Sverak).

However, we wish to emphasize that the trends mentioned in the Introduction can also be found in the above themes.

Finally, it is important to develop at the same time the methods — some of which have been briefly presented in this paper — which are certainly interesting by themselves, and we would like to conclude with a few examples of such developments: (i) H-measures of Tartar [70], Gérard [24] (and the related Wigner measures by Lions and Paul [50], Gérard [25]), (ii) nonlinear partial differential equations in infinite dimensions (and in particular the viscosity solutions approach of Crandall and Lions [15]).

References

[1] H. Andréasson, *A regularity property and strong L^1 convergence to equilibrium for the relativistic Boltzmann equation*, preprint **21**, Chalmers Univ., Göteborg, 1994.

[2] L. Arkeryd and C. Cercignani, *On the convergence of solutions of Enskog equations to solutions of the Boltzmann equation*, Comm. Partial Differential Equations, **14** (1989), 1071–1090.

[3] J. Ball and R. D. James, *Fine phase mixtures as minimizers of energy*, Arch. Rational Mech. Anal., **100** (1987), 13–52.

[4] C. Bardos, F. Golse, and D. Levermore, *Fluid dynamics limits of kinetic equations*, I, J. Statist. Phys., **63** (1991), 323–344; II, Comm. Pure Appl. Math., **46** (1993), 667–753.

[5] M. Bézard, *Régularité L^p précisée des moyennes dans les équations de transport*, preprint.

[6] L. Boltzmann, *Weitere Studien über das Wärmegleichgewicht unter Gasmolekülen.* Sitzungsberichte der Akademie der Wisssenschaften, Vienna, **66** (1972), 275–370. (Trans.: Further studies on the thermal equilibrium of gas molecules, in Kinetic Theory, vol. 2 (S. G. Brush, ed.), Pergamon, Oxford (1966), 88–174).

[7] L. Caffarelli, R. V. Kohn, and L. Nirenberg, *On the regularity of the solutions of Navier-Stokes equations*, Comm. Pure Appl. Math., **35** (1982), 771–832.

[8] T. Carleman, Acta Math., **60** (1933), 91.

[9] T. Carleman, Problèmes mathématiques dans la théorie cinétique des gaz. Notes written by Carleson and Frostman, Uppsala, Almqvist and Wikselles, 1957.

[10] C. Cercignani, The Boltzmann Equation and its Applications. Springer Verlag, Berlin and New York, 1988.

[11] G. Q. Chen, *The theory of compensated compactness and the system of isentropic gas dynamics*, preprint.

[12] R. Coifman, P. L. Lions, Y. Meyer, and S. Semmes, *Compensated compactness and Hardy spaces*, J. Math. Pures Appl. (9), **72** (1993), 247–286.

[13] R. Coifman, R. Rochberg, and G. Weiss, Ann. of Math. (2), **103** (1976), 611–635.

[14] R. Coifman and G. Weiss, *Extensions of Hardy spaces and their use in analysis*, Bull. Amer. Math. Soc., **83** (1977), 579–645.

[15] M. G. Crandall and P. L. Lions, *Hamilton-Jacobi equations in infinite dimensions, I*, J. Funct. Anal., **63** (1985), 379–396; *II*, J. Funct. Anal., **65** (1985), 308–400; *III*, J. Funct. Anal., **68** (1986),214–247 ; *IV*, J. Funct. Anal., **90** (1990), 237–283; *V*, J. Funct. Anal., **97** (1991), 417–465; *VI*, in Evolution Equations, Control Theory and Biomathematics (Ph. Clément and G. Lumer, eds.), Lecture Notes in Pure and Appl. Math. **155**, Dekker, New York, 1994: *VII*, to appear in J. Funct. Anal.

[16] L. Desvillettes, *About the regularizing properties of the non cut-off Kac equation*, preprint, 1994.

[17] R. J. DiPerna, *Convergence of the viscosity method for isentropic gas dynamics*, Comm. Math. Phys., **91** (1983), 27–70.

[18] R. J. DiPerna and P. L. Lions, *On the Cauchy problem for Boltzmann equations: global existence and weak stability*, Ann. of Math. (2), **130** (1989), 312–366.

[19] R. J. DiPerna and P. L. Lions, *Global weak solutions of Vlasov-Maxwell systems*, Comm. Pure Appl. Math., **62** (1989), 729–757.

[20] R. J. DiPerna and P. L. Lions, *Global solutions of Boltzmann's equation and the entropy inequality*, Arch. Rational Mech. Anal., **114** (1991), 47–55.

[21] R. J. DiPerna, P. L. Lions, and Y. Meyer, *L^p-regularity of velocity averages*, Ann. Inst. H. Poincaré Anal. Nonlinéaire, **8** (1991), 271–287.

[22] M. J. Esteban and B. Perthame, *On the modified Enskog equation with elastic or inelastic collisions; Models with spin*, Ann. Inst. H. Poincaré Anal. Nonlinéaire, **8** (1991), 289–398.

[23] C. Fefferman and E. Stein, *H^p spaces of several variables*, Acta Math., **228** (1972), 137–193.

[24] P. Gérard, *Microlocal defect measures*, Comm. Partial Differential Equations, **16** (1991), 1761–1794.

[25] P. Gérard, *Mesures semi-classiques et ondes de Bloch*, in Séminaire EDP, 1990-1991, Ecole Polytechnique, Palaiseau, 1991.

[26] F. Golse, P. L. Lions, B. Perthame, and R. Sentis, *Regularity of the moments of the solutions of a transport equation*, J. Funct. Anal., **76** (1988), 110–125.

[27] F. Golse, B. Perthame, and R. Sentis, *Un résultat pour les équations de transport et application au calcul de la limite de la valeur propre principale d'un opérateur de transport*, C.R. Acad. Sci. Paris Série I, **301** (1985), 341–344.

[28] H. Grad, *Principles of the kinetic theory of gases*, Handbuch der Physik, **12**, Springer Verlag, Berlin (1958), 205–294.

[29] K. Hamdache, *Global existence for weak solutions for the initial boundary value problems of Boltzmann equation*, Arch. Rational Mech. Anal., **119** (1992), 309–353.

[30] F. Hélein, *Régularité des applications faiblement harmoniques entre une surface et une variété riemanienne*, C. R. Acad. Sci. Paris Série I, **312** (1990), 591–596.

[31] D. Hilbert, *Begründung der kinetischen Gastheorie*, Math. Ann. **72** (1912), 562–577.

[32] D. Hoff, *Construction of solutions for compressible, isentropic Navier-Stokes equations in one space dimension with non smooth initial data*, Proc. Roy. Soc. Edinburgh Sect. A, **103** (1986), 301–315.

[33] D. Hoff, *Global existence for 1D compressible, isentropic Navier-Stokes equations with large initial data*, Trans. Amer. Math. Soc., **303** (1987), 169–181.

[34] D. Hoff, *Global well-posedness of the Cauchy problem for non-isentropic gas dynamics with discontinuous initial data*, J. Differential Equations, **95** (1992), 33–73.

[35] R. D. James and D. Kinderlehrer, *Theory of diffusionless phase transformations*, Lecture Notes in Phys. **344**, (M. Rascle, D. Serre, and M. Slemod, eds.), Springer Verlag, Berlin and New York, 1989.

[36] A. V. Kazhikov, *Cauchy problem for viscous gas equations*, Sibirsk. Mat. Zh., **23** (1982), 60–64.

[37] A. V. Kazhikov and V. V. Shelukhin, *Unique global solution with respect to time of the initial boundary value problems for one-dimensional equations of a viscous gas*, J. Appl. Math. Mech., **41** (1977), 273–282.

[38] P. D. Lax, *Hyperbolic systems of conservation laws, II*, Comm. Pure Appl. Math., **10** (1957), 537–566.

[39] P. D. Lax, *Development of singularities of solutions of nonlinear hyperbolic partial differential equations*, J. Math. Phys., **5** (1964), 611–613.

[40] P. D. Lax, *Shock waves and entropy*, in Contributions to Nonlinear Functional Analysis (Zarantonello, ed.), Academic Press, New York, (1973), 603–634.

[41] P. D. Lax, Hyperbolic systems of conservation laws and the mathematical theory of shock waves. CBMS-NSF Regional Conferences Series in Applied Mathematics, **11**, 1973.

[42] J. Leray, *Etude de diverses équations intégrales nonlinéaires et de quelques problèmes que pose l'hydrodynamique*, J. Math. Pures Appl. (9), **12** (1933), 1–82.

[43] J. Leray, *Essai sur les mouvements plans d'un liquide visqueux que limitent des parois*, J. Math. Pures Appl. (9), **13** (1934), 331–418.

[44] J. Leray, *Essai sur le mouvement d'un liquide visqueux emplissant l'espace*, Acta Math., **63** (1934), 193–248.

[45] P. L. Lions, *Existence globale de solutions pour les équations de Navier-Stokes compressibles isentropiques*, C. R. Acad. Sci. Paris Série I, **316** (1993), 1335-1340. *Compacité des solutions des équations de Navier-Stokes compressibles isentropiques*, C. R. Acad. Sci. Paris Série I, **317** (1993), 115–120. *Limites incompressibles et acoustique pour des fluides visqueux compressibles isentropiques*, C. R. Acad. Sci. Paris Série I, **317** (1993), 1197–1202.

[46] P. L. Lions, *Conditions at infinity for Boltzmann's equation*, Comm. Partial Differential Equations, **19** (1994), 335–367.

[47] P. L. Lions, *On Boltzmann and Landau equations*, Phil. Trans. Roy. Soc. London Ser. A, **346** (1994), 191–204.

[48] P. L. Lions, *Compactness in Boltzmann's equation via Fourier integral operators and applications. Parts I–III*, to appear in J. Math. Kyoto Univ., 1994.

[49] P. L. Lions, Mathematical Topics in Fluid Mechanics., to appear in Oxford Univ. Press.

[50] P. L. Lions and T. Paul, *Sur les mesures de Wigner*, Rev. Mat. Iberoamericana, **9** (1993), 553–618.

[51] P. L. Lions, B. Perthame, and P. E. Souganidis, *Existence and compactness of entropy solutions for the one-dimensional isentropic gas dynamics systems*, to appear in Comm. Pure Appl. Math.

[52] P. L. Lions, B. Perthame, and E. Tadmor, *A kinetic formulation of multidimensional scalar conservation laws and related equations*, J. Amer. Math. Soc., **7** (1994), 169–191.

[53] P. L. Lions, B. Perthame, and E. Tadmor, *Kinetic formulation of the isentropic gas dynamics and p-systems*, to appear in Comm. Math. Phys.

[54] A. Majda, Compressible Fluid Flow and Systems of Conservation Laws in Several Space Variables, Springer, Berlin and New York, 1984.

[55] A. Majda, *Mathematical fluid dynamics: The interaction of nonlinear analysis and modern applied mathematics*, in Proceedings of the AMS Centennial Symposium, August 8–12, 1988.

[56] A. Majda, *The interaction of Nonlinear Analysis and Modern Applied Mathematics*, in Proc. Internat. Congress Math., Kyoto, 1990, vol. I, Springer, Berlin and New York, 1991.

[57] J. C. Maxwell, Scientific papers. Vol. 2, Cambridge Univ. Press, Cambridge, 1880 (Reprinted by Dover Publications, New York, 1965).

[58] J. C. Maxwell, *On the dynamical theory of gases*, Phil. Trans. Roy. Soc. London Ser. A, **157** (1886), 49–88.

[59] S. Müller, *A surprising higher integrability property of mappings with positive determinant*, Proc. Amer. Math. Soc., **21** (1989), 245–248.

[60] F. Murat, *Compacité par compensation*, Ann. Scuola Norm. Sup. Pisa Cl. Sci (4), **5** (1978), 489–507; *II*, in Proceedings of the International Meeting on Recent Methods on Nonlinear Analysis (E. De Giorgi, E. Magenes, and U. Mosco, eds.), Pitagora, Bologna, 1979; *III*, Ann. Scuola Norm. Sup. Pisa Cl. Sci (4), **8** (1981), 69–102.

[61] T. Nishida, *Fluid dynamical limit of the nonlinear Boltzmann equation to the level of the compressible equation*, Comm. Math. Phys., **61** (1978), 119–148.

[62] D. H. Phong and E. Stein, *Hilbert integrals, singular integrals and Radon transforms*, Ann. of Math. (2).

[63] D. Serre, *Solutions faibles globales des équations de Navier-Stokes pour un fluide compressible*, C. R. Acad. Sci. Paris Série I, **303** (1986), 629–642.

[64] D. Serre, *Sur l'équation monodimensionnelle d'un fluide visqueux, compressible et conducteur de chaleur*, C. R. Acad. Sci. Paris Série I, **303** (1986), 703–706.

[65] D. Serre, *Variations de grande amplitude pour la densité d'un fluide visqueux compressible*, Phys. D, **48** (1991), 113–128.

[66] E. Stein, *Oscillatory integrals in Fourier analysis*, in Beijing Lectures in Harmonic Analysis (E. Stein, ed.), Princeton Univ. Press, Princeton, NJ (1986), 307–355.

[67] E. Stein and G. Weiss, *On the theory of H^p spaces*, Acta Math., **103** (1960), 25–62.

[68] L. Tartar, *Compensated compactness and applications to partial differential equations*, in Nonlinear Analysis and Mechanics: Heriot-Watt Symposium, vol. 4 (R. J. Knops, ed.), Research Notes in Math., Pitman, London, 1979.

[69] L. Tartar, *The compensated compactness method applied to systems of conservation laws*, in Systems of Nonlinear Partial Differential Equations (J. M. Ball, ed.), NATO ASI Series C III, Reidel, New York, 1983.

[70] L. Tartar, *H-measures, a new approach for studying homogenization, oscillations and concentration effects in partial differential equations*, Proc. Roy. Soc. Edinburgh Sect. A, **115** (1990), 193–230.

[71] S. Ukaï and K. Asano, *The Euler limit and initial layer of the nonlinear Boltzmann equation*, Hokkaido Math. J., **12** (1983), 311–332.

[72] B. Wennberg, *Regularity estimates for the Boltzmann equation*, preprint **2**, 1994, Chalmers Univ., Göteborg.

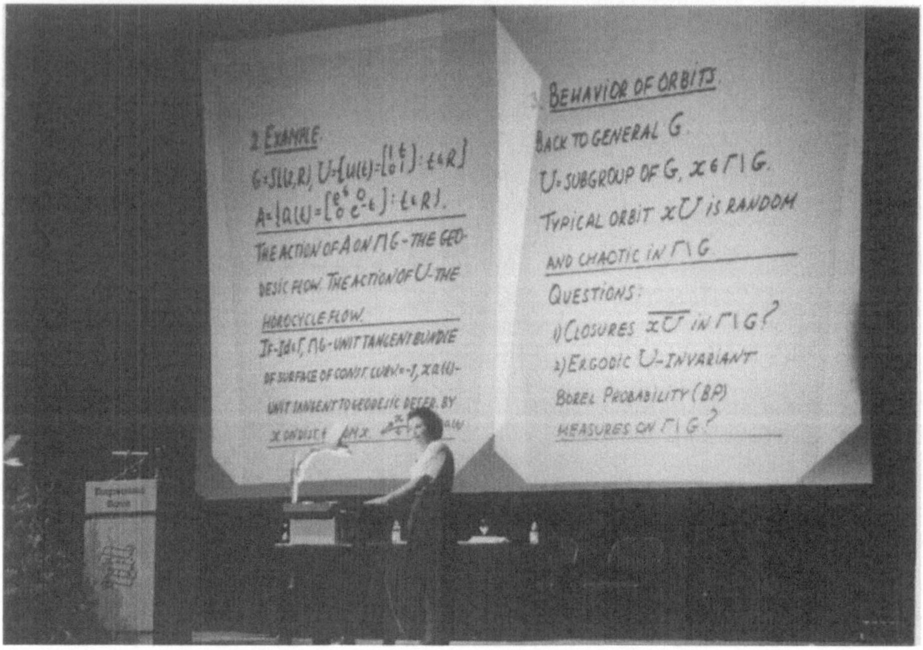

Marina Ratner, first plenary speaker

Interactions Between Ergodic Theory, Lie Groups, and Number Theory

MARINA RATNER

Department of Mathematics
Stanford University
Stanford, CA 94305, USA

and

Department of Mathematics
University of California
Berkeley, CA 94720, USA

Introduction

In this paper we discuss the use of dynamical and ergodic-theoretic ideas and methods to solve some long-standing problems originating from Lie groups and number theory. These problems arise from looking at actions of Lie groups on their homogeneous spaces. Such actions, viewed as dynamical systems, have long been interesting and rich objects of ergodic theory and geometry. Since the 1930s ergodic-theoretic methods have been applied to the study of geodesic and horocycle flows on unit tangent bundles of compact surfaces of negative curvature. From the algebraic point of view the latter flows are examples of semisimple and unipotent actions on finite-volume homogeneous spaces of real Lie groups. It was established in the 1960s through the fundamental work of D. Ornstein that typical semisimple actions are all statistically the same due to their extremal randomness caused by exponential instability of orbits. Their algebraic nature has little to do with the isomorphism problem for such actions: they are measure-theoretically isomorphic as long as their entropies coincide.

In striking contrast, unipotent actions (all having zero entropy), though random and chaotic from a dynamical point of view, were found to be rigidly linked to the algebraic structure of the underlying homogeneous space. In 1981 it was shown by the author that measure theoretic isomorphisms of horocycle flows must be algebraic and imply the isometry of the underlying surfaces. Subsequently, further "rigidities" of an algebraic nature have been found.

While the study of this "rigidity" phenomenon was underway, a powerful impetus came from number theory. Around 1980 Raghunathan made a remarkable observation that the long-standing Oppenheim conjecture on the density of values of irrational quadratic forms at integral points would follow if it were true that closures of orbits of certain unipotent subgroups $\mathbf{U} \subset SL(3,\mathbb{R})$ acting on $SL(3,\mathbb{Z})\backslash SL(3,\mathbb{R})$ were merely orbits of larger groups containing $\mathbf{U}$. The latter result was proved by Margulis in 1986.

Raghunathan's observation led him to propose a general conjecture on orbit closures of unipotent actions. In 1990 it was shown by the author that ergodic-theoretic methods (some of which we developed previously for horocycle flows) can

Proceedings of the International Congress
of Mathematicians, Zürich, Switzerland 1994

be applied to solve this and other related conjectures. This made it possible to answer further number theoretic questions and stimulated subsequent developments in ergodic theory and dynamics of subgroup actions on homogeneous spaces.

This paper consists of six sections. In Sections 1 and 2 we introduce the necessary defintions and state conjectures and results prior to 1990. In Section 3 we state and discuss new results for real Lie groups and in Section 4 we give p-adic and $\mathbb{S}$-arithmetic generalizations of these results. In Section 5 we discuss applications to number theory and in Section 6 applications to ergodic theory and the "rigidity" phenomenon of unipotent actions.

It should be noted that this paper discusses only some of the many important topics that fall under its title.

I thank A. Borel, G. Prasad and D. Witte for their valuable comments on the preliminary version of this paper.

1 Definitions

Let $\mathbf{G}$ be a locally compact second countable topological group, $\mathbf{\Gamma}$ a discrete subgroup of $\mathbf{G}$, and $\mathbf{\Gamma}\backslash\mathbf{G} = \{\mathbf{\Gamma h} : \mathbf{h} \in \mathbf{G}\}$. We shall denote by $\pi : \mathbf{G} \to \mathbf{\Gamma}\backslash\mathbf{G}$ the covering projection $\pi(\mathbf{h}) = \mathbf{\Gamma h}$, $\mathbf{h} \in \mathbf{G}$. The group $\mathbf{G}$ acts on $\mathbf{\Gamma}\backslash\mathbf{G}$ by right translations: $x \to x\mathbf{g}$, $x \in \mathbf{\Gamma}\backslash\mathbf{G}$, $\mathbf{g} \in \mathbf{G}$. We study the dynamics of this action.

Let $\{\mathbf{x}_n\}$ be a sequence in $\mathbf{G}$ and let $\mathbf{e}$ denote the identity element of $\mathbf{G}$. We say that $x_n = \pi(\mathbf{x}_n)$ *cuspidally diverges* in $\mathbf{\Gamma}\backslash\mathbf{G}$ if there are $\mathbf{e} \neq \boldsymbol{\gamma}_n \in \mathbf{\Gamma}$, $n = 1, 2, \ldots$ such that $\mathbf{x}_n^{-1}\boldsymbol{\gamma}_n\mathbf{x}_n \to \mathbf{e}$ as $n \to \infty$. (This means that a left invariant distance between $\mathbf{x}_n$ and $\boldsymbol{\gamma}_n\mathbf{x}_n$ tends to zero as $n \to \infty$ or that the sequence $\{x_n\}$ escapes to the cusps of $\mathbf{\Gamma}\backslash\mathbf{G}$.) For $\mathbf{g} \in \mathbf{G}$ the set

$$\mathcal{D}(\mathbf{g}) = \{x \in \mathbf{\Gamma}\backslash\mathbf{G} : x\mathbf{g}^n \text{ cuspidally diverges as } n \to \infty\}$$

is called the *divergent* set of $\mathbf{g}$. It is clear that if $\mathcal{D}(\mathbf{g}) \neq \emptyset$ for some $\mathbf{g} \in \mathbf{G}$ then $\mathbf{\Gamma}\backslash\mathbf{G}$ is not compact.

The group $\mathbf{\Gamma}$ is called a *lattice* in $\mathbf{G}$ if there is a *finite* $\mathbf{G}$-invariant measure $\nu_{\mathbf{G}}$ on $\mathbf{\Gamma}\backslash\mathbf{G}$. (In this case we shall assume that $\nu_{\mathbf{G}}$ is a *probability* measure, i.e. $\nu_{\mathbf{G}}(\mathbf{\Gamma}\backslash\mathbf{G}) = 1$.) Then a sequence $\{x_n\}$ in $\mathbf{\Gamma}\backslash\mathbf{G}$ cuspidally diverges if and only if it eventually leaves every compact subset of $\mathbf{\Gamma}\backslash\mathbf{G}$ (see [R]).

Now let $\mathbf{U}$ be a subgroup of $\mathbf{G}$ and $x \in \mathbf{\Gamma}\backslash\mathbf{G}$. The set $x\mathbf{U} = \{x\mathbf{u} : \mathbf{u} \in \mathbf{U}\}$ is called the $\mathbf{U}$-orbit of x. A typical orbit $x\mathbf{U}$ in $\mathbf{\Gamma}\backslash\mathbf{G}$ is random and chaotic.

We pose the following questions:

(1) What are the *closures* of orbits $x\mathbf{U}$ in $\mathbf{\Gamma}\backslash\mathbf{G}$?

(2) What are the *ergodic* $\mathbf{U}$-invariant Borel probability measures on $\mathbf{\Gamma}\backslash\mathbf{G}$? (A $\mathbf{U}$-invariant probability measure μ on $\mathbf{\Gamma}\backslash\mathbf{G}$ is ergodic if every $\mathbf{U}$-invariant measurable subset of $\mathbf{\Gamma}\backslash\mathbf{G}$ has μ-measure zero or one.)

Let us give a few natural examples. Suppose $\mathbf{G}$ is a real Lie group, $\mathbf{U} = \{\mathbf{u}(t) : t \in \mathbb{R}\}$ a one-parameter subgroup of $\mathbf{G}$, and $x\mathbf{U}$ a periodic orbit. Then $x\mathbf{U} = \overline{x\mathbf{U}}$ and the normalized length measure on $x\mathbf{U}$ is $\mathbf{U}$-invariant and ergodic.

For a more general example suppose that the closure $\overline{x\mathbf{U}}$ coincides with the orbit of a larger group $\mathbf{H}$ containing $\mathbf{U}$, i.e. $\overline{x\mathbf{U}} = x\mathbf{H}$. In addition, it might happen that $x\mathbf{H}$ is the support of an $\mathbf{H}$-invariant Borel probability measure $\nu_{\mathbf{H}}$ (this happens if and only if $\mathbf{xHx}^{-1} \cap \mathbf{\Gamma}$ is a lattice in $\mathbf{xHx}^{-1}$, $\mathbf{x} \in \pi^{-1}\{x\}$) that is ergodic for the action of $\mathbf{U}$.

These examples motivate the following definitions.

DEFINITION 1. A subset $A \subset \mathbf{\Gamma}\backslash\mathbf{G}$ is called *homogeneous* if there exists a closed subgroup $\mathbf{H} \subset \mathbf{G}$ and a point $x \in \mathbf{\Gamma}\backslash\mathbf{G}$ such that $A = x\mathbf{H}$ and $x\mathbf{H}$ is the support of an $\mathbf{H}$-invariant Borel probability measure $\nu_{\mathbf{H}}$.

We emphasize that this definition of $x\mathbf{H}$ being homogeneous is different from the commonly used one where the existence of a *finite* $\mathbf{H}$-invariant measure on $x\mathbf{H}$ is not required.

DEFINITION 2. A Borel probability measure μ on $\mathbf{\Gamma}\backslash\mathbf{G}$ is *algebraic* if there exist $x \in \mathbf{\Gamma}\backslash\mathbf{G}$ and a closed subgroup $\mathbf{H} \subset \mathbf{G}$ such that $x\mathbf{H}$ is homogeneous and $\mu = \nu_{\mathbf{H}}$.

Equivalently, μ is algebraic if there is $x \in \mathbf{\Gamma}\backslash\mathbf{G}$ such that $\mu(x\mathbf{\Lambda}(\mu)) = 1$, where

$$\mathbf{\Lambda}(\mu) = \{\mathbf{g} \in \mathbf{G} : \text{ the action of } \mathbf{g} \text{ on } \mathbf{\Gamma}\backslash\mathbf{G} \text{ preserves } \mu\}.$$

It is rather exceptional for a subgroup $\mathbf{U}$ to have homogeneous orbit closures or algebraic ergodic measures. However, there are some $\mathbf{U}$ for which this happens. To characterize these $\mathbf{U}$ we need the following definitions.

Let $\mathbf{G}$ be a Lie group over a field κ (where κ is either the real field or a p-adic field) with the Lie algebra $\mathfrak{G}$. For $\mathbf{g} \in \mathbf{G}$ let $\mathrm{Ad}_{\mathbf{g}} : \mathfrak{G} \to \mathfrak{G}$ denote the differential at the identity of the map $\mathbf{h} \to \mathbf{g}^{-1}\mathbf{hg}$, $\mathbf{h} \in \mathbf{G}$. Then $\mathrm{Ad}_{\mathbf{g}}$ (called the adjoint map of $\mathbf{g}$) is a linear automorphism of $\mathfrak{G}$.

It is a fact that there is a neighborhood $\mathfrak{O}$ of zero in $\mathfrak{G}$ such that the exponential map $\exp : \mathfrak{O} \to \mathbf{G}$ is well defined on $\mathfrak{O}$ and maps $\mathfrak{O}$ diffeomorphically onto a neighborhood of $\mathbf{e}$ in $\mathbf{G}$. (When $\kappa = \mathbb{R}$ the map exp is defined on all of $\mathfrak{G}$.) If $\mathbf{x}, \mathbf{y} \in \mathbf{G}$ and $\mathbf{y} = \mathbf{x} \exp v$ for some $v \in \mathfrak{O}$ with $\mathrm{Ad}_{\mathbf{g}^r}(v) \in \mathfrak{O}$ for all $r = 1, \ldots, n$ and some $0 \leq n \in \mathbb{Z}$ then $\mathbf{yg}^r = \mathbf{xg}^r \exp(\mathrm{Ad}_{\mathbf{g}^r}(v))$ for all $r = 1, \ldots, n$. Thus $\mathrm{Ad}_{\mathbf{g}^r}$ characterizes the divergence of $\mathbf{yg}^r$ from $\mathbf{xg}^r$ when r runs from 1 to n.

An element $\mathbf{u} \in \mathbf{G}$ is called Ad-*unipotent* if $\mathrm{Ad}_{\mathbf{u}}$ is a unipotent element of $GL(\mathfrak{n}, \kappa)$, $\mathfrak{n} = \dim \mathfrak{G}$, i.e. every eigenvalue of $\mathrm{Ad}_{\mathbf{u}}$ equals one. Then $\mathrm{Ad}_{\mathbf{u}^r} = \sum_{k=0}^{m}(r^k T_{\mathbf{u}}^k)/k!$ for all $r \in \mathbb{Z}$ and some integer $m \geq 0$, where $T_{\mathbf{u}}$ is a nilpotent endomorphism of $\mathfrak{G}$. This polynomial (in r) form of $\mathrm{Ad}_{\mathbf{u}^r}$ plays a crucial role in all of the results stated below. It shows that Ad-unipotent orbits diverge *polynomially.*

A subgroup $\mathbf{U} \subset \mathbf{G}$ is Ad-*unipotent* if each $\mathbf{u} \in \mathbf{U}$ is Ad-unipotent. A subgroup $\mathbf{U} \subset GL(\mathfrak{n}, \kappa)$ is *unipotent* if each $\mathbf{u} \in \mathbf{U}$ is unipotent. A unipotent $\mathbf{U} \subset GL(\mathfrak{n}, \kappa)$ is Ad-unipotent.

Now let u be an ad-nilpotent element of $\mathfrak{G}$ (this means that the map $\mathrm{ad}_u : \mathfrak{G} \to \mathfrak{G}$, $\mathrm{ad}_u(v) = [v, u]$ is a nilpotent linear transformation of $\mathfrak{G}$). An element $a \in \mathfrak{G}$ is called "*diagonal*" for u if there exists an ad-nilpotent element $u^* \in \mathfrak{G}$ (called an "*opposite*" for u) such that

$$\mathrm{ad}_{u^*}(u) = a, \ \mathrm{ad}_a(u) = -2u, \ \mathrm{ad}_a(u^*) = 2u^*.$$

This terminology is motivated by the fact that u, u^* generate a Lie subalgebra $sl_2(u,a)$ of $\mathfrak{G}$ isomorphic to $sl(2,\kappa)$.

Now let $t \to \mathbf{u}(t)$ be a continuous (hence analytic) homomorphism from κ (as an additive group) to $\mathbf{G}$ with $u = d\mathbf{u}(t)/dt|_{t=0} \neq 0$. The latter condition implies that if κ is a p-adic field then the map $t \to \mathbf{u}(t)$ is one-to-one. We call $\mathbf{U} = \{\mathbf{u}(t) : t \in \kappa\}$ a *one-parameter* subgroup of $\mathbf{G}$ with tangent $u \in \mathfrak{G}$. Then $\mathbf{U}$ is Ad-unipotent if and only if u is ad-nilpotent.

2 Conjectures and Results Prior to 1990

CONJECTURE 1. (Raghunathan's Topological Conjecture) Let $\mathbf{G}$ be a real connected Lie group and $\mathbf{U}$ an Ad-unipotent subgroup of $\mathbf{G}$. Then given any lattice $\mathbf{\Gamma}$ of $\mathbf{G}$ and any $x \in \mathbf{\Gamma}\backslash\mathbf{G}$ the closure of the orbit $x\mathbf{U}$ in $\mathbf{\Gamma}\backslash\mathbf{G}$ is homogeneous.

CONJECTURE 2. (Raghunathan's Measure Conjecture) Let $\mathbf{G}$ and $\mathbf{U}$ be as in Conjecture 1. Then given any lattice $\mathbf{\Gamma}$ in $\mathbf{G}$ every ergodic $\mathbf{U}$-invariant Borel probability measure on $\mathbf{\Gamma}\backslash\mathbf{G}$ is algebraic.

Actually Raghunathan proposed a weaker version of Conjecture 1 and showed its connection with the long-standing Oppenheim conjecture on the density of values of irrational quadratic forms at integral points (see Section 5 below). The latter version as well as Conjecture 2 were stated by Dani [D1] in 1981 for reductive $\mathbf{G}$ and one-parameter $\mathbf{U}$ and by Margulis [M1, Conjectures 2 and 3] in 1986 for general $\mathbf{G}$ and $\mathbf{U}$. (Raghunathan did not propose Conjecture 2. We gave the latter his name because it represents a natural measure-theoretic analogue of his topological conjecture.)

CONJECTURE 3. (Margulis [M1, Conjecture 1], [M2, Conjecture 2]) Let $\mathbf{G}$ be a real connected Lie group and $\mathbf{U}$ a subgroup of $\mathbf{G}$ generated by Ad-unipotent elements of $\mathbf{G}$. Then given any lattice $\mathbf{\Gamma}$ in $\mathbf{G}$ and any $x \in \mathbf{\Gamma}\backslash\mathbf{G}$ the closure of $x\mathbf{U}$ in $\mathbf{\Gamma}\backslash\mathbf{G}$ is homogeneous.

In fact, Margulis proposed a weaker version of this conjecture. Conjecture 3 generalizes Conjecture 1 to a class of subgroups $\mathbf{U}$ much larger than Ad-unipotent subgroups. For example, every connected semisimple $\mathbf{U}$ without compact factors is generated by Ad-unipotent elements of $\mathbf{G}$.

It was shown earlier by Furstenberg [Fu1] and Parry [P1] (see also [AGH]) that Conjectures 1 and 2 hold for one-parameter and one-generator subgroups of *nilpotent* $\mathbf{G}$. Also Starkov [St2] proved Conjecture 1 for one-parameter Ad-unipotent subgroups of *solvable* $\mathbf{G}$ with $\mathbf{\Gamma}$ being an arbitrary closed subgroup of $\mathbf{G}$ such that $\mathbf{\Gamma}\backslash\mathbf{G}$ has finite $\mathbf{G}$-invariant measure. Conjecture 2 for the latter case follows from [St2] and [P1] (in fact, without the assumption of $\mathbf{\Gamma}\backslash\mathbf{G}$ being of finite $\mathbf{G}$-invariant measure).

As for *semisimple* $\mathbf{G}$, Hedlund [H] showed that if $\mathbf{G} = SL(2,\mathbb{R})$ and $\mathbf{\Gamma}\backslash\mathbf{G}$ is compact (in this case $\mathbf{\Gamma}$ is called a uniform lattice in $\mathbf{G}$) then the action of a unipotent one-parameter subgroup $\mathbf{U}$ of $\mathbf{G}$ on $\mathbf{\Gamma}\backslash\mathbf{G}$ is minimal (i.e. every orbit of $\mathbf{U}$ is dense). Subsequently, Furstenberg [Fu2] proved that in this case the action of $\mathbf{U}$ is uniquely ergodic.

It is a fact that one-parameter unipotent subgroups of $\mathbf{G} = SL(2,\mathbb{R})$ are horospherical. A subgroup $\mathbf{U}$ of a Lie group $\mathbf{G}$ is called horospherical if there exists $\mathbf{g} \in \mathbf{G}$ such that

$$\mathbf{U} = \{\mathbf{u} \in \mathbf{G} : \mathbf{g}^{-n}\mathbf{u}\mathbf{g}^{n} \to \mathbf{e} \text{ as } n \to \infty\}$$

where $\mathbf{e}$ denotes the identity element of $\mathbf{G}$.

Generalizing Furstenberg's Theorem, Bowen [Bw], Veech [V] and Ellis and Perrizo [EPe] showed that if $\mathbf{\Gamma}$ is a uniform lattice in a connected semisimple Lie group $\mathbf{G}$ without compact factors then ergodic actions of horospherical subgroups on $(\mathbf{\Gamma}\backslash\mathbf{G}, \nu_{\mathbf{G}})$ are uniquely ergodic. Adapting the method of Furstenberg and Veech, Dani [D1] proved Conjecture 2 when $\mathbf{G}$ is reductive and $\mathbf{U}$ is a maximal horospherical subgroup of $\mathbf{G}$.

As for Conjecture 1, Dani [D3] proved it for horospherical subgroups of reductive $\mathbf{G}$. Also Dani and Margulis [DM2, 3] showed that Conjecture 1 holds for one-parameter unipotent subgroups of $SL(3,\mathbb{R})$.

Dani and Margulis [DM1] proved Conjecture 3 for $\mathbf{G} = SL(3,\mathbb{R})$, $\mathbf{\Gamma} = SL(3,\mathbb{Z})$, and $\mathbf{U} = SO(2,1)^0$.

It should be noted that in 1986 Dani showed [D2, Theorem 3.5] that if $\mathbf{G}$ is a connected semisimple Lie group and $\mathbf{\Gamma}$ a lattice in $\mathbf{G}$ then given $\varepsilon > 0$ there is a compact $K(\varepsilon) \subset \mathbf{\Gamma}\backslash\mathbf{G}$ such that for any $x \in \mathbf{\Gamma}\backslash\mathbf{G}$ and any one-parameter Ad-unipotent subgroup $\mathbf{U} = \{\mathbf{u}(t) : t \in \mathbb{R}\}$ of $\mathbf{G}$ either $\lambda\{t \in [0,T] : x\mathbf{u}(t) \in K(\varepsilon)\} > (1-\varepsilon)T$ for all large T or $x\mathbf{L}$ is homogeneous for some *proper* closed connected subgroup $\mathbf{L}$ of $\mathbf{G}$ containing $\mathbf{U}$. (Here λ denotes the Lebesgue measure on $\mathbb{R}$.) This important result is used in the proofs of Theorems 6 and 8–10 below.

3 New Results (1990 and After)

All Lie groups in this section are assumed to be real, and, unless otherwise stated, the results below are due to the author.

THEOREM 1 (Classification of ergodic invariant measures for Ad-unipotent actions). *Let $\mathbf{G}$ be a connected Lie group and $\mathbf{U}$ an Ad-unipotent subgroup of $\mathbf{G}$. Then given any discrete subgroup $\mathbf{\Gamma}$ (not necessarily a lattice) of $\mathbf{G}$ every ergodic $\mathbf{U}$-invariant Borel probability measure on $\mathbf{\Gamma}\backslash\mathbf{G}$ is algebraic.*

THEOREM 2. *Let $\mathbf{G}$ be a connected Lie group and $\mathbf{U}$ a Lie subgroup of $\mathbf{G}$ of the form $\mathbf{U} = \cup_{i=1}^{\infty}\mathbf{u}_i\mathbf{U}^0$, where $\mathbf{u}_i$ are Ad-unipotent in $\mathbf{G}$, $i = 1, 2, \ldots$, $\mathbf{U}/\mathbf{U}^0$ is finitely generated, and the identity component $\mathbf{U}^0$ is generated by Ad-unipotent elements of $\mathbf{G}$ contained in $\mathbf{U}^0$. Then Theorem 1 holds for $\mathbf{U}$.*

Theorem 1 is stronger than Conjecture 2 and Theorem 2 extends it to groups generated by Ad-unipotent elements.

THEOREM 3 (Orbit closures for Ad-unipotent actions). *Let $\mathbf{G}$ and $\mathbf{U}$ be as in Theorem 1. Then given any lattice $\mathbf{\Gamma}$ in $\mathbf{G}$ and any $x \in \mathbf{\Gamma}\backslash\mathbf{G}$ the closure of the orbit $x\mathbf{U}$ in $\mathbf{\Gamma}\backslash\mathbf{G}$ is homogeneous.*

THEOREM 4. *Let* $\mathbf{G}$ *and* $\mathbf{U}$ *be as in Theorem 2. Then Theorem 3 holds for* $\mathbf{U}$. *Moreover, if* $\mathbf{U}$ *is connected, then for any lattice* $\mathbf{\Gamma}$ *in* $\mathbf{G}$ *and any* $x \in \mathbf{\Gamma}\backslash\mathbf{G}$ *there exists a closed connected subgroup* $\mathbf{H}$ *of* $\mathbf{G}$ *containing* $\mathbf{U}$ *and a one-parameter subgroup* $\mathbf{V}$ *of* $\mathbf{U}$ *Ad-unipotent in* $\mathbf{G}$ *such that* $\overline{x\mathbf{V}} = \overline{x\mathbf{U}} = x\mathbf{H}$ *is homogeneous and* $\mathbf{V}$ *acts ergodically on* $(x\mathbf{H}, \nu_{\mathbf{H}})$, *where* $\nu_{\mathbf{H}}$ *denotes the* $\mathbf{H}$*-invariant Borel probability measure on* $\mathbf{\Gamma}\backslash\mathbf{G}$ *supported on* $x\mathbf{H}$.

Conjecture 1 is implied by Theorem 3 and Theorem 4 extends it to groups generated by Ad-unipotent elements.

THEOREM 5. *Let* $\mathbf{G}$ *be a connected Lie group,* $\mathbf{\Gamma}$ *a discrete subgroup of* $\mathbf{G}$, *and* $x \in \mathbf{\Gamma}\backslash\mathbf{G}$. *Let* $\mathcal{A}_x$ *denote the set of all closed connected* $\mathbf{H} \subset \mathbf{G}$ *such that* $x\mathbf{H}$ *is homogeneous and there is a one-parameter subgroup* $\mathbf{U} \subset \mathbf{H}$ *Ad-unipotent in* $\mathbf{G}$ *acting ergodically on* $(x\mathbf{H}, \nu_{\mathbf{H}})$. *Then* $\mathcal{A}_x$ *is countable.*

We show that in order to prove Theorem 3 for general Ad-unipotent $\mathbf{U}$ it suffices to prove it for one-parameter Ad-unipotent $\mathbf{U}$. But for such $\mathbf{U}$ we have the far stronger Theorem 6 below. To state it we need to introduce a definition.

DEFINITION 3. Let $\mathbf{U} = \{\mathbf{u}(t) : t \in \mathbb{R}\}$ be an arbitrary one-parameter subgroup of $\mathbf{G}$. A point $x \in \mathbf{\Gamma}\backslash\mathbf{G}$ is called *generic* for $\mathbf{U}$ if there exists a closed subgroup $\mathbf{H} \subset \mathbf{G}$ such that $\overline{x\mathbf{U}} = x\mathbf{H}$ is homogeneous and $\frac{1}{t}\int_0^t f(x\mathbf{u}(s))\,ds \xrightarrow[t\to\infty]{} \int_{\mathbf{\Gamma}\backslash\mathbf{G}} f\,d\nu_{\mathbf{H}}$ for every bounded continuous function f on $\mathbf{\Gamma}\backslash\mathbf{G}$.

A similar definition can be given for a one-generator $\mathbf{U} = \{\mathbf{u}^k : k \in \mathbb{Z}\}$ replacing the integral by the sum $\sum_{k=0}^{n-1} f(x\mathbf{u}^k)/n$.

THEOREM 6 (Uniform distribution of Ad-unipotent flows). *Let* $\mathbf{G}$ *be a connected Lie group and* $\mathbf{U}$ *a one-parameter or one-generator* Ad-*unipotent subgroup of* $\mathbf{G}$. *Then given any lattice* $\mathbf{\Gamma}$ *of* $\mathbf{G}$ *every point* $x \in \mathbf{\Gamma}\backslash\mathbf{G}$ *is generic for* $\mathbf{U}$ *and* $\mathbf{U}$ *acts ergodically on* $(\overline{x\mathbf{U}} = x\mathbf{H}, \nu_{\mathbf{H}})$.

This theorem was proved one month before it was conjectured by Margulis at the ICM 1990 in Kyoto, Japan [M2, Conjectures 3 and 4].

Theorem 6 for nilpotent $\mathbf{G}$ follows from [P1] (see also [L]) and for $\mathbf{G} = SL(2,\mathbb{R})$ it was proved earlier by Dani and Smillie [DSm]. Also Shah [S1] proved it for semisimple $\mathbf{G}$ of real rank 1. Their methods are totally different from the author's.

To derive the results stated above we first prove Theorem 1 for one-parameter Ad-unipotent $\mathbf{U}$. Theorem 7 below plays a crucial role in this proof. To state it we introduce the following definition. Let $\mathbf{U} = \{\mathbf{u}(t) = \exp tu : t \in \mathbb{R}\}$ be a one-parameter Ad-unipotent subgroup of $\mathbf{G}$ and assume there is a "diagonal" element $a \in \mathfrak{G}$ for u (see Section 1). Then we call $\mathbf{A} = \{\mathbf{a}(t) = \exp ta : t \in \mathbb{R}\}$ "diagonal" for $\mathbf{U}$ and denote by $SL_2(\mathbf{U}, \mathbf{A})$ the connected subgroup of $\mathbf{G}$ with the Lie algebra $sl_2(u, a)$ (see Section 1). It is clear that $\mathbf{A}$ is "diagonal" for $\mathbf{U}$ if and only if $\mathbf{c}\mathbf{A}\mathbf{c}^{-1}$ is so for every $\mathbf{c} \in \mathbf{C}(\mathbf{U})$ — the centralizer of $\mathbf{U}$ in $\mathbf{G}$.

THEOREM 7. *Let* $\mathbf{G}$ *be a Lie group,* $\mathbf{\Gamma}$ *a discrete subgroup of* $\mathbf{G}$, *and* $\mathbf{U} = \{\mathbf{u}(t) : t \in \mathbb{R}\}$ *a one-parameter* Ad-*unipotent subgroup of* $\mathbf{G}$. *Suppose there is a "diagonal"* $\mathbf{A} = \{\mathbf{a}(t) : t \in \mathbb{R}\}$ *for* $\mathbf{U}$ *in* $\mathbf{G}$ *and let* μ *be an ergodic* $\mathbf{U}$*-invariant Borel probability* *measure on* $\mathbf{\Gamma}\backslash\mathbf{G}$. *Then either* (1) $\mu(\mathcal{D}(\mathbf{a}(t))) = 1$ *for all* $t > 0$ *or* (2) μ *is algebraic and is preserved by* $\mathbf{c}SL_2(\mathbf{U}, \mathbf{A})\mathbf{c}^{-1}$ *for some* $\mathbf{c} \in \mathbf{C}(\mathbf{U})$.

Recall that $\mathcal{D}(\mathbf{g})$, $\mathbf{g} \in \mathbf{G}$, denotes the divergent set of $\mathbf{g}$ (see Section 1).

The central role in the proof of Theorem 7 in [Ra9] is played by a dynamical property of Ad-unipotent actions which we introduced in [Ra8, Theorem 3.1] and called the R-property. It is a consequence of the polynomial divergence of Ad-unipotent orbits. Also it is a generalization of the H-property for horocycle flows introduced in [Ra4] (see Section 6 below).

The R-property states, roughly speaking, that given $0 < \varepsilon < 1$ there exists $0 < \eta(\varepsilon) < 1$ such that if $\mathbf{F}$ is an appropriate sufficiently large open rectangular subset of a closed connected simply connected Ad-unipotent subgroup $\mathbf{U}$ of $\mathbf{G}$ with $\mathbf{e} \in \mathbf{F}$ and $\sup\{d_{\mathbf{G}}(\mathbf{u}, \mathbf{xU}) : \mathbf{u} \in \mathbf{F}\} = \theta$ for some small $\theta > 0$ and some $\mathbf{x} \notin \mathbf{U}$, $d_{\mathbf{G}}(\mathbf{e}, \mathbf{x}) \leq \theta$, then there exists $\mathbf{A} \subset \mathbf{F}$ such that $(1 - \varepsilon)\theta \leq d_{\mathbf{G}}(\mathbf{u}, \mathbf{xU}) \leq \theta$ for all $\mathbf{u} \in \mathbf{A}$ and $\lambda(\mathbf{A}) \geq \eta(\varepsilon)\lambda(\mathbf{F})$, where λ denotes a Haar measure on $\mathbf{U}$ and $d_{\mathbf{G}}$ denotes a left invariant metric on $\mathbf{G}$. Moreover, if $\mathbf{u} \in \mathbf{F}$ and $d_{\mathbf{G}}(\mathbf{u}, \mathbf{xU}) = d_{\mathbf{G}}(\mathbf{u}, \mathbf{ur}(\mathbf{u}))$ for some $\mathbf{r}(\mathbf{u}) \in \mathbf{G}$ with $\mathbf{ur}(\mathbf{u}) \in \mathbf{xU}$ and $d_{\mathbf{G}}(\mathbf{e}, \mathbf{r}(\mathbf{u})) \leq \theta$ then $\mathbf{r}(\mathbf{u})$ is close to the normalizer of $\mathbf{U}$ in $\mathbf{G}$ and this closeness tends to zero as the sides of the rectangular set $\mathbf{F}$ tend to infinity. (When $\mathbf{U} = \{\mathbf{u}(t) : t \in \mathbb{R}\}$ is a one-parameter subgroup of $\mathbf{G}$ we can take $\mathbf{F} = \{\mathbf{u}(t) : 0 \leq t \leq T\}$ for large $T > 0$.)

The rectangular sets $\mathbf{F}$ in the description of the R-property are Fϕlner subsets of $\mathbf{U}$ (see [Ra8]) and the Birkhoff Ergodic Theorem for measure preserving actions of $\mathbf{U}$ holds for averages performed over $\mathbf{F}$.

It should be noted that the R-property and the Birkhoff Ergodic Theorem are *the only* basic facts used in the proof of Theorem 7.

Using Theorem 1 (proved in [Ra8–10]) and Theorem 5 (whose proof is simple [Ra10, Theorem 1.1]) we deduce Theorem 6 [Ra11]. These two proofs (of Theorem 1 and of Theorem 6 from Theorems 1 and 5) are central. Note that Theorem 6 implies Theorem 3 for one-parameter Ad-unipotent $\mathbf{U}$. In [Ra14, Section 8] we outlined (using Theorem 5) how the validity of Theorems 1 and 3 for one-parameter Ad-unipotent $\mathbf{U}$ implies their validity for higher-dimensional connected $\mathbf{U}$ generated by Ad-unipotent elements of $\mathbf{G}$ (see [Ra10] and [Ra11]).

Let us outline the main idea used in [Ra11, Proof of Theorem 2.1] to deduce Theorem 6 from Theorems 1 and 5.

Let $\mathbf{U} = \{\mathbf{u}(t) : t \in \mathbb{R}\}$ be a one-parameter Ad-unipotent subgroup of $\mathbf{G}$. For $x \in \mathbf{\Gamma}\backslash\mathbf{G} = X$ and a sequence $t_n \to \infty$ we denote by μ_n the normalized length measure on the orbit interval $L_n = \{x\mathbf{u}(t) : 0 \leq t \leq t_n\}$, assuming that μ_n is a measure on X supported on L_n. Thus $t_n^{-1} \int_0^{t_n} f(x\mathbf{u}(s))\, ds = \int_X f\, d\mu_n$ for every bounded continuous function f on X. Also the sequence μ_n contains a weak* convergent subsequence (as the closed unit ball in the space of Borel measures on X is weak* compact).

Suppose μ_n converges weak* to μ for some $t_n \to \infty$. Then μ is $\mathbf{U}$-invariant, $\mathrm{Supp}(\mu) \subset \overline{x\mathbf{U}}$ and $\mu(X) = 1$ (by [D2, Theorem 3.5]). Let $\{(C(y), \mu_{C(y)}) : y \in X\}$ be the ergodic decomposition of the action of $\mathbf{U}$ on (X, μ). Here each $\mu_{C(y)}$ is an ergodic $\mathbf{U}$-invariant measure supported on $C(y)$ and μ is the direct integral of the measures $\mu_{C(y)}$, $y \in X$. By Theorem 1 each $\mu_{C(y)}$ is algebraic and $C(y)$

is homogeneous. It follows from Theorem 5 that there exist $y_0 \in X$ and a small $\delta > 0$ such that the set $\Omega = \cup\{C(y) : C(y) = C(y_0)\mathbf{z}, d_{\mathbf{G}}(\mathbf{z}, \mathbf{e}) \leq \delta\}$ has positive μ-measure. It follows from the definition of μ that the proportion of time spent by L_n in every small neighborhood of Ω is close to $\mu(\Omega)$ for all sufficiently large n. Using the polynomial form of $\mathrm{Ad}_{\mathbf{u}(t)}$ (via a version of the R-property for $\mathbf{U}$) we show that this may happen only if $x \in C(y) \subset \Omega$ for some y and $\mu(C(y)) = 1$. Then $\overline{x\mathbf{U}} = C(y)$ is homogeneous and $\mu = \mu_{C(y)}$ is algebraic. Thus there exists $\mathbf{H} \subset \mathbf{G}$ such that $\overline{x\mathbf{U}} = x\mathbf{H}$ is homogeneous and $\mu = \nu_{\mathbf{H}}$. Because this is true for all sequences $t_n \to \infty$ with μ_n weak* convergent, x is generic for $\mathbf{U}$.

Recently, Dani and Margulis [DM4] offered a linearized version of this argument using the action of the adjoint representation of $\mathbf{G}$ on the mth exterior power of $\mathfrak{G}$ with $m = \dim C(y_0)$. Using this version they offered an alternative proof of Theorem 6 (and Theorem 8 below) and showed that the convergence in Theorem 6 (and Theorem 8) is uniform on compact subsets of $\mathbf{\Gamma}\backslash\mathbf{G}$. Also this linearization method is basic for the proofs of Theorems 9, 10, and C2 below.

Now let $\mathbf{U}_n = \{\mathbf{u}_n(t) : t \in \mathbb{R}\}$, $n = 1, 2, \ldots$ and $\mathbf{U} = \{\mathbf{u}(t) : t \in \mathbb{R}\}$ be one-parameter Ad-unipotent subgroups of $\mathbf{G}$. We say that $\mathbf{U}_n \to \mathbf{U}$ if $\mathbf{u}_n(t) \to \mathbf{u}(t)$ for all $t \in \mathbb{R}$.

The argument given above can be applied to derive the following more general version of Theorem 6. (This was pointed out to the author by Marc Burger in December 1990.)

THEOREM 8. *Let $\mathbf{U}_n \to \mathbf{U}$ and $x_n \to x \in \mathbf{\Gamma}\backslash\mathbf{G}$ with $\mathbf{\Gamma}$ being a lattice in $\mathbf{G}$. Suppose that there exists no proper closed connected subgroup $\mathbf{L}$ of $\mathbf{G}$ such that $\mathbf{U} \subset \mathbf{L}$ and $x\mathbf{L}$ is homogeneous. Then*

$$\lim_{n\to\infty} \frac{1}{t_n} \int_0^{t_n} f(x_n\mathbf{u}_n(s))\,ds = \int_{\mathbf{\Gamma}\backslash\mathbf{G}} f\,d\nu_{\mathbf{G}}$$

for every bounded continuous function f on $\mathbf{\Gamma}\backslash\mathbf{G}$ and every sequence $t_n \to \infty$ when $n \to \infty$, where $\nu_{\mathbf{G}}$ denotes the $\mathbf{G}$-invariant Borel probability measure on $\mathbf{\Gamma}\backslash\mathbf{G}$.

Theorem 6 follows from Theorem 8 if we set $\mathbf{U}_n = \mathbf{U}$, $x_n = x$ for all n and use induction on the dimension of $\mathbf{G}$. The main part of the proof of Theorem 8 is given in [Ra14, Section 7].

Now let $\mathbf{\Gamma}$ be a discrete subgroup of $\mathbf{G}$, $X = \mathbf{\Gamma}\backslash\mathbf{G}$, and let $\mathcal{P}(X)$ denote the set of all Borel probability measures on X. Recall that a sequence $\{\mu_n\}$ in $\mathcal{P}(X)$ weak* converges to a measure μ on X if $\int f\,d\mu_n \to \int f\,d\mu$ for every bounded continuous function f on X. Define

$$\mathcal{Q}(X) = \{\mu \in \mathcal{P}(X) : \text{there exists a one-parameter Ad-unipotent subgroup of } \mathbf{G} \text{ that preserves } \mu \text{ and acts ergodically on } (X,\mu)\}.$$

By Theorem 1 every member of $\mathcal{Q}(X)$ is algebraic.

Recently Mozes and Shah proved the following theorem.

THEOREM 9 (Mozes, Shah [MoS]). *Let $\{\mu_n\}$ be a sequence of measures in $\mathcal{Q}(X)$ weak* converging to $\mu \in \mathcal{P}(X)$. Then $\mu \in \mathcal{Q}(X)$. Moreover, there exist $x \in \mathrm{Supp}(\mu)$ and $\mathbf{g}_n \in \mathbf{G}$, $\mathbf{g}_n \to \mathbf{e}$ such that $x\mathbf{g}_n \in \mathrm{Supp}(\mu_n) \subset \mathrm{Supp}(\mu)\mathbf{g}_n$ for all large n.*

The proof of this theorem uses Theorems 1 and 5 and the method of [DM4].

Theorem 9 implies the following extension of Theorem 6 to the case when $\mathbf{\Gamma}$ is not a lattice: if $\mathbf{U}$ is a one-parameter Ad-unipotent subgroup of $\mathbf{G}$ and $\overline{x\mathbf{U}}$ is compact in $\mathbf{\Gamma}\backslash\mathbf{G}$, then x is generic for $\mathbf{U}$. This was conjectured in [Ra12, Conjecture D] and proved there for $\mathbf{G} = SL(2,\mathbb{R})$.

It is clear that if $\{\mathbf{u}(t) : t \in \mathbb{R}\}$ is a one-parameter unipotent subgroup of $GL(n,\mathbb{R})$ then each entry of the matrix $\mathbf{u}(t)$ is a polynomial in t.

Recently, Shah extended Theorem 6 to more general polynomial actions. A map $\boldsymbol{\theta} : \mathbb{R}^k \to SL(n,\mathbb{R})$, $n \in \mathbb{Z}^+$, is called *polynomial* if every entry of the matrix $\boldsymbol{\theta}(t_1,\ldots,t_k) \in SL(n,\mathbb{R})$ is a polynomial in $(t_1,\ldots,t_k) \in \mathbb{R}^k$ and $\boldsymbol{\theta}$ maps the origin to the identity element of $SL(n,\mathbb{R})$.

THEOREM 10 (Shah [S2]). *Let $\boldsymbol{\theta} : \mathbb{R}^k \to SL(n,\mathbb{R})$ be polynomial and let $\mathbf{G}$ be a closed subgroup of $SL(n,\mathbb{R})$ containing $\boldsymbol{\theta}(\mathbb{R}^k)$. Then given any lattice $\mathbf{\Gamma}$ in $\mathbf{G}$ and any $x \in \mathbf{\Gamma}\backslash\mathbf{G}$ there is a closed subgroup $\mathbf{H} \subset \mathbf{G}$ such that $\overline{x\boldsymbol{\theta}(\mathbb{R}^k)} = x\mathbf{H}$ is homogeneous and*

$$\lim_{R\to\infty} \frac{1}{\lambda(B_R)} \int_{B_R} f(x\boldsymbol{\theta}(t))d\lambda(t) = \int_{\mathbf{\Gamma}\backslash\mathbf{G}} f d\nu_{\mathbf{H}}$$

for every bounded continuous function f on $\mathbf{\Gamma}\backslash\mathbf{G}$, where B_R denotes the ball of radius R in $\mathbb{R}^k$ centered at the origin and λ denotes the Lebesgue measure on $\mathbb{R}^k$.

The proof of Theorem 10 uses Theorems 1 and 5 and the method of [DM4]. Shah also showed that if $\boldsymbol{\theta}(t_1,\ldots,t_k) = \boldsymbol{\theta}_1(t_1)\ldots\boldsymbol{\theta}_k(t_k)$ for some polynomial maps $\boldsymbol{\theta}_i : \mathbb{R} \to SL(n,\mathbb{R})$, $i = 1,\ldots,k$, then the conclusion of Theorem 10 holds also for B_n being of the form $[0,T_n^{(1)}] \times \cdots \times [0,T_n^{(k)}]$ with $T_n^{(i)} \to \infty$, $i = 1,\ldots,k$. This implies, in particular, that Theorem 6 holds for higher-dimensional connected simply connected Ad-unipotent $\mathbf{U}$ with averages performed over large rectangular subsets of $\mathbf{U}$ and, in particular, over Fϕlner subsets of $\mathbf{U}$ (see [Ra11]). This gives an affirmative answer to a question raised in [Ra14, Problem 2].

Next we address the following question: Are there subgroups of $\mathbf{G}$ not generated by Ad-unipotent elements of $\mathbf{G}$ for which Theorems 1 and 3 hold? Theorems 11–13 below give an affirmative answer to this question.

Indeed, let $\mathbf{\Gamma}$ be a discrete subgroup of $\mathbf{G}$ and μ a Borel probability measure on $\mathbf{\Gamma}\backslash\mathbf{G}$. Also let $\mathbf{U}$ be a one-parameter Ad-unipotent subgroup of $\mathbf{G}$ and $\mathbf{A}$ be "diagonal" for $\mathbf{U}$. Using Theorem 7 we showed in [Ra10, Proposition 2.1] that if μ is preserved by both $\mathbf{U}$ and $\mathbf{A}$ then μ is preserved by $SL_2(\mathbf{U},\mathbf{A})$. Note that $SL_2(\mathbf{U},\mathbf{A})$ is generated by Ad-unipotent elements of $\mathbf{G}$. This and Theorems 2 and 4 imply the following

THEOREM 11. *Let $\mathbf{G}$ be a connected Lie group and $\mathbf{U}$ a connected subgroup of $\mathbf{G}$ generated by* Ad-unipotent *elements of $\mathbf{G}$. Let $\mathbf{A}_1,\ldots,\mathbf{A}_n$ be "diagonal" for some* *one-parameter* Ad-*unipotent subgroups* $\mathbf{U}_1, \dots, \mathbf{U}_n$ *of* $\mathbf{U}$. *Then Theorems* 1 *and* 3 *hold for the subgroup* $\mathbf{H}$ *of* $\mathbf{G}$ *generated by* $\mathbf{U}$ *and* $\mathbf{A}_1, \dots, \mathbf{A}_n$.

Indeed, if μ is an ergodic $\mathbf{H}$-invariant Borel probability measure on $\mathbf{\Gamma}\backslash\mathbf{G}$ then μ is invariant under the action of the group $\mathbf{H}'$ generated by $\mathbf{U}$ and $SL_2(\mathbf{U}_i, \mathbf{A}_i)$, $i = 1, \dots, n$. Because $\mathbf{H} \subset \mathbf{H}'$, μ is ergodic for $\mathbf{H}'$. Because $\mathbf{H}'$ is generated by Ad-unipotent elements of $\mathbf{G}$, μ is algebraic by Theorem 2. This gives Theorem 1 for $\mathbf{H}$. (This argument was brought to the author's attention by Mozes.) To derive Theorem 3 for $\mathbf{H}$ we show that when $\mathbf{\Gamma}$ is a lattice in $\mathbf{G}$, then $\overline{x\mathbf{H}} = \overline{x\mathbf{H}'}$ for all $x \in \mathbf{\Gamma}\backslash\mathbf{G}$. Hence $\overline{x\mathbf{H}}$ is homogeneous by Theorem 4.

This implies the following

THEOREM 12. *Let* $\mathbf{G}$ *be a connected Lie group and* $\mathbf{G}_1$ *a connected semisimple subgroup of* $\mathbf{G}$ *without compact factors. Let* $\mathbf{H}$ *be a parabolic subgroup of* $\mathbf{G}_1$. *Then Theorems* 1 *and* 3 *hold for* $\mathbf{H}$.

Theorem 12 implies, in particular, that Theorems 1 and 3 hold for the subgroup $\mathbf{H}$ of $\mathbf{G} = SL(n, \mathbb{R})$ consisting of all upper triangular matrices in $\mathbf{G}$.

We say that a subgroup $\mathbf{L}$ of $\mathbf{G}$ is *epimorphic* with respect to $\mathbf{G}$ if for every finite-dimensional representation of $\mathbf{G}$ every vector v fixed by $\mathbf{L}$ is also fixed by $\mathbf{G}$. It is a fact that the group $\mathbf{H}$ described in Theorem 11 is epimorphic with respect to $\mathbf{H}'$ generated by $\mathbf{U}$ and $SL_2(\mathbf{U}_i, \mathbf{A}_i)$, $i = 1, \dots, n$. Recently, Mozes has generalized Theorems 11 and 12 in the following form.

THEOREM 13 (Mozes [Mo2]). *Let* $\mathbf{G}$ *be a connected Lie group and* $\mathbf{L}$ *a subgroup of* $\mathbf{G}$ *epimorphic with respect to a connected semisimple subgroup* $\mathbf{G}_1$ *of* $\mathbf{G}$ *without compact factors. Then Theorem* 1 *holds for* $\mathbf{L}$.

Mozes' proof uses Theorem 2 and a recent result of Bien and Borel [BBo].

PROBLEM. Let $\mathbf{G}$ and $\mathbf{L}$ be as in Theorem 13. Does Theorem 3 hold for $\mathbf{L}$?

In [Ra14] we incorrectly stated that Raghunathan had a counterexample to this problem.

It is a fact that, in general, Theorem 3 does not hold for non-Ad-unipotent one-parameter $\mathbf{U}$. However, using Theorem 3, Starkov proved the following

THEOREM 14 (Starkov [St3]). *Let* $\mathbf{\Gamma}$ *be a lattice in* $\mathbf{G}$ *and* $\mathbf{U}$ *a one-parameter subgroup of* $\mathbf{G}$. *Then the following statements are equivalent:* (1) *for every* $x \in \mathbf{\Gamma}\backslash\mathbf{G}$ *the closure* $\overline{x\mathbf{U}}$ *is a smooth submanifold of* $\mathbf{\Gamma}\backslash\mathbf{G}$; (2) $|\lambda| = 1$ *for every eigenvalue* λ *of* $\mathrm{Ad}_{\mathbf{u}}$ *and every* $\mathbf{u} \in \mathbf{U}$.

Finally we mention that the validity of Theorems 1 and 2 for *discrete* subgroups $\mathbf{\Gamma}$ implies their validity for *arbitrary closed* $\mathbf{\Gamma} \subset \mathbf{G}$. This was shown by Witte in [W3] (see also [St1] for a related result). Witte also showed (in a recent correspondence with the author) that the validity of Theorems 2 and 4 with the assumption of $\mathbf{U}/\mathbf{U}^0$ being finitely generated implies their validity without this assumption (because the assumption holds for the closure of $\mathbf{U}$).

In [Ra10] Theorem 2 is also proved for *disconnected* $\mathbf{G}$ with the additional assumption (which was omitted in [Ra10], though used in the proofs) that $\mathbf{U}/\mathbf{U}^0$

is nilpotent. This assumption automatically holds when $\mathbf{G}$ is connected. (See [W4] for more on the disconnected $\mathbf{G}$ case.)

In closing, we note that the following question remains unanswered.

QUESTION. Do Theorems 2 and 4 hold for *arbitrary* disconnected non-Ad-unipotent subgroups $\mathbf{U}$ of $\mathbf{G}$, generated by Ad-unipotent elements of $\mathbf{G}$?

4 Generalizations to the p-adic and $\mathbb{S}$-Arithmetic Cases

The problem of extending Raghunathan's conjectures to cartesian products of *algebraic* groups over local fields of characteristic zero (this is referred to as the $\mathbb{S}$-arithmetic setting) was raised by Borel and Prasad in [BoPr] (see also [Pr]). They pointed out that the validity of Conjecture 3 for the $\mathbb{S}$-arithmetic case (see Theorem S2 below) would solve the density problem in the Oppenheim conjecture for this case (see Section 5 below).

It turns out that the ideas and methods developed in [Ra8–11] for *real* Lie groups can be applied to prove Conjectures 1–3 for a more general (than the $\mathbb{S}$-arithmetic setting) case, namely, cartesian products of real and p-adic Lie groups. (If κ is a local field of characteristic zero then κ is (isomorphic to) either $\mathbb{R}$, or $\mathbb{C}$ or a finite extension of a p-adic field. Then a Lie group over κ can be viewed as either a real Lie group or a p-adic Lie group.) Also our results allow us to understand the structure of p-adic Lie groups $\mathbf{G}$ that carry discrete subgroups $\mathbf{\Gamma}$ (in particular, lattices) admitting finite Borel measures on $\mathbf{\Gamma}\backslash\mathbf{G}$ preserved by one-parameter subgroups of $\mathbf{G}$ (see Theorem S6 below).

More specifically, let $\mathbb{S}$ be a finite set and for each $\mathrm{s} \in \mathbb{S}$ let $\mathbb{Q}_{\mathrm{s}}$ be either the real field $\mathbb{R}$ or the field of p_{s}-adic numbers for some prime p_{s}. In the latter case we call s ultrametric, otherwise s is called real. The set $\mathbb{S}$ is ultrametric if each $\mathrm{s} \in \mathbb{S}$ is ultrametric.

For $\mathrm{s} \in \mathbb{S}$ let $\mathbf{G}_{\mathrm{s}}$ be a Lie group over $\mathbb{Q}_{\mathrm{s}}$ with the Lie algebra $\mathfrak{G}_{\mathrm{s}}$ and let $\mathbf{G}_{\mathbb{S}} = \prod\{\mathbf{G}_{\mathrm{s}} : \mathrm{s} \in \mathbb{S}\}$ denote the cartesian product of $\mathbf{G}_{\mathrm{s}}$, $\mathrm{s} \in \mathbb{S}$.

Let $\eta : \mathbf{G}_{\mathrm{s}} \to \mathbf{G}_{\mathbb{S}}$ denote the natural embedding of $\mathbf{G}_{\mathrm{s}}$ in $\mathbf{G}_{\mathbb{S}}$ and let $\mathbf{U}_{\mathrm{s}} = \{\mathbf{u}_{\mathrm{s}}(t) : t \in \mathbb{Q}_{\mathrm{s}}\}$ be a one-parameter Ad-unipotent subgroup of $\mathbf{G}_{\mathrm{s}}$. Then $\mathbf{U} = \eta(\mathbf{U}_{\mathrm{s}}) = \{\mathbf{u}(t) = \eta(\mathbf{u}_{\mathrm{s}}(t)) : t \in \mathbb{Q}_{\mathrm{s}}\}$ is called a one-parameter Ad-unipotent subgroup of $\mathbf{G}_{\mathbb{S}}$.

It is a fact (see [Ra15, Theorem 1.1]) that *every* one-parameter subgroup of a p-adic Lie group $\mathbf{G}$ is Ad-unipotent (this was recently proved independently by Lubotzky and Prasad). Also $\mathbf{G}$ is totally disconnected and small neighborhoods of the identity of $\mathbf{G}$ do not generate $\mathbf{G}$. Because of this, $\mathbf{G}$ might contain two *distinct* one-parameter subgroups $\mathbf{U}_1$ and $\mathbf{U}_2$ that have the same tangent (and hence coincide in a neighborhood of $\mathbf{e}$ in $\mathbf{G}$). This motivates the following definitions.

For an ultrametric $\mathrm{s} \in \mathbb{S}$ we call $\mathbf{G}_{\mathrm{s}}$ *Ad-regular* if $\ker \mathrm{Ad}_{\mathbf{G}_{\mathrm{s}}} = \mathbf{Z}(\mathbf{G}_{\mathrm{s}})$, where $\mathrm{Ad}_{\mathbf{G}_{\mathrm{s}}}$ denotes the adjoint representation of $\mathbf{G}_{\mathrm{s}}$ and $\mathbf{Z}(\mathbf{G}_{\mathrm{s}})$ the center of $\mathbf{G}_{\mathrm{s}}$. An Ad-regular $\mathbf{G}_{\mathrm{s}}$ is called *regular* if the orders of all finite subgroups of $\mathbf{G}_{\mathrm{s}}$ do not exceed a constant depending only on $\mathbf{G}_{\mathrm{s}}$.

We show that if two one-parameter subgroups $\mathbf{U}_1 = \{\mathbf{u}_1(t) : t \in \mathbb{Q}_{\mathrm{s}}\}$ and $\mathbf{U}_2 = \{\mathbf{u}_2(t) : t \in \mathbb{Q}_{\mathrm{s}}\}$ of a regular $\mathbf{G}_{\mathrm{s}}$ have the same tangent then $\mathbf{U}_1 = \mathbf{U}_2$ (i.e. $\mathbf{u}_1(t) = \mathbf{u}_2(t)$ for all $t \in \mathbb{Q}_{\mathrm{s}}$).

It is a fact that if κ is a finite extension of $\mathbb{Q}_s$ with an ultrametric s then $GL(n, \kappa)$, $n \in \mathbb{Z}^+$, and its Zariski closed and connected subgroups (viewed as Lie groups over $\mathbb{Q}_s$) are regular.

Also we showed in [Ra15] that if $\mathbf{G}_s$ is a Lie subgroup of a *regular* p_s-adic Lie group then there exists an open subgroup $\mathbf{G}_s^0$ of $\mathbf{G}_s$ such that $\mathbf{G}_s^0$ is *regular* and contains every one-parameter subgroup of $\mathbf{G}_s$. (This implies that if Theorems S1–S6 below hold for $\mathbf{G}_s^0$ in place of $\mathbf{G}_s$ then they hold for $\mathbf{G}_s$. Thus one can reduce these theorems to the case when $\mathbf{G}_s$ is regular for every ultrametric $s \in \mathbb{S}$.)

Henceforth we assume that $\mathbf{G}_s$ is a Lie subgroup of a regular p_s-adic Lie group for every ultrametric $s \in \mathbb{S}$.

THEOREM S1 (Ergodic measures). *Let $\mathbf{H}$ be a closed subgroup of $\mathbf{G}_\mathbb{S}$ and $\mathbf{U}$ a subgroup of $\mathbf{H}$ generated by one-parameter Ad-unipotent subgroups of $\mathbf{G}_\mathbb{S}$. Then given any discrete subgroup $\mathbf{\Gamma}$ of $\mathbf{H}$ every ergodic $\mathbf{U}$-invariant Borel probability measure on $\mathbf{\Gamma}\backslash\mathbf{H}$ is algebraic.*

THEOREM S2 (Orbit closures). *Let $\mathbf{H}$ and $\mathbf{U}$ be as in Theorem 1. Then given any lattice $\mathbf{\Gamma}$ of $\mathbf{H}$ and any $x \in \mathbf{\Gamma}\backslash\mathbf{H}$ the closure $\overline{x\mathbf{U}}$ of the orbit $x\mathbf{U}$ in $\mathbf{\Gamma}\backslash\mathbf{H}$ is homogeneous.*

Theorems S1 and S2 proved in [Ra15, Theorems 1 and 2] extend Theorems 2 and 4 to $\mathbf{G}_\mathbb{S}$. To extend Theorems 6 and 8 we need to introduce the following notation.

Let $\mathbf{H}$ be a closed subgroup of $\mathbf{G}_\mathbb{S}$, $\mathbf{\Gamma}$ a discrete subgroup of $\mathbf{H}$, and $\mathbf{U} = \eta(\mathbf{U}_s) = \{\mathbf{u}(t) : t \in \mathbb{Q}_s\}$, $s \in \mathbb{S}$, a one-parameter Ad-unipotent subgroup of $\mathbf{G}_\mathbb{S}$ contained in $\mathbf{H}$. For $\tau > 0$ let

$$F_s(\tau) = \{t \in \mathbb{Q}_s : |t|_s \leq \tau\},$$

where $|\cdot|_s$ denotes the normalized absolute value on $\mathbb{Q}_s$. When s is ultrametric, $F_s(\tau)$ is a compact open subgroup of $\mathbb{Q}_s$. We denote by λ_s a Haar measure on $\mathbb{Q}_s$.

THEOREM S3 (Uniform Distribution [Ra15, Theorem 3]). *Given any lattice $\mathbf{\Gamma}$ of $\mathbf{H}$ and any $x \in \mathbf{\Gamma}\backslash\mathbf{H}$ there exists a closed subgroup $\mathbf{L}$ of $\mathbf{H}$ such that $\overline{x\mathbf{U}} = x\mathbf{L}$ is homogeneous, $\mathbf{U}$ acts ergodically on $(\overline{x\mathbf{U}} = x\mathbf{L}, \nu_\mathbf{L})$, and*

$$S_f(x,\tau) = \frac{1}{\lambda_s(F_s(\tau))}\int_{F_s(\tau)} f(x\mathbf{u}(t))\, d\lambda_s(t) \to \int_{\mathbf{\Gamma}\backslash\mathbf{H}} f\, d\nu_\mathbf{L} = \nu_\mathbf{L}(f) \text{ as } \tau \to \infty,$$

for every bounded continuous function f on $\mathbf{\Gamma}\backslash\mathbf{H}$.

THEOREM S4 ([Ra15, Theorem 4]). *Let $x_n \to x \in \mathbf{\Gamma}\backslash\mathbf{H}$ with $\mathbf{\Gamma}$ being a lattice in $\mathbf{H}$. Suppose there exists no closed nonopen subgroup $\mathbf{L}$ of $\mathbf{H}$ such that $\mathbf{U} \subset \mathbf{L}$ and $x\mathbf{L}$ is homogeneous. Then there exists an algebraic measure ν on $\mathbf{\Gamma}\backslash\mathbf{H}$ such that $\mathbf{\Lambda}(\nu)$ is an open subgroup of $\mathbf{H}$, $\nu(x\mathbf{\Lambda}(\nu)) = 1$, $\mathbf{U}$ acts ergodically on $(x\mathbf{\Lambda}(\nu), \nu)$, and*

$$\lim_{n\to\infty} S_f(x_n, \tau_n) = \nu(f)$$

for every bounded continuous function f *on* $\mathbf{\Gamma}\backslash\mathbf{H}$ *and every sequence* $\tau_n \to \infty$ *when* $n \to \infty$. *If* $\overline{x\mathbf{U}} = \mathbf{\Gamma}\backslash\mathbf{H}$ *then* ν *is* $\mathbf{H}$*-invariant.*

Recall (see Section 1) that $\mathbf{\Lambda}(\nu) = \{\mathbf{h} \in \mathbf{H} : \text{the action of } \mathbf{h} \text{ on } \mathbf{\Gamma}\backslash\mathbf{H} \text{ preserves } \nu\}$.

Theorem S3 follows from Theorem S4 if we set $x_n = x$ for all n and use induction on the dimension of $\mathbf{H}$.

Note that Theorem 5 has also been extended to $\mathbf{G}_{\mathbb{S}}$ (see [Ra15, Theorem 1.3]).

Recently Margulis and Tomanov [MT1,2] published a particular case of Theorem S1 when for each $\mathrm{s} \in \mathbb{S}$ the group $\mathbf{G}_{\mathrm{s}}$ is the set of κ_{s}-rational points of an algebraic group defined over a local field κ_{s} of characteristic zero. (They also formulated Theorem S3 with $\mathbf{H} = \mathbf{G}_{\mathbb{S}}$ and a weaker version of Theorem S2 for this algebraic case.) As does the author's their proof uses in the most essential way the basic ideas and methods from [Ra8,9] (though they give no specific references to [Ra8,9] in [MT2] and no references to [Ra8,9] at all in [MT1]). In fact, for the most part their proof can be viewed as a translation (with modifications and substantial simplifications possible because $\mathbf{G}_{\mathrm{s}}$ are algebraic) of these ideas and methods to the algebraic group setting. The basic Lemma 7.5 in [MT2] uses the fundamental idea from the proofs of [Ra8, Lemma 4.2], [Ra9, Lemma 3.1] of using the polynomial divergence of Ad-unipotent orbits through the normalizer and the ergodic theorem. Also the results in [MT2, Proposition 6.1] and [MT2, Propositions 6.7 and 8.3] are analogous to [Ra8, Theorem 3.1, Lemma 3.3] and [Ra9, Lemma 3.1].

Next we generalize the notion of a "diagonal" subgroup for a one-parameter Ad-unipotent subgroup $\mathbf{U}_{\mathrm{s}} = \{\mathbf{u}(t) : t \in \mathbb{Q}_{\mathrm{s}}\}$, $u = d\mathbf{u}(t)/dt|_{t=0}$. Suppose there is an "opposite" u^* and a "diagonal" $a = \mathrm{ad}_{u^*}(u)$ for u in $\mathfrak{G}_{\mathrm{s}}$ (see Section 1) and let $\mathbf{A}_{\mathrm{s}}$ be a one-dimensional Lie subgroup of $\mathbf{G}_{\mathrm{s}}$ normalizing $\mathbf{U}_{\mathrm{s}}$ whose Lie algebra is spanned by a.

DEFINITION. The group $\mathbf{A}_{\mathrm{s}}$ is called "*diagonal*" for $\mathbf{U}_{\mathrm{s}}$ if there exists a one-parameter Ad-unipotent $\mathbf{U}^*_{\mathrm{s}} = \{\mathbf{u}^*(t) : t \in \mathbb{Q}_{\mathrm{s}}\}$, $d\mathbf{u}^*(t)/dt|_{t=0} = u^*$ normalized by $\mathbf{A}_{\mathrm{s}}$ such that if we denote by $\mathbf{S} = \langle \mathbf{U}_{\mathrm{s}}, \mathbf{U}^*_{\mathrm{s}} \rangle$ the subgroup of $\mathbf{G}_{\mathrm{s}}$ generated by $\mathbf{U}_{\mathrm{s}}, \mathbf{U}^*_{\mathrm{s}}$ then $\mathbf{A}_{\mathrm{s}} \subset \mathbf{S}$ and $\mathrm{Ad}_{\mathbf{G}_{\mathrm{s}}}$ maps $\mathbf{A}_{\mathrm{s}}$ homomorphically onto the multiplicative one-parameter subgroup $\{\mathfrak{a}(\tau) : \tau \in \mathbb{Q}^*_{\mathrm{s}}\}$ of $\mathrm{Ad}_{\mathbf{G}_{\mathrm{s}}}(\mathbf{S})$ with $d\mathfrak{a}(\tau)/d\tau|_{\tau=1} = \mathrm{ad}_a$. Here $\mathbb{Q}^*_{\mathrm{s}} = \mathbb{Q}_{\mathrm{s}} - \{0\}$.

We write $\mathbf{S} = SL_2(\mathbf{U}_{\mathrm{s}}, \mathbf{A}_{\mathrm{s}})$ and $\mathbf{A}_{\mathrm{s}} = \cup\{\mathbf{A}_{\mathrm{s}}(\tau) : \tau \in \mathbb{Q}^*_{\mathrm{s}}\}$, where

$$\mathbf{A}_{\mathrm{s}}(\tau) = \{\mathbf{a} \in \mathbf{A}_{\mathrm{s}} : \mathrm{Ad}_{\mathbf{a}} = \mathfrak{a}(\tau)\}.$$

Now let $\mathbf{U} = \eta(\mathbf{U}_{\mathrm{s}}) \subset \mathbf{G}_{\mathbb{S}}$. Then we call $\mathbf{A} = \eta(\mathbf{A}_{\mathrm{s}})$ "diagonal" for $\mathbf{U}$ in $\mathbf{G}_{\mathbb{S}}$ and write $SL_2(\mathbf{U}, \mathbf{A}) = \eta(SL_2(\mathbf{U}_{\mathrm{s}}, \mathbf{A}_{\mathrm{s}}))$.

As in the real case, the central role in the proof of Theorem S1 is played by the following Theorem S5 [Ra15, Theorem 6], which generalizes Theorem 7.

THEOREM S5. *Let* $\mathbf{U}$ *be a one-parameter* Ad-*unipotent subgroup of* $\mathbf{G}_{\mathbb{S}}$ *and assume that* $\mathbf{G}_{\mathbb{S}}$ *contains a "diagonal" subgroup* $\mathbf{A}$ *for* $\mathbf{U}$. *Let* $\mathbf{\Gamma}$ *be a discrete subgroup of* $\mathbf{G}_{\mathbb{S}}$ *and* μ *an ergodic* $\mathbf{U}$*-invariant Borel probability measure on* $\mathbf{\Gamma}\backslash\mathbf{G}_{\mathbb{S}}$.

Then either (1) $\mu(\mathcal{D}(\mathbf{a}(\tau))) = 1$ *for every* $\mathbf{a}(\tau) \in \mathbf{A}(\tau)$ *with* $|\tau| > 1$ *or* (2) $\mathbf{c}SL_2(\mathbf{U}, \mathbf{A})\mathbf{c}^{-1} \subset \mathbf{\Lambda}(\mu)$ *for some* $\mathbf{c} \in \mathbf{C}(\mathbf{U})$ *and* μ *is algebraic.*

Recall that $\mathcal{D}(\mathbf{a}(\tau))$ denotes the divergent set of $\mathbf{a}(\tau)$ (see Section 1). It is a fact that if $\mathbb{S}$ is ultrametric then there are no cuspidally divergent sequences in $\mathbf{\Gamma}\backslash\mathbf{G}_{\mathbb{S}}$ (see [Ra13, Proposition 2]). Thus when $\mathbb{S}$ is ultrametric and there is a "diagonal" $\mathbf{A}$ for $\mathbf{U}$, then conclusion (2) holds in Theorem S5. The following theorem shows that the presence of a "diagonal" subgroup for $\mathbf{U}$ is necessary for $\mathbf{U}$ to preserve a finite measure on $\mathbf{\Gamma}\backslash\mathbf{G}_{\mathbb{S}}$.

THEOREM S6. *Assume* $\mathbb{S}$ *is ultrametric. Let* $\mathbf{\Gamma}$ *be a discrete subgroup of* $\mathbf{G}_{\mathbb{S}}$ *and* $\mathbf{U}$ *a one-parameter subgroup of* $\mathbf{G}_{\mathbb{S}}$ *preserving a Borel probability measure on* $\mathbf{\Gamma}\backslash\mathbf{G}_{\mathbb{S}}$. *Then there is a "diagonal" subgroup* $\mathbf{A}$ *for* $\mathbf{U}$ *in* $\mathbf{G}_{\mathbb{S}}$.

COROLLARY S1. *Assume* $\mathbb{S}$ *is ultrametric and* $\mathbf{G}_{\mathbb{S}}$ *admits a lattice. Then for every* $s \in \mathbb{S}$ *and every one-parameter subgroup* $\mathbf{U}_s$ *of* $\mathbf{G}_s$ *there is a "diagonal"* $\mathbf{A}_s$ *in* $\mathbf{G}_s$.

This corollary can be viewed as a generalization of [T, Theorem 3] stating that if an *algebraic* p-adic group $\mathbf{G}$ admits a lattice then $\mathbf{G}$ is reductive.

Finally, we mention that Theorem S5 allows us to extend Theorem 11 to $\mathbf{G}_{\mathbb{S}}$ (see [Ra15, Corollary 3]).

5 Applications to Number Theory

The Oppenheim Conjecture

THEOREM O1 (Margulis). *Let* B *be a real nondegenerate indefinite quadratic form in* n *variables,* $n \geq 3$. *Suppose that the ratio of some two coefficients of* B *is irrational. Then the set of values of* B *at integral points is dense in* $\mathbb{R}$.

This is an equivalent version of the Oppenheim Conjecture proved by Margulis [M1] in 1986. (The original Oppenheim Conjecture asserts that zero is a limit point of $B(\mathbb{Z}^n)$.) In fact, it was Raghunathan who noticed that in order to derive this theorem for $n = 3$ one only needs to prove a weaker version of Theorem 4 for $\mathbf{G} = SL(3,\mathbb{R})$, $\mathbf{\Gamma} = SL(3,\mathbb{Z})$, and $\mathbf{U} = SO(2,1)^0$. This is precisely what Margulis did. (He also observed that Theorem O1 for $n > 3$ can be reduced to the case $n = 3$.) Subsequently he and Dani [DM1,3] showed that the values of B at the primitive elements of $\mathbb{Z}^n$ are dense in $\mathbb{R}$. In 1990–91 Borel and Prasad [BoPr] obtained a remarkable strengthening of this fact, implied by Theorem 4.

THEOREM O2 (Borel, Prasad [BoPr]). *Let* B *be as in Theorem O1. Then given* $c_1, \ldots, c_{n-1} \in \mathbb{R}$ *and* $\varepsilon > 0$ *there are* $x_1, \ldots, x_{n-1} \in \mathbb{Z}^n$ *that are part of a basis in* $\mathbb{Z}^n$ *(and hence are primitive elements of* $\mathbb{Z}^n$*) such that* $|B(x_i) - c_i| < \varepsilon$ *for all* $i = 1, \ldots, n-1$.

In fact, Borel and Prasad [BoPr] have generalized the Oppenheim Conjecture and Theorem O2 to the following more general setting.

Let κ be a number field and $\mathfrak{o}$ the ring of integers of κ. For every normalized absolute value $|\cdot|_v$ on κ, let κ_v be the completion of κ at v. Let $\mathbb{S}$ be a finite set of places of κ containing the set $\mathbb{S}_\infty$ of archimedean ones, $\kappa_{\mathbb{S}}$ the direct sum of the fields κ_{s}, $\mathrm{s} \in \mathbb{S}$, and $\mathfrak{o}_{\mathbb{S}}$ the ring of $\mathbb{S}$-integers of κ (i.e. of elements $x \in \kappa$ with $|x|_v \leq 1$ for all $v \notin \mathbb{S}$).

A quadratic form F on $\kappa_{\mathbb{S}}^n$ is a collection (F_{s}), $\mathrm{s} \in \mathbb{S}$, where F_{s} is a quadratic form on κ_{s}^n. The form is nondegenerate if and only if each F_{s} is nondegenerate. The form is isotropic if each F_{s} is so, i.e. if there exists for each $\mathrm{s} \in \mathbb{S}$ an element $x_{\mathrm{s}} \in \kappa_{\mathrm{s}}^n - \{0\}$ such that $F_{\mathrm{s}}(x_{\mathrm{s}}) = 0$. If s is a real place, this condition is equivalent to F_{s} being indefinite. The form F is said to be *rational* if there exists a unit $\lambda = (\lambda_{\mathrm{s}}) \in \kappa_{\mathbb{S}}^*$ and a form F_0 on κ^n such that $F_{\mathrm{s}} = \lambda_{\mathrm{s}} F_0$ for all $\mathrm{s} \in \mathbb{S}$, and *irrational* otherwise.

THEOREM O3 (Borel, Prasad [BoPr, Theorem A]). *Let F be irrational, nondegenerate, and isotropic, and $n \geq 3$. Then given $\varepsilon > 0$ there exists $x \in \mathfrak{o}_{\mathbb{S}}^n$ such that $0 < |F_{\mathrm{s}}(x)| < \varepsilon$ for all $\mathrm{s} \in \mathbb{S}$.*

THEOREM O4 (Borel, Prasad [BoPr, Corollary 7.9]). *Assume $\mathbb{S} = \mathbb{S}_\infty$ and let F be as in Theorem O3. Let $\lambda_1, \ldots, \lambda_{n-1} \in \kappa_{\mathbb{S}}$. Then for each $j = 1, 2, \ldots$ there are $x_{j,1}, \ldots, x_{j,n-1} \in \mathfrak{o}^n = \mathfrak{o}_{\mathbb{S}}^n$ that are part of a basis of $\mathfrak{o}^n$ over $\mathfrak{o}$ (and hence are primitive elements of $\mathfrak{o}^n$) such that $\lim_{j\to\infty} F(x_{j,i}) = \lambda_i$ for all $i = 1, \ldots, n-1$. In particular, the set of values of F on the primitive elements of $\mathfrak{o}^n$ is dense in $\kappa_{\mathbb{S}}$.*

Theorems O3 and O4 in [BoPr] are deduced by means of Theorem 4, geometry of numbers, and strong approximation in algebraic groups. In [BoPr] Borel and Prasad pointed out that the density of $F(\mathfrak{o}_{\mathbb{S}}^n)$ (and Theorem O4) for non-archimedean $\mathbb{S}$ would follow from the $\mathbb{S}$-arithmetic version of Theorem 4 (see Theorem S2 above). Indeed, the deduction of Theorem O5 below from Theorem S2 is given in [Bo].

THEOREM O5 (Borel, Prasad). *Theorem O4 holds also for non-archimedean $\mathbb{S}$ with $\mathfrak{o}$ replaced by $\mathfrak{o}_{\mathbb{S}}$.*

To illustrate the connection between the Oppenheim conjecture and the orbit closures Theorem 4 let us present the deduction of Theorem O1 from Theorem 4. This deduction is a simplified version of the argument originally given by Raghunathan.

Let B be a quadratic form as specified in Theorem O1. Also let $\mathbf{G} = SL(n, \mathbb{R})$, $\mathbf{\Gamma} = SL(n, \mathbb{Z})$, and $L = \mathbb{Z}^n$ be the lattice of integral points in $\mathbb{R}^n$. Let $L\mathbf{g}$ denote the lattice in $\mathbb{R}^n$ obtained by applying the linear transformation $\mathbf{g} \in \mathbf{G}$ to L. Then $X = \{L\mathbf{g} : \mathbf{g} \in \mathbf{G}\}$ is a set of lattices endowed with the natural Hausdorff topology. Note that if $\boldsymbol{\gamma} \in \mathbf{\Gamma}$ then $L\boldsymbol{\gamma} = L$. This says that we can identify $L\mathbf{g} \in X$ with the coset $\mathbf{\Gamma g} \in \mathbf{\Gamma}\backslash\mathbf{G}$. The identification $L\mathbf{g} \to \mathbf{\Gamma g}$ is a homeomorphism from X onto $\mathbf{\Gamma}\backslash\mathbf{G}$.

Now let $\mathbf{H}$ denote the subgroup of $\mathbf{G}$ preserving the quadratic form B. Then $\mathbf{H}$ is conjugate to $SO(p, q)$, $p+q = n$, $pq \neq 0$, and hence consists of two connected components. Also $\mathbf{H}^0$ is generated by unipotent elements of $\mathbf{G}$ (because $n > 2$). For each $\mathbf{h} \in \mathbf{H}$ the set of values of B on $L\mathbf{h}$ is the same as on L. To prove Theorem O1 it suffices to show that the orbit $L\mathbf{H}^0$ is dense in X or, equivalently, the orbit $z\mathbf{H}^0$ is dense in $\mathbf{\Gamma}\backslash\mathbf{G}$, where $z = \mathbf{\Gamma e}$ and $\mathbf{e}$ denotes the identity element of $\mathbf{G}$.

By Theorem 4 the closure $\overline{z\mathbf{H}^0} = z\mathbf{F}$ is homogeneous for some closed connected subgroup $\mathbf{F} \subset \mathbf{G}$, containing $\mathbf{H}^0$. But the only closed connected subgroups of $\mathbf{G}$ containing $\mathbf{H}^0$ are $\mathbf{G}$ and $\mathbf{H}^0$. Hence either $\mathbf{F} = \mathbf{G}$ or $\mathbf{F} = \mathbf{H}^0$.

We have to show that $\mathbf{F} = \mathbf{G}$, i.e. $\overline{z\mathbf{H}^0} = \mathbf{\Gamma}\backslash\mathbf{G}$. Suppose to the contrary that $\mathbf{F} = \mathbf{H}^0$. Then $\mathbf{H}^0 \cap \mathbf{\Gamma}$ is a lattice in $\mathbf{H}^0$ (by Definition 1) and hence $\mathbf{H} \cap \mathbf{\Gamma}$ is a lattice in $\mathbf{H}$ because $\mathbf{H}/\mathbf{H}_0$ is finite. Because $\mathbf{\Gamma} = SL(n, \mathbb{Z})$, it follows from the Borel density theorem that $\mathbf{H}$ is a $\mathbb{Q}$-subgroup of $\mathbf{G}$. This means that $\mathbf{H}$ is the set of real zeros of some ring of polynomials with rational coefficients. This implies by an elementary argument that B is proportional to a quadratic form with rational coefficients. This contradicts the conditions of Theorem O1 and proves the Theorem.

This proof shows that $L\mathbf{H}$ is dense in X. This is used to prove Theorem O2. Indeed, given $c_1, \dots, c_{n-1} \in \mathbb{R}$ there is a unimodular basis $y_1, \dots, y_n \in \mathbb{R}^n$ such that $B(y_i) = c_i$ for all $i = 1, \dots, n-1$ (because the level surface $B(x) = c$, $x \in \mathbb{R}^n$ is not contained in any hyperplane). Then the $\mathbb{Z}$-span of this basis belongs to X. Hence there are $x_1, \dots, x_n \in L$ and $\mathbf{h} \in \mathbf{H}$ such that $x_1\mathbf{h}, \dots, x_n\mathbf{h}$ are close to $y_1, \dots, y_n$ and $|B(x_i\mathbf{h}) - c_i| < \varepsilon$ for all $i = 1, \dots, n-1$. This implies that $x_1, \dots, x_n$ form a basis in L and $|B(x_i) - c_i| < \varepsilon$ for all $i = 1, \dots, n-1$, because $B(x_i) = B(x_i\mathbf{h})$, $i = 1, \dots, n$. This yields Theorem O2. (This proof is given in [BoPr, Proof of Corollary 7.9].)

Finally we mention the following problem. Let B be a quadratic form as in Theorem O1. Given $0 < a < b$ and $r > 0$, let $E_r(a, b) = \{x \in \mathbb{Z}^n : a \leq |B(x)| < b, \|x\| < r\}$. Then card $E_r(a, b) \to \infty$ when $r \to \infty$. It seems plausible that Theorems 1, 5 can be used to find the asymptotic growth rate for this number. It is believed that card $E_r(a, b) \sim c(a, b)r^{n-2}$ for some $c(a, b) > 0$. A lower bound of this type has already been found by Dani and Margulis in [DM4]. Also recently Eskin and Mozes have informed the author that using the latter lower bound they can prove this asymptotic growth for $n = 4$ and B of the signature (3,1) (and disprove it for the signature (2,2) and the case $n = 3$) and with a modification suggested by Margulis the proof works for any B with $n > 4$.

Counting Integral Points on Homogeneous Varieties

The discussion in this section is related to the following problem recently studied in [DuRuSa] and [EsMc].

Let W be a real finite-dimensional vector space and let $W(\mathbb{Q})(W(\mathbb{Z}))$ denote the set of all vectors in W with rational (integer) coordinates relative to a fixed basis in W. Let $\mathfrak{G}$ be a connected algebraic reductive group defined over $\mathbb{Q}$ and let $\mathfrak{G}(\mathbb{R})$ denote the set of real points of $\mathfrak{G}$, i.e. the set of real zeros of the polynomials defining $\mathfrak{G}$. Similarly, one defines $\mathfrak{G}(\mathbb{Q})$ and $\mathfrak{G}(\mathbb{Z})$. We assume that $\mathfrak{G}$ is homomorphic via a surjective morphism ρ defined over $\mathbb{Q}$ to an algebraic subgroup of $GL(\mathfrak{W})$ defined over $\mathbb{Q}$. (Here $\mathfrak{W}$ denotes the complexification of W.) Then $\mathfrak{G}(\mathbb{R})$ acts linearly on W by $w \to w\mathbf{g} = \rho(\mathbf{g})(w)$, $w \in W$, $\mathbf{g} \in \mathfrak{G}(\mathbb{R})$, and $\mathfrak{G}(\mathbb{Q})$ preserves $W(\mathbb{Q})$.

Now let $V \subset W$ be the set of real points of an affine subvariety of $\mathfrak{W}$ defined over $\mathbb{Q}$. Assume that V has finitely many connected components and $\mathbf{G} = \mathfrak{G}(\mathbb{R})^0$ acts transitively on each of these components. Suppose there exists $v_0 \in V(\mathbb{Z}) =$

$W(\mathbb{Z}) \cap V$ and let $\mathbf{H} = \{\mathbf{g} \in \mathbf{G} : v_0\mathbf{g} = v_0\}$ be the stabilizer of v_0 in $\mathbf{G}$. It is a fact that $\mathbf{H} = \mathfrak{H}(\mathbb{R}) \cap \mathbf{G}$ for some reductive algebraic subgroup $\mathfrak{H}$ of $\mathfrak{G}$ defined over $\mathbb{Q}$. Suppose $\mathfrak{G}$ and $\mathfrak{H}$ do not admit nontrivial characters defined over $\mathbb{Q}$. Then $\mathfrak{G}(\mathbb{Z}) \cap \mathbf{G}$ is a lattice in $\mathbf{G}$ and $\mathfrak{G}(\mathbb{Z}) \cap \mathbf{H}$ a lattice in $\mathbf{H}$. Let $\mathbf{\Gamma}$ be a subgroup of finite index in $\mathfrak{G}(\mathbb{Z}) \cap \mathbf{G}$ whose action on W preserves $W(\mathbb{Z})$.

We denote by $\|\ \ \|$ a norm on W and for $T > 0$ define $B_T = \{w \in W : \|w\| < T\}$. We are interested in the asymptotics of the number of points in $V(\mathbb{Z}) \cap B_T$ as $T \to \infty$. The group $\mathbf{\Gamma}$ acts on $V(\mathbb{Z})$ and it was shown in [BoHC] that $V(\mathbb{Z})$ consists of finitely many $\mathbf{\Gamma}$-orbits. Thus it suffices to know the asymptotics of the number $N(T, V, \mathcal{O})$ of points in $\mathcal{O} \cap B_T$, where $\mathcal{O} = v_0\mathbf{\Gamma}$.

Theorem C1 below recently proved by Eskin, Mozes and Shah [EsMoS] generalizes an earlier result of Duke, Rudnick, and Sarnak [DuRuSa] where an asymptotic of $N(T, V, \mathcal{O})$ was first found. To state the theorem we need the following definition.

Let $\{E_n\}$, $n = 1, 2, \dots$, be an increasing sequence of open subsets of $\mathbf{H}\backslash\mathbf{G} = \cup\{E_n,\ n = 1, 2, \dots\}$ and let $\hat{E}_n$ denote the natural lifting of E_n to $\mathbf{H}^0\backslash\mathbf{G}$. Also let λ denote the $\mathbf{G}$-invariant volume on $\mathbf{H}^0\backslash\mathbf{G}$, p the natural projection from $\mathbf{G}$ onto $\mathbf{H}^0\backslash\mathbf{G}$, and $\mathfrak{H}^0$ the Zariski closure of $\mathbf{H}^0$.

Definition. [EsMoS] The sequence $\{E_n\}$ is said *to be focused* in $\mathbf{H}\backslash\mathbf{G}$ as $n \to \infty$ if there is a compact $\mathbf{C} \subset \mathbf{G}$ and a proper $\mathbb{Q}$-subgroup $\mathfrak{L}$ of $\mathfrak{G}$ containing $\mathfrak{H}^0$ such that

$$\limsup_{n\to\infty} \frac{\lambda(\mathrm{p}((\mathbf{Z_G}(\mathbf{H}^0) \cap \mathbf{\Gamma})\mathbf{LC}) \cap \hat{E}_n)}{\lambda(\hat{E}_n)} > 0\,,$$

where $\mathbf{L} = \mathfrak{L}(\mathbb{R})^0$ and $\mathbf{Z_G}(\mathbf{H}^0)$ denotes the centralizer of $\mathbf{H}^0$ in $\mathbf{G}$.

Theorem C1 [EsMoS]. *Suppose that every $\mathbb{Q}$-subgroup of $\mathfrak{G}$ containing $\mathfrak{H}^0$ is reductive and for every sequence $T_n \uparrow \infty$ the sequence $R_{T_n} = \{\mathbf{Hg} : v_0\mathbf{g} \in B_{T_n}\}$ is not focused in $\mathbf{H}\backslash\mathbf{G}$. Then*

$$N(T, V, \mathcal{O}) \sim \frac{\mathrm{vol}_{\mathbf{H}}((\mathbf{H} \cap \mathbf{\Gamma})\backslash\mathbf{H})}{\mathrm{vol}_{\mathbf{G}}(\mathbf{\Gamma}\backslash\mathbf{G})}\,\mathrm{vol}_{\mathbf{H}\backslash\mathbf{G}}(R_T) \tag{1}$$

as $T \to \infty$, where the volumes in (1) *are induced by a left invariant Riemannian metric on* $\mathbf{G}$.

Corollary C1 [EsMoS]. *Suppose $\mathfrak{H}^0$ is a maximal proper connected $\mathbb{Q}$-subgroup of $\mathfrak{G}$. Then relation* (1) *holds for $N(T, V, \mathcal{O})$ as $T \to \infty$.*

For the particular case when $\mathfrak{H}$ is an affine symmetric subgroup of $\mathfrak{G}$ (i.e. $\mathfrak{H}$ is a fixed point set of an involution of $\mathfrak{G}$ defined over $\mathbb{Q}$) Corollary C1 was proved earlier in [DuRuSa] by other methods (subsequently a simpler proof appeared in [EsMc]).

To give an application of Theorem C1 the authors of [EsMoS] denote by $M_n(\mathbb{Z})$ the set of all $n \times n$ integer matrices and consider the set

$$V_p(\mathbb{Z}) = \{A \in M_n(\mathbb{Z}) : \det(tI - A) = p(t)\}\,,$$

where $p(t)$ denotes a monic polynomial of degree n with integer coefficients irreducible over $\mathbb{Q}$. Theorem C1 implies that

$$N(T, V_p) \sim c_p T^{n(n-1)/2}$$

for some $c_p > 0$, where $N(T, V_p)$ denotes the number of points in $V_p(\mathbb{Z})$ of Hilbert-Schmidt norm less than T.

Theorem C1 is deduced in [EsMoS] from the results on the limit behavior of algebraic measures under translations. More specifically, let $x\mathbf{H}, x = \mathbf{\Gamma e}$ be a homogeneous subset of $\mathbf{\Gamma}\backslash\mathbf{G}$ with $\mathbf{\Gamma}$ being a lattice in a Lie group $\mathbf{G}$ and let $\{\mathbf{g}_i\}_{i\in\mathrm{N}}$ be a sequence in $\mathbf{G}$. We denote by $\nu_{\mathbf{H}}\mathbf{g}_i$ the $\mathbf{g}_i^{-1}\mathbf{H}\mathbf{g}_i$-invariant probability measure supported on $x\mathbf{H}\mathbf{g}_i$. One asks what are the possible weak* limits of the sequence $\{\nu_{\mathbf{H}}\mathbf{g}_i\}_{i\in\mathrm{N}}$ and, in particular, when does the sequence converge to the $\mathbf{G}$-invariant probability measure $\nu_{\mathbf{G}}$?

Using Theorems 1 and 5 and the linearization method of [DM4], Eskin, Mozes, and Shah [EsMoS] showed that if $\mathbf{G}$, $\mathbf{H}$, and $\mathbf{\Gamma}$ satisfy the conditions of Theorem C1 and the sequence $\{\nu_{\mathbf{H}^0}\mathbf{g}_i\}$ weak* converges to a probability measure ν then $\nu = \nu_{\mathbf{L}}\mathbf{c}$ for some $\mathbf{c} \in \mathbf{G}$, where $\nu_{\mathbf{L}}$ is the $\mathbf{L}$-invariant probability measure supported on a homogeneous set $x\mathbf{L}$ with $\mathbf{L} = \mathfrak{L}(\mathbb{R})^0$ for some reductive $\mathbb{Q}$-subgroup $\mathfrak{L}$ of $\mathfrak{G}$ containing $\mathfrak{H}^0$. Also they proved the following

THEOREM C2 [EsMoS]. *Let* $\mathbf{G}$, $\mathbf{H}$, *and* $\mathbf{\Gamma}$ *be as in Theorem C1 and let* $\{E_n\}$ *be an increasing sequence of open subsets of* $\mathbf{H}\backslash\mathbf{G} = \cup\{E_n : n = 1, 2, \dots\}$. *Suppose that* $\{E_n\}$ *is not focused in* $\mathbf{H}\backslash\mathbf{G}$ *as* $n \to \infty$. *Then given any* $\varepsilon > 0$ *there exists an open set* $A \subset \mathbf{H}\backslash\mathbf{G}$ *such that*

$$\liminf_{n\to\infty} \frac{\mathrm{vol}_{\mathbf{H}\backslash\mathbf{G}}(A \cap E_n)}{\mathrm{vol}_{\mathbf{H}\backslash\mathbf{G}}(E_n)} > 1 - \varepsilon$$

and $\{\nu_{\mathbf{H}}\mathbf{g}_i\}$ *weak* converges to* $\nu_{\mathbf{G}}$ *for every sequence* $\{\mathbf{g}_i\}$ *with* $\{\mathbf{H}\mathbf{g}_i\}$ *being a sequence in* A *containing no subsequences convergent in* $\mathbf{H}\backslash\mathbf{G}$.

COROLLARY C2 [EsMoS]. *Suppose* $\mathfrak{H}^0$ *is a maximal connected* $\mathbb{Q}$*-subgroup of* $\mathfrak{G}$ *and let* $\{\mathbf{g}_i\}$ *be a sequence in* $\mathbf{G}$ *such that* $\{\mathbf{H}\mathbf{g}_i\}$ *contains no subsequences convergent in* $\mathbf{H}\backslash\mathbf{G}$. *Then* $\{\nu_{\mathbf{H}}\mathbf{g}_i\}$ *weak* converges to* $\nu_{\mathbf{G}}$.

To deduce Theorem C1 from Theorem C2 one denotes by χ_T the characteristic function of the ball B_T, and for $\mathbf{g} \in \mathbf{G}$ one defines

$$F_T(\mathbf{g}) = \sum\{\chi_T(v_0\boldsymbol{\gamma}\mathbf{g}) : \boldsymbol{\gamma} \in \mathbf{H} \cap \mathbf{\Gamma}\backslash\mathbf{\Gamma}\}.$$

Then F_T is a function on $\mathbf{\Gamma}\backslash\mathbf{G}$, as $F_T(\mathbf{g}) = F_T(\boldsymbol{\gamma}\mathbf{g})$ for all $\boldsymbol{\gamma} \in \mathbf{\Gamma}$. Also $F_T(\mathbf{e}) = N(V, T, \mathcal{O})$. Defining

$$\hat{F}_T(\mathbf{g}) = \frac{\mathrm{vol}(\mathbf{\Gamma}\backslash\mathbf{G})}{\mathrm{vol}(\mathbf{H} \cap \mathbf{\Gamma}\backslash\mathbf{H})\,\mathrm{vol}_{\mathbf{H}\backslash\mathbf{G}}(R_T)} F_T(\mathbf{g})$$

and using Theorem C2, Eskin, Mozes, and Shah showed (following the method of [DuRuSa]) that $\hat{F}_T(\mathbf{g}) \to 1$ weakly and pointwise on $\mathbf{\Gamma}\backslash\mathbf{G}$. In particular, $\hat{F}_T(\mathbf{e}) \to 1$. This yields Theorem C1.

6 Applications to Ergodic Theory and the "Rigidity" Phenomenon of Ad-unipotent Actions

The central problem of ergodic theory is that of classifying measure preserving (m.p.) transformations or flows up to isomorphism.

More precisely, let T and S be two m.p. transformations on probability spaces (X, μ_X) and (Y, μ_Y) respectively. We say that S is a *factor* of T if there is a m.p. ψ from X onto Y such that $\psi(T(x)) = S(\psi(x))$ for μ_X-almost every (a.e.) $x \in X$. If ψ is invertible, then T and S are called *isomorphic* and ψ is called an *isomorphism* between T and S. Similarly, one defines factors and isomorphisms of m.p. flows. One asks what m.p. transformations (or flows) are isomorphic? And what are the possible isomorphisms between T and S?

To approach this problem one looks for properties stable under isomorphisms. There are a number of dynamical properties of this kind, characterizing the degree of randomness of the system. There is also a *numerical* invariant of isomorphism called the *entropy*, which plays an important role in ergodic theory.

The definition of entropy will not be discussed, but we shall only mention that if an element $\mathbf{g}$ of a real Lie group $\mathbf{G}$ acts on $(\mathbf{\Gamma}\backslash\mathbf{G}, \nu_{\mathbf{G}})$, where $\nu_{\mathbf{G}}$ is the $\mathbf{G}$-invariant Borel probability measure on $\mathbf{\Gamma}\backslash\mathbf{G}$, then the entropy of this action is given by $\mathcal{E}(\mathbf{g}) = \sum\{\log|\lambda| : \lambda \text{ is an eigenvalue of } \mathrm{Ad}_{\mathbf{g}} \text{ with } |\lambda| > 1\}$, where the eigenvalues are counted with multiplicities. Thus if $\mathbf{u}$ is Ad-unipotent then $\mathcal{E}(\mathbf{u}) = 0$.

An element $\mathbf{g} \in \mathbf{G}$ is called Ad-semisimple if $\mathrm{Ad}_{\mathbf{g}}$ is diagonalizable over $\mathbb{C}$. The following theorem solves the isomorphism problem for Ad-semisimple actions.

THEOREM E1 (Ornstein, Weiss, Dani). *Let $\mathbf{G}_i$, $i = 1, 2$, be two real connected Lie groups. For each i let $\mathbf{\Gamma}_i$ be a lattice in $\mathbf{G}_i$, $\mathbf{g}^{(i)}$ an* Ad-*semisimple element of $\mathbf{G}_i$ with $\mathrm{Ad}_{\mathbf{g}^{(i)}}$ having at least one eigenvalue $|\lambda| \neq 1$. Suppose that $\mathbf{g}^{(i)}$ acts ergodically on $(M_i = \mathbf{\Gamma}_i\backslash\mathbf{G}_i, \nu_{\mathbf{G}_i})$, $i = 1, 2$. Then the actions of $\mathbf{g}^{(1)}$ and $\mathbf{g}^{(2)}$ are isomorphic if and only if $\mathcal{E}(\mathbf{g}^{(1)}) = \mathcal{E}(\mathbf{g}^{(2)})$.*

This theorem is proved by showing that the actions of $\mathbf{g}^{(1)}$ and $\mathbf{g}^{(2)}$ are isomorphic to Bernoulli shifts and then using Ornstein's isomorphism theorem, which states that Bernoulli shifts with the same entropy are isomorphic. Thus the isomorphism problem for actions of Ad-semisimple elements depends only on the entropy of these actions, hence only on the eigenvalues of $\mathrm{Ad}_{\mathbf{g}^{(1)}}$ and $\mathrm{Ad}_{\mathbf{g}^{(2)}}$. Neither $\mathbf{G}_1$, $\mathbf{G}_2$, nor $\mathbf{\Gamma}_1$, $\mathbf{\Gamma}_2$ play any significant role in the problem. Also one can show that there are uncountably many isomorphisms between isomorphic Ad-semisimple actions.

The following "rigidity" theorem, which can be deduced from our Theorem 1 demonstrates the profoundly different behavior of the actions of Ad-*unipotent* elements.

THEOREM E2 (Rigidity Theorem). *Let $\mathbf{G}_i$ be as above and let $\mathbf{\Gamma}_i$ be a lattice in $\mathbf{G}_i$ containing no nontrivial normal subgroups of $\mathbf{G}_i$, $i = 1, 2$. Let $\mathbf{u}^{(i)}$ be an* Ad-*unipotent element of $\mathbf{G}_i$, $i = 1, 2$. Suppose that the action of $\mathbf{u}^{(1)}$ on $(M_1, \nu_{\mathbf{G}_1})$ is ergodic and there is a m.p. $\psi : (M_1, \nu_{\mathbf{G}_1}) \to (M_2, \nu_{\mathbf{G}_2})$ such that $\psi(x\mathbf{u}^{(1)}) =$*

$\psi(x)\mathbf{u}^{(2)}$ for $\nu_{\mathbf{G}_1}$-a.e. $x \in M_1$. Then there is $\mathbf{c} \in \mathbf{G}_2$ and a surjective homomorphism $\alpha : \mathbf{G}_1 \to \mathbf{G}_2$ such that $\alpha(\mathbf{\Gamma}_1) \subset \mathbf{c}^{-1}\mathbf{\Gamma}_2\mathbf{c}$ and $\psi(\mathbf{\Gamma}_1\mathbf{h}) = \mathbf{\Gamma}_2\mathbf{c}\alpha(\mathbf{h})$ for $\nu_{\mathbf{G}_1}$-a.e. $\mathbf{\Gamma}_1\mathbf{h} \in M_1$. Also α is a local isomorphism whenever ψ is finite to one or $\mathbf{G}_1$ is simple and it is an isomorphism whenever ψ is one-to-one or $\mathbf{G}_1$ is simple with trivial center.

Note that $\mathcal{E}(\mathbf{u}^{(1)}) = \mathcal{E}(\mathbf{u}^{(2)}) = 0$. This theorem says in particular that if the actions of Ad-unipotent elements $\mathbf{u}^{(1)}$ and $\mathbf{u}^{(2)}$ are isomorphic then $\mathbf{G}_1$ must be isomorphic to $\mathbf{G}_2$ and $\mathbf{\Gamma}_1$ to $\mathbf{\Gamma}_2$.

The action of the unipotent group $\mathbf{H} = \{\mathbf{h}(t) = \begin{bmatrix} 1 & t \\ 0 & 1 \end{bmatrix} : t \in \mathbb{R}\}$ on $(M = \mathbf{\Gamma}\backslash\mathbf{G}, \nu_{\mathbf{G}})$, $\mathbf{G} = SL(2,\mathbb{R})$ is called the horocycle flow on M and the action of the diagonal group $\mathbf{A} = \{\mathbf{a}(t) = \begin{bmatrix} e^t & \\ & e-t \end{bmatrix} : t \in \mathbb{R}\}$ the geodesic flow on M.

Theorem E2 for horocycle flows was proved in [Ra2] in 1981. Then using the method of [Ra2] and [Ra4] Witte [W1] extended it to any connected $\mathbf{G}_1, \mathbf{G}_2$ and Ad-unipotent $\mathbf{u}^{(1)}, \mathbf{u}^{(2)}$ and furthermore to any mixing zero entropy affine maps [W2]. (Theorem E2 for nilpotent $\mathbf{G}_1, \mathbf{G}_2$ was proved earlier by Parry [P2].)

The proof in [Ra2] of the rigidity theorem for horocycle flows uses the polynomial divergence of horocycle orbits and the commutation relation with the geodesic flow. Generalizing this method, Feldman and Ornstein [FO] extended the rigidity theorem to horocycle flows on the unit tangent bundles of compact surfaces of variable negative curvature. Also generalizations to higher-dimensional hyperbolic space, to horospherical foliations and to geometrically finite groups were given by Flaminio [Fl], by Witte [W3], and by Flaminio and Spatzier [FlSp].

In fact, Theorem E2 is a consequence of a far stronger "Joinings Theorem" implied by Theorem 1. More specifically, let T and S be as above and let μ be a $T \times S$ invariant probability measure on $X \times Y$. Then μ is called a *joining* of T and S if $\mu(A \times Y) = \mu_X(A)$, $\mu(X \times B) = \mu_Y(B)$ for all measurable subsets $A \subset X$, $B \subset Y$. The joining $\mu_X \times \mu_Y$ is called the *trivial* joining. T and S are called *disjoint* if they have no nontrivial joinings.

It follows from Theorem 1 that every ergodic joining μ of the actions of Ad-unipotent elements $\mathbf{u}^{(1)}$ on $(M_1, \nu_{\mathbf{G}_1})$ and $\mathbf{u}^{(2)}$ on $(M_2, \nu_{\mathbf{G}_2})$ is algebraic. Thus $\mu(x\mathbf{\Lambda}(\mu)) = 1$ for some $x \in M_1 \times M_2$ (see Definition 2). Also $\mathbf{\Lambda}(\mu)$ is a subgroup of $\mathbf{G}_1 \times \mathbf{G}_2$ and the groups $\mathbf{\Lambda}_1(\mu)$ and $\mathbf{\Lambda}_2(\mu)$ defined by

$$\mathbf{\Lambda}_1(\mu) = \{\mathbf{h} \in \mathbf{G}_1 : (\mathbf{h}, \mathbf{e}) \in \mathbf{\Lambda}(\mu)\}$$
$$\mathbf{\Lambda}_2(\mu) = \{\mathbf{h} \in \mathbf{G}_2 : (\mathbf{e}, \mathbf{h}) \in \mathbf{\Lambda}(\mu)\}$$

are closed normal subgroups of $\mathbf{G}_1$ and $\mathbf{G}_2$ respectively. For $\mathbf{c} \in \mathbf{G}_2$ write $\mathbf{\Gamma}_2^{\mathbf{c}} = \{\boldsymbol{\gamma}\mathbf{\Lambda}_2(\mu) : \boldsymbol{\gamma} \in \mathbf{c}^{-1}\mathbf{\Gamma}_2\mathbf{c}\}$ and for $z \in M_1$ let

$$\xi_\mu(z) = \{y \in M_2 : (z, y) \in x\mathbf{\Lambda}(\mu)\}.$$

The set $\xi_\mu(z)$ is called the z-fiber of μ. We showed in [Ra9, Theorem 2] that there is $\mathbf{c} \in \mathbf{G}_2$ and a continuous surjective homomorphism $\alpha : \mathbf{G}_1 \to \mathbf{G}_2/\mathbf{\Lambda}_2(\mu)$ with kernel $\mathbf{\Lambda}_1(\mu)$, $\alpha(\mathbf{u}^{(1)}) = \mathbf{u}^{(2)}\mathbf{\Lambda}_2(\mu)$ such that

$$\xi_\mu(\mathbf{\Gamma}_1\mathbf{h}) = \{\mathbf{\Gamma}_2\mathbf{c}\boldsymbol{\beta}_i\alpha(\mathbf{h}) : i = 1, \dots, n\} \quad (2)$$

for all $\mathbf{h} \in \mathbf{G}_1$, where the intersection $\mathbf{\Gamma}_0 = \alpha(\mathbf{\Gamma}_1) \cap \mathbf{\Gamma}_2^{\mathbf{c}}$ is of finite index in $\alpha(\mathbf{\Gamma}_1)$ and in $\mathbf{\Gamma}_2^{\mathbf{c}}$, $n = |\mathbf{\Gamma}_0 \backslash \alpha(\mathbf{\Gamma}_1)|$ and $\alpha(\mathbf{\Gamma}_1) = \{\mathbf{\Gamma}_0 \boldsymbol{\beta}_i : i = 1, \dots, n\}$. This implies the following

THEOREM E3 [Ra9, Corollary 5] (The Joinings Theorem). *Every ergodic joining μ of the actions of* Ad*-unipotent elements $\mathbf{u}^{(1)}$ and $\mathbf{u}^{(2)}$ is algebraic and the fibers of μ are given by* (2). *If, in addition, $\mathbf{G}_i$ is simple, $i = 1, 2$, and μ is nontrivial, then every fiber of μ is finite and $\mathbf{G}_1$ and $\mathbf{G}_2$ are locally isomorphic.*

Thus if $\mathbf{G}_1$ and $\mathbf{G}_2$ are simple and not locally isomorphic, then the actions of $\mathbf{u}^{(1)}$ and $\mathbf{u}^{(2)}$ are disjoint.

The Joinings Theorem for horocycle flows was proved earlier in [Ra4, Theorem 6]. We showed there that if μ is a nontrivial ergodic joining of the horocycle flows $h_t^{(i)}$ on $M_i = \mathbf{\Gamma}_i \backslash SL(2, \mathbb{R})$, $h_t^{(i)}(x) = x\mathbf{h}(t)$, $x \in M_i$, $i = 1, 2$, then the flow $h_t^{(1)} \times h_t^{(2)}$ on $(M_1 \times M_2, \mu)$ is isomorphic to the horocycle flow on $\mathbf{\Gamma}_0 \backslash SL(2, \mathbb{R})$, where $\mathbf{\Gamma}_0$ is a subgroup of finite index in $\mathbf{\Gamma}_1$ and in $\mathbf{c}^{-1}\mathbf{\Gamma}_2\mathbf{c}$ for some $\mathbf{c} \in SL(2, \mathbb{R})$. This shows that the number of nonisomorphic ergodic joinings of the horocycle flows on M_1 and M_2 is at most countable and if $\mathbf{\Gamma}_1$ is uniform and $\mathbf{\Gamma}_2$ is not or if $\mathbf{\Gamma}_1$ is arithmetic and $\mathbf{\Gamma}_2$ is not then the horocycle flows are disjoint (see [Ra5] for more on this).

The central role in the proof of [Ra4, Theorem 6] is played by a dynamical property of horocycle flows, which we introduced in [Ra4, Definition 1] and called the H-property. It is a consequence of the polynomial divergence of horocycle orbits.

The H-property states that given $0 < \varepsilon < 1$ and $p, N > 0$ there are $\delta(\varepsilon, p, N)$, $\alpha(\varepsilon) \in (0, 1)$ such that if $d_{\mathbf{G}}(\mathbf{x}, \mathbf{e}) < \delta(\varepsilon, p, N)$ for some $\mathbf{x} \in \mathbf{G} = SL(2, \mathbb{R})$ and $\mathbf{x} \notin \mathbf{H}$ then there are $L, T > 0$ with $N < L < T$, $L \geq \alpha(\varepsilon)T$ such that either $d_{\mathbf{G}}(\mathbf{x}\mathbf{h}(t), \mathbf{h}(t+p)) \leq p\varepsilon$ for all $t \in [T-L, T]$ or $d_{\mathbf{G}}(\mathbf{x}\mathbf{h}(t), \mathbf{h}(t-p)) \leq p\varepsilon$ for all $t \in [T-L, T]$.

The H-property was proved in [Ra3, Lemma 2.1]. The latter proof also implies the following more general form of the H-property.

Given small $\theta, \varepsilon \in (0, 1)$ and $N > 1$ there are $\rho(\theta, N), \eta(\varepsilon) \in (0, 1)$ such that if $d_{\mathbf{G}}(\mathbf{x}, \mathbf{e}) < \rho(\theta, N)$ and $\mathbf{x} \notin \mathbf{N_G}(\mathbf{H})$ then there exist $T > N$ and differentiable functions $\sigma(\mathbf{x}, s), \tau(\mathbf{x}, s) : [0, T] \to \mathbb{R}$, $\sigma(\mathbf{x}, 0) = \tau(\mathbf{x}, 0) = 0$ with $\sigma(\mathbf{x}, s)$ increasing in s such that

$$
\begin{aligned}
&d_{\mathbf{G}}(\mathbf{x}\mathbf{h}(\sigma(x,s)), \mathbf{h}(s)\mathbf{a}(\tau(x,s))) < C\theta T^{-1} \text{ for all } s \in [0, T] \\
&\max\{|\tau(\mathbf{x}, s)| : 0 \leq s \leq T\} = |\tau(\mathbf{x}, T)| = \theta \qquad\qquad (*) \\
&|\tau(\mathbf{x}, s) - \tau(\mathbf{x}, T)| < \theta\varepsilon \text{ for all } s \in [(1-\eta(\varepsilon))T, T],
\end{aligned}
$$

where $C > 0$ is a constant. Here $\mathbf{N_G}(\mathbf{H})$ denotes the normalizer of $\mathbf{H}$ in $\mathbf{G} = SL(2, \mathbb{R})$ (it is generated by $\mathbf{A}$ and $\mathbf{H}$).

The first two relations in $(*)$ show, in particular, that for any $\mathbf{M} \subset \mathbf{G} - \mathbf{N_G}(\mathbf{H})$ with $\mathbf{e} \in \overline{\mathbf{M}}$ the group generated by $\mathbf{N_G}(\mathbf{H}) \cap \overline{\{\mathbf{h}_{-s}\mathbf{x}\mathbf{h}_t : \mathbf{x} \in \mathbf{M},\ s, t > 0\}}$ contains $\mathbf{A}$. (It also contains $\mathbf{H}$ by the H-property and hence $\mathbf{N_G}(\mathbf{H})$.) A version of this fact for a more general case was later used by Margulis in [M1, Lemmas 5 and 8].

The H-property was generalized in [W1, the Ratner property] and property $(*)$ was generalized in [Ra8, Theorem 3.1], where it is called the R-property (see Section 3 for a description of the R-property). The latter property plays a crucial role in the proof of Theorem 1. (We showed in [Ra12] how to use property $(*)$ to prove Theorem 1 for $\mathbf{G} = SL(2, \mathbb{R})$.)

Theorem 1 allows one also to show that factors of Ad-unipotent actions on $(\mathbf{\Gamma}\backslash\mathbf{G}, \nu_{\mathbf{G}})$ have simple algebraic form. This was recently done by Witte [W4]. We showed earlier in [Ra3,4] that if S_t is a factor of the horocycle flow on $\mathbf{\Gamma}\backslash SL(2, \mathbb{R})$ then there is a lattice $\mathbf{\Gamma}'$ in $SL(2, \mathbb{R})$ such that $\mathbf{\Gamma} \subset \mathbf{\Gamma}'$ and S_t is isomorphic to the horocycle flow on $\mathbf{\Gamma}'\backslash SL(2, \mathbb{R})$. This implies that the number of nonisomorphic factors of the horocycle flow is finite and if $\mathbf{\Gamma}$ is maximal then there are no non-trivial factors.

It is a fact that there are uncountably many nonisomorphic ergodic joinings and factors of Ad-semisimple actions with positive entropy (this follows from Ornstein's theory of Bernoulli shifts). Also it was shown by Sinai and Bowen-Ruelle that the geodesic flow on $\mathbf{\Gamma}\backslash SL(2, \mathbb{R})$ possesses infinitely many ergodic invariant probability measures that are *not* algebraic. Also there are points x in $\mathbf{\Gamma}\backslash SL(2, \mathbb{R})$ for which closures of geodesic orbits are not manifolds. All these facts put Ad-semisimple actions in striking contrast with the rigid behavior of Ad-unipotent actions discussed in this section and given in Theorems 1, 3, 6, E2, and E3.

The rigidity theorem for the horocycle flows $h_t^{(i)}$ on $(M_i = \mathbf{\Gamma}_i\backslash\mathbf{G}_i, \nu_i)$, $\mathbf{G}_i = SL(2, \mathbb{R})$, $\nu_i = \nu_{\mathbf{G}_i}$, $i = 1, 2$, says that if $h_t^{(1)}$ is isomorphic to $h_t^{(2)}$ then $\mathbf{\Gamma}_1$ is conjugate to $\mathbf{\Gamma}_2$. We ask: Can this "rigidity" be *destroyed* by time changes?

More specifically, let τ be a positive integrable function on M_1 with $\int_{M_1} \tau\, d\nu_1 = 1$. We say that h_t^τ is obtained from $h_t^{(1)}$ by the time change τ if $h_t^\tau(x) = h_{v(x,t)}^{(1)}(x)$ for all $x \in M_1$, $t \in \mathbb{R}$, where $v(x,t)$ is defined by $\int_0^{v(x,t)} \tau(x\mathbf{h}(s)ds) = t$. Then h_t^τ preserves the probability measure $\tau d\nu_1$ on M_1.

We ask: Is there a time change τ such that h_t^τ is isomorphic to $h_t^{(2)}$? If "yes" $h_t^{(1)}$ is called Kakutani equivalent to $h_t^{(2)}$.

Using the Feldman-Katok-Ornstein-Weiss theory of Kakutani equivalence (developed in the 1970s) we showed in [Ra0] that the answer to this question is affirmative. However, we also showed [Ra6] that even very mild smoothness conditions on τ cause the rigidity to persist. Namely, we say that τ is Hölder continuous in the direction of the rotation group

$$\mathbf{R} = \{\mathbf{r}(\theta) = \begin{bmatrix} \cos\theta & \sin\theta \\ -\sin\theta & \cos\theta \end{bmatrix} : \theta \in [-\pi, \pi]\}$$

if $|\tau(x) - \tau(x\mathbf{r}(\theta))| \leq C|\theta|^\alpha$ for some $C, \alpha > 0$ and all $x \in M_1$, $\theta \in [-\pi, \pi]$. We showed in [Ra6] that if h_t^τ is isomorphic to $h_t^{(2)}$ with τ being bounded, measurable, and Hölder continuous in the direction of $\mathbf{R}$, then $\mathbf{\Gamma}_1$ is conjugate to $\mathbf{\Gamma}_2$. Also all isomorphisms between h_t^τ and $h_t^{(2)}$ as well as factors and joinings of h_t^τ have an algebraic form [Ra7].

PROBLEM 1. Are Ad-unipotent ergodic flows on homogeneous spaces of a *general* Lie group $\mathbf{G}$ Kakutani equivalent? In particular, is the flow $h_t^{(1)} \times h_t^{(1)}$ acting on $(M_1 \times M_1, \nu_1 \times \nu_1)$ Kakutani equivalent to the flow $h_t^{(2)} \times h_t^{(2)}$ acting on $(M_2 \times M_2, \nu_2 \times \nu_2)$? (We showed in [Ra1] that $h_t \times h_t$ acting on $(M \times M, \nu \times \nu)$ is not Kakutani equivalent to h_t acting on (M, ν).)

PROBLEM 2. Do the rigidity properties discussed in this section hold for smoothly time-changed Ad-unipotent flows?

Theorem S1 above allows one to extend Theorems E2 and E3 and classify factors of Ad-unipotent flows in the $\mathbb{S}$-arithmetic setting discussed in Section 4. The latter flows represent measure preserving actions of the field $\mathbb{Q}_s$ (as an additive group) on $(\mathbf{\Gamma}\backslash\mathbf{G}_\mathbb{S}, \nu_{\mathbf{G}_\mathbb{S}})$, $s \in \mathbb{S}$, with $\mathbf{\Gamma}$ being a lattice in $\mathbf{G}_\mathbb{S}$ (see Section 4). It would be of interest to develop the ergodic theory of measure preserving actions of the p-adic field $\mathbb{Q}_p$ as an additive group and, in particular, to construct a theory of Kakutani equivalence for these actions. Applying such a theory to p-adic horocycle flows on $\mathbf{\Gamma}\backslash SL(2, \mathbb{Q}_p)$ one can ask questions similar to those discussed in this section for the real case.

To conclude this section we mention that recently Starkov [St4] used Theorem 1 to give an affirmative answer to a question raised by Marcus in [Ma] (see also [M2]). Namely, he showed that if $\mathbf{\Gamma}$ is a lattice in a connected Lie group $\mathbf{G}$ and the action of a one-parameter subgroup $\mathbf{U} \subset \mathbf{G}$ on $(\mathbf{\Gamma}\backslash\mathbf{G}, \nu_\mathbf{G})$ is mixing, then it is mixing of all orders. Marcus [Ma] proved this result for semisimple $\mathbf{G}$ (see also [Mo1]). Starkov's argument exploits Marcus' result and a theorem of Witte [W2, Proposition 2.6] (cf. Corollary 1 in [St4]).

References

[AGH] L. Auslander, L. Green, and F. Hahn, *Flows on homogeneous spaces*, Ann. of Math. Stud. **53** (1963), Princeton Univ. Press, Princeton, NJ.

[BBo] F. Bien and A. Borel, *Sous-groupes épimorphiques de groupes algébriques linéaires I*, C.R. Acad. Sci. Paris **315** (1992).

[Bo] A. Borel, *Values of indefinite quadratic forms at integral points and flows on spaces of lattices*, Bull. Amer. Math. Soc. **32** (1995), 184–204.

[BoHC] A. Borel and Harish-Chandra, *Arithmetic subgroups of algebraic groups*, Ann. of Math. (2) **75** (1962), 485–535.

[BoPr] A. Borel and G. Prasad, *Values of isotropic quadratic forms at S-integral points*, Compositio Math. **83** (1992), 347–372.

[Bw] R. Bowen, *Weak mixing and unique ergodicity on homogeneous spaces*, Israel J. Math. **23** (1976), 267–273.

[D1] S. G. Dani, *Invariant measures and minimal sets of horospherical flows*, Invent. Math. **64** (1981), 357–385.

[D2] ______, *On orbits of unipotent flows on homogeneous spaces*, II, Ergodic Theory Dynamical Systems **6** (1986), 167–182.

[D3] ______, *Orbits of horospherical flows*, Duke Math. J. **53** (1986), 177–188.

[DM1] S. G. Dani and G. A. Margulis, *Values of quadratic forms at primitive integral points*, Invent. Math. **98** (1989), 405–425.

[DM2] ———, *Orbit closures of generic unipotent flows on homogeneous spaces of* $SL(3,\mathbb{R})$, Math. Ann. **286** (1990), 101–128.

[DM3] ———, *Values of quadratic forms at integral points: An elementary approach*, L'enseig. Math. **36** (1990), 143–174.

[DM4] ———, *Limit distributions of orbits of unipotent flows and values of quadratic forms*, Adv. in Sov. Math. **16** (1993), 91–137.

[DSm] S. G. Dani and J. Smillie, *Uniform distribution of horocycle orbits for Fuchsian groups*, Duke Math. J. **5** (1984), 185–194.

[DuRuSa] W. Duke, Z. Rudnick, and P. Sarnak, *Density of integer points on affine homogeneous varieties*, Duke Math. J. **71** (1993), 143–180.

[EPe] R. Ellis and W. Perrizo, *Unique ergodicity of flows on homogeneous spaces*, Israel J. Math. **29** (1978), 276–284.

[EsMc] A. Eskin and C. McMullen, *Mixing, counting and equidistribution in Lie groups*, Duke Math. J. **71** (1993), 181–209.

[EsMoS] A. Eskin, S. Mozes, and N. Shah, *Unipotent flows and counting lattice points on homogeneous varieties*, to appear in Ann. of Math.

[FO] J. Feldman and D. Ornstein, *Semirigidity of horocycle flows over compact surfaces of variable negative curvature*, Ergodic Theory Dynamical Systems **7** (1987), 49–72.

[Fl] L. Flaminio, *An extension of Ratner's rigidity theorem to n-dimensional hyperbolic space*, Ergodic Theory Dynamical Systems **7** (1987), 73–92.

[FlSp] L. Flaminio and R. Spatzier, *Geometrically finite groups, Patherson-Sullivan measures and Ratner's rigidity theorem*, Invent. Math. **99** (1990), 601–626.

[Fu1] H. Furstenberg, *Strict ergodicity and transformations of the torus*, Amer. J. Math. **83** (1961), 573–601.

[Fu2] ———, *The unique ergodicity of the horocycle flow*, in Recent Advances in Topological Dynamics, Lecture Notes in Math. **318**, Springer-Verlag, Berlin and New York (1972), 95–115.

[H] G.A. Hedlund, *Fuchsian groups and transitive horocycles*, Duke Math. J. **2** (1936), 530–542.

[L] E. Lesigne, *Théorèmes ergodiques pour une translation sur nilvariété*, Ergodic Theory Dynamical Systems **9** (1989), 115–126.

[Ma] B. Marcus, *The horocycle flow is mixing of all degrees*, Invent. Math. **46** (1978), 201–209.

[M1] G. A. Margulis, *Discrete subgroups and ergodic theory*, in Number Theory, Trace Formula and Discrete Groups, Symposium in honor of A. Selberg, Oslo 1987, Academic Press, New York (1989), 377–398.

[M2] ———, *Dynamical and ergodic properties of subgroup actions on homogeneous spaces with applications to number theory*, Proc. Internat. Congress Math. (1990), Kyoto, 193–215.

[MT1] G. A. Margulis and G. M. Tomanov, *Measure rigidity for algebraic groups over local fields*, C. R. Acad. Sci. Paris, t. 315, Série I (1992), 1221–1226.

[MT2] ———, *Invariant measures for actions of unipotent groups over local fields on homogeneous spaces*, Invent. Math. **116** (1994), 347–392.

[Mo1] S. Mozes, *Mixing of all orders of Lie groups actions*, Invent. Math. **107** (1992), 235–241.

[Mo2] ———, *Epimorphic subgroups and invariant measures*, Ergodic Theory Dynamical Systems **15** (1995), 1–4.

[MoS] S. Mozes and N. Shah, *On the space of ergodic invariant measures of unipotent flows*, Ergodic Theory Dynamical Systems **15** (1995), 149–159.

[P1] W. Parry, *Ergodic properties of affine transformations and flows on nilmanifolds*, Amer. J. Math. **91** (1969), 757–771.

[P2] ———, *Metric classification of ergodic nilflows and unipotent affines*, Amer. J. Math. **93** (1971), 819–828.

[Pr] G. Prasad, *Ratner's theorem in S-arithmetic setting*, in Workshop on Lie Groups, Ergodic Theory and Geometry, Math. Sci. Res. Inst. Publ. (1992), p. 53.

[R] M. S. Raghunathan, Discrete Subgroups of Lie Groups, Springer-Verlag, Berlin and New York, 1972.

[Ra0] M. Ratner, *Horocycle flows are loosely Bernoulli*, Israel J. Math. **31** (1978), 122–132.

[Ra1] ———, *The Cartesian square of the horocycle flow is not loosely Bernoulli*, Israel J. Math. **34** (1979), 72–96.

[Ra2] ———, *Rigidity of horocycle flows*, Ann. of Math. (2) **115** (1982), 597–614.

[Ra3] ———, *Factors of horocycle flows*, Ergodic Theory Dynamical Systems **2** (1982), 465–489.

[Ra4] ———, *Horocycle flows: Joinings and rigidity of products*, Ann. of Math. (2) **118** (1983), 277–313.

[Ra5] ———, *Ergodic theory in hyperbolic space*, Contemp. Math. **26** (1984), 302–334.

[Ra6] ———, *Rigidity of time changes for horocycle flows*, Acta Math. **156** (1986), 1–32.

[Ra7] ———, *Rigid reparametrizations and cohomology for horocycle flows*, Invent. Math. **88** (1987), 341–374.

[Ra8] ———, *Strict measure rigidity for unipotent subgroups of solvable groups*, Invent. Math. **101** (1990), 449–482.

[Ra9] ———, *On measure rigidity of unipotent subgroups of semisimple groups*, Acta Math. **165** (1990), 229–309.

[Ra10] ———, *On Raghunathan's measure conjecture*, Ann. of Math. (2) **134** (1991), 545–607.

[Ra11] ———, *Raghunathan's topological conjecture and distributions of unipotent flows*, Duke Math. J. **63** (1991), 235–280.

[Ra12] ———, *Raghunathan's conjectures for* $SL(2,\mathbb{R})$, Israel J. Math. **80** (1992), 1–31.

[Ra13] ———, *Raghunathan's conjectures for p-adic Lie groups*, International Mathematics Research Notices, no. 5 (1993), 141–146.

[Ra14] ———, *Invariant measures and orbit closures for unipotent actions on homogeneous spaces*, Geom. Functional Anal (GAFA) **4** (1994), 236–256.

[Ra15] ———, *Raghunathan's conjectures for cartesian products of real and p-adic Lie groups*, Duke Math. J. **77** (1995), 275–382.

[S1] N. Shah, *Uniformly distributed orbits of certain flows on homogeneous spaces*, Math. Ann. **289** (1991), 315–334.

[S2] ———, *Limit distributions of polynomial trajectories on homogeneous spaces*, Duke Math. J. **75** (1994), 711–732.

[St1] A. Starkov, *The reduction of the theory of homogeneous flows to the case of discrete isotropy subgroups*, Dokl. Akad. Nauk **301** (1988), 1328–1331 (Russian).

[St2] ———, *Solvable homogeneous flows*, Math. USSR-Sb. **62** (1989), 243–260.

[St3] ———, *Structure of orbits of homogeneous flows and Raghunathan's conjecture*, Russian Math. Surveys **45** (1990), 227–228.

[St4] ———, *On the mixing of all orders for homogeneous flows*, Dokl. Akad. Nauk **333** (1993), 442–445 (Russian).

[T] T. Tamagawa, *On discrete subgroups of p-adic algebraic groups*, in Arithmetic Algebraic Geometry, Shilling, Harper & Row, New York (1965), 11–17.

[V] W. Veech, *Unique ergodicity of horospherical flows*, Amer. J. Math. **99** (1977), 827–859.

[W1] D. Witte, *Rigidity of some translations on homogeneous spaces*, Invent. Math. **81** (1985), 1–27.

[W2] ———, *Zero entropy affine maps on homogeneous spaces*, Amer. J. Math. **109** (1987), 927–961.

[W3] ———, *Rigidity of horospherical foliations*, Ergodic Theory Dynamical Systems **9** (1989), 191–205.

[W4] ———, *Measurable quotients of unipotent translations on homogeneous spaces*, Trans. Amer. Math. Soc. **345** (1994), 577–594.

Progress on the Four-Color Theorem

PAUL SEYMOUR

Bellcore, 445 South Street
Morristown, NJ 07960, USA

ABSTRACT. The four-color theorem, that every loopless planar graph is 4-colorable, was proved in 1977 by Appel and Haken, but there remain open several conjectured extensions. Here we report on a new (but still computer-based) proof of the four-color theorem itself, and on some progress towards two of the open extensions. We also survey some related results on minors of graphs and on linkless embeddings of graphs in 3-space.

1. Introduction

The four-color theorem (briefly, the 4CT) asserts that every loopless planar graph admits a vertex 4-coloring. (Graphs in this paper are finite, and may have loops or multiple edges. We assume familiarity only with absolutely basic graph theory.) The 4CT was conjectured in about 1850 by F.W.Guthrie, and over the next 125 years became one of the most popular problems of mathematics. Correspondingly, a great many "proofs" of it were proposed, but none of them survived scrutiny until 1977, when Appel and Haken (A&H) gave their famous proof [1], [2], [3].

This was a major triumph. They took the well-established approach of attempting to find an "unavoidable set of reducible configurations" and pushed it through to success, by means of extensive use of a computer, and a novel and elaborate system of "discharging rules" that they invented.

Yet seventeen years have passed, and there still remains some doubt as to whether the theorem is actually true. This implies some serious difficulties with checking the A&H proof, so let me explain what I think these are. First, the most novel feature of the A&H proof, and the most commonly heard criticism of it, is that it makes use of a computer. Evidently, a readable conventional proof would be better, but to me this does not seem such an important issue. Provided the computer calculations can easily be duplicated by the reader, the computer use seems rather like the use of a calculator, and not much more objectionable.

It has to be admitted, however, that the A&H proof does not quite meet this standard; it *is* difficult to duplicate the calculations. The programs themselves are straightforward to write, but getting all the data into the computer is a major headache — it is necessary to input, by hand, descriptions of some 1400 fairly substantial planar graphs. But this would be overcome if the data were available in electronic form, instead of only by figures. If this were the only problem things would not be so bad, someone would have produced the necessary data file and we would have a checkable proof.

Proceedings of the International Congress
of Mathematicians, Zürich, Switzerland 1994

The real difficulty lies in the *conventional* part of the A&H proof, not in the computer part. The noncomputer part of the proof requires such an enormous amount of time and patience that, as far as I am aware, no one has made an independent check of all the details. No major errors have been found in the proof, and the consensus of opinion is that it is probably correct; but, still, it is disturbing that it has not been properly checked.

In the spring of 1993, Neil Robertson, Daniel Sanders, Robin Thomas, and I decided to try to convince ourselves somehow that the theorem really was true. We tried to read the A&H proof, but gave up after a week or so; it became evident that this was not going to work. Instead, we decided it would be easier, and much more fun, to make up our own proof, using the same basic approach as A&H but inventing the details for ourselves. We succeeded, and the new proof turned out to be simpler and more easily checked than the old one. In particular, the necessary data is available in electronic form, and the part analogous to the difficult noncomputer part of the A&H proof has been written in a formal language so that it can be checked by a computer in a few minutes, or by hand (by a very patient reader) in a few months. The other differences are all in the technical details, however, so there is not much point in sketching here what we did — the interested reader should see [16]. For our purposes here, all we need is this: despite all the rumors and doubts, the 4CT is true and provable by the approach that A&H used.

In this paper, we sketch (in Sections 2, 3, and 4) three conjectured generalizations of the 4CT, and some recent progress on two of them. The two new results are both related to the problem of embedding a graph in 3-space so that no subgraph forms a nontrivial link in the sense of knot theory; this is discussed in Section 5. Finally, in Section 6 we sketch a general result on graph excluded minors, related to the two new results.

2. Hadwiger's Conjecture

A graph H is a *minor* of a graph G if H can be obtained from a subgraph of G by contracting edges. The most well-known theorem about minors of graphs is the Kuratowski-Wagner theorem [13], [34], that a graph is planar if and only if it has no K_5 or $K_{3,3}$ minor. (In general, K_n is the complete graph with n vertices; $K_{3,3}$ is the graph with six vertices $a_1, a_2, a_3, b_1, b_2, b_3$, in which each a_i is adjacent to each b_j.) Actually, this is more usually stated in terms of "topological containment", but the two forms are easily interderivable.

In view of this, we can restate the 4CT without mentioning planarity, in the form "Every loopless graph with no K_5 or $K_{3,3}$ minor is 4-colorable." Now K_5 is not 4-colorable, so its presence here is natural, but what about $K_{3,3}$? That of course is 2-colorable, so it looks a little anomalous. What if we leave it out? In 1937, Wagner [34] proved the following lemma:

LEMMA 2.1 *Every 4-connected graph with no K_5 minor is planar.*

From this, it follows that we can indeed leave out $K_{3,3}$, that the 4CT is equivalent to the following:

THEOREM 2.2 *Every loopless graph with no K_5 minor is 4-colorable.*

To prove this equivalence, one direction is clear: certainly (2.2) implies the 4CT, because planar graphs have no K_5 minor. Suppose that the other direction is false, and take a minimal counterexample, that is, a loopless graph with no K_5 minor, not 4-colorable, and as small as possible. It is easy to prove that such a graph must be 4-connected, and hence planar by (2.1), contrary to the 4CT. This proves the equivalence of (2.2) with the 4CT, known as "Wagner's equivalence theorem".

In 1943, Hadwiger [8] proposed the following conjecture:

CONJECTURE 2.3 *For every integer $n \geq 0$, every loopless graph with no K_{n+1} minor is n-colorable.*

(For fixed n, let us call this "HC(n)".) Then HC(0), HC(1), HC(2) are trivial, and HC(3) follows easily from a theorem of Dirac [6] that every nonnull simple graph with no K_4 minor has a vertex of degree ≤ 2. (Actually, HC(3) was proved earlied by Hadwiger [8].) We already saw that HC(4) is equivalent to the 4CT. So the sequence goes

"trivial; trivial; trivial; easy; very very difficult; ...."

What should the next term be? Many people guessed "false"; but it turns out to be true — Robertson, Thomas, and I proved it in [20]. (More precisely, we proved its equivalence with the 4CT — we were not at that time convinced that the 4CT was proved. Actually, this was a strong motivation for the work reported in Section 1.)

At first sight it is surprising that the 4CT, a result about 4-coloring planar graphs, can be equivalent to a result about 5-coloring nonplanar ones. But what happens is rather like the derivation of (2.2) from the 4CT. Let us say a graph is *apex* if for some vertex v (called the *apex*), deleting v from G results in a planar graph. Certainly apex graphs have no K_6 minor, because if one did then deleting the apex would yield a planar graph with a K_5 minor, which is impossible. Moreover, loopless apex graphs are 5-colorable (assuming the 4CT), because four colors suffice for the planar part, and there is still a color left for the apex. We showed, *without* assuming the 4CT, that:

THEOREM 2.4 *Any minimal counterexample to HC(5) is apex.*

(It follows, assuming the 4CT, that there is no minimal counterexample, and so HC(5) is true.) The proof of (2.4) was long and complicated, but it was nothing like the proof of the 4CT, and did not need a computer.

Actually, it is not too difficult to show that every minimal counterexample to HC(5) is 6-connected. There is a conjecture due to Jørgensen [9], still open, that would make the proof of HC(5) even more like the proof of HC(4), by providing an analogue of (2.1). Jørgensen conjectures:

CONJECTURE 2.5 *Every 6-connected graph with no K_6 minor is apex.*

The fact that (2.5) is still open makes it painfully clear that we do not yet really know what graphs with no K_6 minor are like. We shall return to this later.

What about Hadwiger's conjecture in general? I have no idea what a general proof could be like. Proving HC(5) was a lot of work, and I cannot imagine a proof of the general case that uses anything like the same approach. In particular, one of the main results we used was a theorem of Mader [14], the $n = 5$ case of the following:

THEOREM 2.6 *For $n \leq 6$, any simple graph G with at least $n-1$ vertices and with no K_{n+1} minor has at most*

$$(n-1)|V(G)| - n(n-1)/2$$

edges.

(This bound is best possible for each n, as is easily seen.) But for $n > 6$, the inequality of (2.6) is false. For instance, Mader found a simple graph with 10 vertices and 40 edges, with no K_8 minor. (Delete five pairwise nonadjacent edges from K_{10}.) So at least one of our basic tools fails in the general case, and it is not clear what to use instead. The best analogue of (2.6) for general n is the following theorem of Thomason [31]:

THEOREM 2.7 *For every n, every simple graph G with no K_{n+1} minor has at most*

$$Cn(\log n)^{1/2}|V(G)|$$

edges, where C is a constant.

It follows easily from (2.7) that every graph with no K_{n+1} minor is $(C'n(\log n)^{1/2})$-colorable, for some constant C'; but nothing better is known. It is not even known whether every graph with no K_{n+1} minor is $1000n$-colorable.

But we do have one result on HC(n) in general. Before the 4CT was proved, there was speculation as to whether it could be undecidable in some sense. (This would of course mean that there was no counterexample, because a counterexample would make it decidable.) Now that the 4CT is proved, one might still wonder whether, say, HC(6) is undecidable. But it is not.

THEOREM 2.8 *For any fixed n, we can construct N so that if HC(n) is false then it is false for some graph with $< N$ vertices.*

Hence to decide the truth of HC(n) we merely check it for the finitely many graphs with $< N$ vertices. This is a consequence of a general result on graph minors, described in Section 6.

3. Tutte's Conjecture

Some notation: if G is a graph and $X \subseteq V(G)$, we denote by $\delta(X)$ the set of edges of G with one end in X and the other in $V(G) - X$. An edge e is a *bridge* of G if $\delta(X) = \{e\}$ for some X. A graph is *cubic* if every vertex has degree 3. A graph

G is *k-edge-colorable*, where $k \geq 0$ is an integer, if there is a map κ from $E(G)$ into $\{1, \ldots, k\}$ so that for each vertex v, all the edges e incident with v have $\kappa(e)$ different. Tait [30] showed that the 4CT is equivalent to the following:

THEOREM 3.1 *Every bridgeless cubic planar graph is 3-edge-colorable.*

Now "bridgeless" here is necessary; no cubic graph with a bridge is 3-edge-colorable, for trivial reasons. But what about "planar" ? It is not so easy to find *any* cubic bridgeless graph that is not 3-edge-colorable, but one was eventually found, by Petersen [15], the so-called *Petersen graph* (see Figure 1).

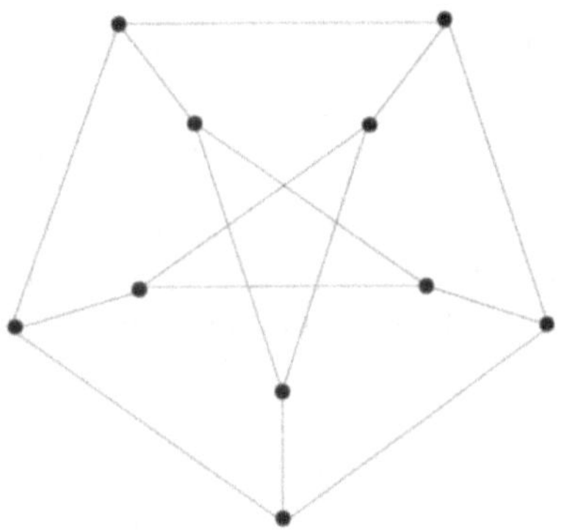

Figure 1. The Petersen graph

In 1966, Tutte [33] proposed the following:

CONJECTURE 3.2 *Every bridgeless cubic graph with no Petersen graph minor is 3-edge-colorable.*

This evidently implies the 4CT, because planar graphs do not have the Petersen graph as a minor. Superficially, this conjecture seems to resemble (2.2), for they are both obtained from a version of the 4CT by replacing the "planar" condition with a condition excluding as a minor the smallest graph not possessing the desired property. But (3.2) seems to be much more difficult than (2.2), even assuming the 4CT, and in particular is still open. Actually, (3.2) is much closer to HC(5) than to HC(4), as we shall see.

Because apex graphs have no Petersen graph minor, (3.2) also implies that:

CONJECTURE 3.3 *Every bridgeless cubic apex graph is 3-edge-colorable.*

There is another specialization of (3.2) of interest, as follows. We say a graph is *doublecross* if it can be drawn in the plane with at most two crossings, so that every crossing is in the boundary of the infinite region in the natural sense. (See Figure 2.)

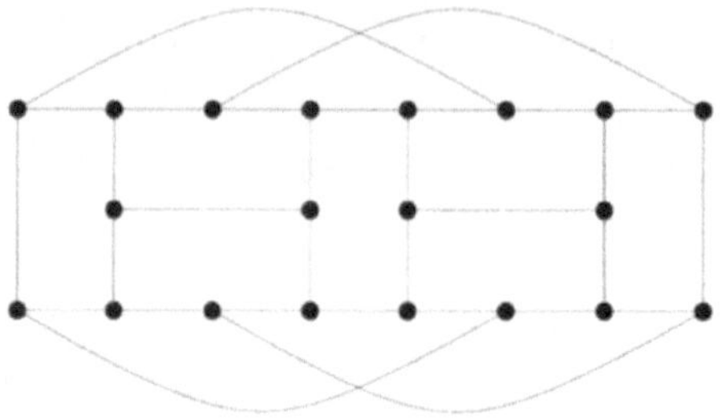

Figure 2. A doublecross graph.

Doublecross graphs have no Petersen graph minors, so (3.2) implies:

CONJECTURE 3.4 *Every bridgeless cubic doublecross graph is 3-edge-colorable.*

It turns out that Tutte's conjecture is equivalent to the conjunction of (3.3) and (3.4). This is a consequence of a very recent result of Robertson, Thomas, and myself [19]. We prove that any minimal counterexample G to (3.2) has the following properties:

(i) Every circuit of G has length at least 5

(ii) $|\delta(X)| \geq 6$ for every $X \subseteq V(G)$ with $|X|, |V(G) - X| \geq 8$.

(This is not particularly difficult, and may well have already been known.) Then we show that:

THEOREM 3.5 *With one exception, every cubic graph satisfying (i) and (ii) with no Petersen graph minor can be converted to an apex or doublecross graph by applying a certain "twisting" operation that does not affect 3-edge-colorability.*

We deduce:

THEOREM 3.6 *If Tutte's conjecture is false, there is a counterexample to it that is either apex or doublecross.*

Consequently, Tutte's conjecture seems to be almost proved. We hope to be able to adapt our proof of the 4CT to prove (3.3) and (3.4), and thereby prove (3.2), but this work is still in progress.

4. Nowhere-Zero Flows

Tutte's conjecture (3.2) is a special case of another conjecture of Tutte about nowhere-zero flows. We have no new results to report about nowhere-zero flows, but survey the topic briefly in this section because it is rather pretty. Also, because the two conjectures of Tutte are often confused, it seems a good idea to clarify the difference between them.

Let G be a graph; direct its edges arbitrarily. (How these edges are directed makes no essential difference to what follows.) Let $k \geq 2$ be an integer. A *nowhere-zero k-flow* in G is a function ϕ with domain $E(G)$, so that

(i) For each edge $e, \phi(e)$ is one of $\pm 1, \pm 2, \ldots, \pm(k-1)$

(ii) For each vertex v, if X and Y denote the set of edges of G with head v and tail v respectively, then

$$\sum(\phi(e) : e \in X) = \sum(\phi(e) : e \in Y).$$

Nowhere-zero flows are of interest basically because of the following. Let G be a graph drawn in the plane, with edges directed arbitrarily. Let κ be a k-region-coloring of G; that is, for each region r, $\kappa(r)$ is one of $1, 2, \ldots, k$, and $\kappa(r_1) \neq \kappa(r_2)$ for any two regions r_1, r_2 both incident with the same edge. For each edge e, define $\phi(e)$ to be $\kappa(r_1) - \kappa(r_2)$, where r_1, r_2 are the regions to the right and left of e respectively. Then it is easy to see that ϕ is a nowhere-zero k-flow in G. Conversely, given a nowhere-zero k-flow in G, it is almost as easy to reconstruct from it a k-region-coloring of G. Thus, nowhere-zero flows give us a way to talk about k-region-colorings of planar graphs *without mentioning regions*, and thereby to extend some of them to nonplanar graphs. (For instance, the elementary theorem "A planar cubic graph is 3-region-colorable if and only if it is bipartite" has a pretty generalization to "A cubic graph has a nowhere-zero 3-flow if and only if it is bipartite.")

Consequently, via planar duality, it follows that another equivalent form of the 4CT is:

THEOREM 4.1 *Every bridgeless planar graph has a nowhere-zero 4-flow.*

(Please note that (4.1) is not just about planar cubic graphs.) Again, the "bridgeless" condition is necessary, because no graph with a bridge has a nowhere-zero k-flow for any k. If we try to drop the "planar" condition, we find ourselves in the same situation as before, because the Petersen graph has no nowhere-zero 4-flow; indeed, it is easy to show that:

THEOREM 4.2 *A cubic graph has a nowhere-zero 4-flow if and only if it is 3-edge-colorable.*

In view of this, Tutte proposed [33]:

CONJECTURE 4.3 *Every bridgeless graph with no Petersen graph minor has a nowhere-zero 4-flow.*

For cubic graphs, this is equivalent to (3.2): but it does not seem possible to reduce (4.3) to the cubic case. It has not been shown that the smallest counterexample to (4.3) is cubic, for instance. Because apex graphs have no Petersen graph minor, a consequence of (4.3) would be:

CONJECTURE 4.4 *Every bridgeless apex graph has a nowhere-zero 4-flow.*

In view of (4.2), it might be thought that (4.4) was close to (3.3), but in fact it seems to me that (4.4) is much stronger. For instance, take a minimal counterexample to (4.4). It is easy to prove that every vertex has degree 3 except for the apex, but the apex might have arbitrary degree. We are going to try to prove (3.3) by adapting our proof of the 4CT, but I have little hope of proving (4.4) this way. The difference is, roughly, that for apex cubic graphs there are only three vertices that are neighbors of the apex, and our proof of the 4CT yields a large (constant) number of vertices in the neighborhood of which a "reducible configuration" can be found; so some of these must be far away from the three bad places. But for (4.4), the number of bad places is unlimited, and I do not see a way to get a reducible configuration avoiding the bad places.

Actually, (4.4) can be shown to be equivalent to the following (unpublished) conjecture of Grötzsch:

CONJECTURE 4.5 *Let G be a planar graph. Then G is 3-edge-colorable if and only if:*

(i) *every vertex of G has degree ≤ 3, and*

(ii) *no subgraph of G has one vertex of degree 2 and all others of degree 3.*

This in turn is a special case of a conjecture of mine [24]:

CONJECTURE 4.6 *Let G be a planar graph, and let $k \geq 0$ be an integer. Then G is k-edge-colorable if and only if:*

(i) *every vertex of G has degree $\leq k$, and*

(ii) *every subgraph of G with an odd number n of vertices has at most $k(n-1)/2$ edges.*

This remains open, for all $k \geq 4$. (Please note that, in all these conjectures, parallel edges are permitted.) The special case of (4.6) for 4-regular graphs when $k = 4$ is particularly interesting; it implies:

CONJECTURE 4.7 *Any 4-connected 4-regular planar graph with an even number of vertices is 4-edge-colorable.*

This might be approachable by some variation on our 4CT proof, but we have not yet tried. (4.7) implies the 4CT, because Kotzig [12] showed that, if G is a connected cubic graph with $|E(G)|$ even, then G is 3-edge-colorable if and only if its line graph $L(G)$ is 4-edge-colorable. ($L(G)$ is defined as the graph with vertex set $E(G)$, in which distinct edges e, f are adjacent if they have a common end in G. The line graph of a planar cubic graph is planar and 4-regular.) As far as I know, (4.7) is strictly stronger than the 4CT (indeed, it seems enormously stronger, but perhaps I am missing something).

Returning to nowhere-zero flows, there is another pretty conjecture of Tutte [33], the following:

CONJECTURE 4.8 *Every bridgeless graph has a nowhere-zero 5-flow.*

(4.8) would imply that, in contrast with the chromatic number of nonplanar graphs, "flow number" cannot be arbitrarily large. This much at least is true; it is shown in [26] that:

THEOREM 4.9 *Every bridgeless graph has a nowhere-zero 6-flow.*

5. Linkless embeddings

In Section 2, we wanted to know "Which graphs have no K_6 minor?", and in Section 3, "Which cubic graphs have no Petersen graph minor?" These two questions are more closely related than they might appear, because of a connection between K_6 and the Petersen graph. By a "Y-Δ operation" we mean the following. Let v be a vertex of a graph G, with degree 3, not incident with any loops or parallel edges. Let its neighbors be x, y, z say. Delete v from G, and add three new edges xy, xz, yz, producing a new graph H. We say H is obtained from G by a *Y-Δ operation*, and G is obtained from H by a *Δ-Y operation.*

These operations are quite powerful; for instance, Steinitz [28], [29], [7] showed that any planar graph can be reduced to the null graph by means of these operations and by deleting loops, multiple edges, and vertices of degree < 2, and suppressing vertices of degree 2. But if we apply these operations repeatedly to K_6 in all possible ways, we obtain a total of only seven graphs, up to isomorphism. One of them is the Petersen graph, and we call the set of seven the *Petersen family.*

A third question we might ask is: "Which graphs have no minor in the Petersen family?" This seems to be quite closely connected with the other two questions. For instance, we can prove that if G is cubic and $|\delta(X)| \geq 5$ for every $X \subseteq V(G)$ with $|X|, |V(G) - X| \geq 3$, then G has a Petersen graph minor if and only if G has a minor in the Petersen family; and I expect there is a similar equivalence between the K_6 question and the Petersen family question for highly connected graphs, though that is not yet shown.

But the Petersen family question has a nice answer, concerned with linkless embeddings. Take an embedding in 3-space of a graph G. It is

(a) *knotless* if every circuit of G bounds a disc in the space

(b) *windless* if for every two disjoint circuits C_1, C_2 of G, the "winding number" (defined in the natural way) of C_1 through C_2 is zero

(c) *linkless* if every set of mutually disjoint circuits forms a "trivial link" in the sense of knot theory; that is, each of the circuits bounds a disc disjoint from the other circuits

(d) *flat* if each circuit of G bounds a disc disjoint from the remainder of G.

(The difference between (a) and (d) should be noted.) Now an embedding can be knotless and yet not windless, and vice versa; but a linkless embedding is both knotless and windless, and seemed to be a nice kind of embedding to investigate. The question arose, "Which graphs have linkless embeddings?" Conway and

Gordon [5] proved that K_6 has no windless embedding; and Sachs [22] independently proved that no graph with a minor in the Petersen family has a windless embedding, and conjectured a strong converse, that every graph with no minor in the Petersen family has a linkless embedding. Böhme [4] suggested that flat embeddings might be even nicer, for every flat embedding is linkless, yet a linkless embedding need not be flat; and he conjectured that any graph that admits a linkless embedding also admits a flat embedding. In [21], we proved that all these conjectures are true. More precisely:

CONJECTURE 5.1 *For a graph G, the following are equivalent:*

(i) *G admits a windless embedding*

(ii) *G admits a linkless embedding*

(iii) *G admits a flat embedding*

(iv) *G has no minor in the Petersen family.*

In some ways, this should be a "deep" theorem. It was certainly difficult enough; the Petersen family is natural from the point of view of graph theory, and flat embeddings are natural from a topological viewpoint (for instance, an embedding is flat if and only if for every subgraph, its complement in 3-space has a free fundamental group). So we are asserting a connection between two, apparently unrelated, natural objects, and such a theorem might be expected to be important. Yet, for our applications to coloring, it was of no help whatsoever. It gives us no idea how to construct the graphs with no K_6 minor (say), because we do not know how to make flat embeddings. Still, I feel that it must be more than a coincidence that the same graphs (K_6 and the Petersen graph) occur here and in the coloring problems.

6. General results

Although this is not an appropriate place to go into the details of how such results as (2.4) and (3.5) can be proved, there is one basic piece of machinery that is used very frequently, and is not as well known as it might be. Let s_1, s_2, t_1, t_2 be vertices of a graph G. When are there two paths of G, pairwise disjoint, from s_1 to t_1 and from s_2 to t_2 respectively? There is one obvious case when such paths do not exist, namely when G can be drawn in the plane with s_1, s_2, t_1, t_2 all on the infinite region, and in the given order; because then clearly any path from s_1 to t_1 will meet any path from s_2 to t_2. The result I want to mention here is that this is essentially the only situation when the paths do not exist. More precisely:

THEOREM 6.1 *Let G, s_1, t_2, s_2, t_2 be as above, and suppose that for every vertex v different from s_1, t_1, s_2, t_2, there are four paths of G from v to s_1, t_1, s_2, t_2 respectively, mutually disjoint except for v. Then exactly one of the following holds:*

(i) *There are two paths of G, pairwise disjoint, from s_1 to t_1 and from s_2 to t_2 respectively*

(ii) *G can be drawn in the plane with s_1, s_2, t_1, t_2 all on the infinite region, and in the given order.*

This was proved for 4-connected graphs by Jung [10], and in the form given by Shiloach [27], Thomassen [32], Kelmans [11] and myself [25], independently. (See also [17] for a simple proof.) This is the basic tool that we use to persuade large parts of wild nonplanar graphs to be planar or almost planar, in the proofs of both (2.4) and (3.5).

(2.4) and (3.5) both tell us that, in some sense, apex graphs are an important way to construct graphs without K_6 and without Petersen graph minors. Can this be made more precise? To clarify the issue, let us replace these little excluded minors by a big one, K_{100} say. How can we construct graphs with no K_{100} minor?

One important construction is as follows. Let G and H both be graphs with no K_{100} minor. Let $g_1, \dots, g_k$ be pairwise adjacent vertices of G, and let $h_1, \dots, h_k$ be pairwise adjacent in H. Take the disjoint union of G and H, and then identify g_i with h_i for $i = 1, \dots, k$, and, if we wish, delete the edges between g_i and g_j, for $1 \leq i < j \leq k$. The graph we produce is called a *clique-sum* of G and H; it is easy to show that it has no K_{100} minor.

But what can we use for building blocks? Certainly graphs with at most 99 vertices have no K_{100} minor, and so do graphs G such that $G \backslash X$ is planar for some set of vertices X with $|X| \leq 95$. Similarly, because K_8 cannot be drawn in a torus, it follows that if $G \backslash X$ can be drawn in a torus for some set X of vertices of size ≤ 92 then G has no K_{100} minor, and so on.

This is not yet very rich, and yet at this point it begins to be difficult to think of other constructions. One important one is to expand the idea of a surface embedding, as follows. A *capacitated surface* Σ is a connected compact 2-manifold with boundary, together with an integer $\lambda(C) \geq 2$ for each component C of the boundary of Σ. By a *drawing* of a graph in Σ, we mean a drawing in the usual sense, with all vertices drawn in $\Sigma - bd(\Sigma)$, with each edge meeting $bd(\Sigma)$ in only finitely many points, and without crossings, except that edges are permitted to cross each other on the boundary of Σ; each point of a boundary component C of Σ may belong to at most $\lambda(C)$ edges of the drawing.

Let us say a capacitated surface Σ *admits* a graph G if there is a graph drawn in Σ with G as a minor. For every Σ, there is a k such that Σ does not admit K_k. For instance, let Σ be a closed disc, with $\lambda(C) = 2$ for the unique boundary component; then Seese and Wessel [23] proved that Σ does not admit K_8.

This gives a slightly more general way to construct graphs with no K_{100} minor. For any integer $k \geq 0$, let $\mathcal{A}(k)$ be the set of all graphs G such that some capacitated surface admits G and not K_k. Then for each graph in $\mathcal{A}(k)$, we have one of finitely many "reasons" why it has no K_k minor; because it is admitted by some Σ that does not admit K_k, and there are only finitely many Σ that do not admit K_k.

Let us denote by $\mathcal{A}(k)+n$ the set of graphs that can be obtained from graphs in $\mathcal{A}(k)$ by adding up to n new vertices, adjacent to anything. Thus, all of

$$\mathcal{A}(100) + 0, \mathcal{A}(99) + 1, \mathcal{A}(98) + 2, \dots$$

are sets of graphs with no K_{100} minor, that in some sense we understand. And we can take repeated clique-sums of these graphs to produce new graphs with no K_{100} minor.

How close are we getting to the complete answer? Probably not very close, if we really want a necessary and sufficient construction for all graphs with no K_{100} minor. But in another sense, these are qualitatively *all* the constructions necessary. Robertson and I [18] proved the following:

THEOREM 6.2 *There exist integers k, n such that every graph with no K_{100} minor is constructible by repeated clique-sums, starting from graphs in $\mathcal{A}(k) + n$.*

(Of course, 100 is not important here; the same is true for any constant.) This had a number of interesting consequences. For instance, (6.2) was the main step in our proof of Wagner's conjecture, that in any infinite set of graphs one of its members is a minor of another. It also was a crucial step in the proof of (2.8), the result implying the decidability of each case of Hadwiger's conjecture. But the main observation to be drawn from it here is that in this area nature is not very inventive. There are capacitated surfaces, there is the trick of adding a few extra vertices, and there are clique-sums; and qualitatively that is the complete list, these suffice to make all the graphs with any fixed minor excluded. Thus, it is perhaps not so surprising that results like (2.4) and (3.5) should be true.

References

[1] K. Appel and W. Haken, *Every planar map is four colorable. Part I. Discharging*, Illinois J. Math. 21 (1977), 429–490.

[2] K. Appel and W. Haken, Every Planar Map Is Four Colorable, Amer. Math. Soc. Contemp. Math. 98 (1989).

[3] K. Appel, W. Haken, and J. Koch, *Every planar map is four colorable. Part II. Reducibility*, Illinois J. Math. 21 (1977), 491–567.

[4] T. Böhme, *On spatial representations of graphs*, Contemporary Methods in Graph Theory, (R. Bodendiek, ed.), BI-Wiss.-Verl. Mannheim, Vienna, Zürich (1990), 151–167.

[5] J. Conway and C. Gordon, *Knots and links in spatial graphs*, J. Graph Theory 7 (1983), 445–453.

[6] G. A. Dirac, *A property of 4-chromatic graphs and some remarks on critical graphs*, J. London Math. Soc. (2) 27 (1952), 85–92.

[7] B. Grünbaum, Convex Polytopes, Pure and Applied Mathematics, vol. XVI (Wiley-Interscience, New York), 1967.

[8] H. Hadwiger, *Über eine Klassifikation der Streckenkomplexe*, Vierteljahrsschr. Naturforsch. Ges. Zürich 88 (1943), 133–142.

[9] L. Jørgensen, *Contractions to K_8*, J. Graph Theory 18 (1994), 431–448.

[10] H. A. Jung, *Eine Verallgemeinerung des n-fachen Zusammenhangs für Graphen*, Math. Ann. 187 (1970), 95–103.

[11] A. K. Kelmans, *Finding special subdivisions of K_4 in a graph*, in Finite and Infinite Sets, Colloq. Janos Bolyai 37 (1981), 487–508.

[12] A. Kotzig, *Z teorie konečných pravidelných grafov tretieho a štvrtého stupna*, Časopis Pěst. Mat. 82 (1957), 76–92.

[13] K. Kuratowski, *Sur le problème des courbes gauches en topologie*, Fund. Math. 15(1930), 271–283.
[14] W. Mader, *Homomorphiesätze für Graphen*, Math. Ann. 178 (1968), 154–168.
[15] J. Petersen, *Die Theorie der regulären Graphen*, Acta Math. 15 (1891), 193–220.
[16] N. Robertson, D. Sanders, P. D. Seymour, and R. Thomas, *The four-colour theorem*, in preparation.
[17] N. Robertson and P. D. Seymour, *Graph minors. IX. Disjoint crossed paths*, J. Combin. Theory Ser. B, 49 (1990), 40–77.
[18] N. Robertson and P. D. Seymour, *Graph minors. XVI. Excluding a non-planar graph*, submitted.
[19] N. Robertson, P. D. Seymour, and R. Thomas, *Tutte's 3-edge-colouring conjecture*, in preparation.
[20] N. Robertson, P. D. Seymour, and R. Thomas, *Hadwiger's conjecture for K_6-free graphs*, Combinatorica 13 (1993), 279–361.
[21] N. Robertson, P. D. Seymour, and R. Thomas, *Sachs' linkless embedding conjecture*, J. Combin. Theory Ser. B, to appear.
[22] H. Sachs, *On spatial representation of finite graphs*, in Finite and Infinite Sets, Colloq. Janos Bolyai 37 (1981).
[23] D. G. Seese and W. Wessel, *Grids and their minors*, J. Combin. Theory Ser. B, 47 (1989), 349–360.
[24] P. D. Seymour, *On multicolourings of cubic graphs, and conjectures of Fulkerson and Tutte*, Proc. London Math. Soc. (3) 38 (1979), 423–460.
[25] P. D. Seymour, *Disjoint paths in graphs*, Discrete Math. 29 (1980), 293–309.
[26] P. D. Seymour, *Nowhere-zero 6-flows*, J. Combin. Theory Ser. B, 30 (1981), 130–135.
[27] Y. Shiloach, *A polynomial solution to the undirected two paths problem*, J. Assoc. Comput. Mach. 27 (1980), 445–456.
[28] E. Steinitz, *Polyeder und Raumeinteilungen*, Enzykl. Math. Wiss. Part 3AB12, Vol. 3 (1922), 1–139.
[29] E. Steinitz and H. Rademacher, Vorlesung über die Theorie der Polyeder, Berlin (1934).
[30] P. G. Tait, *Note on a theorem in geometry of position*, Trans. Roy. Soc. Edinburgh 29 (1880), 657–660.
[31] A. Thomason, *An extremal function for contractions of graphs*, Math. Proc. Cambridge Philos. Soc. 95 (1984), 261–265.
[32] C. Thomassen, *2-linked graphs*, European J. Combin. 1 (1980), 371–378.
[33] W. T. Tutte, *A contribution to the theory of chromatic polynomials*, Canad. J. Math. 6 (1954), 80–91.
[34] K. Wagner, *Über eine Eigenschaft der ebenen Komplexe*, Math. Ann. 114 (1937), 570–590.

Entropy Methods in Hydrodynamic Scaling

S. R. S. VARADHAN*

Courant Institute
251 Mercer Street
New York, NY 10012, USA

1 Introduction

Hydrodynamic scaling is a procedure that attempts rigorously to derive large scale behavior of complex interacting systems from laws governing its evolution that are specified at a smaller scale. The procedure involves statistical averaging over the small scales and can be viewed as part of nonequilibrium statistical mechanics.

The basic example is the classical problem of starting with a Hamiltonian system of interacting particles and deriving from it, after rescaling, the Euler equations of compressible gas dynamics.

Let us consider a Hamiltonian system of N particles moving in a rather large physical space, for instance, the 3-dimensional torus of size ℓ. These particles are governed by a pair interaction $V(x-y)$ between particles. $V(\cdot)$ is an even function that is nonnegative and has support in a fixed compact set independent of ℓ. N and ℓ will be large with $N = \ell^3$ so that the interparticle distance is of order 1 and each particle will typically see only a few particles at any given time. The phase space is $(T_\ell^3 \times R^3)^N$ and the equations of motion for the positions and velocities $(x_i(t), u_i(t))$ are

$$\begin{cases} \dfrac{dx_i^\alpha(t)}{dt} = \dfrac{\partial H}{\partial u_i^\alpha} \\ \dfrac{du_i^\alpha(t)}{dt} = -\dfrac{\partial H}{\partial x_i^\alpha} \end{cases} \tag{1.1}$$

Here i is the particle number, $1 \le i \le N$, and α is the coordinate index, $\alpha = 1, 2, 3$. The Hamiltonian $H(x_1, \ldots, x_N; u_1, \ldots, u_N)$ is, of course, given by

$$H(x; u) = \frac{1}{2}\sum_i \|u_i\|^2 + \frac{1}{2}\sum_{i \neq j} V(x_i - x_j). \tag{1.2}$$

The system has five conserved quantities: the number of particles N, the momenta $\sum u_i^\alpha$, and the total energy H. Suppose we rescale the torus to have size 1 and rescale time by a similar factor ℓ, then quantities of the form

$$\frac{1}{N}\sum J\left(\frac{x_i(\ell t)}{\ell}\right) \tag{1.3}$$

*) Supported by NSF grant DMS-9201222 and ARO grant DAAL03-92-G-0317.

Proceedings of the International Congress
of Mathematicians, Zürich, Switzerland 1994

$$\frac{1}{N}\sum J\left(\frac{x_i(\ell t)}{\ell}\right) u_i^{\alpha}(\ell t) \tag{1.4}$$

and

$$\frac{1}{N}\sum J\left(\frac{x_i(\ell t)}{\ell}\right) e_i(\ell t) \tag{1.5}$$

where

$$e_i = \frac{1}{2}\|u_i\|^2 + \frac{1}{2}\sum_{j:j\neq i} V(x_i - x_j)$$

change in a reasonable manner in t. As N and $\ell \to \infty$ in the manner specified, one should think of (1.3), (1.4), and (1.5) as converging to

$$\int J(y)\rho(y,\tau)\, dy \tag{1.6}$$

$$\int J(y)v^{\alpha}(y,\tau)\rho(y,\tau)\, dy \tag{1.7}$$

and

$$\int J(y)e(y,\tau)\rho(y,\tau)\, dy. \tag{1.8}$$

Here y and τ are rescaled space and time. $\rho(y,\tau)$ is the density at a given point of space time. $\{v^{\alpha}(y,\tau)\}$ are the local fluid velocities and $e(y,\tau)$ is the energy density, related to the temperature at a given point. The equations of gas dynamics in this context are a system of symmetric hyperbolic conservation laws that one can write down for the evolution of ρ, v^{α}, and e. These equations are somewhat different from the usual Euler equation one derives from the Boltzmann equation. In the Boltzmann limit the real density is small; thus, the Euler equation derived from it is linearization in ρ of our equations.

This classical model is deterministic and any randomness has to be in the initial configuration. A precise formulation of the problem has to be done carefully. The goal is to establish some rigorous connection between the Hamiltonian equations on one hand and the Euler equations on the other. Randomness is important because some averaging has to be done with respect to small scales and one needs some information as to how the particles will arrange themselves locally in phase space if we only know their local density, local average velocity, and local temperature. One expects the arrangement to be given by an appropriate Maxwell-Gibbs distribution and formally the equations are derived under that assumption. To justify it, at the least, one needs a reasonable ergodic theory and for that more noise is better.

We will first look at two other models where the evolution is stochastic, say something about these models, and return at the end to our classical model.

The next example is referred to as simple exclusion. The physical space will be the periodic d-dimensional integral lattice of period ℓ. After scaling by a factor of ℓ, this will be viewed as living inside the unit d-torus. We will have a certain number $N = \rho\ell^d$ of particles located at some of the lattice sites. Each site can have at most one particle. A particle at the lattice site x waits for a random exponential waiting time with mean 1 and then picks a random new site x' to which it tries to jump. If the new site already has a particle, then the jump cannot be completed and our particle waits again for a new exponential random time. If the site x' is free, it jumps and starts afresh. The choice of x' is made with probability $\pi(x'-x)$. The probability distribution $\pi(z)$ is of jump sizes, is independent of ℓ, and is assumed to have finite support. All the particles are doing this at the same time and independently of each other. Because of continuous time there will be no ties to resolve.

The only conserved quantity here is the number of particles and we look at the density. We want to study the behavior of

$$\frac{1}{\ell^d}\sum_{I=1}^{M} J\left(\frac{x_i(\tau)}{\ell}\right) \tag{1.9}$$

where $\tau = \ell^\alpha t$. The rescaling of time can be with either $\alpha = 1$ or 2.

The mean $m = \sum z\pi(z)$ plays an important role. If $m \neq 0$, the motion is convective and one has to take $\alpha = 1$. If we think of (1.9) as converging to

$$\int J(y)\rho(\tau, y)\, dy \tag{1.10}$$

then ρ is expected to satisfy the scalar conservation law

$$\frac{\partial \rho}{\partial \rho} + \nabla \cdot m\rho(1-\rho) = 0. \tag{1.11}$$

If $m = 0$, then we need to take $\alpha = 2$. The scaling is diffusive and one then expects ρ to satisfy a nonlinear diffusion equation

$$\frac{\partial \rho}{\partial \tau} = \frac{1}{2}\nabla \cdot a(\rho(\tau, y)\nabla\rho). \tag{1.12}$$

If $\pi(z) = \pi(-z)$; i.e., π is symmetric, then the problem is rather easy and, in fact, one can verify that

$$a(\rho) \equiv a = \sum z \otimes z\ \pi(z) \tag{1.13}$$

is a constant matrix and is, in fact, the covariance matrix of $\pi(\cdot)$. The limiting equation in this case is the linear heat equation.

The third example we will consider is referred to as the Ginzburg-Landau model and is a lattice field model. We again start with the periodic lattice of size ℓ in d dimensions. We scale it down and think of it inside the unit torus. At each lattice site x, we have a variable $\xi(x,t)$ that is real valued and changes in time. The collection $\{\xi(x,t)\}$ is an ℓ^d-dimensional diffusion process and can be described either through a set of stochastic differential equations or though its infinitesimal generator. We will do the former. If $e_1, \dots, e_d$ are the d positive coordinate directions, then $x \pm e_i$ are the neighbors of x.

$$d\xi(x,t) = \sum_{i=1}^{d} [d\eta(x-e_i, x, t) - d\eta(x, x+e_i, t)] \tag{1.14}$$

$$d\eta(x, x+e_i, t) = \frac{1}{2}\left[\phi'(\xi(x,t)) - \phi'(\xi(x+e_i,t))\right] dt + d\beta_{x,x+e_i}(t) \tag{1.15}$$

The equation (1.14) tells us that the way $\xi(x,t)$ changes is by "stuff" coming in or going out along the bonds. We orient the bonds using the positive coordinate directions and the net change is an algebraic sum. Equation (1.15) tells us that the flow along a bond is proportional to some nonlinear gradient modified by white noise. Here $\phi'(\xi)$ is a nonlinear function that satisfies some natural assumptions. We use ϕ', the derivative of $\phi(\xi)$, for convenience. If we again take $\tau = \ell^2 t$ and consider

$$\frac{1}{\ell^d}\sum J\left(\frac{x}{\ell}\right)\xi(x,\tau) \tag{1.16}$$

as an approximation of

$$\int J(y)m(y,\tau)\, dy \tag{1.17}$$

where $m(y,\tau)$ is the limiting "density" of "stuff", then $m(\cdot,\cdot)$ is supposed to satisfy an equation of the form

$$\frac{\partial m}{\partial t} = \frac{1}{2}\Delta\lambda(m)$$

where $\lambda(m)$ is to be determined in terms of ϕ.

We shall look at our three examples in some detail in our next three sections and end with some comments.

2 Ginzburg-Landau Model

It is convenient to write down the infinitesimal generator of our diffusion process on R^{ℓ^d}.

$$\begin{aligned}(\mathcal{L}_\ell F) =& \frac{\ell^2}{2}\sum_{i,x}\left(\frac{\partial}{\partial\xi(x+e_i)} - \frac{\partial}{\partial\xi(x)}\right)^2 F \\ &- \frac{\ell^2}{2}\sum_{i,x}\left[\phi'(\xi(x+e_i)) - \phi'(\xi(x))\right]\left[\frac{\partial}{\partial\xi(x+e_i)} - \frac{\partial}{\partial\xi(x)}\right] F.\end{aligned} \tag{2.1}$$

The factor ℓ^2 appears due to rescaling of time. The object we want to study is

$$G(t) = \frac{1}{\ell^d}\sum J\left(\frac{x}{\ell}\right)\xi(x,t). \tag{2.2}$$

Using the stochastic differential equations one can write

$$dG(t) = A(t)dt + dM(t) \tag{2.3}$$

where

$$\begin{aligned}A(t) &= \mathcal{L}_\ell\left[\frac{1}{\ell^d}\sum J\left(\frac{x}{\ell}\right)\xi(x)\right](t) \\ &\simeq \frac{1}{2\ell^d}\sum (\Delta J)\left(\frac{x}{\ell}\right)\phi'(\xi(x,t))\end{aligned} \tag{2.4}$$

and

$$dM(t) \simeq 0$$

because the Brownian noise terms cancel each other out by averaging. The difficulty now, which is typical of all of these problems, is that due to the nonlinearity of ϕ' the equations do not close and one has to represent

$$\frac{1}{\ell^d} \sum (\Delta J)\left(\frac{x}{\ell}\right) \phi'(\xi(x,t)) \tag{2.5}$$

in terms of $m(x,t)$ as $\ell \to \infty$. These are, of course, defined by (1.16) and (1.17). These require a knowledge of how $\xi(x,t)$ are "distributed" for a given value of "m". If we know this, then the volume average of ϕ' can be replaced by a mean value of ϕ' for a given m.

The evolution governed by $\mathcal{L}_\ell$ is reversible or symmetric with respect to the weight

$$e^{-\sum_x \phi(\xi(x))} = \Phi_\ell(\xi). \tag{2.6}$$

(We normalize ϕ so that $\int e^{-\phi(\xi)}\, d\xi = 1$.) However, the process is not ergodic because

$$m = \frac{1}{\ell^d} \sum_x \xi(x) \tag{2.7}$$

is conserved under our evolution. The conditional distribution $\nu_{m,\ell}(d\xi)$ of $\{\xi(x)\}$, given the average m, are the invariant ergodic pieces. By an "equivalence of ensembles" type theorem we can show

$$\nu_{m,\ell}(d\xi) \to \prod_x e^{-\phi_m(\xi(x))} d\xi(x)$$

and

$$\phi_m(\xi) = \phi(\xi) - \lambda(m)\xi + p(\lambda(m)).$$

Here

$$p(\lambda) = \log \int \exp\left[\lambda\xi - \phi(\xi)\right] d\xi$$

and $\lambda(m)$ solves

$$p'(\lambda(m)) = m$$

and equals

$$\lambda(m) = \frac{d}{dm} \sup_\lambda \left[\lambda m - p(\lambda)\right].$$

One finally verifies that

$$\int \phi'(\xi) e^{-\phi_m(\xi)} d\xi = \lambda(m).$$

This yields our equation

$$\frac{\partial m}{\partial t} = \frac{1}{2} \Delta\lambda(m). \tag{2.8}$$

In order to justify this one has to prove that averages of the form

$$\frac{1}{|B|}\sum_{x\in B}\phi'(\xi(x))$$

are nearly equal to

$$\lambda\left(\frac{1}{|B|}\sum_{x\in B}\xi(x)\right)$$

most of the time with probability nearly 1. The size of the block is important. It should be of size $\ell\epsilon$ with $\epsilon << 1$ but fixed. In [5] with Guo and Papanicolaou we developed a method for handling the problem by using entropy and entropy production as tools. We assume that initially the field $\{\xi(x,0)\}$ is random and is given by a density $f_\ell^0(\xi)\Phi_\ell(\xi)$ satisfying an entropy bound

$$\int f_\ell^0(\xi)\log f_\ell^0(\xi)\Phi_\ell(\xi)\,d\xi \le C\ell^d. \tag{2.9}$$

Such a bound is natural and is satisfied in most cases because one can think of C as the bound for average entropy per site. Then the distribution of $\{\xi(x,t)\}$ will have a density $f_\ell^t(\xi)$ satisfying

$$H_\ell(t) = \int f_\ell^t(\xi)\log f_\ell^t(\xi)\Phi_\ell(\xi)\,d\xi \le H_\ell(0)$$

and, in fact,

$$\frac{dH_\ell(t)}{dt} = -\frac{\ell^2}{2}\int\sum_{i,x}\left(\frac{\partial f_\ell^t}{\partial\xi(x+e_i)} - \frac{\partial f_\ell^t}{\partial\xi(x)}\right)^2\frac{1}{f_\ell^t}\Phi_\ell(\xi)\,d\xi\,dt. \tag{2.10}$$

Because $H_\ell(t)\ge 0$ one gets a trivial bound

$$\int_0^\infty\sum_{i,x}\int\left(\frac{\partial f_\ell^t}{\partial\xi(x+e_i)} - \frac{\partial f_\ell^t}{\partial\xi(x)}\right)^2\frac{1}{f_\ell^t}\Phi_\ell(\xi)\,d\xi\,dt \le 2C\ell^{d-2}.$$

If we fix a finite time T and consider

$$\overline{f}_\ell = \frac{1}{T}\int_0^T f_\ell^t\,dt,$$

by convexity

$$\int\sum_{i,x}\left(\frac{\partial\overline{f}_\ell}{\partial\xi(x+e_i)} - \frac{\partial\overline{f}_\ell}{\partial\xi(x)}\right)^2\frac{1}{\overline{f}_\ell}\Phi_\ell(\xi)\,d\xi \le C_T\ell^{d-2}.$$

We showed in [5] that the above estimate was sufficient to justify the averaging lemma and establish equation (2.8) rigorously. The following theorem was proved.

Assume $f_\ell^0(\xi)$ satisfies the bound (2.8). Assume further that for some density $m_0(y)$ and all smooth test functions $J(\cdot)$

$$\frac{1}{\ell^d}\sum J\left(\frac{x}{\ell}\right)\xi(x) \to \int J(y)m_0(y)\,dy$$

in probability under $f_\ell^0(\xi)\Phi_\ell(\xi)d\xi$. Then for any $t>0$

$$\frac{1}{\ell^d}\sum J\left(\frac{x}{\ell}\right)\xi(x,t) \to \int J(y)m(t,y)\,dy$$

in probability, where $m(t,y)$ solves (2.8) with initial condition $m_0(y)$.

In [14] a modified approach was developed by Yau that has much wider applicability. One could guess that the density at time t should look like

$$g_\ell^t(\xi)\Phi_\ell(\xi) = \frac{1}{Z}\exp\left[\sum \lambda\left(t,\frac{x}{\ell}\right)\xi(x)\right]\Phi_\ell(\xi)$$

where $Z = \exp[\sum p\left(t,\frac{x}{\ell}\right)]$ and $\lambda(t,x) = \lambda(m(t,x))$ and m solves (2.8). In general, the density is $\Phi_\ell(\xi)f_\ell^t(\xi)$ where $f_\ell^t(\xi)$ is obtained by solving the Fokker-Planck equation and is not g_ℓ^t, even if we start off with $f_\ell^0(\xi) = g_\ell^0(\xi)$.

The question is: How close are they?

If we define the specific entropy $s(f_\ell^t, g_\ell^t)$ by

$$s(f_\ell^t, g_\ell^t) = \frac{1}{\ell^d}\int f_\ell^t(\xi)\log\frac{f_\ell^t(\xi)}{g_\ell^t(\xi)}\Phi_\ell(\xi)\,d\xi, \tag{2.11}$$

the theorem of Yau in [14] is that uniformly in $0\le t\le T$,

$$\lim_{\ell\to\infty} s(f_\ell^t, g_\ell^t) = 0.$$

This is enough to justify the hydrodynamic scaling and arrive at the same theorem as in [5]. One assumes more initially but one obtains a stronger conclusion.

The model that we have studied is very special because it is a "gradient" model. While computing

$$dG(t) = d\left(\frac{1}{\ell^d}\sum J\left(\frac{x}{\ell}\right)\xi(x,t)\right)$$

we obtained

$$\begin{aligned}
&\mathcal{L}_\ell\left(\frac{1}{\ell^d}\sum J\left(\frac{x}{\ell}\right)\xi(x)\right)\\
&= -\frac{1}{2}\frac{\ell^2}{\ell^d}\sum\left[J\left(\frac{x+e_i}{\ell}\right) - J\left(\frac{x}{\ell}\right)\right][\phi'(\xi(x+e_i)) - \phi'(\xi(x))]\cdot\\
&\simeq \frac{1}{2\ell^d}\sum(\Delta J)\left(\frac{x}{\ell}\right)\phi'(\xi(x))
\end{aligned}$$

We were able to carry out summation by parts twice. The local "flux" turned out to be

$$\phi'(\xi(x+e)) - \phi'(\xi(x)),$$

which is a gradient and is amenable to another summation by parts. With that the global flux turned out to be

$$\frac{1}{2\ell^d}\sum(\Delta J)\left(\frac{x}{\ell}\right)\xi(x),$$

which is an order 1 quantity.

This is not true in general. To see this let us modify our evolution somewhat. Our operator corresponds to the following Dirichlet form relative to the weight $\Phi_\ell(\xi)$

$$\mathcal{D}_\ell(F) = \frac{\ell^2}{2}\int\sum_{x,i}\left(\frac{\partial F}{\partial\xi(x+e_i)} - \frac{\partial F}{\partial\xi(x)}\right)^2\Phi_\ell(\xi)d\xi.$$

We modify it by

$$\widetilde{\mathcal{D}}_\ell(F) = \frac{\ell^2}{2}\int\sum_{x,i}a(\xi(x+e_i),\xi(x))\left(\frac{\partial F}{\partial\xi(x+e_i)} - \frac{\partial F}{\partial\xi(x)}\right)^2\Phi_\ell(\xi)\,d\xi$$

where $a(\xi_1,\xi_2)$ is a smooth positive function bounded above and below. For simplicity, let us take $d=1$. The operator $\widetilde{\mathcal{L}}_\ell$ is given by

$$\begin{aligned}\widetilde{\mathcal{L}}_\ell F =& \frac{\ell^2}{2}\sum a\left(\xi(x+1),\xi(x)\right)\left(\frac{\partial}{\partial\xi(x+1)} - \frac{\partial}{\partial\xi(x)}\right)^2 F\\ &- \frac{\ell^2}{2}\sum W(\xi(a+1),\xi(x))\left(\frac{\partial}{\partial\xi(x+1)} - \frac{\partial}{\xi(x)}\right)F\end{aligned}$$

where

$$\begin{aligned}W(\xi_1,\xi_2) = &\, a(\xi_1,\xi_2)\left(\phi'(\xi_1)-\phi'(\xi_2)\right)\\ &+ a_2(\xi_1,\xi_2) - a_1(\xi_1,\xi_2)\end{aligned}$$

and

$$a_i(\xi_1,\xi_2) = \frac{\partial}{\partial\xi_i}a(\xi_1,\xi_2)$$

for $i=1,2$.

Now, if we compute

$$\widetilde{\mathcal{L}}_\ell\left(\frac{1}{\ell}\sum J\left(\frac{x}{\ell}\right)\xi(x)\right)$$

we get

$$-\frac{1}{2}\sum J'\left(\frac{x}{\ell}\right)W\left(\xi(x+1),\xi(x)\right). \tag{2.12}$$

The term in (2.12) is a big term. The mean value of W is zero for any given value of m and so averaging produces a meaningless product of zero and infinity.

In [12] we developed a method for handling problems of this kind. We showed that there is a function $D(m)$ such that

$$W(\xi_1, \xi_2) - D(m)\,[\phi'(\xi_1) - \phi'(\xi_2)]$$

is negligible in a somewhat complicated but precise sense. With this modification the correct diffusion equation is

$$\begin{aligned} \frac{\partial m}{\partial t} &= \frac{1}{2}\frac{\partial}{\partial y} D(m(t,y)) \frac{\partial}{\partial y} \lambda(m(t,y)) \\ &= \frac{1}{2}\frac{\partial}{\partial y} \widetilde{D}(m(t,y)) \frac{\partial m}{\partial y} \\ \widetilde{D}(m) &= D(m)\lambda'(m). \end{aligned}$$

For $D(m)$ we provided a variational formula that replaces the traditional Green-Kubo formula. In the Green-Kubo formula the diffusion coefficient $D(m)$ is a space-time integral (sum) of the autocorrelation function in equilibrium, and this is not convenient. Our variational formula is much more convenient and is given explicitly by

$$D(m) = \inf_f E_m\{a(\xi_1, \xi_2)(1 - G_f)^2\}.$$

Here E_m refers to the expectation with respect to the infinite product measure

$$\exp[-\sum \phi_m(\xi(x))] \prod d\xi_x$$

and the infimum is taken over all tame test functions depending on a finite number of coordinates. For each such f, G_f is the gradient defined by

$$(G_f(\xi)) = \sum_x \left(\frac{\partial}{\partial \xi(2)} - \frac{\partial}{\partial \xi(1)}\right) \tau_x f$$

and τ_x is the canonical shift operator on the product space.

In principle one could try to carry out the method of Yau in [14] and one would have to work with a trial function of the form

$$g_N^t(\xi) \simeq \exp\left[\sum \lambda\left(t, \frac{x}{\ell}\right)\xi(x) + \frac{1}{\ell}\sum_x \psi(t, x, \xi)\right]$$

for $g_N^t(\xi)$. One has to choose a suitable corrector ψ in order to carry out the analysis. This has been done very recently by Funaki, Uchiyama, and Yau.

3 Simple Exclusion

We will now return to simple exclusion models. The state of the system can be described by a configuration $\eta = \{\eta(x)\}$, $x \in Z_\ell$, where Z_ℓ is the lattice of integers modulo ℓ. For simplicity, we have taken $d = 1$. $\eta(x) = 1$ if there is a particle at x and 0 otherwise. The generator can be written as

$$(\mathcal{L}_\ell F)(\eta) = \ell^\alpha \sum_{x,x'} \eta(x)(1-\eta(x'))\pi(x'-x)\left[F\left(\eta^{x,x'}\right) - F(\eta)\right]. \tag{3.1}$$

Here $\eta^{x,x'}$ refers to the new configuration obtained when the particle at x jumps to x'. α will be either 1 or 2, depending on the scaling used for time.

The case $\alpha = 1$.

$$\begin{aligned}\mathcal{L}_\ell\left[\frac{1}{\ell}\sum J\left(\frac{x}{\ell}\right)\eta(x)\right] &= \ell\sum\left[J\left(\frac{x'}{\ell}\right) - J\left(\frac{x}{\ell}\right)\right](x'-x)\eta(x)(1-\eta(x'))\\ &\simeq m\int J'(y)\rho(y,t)(1-\rho(y,t))\,dy\end{aligned}$$

where m is the mean $\sum x\ \pi(x)$ of the jump distribution π. The last step is justified because one expects the probability that a site x is occupied to be the local density $\rho(\frac{x}{\ell},t)$, with occupancy of different sites being independent.
This leads to the equation

$$\frac{\partial\rho}{\partial t} + m(\rho(1-\rho))_y = 0 \tag{3.2}$$

for the density ρ. The method of [5] does not work here. The method of relative entropy contained in [14] will work, but needs the solution $\rho(t,x)$ to be smooth. It is known that for most initial data, sooner or later, discontinuities will develop and so the method applies only up to the first shock. There are other coupling methods that establish convergence to the correct weak solution of (3.2). See, for instance, Rezakhanlou [10] for the best results in the case of attractive dynamics.

The case $\alpha = 2$. In the symmetric case; i.e., $\pi(z) = \pi(-z)$, we always have $m = 0$ and we take $\alpha = 2$.

$$\begin{aligned}\mathcal{L}_\ell\frac{1}{\ell}\sum J\left(\frac{x}{\ell}\right)\eta(x) &= \frac{\ell^2}{\ell}\sum\left[J\left(\frac{x'}{\ell}\right) - J\left(\frac{x}{\ell}\right)\right]\eta(x)(1-\eta(x'))\pi(x'-x)\\ &= \frac{\ell^2}{2\ell}\sum\left(J\left(\frac{x'}{\ell}\right) - J\left(\frac{x}{\ell}\right)\right)[\eta(x)-\eta(x')]\,\pi(x'-x)\\ &\sim \frac{1}{2}\sum J''\left(\frac{x}{\ell}\right)\eta(x)\left(\sum z^2\pi(z)\right)\\ &= \frac{D}{2}\sum J''\left(\frac{x}{\ell}\right)\eta(x).\end{aligned}$$

This yields

$$\frac{\partial\rho}{\partial t} = \frac{D}{2}\rho_{xx} \tag{3.3}$$

with

$$D = \sum z^2 \pi(z).$$

The nonlinearity miraculously cancels out and the equations close. No averaging is needed.

If we change the problem by coloring the particles so that some are green and some are red, we can ask how the colors spread. One should compute

$$\mathcal{L}_\ell \left[\frac{1}{\ell} \sum A\left(\frac{x}{\ell}\right) \eta_g(x) + \frac{1}{\ell} \sum B\left(\frac{x}{\ell}\right) \eta_r(x) \right]$$

and proceed from there. Here $\eta_g(x) = 1$ if there is a green particle at x and similarly for $\eta_r(x)$. $\eta_g(x) + \eta_r(x) \leq 1$ due to exclusion. The dynamics is color blind. The analysis is hard because the system turns out to be nongradient. The method of [12] was applied to this situation by Quastel in [9] and he obtained an elliptic system for the pair $\rho_g(t,y)$ and $\rho_r(t,y)$. One first solves for

$$\rho(t,y) = \rho_g(t,y) + \rho_r(t,y)$$

by the heat equation (3.3). Then $\rho_g(t,y)$ is solved by

$$\frac{\partial \rho_g}{\partial t} = \frac{1}{2} \frac{\partial}{\partial y} S(\rho) \frac{\partial}{\partial y} \rho_g - (b(t,y)\rho_g)_y$$

where

$$b(t,y) = \frac{1}{2} \frac{(S(\rho) - D)\rho_y}{\rho}$$

is the pressure due to density gradient. $S(\rho)$ is the self diffusion coefficient determined in [7] and really depends on ρ, with $S(\rho) \to D$ as $\rho \to 0$ as $S(\rho) \to 0$ as $\rho \to 1$. The case $\pi(1) = \pi(-1) = \frac{1}{2}$ is special and $S(\rho) \equiv 0$ in that case.

The case where $\pi(z) \neq \pi(-z)$ but still $m = 0$ is more complex. This is nongradient and nonreversible. The methods of [12] and [9] have to be modified. This was carried out by Xu in [13] who established a limiting equation of the form

$$\frac{\partial \rho}{\partial t} = \frac{1}{2} \frac{\partial}{\partial y} D(\rho) \frac{\partial \rho}{\partial y}$$

with $D(\rho) \geq D = \sum z^2 \pi(z)$. In general, $D(\rho) > D$ for $0 < \rho < 1$ with $D(\rho) \to D$ as $\rho \to 0$ or 1.

Navier-Stokes corrections. Let us return to the case $m \neq 0$, but start very close to equilibrium. We start with a density $\rho(y) = \frac{1}{2} + \frac{1}{\ell} q_0(y)$; i.e.,

$$P[\eta(x) = 1] = \frac{1}{2} + \frac{1}{\ell} q_0\left(\frac{x}{\ell}\right).$$

Now we can rescale with $\alpha = 2$. Recently Esposito, Marra, and Yau [4] have shown that when the dimension $d \geq 3$, the empirical density at the rescaled time t is of the form $\frac{1}{2}+\frac{1}{\ell}q(t,\frac{x}{\ell})$ to within an accuracy $o(\frac{1}{\ell})$, and $q(t,y)$ is given as the solution of a Navier-Stokes type equation; i.e., Burgers' equation with viscosity.

4 Hamiltonian System

Let us return finally to our classical Hamiltonian system. If we start with a random initial configuration given by a density g_N^0 at time $t=0$, and let the configuration evolve by the Hamiltonian motion, one can obtain the density of the configuration at time t by solving the Liouville flow

$$\frac{\partial f_N}{\partial t} = \mathcal{L}_N f_N \qquad f_N = g_N^0 \text{ at } t=0. \tag{4.1}$$

If we are given functions $\rho(y), v^\alpha(y)$, and $T(y)$ on T^3 representing local density, average velocity, and temperature, assuming that those values avoid regions of possible phase transitions, we can associate a density $g_N(y_1,\dots,y_N;u_1,\dots,u_N)$ in the phase space that is a family of slowly varying Maxwell-Gibbs distributions strung together. If we calculate averages like $\frac{1}{N}\sum J(y_i)$ we get $\int J(y)\rho(y)\,dy$ in probability as $N\to\infty$. If we pick ρ, v^α, and T to be, as functions of t and y, a solution of the Euler equation that is quite smooth in some interval, we can use $\rho(t,y), v^\alpha(t,y)$, and $T(t,y)$ to construct a time dependent family $g_N(t,y_1,\dots,y_N;u_1,\dots u_N)$ of such densities.

We would like to establish that the density obtained by the Euler equation; i.e., g_N is close to the density f_N obtained by solving the Hamiltonian or equivalently the Liouville flow. An ideal theorem will say that as $N\to\infty$ the specific (per particle) relative entropy

$$s(f_N,g_N) = \frac{1}{N}\int \log\frac{f_N}{g_N}\cdot f_N \;\; dy\; du \to 0. \tag{4.2}$$

We note that if the specific entropy were to tend to zero, by the usual large deviation estimates, the local density, velocities, and temperature would be the same for f_N and g_N. This would establish the hydrodynamic limit. But we cannot quite prove such a theorem. We have problems at two levels. First there is difficulty with large velocities. This can be overcome by changing the kinetic energy in the Hamiltonian to a function $\phi(u)$ with a bounded gradient instead of $\frac{1}{2}\|u\|^2$. (One choice that will work is the relativistic kinetic energy.) This is a technical point. The more serious problem is the hunger for noise. It is needed to establish some ergodicity. But only very little is needed. We put it in as an additional noisy exchange of velocities between pairs of particles that conserves momenta and energy. The strength of this noisy term is much smaller than the exchange of velocities provided by the Hamiltonian equations. This is then a small second order perturbation of the Liouville operator that does not destroy the conservation laws. The Euler equations are still the same. With (4.1) now replaced by a modified Fokker-Planck equation, one can establish (4.2). The details can be found in our work [8] with Olla and Yau . A key step is that whereas the noise is responsible for keeping the velocitiy distributions locally Maxwellian, for a Hamiltonian dynamics the positions are then shown to satisfy the necessary ergodic behavior with the correct Gibbs distributions.

5 Comments

We have limited our discussion essentially to work that uses entropy-related methods to the study of problems of hydrodynamic scaling. There is considerable work that uses other methods to study similar problems and we have not described them. The monographs [2] and [11] are excellent sources for a much wider collection of material.

We have also not discussed issues of large deviation. In some sense entropy-related methods are intimately related to techniques of large deviation theory and the two often go hand in hand. See, for instance, [6] and [2] for connections to the methods of [5]. As for Hamiltonian systems, there is the earlier work of Boldrighini, Dobrusin, and Sukov [1], which deals with the case of hard rods in one dimension with elastic collision.

References

[1] C. Boldrighini, R. L. Dobrusin and Yu M. Suhov, *One-dimensional hard rod caricatures of hydrodynamics*, J. Stat. Phys. **31** (1983), 577–616.

[2] A. DeMasi and E. Presutti, Mathematical Methods for Hydrodynamical Limits, Lecture Notes in Math., **1501**, Springer Verlag, Berlin-Heidelberg-New York, 1991.

[3] M. D. Donsker and S. R. S. Varadhan, *Large deviations from a hydrodynamic scaling limit*, Comm. Pure. Appl. Math. **42** (1989), 243–270.

[4] R. Esposito, R. Marra and H. T. Yau, *Diffusive limit of asymmetric simple exclusion*, preprint.

[5] M. Z. Guo, G. C. Papanicolaou and S. R. S. Varadhan, *Nonlinear diffusion limit for a system with nearest neighbor interactions*, Comm. Math. Phys. **118** (1988), 31–59.

[6] C. Kipnis, S. Olla and S. R. S. Varadhan, *Hydrodynamics and large deviation for simple exclusion process*, Comm. Pure Appl. Math. **42** (1989), 115–137.

[7] C. Kipnis and S. R. S. Varadhan, *Central limit theorem for additive functionals of reversible Markov processes and application to simple exclusions*, Comm. Math. Phys. **104** (1986), 1–19.

[8] S. Olla, S. R. S. Varadhan and H. T. Yau, *Hydrodynamical limit for a Hamiltonian system with weak noise*, Comm. Math. Phys. **155** (1993), 523–560.

[9] J. Quastel, *Diffusion of color in simple exclusion process*, Comm. Pure Appl. Math. **45** (1992), 623–680.

[10] F. Rezakhanlou, *Hydrodynamical limit for attractive particle systems on* z^d, Comm. Math. Phys. **140** (1991), 417–448.

[11] H. Spohn, Large Scale Dynamics of Interacting Particles, Texts and Monographs in Physics, Springer Verlag, Berlin-Heidelberg-New York, 1991.

[12] S. R. S. Varadhan, *Nonlinear diffusion limit for a system with nearest neighbor interactions, II, Asymptotic Problems in Probability Theory*, in: Stochastic Models and Diffusions on Fractals (K. D. Elworthy and N. Ikeda, eds.), Pitman Res. Notes Math. Ser. **283**, 1991, 75–130.

[13] Lin Xu, Hydrodynamics for asymmetric mean zero simple exclusion, Ph.D. thesis, New York University, 1993.

[14] H. T. Yau, *Relative entropy and the hydrodynamics of Ginzburg-Landau models*, Lett. Math. Phys. **22** (1991), 63–80.

Topology of Discriminants and Their Complements

VICTOR A. VASSILIEV

Independent University of Moscow
P.O. Box 230
Moscow 117463, Russia

and

Steklov Mathematical Institute
Vavilova st. 42
Moscow 117966, GSP-1, Russia

1 Main definitions and examples

The general notion of a discriminant is as follows. Consider any function space $\mathcal{F}$, finite dimensional or not, and some class of singularities S that the functions from $\mathcal{F}$ can take at the points of the issue manifold. The corresponding *discriminant variety* $\Sigma(S) \subset \mathcal{F}$ is the space of all functions that have such singular points. For example, let $\mathcal{F}$ be the space of (real or complex) polynomials of the form

$$x^d + a_1 x^{d-1} + \cdots + a_d, \tag{1}$$

and $S = \{\text{a multiple root}\}$. Then (in the complex case) $\Sigma(S)$ is the zero level set of the usual discriminant polynomial of the coefficients a_i; this is a motivation for the word "discriminant" in the general situation. In the simplest nontrivial case, when $d = 4$, the discriminant variety in the space of real polynomials is ambient diffeomorphic to the direct product of the line $\mathbf{R}^1$ and the "swallowtail", i.e. the surface shown in the lower part of Figure 1a. More generally, we can consider the discriminant Σ_k consisting of all polynomials having roots of multiplicity at least k, or we can consider the space of polynomial systems of the form

$$\begin{array}{c} x^{m_1} + a_1^1 x^{m_1-1} + \cdots + a_{m_1}^1 \\ \cdots\cdots\cdots\cdots\cdots\cdots\cdots\cdots \\ x^{m_k} + a_1^k x^{m_k-1} + \cdots + a_{m_k}^k \end{array} \tag{2}$$

and take for the discriminant the *resultant variety* consisting of all systems that have common roots, etc. In the simplest nontrivial case, when $k = 2$, $m_1 = m_2 = 2$, the real resultant variety is ambient diffeomorphic to the direct product of the line $\mathbf{R}^1$ and the Whitney umbrella, i.e. the surface shown in the lower part of Figure 1b.

Many famous topological spaces can be described as the complements of suitably defined discriminants (or at least are homotopy equivalent to them). Some examples are:

- the classifying spaces of braid groups and, more generally, spaces of polynomials without roots of multiplicity $\geq k$ ($k \geq 2$);
- classical Lie groups;
- spaces of Morse or generalized Morse functions on a manifold M or, more generally, spaces of smooth maps $M \to \mathbf{R}^n$ without complicated singularities;

Proceedings of the International Congress
of Mathematicians, Zürich, Switzerland 1994

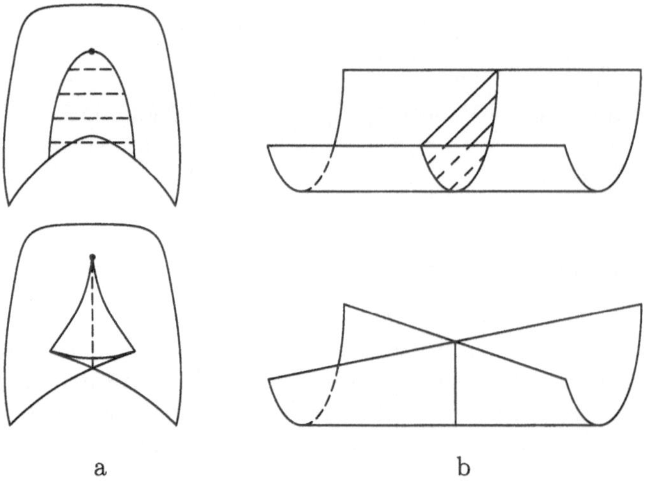

Figure 1 Simplest discriminant and resultant varieties and their simplicial resolutions

- loop spaces ΩY and, more generally, the spaces of maps $X \to Y$ where X is an m-dimensional cell complex and Y an $(m-1)$-connected one;
- spaces of based rational maps $\mathbf{C}P^1 \to \mathbf{C}P^n$ of fixed degree;
- spaces of nonsingular complex manifolds;
- spaces of knots and links in $\mathbf{R}^n$, $n \geq 3$, and many others.

In this paper I describe an easy general method for calculating the cohomologies of all these spaces and, in many cases, also their stable homotopy types. In all the listed cases this method gives new results or at least a simple and uniform way to guess and prove the old ones. Among the known results that appear as very special cases of this general theory and get strong generalizations there are the Adams spectral sequence for the cohomology of loop spaces, the May-Segal formula for the cohomology of stable braid group, the results of Segal and Cohen-Cohen-Mann-Milgram relating the spaces of based rational maps $\mathbf{C}P^1 \to \mathbf{C}P^n$ with the configuration spaces and double loop spaces, the Goresky-MacPherson formula for the homology of complements of plane arrangements, knot and link polynomials, and others. The Snaith and Mahovald splittings for the homology of iterated loop spaces and classical Lie groups also appear naturally in this theory and get an uniform interpretation.

The first reduction

One of the key notions in this construction is the **Spanier-Whitehead duality**. Two topological spaces (having the homotopy types of CW-complexes) are Spanier-Whitehead dual to one another if they have homotopy types of complementary subsets in a sphere of appropriate dimension. (In particular, the homology groups of Spanier-Whitehead dual spaces are related by the Alexander duality.)

An important property of this duality consists in the fact that it is an involution on the space of stable homotopy types: the stable homotopy type of a topological space is completely determined by that of its arbitrary Spanier-Whitehead dual space.

If the function space $\mathcal{F}$ is finite dimensional, then the space Spanier-Whitehead dual to $\mathcal{F} \setminus \Sigma$ is nothing but the one-point compactification $\bar{\Sigma}$ of the discriminant variety, and we can replace the study of all stable homotopical properties (in particular, the homology groups) of $\mathcal{F} \setminus \Sigma$ by the study of this compactification. This reduction, introduced in the work [Arnold 70] (where the topological study of the complements of discriminants was essentially initiated), is very useful because the space of nondiscriminant maps is open and has no evident geometrical structure, whereas the discriminant variety is a naturally stratified set, and, as we shall see, a lot of its topology can by expressed in the terms of this stratification.

The next two steps of the method consist in the use of *simplicial resolutions* of discriminants, and (in the infinite-dimensional case) in the stabilization of the obtained calculations over suitable sequences of finite-dimensional approximating subspaces in $\mathcal{F}$, see [V 87, 89]. We demonstrate these steps in the following simplest case.

2 The main example: real polynomials and functions without multiple roots

Geometrical resolution of the discriminant variety

Denote by P_d the space of *real* polynomials of the form (1) and consider the space $P_d \setminus \Sigma_k$ of polynomials having no roots of multiplicity $\geq k$. As before, we replace the study of $P_d \setminus \Sigma_k$ by the study of its Spanier-Whitehead dual space $\bar{\Sigma}_k$. The homology groups were calculated in [Arnold 89], we give here another calculation that demonstrates our general method.

Namely, we construct a *geometrical resolution* of $\bar{\Sigma}_k$, i.e. a topological space with the same homotopy type but with a more explicit homological structure.

At the first step of this construction we take the *tautological normalization* of the discriminant Σ_k, i.e. the space of all pairs {a point in the line $\mathbf{R}^1$, a polynomial having a k-fold root at this point}. Obviously, this is a smooth manifold diffeomorphic to $\mathbf{R}^{d-k+1}$, and forgetting the first elements of the pairs defines a proper map of it onto Σ_k. A generic point of Σ_k has exactly one preimage under this map, while the number of preimages of an arbitrary discriminant polynomial is equal to its number of geometrically distinct k-fold roots. To get a space homotopy equivalent to the original one we ought to change any such preimage by a contractible space, for instance by inserting a simplex with vertices at the points of this preimage. A precise construction of this resolution is as follows. We fix a *generic* imbedding $I : \mathbf{R}^1 \to \mathbf{R}^N$ of the argument line into a space of a very large dimension. The genericity condition here consists in the claim that the images of no $[d/k]$ distinct points of the line lie in the same $([d/k] - 2)$-dimensional affine subspace. Our resolution will be a subset in the direct product of this space $\mathbf{R}^N$ and the space P_d of all polynomials of the form (1). Namely, for any discriminant polynomial f we take *all* its roots of multiplicity $\geq k$, $z_1, \ldots, z_t$, and consider a simplex in $\mathbf{R}^N \times P_d$, the vertices of which are the points $(I(z_1), f), \ldots, (I(z_t), f)$. The desired resolution space $\sigma \equiv \sigma_k$ is defined as the union of all such simplices. The obvious projection $\mathbf{R}^N \times P_d \to P_d$ defines a proper map of σ onto Σ_k; the extension of this map to a map of the one-point compactifications of these spaces, $\bar{\sigma} \to \bar{\Sigma}_k$, is a homotopy equivalence.

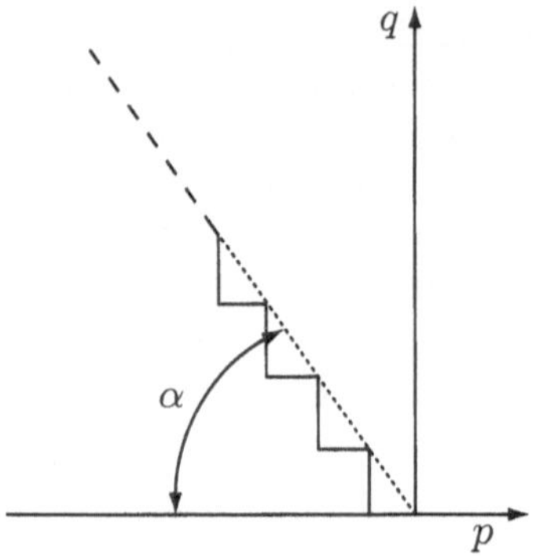

Figure 2 Support of the general cohomological spectral sequence

On the other hand, the space σ (and hence also its one-point compactification $\bar{\sigma}$) has a natural increasing filtration: the term F_p of this filtration consists of all simplices of dimension $\leq p-1$ participating in the previous construction; $F_0 = \{\text{the added point}\}$. The term $E^1_{p,q}$ of the corresponding homological spectral sequence is equal to $\bar{H}_{p+q}(F_p \backslash F_{p-1})$, where $\bar{H}_*$ denotes the Borel-Moore homology. This space $F_p(\bar{\sigma}) \setminus F_{p-1}(\bar{\sigma})$ has a natural structure of a fibre bundle, the base of which is the space $B(\mathbf{R}^1, t)$ of all subsets of cardinality p in the line, and the fibre over such a collection $(z_1, \ldots, z_p)$ is the direct product of an open $(p-1)$-dimensional simplex and an affine space of dimension $d - p \cdot k$, consisting of all polynomials of the form (1) having the k-fold roots at exactly these p points. Hence the space of this bundle is a cell and $E^1_{p,q}$ is equal to $\mathbf{Z}$ if $q = d - p(k-1)$ and $p \leq d/k$ and is trivial for all other p and q. Obviously, this sequence degenerates at the term E^1 and gives immediately the answer.

It is the time now to remember that we are calculating not the homology of the discriminant, but the (Alexander dual to it) cohomology of the complementary space. Therefore it is natural to invert our homological spectral sequence formally into a cohomological one by setting

$$E_1^{p,q} \equiv E^1_{-p,d-q-1}, \tag{3}$$

then its support comes to (the left boundary of) the shaded region in Figure 2 and converges to exactly the cohomology group of the complement of the discriminant. The group $E_1^{p,q}$ of this sequence is equal to $\mathbf{Z}$ if $(k-1)p + q = 0$, $-d/k \leq p \leq 0$, and is equal to 0 for all other p and q. Therefore the group $H^i(P_d \setminus \Sigma_k)$ is free cyclic if i is a multiple of $k-2$ and is no greater than $[d/k](k-2)$, and is trivial for all other i.

Stabilization and jet imbedding

An important property of this *cohomological* spectral sequence is its stabilization when k is fixed and d increases: for any $d' > d$ the corresponding sequences calculating the cohomologies of spaces $P_{d'} \setminus \Sigma_k$, $P_d \setminus \Sigma_k$ naturally coincide in the "stable" domain $\{(p,q) | p \geq -d/k\}$. (Moreover, there is a filtration-preserving homeomorphism between the term $F_{[d/k]}$ of the resolution $\sigma(d')$ of the discriminant $\Sigma_k(d') \subset P_{d'}$ and the direct product $\sigma(d) \times \mathbf{R}^{d'-d}$.) These spaces $P_d \setminus \Sigma_k$, $d \to \infty$, can be considered as finite-dimensional approximations of the space $\mathcal{F} \setminus \Sigma_k$ of smooth functions $\mathbf{R}^1 \to \mathbf{R}^1$ with a fixed behavior at infinity (say, equal identically to 1 outside some compact set) without zeros of multiplicity $\geq k$. Our spectral sequences for increasing d actually converge to a stable spectral sequence calculating the cohomology of the latter space.

Moreover, these sequences prove that $\mathcal{F} \setminus \Sigma_k$ is homologically equivalent to the loop space

$$\Omega(\mathbf{R}^k \setminus 0) \sim \Omega S^{k-1} \tag{4}$$

and this equivalence is induced by the jet extensions mapping a function $f \in \mathcal{F}$ into the map $(f, f', \ldots, f^{(k-1)}) : \mathbf{R}^1 \to \mathbf{R}^k$.

Indeed, the cohomology of the space (4) can be calculated by a similar spectral sequence naturally isomorphic to the previous one starting with the terms E_1.

(Moreover, this jet imbedding induces not only a homology but also homotopy equivalence. The approximation of the space $\mathcal{F} \setminus \Sigma_k$ by the (suitably imbedded) subspaces $P_d \setminus \Sigma_k$ is equivalent (via this homotopy equivalence) to the approximation of the space ΩS^{k-1} by the spaces of piecewise-smooth paths of restricted length, see [Milnor 63].)

All this is a very special case of the general situation described in the next section; in particular the groups $H^*(\mathcal{F} \setminus \Sigma(S))$ defined by arbitrary discriminant varieties can be calculated by appropriate spectral sequences shaped as is shown in Figure 2 with $\tan \alpha = \operatorname{codim}(S)$.

3 The spaces of functions without complicated singularities

Let k be a nonnegative integer, M and N two manifolds. Recall the notation $J^k(M, N)$ (respectively, $J^k_x(M, N)$) for the space of k-jets of smooth maps $M \to N$ (respectively, for the space of k-jets at the point $x \in M$), see for example [Gromov], [AVGL]. Obviously $J^k(M, N)$ is a fibre bundle over M with fibres $J^k_x(M, N)$. Any smooth map $f : M \to N$ defines in an obvious way a section of this bundle, its k-jet extension $j^k(f)$.

DEFINITION. A *singularity class* of maps of m-dimensional manifolds into $\mathbf{R}^n$ is any semialgebraic closed subset of the space $J^k_0(\mathbf{R}^m, \mathbf{R}^n)$ (for some finite k) invariant under the obvious action of the group $\operatorname{Diff}_0(\mathbf{R}^m)$.

If S is such a class, then for any m-dimensional manifold M the corresponding subset $S(M)$ in the jet space $J^k(M, \mathbf{R}^n)$ is well defined: it is a subbundle with fibre S of the jet bundle $J^k(M, \mathbf{R}^n) \to M$. The corresponding discriminant variety $\Sigma(S)$ consists of all smooth maps $M \to \mathbf{R}^n$, the k-jet extensions of which intersect $S(M)$.

THEOREM 1 (The Smale-Hirsch principle for functions without complicated singularities). *Let M be a compact m-dimensional manifold without boundary. If the codimension of the singularity class S in $J^k_0(\mathbf{R}^m, \mathbf{R}^n)$ is at least $m+2$ (or, which is the same, the codimension $\tau(S)$ of the set $\Sigma(S)$ in the space of all smooth maps $M \to \mathbf{R}^n$ is at least 2), then the space of maps without singularities of the class*

S is homology equivalent to the space of all continuous sections of the obvious jet fibration $(J^k(M,\mathbf{R}^n)\setminus S(M))\to M$; this homology equivalence is induced by the k-jet extensions of maps.

(In particular, if $\tau(S)\geq 3$ then by Whitehead's theorem they are not only homology but also homotopy equivalent. In the case when $n=1$ and S consists of the singularities more complicated than the "birth-death" singularity A_2, a similar homotopy equivalence was proved by Igusa (see [Igusa]) up to dimension m and by Eliashberg in all dimensions.)

Indeed, the cohomologies of both spaces can be calculated by the spectral sequences generalizing the one considered above and naturally isomorphic to one another starting with the term E_1. All nontrivial groups $E_1^{p,q}$ of this term lie in the edge $\{(p,q)|p\leq 0,\tau(S)\cdot p+q\geq 0\}$ and have an explicit and "finite-dimensional" description in the terms of the sets S and configuration spaces of M. Namely, consider the obvious fibre bundle $(S(M))^{-p}\to M^{-p}$ with fibre $(S)^{-p}$. Delete from M^{-p} the "diagonale", i.e. the set of all points $(x_1,\dots,x_{-p})\in M^{-p}$ such that $x_i=x_j$ for some $i\neq j$, and delete from $(S(M))^{-p}$ the preimage of this diagonale. The group of permutations of $-p$ elements acts in an obvious way on the base and the space of the resulting bundle. The quotient spaces of these actions also form a fibre bundle with fibre $(S)^{-p}$, the base of which is the configuration space $B(M,-p)$, i.e. the space of all subsets of cardinality $-p$ in M. The space of this bundle will be denoted by $\Lambda_{-p}(M,S)$. Set $\Lambda_0(M,S)\equiv\{\text{a point}\}$ and denote by $\Delta\equiv\Delta(k,m,n)$ the dimension of the space $J_0^k(\mathbf{R}^m,\mathbf{R}^n)$.

THEOREM 2. *In the conditions of Theorem 1, there exists a spectral sequence $E_r^{p,q}$ converging to the cohomology of the space of maps $M\to\mathbf{R}^n$ without the singularities of type S, the term $E_1^{p,q}$ of which is trivial if $p>0$, and for nonpositive p this term is isomorphic to the group*

$$\bar{H}_{-p\Delta-q}(\Lambda_{-p}(M,S),T) \tag{5}$$

where T is a certain local system with fibre $\mathbf{Z}$.

(In particular, $E_1^{p,q}=0$ if $\tau(S)\cdot p+q<0$, and for any integer c there are at most finitely many nontrivial groups $E_1^{p,q}$ with $p+q=c$, thus no problems with the convergence of the spectral sequence appear.) The local system T from (5) is defined as follows.

There are three important elements in the group $H^1(B(M,i),\mathbf{Z}_2)$, the elements W, Alt, and WJ^k. Namely, any loop l in $B(M,i)$ is a simultaneous move of i points in M, and the union of paths swept by these points is a closed curve $[l]$ in M. The class W is defined by identity $\langle W,l\rangle\equiv\langle w_1,[l]\rangle$ where w_1 is the first Stiefel-Whitney class of M. The class Alt takes nontrivial values on the loops that define odd permutations of the i points. Finally, WJ^k is the first Stiefel-Whitney class of the vector bundle over $B(M,i)$, the fibre of which over the collection $(x_1,\dots,x_i)$ is the space $J_{x_1}^k(M,\mathbf{R}^n)\times\cdots\times J_{x_i}^k(M,\mathbf{R}^n)$.

It is easy to calculate that $WJ^k\equiv n\Big(\binom{m+k}{m}\mathrm{Alt}+\binom{m+k}{k-1}W\Big)$.

Consider the local system over $B(M, i)$ locally isomorphic to $\mathbf{Z}$ and changing the orientation after the transportation over a loop l if and only if $\langle(\text{Alt} + WJ^k), l\rangle \neq 0$. The local system T mentioned in (5) is induced from this one by the obvious projection $\Lambda_{-p}(M, S) \to B(M, -p)$.

The construction of this spectral sequence essentially repeats that of the (stable) spectral sequence from Section 2 calculating the cohomology of the space $P_d \setminus \Sigma_k$. First of all, using a generic sequence of finite-dimensional approximating subspaces in the space $\mathcal{F}$ of smooth maps $M \to \mathbf{R}^n$ we can work with this space as with a vector space of a very large finite dimension. (In fact, we construct the spectral sequences calculating the cohomologies of complements of $\Sigma(S)$ in these subspaces and then prove the stabilization of these sequences.)

Then we take the geometrical resolution $\sigma(S)$ of the discriminant variety $\Sigma(S)$. This is a filtered space, the term $F_i \setminus F_{i-1}$ of it is a fibre bundle, the base of which is $B(M, i)$, and the fibre over a point $(x_1, \ldots, x_i) \in B(M, i)$ is a direct product of an open $(i-1)$-dimensional simplex and the space of all smooth maps $M \to \mathbf{R}^n$ having singularities of type S at all points $x_1, \ldots, x_i$. The latter space is obviously diffeomorphic to a direct product of $(S)^i$ and of some affine subspace in $\mathcal{F}$. Thus, the fibre bundle $F_i \setminus F_{i-1} \to B(M, i)$ is a fibred product of three bundles: the bundle of open $(i-1)$-simplices (the first Stiefel-Whitney class of which is equal to Alt), the bundle $\Lambda_i(M, S) \to B(M, i)$, and an affine bundle, the first Stiefel-Whitney class of which is equal to WJ^k. This "proves" formula (5).

REMARK 0. Usually the homology groups of spaces of nonsingular functions (to the calculation of which our spectral sequence is intended) appear as the invariants of the underlying manifold M. However, this spectral sequence itself (especially its higher differentials) provides a much stronger system of invariants than these homology groups.

REMARK 1. If S is invariant also under the obvious action of $\text{GL}(\mathbf{R}^n)$, then instead of the space of maps $M \to \mathbf{R}^n$ in Theorems 1 and 2 we can take the spaces of sections of n-dimensional vector bundles over M (and of the associated bundles with fibres isomorphic to $J_0^k(\mathbf{R}^m, \mathbf{R}^n)$).

REMARK 2. A similar spectral sequence calculates also the cohomology of spaces of functions on manifolds with boundary. Namely, let M be a compact manifold with boundary ∂M, S a singularity class of codimension $\geq m+2$, and ϕ any smooth map of M into $\mathbf{R}^n$ having no singularities of type S in a neighborhood of ∂M. Then the space of functions $M \to \mathbf{R}^n$ coinciding with ϕ close to ∂M and having no singularities of type S is homology equivalent to the space of sections of the bundle $(J^k(M, \mathbf{R}^n) \setminus S(M)) \to M$ coinciding with $j^k(\phi)$ close to ∂M; this homology equivalence is induced by the jet extension map. Indeed, the homology groups of both spaces can be calculated by spectral sequences similar to the previous one and isomorphic to one another starting with the terms E_1. These terms E_1 are described in almost the same way as similar terms in Theorem 2: we need only to define the new space $\Lambda_{-p}(M, S)$ as a fibre bundle with fibre S^{-p} not over $B(M, -p)$ but over $B(M \setminus \partial M, -p)$.

REMARK 3. A similar spectral sequence can be constructed even if $\operatorname{codim}(S) = m+1$ (for example if $S = NM$, the set of non-Morse singularities, and hence $\mathcal{F} \setminus \Sigma(S)$ is the space of Morse functions). In these cases our spectral sequence also provides many cohomology classes of $\mathcal{F} \setminus \Sigma(S)$, but possibly not all of them. The lines $\{p+q = \text{const}\}$ of these sequences may contain infinitely many non-trivial groups, thus the convergence of the sequences is questionable; moreover, generally the 0-dimensional cohomology classes provided by the spectral sequence do not constitute a complete system of invariants of such functions: say, in the case $S = NM$ they are just the polynomials of the numbers of critical points of any Morse index. However, our spectral sequence gives also more delicious invariants. Indeed, the basis higher-dimensional integer homology classes (counted by the dual sequence) are realized by cycles that lie in certain components of the space of Morse functions and thus characterize these components.

Also, the other invariants of the manifold M that appear in our spectral sequences (defined by different S) are of independent interest. For the simplest example, if $S = NM$ the group $E_1^{-1,*}$ coincides (up to a change of grading) with the cohomology of the fibre bundle over M that is the (fibre-wise) join of two bundles: the spherized cotangent bundle of M and the total Grassmann bundle associated with the cotangent bundle of M (the fibre of which over a point $x \in M$ is the disjoint union of all $m+1$ manifolds $G_k(T_x^*M)$, $k = 0, \ldots, m$).

The calculation of columns $E_1^{-i,*}$, $i > 1$, also contains the study of configuration spaces of M and some interaction of them with the Grassmann bundles associated with T^*M. The precise calculation of our spectral sequences (especially of their higher differentials) eventually will allow us to guess many other natural invariant structures and characteristics of M.

4 Spaces of maps of m-dimensional CW-complexes into $(m-1)$-connected complexes

The simplest version of our spectral sequence, corresponding to the case when the forbidden singularity S is defined by the 0-jets (i.e. S is just a subset in $\mathbf{R}^n$), calculates the cohomology of the space of continuous maps of arbitrary m-dimensional CW-complex X into arbitrary m-connected complex Y[1]). (The simplest version of this simplest version, when X is a circle, is the Adams spectral sequence calculating the cohomology of the loop spaces.)

Indeed, we can assume that the complex Y is finite dimensional. We imbed it generically in a sphere S^N (N very large) and choose a subcomplex $\Psi \in S^N$ Spanier-Whitehead dual to it. Let $C\Psi$ be the union of all rays in the Euclidean space $\mathbf{R}^{N+1}$ issuing from the origin and penetrating the unit sphere in the points of Ψ.

We take for the function space $\mathcal{F}$ the space of all continuous maps $X \to \mathbf{R}^{N+1}$ and define the discriminant variety as the space of all maps the images of which intersect the cone $C\Psi$. Because Y is m-connected, the space Ψ can be chosen to

---

1) Note added in proof: As I have learned from F. Cohen's review of [V 92 & 94] (Bull. AMS, October 1994), the same (as an algebraic object) spectral sequence in the case when X is a manifold was constructed by D. Anderson in the early seventies.

be of codimension at least $m+2$ in the sphere, and hence the codimension τ of this discriminant in $\mathcal{F}$ is at least 2.

The space of maps $X \to Y$ is obviously homotopy equivalent to the complement of this discriminant, hence the cohomology of it can be calcucated completely by our spectral sequence. Let us describe this spectral sequence in our particular case.

Again, it lies in the second quadrant in the edge $\{(p,q)|p<0, \tau p+q \geq 0\}$ (see Figure 2), and the term $E_1^{p,q}$ of it is as follows. Let $\Pi(t)$ be the space of a fibre bundle over $B(X,t)$, the fibre of which over a point $\{z_1,\dots,z_t\}$, $z_i \in X$, is the Cartesian product of t examples of $C\Psi$ that are in a correspondence with the points z_i (and permute in the obvious way after the transportations over the loops in the base space $B(X,t)$ that permute these points). Then

$$E_1^{p,q} = \bar{H}_{-p(N+1)-q}(\Pi(-p), \pm\mathbf{Z}^{\otimes N}) \tag{6}$$

where $\pm\mathbf{Z}$ is the local group locally isomorphic to $\mathbf{Z}$ but changing the orientation over the loops defining odd permutations of the points z_i.

Also in the "marginal" case, when X is m-dimensional and Y is $(m-1)$-connected, this spectral sequence provides many homology classes of Y^X, but generally not all of them.

This spectral sequence has also a "relative" version that calculates the cohomology of maps $X \to Y$ coinciding with a fixed map on a subcomplex $K \subset X$. In this case $\Pi(-p)$ in the formula (6) should be defined as a similar fibre bundle over $B((X \setminus K), -p)$.

EXAMPLE 1. If $X = S^1$ and $K = \{\text{a point}\}$, then the relative version of this spectral sequence coincides with the Adams spectral sequence for the loop spaces, cf. [Adams], [Chen].

EXAMPLE 2. If both X and Y are the spheres S^m, S^n, $n > m$, this spectral sequence degenerates at the first term and gives (the homology version of) the Snaith splitting formula

$$H^i(\Omega^m S^n) \simeq \oplus_{t=0}^{\infty} H^{i-t(n-m)}(B(\mathbf{R}^m,t), (\pm\mathbf{Z})^{\otimes(n-m)}). \tag{7}$$

5 Complex analogues: The May-Segal formula for the braid group and multidimensional generalizations

The previous techniques can be applied to holomorphic functions of complex variables: say, the May-Segal formula [Segal 73] for the cohomology of the stable braid group,

$$H^*(Br(\infty)) \simeq H^*(\Omega^2 S^3), \tag{8}$$

can be interpreted in these terms as a simplest version of the complex Smale-Hirsch principle, cf. Theorem 1.

Indeed, the classifying space of the braid group on d strings can be realized as the space $\mathbf{C}P_d \setminus \Sigma$ of complex polynomials (1) without multiple roots. The space $\Omega^2 S^3$ is obviously homotopy equivalent to the space of continuous maps $\mathbf{R}^2 \to \mathbf{R}^4 \setminus 0$ with a fixed behavior at infinity. Any polynomial f of the form (1) defines a map $\mathbf{R}^2 \to \mathbf{R}^4$, its 1-jet extension is given by

$$(f, f') : \mathbf{C}^1 \to \mathbf{C}^2;$$

all these maps (for a given d) have the same asymptotical behavior at infinity, and the condition that f has no double roots coincides with the condition that the image of this map in $\mathbf{C}^2$ does not contain the origin. Thus, we get an imbedding of the space $\mathbf{C}P_d \setminus \Sigma \simeq K(Br(d), 1)$ into a space homotopy equivalent to $\Omega^2 S^3$.

PROPOSITION. *This imbedding induces an isomorphism of cohomology groups in dimensions not exceeding* $[d/2] + 1$.

In particular, for $d \to \infty$ we get the May-Segal formula.

A proof of this proposition is very similar to that of Theorem 1 and is essentially a comparison theorem of two spectral sequences calculating the Borel-Moore cohomology of the standard discriminant $\Sigma \subset \mathbf{C}P_d$ and the discriminant variety in $\Omega^2 \mathbf{R}^4$ defined as the set of double loops $\mathbf{R}^2 \to \mathbf{R}^4$ intersecting the origin.

The multidimensional May-Segal formula

Let $f : (\mathbf{C}^n, 0) \to (\mathbf{C}, 0)$ be a holomorphic function with an isolated singularity at 0, and $F : (\mathbf{C}^n \times \mathbf{C}^l, 0) \to (\mathbf{C}, 0)$ a deformation of f (i.e. $f \equiv F(\cdot, 0)$). F can be considered as a family of functions $f_\lambda : (\mathbf{C}^n, 0) \to (\mathbf{C}, 0)$, $f_\lambda \equiv F(\cdot, \lambda)$, parametrized by the points $\lambda \in \mathbf{C}^l$.

The discriminant variety $\Sigma(F)$ is the set of parameters $\lambda \in \mathbf{C}^l$ such that the corresponding function f_λ has close to the origin in $\mathbf{C}^n$ a critical point with critical value 0 (or, which is the same, the variety $f_\lambda^{-1}(0)$ is singular). The complement of this discriminant variety is the base of several important fibre bundles that appear naturally in integral geometry and mathematical physics, see for example [AVGL], [V 95]. The natural hierarchy in the space of functions and their deformations (defined by the adjacencies of singularities) allows us to define the *stable* (co)homology group of complements of discriminants of isolated singularities of functions $(\mathbf{C}^n, 0) \to (\mathbf{C}, 0)$, see [Arnold 76, 79], [V 87]. (For any fixed dimension i, this stable group can be thought of as just the group $H^i(\mathbf{C}^l \setminus \Sigma(F))$ for a "sufficiently large" deformation F of a "sufficiently complicated" singularity f, see [V 90]. For instance, if $n = 1$, then for a sufficiently complicated singularity we can take any singularity x^d, $d > 2i$, and for the sufficiently large deformation of it the corresponding space of polynomials (1) parametrized by coefficients a_j.)

THEOREM 3 (see [V 87, 90]). *For any n, the stable cohomology ring of complements of discriminants of isolated singularities of functions $(\mathbf{C}^n, 0) \to (\mathbf{C}, 0)$ is naturally isomorphic to the ring $H^*(\Omega^{2n} S^{2n+1})$; this isomorphism is induced by 1-jet extensions.*

6 Complements of resultant and discriminant varieties

Let $\mathbf{C}(m_1, \ldots, m_k)$ be the space of complex polynomial systems (2) and $\Sigma(m_1, \ldots, m_k)$ the resultant subvariety in it. It is easy to see that all spaces $\mathbf{C}(m_1, \ldots, m_k) \setminus$

$\Sigma(m_1,\ldots,m_k)$ with the same k and $\min(m_i)$ are homotopy equivalent to one another, in particular to the space $\mathbf{C}(m,\ldots,m)\setminus\Sigma(m,\ldots,m)$ where $m=\min(m_i)$. Denote the latter space by $\mathbf{C}(m|k)\setminus\Sigma(m|k)$.

THEOREM 4 (see [V 92 & 94]). *The space $\mathbf{C}(m|k)\setminus\Sigma(m|k)$ is stably homotopy equivalent to the space $\mathbf{C}P_{m\cdot k}\setminus\Sigma_k(m\cdot k))$ of all complex polynomials of the form (1) with $d=m\cdot k$ having no roots of multiplicity k. Analogously, the spaces $\mathbf{R}(m|k)\setminus\Sigma(m|k)$ and $P_{m\cdot k}\setminus\Sigma_k(m\cdot k))$ of real systems and polynomials are (nonstably) homotopy equivalent.*

(In the case when $k=2$ this fact was previously established by Cohen, Cohen, Mann, and Milgram, see [CCMM].)

Indeed, the geometrical resolutions of the resultant variety $\Sigma(m|k)$ and of the discriminant $\Sigma_k(m\cdot k)$ are very similar to one another (for illustration, compare the upper parts of Figures 1a, b). The terms $F_i\setminus F_{i-1}$ of their standard filtrations are fibre bundles, the common base of which is the configuration space $B(\mathbf{C}^1,i)$, and the fibres are direct products of a (common) open $(i-1)$-dimensional simplex and a complex affine subspace of codimension $k\cdot i$ in the space $\mathbf{C}(m|k)$ or $\mathbf{C}P_{m\cdot k}$. In particular, the terms $E^1_{p,q}$ of the canonical spectral sequences generated by these filtrations are isomorphic. To prove the stable homotopy equivalence of the one-point compactifications of our resolutions we proceed as follows.

(A) Imbed both $\mathbf{C}P_{m\cdot k}$ and $\mathbf{C}(m|k)$ in $\mathbf{C}(m\cdot k, m\cdot k-1,\ldots,(m-1)\cdot k+1)$: the first imbedding maps a polynomial f into the system $(f,f',\ldots,f^{(k-1)})$ and the second is provided by multiplication of the polynomials of a system (2) by some polynomials with distant and different roots. These imbeddings map the discriminant and resultant varieties into the resultant variety.

(B) Lift these imbeddings $\Sigma_k(m\cdot k)\to\Sigma(m\cdot k,\ldots,(m-1)\cdot k+1)\leftarrow\Sigma(m|k)$ to the maps of their geometrical resolutions. The images of both lifted maps lie in the term F_m of the resolution $\sigma(m\cdot k,\ldots,(m-1)\cdot k+1)$ of $\Sigma(m\cdot k,\ldots,(m-1)\cdot k+1)$.

(C) Prove that both these lifted imbeddings can be extended to the filtration-preserving maps of appropriate multiple (of multiplicity $k(k-1)(2m-1)$) suspensions of the one-point compactifications of $\sigma_k(m\cdot k)$ and $\sigma(m|k)$ onto the one-point compactification of the term F_m of the resolution $\sigma(m\cdot k,\ldots,(m-1)\cdot k+1)$. The isomorphism of terms E^1 of corresponding spectral sequences allows us to prove that (a) there are no topological obstructions for the construction of such maps by induction over our filtration, and (b) both these maps are homotopy equivalences.

7 The knot invariants

For arbitrary $n\geq 3$ consider the space $\mathcal{F}$ of all smooth maps of the circle S^1 (respectively, of a disjoint union of several circles) into $\mathbf{R}^n$ and define a discriminant as the space of such maps that are not the knots (respectively, links), i.e. that have either singularities or self-intersections. Then a spectral sequence similar to the one described above provides the cohomology elements of the space of knots or links, in particular (for $n=3$) the knot and link invariants.

Indeed, consider the space Ψ of all unordered pairs of (not necessarily distinct) points in the circle S^1 and imbed Ψ generically in the space $\mathbf{R}^\Omega$ of immensely large dimension. For any discriminant map $f : S^1 \to \mathbf{R}^n$ consider the simplex $\Delta(f) \subset \mathbf{R}^\Omega$ spanned by the images of all points $(x,y) \in \Psi$ such that either $f(x) = f(y)$ (if $x \neq y$) or $f'(x) = 0$ (if $x = y$). The resolution of the discriminant is a subset in $\mathbf{R}^\Omega \times \mathcal{F}$ equal to the union of simplices of the form $\Delta(f) \times f$ over all discriminant points f. (We ignore here the maps having nondiscrete sets of singular points.) The one-point compactification of the obtained resolution admits a natural increasing filtration that generates (after the standard inversion similar to (3)) the desired spectral sequence. All the quotient spaces F_i/F_{i-1} of this filtration are finite cell complexes (strictly speaking, with infinite-dimensional cells), thus all its terms $E_1^{p,q}$ are finitely generated.

This spectral sequence looks like the one in Figure 2 with $\tan\alpha = n-2$. In particular, for $n \geq 4$ there are no problems with its convergence on any diagonale $\{(p,q)|p+q = \text{const}\}$ and it calculates completely the cohomology of the corresponding knot and link spaces. Moreover, Kontsevich has proved that this spectral sequence (over $\mathbf{Q}$) degenerates at the term E_1, thus the problem of calculating the rational cohomology of spaces of links in $\mathbf{R}^{\geq 4}$ is essentially closed.

CONJECTURE. I suspect that in fact this spectral sequence degenerates in a much stronger "homotopical" sense: namely, that for any i the term F_i of our filtration is (in some precise sense) homotopy equivalent to the wedge $(F_i/F_{i-1}) \vee \ldots \vee (F_2/F_1) \vee F_1$.

In the "marginal" case $n = 3$ it turns out that the invariants that appear from the spectral sequence are stronger than all known polynomial invariants (see [Birman & Lin], [Birman]), and the Milnor's μ-invariants of links ([Bar-Natan 93], [Lin]), in particular distinguish all links up to homotopy equivalence. The problem of completeness of this system of invariants for the isotopy classification is still open.

A majority of the standard notions of the theory developed around these invariants appear naturally in the spectral sequence: say, the groups $E_0^{-i,i}$ of it are generated by the "chord diagrams" with i chords, while the groups $E_0^{-i,i+1}$ are spanned by the "four-term relations" and the "framing independence relations". For further results concerning these invariants see [Kontsevich], [Bar-Natan 92], [Birman], [Arnold 92], [Piunikhin], [Cartier]...[2])

Similarly to Remark 3 of Section 3, besides the "straightforward" invariants of knots (i.e. the 0-dimensional cohomology classes that appear from the terms $E_\infty^{-i,i}$) our spectral sequence provides some more invariants: namely, the basis higher-dimensional integer homology classes of the space of knots or links that are counted in the upper lines of the dual sequence correspond to the cycles that lie in some components of this space and thus characterize these components.

A similar spectral sequence calculates the cohomology of the space of knots or links in an arbitrary 3-dimensional manifold. Of course, all the groups $E_r^{p,q}$ and

2) Note added in proof: Many invariants of this kind can be constructed explicitly from representations of Lie algebras (Bar-Natan, Kontsevich), superalgebras and Yang-Baxter Lie algebras (see A. Vaintrop, *Vassiliev knot invariants and Lie S-algebras*, Math. Research Letters **1** (1994), 579–595). Recently P. Vogel announced that the invariants coming from Lie superalgebras are strictly stronger than those for the semisimple Lie algebras.

all differentials of these sequences provide a system of invariants of the manifolds themselves. It would be very interesting to study these invariants explicitly and to compare them with other known invariants.

8 Invariants of ornaments

A *k-ornament* is a collection of k ordered closed plane curves, no three of which intersect at the same point. (The ornaments are close relatives of the *doodles* of [Fenn & Taylor], which are just the ornaments, the components of which are smooth and have no self-intersections.)

Again a spectral sequence appears that calculates the invariants of such objects: it is based on a resolution of the space of forbidden curves. Many of these invariants have very easy descriptions (but were guessed only by using the general techniques), see [V 94], [Merkov], and the second edition of [V 92 & 94].

The simplest of them are the "index-type invariants": they are the collections of $\binom{k}{2}$ integer-valued functions of k integer variables. Namely, let ϕ be a *generic smooth* k-ornament, i.e. a collection of k ordered immersed circles in $\mathbf{R}^2$ having only transversal double intersections and self-intersections. Let i and j be two naturals, $1 \leq i < j \leq k$. The corresponding function $I_{i,j}(b_1, \ldots, b_k)$ is defined as follows. To any intersection point x of the ith and jth curves we assign in the following way k integer numbers $b_1(x), \ldots, b_k(x)$ and a sign $\sigma(x)$.

If $l \neq i, j$, then $b_l(x)$ is just the number $ind_l(x)$, the index of x with respect to the lth curve. Close to any regular point of the ith curve (in particular, to the intersection point x) the values of the corresponding function $ind_i(\cdot)$ take two neighboring integer values on different sides of the curve. Define the number $b_i(x)$ as the smallest of these values in the neighboring points to x. The number $b_j(x)$ is defined in the same way by means of ind_j. Finally, $\sigma(x)$ is equal to 1 (resp., -1) if the tangent vectors of the ith and jth curves at the point x define a positively (resp., negatively) oriented frame (with respect to a fixed orientation of $\mathbf{R}^2$).

Given a generic smooth k-ornament and k integer numbers $b_1, \ldots, b_k$, define $I_{i,j}(b_1, \ldots, b_k)$ as the number of intersection points x of the i-th and j-th curves of our ornament such that $b_1(x) = b_1, \ldots, b_k(x) = b_k$ and $\sigma(x) = 1$, minus the number of similar points with $\sigma(x) = -1$.

THEOREM 5. *All the functions $I_{i,j}$, $1 \leq i < j \leq k$, are invariants of ornaments.*

Several more delicious invariants were discovered by Merkov, see [Merkov] and Section VI.9 in [V 92 & 94].

These invariants seem to be related to the equations of higher-dimensional simplices in the same way as the knot and link invariants are related to the usual Yang-Baxter equation.

The calculations in this (and the previous) spectral sequence lead to many (partially classical and partially unknown) problems in the modern homological combinatorics, concerning the complexes of connected (multi)graphs, see [V 93, 94], [Björner & Welker].

A parallel classification of plane curves was recently developed by Arnold, see [Arnold 94] and [Aikardi].

9 Complements of affine plane arrangements

Let $V_1, \ldots, V_s$ be a collection of affine planes in $\mathbf{R}^n$ (such collections are called *affine plane arrangements*), and let V be the union of these planes. The topology of spaces $\mathbf{R}^n \setminus V$ in some special cases was studied in [Arnold 69], [Brieskorn] and many other works; in [Goresky & MacPherson] a general formula for the homology groups of these spaces was obtained. The techniques used above also allow us to describe their stable homotopy types.

Indeed, let Λ be any affine plane that is the intersection of several planes V_i. Consider the simplex $\Delta(\Lambda)$, the vertices of which correspond formally to all the planes V_i containing Λ. A face of this simplex, i.e. a collection of such planes, is called *marginal*, if the intersection of these planes is strictly greater than Λ. The quotient space of the simplex $\Delta(\Lambda)$ by the union of all marginal faces will be denoted by $K(\Lambda)$.

THEOREM 6 (see [V 91′, 93], [Ziegler & Živaljević]). *The one-point compactification of the variety V is homotopy equivalent to the wedge of* $\dim(\Lambda)$*-fold suspensions of the spaces $K(\Lambda)$ taken over all planes Λ that are the intersections of some planes of the set $(V_1, \ldots, V_s)$.*

The proof used in [V 91′, 93] consists of two steps: we take the (obvious) geometrical resolution of V and apply to it a version of the stratified Morse theory; see [Goresky & MacPherson].

COROLLARY 1. *The stable homotopy type of the complement $\mathbf{R}^n \setminus V$ of the arrangement V is completely determined by its combinatorial (dimensional) characteristics.*

COROLLARY 2. *The cohomology group $H^i(\mathbf{R}^n \setminus V)$ is isomorphic to* $\oplus \tilde{H}_{n-i-1-\dim(\Lambda)}(K(\Lambda))$ *(the summation over all planes Λ that are the intersections of several planes V_i).*

Indeed, this follows from Theorem 6 by the Alexander duality.

This formula was previously obtained in [Goresky & MacPherson].

10 Homology of classical Lie groups

A few years ago (as J. Milgram informed me in 1992) Mahovald discovered the following splittings of the homology groups of the Lie groups $U(n)$, $O(n)$, and $Sp(n)$ (or, which is the same, of the homotopy equivalent to the groups of all nondegenerate matrices over the complex, real, and quaternion numbers) into the homology groups of the corresponding Grassmannians:

$$H_i(U(n)) \simeq \oplus_{k=0}^{n} H_{i-k^2}(G_k(\mathbf{C}^n)), \tag{9}$$

$$H_i(O(n), \mathbf{Z}_2) \simeq \oplus_{k=0}^{n} H_{i-k(k-1)/2}(G_k(\mathbf{R}^n), \mathbf{Z}_2), \tag{10}$$

$$H_i(O(n)) \simeq \oplus_{k=0}^{n} H_{i-k(k-1)/2}(G_k(\mathbf{R}^n), Or^{\otimes(k-1)}), \tag{11}$$

$$H_i(Sp(n)) \simeq \oplus_{k=0}^{n} H_{i-2k^2-k}(G_k(\mathbf{H}^n)), \tag{12}$$

where Or is the orientation sheaf of the tautological bundle over $G_k(\mathbf{R}^n)$.

These splittings are close relatives of formula (7) and can be guessed automatically with the help of our general techniques (as actually was done in [V 91]). Indeed, they appear from the study of some *conical resolution* of the *determinant varieties*, i.e. of the spaces of all degenerate square matrices. The basis of this resolution is the (suitably topologized) *order complex* (see e.g. [Goresky & MacPherson]) of vector subspaces of the space $\mathbf{K}^n$, $\mathbf{K} = \mathbf{C}, \mathbf{R}$, or $\mathbf{H}$.

Namely, consider the join of n Grassmann manifolds of subspaces in $\mathbf{K}^n$,

$$G_1(\mathbf{K}^n) * \cdots * G_{n-1}(\mathbf{K}^n) * G_n(\mathbf{K}^n), \tag{13}$$

i.e. roughly speaking the union of all simplices, the vertices of which correspond to the points of all these Grassmann manifolds. This join is contractible as $G_n(\mathbf{K}^n)$ is a point. Consider a subset $C\Theta(\mathbf{K}^n)$ in the join (13) consisting of only such simplices that the planes corresponding to their vertices are incident to one another (i.e. form a flag). To any linear subspace $V \subset \mathbf{K}^n$ there corresponds a (contractible) subset $C\Theta(V) \subset C\Theta(\mathbf{K}^n)$, the union of all simplices from the construction of $C\Theta(\mathbf{K}^n)$ all vertices of which correspond to subspaces of V.

Let $\text{Mat}(\mathbf{K}^n)$ be the space of all linear endomorphisms of $\mathbf{K}^n$, and $\text{Det}(\mathbf{K}^n)$ the set of endomorphisms that are not isomorphic. The natural resolution of $\text{Det}(\mathbf{K}^n)$ is constructed as a subspace in $V(\mathbf{K}^n) \times \text{Mat}(\mathbf{K}^n)$ consisting of all pairs (v, A) such that $v \in V(\ker A)$. The splittings (9)–(12) appear automatically in the study of the corresponding spectral sequence.

Here is one more result that also follows from this calculation.

PROPOSITION (see [V 91]). The link of the complex $C\Theta(\mathbf{K}^n)$ (i.e. its intersection with the subjoin $G_1(\mathbf{K}^n) * \cdots * G_{n-1}(\mathbf{K}^n)$ of (13)) is homeomorphic to a sphere of dimension $\binom{n}{2} \dim_{\mathbf{R}} \mathbf{K} + n - 2$.

REMARK. The construction of simplicial resolutions from Sections 2 and 3 also can be reformulated in terms of a certain order complex, the order complex of finite subsets of the issue manifold. (More precisely, if we describe the geometrical resolution in terms of the order complexes, then for every discriminant function f with k singular points the simplices of this order complex that lie over the point f in the space of the resolution form the first barycentric subdivision of the $(k-1)$-dimensional simplex that lies over f in the construction from Section 2.)

11 A homological invariant of rings

Any commutative ring A defines a partially ordered set: the set of its proper ideals. The cohomology ring of the order complex of this poset is obviously an invariant of A.

If A is a finite-dimensional $\mathbf{R}$- or $\mathbf{C}$-algebra, then besides the standard "discrete" topologization the order complex can be supplied with the "Hilbert-scheme" topology. Namely, we take a disjoint union of all Grassmann manifolds $G_1(A), \ldots, G_{\dim A-1}(A)$ of vector subspaces in A, and consider the join $G_1(A) * \ldots * G_{\dim A-1}(A)$ as the union of all simplices the vertices of which correspond to the points of several different Grassmannians. Then we consider only such simplices, all vertices of which correspond to ideals of A that are incident to one another.

The union of these simplices is obviously bijective to the usual order complex of the poset of ideals, but, being supplied with the topology induced from our join, differs from it as a topological space. All the topological characteristics of this space (in particular, the structure of the cohomology ring) are again invariants of A.

A similar invariant also exists if A is a finitely generated graded module over $F[x_0, \ldots, x_n]$. In this case a similar topologization of the order complex of homogeneous ideals can be constructed using the filtration in the space of ideals defined by the lexicographical order of the Hilbert polynomials of corresponding quotient rings, see [Mumford] (first we compare the degrees of the polynomials, if they coincide, then the coefficients of the leading monomials, ...).

The obtained order complexes are a natural means of constructing the resolutions of certain discriminant subsets of the rings: say, in the problem of the rigid isotopy classification of real curves and surfaces it is natural to define a discriminant as the set of polynomials distinguishing singular varieties and to construct a resolution of it as a subset in the product of our order complex and the space of polynomials.

PROBLEM. Let A be a quotient ring $F[x_1, \ldots, x_n]/\{f_1, \ldots, f_m\}$. Is it possible to express the Betti numbers of the corresponding order complex in the terms of some standard characteristics of the polynomials f_i (for example, their Newton polyhedra if the polynomials are "generic")?

Acknowledgment. While preparing this article I was partially supported by grant MQO 000 from the International Science Foundation.

References

[Adams] J.F. Adams, *On the cobar construction*, Colloque de topologie algebrique (Louvain, 1956). – Paris a.o., George Thone, 1957, 81–87.

[Aikardi] F. Aikardi, *Tree-like curves*, in: Singularities and Bifurcations (V. I. Arnold, ed.), Adv. in Sov. Math., **21**, Amer. Math. Soc., Providence, RI, 1994, 1–31.

[Arnold 69] V.I. Arnold, *The cohomology ring of the group of colored braids*, Mat. Zametki (Math. Notes) 1969, **5:2**, 227–231.

[Arnold 70] V.I. Arnold, *On some topological invariants of algebraic functions*, Trudy Moskov. Mat. Obshch. 1970, **21**, 27–46, Engl. transl.: Trans. Moscow Math. Soc. 1970, **21**, 30–52.

[Arnold 76] V.I. Arnold, *Some unsolved problems in the singularity theory*, Trudy Sem. S.L. Soboleva, **1**, Novosibirsk, 1976, 5–15, Engl. transl. in: Singularities, Part I, Proc. Sympos. Pure Math. **40** (Arcata, California), AMS, Providence, RI, 1983.

[Arnold 79] V.I. Arnold, *On some problems in singularity theory*, in: Geometry and Analysis, V.K. Patodi Memorial Volume, Bombay, 1979, 1–10.

[Arnold 89] V.I. Arnold, *The spaces of functions with mild singularities*, Functional Anal. Appl. 1989, **23:3**.

[Arnold 92] V.I. Arnold, *Vassiliev's theory of knots and discriminants*, Proceedings of the First European Congress of Mathematicians, Paris, 1992, Birkhäuser, Basel and Boston, 1994.

[Arnold 94] V.I. Arnold, *Plane curves, their invariants, perestroikas and classifications*, in: Singularities and Bifurcations (V. I. Arnold, ed.), Adv. in Sov. Math. **21**, Amer. Math. Soc., Providence, RI, 1994, 33–91.

[AVGL] V.I. Arnold, V.A. Vassiliev, V.V. Goryunov, and O.V. Lyashko, *Singularities I*, Moscow, VINITI, 1–256. Engl. transl.: Encycl. Math. Sci., **6**, Springer, Berlin and New York, 1993.

[Bar-Natan 92] D. Bar-Natan, On the Vassiliev knot invariants, *Topology*, to appear.

[Bar-Natan 93] D. Bar-Natan, *Vassiliev homotopy string link invariants*, preprint, 1993.

[Birman] Joan S. Birman, *New points of view in knot and link theory*, BAMS, 1993, 253–287.

[Birman & Lin] Joan S. Birman and X.-S. Lin, *Knot polynomials and Vassiliev's invariants*, Invent. Math. 1993, **111**, 225–270.

[Björner & Welker] A. Björner and V. Welker, *The homology of "k-equal" manifolds and related partition lattices*, Report No. 39, 1991/1992, Institut Mittag-Leffler, 1992.

[Brieskorn] E. Brieskorn, *Sur les groupes de tresses (d'apres V.I. Arnold)*, Sém. Bourbaki, 1971/72, No. 401, Lecture Notes in Math. **317**, Springer, Berlin and New York, 1973, 21–44.

[Cartier] P. Cartier, *Construction combinatoire des invariants de Vassiliev-Kontsevich des noeuds*, C.R. Acad. Sci. Paris **316** Série I 1993, 1205–1210.

[Chen] K.T. Chen, *Iterated path integrals*, Bull. of Amer. Math. Soc., 1977, **83:5**, 831–879.

[CCMM] F.R. Cohen, R.L. Cohen, B.M. Mann, and R.J. Milgram, *The topology of rational functions and divisors of surfaces*, Acta Math. 1991, **166**, 163–221.

[CLM] F.R. Cohen, T.J. Lada, and J.P. May, *The homology of iterated loop spaces*, Lecture Notes in Math. **533**, Springer, Berlin and New York, 1976.

[Fenn & Taylor] R. Fenn and P. Taylor, *Introducing doodles*, in: Topology of low-dimensional manifolds (R. Fenn, ed.), Lecture Notes in Math., **722**, Springer, Berlin and New York, 1977, 37–43.

[FHYLMO] P. Freyd, D. Yetter, J. Hoste, W. Lickorish, K. Millet, and A. Ocneanu, *A new polynomial invariant for knots and links*, Bull. Amer. Math. Soc. 1985, **12**, 183–193.

[Goresky & MacPherson] M. Goresky and R. MacPherson, Stratified Morse Theory, Springer, Berlin a.o., 1986.

[Gromov] M. Gromov, Partial Differential Relations, Springer, Berlin a.o., 1986.

[Igusa] K. Igusa, *Higher singularities of smooth functions are unnecessary*, Ann. of Math. (2) 1984, **119**, 1–58.

[Jones] V.F.R. Jones, *Hecke algebra representations of braid groups and link polynomials*, Ann. of Math. (2) 1987, **126**, 335–388.

[Kontsevich] M. Kontsevich, *Vassiliev's knot invariants*, Adv. in Sov. Math., **16**, part 2, Amer. Math. Soc., Providence, RI, 1993, 137–150.

[Lin] X.-S. Lin, *Milnor link invariants are all of finite type*, Columbia Univ., preprint, 1992.

[Merkov] A.B. Merkov, in: Singularities and Bifurcations (V. I. Arnold, ed.), Adv. in Sov. Math., **21**, Amer. Math. Soc., Providence, RI, 1994, 199–211.

[Milgram] R.J. Milgram, *Iterated loop spaces*, Ann. of Math. (2) 1966, **84**, 386–403.

[Milnor 57] J. Milnor, *Isotopy of links*, in: Algebraic geometry and topology, Princeton Univ. Press, Princeton, NJ, 1957, 280–306.

[Milnor 63] J. Milnor, Morse theory, Ann. of Math. Stud., Princeton Univ. Press, Princeton NJ, 1963.

[Mumford] D. Mumford, Algebraic Geometry I. Complex Projective Varieties, Springer-Verlag, Berlin and New York, 1976.

[Piunikhin] S. Piunikhin, *Combinatorial expression for universal Vassiliev link invariant*, Harvard Univ., preprint, March 1993.

[Segal 73] G.B. Segal, *Configuration spaces and iterated loop spaces*, Invent. Math. 1973, **21:3**, 213–221.

[Segal 79] G.B. Segal, *Topology of spaces of rational functions*, Acta Math. 1979, **143**, 39–72.

[Smale] S. Smale, *The classification of immersions of spheres in Euclidean space*, Ann. of Math. (2), 1959, **69:2**, 327–344.

[Snaith] V. Snaith, *A stable decomposition of* $\Omega^n S^n X$, J. London Math. Soc. (2) 1974, 577–583.

[V 87] V.A. Vassiliev, *Stable cohomology of complements of the discriminants of the singularities of smooth functions*, Russian Math. Surveys (Uspekhi) 1987, **42:2**, 219–220 and Current Problems of Math., **33**, Moscow, VINITI, 1988, 3–29; Engl. transl.: J. Soviet Math. 1990, **52:4**, 3217–3230.

[V 89] V.A. Vassiliev, *Topology of spaces of functions without complicated singularities*, Functional Anal. Appl., 1989, **23:4**, 24–36.

[V 90] V.A. Vassiliev, *Topology of complements to discriminants and loop spaces*, in: Theory of Singularities and its Applications (V.I. Arnold ed.), Adv. in Sov. Math., **1**, Amer. Math. Soc., Providence, RI, 1990, 9–21.

[V 90′] V.A. Vassiliev, *Cohomology of knot spaces*, ibid, 23–69.

[V 91] V.A. Vassiliev, *A geometric realization of the homologies of classical Lie groups, and complexes, S-dual to flag manifolds*, Algebra i Analiz **3**, 1991, no. 4, 108–115; Engl. transl. in: St. Petersburg Math. J., 1992, **3**, no. 4, 809–815.

[V 91′] V.A. Vassiliev, *Stable homotopy type of the complement to affine plane arrangement*, preprint, 1991.

[V 92 & 94] V.A. Vassiliev, *Complements of discriminants of smooth maps: Topology and applications.* Revised ed., Transl. of Math. Monographs, **98**, Amer. Math. Soc., Providence RI, 1994.

[V 93] V.A. Vassiliev, *Complexes of connected graphs*, in: Proc. of I.M.Gelfand's seminar, Birkhäuser, Basel, 1993, 223–235.

[V 94] V.A. Vassiliev, *Invariants of ornaments*, in: Singularities and Bifurcations (V. I. Arnold, ed.), Adv. in Sov. Math., **21**, AMS, Providence, RI, 1994, 225–262.

[V 95] V.A. Vassiliev, Ramified integrals, singularities and lacunas, Kluwer Academic Publishers, Dordrecht a.o., 1994, 304 pages.

[Vogel] P. Vogel, Invariants de Vassiliev des noeuds, Sem. Bourbaki, 45-eme annee, 1992–1993, **769**, March 1993.

[Ziegler & Živaljević] G.M. Ziegler and R.T. Živaljević, Homotopy type of arrangements via diagrams of spaces, Report No. 10 (1991/1992), Inst. Mittag-Leffler, 1991.

Free Probability Theory: Random Matrices and von Neumann Algebras

DAN VOICULESCU*

Department of Mathematics, University of California
Berkeley, CA 94720, USA

0 Introduction

Independence in usual noncommutative probability theory (or in quantum physics) is based on tensor products. This lecture is about what happens if tensor products are replaced by free products. The theory one obtains is highly noncommutative: freely independent random variables do not commute in general. Also, at the level of groups, this means instead of $\mathbb{Z}^n$ we will consider the noncommutative free group $F(n) = \mathbb{Z} * \cdots * \mathbb{Z}$ or, looking at the Cayley graphs, a lattice is replaced by a homogeneous tree.

Three different models of free probability theory are provided by convolution operators on free groups, creation and annihilation operators on the Fock space of Boltzmann statistics, and random matrices in the large N limit.

Important problems on the von Neumann algebras of free groups have been solved using free probability techniques, and surprisingly the random matrix model has played a major role in this. In another direction there is a free entropy quantity that goes with free independence.

Concerning connections with other fields we should signal that combinatorial objects (noncrossing partitions, random permutations) have appeared in free probability theory and that random matrices are used in physics.

We have divided our survey into five sections:

(1) Free random variables
(2) Free harmonic analysis
(3) Asymptotic models
(4) Applications to operator algebras
(5) Free entropy.

At the end, an Appendix explains a few basic notions in operator algebras for the reader not conversant in C^*- and W^*-algebras.

1 Free Random Variables

For noncommutative probability spaces, the usual prescription applies: replace the functions on a space by elements of a (possibly noncommutative) algebra. Thus:

*) Work supported in part by Grant from the National Science Foundation

Proceedings of the International Congress
of Mathematicians, Zürich, Switzerland 1994

1.1 Definition. A noncommutative probability space is a unital algebra $\mathcal{A}$ over $\mathbb{C}$ endowed with a linear functional $\phi : \mathcal{A} \to \mathbb{C}$ such that $\phi(1) = 1$. The elements of $\mathcal{A}$ are called random variables and the distribution of a random variable $a \in \mathcal{A}$ is the map $\mu_a : \mathbb{C}[X] \to \mathbb{C}$ given by $\mu_a(P(X)) = \phi(P(a))$.

The above definition is only an algebraic caricature, sufficient for discussing questions such as independence. (Positivity and almost everywhere convergence require additional structure: $\mathcal{A}$ a C^*-algebra and ϕ a state or, even more, a von Neumann algebra with a normal state. In the C^*-case, if $a = a^*$, the distribution functional μ_a extends to a compactly supported probability measure on $\mathbb{R}$.)

Usually independence is modeled on tensor products. The idea of free probability theory is to replace tensor products by free products.

1.2 Definition. A family of subalgebras $1 \in \mathcal{A}_i \subset \mathcal{A}$ $(i \in I)$ in a noncommutative probability space $(\mathcal{A}, \phi)$ is called a *free family of subalgebras* if

$$\phi(a_1 \ldots a_n) = 0$$

whenever $a_j \in \mathcal{A}_{i(j)}$ with $i(j) \neq i(j+1)$ $(1 \leq j < n)$ and $\phi(a_j) = 0$ $(1 \leq j \leq n)$. Families of subsets or of random variables in (A, ϕ) are free if the generated unital subalgebras are free.

As for usual independence, if the free family of subalgebras $\mathcal{A}_i$ $(i \in I)$ generates $\mathcal{A}$, then ϕ is completely determined by the restrictions $\phi|\mathcal{A}_i$ $(i \in I)$. What distinguishes freeness and independence is that free random variables are highly noncommuting.

1.3 Examples. (a) Let $G = \mathop{*}_{i \in I} G_i$ be a free product of groups and let λ be the left regular representation of G on $\ell^2(G)$. Let further W and W_i $(i \in I)$ be the weakly closed subalgebras generated by $\lambda(G)$ and $\lambda(G_i)$ respectively. The von Neumann trace $\tau : W \to \mathbb{C}$ is given by $\tau(T) = \langle T\delta_e, \delta_e \rangle$ where δ_g $(g \in G)$ is the canonical basis of $\ell^2(G)$. Then the W_i $(i \in I)$ are free in (W, τ).

(b) Let $\mathcal{H}$ be a complex Hilbert space and let $\mathcal{TH} = \oplus_{k \geq 0} \mathcal{H}^{\otimes k}$ where $\mathcal{H}^{\otimes 0} = \mathbb{C}1$. Let further $\ell(h)\xi = h \otimes \xi$ be the creation operators on the full Fock space $\mathcal{TH}$ and let $\varepsilon(X) = \langle X1, 1 \rangle$ be the vacuum expectation. If $\mathcal{H}_i$ $(i \in I)$ are mutually orthogonal subspaces of $\mathcal{H}$, then the generated subalgebras $C^*(\ell(\mathcal{H}_i))$ $(i \in I)$ are free in $(C^*(\ell(\mathcal{H})), \varepsilon)$.

1.4 The analogue of the Gaussian law in the free context is the semicircle law, i.e. probability measures on $\mathbb{R}$ with densities having a semiellipse graph: 0 if $|t-a| > R$ and equal to $2\pi^{-1}R^{-2}(R^2 - (t-a)^2)^{\frac{1}{2}}$ if $|t-a| \leq R$. Indeed, we have the following

Free Central Limit Theorem [32]. *If $(f_n)_{n \in \mathbb{N}}$ is a free family of random variables in (A, ϕ) so that $\phi(f_n) = 0$ $(n \in \mathbb{N})$,*

$$\lim_{N \to \infty} N^{-1} \sum_{1 \leq n \leq N} \phi(f_n^2) = 4^{-1}R^2 > 0$$

$$\sup_{n \in \mathbb{N}} |\phi(f_n^k)| < \infty \quad \textit{for all } k \in \mathbb{N}$$

then, if

$$S_N = N^{-\frac{1}{2}} \sum_{1 \le k \le N} f_k$$

the distributions μ_{S_N} *converge pointwise on* $\mathbb{C}[X]$ *to the semicircle law with density* $2\pi^{-1}R^{-2}\,\mathrm{Re}(R^2 - t^2)^{\frac{1}{2}}$.

Convergence in the previous theorem is in a very weak sense. Actually, in the free context convergence to the central limit is much stronger than in usual probability theory (see [1] concerning this superconvergence).

1.5 Roughly speaking, the Gaussian process over a real Hilbert $\mathcal{H}$ space is the process indexed by $\mathcal{H}$, the random variable corresponding to h being $\langle\,\cdot\,, h\rangle : \mathcal{H} \to \mathbb{R}$, when $\mathcal{H}$ is endowed with the Gaussian measure. This is part of the Gaussian functor of second quantization, which takes real Hilbert spaces and contractions to commutative von Neumann algebras with specified trace state and trace- and unit-preserving completely positive maps. The canonical anticommutation relations provide a fermionic analogue. We have found a third such functor, which is the free analogue of these.

THE FREE ANALOGUE OF THE GAUSSIAN FUNCTOR [32]. *If* $\mathcal{H}$ *is a real Hilbert space, let* $\mathcal{H}_{\mathbb{C}}$ *be its complexification and let* $\mathcal{T}\mathcal{H}_{\mathbb{C}}$ *and* $\ell(h)$ *be as in 1.3(b). Let further* $s(h) = 1/2(\ell(h) + \ell(h)^*)$.

(i) *The von Neumann algebra* $\Phi(\mathcal{H})$ *generated by* $s(\mathcal{H})$ *is isomorphic to the* II_1 *factor of a free group on* $\dim \mathcal{H}$ *generators (if* $\dim H > 1$*) and the trace state is given by the vacuum expectation* $\langle \cdot 1, 1\rangle$.

(ii) *If* $T : \mathcal{H}_1 \to \mathcal{H}_2$ *is a contraction there is a unique completely positive map* $\Phi(T) : \Phi(\mathcal{H}_1) \to \Phi(\mathcal{H}_2)$ *such that*

$$(\Phi(T))(X)1 = \mathcal{T}(T_{\mathbb{C}})(X1)\ .$$

The map $\Phi(T)$ *is trace and unit preserving.*

(iii) *If* $(\mathcal{H}_i)_{i\in I}$ *is a family of pairwise orthogonal subspaces in* $\mathcal{H}$ *and* $v(i)$ *are the corresponding inclusions, then* $(\Phi(v(i)))(\Phi(\mathcal{H}_i))_{i\in I}$ *is free in* $\Phi(\mathcal{H})$.

(iv) *Orthogonal vectors correspond to free variables via the map* $s : \mathcal{H} \to \Phi(\mathcal{H})$ *and the distribution of* $s(h)$ *is a centered semicircle law.*

Gaussian processes are obtained by mapping the index set of the process into a Hilbert space and then composing with the Gaussian process over the Hilbert space. Composing with the free analogue (i.e. with $s : \mathcal{H} \to \Phi(\mathcal{H})$) one gets the free analogue of Gaussian processes. Free increments correspond to the requirement of orthogonal increments for the map into the Hilbert space. For instance, Brownian motion corresponds to $\mathcal{H} = L^2(0,\infty)$ and the Hilbert space curve $[0,\infty) \ni t \to \chi_{[0,t)} \in L^2(0,\infty)$. The free analogue of Brownian motion is then obtained by taking $[0,\infty) \ni t \to s(\chi_{[0,t)})$, a possibility used in [28].

1.6 Generalizations of various parts of the free probability context have been studied. We would like to mention here the following two.

(a) Free products with amalgamation over an algebra B [32], [36]. One replaces the complex field $\mathbb{C}$ by an algebra B over $\mathbb{C}$. The noncommutative probability space A is then an algebra containing B as a subalgebra and the expectation ϕ is a $B-B$-bimodule projection $\phi : A \to B$. There is a corresponding definition of B-freeness and the corresponding operator-algebra context has also been studied.

(b) Deformed Cuntz relations [6]. A natural model in which free random variables arise is provided by the creation operators $\ell(h)$ (Example 1.3(b)). They satisfy the Cuntz relations $\ell(h)^*\ell(k) = \langle h, k\rangle I$. A deformation of these relations is

$$\ell(h)^*\ell(k) - \mu\ell(k)\ell(h)^* = \langle h, k\rangle I$$

$\mu \in [-1, 1]$. This provides an interpolation between the three cases $\mu = -1, 0, 1$, which correspond, respectively, to the fermionic, free, and bosonic creation operators.

1.7 Free stochastic integrations. Stochastic integration in the free case has been studied in papers by R. Speicher, K. R. Parthasarathy, B. K. Sinha, F. Fagnola, L. Accardi, and B. Kummerer.

2 Free Harmonic Analysis

2.1 Free convolution. The distribution of the sum of two independent random variables is the (additive) convolution of their distributions. By analogy on $\Sigma = \{f : \mathbb{C}[X] \to \mathbb{C} | f \text{ linear}, f(1) = 1\}$, there are operations $\boxplus$ and $\boxtimes$ called, respectively, additive and multiplicative free convolution so that if a, b are free random variables in some noncommutative probability space then $\mu_{a+b} = \mu_a \boxplus \mu_b$ and $\mu_{ab} = \mu_a \boxtimes \mu_b$ [32]. Because this does not depend on the concrete realizations of the variables with distribution μ_a, μ_b and because the sum of self-adjoint operators is self-adjoint, the product of the unitaries unitary, etc., we have that $\boxplus$ extends to an operation on compactly supported probability measures on $\mathbb{R}$, while $\boxtimes$ defines operations on the compactly supported probability measures on $\mathbb{R}^\times$, $\mathbb{R}_+$, and $\mathbb{T}$. Clearly $\boxplus$ is commutative and actually $\boxtimes$ is also commutative. Moreover, [2] $\boxplus$ extends to an operation on all probability measures on $\mathbb{R}$, while $\boxtimes$ extends to an operation on probability measures on $\mathbb{R}_+$ — which correspond to operations on "unbounded" random variables.

2.2 The linearizing transforms. The computation of free convolution can be done using a linearizing transform. This is like computing the usual convolution of two probability measures using the logarithm of the Fourier transform (which linearizes convolution).

THEOREM [33]. *If $\mu \in \Sigma$ let $G_\mu(z) = z^{-1} + \sum_{n\geq 1} \mu(X^n)z^{-n-1}$ and let $K_\mu(z) \in z^{-1} + \mathbb{C}[[z]]$ be such that $G_\mu(K_\mu(z)) = z$. Then $R_\mu(z) = K_\mu(z) - z^{-1}$ has the property that $R_{\mu_3} = R_{\mu_1} + R_{\mu_2}$ if $\mu_3 = \mu_1 \boxplus \mu_2$.*

If μ is a compactly supported probability measure then G_μ is its Cauchy transform and R_μ is analytic near 0. The linearization result also extends to the case of unbounded supports using analytic functions in angular domains ([2], an intermediate generalization is given in [16]).

A similar linearization result holds for the multiplicative free convolution $\boxtimes$ [34] and also has an analytic function extension to the case of unbounded supports [2].

2.3 *F*-Infinitely divisible laws. A probability measure μ on $\mathbb{R}$ is called F-infinitely divisible if for every $n \in \mathbb{N}$ there is $\mu_{1/n}$ so that $\underbrace{\mu_{1/n} \boxplus \cdots \boxplus \mu_{1/n}}_{n \text{ times}} = \mu$. A family of probability measures $(\mu_t)_{t\geq 0}$ on $\mathbb{R}$ is an F-convolution semigroup if $\mu_{t+s} = \mu_t \boxplus \mu_s$ and μ_t depends continuously on t. There is bijection between F-infinitely divisible measures and F-convolution semigroups. Stationary processes with free increments naturally lead to these definitions.

If $(\mu_t)_{t\geq 0}$ is an F-convolution semigroup, then the Cauchy transforms $G(t,z) = G_{\mu_t \boxplus \gamma}(z)$ for some probability measure γ on $\mathbb{R}$ satisfy the complex quasilinear equation

$$\frac{\partial G}{\partial t} + \frac{\partial G}{\partial z}\,\phi(G) = 0$$

where $\phi(z) = R_{\mu_1}(z)$. In particular, the complex Burger equation

$$\frac{\partial G}{\partial t} + \alpha G\,\frac{\partial G}{\partial z} = 0$$

is the analogue of the heat equation, as $R_\mu(z) = \alpha z$ if μ is a centered semicircle law (which is the free analogue of the Gauss law).

THEOREM. *μ is F-infinitely divisible iff R_μ has an analytic extension to $\{z \in \mathbb{C} | \mathrm{Im}\ z < 0\}$ with values in $\{z \in \mathbb{C} | \mathrm{Im}\ z \leq 0\}$.*

(The case of compactly supported measures is given in [33], the intermediate case of measures with finite variance in [16], and the result in full generality in [2].)

The condition on the imaginary part of $R_\mu(z)$ implies the existence of an integral representation, which makes the above theorem an analogue of the Levy-Khintchine theorem. The analogy goes even further when we remark that the free Poisson distribution defined by

$$\lim_{n\to\infty} \left((1 - \frac{a}{n})\delta_0 + \frac{a}{n}\,\delta_b \right)^{\boxplus n}$$

has the R-function $R(z) = ab(1 - bz)^{-1}$. The free Poisson measure is given by

$$\mu = \begin{cases} (1-a)\delta_0 + \nu & \text{if } 0 \leq a \leq 1 \\ \nu & \text{if } a > 1 \end{cases}$$

where ν has support in $[b(1-\sqrt{a})^2, b(1+\sqrt{a})^2]$ and density $(2\pi bt)^{-1}(4ab^2 - (t - b(1+a))^2)^{\frac{1}{2}}$.

2.4 *F*-Stable laws [2]. Replacing usual convolution by free convolution in the definition of stable laws one defines F-stable laws. F-stable laws were classified in [2], up to taking certain linear combinations, the main types are given by

(i) $R(z) = a, \quad a \in \mathbb{C}, \quad \operatorname{Im} a \leq 0$
(ii) $R(z) = z^{\alpha-1}\operatorname{sign}(\alpha - 1), \quad \alpha \in (0,1) \cup (1,2)$
(iii) $R(z) = \log z$.

As for infinitely divisible laws this runs essentially parallel to the classical case.

The usual Cauchy distribution and the free Cauchy distribution, given by $R(z) = -i$ coincide.

2.5 Multiplicative *F*-infinite divisibility. Infinitely divisible probability measures with respect to the operation $\boxtimes$ on $\mathbb{T}$ were classified in [1] and on $\mathbb{R}_+$ in [2].

Note that the generating function for the measure μ, which is the free analogue of the multiplication Gaussian distribution (i.e. of the log-normal distribution) $\psi(z) = \sum_{n\geq 1} \mu(X^n)z^n$, can be expressed using the generating series for rooted labelled trees

$$\sum_{n\geq 1} \frac{n^{n-1}}{n!} z^n .$$

2.6 Noncrossing partitions. Because the map $\mu \to R_\mu$ linearizes the free convolution it follows that if $R_\mu(z) = \sum_{n\geq 0} R_{n+1}(\mu)z^n$ the coefficients $R_{n+1}(\mu)$ are polynomials in the moments of μ and $R_{n+1}(\mu_1 \boxplus \mu_2) = R_{n+1}(\mu_1) + R_{n+1}(\mu_2)$. The $R_{n+1}(\mu)$ are the free analogues of the cumulants of μ. In [29] it was shown that the formulae giving the free cumulants are entirely analogous to those for the usual cumulants if we replace the lattice of all partitions of $\{1, \ldots, n\}$ by the lattice of noncrossing partitions (i.e. partitions with crossing pairs $\{a, c\}, \{b, c\}$ where $a < b < c < d$ do not lie in different sets of the partition). There are more general such formulae based on noncrossing partitions [29], [30], [20], [21] which characterize freeness of sets of random variables. It seems that the passage from all partitions to the noncrossing partitions is the combinatorial aspect of going from usual independence to free independence.

2.7 Generalizations of the free harmonic analysis. (a) ***B*-free convolution**. Free convolution and its linearization were extended to the context of B-freeness in [36]. A combinatorial approach based on noncrossing partitions to linearization and to the classification of infinitely divisible distributions (with moments of all orders) in the B-free context was developed in [30].

Multiplicative free convolution is no longer commutative for general B and there are nonlinear systems of differential equations that replace linearization [36].

(b) Deformed linearization maps. The linearization map involves certain canonical forms of random variables in creation and annihilation operators on the full Fock-space. Passage to the deformed Cuntz-relation was used to construct deformed free convolution [6] and its linearization map [21].

3 Asymptotic Models

3.1 Gaussian random matrices. The semicircle law that appears in the free central limit theorem also occurs in Wigner's work on the asymptotic distribution of eigenvalues of large Gaussian random matrices [41], [42]. The explanation we found [38] for this coincidence is that large Gaussian random matrices with independent entries give rise asymptotically to free random variables. Moreover, this asymptotic model is the bridge connecting classical and free probability theory. Indeed, independence of matrix-valued random variables is transformed into free independence of the corresponding noncommutative random variables (asymptotically).

The precise statements are as follows.

Let (A_n, ϕ_n) and (A_∞, ϕ_∞) be noncommutative probability spaces and let $(X_{n,i})_{i\in I}$, $n \in \mathbb{N} \cup \{\infty\}$ be in A_n. Then $(X_{n,i})_{i\in I}$ *converges in distribution to* $(X_{\infty,i})_{i\in I}$ if

$$\lim_{n\to\infty} \phi_n(P((X_{n,i})_{i\in I})) \longrightarrow \phi_\infty(P(X_{\infty,i})_{i\in I})$$

for every noncommutative polynomial P in indeterminates indexed by I. In particular the $(X_{n,i})_{i\in I}$ are *asymptotically free* if they converge in distribution to a free family.

A family $(x_i)_{i\in I}$ is called *semicircular* if the x_i have equal centered semicircle distributions and are free. In a C^*-probability space we require in addition that $x_i = x_i^*$.

For asymptotics of random matrices the appropriate (A_n, ϕ_n) are $A_n = \bigcap_{1\le p<\infty} L^p(\Omega, M_n)$ where $(\Omega, d\sigma)$ is some standard probability measure space and

$$\phi_n(X) = \frac{1}{n}\int_\Omega \operatorname{Tr} X(\omega)\, d\sigma(\omega)\ .$$

THEOREM [38]. *Let $Y(\iota, n) = (a(i,j;n,\iota)_{1\le i,j\le n}) \in A_n$ be real random matrices $(\iota \in I)$. Assume $a(i,j;n,\iota) = a(j,i;n,\iota)$ and $\{a(i,j;n,\iota) | 1 \le i \le j \le n,\ \iota \in I\}$ is a family of independent Gaussian $(0, 1/n)$ random variables. Let further $D_n \in A_n$ be a constant diagonal random matrix having a limit distribution as $n \to \infty$. Then $\{Y(\iota,n) | \iota \in I\} \cup \{D_n\}$ is asymptotically free as $n \to \infty$ and $\{Y(\iota,n) | \iota \in I\}$ converges in distribution to a semicircular family.*

3.2 Unitary random matrices. Using polar decomposition (i.e. noncommutative functional calculus) and results of Gromov-Milman on isoperimetric inequalities yields stronger versions of the preceding result for unitary random matrices.

THEOREM [38]. *Given $\varepsilon > 0$ and a nontrivial element*

$$g = g_{i_1}^{k_1} g_{i_2}^{k_2} \dots g_{i_m}^{k_m}$$

$(m \ge 1,\ k_j \ne 0,\ i_s \ne i_{s+1})$ of the free group on p generators, let

$$\Omega_n(g) = \{(u_1, \dots, u_p) \in (U(n))^p | \tau_n(u_{i_1}^{k_1} \dots u_{i_m}^{k_m})| < \varepsilon\}$$

where $\tau_n = n^{-1}\mathrm{Tr}$ *is the normalized trace on* $(n \times n)$*-matrices. Then we have*

$$\lim_{n\to\infty} \mu_n(\Omega_n(g)) = 1$$

where μ_n *is the normalized Haar measure on* $(U(n))^p$.

The preceding theorem has a more general form where a constant diagonal unitary also appears. This implies asymptotic freeness results for random matrices, which as matrix-valued variables are independent and are distributed according to the invariant measures of unitary orbits of self-adjoint matrices (this includes random projections, etc.).

3.3 Further results. Further extensions of the preceding theorems include results for real symmetric and antisymmetric Gaussian random matrices [38], for matrices with fermionic entries [38], and matrices with independent non-Gaussian entries together with a finite-dimensional constant algebra [10]. A generalization of the random matrix result involving representations has been obtained in [4].

In a different direction in [18] freeness results were obtained for independent uniformly distributed random permutation matrices. (Further combinatorial results for words in independent random permutations related to this are given in [19].)

3.4 Applications. Many of the known asymptotic distribution of eigenvalue results for random matrices can be recovered from the asymptotic freeness results. Indeed, many of these are obtained via noncommutative functional calculus from random matrices like those in the preceding theorems. Hence the limit distribution of eigenvalues in the large n limit is the same as the distribution of an element in a certain algebra generated by free random variables, the distribution of which can be computed, in certain cases via free convolution operations.

Related to the asymptotic freeness results for random matrices, it was recently discovered in [5] that free convolution occurs asymptotically in the decomposition into irreducible representations of tensor products of representations of $U(n)$.

Last but not least there are applications to the II_1 factor of free groups, which we shall survey in the next section.

4 Applications to Operator Algebras

Free probability theory and especially asymptotic random matrix realization have led to a surge of new results on the von Neumann algebras of free groups. These recent results will be surveyed here, preceded by some background on II_1-factors.

4.1 II_1**-Factors of discrete groups.** A factor is a von Neumann algebra M with trivial center $Z(M) = \mathbb{C}I$. The factor M is type II_1 if it has a trace-state $\tau : M \to \mathbb{C}$ (which is then unique) and is infinite dimensional. As P ranges over projections in M, $\tau(P)$ takes all values in $[0,1]$, which corresponds to a geometry with subspaces having dimensions in $[0,1]$.

$L(G)$, the von Neumann algebra of the left regular representation $\lambda(G)$, is a II_1-factor iff G has infinite conjugacy classes (i.c.c.). The $L(G)$'s are a rich source of II_1-factors (G will be assumed countable in what follows).

By a deep theorem of Connes [7] all $L(G)$ with amenable G are isomorphic — the hyperfinite II_1-factor. This is the "best" among all II_1-factors; it has a large automorphism group and good finite-dimensional approximation properties, and there are approximately central elements (property Γ of von Neumann). The remarkable properties of the hyperfinite II_1-factor made an in-depth study of its subfactors possible.

At the other extreme are the $L(G)$'s for G with property T of Kazhdan [8], [9]. These II_1-factors have rigidity properties, few automorphisms, no approximation properties, and no approximate center (non-Γ). It is conjectured (by Connes) that isomorphisms among these $L(G)$'s imply isomorphisms of the corresponding groups.

The free group factors $L(F_n)$ ($n = 2, 3, \ldots, \infty$) have intermediate properties: some approximation properties (compact instead of finite-rank) and some properties towards rigidity (non-Γ). Like the hyperfinite II_1-factor, which is related to the fermionic context of the canonical anticommutation relations, the free group factors are related to the free analogue of the Gaussian functor. This could mean that the free group factors are the "best" among the "bad" (i.e. non-Γ) II_1-factors.

4.2 The free probability technique [37] Semicircular and circular systems are the free analogues of, respectively, independent real and complex Gaussian random variables. They provide convenient sets of generators for free group factors. The asymptotic random matrix models based on Gaussian random matrices are the source for many of the properties of circular and semicircular systems.

A system of self-adjoint random variables $(s_i)_{i\in I}$ is semicircular if the s_i's are free and have identical centered semicircle distributions. Similarly, $(c_i)_{i\in I}$ is circular if $(\mathrm{Re}\ c_i)_{i\in I} \cup (\mathrm{Im}\ c_i)_{i\in I}$ is semicircular.

A block of a Gaussian random matrix, being a matrix of the same kind, implies that if $p = p^* = p^2$ is free with respect to a semicircular system $(s_i)_{i\in I}$ then the compression $(ps_ip)_{i\in I}$ is semicircular in $(pAp, \phi(p)^{-1}\phi(\cdot))$. In the polar decomposition $c = u|c|$ of a circular element, u and $|c|$ are free. Cutting and pasting blocks of Gaussian random matrices have analogues for circular and semicircular systems. For instance, if $(c_{i,j;s})_{1\le i,j\le n, s\in S}$ is circular, then the matrices $X_s = \sum_{1\le i,j\le n} c_{i,j;s} \otimes e_{ij}$, $s \in S$, form a circular system.

I introduced this free probability technique and used it to obtain results on free group factors in [37]; the applications to free group factors were subsequently carried much further by Radulescu and Dykema.

4.3 The fundamental group $\mathcal{F}(L(F(\infty)))$. If M is a II_1-factor and $p = p^2 \in M$ the isomorphism class of pMp depends only on $\lambda = \tau(p)$ and is denoted M_λ. The fundamental group $\mathcal{F}(M)$ [17] consists of those $\lambda \in (0, 1]$ such that $M_\lambda \sim M$ and their inverses. For the hyperfinite II_1-factor R, $\mathcal{F}(R) = (0, \infty)$. By a result of Connes $\mathcal{F}(L(G))$ is countable if G is an i.c.c. group with property T.

THEOREM. $\mathcal{F}(L(F(\infty))) = (0, \infty)$.

That $\mathcal{F}(L(F(\infty))) \supset \mathbb{Q} \cap (0, \infty)$ was proved in [37] by me, the complete result was then obtained by Radulescu [23].

4.4 The compressions $(L(F(n)))_\lambda$

THEOREM [37]. $(L(F(n)))_{1/N} \sim L(F(N^2(n-1)+1))$, $N \in \mathbb{N}$, $n = 2, 3, \ldots, \infty$.

The preceding result, a first application of the free probability technique, was extended in several directions.

THEOREM [24], [25]. *If* $p, q \in \mathbb{N}$, $2 \leq p < q$, *and* $\lambda = (p-1)^{\frac{1}{2}}(q-1)^{-\frac{1}{2}}$, *then*

$$(L(F(p)))_\lambda = L(F(q)) \ .$$

Building on this, Dykema [12] and Radulescu [25] (independently) defined *interpolated free group factors* $L(F(s))$, $s > 1$, $s \in \mathbb{R}$, satisfying the formula in the preceding theorem for arbitrary real $q > p > 1$. Moreover for arbitrary real $p > 1$, $q > 1$,

$$L(F(p)) * L(F(q)) \sim L(F(p+q)) \ .$$

4.5 Free products. A few preliminary results [37], [10] identifying certain free product von Neumann algebras with free group factors were greatly extended by Dykema [11]. If A, B are injective separable von Neumann algebras with specified faithful normal trace-states and if $A * B$ is a factor, then it is isomorphic to one of the interpolated free group factors $L(F(s))$. Moreover, formulae for the parameter s are given in [11]. A further generalization is given in [13].

4.6 Subfactors. Radulescu has shown in [25] that $L(F(\infty))$ has subfactors of all allowable Jones indices < 4, i.e. the numbers $4\cos^2 \frac{\pi}{n}$ of [15]. The proof involves random matrices and results of [22] on constructing subfactors via amalgamated free products. Note that the fundamental group of $L(F(\infty))$ being $(0, \infty)$ implies the existence of subfactors of indices ≥ 4.

4.7 The isomorphism problem. The question of whether the free group factors $L(F(m))$ are isomorphic or not for different values of m is still unresolved (this problem appears on Kadison's Baton Rouge problem list).

4.8 Type III factors. In [26] Radulescu showed that the free product of $L(\mathbb{Z})$ with the (2×2) matrix algebra endowed with a nontracial state is a type III factor and that its core is isomorphic to $L(F_\infty) \otimes B(H)$. Further results on free product type III factors were obtained by Barnett and Dykema.

4.9 Quasitraces. Uses of semicircular systems have not been confined to W^*-algebra questions. A surprising application of semicircular systems appears in Haagerup's solution of the quasitraces problem for exact C^*-algebras [14].

5 Free Entropy [39]

5.1 The definition of free entropy. In classical probability theory, the entropy of an n-tuple $f = (f_1, \dots, f_n)$ of random variables is given by

$$S(f) = -\int_{\mathbb{R}^n} p(t) \log p(t)\, dt$$

where p is the density of the distribution of $(f_1, \dots, f_n)$. To define a free entropy $\chi(X_1, \dots, X_n)$ for an n-tuple of self-adjoint random variables in a tracial W^*-probability space, we had to go back to Boltzmann's $S = k \log W$ (i.e., roughly, the entropy is proportional to the logarithm of the measure of a set of microstates) and take into account that independence of random matrices gives rise asymptotically to freeness. This means we will choose approximating microstates $\Gamma_R(X_1, \dots, X_n; m, k, \varepsilon)$ to be the sets of $(A_1, \dots, A_n) \in (\mathcal{M}_k^{sa})^n$ so that

$$|\tau(X_{i_1} \dots X_{i_p}) - k^{-1}\mathrm{Tr}(A_{i_1} \dots A_{i_p})| < \varepsilon$$

for all $1 \leq p \leq m$, $(i_1, \dots, i_p) \in \{1, \dots, n\}^p$ and $\|A_j\| \leq R$, $1 \leq j \leq n$. With vol denoting the volume on $(\mathcal{M}_k^{sa})^n$ for the scalar product defined by the trace Tr, we take

$$\limsup_{k \to \infty} (k^{-2} \log \mathrm{vol}\, \Gamma_R(X_1, \dots, X_n; m, k, \varepsilon) + \frac{n}{2} \log k)$$

and then define $\chi(X_1, \dots, X_n)$ to be

$$\sup_{R>0}\ \inf_{m \in \mathbb{N}}\ \inf_{\varepsilon > 0}$$

of that quantity.

Note that a similar definition for the classical entropy is possible, taking instead of all matrices $\mathcal{M}_k$ only the diagonal ones.

5.2 Properties of free entropy

(1) For one variable X with distribution μ,

$$\chi(X) = \iint \log|s - t|\, d\mu(s)\, d\mu(t) + \tfrac{3}{4} + \tfrac{1}{2}\log 2\pi$$

(2) $\chi(X_1, \dots, X_n) \leq \frac{n}{2} \log(2\pi e n^{-1} C)$ where $C^2 = \tau(X_1^2 + \dots + X_n^2)$.

(3) If $(X_1^{(p)}, \dots, X_n^{(p)})$ converge strongly to $(X_1, \dots, X_n)$ then

$$\limsup_{p \to \infty} \chi(X_1^{(p)}, \dots, X_n^{(p)}) \leq \chi(X_1, \dots, X_n)$$

(4) $\chi(X_1, \dots, X_{m+n}) \leq \chi(X_1, \dots, X_m) + \chi(X_{m+1}, \dots, X_{m+n})$

(5) If $X_1, \dots, X_n$ are free, then $\chi(X_1, \dots, X_n) = \chi(X_1) + \dots + \chi(X_n)$.

(6) Let $F = (F_1, \dots, F_n)$ where F_j are noncommutative power series in n indeterminates. Under suitable convergence assumptions and the existence of an inverse (with respect to composition) of the same kind,

$$\chi(F(X_1, \dots, X_n)) = \chi(X_1, \dots, X_n) + \log|\mathcal{J}|$$

where $|\mathcal{J}|$ (the "positive Jacobian") is the Kadison-Fuglede positive determinant of the differential $DF(X_1, \dots, X_n)$ viewed as an element of $M \otimes M^{\mathrm{op}} \otimes \mathcal{M}_n$.

5.3 The free analogue of Fisher's information measure. By analogy with the classical case, the free analogue of the Fisher information measure is

$$\Phi(X) = \lim_{\varepsilon \downarrow 0} \varepsilon^{-1}(\chi(X + \sqrt{\varepsilon} S) - \chi(X))$$

where S is $(0,1)$-semicircular and X, S are free. If $d\mu(t) = v(t)dt$ (μ the distribution of X) then

$$\Phi(X) = \tfrac{2}{3} \int v^3(t)\, dt\ .$$

The free analogue of the Cramér-Rao inequality is

$$\left(\int v^3(t)\, dt\right)\left(\int t^2 v(t)\, dt\right) \geq \frac{3}{4\pi^2}$$

where $v \in L^1 \cap L^3$ is a probability density. Equality holds iff v is a centered semicircle law.

5.4 The free entropy dimension. The entropy being a kind of normalized (logarithm of) volume, one may imitate the idea of the Minkowski content and define a dimension quantity from the asymptotic of volumes of ε-neighborhoods. This is realized via a free semicircular perturbation. The free entropy dimension is

$$\delta(X_1, \ldots, X_n) = n + \limsup_{\varepsilon \to 0} \frac{\chi(X_1 + \varepsilon S_1, \ldots, X_n + \varepsilon S_n)}{|\log \varepsilon|}$$

where $(S_1, \ldots, S_n)$ and $(X_1, \ldots, X_n)$ are free and $(S_1, \ldots, S_n)$ is a semicircular system.

(1) $\delta(X_1, \ldots, X_n) \leq n$ and it is ≥ 0 if $X_1, \ldots, X_n$ can be realized in a free group factor $L(F_m)$.

(2) $\delta(X_1, \ldots, X_{p+q}) \leq \delta(X_1, \ldots, X_p) + \delta(X_{p+1}, \ldots, X_{p+q})$.

(3) If $X_1, \ldots, X_n$ are free then $\delta(X_1, \ldots, X_n) = \delta(X_1) + \cdots + \delta(X_n)$.

(4) If μ is the distribution of X, $\delta(X) = 1 - \sum_{t \in \mathbb{R}} (\mu(\{t\}))^2$.

5.5 Free entropy dimension and smooth changes of generators

THEOREM. *If $X_1, \ldots, X_n$ and $Y_1, \ldots, Y_m$ are semicircular generators of the same W^*-algebra M and if $Y_1, \ldots, Y_m$ are "smooth noncommutative functions of $X_1, \ldots, X_n$" then $n \geq m$.*

Here Y_j is a smooth noncommutative function of $(X_1, \ldots, X_n)$ if

$$d_2(Y_j, W^*(X_1 + \varepsilon S_1, \ldots, X_n + \varepsilon S_n)) = O(\varepsilon^s) \quad \text{for all } s < 1$$

where d_2 is the 2-norm distance defined by the trace $(S_1, \ldots, S_n)$ semicircular and free with respect to $(X_1, \ldots, X_n)$. For instance, elements obtained via suitably convergent noncommutative power series are smooth.

Note that if "smooth" could be replaced by "Borel" the corresponding result would imply $m \neq n \Rightarrow L(F_m)$ nonisomorphic to $L(F_n)$. In particular, the same conclusion, concerning the isomorphism problem of free group factors, would be reached, from an affirmative answer to the

SEMICONTINUITY PROBLEM. *If* $(X_1^{(p)},\dots,X_n^{(p)})$ *converges strongly to* $(X_1,\dots,X_n)$ *does it follow that* $\liminf_{p\to\infty}\delta(X_1^{(p)},\dots,X_n^{(p)})\geq\delta(X_1,\dots,X_n)$?

The explicit formula for δ in case $n=1$ implies an affirmative answer.

Note also that if $X_1,\dots,X_n$ are free and generate a factor, then comparing $\delta(X_1,\dots,X_n)$ with the results of [11] we have

$$W^*(X_1,\dots,X_n)\simeq L(F(\delta(X_1,\dots,X_n)))\ .$$

Appendix: Operator Algebra Glossary

C^*-algebras are involutive Banach algebras isomorphic to norm-closed sub-algebras of the algebra of all bounded operators on some complex Hilbert space $B(\mathcal{H})$ and which together with an operator T contain its adjoint T^*.

A functional $\phi: A\to\mathbb{C}$ (A a C^*-algebra) is a *state* if $\|\phi\|=1$ and ϕ is positive, i.e. $\phi(a^*a)\geq 0$ for all $a\in A$. By a theorem of Gel'fand-Naimark *commutative* ***C^*-algebras*** are precisely the algebras of continuous functions $C_0(X)$ vanishing at infinity on some locally compact space — states are Radon probability measures on X.

A *von Neumann algebra* M (or W^*-*algebra*) is a $*$-subalgebra of $B(\mathcal{H})$ that contains the identity and is closed in the weak operator topology, i.e. if x_i is a net of operators in M and $\langle x_i,h,k\rangle\to\langle xh,k\rangle$ for some $x\in B(\mathcal{H})$ and all $h,k\in\mathcal{H}$, then x is in M.

A functional $\tau: A\to\mathbb{C}$ on an algebra is a ***trace*** if $\tau(ab)=\tau(ba)$ for all $a,b\in A$.

References

[1] H. Bercovici and D. Voiculescu, *Levy-Hincin type theorems for multiplicative and additive-free convolution*, Pacific J. Math. **153** (1992), no. 2, 217–248.

[2] H. Bercovici and D. Voiculescu, *Free convolution of measures with unbounded support*, Indiana Univ. Math. J. **42**, no. 3 (1993), 733–773.

[3] H. Bercovici and D. Voiculescu, *Superconvergence to the central limit and failure of the Cramer theorem for free random variables*, preprint 1994.

[4] P. Biane, *Permutation model for semicircular systems and quantum random walks*, preprint 1993.

[5] P. Biane, *Representations of unitary groups and free convolution*, Publ. RIMS Kyoto Univ. **31** (1995), 63–79.

[6] M. Bozejko and R. Speicher, *An example of generalized Brownian motion*, Comm. Math. Phys. **137** (1991), 519–531; II in Quantum Probab. & Related Topics **VII** (1990), 67–77.

[7] A. Connes, *Classification of injective factors, Ann. of Math.* (2) **104** (1976), 73–115.

[8] A. Connes, *A factor of type II_1 with countable fundamental group*, J. Operator Theory **4** (1980), 151–153.

[9] A. Connes and V. F. R. Jones, *Property T for von Neumann algebras*, Bull. London Math. Soc. **17** (1985), 57–62.

[10] K. J. Dykema, *On certain free product factors via an extended matrix model*, J. Funct. Anal. **112** (1993), 31–60.

[11] K. J. Dykema, *Free products of hyperfinite von Neumann algebras and free dimension*, Duke Math. J. **69** (1993), 97–119.

[12] K. J. Dykema, *Interpolated free group factors*, Pacific J. Math. **163** (1994), 123–135.

[13] K. J. Dykema, *Amalgamated free products of multi-matrix algebras and a construction of subfactors of a free group*, preprint 1994.

[14] U. Haagerup, *Quasitraces on exact C^*-algberas are traces*, preprint 1991.

[15] V. F. R. Jones, *Index of subfactors*, Invent. Math. **72** (1983), 1–25.

[16] H. Maassen, *Addition of freely independent random variables*, J. Funct. Anal. **106** (1992), 409–438.

[17] F. J. Murray and J. von Neumann, *On rings of operators, IV*, Ann. of Math. (2) **44** (1943), 716–808.

[18] A. Nica, *Asymptotically free families of random unitaries in symmetric groups*, Pacific J. Math. **157** no. 2, (1993), 295–310.

[19] A. Nica, *Expected number of cycles of length c of a free word in k permutations*, preprint.

[20] A. Nica, *R-transforms of free joint distributions, and non-crossing partitions*, to appear in J. of Funct. Anal.

[21] A. Nica, *A one-parameter family of transforms linearizing convolution laws for probability distributions*, Commun. Math. Phys. **168** (1995), 187–207.

[22] S. Popa, *Markov traces on universal Jones algebras and subfactors of finite index*, Invent. Math. **111** (1993), 375–405.

[23] F. Radulescu, *A one-parameter group of automorphism of $L(F_\infty) \otimes B(H)$ scaling the trace*, C. R. Acad. Sci. Paris, t.**314**, Serie I (1992), 1027–1032.

[24] F. Radulescu, *Stable equivalence of the weak closures of free groups convolution algebras, Comm. Math. Phys.* **156** (1993), 17–36.

[25] F. Radulescu, *Random matrices, amalgamated free products and subfactors of the von Neumann algebra of a free group*, Invent. Math. **115**, no. 2 (1994), 347–389.

[26] F. Radulescu, *A type III_λ factor with core isomorphic to the free group von Neumann algebra of a free group tensor $B(H)$*, preprint.

[27] F. Radulescu, *On the von Neumann algebra of Toeplitz operators with automorphic symbol*, preprint 1993.

[28] R. Speicher, *A new example of independence and 'white noise'*, Probab. Theory Related Fields **84** (1990), 141–159.

[29] R. Speicher, *Multiplicative functions on the lattice of non-crossing partitions and free convolution*, Math. Ann. **298** (1994), 611.

[30] R. Speicher, *Combinatorial theory of the free product with amalgamation and operator-valued free probability theory*, Habilitationsschrift, Universität Heidelberg, 1994.

[31] E. Störmer, *Entropy of some automorphisms of the II_1-factors of free groups in infinite number of generators*, preprint.

[32] D. Voiculescu, *Symmetries of some reduced free product C^*-algebras*, in Lecture Notes in Math., vol. **1132** Springer-Verlag, Berlin and New York (1985), 556–588.

[33] D. Voiculescu, *Addition of certain non-commuting random variables*, J. Funct. Anal. **66** (1986), 323–346.

[34] D. Voiculescu, *Multiplication of certain non-commuting random variables*, J. Operator Theory **18** (1987), 223–235.

[35] D. Voiculescu, *Dual algebraic structures on operator algebras related to free products*, J. Operator Theory **17** (1987), 85–98.

[36] D. Voiculescu, *Operations on certain noncommutative operator-valued random variables*, preprint, Berkeley, 1992.

[37] D. Voiculescu, *Circular and semicircular systems and free product*, in Progr. Math. vol. **92** (1990), 45–60, Birkhäuser, Boston, MA.

[38] D. Voiculescu, *Limit laws for random matrices and free products*, Invent. Math. **104** (1991), 201–220.

[39] D. Voiculescu, *The analogues of entropy and of Fisher's information measure in free probability theory, I*, Comm. Math. Phys. **155** (1993), 71–92; II, Invent. Math., **118** (1994), 411–440.

[40] D. Voiculescu, K. J. Dykema, and A. Nica, *Free random variables*, CRM Monograph Series, vol. **I**, American Mathematical Society, Providence, RI, 1992.

[41] E. Wigner, *Characteristic vectors of bordered matrices with infinite dimension*, Ann. of Math. (2) **62** (1955), 548–564.

[42] E. Wigner, *On the distribution of the roots of certain symmetric matrices*, Ann. of Math. **67** (1958), 325–327.

Andrew Wiles, last plenary speaker

Modular Forms, Elliptic Curves, and Fermat's Last Theorem

ANDREW WILES

Mathematics Department, Princeton University
Princeton, NJ 08544, USA

The equation of Fermat has undoubtedly had a far greater influence on the development of mathematics than anyone could have imagined. After 1847 most serious mathematical approaches to the problem followed the line introduced by Kummer. This approach involved a detailed analysis of the ideal class groups of cyclotomic fields.

The class number formulas used in Kummer's theory are refinements of the well-known class number formula of Dirichlet of 1838. To recall a special case, if q is a prime with $q \equiv 3 \bmod 4$ and $q \neq 3$, then the class number of $\mathbf{Q}(\sqrt{-q})$ is $(B - A)/q$ where

$$\begin{aligned} B &= \Sigma \text{ quadratic nonresidues mod } q \\ A &= \Sigma \text{ quadratic residues mod } q. \end{aligned}$$

Such formulas can be recast in the language of Galois modules as follows. If $M \simeq (\mathbf{Q}_p/\mathbf{Z}_p)(\chi)$ with a $\mathrm{Gal}(\bar{\mathbf{Q}}/\mathbf{Q})$-action we define $h(M)$ by

$$h(M) = \# \ker : H^1(\mathbf{Q}, M) \longrightarrow \prod_{\ell} H^1(\mathbf{Q}_\ell^{unr}, M).$$

For a general p-divisible M we need to modify the condition at p, but for the example given above we can just take $M_p = (\mathbf{Q}_p/\mathbf{Z}_p)(\chi)$ where χ is the quadratic character of $\mathbf{Q}(\sqrt{-q})$. Then knowing the class number of $\mathbf{Q}(\sqrt{-q})$ is equivalent to determining $h(M_p)$ for all p.

However, despite considerable progress on such problems, no convincing conjectures appeared that were strong enough to imply Fermat's Last Theorem. Ultimately, the solution came from a quite different source, although it did also rely in part on a generalized class number formula of the above type.

We begin with a rather special but influential example from the work of Eichler (1954). Let E be the elliptic curve: $y^2 + y = x^3 - x^2 - 10x - 20$. Let N_p denote the number of solutions of this (affine) equation mod p. Consider the modular form

$$\begin{aligned} f &= q \prod_{n=1}^{\infty} (1 - q^n)^2 (1 - q^{11n})^2 \\ &= \eta(z)^2 \eta(11z)^2 = q - 2q^2 - q^3 + 2q^4 + q^5 + q^6 - 2q^7 + \cdots \end{aligned}$$

Proceedings of the International Congress
of Mathematicians, Zürich, Switzerland 1994

Then f is a modular form on $\Gamma_0(11)$ and if we write $f = \Sigma a_n q^n$ then Eichler showed that $a_p = p - N_p$ for each prime $p \neq 11$. The L-function of E is thus given by the Mellin transform of f, $L(f,s) = \Sigma a_n n^{-s}$. We call an elliptic curve modular if there is a modular form with this property.

In the 1960s Shimura, building on the work of Eichler, showed how to associate an elliptic curve over $\mathbf{Q}$ to any newform of weight 2 on $\Gamma_0(N)$ that has rational Fourier coefficients. One therefore has a triangle

Newforms of weight 2 on $\Gamma_0(N)$ with rational Fourier coefficients	Shimura $\xrightarrow{\quad r \quad}$	Elliptic curves over $\mathbf{Q}$ up to isogeny
$\searrow s$		$t \swarrow$
	Compatible systems of ℓ-adic representations	

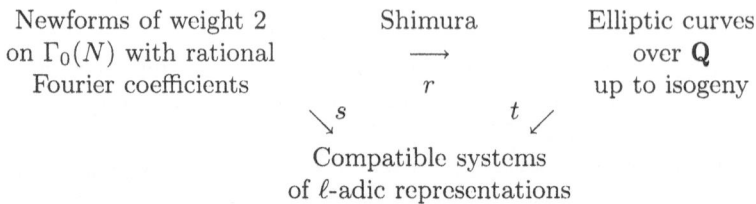

where s is simply defined as the composite. Serre in special cases and Faltings in general later proved that t was injective. However, the fundamental question posed in [We], which was apparently first raised in an imprecise form by Taniyama and in a precise form by Shimura, was whether r was surjective. In other words, is every elliptic curve over $\mathbf{Q}$ modular?

During the next 25 years, the triangle was enormously developed. The map r was studied in great generality under the name of Shimura varieties and the map s was also studied in great generality as part of the Langlands programme. One significant advance was in the analogue of s for weight one. Here the image is more naturally replaced by 2-dimensional complex representations. The construction of these was given in [D-S] and the crucial converse theorem for representations with soluble image was proved by Langlands in [L] and completed by Tunnell in [T]. However, the original problem on the surjectivity of r remained untouched.

In 1985, Frey suggested a completely new approach to Fermat's Last Theorem. If $p \geq 5$ is an odd prime and $a^p + b^p = c^p$ were a solution to Fermat's equation then he proposed showing that the elliptic curve $E: y^2 = x(x-a^p)(x+b^p)$ could not be modular. To exploit this idea Serre formulated a conjecture on Galois representations which applied to the Galois module $E[p]$ of p-division points implied that this curve could not be modular. Ribet then proved Serre's conjecture in the summer of 1986 (see [R]).

It remained to prove that elliptic curves over $\mathbf{Q}$ are modular, or less generally that all semistable elliptic curves over $\mathbf{Q}$ are modular, as those considered by Frey would necessarily be of this form. The main theorems of [Wi] are:

THEOREM 1 *Every semistable elliptic curve over* $\mathbf{Q}$ *is modular.*

THEOREM 2 *(Fermat's Last Theorem) If* $a^p + b^p = c^p$ *with* a, b, c *in* $\mathbf{Q}$, *then* $abc = 0$.

To set the stage for the proof in [Wi] one begins by replacing the problem on elliptic curves with a problem on Galois representations. Thus instead of considering the map r we consider the map s. We consider the extension of the map

s covering forms on $\Gamma_1(N)$ and restrict its image to consider only ℓ-adic representations for a single choice of ℓ.

The proof itself begins with the crucial observation that for any elliptic curve E over $\mathbf{Q}$, $\rho_{E,3}$ is modular where $\rho_{E,3}$ is the representation on the 3-division points of E. This is immediate from the theorem of Langlands and Tunnell referred to above, although we actually want to know that it is also the reduction of the representation associated to a form of weight 2. The proof then proceeds by showing that, under the hypothesis that $\rho_{E,3}$ is irreducible, every suitable lifting to a $\mathrm{GL}_2(\mathbf{Z}_3)$ representation is modular; in other words, it is in the image of the extended map s. This is part of a more general theory describing conditions under which 2-dimensional ℓ-adic representations that are liftings of modular mod ℓ representations should themselves be modular. The key ingredient in proving these results is the forging of a surprising link with a certain generalized class number formula. The link is made using some new arguments from commutative algebra as well as an elaborate study of the properties of modular forms. The commutative algebra enters in trying to relate two rings, one arising from the theory of deformations of Galois representations and the other from the theory of modular forms.

The theorem of Langlands and Tunnell permits one to choose a modular lifting of $\rho_{E,3}$ and it is the adjoint of this representation to which the class number formula is attached in the manner described earlier. The solution to this class number problem is based on duality theorems in Galois cohomology and on a construction using Hecke rings, which was inspired by a variant of Iwasawa theory (see [Wi] and [T-W]). These arguments only work at the moment when $\rho_{E,3}$ is irreducible. To include the other semistable curves we use a different argument involving families of elliptic curves with the same representation on their 5-division points.

At the time of the congress, one step in the argument was not complete, but it was completed a few weeks afterwards. For a fuller account of the proof we refer to [Wi] and [T-W].

References

[D-S] Deligne, P. and Serre, J-P., *Formes modulaires de poids 1*, Ser 7, *Ann. Sci. École Norm. Sup.* **4** (1974), 507–530.

[L] Langlands, R., *Base change for GL*(2), Ann. of Math Stud., **96** (1980), Princeton University Press, Princeton, NJ.

[R] Ribet, K., *On modular representations of Gal*($\bar{\mathbf{Q}}/\mathbf{Q}$) *arising from modular forms*, Invent. Math. **100** (1990), 431–476.

[T] Tunnell, J., *Artin's conjecture for representations of octahedral type*, Bull. Amer. Math. Soc. **5** (1981), 173–175.

[T-W] Taylor, R. and Wiles, A., *Ring-theoretic properties of certain Hecke algebras*, Ann. of Math. **142** (1995), 553–572.

[We] Weil, A., *Über die Bestimmung Dirichletscher Reihen durch Funktionalgleichungen*, Math. Ann. **168** (1967), 149–156.

[Wi] Wiles, A., *Modular elliptic curves and Fermat's Last Theorem*, Ann. of Math. **142** (1995), 443–551.

Recent Developments in Dynamics

JEAN-CHRISTOPHE YOCCOZ

Lab. Math., Bâtiment 425
Université Paris-Sud Orsay
F-91405 Orsay Cedex, France

1 Introduction

1.1 Broadly speaking, the goal of the theory of dynamical systems is, as it should be, to understand *most* of the dynamics of *most* systems.

The dynamical systems that we will consider in this survey are smooth maps f from a smooth manifold M to itself; the time variable then runs amongst non-negative integers.

Frequently, we will also assume that the map f is a diffeomorphism, allowing the time variable to take all integer values. We could also consider smooth flows on M, with a real time variable: the ideas and concepts are pretty much the same in this case.

Given two dynamical systems $f: M \to M$ and $g: N \to N$, a morphism from f to g is a smooth map $h: M \to N$ such that $g \circ h = h \circ f$, in other words the diagram

$$\begin{array}{ccc} M & \xrightarrow{\ f\ } & M \\ {\scriptstyle h}\downarrow & & \downarrow{\scriptstyle h} \\ N & \xrightarrow[\ g\ ]{} & N \end{array}$$

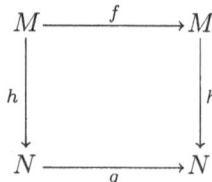

is commutative.

When h is a diffeomorphism, we will say that f and g are (smoothly) *conjugated.* When h is an embedding, f is a *subsystem* of g. When h is a submersion, f is an *extension* of g, and g is a *factor* of f.

The ultimate goal of the theory should be to classify dynamical systems up to conjugacy. This can be achieved for some classes of simple systems [PY1]; but even for (say) smooth diffeomorphisms of the two-dimensional torus, such a goal is totally unrealistic. Hence we have to settle to the more limited, but still formidable, task to understand most of the dynamics of most systems.

The word "most" in the last sentence may assume both a topological and metrical meaning. From a topological point of view, it means open and dense, or

Proceedings of the International Congress
of Mathematicians, Zürich, Switzerland 1994

more frequently G_δ-dense (i.e. countable intersection of open and dense); from a metrical point of view, we would like to understand the trajectories of Lebesgue for almost every point of the system; when considering a smoothly parametrized family of maps or diffeomorphisms, we would also like to understand the dynamics for almost all values of the parameter.

1.2 The dynamical features that we are able to understand fall into two classes, hyperbolic dynamics and quasiperiodic dynamics; it may very well happen, especially in the conservative case, that a system exhibits both hyperbolic and quasiperiodic features.

I will not try to give a precise definition of what is hyperbolic or quasiperiodic: actually, we seek to extend these concepts, keeping a reasonable understanding of the dynamics, in order to account for as many systems as we can. The big question is then: Are these concepts sufficient to understand most systems?

1.3 The prototype of a quasiperiodic dynamical system is a translation T in a compact abelian group G; typically, G is the n-dimensional torus $\mathbf{T}^n = \mathbf{R}^n/\mathbf{Z}^n$, but the additive group $\mathbf{Z}_p$ of p-adic integers (or more generally any profinite abelian group) is also relevant.

Every translation commutes with T, hence is a symmetry of the dynamics of T: this makes the dynamics homogeneous, with a group of symmetries acting transitively. Another significant feature is that the family of iterates of T is equicontinuous; the topological entropy of T is zero.

Finally, the Haar measure on G is invariant under T, and the unitary operator $\varphi \mapsto \varphi \circ T$ of $L^2(G)$ induced by T has a discrete spectrum.

1.4 As prototypes of hyperbolic dynamical systems, we will consider two examples.

The first one is the Bernoulli shift σ on the profinite abelian group $\Sigma = \{0,1\}^{\mathbf{Z}}$, defined by

$$\sigma\left((x_i)_{i\in\mathbf{Z}}\right) = (y_i)_{i\in\mathbf{Z}}, \qquad y_i = x_{i+1}.$$

For the second one, we consider a matrix $A \in \mathrm{GL}(n,\mathbf{Z})$ which is hyperbolic, i.e. no eigenvalue has modulus one. Such a matrix induces an automorphism of $\mathbf{T}^n$, which is a typical example of Anosov diffeomorphism.

Let us consider some significant features of the dynamics (in both examples).

Perhaps the most important is the *shadowing property*: define an ε-pseudo orbit as a sequence $(z_i)\_i \in \mathbf{Z}$ such that $d(f(z_i), z_{i+1}) < \varepsilon$ for all i; then, for given $\delta > 0$, there exists $\varepsilon > 0$ such that every ε-pseudo orbit $(z_i)\_i \in \mathbf{Z}$ is "shadowed" by a true orbit $(w_i)\_i \in \mathbf{Z}$ in the sense that $d(w_i, z_i) < \delta$ for all i.

A counterpart of the shadowing property is the *expansivity property*: there exists $\varepsilon_o > 0$ such that

$$\operatorname*{Sup}_n d(f^n x, f^n y) \geq \varepsilon_o$$

for all distinct x, y: this makes the shadowing orbit unique (for δ small enough) and is in contrast with the equicontinuity of iterates of the quasiperiodic case.

In both examples, the topological entropy is strictly positive. As automorphisms of compact abelian groups, the two examples preserve the Haar measure; the corresponding unitary operators have a Lebesgue spectrum.

2 Quasiperiodic dynamics

2.1 Before giving some specific results, let us begin with a broad overview.

There are three approaches to quasiperiodic dynamics that have been very fruitful.

The first one is the function-theoretical approach, dealing with the stability of diophantine quasiperiodic motions. This includes the so-called KAM-theory, and techniques where functional equations are solved via Newton's method (combined with smoothing operators) or implicit function theorems in Fréchet spaces (which are conceptual analogues of Newton's method). In several special but important contexts, Herman [H2] has been able to solve the functional equations via the Schauder-Tichonoff fixed point theorem.

Finally, Rüssmann [Ru2] has announced the proof of several KAM-theorems relying only on the standard fixed point theorem.

In the symplectic context, the variational approach has also been quite successful; there is a huge number of results related to the existence of periodic orbits. We will present briefly the pionneering work of Mather on quasiperiodic dynamics in this context.

The last approach to quasiperiodic phenomena is more geometric, and frequently coined as "renormalization". Roughly speaking, the combinatorics of the recurrence are unravelled in an infinite sequence of simple successive steps, each of them involving a change of scales both in time and space. Typically, for a circle diffeomorphism f, with irrational rotation number α having convergents $(p_n/q_n)\_n \geq 0$, two successive iterates f^{q_n}, $f^{q_{n+1}}$ give rise to a circle diffeomorphism f_n, which is the "nth-renormalization" of f (Herman, Yoccoz). Sullivan has developed this approach when the recurrence is combinatorially described as a translation in a profinite abelian group.

2.2 Let us consider a holomorphic germ $f(z) = \lambda z + O(z^2)$, $\lambda \in \mathbf{C}^*$, in one complex variable.

We are interested in the dynamics near the fixed point 0, when the eigenvalue λ has modulus 1 but is not a root of unity; we write $\lambda = e^{2\pi i \alpha}$, with irrational $\alpha \in (0,1)$.

It is convenient to assume some normalization on f; we will consider the class S_α of germs as above that are defined and univalent in the unit disk $\mathbf{D}$.

The germ f is always formally linearizable: there exists a unique formal power series $h_f(z) = z + O(z^2)$ satisfying $h_f \circ R_\lambda = f \circ h_f$, where $R_\lambda: z \to \lambda z$ is the linear part of f.

Consider $V_f = \operatorname{int}\left(\bigcap_{n\geq 0} f^{-n}(\mathbf{D})\right)$; it is easy to see that $0 \in V_f$ if and only if h_f is convergent, and that in this case there exists $r_f > 0$ such that the restriction of h_f to $\{|z| < r_f\}$ is a conformal representation of the component U_f of 0 in V_f. Actually, when $\overline{U}_f \subset \mathbf{D}$, r_f is the radius of convergence of h_f.

Let us define

$$r(\alpha) = \inf_{S_\alpha} r_f,$$

and denote by $(p_n/q_n)\_n \geq 0$ the convergents of α.

Siegel [Si] proved in 1942 that $r(\alpha) > 0$ as soon as the diophantine condition $\operatorname{Log} q_{n+1} = O(\operatorname{Log} q_n)$ holds; he achieved this first breakthrough through small divisors problems by a direct estimation of the coefficients of h_f. Later, Brjuno [Br], through a refinement of Siegel's estimates, proved that if

$$\text{(B)} \qquad \Phi(\alpha) = \sum_{n\geq 0} q_n^{-1} \operatorname{Log} q_{n+1} < +\infty,$$

then $r(\alpha) > 0$ and even $\operatorname{Log} r(\alpha) > 2\Phi(\alpha) - c$ (for some $c > 0$ independent of α). See also [C].

Using a "renormalization" approach based on a geometric construction, I gave a new proof of the Siegel-Brjuno theorem and proved the converse ([Y5], [Y4]).

THEOREM. (1) *If* $\Phi(\alpha) < +\infty$, *then*

$$|\operatorname{Log} r(\alpha) + \Phi(\alpha)| < c,$$

for some $c > 0$ *independent of* α.

(2) *If* $\Phi(\alpha) = +\infty$, *the quadratic polynomial* $P_\lambda(z) = \lambda z + z^2$ *is not linearizable: every neighborhood of 0 contains a periodic orbit, distinct from 0.*

Actually, one first constructs a nonlinearizable germ with this property, and then shows that the same holds for the quadratic polynomial, via Douady-Hubbard's theory of quadratic-like maps.

Significant progress has been achieved by Perez-Marco [PM2], [PM3] in the understandings of the dynamics in the nonlinearizable case. He first showed that for a germ $f \in S_\alpha$ that is not linearizable and has no periodic orbit in $\mathbf{D}$ (except for 0) to exist, it is necessary and sufficient that

$$\sum_{n\geq 0} q_n^{-1} \operatorname{Log}\operatorname{Log} q_{n+1} = +\infty.$$

He also defines "degenerate" Siegel disks as follows: assuming f to be univalent in a neighborhood of $\overline{\mathbf{D}}$, the connected component K_f of 0 in $\bigcap_{\mathbf{Z}} f^{-n}(\overline{\mathbf{D}})$ is a full, compact, connected, invariant subset of $\overline{\mathbf{D}}$ that meets S^1. When α satisfies the diophantine condition (H) (see 2.3), one has just $K_f = \overline{U}_f$.

These invariant sets provide a rich connection with the theory of analytic circle diffeomorphisms; if $k\colon \mathbf{H}/\mathbf{Z} \to \mathbf{C} - K_f$ is a conformal representation, the map $g = k^{-1} f\, k$ is defined in some strip $\{0 < \operatorname{Im} z < \delta\}$ and extends by Schwarz's reflection principle to a circle diffeomorphism with the same rotation number as f.

2.3 Let us now consider analytic circle diffeomorphisms. For $\delta > 0$, define $B_\delta = \{z \in \mathbf{C}/\mathbf{Z}, |\operatorname{Im} z| < \delta\}$. For irrational $\alpha \in \mathbf{R}/\mathbf{Z}$, let $S_\alpha(\delta)$ be the set of orientation preserving analytic diffeomorphisms f of $\mathbf{R}/\mathbf{Z}$ with rotation number α that extend to a univalent map from B_δ to $\mathbf{C}/\mathbf{Z}$.

By Denjoy's theorem, f is conjugated to the translation $R_\alpha: z \mapsto z+\alpha$ on the circle by a homeomorphism h_f of $\mathbf{R}/\mathbf{Z}$ (uniquely defined if we require $h_f(0)=0$). As for germs, h_f is analytic if and only if the circle $\mathbf{R}/\mathbf{Z}$ is contained in the interior of $\bigcap_{n\geq 0} f^{-n}(B_\delta)$. There are two kinds of results, depending on whether we assume or not that f is near the translation R_α; the breakthroughs (under more restrictive arithmetic conditions) are respectively due to Arnold (1960) and Herman (1976). We state the results in their final form before some comments.

THEOREM 1. (Arnold [A], Rüssmann, Yoccoz [Y6]) *Assume that* $\Sigma\, q_n^{-1} \operatorname{Log} q_{n+1} < +\infty$. *There exists* $\varepsilon = \varepsilon(\alpha,\delta)$ *such that, if*

$$\|f - R_\alpha\|_{C^o(B_\delta)} < \varepsilon(\alpha,\delta),$$

then h_f *is analytic. Moreover, the diophantine condition is optimal.*

THEOREM 2. (Herman, Yoccoz) *Assume that the rotation number satisfies the diophantine condition (H) below. Then* h_f *is analytic. Moreover, the diophantine condition is optimal.*

The arithmetic condition (H)

Assume that $0<\alpha<1$ and define $\alpha_o=\alpha$, $\alpha_n=\{\alpha_{n-1}^{-1}\}$ for $n\geq 1$. For $m\geq n\geq 0$, define inductively $\Delta(m,n)$ as follows:

$$\Delta(n,n)=0, \qquad \forall n\geq 0$$

$$\Delta(m+1,n)=\begin{cases}\exp\Delta(m,n) & \text{if } \Delta(m,n)\leq \operatorname{Log}\alpha_m^{-1}\\ \alpha_m^{-1}(\Delta(m,n)-\operatorname{Log}\alpha_m^{-1}+1) & \text{if } \Delta(m,n)\geq \operatorname{Log}\alpha_m^{-1}\end{cases}$$

Then α satisfies (H) if for every $n\geq 0$ we have $\Delta(m,n)\geq \operatorname{Log}\alpha_m^{-1}$ for $m\geq m_o=m_o(n)$.

The set of numbers α satisfying (H) is a $F_{\sigma\delta}$ set (a countable intersection of F_σ sets) but neither a F_σ or a G_δ set (this explains why the definition has to be complicated). Numbers α such that

$$\operatorname{Log} q_{n+1} = 0\left((\operatorname{Log} q_n)^c\right), \quad \text{for some } c>0$$

satisfy (H). On the other hand, condition (H) is strictly stronger than condition (B). Indeed, for numbers $\alpha = 1/(a_1+1/(a_2+\cdots$ such that

$$a_i \leq a_{i+1} \leq \exp(a_i)$$

condition (B) is always fullfilled; on the other hand, defining $b_o=0$, $b_n=\exp(b_{n-1})$, the number α satisfies (H) if and only if, for any $k\geq 0$, we have $a_{m+k}\leq b_m$ for m large enough; for instance, if $a_{i+1}\geq \exp(a_i^\theta)$, for some $\theta\in(0,1)$, α does not satisfy (H).

Conditions (B) and (H) are closely related: let $\mathcal{H}_o$ be the set of irrational α such that $\Delta(m, 0) \geq \text{Log } \alpha_m^{-1}$ for large m; then α satisfies (H) if and only if its orbit under $\text{GL}(2, \mathbf{Z})$ is contained in $\mathcal{H}_o$, whereas it satisfies (B) if and only if its orbit meets $\mathcal{H}_o$.

REMARKS. (1) The fact that the optimal arithmetic condition is not the same in the local and global conjugacy theorems is in strong contrast with the smooth (C^∞) case; then in both theorems, the optimal arithmetic condition is the standard one

$$\text{Log } q_{n+1} = O(\text{Log } q_n)$$

(Moser [Mo2], Herman [H1], Yoccoz [Y2]).

(2) Another important difference between the smooth and analytic cases is that the effect of good rational approximations is cumulative in the analytic case, but not in the smooth case. Another way to see this difference is to observe that the arithmetic condition in the smooth case is given by the linearized equation, whereas both conditions (H) and (B) do not appear naturally when looking at linear difference equations.

(3) All known proofs ([H1], [Y2], [KO1], [KO2], [KS]) of *global* conjugacy theorems (smooth or analytic) are based on a renormalization scheme that relies in an essential way on the relationship between the good rational approximations of the rotation number (given by the continued fraction).

2.4 When several frequencies are involved, KAM techniques are available, but they do not give as much geometric insight as we would like to have. One would like to have some geometric renormalization scheme as above, but the problem, of a purely arithmetical nature, is then to understand thoroughly the relationships between good rational approximations.

Here is a test case. Consider the following two theorems.

THEOREM 1. (Arnold [A], Moser [Mo2]) *Let $\alpha = (\alpha_1, \ldots, \alpha_n) \in \mathbf{T}^n$ satisfy the diophantine condition: $\exists \gamma > 0$, $\tau \geq 0$ s.t.*

$$|\langle k, \alpha \rangle + k_o| \geq \gamma ||k||^{-n-\tau}$$

for all $(k_o, k_1 \ldots k_n) \in \mathbf{Z}^{n+1} - \{0\}$.

There exists $\varepsilon = \varepsilon(\alpha)$ and $k = k(\tau)$ such that if f is a smooth diffeomorphism of $\mathbf{T}^n$ satisfying

$$||f - R_\alpha||_{C^k} < \varepsilon,$$

then there exists a (small) translation R_λ and a smooth diffeomorphism h such that

$$f = R_\lambda \circ h \circ R_\alpha \circ h^{-1}.$$

THEOREM 2. (Moser [Mo3]) *Let* $(\alpha_1, \dots, \alpha_n) \in \mathbf{T}^n$ *satisfy the diophantine condition:* $\exists \gamma > 0,\ \tau \geq 0$ *s.t.*

$$|k_o\alpha_i - k_i| \geq \gamma ||k||^{-\frac{1}{n}-\tau}, \quad 1 \leq i \leq n$$

for all $(k_o, \dots, k_n) \in \mathbf{Z}^{n+1} - \{0\}$.

There exists $\varepsilon = \varepsilon(\alpha)$ *and* $k = k(\tau)$ *such that if* $f_1, \dots, f_n$ *are smooth commuting diffeomorphisms of* $\mathbf{T}^1$ *satisfying*

$$\|f_i - R_{\alpha_i}\|_{C^k} < \varepsilon, \quad \rho(f_i) = \alpha_i$$

then there exists a smooth diffeomorphism h *such that*

$$f_i = h\, R_{\alpha_i}\, h^{-1}, \quad 1 \leq i \leq n.$$

PROBLEM 1: Prove Theorem 2 without assuming that f_i is close to R_{α_i}.

PROBLEM 2: Find the *optimal* arithmetical conditions in Theorem 1 and Theorem 2 in the *analytic* case.

The first problem should be easier: diophantine conditions in smooth small divisors problems tend to be more "stable" than in analytic ones.

2.5 Codimension 1 invariant tori

The fundamental result of Moser [Mo1] on the existence of invariant curves for near integrable area-preserving twist diffeomorphisms of the annulus was first generalized by Rüssmann as a "translated curve" theorem (removing the area-preserving hypothesis) [Ru1], [H3]. This has recently been further generalized to higher dimensions as follows (by Cheng-Sun [CS] and Herman [H6]).

Let L be a smooth orientation preserving diffeomorphism of $\mathbf{T}^n \times \mathbf{R}$ such that $L(\mathbf{T}^n \times \{0\}) = \mathbf{T}^n \times \{0\}$ and the restriction of L to $\mathbf{T}^n \times \{0\}$ is a translation; let also $\alpha \in \mathbf{T}^n$ satisfy the standard diophantine condition (see Theorem 1 in 2.4).

Then, if F is a smooth diffeomorphism of $\mathbf{T}^n \times \mathbf{R}$ close enough to L, there exists a translation R in $\mathbf{T}^n \times \mathbf{R}$ such that $R \circ F$ leaves invariant a codimension one torus T, going through the origin, C^∞-close to $\mathbf{T}^n \times \{0\}$, and $R \circ F/T$ is smoothly conjugated to the given translation R_α.

Herman has derived important consequences of this result.

The first is the failure of the quasi-ergodic hypothesis. The ergodic (resp. quasi-ergodic) hypothesis states that the generic Hamiltonian flow is ergodic (resp. has a dense orbit) on the generic (compact, connected) energy surface. The classical KAM-theorems provide for open sets of Hamiltonian flows a set of positive measure (on each energy surface) made of diophantine invariant tori; hence the ergodic hypothesis fails. Herman has discovered a rigidity property of the rotation number in the symplectic context that guarantees a similar phenomenon: there exist a nonempty open set of Hamiltonian flows and energy values for which the energy surface contains a Cantor set of *codimension one* diophantine invariant tori; the orbits "between" the tori are thus constrained to stay there.

Another important consequence is the failure of a conjecture of Pesin: Herman shows that on any manifold M (of dimension ≥ 3) there exists a nonempty open set of volume-preserving diffeomorphisms whose Lyapunov exponents are all 0 on a set of positive volume. In dimension 2, this follows from Moser's twist theorem.

2.6 In the symplectic context, Mather has been pioneering the study of quasiperiodic motions through a variational approach. In one degree of freedom, we now have (due to Aubry [AD], Mather [Ma1], Le Calvez [LeC1], ...) a fairly satisfactory theory of Aubry-Mather Cantor sets. In more degrees of freedom, Mather has obtained a yet partial generalization that seems quite promising in understanding somewhat Arnold diffusion ([Ma3], [Ma2]).

Let me explain this very roughly for discrete time (diffeomorphisms). Consider an integrable diffeomorphism L of $\mathbf{T}^n \times \mathbf{R}^n = T^*\mathbf{T}^n$:

$$L(\theta, r) = (\theta + \nabla\ell(r), r),$$

with strictly convex ℓ superlinear at ∞.

To each invariant lagrangian torus $\{r = r_o\}$, we can associate the cohomology class $r_o \in H^1(\mathbf{T}^n, \mathbf{R})$ and the rotation number $\alpha = \nabla\ell(r_o)$ that belongs in a natural way to $H_1(\mathbf{T}^n, \mathbf{R})$.

Let now $F\colon (\theta, r) \mapsto (\Theta, R)$ be an exact symplectic diffeomorphism close to L. Writing

$$\Sigma\, R_i\, d\Theta_i - \Sigma\, r_i\, d\theta_i = dH(\theta, \Theta)$$

we obtain the generating function H of F, defined on $\mathbf{R}^n \times \mathbf{R}^n$ and satisfying

$$H(\theta + k, \Theta + k) = H(\theta, \Theta), \quad k \in \mathbf{Z}^n.$$

Given an invariant measure μ with compact support, we transport it via $(\theta, r) \mapsto (\theta, \Theta)$ to the diagonal quotient $(\mathbf{R}^n \times \mathbf{R}^n)/\mathbf{Z}^n$ and consider, for $\omega \in \mathbf{R}^n = H^1(\mathbf{T}^n, \mathbf{R})$, the ω-action:

$$A_\omega(\mu) = \int [H(\theta, \Theta) - \langle \omega, \Theta - \theta \rangle]\, d\mu.$$

The invariant measure is *minimal* if it minimizes the ω-action (amongst invariant measures, or equivalently amongst all measures) for some cohomology class ω.

On the other hand, to any invariant measure, one can associate a rotation number

$$\alpha(\mu) = \int (\Theta - \theta)\, d\mu \in \mathbf{R}^n = H_1(\mathbf{T}^n, \mathbf{R}).$$

Then μ is minimal if and only if it minimizes the action A_o amongst all invariant measures with the same rotation number.

For any ω, there exist ω-minimal measures; there also exist minimal measures with any given rotation number α. The correspondence between α and ω is realized by Legendre transform (in a nonsmooth, nonstrictly convex context).

The support of such a minimal measure is an invariant torus in the integrable case and shares in the general case some properties of Aubry-Mather sets: in particular, it is the *graph* of the restriction to a closed subset of a *Lipschitz* map from $\mathbf{T}^n$ to $\mathbf{R}^n$.

The key point for further progresses is to understand the "shadowing" properties of these minimal measures. With one degree of freedom, Mather has proved (see also Le Calvez) that there are no obstructions except for the obvious ones: if $(\Lambda_n)_n \in \mathbf{Z}$ is a sequence of Aubry-Mather sets, *not separated by an invariant curve*, there exists an orbit coming successively (in the prescribed order) close to each of the Λ_i. In more degrees of freedom, invariant tori do not separate and there is no obvious obstruction preventing an orbit to come successively close to the supports of any given sequence of minimal measures (for a generic diffeomorphism). Mather has a partial result in this direction.

2.7 Renormalization theory for quadratic polynomials

The Aubry-Mather sets and minimal measures we have just discussed are important generalizations of the classical KAM quasiperiodic motions. Another nonstandard "generalization" is provided by the dynamics of infinitely renormalizable quadratic polynomials.

The key tool is the Douady-Hubbard theory of quadratic-like maps [DH2], i.e. ramified covering $f: U \to U'$ of degree 2, with U, U' simply connected and $U \subset\subset U'$. Such a map is quasiconformally conjugated to a quadratic polynomial, its filled-in Julia set is $K_f = \bigcap_{n \geq 0} f^{-n}(U')$.

An integer $n \geq 2$ is a renormalization period for the quadratic polynomial $P_c: z \mapsto z^2 + c$ if there exist open neighborhoods $U_n \subset\subset U'_n$ of 0 such that $P_c^n: U_n \to U'_n$ is quadratic-like with connected filled-in Julia set. The quadratic polynomial is infinitely renormalizable if the set $\mathcal{N} = \{n_1 < n_2 < \dots\}$ of its renormalization periods is infinite; then n_k divides n_{k+1} (we write $n_{k+1} = p_{k+1} n_k$, $p_1 = n_1$); let $f_o = P_c$ and $f_k = P_c^{n_k}/U_{n_k}$. Then f_{k+1} is the restriction of $f_k^{p_k}$ to the smaller domain $U_{n_{k+1}}$; it is called the renormalization of f_k. (See Figure 1).

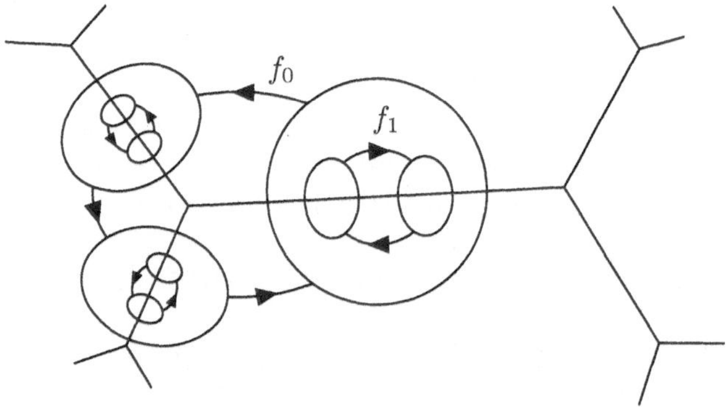

Figure 1

In the study of the dynamics of rational maps, a key point is to understand the geometry and the dynamics of the post-critical set:

$$P(f) = \overline{\{f^n(c), n \geq 1, c \text{ critical value}\}}.$$

In our case, $P(f_o)$ is contained in the fully invariant compact set

$$K_\infty = \bigcap_{k\geq 0} \bigcup_{0\leq j<n_k} f_o^j(K_{f_k}).$$

We have a natural continuous surjective map

$$K_\infty \xrightarrow{p} \mathbf{Z}_{\mathcal{N}} = \lim_{\leftarrow} \mathbf{Z}/n_k\mathbf{Z}$$

onto the profinite abelian group of $\mathcal{N}$-adic integers, the dynamics on $\mathbf{Z}_{\mathcal{N}}$ being translation by 1; the post-critical set is sent to the set of positive integers. The map p is known to be a homeomorphism for real c (Sullivan), but it is also known that it is not always injective.

PROBLEM: Find a necessary and sufficient condition on the combinatorics for p to be a homeomorphism.

Very little is known in the general case (with complex c). On the other hand, a beautiful approach pioneered by Sullivan [Su1], [Su2] has been fruitful in important particular cases. The general strategy is the following; one first constructs, from an infinitely renormalizable quadratic-like map, a geometric object that is a compact set laminated by Riemann surfaces (Riemann lamination). The dynamical properties of the initial quadratic-like map correspond to some properties of the complex geometry of this Riemann lamination. Such laminations have, as usual Riemann surfaces, a Teichmüller space; the renormalization operator (from f_k to f_{k+1}) corresponds to a map between such Teichmüller spaces and we are led to study the *dynamics* of this new map (at the "parameter" level). This map does not increase the Teichmüller distance, and the central problem is to understand to which extent it is contracting. There are partial results in this direction by Sullivan (for real c, with $(p_k)\_k \geq 1$ bounded) and McMullen (under a potentially more general geometric assumption) [McM].

3 Hyperbolic dynamics

3.1 Before we discuss some recent developments, we recall some "classical" hyperbolic dynamics, as developped by Anosov, Sinaï, Smale, Palis, ... in the 1960s [Bo], [Sm], [Sh], [Y9].

The central concept is that of a *basic set*: if f is a smooth diffeomorphism of a manifold M, a basic set of f is a *compact, invariant* subset K of M that is *transitive* (f/K has a dense orbit), *locally maximal* (K is the maximal invariant set in an open neighborhood), and *hyperbolic*: the tangent bundle $E = TM/K$ admits an invariant splitting $E = E^s \oplus E^u$ in a stable subbundle E^s uniformly contracted by Tf and an unstable subbundle E^u uniformly contracted by Tf^{-1}.

The dynamics on a basic set are fairly well understood (and completely so when dim $M = 2$); in particular, the existence of Markov partitions allows us to reduce the study of periodic orbits, invariant measures, ... to the same problems in symbolic dynamics, i.e. subshifts of a finite type on a finite alphabet.

The existence of a basic set K for a diffeomorphism is a semilocal property: it only involves the dynamics of f near K. One gets to more global properties (Anosov diffeomorphisms, Axiom A diffeomorphisms, ...) if one asks that some big invariant subset, carrying "most" of the dynamical properties of f, is hyperbolic.

For instance the *chain recurrent set* $C(f)$ of a smooth diffeomorphism of a compact manifold is the locus of points that are periodic for some arbitrarily small C^o-perturbation of f. Let us say that f is *uniformly hyperbolic* if $C(f)$ is hyperbolic.

It can be proven that $C(f)$ is then a finite union of disjoint basic sets. Uniformly hyperbolic diffeomorphisms form an open subset of $\mathrm{Diff}^\infty(M)$ (it is even open in the C^1-topology), and they are *stable* [R], [Ro]: two C^1-close uniformly hyperbolic diffeomorphisms are topologically conjugated on a neighborhood of their respective chain recurrent sets. Actually, a deep theorem of Mañé (extended by Palis) states that the converse is also true: a diffeomorphism that is C^1 stable in this sense is uniformly hyperbolic [M3].

It was hoped at some point that such globally hyperbolic diffeomorphisms could account, at least in the dissipative case, for most diffeomorphisms. This was shown to be too optimistic when Newhouse [N1], [N2] discovered in the 1970s that there exist open sets of diffeomorphisms that exhibit generically infinitely many attractive periodic orbits (this is not compatible to any global uniform hyperbolic behaviour). Nevertheless, uniformly hyperbolic diffeomorphisms still constitute a good starting point from which one can bifurcate and study more complicated diffeomorphisms. Also, there are many important classes of diffeomorphisms that are not uniformly hyperbolic, but that admit many basic sets that together should carry a lot of information on the dynamics.

3.2 The conceptual apparatus to study weaker forms of hyperbolicity is based on Oseledets' theorem (1968) [O] and Pesin's theory (1976) [Pe1], [Pe2], [FHY]. Oseledets theorem, itself based on a subadditive ergodic theorem, asserts the existence of Lyapunov exponents of a diffeomorphism on a (Borel) set of points that has full measure with respect to all invariant measures. From this starting point, Pesin then constructed the stable and unstable "foliations" associated to the nonzero exponents and proved the crucial fact that they are absolutely continuous.

How frequently are all (or some) of the Lyapunov exponents of a non-uniformly hyperbolic diffeomorphism different from 0? I have mentioned above Herman's theorem (see 2.5), which indicates that we cannot be too optimistic. On the other hand, there have been several breakthroughs showing that it tends to happen with positive measure in the parameter space.

3.3 The first crucial step in this direction is Jakobson's theorem (1981) [J]. He considers real quadratic polynomials $P_c(x) = x^2 + c$, for c in some subset $A_\varepsilon \subset [-2, -2 + \varepsilon]$, whose relative measure tends to 1 as ε goes to zero. For such a parameter, let α be the negative fixed point of P_c and $I = (\alpha, -\alpha)$; he constructs a countable partition $I = \bigcup I_i$ (mod 0) into disjoint open intervals and a map $T: \bigcup I_i \to I$ that is *uniformly expanding* (with bounded distortion) and whose restriction to each I_i is an iterate $P_c^{k_i}$ realizing a diffeomorphism onto I (see Figure 2).

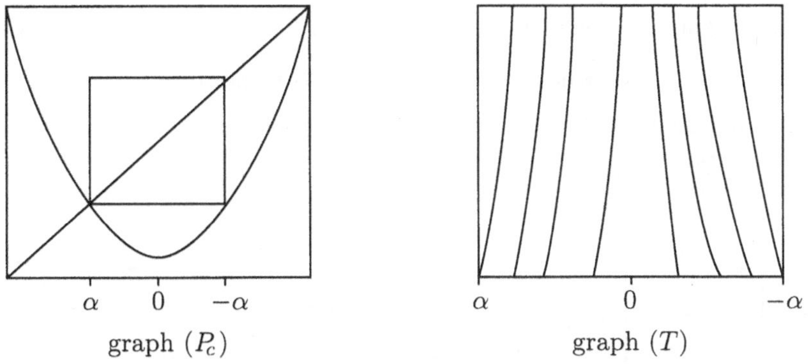

Figure 2

An important point is that although the k_i are not bounded, the measures of the I_i for which $k_i \geq k$ is exponentially small with k. From the existence of such a map T, it is easy to deduce that P_c has an ergodic invariant measure that is equivalent to Lebesgue measure on the (real) Julia set, and that the corresponding Lyapunov exponent is positive.

This kind of result has since been extended in several directions. Rees [Re] has proved that a similar statement holds for an holomorphic family of rational maps (see also [Be]). Jakobson and Swiatek have extended the set of values of c for which the map T is constructed (putting no restriction on the k_i's) [JS].

For *complex* quadratic polynomials whose all periodic orbits are repulsive and that are not infinitely renormalizable, I proved that the dynamics are still sufficiently expanding to guarantee that the Julia sets are locally connected, as a consequence of a (very weak) self-similarity property [Y10]. Lyubich [Ly1] then went on to prove that such a Julia set has measure 0. These results are related to previous work of Branner-Hubbard [BH] and McMullen [McM] on complex cubic polynomials.

3.4 Another very important breakthrough, going to higher dimensions, was achieved by Benedicks-Carleson (1989). They consider Hénon's family [He] of polynomial diffeomorphisms of the plane:

$$H_{a,b}(x, y) = (x^2 + a - y, bx).$$

The parameter b is the constant value of the Jacobian; it is fixed and very small. The parameter a belongs to a subset $A_\varepsilon \subset [-2+\varepsilon, -2+2\varepsilon]$ of relative large measure, with $0 < |b| \ll \varepsilon \ll 1$. For such parameters, the rectangle $U = \{(x, y), |x| < 2 - \frac{3}{4}\varepsilon, |y| < 3b\}$ satisfies $H_{a,b}(U) \subset\subset U$, and one wants to describe the "attractor"

$$\Lambda_{a,b} = \bigcap_{n>0} H^n_{a,b}(U).$$

What emerges from Benedicks-Carleson's study [BC2], together with more recent work of Benedicks-Young [BY] and Jakobson-Newhouse [JN] is the following structure (see Figure 3): one can construct an open subrectangle $V \subset U$, a countable family of disjoint subrectangles $V_i \subset V$ such that $\bigcup_i V_i$ "essentially" covers $V \cap \Lambda$, and a map $T: \bigcup_i V_i \longrightarrow V$, whose restriction to V_i is some iterate H^{k_i} of H, and that is *uniformly* hyperbolic; the k_i's are not bounded, but they take big values on very small sets.

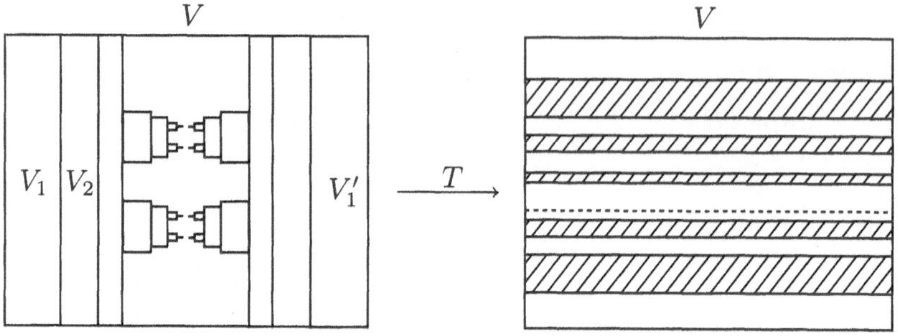

Figure 3

From there, one constructs a nice Sinaï-Bowen-Ruelle invariant measure on Λ; it describes the asymptotics of a positive *Lebesgue* measure set of orbits in U, and the Lyapunov exponents with respect to that measure are nonzero. One also recovers many classical properties of uniformly hyperbolic attractors (it is easy to see that Λ cannot be uniformly hyperbolic; in fact, there is a dense subset of Λ where the stable and unstable manifolds are tangent).

One should note that the admissible set A_ε of values of the parameter a has an empty interior; also, two distinct values of a give rise to attractors that admit the same qualitative description, but are definitely not conjugated. This is in strong contrast to the uniformly hyperbolic case.

The kind of phenomenon that we have tried to describe is not particular to the Hénon family. A first extension of these results, extremely important for applications (see below), was given by Mora-Viana [MV], who introduced the concept of "Hénon-like" families. More recently, Viana [V] proved similar results for some families in higher dimensions, for instance skew products:

$$T: \mathbf{T}^2 \times \mathbf{R}^2 \longrightarrow \mathbf{T}^2 \times \mathbf{R}^2$$
$$T(\theta, (x, y)) = (A(\theta), H_{a(\theta),b}(x, y)),$$

where A is an Anosov diffeomorphism of $\mathbf{T}^2$ (for instance linear hyperbolic) and a is a Morse function on $\mathbf{T}^2$ (subjected to some conditions). In this context, because of the uniform hyperbolicity on the base, it is no more necessary to exclude parameters.

3.5 What we would like to do in the next few years is to obtain a conceptual theory of "weakly hyperbolic basic sets" (including of course the striking examples considered above). For a smooth diffeomorphism f of a manifold M, such a "weakly hyperbolic basic set" should again be a *compact, invariant, transitive, locally maximal* subset K of M satisfying moreover some kind (?) of *weak hyperbolicity* condition. Let me speculate, based on the examples above, on what could be some aspects of this theory.

(1) One would be able to cover "most" of K with a countable family of disjoint open sets V_i and define on $\bigcup V_i$ a *uniformly hyperbolic* map T whose restriction to V_i is some iterate f^{k_i} of f;

(2) One would thus obtain K as the limit (for the Hausdorff distance on compact sets) of an increasing sequence of (uniformly hyperbolic) basic sets

$$K_n = \bigcap_{m\in\mathbf{Z}} f^{-m}\left(\overline{\bigcup_{k_i\leq n} V_i}\right);$$

(3) One should be able to construct some kind of Sinaï-Bowen-Ruelle invariant measure, whose "restriction" to most unstable manifolds would be absolutely continuous with respect to some Hausdorff measure on the unstable manifold. The Lyapunov exponents with respect to this measure would be nonzero;

(4) There would exist some "infinite Markov partition" (as the V_i above) allowing a description by symbolic dynamics (with an infinite alphabet).

4 Parameter space

4.1 I want to discuss now "how many" dynamical systems we are able to understand.

Let us start with a "test case", the family of quadratic (real or complex) polynomials $P_c\colon x \mapsto x^2 + c$, where only one (real or complex) parameter c is involved.

If the critical point 0 escapes under iteration to infinity or converges to some attractive periodic orbit, the dynamics on J_c are *uniformly hyperbolic* (expanding) and stable; such parameters c form an open set U_{hyp}.

If there is a periodic orbit with eigenvalue λ of modulus 1, the parameter c is determined (algebraically) by the period and λ (up to a finite number of choices); the dynamics on the Julia set is of "weak-hyperbolic" type when λ is a root of unity, whereas quasiperiodic features are dominant when λ is not a root of unity (see 2.2).

We are left with the case where all periodic orbits are repulsive, but the critical point belongs to the Julia set J_c (preventing it to be uniformly hyperbolic); it is here natural to discuss separately the cases where P_c is infinitely renormalizable and it is not.

In the non-infinitely renormalizable case, the dynamics on the Julia set still exhibit some (very weak) form of hyperbolicity (see 3.3). As a consequence, I proved that such parameters are *rigid*, i.e. they are determined by the combinatorics of the Julia set [Y10].

In the infinitely renormalizable case, the dynamics on the Julia set exhibit many quasiperiodic features (see 2.7). Swiatek [Sw] and Lyubich [Ly2] have proved that for *real* quadratic polynomials, these parameters are rigid (in the real sense). As a consequence, the open set $U_{\text{hyp}} \cap \mathbf{R}$ of (uniform) hyperbolicity is dense in $\mathbf{R}$. Actually, one would expect that even the following stronger statement should be true: for almost all $c \in \mathbf{R}$, either $c \in U_{\text{hyp}}$ or there exists (as in Jakobson's theorem) an invariant measure on $J_c \cap \mathbf{R}$, absolutely continuous with respect to Lebesgue measure, for which the Lyapunov exponent is nonzero.

On the other hand, although there is a partial result by Lyubich, it is not known whether infinitely renormalizable parameters are rigid in the complex sense. This is the missing step in the Douady-Hubbard's conjecture that the Mandelbrot set is locally connected; actually, assuming this rigidity, we would have from Douady-Hubbard [DH1] and Thurston a complete *topological* description of the Mandelbrot set (whereas we have only a *combinatorial* one at the moment).

4.2 In higher dimensions, such as for instance for diffeomorphisms of surfaces, we are still very far from having a near complete understanding of the "parameter space".

Nevertheless, the results that we have discussed above and others in the same line have led to a change of point of view in looking at these problems.

The classical, uniformly hyperbolic, basic sets have a strong stability property known as hyperbolic continuation: a nearby diffeomorphism admits a basic set close to the original one and the dynamics on the two basic sets are conjugated by a homeomorphism close to the identity. The "parameter space", for instance the space of all smooth diffeomorphisms of the given compact manifold, was in the 1960s and 1970s mostly considered from a topological point of view; one was looking for dynamical features appearing on some open set, or some G_δ set (dense into some open set).

Although this point of view remains important, properties of "weakly hyperbolic basic sets" such as Hénon-like attractors have given a strong impetus on the measure-theoretic point of view of the parameter space: typically, in a generic parameter family of diffeomorphisms, one expects to meet these weakly hyperbolic features on a F_σ subset of the parameter space (because one needs to exclude parameters), but one that has positive Lebesgue measure.

4.3 I would like here to emphasize the importance of the Hénon family, or rather of the Hénon-like families introduced by Mora and Viana (see 3.4), for the study of surface diffeomorphisms.

Consider a smooth diffeomorphism f_o of a surface M that exhibits a homoclinic tangency: this means that f_o has a fixed saddle point p such that the stable manifold $W^s(p)$ and the unstable manifold $W^u(p)$ are tangent along a homoclinic orbit $(f^n(q))_{n \in \mathbf{Z}}$. This is a codimension one phenomenon.

Assume that the tangency is quadratic and consider a one-parameter family of diffeomorphisms $(f_t)_{\text{-}|t| < \varepsilon}$ unfolding generically the tangency.

One would like to understand the orbits under f_t, $|t|$ small, which remain in an appropriately small neighborhood of the orbit of q under f_o. To do this, the first step is to compute the return map R_t into some small neighborhood V of q; it is a disjoint union $V = \bigcup_{n \geq n_o} V_n$; R_t is equal to f_t^n on V_n and is a Hénon-like family of approximately constant Jacobian. The striking conclusion is that every dynamical feature exhibited by the Hénon family, or Hénon-like families, actually occurs in a generic one-parameter family near homoclinic bifurcations. For instance, if the fixed point p is dissipative, the Jacobian of these Hénon-like families will be very small and we will get for a positive measure set of parameters Hénon-like attractors. Let me recall in this context an older result of Newhouse: there are arbitrarily close to 0 intervals of parameter values in which for generic t the diffeomorphism f_t has *infinitely many* periodic orbits; still we suspect (but we don't know) that this only happens for a set of t of Lebesgue measure zero.

Homoclinic bifurcations are by no means the only codimension one bifurcations where Hénon-like families occur; another such example, studied by Diaz, Rocha, and Viana, is the critical saddle-node bifurcation [DRV1], [DRV2].

4.4 Palis has proposed a general program to study the dynamics of (non-uniformly hyperbolic) diffeomorphisms (of compact surfaces, to begin with). He suggests that one should look first to a *dense* subset in the space of non-uniformly hyperbolic diffeomorphisms for which we have at least some grasp of the dynamics; he conjectures actually that homoclinic tangencies could be such a subset. Then one should consider generic parametrized families through these "simple" diffeomorphisms, and study which dynamical features are "persistent" in the measure theoretical sense, i.e. they occur on sets of parameters of positive measure or even relative positive density near the initial diffeomorphism.

Starting with Newhouse-Palis-Takens [NPT] and Palis-Takens ([PT1], [PT2]), there has been a great deal of effort and results related to the study of homoclinic bifurcations, which give rise to an extremely rich number of complicated phenomena; still we are quite far from a satisfactory understanding of f. One very interesting feature of these results, when the fixed saddle point belongs to some (uniformly hyperbolic) basic set, is the subtle relationship between the geometry of the basic set (Hausdorff dimension, thickness, ...) and the dynamics near the bifurcating parameter [PY2].

4.5 Still there is a very central difficulty in carrying out Palis's program, a difficulty that occurs at very many places in dynamical systems, the so-called closing-lemma problem. Pugh's closing lemma [Pu] asserts that if p is a recurrent point for a smooth diffeomorphism f, then one can perturb f *in the C^1-topology* in order for p to become periodic. See also [M1].

We still have no idea whether it is possible to achieve the same goal by a C^2 (or even $C^{1+\varepsilon}$) perturbation. In particular, we still don't know whether the C^2 diffeomorphisms of $\mathbf{T}^2$ that have a periodic orbit form a C^2-dense subset of $\mathrm{Diff}^2(\mathbf{T}^2)$! Guttierez [G] has constructed an example (on the noncompact surface $\mathbf{T}^2 - \{0\}$) that indicates that the localized perturbations used by Pugh in the C^1-case cannot be sufficient in the C^2-case. Also Herman [H4], [H5] has constructed a Hamiltonian flow on a compact symplectic manifold, for which no periodic orbit can be created by smooth perturbations of the Hamiltonian (because of KAM-theory and a symplectic rigidity of the rotation number).

In a similar vein, recent results of Herman suggest that it would be very interesting to know the answer to the following question: Let C be the set of smooth diffeomorphisms (of a compact manifold M) that are of finite order on a nonempty open set (depending on the diffeomorphism); what is the closure of C in the C^∞-topology?

References

[A] V. I. Arnold, *Small denominators I: on the mapping of a circle into itself*, Izv. Akad. Nauk., serie Math. 25, 21–86 (1961) = Trans. Am. Math. Soc., 2nd series, 46, 213–284.

[AD] S. Aubry and P. Y. Le Daeron, *The discrete Frenkel-Kantorova model and its extensions*, Phys. D, 8, 381–422 (1983).

[BC1] M. Benedicks and L. Carleson, *On iterations of* $1-ax^2$ *on* $(-1,1)$, Ann. of Math. (2), 122, 1–25 (1985).

[BC2] M. Benedicks and L. Carleson, *The dynamics of the Hénon map*, Ann. of Math. (2) 133, 73–169 (1991).

[BY] M. Benedicks and L. S. Young, *SBR-measures for certain Hénon maps*, Invent. Math., 112-3, 541–576 (1993).

[Be] J. Bernard, *Sur la dynamique des exemples de Lattès*, thèse Univ. Paris-Sud, (1994).

[Bo] R. Bowen, Equilibrium states and the ergodic theory of Anosov diffeomorphisms, Lecture Notes in Math. 470, (1975), Springer-Verlag, Berlin and New York.

[BH] B. Branner and J. H. Hubbard, *The iteration of cubic polynomials*, Acta Math., 160, 143–206 (1988); 169, 229–325 (1992).

[Br] A. D. Brjuno, *Analytical form of differential equations*, Trans. Moscow Math. Soc. 25, 131–288 (1971); 26, 199–239 (1972).

[CS] C. Q. Cheng and Y. S. Sun, *Existence of invariant tori in three-dimensional measure preserving mappings*, Celestial Mech. 47, 275–293 (1990).

[C] T. M. Cherry, *A singular case of iteration of an analytic function; a contribution to the small divisor problem*, in Nonlinear Problems of Engineering, ed. W. Ames, Acad. Press, New York and San Diego, CA, 29–50 (1964).

[DRV1] L. J. Diaz, J. Rocha, and M. Viana, *Saddle-node cycles and prevalence of strange attractors*, preprint IMPA 1993.

[DRV2] L. J. Diaz, J. Rocha, and M. Viana, *Global strange attractors for dissipative maps of the annulus*, in preparation.

[DH1] A. Douady and J. H. Hubbard, *Étude dynamique des polynômes complexes*, Publ. Math. Orsay 84–82 (1984).

[DH2] A. Douady and J. H. Hubbard, *On the dynamics of polynomial-like mappings*, Ann. Sci. École Norm. Sup. (4), 18, 287–343 (1985).

[FHY] A. Fathi, M. Herman, and J. C. Yoccoz, *A proof of Pesin's stable manifold theorem dans Geometric dynamics*, Proceedings of the Symposium in Rio de Janeiro 1981, Springer Lecture Notes no. 1007.

[G] C. Gutierrez, *A counter-example to a* C^2 *closing lemma*, Ergodic Theory Dynamical Systems 7, 509–530 (1987).

[He] M. Hénon, *A two-dimensional mapping with a strange attractor*, Comm. Math. Phys. 50, 69–77 (1976).

[H1] M. R. Herman, *Sur la conjugaison differentiable des difféomorphismes du cercle à des rotations*, Publ. Math. I.H.E.S. 49, 5–233 (1979).

[H2] M. R. Herman, *Simple proofs of local conjugacy theorems for diffeomorphisms of the circle with almost every rotation number*, Bol. Soc. Brasil. Mat., 16, 45–83 (1985).

[H3] M. R. Herman, *Sur les courbes invariantes par les difféomorphismes de l'anneau*, Astérisque, vol. 1, 103–104 (1983), vol. 2, 144 (1986).

[H4] M. R. Herman, *Exemples de flots hamiltoniens dont aucune perturbation en topologie C^∞ n'a d'orbites périodiques sur un ouvert de surfaces d'énergie*, C.R. Acad. Sci. Paris, t. 321, Série I, 989–994 (1991).

[H5] M. R. Herman, *Differentiabilité optimale et contre-exemples à la fermeture en topologie C^∞ des orbites récurrentes des flots hamiltoniens*, C.R. Acad. Sci. Paris, t. 313, Série I, 49–51 (1991).

[H6] M. R. Herman, *Théorème des tores translatés et quelques applications à la stabilité topologique des systèmes dynamiques consrvatifs*, en preparation.

[J] M. V. Jakobson, *Absolutely continuous invariant measures for one-parameter families of one-dimensional maps*, Comm. Math. Ph. 81, 39–88 (1981).

[JN] M. V. Jakobson and S. Newhouse, *Strange attractors in strongly dissipative surface diffeomorphisms*, in preparation.

[JS] M. V. Jakobson and G. Swiatek, *Metric properties of non-renormalizable S-unimodal maps*, preprint (1994).

[KO1] Y. Katznelson and D. Ornstein, *The differentiability of conjugation of certain diffeomorphisms of the circle*, Ergodic Theory Dynamical Systems 9, 643–680 (1989).

[KO2] Y. Katznelson and D. Ornstein, *The absolute continuity of conjugation of certain diffeomorphisms of the circle*, Ergodic Theory Dynamical Systems 9, 681–690 (1989).

[KS] K. M. Khanin and Ya. G. Sinai, *A new proof of M. Herman's theorem*, Comm. Math. Phys. 112, 89–101 (1987).

[LeC1] P. Le Calvez, *Propriétés dynamiques des zones d'instabilité*, Ann. Sci. École Norm. Sup. (4), 20, 443–464 (1987).

[LeC2] P. Le Calvez, *Étude topologique des applications deviant la verticale*, Ensaios Mat., vol. 2, 1990.

[Ly1] M. Lyubich, *On the Lebesgue measure of the Julia set of a quadratic polynomial*, preprint Stonybrook (1991).

[Ly2] M. Lyubich, *Geometry of quadratic polynomials: Moduli, rigidity and local connectivity*, preprint Stonybrook (1993).

[M1] R. Mañé, *An ergodic closing lemma*, Ann. of Math. (2) 116, 503–540 (1982).

[M2] R. Mañé, Ergodic Theory and Differentiable Dynamics, Springer-Verlag, Berlin and New York, 1987.

[M3] R. Mañé, *A proof of the C^1 stability conjecture*, Publ. Math. I.H.E.S. 66, 161–210 (1988).

[Ma1] J. Mather, *Existence of quasiperiodic orbits for twist diffeomorphisms of the annulus*, Topology, 21, 457–467 (1982).

[Ma2] J. Mather, *Minimal measures*, Comm. Math. Helv., 64, 375–394 (1989).

[Ma3] J. Mather, *Action minimizing invariant measures for positive definite Lagrangian systems*, Math. Z., 207, 169–207 (1991).

[McM] C. McMullen, *Renormalization and 3-manifolds which fiber over the circle*, preprint (1994).

[dMvS] W. de Melo and S. van Strien, One Dimensional Dnamics, Springer-Verlag, Berlin and New York, 1993.

[MV] L. Mora and M. Viana, *Abundance of strange attractors*, Acta Math. 171, 1–71 (1993).

[Mo1] J. Moser, *A rapidly convergent iteration method, part II*, Ann. Scuola Norm. Sup. Pisa Cl. Sci. (3), 20, 499–535 (1966).

[Mo2] J. Moser, *On invariant curves of area preserving mappings of an annulus*, Nachr. Akad. Wiss. Göttingen, Math.-Phys. Kl., 1–20 (1962).

[Mo3] J. Moser, *On commuting circle mappings and simultaneous diophantine approximations*, Math. Z. 205, 105–121 (1990).

[N1] S. Newhouse, *Diffeomorphisms with infinitely many sinks*, Topology, 13, 9–18 (1974).

[N2] S. Newhouse, *The abundance of wild hyperbolic sets and non smooth stable sets of diffeomorphisms*, Publ. Math. I.H.E.S. 50, 101–151 (1979).

[NPT] S. Newhouse, J. Palis, and F. Takens, *Bifurcations and stability of families of diffeomorphisms*, Publ. Math. I.H.E.S. 57, 5–71 (1983).

[O] V. Oseledec, *A multiplicative ergodic theorem: Lyapunov characteristic numbers for dynamical systems*, Trans. Moscow Math. Soc. 19, 197– 231 (1968).

[P] J. Palis, *On the* $C^1 - \Omega$*-stability conjecture*, Publ. Math. I.H.E.S. 66, 211–215 (1988).

[PdM] J. Palis and W. de Melo, Geometric Theory of Dynamical Systems, Springer-Verlag, Berlin and New York, 1982.

[PT1] J. Palis and F. Takens, *Cycles and measure of bifurcation sets for two-dimensional diffeomorphisms*, Invent. Math. 82, 397–422 (1985).

[PT2] J. Palis and F. Takens, *Hyperbolicity and the creation of homoclinic orbits*, Ann. of Math. (2) 125, 337–374 (1987).

[PT3] J. Palis and F. Takens, Hyperbolicity and Sensitive Chaotic Dynamics, Cambridge Univ. Press, London, 1993.

[PY1] J. Palis and J. C. Yoccoz, *Differentiable conjugacy of Morse Smale diffeomorphisms*, Bol. Soc. Brasil. Mat. 25, 25–48 (1990).

[PY2] J. Palis and J. C. Yoccoz, *Homoclinic tangencies for hyperbolic sets of large Hausdorff dimension*, Acta Math. 172, 92–136 (1994).

[PM1] R. Perez-Marco, *Solution complète au problème Siegel de linearisation d'une application holomorphe au voisinage d'un point fixe*, Sem. Bourbaki no. 753, Astérisque, 206 (1992).

[PM2] R. Perez-Marco, *Sur les dynamiques holomorphes non linéarisables et une conjecture de V. I. Arnold*, Ann. Sci. École Norm. Sup. (4), 26, 565–644 (1993).

[PM3] R. Perez-Marco, *Non-linearizable dynamics having an noncountable number of symmetries*, to appear in Invent. Math.

[PM4] R. Perez-Marco, *Fixed point and circle maps*, in préparation.

[Pe1] Y. Pesin, *Families of invariant manifolds corresponding to non- zero characteristic exponents*, Math. USSR-Izv. 10, 1261–1305 (1976).

[Pe2] Y. Pesin, *Characteristic Lyapunov exponents and ergodic theory*, Russian Math. Surveys 32, 55–114 (1977).

[Pu] C. Pugh, *The closing lemma*, Amer. J. Math. 89, 956–1009 (1967).

[Re] M. Rees, *Positive measure sets of ergodic rational maps*, Ann. Sci. École Norm. Sup. (4), 19, 383–407 (1986).

[R] J. Robbin, *A structural stability theorem*, Ann. of Math. (2) 94, 447–493 (1971).

[Ro] C. Robinson, *Structural stability of vector fields*, Ann. of Math. (2) 99, 154–175 (1974).

[Ru1] H. Rüssmann, *Kleine Nenner I: Über invariante Kurven differenzierbarer Abbildungen eines Kreisringes*, Nachr. Akad. Wiss. Göttingen Math.-Phys. Kl., 67–105 (1970).

[Ru2] H. Rüssmann, *Non-degeneracy in the perturbation theory of integrable dynamical systems*, in Number Theory and Dynamical Systems, Dodson-Vickers (eds), L.N. 134, Lond. Math. Soc., 1–18 (1989).

[Sh] M. Shub, *Stabilité globale des systèmes dynamiques*, Astérisque 50 (1978).

[Si] C. L. Siegel, *Iteration of analytic functions*, Ann. of Math. (2) 43, 807–812 (1942).

[Sm] S. Smale, *Differentiable dynamical systems*, Bull. Amer. Math. Soc. 73, 747–817 (1967).

[Su1] D. Sullivan, *Bounded structure of infinitely renormalizable mappings*, in Universality and Chaos, ed. P. Cvitanovic, 2nd edition, Adam Hilger, 1989.

[Su2] D. Sullivan, *Bounds, quadratic differentials and renormalization conjectures*, in AMS Centennial Publications, vol. 2, Mathematics in the 21st Century, 1992.

[Sw] G. Swiatek, *Hyperbolicity is dense in the real quadratic family*, preprint Stonybrook, 1992.

[T] F. Takens, *Homoclinic bifurcations*, Proc. Internat. Congress Math., Berkeley (1986), 1229–1236.

[V] M. Viana, *Strange attractors exhibiting multidimensional expansion*, in preparation.

[Y1] J. C. Yoccoz, *Il n'y a pas de contre-exemple de Denjoy analytique*, C.R. Acad. Sci. Paris Série I, t.298 (1984), 141–144.

[Y2] J. C. Yoccoz, *Conjugaison différentiable des difféomorphismes du cercle dont le nombre de rotation vérifie une condition diophantienne*, Ann. Sci. École Norm. Sup. (4), t.17 (1984), 333–359.

[Y3] J. C. Yoccoz, *Centralisateurs et conjugaison différentiable des difféomorphismes du cercle*, Thèse d'Etat, Juin 1985, Université de Paris-Sud.

[Y4] J. C. Yoccoz, *Théorème de Siegel, nombres de Brjuno et polynômes quadratiques*, manuscrit (1987).

[Y5] J. C. Yoccoz, *Linéarisation des germes de difféomorphismes holomorphes de* $(\mathbf{C}, 0)$, C.R. Acad. Sci. Paris Série I, t.306 (1988), 55–58.

[Y6] J. C. Yoccoz, *Conjugaison analytique des difféomorphismes du cercle*, manuscrit (1989).

[Y7] J. C. Yoccoz, *Polynômes quadratiques et attracteur de Hénon*, Sém. Bourbaki no. 734, Astérisque (1990).

[Y8] J. C. Yoccoz, *Travaux de Herman sur les tores invariants*, Sém. Bourbaki no. 754, fevrier 1992.

[Y9] J. C. Yoccoz, *Hyperbolic dynamics*, cours donné à Hilleröd 1993, basé sur des notes de Workshop on Dynamical Systems, Int. Center Theor. Physics, Trieste, 1991, à paraître chez Kluwer.

[Y10] J. C. Yoccoz, *Vers la connexité locale du lieu de connexité des polynômes quadratiques*, en préparation.

Invited Forty-Five Minute Addresses at the Section Meetings

On the Size of Quotients by Definable Equivalence Relations

ALAIN LOUVEAU

CNRS, Equipe d'Analyse
Université Paris 6
Tour 46-4e, 4 Place Jussieu
F-75252 Paris Cedex 05, France

Introduction

The study of simply definable equivalence relations, and in particular of Borel and analytic ones, on Polish spaces, has attracted a lot of attention in descriptive set theory since 1970, when the first fundamental result in the subject, Silver's theorem, was proved.

One motivation has certainly been the conjecture of Vaught in model theory, one of the oldest still open problems in mathematical logic, which can be interpreted as a question about certain analytic equivalence relations. Another motivation is that these objects are extremely common in many different fields of mathematics, and the answers to basic facts about them are certainly desirable. A third motivation is that the progress made on the tools and techniques of what is now called "effective" descriptive set theory, which is instrumental in many results in the subject, made it possible to hope for nontrivial results.

Since the 1970s, the study of these equivalence relations has been developed in many directions. This article will only consider one central question: Given a Polish space X, and a simply definable equivalence relation E on X, what can be said about the size of the quotient set X/E of equivalence classes?

In the rest of this paper, X will always denote a Polish space, i.e. a topological space homeomorphic to a separable complete metric space. And as all uncountable Polish spaces are Borel isomorphic, and we will work up to Borel isomorphism anyway, one can think of X as being the Cantor space $C = \{0,1\}^{\mathbb{N}}$. (One could also consider more general domains, like analytic or coanalytic ones, but simple manipulations usually allow us to reduce the questions to Polish domains.) It will be understood that X is the domain of an equivalence relation E. And as E is a subset of the Polish space X^2, definability properties of E make sense in that space. This paper will focus on the Borel equivalence relations and on the analytic ones. (A set is *analytic* in a Polish space if it is a continuous image of a Borel set. It is *coanalytic* if its complement is analytic.) What about the "size" of the quotient X/E? The usual set-theoretic notion is that of cardinality, and it is historically the first that has been studied, again in relation with Vaught's conjecture. In the first section of this paper, I will present the dichotomy results

Proceedings of the International Congress
of Mathematicians, Zürich, Switzerland 1994

of Silver (which imply that for Borel E, the quotient is either countable or of cardinality the continuum) and Burgess (that for analytic E, and regardless of the exact value of the continuum, the only other possibility is $\aleph_1$). I will also relate these results to Vaught's conjecture.

In a second section, I will introduce and study another notion of "size" for quotients, more in the spirit of descriptive set theory, where the one-to-one maps between quotients that compare their size are now assumed to be "definable" (i.e. to come from definable maps on the domains). This leads to a partial ordering between equivalence relations, first introduced by Friedman and Stanley, for which it is still possible to get dichotomy results in the Borel case (Harrington-Kechris-Louveau) and in the analytic case (Becker, Hjorth-Kechris), which I will discuss, together with some other properties of this ordering.

This article is not a research paper (in particular, no proofs are given, and it is a bit of a pity, for proofs convey often more of the flavor of the subject than the results). It is also not a survey paper (even for the narrow problem it considers, this would need much more space), but a mere introduction to the subject, with a chosen sample of results (many important works, especially about the Vaught conjecture and its generalization, the topological Vaught conjecture, are not mentioned). For the reader interested in the subject, there should soon be available a monograph, by Kechris ([K]), called "Lectures on definable group actions and equivalence relations", that will contain the material presented here, and much, much more.

1 Cardinality of quotients

The following result of Silver gives the possible cardinalities of quotients X/E, when E is a Borel, or even a coanalytic, equivalence relation on a Polish space X.

THEOREM (SILVER [SI]). *Let E be a coanalytic equivalence relation on the Polish space X. Then either*

(i) X/E is countable

or

(ii) There is in X a copy of the Cantor space C consisting of pairwise E-inequivalent elements (so in particular X/E has cardinality the continuum).

A subset of X that picks at most one point in each equivalence class of E is called a *partial transversal*, and a *transversal* if in addition it meets all classes. So Silver's theorem asserts that if X/E is uncountable, it admits a partial transversal that is a perfect set homeomorphic to C. One then says that E has *perfectly many* classes. The question of the existence of well-behaved (total) transversals will be discussed in the next section.

Silver's result is reminiscent of the classical perfect set theorem for analytic sets — which indeed can be viewed as a particular case of it. But it is much harder to prove. Silver's original proof is both difficult and metamathematical. A later simpler proof by Harrington (unpublished) had a considerable influence on the subject because it brought into it the "effective" techniques borrowed from the theory of definability on the integers, hence based ultimately on computability theory, that are now central to the subject.

Silver's result was the starting point of a series of investigations about the number of classes of definable equivalence relations. The next result takes care of the analytic case.

THEOREM (BURGESS [B]). *Let E be an analytic equivalence relation on the Polish space X. Then either*
(i) X/E has cardinality at most $\aleph_1$
or
(ii) E has perfectly many classes.

This result says that, compared to the Borel case, the analytic case adds at most one possibility, namely that of $\aleph_1$ many, but not perfectly many, equivalence classes. Such analytic equivalence relations do exist: it is easy to give the set LO of linear orders with domain $\mathbb{N}$ a Polish topology. For this topology, the following relation between linear orders is analytic: two orders are equivalent if they are well-orders of the same length, or if both are not well-orders. This relation does not have perfectly many classes, and its quotient space is isomorphic to the set ω_1 of countable ordinals.

After the results of Silver and Burgess, investigations have been pursued in two main directions. The first one concerns even more complicated definable equivalence relations. For this line of research, see the papers of Harrington-Sami [HSa] and Harrington-Shelah [HSh].

Another line of research is concerned with the possibility of extending Silver's dichotomy result to interesting subclasses of analytic equivalence relations, in particular in relation with Vaught's conjecture in model theory.

Vaught's original conjecture is the statement that any first-order theory (in a countable language) has either countably many or else continuum-many nonisomorphic countable models.

As in the case of linear orders, it is possible to equip the set of models of a given first-order theory that have domain $\mathbb{N}$ with a natural Polish topology, for which the isomorphism relation becomes analytic, but not always Borel. So Silver's result does not apply to this situation, and the result of Burgess is inconclusive.

On the other hand, the isomorphism equivalence relations that arise in model theory are not arbitrary analytic equivalence relations. They are a particular case of the following situation: Suppose we are given a Polish group G (i.e. a topological group that is a Polish space), and a Borel action $\alpha : G \times X \to X$ of G on X. One can then consider the associated orbit equivalence relation E_α on X defined by

$$xE_\alpha y \leftrightarrow \exists g \in G \ \alpha(g, x) = y$$

(the model theory case corresponds to the natural action of the symmetric group $S_\infty(\mathbb{N})$ of permutations of $\mathbb{N}$ on the space of models with domain $\mathbb{N}$).

Orbit equivalence relations of Borel actions of Polish groups are analytic (and again not always Borel). The natural generalization in descriptive set theory of Vaught's conjecture is the so-called topological Vaught conjecture: "For any Borel action of a Polish group on a Polish space, the orbit equivalence relation has either countably or perfectly many classes."

A lot of work has been done on the two conjectures, both in model theory and in descriptive set theory, and it is impossible to give here a fair account of the known results.

In model theory, Vaught's conjecture has been established for many theories, in particular among the theories for which the classification tools apply. The best results at present are that Vaught's conjecture holds for the so-called ω-stable theories (Harrington-Makkai-Shelah [HMS]), and the superstable theories of finite rank (Buechler [Bu]). The conjecture is also known for some specific theories, trees for example (Steel [St]).

In descriptive set theory, the topological Vaught conjecture is known for various acting groups (the locally compact ones, the abelian ones), and for orbit equivalence relations with special (rather technical) properties (see Steel [St]). Still, both conjectures are open.

2 An alternative notion of size for quotients

DEFINITION. *Let E and F be two equivalence relations on the Polish spaces X and Y, respectively. A reduction of E to F is a map $f : X \to Y$ that satisfies, for all x and y in the space X,*

$$xEy \leftrightarrow f(x)Ff(y).$$

We say that E is Borel reducible to F, or F Borel reduces E, written $E \preceq F$, if there exists a Borel reduction of E to F. The relations E and F are Borel bireducible, written $E \cong F$, if both $E \preceq F$ and $F \preceq E$.

Note that a reduction f of E to F induces a quotient map $f^* : X/E \to Y/F$, which is one-to-one. So the idea of Borel reducibility is to compare the sizes of quotients not by arbitrary one-to-one maps, as is done in cardinality theory, but by one-to-one maps that are induced by Borel functions. And heuristically the class of equivalence relations that are Borel bireducible with a given E is a measure of the "Borel size" of the quotient space X/E.

The notion of Borel reducibility was first introduced, in the context of model theory, by Friedman and Stanley [FS]. It corresponds there to the search, for countable models of a given theory, of invariants that describe the models up to isomorphism, and is used in [FS] as a way of comparing theories.

What are the possible "Borel sizes" of quotients? Let us first consider the case of Borel equivalence relations.

First, Silver's dichotomy result can be interpreted in terms of the ordering $\preceq$. Denote by $\Delta(A)$, for any set A, the relation of equality on A. Then Silver's result asserts that for any Borel E, either $E \preceq \Delta(\mathbb{N})$, or else $\Delta(C) \preceq E$, where C is the Cantor space. So among uncountable Borel quotients, C has the smallest "Borel size".

The equivalence relations that are Borel reducible to equality on C are called *smooth*. They can alternatively be characterized as those Borel equivalence relations that admit a countable Borel separating family, i.e. a sequence (B_n) of Borel E-invariant sets such that

$$xEy \leftrightarrow \forall n(x \in B_n \leftrightarrow y \in B_n)$$

and also as those for which the σ-algebra of E-invariant Borel sets is countably generated.

Moreover, using a selection theorem of Jankov and von Neumann, smooth equivalence relations admit C-measurable transversals. (In a Polish space, a set is *C-measurable* if it is in the smallest σ-algebra containing the open sets and closed under Suslin operation. These sets are more general than the analytic and coanalytic ones, but still well behaved, in particular, universally measurable.)

Closed, and even G_δ equivalence relations are smooth. But it is well known that there are simple F_σ equivalence relations with no universally measurable transversals, hence that are not smooth, e.g. Vitali's famous example of equality mod $\mathbb{Q}$ on the space $\mathbb{R}$. Here we will consider another example (which happens to be Borel bireducible to Vitali's example), the relation E_0 of eventual equality on the Cantor space C, given by

$$\alpha E_0 \beta \leftrightarrow \exists n \forall m \geq n \ \alpha(m) = \beta(m).$$

This F_σ equivalence relation is not smooth, hence C/E_0 represents a new quotient, of "Borel size" bigger than C. The next dichotomy result, which extends earlier results of Glimm [G] and Effros [E1], [E2], says that it is indeed the smallest Borel quotient above C.

THEOREM (HARRINGTON-KECHRIS-LOUVEAU [HKL]). *Let E be a Borel equivalence relation on the Polish space X. Then either*
(i) E is smooth, i.e. E is Borel reducible to $\Delta(C)$,
or
(ii) $E_0 \preceq E$, in fact there is a one-to-one and continuous reduction of E_0 to E.

A consequence of this result is the following: Say that a probability measure μ on X is *ergodic* for E if every Borel E-invariant set has μ-measure 0 or 1, and *nonatomic* for E if every E-class has μ-measure 0. For example, $\Delta(C)$ admits no ergodic nonatomic measures, but the usual Lebesgue measure on C is both ergodic and nonatomic for E_0. It is easy to check that ergodicity and nonatomicity are preserved under images by Borel (or even universally measurable) reductions. Hence the previous result implies that, for a Borel equivalence relation, smoothness is indeed equivalent to the existence of a C-measurable transversal (this is the way smoothness is proved for G_δ equivalence relations), and also that the relation $E_0 \preceq E$ is equivalent to the existence of an ergodic and nonatomic probability measure for E.

Equivalence relations that are Borel reducible to E_0 are called *hyperfinite*. Among the Borel equivalence relations with countable equivalence classes, they can be characterized as those induced by a Borel automorphism of X, i.e. as the orbit equivalence relations of Borel $\mathbb{Z}$-actions. So they are the ones considered in ergodic theory (although in a different context, as the domain is a measure space, and the automorphism is usually assumed to be nonsingular with respect to the measure). Hyperfinite equivalence relations with countable equivalence classes have been extensively studied by Dougherty, Jackson, and Kechris, in a paper [DJK] that contains in particular a complete classification up to Borel isomorphism.

However, there is no known dichotomy result, similar to the Harrington-Kechris-Louveau result, that would separate the hyperfinite equivalence relations from the other ones, and some basic questions about them are open. For example, it is not known exactly which countable groups have the property that all their Borel actions have hyperfinite orbit equivalence relations. It holds when the group is $\mathbb{Z}^n$ (Weiss) or more generally is a finitely generated group of polynomial growth (Jackson-Kechris-Louveau), and it does not hold if the group is not amenable (in analogy with the measure case, amenability should be the right conjecture).

Above E_0, the partial order of Borel reducibility between Borel equivalence relations is not very well understood, and the known results indicate that it has a rather complicated structure.

First, it has a cofinal sequence of length ω_1, but no maximum element (Friedman). In fact, if one defines, for a Borel equivalence relation E on X, the Borel relations E^+ and E^* on $X^{\mathbb{N}}$ by

$$(x_n)_{n\in\mathbb{N}}\ E^+\ (y_n)_{n\in\mathbb{N}} \leftrightarrow \{x_n/E : n \in \mathbb{N}\} = \{y_n/E : n \in \mathbb{N}\},$$

where x/E denotes the E-class of x, and

$$(x_n)_{n\in\mathbb{N}}\ E^*\ (y_n)_{n\in\mathbb{N}} \leftrightarrow \exists n\ \forall m \geq n\ x_n E y_n,$$

then for E with at least two classes, both E^+ (Friedman) and E^* (Louveau) are strictly bigger than E in the ordering $\preceq$. These two "jump operators" are also incomparable (e.g. the relations $\Delta(C)^+$ and $\Delta(C)^*$ are $\preceq$-incomparable).

Another result (Kechris-Louveau) is that for any Borel E strictly above E_0, there is always another Borel E' that is $\preceq$-incomparable to it. It follows that there can be no similar dichotomy results above E_0, at least for the whole class of Borel equivalence relations. Finally (in this list of negative results), Louveau and Velickovic [LV] have shown that the partial order of almost inclusion between subsets of $\mathbb{N}$ can be embedded, as a partial order, in the order $\preceq$ between Borel equivalence relations. Hence so does any partial order of size at most $\aleph_1$, and there is no hope to describe in a reasonable way the various possible "Borel sizes" of Borel quotients.

The ongoing research on this partial ordering focuses on some important subclasses of Borel equivalence relations, like the ones that have countable equivalence classes. For more on the subject, we refer the reader to the forthcoming monograph of Kechris [K].

Let us consider now the analytic equivalence relations. First, one should relax a bit the notion of definable reducibility used to compare the quotients, and consider, say, C-measurable reductions instead of Borel ones. Even then, the Harrington-Kechris-Louveau result does not extend to the analytic case.

Consider for example, in model theory, the relation E of isomorphism between countable abelian torsion groups. These groups can be classified, up to isomorphism, by invariants called the Ulm invariants. These invariants can be considered as transfinite sequences of 0's and 1's of countable length, i.e. elements of $\{0, 1\}^{<\omega_1}$. If one considers only groups with Ulm invariant of fixed bounded length, isomorphism is smooth on them. This prevents E_0 from being reducible to the isomorphism in any reasonable way. But the length of the Ulm invariants cannot be bounded below ω_1, and this prevents E from being reducible to $\Delta(C)$ in any reasonable way.

The following very recent and unpublished result says that the situation above is in fact typical. To state the result, note that, like the set ω_1, the set $\{0,1\}^{<\omega_1}$ can be realized as the quotient of the Cantor space C by some analytic equivalence relation, call it E_1. One then has

THEOREM (BECKER, HJORTH-KECHRIS). *Let E be the orbit equivalence relation of the Borel action of a Polish group on the Polish space X. Then either*
(i) E is reducible, via a C-measurable reduction, to E_1,
or
(ii) E_0 is continuously reducible to E.

This result could have interesting consequences in model theory. It suggests that some of the theories with continuum many nonisomorphic models, which are usually considered unclassifiable, might still admit some kind of classification, analogous to the Ulm invariants for abelian torsion groups, when E_0 is not reducible to the isomorphism relation. But work has to be done, first to understand what it means for a theory that E_0 is not reducible to its isomorphism relation, and secondly in order to replace the abstract invariants of case (i) of the theorem by concrete ones in specific situations.

Let us mention finally that a weak version of the above theorem, valid for arbitrary analytic equivalence relations, has also been established by Hjorth and Kechris, using a strong set-theoretic assumption.

References

[Bu] S. Buechler, *Vaught's conjecture for superstable theories of finite rank*, preprint.

[B] J.P. Burgess, *Equivalences generated by families of Borel sets*, Proc. Amer. Math. Soc. **69** (1978), 323–326.

[DJK] R. Dougherty, S. Jackson, and A.S. Kechris, *The structure of hyperfinite Borel equivalence relations*, Trans. Amer. Math. Soc. **341** (1994), 193–225.

[E1] E.G. Effros, *Transformation groups and C^*-algebras*, Ann. of Math. (2), **81** (1965), 38–55.

[E2] E.G. Effros, *Polish transformation groups and classification problems*, in: General Topology and Modern Analysis (Rao and McAuley, eds.), Academic Press, San Diego, CA (1980), 217–227.

[FS] H. Friedman and L. Stanley, *A Borel reducibility theory for classes of countable structures*, J. of Symbolic Logic **54** (1989), 894–914.

[G] J. Glimm, *Locally compact transformation groups*, Trans. Amer. Math. Soc. **101** (1961), 124–138.

[HKL] L.A. Harrington, A.S. Kechris, and A. Louveau, *A Glimm-Effros dichotomy for Borel equivalence relations*, J. of the Amer. Math. Soc. **3** (1990), 903–928.

[HMS] L.A. Harrington, M. Makkai, and S. Shelah, *A proof of Vaught's conjecture for ω-stable theories*, Israel J. Math. **49** (1984), 259–280.

[HSa] L.A. Harrington and R. Sami, *Equivalence relations, projective and beyond*, in: Logic Colloquium'78, Studies in Logic and the Foundation of Mathematics 97, North-Holland (1979), 247–264.

[HSh] L.A. Harrington and S. Shelah, *Counting equivalence classes for co-κ-Suslin equivalence relations*, in: Logic Colloquium'80, Studies in Logic and the Foundation of Mathematics **108**, North-Holland, Amsterdam (1982), 147–152.

[K] A. S. Kechris, *Lectures on definable group actions and equivalence relations*, circulating manuscript.

[LV] A. Louveau and B. Velickovic, *A note on Borel equivalence relations*, Proc. Amer. Math. Soc. **120** (1994), 255–259.

[Si] J. Silver, *Counting the number of equivalence classes of Borel and coanalytic equivalence relations*, Ann. of Math. Logic **18** (1980), 1–28.

[St] J.R. Steel, *On Vaught's conjecture*, In: Cabal Seminar 76–77 (Kechris and Moschovakis, eds.), Lecture Notes in Math. **689**, Springer-Verlag, Berlin and New York (1978), 193–208.

Model Theory, Differential Algebra, and Number Theory

ANAND PILLAY

University of Notre Dame
Notre Dame, IN 46556, USA

1 Introduction

The past few years have seen some new connections developing between model theory (a branch of mathematical logic) and both algebra and number theory. Part of the novelty, for the model theorists, is that this work depends on the techniques, machinery, and point of view of *stability theory*, whose use, up to now, has been largely, but not exclusively, confined to problems of *pure* model theory (such as classifying first order theories and their models).

For us here, differential fields (and more generally, fields with operators) will serve as objects of study, as well as intermediaries between model theory and diophantine geometry. The connection between differential algebra and diophantine geometry goes back at least to Manin's proof [M] of the Mordell conjecture for function fields. Lang made a conjecture (discussed in Section 4) that generalized Mordell's conjecture. Buium [B1], [B2] and Hrushovski [H] (separately) used differential algebra to prove the function field version of Lang's conjecture. Hrushovski's proof uses in addition some model-theoretic methods and theorems and generalizes to the positive characteristic case (where the model theory of separably closed fields replaces that of differentially closed fields).

On the other hand, differential algebra (the study of fields equipped with derivations, and the resulting *algebraic* study of solutions of algebraic differential equations) also has a relatively long history. The theory was developed by Ritt and Kolchin. An important achievement of Kolchin [K1] was his "differential Galois theory of strongly normal extensions", which generalized the theory of Picard-Vessiot extensions, and was moreover one of the starting points of the modern theory of algebraic groups; algebraic groups are precisely the Galois groups of "strongly normal" extensions. In Section 3, we present a further generalization of Kolchin's theory, where the Galois groups are precisely "finite-dimensional differential algebraic groups".

The general theory of differential algebraic groups (which can be thought of either as group objects in differential algebraic geometry, or as groups definable in differentially closed fields) was developed by Cassidy [C1], [C2] and then by Kolchin [K2]. Finite-dimensional such groups have already been mentioned in connection with differential Galois theory, but also play a fundamental role in the aforementioned diophantine questions. In Section 2 we give a classification of the finite-dimensional "simple" differential algebraic groups (continuing the work of

Proceedings of the International Congress
of Mathematicians, Zürich, Switzerland 1994

Cassidy and Buium). This leads also to a solution of the problem of how many countable differentially closed fields of characteristic 0 there are. In Section 2 we also give solutions to some problems stated by Kolchin [K2] concerning arbitrary (not necessarily finite-dimensional) differential algebraic groups, such as their embeddability in algebraic groups.

It is now time to mention model theory. The reader is referred to Hodges' recent book [Ho] and also [P] for the basic ingredients of first order logic, model theory, and basic stability theory. [Pi1] and [Pi2] are recommended for stability theory. Many of the basic notions of stability theory and stable group theory also appear in the survey article [Pi3]. Model theory studies, among other things, the family of first order definable sets in a given structure M. Under various hypotheses (such as stability) on the structure M, an array of tools has been developed, largely by Shelah, for understanding these definable sets and their mutual interaction. In particular, there is a highly developed theory of *independence, orthogonality*, and *stable groups*, all of which play important roles in the applications. A basic point is that differentially closed fields (introduced below) are ω-stable structures, and all the techniques apply. Separably closed fields are stable. The model completion of the theory of fields with an automorphism, although unstable, turns out to have a manageable model theory.

If M is a model of a stable theory, $A \subseteq M$, and $\mathbf{a}, \mathbf{b}$ are tuples from M, the theory of forking allows us to give sense to the statement "**a** is independent from **b** over A". If also M is ω-stable, then definable sets, and also types (intersections of definable sets) can be assigned an ordinal valued "dimension" (often called a "rank") in such a way that **a** is independent from **b** over A iff $\dim(\text{tp}(\mathbf{a}/A\mathbf{b})) = \dim(\text{tp}(\mathbf{a}/A))$. Among such ranks we shall use Morley-rank (for definable sets and types) and U-rank (for types and definable groups). So if X is a definable set in M (say $X \subseteq M^n$), defined with parameters in $A \subseteq M$, we will be able to speak of a point $\mathbf{a} \in X$ being a *generic* point of X over A (namely if Morley-rank(X) = Morley-rank$(\text{tp}(\mathbf{a}/A))$). If M happens to be an algebraically closed field $(K, +, .)$, then the definable sets in M are precisely the *constructible* sets (finite Boolean combinations of affine algebraic sets), and Morley-rank = algebro-geometric dimension. So in this sense the model-theoretic context generalizes the Weil point of view and language of algebraic geometry. Partly for this reason we will follow Weil's language of varieties, rather than the language of schemes, in any algebraic-geometrical considerations.

It is worth recalling the definition of Morley-rank. If M is a sufficiently saturated structure and X is a set definable (with parameters) in M, then Morley-rank$(X) \geq \alpha + 1$ if there is a set $\{Y_i : i < \omega\}$ of pairwise disjoint nonempty definable sets in M such that for all $i < \omega$, $Y_i \subseteq X$ and Morley-rank$(Y_i) \geq \alpha$. If Morley-rank$(X) = \alpha$, then X is said to have multiplicity 1 if X cannot be partitioned into two disjoint definable sets, each of Morley-rank α. A *strongly minimal* set is defined to be a definable set of Morley-rank 1 and multiplicity 1. The strongly minimal set D is said to be *locally modular* if (after naming parameters) for any **a**, **b** from D, **a** is independent from **b** over $acl(\mathbf{a}) \cap acl(\mathbf{b})$.

Finally in this introduction we introduce and give the basic properties of differentially closed fields. All the differential algebras we will mention here will be in characteristic 0. An (ordinary) differential field is a field F equipped with a derivation $\delta: F \to F$. There is a natural notion of a differential polynomial $P(X_1, \ldots, X_n)$ over F, in the differential indeterminates $X_1, \ldots, X_n$. (P will be simply an ordinary polynomial over F in indeterminates $\delta^j(X_i)$, for $j < \omega$ and $i = 1, \ldots, n$.). If $n = 1$, then we can speak of the *order* of $P(X)$ (the greatest m such that $\delta^m(X)$ occurs in P). The differential field F is said to be *differentially closed* if whenever $P(X)$, $Q(X)$ are differential polynomials over F, with order$(Q) <$ order(P), then $P(X) = 0$ and $Q(X) \neq 0$ has a solution in F. DCF_0 denotes the theory of differentially closed fields (of characteristic 0), in the language with symbols for $+, ., 0, 1, \delta$ (namely the theory whose axioms are those for differential fields, together with those expressing the existence of solutions to the above systems). See Poizat's book [P], or Marker's survey paper [Mar] for the following facts, and also for attributions.

FACT 1.1. DCF_0 *is consistent, complete, has quantifier elimination, and is ω-stable.*

A similar result holds for "partial" differential fields, namely fields equipped with a fixed finite set of commuting derivations. All our results generalize to this context.

FACT 1.2. DCF_0 *has elimination of imaginaries.*

This means that if F is a model of DCF_0, $X \subseteq F^n$ is a definable set, and E is a definable equivalence relation on X, then there is some definable set $Y \subseteq F^k$ (for some k) and a definable bijection between X/E and Y.

If U is a κ-saturated model of DCF_0 of cardinality κ (for some large κ) then U is a "universal domain" for differential algebraic geometry, in the sense of Kolchin. In particular, if F is a differential subfield of U of cardinality $< \kappa$ and L is a differential field extension of F of cardinality $\leq \kappa$, then there is an embedding of L into U over F. Quantifier elimination of DCF_0 implies that if F is a differential subfield of U and F is differentially closed then F is an *elementary* substructure of U.

If F is a differential subfield of U and $A \subseteq U$, then $F\langle A\rangle$ denotes the differential field generated by F and A. The notion of independence (coming from stability) takes the following form here: if F is a differential subfield of U then $\mathbf{a}$ is independent from $\mathbf{b}$ over F just if $F\langle\mathbf{a}\rangle$ is *algebraically disjoint* from $F\langle\mathbf{b}\rangle$ over F.

In the next section we will work inside such a universal domain U. An important feature of DCF_0 (compared with say the theory of algebraically closed fields) is that, from the point of view of Morley-rank, definable sets may be infinite-dimensional. For example U itself has Morley-rank ω. On the other hand C_U, the field of constants of U, is a definable set of Morley-rank 1.

Now U, as an algebraically closed field, is also a universal domain for algebraic geometry. We will be concerned with both sets definable in the structure $(U, +, ., \delta)$, and sets definable in the "reduct" $(U, +, .)$. We will call the former δ-definable sets, and the latter f-definable sets. Of course any f-definable set is δ-definable, but not vice versa. We will identify an algebraic variety V over U with its set of U-rational points, which will be an f-definable set. Although V has finite Morley-rank in the structure $(U, +, .)$, it will have (if it is not trivial) infinite Morley-rank as a δ-definable set. It is important to note that if X is δ-definable and $X \subseteq (C_U)^n$ then X is definable in the structure $(C_U, +, .)$. When we say that the definable set X is *defined over* A ($A \subseteq U$), we mean that X is defined by a formula (of the relevant kind) whose parameters are from A.

2 Differential algebraic groups

We could introduce differential algebraic groups, by first introducing the category of δ-varieties and δ-morphisms (analogous to abstract algebraic varieties and morphisms), and then defining differential algebraic groups to be group objects in this category. It is easier, however, to consider differential algebraic groups as simply δ-definable groups (groups G whose underlying set and group operation are definable sets in $(U, +, ., \delta)$). The identification of these two classes of groups is actually a theorem [Pi4]. When it comes to fields of definition, however, this identification is a more subtle matter, and is treated in [Pi5]. In a similar fashion we will identify algebraic groups with f-definable groups.

Examples

2.1. If G is an algebraic group defined over C_U, then $G(C_U)$, the set of C_U-rational points of G, is a δ-definable group.

2.2. If G is an algebraic subgroup of $\mathrm{GL}_n(U)$ (so a linear algebraic group), then we can give $\{(A, A') \colon A \in G\}$ the structure of a δ-definable group by setting $(A, A').(B, B') = (A.B, (A.B)')$. (Here $(a_{ij})'$ means (a'_{ij}).)

This group can be considered naturally as a δ-definable subgroup of $\mathrm{GL}_{2n}(U)$.

2.3. Let A be an Abelian variety over U. Manin [M] shows, using generalized "Picard-Fuchs" equations, that there is some nontrivial δ-definable homomorphism f from A into some vector group U^n. The kernel of f is then a δ-definable group, which contains $\mathrm{Tor}(A)$. A particular case of this is given in Example 3.7.

The following answers a question of Kolchin [K2], and is proved using ideas from stable group theory. The key point in the proofs is a Weil theorem for pro-algebraic varieties and pro-algebraic groups.

THEOREM 2.4 [Pi5]. *If G is a δ-definable group then there is a δ-definable embedding of G into an algebraic group.*

In fact the proof also yields information on the rather delicate matter of fields of definition, and shows that if the δ-group G is defined over F in the Kolchin sense (or model-theoretic sense) then G is defined over F in the δ-variety sense.

Let us say that the δ-group G is *linear* if there is a δ-definable embedding of G into some $\mathrm{GL}_n(U)$, and of Abelian type if there is a δ-definable embedding of G into some Abelian variety. A δ-definable group is said to be δ-connected if it has no proper δ-definable subgroup of finite index.

COROLLARY 2.5. *If G is a δ-connected δ-definable group then G has a (unique) maximal normal linear δ-connected δ-definable subgroup N, and G/N is of Abelian type.*

A δ-definable set X is said to be *finite dimensional* if Morley-rank$(X) < \omega$. Suppose X is defined over F (namely defined by a formula with parameters in F). Then X will be finite dimensional iff for any $\mathbf{a} \in X$ (or even for generic $\mathbf{a} \in X$ over F) $F\langle \mathbf{a} \rangle$ has finite transcendence degree over F. The δ-definable groups in Example 2.2 are infinite dimensional. Concerning Example 2.3, there will always be some f such that Ker(f) is finite dimensional.

We will call the δ-definable group G *δ-simple*, if G contains no proper infinite normal δ-definable subgroup. If G is noncommutative, then (from the theory of superstable groups) G will be δ-simple just if $Z(G)$ is finite and $G/Z(G)$ is simple as an abstract group. A δ-simple group will (by Corollary 2.5) be either linear or of Abelian type. So a noncommutative δ-simple group will be linear. The following was proved by Cassidy, but another proof appears in [Pi6].

2.6. *Let G be a finite-dimensional linear δ-simple δ-definable group. Then there is a (linear) algebraic group H defined over C_U such that G is isomorphic (by a δ-definable isomorphism) to $H(C_U)$. (In particular G is δ-definably isomorphic to a group definable in $(C_U, +, .)$.)*

A consequence is:

2.7 [S]. *If K is a finite-dimensional δ-definable field, then F is δ-definably isomorphic to the field C_U.*

Fact 2.6 implies that there are no exotic simple noncommutative finite-dimensional groups in the context of differential algebraic geometry. The case of δ-simple δ-definable groups of Abelian type presents a different picture. It is clear that any such group G δ-definably embeds in a *simple* Abelian variety A (namely one without proper sub-abelian varieties). Also, as Buium [B1] proves:

LEMMA 2.8. *Let A be a simple Abelian variety over U. Then A contains a* unique *δ-simple δ-definable subgroup G. G is finite dimensional and contains* Tor(A). *(In fact G will be the kernel of some δ-definable homomorphism of A into some U^m.)*

LEMMA 2.9. *Let A_1, A_2 be simple Abelian varieties and G_1, G_2 δ-simple δ-definable subgroups of A_1, A_2 respectively. Then G_1 is δ-definably isomorphic to G_2 if and only if A_1 is f-definably (birationally) isomorphic to A_2.*

The following brings in the key notion of orthogonality. If D_1, D_2 are definable sets in a saturated model of an ω-stable theory, and both have multiplicity 1, then they are said to be *orthogonal* if for any set A of parameters over which D_1, D_2 are defined, and for a, b generic points of D_1, D_2 respectively over A, a is independent from b over A. The following is proved in [HS], using the deep result of [HZ], concerning strongly minimal sets with a nice "Zariski topology".

THEOREM 2.10. *Let D be a strongly minimal δ-definable set. Then either D is nonorthogonal to C_U or D is locally modular.*

Using Theorem 2.10, and the structure of "1-based" or "locally modular" groups (from [HPi]), one obtains:

THEOREM 2.11.

(i) Suppose A is a simple Abelian variety over U that is not birationally isomorphic to one defined over C_U. Let G be the unique δ-simple δ-definable subgroup of A. Then any δ-definable subset of G^n is a finite Boolean combination of cosets of δ-definable subgroups.

(ii) Let G_1, G_2 be δ-definable δ-simple (finite-dimensional) groups, both of which are orthogonal to C_U. Then G_1 is nonorthogonal to G_2 iff there is a δ-definable isogeny from G_1 to G_2 (iff the corresponding simple abelian varieties are isogenous).

The same conclusion as in Theorem 2.11(i) holds if G is any almost strongly minimal, δ-connected δ-definable group that is orthogonal to C_U.

Lemma 2.8 and Theorem 2.11 (ii) (together with the existence of elliptic curves that do not descend to C_U) are used to give a fast proof of:

THEOREM 2.12 [HS]. *There are continuum many countable models of* DCF_0 *(up to isomorphism).*

Another proof, not making use of Theorems 2.10 and 2.11 appears in [Pi6].

3 Differential Galois theory

Picard [Pic] initiated the Galois theory of linear differential equations. Let $\mathbf{C}$ denote the complex numbers. Let $\tilde{\mathbf{C}}$ denote the differential field of meromorphic functions on $\mathbf{C}$ (where δ is d/dz). Let $a_0(z), \ldots, a_{n-1}(z)$ be polynomials over $\mathbf{C}$. Consider for example the nth order homogeneous linear differential equation:

$$d^n f/dz^n + a_{n-1}(z)(d^{n-1}f/dz^{n-1}) + \cdots + a_0(z)f = 0.$$

The set of everywhere holomorphic solutions forms an n-dimensional vector space over $\mathbf{C}$. Let $f_1(z), \ldots, f_n(z)$ be a basis for this space. Let $F = \mathbf{C}(z)$, and L be the differential subfield of $\tilde{\mathbf{C}}$ generated by $\mathbf{C}(z, f_1(z), \ldots, f_n(z))$. So $F < L$ and the field of constants of both F and L is $\mathbf{C}$. Let G be the group of automorphisms of the differential field L that fix F pointwise. For each $g \in G$, there are $x_{ji}(g)$ in $\mathbf{C}$ such that $g.f_i = \sum x_{ji}(g)f_j$. Then it turns out that the set of matrices $(x_{ji}(g))$, $g \in G$, forms an *algebraic* subgroup of $\mathrm{GL}_n(\mathbf{C})$, isomorphic to G. In fact it appears that this is the first time the expression "algebraic group" was used.

Kolchin generalized the theory in the framework of differential algebra as follows. Let $F < L$ be differential fields. Assume L has finite transcendence degree over F and is finitely generated over F as a differential field. We consider both F, L as differential subfields of the universal domain U. Assume also that $C_F = C_L$, and is algebraically closed. L is called a *strongly normal* extension of F if, for *any* (differential) embedding σ of L in U that fixes F pointwise, $\langle L, \sigma(L)\rangle \subseteq \langle L, C_U\rangle$. Let G be the group of automorphisms of (the differential field) L that fix F pointwise. Kolchin proves that there is an algebraic group H defined over C_F

such that G is canonically isomorphic (in a manner that we make precise below) to $H(C_F)$. Here H is an abstract algebraic group (and so could be, for example, an Abelian variety). Note that the group $H(C_U)$ is a δ-definable group of finite dimension, which Kolchin calls the Galois group associated to the extension L/F. Moreover Kolchin shows that for any differential field F_0 such that $C = C_{F_0}$ is algebraically closed, and for any algebraic group H over C, there are $F_0 < F < L$, with $C = C_F = C_L$ such that L is a strongly normal extension of F with Galois group $H(C_U)$. In addition to this we have the usual Galois correspondences holding. Here I will point out a generalization of this theory, where the Galois groups can now be arbitrary finite-dimensional δ-definable groups. The remaining results of this section appear in [Pi7] and [Pi8].

It is convenient to make use of "differential closures" of differential fields. If F is a differential subfield of U, then ω-stability implies that there is a *prime model* over F, namely there is a differentially closed field $\hat{F} < U$ that contains F and can be (elementarily) embedded over F in *any* differentially closed field $F_2 < U$ that contains F. $\hat{F}$ is called the differential closure of F, and is unique up to F-isomorphism, but is not in general unique as a differential subfield of U. (This is the analogue of the *algebraic closure* of a field.) $\hat{F}$ has the feature that for any tuple $\mathbf{a}$ from $\hat{F}$, there is a single formula $\varphi(\mathbf{x})$ over F true of $\mathbf{a}$ such that for any tuple $\mathbf{b}$ from U, $\varphi(\mathbf{x})$ is true of $\mathbf{b}$ in U iff there is an automorphism σ of U that fixes F pointwise and takes $\mathbf{a}$ to $\mathbf{b}$. In the usual terminology, $\varphi(\mathbf{x})$ isolates the type of $\mathbf{a}$ over F.

DEFINITION 3.1. Suppose that F is a differential field, $\hat{F}$ is a fixed copy of the prime model over F, and X is a δ-definable set in U, defined by a formula with parameters in F. Suppose also that $X(F) = X(\hat{F})$ (namely all points of X in $\hat{F}$ have all coordinates in F). Let L be a differential field such that $F < L < \hat{F}$ and L is finitely generated over F. We will then call L an *X-strongly normal extension of F*, if for every embedding σ of L in U over F, $\langle L, \sigma(L)\rangle \subseteq \langle L, X\rangle$.

REMARK 3.2. Suppose L is an X-strongly normal extension of F. Suppose $L = F\langle\mathbf{a}\rangle$. The type of $\mathbf{a}$ over F is isolated by a formula $\varphi(\mathbf{x})$ in U. Let S be the set of solutions of φ in U. Recall that L is a subfield of $\hat{F}$. It follows that $S(\hat{F})$ (the set of points of S all of whose coordinates are in $\hat{F}$) is contained in L. For, by the compactness theorem, there is some δ-definable function f (defined over F) such that for any $\mathbf{c} \in S$, there is $\mathbf{b} \subseteq X$ such that $\mathbf{c} = f(\mathbf{a}, \mathbf{b})$. This sentence is then true in the elementary substructure $\hat{F}$ of U. But $X(\hat{F})$ is contained in F. Thus for any $\mathbf{c}$ in $S(\hat{F})$, $\mathbf{c} = f(\mathbf{a}, \mathbf{b})$ for some $\mathbf{b}$ in F. So (as DCF_0 has quantifier elimination) $\mathbf{c}$ is contained in L. This shows also that $\mathrm{Aut}(L/F)$ acts on $S(\hat{F})$, and any $\sigma \in \mathrm{Aut}(L/F)$ is even determined by its action on $S(\hat{F})$.

THEOREM 3.3. *Let L be an X-strongly normal extension of F. Suppose that $L = F\langle\mathbf{a}\rangle$. Let $\varphi(\mathbf{x})$ isolate the type of $\mathbf{a}$ over F. Let S be the set of realizations of φ in U. Then there is a finite-dimensional δ-definable group H such that*

(i) H is defined over F, and $H \subseteq \mathrm{dcl}(X)$,

(ii) H acts δ-definably (over L) and regularly on S.

(iii) *$H(\hat{F})$ is isomorphic to* $\mathrm{Aut}(L/F)$*, as groups acting on $S(\hat{F})$. (See Remark 3.2.)*
(iv) *If F is algebraically closed, or even relatively algebraically closed in L, then H is δ-connected.*

REMARK 3.4.
(i) When we say that $H \subseteq \mathrm{dcl}(X)$ we mean that there is some δ-definable function from a δ-definable subset of X^m onto H.
(ii) As H acts regularly on S (and the action is δ-definable over $\hat{F}$), $H(\hat{F})$ acts regularly on $S(\hat{F})$. So Theorem 3.3 (iii) makes sense. Often we say that the Galois group of L over F *is* H.
(iii) Again one has a Galois correspondence for X-strongly normal extensions: if L is an X-strongly normal extension, and H is the differential algebraic group given by Theorem 3.3, then there is a Galois correspondence between intermediate differential fields: $F < K < L$ and δ-definable subgroups of H that are defined over $\hat{F}$.
(iv) So actually, L will be an H-strongly normal extension of F.

By a *generalized* strongly normal extension of F we mean an X-strongly normal extension of F, for some F-definable set X. The following gives a canonical form for generalized strongly normal extensions (if F is algebraically closed), generalizing Kolchin's theorem [K1] on G-primitive elements. It amounts to the triviality of a certain "rational first cohomology group". Remember by Theorem 2.4 that any δ-definable group defined over F can be δ-definably embedded into an algebraic group.

THEOREM 3.5. *Let L be an X-strongly normal extension of F, where F is algebraically closed. Let H be as in Theorem 3.3. Let H_1 be any connected algebraic group that is defined over F and into which H F-definably embeds. Assume $H < H_1$. Then there is $\alpha \in H_1(L)$, and an isomorphism $h : \mathrm{Aut}(L/F) \to H(F)$, such that*
(i) *$L = K\langle\alpha\rangle$, and*
(ii) *for any $\sigma \in \mathrm{Aut}(L/F)$ $\sigma(\alpha) = h(\sigma)^{-1}.\alpha$ (where the last product is in the sense of H_1).*

THEOREM 3.6. *Let F_0 be any differential field. Let H be any finite-dimensional differential algebraic group defined over F_0. Then there are $F_0 < F < L$, such that L is a generalized strongly normal extension of F, and H is the Galois group of L over F.*

The proof of Theorem 3.6 uses the embeddability of H in some algebraic group and also the fact that types of finite U-rank (such as generic types of finite-dimensional definable groups) are *orthogonal* to types of U-rank $\omega.m$ (such as generic, in the DCF_0-sense, types of algebraic groups).

EXAMPLE 3.7. Assume that $\mathbf{C}$, the field of complex numbers, is C_U. Let $t \in U$ be such that t' (short for $\delta(t)$) $= 0$. Let A be the elliptic curve whose equation (in affine coordinates) is $y^2 = x(x-1)(x-t)$. (A can be thought of as a family of elliptic curves in $\mathbf{C}$, parametrized by t.) Manin [M] constructs a δ-definable homomorphism μ from A onto the algebraic group $(U,+)$, whose kernel is defined (according to [B1]) by $-y^3-2(2t-1)(x-t)^2x'y+2t(t-1)(x-t)^2(x''y-2x'y')=0$.

Let $G = ker(\mu)$. G is a finite-dimensional δ-definable group, defined over $\mathbf{C}(t)$. Let α be a δ-generic point of A over $\mathbf{C}$. Let $d=\mu(\alpha)$. Let $F = acl(\mathbf{C}(t)\langle d\rangle)$ $(= acl(\mathbf{C}(t)(d)))$, and $L = F\langle\alpha\rangle$. Then L is a G-strongly normal extension of F with Galois group G. So G is in a sense the Galois group of the differential system "$(x,y)\in A$ and $\mu((x,y))=d$".

An open question is whether a superstable differential field must be differentially closed (i.e. a model of DCF_0). In [PiS] it was shown that superstable fields have no proper strongly normal extensions. An extension of the proof shows that they have no proper generalized strongly normal extensions. Using this together with the classification of nontrivial strongly minimal δ-definable sets, obtained from the results of Section 2, we obtain the following statement (using the language of stability theory):

THEOREM 3.8. *Let F be a differential field whose theory is superstable. Let $\hat{F}$ be a copy of the prime model over F in the sense of* DCF_0. *(So $\hat{F}\models \mathrm{DCF}_0$.) Work now in U. Let Φ be the family of strongly minimal formulas that are over F and have nontrivial pregeometries. Let $\Theta(x)$ be a formula over F that is Φ-analyzable. Then all solutions of Θ that are in $\hat{F}$ are already in F.*

4 A conjecture of Lang

Thirty years ago or so, Lang formulated the following conjecture (often called the Mordell-Lang conjecture, which we state here for semi-Abelian varieties, namely extensions of Abelian varieties by algebraic tori):

LANG CONJECTURE. *Let A be a semi-Abelian variety over $\mathbf{C}$. Let X be a subvariety of A. Let $\Gamma < A$ be the divisible hull of some finitely generated subgroup of $A(\mathbf{C})$. Then $X\cap\Gamma$ is a finite union of cosets.*

See [L] for a thorough discussion and more background. The conjecture implies Mordell's conjecture on the finiteness of the set of rational points on curves (defined over $\mathbf{Q}$) of genus >1. (The implication can be seen by embedding the curve in its Jacobian variety.) The full Lang conjecture has recently been proved by McQuillan. But the past few years have also seen proofs of the geometric (or function field) version, using surprisingly general methods.

Let me first mention that Lang's conjecture is equivalent to a purely model-theoretic statement. First, if T is a stable theory and $\theta(x)$ is a formula over $\emptyset$, then $\theta(x)$ is said to be *1-based* if for any model M of T, and any algebraically closed subsets X,Y of $(\theta^M)^{eq}$, X is independent from Y over $X\cap Y$ (in the sense of forking).

DEFINITION 4.1. Suppose K is an algebraically closed field. Let A be a commutative algebraic group over K, and let Γ be a subgroup of $A(K)$. We say that (K,A,Γ) is of *Lang-type* if for any subvariety X of $A(K)$, $X\cap\Gamma$ is a finite union of cosets.

So Lang's conjecture can be restated as:

If A is a semi-Abelian variety over $\mathbf{C}$ *and* Γ *is a subgroup of* $\mathbf{A}$ *which is the divisible hull of a finitely generated subgroup, then* $(\mathbf{C}, A, \Gamma)$ *is of Lang-type.*

PROPOSITION 4.2. (K, A, Γ) *is of Lang-type if and only if the complete theory of the structure* $(K, +, ., a, \Gamma)_{a\in K}$ *is stable and the formula "*$X \in \Gamma$*" is 1-based.*

The following result of Hrushovski [H] proves the "geometric" or "function field" version of Lang's conjecture in *all characteristics*.

THEOREM 4.3. *Let* $k < K$ *be fields, where* k *is algebraically closed. Let* A *be a semi-Abelian variety defined over* K*. Let* X *be a subvariety of* A *defined over* K*. Let* Γ *be a subgroup of* $A(acl(K))$ *that is contained in either the divisible hull of a finitely generated subgroup of* A *(in the characteristic* 0 *case) or the "prime to* p*"-divisible hull of a finitely generated subgroup of* A *(in the characteristic* p *case). Assume that* $X \cap \Gamma$ *is Zariski dense in* X*. Then there are a semi-Abelian subvariety* B *of* A*, a semi-Abelian variety* A_0 *defined over* k*, and a subvariety* X_0 *of* A_0 *also defined over* k*, and a rational homomorphism* h *from* B *onto* A_0 *such that* X *is a translate of* $h^{-1}(X_0)$.

In the characteristic 0 case, the proof uses the results from Section 2, the key idea being to replace Γ by a finite-dimensional δ-definable subgroup of A. We give a sketch of the proof. We may assume K is also algebraically closed. We identify A, X with $A(K)$, $A(X)$ respectively. First enrich K by adjoining a derivation δ such that $(K, +, ., \delta)$ is a model of DCF_0 and $k = C_K$. Work in the structure $(K, +, ., \delta)$, which we may assume to be saturated. Using essentially Lemma 2.8, find a δ-definable, finite-dimensional subgroup G of A that contains Γ. So $X \cap G$ is Zariski dense in X. Quotienting out by the stabilizer of X (in the sense of algebraic groups) we may suppose this stabilizer of X to be finite. Applying some "finite Morley-rank group theory" we can find a δ-definable connected subgroup H of G such that (up to a translation of X) $X \cap H$ is Zariski dense in X, H is a sum of almost strongly minimal groups, and $\mathrm{Stab}(X \cap H)$ (in the sense of model theory) is finite. The results of Section 2 imply that H is the sum of two definable groups H_1 and H_2, where H_1 satisfies the conclusion of Theorem 2.11(i), and H_2 is δ-definably isomorphic to the k-rational points of a semi-Abelian variety A_0 defined over k. If $H_1 \neq \{0\}$, then one concludes that $\mathrm{Stab}(X \cap H)$ is infinite, a contradiction. So $H = H_2$. The δ-definable isomorphism of H_2 with $A_0(k)$ lifts to a rational isomorphism h of some semi-Abelian subvariety B of A with A_0. If X_0 is $h(X)$, then X_0 is defined over k too. This completes the sketch.

In the characteristic p case, one works in a separably closed field K of some suitable finite, nonzero, Ersov invariant, such that $k = \bigcap\{F^{p^n} : n < \omega\}$. The theory of the field K is stable, but not superstable. Nevertheless there is a theory of "finite-dimensional" infinitely definable sets and groups in K. The finite-dimensional ∞-definable subgroup of A that plays the role of G (from the characteristic 0 case) is $\bigcap_n (p^n(A(K)))$. From here on the proof parallels the characteristic 0 case (of course using suitable analogues, such as those in [Me], of the results from Section 2).

References

[B1] A. Buium, Differential algebraic groups of finite dimension, Lecture Notes in Math., **1506**, Springer-Verlag, Berlin and New York, 1992.

[B2] A. Buium, *Intersections in jet spaces and a conjecture of S. Lang*, Ann. of Math. (2) **136** (1992), 557–567.

[C1] P. Cassidy, *Differential algebraic groups*, Amer. J. Math. **94** (1972), 891–954.

[C2] P. Cassidy, *The classification of semisimple differential algebraic groups*, J. of Algebra **121** (1989), 169–238.

[Ho] W. A. Hodges, *Model theory*, Encyclopaedia of Mathematics and its Applications, Cambridge University Press, Cambridge, 1993.

[H] E. Hrushovski, *The Mordell-Lang conjecture for function fields*, J. Amer. Math. Soc., to appear.

[HPi] E. Hrushovski and A. Pillay, *Weakly normal groups*, Logic Colloquium **85**, (1987), North-Holland, Amsterdam.

[HS] E. Hrushovski and Z. Sokolovic, *Minimal subsets of differentially closed fields*, preprint 1994.

[HZ] E. Hrushovski and B. Zilber, *Zariski geometries*, Bulletin of Amer. Math. Soc., **28** (1993), 315–323.

[K1] E. R. Kolchin, Differential Algebra and Algebraic Groups, Academic Press, New York, 1973.

[K2] E. R. Kolchin, Differential Algebraic Groups, Academic Press, New York, 1973.

[L] S. Lang, *Number theory III: Diophantine geometry*, Encyclopaedia of Mathematical Sciences, Springer-Verlag, Berlin and New York, 1991.

[M] Yu. Manin, *Rational points of algebraic curves over function fields*, Amer. Math. Soc. Transl. Ser. (2) **59** (1966), 189–234.

[Mar] D. Marker, *Model theory of differential fields*, in Model Theory of Fields (by D. Marker, ed.) Lect. Notes in Log., Springer-Verlag, Berlin and New York, to appear.

[Me] M. Messmer, *Groups and fields interpretable in separably closed fields*, Trans. Amer. Math. Soc., **344** (1994), 361–377.

[Pic] E. Picard, *Equations differentielles lineares et les groupes algebriques de transformations*, Oevres II, 117–131.

[Pi1] A. Pillay, An Introduction to Stability Theory, Oxford University Press, London and New York, 1983.

[Pi2] A. Pillay, Geometrical Stability Theory, Oxford University Press, in preparation.

[Pi3] A. Pillay, *Model theory, stability theory and stable groups*, in The Model Theory of Groups, (A. Nesin and A. Pillay, eds.), University of Notre Dame Press, Notre Dame, IN, 1989.

[Pi4] A. Pillay, *Differential algebraic group chunks*, J. Symb. Log. **55** (1990), 1138–1142.

[Pi5] A. Pillay, *Some foundational questions on differential algebraic groups*, 1994, submitted.

[Pi6] A. Pillay, *Differential algebraic groups and the number of countable differentially closed fields*, in Model Theory of Fields, Lecture Notes in Logic, Springer-Verlag, Berlin and New York, to appear.

[Pi7] A. Pillay, *Differential Galois theory I*, submitted.

[Pi8] A. Pillay, *Differential Galois theory II*, submitted.

[PiS] A. Pillay and Z. Sokolovic, *Superstable differential fields*, J. Symbolic Logic **57** (1992), 97–108.

[P] B. Poizat, *Cours de théorie des Modèles*, Nur al-Mantiq wal-Ma'rifah, Villeurbanne, 1985.

[S] Z. Sokolovic, *Model theory of differential fields*, Ph.D. thesis, University of Notre Dame, Notre Dame, IN, 1992.

Logic and Complexity: Independence results and the complexity of propositional calculus

Pavel Pudlák

Mathematical Institute
Academy of Sciences of Czech Republic
11567 Praha 1, Žtná 25, Czech Republic

1. Introduction

The problem of whether $\mathcal{P} = \mathcal{NP}$ is generally recognized as one of the most important problems in contemporary mathematics. It is one of the many problems in complexity theory that have resisted for years all attempts to solve them. The problem $\mathcal{P} = \mathcal{NP}$? originated in logic and thus there were some hopes that logic would help to solve it. Naturally the question of whether $\mathcal{P} = \mathcal{NP}$? or a similar problem is independent from theories used as foundations of parts of mathematics, e.g. Peano arithmetic, also was brought up. Later more researchers started to use finite combinatorics and algebra in this field. This approach has been quite successful in solving some restricted versions of the problems, but the fundamental problems remain open.

Also it turned out that it is very difficult to prove independence of sentences of such a form from Peano arithmetic; therefore the attention of many logicians has focused on much weaker theories, called *bounded arithmetic*. For such theories it is possible to prove interesting connections with problems in complexity theory and at least some weak independence results. In particular it is also possible to reduce independence results to combinatorial questions. In complexity theory such reductions use boolean circuits; however, in logic the connection is based on propositional calculus. Thus, a new field has emerged that combines proof theory and model theory with techniques and results from complexity theory and finite combinatorics.

In this paper we want to explain these relationships between logic and complexity theory using some recent results in this field. These results concern provability and mutual dependence of counting principles. We think that this will give you a better idea of the methods used in this field than a general survey of the whole area.

We shall start with an algebraic problem, which may be more understandable for mathematicians not working in logic or complexity theory and which incidentally arose in an independence proof for the counting principles. Then we shall explain some basic concepts used in this field — propositional proof systems and bounded arithmetic. Finally we shall describe the results on the independence of counting principles and state an important open problem.

Proceedings of the International Congress
of Mathematicians, Zürich, Switzerland 1994

2. An example

Let us consider the problem of deciding whether a system of algebraic equations has a solution. Let a field F be given and let

$$\begin{aligned} f_1(x_1,\ldots,x_n) &= 0 \\ &\vdots \\ f_m(x_1,\ldots,x_n) &= 0 \end{aligned} \tag{1}$$

be a system of equations with $f_1,\ldots,f_m$ polynomials over F. We are not interested in solving the equations, we only want to know whether there is a solution. In order to simplify the exposition, let us assume for a moment that F is a finite field. Then the problem is $\mathcal{NP}$-complete, even if each f_i is a polynomial of degree at most 2. This means that any solution to the problem can be described in polynomially many bits and verified in polynomial time (i.e. the problem is in $\mathcal{NP}$) and any such problem (i.e. any problem in $\mathcal{NP}$) can be reduced by a polynomial algorithm to this problem. The first property is trivial. The second property implies that there is no polynomial time algorithm for it, if $\mathcal{P} \neq \mathcal{NP}$.

A general concept of a proof [9] is based on the following simple condition: potential proofs are codable as finite strings over a fixed finite alphabet, and there is a polynomial time procedure to test the validity of a proof. In this sense every $\mathcal{NP}$-problem has an associated proof system. In the case of algebraic equations, we can simply proclaim that a solution to a system of equations is a proof that the equations are solvable. The property that characterizes $\mathcal{NP}$-problems is the fact that they have proofs of *sizes bounded by a polynomial in the sizes of inputs*. For example, a solution of (1) in F is just a string $a \in F^n$, hence it is polynomially bounded in the size of the sequence $f_1,\ldots,f_m$.

Now we know that it is unlikely that we find a polynomial time algorithm for deciding solvability of algebraic equations. We also know that we can easily find polynomial size proofs that they are solvable, if they have a solution in a finite field. But can we find polynomial size proofs that they are *unsolvable*, if they are unsolvable? Does this hold for all $\mathcal{NP}$-problems? Because polynomial size proofs define the class $\mathcal{NP}$, this is just the question, whether $\mathcal{NP}$ is closed under complements, i.e. whether $\mathcal{NP} = co\mathcal{NP}$. It follows from the $\mathcal{NP}$-completeness of the problem of solvability of algebraic equations in finite fields that the general problem is equivalent to this special case.

A nice property of our example of an $\mathcal{NP}$-complete problem is that we have a natural proof system for its complement. This is the basic result known as *Hilbert's Nullstellensatz* (weak form):

THEOREM 1 *The system (1) does not have a solution in $\bar{F}$ (the algebraic closure of F), iff 1 is in the ideal generated by $f_1,\ldots,f_m$.*

Note that the fact that the solutions are considered in the algebraic closure of F does not cause any problems, if we are interested only in solutions in a *finite* field F, as we can add equations $\Pi_{a\in F}(x_i - a) = 0$, $i = 1,\ldots,m$, which ensure that any solution must be in F.

Let us state this theorem more explicitly. We can prove that (1) does not have a solution by finding other polynomials $g_1, \ldots, g_m \in F[x_1, \ldots, x_n]$ such that

$$\sum_i g_i f_i = 1 \tag{2}$$

holds in $F[x_1, \ldots, x_n]$. Thus, polynomials $g_1, \ldots, g_m$ are a *proof that (1) does not have a solution.* Hilbert's Nullstellensatz is a completeness result for this "calculus", it says that there is such a proof if (1) is unsolvable. (As we want to be able to test the condition (2) in polynomial time, we assume that all polynomials are represented as sums of monomials.)

There are several results about the degrees of polynomials $g_1, \ldots, g_m$ in (2). In particular, it is very easy to give an example of constant degree polynomials $f_1, \ldots, f_m$ such that, for any polynomials $g_1, \ldots, g_m$ satisfying (2), at least one of them has an exponentially large degree [6] (the example is due to Masser and Philippone). We shall show below that a certain lower bound on the degrees can be used to obtain results about stronger systems, including first order theories. The set of equations is defined below, see (5); presently we have only a very small nonlinear lower bound, and the proof is quite involved.

3. Propositional calculus

The above calculus is not a typical logical proof system; it does not talk about formulas and the proof is not based on successive derivation of small pieces (though we can reformulate it in this way). The most important types of proof systems for propositional logic are the so-called Frege systems and extended Frege systems. We shall briefly describe them in this section.

We consider propositional formulas in a fixed finite basis, say $\{\wedge, \vee, \neg\}$, with propositional variables $p_1, p_2, \ldots$. A *Frege system* is determined by a finite set of *rules* that are sound and enable us to derive all tautologies. A rule may have zero premises, in which case it is called an *axiom schema;* an example is the law of excluded middle

$$\varphi \vee \neg\varphi$$

The *cut rule*, which allows us to derive $\varphi \vee \chi$ from $\varphi \vee \psi$ and $\neg\psi \vee \chi$, usually written as

$$\frac{\varphi \vee \psi \quad \neg\psi \vee \chi}{\varphi \vee \chi}$$

is an example of a rule with two premises. Let us remark that the particular choice of rules is not so important, as it can change the minimal length of a proof at most polynomially [9].[1]

The propositional proof systems that are used in textbooks on logic are in most cases Frege systems.

An *extended Frege system* is a Frege system augmented with a special rule, the *extension rule*, which allows us to abbreviate a formula by a variable. (Namely, we are allowed to introduce a sentence expressing that a formula is equivalent to

[1]This is not quite so in the *sequent calculi,* where the presence of the cut rule plays an important role.

a propositional variable, provided that the variable does not occur in the previous part of the proof nor in the conclusion.) It seems that because of this rule some tautologies may have much shorter proofs in extended Frege systems than in Frege systems, but we are not able to prove it.

In both cases a *proof* is a sequence of formulas that are either axioms or follow from the previous ones by a rule.

The main question about these proof systems is whether there is a polynomial $p(x)$ such that every tautology τ has a proof of length $p(|\tau|)$, where $|\tau|$ denotes the length of τ as a string in a finite alphabet. Because the set of tautologies is a complement of an $\mathcal{NP}$-complete set, this is not true, unless $\mathcal{NP} = co\mathcal{NP}$. However already Frege systems are quite strong, so we are not able to prove even such consequences of the conjecture $\mathcal{NP} \neq co\mathcal{NP}$.

Exponential lower bounds can be proved under the restriction of *bounded depth.* By the depth of a formula in the basis $\{\wedge, \vee, \neg\}$ we mean the number of alternations of the connectives $\wedge, \vee, \neg$. For instance a conjunction of an arbitrary number of propositional variables is of depth 1, the negation of such a formula and disjunctions of such formulas are of depth 2, etc. Let a particular Frege system be given. For each positive integer k we can consider a proof system obtained by restricting the Frege system to formulas of depth k.

The techniques used for lower bound proofs for bounded depth Frege systems come from boolean circuit complexity theory. The two subjects — propositional proofs and boolean circuits — are very much related. Nevertheless, applications of methods from boolean complexity require additional work. (One should realize that a Frege proof is a sequence of tautologies, which are formulas that define boolean functions constantly equal to 1, thus we cannot simply transfer results on the difficulty of specific boolean functions.)

4. Bounded arithmetic

We turn now to some first order theories that are important in this field. Let us first recall the classical theory Peano arithmetic used to formalize number theory. This is a theory that axiomatizes a part of the true sentences about nonnegative integers with 0, 1, $+$, $\cdot$, and $\leq$ as primitive notions. The axioms consist of 8 axioms describing some basic properties of the operations, e.g. the inductive clause for addition

$$x + (y + 1) = (x + y) + 1,$$

and induction axioms for all formulas in this language. This theory is adequate for the formalization of most mathematical reasonings about numbers and finite combinatorics, except for those involving very fast growing functions. (The theory does not contain the concept of a set as a primitive notion, but finite sets can be coded by numbers.) As the interesting problems in complexity are statements that do not contain fast growing functions, it is impossible to prove their independence from Peano arithmetic with the current methods.

Thus, the interest of many logicians gradually switched to weaker theories. A typical theory of the family bounded arithmetic is defined roughly as follows. We choose a complexity class $\mathcal{C}$ and a set of arithmetical formulas Φ whose formulas define just the sets in $\mathcal{C}$. To this end we may need to extend the standard language of arithmetic $0, 1, +, \cdot$ by more function and predicate symbols. Then the axioms of the theory T are basic axioms about the relations and functions of the language plus induction axioms for all formulas in Φ.

Let us stress that our motivation for studying bounded arithmetic is not based on some philosophical approach to foundations of mathematics like constructivism. The real importance of bounded arithmetic stems from the fact that we can *prove* some dependence between problems in complexity theory and these theories.

Here is such an important connection. Most problems in complexity theory are stated as the question of whether two complexity classes $\mathcal{C}_1$ and $\mathcal{C}_2$ are the same, where $\mathcal{C}_1 \subseteq \mathcal{C}_2$. If we have theories T_1 and T_2 corresponding to these complexity classes, we can ask whether $T_1 = T_2$. The answers to these two questions can be independent, but at least one thing is clear (provided that we consider natural formalizations of the classes $\mathcal{C}_1, \mathcal{C}_2$): *if we prove that $T_1 \neq T_2$, then it is not provable in T_1 that $\mathcal{C}_1 = \mathcal{C}_2$.*

We shall mention, in particular, two of the many theories of bounded arithmetic. The first one, introduced by Parikh [15] and denoted by $I\Delta_0$, is obtained from Peano arithmetic simply by restricting the induction to bounded formulas. A *bounded (arithmetical) formula* is a formula where each quantifier is followed by an inequality limiting the range of the quantified variable; e.g. the following formula $\varphi(x)$, which defines composite numbers, is bounded:

$$\exists y,\ y \leq x\ \exists z,\ z \leq x\ (x = y \cdot z \wedge y < x \wedge z < x).$$

The class of bounded formulas is denoted by Δ_0. The class of sets that bounded formulas define is the so-called *linear time hierarchy* (which is equivalently defined as the class of sets accepted by alternating Turing machines with constant number of alternations in linear time).

The second theory is denoted by S^1_2. In this theory induction is postulated for formulas that define sets in $\mathcal{NP}$. This theory uses a richer language and the induction axioms are slightly modified; we shall not go into details and refer the reader to [7], [10], [12].

An important property of these theories is that they cannot define fast growing functions, e.g. they do not prove that for every x there exists 2^x.

We need the following modification of $I\Delta_0$. Let R be a predicate symbol. Consider all arithmetical formulas that may also contain R. Then we can extend the concept of bounded formula to these formulas; let us denote this class of formulas by $\Delta_0(R)$. Finally we extend also the theory $I\Delta_0$ to $I\Delta_0(R)$ by extending the induction schema to all formulas in $\Delta_0(R)$. We can think of R as a free second order variable. As we do not specify interpretation of R, proving independence results is easier, but still requires quite ingenious arguments.

5. Counting principles

We shall show further connections between the concepts introduced above using a concrete example of counting principles.

Many important problems in complexity theory are connected with counting, especially with counting modulo a fixed number. In boolean complexity it has been proved that bounded depth polynomial size circuits with gates computing modulo a prime p cannot compute the boolean function that computes the number of ones modulo a different prime q. Other problems are connected with counting the number of accepting computations of a nondeterministic Turing machine. A related problem concerns bounded formulas. Suppose we use not only the usual quantifiers $\forall$ and $\exists$, but also a counting quantifier, saying that *the number of x's satisfying φ is divisible by p*. Then we can ask if the class of bounded formulas augmented with such a counting quantifier defines the same class of sets as bounded formulas without counting quantifiers. By the above schema, there is a corresponding question, if $I\Delta_0$ is equivalent to its version with counting quantifiers. We shall relax the problem by asking about the provability of certain consequences of such theories, which we call *counting principles.*

Let $q \geq 2$. The counting principle $Count_q$ is the statement that the set $\{0, \ldots, n-1\}$ cannot be decomposed into blocks of size q, for every n not divisible by q. In bounded arithmetic with an extra relation we can express this principle as follows. We take a q-ary relation symbol R. For a given x we shall think of R as encoding q element subsets $\{x_1, \ldots, x_q\}$ where $x_1 < \cdots < x_q < x$ and $R(x_1, \ldots, x_q)$. Then the statement that R encodes a partition of $\{0, \ldots, x-1\}$ is a $\Delta_0(R)$ formula; let us denote it by $Part_q(x, R)$. Hence the counting principle $Count_q$ can be expressed by

$$\forall x(\forall y \leq x(q \cdot y \neq x) \rightarrow \neg Part_q(x, R)). \tag{3}$$

We shall denote this formula by $Count_q(R)$. This formula is $\Pi_1(R)$ (universally quantified $\Delta_0(R)$), so it has the same logical complexity as the statement of $\Delta_0(R)$-induction. Therefore it is natural to ask if it is derivable from $\Delta_0(R)$-induction. We can consider the theory axiomatized by 8 basic axioms together with $Count_q(\Delta_0(R))$ (schema of counting modulo q for all bounded formulas with R) instead of $I\Delta_0(R)$. Then the above problem is equivalent to the question of whether this theory is contained in $I\Delta_0(R)$.

Let us now consider propositional calculus. Let n be fixed. Take propositional variables p_e indexed by q-element subsets of $\{0, \ldots, n-1\}$; we shall use the variables to code sets of q-element subsets, where p_e is true, if the subset e is present. Then we can express the counting principle for this particular n by

$$\bigvee_{v<n} \bigwedge_{v \in e} \neg p_e \quad \vee \quad \bigvee_{e,f;\ e \perp f} (p_e \wedge p_f) \tag{4}$$

where $e \perp f$ abbreviates the conjunction $e \cap f \neq \emptyset \wedge e \neq f$. We denote this formula by $Count_q^n$. Of course, this is a tautology iff n is not divisible by q.

We can also express the same counting principle using algebraic equations as the unsolvability of the following set of equations:

$$\left(\sum_{e;\ v \in e} x_e \right) - 1 = 0, \ \text{for } v < n, \tag{5}$$
$$x_e \cdot x_f = 0, \ \text{for } e \perp f.$$

Again $x_e = 1$ expresses that the subset e is in the alleged partition.

One of the main motivations for studying the length of proofs in propositional calculus is its relation to provability in bounded arithmetic. Paris and Wilkie [16] showed that provability of a $\Pi_1(R)$ sentence φ in $I\Delta_0(R)$ implies that certain tautologies have polynomial size proofs in a bounded depth Frege system. These are obtained from numeric instances of the universal sentence φ. In order to simplify the exposition, we shall state their theorem only for the special case of counting principles.

THEOREM 2 *Suppose $Count_q(R)$ is provable in $I\Delta_0(R)$. Then the propositions $Count_q^n$, for n not divisible by q, have depth k and polynomial size proofs in a Frege system.*

This reduction has actually been used. Ajtai [1], [2] proved superpolynomial lower bounds on the lengths of proofs in bounded depth Frege systems for such formulas and for the related pigeon hole principle.

THEOREM 3 *[2] There are no polynomial size bounded depth Frege proofs of $Count_q^n$, for n not divisible by q.*

COROLLARY 4 *$Count_q(R)$ is not provable in $I\Delta_0(R)$.*

A similar relation holds between bounded arithmetic S_2^1 and extended Frege systems, namely the provability of certain universal sentences in S_2^1 implies polynomial upper bounds on the length of some tautologies in extended Frege systems [8] (this was done first for an equational theory PV and later extended to S_2^1 [7]). However such proof systems are quite strong, so we are not able to prove any superpolynomial lower bounds for them at present.

Let us note that there are more such relations, see [14], [12]. When working on the complexity of propositional calculus it is necessary to know about these relations, as one can more easily find proofs in bounded arithmetic than construct them in particular propositional proof systems. This has actually helped to refute some conjectures about the lengths of propositional proofs.

6. Independence of counting principles

Probably the most interesting and deepest results in this field deal with the independence of counting principles. In the same way as above we can ask the question either in bounded arithmetic or in propositional calculus. In bounded arithmetic it is the question if, for given q and r, one can prove $Count_q(R)$ from $I\Delta_0(R) + Count_r(\Delta_0(R))$. In the latter theory we assume the counting principle $Count_r$ for all bounded formulas that one can construct from R. Using the same method as in Theorem 2, one can reduce this problem to propositional calculus. There the question is: Are there polynomial size bounded depth Frege proofs of $Count_q^n$, for n not divisible by q, using instances of $Count_r^m$, for m not divisible by r, as assumptions?

These problems have been completely resolved by two people and one group working, more or less, independently. The main progress was done by Ajtai [4], who proved the independence for q and r different primes. He applied symmetric group representation theory over finite fields to prove a general theorem about symmetric systems of linear equations, which he then applied to solve a key combinatorial problem in his proof of the independence of counting principles. His theorem states, roughly speaking, that every solvable *symmetric* system of linear equations over GF_p, p prime, has in a certain sense *regular* solutions.

In [5], [17] different methods were used and a complete analysis of the possible cases was accomplished also for composite numbers q and r, which is:

THEOREM 5 *There are polynomial size bounded depth Frege proofs of $Count_q^n$, for n not divisible by q, using instances of $Count_r^m$, for m not divisible by r, if and only if all prime factors of q are prime factors of r.*

As a corollary we get that the same relation holds in bounded arithmetic.

I shall briefly describe the solution of the group in which I participated [5]. Our proof is based on a reduction to a lower bound on the degree of polynomials g_i in the Nullstellensatz (see equation (2)), where the system of algebraic equations (1) is (5).

Let q and n be given, n not divisible by q. Let us first suppose that q is a prime and F is a field of characteristic p. Then the equations (5) are not solvable and we can find polynomials g_v and $g_{e,f}$ such that

$$\sum_{v<n} g_v \left(\left(\sum_{e;v\in e} x_e \right) - 1 \right) + \sum_{e\perp f} g_{e,f} x_e x_f = 1. \tag{6}$$

If $p = q$ one can find such polynomials of degree ≤ 1 (take $g_v = -((\sum_{e;v\in e} x_e) + 1)/n$ and $g_{e,f} = |e \cap f|/n$.) This is not true, if $p \neq q$; we have the following result, which can be interpreted as a lower bound on the complexity of proofs of $Count_q^n$ in the Nullstellensatz proof system over a field of characteristic r.

LEMMA 6 *Let F be a field of characteristic r and suppose that q is a different prime. Then the degrees of polynomials g_v and $g_{e,f}$ satisfying (6) cannot be bounded by any constant.*

In the proof of Theorem 5 we reduce the lower bound on the size of bounded depth Frege proofs to the lower bound on the degree of polynomials g_v and $g_{e,f}$ in Lemma 6 in a little stronger form (the above lemma suffices for q and r different primes). The reduction is not as direct as when we reduce provability in bounded arithmetic to the lengths of proofs in propositional calculus. It is based on a lemma of the type that is known in complexity theory as a *switching lemma* [11]. This lemma enables us to reduce bounded depth formulas to constant depth decision trees by fixing some variables.

Finally, let us note that Lemma 6 can be also proved using Ajtai's theorem on symmetric systems of linear equations. Our proof is combinatorial, based on the Ramsey theorem, and thus can be applied also in the case when r is not a prime.

7. Conclusions

From what we have presented it is clear that we are able to prove independence results only for very weak systems. Also the sentences that we consider are trivial combinatorial statements. This is the present situation and we can only add that the proofs of these results are not easy and that a similar situation holds in complexity theory. For instance it is not proved for any nonrestricted model of computation that multiplication is harder than addition, though experience confirms it strongly. By understanding the power of counting we get closer to the solution of such problems.

We shall conclude this paper by stating a fundamental problem, which shows another possible way that logic could help complexity theory.

Consider the problem of whether $\mathcal{NP} = co\mathcal{NP}$, i.e. if the complement of every $\mathcal{NP}$ set is also in $\mathcal{NP}$. As we noted above, this problem is equivalent to the problem of whether there is a propositional proof system in which all tautologies have polynomial size proofs. For all we know, such a system could be just a Frege system, i.e. the most usual system.

We have mentioned that there is a relation between S^1_2 and extended Frege systems, which is similar to the relation between $I\Delta_0(R)$ and bounded depth Frege systems. More importantly, extended Frege systems are the strongest propositional proof systems whose soundness S^1_2 can prove. It follows that if S^1_2 proves that $\mathcal{NP} = co\mathcal{NP}$, then it must actually prove that each tautology has a polynomial size proof in an extended Frege system.

Hence, to prove that S^1_2 does not prove $\mathcal{NP} = co\mathcal{NP}$, it suffices to prove superpolynomial lower bounds on the lengths of proofs in extended Frege systems. Such a lower bound would give us models of arithmetic very similar to the natural numbers, where $\mathcal{NP} \neq co\mathcal{NP}$. At present this seems to be very hard (even for ordinary Frege systems), but it should be easier than proving superpolynomial lower bounds for *all proof systems*, which is needed for a proof of $\mathcal{NP} \neq co\mathcal{NP}$.

Acknowledgments. I want to thank Sam Buss, Jan Krajíček, Russell Impagliazzo, and Jiří Sgall for their comments on a draft of this paper.

References

[1] Ajtai, M., *The complexity of the pigeonhole principle*, in Proc. 29-th IEEE Annual Symp. on Foundation of Computer Science, 1988, pp. 346–355, also to appear in Combinatorica.

[2] Ajtai, M., *Parity and the pigeonhole principle*, in Feasible Mathematics, Eds. S. R. Buss and P. J. Scott, Birkhäuser, Basel and Boston, MA 1990, pp. 1–24.

[3] Ajtai, M., Symmetric systems of linear equations modulo p, Israel J. of Math., to appear.

[4] Ajtai, M., *The independence of the modulo p counting principles*, Proc. 26-th ACM Symp. on Theory of Computing, 1994, pp. 402–411.

[5] Beame, P., Impagliazzo, R., Krajíček, J., Pitassi, T., and Pudlák, P., *Lower bounds on Hilbert's Nullstellensatz and propositional proofs*, preliminary version to appear in Proc. 35-th Symp. on Foundation of Computer Science, 1994, also to appear in Proc. of London Math. Society.

[6] Brownawell, W. D., *Bounds for the degrees in the Nullstellensatz*, Ann. Math. 126, 1987, pp. 577–591.

[7] Buss, S.R., Bounded Arithmetic, Bibliopolis, Naples, 1986.

[8] Cook, S. A., *Feasibly constructive proofs and the propositional calculus*, in Proc. 7-th ACM Symp. on Theory of Computing, 1975, pp. 83–97.

[9] Cook, S. A., and Reckhow, A. R., *The relative efficiency of propositional proof systems*, J. Symbolic Logic 44(1), 1979, pp. 36–50.

[10] Hájek, P., and Pudlák, P., Metamathematics of First-order Arithmetic, Perspect. Math. Logic, Springer-Verlag, Berlin and New York, 1993.

[11] Håstad, J., *Computation limits of small depth circuits*, ACM dissertation award, MIT Press, Cambridge, MA, 1986.

[12] Krajíček, J., Bounded arithmetic, propositional calculus and complexity theory, Cambridge Univ. Press, London and New York, to appear.

[13] Krajíček, J., and Pudlák, P., *Propositional proof systems, the consistency of first-order theories and the complexity of computations*, J. Symbolic Logic 54, 1989, pp. 1063–1079.

[14] Krajíček, J., and Pudlák, P., *Quantified propositional calculi and fragments of bounded arithmetic*, Z. Math. Logik Grundlag. Math. 36, 1989, pp. 29–46.

[15] Parikh, R., *Existence and feasibility in arithmetic*, J. Symbolic Logic 36, 1971, pp. 494–508.

[16] Paris, J. B., and Wilkie, A. J., *Counting problems in bounded arithmetic*, in Methods in Mathematical Logic, Lecture Notes in Math., Springer, Berlin and New York, 1130, 1985, pp. 317–340.

[17] Riis, S., *Count(q) does not imply Count(p)*, preprint, 1994, submitted to Ann. Pure Appl. Logic.

The Path Model for Representations of Symmetrizable Kac-Moody Algebras

PETER LITTELMANN*

Mathematisches Institut
Universität Basel
Rheinsprung 21
4051 Basel, Switzerland

and

IRMA
Université Louis Pasteur
7 rue Réné Descartes
F-67084 Strasbourg Cedex, France

In the theory of finite-dimensional representations of complex reductive algebraic groups, the group $GL_n(\mathbb{C})$ is singled out by the fact that besides the usual language of weight lattices, roots, and characters, there exists an additional important combinatorial tool: the Young tableaux. For example, the sum over the weights of all tableaux of a fixed shape is the character of the corresponding representation, and the Littlewood-Richardson rule describes the decomposition of tensor products of $GL_n(\mathbb{C})$-modules purely in terms of the combinatoric of these Young tableaux. The advantage of this type of formula is (for example compared to Steinberg's formula to decompose tensor products) that there is no cancellation of terms. This makes it much easier (and sometimes even possible) to prove for example that certain representations occur in a given tensor product.

To construct objects like the tableaux in a more general setting, consider the weight lattice X of a complex symmetrizable Kac-Moody algebra $\mathfrak{g}$, and denote by Π the set of all piecewise linear paths $\pi : [0,1]_{\mathbb{Q}} \to X_{\mathbb{Q}}$ starting in 0 and ending in an integral weight. We associate to a simple root α linear operators e_α and f_α on the free $\mathbb{Z}$-module $\mathbb{Z}\Pi$ spanned by Π. Let $\mathcal{A} \subset \mathrm{End}_{\mathbb{Z}}\,\mathbb{Z}\Pi$ be the subalgebra generated by these operators.

Fix $\pi \in \Pi$ such that the image is completely contained in the dominant Weyl chamber. The $\mathcal{A}$-module $\mathcal{A}\pi \subset \mathbb{Z}\Pi$ generated by π is a "model" for the irreducible, integrable representation V_λ of $\mathfrak{g}$ of highest weight $\lambda = \pi(1)$: for example, the sum over the endpoints of all paths in $\mathcal{A}\pi$ is the character of V_λ, and the Littlewood-Richardson rule can be generalized in a straightforward way. As an application, one gets a purely combinatorial proof of the P–R–V conjecture.

So the paths can be viewed in a natural way as a generalization of Young tableaux to the setting of symmetrizable Kac-Moody algebras. Though the theory of the path modules is completely independent of the theory of quantum groups, the path modules are strongly related to the crystal graph of representations of the q-analog of the enveloping algebra of $\mathfrak{g}$. In fact, they can be viewed as a geometric realization of the theory of crystals of representations.

---

*) Supported by Schweizerischer Nationalfonds.

Proceedings of the International Congress
of Mathematicians, Zürich, Switzerland 1994

1 Review on the $GL_n(\mathbb{C})$-case

Let T be the maximal torus of $GL_n(\mathbb{C})$ of diagonal matrices, and denote by $\epsilon_i : T \to \mathbb{C}^*$ the projection of a diagonal matrix onto its ith entry. The irreducible polynomial representations of $GL_n(\mathbb{C})$ are in bijection with the dominant weights X^+ of the form $\lambda = p_1\epsilon_1 + \cdots + p_n\epsilon_n$, where $p_1 \geq p_2 \geq \ldots \geq p_n \geq 0$. A *Young diagram* of shape $\lambda \in X^+$ is a left-justified sequence of rows of boxes such that there are p_1 boxes in the first row, p_2 boxes in the second row, etc. By a *Young tableau* $\mathcal{T}$ of shape λ we mean a filling of the boxes of the diagram with numbers $\{1, \ldots, n\}$. $\mathcal{T}$ is called *semi-standard* if the entries are strictly increasing in the columns and nondecreasing in the rows.

A word $\mathbf{w}$ in the alphabet $\{1, \ldots, n\}$ is a finite sequence $\mathbf{w} = i_1 i_2 \ldots i_s$ with $1 \leq i_1, \ldots, i_s \leq n$. Its weight is the sum $\nu(\mathbf{w}) := \epsilon_{i_1} + \cdots + \epsilon_{i_s}$, and it is called λ-dominant for some $\lambda \in X^+$ if $\lambda + \epsilon_{i_1}$, $\lambda + \epsilon_{i_1} + \epsilon_{i_2}$, $\ldots$, are all in X^+. For a tableau $\mathcal{T}$ let $\mathbf{w}_\mathcal{T}$ be the word obtained by reading the entries from top to bottom, right to left: If $\mathcal{T}$ is of shape $(3, 2)$ having in the first row the entries 1 2 4 and in the second row 3 6, then $\mathbf{w}_\mathcal{T} = 4\,2\,1\,6\,3$. The weight $\nu(\mathcal{T})$ is the weight of $\mathbf{w}_\mathcal{T}$, and $\mathcal{T}$ is called λ-dominant if $\mathbf{w}_\mathcal{T}$ is λ-dominant.

THEOREM 1. *For $\lambda \in X^+$ let V_λ be the irreducible $GL_n(\mathbb{C})$-representation.*

(a) Character formula: $\operatorname{Char} V_\lambda = \sum e^{\nu(\mathcal{T})}$, *where the sum runs over all semi-standard Young tableaux of shape λ.*

(b) Littlewood-Richardson rule: For $\lambda, \mu \in X^+$ the tensor product $V_\lambda \otimes V_\mu$ decomposes into the direct sum: $V_\lambda \otimes V_\mu \simeq \bigoplus V_{\lambda+\nu(\mathcal{T})}$, where the sum runs over all semi-standard Young tableaux of shape μ that are λ-dominant.

For more details on these classical results we refer to [3], [20], [27]. The Littlewood-Richardson rule was first stated in [17] (without proof), it seems that the first complete proof was given in [25] (the proofs in [16], [24] are not complete). For various generalizations of Young tableaux, Gelfand-Zetlin patterns and related topics we refer to [1], [2], [8], [9], [11], [12], [13], [22], [23], [26] (this list is far from being complete). To the best of my knowledge, the (conjectural) definition of a "Young tableau" in [12] is the first that is independent of the type of the Lie algebra (see Section 8).

2 The paths

Let now X be the weight lattice of a symmetrizable Kac-Moody algebra, and set $X_\mathbb{Q} := X \otimes_\mathbb{Z} \mathbb{Q}$.

DEFINITION. *A piecewise linear path in $X_\mathbb{Q}$ is a piecewise linear, continuous map $\pi : [0,1]_\mathbb{Q} \to X_\mathbb{Q}$ of the interval $[0,1]_\mathbb{Q} := \{x \in \mathbb{Q} \mid 0 \leq x \leq 1\}$ into $X_\mathbb{Q}$. We consider two paths π, η as identical if there exists a piecewise linear, nondecreasing, continuous, surjective map $\phi : [0,1]_\mathbb{Q} \to [0,1]_\mathbb{Q}$ such that $\pi = \eta \circ \phi$. Denote by Π the set of all piecewise linear paths such that $\pi(0) = 0$ and $\pi(1) \in X$, and let $\mathbb{Z}\Pi$ be the free $\mathbb{Z}$-module with basis Π.*

EXAMPLE. (*i*) For $\lambda \in X_{\mathbb{Q}}$ set $\pi_\lambda(t) := t\lambda$, then $\pi_\lambda \in \Pi \Leftrightarrow \lambda \in X$.

(*ii*) Let π_1, π_2 be two piecewise linear paths starting in 0. By $\pi := \pi_1 * \pi_2$ we mean the path defined by

$$\pi(t) := \begin{cases} \pi_1(2t), & \text{if } 0 \le t \le 1/2; \\ \pi_1(1) + \pi_2(2t-1), & \text{if } 1/2 \le t \le 1. \end{cases}$$

(*iii*) For a word $\mathbf{w} = \lambda_1 \ldots \lambda_s$ in the alphabet $X_{\mathbb{Q}}$ (a finite sequence of rational weights) set $\pi_{\mathbf{w}} := \pi_{\lambda_1} * \cdots * \pi_{\lambda_s}$. The path connects successively the weights $0, \lambda_1, \lambda_1 + \lambda_2$, etc. Of course, $\pi_{\mathbf{w}} \in \Pi \Leftrightarrow \lambda_1 + \ldots + \lambda_s \in X$. Note that any path in Π is (up to reparameterization) of this form.

(*iv*) Suppose $\mathfrak{g} = \mathfrak{gl}_n(\mathbb{C})$ and $\mathbf{w} = i_1 \ldots i_s$ is a word as in Section 1. Denote by $\mathbf{w}$ also the path associated to the word $\epsilon_{i_1} \ldots \epsilon_{i_s}$ as in (*iii*). In this way we can consider the free $\mathbb{Z}$-module $\mathbb{Z}\mathcal{W}$ spanned by the words as a subset of $\mathbb{Z}\Pi$.

3 The root operators

Suppose $M \subset \mathbb{Z}\Pi$ is a $\mathbb{Z}$-submodule having a set of paths $B_M \subset \Pi$ as a $\mathbb{Z}$-basis. By the character of M we mean the formal sum: $\operatorname{Char} M := \sum_{\pi \in B_M} e^{\pi(1)}$.

EXAMPLE. For $\mathfrak{g} = \mathfrak{gl}_n(\mathbb{C})$ and $\lambda \in X^+$ let $M_\lambda \subset \mathbb{Z}\Pi$ be spanned by the paths $\mathbf{w}_{\mathcal{T}}$, $\mathcal{T}$ semi-standard of shape λ. Then $\operatorname{Char} M_\lambda = \operatorname{Char} V_\lambda$ (Theorem 1).

To construct such submodules in the setting of symmetrizable Kac-Moody algebras, we associate to every simple root two linear operators $e_\alpha, f_\alpha \in \operatorname{End}_{\mathbb{Z}} \mathbb{Z}\Pi$. For a piecewise linear path π let $s_\alpha(\pi)$ be the path defined by $s_\alpha(\pi)(t) := s_\alpha(\pi(t))$. Choose $\pi \in \Pi$ and fix a simple root α, let $\alpha^\vee$ be its coroot. Denote by h_α the function: $h_\alpha : [0,1]_{\mathbb{Q}} \to \mathbb{Q}$, $t \mapsto \langle \pi(t), \alpha^\vee \rangle$, and let m_α be the minimal value attained by the function h_α.

If $m_\alpha \le -1$, then fix $t_1 \in [0,1]_{\mathbb{Q}}$ minimal such that $h_\alpha(t_1) = m_\alpha$ and let $t_0 \in [0, t_1]_{\mathbb{Q}}$ be maximal such that $h_\alpha(t) \ge m_\alpha + 1$ for $t \in [0, t_0]_{\mathbb{Q}}$. We "cut" now the path between t_0 and t_1 into smaller pieces:

Choose $t_0 = s_0 < s_1 < \cdots < s_r = t_1$ such that either

(1) $h_\alpha(t) \ge h_\alpha(s_{i-1}) = h_\alpha(s_i)$ for $t \in [s_{i-1}, s_i]_{\mathbb{Q}}$;

(2) or h_α is strictly decreasing on $[s_{i-1}, s_i]_{\mathbb{Q}}$ and $h_\alpha(t) \ge h_\alpha(s_{i-1})$ for $t \le s_{i-1}$.

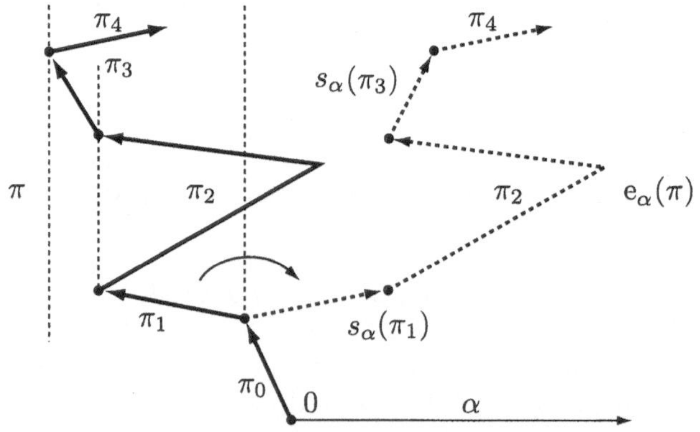

Figure 1 The part of $e_\alpha(\pi)$ different from π is drawn as a dashed line.

Let π_i be the part of π between s_{i-1} and s_i, or more precisely: set $s_{-1} := 0$ and $s_{r+1} := 1$, and denote by π_i the path defined by

$$\pi_i(t) := \pi(s_{i-1} + t(s_i - s_{i-1})) - \pi(s_{i-1}), \quad i = 0, \dots, r+1.$$

It is clear that $\pi = \pi_0 * \pi_1 * \dots * \pi_{r+1}$.

DEFINITION. If $m_\alpha \leq -1$, then $e_\alpha \pi := \pi_0 * \eta_1 * \eta_2 * \dots * \eta_r * \pi_{r+1}$, where $\eta_i = \pi_i$ if the function h_α behaves on $[s_{i-1}, s_i]_{\mathbb{Q}}$ as in (1), and $\eta_i = s_\alpha(\pi_i)$ if the function h_α behaves on $[s_{i-1}, s_i]_{\mathbb{Q}}$ as in (2). If $m_\alpha > -1$, then we set $e_\alpha \pi := 0$. See Figure 1.

The definition of the operator f_α is similar. Let $t_0 \in [0,1]_{\mathbb{Q}}$ be maximal such that $h_\alpha(t_0) = m_\alpha$. If $h_\alpha(1) - m_\alpha \geq 1$, then fix $t_1 \in [t_0, 1]_{\mathbb{Q}}$ minimal such that $h_\alpha(t) \geq m_\alpha + 1$ for $t \in [t_1, 1]_{\mathbb{Q}}$.

Choose $t_0 = s_0 < s_1 < \dots < s_r = t_1$ such that either

(1) $h_\alpha(t) \geq h_\alpha(s_i) = h_\alpha(s_{i-1})$ for $t \in [s_{i-1}, s_i]_{\mathbb{Q}}$;

(2) or h_α is strictly increasing on $[s_{i-1}, s_i]_{\mathbb{Q}}$ and $h_\alpha(t) \geq h_\alpha(s_i)$ for $t \geq s_i$.

As above, we write now π as $\pi = \pi_0 * \pi_1 * \dots * \pi_{r+1}$, where the paths π_i are defined by $\pi_i(t) := \pi(s_{i-1} + t(s_i - s_{i-1})) - \pi(s_{i-1})$.

DEFINITION. If $h_\alpha(1) - m_\alpha \geq 1$, then $f_\alpha \pi := \pi_0 * \eta_1 * \eta_2 * \dots * \eta_r * \pi_{r+1}$, where $\eta_i = \pi_i$ if the function h_α behaves on $[s_{i-1}, s_i]_{\mathbb{Q}}$ as in (1), and $\eta_i = s_\alpha(\pi_i)$ if h_α behaves on $[s_{i-1}, s_i]_{\mathbb{Q}}$ as in (2). If $h_\alpha(1) - m_\alpha < 1$, then we set $f_\alpha \pi := 0$.

EXAMPLE. **a)** Suppose that $\mathfrak{g} = \mathfrak{sl}_3$ and β is the highest root. The paths obtained from $\pi_\beta : t \mapsto t\beta$ by applying the operators f_α, e_α are the paths $\pi_\gamma(t) := t\gamma$, where γ is an arbitrary root, and for the two simple roots the paths $\eta_\alpha := \pi_{-\alpha/2} * \pi_{\alpha/2}$. See Figure 2.

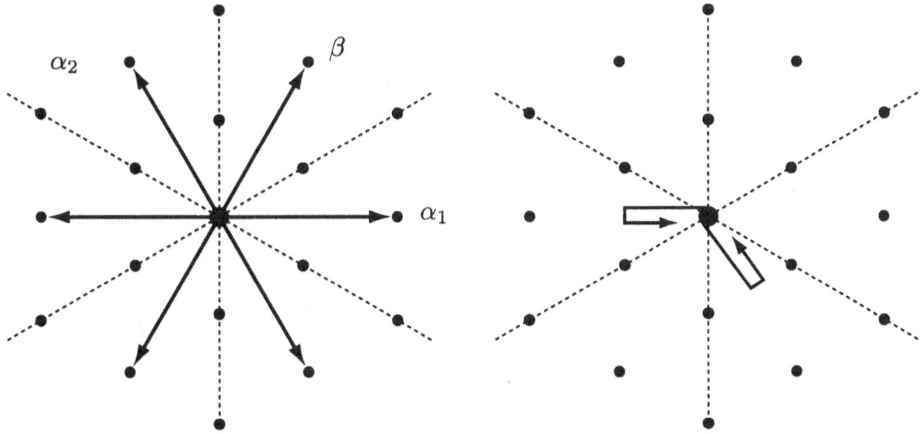

Figure 2 The paths generated by π_β.

(b) Suppose $\mathfrak{g} = \mathfrak{gl}_n$. Let $\mathbf{w} \in \mathbb{Z}\mathcal{W} \subset \mathbb{Z}\Pi$ be a word (path), and let α be the simple root $\epsilon_i - \epsilon_{i+1}$. It is easy to see that if $f_\alpha \mathbf{w} \neq 0$ (respectively $e_\alpha \mathbf{w} \neq 0$), then the operator f_α just replaces some i in $\mathbf{w}$ by an $i+1$ (respectively some $i+1$ by an i). In particular, $\mathbb{Z}\mathcal{W}$ is stable with respect to the root operators.

4 Some simple properties of the root operators

The following properties of the paths are easy to prove [14], [15]:

LEMMA 1. *Let α be a simple root and suppose $\pi \in \Pi$.*

(i) If $e_\alpha \pi \neq 0$, then $e_\alpha(\pi)(1) = \pi(1) + \alpha$, and if $f_\alpha(\pi) \neq 0$, then $f_\alpha(\pi)(1) = \pi(1) - \alpha$.

(ii) If $e_\alpha(\pi) \neq 0$, then $f_\alpha e_\alpha(\pi) = \pi$, and if $f_\alpha(\pi) \neq 0$, then $e_\alpha f_\alpha(\pi) = \pi$.

(iii) Let π^* be the dual path; i.e., $\pi^*(t) := \pi(1-t) - \pi(1)$. Then $(f_\alpha \pi)^* = e_\alpha(\pi^*)$ and $(e_\alpha \pi)^* = f_\alpha(\pi^*)$.

(iv) Let n be maximal such that $f_\alpha^n(\pi) \neq 0$ and let m be maximal such that $e_\alpha^m(\pi) \neq 0$. Then $\langle \pi(1), \alpha^\vee \rangle = n - m$.

(v) For $k \in \mathbb{N}$ let $k\pi$ be the path obtained by stretching π: $(k\pi)(t) := k\pi(t)$. Then $k(f_\alpha \pi) = f_\alpha^k(k\pi)$ and $k(e_\alpha \pi) = e_\alpha^k(k\pi)$.

(vi) Set $X_\alpha := \sum_{i \geq 1} e_\alpha^i f_\alpha^{i-1}$, $Y_\alpha := \sum_{i \geq 1} f_\alpha^i e_\alpha^{i-1}$, and $H_\alpha := \sum_{i \geq 1} e_\alpha^i f_\alpha^i - f_\alpha^i e_\alpha^i$. The Lie algebra in $\mathrm{End}_\mathbb{Z} \mathbb{Z}\Pi$ generated by X_α, Y_α, and H_α is isomorphic to $\mathfrak{sl}_2(\mathbb{Z})$.

5 Concatenation of integral modules

Denote by $\mathcal{A} \subset \mathrm{End}_\mathbb{Z} \Pi$ the algebra generated by the operators e_α and f_α. Let $M \subset \mathbb{Z}\Pi$ be an $\mathcal{A}$-stable submodule. We call M an *integral $\mathcal{A}$-module* if the set of paths $B := M \cap \Pi$ contained in M forms a $\mathbb{Z}$-basis for M, and if for all $\pi \in B$ and all simple roots α the minimum of the function $h_\alpha : t \mapsto \langle \pi(t), \alpha^\vee \rangle$ is an integer.

If M, N are integral $\mathcal{A}$-modules, then let $M * N$ be the submodule of $\mathbb{Z}\Pi$ spanned by the concatenation $\pi * \eta$ of all paths $\pi \in M$ and $\eta \in N$. The integrality condition on the function h_α implies [14], [15]:

LEMMA 2. *If M, N are integral $\mathcal{A}$-modules, then $M * N$ is an integral $\mathcal{A}$-module, and for $\pi \in M \cap \Pi$ and $\eta \in N \cap \Pi$ one has*

$$f_\alpha(\pi * \eta) = \begin{cases} (f_\alpha \pi) * \eta, & \textit{if } \exists\ n \geq 1 \textit{ such that } f_\alpha^n \pi \neq 0 \textit{ but } e_\alpha^n \eta = 0; \\ \pi * (f_\alpha \eta), & \textit{else.} \end{cases}$$

and $e_\alpha(\pi * \eta) = \pi * (e_\alpha \eta)$ if $\exists\ n \geq 1$ such that $e_\alpha^n \eta \neq 0$ but $f_\alpha^n \pi = 0$, and $e_\alpha(\pi * \eta) = (e_\alpha \pi) * \eta$ else.

6 Continuity

For $\pi_1, \pi_2 \in \Pi$ fix a parameterization. With respect to this parameterization we set:

$$d(\pi_1, \pi_2) := \max\{ |\langle \pi_1(t) - \pi_2(t), \alpha^\vee \rangle| \mid \alpha \text{ simple}, t \in [0,1]_\mathbb{Q} \}.$$

Denote by $\mathfrak{c}$ the maximum $\max\{ |\langle \alpha, \gamma^\vee \rangle| \mid \alpha, \gamma \text{ simple roots } \}$. An important property of the operators is that they are "continuous" with respect to $d(\cdot,\cdot)$. But note that $d(\pi_1, \pi_2)$ depends on the parameterization (see [15]).

PROPOSITION 1. *Suppose $d(\pi_1, \pi_2) < \epsilon$. If $f_\alpha \pi_1, f_\alpha \pi_2 \neq 0$ (respectively $e_\alpha \pi_1, e_\alpha \pi_2 \neq 0$), then $d(f_\alpha \pi_1, f_\alpha \pi_2) < 3\mathfrak{c}\epsilon$ (respectively $d(e_\alpha \pi_1, e_\alpha \pi_2) < 3\mathfrak{c}\epsilon$).*

7 The main results

Denote by $\Pi^+ \subset \Pi$ the subset of paths such that the image is completely contained in the dominant Weyl chamber. The structure of the modules $\mathcal{A}\pi$ generated by a path $\pi \in \Pi^+$ is described in the following theorems [15]:

MAIN THEOREM. *Suppose* $\pi, \pi_1, \pi_2 \in \Pi^+$.

(a) *Integrality:* $\mathcal{A}\pi$ *is an integral* $\mathcal{A}$*-module; i.e.,* $B_\pi := \mathcal{A}\pi \cap \Pi$ *is a* $\mathbb{Z}$*-basis for* $\mathcal{A}\pi$*, and the minimum of the function* $h_\alpha : t \mapsto \langle \eta(t), \alpha^\vee \rangle$ *is an integer for all* α *simple,* $\eta \in B_\pi$.

(b) *Highest weight module:* $\mathcal{A}\pi \cap \Pi^+ = \{\pi\}$, $\mathcal{A}\pi = \sum \mathbb{Z} f_{\alpha_1} \cdots f_{\alpha_s}\pi$*, and* π *is the only path in* $\mathcal{A}\pi$ *such that* $e_\alpha \pi = 0$ *for all simple roots.*

(c) *Isomorphism: If* $\pi_1(1) = \pi_2(1)$*, then* $\mathcal{A}\pi_1 \simeq \mathcal{A}\pi_2$.

(d) *Weyl group: The action of the simple reflections* s_α *on* $\mathbb{Z}\Pi$ *defined by:*

$$s_\alpha(\eta) := \begin{cases} f_\alpha^p(\eta), & \text{if } p := \langle \eta(1), \alpha^\vee \rangle \geq 0, \\ e_\alpha^p(\eta), & \text{if } -p := \langle \eta(1), \alpha^\vee \rangle < 0, \end{cases}$$

extends to an action of the Weyl group W *on* $\mathbb{Z}\Pi$ *such that the* $\mathcal{A}$*-modules* $\mathcal{A}\pi$ *are stable with respect to this action and* $w(\eta)(1) = w(\eta(1))$.

WEYL CHARACTER FORMULA FOR $\mathcal{A}$-MODULES. *Fix* $\rho \in X$ *such that* $\langle \rho, \alpha^\vee \rangle = 1$ *for all simple roots, and suppose* $\pi \in \Pi^+$*. Then*

$$\sum_{\sigma \in W} \operatorname{sgn}(\sigma) e^{\sigma(\rho)} \operatorname{Char} \mathcal{A}\pi = \sum_{\sigma \in W} \operatorname{sgn}(\sigma) e^{\sigma(\rho+\lambda)}.$$

In particular, $\operatorname{Char} \mathcal{A}\pi$ *is equal to the character of the irreducible, integrable* $\mathfrak{g}$*-module* V_λ *of highest weight* $\lambda := \pi(1)$.

CONCATENATION OF $\mathcal{A}$-MODULES. *Suppose* $\pi_1, \pi_2 \in \Pi^+$*. Then*

$$\mathcal{A}\pi_1 * \mathcal{A}\pi_2 = \bigoplus \mathcal{A}(\pi_1 * \eta),$$

where the sum runs over all paths $\eta \in \mathcal{A}\pi_2$ *such that* $\pi_1 * \eta \in \Pi^+$.

Because $\operatorname{Char} \mathcal{A}\pi_1 * \mathcal{A}\pi_2 = \operatorname{Char} \mathcal{A}\pi_1 \operatorname{Char} \mathcal{A}\pi_2 = \operatorname{Char} V_{\pi_1(1)} \otimes V_{\pi_2(1)}$, we get as an immediate consequence:

GENERALIZED LITTLEWOOD-RICHARDSON RULE. *For dominant weights* λ, μ *let* $\pi_1, \pi_2 \in \Pi^+$ *be such that* $\pi_1(1) = \lambda$ *and* $\pi_2(1) = \mu$*. Then the tensor product of the integrable, irreducible representations* V_λ, V_μ *of* $\mathfrak{g}$ *of highest weight* λ, μ *is isomorphic to the direct sum*

$$V_\lambda \otimes V_\mu \simeq \bigoplus V_{\lambda+\eta(1)},$$

where the sum runs over all paths $\eta \in \mathcal{A}\pi_2$ *such that* $\pi_1 * \eta \in \Pi^+$.

EXAMPLE. Suppose $\mathfrak{g} = \mathfrak{gl}_n$ and $\lambda \in X^+$, and we write again $\mathbf{w}$ for the path associated to the word. Let $\mathcal{T}_0$ be the tableau of shape λ having only 1's in the first row, 2's in the second row, etc. One sees easily that $\mathbf{w}_{\mathcal{T}_0}$ is completely contained in the dominant Weyl chamber, and the module $\mathcal{A}\mathbf{w}_{\mathcal{T}_0}$ has as basis the paths $\mathbf{w}_{\mathcal{T}}$, where $\mathcal{T}$ is a semi-standard Young tableau of shape λ. In this way we get the Young tableaux and the classical Littlewood-Richardson rule as a special case of the theory of path modules.

8 An example: The Lakshmibai-Seshadri paths

Let V_λ be the simple integrable module of a symmetrizable Kac-Moody algebra of highest weight λ. Denote by $\pi_\lambda : t \to t\lambda$ the path that connects the origin with λ by a straight line.

We are going to describe the basis $\mathcal{A}\pi_\lambda \cap \Pi$ of the $\mathcal{A}$-module generated by π_λ: the Lakshmibai-Seshadri paths. The definition given here is a "translation" of the definition in [12] into the language of paths.

Let W_λ be the stabilizer of λ, and let "$\leq$" be the Bruhat order on W/W_λ. We identify a pair $\pi = (\underline{\tau}, \underline{a})$ of sequences:

- $\underline{\tau} : \tau_1 > \tau_2 > \ldots > \tau_r$ is a sequence of linearly ordered cosets in W/W_λ and
- $\underline{a} : a_0 := 0 < a_1 < \ldots < a_r := 1$ is a sequence of rational numbers with the path:

$$\pi(t) := \sum_{i=1}^{j-1}(a_i - a_{i-1})\tau_i(\lambda) + (t - a_{j-1})\tau_j(\lambda) \quad \text{for} \quad a_{j-1} \leq t \leq a_j.$$

Note that $\lambda - \pi(1) = (\lambda - \tau_r(\lambda)) + \sum_{i=1}^{r-1} a_i(\tau_{i+1}(\lambda) - \tau_i(\lambda))$, so if the a_i are chosen such that the $a_i(\tau_{i+1}(\lambda) - \tau_i(\lambda))$ are still in the root lattice, then $\pi(1) \in X$. To ensure this, we introduce now the notion of an a-chain. Let $l(\cdot)$ be the length function on W/W_λ and denote by $\beta^\vee$ the coroot of a positive real root β:

Let $\tau > \sigma$ be two elements of W/W_λ and let $0 < a < 1$ be a rational number. By an *a-chain* for the pair (τ, σ) we mean a sequence of cosets in W/W_λ:

$$\kappa_0 := \tau > \kappa_1 := s_{\beta_1}\tau > \kappa_2 := s_{\beta_2}s_{\beta_1}\tau > \cdots > \kappa_s := s_{\beta_s} \cdot \ldots \cdot s_{\beta_1}\tau = \sigma,$$

where $\beta_1, \ldots, \beta_s$ are positive real roots and $l(\kappa_i) = l(\kappa_{i-1}) - 1$, $a\langle \kappa_i(\lambda), \beta_i^\vee \rangle \in \mathbb{Z}$ for all $i = 1, \ldots, s$.

DEFINITION. A pair $(\underline{\tau}, \underline{a})$ is called a *Lakshmibai-Seshadri* path of shape λ if for all $i = 1, \ldots, r-1$ there exists an a_i-chain for the pair (τ_i, τ_{i+1}).

EXAMPLE. For $\sigma \in W/W_\lambda$ let $\pi_{\sigma(\lambda)}$ be the path $t \mapsto t\sigma(\lambda)$ that connects 0 with $\sigma(\lambda)$ by a straight line. Then $\pi_{\sigma(\lambda)}$ is the Lakshmibai-Seshadri path $(\sigma; 0, 1)$.

THEOREM 2. [14] *The $\mathcal{A}$-module $\mathcal{A}\pi_\lambda$ generated by the path π_λ has as basis the set of all Lakshmibai-Seshadri paths $(\underline{\tau}, \underline{a})$ of shape λ.*

For $\tau \in W$ let $\mathcal{A}\pi_\lambda(\tau)$ be the subset of all Lakshmibai-Seshadri paths of shape λ such that $\tau_1 \leq \tau$, and denote by Λ_α the Demazure operator on $\mathbb{Z}[X]$:

$$\Lambda_\alpha(e^\mu) := \frac{e^{\mu+\rho} - e^{s_\alpha(\mu+\rho)}}{1 - e^{-\alpha}} e^{-\rho}.$$

DEMAZURE TYPE CHARACTER FORMULA. [14] *For any reduced decomposition* $\tau = s_{\alpha_1} \dots s_{\alpha_r}$ *one has* $\Lambda_{\alpha_1} \circ \dots \circ \Lambda_{\alpha_r}(e^\lambda) = \sum_{\eta \in \mathcal{A}\pi_\lambda(\tau)} e^{\eta(1)}$.

9 The P-R-V conjecture

Consider the tensor product $V_\lambda \otimes V_\mu$ of two simple, integrable $\mathfrak{g}$-modules of highest weight λ and μ. The Parthasarathy-Ranga-Rao-Varadarajan conjecture states:

THEOREM 3. *If* $\sigma, \tau \in W$ *are such that* $\nu := \tau(\lambda) + \sigma(\mu)$ *is a dominant weight, then the module* V_ν *occurs in* $V_\lambda \otimes V_\mu$.

Proofs of the conjecture have been given independently in [10] and [21]. Using the generalized Littlewood-Richardson rule, one can give a simple purely combinatorial rule. Figure 3 shows the idea of the proof given in [14]:

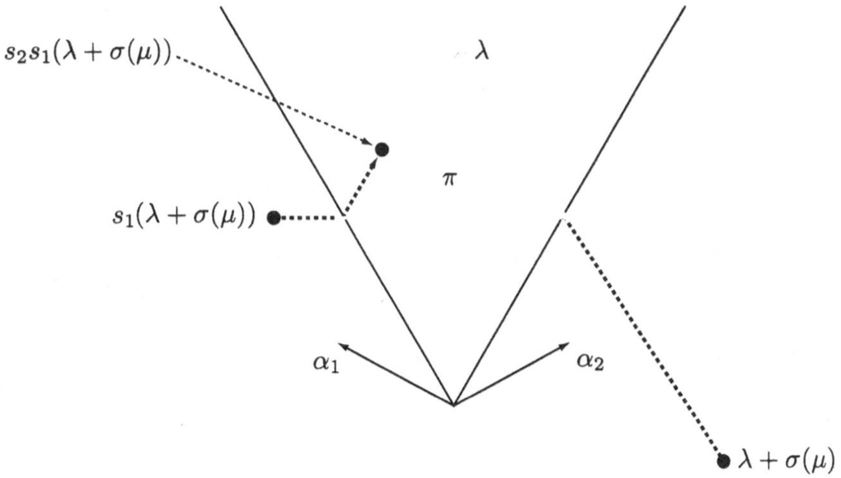

Figure 3

For $\nu' \in X$ let $[\nu'] \in X^+$ be the unique dominant element in the Weyl group orbit $W \cdot \nu'$. Suppose now $\nu := \tau(\lambda) + \sigma(\mu)$ is a dominant weight, then for $\kappa := \tau^{-1}\sigma$ we have $\nu = [\lambda + \kappa(\mu)]$. Starting with the path $\pi_{\kappa(\mu)} = (\kappa; 0, 1)$, Figure 3 shows how to construct inductively a Lakshmibai-Seshadri path $\pi = (\underline{\tau}, \underline{a})$ of shape μ such that $\pi_\lambda * \pi \in \Pi^+$ and $\nu = \lambda + \pi(1)$, which proves the theorem.

10 The paths and the crystal graph

Let $U_q(\mathfrak{g})$ be the q-analog of the enveloping algebra of $\mathfrak{g}$ (see for example [5], [19]). Kashiwara [6] and Lusztig [18] gave decomposition formulas for tensor products of $\mathfrak{g}$-representations using the theory of representations of $U_q(\mathfrak{g})$.

For $\pi \in \Pi^+$ let $\mathcal{G}(\pi)$ be the oriented, colored graph having as points the elements of the basis $\mathcal{A}\pi \cap \Pi$, and put an arrow $\pi_1 \xrightarrow{\alpha} \pi_2$ with color α if and only if $f_\alpha(\pi_1) = \pi_2$. Though the theory of the path modules is independent of the theory of quantum groups (the proofs of the properties of the path modules involve only combinatorics of Weyl groups), the following connection has been found (independently) by Joseph [5], Kashiwara [7], and Lakshmibai.

THE CRYSTAL GRAPH. *For $\pi \in \Pi^+$ set $\lambda := \pi(1)$. The graph $\mathcal{G}(\pi)$ is isomorphic to the crystal graph of the irreducible representation V_λ of highest weight λ of the q-analog $U_q(\mathfrak{g})$ of the enveloping algebra of $\mathfrak{g}$.*

11 The representation ring

Denote by $\mathbb{Z}\Pi_0 := \mathcal{A}\Pi^+ \subset \mathbb{Z}\Pi$ the $\mathcal{A}$-submodule generated by the elements of Π^+, and set $\Pi_0 := \mathbb{Z}\Pi_0 \cap \Pi$. By the decomposition formula, $\mathbb{Z}\Pi_0$ is a subalgebra of $\mathbb{Z}\Pi$.

We say $\pi, \eta \in \Pi_0$ are equivalent if and only if there exist $\pi^+, \eta^+ \in \Pi^+$ such that $\pi^+(1) = \eta^+(1)$, $\pi \in \mathcal{A}\pi^+$, and $\eta \in \mathcal{A}\eta^+$, and the isomorphism $\mathcal{A}\pi^+ \to \mathcal{A}\eta^+$, $a\pi^+ \mapsto a\eta^+$ maps π onto η.

PROPOSITION. *The concatenation on $\mathbb{Z}\Pi_0$ induces on $\mathbb{Z}\Pi_0/$"$\sim$" in a canonical way the structure of a $\mathbb{Z}$-algebra, and $\mathbb{Z}\Pi_0/$"$\sim$" is isomorphic to $\bigoplus_{\lambda \in X^+} \mathcal{A}\pi_\lambda$ as an $\mathcal{A}$-module. Further, for $\lambda, \mu \in X^+$ one has*

$$\mathcal{A}\pi_\lambda * \mathcal{A}\pi_\mu \equiv \oplus \mathcal{A}\pi_{\nu+\eta(1)} \mod \text{"}\sim\text{"},$$

where the sum runs over all $\eta \in \mathcal{A}\pi_\mu$ such that $\pi_\lambda * \eta \in \Pi^+$.

So one can hence view $\mathbb{Z}\Pi_0/$"$\sim$" as a model for the representation ring of $\mathfrak{g}$. Note that for $\mathfrak{g} = \mathfrak{sl}_n$ this algebra is isomorphic to the monoide plactique of Lascoux and Schützenberger [25]. Here we define the multiplication on the semi-standard Young tableaux as follows: let (as before) $\mathbf{w}_T$ be the path associated to a tableau. Then $T \cdot T' := T''$, where T'' is the unique semi-standard Young tableau such that $\mathbf{w}_T * \mathbf{w}_{T'} \equiv \mathbf{w}_{T''} \mod$ "$\sim$".

12 Some idempotents

Let $\mathbb{Z}\Pi_0$ be as above. Denote by $\mathcal{A}_0$ the subalgebra of $\mathrm{End}_\mathbb{Z}\,\mathbb{Z}\Pi_0$ generated by the restrictions $e_\alpha|_{\mathbb{Z}\Pi_0}, f_\alpha|_{\mathbb{Z}\Pi_0} \in \mathrm{End}_\mathbb{Z}\,\mathbb{Z}\Pi_0$. We consider here for simplicity only the case where $\mathfrak{g}$ is a finite-dimensional, semi-simple Lie algebra. For a dominant weight λ set $\lambda_\alpha := \langle \lambda, \alpha^\vee \rangle$, and denote by $p_\lambda \in \mathcal{A}_0$ the element

$$p_\lambda := \prod_{\alpha\ simple} e_\alpha^{\lambda_\alpha} f_\alpha^{\lambda_\alpha} (1 - e_\alpha^{\lambda_\alpha+1} f_\alpha^{\lambda_\alpha+1})(1 - f_\alpha e_\alpha).$$

Note that $p_\lambda \eta = \eta$ if $\eta \in \Pi^+$ and $\eta(1) = \lambda$, and $p_\lambda \eta = 0$ else. It follows that p_λ is an idempotent and $p_\lambda a p_\mu = 0$ for all $a \in \mathcal{A}_0$ and $\lambda \neq \mu$. Denote by $\mathcal{A}_\lambda$ the ideal $\mathcal{A}_0 p_\lambda \mathcal{A}_0$, and let λ^* be the highest weight of $\mathrm{Hom}_\mathbb{C}(V_\lambda, \mathbb{C})$.

THEOREM 4. *(a) As an* $(\mathcal{A}_0 \times \mathcal{A}_0)$*-bimodule, the ideal* $\mathcal{A}_\lambda$ *is isomorphic to the tensor product* $\mathcal{A}_0\pi_\lambda \otimes \mathcal{A}_0\pi_{\lambda^*}$*, and the representation map* $\mathcal{A}_0 \longrightarrow \mathrm{End}_\mathbb{Z}\, \mathcal{A}_0\pi_\lambda$*,* $a \mapsto a|_{\mathcal{A}_0\pi_\lambda}$*, induces an isomorphism of algebras* $\mathcal{A}_\lambda \simeq \mathrm{End}_\mathbb{Z}\, \mathcal{A}_0\pi_\lambda$.

(b) Set $U := \bigoplus_{\lambda\in X^+} \mathcal{A}_0\pi_\lambda$*. The representation* $\rho : \mathcal{A}_0 \longrightarrow \mathrm{End}_\mathbb{Z}\, U$*,* $a \mapsto a|_U$*, is faithful and induces an isomorphism:* $\bigoplus_{\lambda\in X^+} \mathcal{A}_\lambda \longrightarrow \bigoplus_{\lambda\in X^+} \mathrm{End}_\mathbb{Z}\, \mathcal{A}_0\pi_\lambda$.

(c) Let $\mathcal{F} \subset \mathcal{A}_0$ *be the subalgebra generated by the* f_α *and let* $\mathcal{E} \subset \mathcal{A}_0$ *be the subalgebra generated by the* e_α*. For any* $a \in \mathcal{A}_0$ *and* $v \in \mathbb{Z}\Pi_0$ *there exists an* $\mathcal{A}$*-stable submodule* $M \subset \mathbb{Z}\Pi_0$ *such that* $v \in M$*, and there exist* $F_\lambda \in \mathcal{F}$ *and* $E_\lambda \in \mathcal{E}$ *such that* $a|_M = \sum F_\lambda p_\lambda E_\lambda$ *for a finite number of* $\lambda \in X^+$.

References

[1] A. Berele, *A Schensted-type correspondence for the symplectic group*, J. Combin. Theory Ser. A, **43** (1986), 320–328.

[2] A. D. Berenstein and A. V. Zelevinsky, *Triple multiplicities for* $\mathfrak{sl}_{r+1}$ *and the spectrum of the exterior algebra of the adjoint representation*, J. Alg. Combinatorics, **1** (1992), 7–22.

[3] W. Fulton and J. Harris, Representation theory, Graduate Texts in Math., Springer Verlag, Berlin and New York, (1991).

[4] I. M. Gelfand and M. L. Zetlin, *Finite dimensional representations of the group of unimodular matrices*, Dokl. Akad. Nauk. USSR, **71** (1950), 825–828.

[5] T. Joseph, Quantum Groups and Their Primitive Ideals, Springer Verlag, Berlin and New York, to appear.

[6] M. Kashiwara, *Crystalizing the q-analogue of Universal Enveloping algebras*, Commun. Math. Phys., **133** (1990), 249–260.

[7] M. Kashiwara, Crystal bases of modified quantized enveloping algebras, RIMS, **917** (1993).

[8] R. C. King, Weight multiplicities for the classical groups, Group Theoretical methods in Physics, Lecture Notes in Phys., (Janner, Jannssen, and Boon, eds.), Springer Verlag, Berlin and New York, **50** (1976).

[9] K. Koike and I. Terada, *Young diagrammatic methods for the restriction of representations of complex classical Lie groups to reductive subgroups of maximal rank*, Adv. in Math., **79** (1990), 104–135.

[10] S. Kumar, *Proof of the Parthasarathy-Ranga-Rao-Varadarajan Conjecture*, Invent. Math., **93** (1988), 117–130.

[11] V. Lakshmibai and C. S. Seshadri, *Geometry of G/P V*, J. Algebra, **100** (1986), 462–557.

[12] V. Lakshmibai and C. S. Seshadri, *Standard monomial theory*, in: Proceedings of the Hyderabad Conference on algebraic groups, Manoj Prakashan, Madras, (1991), 279–323.

[13] P. Littelmann, *A generalization of the Littlewood-Richardson rule*, J. Algebra, **130** (1990), 328–368.

[14] P. Littelmann, *A Littlewood-Richardson type rule for symmetrizable Kac-Moody algebras*, Invent. Math., **116** (1994), 329–346.

[15] P. Littelmann, *Paths and root operators in representation theory*, preprint, to appear in: Ann. of Math. (2) (1993).

[16] D. E. Littlewood, The Theory of Group Characters and Matrix Representations of Groups, Oxford University Press, London, (1950).

[17] D. E. Littlewood and A. R. Richardson, *Group characters and algebra*, Philos. Trans. Roy. Soc. London Ser. A, **233** (1934), 99–141.

[18] G. Lusztig, *Canonical bases arising from quantized enveloping algebras II*, Progr. Theoret. Phys., **102** (1990), 175–201.

[19] G. Lusztig, Introduction to quantum groups, Birkhäuser Verlag, Boston, Progr. Math., **110** (1993).

[20] I. G. Macdonald, Symmetric Functions and Hall polynomials, Oxford University Press, London, (1979).

[21] O. Mathieu, *Construction d'un groupe de Kac-Moody et applications*, Compos. Math., **69** (1989), 37–60.

[22] T. Nakashima, Crystal base and a generalization of the Littlewood-Richardson rule for the classical Lie algebras, preprint, RIMS, **783** (1991).

[23] R. A. Proctor, *A Schensted algorithm which models tensor representations of the orthogonal group*, Canad. J. Math., **42** (1990), 28–49.

[24] G. B. Robinson, *On the representation of the symmetric group*, Amer. J. Math., **69** (1938), 745–760.

[25] M. P. Schützenberger, La correspendance de Robinson, Combinatoire et représentation du groupe symmetrique, Lecture Notes in Math., Springer Verlag, Berlin and New York, **579** (1976).

[26] S. Sundaram, *Orthogonal tableaux and an insertion algorithm for* $SO(2n+1)$, J. Combin. Theory Ser. A, **53** (1990), 239–256.

[27] H. Weyl, The classical groups, their invariants and representations, Princeton University Press, Princeton, (1946).

Subgroup Growth

ALEXANDER LUBOTZKY

Institute of Mathematics, Hebrew University
Jerusalem, 91904, Israel

Dedicated to the memory of S. A. Amitsur

1 Introduction

Let Γ be a group generated by a finite set S. Denote by $b_n^S(\Gamma)$ the number of elements of Γ of length at most n with respect to $S \cup S^{-1}$. The word growth of Γ, i.e., the growth of the sequence $b_n^S(\Gamma)$ has received considerable attention following the observation that it has some geometric meaning (see [Gr] and the references therein).

Here we look at a different growth — *the subgroup growth*. Let $\sigma_n(\Gamma) =$ the number of subgroups of Γ of index at most n and $a_n(\Gamma) = \sigma_n(\Gamma) - \sigma_{n-1}(\Gamma)$. The study of $\sigma_n(\Gamma)$ was started in 1949 by Hall, who computed $\sigma_n(\Gamma)$ for $\Gamma = F_r$, the free group on r generators. But it was only in the last decade that the research of this concept blossomed when connections were found to problems in finite groups, combinatorics, algebraic and arithmetic groups, p-adic Lie groups, number theory, model theory, complex functions, Igusa's zeta functions, and more.

Subgroup growth can be thought of as a chapter in "asymptotic group theory" where we study infinite groups via the asymptotic behavior of some of their invariants. As will be seen below it leads naturally to another direction of asymptotic group theory where invariants of infinite families of *finite* groups are studied. The interplay between the finite and the infinite turns out to be extremely useful.

A related development is the resurgence of interest in the last decade in the subject of "finiteness properties of infinite groups". This topic deserves an independent survey. In Section 7 we only touch on it briefly and give some references.

We will try here to give a brief description of the main results achieved so far and some indication on the problems for future research. Other survey papers are [L4], [MS2], [dS2].

In Section 2 we will present the theorem that characterizes the groups with polynomial subgroup growth (PSG-groups for short), i.e., those Γ for which $\sigma_n(\Gamma) = O(n^c)$ for some constant c. This is probably the major theorem in this topic so far. We give some hints of its proof as they shed light on the connection between subgroup growth and other topics, such as the classification of finite simple groups, p-adic Lie groups, strong approximation results for arithmetic groups, and some number theory.

Proceedings of the International Congress
of Mathematicians, Zürich, Switzerland 1994

The proof of the PSG-theorem calls attention to the problem of counting the *congruence subgroups* of arithmetic groups. This problem is addressed in Section 3, where estimates of the growth of congruence subgroups are given. This is $n^{c\log n/\log\log n}$ for characteristic zero arithmetic groups (e.g., $SL_r(\mathbb{Z})$) and slightly higher for characteristic $p > 0$ (e.g., $SL_r(\mathbb{F}_p[t])$). In case the arithmetic groups satisfy the *congruence subgroup property* (CSP) (e.g. $r \geq 3$), the congruence subgroups are all the finite index subgroups and so we have groups with $\sigma_n(\Gamma) \sim n^{c\log n/\log\log n}$, i.e., groups with subgroup growth that is strictly between polynomial and exponential. It turns out that the congruence subgroup property can be characterized by means of subgroup growth. The latter has a purely group theoretic meaning — so the congruence subgroup problem can now be asked for nonarithmetic lattices in semi-simple groups. Some conjectures and some very partial results on subgroup growth of fundamental groups of hyperbolic groups are presented.

Another application of the counting of congruence subgroups is that for finitely generated linear groups Γ, $\sigma_n(\Gamma)$ grows either polynomially or at least as fast as $n^{c\log n/\log\log n}$. For profinite groups there is no such gap, but for pro-p groups G the gap is even larger: $\sigma_n(G) \geq n^{c\log n}$ if not polynomial. The question of existence of a "gap" for general finitely generated groups is still open. There are however (nonlinear) groups with growth $n^{c\log n/(\log\log n)^2}$ and it might be that this is the minimum possible for discrete non PSG-groups. The gap problem is discussed in Section 4.

In Section 5, we turn to nilpotent groups Γ. They have polynomial subgroup growth and the zeta function

$$\zeta_\Gamma(s) = \sum_{n=1}^{\infty} a_n(\Gamma)n^{-s} \qquad \text{where} \qquad a_n(\Gamma) = \sigma_n(\Gamma) - \sigma_{n-1}(\Gamma)$$

has nontrivial domain of convergence. Moreover, it has Euler factorization and the local p-factors are rational functions of p^{-s}. One is tempted to try to develop a theory of zeta functions for nilpotent groups, in analogy with the Dedekind ζ-functions of number fields. Only very partial results on analytic continuation and functional equations are known. In Section 6, we briefly describe some of the detailed information known about $\sigma_n(\Gamma)$ for virtually free groups Γ, and we hint at the connection to some combinatorial problems.

2 Groups of polynomial subgroup growth

A group Γ is said to have *polynomial subgroup growth* (a PSG-group) if there exists c such that $a_n(\Gamma) \leq n^c$ for every n. Here is the main result on such groups.

THEOREM 2.1. [LMS], [MS1], [LM2], [S] *Let Γ be a finitely generated, residually finite group. Then Γ has polynomial subgroup growth if and only if Γ is virtually solvable of finite rank.*

Recall that a group is virtually solvable if it has a finite index solvable subgroup. It is of finite rank if there is a bound on the number of generators of its finitely generated subgroups.

The difficult part of Theorem 2.1 is, of course, the part saying that a PSG-group is virtually solvable. This is done in a sequence of steps, which we describe briefly:

The core of the proof is the linear case. Assume Γ is a PSG-subgroup of $GL_n(\mathbb{C})$ for some n and Γ is not virtually solvable. Some results on linear and algebraic groups are used to deduce that Γ has a representation Π into $GL_m(\mathbb{Q})$ for some m with a non virtually solvable image. Moreover, the Zariski closure G of $\Pi(\Gamma)$ can be assumed to be a semi-simple, connected, simply connected algebraic subgroup of GL_m. Hence $\Pi(\Gamma)$ is a subgroup of an S-arithmetic group $G(\mathbb{Z}_S)$. Strong approximation results for subgroups of S-arithmetic groups imply that $\sigma_n(\Gamma)$ grows at least as fast as $\gamma_n(G(\mathbb{Z}_S))$. The latter is the number of *congruence* subgroups of $G(\mathbb{Z}_S)$ of index at most n. A congruence subgroup of Γ is one containing $\Gamma(m) = \Gamma \cap \mathrm{Ker}(GL_n(\mathbb{Z}) \to GL_n(\mathbb{Z}/m\mathbb{Z}))$ for some $0 \neq m \in \mathbb{Z}$. The Prime Number Theorem is then used to count congruence subgroups and to show that their growth is not polynomial. (A more precise counting of congruence subgroups is described in Theorem 3.2 below.) This proves the linear case in characteristic zero.

Theorem 2.1 is extended to residually-p groups in the following way. A residually-p group Γ is characterized by the fact that Γ is embedded in its pro-p completion $\Gamma_{\hat{p}}$. If Γ is PSG so is $H = \Gamma_{\hat{p}}$. In addition we have the following theorem, which is of independent interest:

THEOREM 2.2. [LM2] *Let H be a finitely generated pro-p group. Then H is a PSG group if and only if it is p-adic analytic.*

The theory of p-adic analytic groups, i.e., groups that are Lie groups over $\mathbb{Q}_p$, the field of p-adic numbers, was studied in detail by Lazard. For a more modern treatment (via the notion of powerful p-groups) and references see [DDMS]. A compact p-adic analytic group is linear. Thus Theorem 2.2 and the remark proceeding it imply that a PSG residually-p group is linear over $\mathbb{C}$. Theorem 2.1 is therefore proved for residually-p groups that include linear groups in characteristic p. The linearity argument described above is a prototype of a linearization method, initiated in [L2], that uses p-adic analytic groups as a tool to ensure linearity of various groups.

The case of general Γ in Theorem 2.1 is carried out in the following way. Using the classification of the finite simple groups one analyzes the possible chief factors of $\hat{\Gamma}$, the profinite completion of Γ. The nonabelian simple ones should have bounded multiplicity and should have bounded degree linear representations. This, with the linear case proved below, reduces the proof to residually solvable groups. Some results and arguments from the theory of infinite solvable groups are then used to reduce to the residually-p case, which was treated before.

The same proof can be presented via different patterns. The reader is referred to other surveys like [DDMS], [LMS], and [L4].

We finally define for a PSG-group the invariant: $\alpha(\Gamma) = \limsup_{n\to\infty} \frac{\log \sigma_n(\Gamma)}{\log n}$. Very little is known about $\alpha(\Gamma)$. It is not even known whether or not it is always a rational number (this is so for pro-p groups ([dS1])), not even for nilpotent groups Γ.

3 Counting congruence subgroups and intermediate subgroup growth

The proof of the PSG-Theorem (2.1) called attention to counting congruence subgroups in arithmetic subgroups of semi-simple groups. The proof there required only that this growth not be polynomial, a fact that follows quickly from a weak version of the prime number theorem. Deeper versions of the prime number theorem combined with group theoretical methods from [P] can give more precise estimates. (For simplicity we present here weak forms of the actual results.)

DEFINITION 3.1. Let $f : \mathbb{N} \to \mathbb{R}$ be a function and Γ a group. Γ is said to have subgroup growth of type f if there exist $a, b \in \mathbb{R}_+$ such that $\sigma_n(\Gamma) \leq f(n)^a$ for every n and $\sigma_n(\Gamma) \geq f(n)^b$ for infinitely many n. *We will write* $\sigma_n(\Gamma) \sim f(n)$.

THEOREM 3.2. [L3] *Let G be a simple Chevalley subgroup of GL_n, and $\Gamma = G(\mathbb{Z})$. Then $\gamma_n(\Gamma) \sim n^{\log n/\log\log n}$ where $\gamma_n(\Gamma)$ is the number of congruence subgroups of Γ of index at most n.*

Interestingly enough, for global fields of characteristic $p > 0$ the type of growth is different as is the method of proof (which is influenced by [Sh1]).

THEOREM 3.3. [L3] *Let G be a simple Chevalley subgroup of GL_n and $\Gamma = G(\mathbb{F}_p[t])$ (with G not of type A_1 or C_n if $p = 2$). Then there exist two constants a and b such that for every n,*

$$n^{a\log n} \leq \gamma_n(\Gamma) \leq n^{b\log^2 n}$$

where $\gamma_n(\Gamma)$ is the number of congruence subgroups of index at most n.

Note that in the characteristic p case we do not have a complete answer. It is however different from characteristic zero.

In any event, if Γ has the congruence subgroup property, e.g., $\Gamma = SL_3(\mathbb{Z})$ or $\Gamma = SL_3(\mathbb{F}_p[t])$ where every finite index subgroup is a congruence subgroup, we can conclude:

COROLLARY 3.4. *There exist finitely generated groups of intermediate subgroup growth.*

The above-mentioned examples are of growth that is just slightly faster than polynomial. Other types of examples were constructed using arguments from commutative algebra and the geometry of numbers:

THEOREM 3.5. [SS] *For every $d > 1$ there exist a finitely presented meta-abelian group Γ and constants $b, c > 1$ such that $b^{n^{1/d}} \leq \sigma_n(\Gamma) \leq c^{n^{1/d}}$ for all large n.*

It is widely open as to what the possible types of subgroup growth are for finitely generated discrete (or pro-p) groups.

The congruence subgroup property is connected to subgroup growth in an even deeper way. Assume G is a semi-simple, simply connected, connected $\mathbb{Q}$-algebraic subgroup of GL_n such that $\Gamma = G(\mathbb{Z})$ is infinite and $G(\mathbb{Q})$ has the standard description of normal subgroups (cf. [PR], [L3]). $\Gamma = G(\mathbb{Z})$ is said to have the congruence subgroup property (CSP) if $\mathrm{Ker}(\widehat{G(\mathbb{Z})} \to G(\hat{\mathbb{Z}}))$ is finite.

THEOREM 3.6. $\Gamma = G(\mathbb{Z})$ *has the congruence subgroup property if and only if* $\sigma_n(\gamma) = O_\varepsilon(n^{\varepsilon \log n})$ *for every* $\varepsilon > 0$.

So, if CSP fails there are many more noncongruence subgroups than congruence subgroups. Moreover, the Theorem says that an arithmetic property like CSP can be characterized by the subgroup growth, which is a purely group theoretical property. This enables one to pose the congruence subgroup problem for groups without arithmetic structure, in particular for nonarithmetic lattices in a semi-simple Lie group H. One of the features of Serre's conjecture on CSP [Se], is that the CSP depends on H and not on Γ. It is natural to extend this conjecture to nonarithmetic lattices and to suggest that fundamental groups of hyperbolic manifolds have fast subgroup growth. Only partial results are known, e.g.:

PROPOSITION 3.7. [L1], [L3] *Let* M *be a 3-dimensional hyperbolic manifold of finite volume. Then for some* $\varepsilon > 0$, $\sigma_n(\Pi_1(M)) \geq n^{\varepsilon \log n}$ *for infinitely many* n*'s.*

It can also be shown that the arithmetic lattices treated in [M] and the nonarithmetic lattices in [GP] have super-exponential subgroup growth.

On the other hand, it is natural to conjecture (and it is supported by the known examples) that discrete groups with Kazhdan property (T) have modest subgroup growth. No nontrivial bound is known at this point.

4 The gap problem

A byproduct of the PSG-Theorem (2.1), the precise counting of congruence subgroups (Theorem 3.2), and arguments from the geometry of numbers from [SS] is:

THEOREM 4.1. [L3] *Let* Γ *be a finitely generated linear group. Then either* Γ *is a PSG-group or* $\sigma_n(\Gamma) \geq n^{c \log n/ \log\log n}$ *for some* $c > 0$ *and infinitely many* n*'s.*

This joins a result of similar flavor (but with a different bound!) for pro-p groups.

THEOREM 4.2. [Sh1] *Let* G *be a finitely generated pro-p group. Then either* G *is PSG or* $\sigma_n(G) \geq n^{c \log n}$ *for some* $c > 0$ *and infinitely many* n*'s.*

On the other hand for profinite groups G, there is no such gap.

THEOREM 4.3. [Sh2] *For every function* $f : \mathbb{N} \to \mathbb{N}$ *such that* $f(1) = 1$ *and* $\log f(n)/\log n$ *goes to infinity, there exists a finitely generated non PSG-profinite group* G *satisfying* $\sigma_n(G) \leq f(n)$ *for all* n.

Theorems 4.1 and 4.2 are best possible in the sense that linear (resp. pro-p) groups exist with $\sigma_n(\Gamma) \sim n^{\log n/ \log\log n}$ (resp. $\sigma_n(G) \sim n^{\log n}$). But it is not known whether or not any gap exists for general finitely generated discrete groups. Recently, it was shown that the lower bound for linear groups given in Theorem 4.1 can be beaten.

THEOREM 4.4. [LPSh] *There exist finitely generated groups whose subgroup growth is of type* $n^{\log n/(\log\log n)^2}$.

We are left with:

OPEN PROBLEM. Is there a nontrivial lower bound on the subgroup growth of non-PSG finitely generated groups? Is it $n^{\log n/(\log\log n)^2}$?

5 Nilpotent groups and zeta functions

For word growth the numbers $b_n^S(\Gamma)$ defined in Section 1 are of little interest; their growth is the focus of research. With subgroup growth the situation is different: the numbers $a_n(\Gamma)$ (= the number of subgroups of index n) are of intrinsic interest by themselves. In [GSS] the function $\zeta_\Gamma(s) = \sum_{n=1}^{\infty} a_n(\gamma)n^{-s}$ was associated with Γ and its properties were studied when Γ is a nilpotent group.

THEOREM 5.1. [GSS] *Let Γ be a finitely generated nilpotent group and $\zeta_\Gamma(s) = \sum a_n(\Gamma)n^{-s}$. Then:*
(1) *$a_n(\Gamma)$ grows polynomially and so*

$$\alpha(\Gamma) = \limsup \frac{\log \sigma_n(\Gamma)}{\log n} < \infty.$$

(2) *$\zeta_\gamma(s)$ is convergent for $\mathrm{Re}(s) > \alpha(\Gamma)$.*
(3) *Euler factorization: $\zeta_\Gamma(s) = \prod_{p \text{ prime}} \zeta_{\Gamma,p}(s)$ where $\zeta_{\Gamma,p}(s) = \sum_{i=0}^{\infty} a_{p^i}(\Gamma)p^{-is}$.*
(4) *For every prime p, $\zeta_{\Gamma,p}(s)$ is a rational function of p^{-s}.*

The first three parts of Theorem 5.1 are easy — but not the fourth. Its proof is based on a presentation of $\zeta_{\Gamma,p}(s)$ as a p-adic integral. This p-adic integral is rational by the main result of [D]. The proof of the latter is based on model theoretic considerations.

Of course, the dream is to establish a theory of zeta functions for nilpotent groups in analogy to Dedekind zeta functions of number fields. The many examples computed in [GSS] are quite encouraging — but very little is known (see [dS2]). The computations suggest also a Čebotarev type theorem. Recently it was shown in [dSL] that an analogous local zeta function $\zeta^{\wedge}_{\Gamma,p}(s)$ satisfies, in some cases, a kind of functional equation. Here, $\zeta^{\wedge}_{\Gamma}(s) = \sum a^{\wedge}_n(\Gamma)n^{-s}$ where $a^{\wedge}_n(\Gamma)$ is the number of subgroups of index n in Γ whose profinite completion is isomorphic to $\hat{\Gamma}$. This $\zeta^{\wedge}_{\Gamma,p}(s)$ turns out to be an Igusa zeta function computed as an integral over some p-adic algebraic group [I]. On the other hand in [dS2] some examples are shown where $\zeta^{\wedge}_{\Gamma}$ do not have a meromorphic continuation. Clearly only the tip of the iceberg has been revealed so far. Only the future can say where this theory will lead us.

A far-reaching generalization of Theorem 5.1(4) is the following:

THEOREM 5.2. [dS1] *Let G be a compact p-adic analytic group; then $\zeta_{G,p}(s)$ is a rational function of p^{-s}.*

Although many examples were computed with G nilpotent, the only semi-simple groups G for which $\zeta_{G,p}(s)$ has been computed are the principal congruence subgroups of $SL_2(\mathbb{Z}_p)$ [Il].

6 Virtually free groups and permutational representations

As mentioned above the number of subgroups of index n, $a_n(\Gamma)$, in $\Gamma = F_r$, the free group on r generators was given (recursively) in [H]. The basic observation is that $a_n(\Gamma) = \frac{1}{(n-1)!}|\mathrm{Hom}_t(\Gamma, S_n)|$ where $\mathrm{Hom}_t(\Gamma, S_n)$ is the number of transitive actions of Γ on the set $\{1, \ldots, n\}$. For $\Gamma = F_r$, most actions are transitive so $|\mathrm{Hom}_t(F_r, S_n)| \sim |\mathrm{Hom}(F_r, S_n)| = (n!)^r$ and hence $a_n(F_r) \sim n(n!)^{r-1}$, which is super-exponential. For general groups it is not easy to estimate $|\mathrm{Hom}(\Gamma, S_n)|$ (or $|\mathrm{Hom}_t(\Gamma, S_n)|$).

Detailed study for Γ virtually free groups was carried out in [Mu] (and the references therein). For example if Γ is a free product of two finite groups $\Gamma = H_1 * H_2$, then $|\mathrm{Hom}(\Gamma, S_n)| = \prod_{i=1}^{2} |\mathrm{Hom}(H_i, S_n)|$. The problem of estimating $|\mathrm{Hom}(H, S_n)|$ for a finite group H was solved in [Mu] and the following Theorem is deduced:

THEOREM 6.1. [Mu] *Let $\Gamma = *_{i=1}^{s} G_i$ be a free product of $2 \le s < \infty$ nontrivial finite groups of orders $m_1, \ldots, m_s$ respectively. If $s = 2$ assume that not both G_1 and G_2 are cyclic of order 2. Then*

$$a_n(\Gamma) \sim L_\Gamma \cdot \Phi_\Gamma(n) \quad \text{as } n \to \infty$$

where

$$L_\Gamma = (2\pi m_1 * \cdots * m_s)^{-1/2} \exp\left(- \sum_{\{i \mid 2 \mid m_i\}} \frac{(a_{m_i/2}(G_i))^2}{2m_i} \right)$$

$$\Phi_\Gamma(n) = n^{-h(\Gamma)n} \exp\left(h(\Gamma)n + \sum_{i=1}^{s} \sum_{\substack{d_i \mid m_i \\ d_i < m_i}} \frac{a_{d_i}(G_i)}{d_i} n^{d_i/m_i} + \frac{1}{2}\log n \right)$$

and

$$h(\Gamma) = \text{Euler characteristic of } \Gamma = \frac{1 - (m_1 - 1) \cdot \ldots \cdot (m_s - 1)}{m_1 \cdot \ldots \cdot m_s}.$$

This applies in particular for the modular group $PSL_2(\mathbb{Z}) \simeq \mathbb{Z}/2\mathbb{Z} * \mathbb{Z}/3\mathbb{Z}$. Of course the subgroup growth of $PSL_2(\mathbb{Z})$ is super-exponential, whereas the growth of the congruence subgroups is much slower (see Theorem 3.2), so "most" subgroups are noncongruence as Theorem 3.6 says. It even suggests that a much stronger form of Theorem 3.6 might be true. Compare Proposition 3.7 and the remark preceding it.

7 Finiteness conditions on infinite groups

Subgroup growth can be viewed as part of the general theory of "finiteness conditions on infinite groups". This theory was a main theme of group theory in the first half of the century. There have been various conjectures, all of the following type: Assume a group G satisfies a set of conditions P, then G is finite (or solvable or virtually solvable). The theory had only partial success. The examples of Tarski monsters by Olshanski and Rips explain why: these groups are finitely generated where every proper subgroup is of order p (for a fixed prime p) and still they are infinite groups. They are, therefore, of finite exponent, finite rank, etc. and yet not virtually solvable.

In recent years it has been shown that under the assumption of residual finiteness, some of these old conjectures become theorems. The most notable example is of course the restricted Burnside problem, which is nothing more than the Burnside problem for residually finite groups:

THEOREM 7.1. [Z1] *Let Γ be a finitely generated residually finite group of finite exponent (i.e. for some m, $x^m = 1$, for every $x \in \Gamma$). Then Γ is finite.*

Similarly for finite rank instead of finite exponent:

THEOREM 7.2. [LM1] *Let Γ be a finitely generated residually finite group of finite rank (i.e. for some m, every finitely generated subgroup of Γ is generated by at most m elements). Then Γ is virtually solvable.*

Another result of a similar flavor is:

THEOREM 7.3. [W], [Z2] *A finitely generated residually finite n-Engel group (i.e. for every x and y in Γ, $[x, y, y, \dots, y] = 1$ where y occurs n times) is nilpotent.*

These developments were made possible by the accumulation of four theories:
(1) The classification of finite simple groups,
(2) The remarkable progress made in Lie algebras techniques,
(3) The use of p-adic analytic pro-p groups, and
(4) The use of the theories of linear/algebraic/arithmetic groups.

It seems that the future will bring many more exciting results in these directions. Some connections of finiteness conditions with arithmetic groups have already surfaced (cf. [PR], [L3]).

References

[D] J. Denef, *The rationality of the Poincaré series associated to the p-adic points on a variety*, Invent. Math. **77** (1984) 1–22.

[DDMS] J. D. Dixon, M. P. F. du Sautoy, A. Mann, and D. Segal, *Analytic pro-p groups*, LMS Lecture Notes **157**, Cambridge (1991).

[dS1] M. P. F. du Sautoy, *Finitely generated groups, p-adic analytic groups and Poincaré series*, Ann. of Math. (2) **137** (1993), 639–670.

[dS2] M. P. F. du Sautoy, *Zeta functions of groups: A survey*, in preparation.

[dSL] M. P. F. du Sautoy and A. Lubotzky, *Functional equations and uniformity for local zeta functions of nilpotent groups*, American J. of Math. (1996), to appear.

[Gr] R. I. Grigorchuk, *On growth in group theory*, Proc. Internat. Congress Math., Kyoto, Japan, 1990, 325–338.

[GP] M. Gromov and I. Piatetski-Shapiro, *Non arithmetic groups in Lobachevsky spaces*, Publ. Math. IHES **66** (1988) 93–103.

[GSS] F. J. Grunewald, D. Segal, and G. C. Smith, *Subgroups of finite index in nilpotent groups*, Invent. Math. **93** (1988), 185–223.

[H] M. Hall, *Subgroups of finite index in free groups*, Canad. J. Math. **1** (1949), 187–190.

[I] J. I. Igusa, *Universal p-adic zeta functions and their functional equations*, Amer. J. Math. **111** (1989), 671–716.

[Il] I. Ilani, *Counting subgroups of finite index in analytic pro-p groups and Zeta functions of groups*, Hebrew-University thesis, 1992 (in Hebrew).

[L1] A. Lubotzky, *Group presentation, p-adic analytic groups and lattices in $SL_2(\mathbb{C})$*, Ann. of Math. (2), **118** (1983) 115–130.

[L2] A. Lubotzky, *A group theoretic characterization of linear groups*, J. Algebra, **113** (1988) 207–214.

[L3] A. Lubotzky, *Subgroup growth and congruence subgroups*, Invent. Math., **119** (1995) 267–295.

[L4] A. Lubotzky, *Counting finite index subgroups*, Proc. Groups, Galway/St. Andrews 1993, Lecture Notes LMS, **212**, 368–404, Cambridge 1995.

[LM1] A. Lubotzky and A. Mann, *Residually finite groups of finite rank*, Math. Proc. Cambridge Philos. Soc. **106** (1989), 385–388.

[LM2] A. Lubotzky and A. Mann, *On groups of polynomial subgroup growth*, Invent. Math. **104** (1991), 521–533.

[LMS] A. Lubotzky, A. Mann, and D. Segal, *Finitely generated groups of polynomial subgroup growth*, Israel J. Math. **82** (1993), 363–371.

[LPSh] A. Lubotzky, L. Pyber, and A. Shalev, *Discrete groups of slow subgroup growth*, Israel J. of Math., 1996, to appear.

[MS1] A. Mann and D. Segal, *Uniform finiteness conditions in residually finite groups*, Proc. London Math. Soc. (3) **61** (1990), 529–545.

[MS2] A. Mann and D. Segal, *Subgroup growth: Current developments*, Proc. of Infinite Groups **94** (Ravelo), de Gruyter, to appear.

[M] J. Millson, *On the first Betti number of a constant negatively curved manifold*, Ann. of Math. (2) **104** (1976), 235–247.

[Mu] T. Müller, *Finite group actions, subgroups of finite index in free products and asymptotic expansions of $e^{P(z)}$*, to appear.

[PR] V. Platonov and A. Rapinchuk, *Abstract characterization of arithmetic groups with the congruence subgroup property*, Dokl. Akad. USSR **319** (1991), 1322–1327.

[P] L. Pyber, *Enumerating finite groups of given order*, Ann. of Math. (2) **137** (1993), 203–220.

[S] D. Segal, *Subgroups of finite index in soluble groups I*, Proc. Groups St. Andrews 1985, LMS Lecture Notes **121**, 307–314, Cambridge (1986).

[SS] D. Segal and A. Shalev, *Groups with fractionally exponential subgroup growth*, J. Pure Appl. Algebra **88** (1993), 205–223.

[Se] J. P. Serre, *Le problème des groupes de congruences pour SL_2*, Ann. of Math. (2) **92** (1970), 489–527.

[Sh1] A. Shalev, *Growth functions, p-adic analytic groups and groups of finite coclass*, J. London Math. Soc. **46** (1992), 111–122.

[Sh2] A. Shalev, *Subgroup growth and sieve methods*, preprint.

[W] J. S. Wilson, *Two generator conditions for residually finite groups*, Bull. London Math. Soc. **23** (1991) 239–248.

[Z1] E. I. Zel'manov, *On the restricted Burnside problem*, Proc. Internat. Congress Math., Kyoto, Japan, 1990, vol. 1 pp. 395–402, Springer-Verlag, 1991.

[Z2] E. I. Zel'manov, *On some problems of group theory and Lie algebras*, Math. USSR-Sb. **66** (1990) 159–168.

Constructions polynomiales et théorie de Galois

JEAN-FRANÇOIS MESTRE

UFR de Math.
Université Paris 7
2 place Jussieu, F-75251 Paris Cedex 05, France

1. Introduction

Soit G un groupe fini. Existe-t-il une extension de $\mathbf{Q}$ (ou plus généralement une infinité d'extensions de $\mathbf{Q}$, ou plus généralement encore une extension régulière de $\mathbf{Q}(t)$) galoisienne de groupe de Galois G?

Lorsque G est extension centrale d'un groupe H par $\mathbf{Z}/2\mathbf{Z}$, et que l'on connaît une extension régulière K de $\mathbf{Q}(t)$ de groupe de Galois H, l'existence d'une extension quadratique M de K telle que $M/\mathbf{Q}(t)$ soit galoisienne de groupe de Galois G est reliée au fait qu'un certain élément α_K de $\mathrm{Br}_2(\mathbf{Q}(t))$ associé à l'extension K est nul [1].

Dans la section 1, nous considérons le cas où H est le groupe alterné A_n ($n \geq 4$), et $G = \tilde{A}_n$ l'unique extension centrale non triviale de A_n par $\mathbf{Z}/2\mathbf{Z}$. Pour n impair, nous construisons des extensions K de $\mathbf{Q}(t)$ de groupe de Galois A_n, dont les ordres des groupes d'inertie sont impairs (en fait égaux à 3). L'invariant α_K est alors constant. Comme, pour $t = \infty$, ces extensions se spécialisent en des algèbres étales scindées $\mathbf{Q}^n$, l'invariant α_K est nul, et le groupe $\tilde{A}_n$ est groupe de Galois d'une extension régulière de $\mathbf{Q}(t)$. Il est facile de voir que le cas n pair se déduit du cas n impair, d'où le théorème:

THÉORÈME 1 *Si n est un entier ≥ 4, le groupe $\tilde{A}_n$ est groupe de Galois d'une extension régulière de $\mathbf{Q}(t)$.*

Dans la section 2, nous traitons le cas où $H = \mathrm{PSL}_2(F_7)$ et $G = \mathrm{SL}_2(\mathbf{F}_7)$. Dans ce cas, LaMacchia a construit des extensions régulières de $\mathbf{Q}(x)$ de groupe de Galois $\mathrm{PSL}_2(\mathbf{F}_7)$ telles que l'invariant $w \in \mathrm{Br}_2(\mathbf{Q}(x))$ associé a 4 points de ramification d'indice pair, et qui, pour certaines valeurs de t, s'annule.

Le théorème

THÉORÈME 2 *Le groupe $\mathrm{SL}_2(\mathbf{F}_7)$ est groupe de Galois d'une extension régulière de $\mathbf{Q}(t)$*

apparaît alors comme un corollaire du théorème suivant:

[1] Plus précisément, si β est l'élément de $H^2(H, \mathbf{Z}/2\mathbf{Z})$ associé à G, et $\rho : \mathrm{Gal}(\overline{\mathbf{Q}(t)}/\mathbf{Q}(t)) \to H$ le morphisme associé à l'extension K, on a $\alpha_K = \rho^*(\beta)$.

Proceedings of the International Congress
of Mathematicians, Zürich, Switzerland 1994

THÉORÈME 3 *Soit k un corps de caractéristique différente de 2, et α un élément de $\mathrm{Br}_2(k(x))$, tel que l'élément $\alpha(\infty) \in \mathrm{Br}_2(k)$ obtenu par spécialisation de x en l'infini est nul, et dont la somme des degrés des pôles est ≤ 4. Il existe alors une fraction rationnelle $x = f(t)$ non constante, telle que $f(\infty) = \infty$, et telle que l'élément $f^*(\alpha)$ de $\mathrm{Br}_2(k(t))$ est nul.*

Ce théorème (ainsi que le théorème analogue lorsque la somme des degrés des pôles est ≤ 5) résulte de constructions polynomiales élémentaires.

Dans la section 3, nous montrons comment ces mêmes constructions permettent de construire une courbe elliptique sur $\mathbf{Q}(t)$ de rang 12 (et donc une infinité de courbes elliptiques sur $\mathbf{Q}$ de rang 12).

Les résultats ci-dessus étant soit parus, soit à paraître prochainement, nous avons choisi ici d'insister sur la description des constructions polynomiales qui ont permis de les démontrer.

2. Le groupe $\tilde{A}_n$ est groupe de Galois d'une extension régulière de $\mathbf{Q}(t)$

Le point crucial de la démonstration (*cf.* [2]) est la construction suivante:

Soit n un entier impair. Il existe un polynôme $H \in \mathbf{Q}[X_1, \dots, X_n]$ tel que tout élément $P = x^n + a_1x^{n-1} + \dots + a_n$ de $\mathbf{Q}[x]$ avec $H(a_1, \dots, a_n) \neq 0$ vérifie les propriétés suivantes:

1) Il existe deux polynômes Q et R de degré $\leq n-1$, premiers avec P, vérifiant

$$P'Q - PQ' = R^2.$$

2) Le polynôme R a ses racines distinctes.
3) Soit t une nouvelle indéterminée. Le discriminant (par rapport à la variable x) du polynôme $F_t = P - tQ$ est égal à $S(t)^2D$, où D est le discriminant de P et où S est un polynôme en t de degré $n-1$ ayant ses racines distinctes.
4) Si a est une racine de S, le polynôme F_a a une racine triple et $n-3$ racines simples.

L'existence de Q et R vérifiant $P'Q - PQ' = R^2$ résulte du fait que cette égalité est équivalente à un système de n équations linéaires à n inconnues, dont la matrice associée est antisymétrique (et a donc un noyau non nul pour n impair) ([2], p. 485). Les points 2), 3), 4) sont techniques, et en fait résultent de ce qu'ils sont vérifiés dans le cas particulier où $P(x) = x^n - x$ ([2], p. 486).

Soit à présent un polynôme $P \in \mathbf{Q}[x]$ de degré n impair dont les coefficients n'annulent pas H, et dont les racines sont simples et appartiennent à $\mathbf{Q}$. Soit K l'extension de $\mathbf{Q}(t)$ engendrée par les racines du polynôme F_t. Il n'est pas difficile de montrer que son groupe de Galois est le groupe alterné A_n, que les valeurs de ramification sont les $n-1$ racines t_i du polynôme S, et que les groupes d'inertie sont des 3-cycles, donc d'ordre impair. Par suite, l'obstruction $\alpha \in \mathrm{Br}_2(k(t))$ est constante. Pour $t = \infty$, l'extension K se spécialise en l'algèbre étale $\mathbf{Q}^n$, dont l'invariant α associé est nul. Donc $\alpha = 0$, et il existe une extension quadratique M de K dont le groupe de Galois est $\tilde{A}_n$.

REMARQUE 1. La méthode ci-dessus permet également de construire, pour n impair, une infinité d'extensions de $\mathbf{Q}$ totalement réelles de groupe de Galois $\tilde{A}_n$: en effet, la forme $\mathrm{Tr}(x^2)$ de l'extension $K/\mathbf{Q}(t)$ est indépendante de t, et est donc équivalente à la forme $\mathrm{Tr}(x^2)$ de l'algèbre $\mathbf{Q}[x]/(P)$. Si P est scindé sur $\mathbf{Q}$, cette forme est définie positive, d'où le résultat.

REMARQUE 2. Par des constructions analogues, on peut prouver que les groupes $6A_6$ et $6A_7$ (dans les notations de l'Atlas de Conway) sont groupes de Galois d'une extension régulière de $\mathbf{Q}(t)$.

3. Le groupe $\mathrm{SL}_2(\mathbf{F}_7)$ est groupe de Galois d'une extension régulière de $\mathbf{Q}(t)$

Dans [1], LaMacchia a construit des familles d'extensions $K_{a,b}$ régulières de $\mathbf{Q}(a,b)$ de groupe de Galois $\mathrm{PSL}_2(\mathbf{F}_7)$, où a et b sont deux indéterminées.

Lorsque $a = 4$ et $b = 1$, Malle a prouvé que l'extension de $\mathbf{Q}$ obtenue par spécialisation admet une extension quadratique galoisienne sur $\mathbf{Q}$ de groupe de Galois $\mathrm{SL}_2(\mathbf{F}_7)$. Considérons donc l'extension $K_{4,b}$ de $\mathbf{Q}(b)$. On peut montrer qu'elle est ramifiée de type $(3,3)$ en l'infini, et ramifiée d'ordre 2 en 4 autres points. L'invariant $\alpha \in Br_2(\mathbf{Q}(b))$ est donc tel que sa spécialisation en $b = 1$ est nulle, et que la somme des degrés de ses pôles est égale à 4. Le théorème 2 est donc conséquence du théorème 3 ci-dessus.

Soit donc α un élément de $\mathrm{Br}_2(k(x))$, où k est un corps de caractéristique différente de 2, dont la spécialisation en $+\infty$ est nulle. À un tel élément sont associés le polynôme unitaire p produit des pôles de α et le "résidu" de α, qui est un élément $r \in A^*/A^{*2}$, où $A = k[x]/(p)$ (*cf.* [5]).

On va voir que le théorème 3 est une conséquence facile du lemme élémentaire suivant:

LEMME. *Soit p un élément de $k[x]$ de degré $2n$, où n est un entier ≥ 1, et a le coefficient de son terme de degré $2n$. Il existe un unique polynôme $g \in k[x]$ de degré n et de terme dominant a tel que $ap - g^2$ soit de degré $\leq n-1$.*

Il suffit de prendre pour g la partie polynomiale du développement asymptotique de $(ap)^{1/2}$.

La démonstration du théorème 3 est alors immédiate: si t est une nouvelle indéterminée, et si R est un représentant de degré 3 de r (il en existe toujours), d'après le lemme, le polynôme $p + t^2R = g^2 + A(t)x - B(t)$, où A et B sont des éléments de $k(t)$. On peut montrer que B est de degré 8 et A est non nul de degré ≤ 6; le changement de variable $x = B(t)/A(t)$ répond alors à la question: il est tel que $f^*(\alpha) = 0$ et $f(\infty) = \infty$).

Le lemme précédent permet de prouver facilement des assertions voisines, par exemple:

THÉORÈME 4 *Soit k un corps de caractéristique distincte de 2, $p \in k[x]$ un polynôme unitaire de degré 8, $c \in k$, et α l'élément de $\mathrm{Br}_2(k(x))$ égal à (p,c). Il existe un changement de variable non constant $x = f(t)$, où $f \in k(t)$ vérifie $f(\infty) = \infty$, tel que $f^*(\alpha) = 0$.*

La démonstration complète de ce résultat fera l'objet d'un article ultérieur. Nous donnons ici la preuve du théorème ci-dessus lorsque le coefficient de degré 7 de p est nul (ce à quoi on peut toujours se ramener) et lorsque p n'est pas pair (i.e. $p(-x) \neq p(x)$).

Dans ce cas, d'après le lemme, il existe g de degré 4 et r de degré ≤ 3 tels que $p = g^2 + r$; si t est une nouvelle indéterminée, et si $G = g + ct^2/2$, on a donc $p = G^2 - cu^2G + r + c^2u^4/4$. Si l'on applique à présent le lemme au polynôme $cu^2G - r - c^2u^4/4$, on voit que $cu^2G - r - c^2u^4/4 = cH^2 + A(t)x - B(t)$; puisque p n'est pas pair, le polynôme A n'est pas nul. La fraction rationnelle $f = B/A$ répond alors à la question (Et on a même une solution explicite de l'équation $X^2p(f(t)) + Y^2c = Z^2$, avec $(X, Y, Z) = (1, H(f(t)), G(f(t)))$.)

Par une méthode légèrement différente (qui pourrait se ramener à l'application du lemme ci-dessus, mais ce serait artificiel), on peut également montrer le théorème suivant:

THÉORÈME 5 *Soit k un corps de caractéristique différente de 2, et α un élément de $\mathrm{Br}_2(k(x))$ dont la somme des degrés des pôles est ≤ 5, et dont la spécialisation en ∞ est nulle. Il existe une fraction rationnelle $x = f(t)$, de degré ≤ 16, telle que $f(\infty) = \infty$ et $f^*(\alpha) = 0$.*

Nous donnons ici les points principaux permettant de démontrer ce théorème, la démonstration détaillée faisant l'objet d'une publication ultérieure.

Soit p le polynôme unitaire produit des pôles de α, et r un polynôme de degré ≤ 4 représentant le résidu de α. Posons $U = r + ax + b$, $V = v0 + v1x + tx^2$, et $R = R_4x^4 + \ldots + R_0$ le reste de $U^2 - rV^2 \bmod p$. Il est clair que le système $R_4 = R_3 = 0$ est linéaire en a et b; lorsqu'on le résout, on voit que a est de degré 1 en v_0, et que le coefficient de degré 1 de a par rapport à v_0 est linéaire en v_1. On l'annule, et a et v_1 sont donc à présent des éléments de $k(t)$. On remarque alors que R_2 est un polynôme de degré 1 en v_0 (à coefficients dans $k(t)$). On résout donc $R_2 = 0$ par rapport à v_0; par suite, a, b, v_0, v_1 sont à présent des éléments de $k(t)$, et $R = R_1(t)x + R_0(t)$, où R_0 (resp. R_1) est de degré 16 (resp. 14) en t. Par suite, $U^2 - rV^2 \equiv R_1(t)x + R_0(t) \bmod p$, et le lemme 1.2 de [5] permet de conclure: la fraction rationnelle non constante $x = f(t) = -R_0/R_1$ est telle que $f(\infty) = \infty$ et l'on a $f^*(\alpha) = 0$.

4. Courbes elliptiques de rang élevé

Le lemme de la section précédente permet également de construire des courbes elliptiques de rang élevé: soit p un élément de $\mathbf{Q}(t)$ unitaire de degré 12, dont les racines sont distinctes et appartiennent à $\mathbf{Q}$. D'après le lemme, il existe g de degré 6 et r de degré ≤ 5 tels que $p = g^2 - r$. Supposons que r soit de degré 4; la courbe de genre 1 d'équation $y^2 = r(x)$ possède alors les 12 points rationnels $P_i = (x_i, g(x_i))$, où les x_i sont les racines de p.

Une méthode efficace pour trouver de tels polynômes p (tels que r soit de degré 4) est la suivante: soit $q = (x - a_1) \ldots (x - a_6)$ un polynôme de degré 6 scindé sur $\mathbf{Q}$, soit t une indéterminée, et soit $p(x) = q(x - t)q(x + t)$. Le polynôme

r est alors divisible par t^2, et si $R = r/t^2$, le coefficient de degré 5 de R est un polynôme $s(a_1, \ldots, a_6)$ en les a_i indépendant de t.

Par suite, si $s(a_1, \ldots, a_6) = 0$, on obtient une courbe de genre 1 sur $\mathbf{Q}(t)$, d'équation $y^2 = R(x)$, et possédant 12 points rationnels. On en choisit un comme origine, et la courbe elliptique ainsi obtenue est en général de rang 11 sur $\mathbf{Q}(t)$.

Pour obtenir des courbes de rang 12, on remarque que le coefficient de degré 4 de R est de la forme $Pt^2 + Q$, où P et Q sont des polynômes en les a_i. Si, pour certains choix de $(a_1, \ldots, a_6)$ annulant s, la conique (en (t, z)) $Pt^2 + Q = z^2$ a des points, on la paramétrise par $t = f(w)$: les points à l'infini de la courbe $y^2 = R(x)$, définie sur $\mathbf{Q}(w)$, sont alors rationnels sur $\mathbf{Q}(w)$, et l'on obtient ainsi en général une courbe de rang 12 sur $\mathbf{Q}(w)$ ([3]). Comme l'invariant modulaire de cette courbe est non constant, on obtient par spécialisation du paramètre w une infinité de courbes elliptiques sur $\mathbf{Q}$ de rang 12, deux à deux non isomorphes.

On peut donc espérer que par spécialisation on obtienne des courbes elliptiques sur $\mathbf{Q}$ de rang > 12. C'est ce qu'ont fait notamment Fermigier (qui a obtenu des courbes de rang ≥ 19) et Nagao (qui a obtenu une courbe de rang ≥ 21).

De façon plus surprenante, Nagao a remarqué ([6]) qu'en considérant la courbe obtenue en prenant $(a_1, a_2, a_3, a_4, a_5, a_6) = (148, 116, 104, 57, 25, 0)$, on obtient un treizième point rationnel sur $\mathbf{Q}(t)$, d'abscisse $(t + 103)/15$, et indépendant des précédents. De plus, la conique $Pt^2 + Q = z^2$ a un point (car ici $P = 1$), donc il obtient une courbe elliptique sur $\mathbf{Q}(w)$ (d'invariant non constant) de rang ≥ 13 (et donc une infinité de courbes elliptiques deux à deux non isomorphes, définies sur $\mathbf{Q}$, et de rang ≥ 13).

En fait, on peut trouver une famille à deux paramètres (u, v) de telles courbes: Soit $(a_1, a_2, a_3, a_4, a_5, a_6) =$

$$(u^3v^2 - 2\,u^2v^3 + uv^4 - u^4 - 2\,u^2v^2 - 2\,uv^3 + v^4 + u^2v + uv^2 - v^3 + u^2 + 2\,uv + v^2 - v,$$
$$u^4v - 2\,u^3v^2 + u^2v^3 + u^4 - 2\,u^3v - 2\,u^2v^2 - v^4 - u^3 + u^2v + uv^2 + u^2 + 2\,uv + v^2 - u,$$
$$u^4 + u^4v - u^3v^2 - 2\,u^3v - 2\,u^3 + u^2v^3 + u^2v^2 + u^2 - 2\,u^2v + 2\,uv^3 - uv^4 + uv^2 + v^3 - v,$$
$$u^3 - u^4v - 2\,u^2v + u^2v^2 + u^2v^3 - 2\,u^2 - 2\,uv + uv^2 + 2\,uv^3 + u - v^2 + v^3 - v^4 + v,$$
$$u^3v^2 - u^4v + 2\,u^3v + u^3 + u^2v^2 - u^2v^3 + u^2v - 2\,uv^3 - 2\,uv^2 - u + uv^4 - 2\,v^3 + v^2 + v^4,$$
$$-a_1 - a_2 - a_3 - a_4 - a_5).$$

On a alors $s(a_1, \ldots, a_6) = 0$, et la courbe elliptique $y^2 = R(x)$, définie sur $\mathbf{Q}(u, v)(t)$, contient un treizième point rationnel sur $\mathbf{Q}(u, v)(t)$, indépendant des précédents, à savoir le point d'abscisse $(A + Bt)/(u^2 + v^2 + 1)$, où

$$A = 3\,u^3v^2 + 2\,u^4v + uv - 4\,v^3u - 3\,v^2u^2 + 3\,v^3u^2 - 4\,u^3v + u + v - u^6 + v^3 - 3\,v^4$$
$$+2\,v^5 - v^6 + 3\,u^2v + 3\,v^2u + 2\,v^4u - 3\,u^4 + u^3 + 2\,u^5 + u^5v^2 + u^5v - u^4v^3 - 2\,u^4v^2$$
$$-u^3v^4 - 4\,u^3v^3 + u^2v^5 - 2\,u^2v^4 + uv^5$$

et

$$B = -u^2 - v^2 + 2\,u + 2\,v + 1.$$

De plus, le terme de degré 4 de R est de la forme $R_4 = C(u,v)^2t^2 + D(u,v)$, où $C(u,v)$ et $D(u,v)$ sont des éléments de $\mathbf{Q}[u,v]$. donc en paramétrant la conique (en (t,z)) $R_4 = z^2$ par $t = f(w)$, on obtient une famille à deux paramètres (u,v) de courbes elliptiques sur $\mathbf{Q}(w)$ de rang ≥ 13. On retrouve la courbe de Nagao en prenant $(u,v) = (2,5)$.

REMARQUE. Le même type de constructions permet de construire une infinité de courbes elliptiques sur $\mathbf{Q}$, non deux à deux $\mathbf{Q}$-isomorphes, d'invariant modulaire égal à 0 (resp. 1728), et dont le rang est ≥ 6 (resp. ≥ 4) ([4]).

References

[1] M. E. LaMacchia, *Polynomials with Galois group* $PSL(2,7)$, Commun. Alg. **8** (1980), 983–992.

[2] J.-F. Mestre, *Extensions régulières de* $\mathbf{Q}(t)$ *de groupe de Galois* $\tilde{A}_n$, J. of Alg. **131** (1990), 483–495.

[3] ———, *Courbes elliptiques de rang 12 sur* $\mathbf{Q}(t)$, C. R. Acad. Sci. Paris **313** (1991), 171–174.

[4] ———, *Rang de courbes elliptiques d'invariant donné*, C. R. Acad. Sci. Paris **314** (1992), 919–922.

[5] ———, *Annulation, par changement de variable, d'éléments de* $\mathrm{Br}_2(k(x))$ *ayant quatre pôles*, C. R. Acad. Sci. Paris **319** (1994), 529–532.

[6] K. Nagao, *An example of Elliptic Curve over* $\mathbf{Q}(T)$ *with Rank* ≥ 13, Proc. of the Japan Acad. **70** (1994), 152–153.

Study of Quadratic Forms — Some Connections with Geometry

RAMAN PARIMALA

School of Mathematics
Tata Institute of Fundamental Research
Homi Bhabha Road, Bombay 4000 05, India

Dedicated to my teacher Professor R. Sridharan on his 60th birthday.

ABSTRACT. Let X be an algebraic variety over a field k of characteristic not 2. A *quadratic space on* X is a locally free sheaf $\mathcal{E}$ on X together with a self-dual isomorphism $q : \mathcal{E} \to \mathcal{E}^\vee$. In this article we outline some recent developments concerning the stable and nonstable study of quadratic spaces over algebraic varieties. Although this study borrows tools from algebra and geometry, it yields in return new insights into certain seemingly unrelated questions in algebra and geometry.

1 Quadratic spaces over the affine plane

The nonstable study of quadratic spaces acquired an impetus with the solution of Serre's conjecture on the triviality of algebraic vector bundles on the affine space by Quillen and Suslin. In [OS], Ojanguren and Sridharan constructed nonfree, rank one projective modules over $D[X, Y]$, where D is any noncommutative division ring. There was a classification [PS1] of nonfree projective left ideals of $\mathbb{H}[X, Y]$, where $\mathbb{H}$ denotes the algebra of real quaternions, in terms of certain 2×2 hermitian matrices, modulo "hermitian" equivalence. This led to the construction [P1] of an explicit family of indecomposable rank 4 quadratic spaces over $\mathbb{R}[X, Y]$, thereby giving a negative answer to an analogue of Serre's conjecture for orthogonal bundles over $\mathbb{A}^2_k$.

Given a quadratic space $(\mathcal{E}, q)$ over $\mathbb{A}^n_k$, there is a quadratic space q_0 over k such that the form on the fibre of $(\mathcal{E}, q)$ at any point of $\mathbb{A}^n_k$ is isometric to q_0. We call q_0 the *form on the fibre* of $(\mathcal{E}, q)$. We say that a space over $\mathbb{A}^n_k$ is *isotropic* if the form on the fibre is isotropic. This is equivalent to the form being isotropic generically. The indecomposable spaces over $\mathbb{R}[X, Y]$ mentioned above are anisotropic; indeed they have $\langle 1, 1, 1, 1 \rangle$ as the form on the fibre. It was shown by Ojanguren [O] and independently by Kopeiko and Suslin [KS] that any isotropic quadratic space over $\mathbb{A}^n_k$ is extended from k. Thus, the obstruction to the quadratic spaces on $\mathbb{A}^n_k$ being extended from k lies in the existence of anisotropic quadratic spaces over k. That this is the precise obstruction to the extendibility question follows from the theorem [P2] below.

Proceedings of the International Congress
of Mathematicians, Zürich, Switzerland 1994

THEOREM. *Let k be a field of characteristic not 2 and q_0 an anisotropic quadratic space of rank at least 3 over k. Then there exists an infinite family of indecomposable quadratic spaces over $\mathbb{A}_k^2$ with q_0 as the fibre at any rational point.*

Raghunathan [R] proved the following theorems extending these results for principal G-bundles on $\mathbb{A}_k^2$, where G is a connected reductive algebraic group over k.

THEOREM A. *Assume that if $[G,G]$ is nontrivial, then every one of its k-simple components is isotropic over k. Let P be a G-bundle on $\mathbb{A}_k^n$ such that P becomes trivial under base change to $\mathbb{A}_{k_s}^n$. Assume further that over some (hence any) k-point of $\mathbb{A}^n, P$ is trivial. Then P is trivial.*

THEOREM B. *Assume that G is connected anisotropic absolutely almost simple and is not of type F_4 or G_2. Then there are infinitely many mutually nonisomorphic G-bundles P on $\mathbb{A}_k^2$ that become trivial over $\mathbb{A}_{k_s}^2$, such that P do not admit a reduction of structure group to any proper connected reductive subgroup of G.*

A rational construction of Cayley algebras over any affine scheme [KPS4] leads to examples of nontrivial G_2-bundles on $\mathbb{A}_k^2$, provided k admits a Cayley division algebra.

It was shown in [KPS1] that anisotropic quadratic spaces over $\mathbb{A}_k^2$ admit a *unique* extension to the projective plane $\mathbb{P}_k^2$. This reduces the problem of classification of quadratic spaces on $\mathbb{A}_k^2$ to a corresponding problem over $\mathbb{P}_k^2$, where there is an abundance of vector bundle techniques available. An adaptation of methods of Barth and Hulek [BaH] yields a classification [OPS2] of anisotropic "s-stable" quadratic spaces over $\mathbb{P}_k^2$ in terms of certain orthogonal equivalence classes of triples of skew symmetric matrices. Thus, the classification problem of quadratic spaces on $\mathbb{A}_k^2$ is reduced to a problem in linear algebra. This was used in the construction of certain large rank indecomposable quadratic spaces over $\mathbb{A}_{\mathbb{R}}^2$ [OPS1] and in the classification of rank 4 quadratic spaces over $\mathbb{A}_{\mathbb{R}}^2$ with prescribed "Chern classes" [OPS2].

2 Pfaffians and discriminants

There has been a systematic study due to Knus-Ojanguren-Parimala-Sridharan of low rank quadratic spaces over arbitrary commutative rings via Clifford algebras. Let R be a commutative ring in which 2 is invertible. Let A be an Azumaya algebra over R of degree n; i.e., for some faithfully flat extension S of R, $S \otimes_R A \simeq M_n(S)$. Let P be a projective A-module of rank one. In [KOS], there is a functorial assignment to P of a projective R-module $Nrd\,P$ of rank one and a polynomial map $Nrd : P \to Nrd\,P$ of degree n. If $P = A$, then $Nrd\,A = R$ and $Nrd : A \to R$ is the usual reduced norm. If degree $A = 2$, $Nrd : P \to Nrd\,P$ is a quadratic map. If $Nrd\,P$ is free, a choice of a generator for $Nrd\,P$ yields a quadratic space (P, Nrd) of rank 4 and trivial discriminant. In [KOS], it is proved that every rank 4 quadratic space of trivial discriminant arises this way.

Let A be an Azumaya algebra of degree $2n$ over R which is 2-torsion in the Brauer group of R. Let $\varphi : A \otimes_R A \to End_R P$ be an isomorphism of R-algebras, P being a finitely generated projective R-module. The "switch" map $x \otimes y \to y \otimes x$ on $A \otimes_R A$ is given by inner conjugation by a "canonical" unit $u \in (A \otimes A)^*$ with $u^2 = 1$. Let $\varphi(u) = \psi$. We call ψ the "module involution" on P. Let $Alt\, P = \{x - \psi(x), x \in P\}$ be the R-module of ψ-*alternating* elements in P. To the triple (A, P, φ) is associated in [KPS2] functorially a projective R-module $Pf\, P$ of rank one and a polynomial map $Pf : Alt\, P \to Pf\, P$ of degree n. Further, there is a natural pairing

$$\delta : Pf\, P \otimes Pf\, P \to Nrd\, P$$

such that for $x \in Alt\, P$, $\delta\big(Pf(x) \otimes Pf(x)\big) = Nrd\, x$ [PS3]. If $A = M_{2n}(R)$, $\varphi : A \otimes_R A \to End_R A$ is the map $x \otimes y \to (z \to xzy^t)$, $Alt\, A = Alt_{2n}(R)$, the subset of $M_{2n}(R)$ of alternating matrices, $Pf\, A \simeq R$ and $Pf : Alt_{2n}(R) \to R$ is the classical pfaffian; the pairing $\delta : R \otimes R \to R$ is the multiplication and the formula above simply says that the square of the pfaffian of an alternating matrix is its determinant. Let degree $A = 4$ and $Pf\, P$ be free. Then, a choice of a generator for $Pf\, P$ yields a quadratic space $Pf : Alt\, P \to R$ of rank 6 and of trivial discriminant. In [KPS2], it was shown conversely that every rank 6 quadratic space of trivial discriminant arises this way.

Let A be an Azumaya algebra of degree $2n$ with an *orthogonal* involution τ; i.e., τ splits as $X \to X^t$ in a faithfully flat splitting of A. We have an isomorphism $\varphi_\tau : A \otimes_R A \to End_R A$, defined by $x \otimes y \mapsto \big(z \to xz\tau(y)\big)$. The module involution ψ on A coincides with τ and $Alt\, A = \{x - \tau x, x \in A\}$. Since $Nrd\, A = R$, $\delta : Pf(A) \otimes Pf(A) \to R$ gives a discriminant module over R, which was called the *pfaffian discriminant* of (A, τ) in [KPS2]. If A is a central simple algebra over a field k with an involution τ, the discriminant defined above for (A, τ) coincides with the one defined by Jacobson [J] and Tits [T]. The discriminant in this case coincides with the square class of any unit u such that $\tau u = -u$. In [KPS3] the following theorem was proved.

THEOREM. *A rank 16 Azumaya algebra with an orthogonal involution splits into a tensor product of involutions on quaternion subalgebras if and only if the discriminant of the involution is trivial.*

This criterion for decomposability of involutions on rank 16 algebras is interesting already for fields where examples of indecomposable involutions have been constructed by Amitsur-Rowen-Tignol [AmRT].

Subsequently the determination of possible discriminants on a given division algebra D with deg $D \geq 4$ over a field k has become a problem of wider interest. It was not a priori clear whether one could construct, on a given algebra D, an orthogonal involution that had a trivial discriminant or, for that matter, one that had a nontrivial discriminant.

The following theorem [PSSu] gives a complete answer to this question.

THEOREM. *Let D be a central division algebra over k of degree at least 4, whose class in the Brauer group of k is 2-torsion. Then every element of D is symmetric with respect to an orthogonal involution of discriminant one. In particular, the group of reduced norms of D^* modulo k^{*2} coincides with the set of discriminants of orthogonal involutions on D.*

Yanchevskii [Y] uses this theorem to construct indecomposable involutions on division algebras D for which $K_1 \operatorname{Spin} D$ is nontrivial.

In terms of the invariants of hermitian forms over division algebras with involutions like the discriminant and the Clifford invariant for involutions of orthogonal type, Bayer-Flückiger and Parimala [BP] have recently obtained a classification result for hermitian forms over division algebras with involutions over fields of characteristic not 2 and of cohomological dimension ≤ 2. This has led [BP] to an affirmative solution of a conjecture of Serre, for groups of classical type.

THEOREM. *Let k be a perfect field of characteristic not 2 and of cohomological dimension at most 2. Let G be a semisimple, simply connected linear algebraic group defined over k. If G is of classical type, then $H^1(k, G) = \{1\}$.*

The validities of the conjecture for $G = SL_{1,D}$, for D a finite dimensional central division algebra over k and for $G = \operatorname{Spin} q$, for a quadratic form q over k, are due to Merkurjev-Suslin [S] and Merkurjev respectively.

3 Invariants for quadratic spaces — Witt groups

For quadratic spaces over algebraic varieties, we have the "classical" invariants, namely the *rank* (modulo 2), which has values in $H^0_{\text{et}}(X, \mu_2)$, the *discriminant* with values in $H^1_{\text{et}}(X, \mu_2)$, and the *Clifford invariant* (cf. [PSr], [KPS4]) with values in $H^2_{\text{et}}(X, \mu_2)$. Whereas the first two invariants are surjective onto the cohomology, it is interesting to analyse the image of the Clifford invariant. The following two theorems relate this question for a curve to some purely geometric questions concerning the curve. We begin with a definition.

Let $x_0 \in X(k)$. The curve X is said to have *extension property* (for quadratic spaces with respect to x_0) if every quadratic space over $X \backslash x_0$ extends to a quadratic space over X. We have the following theorems:

THEOREM. [PS2] *Let X be a smooth projective curve over a local field of characteristic not 2. Suppose $X(k) \neq \emptyset$. Then the Clifford invariant map is surjective if and only if X has the extension property.*

THEOREM. [PSc] *Let X be a smooth, projective hyperelliptic curve of genus at least 2 over a local field k of characteristic not 2. Suppose $X(k) \neq \emptyset$. Then the following are equivalent:*

(1) *X has the extension property.*

(2) *The canonical line bundle Ω_X is a square in* $\operatorname{Pic} X$.

(3) *The genus of X is odd or the genus of X is even and X satisfies one of the following conditions for the double covering $\pi : X \to \mathbb{P}^1$.*
  (a) *π has a ramification point of odd degree;*
  (b) *All ramification points of π have even degree and there is a quadratic extension of k that is contained in the residue fields of all ramification points of π.*

The criterion (3) leads to examples of hyperelliptic curves over local fields where the Clifford invariant map is not surjective.

For any affine variety X, there are also invariants e_i for quadratic spaces with values in certain quotients of the K-groups, $K_i(X)$ for $1 \leq i \leq 3$, e_i being defined for spaces for which the previous invariants vanish. The invariant e_2 has been defined by Giffen [G]. This is a refinement of the Clifford invariant in the following sense: The Grothendieck equivariant Chern class map $c_{22} : K_2(X) \to H^2_{\text{et}}(X, \mu_2)$ maps the class of the e_2 invariant to the class of the Clifford invariant [OPS3]. The invariant e_3 has recently been constructed by Barge-Ojanguren [BaO] and is related to the Arason invariant for fields [A].

The stable theory of quadratic spaces over algebraic varieties was initiated by Knebusch, who defined the *Witt group* $W(X)$ of a variety X following the classical notion of Witt groups of quadratic forms over fields. This group is the quotient of the Grothendieck group of isometry classes of quadratic spaces on X with respect to orthogonal sum, modulo the subgroup generated by *metabolic* spaces. Metabolic spaces are those spaces $(\mathcal{E}, q)$ that admit a sub-bundle $\mathcal{V}$ with $\mathcal{V} = \mathcal{V}^{\perp}$. The following are a few computations for the Witt groups of varieties over arbitrary fields: Witt group of the projective space due to Arason [A], of conics [P3], and hyperelliptic curves due to Parimala-Sujatha [PSj], Shick [Sh], and Arason-Elman-Jacob [AEJ2].

A general method of studying the Witt group of a smooth variety is through the graded group associated to the filtration induced by the filtration of the Witt group of the function field by powers of the fundamental ideal of even rank forms. Whether the graded Witt group is isomorphic to the graded Galois cohomology group is a wide-open question even for fields, posed by Milnor [Mi]. In [P4], the graded Witt group of X is related to the graded *unramified cohomology group* of X provided the Milnor conjecture is valid for the function field of X and the residue fields of X at codimension one points. This was achieved by using a flasque resolution due to Bloch-Ogus [BOg] of the Zariski sheaf $\mathcal{H}^n$ associated to the presheaf $U \to H^n_{\text{et}}(U, \mu_2)$. This association to the graded Witt group of the graded unramified cohomology group has since been used widely in the study of the "unramified" Witt group of a smooth variety ([Sj], [CSj], [SaSj]) and is also crucial in the analysis of the finiteness questions of Witt groups of real varieties, which we discuss in the next paragraph.

4 Some connections with geometry

The study of Witt groups is especially interesting for real varieties where there is a subtle interplay between the geometry of X and the real topology of $X(\mathbb{R})$. In this connection, Knebusch [Kn2] raised the question whether the Witt group of a smooth variety over $\mathbb{R}$ is finitely generated. Finite generation of $W(X)$ was proved for dim $X = 1$ by Knebusch himself [Kn1] and for dim $X = 2$ by Ayoub [Ay]. For smooth affine 3-folds, we have the following theorem [P4], which relates the question of finite generation of $W(X)$ to one purely concerned with the geometry of X.

THEOREM. *Let X be a smooth affine 3-fold over $\mathbb{R}$. Then $W(X)$ is finitely generated if and only if $CH^2(X)/2$ is finite, $CH^2(X)$ denoting the group of codimension 2 cycles on X, modulo rational equivalence.*

In fact, if $CH^2(X_{\mathbb{C}})/2$ is finite, then $CH^2(X)/2$ is finite. However, finiteness of $CH^2(Y)/2$ for smooth 3-folds over $\mathbb{C}$ is in general an open question which is known to be true only in some special cases; e.g. where Y is unirational over $\mathbb{C}$, or where Y is a conic bundle over a smooth projective variety ([P5], Appendix).

The proof of the above theorem uses the following results:

(1) Separation of connected components of $X(\mathbb{R})$ by signatures, due to Mahé [Ma].
(2) Relationship between the graded Witt ring and the graded cohomology ring for function fields of real 3-folds, due to Arason, Elman, and Jacob [AEJ1].
(3) Finiteness of certain *unramified cohomology groups* due to Colliot-Thélène and Parimala [CP], which we discuss in the next paragraph.

Let $\Gamma(X, \mathcal{H}^n)$ denote the group of sections of the Zariski sheaf $\mathcal{H}^n$ associated to the presheaf $U \mapsto H^n_{\text{et}}(U, \mu_2)$. By results of Bloch and Ogus [BOg], $\Gamma(X, \mathcal{H}^n)$ is a birational invariant for smooth projective varieties. To prove the theorem above, one required the finiteness of $\Gamma(X, \mathcal{H}^{d+1})$ for smooth real varieties X of dimension d. If dim $X = 1$, $\Gamma(X, \mathcal{H}^2) \simeq \text{Br}(X)$, the "unramified" Brauer group of X. For smooth quasiprojective curves over $\mathbb{R}$, Witt [W] proved that $Br(X) \simeq (\mathbb{Z}/2)^s$, s denoting the number of connected components of $X(\mathbb{R})$. The following theorem [CP] provides a generalization of Witt's theorem to higher dimensional varieties.

THEOREM. *Let X be a smooth variety over $\mathbb{R}$ of dimension d. For any integer $n \geq d+1$, we have an isomorphism*

$$H^0(X, \mathcal{H}^{d+1}) \xrightarrow{\sim} (\mathbb{Z}/2)^s ,$$

where s denotes the number of connected components of $X(\mathbb{R})$ for the real topology.

This result has in turn been extended to varieties with possible singularities by Colliot-Thelène and Scheiderer ([CSc]). In fact, they show that for any variety X of dimension d over $\mathbb{R}$, and for $n \geq d+1$, $i \geq 0$,

$$H^i_{\text{Zar}}(X, \mathcal{H}^n) \simeq H^i(X(\mathbb{R}), \mathbb{Z}/2) .$$

The obstruction to the finite generation of $W(X)$ mentioned above for real 3-folds vanishes if the cycle map

$$c\ell : CH^2(X)/2 \longrightarrow H^4_{\text{et}}(X, \mu_2)$$

is injective, in view of the fact that the étale cohomology groups $H^n_{et}(X,\mu_2)$ are finite for smooth varieties over $\mathbb{R}$. It was shown in [CP] that if X is a smooth projective surface over $\mathbb{R}$, this map is indeed injective. The question of the injectivity of the cycle map for arbitrary smooth projective surfaces with rational points has been of general interest and has been open for a while.

There are examples of surfaces without rational points over $\mathbb{Q}_p(t)$ for which the cycle map is not injective (cf. [C], Remark 7.6.1). Recently, Parimala and Suresh [PSu] have constructed examples of smooth conic fibrations of hyperelliptic curves with rational points over $\mathbb{Q}_3$ for which the cycle map is not injective.

For a smooth projective variety X over any field k, there is a pairing

$$CH_0(X) \times H^2_{\mathrm{et}}(X,\mathbb{G}_m) \to H^2_{et}(k,\mathbb{G}_m) = \mathrm{Br}(k)\,,$$

defined by

$$\langle p,\xi\rangle = \mathrm{cores}_{k(p)/k}(\xi_p)\,,$$

for a closed point p of X and $\xi \in H^2_{\mathrm{et}}(X,\mathbb{G}_m)$, where ξ_p denotes the restriction of ξ to the fibre at p and $\mathrm{cores}_{k(p)/k} : \mathrm{Br}(k(p)) \to \mathrm{Br}(k)$ is the corestriction homomorphism. Let

$$\alpha : CH_0(X) \to \mathrm{Hom}((H^2_{\mathrm{et}}(X,\mathbb{G}_m),\mathrm{Br}(k))$$

be the map induced by the above pairing.

Let k be a p-adic field. If $\dim X = 1$, Lichtenbaum [L] shows that α is injective. Let X be a smooth projective surface over k. One can show that the kernel of α modulo 2 is contained in the kernel of the cycle map. The examples in [PSu] show that for some conic fibrations of curves over p-adic fields, $\ker\alpha$ is in general not zero. Thus, the cohomological Brauer group fails to detect zero cycles modulo rational equivalence, for surfaces over p-adic fields.

We observe that Saito in [Sa] has shown that for a smooth projective surface over a p-adic field with $H^2(X,\mathcal{O}_X) = 0$ and for which the Albanese of X has good reduction, the map α is an injection. The examples in [PSu] show that the condition of good reduction for Alb_X in Saito's result is indeed essential.

References

[AmRT] S.A. Amitsur, L.H. Rowen, and J.P. Tignol, *Division algebras of degrees* 4 *and* 8 *with involution*, Israel J. Math. **33** (1979), 133–148.

[A] J.Kr. Arason, *Der Wittring projektiver Räume*, Math. Ann. **253** (1980), 205–212.

[AEJ1] J.Kr. Arason, R. Elman, and B. Jacob, *The graded Witt ring and Galois cohomology*, Can. Math. Soc. Conf. Proc. **4** (1984), 17–50.

[AEJ2] J.Kr. Arason, R. Elman, and B. Jacob, *On generators for the Witt ring*, Contemp. Math. **155** (1994), 247–269.

[Ay] G. Ayoub, *Le groupe de Witt d'une surface réelle*, Comm. Math. Helv. **62** (1987), 74–105.

[BaO] J. Barge, and M. Ojanguren, *Sur le troisième invariant d'une forme quadratique*, preprint 1994.

[BaH] W. Barth, and K. Hulek, *Monads and moduli of vector bundles*, Manuscripta Math. **25** (1978), 323–347.

[BP] E. Bayer-Flückiger, and R. Parimala, *Galois cohomology of the classical groups over fields of cohomological dimension* ≤ 2, to appear in Invent. Math.

[BOg] S. Bloch, A. Ogus, *Gersten's conjecture and the homology of schemes*, Ann. Sci. École Norm. Sup., 4e série **7** (1974), 181–202.

[C] J.-L. Colliot-Thélène, *Cycles algébriques de torsion et K-théorie algébrique*, Arithmetic Algebraic Geometry, Trento 1991, 1–49, SLN 1553.

[CP] J.-L. Colliot-Thélène, and R. Parimala, *Real components of algebraic varieties and étale cohomology*, Invent. Math. **101** (1990), 81–99.

[CSc] J.-L. Colliot-Thélène, and C. Scheiderer, *Zero cycles and cohomology of real algebraic varieties*, preprint 1994.

[CSj] J.-L. Colliot-Thélène, and R. Sujatha, *Unramified Witt groups of real anisotropic quadrics*, Proc. Symp. Pure Math. **58**, Part II (1995), 127–147.

[G] C.H. Giffen, *Hasse-Witt invariants for* (α, u) *reflexive forms and automorphisms I: Algebraic* K_2*-valued Hasse-Witt invariants*, J. Algebra **44** (1977), 434–456.

[J] N. Jacobson, *Clifford algebras for algebras with involution of type D*, J. Algebra **1** (1964), 288–300.

[Kn1] M. Knebusch, *On algebraic curves over real closed fields II*, Math. Z. **151** (1976), 189–205.

[Kn2] M. Knebusch, *Some open problems*, in: Conference on Quadratic Forms, Queen's Papers in Pure and Appl. Math. **46** (1977), Kingston, Ontario, 361–370.

[KOS] M.-A. Knus, M. Ojanguren, and R. Sridharan, *Quadratic forms and Azumaya algebras*, J. Reine Angew. Math. **303/304** (1978), 231–248.

[KPS1] M.-A. Knus, R. Parimala, and R. Sridharan, *Non-free projective modules over* $\mathbb{H}[X, Y]$ *and stable bundles over* $\mathbb{P}_2(\mathbb{C})$, Invent. Math. **65** (1981), 13–27.

[KPS2] M.-A. Knus, R. Parimala, and R. Sridharan, *A classification of rank* 6 *quadratic spaces via Pfaffians*, J. Reine Angew. Math. **398** (1989), 187–218.

[KPS3] M.-A. Knus, R. Parimala, and R. Sridharan, *Pfaffians, central simple algebras and similitudes*, Math. Z. **206** (1991), 589–604.

[KPS4] M.-A. Knus, R. Parimala, and R. Sridharan, *Compositions and triality*, J. Reine. Angew. Math. **457** (1994), 45–70.

[KS] V.I. Kopeiko, and A.A. Suslin, *Quadratic modules over polynomial rings*, J. Sov. Math. **17** (1981), 2024–2031.

[L] S. Lichtenbaum, *Duality theorems for curves over p-adic fields*, Invent. Math. **7** (1969), 120–136.

[Ma] L. Mahé, *Signatures et composantes connexes*, Math. Ann. **260** (1982), 191–210.

[Mi] J. Milnor, *Algebraic K-theory and quadratic forms*, Invent. Math. **9** (1970), 318–344.

[O] M. Ojanguren, *Formes quadratiques sur les algèbres de polynômes*, C.R. Acad. Sci. Paris, Sér. A **287** (1978), 695–698.

[OPS1] M. Ojanguren, R. Parimala, and R. Sridharan, *Indecomposable quadratic bundles of rank* 4n *over the real affine plane*, Invent. Math. **71** (1983), 648–653.

[OPS2] M. Ojanguren, R. Parimala, and R. Sridharan, *Anisotropic quadratic spaces over the plane*, in: Vector bundles on algebraic varieties, Bombay 1984, OUP 1987, 465–489.

[OPS3] M. Ojanguren, R. Parimala, and R. Sridharan, *Ketu and the second invariant of a quadratic space*, K-theory **7** (1993), 501–515.

[OS] M. Ojanguren, and R. Sridharan, *Cancellation of Azumaya algebras*, J. Algebra **18** (1971), 501–505.

[P1] R. Parimala, *Failure of a quadratic analogue of Serre's conjecture*, Amer. J. Math. **100** (1978), 913–924.

[P2] R. Parimala, *Indecomposable quadratic spaces over the affine plane*, Adv. in Math. **62** (1986), 1–6.

[P3] R. Parimala, *Witt groups of conics, elliptic and hyperelliptic curves*, J. Number Theory **28** (1988), 69–93.

[P4] R. Parimala, *Witt groups of affine 3-folds*, Duke Math. J. **57** (1989), 947–954.

[P5] R. Parimala, *Witt groups vis-à-vis Chow groups*, in: Geometry, Bombay 1990, NBHM (1993), 149–154.

[PSc] R. Parimala, and W. Scharlau, *On the canonical class of a curve and extension property for quadratic forms*, Contemp. Math. **155** (1994), 339–350.

[PS1] R. Parimala, and R. Sridharan, *Projective modules over polynomial rings over division rings*, J. Math. Kyoto Univ. **15** (1975), 129–148.

[PS2] R. Parimala, and R. Sridharan, *Graded Witt rings and unramified cohomology*, K-Theory **6** (1992), 29–44.

[PS3] R. Parimala, and R. Sridharan, *Reduced norms and Pfaffians via Brauer-Severi schemes*, Contemp. Math. **155** (1994), 351–363.

[PSSu] R. Parimala, R. Sridharan, and V. Suresh, *A question on the discriminants of involutions of central division algebras*, Math. Ann. **297** (1993), 575–580.

[PSr] R. Parimala, and V. Srinivas, *Analogues of the Brauer group for algebras with involution*, Duke Math. J. **66** (1992), 207–237.

[PSj] R. Parimala, and R. Sujatha, *Witt group of hyperelliptic curves*, Comm. Math. Helv. **65** (1990), 559–580.

[PSu] R. Parimala, and V. Suresh, *Zero cycles on quadric fibrations: finiteness theorems and the cycle map*, to appear in Invent. Math.

[R] M.S. Raghunathan, *Principal bundles on affine space and bundles on the projective line*, Math. Ann. **285** (1989), 309–332.

[Sa] S. Saito, *A conjecture of Bloch and Brauer groups of surfaces over p-adic fields*, preprint 1990.

[SaSj] S. Saito, and R. Sujatha, *Finiteness theorems for cohomology of surfaces over p-adic fields and an application to Witt groups*, Proc. Symp. Pure Math. **58**, Part II (1995), 403–415.

[Sh] J. Shick, *Witt groups of function fields of hyperelliptic curves*, Comm. Algebra **21** (4) (1993), 1371–1388.

[Sj] R. Sujatha, *Witt groups of real projective surfaces*, Math. Ann. **288** (1990), 89–101.

[S] A.A. Suslin, *Algebraic K-theory and the norm residue homomorphism*, J. Soviet Math. **30** (1985), 2556–2611.

[T] J. Tits, *Formes quadratiques, groupes orthogonaux et algèbres de Clifford*, Invent. Math. **5** (1968), 19–41.

[W] E. Witt, *Zerlegung reeller algebraischer Funktionen in Quadrate, Schiefkörper über reellem Funktionenkörper*, J. Reine Angew. Math. **171** (1934), 4–11.

[Y] V.I. Yanchevskii, *Symmetric and skew-symmetric elements of involutions, associated groups and the problem of decomposability of involutions*, Proc. Symp. Pure Math. **58**, Part II (1995), 431–444.

Invariant Differential Operators

GERALD W. SCHWARZ

Department of Mathematics
Brandeis University
PO Box 9110, Waltham, MA 02254-9110, USA

0 Introduction

All varieties we consider will be irreducible, algebraic, and defined over our base field $\mathbb{C}$.

Let Z be an affine variety, and set $A := \mathcal{O}(Z)$. Let E denote a vector bundle over Z or a coherent sheaf of $\mathcal{O}_Z$-modules, with global sections $M := \Gamma(Z, E)$. Then we define the algebra of (algebraic) differential operators on M and E as follows: If $P \in \operatorname{End}_{\mathbb{C}}(M)$ and $a \in A$, then $[P, a]$ denotes the usual commutator: $[P, a](m) = P(am) - a(P(m))$, $m \in M$. Define $D_A^n(M) = 0$ for $n < 0$, and for $n \geq 0$ inductively define:

$$D_A^n(M) = \{P \in \operatorname{End}_{\mathbb{C}}(M) : [P, a] \in D_A^{n-1}(M) \text{ for all } a \in A\}.$$

Clearly, $D_A^0(M) = \operatorname{End}_A(M)$. Note that $D_A^n(M) \subseteq D_A^{n+1}(M)$ for all n, and we define $D_A(M) := \bigcup D_A^n(M)$. Now we set $\mathcal{D}_E^n(Z) := D_A^n(M)$ and $\mathcal{D}^n(Z) := D_A^n(A)$, and similarly for $\mathcal{D}_E(Z)$ and $\mathcal{D}(Z)$. We call $\mathcal{D}_E^n(Z)$ (resp. $D_A^n(M)$) the *differential operators on E (resp. M) of order at most n*, $\mathcal{D}_E(Z)$ (resp. $D_A(M)$) the *algebra of differential operators on E (resp. M)*, and $\mathcal{D}(Z)$ (resp. $D_A(A)$) the *algebra of differential operators on Z (resp. A)*. Note that $\mathcal{O}(Z)$ acts on $\mathcal{D}_E^n(Z)$, etc. by left multiplication, making $\mathcal{D}_E^n(Z)$, etc. into left $\mathcal{O}(Z)$-modules.

PROPOSITION (see [Sc1, Section 3]). *Let Z and E be as above. Then*

(1) *$\mathcal{D}_E^n(Z)$ is finitely generated for all n.*
(2) *If $P \in \mathcal{D}_E^n(Z)$ and $Q \in \mathcal{D}_E^m(Z)$, then $Q \circ P \in \mathcal{D}_E^{n+m}(Z)$. If $E = \mathcal{O}_Z$, then $[Q, P] := Q \circ P - P \circ Q \in \mathcal{D}^{n+m-1}(Z)$.*

From the filtrations $\{\mathcal{D}_E^n(Z)\}$ we obtain associated graded algebras $\operatorname{gr} \mathcal{D}_E(Z)$, and $\operatorname{gr} \mathcal{D}(Z)$ is commutative by (2) above.

If Z is smooth, then $\operatorname{gr} \mathcal{D}(Z)$ is a finitely generated domain [Bj], hence $\mathcal{D}(Z)$ is a finitely generated domain, left and right Noetherian. Moreover, if E is a vector bundle over Z, then $\operatorname{gr} \mathcal{D}_E(Z)$ is a finite $\operatorname{gr} \mathcal{D}(Z)$-module. Finally, $\mathcal{D}(Z)$ is a simple algebra. If Z is not smooth, all of these properties can fail [BGG]. It seems to be very difficult to determine the properties of $\mathcal{D}(Z)$ in the singular case.

We consider the case of quotient singularities: let X be an affine G-variety, where G is reductive. Then $\mathcal{O}(X)^G$ is finitely generated, corresponding to an affine

Proceedings of the International Congress
of Mathematicians, Zürich, Switzerland 1994

variety $X/\!\!/G$, and we have a surjection $\pi_X : X \to X/\!\!/G$ dual to the inclusion $\mathcal{O}(X)^G \subset \mathcal{O}(X)$ [Kr, II.3.2]. Recent work ([Ka], [Le], [LS1], [Mu], [MuV], [Sc1], [Sc2], [VdB1], [VdB3]) has given support to the following.

CONJECTURE 1. *Let X be a smooth affine G-variety, where G is reductive. Then* $\operatorname{gr}\mathcal{D}(X/\!\!/G)$ *is finitely generated.*

CONJECTURE 2. *Let X and G be as above. Then $\mathcal{D}(X/\!\!/G)$ is simple.*

We also consider the properties of algebras $\operatorname{gr}\mathcal{D}_{\mathcal{E}}(X/\!\!/G)$, where $\mathcal{E}$ is the sheaf of $\mathcal{O}_{X/\!\!/G}$-modules corresponding to the G-invariant sections of a G-vector bundle E over X. Note that Conjectures 1 and 2 are true if they are locally true on X, i.e., if they are true for every variety ${\pi_X}^{-1}(U)$ for U affine open in $X/\!\!/G$ ([Sc1], [VdB3]). Luna's slice theorem ([Lu1], [Sl]) then allows us to reduce to the case of representations (of all the subgroups of G that are isotropy groups of closed orbits in X). Moreover, if E is a G-vector bundle over X, then we can similarly reduce questions about $\mathcal{D}_{\mathcal{E}}(X/\!\!/G)$ to cases of the form $X = V$ and $E = \Theta_W := V \times W \to V$, where V and W are G-modules. In other words, it is sufficent to consider differential operators on algebras $\mathcal{O}(V)^G$ and on $\mathcal{O}(V)^G$-modules of covariants $\operatorname{Mor}(V, W)^G$. From now on V and W will be G-modules and E will denote Θ_W.

Let $P \in \mathcal{D}^n(V)^G \simeq \mathcal{D}^n(\mathcal{O}(V))^G$. Then $P|_{\mathcal{O}(V)^G} \in \mathcal{D}^n(\mathcal{O}(V)^G)$, hence we have an element $(\pi_V)_*(P) \in \mathcal{D}^n(V/\!\!/G)$. If $(\pi_V)_*(\mathcal{D}^n(V)^G) = \mathcal{D}^n(V/\!\!/G)$ for every $n \geq 0$ (equivalently, if $\operatorname{gr}(\pi_V)_* : \operatorname{gr}(\mathcal{D}(V)^G) \to \operatorname{gr}\mathcal{D}(V/\!\!/G)$ is surjective) we say that $(\pi_V)_*$ is *graded surjective*. As $\operatorname{gr}\mathcal{D}(V) \simeq \mathcal{O}(V \oplus V^*)$, $\operatorname{gr}\mathcal{D}(V)^G \simeq \mathcal{O}(V \oplus V^*)^G$ is finitely generated. Hence, $\operatorname{gr}\mathcal{D}(V/\!\!/G)$ *is finitely generated whenever $(\pi_V)_*$ is graded surjective.*

Similarly, we can define $\pi_{V,E} : \mathcal{D}_E(V)^G \to \mathcal{D}_{\mathcal{E}}(V/\!\!/G)$, where $\operatorname{gr}\mathcal{D}_E(V)^G$ is finitely generated over $\operatorname{gr}\mathcal{D}(V)^G$. We say that $\pi_{V,E}$ *is graded surjective* if $\operatorname{gr}\pi_{V,E} : \operatorname{gr}\mathcal{D}_E(V)^G \to \operatorname{gr}\mathcal{D}_{\mathcal{E}}(V/\!\!/G)$ is surjective. *If $\pi_{V,E}$ and $(\pi_V)_*$ are graded surjective, then* $\operatorname{gr}\mathcal{D}_{\mathcal{E}}(V/\!\!/G)$ *is a finite* $\operatorname{gr}\mathcal{D}(V/\!\!/G)$*-module* in a natural way [Sc1, Section 3].

Let $\mathcal{K}^n(V)$ denote the elements in $\mathcal{D}^n(V)$ that act trivially on $\mathcal{O}(V)^G$. Set $\mathcal{K}(V) = \bigcup \mathcal{K}^n(V)$. Then $\mathcal{K}(V)^G$ is the kernel of $(\pi_V)_* : \mathcal{D}(V)^G \to \mathcal{D}(V/\!\!/G)$, and analogously for $\mathcal{K}^n(V)^G$. Similarly, one defines $\mathcal{K}_E(X) = \bigcup \mathcal{K}_E^n(X)$, where $\mathcal{K}_E^n(X)^G$ is the kernel of $\pi_{X,E} : \mathcal{D}_E^n(V)^G \to \mathcal{D}_{\mathcal{E}}^n(V/\!\!/G)$.

In order to show that $(\pi_V)_*$ is (graded) surjective, it is obviously very useful to have a good description of $\mathcal{K}^n(V)$. There is an obvious subspace of $\mathcal{K}^n(V)$: let $X \in \mathfrak{g} :=$ Lie algebra of G. Let $\tau(X)$ denote the action of X on $\mathcal{O}(V)$ (as derivations). Then $\tau(\mathfrak{g})$ annihilates $\mathcal{O}(V)^G$, hence $\mathcal{D}^{n-1}(V)\tau(\mathfrak{g}) \subset \mathcal{K}^n(V)$. It is natural to pose the following questions:

QUESTIONS 3. *Let V be a G-module.*

(1) *What are sufficient conditions for $(\pi_V)_*$ to be graded surjective?*
(2) *What are necessary conditions for $(\pi_V)_*$ to be graded surjective?*
(3) *Is it possible for $(\pi_V)_*$ to be surjective without being graded surjective?*
(4) *When is there equality in the inclusion $\mathcal{D}^{n-1}(V)\tau(\mathfrak{g}) \subset \mathcal{K}^n(V)$?*
(5) *When is there equality in the inclusion $(\mathcal{D}^{n-1}(V)\tau(\mathfrak{g}))^G \subset \mathcal{K}^n(V)^G$?*

There are analogues of all the questions above with $(\pi_V)_*$ replaced by $\pi_{V,E}$, $\mathcal{K}^n(V)^G$ replaced by $\mathcal{K}^n_E(V)^G$, and $\tau(\mathfrak{g})$ replaced by $\tau_E(\mathfrak{g})$, where $\tau_E(X)$ is the action of $X \in \mathfrak{g}$ as a differential operator of order 1 on $\Gamma(V, E)$.

In the next section we will deal with Conjecture 1 and Question 3. Conjecture 2 is reported on in Section 2, and we consider an interesting analogue of Question 3(3) in Section 3, where we deal with differential operators on adjoint representations.

1 On Conjecture 1

We first define some properties of the G-module V that figure in necessary conditions and sufficient conditions for $(\pi_V)_*$ (or $\pi_{V,E}$) to be graded surjective.

Let X be an affine G-variety. Let H be a minimal element (with respect to set inclusion) of $\{G_x : Gx \text{ is closed in } X\}$. Then H is called a *principal isotropy group of* X, and all the principal isotropy groups are conjugate. One calls $X' := \{x \in X : Gx \text{ is closed and } G_x \text{ is principal}\}$ the *principal orbits of* X. We say that X *has FPIG* when the principal isotropy groups of X are finite. Set $X_{\text{pr}} = {\pi_X}^{-1}(\pi_X(X'))$. If X has FPIG (which we normally assume), then $X' = X_{\text{pr}}$.

Define $X_{(n)} = \{x \in X : \dim G_x = n\}$, and define $\operatorname{mod}(X, G)$, the *modularity of* (X, G), to be $\sup_n\{\dim X_{(n)} - \dim G + n\}$ (see [Vi]). Define $\mathrm{d}(X, G)$ to be the transcendence degree of $\mathcal{Q}(X)^G$, where $\mathcal{Q}(X)$ denotes the field of rational functions on X.

REMARKS 4. (1) By a theorem of Rosenlicht, $\mathrm{d}(X, G) = \dim X - \sup_x \dim Gx$.

(2) If X has FPIG, then $\mathrm{d}(X, G) = \dim X/\!\!/G = \dim X - \dim G$.

(3) $\mathrm{d}(X, G) \leq \operatorname{mod}(X, G)$.

DEFINITIONS 5. Let $k \geq 0$. Then

(1) V is *k-modular* if V has FPIG and $\operatorname{mod}(V \setminus V_{(0)}, G) + k \leq \dim V/\!\!/G$,
(2) V is *k-principal* if $\operatorname{codim} V \setminus V_{\text{pr}} \geq k$, and
(3) V is *k-large* if it is k-modular and k-principal.

REMARKS 6. Let V have FPIG.

(1) V is k-modular if and only if $\operatorname{codim} V_{(n)} \geq n + k$; $n = 1, 2, \ldots, \dim G$.
(2) V is k-large if and only if $\operatorname{mod}(V \setminus V_{\text{pr}}, G) + k \leq \dim V/\!\!/G$.

We say that V is *coregular* if $\mathcal{O}(V)^G$ is a polynomial ring, equivalently, if $V/\!\!/G$ is smooth. Regarding the surjectivity of $(\pi_V)_*$ and $\pi_{V,E}$ we then have the following.

THEOREM 7 [Sc1], [Sc2]. *(1) If V is 2-large, then $(\pi_V)_*$ and $\pi_{X,E}$ are graded surjective.*

(2) *Suppose that G is semisimple. Consider G-modules V such that $\operatorname{Ker}(G \to \operatorname{GL}(U))$ is finite for each nonzero irreducible G-submodule U of V. Then, up to isomorphism, all but finitely many V are 2-large.*

(3) *Suppose that G is simple and $V^G = (0)$. Then, up to isomorphism, all but finitely many V are 2-large. Moreover, if V is irreducible, then either V is coregular or $(\pi_V)_*$ is graded surjective.*

THEOREM 8 [Sc1]. *Suppose that* $(\pi_V)_*$ *is surjective.*

(1) *If* G *is finite, or* G^0 *is semisimple or a torus, then* V *is 2-principal.*
(2) *If* G^0 *is a torus, then* $(\pi_V)_*$ *is graded surjective. If, in addition,* V *has FPIG, then* V *is 2-large.*
(3) *The smooth points of* $V/\!\!/G$ *are exactly the principal orbits* $(V/\!\!/G)_{\text{pr}}$. *In particular, if* V *is coregular, then* $0 \in V$ *is a principal orbit. Hence,* V *is fix pointed, i.e.,* $V = V^G \oplus V_1$, *where* $\mathcal{O}(V_1)^G = \mathbb{C}$.

COROLLARY 9. (1) *If* G *is finite, then* $(\pi_V)_*$ *is surjective iff it is graded surjective iff* V *is 2-principal iff* V *is 2-large iff* $G \to \mathrm{GL}(V)$ *contains no pseudoreflections* (Kantor [Ka]).

(2) *If* G^0 *is a torus, then* $(\pi_V)_*$ *is surjective iff it is graded surjective iff* V *is 2-principal* (Musson [Mu]). *Moreover, if* V *has FPIG, then* V *is 2-principal iff it is 2-large.*
(3) *If* G *is semisimple, then, up to isomorphism, there are only finitely many coregular* G*-modules* V *such that* $V^G = (0)$ [Pop], [Go], [Kn].

Using some tricks one obtains from the above:

COROLLARY 10. *Conjecture 1 holds in the following cases.*

(1) V *is 2-large.*
(2) V *is coregular.*
(3) G *is commutative.*
(4) G *is finite.*
(5) G *is simple and* V *is irreducible.*

The situation for the G-vector bundle analogue of Conjecture 1 is more complicated. Parts (1), (4), and (5) above hold. Part (2) holds if you add the condition that π_V is equidimensional (we say that V is *cofree* in this case, because the condition is equivalent to $\mathcal{O}(V)$ being a free $\mathcal{O}(V)^G$-module). As stated, however, parts (2) and (3) fail:

EXAMPLE 11. Let ν_j denote the irreducible representation of $G = \mathbb{C}^*$ of weight j. Set $V = \nu_1 \oplus \nu_{-1} \oplus \nu_{-1}$, $W = \nu_1$. Then V is coregular and 1-large, G is commutative, yet $\operatorname{gr}\mathcal{D}_{\mathcal{E}}(V/\!\!/G)$ (which is commutative) is not finitely generated [Sc1, 3.27].

EXAMPLE 12. Let $G = \mathrm{SL}_n$, $V = (n+1)\mathbb{C}^n$ ($(n+1)$ copies of the standard representation on $\mathbb{C}^n$), and $W = \mathbb{C}^n$, $n \geq 2$. Then V is coregular with quotient $\mathbb{C}^{n+1}$, but $\mathcal{D}_{\mathcal{E}}(V/\!\!/G)$ is not left Noetherian. Hence $\operatorname{gr}\mathcal{D}_{\mathcal{E}}(V/\!\!/G)$ is not finitely generated over any finitely generated commutative algebra [Sc1, 3.28].

We return to considering "ordinary" differential operators. Coregularity is a property of "small" G-modules, and 2-largeness is true for modules that are "sufficiently large." In between there is a "gray area" of modules that are neither 2-large nor coregular. We have no general tools to determine whether or not Conjecture 1 holds in these cases. Examples are some of the SL_n-modules of the form $k\mathbb{C}^n \oplus l(\mathbb{C}^n)^*$, $k + l < 2n$. However, there are cases where there is no "gray area."

THEOREM 13. *V is always coregular or 2-large in the following cases.*

(1) $G = \mathrm{SL}_2$ [Sc1, 11.9].

(2) $(V, G) = (k\mathbb{C}^n \oplus l(\mathbb{C}^n)^*, \mathrm{GL}_n)$, $(k\mathbb{C}^n, \mathrm{O}_n)$, $(k\mathbb{C}^n, \mathrm{SO}_n)$ *or* $(k\mathbb{C}^{2n}, \mathrm{Sp}_{2n})$; k, $l \geq 0$, $n \geq 1$ [LS1].

(3) $G = G_2$ *(resp.* Spin_7*), and V is a direct sum of copies of the irreducible 7-dimensional (resp. 8-dimensional) module* [Sc1, 11.21].

Regarding Question 3(3) little is known. If V is 1-large, then one can show that $(\pi_V)_*$ is surjective iff it is graded surjective, and similarly for $\pi_{V,E}$, provided that the principal isotropy groups of V act trivially on W (otherwise there are counterexamples [Sc1, 5.4]).

EXAMPLE 14. Let $V = k\mathbb{C}^2$ have the diagonal action of SL_2. Then V is $(k-2)$-large. When $k = 2$, $\mathcal{K}(V)^G \neq (\mathcal{D}(V)\tau(\mathfrak{g}))^G$. When $k = 3$, V is 1-large but $(\pi_V)_*$ is not surjective (because V is coregular). For $k \geq 4$, $(\pi_V)_*$ is graded surjective (and V is not coregular!).

2 On Conjecture 2

Let V and W be G-modules, and, as usual, let E denote Θ_W. The question is whether or not $\mathcal{D}_\mathcal{E}(V/\!\!/G)$ is simple. If $\pi_{V,E}$ is surjective, this is the same as asking if $\mathcal{K}_E(V)^G$ is a maximal 2-sided ideal in $\mathcal{D}_E(V)^G$. Of course, $\mathcal{D}(V/\!\!/G) = \mathcal{D}_\mathcal{E}(V/\!\!/G)$ when W is the trivial one-dimensional G-module.

Suppose that G is finite, and set $H = \{g \in G : g$ acts trivially or as a reflection on $V\}$. Then $V' := V/\!\!/H$ is a G/H-module without reflections, and there is a G/H-module W' such that $\Gamma(V, E)^H \simeq \Gamma(V', E')$ as $\mathcal{O}(V')$ and G/H-module, where $E' := V' \times W' \to V'$. Hence we can reduce to the case that G acts effectively on V and contains no reflections.

THEOREM 15. *Suppose that V and W are G-modules, where G is finite and acts faithfully and without reflections on V. Then*

(1) $\pi_{V,E}$ *is an isomorphism.*

(2) $\mathcal{D}_E(V)^G \simeq \mathcal{D}_\mathcal{E}(V/\!\!/G)$ *is simple.*

Proof. It is easy to see that $\pi_{V,E}$ is injective, because $\pi_{V,E} : \mathcal{D}_E(V_{\mathrm{pr}})^G \xrightarrow{\sim} \mathcal{D}_\mathcal{E}(V_{\mathrm{pr}}/\!\!/G)$ is an isomorphism. As V is 2-large, $\pi_{V,E}$ is also surjective, giving (1). For (2), one can modify the proof of [Wa] or apply [Mo, Corollary 2.6]. In the latter case one has to show that G consists of "outer" automorphisms of the simple algebra $\mathcal{D}_E(V)$, which follows from the fact that G acts faithfully. □

For tori, we have the work of Van den Bergh [VdB1] and Musson and Van den Bergh [MuV]. They considered differential operators on modules of covariants of torus representations. I add the assumption of 2-largeness to their hypotheses, so that we are talking about differential operators on the quotient.

THEOREM 16 [MuV]. *Let G be a torus and V a G-module that is 2-large. Let $\eta_1, \ldots, \eta_n$ be the weights of V ($\dim V = n$). Suppose that W is an irreducible*

G-module with weight $\sum a_i\eta_i$, $a_i \in \mathbb{Q}$. Then $\mathcal{D}_{\mathcal{E}}(V/\!\!/G)$ is simple if $-1 < a_i \leq 0$ for all i. In particular, $\mathcal{D}(V/\!\!/G)$ is simple.

In case G^0 is not a torus results are even fewer and farther between. Levasseur and Stafford established simplicity in the case of the "classical" representations of GL_n, O_n, and Sp_n [LS1]. Of course, there is nothing to prove in the coregular case, because the Weyl algebras $\mathcal{D}(\mathbb{C}^n)$ are simple. However, in general, there are not many W such that $\mathcal{D}_{\mathcal{E}}(V/\!\!/G)$ is simple:

THEOREM 17. *Let V be 2-large.*

(1) *If $\mathcal{D}_{\mathcal{E}}(V/\!\!/G)$ is simple, then $\Gamma(V,E)^G$ is Cohen-Macaulay* [VdB1].
(2) *Suppose that G acts faithfully on V. Then there are only finitely many W, up to isomorphism and addition of trivial factors, such that $\Gamma(V,E)^G$ is Cohen-Macaulay* (Brion[Br]). *Hence there are only finitely many W for which $\mathcal{D}_{\mathcal{E}}(V/\!\!/G)$ is simple.*

Here is an example where everything works out nicely: Let $G = \mathrm{SL}_2$, and let $R_j = \mathcal{S}ym^j(\mathbb{C}^2)$ denote the irreducible representation of G of dimension $j+1$. We have the following result of Van den Bergh [VdB2], [VdB3]:

THEOREM 18. *Let $V = \sum_{i=1}^e R_{k_i}$, where each k_i is odd. Set $s = \sum_{i=1}^e \dfrac{(k_i+1)^2}{4}$. We assume that V is not coregular, hence V is 2-large. Let $W = R_m$. Then*

(1) *$\mathcal{D}_{\mathcal{E}}(V/\!\!/G) \simeq \mathcal{D}_E(V)^G/\mathcal{K}_E(V)^G$ is simple iff $m < s+2$.*
(2) *$\Gamma(V,E)^G$ is Cohen-Macaulay iff $m < s+2$.*

3 Differential operators on Lie algebras

Let G be connected reductive with maximal torus T. Let $\mathfrak{g}$ (resp. $\mathfrak{t}$) denote the Lie algebra of G (resp. T) and let $\mathcal{W}$ denote the Weyl group. Long ago, Harish-Chandra [HC1], [HC2] constructed a map $\delta : \mathcal{D}(\mathfrak{g})^G \to \mathcal{D}(\mathfrak{t})^W$ with the following properties:

($\delta 1$) δ is an algebra homomorphism.
($\delta 2$) On $\mathcal{O}(\mathfrak{g})^G$, δ is the isomorphism given by restriction $\mathcal{O}(\mathfrak{g})^G \xrightarrow{\sim} \mathcal{O}(\mathfrak{t})^{\mathcal{W}}$.
($\delta 3$) On $\mathcal{S}ym(\mathfrak{g})^G$ (considered as the invariant constant coefficient differential operators on $\mathfrak{g}$), δ is the isomorphism $\mathcal{S}ym(\mathfrak{g})^G \xrightarrow{\sim} \mathcal{S}ym(\mathfrak{t})^{\mathcal{W}}$ induced by the canonical projection $\mathfrak{g} \to \mathfrak{t}$.
($\delta 4$) The kernel of δ is $\mathcal{K}(\mathfrak{g})^G$.
($\delta 5$) δ is surjective.

There are several other properties of δ, but the ones above are the most important. The construction of δ is quite simple, but proving the above properties is quite arduous and, in Harish-Chandra's case, rather analytic in nature. Also, ($\delta 5$) is only recent. It is a corollary of the following theorem of Levasseur and Stafford [LS2]:

THEOREM 19. *Let H be a finite group and U an H-module. Then $\mathcal{D}(U)^H$ is generated by $\mathcal{O}(U)^H$ and $\mathcal{S}ym(U)^H$.*

Here is a slight reformulation of Harish-Chandra's construction of δ: Let $\mathfrak{t}'$ denote the principal $\mathcal{W}$-orbits in $\mathfrak{t}$, i.e., those with trivial isotropy. Then $\pi_{\mathfrak{t}} : \mathfrak{t}' \to \mathfrak{t}'/\!\!/\mathcal{W}$ is a covering map, so $(\pi_{\mathfrak{t}})_*$ induces an isomorphism of $\mathcal{D}(\mathfrak{t}')^{\mathcal{W}}$ and $\mathcal{D}(\mathfrak{t}'/\!\!/\mathcal{W})$. Define δ' by the following commutative diagram:

$$\begin{array}{ccc} \mathcal{D}(\mathfrak{g})^G & \xrightarrow{\delta'} & \mathcal{D}(\mathfrak{t}')^W \\ \big\downarrow (\pi_{\mathfrak{g}})_* & & \big\uparrow (\pi_{\mathfrak{t}})_*^{-1} \\ \mathcal{D}(\mathfrak{g}/\!\!/G) & \xrightarrow{\text{incl.}} & \mathcal{D}(\mathfrak{t}'/\!\!/\mathcal{W}) \end{array}$$

Then δ' is clearly a homomorphism. Let ρ denote the product of linear functionals that define the reflection hyperplanes in $\mathfrak{t}$. Define δ to be the composition of differential operators $m_\rho \circ \delta' \circ m_{\rho^{-1}}$, where m_ρ denotes multiplication by ρ. Considered as morphisms from $\mathcal{D}(\mathfrak{g})^G \to \mathcal{D}(\mathfrak{t}')^{\mathcal{W}}$, δ and δ' satisfy ($\delta 1$), ($\delta 2$), and ($\delta 4$). One needs "only" to show ($\delta 3$) and that $\operatorname{Im}\delta \subset \mathcal{D}(\mathfrak{t})^{\mathcal{W}}$.

Let $P \in \mathcal{D}(\mathfrak{g})^G$. Then $\delta(P) = \sum_\alpha a_\alpha \partial^\alpha$, where $\delta(P)$ lies in $\mathcal{D}(\mathfrak{t})^{\mathcal{W}}$ iff the rational functions a_α are polynomial. If one of the a_α is not polynomial, then it must have poles along the reflection hyperplanes in $\mathfrak{t}$. However, the hyperplanes correspond to copies of $\mathrm{SL}_2 \subset G$, and using Luna's slice theorem (there is a bit of work here), one can reduce the problem to the case of SL_2, where it is an easy calculation to see that $\operatorname{Im}\delta = \mathcal{D}(\mathfrak{t})^{\mathcal{W}}$.

It remains to show ($\delta 3$), and we can clearly reduce to the case where $\mathfrak{g}$ is simple. We construct an action of SL_2 on $\mathcal{D}(\mathfrak{g})$: let q denote the quadratic invariant of $\mathfrak{g}$ and let $Q \in \mathcal{S}ym^2(\mathfrak{g})^G$ denote the corresponding constant coefficient operator. Then the commutator $h := [q, Q]$ is easily seen to be $aE + b$, where E is the Euler operator and $a, b \in \mathbb{C}$. Adjusting coefficients, one can arrange that $e := q$, $f := Q$, and h form a simple algebra of type $\mathfrak{sl}_2$. Because q and Q act ad-nilpotently on $\mathcal{D}(\mathfrak{g})$, the $\mathfrak{sl}_2$-action integrates to an action of SL_2.

Now $\delta(q) = q'$ is the quadratic generator in $\mathcal{O}(\mathfrak{t})^{\mathcal{W}}$, and hopefully, $\delta(Q) = Q'$ is a generator of $\mathcal{S}ym^2(\mathfrak{t})^{\mathcal{W}}$. From the construction of δ, $0 \neq Q'$ has order 2, and it sends elements of $\mathcal{O}(\mathfrak{t})^{\mathcal{W}}$ of degree k to elements of degree $k-2$. As the coefficients of Q' are regular, it can only be a constant coefficient differential operator, i.e., $Q' \in \mathcal{S}ym^2(\mathfrak{t})^{\mathcal{W}}$. Because $\operatorname{ad} q$, $\operatorname{ad} Q$, and $\operatorname{ad}[q, Q]$ define an action of $\mathfrak{sl}_2$ on $\mathcal{D}(\mathfrak{g})$, the same is true for $\operatorname{ad} q'$, $\operatorname{ad} Q'$, and $\operatorname{ad}[q', Q']$ on $\mathcal{D}(\mathfrak{t})$. By construction, δ is equivariant with respect to the two $\mathfrak{sl}_2$ and SL_2-actions. The generator of the Weyl group of SL_2 interchanges $\mathcal{O}(\mathfrak{g})^G$ and $\mathcal{S}ym(\mathfrak{g})^G$ (and $\mathcal{O}(\mathfrak{t})^{\mathcal{W}}$ and $\mathcal{S}ym(\mathfrak{t})^{\mathcal{W}}$), and ($\delta 3$) follows from ($\delta 2$).

Wallach and Hunziker [WaH] have a purely Lie algebra theoretic construction of δ. In their approach, properties ($\delta 1$), ($\delta 2$), ($\delta 3$), and ($\delta 5$) are immediate. The difficult part is to establish ($\delta 4$). They do this by showing that δ annihilates $K' := (\mathcal{D}(\mathfrak{g}) \cdot \tau(\mathfrak{g}))^G$, and then showing that every element $P \in \mathcal{K}(\mathfrak{g})^G$ can be multiplied by an invariant $h \neq 0$ so that it lands in K' [Wa]. Because δ is a homomorphism, $\delta(hP) = 0$ implies that $h|_{\mathfrak{t}}$ annihilates $\delta(P)$. Hence, $\delta(P) = 0$.

Finally, we consider connections to our previous questions. First of all, *does* $\mathcal{K}(\mathfrak{g})^G$ *equal* $(\mathcal{D}(\mathfrak{g})\tau(\mathfrak{g}))^G$? This has been recently established by Levasseur and Strafford [LS3], using noncommutative methods. Secondly, *is the surjective homomorphism* $\delta : \mathcal{D}(\mathfrak{g})^G \to \mathcal{D}(\mathfrak{t})^{\mathcal{W}}$ *graded surjective*? Identifying $\mathfrak{g}^*$ and $\mathfrak{t}^*$ with $\mathfrak{g}$ and $\mathfrak{t}$

via the Killing form, we are asking if the restriction map $\sigma : \mathcal{O}(\mathfrak{g}\oplus\mathfrak{g})^G \to \mathcal{O}(\mathfrak{t}\oplus\mathfrak{t})^{\mathcal{W}}$ is surjective. By a theorem of Richardson [Ri], the closure of $G\cdot(\mathfrak{t}\oplus\mathfrak{t})$ in $\mathfrak{g}\oplus\mathfrak{g}$ is the commuting variety $\mathcal{C} := \{(A,B) \in \mathfrak{g}\oplus\mathfrak{g} : [A,B] = 0\}$. By a theorem of Luna [Lu2], σ is surjective iff $\mathcal{C}/\!\!/G \subset (\mathfrak{g}+\mathfrak{g})/\!\!/G$ is normal. For example, if one knew that $\mathcal{C}$ were normal, then σ would be onto. However, normality has only been established in a few cases by calculation.

References

[BGG] I. N. Bernstein, I. M. Gel'fand, and S. I. Gel'fand, *Differential operators on the cubic cone*, Russian Math. Surveys **27** (1972), 169–174.

[Bj] J.-E. Björk, Rings of Differential Operators, North-Holland, Amsterdam, 1979.

[Br] M. Brion, *Sur les modules de covariants*, Ann. Sci. École Norm. Sup. (4) **26** (1993), 1–21.

[Go] N. Gordeev, *Coranks of elements of linear groups and the complexity of algebras of invariants*, Leningrad Math. J. **2** (1991), 245–267.

[HC1] Harish-Chandra, *Differential operators on a semisimple Lie algebra*, Amer. J. Math. **79** (1957), 87–120.

[HC2] ———, *Invariant differential operators and distributions on a semisimple Lie algebra*, Amer. J. Math. **86** (1964), 534–564.

[Ka] J.-M. Kantor, *Formes et opérateurs différentiels sur les espaces analytiques complexes*, Bull. Soc. Math. France, Mémoire **53** (1977), 5–80.

[Kn] F. Knop, *Über die Glattheit von Quotientenabbildungen*, Manuscripta Math. **56** (1986), 410–427.

[Kr] H. Kraft, Geometrische Methoden in der Invariantentheorie, Vieweg-Verlag, Braunschweig, 1985.

[Le] T. Levasseur, *Relèvements d'opérateurs différentiels sur les anneaux d'invariants*, Operator Algebras, Unitary Representations, Enveloping Algebras and Invariant Theory, Progr. Math., vol. 92, Birkhäuser, Boston, MA, 1990, pp. 449–470.

[LS1] T. Levasseur and J. T. Stafford, Rings of differential operators on classical rings of invariants, vol. 412, 1989.

[LS2] ———, *Invariant differential operators and an homomorphism of Harish-Chandra*, J. Amer. Math. Soc. **8** (1995), 365–372.

[LS3] ———, *The kernel of an homomorphism of Harish-Chandra*, Ann. Sci. École Norm. Sup., to appear.

[Lu1] D. Luna, *Slices étales*, Bull. Soc. Math. France, Mémoire **33** (1973), 81–105.

[Lu2] ———, *Adhérences d'orbite et invariants*, Invent. Math. **29** (1975), 231–238.

[Mo] Susan Montgomery, Fixed rings of finite automorphism groups of associative rings, Lecture Notes in Math. **818**, Springer-Verlag, Berlin and New York, 1980.

[Mu] I. Musson, *Rings of differential operators on invariant rings of tori*, Trans. Amer. Math. Soc. **303** (1987), 805–827.

[MuV] I. Musson and M. Van den Bergh, *Rings of global differential operators on toric varieties and related topics*, in preparation.

[Pop] V. Popov, *A finiteness theorem for representations with a free algebra of invariants*, Math. USSR-Izv. **20** (1983), 333–354.

[Ri] R. W. Richardson, *Commuting varieties of semisimple Lie algebras and algebraic groups*, Compositio Math. **38** (1979), 311–327.

[Sc1] G. W. Schwarz, *Lifting differential operators from orbit spaces*, Ann. Sci. École Norm. Sup. **28** (1995), 253–306.

[Sc2] ———, *Differential operators on quotients of simple groups*, J. Algebra **169** (1995), 248–273.

[Sl] P. Slodowy, *Der Scheibensatz für algebraische Transformationsgruppen*, Algebraic Transformation Groups and Invariant Theory, DMV Sem. **13**, Birkhäuser Verlag, Basel and Boston, 1989, pp. 89–113.

[VdB1] M. Van den Bergh, *Differential operators on semi-invariants for tori and weighted projective spaces*, Séminaire Malliavin (1990), Lecture Notes in Math. **1478**, Springer-Verlag, Berlin and New York, 1992, pp. 255–272.

[VdB2] ———, *A converse to Stanley's conjecture for* SL_2, Proc. Amer. Math. Soc., to appear.

[VdB3] ———, *Some rings of differential operators for* SL_2*-invariants are simple*, in preparation.

[Vi] E. B. Vinberg, *Complexity of actions of reductive groups*, Functional Anal. Appl. **20** (1986), 1–11.

[Wa] N. R. Wallach, *Invariant differential operators on a reductive Lie algebra and Weyl group representations*, J. Amer. Math. Soc. **6** (1993), 779–816.

[WaH] N. R. Wallach and M. Hunziker, *On the Harish-Chandra homomorphism of invariant differential operators on a reductive Lie algebra*, preprint.

Algebraic K-Theory and Motivic Cohomology

ANDREI SUSLIN*

Saint Petersburg Branch of the
Steklov Math. Institute (POMI)
Fontanka 27, St. Petersburg 191011
Russia

and

Department of Mathematics
Northwestern University
Evanston, IL 60208
USA

The general idea of motivic cohomology as a universal cohomology theory on the category of schemes goes back to Grothendieck. But it was not until 1982 that this general idea got a precise form. Around that time Beilinson formulated his famous conjectures. Beilinson conjectured that for all $q \geq 0$ there should exist complexes of sheaves $\mathbb{Z}(q)$ on the big Zariski site of regular schemes that should satisfy (among others) the following properties:

(1) $\mathbb{Z}(0) = \mathbb{Z}, \mathbb{Z}(1) = \mathcal{O}^*[-1]$

(2) **Vanishing conjecture:** for $q > 0$ the complex $\mathbb{Z}(q)$ is acyclic outside $[1, q]$, the sheaf $H^q(\mathbb{Z}(q))$ coincides with the sheaf K_q^M of Milnor K-groups

(3) **Relationship to algebraic K-theory:** there exists a spectral sequence

$$E_2^{p,q} = H_{\mathcal{M}}^{p-q}(X, \mathbb{Z}(-q)) \Longrightarrow K_{-p-q}(X)$$

that is split up to standard factorials by means of Chern classes.
Here and below we denote by $H_{\mathcal{M}}^*(X, \mathbb{Z}(q))$ the Zariski hypercohomology of X with coefficients in the complex $\mathbb{Z}(q)$

(4) **Relationship to etale cohomology:** set $\mathbb{Z}/l(q) = \mathbb{Z}(q) \otimes^L \mathbb{Z}/l$, then, restricting to the subcategory of schemes over $Spec\ \mathbb{Z}[1/l]$, we have a functorial quasiisomorphism $\mathbb{Z}/l(q) = \tau_{\leq q} R\pi_*(\mu_l^{\otimes q})$ where $\pi : (Sch)_{\text{et}} \to (Sch)_{\text{Zar}}$ is the canonical morphism of sites and $\tau_{\leq q}$ denotes the degree q truncation of the complex.

Several approaches to the construction of motivic complexes and motivic cohomology were proposed during the last years — see [B1], [FL], [FG], [G], [L], [V1], [V2], [V3]. All these approaches are based on the theory of algebraic cycles. We'll discuss below two of these constructions. Even though a significant part of the theory may be applied in a more general situation of arbitrary noetherian schemes we always restrict ourselves to the category of schemes of finite type over a field.

---

*) The author was supported in part by Grant MOG 000 from the International Science Foundation.

Proceedings of the International Congress
of Mathematicians, Zürich, Switzerland 1994

1 The triangulated category of motives

The general framework for the development of the theory of motives was laid down in a series of works of Voevodsky [V1], [V2], [V3], [FV]. Fix a field F and denote by Sch/F (resp. Sm/F) the category of schemes of finite type over the field F (resp. the category of smooth schemes of finite type over F). Any "reasonable" cohomology theory should be a contravariant functor $\mathcal{H} : Sch/F \to Ab$, satisfying the property $\mathcal{H}(X \coprod Y) = \mathcal{H}(X) \oplus \mathcal{H}(Y)$. Moreover $\mathcal{H}$ should be equipped with transfer homomorphisms $Tr_{X/S} : \mathcal{H}(X) \to \mathcal{H}(S)$ defined (at least) for finite surjective morphisms $f : X \to S$ with X integral and S smooth and irreducible. Finally $\mathcal{H}$ should be homotopy invariant: $\mathcal{H}(X \times \mathbb{A}^1) = \mathcal{H}(X)$ for any $X \in Sch/F$. This observation leads to the following definition:

DEFINITION 1.1 [SV1]. *A presheaf with transfers on the category Sch/F is a contravariant functor $\mathcal{H} : Sch/F \to Ab$, satisfying the property $\mathcal{H}(X \coprod Y) = \mathcal{H}(X) \oplus \mathcal{H}(Y)$, and equipped with transfer homomorphisms $Tr_{X/S} : \mathcal{H}(X) \to \mathcal{H}(S)$ defined for finite surjective morphisms $X \to S$ with X irreducible and S smooth and irreducible. These transfer homomorphisms should satisfy the following compatibility properties:*

(1.1.1) *If $f : X \to S$ is an isomorphism, then $Tr_{X/S}$ coincides with $(f^{-1})^*$.*

(1.1.2) *Let $S' \to S$ be a morphism of smooth irreducible schemes. Set $X' = X \times_S S'$, denote components of X' by X'_i, and by n_i denote the corresponding multiplicities. Then the following diagram commutes:*

$$\begin{array}{ccc} \mathcal{H}(X) & \longrightarrow & \coprod \mathcal{H}(X'_i) \\ {\scriptstyle Tr_{X/S}}\downarrow & & \downarrow{\scriptstyle \sum n_i \cdot Tr_{X'_i/S'}} \\ \mathcal{H}(S) & \longrightarrow & \mathcal{H}(S') \end{array}$$

In many cases it is preferable to work with smooth schemes only. For a pair X, Y of smooth schemes over F denote by $Cor(X,Y)$ the free abelian group generated by integral closed subschemes $Z \subset X \times Y$ that are finite and surjective over a component of X. Elements of $Cor(X,Y)$ are called finite correspondences from X to Y. One defines easily the composition homomorphism $Cor(Y,T) \times Cor(X,Y) \to Cor(X,T) \quad (\beta \times \alpha \mapsto \beta \circ \alpha)$. In this way we obtain an additive category $SmCor/F$ whose objects are smooth schemes over F and $Hom_{SmCor/F}(X,Y) = Cor(X,Y)$. Associating to a morphism $X \to Y$ its graph we get a canonical functor $Sm/F \to SmCor/F$.

DEFINITION 1.2 [V3]. *A presheaf with transfers on the category Sm/F is a contravariant additive functor $\mathcal{H} : SmCor/F \to Ab$. We say that $\mathcal{H}$ is a Zariski (Nisnevich, etale ...) sheaf with transfers on the category Sm/F if the composed functor $Sm/F \to SmCor/F \xrightarrow{\mathcal{H}} Ab$ is a sheaf in the corresponding topology.*

The Nisnevich topology is very convenient when dealing with presheaves with transfer in view of the following lemma, which is false in the Zariski topology.

LEMMA 1.3 [V3]. *Let $\mathcal{F}$ be a presheaf with transfers on the category Sm/F and let $\mathcal{F}^{\sim}_{\mathrm{Nis}}$ denote the associated sheaf in the Nisnevich topology. Then $\mathcal{F}^{\sim}_{\mathrm{Nis}}$ has a canonical structure of a Nisnevich sheaf with transfers.*

This lemma implies in particular that the category $Shv_{\mathrm{Nis}}(SmCor/F)$ of Nisnevich scheaves with transfer is abelian.

For any scheme $X \in Sch/F$ denote by $L(X)$ a presheaf with transfers on the category Sm/F given by the formula:

$L(X)(S) =$ the free abelian group generated by closed integral subschemes $Z \subset X \times S$ that are finite and surjective over a component of S.
One verifies easily that $L(X)$ is a Nisnevich sheaf with transfers.

PROPOSITION 1.4 [V3]. *Let X be a smooth scheme over F and let K be a bounded from above complex of Nisnevich scheaves with transfers. Then for any $i \in \mathbb{Z}$ we have a canonical isomorphism*

$$Hom_{D^-(Shv_{\mathrm{Nis}}(SmCor/F))}(L(X), K[i]) = H^i_{\mathrm{Nis}}(X, K)\,.$$

For any $X \in Sm/F$ and any presheaf with transfers $\mathcal{F}$ define a new presheaf with transfers $\mathcal{H}om(X, \mathcal{F})$ by means of the formula $\mathcal{H}om(X, \mathcal{F})(S) = \mathcal{F}(X \times S)$. Note that $\mathcal{H}om(X, \mathcal{F})$ is a Nisnevich (Zariski, etale ...) sheaf provided that $\mathcal{F}$ is. A presheaf $\mathcal{F}$ is called homotopy invariant if the canonical homomorphism $\mathcal{F} \to \mathcal{H}om(\mathbb{A}^1, \mathcal{F})$ is an isomorphism. A presheaf $\mathcal{F}$ is called contractible if there exists a presheaf homomorphism $\phi : \mathcal{F} \to \mathcal{H}om(\mathbb{A}^1, \mathcal{F})$ such that $i_0\phi = 0, i_1\phi = id$, where $i_0, i_1 : \mathcal{H}om(\mathbb{A}^1, \mathcal{F}) \to \mathcal{F}$ are homomorphisms defined by points $0, 1 \in \mathbb{A}^1$ respectively. One checks immediately that the Nisnevich (Zariski, etale ...) sheaf associated with a contractible presheaf is again contractible. The corresponding statement for homotopy invariant sheaves is much more difficult.

THEOREM 1.5 [V2]. *Assume that the field F is perfect. Let $\mathcal{F}$ be a homotopy invariant presheaf with transfers on the category Sm/F, then*

1. *the sheaf $\mathcal{F}^{\sim}_{\mathrm{Zar}}$ is strictly homotopy invariant, i.e. $H^i_{\mathrm{Zar}}(X, \mathcal{F}^{\sim}_{\mathrm{Zar}}) = H^i(X \times \mathbb{A}^1, \mathcal{F}^{\sim}_{\mathrm{Zar}})$ for all i and all $X \in Sm/F$.*
2. *the sheaf $\mathcal{F}^{\sim}_{\mathrm{Nis}}$ coincides with $\mathcal{F}^{\sim}_{\mathrm{Zar}}$, moreover for any $X \in Sm/F$ we have: $H^*_{\mathrm{Zar}}(X, \mathcal{F}^{\sim}_{\mathrm{Zar}}) = H^*_{\mathrm{Nis}}(X, \mathcal{F}^{\sim}_{\mathrm{Nis}})$.*

Define the triangulated category of motives over a field F $DM(F)$ to be the full subcategory of $D^-(Shv_{\mathrm{Nis}}(SmCor/F))$ consisting of complexes with homotopy invariant cohomology sheaves.

Let $\Delta^{\cdot}$ be the standard cosimplicial scheme: $\Delta^n = Spec\ F[T_0, \dots, T_n]/(T_0 + \dots + T_n - 1)$. For any presheaf with transfers $\mathcal{F}$ denote by $\underline{C}_*(\mathcal{F})$ a simplicial presheaf given by the formula $\underline{C}_n(\mathcal{F}) = \mathcal{H}om(\Delta^n, \mathcal{F})$ and by $h_i(\mathcal{F})$ denote the ith homology presheaf of the complex $\underline{C}_*(\mathcal{F})$. Let further $C_*(\mathcal{F})$ be the complex of global sections of the complex of presheaves $\underline{C}_*(X)$ and $H^{sing}_i(\mathcal{F})$ be the group of global sections of the presheaf $h_i(\mathcal{F})$, i.e. $H^{sing}_i(\mathcal{F}) = H_i(C_*(\mathcal{F}))$. For any $n > 0$ define singular homology and cohomology of $\mathcal{F}$ with finite coefficients $\mathbb{Z}/n$ via the formulae:

$$H^{sing}_i(\mathcal{F}, \mathbb{Z}/n) = H_i(C_*(\mathcal{F}) \otimes^L \mathbb{Z}/n), \quad H^i_{sing}(\mathcal{F}, \mathbb{Z}/n) = H^i(RHom(C_*(\mathcal{F}), \mathbb{Z}/n)\,.$$

Passing to the cohomology notations (i.e. setting $\underline{C}_n(\mathcal{F}) = \underline{C}^{-n}(\mathcal{F})$) we get a functor $\underline{C}^* : D^-(Shv_{\mathrm{Nis}}(SmCor/F)) \to D^-(Shv_{\mathrm{Nis}}(SmCor/F))$. An immediate verification shows that presheaves $h_i(\mathcal{F})$ are homotopy invariant. Theorem 1.5 shows now that the image of the functor $\underline{C}^*$ lies in $DM(F)$. Moreover we have the following result:

THEOREM 1.6 [V3]. *The functor $\underline{C}^*$ induces an equivalence between the localization of $D^-(Shv_{\mathrm{Nis}}(SmCor/F))$ with respect to a thick subcategory of complexes quasiisomorphic to complexes consisting of contractible sheaves and the category $DM(F)$.*

For any scheme X define its motive $M(X)$ as the image of $L(X)$ in $DM(F)$ under the action of the localizing functor $\underline{C}^*$. The category $DM(F)$ has a canonical tensor structure characterized by the property $M(X) \otimes M(Y) = M(X \times Y)$. The closed embedding $SpecF \overset{0}{\hookrightarrow} \mathbb{P}^1$ and the projection $\mathbb{P}^1 \to SpecF$ give a splitting $M(\mathbb{P}^1) = M(SpecF) \oplus \widetilde{M}(\mathbb{P}^1)$. Define the Tate motive $\mathbb{Z}(1)$ as $\widetilde{M}(\mathbb{P}^1)[-2]$. Define further $\mathbb{Z}(n)$ as the nth tensor power of $\mathbb{Z}(1)$. One checks easily that the motive of the projective space $\mathbb{P}^n$ splits into the direct sum of Tate motives $M(\mathbb{P}^n) = \coprod_{i=0}^n \mathbb{Z}(n)[2n]$ so that essentially the study of Tate motives $\mathbb{Z}(n)$ is equivalent to that of motives $M(\mathbb{P}^n)$ corresponding to projective spaces.

Now we can define motivic cohomology of any scheme X via the formula

$$H^i_{\mathcal{M}}(X, \mathbb{Z}(n)) = Hom_{DM(F)}(M(X), \mathbb{Z}(n)[i])$$

If X is smooth, then using Proposition 1.4 and Theorems 1.5 and 1.6 we conclude that

$$\begin{aligned} H^i_{\mathcal{M}}(X, \mathbb{Z}(n)) &= Hom_{D^-(Shv_{\mathrm{Nis}}(SmCor/F))}(L(X), \mathbb{Z}(n)[i]) \\ &= H^i_{\mathrm{Nis}}(X, \mathbb{Z}(n)) = H^i_{\mathrm{Zar}}(X, \mathbb{Z}(n)). \end{aligned}$$

Thus, this definition of motivic cohomology fits into the picture predicted by the Beilinson cojecture.

2 h-topologies

Another important innovation of Voevodsky was the introduction of several new topologies on the category of schemes — h-topologies. Roughly speaking, the h-topology is obtained by declaring all proper surjective morphisms to be coverings (all finite surjective morphisms in the case of qfh-topology), more precisely we have the following result:

PROPOSITION 2.1 [V1]. *The h-topology is stronger than the qfh-topology, which is stronger than the etale topology. Every finite surjective morphism is a qfh-covering, every proper surjective morphism is an h-covering. Furthermore, let $S \in Sch/F$ be a normal connected scheme. Then*

(1) *Every qfh-covering of S has a refinement of the form $(Y_i \to S)_{i \in I}$, where $Y \to S$ is the normalization of S in a finite normal extension of the field $F(S)$ and $(Y_i \to Y)_{i \in I}$ is a Zariski open covering of Y.*

(2) *Every h-covering of S has a refinement of the form $(Y_i \to S)_{i\in I}$, where $(Y_i \to Y)_{i\in I}$ is a Zariski open covering of the scheme Y, $Y \to Z$ is the normalization of the scheme Z in a finite normal extension of the field $F(Z)$ and $Z \to S$ is a blow up of a closed subscheme of S.*

Even though the h-topology is stronger than the etale topology the corresponding cohomology groups coincide in many cases. For example we have the following result:

THEOREM 2.2 [SV1]. *For any scheme $S \in Sch/F$ and any integer $n > 0$ we have:*

$$H^*_{et}(S, \mathbb{Z}/n) = H^*_{qfh}(S, \mathbb{Z}/n) = H^*_h(S, \mathbb{Z}/n)\,.$$

The following important rigidity theorem is proved in [SV1]. The proof is similar to the proof of the rigidity theorem in algebraic K-theory — see [S1], but at a certain moment one has to use resolution of singularities, which at the moment is known in the characteristic zero case only.

THEOREM 2.3 [SV1]. *Assume that F is an algebraically closed field of characteristic zero. Assume further that $\mathcal{F}$ is a homotopy invariant presheaf with transfers on the category Sch/F. Then for any $n > 0$ we have canonical isomorphisms*

$$Ext^*_{et}(\mathcal{F}^{\sim}_{et}, \mathbb{Z}/n) = Ext^*_{qfh}(\mathcal{F}^{\sim}_{qfh}, \mathbb{Z}/n) = Ext^*_h(\mathcal{F}^{\sim}_h, \mathbb{Z}/n) = Ext^*_{Ab}(\mathcal{F}(F), \mathbb{Z}/n)\,.$$

Analyzing two hyperhomology spectral sequences associated with the complex $\underline{C}_*(\mathcal{F})$ we derive from theorem 2.3 the following result:

THEOREM 2.4 [SV1]. *Let F be an algebraically closed field of characteristic zero. Let further $\mathcal{F}$ be a presheaf with transfers on the category Sch/F. Then both arrows in the diagram*

$$C_*(\mathcal{F}) = C_*(\mathcal{F}^{\sim}_{qfh}) \hookrightarrow \underline{C}_*(\mathcal{F}^{\sim}_{qfh}) \hookleftarrow \mathcal{F}^{\sim}_{qfh}$$

induce isomorphisms on $Ext^*(-, \mathbb{Z}/n)$. In particular, for any $n > 0$ we have canonical isomorphisms

$$H^*_{sing}(\mathcal{F}, \mathbb{Z}/n) = Ext^*_{qfh}(\mathcal{F}^{\sim}_{qfh}, \mathbb{Z}/n) = Ext^*_h(\mathcal{F}^{\sim}_h, \mathbb{Z}/n).$$

3 Sheaves of equidimensional cycles

Every qfh-sheaf has canonical transfers [SV1]. The Nisnevich sheaf $L(X)$ introduced in the first section was defined on the category Sm/F only, however it admits an extension to the category Sch/F. To do so one has to define a pullback of a cycle finite and surjective over the base scheme S in the case when S is not necessarily regular. This can be done using elementary Galois theory (see [SV1]). It turns out however that multiplicities of components are no longer integers, but might have the exponential characteristic of F in the denominator. Thus, denoting the exponential characteristic of F by p we get a qfh-sheaf $L(X)[1/p]$. Moreover this sheaf coincides with the free qfh-sheaf of $\mathbb{Z}[1/p]$-modules $\mathbb{Z}[1/p](X)$ generated by X. Applying Theorem 2.4 to the qfh-sheaf $\mathbb{Z}_{qfh}(X)$ we get the following corollary:

COROLLARY 3.1 [SV1]. *Let F be an algebraically closed field of characteristic zero, then for any $X \in Sch/F$ we have canonical isomorphisms*

$$H^*_{sing}(X, \mathbb{Z}/n) = H^*_{qfh}(X, \mathbb{Z}/n) = H^*_{et}(X, \mathbb{Z}/n).$$

For schemes over $\mathbb{C}$ we can do even better. Note that the simplicial abelian group $C_*(X)$ coincides with $Hom(\Delta^{\cdot}, \coprod_{d=0}^{\infty} S^d(X))^+$, where $S^d(X)$ is the dth symmetric power of X and $+$ stands for the group completion. By restricting the algebraic maps $\Delta^n \to S^d(X)$ to the usual topological simplex $\Delta^n_{top} \subset \Delta^n$ we get a canonical morphism of simplicial abelian groups

$$Hom(\Delta^{\cdot}, \coprod_{d=0}^{\infty} S^d(X))^+ \longrightarrow Hom_{top}(\Delta^{\cdot}_{top}, \coprod_{d=0}^{\infty} S^d(X(\mathbb{C})))^+$$

and hence the induced homomorphism on the homotopy groups $H^{sing}_*(X, \mathbb{Z}/n) \to H_*(X(\mathbb{C}), \mathbb{Z}/n)$.

COROLLARY 3.2 [SV1]. *The canonical homomorphism*

$$H^{sing}_*(X, \mathbb{Z}/n) \to H_*(X(\mathbb{C}), \mathbb{Z}/n)$$

is an isomorphism for any quasiprojective scheme $X \in Sch/\mathbb{C}$.

The sheaf $\mathbb{Z}[1/p]_{qfh}(X)$ is just one example of the sheaves of equidimensional cycles. The general construction of sheaves of relative cycles is given in [SV2]. In particular for any scheme $X \in Sch/F$ and any $t \geq 0$ one can consider a qfh-sheaf $z_t(X)$, which is characterized by the following property: for any normal irreducible scheme S one has $z_t(X)(S) =$ a free $\mathbb{Z}[1/p]$-module generated by closed reduced irreducible subschemes $Z \subset X \times S$ which are equidimensional of relative dimension t over S.

Note that the sheaf $z_0(X)$ coincides with $\mathbb{Z}[1/p]_{qfh}(X)$ when X is proper over F. If $X \in Sch/F$ is any separated scheme then one can choose an open embedding $X \hookrightarrow \overline{X}$ with $\overline{X}$ proper. Let X_∞ denote $\overline{X} \setminus X$ considered as a closed reduced subscheme of $\overline{X}$. The sequence of qfh-sheaves

$$0 \longrightarrow z_0(X_\infty) \longrightarrow z_0(\overline{X}) \longrightarrow z_0(X) \longrightarrow 0$$

is left-exact but not right-exact. However it becomes right-exact if one replaces all qfh-sheaves involved by the associated h-sheaves — see [SV2]. Using this fact and Theorem 2.2 we obtain

PROPOSITION 3.3. *$Ext^*_{qfh}(z_0(X), \mathbb{Z}/n) = Ext^*_h(z_0(X)^{\sim}_h, \mathbb{Z}/n) = H^*_c(X, \mathbb{Z}/n)$, where H^*_c stands for etale cohomology with compact supports.*

Applying Theorem 2.4 to the sheaf $z_0(X)$ and using Proposition 3.3 we get the following result:

THEOREM 3.4. *Let F be an algebraically closed field of characteristic zero and let $X \in Sch/F$ be any separated scheme. Then $H^{sing}_*(z_0(X), \mathbb{Z}/n)^\# = H^*_c(X, \mathbb{Z}/n)$, where $\#$ stands for dual (i.e. $Hom(-, \mathbb{Q}/\mathbb{Z})$).*

4 Higher Chow groups

Higher Chow groups introduced by Bloch in 1985 [B1] give an alternative approach to motivic cohomology. In this way we won't be able to construct the motivic complexes $\mathbb{Z}(n)$ but at least we get a rather direct approach to the computation of motivic cohomology groups $H^*_{\mathcal{M}}(X)$. What is more important, so far higher Chow groups give the only known construction of the spectral sequence relating motivic cohomology to algebraic K-theory.

Let $X \in Sch/F$ be any equidimensional scheme. Denote by $Z^q(X,n)$ the group of cycles of codimension q in $X \times \Delta^n$ that intersect properly all faces $X \times \Delta^m \subset X \times \Delta^n$. $Z^q(X,-)$ is a simplicial abelian group and higher Chow groups $CH^q(X,n)$ are defined as its homotopy groups (i.e. homology of the corresponding complex) — see [B1]. It's evident from the definition that $CH^q(X,0)$ coincides with the group $CH^q(X)$ of codimension q cycles modulo rational equivalence. The following is a list of further properties of higher Chow groups:

(4.1.1) **Homotopy invariance** (see [B1]). $CH^q(X,n) = CH^q(X \times \mathbb{A}^1, n)$

(4.1.2) **Functorial behaviour** (see [B1], [B2]). Any morphism $f : X \to Y$ of smooth equidimensional quasiprojective varieties induces homomorphisms $f^* : CH^q(Y,n) \to CH^q(X,n)$. Every proper morphism of equidimensional schemes $f : X \to Y$ induces homomorphisms $f_* : CH^q(X,n) \to CH^{q-d}(Y,n)$, where $d = \dim X - \dim Y$.

(4.1.3) **Higher Chow groups of codimension** ≤ 1 (see [B1]). For any smooth equidimensional scheme X we have:

$$CH^0(X,n) = \begin{cases} H^0(X,\mathbb{Z}) & n = 0 \\ 0 & n > 0 \end{cases}$$

$$CH^1(X,n) = \begin{cases} H^1(X,\mathcal{O}^*) & n = 0 \\ H^0(X,\mathcal{O}^*) & n = 1 \\ 0 & n > 1 \end{cases}$$

(4.1.4) **Localization exact sequence** (see [B2]). Let X be an equidimensional quasi-projective scheme and let $Y \subset X$ be a closed subscheme of pure codimension d. Then denoting the open subscheme $X \setminus Y$ by U we have a functorial long exact sequence

$$\cdots \to CH^q(Y,n) \to CH^{q+d}(X,n) \to CH^{q+d}(U,n) \to CH^q(Y,n-1) \to \cdots$$

(4.1.5) **Relations to Milnor** K**-theory** (see [NS])

$$CH^q(Spec\ F,n) = \begin{cases} 0 & n < q \\ K_q^M(F) & n = q \end{cases}$$

The following important result due to Bloch and Lichtenbaum relates higher Chow groups to algebraic K-theory:

THEOREM 4.2 [BL]. *For any field F there exists a natural spectral sequence of cohomological type*

$$E_2^{p,q} = CH^{-q}(Spec\ F, -p-q) \Longrightarrow K_{-p-q}(F)\,.$$

Higher Chow groups are related to motivic cohomology according to the following theorem of Voevodsky [V3].

THEOREM 4.3. *Let $X \in Sch/F$ be an equidimensional scheme over a field F of characteristic zero. Then higher Chow groups of X coincide with motivic Borel-Moore homology of X, so that if X is smooth then higher Chow groups coincide with motivic cohomology of X, more precisely: $CH^q(X,n) = H_{\mathcal{M}}^{2q-n}(X,\mathbb{Z}(q))$.*

Theorems 4.2 and 4.3 give a desired spectral sequence relating motivic cohomology to algebraic K-theory, at least in the case of fields. A different approach to the construction of the spectral sequence relating motivic cohomology to algebraic K-theory was suggested by Grayson [G]. The spectral sequence constructed by Grayson has extremely nice functorial properties. Its second term resembles motivic cohomology. Whether it coincides with motivic cohomology or not remains unclear.

Computation of higher Chow groups with finite coefficients for varieties over an algebraically closed field of characteristic zero is fullfilled using the following result:

THEOREM 4.4 [S2]. *Denote by $Z^q_{\text{equi}}(X,n)$ the subgroup of $Z^q(X,n)$ generated by cycles $Z \subset X \times \Delta^n$ that are equidimensional (of relative dimension* $\dim\ X - q$ *over Δ^n. Assume that X/F is an affine equidimensional variety and $q \leq \dim\ X$; then the evident embedding of complexes $Z^q_{\text{equi}}(X,-) \hookrightarrow Z^q(X,-)$ is a quasiisomorphism. The same statement holds for any quasiprojective scheme X if F is a field of characteristic zero.*

COROLLARY 4.5. *In conditions of Theorem 4.4 the group $CH^q(X,n)$ coincides with $H_n^{sing}(z_{\dim X - q}(X))$.*

Assume that X is an affine equidimensional variety over an algebraically closed field of characteristic zero F and set $d = \dim\ X$. According to Corollary 4.5 the higher Chow groups of X with finite coefficients $CH^d(X,n;\mathbb{Z}/m)$ coincide with $H_n^{sing}(z_0(X),\mathbb{Z}/m)$. Dual to the last group coincides with $H^n_{sing}(z_0(X),\mathbb{Z}/m) = H^n_c(X,\mathbb{Z}/m)$ and because this group is finite we conclude that $CH^d(X,n;\mathbb{Z}/m) = H^n_c(X,\mathbb{Z}/m)^\#$. Using the localization exact sequence this computation generalizes to any quasiprojective variety. Furthermore, for any $q \geq d$ we may apply the previous computation to $X \times \mathbb{A}^{q-d}$. Using the homotopy invariance of higher Chow groups we finally come to the following theorem:

THEOREM 4.6 [S2]. *Let $X \in Sch/F$ be an equidimensional quasiprojective scheme over an algebraically closed field of characteristic zero F. Set $d = \dim\ X$. Then for any $q \geq d$ we have:*

$$CH^q(X,n;\mathbb{Z}/m) = H_c^{2(d-q)+n}(X,\mathbb{Z}/m(d-q))^\#\,.$$

If furthermore X is smooth then this formula simplifies to

$$CH^q(X,n;\mathbb{Z}/m) = H^{2q-n}_{et}(X,\mathbb{Z}/m(q)).$$

Let $X \in Sch/\mathbb{C}$ be a smooth variety of dimension ≤ 2. All higher Chow groups of X may be expressed in terms of etale cohomology: if $q \leq 1$ we may use (4.1.3) and if $q \geq 2$ we may use Theorem 4.6. Because both higher Chow groups and etale cohomology respect filtered limits the same relationship holds for the function field $\mathbb{C}(X)$. Consider now the spectral sequence of Bloch and Lichtenbaum

$$E_2^{p,q} = CH^{-q}(Spec\ \mathbb{C}(X), -p-q; \mathbb{Z}/m) \Longrightarrow K_{-p-q}(\mathbb{C}(X), \mathbb{Z}/m).$$

All differentials in this spectral sequence are zero by dimension considerations. Using further easy multiplicative properties of the spectral sequence we get the following canonical direct sum decompositions:

$$\begin{aligned} K_0(\mathbb{C}(X),\mathbb{Z}/m) &= H^0_{et}(\mathbb{C}(X),\mathbb{Z}/m) \\ K_{2k}(\mathbb{C}(X),\mathbb{Z}/m) &= H^0_{et}(\mathbb{C}(X),\mathbb{Z}/m(k)) \oplus H^2_{et}(\mathbb{C}(X),\mathbb{Z}/m(k+1)), \quad k>0 \\ K_{2k+1}(\mathbb{C}(X),\mathbb{Z}/m) &= H^1_{et}(\mathbb{C}(X),\mathbb{Z}/m(k+1)). \end{aligned}$$

From these computations we conclude that Bott multiplication $\beta : K_i(\mathbb{C}(X),\mathbb{Z}/m) \to K_{i+2}(\mathbb{C}(X),\mathbb{Z}/m)$ is an isomorphism in degrees $i > 0$ (in all degrees if X is a curve). Using the localization exact sequence in algebraic K-theory we derive immediately from the previous remark that Bott multiplication $\beta : K'_i(X,\mathbb{Z}/m) \to K'_{i+2}(X,\mathbb{Z}/m)$ is an isomorphism for $i \geq \dim\ X$ for any irreducible scheme X of dimension ≤ 2. According to the theorem of Thomason [Th] this implies the following theorem:

Theorem 4.7 (Quillen-Lichtenbaum conjecture for curves and surfaces). *Let $X \in Sm/\mathbb{C}$ be a smooth variety of dimension ≤ 2. Then the canonical homomorphism $K_i(X,\mathbb{Z}/m) \to K_i^{top}(X(\mathbb{C}),\mathbb{Z}/m)$ is an isomorphism when $i \geq \dim\ X$.*

References

[B1] S. Bloch, *Algebraic cycles and higher K-theory*, Adv. in Math. **61** (1986), 267–304.
[B2] S. Bloch, *The moving lemma for higher Chow groups*, J. Alg. Geom., to appear.
[BL] S. Bloch and S. Lichtenbaum, *A spectral sequence for motivic cohomology*, preprint 1994.
[FG] E. Friedlander and O. Gabber, *Cycle spaces and intersection theory*, in Topological Methods in Modern Mathematics (1993), pp. 325–370.
[FL] E. Friedlander and H. B. Lawson, *A theory of algebraic cocycles*, Ann. of Math. (2) **136** (1992), 361–428.
[FV] E. Friedlander and V. Voevodsky, *Bivariant cycle cohomology*, preprint 1994.
[G] D. Grayson, *Weight filtration via commuting automorphisms*, preprint 1993.

[L] S. Lichtenbaum, *Suslin homology and Deligne 1-motives*, preprint 1992.

[NS] Yu. Nesterenko and A. Suslin, *Homology of the general linear group over a local ring and Milnor K-theory*, Izv. Akad. Nauk SSSR **53** (1989), 121–146; English translation in Soviet Math. Iz. (vuz).

[S1] A. Suslin, *Algebraic K-theory of fields*, in Proceedings of ICM-86 Berkeley, pp. 222–244.

[S2] ———, *Higher Chow groups and etale cohomology*, preprint 1993.

[SV1] A. Suslin and V. Voevodsky, *Singular homology of abstract algebraic varieties*, Invent. Math., to appear.

[SV2] ———, *Relative cycles and Chow sheaves*, preprint 1994.

[Th] R. W. Thomason, *Algebraic K-theory and etale cohomology*, Ann. Sci. École Norm. Sup. (4) **13** (1985), 437–552.

[V1] V. Voevodsky, *Homology of schemes and covariant motives*, Harvard Univ. Ph.D. thesis 1992.

[V2] ———, *Homology of schemes II*, preprint 1993.

[V3] ———, *Homology of schemes III*, preprint 1994.

Modules of Covariants

MICHEL VAN DEN BERGH

Limburgs Universitair Centrum, Dept. WNI, Univ. Campus
B-3590 Diepenbeek, Belgium

1. The Hochster-Roberts theorem

Let G be a reductive group over $\mathbb{C}$ and let W be a finite-dimensional representation of W. Set $R = SW$, the symmetric algebra of W. Then it follows from the famous Hochster-Roberts theorem [7] that R^G is Cohen-Macaulay. In this setting this means that there exist homogeneous $r_1, \dots, r_h \in R^G$ such that

- The $\mathbb{C}$-algebra R_0 generated by $r_1, \dots, r_h$ is a polynomial algebra.
- R^G is a finitely generated free R_0-module.

Hence one may view this result as giving, at least in principle, a unique representation for each invariant. Indeed if $m_1, \dots, m_p$ is a basis of R^G over R_0 then each $f \in R^G$ may be written uniquely as $\sum P_i(r_1, \dots, r_h)m_i$ where the P_i are polynomials in h variables.

The original Hochster-Roberts theorem had a long and complicated proof, based upon reduction mod p. Subsequent simplifications by Kempf yielded the following result.

THEOREM 1.1. [9] *Assume that there is a pure homomorphism $S \longrightarrow T$ between commutative rings of finite type over a field. Then if T is regular, S is Cohen-Macaulay.*

Pure means that for any $M \in S$-mod the induced map $M \rightarrow T \otimes_S M$ is injective. This happens for example if S is a direct summand of T as S-module. This is the case relevant for invariant theory. Indeed, the Reynolds operator defines an R^G-linear splitting for the inclusion $R^G \hookrightarrow R$.

Nowadays there are short proofs of stronger versions of the Hochster-Roberts theorem. One of these is due to Boutot [2]. He proves that if T/S is pure and T has rational singularities then S has rational singularities. This implies the Hochster-Roberts theorem as there are implications

$$\text{regular} \Rightarrow \text{rational singularities} \Rightarrow \text{Cohen-Macaulay}.$$

Boutot's proof, which is very short, uses the deep Grauert-Riemenschneider vanishing theorem, whose proof uses analysis.

The author is a senior researcher at the NFWO.

Proceedings of the International Congress
of Mathematicians, Zürich, Switzerland 1994

Another new proof for the Hochster-Roberts theorem is due to Hochster and Huneke [6] who prove that "F-regularity" is preserved under pure maps. Again this implies the original Hochster-Roberts theorem as

$$\text{regular} \Rightarrow \text{F-regular} \Rightarrow \text{Cohen-Macaulay}.$$

This new approach is still based upon reduction mod p. For a self-contained summary of the proof of Hochster and Huneke, see [11].

One has the impression that the Hochster-Roberts theorem is a very subtle result. This feeling is strengthened when one considers the obvious generalization to covariants which turns out to fail rather drastically (Theorem 4.3). Therefore I would propose the following question.

QUESTION 1.2. Does there exist a proof for the Hochster-Roberts theorem that does not use reduction mod p and that is completely algebraic (and hence does not use results such as the Grauert-Riemenschneider vanishing theorem) ?

In the setting of invariant theory the answer is known to be yes in the case that G is a torus [5], [16] and in the case that W is sufficiently big [18].

2. Covariants

We let G, W be as before and we let U be another finite-dimensional G-representation. Then the module of covariants associated to U is defined as $(R \otimes_{\mathbb{C}} U)^G$. It is easy to see that this is a finitely generated R^G-module. The definition is clearly compatible with direct sums, so henceforth we assume that U is irreducible. An element of some $(R \otimes U)^G$ is called a covariant. It corresponds to an equivariant polynomial map $W^* \to U$.

The notion of covariants stems from the classical invariant theory of forms. In modern terms this would correspond to $G = SL_n(\mathbb{C})$, $W = (S^u\mathbb{C}^n)^*$, $U = S^v\mathbb{C}^n$. One of the reasons that classical invariant theorists studied covariants rather than just invariants is probably that these allowed for greater flexibility. Indeed the forms themselves represent canonical covariants, and from these other covariants can be derived using equivariant differential operators such as transvectants.

More recently the interest in modules of covariants has revived because of their connection with the theory of inhomogeneous linear diophantine equations [16] and with the theory of PI-algebras and trace rings [13].

Given the Hochster-Roberts theorem, and the fact that $R \otimes U$ is a free R-module, it is rather natural to pose the following question.

QUESTION 2.1. Is $(R \otimes U)^G$ a Cohen-Macaulay R^G-module?

We are asking whether $(R \otimes U)^G$ is a finitely generated free module over a polynomial ring in R^G. It is rather easy to see that the answer to this question is "no" in general.

EXAMPLE 2.2. Let $G = T = \mathbb{C}^*$, a one-dimensional torus and let $\chi : T \to \mathbb{C}^*$: $\alpha \mapsto \alpha$ be the identity character. Define $U = L_{\chi^{-1}}$, $W = L_\chi \oplus L_\chi \oplus L_{\chi^{-1}}$ where L_χ is the one-dimensional T-representation associated to χ. Then $R = \mathbb{C}[x, y, z]$, $M \overset{\text{def}}{=} U \otimes R = \mathbb{C}[x, y, z]$ where T acts on R and M as follows: take $\alpha \in T$, $f \in R$,

$g \in M$. Then $\alpha \cdot f = f(\alpha x, \alpha y, \alpha^{-1} z)$, $\alpha \cdot g = \alpha^{-1} f(\alpha x, \alpha y, \alpha^{-1} z)$ $R^T = \mathbb{C}[xz, yz] \cong \mathbb{C}[s,t]$, $M^T = xR^T + yR^T = z^{-1}(xzR^T + yzR^T) \cong (s,t) \subset \mathbb{C}[s,t]$. So M^T is not free over R^T, which is a polynomial ring, and hence M^T is not Cohen-Macaulay.

3. Isotypical components

There is a direct sum decomposition $R = \oplus R^G_\chi$ where the sum is over all irreducible characters of G and where R^G_χ denotes the sum of all irreducible subrepresentations of R with character χ. Let χ be the character of an irreducible representation U. Then it is easily seen that $R^G_\chi = (R \otimes U^*)^G \otimes U$ so that the Cohen-Macaulayness of $(R \otimes U^*)^G$ is equivalent to that of R^G_χ.

4. Some basic results

In this section we indicate the extremal situations with regard to Question 2.1. Indeed, we will show that sometimes all modules of covariants for a given (G, W) are Cohen-Macaulay, but that usually this is only true for a finite number of them. The latter result was proved by Brion in [3]. We give a somewhat weaker version with a simple proof.

THEOREM 4.1. [12, Section 5.1] *Every R^G_χ is Cohen-Macaulay if and only if the quotient map from* Spec R *to* Spec R^G *is equidimensional (that is, the generic fiber has the same dimension as the zero-fiber).*

Proof. Assume that Spec $R \to$ Spec R^G is equidimensional. Let $R_0 \subset R^G$ be a graded polynomial ring such that R^G/R_0 is finite. Then SpecR$\to$ Spec R_0 is an equidimensional map between smooth varieties, and hence is flat by [1]. Because R^G_χ is a direct summand of R this implies that R^G_χ is flat over R_0, and because R^G_χ is in addition finitely generated this yields that R^G_χ/R_0 is free. The converse is proved by reversing the above argument. □

REMARK 4.2.
- The above result applies when G is finite.
- If G is semisimple then, up to trivial summands, there are only a finite number of W such that Spec $R \to$ Spec R^G is equidimensional [12, Section 5.9].
- If Spec $R \to$ Spec R^G is equidimensional and G is connected then it is conjectured that R^G is itself a polynomial ring. This would imply that all modules of covariants are actually free, and hence R/R^G would also be free.

Now we state a weak version of Brion's finiteness result. For the full version we refer to [3, Section 4.2]. Define $X =$ Spec $R(\cong W^*)$ and $X_s = \{x \in X \mid G_x = \text{triv}\}$.

THEOREM 4.3. *Assume that the generic orbit in X is closed and that* $\operatorname{codim}(X - X_s, X) \geq 2$. *Then there are only a finite number of irreducible U such that $(R \otimes U)^G$ is Cohen-Macaulay.*

Proof. Let $R_0 \subset R^G$ be a graded polynomial ring such that R^G/R_0 is finite. Define $\Omega = \operatorname{Hom}(R^G, R_0)$. This is in a natural way a graded R^G-module that is isomorphic to the dualizing module of R^G, suitably shifted. It follows from [10] that $\Omega = (R \otimes L_\chi)^G(q)$ for some $\chi \in \operatorname{Hom}(G, \mathbb{C}^*)$ and some $q \in \mathbb{N}$.

Let $M = (R \otimes U)^G$. Then $\tilde{M} = \operatorname{Hom}_{R_0}(M, R_0)$ is the (shifted) Cohen-Macaulay dual of M. We have

$$\begin{aligned} \tilde{M} &= \operatorname{Hom}_{R_0}(M, R_0) \\ &= \operatorname{Hom}_{R^G}(M, \Omega) \\ &= \operatorname{Hom}_{R^G}((R \otimes U)^G, (R \otimes L_\chi)^G(q)) \\ &= (R \otimes U^* \otimes L_\chi)^G(q) . \end{aligned} \tag{1}$$

The last equality is true on $X_s /\!\!/ G$ by descent and because of the fact that the quotient map $X_s \to X_s /\!\!/ G$ is a principal G-bundle (this follows from the Luna slice theorem). Then this equality is true on $X /\!\!/ G$ because $X \to X /\!\!/ G$ contracts no divisor and hence all modules of covariants are reflexive [3, Section 1.3].

The only thing we actually need from (1) is that the dual of an arbitrary module of covariants lives in degree $\geq -q$.

Assume now that M is Cohen-Macaulay. Then M is graded free over R_0. So as a graded R_0-module

$$M = \bigoplus_{i=1}^{n} R_0(-a_i)$$

for suitable $a_i \in \mathbb{N}$. Because $\tilde{M} = \operatorname{Hom}_{R_0}(M, R_0) = R_0(a_i)$ lives in degrees $\geq -q$ we deduce that $a_i \leq q$, from which it follows that M contains an element of degree $\leq q$. Hence U^* is contained in $\oplus_{i=0}^{q} S^i W$. So there are only a finite number of possibilities. □

REMARK 4.4.
- In [19], in the case where G is a torus, necessary and sufficient conditions are given for the existence of only a finite number of Cohen-Macaulay modules of covariants.
- The proof of Theorem 4.3 yields a necessary condition for $(R \otimes U)^G$ to be Cohen-Macaulay. This condition is in general far from sufficient. However consider the following case [3] $G = \mathrm{Sl}_n(\mathbb{C})$, $W = (\mathbb{C}^n)^{n+1}$. Then the hypothesis of Theorem 4.3 is satisfied, and furthermore R^G is a polynomial ring. Thus, we may take $q = 0$. Then it follows that the only U for which $(R \otimes U)^G$ is Cohen-Macaulay is the trivial representation.
- If G is semisimple and connected, then, up to trivial summands, there are only a finite number of W for which the hypothesis of Theorem 4.3 is not satisfied.

5. The torus case

In this section we assume that $G = T = (\mathbb{C}^*)^s$ is an s-dimensional torus. Let $\gamma_1, \gamma_2, \ldots, \gamma_d \in X(T) = \operatorname{Hom}(T, \mathbb{C}^*)$ be the weights of W. So $R = \mathbb{C}[x_1, \ldots, x_d]$ with, for $z \in T$, $z \cdot x_i = \gamma_i(z) x_i$. Let $\chi \in X(T)$. Then there exist $\Phi = (\phi_{ij})_{ij}$, $\alpha = (\alpha_i)_i$ such that $\gamma_i(z_1, \ldots, z_s) = z_1^{\phi_{1i}} \cdots z_s^{\phi_{si}}$, $\chi(z_1, \ldots, z_s) = z_1^{\alpha_1} \cdots z_s^{\alpha_s}$ Then it follows that

$$\begin{aligned} R^T &= \oplus_{\Phi\beta = 0, \beta \geq 0} \mathbb{C} x^{\beta_1} \cdots x_d^{\beta_d} \subset R \\ R_\chi^T &= \oplus_{\Phi\beta = \alpha, \beta \geq 0} \mathbb{C} x^{\beta_1} \cdots x_d^{\beta_d} \subset R . \end{aligned}$$

So R^T, R^T_χ have a basis that is indexed by the solutions of a system of respectively homogeneous and inhomogeneous linear diophantine equations. This observation was used by Stanley to make a connection between combinatorics and commutative algebra [16].

Because R is $\mathbb{Z}^d$-graded in the natural way, and this grading is compatible with the T-action on R, we may define the $\mathbb{Z}^d$-graded Hilbert series of R^T_χ,

$$F(R^T_\chi, u) = \sum_{\beta\in\mathbb{Z}^d} \dim(R^T_\chi)_\beta u_1^{\beta_1}\cdots u_d^{\beta_d} = \sum_{\Phi\beta=\alpha} u_1^{\beta_1}\cdots u_d^{\beta_d}\,.$$

Thus, $F(T^T_\chi, u)$ is the generating function for the solutions to $\Phi\beta = \alpha$. It is easily seen to be a rational function.

The Hochster-Roberts theorem, and also the proof of Brion's result, suggest that in order for R^T_χ to be Cohen-Macaulay, χ should in some sense be small compared to W. An appropriate notion of smallness was found by Stanley.

Definition 5.1. A character $\chi \in X(T)$ is strongly critical for (T, W) if and only if $\chi = \sum_{i=1}^d \mu_i\gamma_i$ in $X(T)\otimes_\mathbb{Z}\mathbb{R}$, with $\mu_i \in]-1, 0]$.

Theorem 5.2. [15] *Assume that χ is strongly critical for (T, W). Then*

(1) *R^T_χ is Cohen-Macaulay.*
(2) *Assume that the generic orbit in W^* is closed. Then the following identity holds for the $\mathbb{Z}^d$-graded Hilbert series of R^T_χ (the "Monster reciprocity theorem")*

$$F(R^T_\chi, u^{-1}) = \pm u_1\cdots u_d F(R^T_{-\chi-\sum\gamma_i}, u)\,. \tag{2}$$

The reciprocity theorem implies "reciprocity" results for the solutions of $\Phi\beta = \alpha$ [14].

Stanley proved Theorem 5.2 (2) first in [14] by writing $F(R^T_\chi, u)$ as an integral over a complex torus (the "Molien-Weyl" formula). Then repeated application of the residue theorem yields the result.

In [15], [16] the results 5.2 (1) and 5.2 (2) are proved using commutative algebra. It turns out that 5.2 (1) implies 5.2 (2).

Remark 5.3. The terminology of strongly critical weights is mine, but the notion was introduced by Stanley in [15] as a special case of so-called critical weights, a slightly more general concept.

6. Strongly critical characters in the general case

In this section we let G be general again and we let $T \subset G$ be a maximal torus. Let $\chi : G \to \mathbb{C}$ be an irreducible character and let $\chi \mid T = \chi_1 + \cdots + \chi_l$ where the χ_i are characters of T. Let Ω be the set of roots of G.

Definition 6.1. χ is strongly critical for (G, W) if and only if, for all $S \subset \Omega$ and for all $i \in \{1, \ldots d\}$, it is true that

$$\chi_i - \sum_{\rho\in S}\rho \tag{3}$$

is strongly critical for (T, W).

Again the strongly critical characters are those that are sufficiently close to the trivial representation.

The motivation for Definition 6.1 is the following result by Stanley.

THEOREM 6.2. [15] *Assume that χ is strongly critical for (G,W) and that the generic T-orbit in W^* is closed. Then the $\mathbb{Z}$-graded Hilbert series of R^G_χ has the property*

$$F(R^G_\chi, u^{-1}) = \pm u^d F(R^G_\psi, u) \tag{4}$$

where $\psi(g) = \chi(g^{-1})\lambda(g^{-1})$, with λ the character of $\Lambda^d W$ (so in the notation of Section 5, $\lambda = \sum \gamma_i$).

Proof. We will only sketch the proof. If we expand the integrand in the Molien-Weyl formula for $F(R^G_\chi, u)$, then we see that it is a sum of $F(R^T_\mu, u)$ where the μ's are weights of the form (3). Then (2) applies to each of these terms, and one obtains (4) by summing. □

If one compares Theorem 5.2 with Theorem 6.2 then one is naturally lead to the following conjecture.

CONJECTURE 6.3. If χ is strongly critical for (G,W) then R^G_χ is Cohen-Macaulay.

This conjecture was proved by the author under some reasonable additional hypotheses [18]. For example it is true if G is semisimple, and the generic orbit of W^* has a finite stabilizer.

7. Local cohomology

Let us specialize to our setting the standard local cohomology criterion for a module to be Cohen-Macaulay.

LEMMA 7.1. *R^G_χ is Cohen-Macaulay if and only if $H^i_{(R^G)^+}(R^G_\chi) = 0$ for $i = 0,\dots,h-1$, where h is the Krull dimension of R^G.*

There are many equivalent definitions of local cohomology. One of them is as follows. Choose $f_1,\dots,f_k$ homogeneous such that $\mathrm{rad}(f_1,\dots,f_k) = (R^G)^+$. Then $H^*_{(R^G)^+}(R^G_\chi)$ is the homology of the complex

$$0 \to R^G_\chi \to \bigoplus_i (R^G_\chi)_{f_i} \to \bigoplus_{j>i} (R^G_\chi)_{f_i f_j} \to \cdots \to (R^G_\chi)_{f_1\cdots f_k} \to 0 \tag{5}$$

with standard alternating boundary maps. Suppose now that $G = T$ is a torus. Then Stanley found a combinatorial way to describe (5). Because T is a torus we may choose the $f_1,\dots,f_k$ to be monomials and then (5) is compatible with the $\mathbb{Z}^d$-grading on R. Furthermore if we fix $\beta \in \mathbb{Z}^d$ and we restrict (5) to degree β then it is easy to see that one obtains the chain complex of a simplicial complex. Then further results in algebraic topology are used to study these simplicial complexes. This eventually leads to a precise criterion for R^T_χ to be Cohen-Macaulay, and also to the proof of Theorem 5.2.

As beautiful as this method is, it has the drawback of being restricted to the torus case. Our goal is to study the general case; thus, a different approach is needed.

Let G be general again. Comparing (5) with the complex

$$L^{\cdot}: \quad 0 \to R \to \bigoplus_{i} R_{f_i} \to \bigoplus_{j>i} R_{f_i f_j} \to \cdots \to (R^G_\chi)_{f_1\cdots f_k} \to 0$$

we find that $H^*_{(R^G)^+}(R^G_\chi) = H^*(L^{\cdot})^G_\chi = H^*_I(R)^G_\chi$ where $I = \operatorname{rad}(f_1, \dots, f_k)$. This formula was stated in [9]. It essentially says that $H^*_I(R)$ encodes the information about the Cohen-Macaulay property for all modules of covariants. Of course, given the difficulty of computing $H^*_I(R)$ for an arbitrary ideal $I \subset R$, this doesn't appear to be much progress. Things become clearer when we go to the associated varieties. Put $X \cong \operatorname{Spec} R(\cong W^*)$. Then we have $V(I) = X^u$ where X^u is the so-called null-cone or unstable locus in X. That is

$$X^u = \{x \in X \mid o \in \overline{Gx}\}.$$

So we have $H^*_I(R) = H^*_{X^u}(X, \mathcal{O}_X)$. This leads us to the following more general formulation of our original question.

QUESTION 7.2. What is the structure of $H^*_{X^u}(X, \mathcal{O}_X)$ as G-module?

At this stage, it is useful to make another generalization, which turns out to clarify matters. Let $\mathcal{D}_X$ be the sheaf of differential operators on X. Then it is standard that the (sheafifications of) $H^*_{X^u}(X, \mathcal{O}_X)$ are holonomic $\mathcal{D}_X$-modules with regular singularities. Thus we ask

QUESTION 7.3. What is the structure of $H^*_{X^u}(X, \mathcal{O}_X)$ as $(G, \mathcal{D}_X)$-module ?

A very useful fact is that if G is connected then the G-structure on $H^*_{X^u}(X, \mathcal{O}_X)$ is encoded in the $\mathcal{D}_X$-structure [20, Section 3.1].

8. Geometric invariant theory

Insight into the structure of X^u is provided by geometric invariant theory. Let $T \subset B \subset G$ be a maximal torus and a Borel subgroup of G. Let $\lambda \in Y(T)$ be a one-parameter subgroup of T. Define

$$X_\lambda = \{x \in X \mid \lim_{t \to 0} \lambda(t)x = 0\}$$
$$P_\lambda = \{g \in G \mid \lim_{t \to 0} \lambda(t)g\lambda(t)^{-1} \text{ exists}\}.$$

It is easy to see that X_λ is a linear subspace of X and $P_\lambda X_\lambda = X_\lambda$. In addition it was shown by Kempf [8] that P_λ is a parabolic subgroup of G.

Now the Hilbert-Mumford theorem yields that

$$X^u = \bigcup_{P_\lambda \supset B} GX_\lambda .$$

Furthermore the projection map $G \times^{P_\lambda} X_\lambda \to GX_\lambda$ is usually a resolution of singularities.

9. The torus case revisited

Here we indicate how one may answer Question 7.3 in the torus case. The notation is as in Section 5 and Section 7. By Section 8: $X^u = \bigcup_{\lambda \in Y(T)} X_\lambda$. Let $\gamma_1, \dots, \gamma_d$ be the weights of W, with corresponding weight vectors $w_1, \dots, w_d$. Then $X_\lambda \subset X(\cong W^*)$ is the linear span of those w_i^* for which $\langle \lambda, \gamma_i \rangle < 0$, where $\langle\ ,\ \rangle$ denotes the canonical pairing between $Y(T)$ and $X(T)$. This description of X_λ still makes sense if $\lambda \in Y(T)_{\mathbb{R}}$, so we extend the definition of X_λ accordingly. If $U \subset Y(T)_{\mathbb{R}}$ then we put $X_U = \bigcup_{\lambda \in U} X_\lambda$. If $B \subset Y(T)_{\mathbb{R}}$ is the open unit ball for some norm on $Y(T)_{\mathbb{R}}$ then with this notation $X^u = X_B$. The key lemma is the following.

LEMMA 9.1. *Assume that $U_1, U_2 \subset Y(T)_{\mathbb{R}}$ are convex and that $U_1 \cup U_2$ is also convex. Then*

$$X_{U_1} \cap X_{U_2} = X_{U_1 \cap U_2}\,.$$

Lemma 9.1 leads to a Mayer-Victoris sequence

$$\to H^i_{X_{U_1\cap U_2}}(X, \mathcal{O}_X) \to H^i_{X_{U_1}}(X, \mathcal{O}_X) \oplus H^i_{X_{U_2}}(X, \mathcal{O}_X) \to H^i_{X_{U_1\cup U_2}}(X, \mathcal{O}_X) \to$$

This observation makes it plausible that by "cutting up" B in small pieces, the computation of $H^*_{X^u}(X, \mathcal{O}_X)$ may somehow be reduced to that of $H^*_{X_\sigma}(X, \mathcal{O}_X)$ for $\sigma \subset B$ so small that there exist $\lambda \in \sigma$ such that $X_\lambda = X_\sigma$. The appropriate technical tool to do this is to use a certain spectral sequence which is shown to degenerate [19].

Now we will state the main result from [19] but first we have to introduce some more notation. We set $s = \dim T$ and if $\lambda, \mu \in Y(T)_{\mathbb{R}}$ then we set $\lambda \sim \mu$ if $X_\lambda = X_\mu$. We let Λ be a set of representants for $B/\sim$ and if $\lambda \in \Lambda$ then we set $B_\lambda = \{\mu \in B \mid \mu \sim \lambda\}$ and $\Phi_\lambda = \overline{B_\lambda} - B_\lambda$. It is easy to see that Φ_λ may be given the structure of a finite CW-complex.

THEOREM 9.2. *There exists a filtration on $H^i_{X^u}(X, \mathcal{O}_X)$ as $(T, \mathcal{D}_X)$-module such that*

$$\operatorname{gr} H^i_{X^u}(X, \mathcal{O}_X) = \oplus_{\lambda \in \Lambda} H^{d_\lambda}_{X_\lambda}(X, \mathcal{O}_X)^{\oplus u_{i,\lambda}}$$

where $u_{i,\lambda} = \dim \tilde{H}^{i+s-d_\lambda-1}(\Phi_\lambda)$, $d_\lambda = \operatorname{codim}(X_\lambda, X)$.

To apply the above theorem, one needs to know the T-action on $H^{d_\lambda}_{X_\lambda}(X, \mathcal{O}_X)$ but this can be very easily computed [19]. This yields a description of the T-structure of $H^*_{X^u}(X, \mathcal{O}_X)$ from which we can obtain the local cohomology modules $H^*_{(R^T)^+}(R^T_\chi)$. Inspection reveals that there should be a connection between the Φ_λ and the simplicial and polyhedral complexes introduced by Stanley, but the nature of this connection is not at all clear.

EXAMPLE 9.3. Let $T = (\mathbb{C}^*)^2$. Then $X(T), Y(T)$ may be identified with $\mathbb{Z}^2$. Take

$$W = L^{\oplus 2}_{(-1,0)} \oplus L^{\oplus 2}_{(0,1)} \oplus L^{\oplus 2}_{(1,1)} \oplus L^{\oplus 2}_{(1,-1)}\,.$$

In Figure 1 we have given a graphical representation of this example. The weights of W have been indicated by fat dots. Hence there are four fat dots, each representing a weight of multiplicity two. The interior of the white area represents those χ such that R^T_χ is Cohen-Macaulay. In particular there are only a finite

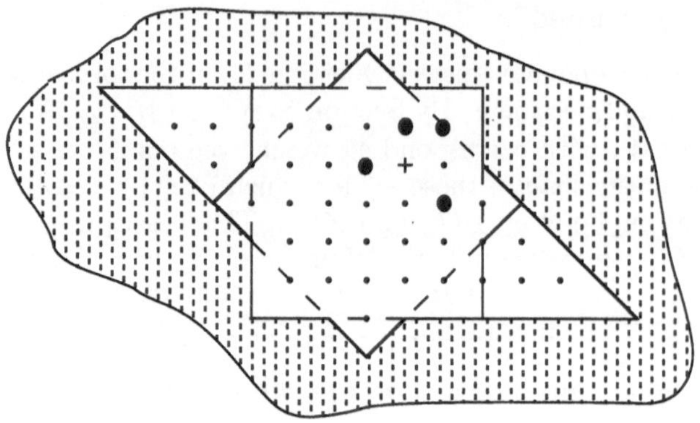

FIGURE 1

number of such weights, in accordance with Theorem 4.3. The interior of the region bounded by the dashed lines is the strongly critical weights. Because this is not all of the white area, we see that the converse to 5.2 (1) is false.

10. The general case

Theorem 9.2 may to some extent be generalized to arbitrary G. Let the notation be as in Section 7 and Section 8. We need the following conditions.

CONDITION (*).

- If $\lambda, \mu \in Y(T)$ such that $X_\lambda \neq X_\mu$ then $GX_\lambda \neq GX_\mu$.
- If $\lambda \in Y(T)$ then there exist $\mu \in Y(T)$ with $X_\mu = X_\lambda$ such that the projection map $G \times^{P_\mu} X_\mu \to GX_\mu$ is small. (A map $\pi : Y \to X$ is said to be small if for any $n > 0$, $\operatorname{codim}\{y \in Y \mid \dim \pi^{-1} y \geq n\} > 2n$.)

Then we have [20]:

THEOREM 10.1. *Assume that condition (*) holds. Then there is a filtration on* $H^i_{X^u}(X, \mathcal{O}_X)$ *as* $(G, \mathcal{D}_X)$*-module such that*

$$\operatorname{gr} H^i_{X^u}(X, \mathcal{O}_X) = \oplus_{\lambda \in Y(T)} \mathcal{L}(GX_\lambda, X)^{\oplus u_{i,\lambda}}$$

where $u_{i,\lambda} \in \mathbb{N}$ *and* $\mathcal{L}(GX_\lambda, X)$ *is the unique simple subquotient of* $H^*_{GX_\lambda}(X, \mathcal{O}_X)$ *with support equal to* GX_λ.

REMARK 10.2.

(1) As in Theorem 9.2 one may give explicit formulas for the $u_{i,\lambda}$ in terms of the reduced cohomology groups of certain finite CW-complexes.
(2) The G-structure of $\mathcal{L}(GX_\lambda, X)$ may be explicitly computed.

11. Some examples

We conclude with a few examples that can be computed using Theorem 10.1 and Remark 10.2.

First let $G = \mathrm{Sl}_2(\mathbb{C})$ and $W = \oplus_{i=1}^e S^{d_i}\mathbb{C}^2$ with all $d_i > 0$. Define for $n \geq 0$

$$s^{(n)} = \begin{cases} \dfrac{(n+1)^2}{4} & \text{if } n \text{ is odd} \\ \dfrac{n(n+2)}{4} & \text{if } n \text{ is even} \end{cases}$$

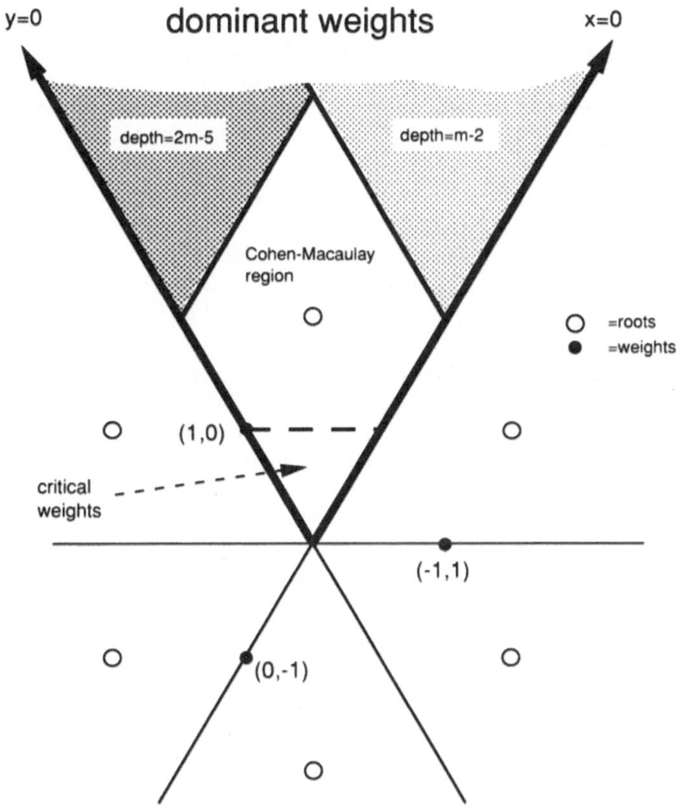

FIGURE 2

and set $s = \sum s^{(d_i)}$. It follows from Stanley's criterion that $(R \otimes S^u V)^G$ is Cohen-Macaulay if $u < s - 2$. In this case it turns out that the converse is almost true. We separate the following cases:

(A) $W = \mathbb{C}^2, S^2\mathbb{C}^2, \mathbb{C}^2 \oplus \mathbb{C}^2, \mathbb{C}^2 \oplus S^2\mathbb{C}^2, S^2\mathbb{C}^2 \oplus S^2\mathbb{C}^2, S^3\mathbb{C}^2, S^4\mathbb{C}^2$;
(B) All d_i are even and u is odd;
(C) All other cases.

THEOREM 11.1. [4], [17], [20] *In case (A) the module $(R \otimes S^u\mathbb{C}^2)^G$ will always be Cohen-Macaulay. In case (B) $(R \otimes S^u\mathbb{C}^2)^G = 0$. In case (C) $(R \otimes S^u\mathbb{C}^2)^G$ is Cohen-Macaulay if and only if Stanley's criterion holds.*

Now let us look at the case $G = \mathrm{Sl}_3(\mathbb{C})$, $W = (\mathbb{C}^3)^n$. Let ω_1, ω_2 be the fundamental weights and let χ be a character of G with corresponding highest weight $x\omega_1 + y\omega_2$, $x \geq 0$, $y \geq 0$. Then R_χ^G is Cohen-Macaulay if $x < m - 3$, $y < m - 3$. On the other hand χ is strongly critical for (G, W) if $x + y + 4 < m$. This is represented graphically in Figure 2 which is for $m = 5$.

References

[1] A. Altman and S. Kleiman, Introduction to Grothendieck duality, Lecture Notes in Math., **146**, Springer-Verlag, Berlin and New York, 1970.

[2] J. F. Boutot, *Singularités rationelles et quotients par les groupes réductifs*, Invent. Math. **88** (1987), 65–68.

[3] M. Brion, *Sur les modules de covariants*, to appear in Ann. Sci. École Norm. Sup. (4).

[4] B. Broer, *On the generating functions associated to a system of binary forms*, Indag. Math., (New Ser.) **1** (1990), no. 1, 15–25.

[5] M. Hochster, *Rings of invariants of tori, Cohen-Macaulay rings generated by monomials, and polytopes*, Ann. of Math. (2) **96** (1972), 318–337.

[6] M. Hochster and C. Huneke, *Tight closure, invariant theory and the Briançon-Skoda theorem*, J. Amer. Math. Soc. **3** (1990), 31–116.

[7] M. Hochster and J. Roberts, *Rings of invariants of reductive groups acting on regular rings are Cohen-Macaulay*, Adv. in Math. **13** (1974), 115–175.

[8] G. Kempf, *Instability in invariant theory*, Ann. of Math. (2) **108** (1978), 299.

[9] ———, *The Hochster-Roberts theorem in invariant theory*, Michigan Math. J. **26** (1979), 19–32.

[10] F. Knop, *Der kanonische Modul eines Invariantenrings*, J. Algebra **127** (1989), 40–54.

[11] ———, *Die Cohen-Macaulay-Eigenschaft von Invariantenringen*, unpublished, 1990.

[12] V. L. Popov, Groups, generators, syzygies and orbit in invariant theory, Transl. Math. Monographs, vol. 100, Amer. Math. Soc., Providence, RI, 1993.

[13] C. Procesi, *Invariant theory of $n \times n$-matrices*, Adv. in Math. **19** (1976), 306–381.

[14] R. P. Stanley, *Combinatorial reciprocity theorems*, Adv. in Math. **14** (1974), 194–253.

[15] ———, *Combinatorics and invariant theory*, Proc. Sympos. Pure Math. **34** (1979), 345–355.

[16] ———, *Linear diophantine equations and local cohomology*, Invent. Math. **68** (1982), 175–193.

[17] M. Van den Bergh, *A converse to Stanley's conjecture for* Sl_2, to appear in Proc. Amer. Math. Soc.

[18] ———, *Cohen-Macaulayness of modules of covariants*, Invent. Math. **106** (1991), 389–409.

[19] ———, *Cohen-Macaulayness of semi-invariants for tori*, Trans. Amer. Math. Soc. **336** (1993), no. 2, 557–580.

[20] ———, *Local cohomology of modules of covariants*, to appear, 1993.

Bounds for L-Functions

JOHN B. FRIEDLANDER

Scarborough College
University of Toronto
Toronto, Ontario
Canada M5S 1A1

This paper is a report on joint work, in part with Bombieri, and in part with Duke and Iwaniec. We shall consider a number of different kinds of L-functions $L(s)$. All of these will, however, share the following properties:

(P1) In the half plane $\sigma = \operatorname{Re} s > 1$, the function $L(s)$ is given by an absolutely convergent Dirichlet series

$$L(s) = \sum_{n \geq 1} a_n n^{-s}$$

with coefficients a_n satisfying $a_1 = 1$ and $a_n \ll n^{o(1)}$.

(P2) The function $L(s)$ has a meromorphic continuation to the whole complex plane, has (at most) finitely many poles, and satisfies a functional equation

$$\Phi(s)L(s) = \overline{\Phi}(1-s)\overline{L}(1-s), \quad \text{where} \quad \Phi(s) = wQ^s \prod_{j=1}^{J} \Gamma(\lambda_j s + \mu_j)$$

where $|w| = 1$, $Q > 0$, $\lambda_j > 0$, and $\operatorname{Re} \mu_j \geq 0$. By $\bar{f}(s)$ we mean $\overline{f(\bar{s})}$.

It follows by a well-known argument (cf. [T]) that the number $N(T; L)$ of nontrivial zeros (that is, those not located at the poles of the Γ factors) of $L(s)$ satisfying $0 < t = \operatorname{Im} s \leqq T$, is given asymptotically by

$$N(T; L) = \frac{T}{2\pi} \log(C_L T^{2\Lambda}) + O(\log T), \tag{1}$$

where $\Lambda = \sum_{j=1}^{J} \lambda_j$ and $C_L = c_\Gamma Q^2$ with $c_\Gamma = \prod_j \lambda_j^{2\lambda_j}$.

The choice of the parameter Q and the gamma factors in the above decomposition of $\Phi(s)$ is not quite uniquely determined, due to the multiplication formula for the gamma function. However, Λ and C_L are uniquely given by $L(s)$ so that the order of magnitude of Q is almost prescribed. Throughout we allow implied constants to depend on the λ_j, μ_j. Once these are fixed, Q is determined by L.

Proceedings of the International Congress
of Mathematicians, Zürich, Switzerland 1994

The basic problem with which we shall be concerned is that of obtaining upper bounds for $|L(s)|$. It follows already from (P1) that for each $\delta > 0$ we have $L(s) \ll 1$ uniformly for $\sigma \geq 1 + \delta$. It then follows from the functional equation (P2) that for $\sigma \leq -\delta$ we have, uniformly in any fixed vertical strip,

$$L(s) \ll \left(Q\tau^{\Lambda}\right)^{1-2\sigma},$$

where $\tau = \max\{1, |t|\}$.

These bounds are best possible and the more interesting (and important) problem is to obtain bounds in the critical strip $0 \leq \sigma \leq 1$. From the Phragmén-Lindelöf convexity principle and the above bounds it follows that we have there

$$L(s) \ll_{\varepsilon} \left(Q\tau^{\Lambda}\right)^{1-\sigma+\varepsilon}; \tag{2}$$

we shall refer to this as the convexity bound. Our goal is to improve the exponent $1 - \sigma$. The best we can hope for is

$$L(s) \ll_{\varepsilon} \left(Q\tau^{\Lambda}\right)^{\varepsilon}, \tag{3}$$

valid for $\sigma \geq \frac{1}{2}$. One can't hope for such a bound for $\sigma < \frac{1}{2}$, because if this were true at say $\sigma = \frac{1}{4}$, then the functional equation would imply too good a bound on $\sigma = \frac{3}{4}$. By convexity, (3) gives $L(s) \ll_{\varepsilon} \left(Q\tau^{\Lambda}\right)^{1-2\sigma+\varepsilon}$ for $0 \leq \sigma \leq \frac{1}{2}$ and the bound (3) holds for fixed $\sigma > \frac{1}{2}$ once it is known to hold on $\sigma = \frac{1}{2}$.

Without losing anything essential in the generality of the problem we shall restrict to $\sigma = \frac{1}{2}$. The bound (3) is the "generalized Lindelöf hypothesis". It follows rather easily from the Generalized Riemann Hypothesis (the assumption that the nontrivial zeros of the L-function are on the central line $\sigma = \frac{1}{2}$) but may be true more generally. Note that the Generalized Riemann Hypothesis is not true for every L-function satisfying (P1), (P2).

There is a second, essentially equivalent, way of deducing the convexity bound (2). Given an L-function satisfying (P1), the Dirichlet polynomial $D = \sum_{n \leq x} a_n n^{-s}$ offers a reasonable approximation to L in the region $\sigma > 1$. Once we are given also the property (P2), then the approximation continues to hold in the critical strip as long as we stay away from the poles and as long as x is reasonably large.

If we allow the replacement of D by other Dirichlet polynomials $D_x(s) = \sum_{n \leq x} a_n(x) n^{-s}$, in particular those that give smoother truncations of the Dirichlet series, then it is possible to approximate $L(s)$ more closely, but still with the same restriction on the size of x, namely that $x > \left(Q\tau^{\Lambda}\right)^{2+\varepsilon}$. By taking advantage of the functional equation it is possible to use shorter polynomials in an approximation to $L(s)$ (called the approximate functional equation) of the form

$$L(s) = \sum_{n \leq x} a_n n^{-s} + \chi(s) \sum_{n \leq y} \bar{a}_n n^{s-1} + \text{ small}, \tag{4}$$

where $\chi(s) = \overline{\Phi}(1-s)\Phi(s)^{-1}$ and we now only require that $xy > \left(Q\tau^{\Lambda}\right)^{2+\varepsilon}$.

Choosing $s = \frac{1}{2} + it$ and $x = y$ we may, using only trivial bounds, recover the convexity bound (2). This seems a more complicated way to achieve the same result but, with (4) as a starting point, we may hope to develop nontrivial estimates for the Dirichlet polynomials involved and thereby sharpen this bound. Thus, we have broken up the problem of bounding $L(s)$ into two relatively independent subproblems:

(A) Finding good approximations to L functions by short Dirichlet polynomials.

(B) Finding techniques to detect cancellation in and give nontrivial bounds for Dirichlet polynomials.

(A) Dirichlet polynomial approximation

In joint work with Bombieri we have been studying the admissible lengths of such approximations. More precisely, consider Dirichlet polynomials $D_x(s)$ with $a_1(x)$ bounded away from zero and $a_n(x) \ll n^{o(1)}$ and let $I(\sigma, T)$ denote the interval $\{\sigma + it, T \leq t \leq (1+\varepsilon)T\}$. For fixed σ there are always on $I(\sigma, T)$ such approximations with

$$L(s) = D_x(s) + O\left(T^{-\varepsilon}\right) \tag{5}$$

(and in fact with much sharper error term) as long as we have $x > (1+2\varepsilon)\, C_L T^{2\Lambda}$. In case $\sigma > \frac{1}{2}$ the problem of the existence of very much shorter approximations is, in a rough sense, equivalent to the Lindelöf Hypothesis, but for $\sigma < \frac{1}{2}$ it is possible to show that, under the assumption of a very weak zero density estimate, one really cannot find approximations shorter than the above. More precisely we have:

THEOREM 1. *Assume that $L(s)$ satisfies* (P1), (P2) *and that, for each fixed $\sigma > \frac{1}{2}$, the number $N(\sigma, T)$ of zeros $\varrho = \beta + i\gamma$ with $\beta \geq \sigma$, $0 < \gamma \leq T$ satisfies $N(\sigma, T) = o(T \log T)$. Suppose that we have for some $\varepsilon, \varepsilon' > 0$ the approximation* (5) *on $I\left(\frac{1}{2} - \varepsilon', T\right)$. Then $x > T^{2\Lambda - o(1)}$.*

The exponent 2Λ in Theorem 1 is optimal but if one is willing to strengthen the assumptions a little, then one can derive a more precise conclusion.

THEOREM 2. *Assume that L satisfies* (P1), (P2) *and, for each fixed $\sigma > \frac{1}{2}$, the stronger density bound $N(\sigma, T) = o\left(T(\log T)^{-1}\right)$. Suppose that the approximation* (5) *holds on $I(-\varepsilon', T)$. Then $x > (1 - o(1))\, C_L T^{2\Lambda}$, where C_L is given in* (1).

Here not only the exponent but also the constant C_L is optimal for every L function satisfying (P1) and (P2).

The basic idea behind the proofs of these theorems is to compare in a large rectangle the number of zeros (perhaps with suitable weights) of the function $L(s)$ with that of the Dirichlet polynomial $D_x(s)$. These should be nearly equal if the approximation is good. One can use (1) to give an asymptotic formula for the former. One may expect, on the other hand, that it is not possible for a Dirichlet polynomial to have too many zeros and that one can give an upper bound for the number of these, which is too small in the case where x is too small. The following result of this type provides a bound that is sharp and suffices to yield the above theorems.

PROPOSITION. *Let $N(\alpha, T, T+H; D_x)$ denote the number of zeros of $D_x(s)$ satisfying $\sigma \geq \alpha$, $T \leq t \leq T+H$, where $H \leq T$. Then, uniformly for $-H < \alpha < -1$, we have*

$$N(\alpha, T, T+H; D_x) \leq \frac{H}{2\pi} \log x + O\big(|\alpha|^{\frac{1}{2}} H^{\frac{1}{2}} \log x\big).$$

In Theorems 1 and 2 we have treated the L-function as fixed and dealt with the problem only in its T-aspect. One may also consider Q as a variable and it would be of interest to have a Q-analogue to these results that shows that the above approximations are still optimal. This is definitely possible under certain conditions, if not in general.

In a different direction, it would seem natural to conjecture that the known approximate functional equations are also optimal and that in any approximation such as (4) we must necessarily have $xy > (1-o(1))\, C_L T^{2\Lambda}$. Here however the method used to prove Theorems 1 and 2 appears to break down and a new idea is wanted.

(B) Cancellation in Dirichlet polynomials

Although the results of the previous section are rather precise and shed light on the problem (A), they do not help us to improve the convexity bound (2) and we are left with the problem of finding some cancellation in the sums occurring in the approximate functional equation.

The considerations of the previous sections and, in particular, the convexity bound (2) apply rather generally to the L-functions that we have been considering. By contrast, the methods that have been used to improve upon this bound have, until recently, been of a more specialized nature.

(i) Riemann zeta function

In the case of this, the prototypical example, the first progress was made in the years around 1915–1920 based on ideas of Weyl [W], some of which were discovered independently by Hardy and Littlewood, and on a somewhat different method of van der Corput, cf. [T, Chapter V].

Weyl's method proceeds as follows. By the approximation we are led to study sums $\sum_n n^{-\frac{1}{2}-it}$ and, by partial summation, bounds for these may be inferred from bounds for sums $\sum_n n^{-it} = \sum_n e\left(-\frac{t}{2\pi}\log n\right)$, where, as usual, $e(u) = e^{2\pi i u}$. By Taylor's theorem such sums may be approximated by sums $\sum_n e(tf(n))$, where f is a polynomial with real coefficients.

In the case of a linear polynomial this is just the sum of a geometric progression. A nontrivial bound for the general case may be deduced from this, by induction on the degree, using the identity

$$\Big|\sum_n e(tf(n))\Big|^2 = \sum_{n_1}\sum_{n_2} e(t(f(n_1)-f(n_2))) = \sum_h \sum_n e(t(f(n+h)-f(n))).$$

Here the diagonal terms $h = 0$ can only be estimated trivially but are few in number, thus giving a small contribution to the total, whereas, for each fixed $h \neq 0$, the polynomial $f(n+h) - f(n)$ has had its degree reduced by one.

The van der Corput method is somewhat different, using approximations to the above exponential sums by exponential integrals, but both rely on the continuous nature of the function $\log n$ and this limits their applicability to L-functions with relatively simple coefficients. On the other hand, for the Riemann zeta function itself the above ideas form only the beginnings of a topic that has been extensively developed. The most recent breakthrough, due to Bombieri and Iwaniec [BI] and carried forward by Huxley, partly in joint work with Watt, was reported in [H] in the proceedings of the previous ICM in Kyoto.

(ii) Dirichlet L-functions

In the case where one considers a single fixed Dirichlet L-function the problem reduces to (a very small modification of) that for $\zeta(s)$. In the case where one considers the whole class of Dirichlet L-functions and wants a bound explicit in its dependence on Q (and herein lie the most interesting applications), then the problem becomes new, with additional difficulties.

Henceforth we shall consider the case where t is fixed and Q is the only variable. This isolates the problem of most central interest, although there still remains then the not insignificant problem of getting "hybrid" bounds that are simultaneously strong in their dependence on both variables. In this case we have $Q = (q/\pi)^{\frac{1}{2}}$, where q is the conductor of χ. Because s is taken to be fixed we may as well (by partial summation) take our Dirichlet polynomial at $s = 0$ and hence the relevant sum is the character sum $S(N) = \sum_{n \leq N} \chi(n)$. With this normalization a bound $S(N) \ll N^{\frac{1}{2}} q^{\alpha+\varepsilon}$ leads by partial summation to a bound $L(s,\chi) \ll q^{\alpha+\varepsilon}$ on $\sigma = \frac{1}{2}$. The first nontrivial bound for $S(N)$, the Pólya-Vinogradov inequality [P], [V], gives $S(N) \ll q^{\frac{1}{2}} \log q$, sufficing only to recover the convexity bound

$$L\left(\tfrac{1}{2} + it\right) \ll_{\varepsilon,t} q^{\frac{1}{4}+\varepsilon}.$$

Actually the factor q^{ε} can be sharpened; we ignore throughout improvements that are smaller than a fixed power. The problem of improving these bounds for $S(N)$ and consequently for $L(s,\chi)$ was first successfully accomplished by Burgess. Amongst a number of character sum bounds, he showed that $S(N) \ll N^{\frac{1}{2}} q^{\frac{3}{16}+\varepsilon}$ from which one deduces [B] that, for fixed $s = \frac{1}{2} + it$,

$$L(s,\chi) \ll q^{\frac{3}{16}+\varepsilon},$$

which is still the best known bound. Like the earlier methods for studying the t aspect, this one is somewhat special, making essential use of the special structure of the coefficients $a_n = \chi(n)$.

Roughly speaking, one may argue as follows. If $1 \leq a \leq A$, $1 \leq b \leq B$, and if AB is small compared to N, then $S(N)$ is approximated by $\sum_{n \leq N} \chi(n + ab)$ and hence, doing this for each a, b, $S(N)$ is essentially bounded by

$$\begin{aligned}(AB)^{-1} \Big|\sum_a \sum_b \sum_n \chi(n+ab)\Big| &\leq (AB)^{-1} \sum_a \sum_n \Big|\sum_b \chi(\bar{a}n + b)\Big| \\ &= (AB)^{-1} \sum_{y \bmod q} \nu(y) \Big|\sum_b \chi(y+b)\Big|,\end{aligned}$$

where $\bar{a}$ is the multiplicative inverse of a mod q, and $\nu(y)$ is the number of pairs (a, n) with $\bar{a}n \equiv y \pmod{q}$. Applying Hölder's inequality to this we reduce to the problem of bounding the sum $\sum_{y(q)} \nu^2(y)$, which counts solutions of a simple congruence and may be done in elementary fashion, and bounding

$$\sum_{y(q)} \Big|\sum_{b} \chi(y+b)\Big|^4 = \sum_{b_1} \cdots \sum_{b_4} \sum_{y} \chi\left(\frac{(y+b_1)(y+b_2)}{(y+b_3)(y+b_4)}\right).$$

For this last sum over y, Weil's theorem provides a sharp bound and, with optimal choice of A, B, this yields the above Burgess estimate.

(iii) More recondite L-functions

In the following sections we shall describe some of the workings behind a new approach that we have been developing in a series of joint works with Duke and Iwaniec. This leads to improvements in the convexity bound in the above-mentioned cases (albeit with weaker exponents than the best previously known; see [FI] for the case of Dirichlet characters) but also applies more generally to a number of L-functions for which the earlier methods do not.

For the most part this method has been applied to the automorphic L-functions arising in connection with the modular group and its congruence subgroups. In the first place [DFI$_1$], one may take a fixed modular form, for example, a holomorphic cusp form g of weight k for the modular group

$$g(z) = \sum_{n=1}^{\infty} g_n n^{(k-1)/2} e(nz),$$

where here the coefficients have been normalized so that $g_n \ll n^{\varepsilon}$. We consider the twist of g by a primitive Dirichlet character χ mod q and the associated L-function

$$L_g(s,\chi) = \sum_{n=1}^{\infty} g_n \chi(n) n^{-s}.$$

Here one has a functional equation proven by Hecke that is, due to the above normalization, of type (P2) and with, say, $Q = q/\pi$. We then have the estimate of [DFI$_1$]:

THEOREM 3. *With L_g as above,*

$$L_g\left(\tfrac{1}{2} + it, \chi\right) \ll q^{\frac{5}{11}+\varepsilon}. \tag{6}$$

This improves the convexity bound, which would have $\frac{1}{2}$ in place of $\frac{5}{11}$.

Rather than twists of a single cusp form, a natural alternative [DFI$_2$] is to consider the space of holomorphic cusp forms of weight k on the congruence group $\Gamma_0(q)$ and to study, for any individual one of these, the dependence of the corresponding L-function L_f on the level q. We may again normalize the Fourier coefficients $a_f(n)$ so that the Deligne-Ramanujan bound $a_f(n) \ll n^{\varepsilon}$ holds and so that we have a functional equation of type (P2) with, say, $Q = q^{\frac{1}{2}}/2\pi$. In [DFI$_2$] the convexity exponent of $\frac{1}{4}$ is improved to:

THEOREM 4. *For any normalized cusp form f of weight $k \geq 2$ that is a new form on $\Gamma_0(q)$, we have*

$$L_f\left(\tfrac{1}{2}+it\right) \ll q^{\frac{1}{4}-\delta},$$

where $\delta = \frac{1}{192}$.

For $k = 2$ the above includes the L-functions of modular elliptic curves over $\mathbb{Q}$.

Other examples to which the method applies have been considered in [I$_2$] and [IS]. The final example [DFI$_3$] we mention does not begin with $SL(2)$, although the study of the problem leads there soon enough. In this case we consider the imaginary quadratic field $\mathbb{Q}(\sqrt{-D})$ and the dependence on D of the L-functions attached to the characters χ of the class group. In the case of the principal character this is just the Dedekind zeta function of the field, and in any case where χ is real it may be, by a theorem of Kronecker, expressed in terms of Dirichlet L-functions to which the Burgess estimate may be applied. In the general case we are now able to give a subconvexity bound $D^{\frac{1}{4}-\delta}$ with $\delta = 1/1156$, if D is free of large prime factors as in the work of Graham and Ringrose [GR] or for more general discriminants, but then only subject to the assumption of any one of a number of unproven hypotheses. This is due to a serious auxiliary problem that arises in this case; we shall return to this later.

(iv) A new method

We sketch some of the main ideas involved in the estimates of the previous section. As before the problem may be reduced by approximation to the estimation of coefficient sums of the form

$$S_{\boldsymbol{b}}(M) = \sum_{m\sim M} b(m).$$

Here $m \sim M$ means that m runs over the interval $M \leq m \leq 2M$ and the $b(m)$ are either the coefficients a_m of the L-function or else some close relatives. Thus, it may be convenient for reasons of convergence to take $b(m) = a_m h(m)$, where $h(m)$ is a smooth weight function. Alternatively, it will often be more convenient to find a bound for an approximation to some power $L^{\kappa}(s)$ rather than to $L(s)$ itself.

In order for our method to work we need to find some family $\mathcal{F}$ of "companions" to our sequence $\boldsymbol{b}$ of coefficients. In the above examples these companions come from the other characters of the relevant group or, in the case of the space of cusp forms, the other members of an orthogonal basis. We have need of a mean value theorem in the form of an upper bound for a sum of the type

$$\mathcal{B} = \frac{1}{F}\sum_{f\in\mathcal{F}}\Big|\sum_{n\sim N}\alpha_n b_f(n)\Big|^2 \tag{7}$$

with fairly general complex coefficients α_n and $F = |\mathcal{F}|$ being the number of family members. To establish such a mean value theorem we require some orthogonality properties within the family $\mathcal{F}$. Expanding (7) we have

$$\mathcal{B} = \sum_{n_1}\sum_{n_2}\alpha_{n_1}\bar{\alpha}_{n_2}\Delta(n_1,n_2),$$

where $\Delta(n_1, n_2) = \frac{1}{F}\sum_{f\in\mathcal{F}} b_f(n_1)\overline{b_f(n_2)}$. If we assume that the sequences b_f have been normalized so that $b_f(n) \ll_\varepsilon n^\varepsilon$, then we expect (and rather generally get) a bound

$$\mathcal{B} \ll \left(1 + F^{-1}N\right) \|\alpha\|^2 (FN)^\varepsilon\,, \tag{8}$$

where $\|\alpha\|$ denotes the ℓ_2-norm. Here the first term on the right arises from the diagonal terms $n_1 = n_2$ combined with the trivial bound for Δ while the second term arises from the remaining terms and Cauchy's inequality and assumes that, for $n_1 \neq n_2$, the orthogonality gives almost complete cancellation in Δ.

The bound (8) is the best that one can hope for in general, but for certain choices of the sequence α_n it can be improved using more sophisticated techniques. Because there is no cancellation coming from the diagonal terms, any hope of improvement must come from the others and so we want N to be larger than q and we need a nontrivial treatment of the off-diagonal terms. One way to achieve this and to simultaneously retain the special sum $\sum_n b_n$, which it was our goal to estimate, is to introduce the modified mean value

$$\mathcal{B}_{\mathbf{c}} = \frac{1}{F}\sum_{f\in\mathcal{F}} \Big|\sum_{\ell\sim L} c_\ell b_f(\ell)\Big|^2 \Big|\sum_{m\sim M} b_f(m)\Big|^2\,, \tag{9}$$

where the complex coefficients c_ℓ are arbitrary.

In case the b_f are completely multiplicative then this is actually a special case of the previous sum with $\alpha = \boldsymbol{c} * 1$ but, in general, it is slightly different. Sums of this type have been used in the study of the zeros of the zeta function. There the sum $\mathcal{A} = \sum_{\ell\sim L} c_\ell b_f(\ell)$, is called a "mollifier" and the coefficients c_ℓ are chosen so that it imitates the inverse of the target function and thereby acts to mollify or smooth out some of the irregularities in its behavior. Here we intend to choose the coefficients c_ℓ to achieve the opposite effect so that $\mathcal{A}$ behaves as an "amplifier" increasing the contribution of the target function $\boldsymbol{b}$ within the family. If, as happens in all of our examples, we have $F = Q^{\kappa+o(1)}$ for some integer κ, then, with the trivial amplifier $\mathcal{A} = 1$, the estimate (8) is just sufficient to recover the convexity bound, provided that we take our coefficient sums to approximate the function $L^\kappa(s)$ rather than $L(s)$, in order to balance the two terms on the right-hand side of (8).

In the general case we may hope to attain a bound for $\mathcal{B}_{\mathbf{c}}$ of the shape:

$$\mathcal{B}_{\mathbf{c}} \ll \left(1 + L^A F^{-\delta}\right) \|\boldsymbol{c}\|^2 M (FLM)^\varepsilon. \tag{10}$$

Here A is a large positive constant, hopefully not too large, and δ is a small positive constant, hopefully not too small.

We may choose the coefficients c_ℓ by $c_\ell = \overline{b(\ell)}$, where $\boldsymbol{b}$ is the target function, to amplify the contribution to $\mathcal{B}_{\mathbf{c}}$ coming from this term. Presumably the contribution of the terms coming from the companion functions is, by orthogonality, small, but we do not need to quantify this as we may by positivity drop all of these, getting the lower bound

$$\mathcal{B}_{\mathbf{c}} \gg F^{-1}\Big(\sum_{\ell\sim L} |b(\ell)|^2\Big)^2 \Big|\sum_{m\sim M} b(m)\Big|^2 \tag{11}$$

and, combining this with (10),

$$\sum_{m\sim M} b(m) \ll \left(1 + L^{\frac{A}{2}} F^{-\frac{\delta}{2}}\right) \|\mathbf{c}\|^{-1} F^{\frac{1}{2}} M^{\frac{1}{2}} (FLM)^{\varepsilon}. \tag{12}$$

Here, in comparison with the convexity bound that follows from the mean value estimate (8), we have saved a factor $\|\mathbf{c}\|$ that we should like to be large. This leads to two distinct problems, the first being that we would like the size of L to be large. This is obviously constrained by the exponents A and δ that we can achieve in the mean value bound (10). The second problem concerns the size of the coefficients c_ℓ. In the case of Dirichlet L-functions we have $|c_\ell| \gg |\overline{\chi}(\ell)|$, which has modulus one when $(\ell, q) = 1$ so that for $L = q^\eta$ with any fixed η we have $\|\mathbf{c}\| \gg L^{\frac{1}{2}} q^{-\varepsilon}$. We should like to have a similar bound more generally but, in practice, this sometimes proves to be a problem.

Nevertheless, it is clear that, provided we have (12) for the sums $\sum_{m\sim M} b(m)$ approximating L^κ with some $A > 0$, $\delta > 0$, and provided $\|\mathbf{c}\| \gg L^\varphi$ for $L = F^\eta$ with some $\varphi > 0$, $0 < \eta < \delta A^{-1}$, then we get a bound $L\left(\frac{1}{2} + it\right) \ll Q^{\frac{1}{2}-\theta}$ for some $\theta > 0$, and this improves the convexity estimate.

(v) The mean value theorem

An upper bound for the sum $\mathcal{B}_{\mathbf{c}}$ given in (9) may be quickly deduced from a bound for the sum

$$B_\ell = \frac{1}{F} \sum_{f\in\mathcal{F}} b_f(\ell) \Big|\sum_m b_f(m)\Big|^2 \tag{13}$$

as long as there is some multiplicativity that allows the expression of $b_f(\ell_1)\overline{b_f(\ell_2)}$ in terms of $b_f(\ell)$. In the cases considered this may be achieved by the multiplicativity of Dirichlet (or class group) characters or of Hecke eigenvalues.

The sum B_ℓ looks a lot more like the sum considered in the similar mean value (7) and we may expand in the same way getting $B_\ell = \sum\limits_{m_1}\sum\limits_{m_2} \Delta(m_1, m_2)$, where now

$$\Delta(m_1, m_2) = \frac{1}{F} \sum_{f\in\mathcal{F}} b_f(\ell) b_f(m_1) \overline{b_f(m_2)}.$$

In the case of complete multiplicativity this is the same as before, and in the other cases considered it is not too radical a departure therefrom.

One now derives the mean value theorem from the orthogonality properties of $\mathcal{F}$. This reduces to an equidistribution problem of some sort, but the proofs are somewhat long and complicated and it is difficult to describe them briefly or in a unified fashion.

(vi) The amplifier

As already mentioned, in the case where the coefficients c_ℓ may be chosen essentially to be $\overline{\chi}(\ell)$ for χ a Dirichlet character we have $|c_\ell| \gg 1$ for $(\ell, q) = 1$ and $\|\mathbf{c}\|$ has a good lower bound. This includes the case of Dirichlet L-functions and also the case of the twists of a single automorphic L-function. In the case [DFI$_2$], where c_ℓ are chosen as Hecke eigenvalues, one can make use of a trick. The fact that for such λ_f we have $\lambda_f(\ell)^2 - \lambda_f(\ell^2) = 1$ for primes $\ell \nmid q$ means that we cannot have $|\lambda_f|$ small at both ℓ and ℓ^2 and hence $\|\mathbf{c}\|^2 \gg L^{\frac{1}{2}}(\log L)^{-1}$.

In the case of class group L-functions, a more serious problem emerges. It turns out that, choosing say c_ℓ to be $\overline{\chi}$ summed over ideals of norm ℓ, we require, in order to get a nontrivial result, a contribution coming from ideals made up of primes of first degree. But because L is to be a small power of D it is a well-known unsolved problem to show the existence of such ideals. This leads to an interesting question but also to the conditional nature of some of the results in connection with this type of L-function.

(C) Equidistribution

The improvement in Theorem 3 of $\frac{1}{2}$ to $\frac{5}{11}$, or indeed to $\frac{1}{2} - \delta$ for any positive δ, has a significant application. Via the Shimura correspondence the form g is related to a certain half integral weight form on $\Gamma_0(4)$ and a well-known theorem of Waldspurger expresses, for $-q$ a fundamental discriminant, the Fourier coefficient $c(q)$ of the latter in terms of the special value $L_g(\frac{1}{2}, \chi_q)$, where χ_q is the real primitive character $\chi_q(n) = (\frac{-q}{n})$.

The bound in Theorem 3 then translates into a bound $c(q) \ll q^{\frac{5}{22}+\varepsilon}$, which is sufficient to settle the celebrated Linnik problem of the equidistribution of integer points on the sphere. See the paper [D] for more details and for the original solution of the Linnik problem.

The above example illustrates a phenomenon that is present in other circumstances. An older example of this in relation to Dirichlet characters may be found in [BS] and a number of applications coming from L-functions of higher degree may be found in [S]. The achievement of a subconvexity bound is roughly equivalent to the fact that "something or other is equidistributed" and thus can have application to a problem that superficially appears unrelated.

The original solution of the Linnik problem depends on estimates of Iwaniec [I$_1$] for certain sums of Salié sums. Recently, with Duke and Iwaniec [DFI$_4$] we have established the equidistribution of angles of Salié sums. This follows as a consequence of the following theorem establishing the equidistribution of roots of a quadratic congruence as the modulus runs through the primes.

THEOREM 5. *Let $f(x) = ax^2+bx+c$ have integer coefficients and negative discriminant $b^2 - 4ac$. Let $0 \le \alpha < \beta \le 1$ be fixed and let $S(\alpha, \beta; P)$ denote the number of those solutions x of the congruence $f(x) \equiv 0 \pmod p$ for which $\alpha p \le x \le \beta p$, summed over all primes $p < P$. Then*

$$S(\alpha, \beta; P) = \{\beta - \alpha + o(1)\} S(0, 1; P).$$

References

[BS] A. Baker and A. Schinzel, *On the least integers represented by the genera of binary quadratic forms*, Acta Arith. **28** (1971), 137–166.

[BF] E. Bombieri and J. B. Friedlander, *Dirichlet polynomial approximations to zeta functions*, Ann. Scuola Norm. Sup. Pisa Cl. Sci (4), to appear.

[BI] E. Bombieri and H. Iwaniec, *On the order of* $\zeta(\frac{1}{2}+it)$, Ann. Scuola Norm. Sup. Pisa Cl. Sci. (4) **13** (1986), 449–472.

[B] D. A. Burgess, *On character sums and L-series II*, Proc. London Math. Soc. **13** (1963), 524–536.

[D] W. Duke, *Hyperbolic distribution problems and half-integral weight Maass forms*, Invent. Math. **92** (1988), 73–90.

[DFI$_1$] W. Duke, J. B. Friedlander, and H. Iwaniec, *Bounds for automorphic L-functions*, Invent. Math. **112** (1993), 1–8.

[DFI$_2$] ———, *Bounds for automorphic L-functions, II*, Invent. Math. **115** (1994), 219–239.

[DFI$_3$] ———, *Class group L-functions*, to appear in Duke J. Math.

[DFI$_4$] ———, *Equidistribution of roots of a quadratic congruence to prime moduli*, Ann. of Math. (2), **141** (1995), 423–441.

[FI] J. Friedlander and H. Iwaniec, *A mean-value theorem for character sums*, Michigan Math. J. **39** (1992), 153–159.

[GR] S. W. Graham and C. J. Ringrose, *Lower bounds for least quadratic non-residues*, Analytic Number Theory, Proc. Bateman Conf., ed. B. C. Berndt et al., Progr. Math., vol. 85, Birkhäuser, Boston, MA, 1990, pp. 269–309.

[H] M. N. Huxley, *Area, lattice points and exponential sums*, Proc. ICM Kyoto 1990, vol. 1, Springer, Tokyo, 1991, pp. 413–417.

[I$_1$] H. Iwaniec, *Fourier coefficients of modular forms of half-integral weight*, Invent. Math. **87** (1987), 385–401.

[I$_2$] ———, *The spectral growth of automorphic L-functions*, J. Reine Angew. Math. **428** (1992), 139–159.

[IS] H. Iwaniec and P. Sarnak, *L^∞ norms of eigenfunctions of arithmetic surfaces*, Ann. Math. (2), **141** ((1995), 301–320.

[P] G. Polya, *Über die Verteilung der quadratischen Reste und Nichtreste*, Königl. Ges. Wiss. Gött. Nachr. (1918), 21–29.

[S] P. Sarnak, Arithmetic quantum chaos, First Annual R.A. Blyth Lectures, University of Toronto, Canada, preprint, 1993.

[T] E. C. Titchmarsh, The Theory of the Riemann Zeta-Function, 2nd ed., revised by D.R. Heath-Brown, Oxford University Press, Oxford, 1986.

[V] I. M. Vinogradov, On the distribution of power residues and non-residues, Selected works, Springer-Verlag, Berlin and New York, 1985.

[W] H. Weyl, *Über die Gleichverteilung von Zahlen mod Eins*, Math. Ann. **77** (1916), 313–352.

Polylogarithms in Arithmetic and Geometry

ALEXANDER B. GONCHAROV*

Department of Mathematics, MIT
Cambridge MA 02139, USA

The classical polylogarithms were invented in the correspondence of Leibniz with Bernoulli in 1696 [Lei]. They are defined by the series

$$Li_n(z) = \sum_{k=1}^{\infty} \frac{z^k}{k^n} \qquad |z| < 1$$

and continued analytically to a covering of $\mathbb{C}P^1 \backslash \{0, 1, \infty\}$:

$$Li_n(z) := \int_0^z Li_{n-1}(t) \frac{dt}{t}, \qquad Li_1(z) = -\log(1-z).$$

1. The dilogarithm

This was studied by Spence, Abel, Kummer, Lobachevsky, Rogers, Ramanujan, ... [L]. The main discovery was that the dilogarithm satisfies many functional equations. For example Rogers' version of the dilogarithm $L_2(x) := Li_2(x) + \frac{1}{2}\log(x)\log(1-x) - \frac{\pi^2}{6}$ for $1 > x > y > 0$ satisfies the relation

$$L_2(x) - L_2(y) + L_2(y/x) - L_2(\frac{1-x^{-1}}{1-y^{-1}}) + L_2(\frac{1-x}{1-y}) = 0. \qquad (1)$$

After a century of neglect the dilogarithm appeared twenty years ago in works of Gabrielov-Gelfand-Losik [GGL] on a combinatorial formula for the first Pontryagin class, Bloch on K-theory and regulators [Bl1], and Wigner on Lie groups.

The dilogarithm has a single-valued cousin: the Bloch-Wigner function

$$\mathcal{L}_2(z) := \operatorname{Im} Li_2(z) + \arg(1-z)\log|z|.$$

Let $r(x_1, \dots, x_4)$ be the cross-ratio of four distinct points on $\mathbb{C}P^1$. Then

$$\sum_{i=0}^{4} (-1)^i \mathcal{L}_2(r(z_0, \dots, \hat{z}_i, \dots, z_4)) = 0 \qquad z_i \in \mathbb{C}P^1. \qquad (2)$$

If $(z_1, \dots, z_5) = (\infty, 0, 1, x, y)$ the arguments here are the same as in (1).

Choose $x \in \mathbb{C}P^1$. Then (2) just means that the function $c_3(g_0, \dots, g_3) := \mathcal{L}_2(r(g_0 x, \dots, g_3 x))$, where $g_i \in GL_2(\mathbb{C})$ and $g_i x \neq g_j x$, is a *measurable* 3-cocycle on $GL_2(\mathbb{C})$ (Wigner).

The function $\log|x|$ is characterized by its functional equation $\log|xy| = \log|x| + \log|y|$. The 5-term equation plays a similar role for the dilogarithm: any

* The author gratefully acknowledges a travel grant from ICM-94.

Proceedings of the International Congress
of Mathematicians, Zürich, Switzerland 1994

measurable function $f(z)$, $z \in \mathbb{C}$ satisfying (2) is proportional to $\mathcal{L}_2(z)$ [Bl1]. Moreover, any functional equation for $\mathcal{L}_2(z)$ is a formal consequence of (2).

Let us show how to prove (2). For a set X denote by $\mathbb{Z}[X]$ the free abelian group generated by symbols $\{x\}$, $x \in X$. Let F be a field. Consider the homomorphism [Bl1]

$$\tilde{\delta}_2 : \mathbb{Z}[F^*\backslash 1] \longrightarrow \Lambda^2 F^*, \quad \{x\} \longmapsto (1-x) \wedge x\,.$$

Let $R_2(F)$ be the subgroup of $\mathbb{Z}[F^*\backslash 1]$ generated by the elements $\sum(-1)^i\{r(z_0, \ldots, \hat{z}_i, \ldots, z_4)\}$ where $z_i \neq z_j \in P^1_F$. One can check that $\tilde{\delta}_2(R_2(F)) = 0$. This together with the formula

$$d\mathcal{L}_2(z) = -\log|1-z| d\arg|z| + \log|z| d\arg(1-z)$$

(which reminds us of the definition of $\tilde{\delta}_2$) implies (2).

Setting $B_2(F) := \mathbb{Z}[F^*\backslash 1]/R_2(F)$ we get the Bloch complex [DS], [S1]

$$B_2(F) \xrightarrow{\delta_2} \Lambda^2 F^*, \quad \{x\} \mapsto (1-x) \wedge x\,. \tag{3}$$

By the Matsumoto theorem $\mathrm{Coker}\tilde{\delta}_2 = K_2(F)$. Suslin proved [S1] that $\mathrm{Ker}\delta_2 \otimes \mathbb{Q} = K_3^{\mathrm{ind}}(F)_{\mathbb{Q}}$. Here $K_3^{\mathrm{ind}}(F)$ is the cokernel of the multiplication $K_1(F)^{\otimes 3} \to K_3(F)$, and the subscript $\mathbb{Q}$ means tensoring with $\mathbb{Q}$.

If $F = \mathbb{C}$ any real-valued function $f(z)$ defines a homomorphism $\tilde{f} : \mathbb{Z}[\mathbb{C}] \longrightarrow \mathbb{R}$, $\{z\} \longmapsto f(z)$. Thanks to (2) we have a homomorphism $\tilde{\mathcal{L}}_2 : B_2(\mathbb{C}) \to \mathbb{R}$. Combined with the above homomorphism $K_3(\mathbb{C}) \to \mathrm{Ker}\delta_2$ it gives an explicit formula for the Borel regulator $K_3(\mathbb{C}) \to \mathbb{R}$ and hence [Bo2] a formula for $\zeta_F(2)$ for any number field F (see Section 5 below).

Let $\mathcal{H}^3$ be the 3-dimensional hyperbolic space. Denote by $I(z_0, \ldots, z_3)$ the ideal geodesic symplex with vertices at points $z_0, \ldots, z_3$ of the absolute $\partial\mathcal{H}^3 = \mathbb{C}P^1$. Then $\mathrm{vol}\, I(z_0, \ldots, z_3) = 3/2\mathcal{L}_2(r(z_0, \ldots, z_3))$ (Lobachevsky).

Any complete hyperbolic 3-fold of finite volume V^3 can be represented as a formal sum of ideal geodesic simplices. So $\mathrm{vol}\, V^3 = 3/2 \sum \mathcal{L}_2(x_i)$. It turns out the condition "V^3 is a manifold" implies $\delta_2 \sum\{x_i\} = 0$. (Thurston, [DS], [NZ]).

At first glance many features of this picture seem special for the dilogarithm. For example the classical n-logarithms are functions of just one variable, but for $n > 2$, GL_n does not act on P^1, $\partial\mathcal{H}^n$ is no longer a complex manifold, and so on. In this lecture I will explain how most of these facts about the dilogarithm are generalized to the trilogarithm and outline what should happen in general.

2. The trilogarithm and $\zeta_F(3)$ [G2]

A single-valued version of $Li_3(z)$ is

$$\mathcal{L}_3(z) := \mathrm{Re}\Big(Li_3(z) - Li_2(z) \cdot \log|z| + \frac{1}{3} Li_1(z) \cdot \log^2|z|\Big)\,.$$

Denote by $\{x\}_2$ the image of $\{x\}$ in $B_2(F)$. Set

$$\mathbb{Z}[F^*] \xrightarrow{\delta_3} B_2(F) \otimes F^*, \qquad \delta_3 : \{x\} \mapsto \{x\}_2 \otimes x, \quad \{1\} \mapsto 0\,. \tag{4}$$

Let F be a number field with r_1 real and r_2 complex places, $\{\sigma_j\}$ be the set of all possible embeddings $F \hookrightarrow \mathbb{C}$ numbered so that $\sigma_{r_1+k} = \overline{\sigma_{r_1+r_2+k}}$, and d_F be the discriminant of F. For $x \in \mathbb{Z}[F^*]$ one gets numbers $\tilde{\mathcal{L}}_3(\sigma_j(x))$ defined via the composition $\mathbb{Z}[F^*] \xrightarrow{\sigma_j} \mathbb{Z}[\mathbb{C}^*] \xrightarrow{\tilde{\mathcal{L}}_3} \mathbb{R}$.

THEOREM 1. *For any number field F there exist elements $y_1, \dots, y_{r_1+r_2} \in \mathrm{Ker}\delta_3 \otimes \mathbb{Q} \subset \mathbb{Q}[F^*]$ such that*

$$\zeta_F(3) = \pi^{3r_2} d_F^{-\frac{1}{2}} \det |\tilde{\mathcal{L}}_3(\sigma_j(y_i))| , \quad (1 \le i, j \le r_1 + r_2) . \tag{5}$$

It was conjectured by Zagier, who gave many numerical examples [Z1]. Here is one of them:

$$\zeta_{\mathbb{Q}(\sqrt{5})}(3) = \frac{24}{25\sqrt{5}} \cdot \mathcal{L}_3(1) \cdot \Big(\mathcal{L}_3(\frac{1+\sqrt{5}}{2}) - \mathcal{L}_3(\frac{1-\sqrt{5}}{2})\Big) .$$

If $\alpha := \frac{1+\sqrt{5}}{2}$ and $\bar{\alpha} := \frac{1-\sqrt{5}}{2}$ then $\alpha \cdot \bar{\alpha} = -1, \alpha + \bar{\alpha} = 1$, so $\{\alpha\}_2 \otimes \alpha - \{\bar{\alpha}\}_2 \otimes \bar{\alpha} = (\{\alpha\}_2 + \{1-\alpha\}_2) \otimes \alpha = 0$ modulo torsion because $6 \cdot (\{x\}_2 + \{1-x\}_2) \in R_2(F)$.

Let $\Delta : G \to G \times G$ be the diagonal map. An element $x \in H_n(G)$ is called primitive if $\Delta_*(x) = x \otimes 1 + 1 \otimes x$. For any field F one can define $K_n(F)_{\mathbb{Q}}$ as the subspace of primitive elements in $H_n(GL(F), \mathbb{Q})$.

Let $H_c^*(G, \mathbb{R})$ be a continous cohomology of a Lie group G. It is known that $H_c^*(GL(\mathbb{C}), \mathbb{R}) = \Lambda_{\mathbb{R}}^*(c_1, c_3, \dots)$ where $c_{2n-1} \in H_c^{2n-1}(GL(\mathbb{C}), \mathbb{R})$ are certain classes. For example $c_1(g_1, g_2) = \log |\det g_1^{-1} g_2|$. Considered as a functional on homology c_{2n-1} induces a map $r_{\mathbb{C}}(n) : K_{2n-1}(\mathbb{C})_{\mathbb{Q}} \to \mathbb{R}$. It is called the Borel regulator [Bo2]. Let F be a number field. Then the image of the composition

$$r(n) : K_{2n-1}(F) \longrightarrow \oplus_{\mathrm{Hom}(F,\mathbb{C})} K_{2n-1}(\mathbb{C})_{\mathbb{Q}} \overset{r_{\mathbb{C}}(n) \otimes \mathbb{R}(n-1)}{\longrightarrow} \mathbb{Z}^{\mathrm{Hom}(F,\mathbb{C})} \otimes \mathbb{R}(n-1)$$

is invariant under the complex conjugation. So we get a regulator map $r(n) : K_{2n-1}(F) \longrightarrow \mathbb{R}(n-1)^{d_n}$. Here $d_n = r_1 + r_2$ for odd n and r_2 for even. We will use notation $a \sim b$ if $a/b \in \mathbb{Q}^*$. Borel proved that $r(n)(K_{2n-1}(F))$ is a lattice with covolume $\sim d_F^{1/2} \zeta_F(n)(\pi i)^{-n d_{n-1}}$.

The proof of our theorem is based on an explicit description of the regulator $K_5(\mathbb{C}) \to \mathbb{R}$ by means of the trilogarithm $\mathcal{L}_3$. The key step is a formula for a measurable 5-cocycle of $GL(\mathbb{C})$ representing the class c_5. For $GL_3(\mathbb{C})$ it appears as follows.

Let V^3 be a 3-dimensional vector space over F. Choose a volume form $\omega \in \wedge^3(V^3)^*$. For 6 vectors $l_1, \dots, l_6$ in generic position in V^3 set $\Delta(l_i, l_j, l_k) := \langle \omega, l_i \wedge l_j \wedge l_k \rangle \in F^*$. Let $\mathrm{Alt}_6 f(l_1, \dots, l_6) := \sum_{\sigma \in S_6} (-1)^{|\sigma|} f(l_{\sigma(1)}, \dots, f_{\sigma(6)})$. Set

$$r_3(l_1, \dots, l_6) := \mathrm{Alt}_6 \left\{ \frac{\Delta(l_1, l_2, l_4)\Delta(l_2, l_3, l_5)\Delta(l_3, l_1, l_6)}{\Delta(l_1, l_2, l_5)\Delta(l_2, l_3, l_6)\Delta(l_3, l_1, l_4)} \right\} \in \mathbb{Z}[F^*] . \tag{6}$$

$r_3(l_1, \dots, l_6)$ clearly does not depend on the lengths of vectors l_i and so is a generalized cross-ratio of 6 points on the projective plane.

THEOREM 2. *(a) For any 7 points $(m_1, \dots, m_7)$ in generic position in $\mathbb{C}P^2$*

$$\sum_{i=1}^{7} (-1)^i \tilde{\mathcal{L}}_3(r_3(m_1, \dots, \hat{m}_i, \dots, m_7)) = 0 .$$

(b) Choose $x \in \mathbb{C}P^2$. Then the function $c_5(g_0, \dots, g_5) := \tilde{\mathcal{L}}_3(r_3(g_0x, \dots, g_5x))$ defined for $g_i \in GL_3(\mathbb{C})$ such that $(g_0x, \dots, g_5x)$ is in general position is a measurable 5-cocycle representing a nontrivial cohomology class of the group $GL_3(\mathbb{C})$.

3. The trilogarithm and algebraic K-theory [G3]

Let $R_3(F)$ be the subgroup of $\mathbb{Z}[F^*]$ generated by the elements $\{x\} + \{x^{-1}\}$ and $\sum_{i=1}^{7}(-1)^i r_3(m_1, \dots, \hat{m}_i, \dots, m_7)$ where $(m_1, \dots, m_7)$ run through all generic configurations of 7 points in P_F^2. Then $\delta_3 R_3(F) = 0$. Let $B_3(F)$ be the quotient of $\mathbb{Z}[F^*]$ by $R_3(F)$. We get a complex

$$B_F(3): \qquad B_3(F) \xrightarrow{\delta_3} B_2(F) \otimes F^* \xrightarrow{\delta_2 \wedge id} \Lambda^3 F^*$$

placed in degrees [1,3]. Here δ_2, δ_3 were defined in (3) and (4).

According to [S2], $H_n(GL_n(F), \mathbb{Q}) = H_n(GL(F), \mathbb{Q})$. Let

$$K_n^{(i)}(F)_{\mathbb{Q}} := K_n(F)_{\mathbb{Q}} \cap \mathrm{Im}\Big(H_n(GL_{n-i}(F), \mathbb{Q}) \to H_n(GL_n(F), \mathbb{Q})\Big)$$

be the rank filtration. Denote by $K_n^{[i]}(F)_{\mathbb{Q}}$ its graded quotients.

THEOREM 3. *There are canonical maps $K_{6-i}^{[3-i]}(F)_{\mathbb{Q}} \longrightarrow H^i(B_F(3) \otimes \mathbb{Q})$.*

They should be isomorphisms. This is known for $i = 3$ [S2].

4. Classical polylogarithms and motivic complexes [G1]

The following single-valued version of $Li_n(z)$ was invented by Zagier [Z1].

$$\mathcal{L}_n(z) \quad := \quad \begin{matrix} \mathrm{Re} & (n: \text{ odd}) \\ \mathrm{Im} & (n: \text{ even}) \end{matrix} \left(\sum_{k=0}^{n-1} \beta_k \log^k |z| \cdot Li_{n-k}(z) \right), \quad n \geq 2.$$

It is continuous on $\mathbb{C}P^1$. Here $\frac{2x}{e^{2x}-1} = \sum_{k=0}^{\infty} \beta_k x^k$. A Hodge-theoretic interpretation of $\mathcal{L}_n(z)$ was given by Beilinson and Deligne [BD1].

Let us define inductively for each $n \geq 1$ a subgroup $\mathcal{R}_n(F) \subset \mathbb{Z}[P_F^1]$, which for $F = \mathbb{C}$ will be the subgroup of *all* functional equations for $\mathcal{L}_n(z)$.

Set $\mathcal{B}_n(F) := \mathbb{Z}[P_F^1]/\mathcal{R}_n(F)$. Set $\mathcal{R}_1(F) := (\{x\}+\{y\}-\{xy\}; \{0\}; \{\infty\})$.Then $\mathcal{B}_1(F) = F^*$. Let $\{x\}_n$ be the image of $\{x\}$ in $\mathcal{B}_n(F)$. Consider homomorphisms

$$\mathbb{Z}[P_F^1] \xrightarrow{\delta_n} \begin{cases} \mathcal{B}_{n-1}(F) \otimes F^* & : \ n \geq 3 \\ \Lambda^2 F^* & : \ n = 2 \end{cases} \tag{7}$$

$$\delta_n : \{x\} \mapsto \begin{cases} \{x\}_{n-1} \otimes x & : \ n \geq 3 \\ (1-x) \wedge x & : \ n = 2 \end{cases} \qquad \delta_n : \{\infty\}, \{0\}, \{1\} \mapsto 0. \tag{8}$$

Set $\mathcal{A}_n(F) := \mathrm{Ker}\, \delta_n$. Any element $\alpha(t) = \Sigma n_i \{f_i(t)\} \in \mathbb{Z}[P_{F(t)}^1]$ has a specialization $\alpha(t_0) := \Sigma n_i \{f_i(t_0)\} \in \mathbb{Z}[P_F^1]$ at each point $t_0 \in P_F^1$.

DEFINITION 4. *$\mathcal{R}_n(F)$ is generated by elements $\{\infty\}$, $\{0\}$, and $\alpha(0) - \alpha(1)$ where $\alpha(t)$ runs through all elements of $\mathcal{A}_n(F(t))$.*

One can show that $\delta_n \mathcal{R}_n(F) = 0$ [G1, 1.16]. So we get homomorphisms

$$\delta_n : \mathcal{B}_n(F) \longrightarrow \mathcal{B}_{n-1}(F) \otimes F^*, \quad n \geq 3; \quad \delta_2 : \mathcal{B}_2(F) \longrightarrow \Lambda^2 F^*$$

and finally the following complex $\Gamma(F, n)$:

$$\mathcal{B}_n \xrightarrow{\delta} \mathcal{B}_{n-1} \otimes F^* \xrightarrow{\delta} \mathcal{B}_{n-2} \otimes \Lambda^2 F^* \xrightarrow{\delta} \cdots \xrightarrow{\delta} \mathcal{B}_2 \otimes \Lambda^{n-2} F^* \xrightarrow{\delta} \Lambda^n F^*$$

where $\delta : \{x\}_p \otimes \bigwedge_{i=1}^{n-p} y_i \to \delta_p(\{x\}_p) \wedge \bigwedge_{i=1}^{n-p} y_i$ has degree $+1$ and $\mathcal{B}_n \equiv \mathcal{B}_n(F)$ placed in degree 1. One can prove that $\tilde{\mathcal{L}}_n\Big(\mathcal{R}_n(\mathbb{C})\Big)$ (see [G2, Theorem 1.13].

CONJECTURE 5. *Let $f(z)$ be a measurable function such that $\tilde{f}(\mathcal{R}_n(\mathbb{C})) = 0$. Then* $f(z) = \lambda_0 \mathcal{L}_n(z) + \lambda_1 \mathcal{L}_{n-1}(z) \log|z| + \cdots + \lambda_{n-2} \mathcal{L}_2(z)(\log|z|)^{n-2}$, $\lambda_i \in \mathbb{C}$.

This is true for $n = 2$ [Bl1] and $n = 3$ (unpublished).

Let γ be the Adams filtration on $K_n(F)_{\mathbb{Q}}$. Hypothetically it is opposite to the rank filtration. For number fields $gr_n^{\gamma} K_m(F)_{\mathbb{Q}} = 0$ if $m \neq 2n-1$.

CONJECTURE A. *(a) For any field F one has $H^i\Gamma(F, n) \otimes \mathbb{Q} = gr_n^{\gamma} K_{2n-i}(F) \otimes \mathbb{Q}$.*

(b) The composition $gr_n^{\gamma} K_{2n-1}(\mathbb{C})_{\mathbb{Q}} \to H^1\Gamma(\mathbb{C}, n)_{\mathbb{Q}} \to \mathbb{R}$ is a nonzero rational multiple of the Borel regulator.

For number fields the isomorphism $K_{2n-1}(F)_{\mathbb{Q}} = \operatorname{Ker} \delta_n$ was conjectured (slightly differently, without the complexes $\Gamma(F, n)$) by Zagier [Z1].

$H^n\Gamma(F, n) = K^M(F)$ by definition. So we get a hypothetical description of Quillen's K-groups by symbols that generalizes Milnor's approach to K-theory:

$$K_m(F)_{\mathbb{Q}} \stackrel{?}{=} \oplus_n H^{2n-m}(\Gamma(F, n) \otimes \mathbb{Q}) . \tag{9}$$

This suggests that $\Gamma(F, n) \otimes \mathbb{Q}$ should be the weight n motivic complex conjectured by Beilinson and Lichtenbaum [B1], [Li]. Another approach is shown in [Bl2].

For a compact smooth i-dimensional variety X over $\mathbb{Q}$ Beilinson defined a regulator map to Deligne cohomology [B1] $r_{\mathrm{Be}} : gr_n^{\gamma} K_{2n-i}(X) \longrightarrow H^i_{\mathcal{D}}(X/\mathbb{R}, \mathbb{R}(n))$.

A model $X_{\mathbb{Z}}$ of X over $\mathbb{Z}$ defines a subgroup $gr_n^{\gamma} K_{2n-i}(X_{\mathbb{Z}}) \subset gr_n^{\gamma} K_{2n-i}(X)$. For $n > i+1$ they should coincide. Beilinson conjectured that $r_{\mathrm{Be}}(gr_n^{\gamma} K_{2n-i}(X_{\mathbb{Z}}))$ is a lattice whose covolume with respect to the natural $\mathbb{Q}$-structure provided by $H^i_{\mathcal{D}}(X/\mathbb{R}, \mathbb{Q}(n))$ up to a standard factor coincides with the value of L-function $L(h^i(X), s)$ at $s = i$. Unfortunately, the definition of the regulator is rather implicit.

Conjecture A, together with Beilinson's conjecture on regulators [B2], should give explicit formulas for special values of the L-functions of varieties over number fields in terms of *classical* polylogarithms. Below two examples are discussed: ζ-functions of number fields and L-functions of elliptic curves.

5. Zagier's conjecture

Conjecture A (b) and Borel's theorem [Bo2] lead to

CONJECTURE 6. *For any number field F there exist elements $y_1, \ldots y_{d_n} \in \operatorname{Ker}\delta_n \otimes \mathbb{Q} \subset \mathcal{B}_n(F)_{\mathbb{Q}}$ such that*

$$\zeta_F(n) = \pi^{d_{n-1} \cdot n} d_F^{-\frac{1}{2}} \det|\tilde{\mathcal{L}}_n(\sigma_j(y_i))| , \quad (1 \leq i, j \leq d_n) . \tag{10}$$

This was stated in [Z1]. The case $n = 2$, essentially proved in [Z2], follows from the Borel theorem and the results of Bloch [Bl1] and Suslin [S2] (see Section 1); for a simpler proof see in Section 2 of [G1]. It is not proved for $n > 3$.

THEOREM 7. *For any number field F there is a map $l_n : \mathrm{Ker}\,\delta_n \otimes \mathbb{Q} \to K_{2n-1}(F)_{\mathbb{Q}}$ such that for any $\sigma : F \hookrightarrow \mathbb{C}$ one has $r_{\mathbb{C}}(n)(\sigma \circ l_n(y)) = \tilde{\mathcal{L}}_n(\sigma(y))$.*

This was proved by Beilinson and Deligne [BD2] and later De Jeu [J]. It can be deduced from the existence of the triangulated category of mixed Tate motives over Spec(F) constructed by Levine [Le] and Voevodsky [V], see [BD1].

6. Motivic complexes for curves [G1]

Let K be a field with a discrete valuation v, the residue field k_v, and the group of units U. Let $u \to \bar{u}$ be the projection $U \to k_v^*$. Choose a uniformizer π. There is a homomorphism $\theta : \Lambda^n F^* \longrightarrow \Lambda^{n-1} F_v^*$ uniquely defined by the following properties ($u_i \in U$):

$$\theta\,(\pi \wedge u_1 \wedge \cdots \wedge u_{n-1}) = \bar{u}_1 \wedge \cdots \wedge \bar{u}_{n-1}; \qquad \theta\,(u_1 \wedge \cdots \wedge u_n) = 0\,.$$

It is clearly independent of π. Let us define a homomorphism $s_v : \mathbb{Z}[P_K^1] \longrightarrow \mathbb{Z}[P_{k_v}^1]$ setting $s_v\{x\} = \{\bar{x}\}$ if x is a unit and 0 otherwise. It induces a homomorphism $s_v : \mathcal{B}_m(K) \longrightarrow \mathcal{B}_m(k_v)$. Set

$$\partial_v := s_v \otimes \theta : \mathcal{B}_m(K) \otimes \Lambda^{n-m} K^* \longrightarrow \mathcal{B}_m(k_v) \otimes \Lambda^{n-m-1} k_v^*\,.$$

This defines a morphism of complexes $\partial_v : \Gamma(K, n) \longrightarrow \Gamma(k_v, n-1)[-1]$. Let X be a regular curve over a field F and F_x be the residue field of a point $x \in X$. Let us define the motivic complex $\Gamma(X, n)$ as follows:

$$\begin{array}{ccccc}
\mathcal{B}_n(F(X)) & \xrightarrow{\delta} & \mathcal{B}_{n-1}(F(X)) \otimes F(X)^* \xrightarrow{\delta} \ldots & \xrightarrow{\delta} & \Lambda^n F(X)^* \\
 & & \downarrow \coprod_x \partial_x & & \downarrow \coprod_x \partial_x \\
 & & \coprod_{x \in X^1} \mathcal{B}_{n-1}(F_x) \quad \xrightarrow{\delta} \ldots & \xrightarrow{\delta} & \coprod_{x \in X^1} \Lambda^{n-1} F_x^*
\end{array} \qquad (11)$$

Here the group $\mathcal{B}_n(F(X))$ is placed in degree 1.

CONJECTURE 8. *For a regular curve X one has*

$$H^i(\Gamma(X, n) \otimes \mathbb{Q}) = gr_n^{\gamma} K_{2n-i}(X)_{\mathbb{Q}}.$$

7. Explicit formulas for regulators in the case of curves [G6]

Let me recall that, for a curve X over $\mathbb{R}$ and $n > 1$,

$$H^2_{\mathcal{D}}(X/\mathbb{R}, \mathbb{R}(n)) = H^1(X, \mathbb{R}(n-1))^+$$

where "+" means invariants of the complex conjugation acting both on $X(\mathbb{C})$ and $\mathbb{R}(n-1)$. Beilinson's regulator for curves over $\mathbb{Q}$ is a homomorphism

$$r_{\mathrm{Be}}(n) : K_{2n-2}(X)_{\mathbb{Q}} \longrightarrow H^2_{\mathcal{D}}(X/\mathbb{R}, \mathbb{R}(n))\,.$$

The cup product with $\omega \in \Omega^1(\bar{X})$ identifies $H^1(\bar{X}, \mathbb{R}(n-1))$ with $H^0(\bar{X}, \Omega^1)^{\vee}$. Thus, we will view elements of $H^2_{\mathcal{D}}(\bar{X}/\mathbb{R}, \mathbb{R}(n))$ as functionals on $H^0(\bar{X}, \Omega^1)^{\vee}$.

Set $\alpha(f, g) := \log|f| d\log|g| - \log|g| d\log|f|$.

THEOREM 9. *Let X be a curve over $\mathbb{Q}$. Then for each element $\gamma_{2n-2} \in K_{2n-2}(X)$, $n = 3, 4$, there are rational functions $f_i, g_i \in \mathbb{Q}(X)^*$ such that $\sum_i \{f_i\}_{n-1} \otimes g_i$ is a 2-cocycle in (11) and for any $\omega \in \Omega^1(X)$ one has ($a_n, b_n \in \mathbb{Q}^*$):*

$$\int_{X(\mathbb{C})} r_{\mathrm{Be}}(n)(\gamma_{2n-2}) \quad \wedge \quad \omega = a_n \cdot \sum_i \int_{X(\mathbb{C})} \mathcal{L}_{n-1}(f_i)\, d\log|g_i| \wedge \omega \tag{12}$$

$$= \quad b_n \cdot \sum_i \int_{X(\mathbb{C})} \log|g_i| \log^{n-3}|f_i| \alpha(1-f_i, f_i) \wedge \omega . \tag{13}$$

For $n = 2$ this is the famous symbole modéré of Beilinson and Deligne. Hypothetically (12) should be true for all n.

EXAMPLE. For $n = 3$ the condition "$\sum_i \{f_i\}_2 \otimes g_i$ is a 2-cocycle in (11)" means that $\sum_i (1-f_i) \wedge f_i \wedge f_i = 0$ in $\Lambda^3 \mathbb{Q}(X)^*$ and $\sum_i v_x(g_i)\{f_i(x)\}_2 = 0$ in $\mathcal{B}_2(\bar{\mathbb{Q}})$ for any $x \in X(\bar{\mathbb{Q}})$. Here v_x is the valuation defined by a point x.

8. Special values of L-functions of elliptic curves [G6]

Let E be an elliptic curve $/\mathbb{Q}$ and $\Gamma := H_1(E(\mathbb{C}), \mathbb{Z})$. A holomorphic 1-form ω defines an embedding $\Gamma \hookrightarrow \mathbb{C}$ together with an isomorphism $E(\mathbb{C}) = \mathbb{C}/\Gamma = \Gamma \otimes \mathbb{R}/\Gamma$. The intersection pairing $\Gamma \times \Gamma \to \mathbb{Z}(1)$ provides a pairing $(\cdot,\cdot) : E(\mathbb{C}) \times \Gamma \longrightarrow U(1) \subset \mathbb{C}^*$. If $\Gamma = \mathbb{Z}u + \mathbb{Z}v \subset \mathbb{C}$ with $\mathrm{Im}(u/v) > 0$, then

$$(z, \gamma) = \exp A(\Gamma)^{-1}(z\bar{\gamma} - \bar{z}\gamma) \quad \text{where} \quad A(\Gamma) = \frac{1}{2\pi i}(\bar{u}v - u\bar{v}).$$

Consider the generalized Eisenstein-Kronecker series ($\gamma_i \in \Gamma$)

$$K_n(x,y,z) := \sum_{\gamma_1+\dots+\gamma_n=0}{}' \frac{(x,\gamma_1)(y,\gamma_2+\dots+\gamma_{n-1})(z,\gamma_n)(\bar{\gamma}_n - \bar{\gamma}_{n-1})}{|\gamma_1|^2|\gamma_2|^2\dots|\gamma_n|^2}, \quad n \geq 3.$$

They are invariant under the shift $(x,y,z) \to (x+t, y+t, z+t)$ and so live actually on $E(\mathbb{C}) \times E(\mathbb{C})$. For $n = 2$ set $K_2(x,y,z) := \sum'_\gamma \frac{(x-z,\gamma)}{|\gamma|^2\gamma}$.

Let $\omega \in H^0(E, \Omega^1_{E/\mathbb{Q}})$, $\Omega = \int_{E(\mathbb{R})} \omega$ be the real period of E. Let f_E be the conductor of E.

CONJECTURE 10. *(a) Let E be an elliptic curve $/\,\mathbb{Q}$ and $n \geq 3$. Then there exist functions $f_i, g_i \in \mathbb{Q}(E)^*$ such that $\sum_i \{f_i\}_{n-1} \otimes g_i$ is a 2-cocycle in (11) and*

$$q \cdot L(E,n) = \left(\frac{2\pi A(\Gamma)}{f_E}\right)^{n-1} \Omega \cdot \sum_i K_n(x_i, y_i, z_i) \tag{14}$$

where x_i, y_i, z_i are divisors of $f_i, g_i, 1-f_i$, and $q \in \mathbb{Q}^*$.
(b) For any $f_i, g_i \in \mathbb{Q}(E)^*$ as above formula (14) holds with (possibly 0) $q \in \mathbb{Q}$.

For $n = 2$ (14) is Bloch's conjecture [Bl1] and for $n = 3$ it was conjectured (slightly differently, using Massey products) by Deninger [Den]. A conjecture for any elliptic curve over a number field involves determinants with entries $K_n(x,y,z)$.

THEOREM 11. *Conjecture 10 holds for modular elliptic curves over $\mathbb{Q}$ in the cases $n = 3$ and $n = 4$.*

The proof uses Theorem 3, a similar result in weight 4, Theorem 9, and the weak Beilinson's conjecture for modular curves proved in [B3]. For example, for $n = 3$ we get the formula

$$L(E,3) \sim \Big(\frac{2\pi A(\Gamma)}{f_E}\Big)^2 \Omega \cdot \sum_i \sum_{\gamma_1+\gamma_2+\gamma_3=0}' \frac{(x_i,\gamma_1)(y_i,\gamma_2)(z_i,\gamma_3)}{|\gamma_1|^2|\gamma_2|^2|\gamma_3|^2}.$$

In a similar conjecture about $L(S^nE, n+1)$ determinants appear whose entries are the classical Kronecker-Eisenstein series $\sum_{\gamma\in\Gamma} \frac{(x-y,\gamma)}{\gamma^a\bar{\gamma}^b}$ $(a+b=2n+1)$. Their motivic interpretation was given in [BL]. One should have it also for functions $K_n(x,y,z)$, and, more generally, for functions needed to compute $L(S^nE, m)$.

9. Motivic Lie algebra $L(F)_\bullet$ [G2]

Beilinson conjectured [B1], [BD2] that for any field F there should exist a Tannakian (i.e. abelian, with inner Hom and $\otimes$, ...) category $\mathcal{M}_T(F)$ of mixed Tate motives over F. It is generated (as tensor category) by an invertible object $\mathbb{Q}(1)_\mathcal{M}$ (the Tate motive). For any $n \in \mathbb{Z}$ set $\mathbb{Q}(n)_\mathcal{M} := \mathbb{Q}(1)_\mathcal{M}^{\otimes n}$. The crucial axiom is:

$$\mathrm{Ext}^i_{\mathcal{M}_T(F)}(\mathbb{Q}(0)_\mathcal{M}, \mathbb{Q}(n)_\mathcal{M}) \stackrel{?}{\cong} gr^n_\gamma K_{2n-i}(F)_\mathbb{Q}\,. \tag{15}$$

Any object M of this category carries canonical increasing weight filtration $W_\bullet M$ such that $gr^W_{2k}M = \oplus\mathbb{Q}(-k)_\mathcal{M}$ and $gr^W_{2k-1}M = 0$. There is a canonical fiber functor ω from $\mathcal{M}_T(F)$ to the category of finite-dimensional graded $\mathbb{Q}$-vector spaces: $\omega(M) := \oplus \mathrm{Hom}(\mathbb{Q}(-k)_\mathcal{M}, gr^W_{2k}M)$. Let $U(F)_\bullet := \mathrm{End}\,\omega$ be the space of all endomorphisms of the functor ω. It is a graded (pro) Hopf algebra over $\mathbb{Q}$.

Let $L(F)_\bullet$ be the Lie algebra of all derivations of ω. It is naturally graded: $L(F)_\bullet = \oplus_{n\geq 1}L(F)_{-n}$ and $U(F)_\bullet$ is its universal enveloping algebra. The functor ω is an equivalence of the category $\mathcal{M}_T(F)$ with the category of finite-dimensional graded modules over $L(F)_\bullet$.

The degree n part of the cochain complex $(\Lambda^\bullet(L(F)^\vee_\bullet), \partial)$ of the Lie algebra $L(F)_\bullet$ forms a subcomplex (here $V^\vee$ is dual to V, and $L^\vee_{-n}$ is in degree 1):

$$L^\vee_{-n} \xrightarrow{\partial} \cdots \xrightarrow{\partial} L^\vee_{-2} \otimes \Lambda^{n-2}L^\vee_{-1} \xrightarrow{\partial} \Lambda^n L^\vee_{-1} \tag{16}$$

Its cohomology is predicted by formula (15). Moreover it should be quasiisomorphic to the weight n motivic complex for Spec(F): (15) provides its key property. So Conjecture A suggests that it should be quasiisomorphic to our complex $\Gamma(F,n)$.

One can argue (see Section 13 below) that one should have canonical injective homomorphisms $l_n : \mathcal{B}_n(F) \hookrightarrow L(F)^\vee_{-n}$. They should be isomorphisms for $n \leq 3$. However complex (16) is much bigger then $\Gamma(F,n)$. For example, for $n = 4$ in degree 2 of (16) appears $\Lambda^2L^\vee_{-2}(F) \stackrel{?}{=} \Lambda^2\mathcal{B}_2(F)$, which is absent in $\Gamma(F,4)$.

It turns out that Conjecture A is essentially equivalent to the following conjecture about the structure of the Lie algebra $L(F)_\bullet$. Set $I_\bullet := \oplus^\infty_{n=2}L(F)_{-n}$ and let $H^1_{(n)}(I(F)_\bullet)$ be the degree n part of $H^1(I(F)_\bullet)$.

CONJECTURE B. (a) *$I(F)_\bullet$ is a free graded pro-Lie algebra.*

(b) $H^1_{(n)}(I(F)_\bullet) = \mathcal{B}_n(F)_\mathbb{Q}$ for $n \geq 2$, i.e. *$I(F)_\bullet$ is generated as a graded pro-Lie algebra by the spaces $\mathcal{B}_n(F)^\vee$ of degree $-n$.*

(c) *The action of* $L_\bullet/I_\bullet = F_{\mathbb{Q}}^{*\vee}$ on $H_1^{(n)}(I(F)_\bullet) = \mathcal{B}_n(F)_{\mathbb{Q}}^{\vee}$ *coming from the extension* $0 \to H_1(I_\bullet) \to L_\bullet/[I_\bullet, I_\bullet] \to L_\bullet/I_\bullet \to 0$ *is described by the homomorphism dual to* $\delta_n : \mathcal{B}_n(F)_{\mathbb{Q}} \to \mathcal{B}_{n-1}(F)_{\mathbb{Q}} \otimes F^*$.

Assuming Conjecture B it is easy to see that the Hochschild-Serre spectral sequence for $H^*_{(n)}(L(F)_\bullet)$ with respect to the ideal $I_\bullet$ reduces exactly to the complex $\Gamma(F,n)$. Indeed thanks to (a) and (b) we have

$$E_1^{p,q} = C^p(L_\bullet/I_\bullet, H^q_{(n-p)}(I_\bullet)) = \begin{cases} \Lambda^p F_{\mathbb{Q}}^* \otimes \mathcal{B}_{n-p}(F)_{\mathbb{Q}} : & q = 1 \\ \Lambda^n F_{\mathbb{Q}}^* & : q = 0, n = p \\ 0 & : \text{otherwise} \end{cases}$$

and the differentials coincide with the ones in $\Gamma(F,n)$ because of (c).

10. Framed mixed Tate motives and $U(F)_\bullet$ [BMS], [BGSV]

A mixed $\mathbb{Q}$-Hodge structure H is called a Hodge-Tate structure if all the quotients $gr_\bullet^W H$ are of Hodge type (p,p). It is an n-framed Hodge-Tate structure if supplied with nonzero vectors $v \in gr_{2n}^W H$ and $f \in (gr_0^W H)^*$.

Consider the finest equivalence relation on the set of all n-framed Hodge-Tate structures for which $H_1 \sim H_2$ if there is a morphism of mixed Hodge structures $H_1 \to H_2$ respecting the frames. Let $\mathcal{H}_n$ be the set of equivalence classes. It has an abelian group structure: $(H; v, f) \oplus (H'; v', f') := (H \oplus H'; (v, v'), f + f')$ and $\mathbb{Q}(0) \oplus \mathbb{Q}(n)$ represents zero. Set $\mathcal{H}_0 := \mathbb{Z}$. The tensor product of mixed Hodge structures induces the commutative multiplication $\mu : \mathcal{H}_k \otimes \mathcal{H}_\ell \to \mathcal{H}_{k+\ell}$. A comultiplication $\nu = \bigoplus_k \nu_{k,n-k} : \mathcal{H}_n \to \bigoplus_k \mathcal{H}_k \otimes \mathcal{H}_{n-k}$ is defined as follows. Let $\{e_j\}$ and $\{e^j\}$ be dual bases in $gr_{2k}^W H_{\mathbb{Q}}$ and $gr_{2k}^W H_{\mathbb{Q}}^*$. Set $\nu_{k,n-k}((H; v, f)) := \sum_j (H; v, e^j) \otimes (H; e_j, f)$.

Then $\mathcal{H}_\bullet := \oplus \mathcal{H}_n$ is a commutative graded Hopf algebra.

Similarly the equivalence classes of n-framed objects in the category $\mathcal{M}_T(F)$ form a commutative graded Hopf algebra $\mathcal{M}_\bullet$. It maps to $U(F)_\bullet^\vee$: the value of the functional defined by $(\omega(M), v, f)$ on $A \in \operatorname{End}\omega$ is $< f, Av >$. *This map is an isomorphism of Hopf algebras.* In particular,

$$\operatorname{Ker}\Big(U(F)^\vee_{-n} \xrightarrow{\Delta} \oplus_k U(F)^\vee_{-(n-k)} \otimes U(F)^\vee_{-k}\Big) \overset{?}{\cong} gr_n^\gamma K_{2n-1}(F)_{\mathbb{Q}}\,. \qquad (17)$$

A variation of mixed Hodge structures over $\mathbb{C}P^1 \setminus \{0, 1, \infty\}$ related to the classical n-logarithm (see [BD1]) was discovered in the 1970s by Deligne.

It seems that any example of variation of framed mixed Tate motives should be of great interest; the corresponding Hodge periods deserve to be called polylogarithms (don't confuse them with the *classical* polylogarithms!). Below I discuss two such examples where periods are hyperlogarithms and volumes of noneuclidian geodesic simplices. Another example is shown in [BGSV].

11. Hyperbolic geometry [G4]

THEOREM 12. *Let* V^5 *be a 5-dimensional complete hyperbolic manifold of finite volume. Then there are algebraic numbers* $z_i \in \bar{\mathbb{Q}}^*$ *such that*

$$\sum_i \{z_i\}_2 \otimes z_i = 0 \text{ in } B_2(\bar{\mathbb{Q}}) \otimes \bar{\mathbb{Q}}^* \quad \text{and} \quad \operatorname{vol}(V^5) = \sum_i \mathcal{L}_3(z_i)\,.$$

A similar result for hyperbolic 3-manifolds was proved in [DS] and [NZ].

CONJECTURE 13. *Let V^{2n-1} be a $(2n-1)$-dimensional complete hyperbolic manifold of finite volume. Then there are algebraic numbers $z_i \in \bar{\mathbb{Q}} \subset \mathbb{C}$ such that $(n \geq 3)$ $\delta_n(\sum_i \{z_i\}_n) = 0$ and $\mathrm{vol}(V^{2n-1}) = \sum_i \mathcal{L}_n(z_i)$.*

A geodesic simplex M in the hyperbolic space $\mathcal{H}^m$ defines a mixed Tate motive. Indeed, in the Klein model $\mathcal{H}^m$ is the interior of a ball in $\mathbb{R}^m$ and geodesics are straight lines. So a geodesic simplex is the usual one inside the absolute: sphere Q.

Let us complexify and compactify this picture. We will get $\mathbb{CP}^m$ together with a quadric Q (the absolute) and a collection of hyperplanes $M = (M_1, \dots, M_{m+1})$ ($(n-1)$-faces of a geodesic simplex). $H^m(\mathbb{C}P^m \setminus Q, M)$ is a Hodge-Tate structure.

Let $m = 2n-1$ and $\tilde{Q}(x) = 0$ is a quadratic equation of Q. Set

$$\omega_Q := \pm \frac{\sqrt{\det \tilde{Q}}}{(2\pi i)^n} \frac{\sum_{i=o}^{2n-1}(-1)^i x_i dx_0 \wedge \dots \hat{dx_i} \wedge dx_{2n-1}}{\tilde{Q}(x)^n}.$$

The sign depends on the choise of a generator in the primitive part of $H^{n-1}(Q, \mathbb{Z})$. It is provided by the orientation of $\mathcal{H}^{2n-1}$. The simplex M defines a chain Δ_M representing a generator in $H_{2n}(\mathbb{C}P^{2n-1}, M)$. Then $\mathrm{vol}(M) = \int_{\Delta_M} \omega_Q$.

The scissor congruence group $\mathcal{P}(\mathcal{H}^m)$ is an abelian group generated by pairs $[M, \alpha]$ where M is an oriented geodesic simplex and α is an orientation of $\mathcal{H}^m$. The relations are: $[M, \alpha] = [M_1, \alpha] + [M_2, \alpha]$ if $M = M_1 \cup M_2$; $[M, \alpha]$ changes sign if we change orientation of M or α, and $[M, \alpha] = [gM, g\alpha]$ for any $g \in O(m, 1)$. Similarly, there are the spherical scissor congruence groups $\mathcal{P}(S^m)$.

The volume provides homomorphisms $\mathcal{P}(\mathcal{H}^m) \to \mathbb{R}$ and $\mathcal{P}(S^m) \to \mathbb{R}/\mathbb{Z}$.

We have a vector $[\omega_Q]$ in $H^{2n-1}(\mathbb{C}P^{2n-1} \backslash Q) = gr^W_{2n} H(Q, M)$ and a functional $[\Delta_M]$ on $H^{2n-1}(\mathbb{C}P^{2n-1}, M) = gr^W_0 H(Q, M)$. So we get an n-framed Hodge-Tate structure associated with $[M, \alpha]$. This construction defines a homomorphism of groups $\mathcal{P}(\mathcal{H}^{2n-1}) \to \mathcal{H}_n$ and similarly $\mathcal{P}(S^{2n-1}) \to \mathcal{H}_n$.

One can show that $\mathcal{P}(\mathcal{S}^{2k}) = 0$. Let us define the Dehn invariant

$$\mathcal{P}(\mathcal{H}^{2n-1}) \xrightarrow{D^h_n} \oplus_k \mathcal{P}(\mathcal{H}^{2k-1}) \otimes \mathcal{P}(S^{2(n-k)-1}).$$

Each $(2k-1)$-face A of M is a hyperbolic simplex $h(A)$. In the orthogonal plane $A^\perp$ M cuts a spherical simplex $s(A)$. Choose orientations α_A and β_A of A and $A^\perp$ such that $\alpha_A \otimes \beta_B = \alpha$. Then $D^h_n([M, \alpha]) := \sum_A [h(A), \alpha_A] \otimes [s(A), \beta_A]$.

THEOREM 14. *The following diagram is commutative:*

$$\begin{array}{ccc} \mathcal{P}(\mathcal{H}^{2n-1}) & \xrightarrow{D^h_n} & \oplus_k \mathcal{P}(\mathcal{H}^{2k-1}) \otimes \mathcal{P}(S^{2(n-k)-1}) \\ \downarrow & & \downarrow \\ \mathcal{H}_n & \xrightarrow{\nu} & \oplus_k \mathcal{H}_k \otimes \mathcal{H}_{n-k} \end{array}$$

A similar motivic interpretation has the spherical Dehn invariant

$$D^s_n : \mathcal{P}(S^{2n-1}) \longrightarrow \oplus_k \mathcal{P}(\mathcal{S}^{2k-1}) \otimes \mathcal{P}(S^{2(n-k)-1}).$$

So (17) leads to

CONJECTURE 15. *There are canonical injective homomorphisms*

$$\mathrm{Ker}\, D_n^h \otimes \mathbb{Q} \hookrightarrow [gr_n^\gamma K_{2n-1}(\mathbb{C}) \otimes \mathbb{Q}(n)]^- \qquad \mathrm{Ker}\, D_n^s \otimes \mathbb{Q} \hookrightarrow [gr_n^\gamma K_{2n-1}(\mathbb{C}) \otimes \mathbb{Q}(n)]^+$$

whose compositions with Beilinson's regulator coincide with the volume homomorphisms.

For $n = 2$ these homomorphisms exist and are isomorphisms by the results of [D], [DS], and [S1].

Each complete hyperbolic $(2n-1)$-manifold can be cut on geodesic simplices and so produces an element in $\mathcal{P}(\mathcal{H}^{2n-1})$. Its Dehn invariant is equal to zero. So Conjecture 13 follows from Conjectures 15 and A.

12. Hyperlogarithms [G5]

These were considered by Kummer [Ku], Poincaré, Lappo-Danilevsky, etc. We define them as the following iterated integrals:

$$\Psi_{m_1,\dots,m_l}(a_1,\dots,a_l) := \int_0^1 \underbrace{\frac{dt}{t-a_1} \circ \frac{dt}{t} \circ \dots \circ \frac{dt}{t}}_{m_1 \text{ times}} \circ \dots \circ \underbrace{\frac{dt}{t-a_l} \circ \frac{dt}{t} \circ \dots \frac{dt}{t}}_{m_l \text{ times}} .$$

This formula means the following. Let $n := m_1 + \dots + m_l$ and

$$\begin{aligned}\Delta := \{(t_1,\dots,t_n) \subset \mathbb{R}^n | 0 &\leq t_1 - a_1 \leq t_2 \leq \dots \leq t_{m_1} \leq t_{m_1+1} - a_2 \\ &\leq t_{m_1+2} \dots \leq t_{m_l}\}.\end{aligned}$$

Let L be a coordinate simplex in $\mathbb{C}P^n$ related to coordinates $(t_0 : \dots : t_n)$ and $\omega_L := \frac{dt_1}{t_1} \wedge \dots \wedge \frac{dt_n}{t_n}$. Then $\Psi_{m_1,\dots,m_l}(a_1,\dots,a_l) = \int_\Delta \omega_L$.

Let M be a collection of all the hyperplanes corresponding to codimension 1 faces of Δ. Then $H(L,M) := H^n(\mathbb{C}P^n \backslash L, M)$ is a Hodge-Tate structure. It has canonical n-framing: $[\omega_L]$ is a vector in $H^n(\mathbb{C}P^n \backslash L) = gr_{2n}^W H(L,M)$ and Δ produces a class $[\Delta] \in H_n(\mathbb{C}P^n, M) = gr_0^W H(L,M)$. So we get an element $\Psi^{\mathcal{H}}_{m_1,\dots,m_l}(a_1,\dots,a_l) \in \mathcal{H}_n$. According to the general philosophy *a mixed Hodge structure in the cohomology of a (simplicial) variety is a realization of a mixed motive.* So we should have an n-framed mixed Tate motive $\Psi^{\mathcal{M}}_{m_1,\dots,m_l}(a_1,\dots,a_l)$.

More generally, if F is a field and $a_i \in F^*$ one should also have an n-framed mixed Tate motive $\Psi^{\mathcal{M}}_{m_1,\dots,m_l}(a_1,\dots,a_l)$ related to $H^n(P^n_F \backslash L, M)$.

There is a remarkable power series expansion of the hyperlogarithms. Namely, consider *multiple* polylogarithms (which were also thought of by Zagier)

$$\Phi_{m_1,\dots,m_l}(x_1,\dots,x_l) := \quad (-1)^l \sum_{0<k_1<k_2<\dots<k_l} \frac{x_1^{k_1} x_2^{k_2} \dots x_l^{k_l}}{k_1^{m_1} k_2^{m_2} \dots k_l^{m_l}} .$$

THEOREM 16. [G5]. *Suppose* $|a_i/a_{i-1}| < 1$. *Then*

$$\Psi_{m_1,\dots,m_l}(a_1,\dots,a_l) = \Phi_{m_1,\dots,m_l}(\frac{a_2}{a_1}, \frac{a_3}{a_2}, \dots, \frac{1}{a_l}).$$

In particular, $\zeta(m_1,\dots,m_l) := \Phi_{m_1,\dots,m_l}(1,1,\dots,1)$ are the multiple zeta values of Euler [E], rediscovered and studied in [Z3]; see also [Dr] and [Ko].

CONJECTURE 17. *(a) Any n-framed mixed Tate motive over F is a sum of hyperlogarithmic ones* $\Psi^{\mathcal{M}}_{m_1,\dots,m_l}(a_1,\dots,a_l)$, *where* $n = m_1 + \cdots + m_l$; $a_i \in F^*$.

(b) Any n-framed mixed Tate motive over Spec($\mathbb{Z}$) *is a sum of motivic multiple zeta's* $\zeta^{\mathcal{M}}(m_1,\dots,m_l)$.

The first part of the conjecture is motivated by the following

PROPOSITION 18. (Universality of hyperlogarithms) *Any iterated integral* $F(z) = \int_x^z \omega_1 \circ \cdots \circ \omega_n$ *of rational 1-forms* ω_i *on a rational variety X is a sum of hyperlogarithms, i.e. there exist* $f_j^{(i)}(z) \in \mathbb{C}(X)^*$ *such that*

$$F(z) = \sum_i \Psi_{m_1^{(i)},\dots,m_{l(i)}^{(i)}}(f_1^{(i)}(z),\dots,f_l^{(i)}(z)) \quad + \quad C \qquad (C \textit{ is a constant}).$$

13. Motivic interpretation of the "weak" part of Conjecture A

For any $a \in F^*$ the n-framed mixed Tate motive $\Psi_n^{\mathcal{M}}(a^{-1})$ (corresponding to $Li_n(a)$) provides a homomorphism $\tilde{l}_n : \mathbb{Z}[F^*] \to U(F)^{\vee}_{-n}$. Denote by l_n its composition with the canonical projection $U(F)^{\vee}_{-n} \to L(F)^{\vee}_{-n}$.

One should have $l_n(\mathcal{R}_n(F)) = 0$, so $l_n : \mathcal{B}_n(F) \to L(F)^{\vee}_{-n}$. It turns out that $\partial(l_n\{a\}) = l_{n-1}\{a\} \wedge a$ (we identified $L(F)^{\vee}_{-1}$ with $F^*_{\mathbb{Q}}$). Therefore *homomorphisms* $\{l_i\}$ *provide a canonical homomorphism of the complex* $\Gamma(F,n)$ *to the complex (16).* Using (15) we get canonical maps $H^i(\Gamma(F,n) \otimes \mathbb{Q}) \to gr^{\gamma}_n K_{2n-i}(F)_{\mathbb{Q}}$.

14. The quantum dilogarithm [FK]

Mixed Tate motives give the best explanation *all* of the different appearances of the dilogarithm discussed above. However recently the dilogarithm appeared in conformal field theory and exactly solvable problems of statistical mechanics. Here is one example.

Let $\Psi(x) := \prod_{n=1}^{\infty}(1 - xq^n)$, $|q| < 1$. Then for $q = \exp(\epsilon)$, $\mathrm{Im}(\epsilon) < 0$.

$$\Psi(x) = \frac{1}{\sqrt{1-x}} \exp(Li_2(x)/\epsilon)(1 + O(\epsilon)), \quad \epsilon \to 0.$$

THEOREM 19. ([FK]) *Suppose* $\hat{U}$ *and* $\hat{V}$ *satisfy* $\hat{U}\hat{V} = q\hat{V}\hat{U}$. *Then*

$$\Psi(\hat{V})\Psi(\hat{U}) = \Psi(\hat{U})\Psi(-\hat{U}\hat{V})\Psi(\hat{V})$$

and in the classical limit we get the 5-term relation for the Rogers dilogarithm.

Acknowledgment. I am grateful to Professor A. Borel for many useful remarks about a preliminary version of the paper.

References

[B1] Beilinson, A. A., *Height pairings between algebraic cycles*, Lecture Notes in Math. 1289, (1987), 1–26.

[B2] Beilinson, A. A., *Higher regulators and values of L-functions*, VINITI 24 (1984), 181–238.

[B3] Beilinson, A. A., *Higher regulators for modular curves*, Cont. Math. vol. 55, (1987), 1–34.

[BD1] Beilinson, A. A. and Deligne, P., *Interpretation motivique de la conjecture de Zagier* in Sympos. Pure Math. vol. 55, part 2, (1994) 23–41.

[BD2] Beilinson, A. A. and Deligne, P., *Polylogarithms and regulators*, in preparation.

[BL] Beilinson, A. A. and Levin, A., *Elliptic polylogarithms* in Sympos. Pure Math. vol. 55, part 2, (1994) 101–156.

[BMS] Beilinson, A. A.; MacPherson, R.; and Schechtman, V. V., *Notes on motivic cohomology*, Duke Math. J. 54 (1987), 679–710.

[BGSV] Beilinson, A. A.; Goncharov, A. B.; Schechtman, V. V.; and Varchenko, A. N., *Projective geometry and algebraic K-theory*, Algebra and Analysis (1990) no. 3, 78–131.

[Bl1] Bloch, S., Higher regulators, algebraic K-theory and zeta functions of elliptic curves, Lecture Notes U.C. Irvine, (1977).

[Bl2] Bloch, S., *Algebraic cycles and higher K-theory*, Adv. in Math. (1986), vol. 61, 267–304.

[BK] Bloch, S. and Kriz, I., *Mixed Tate motives*, preprint, 1992.

[Bo1] Borel, A., *Stable real cohomology of arithmetic groups*, Ann. Sci. École Norm. Sup., (4) 7 (1974), 235–272.

[Bo2] Borel, A., *Cohomologie de SL_n et valeurs de fonctions zêta aux points entiers*, Ann. Scuola Norm. Sup. Pisa Cl. Sci. (4) (1977), 613–636.

[Den] Deninger, C., *Higher order operations in Deligne cohomology*, Invent. Math. 120, no. 2, (1995), 289–316.

[Dr] Drinfeld, V. G., *On quasi-triangular quasi-Hopf algebras and some groups related to Gal*($\bar{\mathbb{Q}}/\mathbb{Q}$), Algebra and Analysis (1991).

[D] Dupont, J., *Homology of flag complexes* Osaka Math. J. 5, (1982), 599–641.

[DS] Dupont, J. and Sah, C.-H., *Scissors congruences II*, J. Pure Appl. Algebra vol. 25 (1982), 159–195.

[E] Euler, L., Opera Omnia, Ser. 1, Vol XV, Teubner, Berlin 1917, 217–267.

[FK] Faddeev, L. D. and Kashaev, R. M., Quantum dilogarithm, preprint, 1993.

[GGL] Gabrielov, A. M.; Gelfand, I. M.; and Losik, M. V., *Combinatorial computation of characteristic classes*, Functional Anal. Appl. vol. 9, no. 2 (1975), 5–26.

[GM] Gelfand, I. M. and MacPherson, R., *Geometry in Grassmannians and a generalisation of the dilogarithm*, Adv. in Math. 44 (1982), 279–312.

[G1] Goncharov, A. B., *Geometry of configurations, polylogarithms and motivic cohomology*, Adv. in Math. vol. 114, no. 2, (1995), 197–318.

[G2] Goncharov, A. B., *Polylogarithms and motivic Galois groups*, Sympos. Pure Math. vol. 55, part 2, (1994), 43–97.

[G3] Goncharov, A. B., *Explicit construction of characteristic classes*, Adv. Sov. Math. (1993), vol. 16, 169–210.

[G4] Goncharov, A. B., *Volumes of hyperbolic manifolds*, preprint, 1993.

[G5] Goncharov, A. B., *Multiple ζ-values, hyperlogarithms and mixed Tate motives*, preprint, 1993.

[G6] Goncharov, A. B., *Special values of Hasse-Weil L-functions and generalized Eisenstein-Kronecker series*, preprint, 1994.

[HM] Hain, R. and MacPherson, R., *Higher Logarithms*, Ill. J. Math. vol. 34, (1990) no. 2, 392–475.

[HaM] Hanamura, M. and MacPherson, R., *Geometric construction of polylogarithms*, Duke Math. J. 70, no. 3, (1993), 481–517.

[J] De Jeu, R., *Zagier's conjecture and wedge complexes in algebraic K-theory*, preprint, Utrecht University, 1993.

[Ko] Kontsevich, M., private communication.

[K] Kummer, E. E., *Über die Transzendenten, welche aus wiederholten Integrationen rationaler Formeln entstehen*, J. Reine Angew. Math. 21 (1840), 74–90.

[Le] Levine, M., *The derived motivic category*, preprint, 1993.

[L] Lewin, L., Dilogarithms and Associated Functions, North-Holland, 1981.

[Lei] Gerhardt, C. I. (ed)., G. W. Leibniz, Mathematische Schriften III/1 pp. 336–339 Georg Olms Verlag, Hildesheim and New York, 1971.

[Li] Lichtenbaum, S., *Values of zeta functions at non-negative integers*, Lecture Notes in Math., 1068, Springer-Verlag, Berlin and New York, (1984), 127–138.

[NZ] Neumann, W. and Zagier, D., *Volumes of hyperbolic 3-folds*, Topology 24 (1985), 307–331.

[S1] Suslin, A. A., *K_3 of a field and Bloch's group*, Proc. Steklov Inst. (1991), Issue 4.

[S2] Suslin, A. A., *Homology of GL_n, characteristic classes and Milnor's K-theory*, Lecture Notes in Math. 1046 (1989), 357–375.

[V] Voevodsky, V., *Triangulated categories of motives over a fileld*, preprint, 1994.

[Y] Yang, R., Ph.D. thesis, Univ. of Washington, Seattle (1991).

[Z1] Zagier, D., *Polylogarithms, Dedekind zeta functions and the algebraic K-theory of fields*, Arithmetic Algebraic Geometry (G.v.d. Geer, F. Oort, and J. Steenbrink, eds.), Prog. Math., vol. 89, Birkhäuser, Basel and Boston, (1991), pp. 391–430.

[Z2] Zagier, D., *Hyperbolic manifolds and special values of Dedekind zeta functions*, Invent. Math. 83 (1986), 285–301.

[Z3] Zagier, D., *Special values of L-functions*, Proc. First European Congress of Mathematicians in Paris, (1992).

Unexpected Irregularities in the Distribution of Prime Numbers

ANDREW GRANVILLE*

Department of Mathematics
University of Georgia
Athens, Georgia 30602, USA

In 1849 the Swiss mathematician ENCKE wrote to GAUSS, asking whether he had ever considered trying to estimate $\pi(x)$, the number of primes up to x, by some sort of "smooth" function. On Christmas Eve 1849, GAUSS replied that "*he had pondered this problem as a boy*" and had come to the conclusion that "*at around x, the primes occur with density* $1/\log x$." Thus, he concluded, $\pi(x)$ could be approximated by

$$\mathrm{Li}(x) := \int_2^x \frac{dt}{\log t} = \frac{x}{\log x} + \frac{x}{\log^2 x} + O\left(\frac{x}{\log^3 x}\right).$$

Comparing GAUSS's guess to the best data available today (due to DELEGLISE and RIVAT), we have:

x	$\pi(x)$	$[\mathrm{Li}(x) - \pi(x)]$
10^8	5761455	754
10^9	50847534	1701
10^{10}	455052511	3104
10^{11}	4118054813	11588
10^{12}	37607912018	38263
10^{13}	346065536839	108971
10^{14}	3204941750802	314890
10^{15}	29844570422669	1052619
10^{16}	279238341033925	3214632
10^{17}	2623557157654233	7956589
10^{18}	24739954287740860	21949555

The number of primes, $\pi(x)$, up to x.

This data certainly seems to support GAUSS's prediction, because $\mathrm{Li}(x) - \pi(x)$ appears to be no bigger than a small power of $\pi(x)$. In 1859, RIEMANN, in a now famous memoir, illustrated how the question of estimating $\pi(x)$ could be

*) The author is an Alfred P. Sloan Research Fellow; and is also supported, in part, by the National Science Foundation.

Proceedings of the International Congress
of Mathematicians, Zürich, Switzerland 1994

turned into a question in analysis: Define $\zeta(s) := \sum_{n\geq 1} n^{-s}$ for $\mathrm{Re}(s) > 1$, and then analytically continue $\zeta(s)$ to the rest of the complex plane. We have, for sufficiently large x,

$$x^{b-\varepsilon} \ll \max_{y\leq x} |\pi(y) - \mathrm{Li}(y)| \ll x^{b+\varepsilon} \text{ where } b := \sup_{\zeta(\beta+i\gamma)=0} \beta. \tag{1}$$

The (as yet unproven) *Riemann Hypothesis* (RH) asserts that $b = 1/2$ (in fact that $\beta = 1/2$ whenever $\zeta(\beta+i\gamma) = 0$ with $0 \leq \beta \leq 1$) leading to the sharp estimate

$$\pi(x) \;=\; \mathrm{Li}(x) + O(x^{1/2} \log x)\ . \tag{2a}$$

It was not until 1896 that HADAMARD and DE LA VALLÉE-POUSSIN independently proved that $\beta < 1$ whenever $\zeta(\beta + i\gamma) = 0$, which implies *the prime number theorem*: that is, GAUSS's prediction that

$$\pi(x) \;\sim\; \mathrm{Li}(x) \;\sim\; \frac{x}{\log x}.$$

In 1914, LITTLEWOOD showed, unconditionally, that

$$\pi(x) - \mathrm{Li}(x) = \Omega_{\pm}\left(x^{1/2}\, \frac{\log\log\log x}{\log x}\right), \tag{2b}$$

the first proven "irregularities" in the distribution of primes.[1]

As GAUSS's vague "density assertion" was so prescient, CRAMÉR [4] decided, in 1936, to interpret GAUSS's statement more formally in terms of probability theory, to try to make further predictions about the distribution of prime numbers: let $Z_2, Z_3, \ldots$ be a sequence of independent random variables with

$$\mathrm{Prob}(Z_n = 1) \;=\; \frac{1}{\log n} \quad \text{and} \quad \mathrm{Prob}(Z_n = 0) \;=\; 1 - \frac{1}{\log n}\ .$$

Let S be the space of sequences $T = z_2, z_3, \ldots$ and for each $x \geq 2$ define

$$\pi_T(x) \;=\; \sum_{2\leq n\leq x} z_n\ .$$

The sequence $P = \pi_2, \pi_3, \ldots$, where $\pi_n = 1$ if and only if n is prime, belongs to S. CRAMÉR wrote: "*In many cases it is possible to prove that,* with probability *1, a certain relation R holds for sequences in S ... Of course we cannot in general conclude that R holds for the particular sequence P, but results suggested in this way may sometimes afterwards be rigorously proved by other methods.*" For example CRAMÉR was able to show, with probability 1, that

$$\max_{y\leq x} |\pi_T(y) - \mathrm{Li}(y)| \sim \sqrt{2x \cdot \frac{\log\log x}{\log x}},$$

[1] $f(x) = \Omega_{\pm}(g(x))$ means that there exists a constant $c > 0$ such that $f(x_+) > cg(x_+)$ and $f(x_-) < -cg(x_-)$ for certain arbitrarily large values of $x_{\pm}$.

which corresponds well with the estimates in (2); and if true for $T = P$ implies RH, by (1).

GAUSS's assertion was really about primes in short intervals, and so is best applied to $\pi(x+y)-\pi(x)$, where y is "small" compared to x. The binomial random variables Z_n are, more-or-less, the same for all integers n in such an interval. If we take $y = \lambda \log x$ so that the "expected" number of primes λ in the interval is fixed then we would expect that the number of primes in such intervals should follow a Poisson distribution. Indeed, we can prove that for any fixed $\lambda > 0$ and integer $k \geq 0$, we have

$$\#\{\text{integers } x \leq X : \quad \pi_T(x+\lambda \log x) - \pi_T(x) = k\} \sim e^{-\lambda}\frac{\lambda^k}{k!}X \tag{3}$$

as $X \to \infty$, with probability 1 for $T \in S$. In 1976, GALLAGHER [11] showed that this holds for the sequence of primes (that is, for P) under the assumption of a reasonable "uniform" version of HARDY AND LITTLEWOOD's *Prime k-tuplets conjecture* [14]. This conjecture is the case where we take each $f_j(x)$ to be a linear polynomial in SCHINZEL [22]

HYPOTHESIS H. *Let $F = \{f_1(x), f_2(x), \ldots, f_k(x)\}$ be a set of irreducible polynomials with integer coefficients. Then the number of integers $n \leq x$ for which each $|f_j(n)|$ is prime is*

$$\pi_F(x) = \{C_F + o(1)\}\frac{x}{\log|f_1(x)|\log|f_2(x)|\ldots\log|f_k(x)|}$$

$$\text{where} \quad C_F = \prod_{p \text{ prime}} \left(1 - \frac{\omega_F(p)}{p}\right) \Big/ \left(1-\frac{1}{p}\right)^k,$$

and $\omega_F(p)$ counts the number of integers n, in the range $1 \leq n \leq p$, for which $f_1(n)f_2(n)\ldots f_k(n) \equiv 0 \pmod p$.[2,3]

Estimates analogous to (2a) should hold for the number of primes in intervals of various lengths, if we believe that what almost always occurs in S should also hold for P. Specifically, if $10\log^2 x \leq y \leq x$ then

$$\pi_T(x+y) - \pi_T(x) \quad = \quad \mathrm{Li}(x+y) - \mathrm{Li}(x) + O(y^{1/2}) \tag{4}$$

with probability 1 for $T \in S$. In 1943 SELBERG [23] showed that primes do, on the whole, behave like this by proving, under the assumption of RH, that

$$\pi(x+y) - \pi(x) \sim \frac{y}{\log x} \tag{5}$$

for "almost all" integers x, provided $y/\log^2 x \to \infty$ as $x \to \infty$.

MONTGOMERY [17] has shown that one can deduce estimates about primes in short intervals by understanding local distribution properties of the zeros of $\zeta(s)$:

[2]Elementary results on prime ideals guarantee that the product defining C_F converges if the primes are taken in ascending order.

[3]The asymptotic formula proposed here for $\pi_F(x)$ has a "local part" C_F, which has a factor corresponding to each rational prime p, and an "analytic part" $x/\prod_i \log|f_i(x)|$. This reminds one of formulae that arise when counting points on varieties.

PAIR CORRELATION CONJECTURE (PC). *Assume* RH. *For any fixed* $\alpha > 0$, *the average number of zeros* $1/2 + i\gamma'$ *of* $\zeta(s)$ *"close" to a given zero* $1/2 + i\gamma$, *that is with* $\gamma < \gamma' \le \gamma + 2\pi\alpha/\log(|\gamma| + 1)$, *is*

$$\sim \int_0^\alpha \left\{1 - \left(\frac{\sin \pi u}{\pi u}\right)^2\right\} du. \tag{6a}$$

Inspired by the work of MONTGOMERY and others, GOLDSTON [12] showed that (6a) holds for every fixed $\alpha > 0$ *if and only if* for every fixed $\beta > 0$,

$$\int_T^{T^{1+\beta}} \left(\psi\left(x + \frac{x}{T}\right) - \psi(x) - \frac{x}{T}\right)^2 \frac{dx}{x^2} \sim \beta \frac{\log^2 T}{T}, \tag{6b}$$

where $\psi(x) := \sum_{p^m \le x} \log p$. Because (6b) is predicted by CRAMÉR's model, thus so is PC. DYSON predicted an analogous density function for the correlation of n-tuples of zeros of $\zeta(s)$,[4] which presumably may be shown to be equivalent to estimates for primes in short intervals, and thus be predicted by CRAMÉR's model.

CRAMÉR's model does seem to accurately predict what we already believe to be true about primes for more substantial reasons.[5] To be sure, one can find small discrepancies[6] but the probabilistic model usually gives one a strong indication of the truth. CRAMÉR made one conjecture, based on his model, that does not seem to be attackable by other methods: if $p_1 = 2 < p_2 = 3 < p_3 = 5 < \ldots$ is the sequence of prime numbers then

$$\max_{p_n \le x} (p_{n+1} - p_n) \sim \log^2 x.$$

This statement (or the weaker $O(\log^2 x)$) is known as "*Cramér's conjecture*"; there is some computational evidence to support it:

p_n	$p_{n+1} - p_n$	$(p_{n+1} - p_n)/\log^2 p_n$
31397	72	.6715
370261	112	.6812
2010733	148	.7025
20831323	210	.7394
25056082087	456	.7953
2614941710599	652	.7975
19581334192423	778	.8177

Record-breaking gaps between primes, up to 10^{14}

[4]Based somewhat surprisingly on the fact that (6a) also describes the distribution of eigenvalues of random Hermitian matrices, which arise in physics as models for various naturally occurring quanta. The "Fourier transforms" of these pair correlation and n-tuple correlation conjectures for the zeros of $\zeta(s)$ are now known to hold in a natural restricted range, under the assumption of RH (see [17] and [21]).

[5]Though HARDY AND LITTLEWOOD [14] remarked thus on probabilistic models: "*Probability is not a notion of pure mathematics, but of philosophy or physics*".

[6]As has been independently pointed out to me by SELBERG, MONTGOMERY, and PINTZ: for example, PINTZ noted that the mean square of $|\psi(y) - y|$ for $y \le x$, is $\gg x^{2b-\varepsilon}$ (with b as in (1)), in fact $\asymp x$ assuming RH, whereas the probabilistic model predicts $\asymp x \log x$.

In 1985 MAIER [16] surprisingly proved that, despite SELBERG showing (5) holds "almost all" the time when $y = \log^B x$ (assuming RH) for fixed $B > 2$, it cannot hold all of the time for such y. This not only radically contradicts what is predicted by the probabilistic model, but also what most researchers in the field had believed to be true, whether or not they had faith in the probabilistic model. Specifically, MAIER showed the existence of a constant $\delta_B > 0$ such that for occasional, but arbitrarily large, values of x_+ and x_-,

$$\begin{aligned} & \pi(x_+ + \log^B x_+) - \pi(x_+) > (1+\delta_B)\log^{B-1} x_+ , \\ \text{and} \quad & \pi(x_- + \log^B x_-) - \pi(x_-) < (1-\delta_B)\log^{B-1} x_- . \end{aligned} \tag{7}$$

Outline of the Proof. There are $\sim e^{-\gamma}x/\log z$ integers $\leq x$, all of whose prime factors are $> z$, provided z is not too large. Among these we have all but $\pi(z)$ of the primes $\leq x$, and so the probability that a randomly chosen such integer is prime is $\sim e^{\gamma}\log z/\log x$. Thus, in a specific interval $(x, x+y]$ we should "expect" $\sim e^{\gamma}\Phi\log z/\log x$ primes, where Φ is the number of integers in the interval that are free of prime factors $\leq z$. Now if we can select our interval so that $\Phi \not\sim e^{-\gamma}y/\log z$ then our new prediction is not the same as that in (5).

If x is divisible by $P = \prod_{p\leq z} p$ then

$$\Phi = \Phi(y,z) := \#\{1 \leq n \leq y : \ p|n \Rightarrow p > z\} \sim \omega(u)\frac{y}{\log z}$$

for $y = z^u$ with u fixed (see [3]), where $\omega(u) = 0$ if $0 < u < 1$ and satisfies the differential-delay equation $u\omega(u) = 1 + \int_1^{u-1}\omega(t)\,dt$ if $u \geq 1$. Obviously $\lim_{u\to\infty}\omega(u) = \mathrm{e}^{-\gamma}$. IWANIEC showed that $\omega(u) - \mathrm{e}^{-\gamma}$ oscillates, crossing zero either once or twice in every interval of length 1. Thus, if we fix $u > B$, chosen so that $\omega(u) > \mathrm{e}^{-\gamma}$ or $< \mathrm{e}^{-\gamma}$ (as befits the case of (7)), select $y = \log^B x$ and $z = y^{1/u}$, and "adjust" x so that it is divisible by P, then we expect (5) to be false.

To convert this heuristic into a proof, MAIER considered a progression of intervals of the form $(rP, rP+y]$, with $R \leq r < 2R$ for a suitable value of R. Visualizing this as a "matrix", with each such interval represented by a different row, we see that the primes in the matrix are all contained in those columns j for which $(j, P) = 1$.

$RP+1$	$RP+2$	$RP+3$	$\cdots$	$RP+y$
$(R+1)P+1$	$(R+1)P+2$	$(R+1)P+3$	$\cdots$	$(R+1)P+y$
$(R+2)P+1$	$(R+2)P+2$	$\cdot$	$\cdot$	$\vdots$
$(R+3)P+1$	$\vdots$	(i,j)th entry : $(R+i)P+j$	$\vdots$	$\vdots$
$\vdots$	$\vdots$	$\cdot$	$\vdots$	$\vdots$
$(2R-1)P+1$	$(2R-1)P+2$	$\cdots$	$\cdots$	$(2R-1)P+y$

The "Maier matrix" for $\pi(x+y)-\pi(x)$

Now, for any integer $q > 1$, the primes are roughly equi-distributed amongst those arithmetic progressions $a \pmod q$ with $(a, q) = 1$: in fact up to x we expect that the number of such primes

$$\pi(x; q, a) \sim \frac{\pi(x)}{\phi(q)}. \tag{8}$$

If so, then the number of primes in the jth column, when $(j, P) = 1$, is

$$\pi(2X; P, j) - \pi(X; P, j) \sim \frac{1}{\phi(P)} \frac{X}{\log X} \quad \text{where } X = RP.$$

To get the total number of primes in the matrix we sum over all such j, and then we can deduce that, on average, a row contains

$$\sim \frac{\Phi(y, z)}{R} \frac{1}{\phi(P)} \frac{RP}{\log RP} \sim \omega(u) \frac{y}{\log z} \frac{P}{\phi(P)} \frac{1}{\log RP} \sim e^{\gamma} \omega(u) \frac{y}{\log RP}$$

primes. MAIER's result follows provided we can prove a suitable estimate in (8). □

In general, it is desirable to have an estimate like (8) when x is not too large compared to q. It has been proved that (8) holds uniformly for

(i) All $q \leq \log^B x$ and all $(a, q) = 1$, for any fixed $B > 0$ (SIEGEL-WALFISZ).
(ii) All $q \leq \sqrt{x}/\log^{2+\varepsilon} x$ and all $(a, q) = 1$, assuming GRH.[7] In fact (8) then holds with error term $O\left(\sqrt{x}\log^2(qx)\right)$.
(iii) Almost all $q \leq \sqrt{x}/\log^{2+\varepsilon} x$ and all $(a, q) = 1$ (BOMBIERI-VINOGRADOV).[8]
(iv) Almost all $q \leq x^{1/2+o(1)}$ with $(q, a) = 1$, for fixed $a \neq 0$ (BOMBIERI-FRIEDLANDER-IWANIEC, FOUVRY).
(v) Almost all $q \leq x/\log^{2+\varepsilon} x$ and almost all $(a, q) = 1$ (BARBAN-DAVENPORT-HALBERSTAM, MONTGOMERY, HOOLEY).

Thus, when GRH is true, we get a good enough estimate in (8) with $R = P^2$ to complete MAIER's proof. However MAIER, in the spirit of the BOMBIERI-VINOGRADOV theorem, showed how to pick a "*good*" value for P (see [8, Prop. 2]), so that (8) is off by, at worst, an insignificant factor when R is a large, but fixed, power of P (thus proving his result unconditionally).

In [15], HILDEBRAND AND MAIER extended the range for y in the proof above, establishing that there are arbitrarily large values of x for which (4) fails to hold for some $y > \exp\left((\log x)^{1/3-\varepsilon}\right)$; and, assuming GRH, for some $y > \exp\left((\log x)^{1/2-\varepsilon}\right)$. Moreover, they show that such intervals $(x, x + y]$ occur within every interval $[X, 2X]$.[9]

It is plausible that (5) holds uniformly if $\log y/\log\log x \to \infty$ as $x \to \infty$; and that (4) holds uniformly for $T = P$ if $y > \exp\left((\log x)^{1/2+\varepsilon}\right)$ (at least, we can't

[7] The *Generalized Riemann Hypothesis* (GRH) states that if $\beta + i\gamma$ is a zero of *any* Dirichlet L-function then $\beta \leq 1/2$.

[8] This result is often referred to as "GRH on average".

[9] A far better localization than those obtained in any proof of (2b).

disprove these statements as yet). We conjecture, presumably safely, that (4) and (5) hold uniformly when $y > x^{\varepsilon}$.

One can show that there are more than $x/\exp((\log x)^{c_B})$ integers $x_\pm \leq x$ satisfying the unexpected inequalities in (7). Although this may not be enough to upset (6b), it surely guarantees that the error term there will not be as small as might have been hoped. Thus, we should not expect the pair correlation conjecture to hold with as much uniformity as might have been believed. Evidence that this is so may be seen in the computations represented by [19, Figures 2.3.1,2,3]: the pair correlation function for the nearest 10^6 zeros to the 10^{12}th zero fits well with (6a) for $\alpha < .8$ and then has larger amplitude for $.8 < \alpha < 3$; also, the pair correlation function for the nearest 8×10^6 zeros to the 10^{20}th zero fits well with (6a) for $\alpha < 3$ and then has larger amplitude for $3 < \alpha < 5$.

MAIER's work suggests that CRAMÉR's model should be adjusted to take into account divisibility of n by "small" primes.[10] It is plausible to define "small" to mean those primes up to a fixed power of $\log n$. Then we are led to conjecture that there are infinitely many primes p_n with $p_{n+1} - p_n > 2\mathrm{e}^{-\gamma} \log^2 p_n$, contradicting CRAMÉR's conjecture![11]

If we analyze the distribution of primes in arithmetic progressions using a suitable analogue of CRAMÉR's model, then we would expect (8), and even

$$\pi(x;q,a) = \frac{\pi(x)}{\phi(q)} + O\left(\left(\frac{x}{q}\right)^{1/2} \log(qx)\right) , \tag{8'}$$

to hold uniformly when $(a,q) = 1$ in the range

$$q \leq Q = x/\log^B x , \tag{9}$$

for any fixed $B > 2$. However the method of MAIER is easily adapted to show that neither (8) nor (8′) can hold in at least part of the range (9): for any fixed $B > 0$ there exists a constant $\delta_B > 0$ such that for any modulus q, with "not too many small prime factors", there exist arithmetic progressions $a_\pm \pmod q$ and values $x_\pm \in [\phi(q)\log^B q, 2\phi(q)\log^B q]$ such that

$$\pi(x_+;q,a_+) > (1+\delta_B)\frac{\pi(x_+)}{\phi(q)} \quad \text{and} \quad \pi(x_-;q,a_-) < (1-\delta_B)\frac{\pi(x_-)}{\phi(q)} . \tag{10}$$

The proof is much as before, though now using a modified "Maier matrix":

[10] One has to be careful about the meaning of "small" here, because if we were to take into account the divisibility of n by all primes up to $\sqrt{n}$, then we would conclude that there are $\sim \mathrm{e}^{-\gamma} x/\log x$ primes up to x.

[11] It is unclear what the "correct conjecture" here should be because, to get at it with this approach, we would need more precise information on "sifting limits" than is currently available.

RP	$RP+q$	$RP+2q$	$\cdots$	$RP+yq$
$(R+1)P$	$(R+1)P+q$	$(R+1)P+2q$	$\cdots$	$(R+1)P+yq$
$(R+2)P$	$(R+2)P+q$	$\cdot$	$\cdot$	$\vdots$
$(R+3)P$	$\vdots$	(i,j)th entry : $(R+i)P+jq$	$\vdots$	$\vdots$
$\vdots$	$\vdots$	$\cdot$	$\vdots$	$\vdots$
$(2R-1)P$	$(2R-1)P+q$	$\cdots$	$\cdots$	$(2R-1)P+yq$

The Maier matrix for $\pi(yq;q,a)$

The BOMBIERI-VINOGRADOV theorem is usually stated in a stronger form than above: for any given $A>0$, there exists a value $B=B(A)>0$ such that

$$\sum_{q\le Q}\max_{(a,q)=1}\max_{y\le x}\left|\pi(y;q,a)-\frac{\pi(y)}{\phi(q)}\right| \ll \frac{x}{\log^A x} \tag{11}$$

where $Q=\sqrt{x}/\log^B x$. It is possible [6] to take the same values of R and P in the Maier matrix above for many different values of q, and thus deduce that there exist arbitrarily large values of a and x for which

$$\left|\sum_{\substack{Q\le q\le 2Q\\(q,a)=1}}\left\{\pi(x;q,a)-\frac{\pi(x)}{\phi(q)}\right\}\right| \gg x; \tag{12}$$

thus refuting the conjecture that for any given $A>0$, (11) should hold in the range (9) for some $B=B(A)>0$. In [10] we showed that (11) even fails with

$$Q \;=\; x/\exp\left((A-\varepsilon)(\log\log x)^2/(\log\log\log x)\right).$$

We also showed that (8′) cannot hold for every integer a, prime to q, for

(i) Any $q\ge x/\exp\left((\log x)^{1/5-\varepsilon}\right)$.
(ii) Any $q\ge x/\exp\left((\log x)^{1/3-\varepsilon}\right)$ that has $<1.5\log\log\log q$ distinct prime factors $<\log q$.[12]
(iii) Almost any $q\in(y,2y]$, for any $y\ge x/\exp\left((\log x)^{1/2-\varepsilon}\right)$.

Moreover, under the assumption of GRH we can improve the values 1/5 and 1/3 in (i) and (ii), respectively, to 1/3 and 1/2.

It seems plausible that (8) holds uniformly if $\log(x/q)/\log\log q\to\infty$ as $q\to\infty$; and that (11) holds uniformly for $Q<x/\exp\left((\log x)^{1/2+\varepsilon}\right)$. At least we can't disprove these statements as yet, though we might play it safe and conjecture only that they hold uniformly for $q,Q<x^{1-\varepsilon}$.

Notice that in the proof described above, the values of a increase with x, leaving open the possibility that (8) might hold uniformly for all $(a,q)=1$ in the

[12]Which, by the TURAN-KUBILIUS inequality, includes "almost all" integers.

range (9) if we fix a.[13] However, in [8] we observed that when a is fixed one can suitably modify the Maier matrix, by forcing the elements of the second column to all be divisible by P:

$$\begin{array}{|lcccc|}
\hline
1 & 1+(RP-1) & 1+2(RP-1) & \cdots & 1+y(RP-1) \\
1 & 1+((R+1)P-1) & 1+2((R+1)P-1) & \cdots & 1+y((R+1)P-1) \\
1 & 1+((R+2)P-1) & \cdot & \cdot & \vdots \\
1 & \vdots & \boxed{\begin{array}{c}(i,j)\text{th entry}: \\ 1+j((R+i)P-1)\end{array}} & \vdots & \vdots \\
\vdots & \vdots & \cdot & \vdots & \vdots \\
1 & 1+((2R-1)P-1) & \cdots & \cdots & 1+y((2R-1)P-1) \\
\hline
\end{array}$$

The Maier matrix for $\pi(yq;q,1)$

Notice that the jth column here is now part of an arithmetic progression with a varying modulus, namely $1-j \pmod{jP}$. With this type of Maier matrix we can deduce that, for almost all $0<|a|<x/\log^B x$ (including all fixed $a\neq 0$), there exist $q\in(x/\log^B x, 2x/\log^B x]$, coprime to a, for which (8) does not hold. However (8) cannot be false too often (like in (12)), because this would contradict the BARBAN-DAVENPORT-HALBERSTAM theorem. So for which a is (8) frequently false? It turns out that the answer depends on the number of prime factors of a: In [9], extending the results of [2], we show that for any given $A>1$ there exists a value $B=B(A)>0$ such that, for any $Q\leq x/\log^B x$ and any integer a that satisfies $0<|a|<x$ and has $\ll \log\log x$ distinct prime factors,[14] we have

$$\left|\sum_{\substack{Q\leq q\leq 2Q\\ (q,a)=1}}\left\{\pi(x;q,a)-\frac{\pi(x)}{\phi(q)}\right\}\right| \ll \frac{x}{\log^A x}. \tag{13}$$

On the other hand, for every given $A,B>0$, there exists $Q\leq x/\log^B x$ and an integer a that satisfies $0<|a|<x$ and has $<(\log\log x)^{6/5+\varepsilon}$ distinct prime factors, for which (13) does not hold (and assuming GRH we may replace $6/5+\varepsilon$ here by $1+\varepsilon$).

Finding primes in $(x,x+y]$ is equivalent to finding integers $n\leq y$ for which $f(n)$ is prime, where $f(t)$ is the polynomial $t+x$. Similarly, finding primes $\leq x$ that belong to the arithmetic progression $a \pmod q$ is equivalent to finding integers $n\leq y:=x/q$ for which $f(n)$ is prime, where $f(t)$ is the polynomial $qt+a$. Define the *height* $h(f)$ of a given polynomial $f(t)=\sum_i c_i t^i$ to be $h(f):=\sqrt{\sum_i c_i^2}$. In the cases above, in which the degree is always 1, we proved that we do not always get the asymptotically expected number of prime values $f(n)$ with $n\leq y=\log^B h(f)$,

[13] Which would be consistent with the BARBAN-DAVENPORT-HALBERSTAM theorem.

[14] Which includes almost all integers a once the inexplicit constant here is >1 (by a famous result of HARDY AND RAMANUJAN).

for any fixed $B > 0$. In [7] we showed that this is true for polynomials of arbitrary degree d, which is somewhat ironic because it is not known that *any* polynomial of degree ≥ 2 takes on infinitely many prime values, nor that the prime values are ever "well-distributed". NAIR AND PERELLI [18] showed that some of the polynomials $F_R(n) = n^d + RP$ attain more than, and others attain less than, the number of prime values expected in such a range, by considering the following Maier matrix:

$$\begin{array}{|ccccc|}
\hline
F_R(1) & F_R(2) & F_R(3) & \cdots & F_R(y) \\
F_{R+1}(1) & F_{R+1}(2) & F_{R+1}(3) & \cdots & F_{R+1}(y) \\
F_{R+2}(1) & F_{R+2}(2) & \cdot & \cdot & \vdots \\
F_{R+3}(1) & \vdots & \boxed{\begin{array}{c}(i,j)\text{th entry}: \\ F_{R+i}(j)\end{array}} & \vdots & \vdots \\
\vdots & \vdots & \cdot & \vdots & \vdots \\
F_{2R-1}(1) & F_{2R-1}(2) & \cdots & \cdots & F_{2R-1}(y) \\
\hline
\end{array}$$

The Maier matrix for $\pi_F(y)$

Notice that the jth column here is part of the arithmetic progression $j^d \pmod P$.

Using Maier matrices it is possible to prove "bad equi-distribution" results for primes in other interesting sequences, such as the values of binary quadratic forms, and of prime pairs. For example, if we fix $B > 0$ then, once x is sufficiently large, there exists a positive integer $k \leq \log x$ such that there are at least $1 + \delta_B$ times as many prime pairs $p, p + 2k$, with $x < p \leq x + \log^B x$, as we would expect from assuming that the estimate in Hypothesis H holds uniformly for $n \ll \log^B h((t + x)(t + (x + 2k)))$.

We have now seen that the asymptotic formula in Hypothesis H fails when x is an arbitrary fixed power of $\log h(F)(:= \sum_i \log h(f_i))$, for many different non-trivial examples F. Presumably the asymptotic formula *does* hold uniformly as $\log x / \log\log h(F) \to \infty$. However, to be safe, we only make the following prediction:

CONJECTURE. *Fix $\varepsilon > 0$ and positive integer k. The asymptotic formula in Hypothesis H holds uniformly for $x > h(F)^\varepsilon$ as $h(F) \to \infty$.*

Our work here shows that the "random-like" behavior exhibited by primes in many situations does not carry over to *all* situations. It remains to discover a model that will always accurately predict how primes are distributed, as it seems that minor modifications of CRAMÉR's model will not do. We thus agree that:

> *"It is evident that the primes are randomly distributed but, unfortunately, we don't know what 'random' means."* — R.C. VAUGHAN (February 1990).

Final remarks

Armed with MAIER's ideas it seems possible to construct incorrect conclusions from, more or less, any variant of CRAMÉR's model. This flawed model may still be used to make conjectures about the distribution of primes, but one should be very cautious of such predictions!

There are no more than $O(x^2/\log^{3B} x)$ arithmetic progressions $a \pmod q$, with $1 \leq a < q < x/\log^B x$ and $(a,q) = 1$, for which (8) fails, by the BARBAN-DAVENPORT-HALBERSTAM theorem. However, our methods here may be used to show that (8) does fail for more than $x^2/\exp((\log x)^\varepsilon)$ such arithmetic progressions.

Maier's matrix has been used in other problems too: KONYAGIN recently used it to find unusually large gaps between consecutive primes. MAIER used it to find long sequences of consecutive primes, in which there are longer than average gaps between each pair. SHIU has used it to show that every arithmetic progression $a \pmod q$ with $(a,q) = 1$ contains arbitrarily long strings of consecutive primes.

BALOG [1] has recently shown that the prime k-tuplets conjecture holds "on average" (in the sense of the BOMBIERI-VINOGRADOV Theorem).[15]

As we saw in the table above, $\text{Li}(x) > \pi(x)$ for all $x \leq 10^{18}$. However, (2b) implies that this inequality does not persist for ever; indeed, it is reversed for some $x < 10^{370}$ (TE RIELE). Recently, however, RUBINSTEIN AND SARNAK [20][16] showed that it does hold more often than not, in the sense that there exists a constant $\delta \approx 1/(4 \cdot 10^6)$ such that the (logarithmically scaled) proportion of x for which $\pi(x) > \text{Li}(x)$ exists and equals δ. Such biases may also be observed in arithmetic progressions, in that there are "more" primes belonging to arithmetic progressions that are quadratic nonresidues than those that are quadratic residues. In particular they prove that $\pi(x; 4, 3) > \pi(x; 4, 1)$ for a (logarithmically scaled) proportion $0.9959\ldots$ of the time.

Delicate questions concerning the distribution of prime numbers still seem to be very mysterious. It may be that by taking into account divisibility by small primes we can obtain a very accurate picture; or it may be that there are other phenomena, disturbing the equi-distribution of primes, that await discovery ...

"Mathematicians have tried in vain to discover some order in the sequence of prime numbers but we have every reason to believe that there are some mysteries which the human mind will never penetrate."

— L. EULER (1770).

Acknowledgments: I'd like to thank Red Alford, Nigel Boston, John Friedlander, Dan Goldston, Ken Ono, Carl Pomerance, and Trevor Wooley for their helpful comments on earlier drafts of this paper.

---

[15] Which was improved to the exact analogue of (11) by MIKAWA for $k = 2$, and by KAWADA for all $k \geq 1$.

[16] All of their results are proved assuming appropriate conjectures such as RH, GRH, and that the zeros of the relevant L-functions are linearly independent over $\mathbb{Q}$.

References

[1] A. Balog, *The prime k-tuplets conjecture on average*, Analytic Number Theory (B.C. Berndt, H.G. Diamond, H. Halberstam, A. Hildebrand, eds.), Basel and Birkhäuser, Boston, 1990, pp. 165–204.

[2] E. Bombieri, J. B. Friedlander, and H. Iwaniec, *Primes in arithmetic progressions to large moduli III*, J. Amer. Math. Soc. **2** (1989), 215–224.

[3] A. A. Buchstab, *On an asymptotic estimate of the number of numbers of an arithmetic progression which are not divisible by relatively small prime numbers*, Mat. Sb. **28/70** (1951), 165–184 (Russian).

[4] H. Cramér, *On the order of magnitude of the difference between consecutive prime numbers*, Acta Arith. **2** (1936), 23–46.

[5] H. Davenport, Multiplicative Number Theory (2nd ed.), Springer-Verlag, Berlin and New York, 1980.

[6] J. B. Friedlander and A. Granville, *Limitations to the equi-distribution of primes I*, Ann. of Math. (2) **129** (1989), 363–382.

[7] ———, *Limitations to the equi-distribution of primes IV*, Proc. Royal Soc. London Ser. A **435** (1991), 197–204.

[8] ———, *Limitations to the equi-distribution of primes III*, Compositio Math. **81** (1992), 19–32.

[9] ———, *Relevance of the residue class to the abundance of primes*, Proc. Amalfi Conf. on Analytic Number Theory (E. Bombieri, A. Perelli, S. Salerno, U. Zannier, eds.), Salerno, Italy, 1993, pp. 95–104.

[10] J. B. Friedlander, A. Granville, A. Hildebrand, and H. Maier, *Oscillation theorems for primes in arithmetic progressions and for sifting functions*, J. Amer. Math. Soc. **4** (1991), 25–86.

[11] P. X. Gallagher, *On the distribution of primes in short intervals*, Mathematika **23** (1976), 4–9.

[12] D. A. Goldston, *On the pair correlation conjecture for zeros of the Riemann zeta-function*, J. Reine Angew. Math. **385** (1988), 24–40.

[13] A. Granville, *Harald Cramér and the distribution of prime numbers*, Scand. Actuarial J. **2** (1995), to appear.

[14] G. H. Hardy and J. E. Littlewood, *Some problems on partitio numerorum III On the expression of a number as a sum of primes*, Acta Math. **44** (1923), 1–70.

[15] A. Hildebrand and H. Maier, *Irregularities in the distribution of primes in short intervals*, J. Reine Angew. Math. **397** (1989), 162–193.

[16] H. Maier, *Primes in short intervals*, Michigan Math. J. **32** (1985), 221–225.

[17] H. L. Montgomery, *The pair correlation of zeros of the zeta function*, Proc. Sympos. Pure Math., vol. 24, Amer. Math. Soc., Providence, RI, 1973, pp. 181–193.

[18] M. Nair and A. Perelli, *On the prime ideal theorem and irregularities in the distribution of primes*, Duke Math. J. **77** (1995), 1–20.

[19] A. M. Odlyzko, *The 10^{20}th zero of the Riemann zeta function and 70 million of its neighbors, to appear.*

[20] M. Rubinstein and P. Sarnak, *Chebyshev's bias, to appear.*

[21] Z. Rudnick and P. Sarnak, *The n-level correlations of the zeros of the zetafunction*, C.R. Acad. Sci. Paris, to appear.

[22] A. Schinzel and W. Sierpiński, *Sur certaines hypothèses concernant les nombres premiers*, Acta Arith. **4** (1958), 185–208.

[23] A. Selberg, *On the normal density of primes in small intervals and the difference between consecutive primes*, Arch. Math. Naturvid. J. **47** (1943), 87–105.

Fonctions L p-adiques

BERNADETTE PERRIN-RIOU

Université Paris-Sud
Bât. 425, F-91405 Orsay Cedex, France

Ce qui suit est une présentation d'un travail que l'on peut trouver dans [P92], [P93], [P94], [Pa], [Pb]. Le but principal est d'attacher à toute représentation p-adique galoisienne V:

- un module de fonctions L p-adiques défini par utilisation de cohomologie galoisienne non ramifiée en dehors d'un nombre fini de places et de la théorie de Fontaine;
- une fonction L p-adique lorsque V est la réalisation p-adique d'un motif M par interpolation des valeurs de la fonction L de M (conjecturalement);

et de les relier (conjecturalement encore).
Nous insistons ici sur l'étude locale préliminaire dans laquelle une sorte de logarithme étendu aux modules d'Iwasawa locaux est défini et nous en tirons au passage quelques conséquences sur la continuité des logarithmes de Bloch-Kato.

On fixe un nombre premier p impair. On fixe une clôture algébrique $\overline{\mathbb{Q}}$ (resp. $\overline{\mathbb{Q}}_p$) de $\mathbb{Q}$ (resp. $\mathbb{Q}_p$) et un plongement de $\overline{\mathbb{Q}}$ dans $\overline{\mathbb{Q}}_p$. Si K est une extension de $\mathbb{Q}$ (resp. de $\mathbb{Q}_p$) contenue dans $\overline{\mathbb{Q}}$ (resp. dans $\overline{\mathbb{Q}}_p$), on note G_K le groupe de Galois de $\overline{\mathbb{Q}}$ (resp. $\overline{\mathbb{Q}}_p$) sur K. Si M est un G_K-module topologique, on note $H^1(K, M) = H^1(G_K, M)$ les groupes de cohomologie continue.

1. Etude locale

1.1. Exponentielle et logarithme de Bloch-Kato.

1.1.1. Soit V une représentation p-adique de $G_{\mathbb{Q}_p}$, c'est-à-dire un $\mathbb{Q}_p$-espace vectoriel de dimension finie d muni d'une action linéaire et continue de $G_{\mathbb{Q}_p}$, que l'on suppose cristalline; on note $\mathbf{D}_p(V)$ le φ-module filtré associé par la théorie de Fontaine: c'est un $\mathbb{Q}_p$-espace vectoriel de même dimension que V, muni d'un endomorphisme bijectif φ et d'une filtration $\mathrm{Fil}^i\, \mathbf{D}_p(V)$ décroissante, exhaustive et séparée: ainsi, $\mathbf{D}_p(V) = (B_{cris} \otimes V)^{G_{\mathbb{Q}_p}}$ (pour la définition des anneaux B_{cris}, B_{dR}, A_{cris}, voir [Bu]).

Proceedings of the International Congress
of Mathematicians, Zürich, Switzerland 1994

1.1.2. EXEMPLES. (i) Si $\mathbb{Z}_p(1)$ est le module de Tate associé aux racines de l'unité d'ordre une puissance de p et $V = \mathbb{Q}_p \otimes \mathbb{Z}_p(1)$, $\mathbf{D}_p(V)$ est $\mathbb{Q}_p$, φ agit par multiplication par p^{-1}, la filtration est donnée par $\mathrm{Fil}^{-1}\mathbf{D}_p(V) = \mathbf{D}_p(V)$, $\mathrm{Fil}^0\mathbf{D}_p(V) = 0$. On note $\mathbb{Q}_p(k) = \mathbb{Q}_p(1)^{\otimes k}$ si $k > 0$, $= \mathrm{Hom}_{\mathbb{Q}_p}(\mathbb{Q}_p(-k), \mathbb{Q}_p)$ si $k < 0$. Si χ est le caractère cyclotomique, l'action de Galois sur $\mathbb{Q}_p(k)$ est donc donnée par χ^k.

(ii) Si V est la représentation p-adique associée à une variété abélienne de dimension g définie sur $\mathbb{Q}_p$ et ayant bonne réduction, $\mathbf{D}_p(V)$ est essentiellement son module covariant de Dieudonné. La filtration est donnée par $\mathrm{Fil}^{-1}\mathbf{D}_p(V) = \mathbf{D}_p(V)$, $\mathrm{Fil}^0\mathbf{D}_p(V)$ de dimension g, $\mathrm{Fil}^1\mathbf{D}_p(V) = 0$.

(iii) Si X est une variété projective lisse définie sur $\mathbb{Q}_p$, $V = \mathbb{Q}_p \otimes H^i_{\text{ét}}(X)$.

(iv) Si V est une représentation p-adique, on pose $V(k) = V \otimes \mathbb{Q}_p(k)$. On identifie dans ce texte les $\mathbb{Q}_p$-espaces vectoriels $\mathbf{D}_p(V)$ et $\mathbf{D}_p(V(k))$, la lettre φ désignant alors l'endomorphisme de $\mathbf{D}_p(V)$.

1.1.3. On a la suite exacte de $G_{\mathbb{Q}_p}$-modules

$$0 \to \mathbb{Q}_p \to B_{cris} \to B_{cris} \oplus B_{dR}/\mathrm{Fil}^0 B_{dR} \to 0 \tag{1}$$

où la seconde application est donnée par $b \mapsto ((1-\varphi)b, b \bmod \mathrm{Fil}^0 B_{dR})$. De plus, cette suite exacte est scindée en tant que suite exacte d'applications continues. Soit K une extension finie de $\mathbb{Q}_p$. En tensorisant la suite (1) par V et en prenant la suite exacte de cohomologie, on obtient la suite exacte

$$\begin{array}{rcl} 0 \to H^0(K,V) \to \mathbf{D}_p(V)_{K_0} & \to & \mathbf{D}_p(V)_{K_0} \oplus t_V(K) \\ & \to & H^1(K,V) \to H^1(K, B_{cris} \otimes V) \end{array} \tag{2}$$

où K_0 est la plus grande extension de $\mathbb{Q}_p$ non ramifiée contenue dans K et $\mathbf{D}_p(V)_{K_0} = K_0 \otimes \mathbf{D}_p(V)$, où le K-espace vectoriel $t_V(K) = ((B_{dR}/\mathrm{Fil}^0 B_{dR}) \otimes V)^{G_K} = K \otimes t_V$ avec $t_V = \mathbf{D}_p(V)/\mathrm{Fil}^0\mathbf{D}_p(V)$ est appelé espace tangent de V sur K ; $\mathbf{D}_p(V)_{K_0} \to \mathbf{D}_p(V)_{K_0} \oplus t_V(K)$ est donnée par $x \mapsto ((1-\varphi)x, x \bmod \mathrm{Fil}^0\mathbf{D}_p(V))$. Bloch et Kato ont noté $H^1_f(K,V)$ le noyau de $H^1(K,V) \to H^1(K, B_{cris} \otimes V)$. La suite exacte (2) devient:

$$0 \to H^0(K,V) \to \mathbf{D}_p(V)_{K_0} \to \mathbf{D}_p(V)_{K_0} \oplus t_V(K) \to H^1_f(K,V) \to 0\,. \tag{3}$$

On appelle exponentielle de Bloch-Kato l'application $t_V(K) \to H^1_f(K,V)$ qui s'en déduit et on la note $\exp_{V,K}$. Lorsque $\mathbf{D}_p(V)_{K_0}^{\varphi=1}$ est nul, cette application est bijective, on note alors $\log_{V,K}$ l'application réciproque.

1.1.4. EXEMPLES. (i) On a $\varphi = 1$ sur $\mathbf{D}_p(\mathbb{Q}_p)$, si K^{nr} est la plus grande extension non ramifiée de K, on a $H^1_f(K, \mathbb{Q}_p) = \mathrm{Hom}_{\mathbb{Z}_p}(\mathrm{Gal}(K^{nr}/K), \mathbb{Q}_p)$. On a $t_{\mathbb{Q}_p(1)}(K) = K$, $H^1_f(K, \mathbb{Q}_p(1)) = \mathbb{Q}_p \otimes U_K$ où U_K est le groupe des unités de K, l'exponentielle est l'exponentielle p-adique classique.

(ii) Si V est attachée à une variété abélienne A ayant bonne réduction en p, $H^1_f(K,V)$ est naturellement isomorphe à $\mathbb{Q}_p \otimes A(K)$ par la théorie de Kummer sur la variété abélienne, l'espace tangent t_V est l'espace tangent usuel et $\exp_{V,K}$ l'application exponentielle classique.

(iii) Supposons que $\mathrm{Fil}^0 \mathbf{D}_p(V) = 0$ et que p^{-1} n'est pas valeur propre de φ sur $\mathbf{D}_p(V)_{K_0}$. Alors, l'application $\log_{V,K}$ est un isomorphisme de $H^1(K,V) = H^1_f(K,V)$ sur $K \otimes \mathbf{D}_p(V)$.

1.1.5. Le cup-produit induit une dualité $H^1(K,V) \times H^1(K,V^*(1)) \to \mathbb{Q}_p$ où $V^*(1) = \mathrm{Hom}_{\mathbb{Q}_p}(V,\mathbb{Q}_p(1))$.[1] Bloch et Kato ont montré que le sous-espace vectoriel $H^1_f(K,V)$ est l'orthogonal de $H^1_f(K,V^*(1))$. D'autre part, dans la dualité naturelle $\mathbf{D}_p(V) \times \mathbf{D}_p(V^*(1)) \to \mathbb{Q}_p$, le sous-espace $\mathrm{Fil}^0 \mathbf{D}_p(V)$ est l'orthogonal de $\mathrm{Fil}^0 \mathbf{D}_p(V^*(1))$. La transposée de $t_{V^*(1)}(K) \to H^1_f(K,V^*(1))$ pour cette dualité donne une application $\lambda_V = \lambda_{V,K}\colon H^1(K,V)/H^1_f(K,V) \to K \otimes \mathrm{Fil}^0 \mathbf{D}_p(V)$.

1.2. Quelques problèmes. On note T un réseau de V stable par $G_{\mathbb{Q}_p}$ et M un réseau de $\mathbf{D}_p(V)$ que l'on munit de la filtration induite.

1.2.1. *Problème A.* Les constructions qui ont été décrites pour V ont bien sûr un sens pour $V(k) = V \otimes \mathbb{Q}_p(k)$ pour tout entier k. Remarquons que pour $k \gg 0$, par exemple $k > k_0$, $\mathrm{Fil}^0 \mathbf{D}_p(V(k)) = 0$. On peut se poser le problème de la continuité de $\log_{V(k),\mathbb{Q}_p}$ en fonction de k de la manière suivante: On pose $T(k) = T \otimes \mathbb{Z}_p(k)$. Prenons deux entiers $k > k_0$ et $k' > k_0$ congrus modulo $(p-1)p^n$. Disons que $x \in H^1(\mathbb{Q}_p,T(k))$ et $x \in H^1(\mathbb{Q}_p,T(k'))$ sont congrus modulo p^{n+1} si leurs images respectives dans $H^1(\mathbb{Q}_p,T(k)/p^{n+1}T(k))$ et dans $H^1(\mathbb{Q}_p,T(k')/p^{n+1}T(k'))$ se correspondent par l'isomorphisme naturel $T(k)/p^{n+1}T(k) \cong T(k')/p^{n+1}T(k')$. La question est alors: si x et x' comme ci-dessus sont congrus modulo p^{n+1}, que peut-on dire de $\log_{V(k),\mathbb{Q}_p}(x)$ et de $\log_{V(k'),\mathbb{Q}_p}(x')$?

1.2.2. *Problème B.* Soit ω un élément non nul de $det_{\mathbb{Q}_p} t_V(K)$; on introduit un nombre $Tam_{\omega,K}(T)$ défini de la manière suivante:

$$\text{si} \quad det_{\mathbb{Q}_p} t_V(K) \cong det_{\mathbb{Q}_p} H^1_f(K,V) \otimes (det_{\mathbb{Q}_p} H^0(K,V))^*$$

est l'isomorphisme (défini au signe près) déduit de la suite exacte (3), $Tam_{\omega,K}(T)$ est l'élément de $p^{\mathbb{Z}}$ vérifiant

$$Tam_{\omega,K}(T)^{-1} \omega \mathbb{Z}_p = det_{\mathbb{Z}_p} H^1_f(K,T) \otimes (det_{\mathbb{Z}_p} H^0(K,T))^* \ .$$

Ainsi, lorsque 1 n'est pas valeur propre de φ et si $L_p(V,1) = det(1-\varphi|\mathbf{D}_p(V))^{-1}$, $L_p(V,1)^{-1} Tam_{\omega,K}(T)$ est, à une unité près, le déterminant de

$$\exp_{V,K} : t_V(K) \to H^1_f(K,V)$$

dans la base ω et dans une base du $\mathbb{Z}_p$-module $det_{\mathbb{Z}_p} H^1_f(K,T)$.

Que peut-on dire de ces nombres et de l'image de $H^1_f(K,T)$ dans $K \otimes \mathbf{D}_p(V)$ lorsque K varie le long de l'extension cyclotomique ? Que peut-on dire par exemple de $Tam_{\omega,K}(T)/Tam_{\omega',K}(T^*(1))$?

Donnons tout de suite une réponse conjecturale à cette dernière question (celle-ci s'étend au cas où V est semi-stable à condition d'introduire les facteurs ϵ de V). Posons $h_i(V) = dim_{\mathbb{Q}_p} \mathrm{Fil}^i \mathbf{D}_p(V)/\mathrm{Fil}^{i+1} \mathbf{D}_p(V)$ et $t_H(V) = \sum_i i h_i(V)$. Notons e_T une base du $\mathbb{Z}_p$-module (de rang 1) $(t^{t_H(V)} A_{cris} \otimes det_{\mathbb{Z}_p} T)^{G_{\mathbb{Q}_p}}$ où t est

[1] Si W est un $\mathbb{Q}_p$-espace vectoriel, on pose $W^* = \mathrm{Hom}_{\mathbb{Q}_p}(W,\mathbb{Q}_p)$, si U est un $\mathbb{Z}_p$-module de type fini, on pose $U^* = \mathrm{Hom}_{\mathbb{Z}_p}(U,\mathbb{Z}_p)$; si ω est une base, on note ω^{-1} la base duale.

un générateur de l'image de $\mathbb{Z}_p(1)$ dans B_{cris}. Posons $\Gamma^*(j) = (j-1)!$ si $j > 0$ et $\Gamma^*(j) = (-1)^j(-j)!^{-1}$ si $j \leq 0$.

1.2.3. CONJECTURE. *Pour $\omega \in det_{\mathbb{Q}_p} t_V(K)$, $\omega' \in det_{\mathbb{Q}_p} t_{V^*(1)}(K)$ tels que $\omega \otimes \omega'^{-1}$ engendre $(det_{\mathbb{Z}_p} \mathcal{O}_K)^{\otimes d} \otimes \mathbb{Z}_p e_T^{\otimes [K:\mathbb{Q}_p]}$ vu comme sous-module de $det_{\mathbb{Q}_p}(K \otimes \mathbf{D}_p(V))$,*

$$\frac{Tam_{\omega,K}(T)}{Tam_{\omega',K}(T^*(1))} \sim d_K^{-t_H(V^*(1))} \prod_i \Gamma^*(-i)^{-h_i(V)[K:\mathbb{Q}_p]} \tag{4}$$

où d_K est le discriminant de K sur $\mathbb{Q}_p$.[2]

La formule (4) a été montrée par Bloch et Kato pour $V = \mathbb{Q}_p(r)$ et K non ramifié sur $\mathbb{Q}_p$. Elle est vraie aussi pour les représentations p-adiques ordinaires et K non ramifié. On notera $M(T)$ un réseau de $\mathbf{D}_p(V)$ tel que $det_{\mathbb{Z}_p} M(T) = (t^{t_H(V)} A_{cris} \otimes det_{\mathbb{Z}_p} T)^{G_{\mathbb{Q}_p}}$.

1.2.4. *Problème C.* L'application λ_V est extrêmement importante. Elle mesure le défaut d'un élément de $H^1(\mathbb{Q}_p, T)$ à être dans $H^1_f(\mathbb{Q}_p, T)$ et intervient dans les problèmes de "loi explicite de réciprocité" [3]. Comment la calculer?

1.3. Résultats.

1.3.1. Soient $K_n = \mathbb{Q}_p(\mu_{p^{n+1}})$, $G_\infty = \mathrm{Gal}(K_\infty/\mathbb{Q}_p)$, $\chi : G_\infty \cong \mathbb{Z}_p^\times$ le caractère cyclotomique. Si X est une indéterminée, on fait agir G_∞ sur $1+X$ par $\tau.(1+X) = (1+X)^{\chi(\tau)}$ et on prolonge cette action par linéarité et continuité à $\Lambda = \mathbb{Z}_p[[G_\infty]]$. D'autre part, on fait agir φ sur $\mathbb{Q}_p[[X]]$ par $\varphi(1+X) = (1+X)^p$. Fixons un générateur (multiplicatif) $\epsilon = (\zeta_n)$ de $\mathbb{Z}_p(1)$. Notons $\tilde{\Delta}_j$ le composé de l'évaluation en χ^j et de la projection $\mathbf{D}_p(V) \to \mathbf{D}_p(V)/(1-p^j\varphi)$ et $\tilde{\Delta} = \oplus \tilde{\Delta}_j$. Si $g \in \Lambda \otimes \mathbf{D}_p(V)$ vérifie $\tilde{\Delta}(g) = 0$ (il n'y a qu'un nombre fini de conditions), on montre qu'il existe un élément $G \in \mathbb{Q}_p[[X]] \otimes_{\mathbb{Q}_p} \mathbf{D}_p(V)$ convergeant sur le disque unité ouvert, tel que $(1-\varphi)G = g.(1+X)$ (φ agit ici de manière diagonale). On pose $\Xi_{n,V}(g) = (p \otimes \varphi)^{-(n+1)}(G)(\zeta_n - 1) \in (K_n \otimes_{\mathbb{Q}_p} \mathbf{D}_p(V))/\mathbf{D}_p(V)^{\varphi=1}$.

Notons $Z^1_\infty(T)$ la limite projective relativement aux applications de corestriction des $H^1(K_n, T)$. C'est un Λ-module de rang d. Son sous-Λ-module de torsion[4] est isomorphe à $T^{G_{K_\infty}}$. On définit un isomorphisme, noté Tw_ϵ: $x \mapsto x \otimes \epsilon$, de $Z^1_\infty(T(j))$ dans $Z^1_\infty(T(j+1))$ et induit par les isomorphismes $H^1(K_n, T_n(j)) \to H^1(K_n, T_n(j)) \otimes \mu_{p^{n+1}} = H^1(K_n, T_n(j+1))$, avec $T_n = T/p^{n+1}T$.

Définissons $\mathcal{H}(G_\infty)$ comme le tensorisé avec $\mathbb{Z}_p[\mathrm{Gal}(\mathbb{Q}_p(\mu_p)/\mathbb{Q}_p)]$ de la sous-algèbre de $\mathbb{Q}_p[[\gamma - 1]]$ formée des $f = \sum_n a_n(\gamma-1)^n$ avec $\sup_{n>0} \frac{|a_n|}{n^r} < \infty$ pour un r convenable (ici, γ est un générateur topologique de $\mathrm{Gal}(K_\infty/\mathbb{Q}_p(\mu_p))$). Soit $\mathcal{K}(G_\infty)$ l'anneau total des fractions de $\mathcal{H}(G_\infty)$. Notons Tw l'application de twist induite par $\gamma \mapsto \chi(\gamma)\gamma$.

---

[2] $a \sim b$ si $a = ub$ avec u unité p-adique

[3] voir aussi les démonstrations "à la Kolyvagin".

[4] pour chacune des composantes relativement à l'action de $\mathrm{Gal}(K_0/\mathbb{Q}_p)$

1.3.2. THÉORÈME. *Pour tous entiers h et j, il existe un unique Λ-homomorphisme injectif*

$$\Omega^{\epsilon}_{V(j),h} : (\Lambda \otimes \mathbf{D}_p(V(j)))^{\tilde{\Delta}=0} \to \mathcal{K}(G_\infty) \otimes_\Lambda (Z^1_\infty(T(j))/T(j)^{G_{K_\infty}})$$

tel que: (i) *pour tout entier* $h \geq 1$, *supérieur à la longueur de la filtration de Hodge de* V *et tel que* $\mathrm{Fil}^{-h}\,\mathbf{D}_p(V) = \mathbf{D}_p(V)$ *et pour tout entier* j *tel que* $\mathrm{Fil}^0\,\mathbf{D}_p(V(j)) = 0$, $\Omega^{\epsilon}_{V(j),h+j}$ *est à valeurs dans* $\mathcal{H}(G_\infty) \otimes_\Lambda (Z^1_\infty(T(j))/T(j)^{G_{K_\infty}})$; *pour* $n \geq 0$, *le diagramme suivant est alors commutatif*

$$\begin{array}{ccc} (\Lambda \otimes \mathbf{D}_p(V(j)))^{\tilde{\Delta}=0} & \longrightarrow & \mathcal{K}(G_\infty) \otimes_\Lambda (Z^1_\infty(T(j))/T(j)^{G_{K_\infty}}) \\ \Xi_{n,V(j)} \downarrow & & \downarrow \\ K_n \otimes \mathbf{D}_p(V(j)) & \xrightarrow{(h+j-1)!exp_{K_n,V(j)}} & H^1(K_n, V(j)) \end{array}$$

où l'application verticale de droite se déduit de la projection naturelle;
(ii) *pour tous entiers* j *et* h, *on a*

$$Tw_\epsilon \circ \Omega^{\epsilon}_{V(j),h+j} \circ Tw = -\Omega^{\epsilon}_{V(j+1),h+j+1} \;;$$

(iii) *pour tous entiers* j *et* h, $\Omega^{\epsilon}_{V(j),h+1} = l_h \Omega^{\epsilon}_{V(j),h}$ *avec* $l_h = h - \log\gamma/\log\chi(\gamma)$ *et* γ *élément de* G_∞ *d'ordre infini.*

1.3.3. On en déduit une famille d'isomorphismes de $\mathcal{K}(G_\infty)$-modules

$$\Omega^{\epsilon}_{V,h} : \mathcal{K}(G_\infty) \otimes (\Lambda \otimes \mathbf{D}_p(V(j))) \to \mathcal{K}(G_\infty) \otimes_\Lambda Z^1_\infty(T(j)) \;,$$

que l'on peut voir comme une sorte d'exponentielle au niveau des modules d'Iwasawa. Notons $\delta_h(V)$ (resp. $\delta_{\mathbb{Z}_p,h}(V)$) le sous-$\mathbb{Q}_p \otimes \Lambda$-module (resp. sous-Λ-module) de $\mathcal{K}(G_\infty)$ engendré par le déterminant de $\Omega^{\epsilon}_{V,h}$ dans des bases des $\mathbb{Q}_p \otimes \Lambda$-modules $\Lambda \otimes det_{\mathbb{Q}_p}\mathbf{D}_p(V)$ et $\mathbb{Q}_p \otimes det_\Lambda Z^1_\infty(T) \otimes (det_\Lambda Z^2_\infty(T))^{*}$[5] (resp. des Λ-modules $\Lambda \otimes det_{\mathbb{Z}_p} M(T)$ et $det_\Lambda Z^1_\infty(T) \otimes (det_\Lambda Z^2_\infty(T))^*$). [6]

1.3.4. THÉORÈME. $\prod_{j>-h} l_{-j}^{-dim_{\mathbb{Q}_p} \mathrm{Fil}^j\,\mathbf{D}_p(V)} \delta_h(V)$ *est indépendant de* h *et contenu dans* $\mathbb{Q}_p \otimes \Lambda$.

1.3.5. CONJECTURE. $\tilde{\delta}_{\mathbb{Z}_p}(V) = \prod_{j>-h} l_{-j}^{-dim_{\mathbb{Q}_p} \mathrm{Fil}^j\,\mathbf{D}_p(V)} \delta_{\mathbb{Z}_p,h}(V)$ *est égal à* Λ.

La conjecture est indépendante du choix de T.

---

[5]Ici, * désigne $\mathrm{Hom}_\Lambda(-,\Lambda)$

[6]Nous emploierons désormais le langage des déterminants pour les Λ-modules: rappelons que pour un Λ-module M de type fini et de Λ-torsion, $det_\Lambda M$ se plonge canoniquement dans l'anneau total des fractions de Λ et est égal à $f^{-1}\Lambda$ où f est une série caractéristique de M.

1.3.6. PROPOSITION. *Si* $\mathrm{Fil}^0\mathbf{D}_p(V)=0$ *ou* $\mathbf{D}_p(V)$ *et si 1 et* p^{-1} *ne sont pas valeurs propres de* φ*, avec les notations de 1.2.3,*

$$\prod_i \Gamma^*(-i)^{h_i(V)[K_n:\mathbb{Q}_p]} d_{K_n}^{t_H(V^*(1))} Tam_{\omega,K_n}(T)/Tam_{\omega',K_n}(T^*(1))$$

est égal (à une unité près) à $\prod_\eta \eta(f)$ *où* f *est un générateur de* $\tilde{\delta}_{\mathbb{Z}_p}(V)$ *et où* η *parcourt les caractères de* $\mathrm{Gal}(K_n/\mathbb{Q}_p)$.

Ainsi, la conjecture 1.3.5 implique la conjecture 1.2.3. D'autre part, à condition de supposer vraie la conjecture 1.3.10 qui suit, on peut enlever l'hypothèse sur Fil^0.

Supposons $\mathrm{Fil}^0\mathbf{D}_p(V)=0$. Alors, $Tam_{\omega',K_n}(T^*(1))$, pour ω' base canonique, est le cardinal du quotient de $(V^*(1)/T^*(1))^{G_{K_n}}$ par sa partie divisible et est borné par rapport à n. On déduit alors de 1.3.4 que, si ω_n est une base de $\det_{\mathbb{Z}_p}(\mathcal{O}_{K_n}\otimes M(T))$,

$$Tam_{\omega_n,K_n}(T)\sim d_{K_n}^{-t_H(V^*(1))}\prod_i \Gamma^*(-i)^{-h_i(V)p^n(p-1)}\prod_\eta \eta(f)p^c$$

(c constant pour n assez grand). Ainsi, si 1.3.5 est vraie,

$$Tam_{\omega,K_n}(T)\sim p^{-p^n(p-1)(n-\frac{1}{p-1})t_H(V^*(1))}\prod_i \Gamma^*(-i)^{-h_i(V)p^n(p-1)}p^c$$

où c est constant pour n assez grand. Sans la conjecture 1.3.5, il faut rajouter $p^{\mu p^{n+1}+\lambda n}$ où μ et λ sont des constantes.

1.3.7. EXEMPLE. Pour $\mathbb{Q}_p(r)$, la conjecture 1.3.5 est démontrée. Pour ω_n base de $M(\mathbb{Z}_p(r))\otimes det_{\mathbb{Z}_p}\mathcal{O}_{K_n}$ et $r>0$, on a $Tam_{\omega_n,K_n}(\mathbb{Z}_p(r))\sim p^c d_{K_n}^{1-r}(r-1)!^{-p^n(p-1)}$.

1.3.8. Revenons au problème A et prenons h comme dans le théorème 1.3.2. Soit $x\in Z^1_\infty(T)$. On déduit du théorème précédent qu'il existe un unique élément $\text{Ł}_h(x)$ de $\mathcal{K}(G_\infty)\otimes\mathbf{D}_p(V)$ tel que pour tout entier $j\gg 0$,

$$\chi^{-j}(\text{Ł}_h(x))=(-1)^j P_j(\varphi)\frac{\log_{V(j),\mathbb{Q}_p}P(x\otimes\epsilon^{\otimes j})}{(h+j-1)!} \tag{5}$$

où $P(x\otimes\epsilon^{\otimes j})$ désigne la projection de $x\otimes\epsilon^{\otimes j}$ dans $H^1(\mathbb{Q}_p,V(j))$ et où

$$P_j(\varphi)=(1-p^{-j}\varphi)(1-p^{j-1}\varphi^{-1})^{-1}.$$

Les propriétés d'analycité de $\text{Ł}_h(x)$ ont des conséquences sur la fonction $j\mapsto\chi^{-j}(\text{Ł}_h(x))$, en particulier sur sa continuité. Remarquons que $\text{Ł}_h(x)$ est entièrement déterminée (par continuité) par ses valeurs aux entiers $j\gg 0$. Rappelons que M est un réseau fixé de $\mathbf{D}_p(V)$. On déduit des théorèmes précédents que pour tout $k\in\mathbb{Z}_p$, sauf un nombre fini, il existe une constante C telle que, pour n assez grand et $j\equiv j'\equiv k \bmod p^n(p-1)$, $j>k_0$, $j'>k_0$, si $x\in H^1(\mathbb{Q}_p,T(j))$ et $x'\in H^1(\mathbb{Q}_p,T(j'))$ sont congrus modulo p^{n+1}, $P_j(\varphi)\frac{\log_{V(j),\mathbb{Q}_p}(x)}{(h+j-1)!}$ et $P_{j'}(\varphi)\frac{\log_{V(j'),\mathbb{Q}_p}(x')}{(h+j'-1)!}$ sont congrus modulo $C^{-1}p^nM$. Si la conjecture 1.3.5 est vraie, l'ensemble exceptionnel est contenu dans la réunion des k tels que $V(k)^{G_{\mathbb{Q}_p}}$ ou $V(1-k)^{G_{\mathbb{Q}_p}}$ est non nul et des entiers compris entre $-h$ et j_0. Sinon il faut aussi enlever les zéros d'un générateur de $\tilde{\delta}_{\mathbb{Z}_p}(V)$.

1.3.9. Soit $<,>_V$ l'application $Z^1_\infty(T) \times Z^1_\infty(T^*(1)) \to \Lambda$ qui est Λ-sesquilinéaire (relativement à $\iota : \Lambda \to \Lambda$ induit par $\tau \mapsto \tau^{-1}$ pour $\tau \in G_\infty$) et induit par le cup-produit: $< x, y >_V = \lim_{n\to\infty} \sum_{\tau\in \mathrm{Gal}(K_n/\mathbb{Q}_p)} (\tau^{-1}x_n \cup y_n)\tau$. Notons $[\ ,\ ]_{\mathbf{D}_p(V)}$ la dualité naturelle $\mathbf{D}_p(V) \times \mathbf{D}_p(V^*(1)) \to \mathbb{Q}_p$. On note de la même manière les formes obtenues par extension des scalaires à $\mathcal{H}(G_\infty)$.

1.3.10. CONJECTURE. *Pour tous* $x \in (\Lambda \otimes \mathbf{D}_p(V))^{\tilde{\Delta}=0}$, $y \in (\Lambda \otimes \mathbf{D}_p(V^*(1)))^{\tilde{\Delta}=0}$,

$$(-1)^{h-1} < \Omega^{\epsilon}_{V,h}(x), \Omega^{\epsilon^{-1}}_{V^*(1),1-h}(y^\iota) >_V = [x,y]_{\mathbf{D}_p(V)} \ . \qquad (6)$$

1.3.11. En prenant la valeur de (6) au caractère unité et en utilisant le (iii) du théorème 1.3.2, on obtient une formule que l'on peut voir comme une description de l'application $\lambda_{V(-r)}$ et qui pour $\mathbb{Q}_p$ redonne la loi explicite de réciprocité de Bloch-Kato. Soit r un entier tel que $\mathrm{Fil}^0\, \mathbf{D}_p(V(r)) = 0$ et $\alpha \in \mathbf{D}_p(V(r))$. Pour $h \gg 0$ et $r_0 > h - r$, $\mathrm{Fil}^0\, \mathbf{D}_p(V^*(1+r_0)) = 0$, si $y \in H^1(\mathbb{Q}_p, V^*(1-r))$, il existe $g_h \in \mathcal{K}(G_\infty) \otimes \mathbf{D}_p(V^*(1+r_0))$ tel que $P(\Omega^{\epsilon}_{V^*(1+r_0),h}(g_h) \otimes \epsilon^{\otimes(-r_0-r)}) = y$. Soit G_h tel que $(1-\varphi)G_h = g_h(1+x)$. Alors,

$$< exp_{V(r),\mathbb{Q}_p}(\alpha), y > = \pm \frac{[\alpha, D^{r+r_0}(G_h)(0)]_{\mathbf{D}_p(V)}}{(r+r_0-h)!} \ .$$

On en déduit facilement une formule pour $\lambda_{V^*(1-r)}$.

1.3.12. Une autre conséquence de (6) est que pour $r < -h$, $\mathrm{Fil}^r\, \mathbf{D}_p(V) = \mathbf{D}_p(V)$,

$$\chi^{-r}(\text{Ł}_h(x)) = \pm(1-p^{-r}\varphi)(1-p^{r-1}\varphi^{-1})^{-1} \frac{\lambda_{V(r),\mathbb{Q}_p}(P(x \otimes \epsilon^{\otimes r}))}{\Gamma^*(h+r)!} \ .$$

Ainsi, le membre de droite est une limite pour j tendant p-adiquement vers r et tendant vers l'infini dans $\mathbb{R}$, d'expressions du type (5).

2. Etude globale

A partir de maintenant, V est une représentation p-adique de $G_{\mathbb{Q}}$ non ramifiée en dehors d'un ensemble fini S de nombres premiers (avec $p \in S$) et cristalline en p, on désigne par T un réseau de V stable par $G_{\mathbb{Q}}$. On pose $F_n = \mathbb{Q}(\mu_{p^{n+1}})$, $G_\infty = \mathrm{Gal}(F_\infty/\mathbb{Q})$ et $\Lambda = \mathbb{Z}_p[[G_\infty]]$. Si M est un $\mathbb{Z}_p[\mathrm{Gal}(\mathbb{Q}(\mu_p)/\mathbb{Q})]$-module, on note $M_\pm$ le sous-module sur lequel la conjuguaison complexe agit par $\pm$. Si F est une extension de $\mathbb{Q}$ contenue dans $\overline{\mathbb{Q}}$, on note $G_{S,F}$ le groupe de Galois de la plus grande extension de F non ramifiée en dehors de S sur F.

2.1. Point de vue arithmétique.

2.1.1. Soit $H^i_{\infty,S}(T)$ la limite projective des $H^i(G_{S,F_n}, T)$ relativement aux applications de corestriction. Ces Λ-modules de type fini sont nuls pour i différent de 1 et 2. On montre que $rg_{\Lambda_\pm} H^1_{\infty,S}(T) - rg_{\Lambda_\pm} H^2_{\infty,S}(T) = d_{-\pm}(V)$ où $d_+(V) = dim_{\mathbb{Q}_p} V^{\mathrm{Gal}(\mathbb{C}/\mathbb{R})}$ et $d_-(V) = d - d_+(V)$.

2.1.2. CONJECTURE. *Le Λ-module $H^2_{\infty,S}(T)$ est de torsion.*

Dans les cas classiques, cette conjecture est ce qu'on appelle la conjecture faible de Leopoldt. Elle est équivalente à la nullité de $H^2(G_{S,F_\infty}, V/T)$. Elle est démontrée pour $V = \mathbb{Q}_p(r)$, pour les représentations associées aux courbes elliptiques à multiplication complexe, définies sur $\mathbb{Q}$ et ayant bonne réduction en p à quelques exceptions près (Rubin, [Mc]). Les résultats de finitude de Flach impliquent cette conjecture pour le carré symétrique d'une forme modulaire de poids 2 (et pour certains espaces propres relativement à $\mathrm{Gal}(\mathbb{Q}(\mu_p)/\mathbb{Q})$). De manière un peu anecdotique, on peut aussi montrer que les conjectures à la Bloch-Kato sur l'ordre du zéro de la fonction L énoncées comme en [FP94] impliquent cette conjecture. Enfin, des critères vérifiables numériquement peuvent être donnés. Par exemple, Rubin et Silverberg [RS] ont calculé des polynômes $a_D(t)$ et $b_D(t)$, $c_D(t)$ tels que pour tout $t \in \mathbb{Q}$, si $E_t^{(D)}$ est la courbe elliptique $y^2 = x^3 + a_D(t)x^2 + b_D(t)x + c_D(t)$, le $G_{\mathbb{Q}}$-module $E_t^{(D)}[5]$ soit isomorphe à $\mathcal{E}[5]$ où $\mathcal{E}$ est la courbe d'équation $y^2 = x^3 - Dx$. On peut alors montrer en utilisant les calculs numériques de [BGS84] que si $D = 1, -1, -2, 3, 9, -27, \ldots$ et $t \in \mathbb{Z}_{(5)}$, $Dt^2 \not\equiv 3 \bmod 5$, la conjecture de Leopoldt faible pour la courbe $E_t^{(D)}$ et $p = 5$ est vraie.

2.1.3. Nous supposons désormais la conjecture 2.1.2 vraie. Si Σ est un ensemble fini de places de $\mathbb{Q}$, on pose $Z^i_{\infty,\Sigma}(T) = \varprojlim \oplus_{w|v\in\Sigma} H^i(F_{n,w}, T)$ et $\Delta^{glob}_{\infty,S}(T) = det_\Lambda H^2_{\infty,S}(T) \otimes det_\Lambda H^1_{\infty,S}(T)^*$. Posons $\Delta^{loc}_{\infty,S}(T) = det_\Lambda Z^2_{\infty,S}(T) \otimes det_\Lambda Z^1_{\infty,S}(T)^*$ et $\Delta_{\infty,S}(T)^* = \Delta^{loc}_{\infty,S}(T) \otimes \Delta^{glob}_{\infty,S}(T)^*$.

Donnons deux définitions possibles du module des fonctions L p-adiques attachées à V:

2.1.4. DÉFINITION. *Si $\omega = \omega_{glob} \otimes \omega_{loc}^{-1}$ est un élément de $\Delta_{\infty,S}(T)^*$ avec $\omega_{glob} \in \Delta^{glob}_{\infty,S}(T)$ et $\omega_{loc} \in \Delta^{loc}_{\infty,S}(T)$ et si $\delta \in \wedge^{d_\pm(V)} \mathbf{D}_p(V)$, on note $\lambda_{V,S,h,\pm}(\omega)(\delta)$ l'élément de $\mathcal{K}(G_\infty)_\pm$ défini par*

$$\lambda_{V,S,h,\pm}(\omega)(\delta)\omega_{loc,\pm}^{-1} = \wedge^{d_\pm(V)}(\Omega^\epsilon_{V,h,\pm})(\delta) \wedge \omega_{glob,\pm}^{-1} \tag{7}$$

Le sous-$\Lambda_\pm$-module de $\mathrm{Hom}_{\mathbb{Q}_p}(\wedge^{d_\pm(V)}\mathbf{D}_p(V), \mathcal{K}(G_\infty)_\pm)$ engendré par l'image de $\Delta_{\infty,S}(T)^*$ est noté $\mathbb{I}_{arith,\{\infty,p\},h}(T)_\pm$.

2.1.5. DÉFINITION. *Le sous-Λ-module de $\mathrm{Hom}_{\mathbb{Q}_p}(\wedge^{d_\pm(V)}\mathbf{D}_p(V), \mathcal{K}(G_\infty)_\pm)$ défini pour $\delta \in \wedge^{d_\pm(V)}\mathbf{D}_p(V)$ par*

$$\begin{aligned} \mathbb{I}_{arith,\{\infty,p\},h}(T)_\pm(\delta) det_{\mathbb{Z}_p} M(T) &= \prod_{j>-h} l_{-j}^{dim_{\mathbb{Q}_p} \mathrm{Fil}^j \mathbf{D}_p(V)} \\ det_{\Lambda_\pm} Z^2_{\infty,S-\{p\}}(T)_\pm &\delta \wedge (\wedge^{d-\pm(V)}(\Omega^\epsilon_{V,h,\pm})^{-1}(\Delta^{glob}_{\infty,S}(T)^*)) . \end{aligned} \tag{8}$$

est noté $\mathbb{I}_{arith,\{\infty,p\},h}(T)_\pm$.

2.1.6. Si la conjecture 1.3.10 est vraie, ces deux définitions coïncident. Dans la suite, nous laissons l'ambiguïté. Ces modules généralisent la notion de "série caractéristique d'un module de Selmer" dans le cas où V est ordinaire en p. Dans les cas particuliers, il arrive qu'un choix de h s'impose. La dépendance en h est de toute façon simple et se déduit du théorème 1.3.2, (iii).

2.2. Point de vue analytique. Nous arrivons dans le domaine purement conjectural. Les énoncés s'appuient sur les conjectures de Beilinson et Deligne sur les valeurs spéciales des fonctions L des motifs. Ce qui suit tient de la loi-cadre. Nous renvoyons à [FP91] et [FP94] entre autres pour des bases plus précises et à [Pb] pour des "énoncés" plus complets.

2.2.1. Soit M un motif donné par ses réalisations de de Rham M_{dR}, de Betti M_B et l-adiques M_l pour tout l. Rappelons que M_{dR} est un $\mathbb{Q}$-espace vectoriel muni d'une filtration $\mathrm{Fil}^i\, M_{dR}$, M_B est un $\mathbb{Q}$-espace vectoriel muni d'une involution c (on note $M_B^{\pm}$ les parties $\pm$ pour cette involution), M_l est une représentation l-adique de $G_{\mathbb{Q}}$. Nous supposons que $V = M_p$ vérifie les hypothèses des paragraphes précédents et est en particulier cristalline en p. On a des isomorphismes de comparaison $\mathbb{C}\otimes M_{dR} \cong \mathbb{C}\otimes M_B, B_{cris}\otimes M_{dR} \cong B_{cris}\otimes M_p$ et $\mathbb{Q}_p\otimes M_{dR} \cong \mathbf{D}_p(M_p)$[7] compatibles avec les différentes structures supplémentaires. On note $\mathbb{Q}$ le motif trivial, $\mathbb{Q}(1)$ le motif de Tate et $\mathbb{Q}(j) = \mathbb{Q}(1)^{\otimes j}$ si $j > 0$ et $\mathrm{Hom}(\mathbb{Q}(-j), \mathbb{Q})$ si $j \leq 0$. Pour tout entier j, on définit le motif $M(j)$ twist de M par $\mathbb{Q}(j)$.

Fixons une $\mathbb{Z}$-structure $\mathcal{M}$ sur M, c'est-à-dire un réseau $\mathcal{M}$ de M_B tel que $\mathbb{Z}_l \otimes \mathcal{M}$ vu dans M_l soit stable par $G_{\mathbb{Q}}$. On choisit une base de la $\mathbb{Z}$-structure canonique $\mathbb{Z}(1)$ de $\mathbb{Q}(1)$. On fixe sur $\mathcal{M}$ une orientation c'est-à-dire des bases $\omega_{\mathcal{M}}^+$ de $det_{\mathbb{Z}}\mathcal{M}^+$ et $\omega_{\mathcal{M}(1)}^+$ de $det_{\mathbb{Z}}\mathcal{M}(1)^+$. On en déduit alors pour tout j une base $\omega_{\mathcal{M}(j)}^+$ de $det_{\mathbb{Z}}\mathcal{M}(j)^+$.

Soit j un entier tel que $\mathrm{Fil}^{-1}\, M_{dR}(j) = 0$. **Conjecturalement**, on définit un $\mathbb{Q}$-espace vectoriel $H^1_f(\mathbb{Q}, M(j))$ de dimension $d_-(M(j))$ et une application $\mathbb{Q}$-linéaire $H^1_f(\mathbb{Q}, M(j)) \to H^1_f(\mathbb{Q}, M(j)_p)$ qui devient un isomorphisme lorsqu'on tensorise par $\mathbb{Q}_p$; on définit un isomorphisme $\mathbb{R}\otimes H^1_f(\mathbb{Q}, M(j)) \to \mathbb{R}\otimes \mathcal{M}_{dR}(j)/M_B(j)^+$. D'où une application

$$det_{\mathbb{Q}}H^1_f(\mathbb{Q}, M(j)) \otimes (det_{\mathbb{Q}}M_{dR}(j))^* \otimes det_{\mathbb{Q}}M_B(j)^+ \to \mathbb{R}\ .$$

Si $\omega \in (det_{\mathbb{Q}}M_{dR}(j))^* \otimes det_{\mathbb{Q}}H^1_f(\mathbb{Q}, M(j))$, on note $Per_{\infty,\mathcal{M}(j)}(\omega)$ l'image de $\omega \otimes \omega_{\mathcal{M}(j)}^+$ par cet isomorphisme. De même, en utilisant l'application exponentielle de $M(j)_p$, on définit une application

$$det_{\mathbb{Q}_p}H^1_f(\mathbb{Q}, M(j)_p) \otimes (det_{\mathbb{Q}_p}\mathbf{D}_p(M(j)))^* \otimes \wedge^{d_+(M(j))}\mathbf{D}_p(M(j)) \to \mathbb{Q}_p\ .$$

Si $\omega \in (det_{\mathbb{Q}_p}\mathbf{D}_p(M(j)))^* \otimes det_{\mathbb{Q}_p}H^1_f(\mathbb{Q}, M(j)_p)$ et si $\delta \in \wedge^{d_+(M(j))}\mathbf{D}_p(M(j))$, on note $Per_{p,M(j)_p}(\omega)(\delta)$ l'image de $\omega\otimes\delta$ par cet isomorphisme. Ainsi, $Per_{p,M(j)_p}(\omega)$ appartient à $\mathrm{Hom}_{\mathbb{Q}_p}(\wedge^{d_+(M(j))}\mathbf{D}_p(M(j)), \mathbb{Q}_p)$.

L'élément $Per_{\infty,\mathcal{M}(j)}(\omega)^{-1} \otimes Per_{p,M(j)_p}(\omega)(\delta)$ de $\mathbb{R}\otimes\mathbb{Q}_p$ ne dépend pas de $\omega \in (det_{\mathbb{Q}}M_{dR}(j))^* \otimes det_{\mathbb{Q}}H^1_f(\mathbb{Q}, M(j))$. Un choix possible pour δ est un élément

[7] On pose $\mathbf{D}_p(M) = \mathbf{D}_p(M_p)$

de $det_{\mathbb{Z}_p}(\mathbb{Z}_p \otimes \mathcal{M}^+)$, qui dépend du choix d'une conjugaison complexe, ce qui n'est pas toujours naturel dans le cadre p-adique. Cela explique la dissymétrie entre les définitions complexe et p-adique.[8]

2.2.2. Au motif M est associée une fonction L: on note $\mathbf{L}^{\infty}_{\{\infty,p\}}(M,s) = \prod_{l\neq p} L_l(M,s)$ la fonction incomplète en p. Les conjectures de Beilinson prédisent que, avec les notations précédentes,

$$\frac{\mathbf{L}^{\infty}_{\{\infty,p\}}(M,j)}{Per_{\infty,\mathcal{M}(j)}(\omega)} \in \mathbb{Q}.$$

2.2.3. CONJECTURE. *Il existe un unique générateur* $\mathbf{L}^{p}_{\{\infty,p\},h}(\mathcal{M})$ *du* Λ*-module* $\mathbb{I}_{arith,\{\infty,p\},h}(\mathcal{M}_p)$ *tel que pour tout entier* $j \gg 0$, *on a l'égalité d'éléments de* $\mathrm{Hom}_{\mathbb{Q}_p}(\wedge^{d_+(M(j))}\mathbf{D}_p(M_p),\mathbb{Q}_p)$

$$\begin{aligned}&\wedge^{d_+(M(j))}((1-p^j\varphi)^{-1}(1-p^{-j-1}\varphi^{-1}))\chi^{-j}\mathbf{L}^{p}_{\{\infty,p\},h}(\mathcal{M})\\ =&(h+j-1)!^{d_+(M(j))}2^{-d_+(M(j))}\frac{\mathbf{L}^{\infty}_{\{\infty,p\}}(M(j),0)}{Per_{\infty,\mathcal{M}(j)}(\omega)}Per_{p,M(j)_p}(\omega)\end{aligned} \quad (9)$$

pour $\omega \in (det_{\mathbb{Q}}M_{dR}(j))^* \otimes det_{\mathbb{Q}}H^1_f(\mathbb{Q},M(j))$.

Cette conjecture, d'ailleurs totalement inabordable car elle présuppose déjà les conjectures de Beilinson, comporte en fait une partie d'interpolation (existence, pour h assez grand, d'un élément de $\mathrm{Hom}_{\mathbb{Q}_p}(\wedge^{d_+(M(j))}\mathbf{D}_p(M_p),\mathcal{H}(G_\infty))$ vérifiant les équations (9)) et une partie "conjecture principale" (lien entre le module arithmétique et la fonction d'interpolation analytique). D'autre part, des conjectures sur l'ordre du zéro de $\mathbf{L}^{p}_{\{\infty,p\},h}(\mathcal{M})$ et sur le terme dominant en χ^j pour tout j peuvent être formulées ([Pb]). Ces conjectures que l'on peut aussi voir comme un analogue p-adique des conjectures à la Bloch-Kato sont essentiellement démontrées pour le module des fonctions L p-adiques de M_p.

References

[BGS84] D. Bernardi, C. Goldstein et N. Stephens, *Notes p-adiques sur les courbes elliptiques*, J. Reine Angew. Math. **351** (1984), 129–170.

[BP93] D. Bernardi et B. Perrin-Riou, *Variante p-adique de la conjecture de Birch et Swinnerton-Dyer (le cas supersingulier)*, C.R. Acad. Sci. Paris, **317**, série I (1993), 227–232.

[BK90] S. Bloch et K. Kato, *L functions and Tamagawa numbers of motives*, dans The Grothendieck Festschrift, vol. 1, Prog. in Math. **86**, Birkhäuser, Boston, 1990, 333–400.

[C] P. Colmez, *Fonctions zêta p-adiques en* $s=0$, prépublication DMI, ENS (1994), LMENS 94–11.

[Bu] *Périodes p-adiques*, (tenu à Bures en 1988), Astérisque **223** (1994), (J.-M. Fontaine, éd.).

[FP91] J.-M. Fontaine et B. Perrin-Riou, *Autour des conjectures de Bloch et Kato*, C.R. Acad. Sci. Paris, **313**, série I (1991), *I - Cohomologie galoisienne*, 189-196, *II - Structures motiviques f-closes*, 349–356; *III - Le cas général*, 421–428.

[8] voir [C] pour une discussion à ce propos.

[FP94] J.-M. Fontaine et B. Perrin-Riou, *Autour des conjectures de Bloch et Kato: cohomologie galoisienne et valeurs de fonctions L*, dans Motives, (U. Jannsen, S. Kleiman, J.-P. Serre, éds), Proc. Symp. Pure Math. **5** (1994), 599–706.

[Mc] McCornell, *On the Iwasawa theory of CM elliptic curves at supersingular primes*, prépublication (1993).

[P92] B. Perrin-Riou, *Théorie d'Iwasawa des représentations p-adiques: le cas local*, C.R. Acad. Sci. Paris, **315**, série I (1992), 629–632.

[P93] B. Perrin-Riou, *Fonctions L p-adiques d'une courbe elliptique et points rationnels*, Ann. Inst. Fourier, **43**, (1993), 945–995.

[P94] B. Perrin-Riou, *Théorie d'Iwasawa des représentations p-adiques sur un corps local*, Invent. Math. **115** (1994), 81–149.

[Pa] B. Perrin-Riou, *La fonction de Kubota-Leopoldt*, dans Arithmetic Geometry (N. Childress et J. Jones, eds), Contemp. Math. **174** (1994), 65–93.

[Pb] B. Perrin-Riou, *Fonctions L p-adiques des représentations p-adiques*, Astérisque **229** (1995).

[RS] K. Rubin et A. Silverberg, *Families of elliptic curves with constant mod p representations*, prépublication (1994).

The Role of Smooth Numbers in Number Theoretic Algorithms

CARL POMERANCE*

The University of Georgia
Athens, Georgia 30602, USA

1 Introduction

A *smooth number* is a number with only small prime factors. In particular, a positive integer is *y-smooth* if it has no prime factor exceeding y. Smooth numbers are a useful tool in number theory because they not only have a simple multiplicative structure, but are also fairly numerous. These twin properties of smooth numbers are the main reason they play a key role in almost every modern integer factorization algorithm. Smooth numbers play a similar essential role in discrete logarithm algorithms (methods to represent some group element as a power of another), and a lesser, but still important, role in primality tests.

In this article we shall survey some of the more interesting theoretical and practical algorithms for factoring, computing discrete logarithms, and primality testing, and will especially highlight the central role of smooth numbers in the subject.

2 A "fundamental lemma"

We begin with a problem that does not appear to have anything to do with our main topic. We shall first see that smooth numbers play an essential role in both the theoretical and algorithmic solution of the problem. We next shall show how the problem is the key ingredient in a robust class of factoring algorithms.

Suppose we choose integers independently and with uniform distribution in the interval $[1, x]$. How many should we choose so that almost surely some nonempty subset of our choices will have a product that is a square? The answer depends on the function $\exp(\sqrt{\ln x \ln\ln x})$, which we shall abbreviate as $L(x)$.

Lemma 2.1. *Let ε be an arbitrarily small positive number. If we choose $L(x)^{\sqrt{2}+\varepsilon}$ integers from $[1, x]$ (independently and with uniform distribution), then as $x \to \infty$, the probability tends to 1 that some nonempty subset product is a square, whereas if we choose $L(x)^{\sqrt{2}-\varepsilon}$ integers, the probability tends to 0.*

A proof of the first statement, which is considerably easier than a proof of the second, is implicit in [BLP, Theorem 10.1], and explicit in [P4, Proposition 4.1].

*) Supported in part by a grant from the National Science Foundation.

Proceedings of the International Congress
of Mathematicians, Zürich, Switzerland 1994

A proof of the entire result will be given in a forthcoming paper of the author. For our purposes it will be interesting to see why Lemma 2.1 is true. In addition, there is an algorithmic problem implicit in Lemma 2.1. Namely, if you are actually choosing the random integers and want to explicitly find a subset product that is a square, what is an efficient way to do this?

If we choose a number $n \in [1, x]$ with a large prime factor p, then it is unlikely that p^2 divides n and it will probably be a long wait before we ever see another number $m \in [1, x]$ divisible by p. Thus, it is unlikely that we can use n in a subset product that is a square. That is, the numbers that we can potentially use in the subset product are smooth numbers. Let y be some positive number, which we shall specify shortly. If we have chosen more y-smooth numbers than there are primes up to y, then some nonempty subset of these numbers has a product that is a square. This follows from a simple linear algebra argument. Each y-smooth integer n has an *exponent vector* $\mathbf{v}(n)$ of length the number of primes up to y. Indeed, if $p \le y$ is prime, then the coordinate in $\mathbf{v}(n)$ corresponding to p is the exponent on p in the prime factorization of n. Let $\pi(y)$ denote the number of primes up to y. So if we have more than $\pi(y)$ of these exponent vectors, they must be linearly dependent. In particular, they are dependent over the field $\mathbf{F}_2$ of two elements. A dependency here is represented as a nonempty subset sum being the 0-vector, which corresponds exactly to the corresponding subset product being a square.

Let $\psi(x, y)$ denote the number of y-smooth integers up to x. Then the expected number of choices of random integers in $[1, x]$ to find one y-smooth number is $x/\psi(x, y)$, so that the expected number of choices to find $\pi(y)+1$ such y-smooths is $x(\pi(y) + 1)/\psi(x, y)$. We thus wish to choose y as a function of x so as to minimize this expression. It turns out that the optimal value of y is about $L(x)^{\sqrt{2}/2}$, and the resulting expected number is about $L(x)^{\sqrt{2}}$. This is how the upper bound in Lemma 2.1 is shown.

This proof sketch also serves to suggest how the algorithmic problem implicit in Lemma 2.1 may be efficiently solved. Namely, for $y = L(x)^{\sqrt{2}/2}$, test each choice of a number $n \in [1, x]$ to see if it is y-smooth, discarding those that are not. When $\pi(y) + 1$ successes n have been found, create the exponent vectors $\mathbf{v}(n)$, and with Gaussian elimination over $\mathbf{F}_2$, find a subset product that is a square.

The lower bound in Lemma 2.1 shows us that we cannot do substantially better; that is, smooth numbers are essentially forced upon us. The proof is tricky, but not especially deep, the idea being that for each fixed k, almost surely a subset whose product is a square will contain a number with kth largest prime factor maximal over the subset and k other numbers in the subset, each divisible by a different one of these k large primes. Further, a calculation shows that it is unlikely that we will see such a $(k + 1)$-tuple if we only choose $L(x)^{\sqrt{2k/(k+1)}-\varepsilon}$ numbers. Because k may be taken arbitrarily large, we get our result.

The upper and lower bound calculations in Lemma 2.1 depend on an estimate for the number $\psi(x, y)$ of y-smooths up to x: for any fixed, positive real number a, we have $\psi(x, L(x)^a) = xL(x)^{-1/(2a)+o(1)}$ as $x \to \infty$, see [CEP]. For more on the distribution and application of smooth numbers, see [HT] and [V].

3 Combinations of congruences

In this section we shall discuss the connection between Lemma 2.1 and factoring. An old factoring method due to Fermat is to represent the number to be factored as the difference of two squares. For example, one can verify mentally that $8051 = 90^2 - 7^2$, so that $8051 = 83 \times 97$. The problem with this method is that it is often very difficult to find two squares that work.

Instead, one may search for two squares whose difference is a multiple of the number to be factored. If $u^2 \equiv v^2 \bmod n$ and $u \not\equiv \pm v \bmod n$, then the greatest common factor of $u - v$ and n, which may be computed rapidly via Euclid's algorithm, is a nontrivial factor of n.

Assume n is an odd composite that is not a power. Suppose we choose a random integer $A \in [1, n]$ and compute the least positive residue $Q \equiv A^2 \bmod n$. Then Q is "close" to being a random integer in $[1, n]$. If Lemma 2.1 is applicable, we would expect to find a set of such numbers Q, with their product a square, after taking about $L(n)^{\sqrt{2}}$ values of A. Multiplying the corresponding congruences $Q \equiv A^2 \bmod n$ would thus give rise to a congruence of the shape $u^2 \equiv v^2 \bmod n$, from which we would have a chance of factoring n. (It is not certain that such a congruence could split n because it may be that $u \equiv \pm v \bmod n$.)

I have just described the "random squares" factorization method of Dixon. It can be proved that the likelihood of Q being smooth is about the same as a uniformly distributed random integer in $[1, n]$, so that this step in the above heuristic method can be made rigorous. It is also not hard to show that the final congruence is nontrivial with probability at least $1/2$, so the random squares method is a completely rigorous probabilistic factoring algorithm. With special subroutines to determine if the numbers Q are smooth and to do the linear algebra over $\mathbf{F}_2$, the expected running time for the random squares method is $L(n)^{\sqrt{2}+o(1)}$, see [P3].

A simple, but crucial fact about smooth numbers is that large numbers are less likely to be y-smooth than small numbers. In the random squares method we are presented with a stream of random auxiliary numbers Q that we examine for smoothness, discarding the majority that are not, and stopping when we have found sufficiently many that are smooth. If we could alter the stream so that the numbers Q are smaller, then each would have a better chance of being smooth, and we would not have to examine so many.

One simple way to make Q smaller is to replace it with $Q - n$ if it exceeds $n/2$; that is, to use the residue closest to 0 rather than the least positive residue. To make a square, we now would also have to worry about the sign of the product, but this can be easily handled by adding one extra coordinate to the exponent vectors $\mathbf{v}(Q)$ to represent the sign of Q. However, reducing the size from n to $n/2$ is not sufficient to substantially affect the complexity estimate.

If A_i/B_i is the ith convergent in the continued fraction expansion of $\sqrt{n}$, then the residue Q_i of $A_i^2 \bmod n$ that is closest to 0 satisfies $|Q_i| < 2\sqrt{n}$. Further, the numbers A_i, Q_i are easy to find by a simple recursive procedure. If it could be shown that the numbers Q_i are sufficiently random, then indeed we would have a significant improvement over the random squares method, with a complexity of $L(n)^{1+o(1)}$.

In some cases it can be shown that the numbers Q_i are definitely *not* sufficiently random. For example, if the period of the (periodic) continued fraction for $\sqrt{n}$ is too short, then the pairs A_i, Q_i may begin repeating before we have found enough smooth values of Q_i. However, for most numbers n this phenomenon does not occur, and even when it does, looking at the continued fraction for $\sqrt{an}$ for a small integer a seems to solve the problem.

This method, due to Brillhart and Morrison [MB], is known as the continued fraction method. It completely majorizes the random squares method. However, no one has *proved* that it is likely to work. Of course, the number n we are trying to factor does not *know* this! The continued fraction algorithm, like all modern, practical factoring algorithms, does not have a rigorous complexity analysis. However, heuristic analyses help us to compare various methods, and to see which may be worthy of further tinkering.

The fastest factoring algorithm that has been rigorously analyzed is the class group method (see [LP]). This method uses the group of primitive binary quadratic forms with discriminant either $-n$, or a small multiple of $-n$. Smooth numbers play a key role here as well. The algorithm generates random forms (Q, R, S) in the class group by looking at random words on a small generating set. Corresponding to the prime factorization of Q, we get a factorization of (Q, R, S) into corresponding "prime forms". By accepting only those cases where Q is smooth, we collect relations between the generating forms and the prime forms. When enough such relations are collected, we can use them, again via a linear algebra step over $\mathbf{F}_2$, to construct an "ambiguous form", namely one whose square is the identity. From Gauss, we may use such forms to factor the discriminant, which is exactly what we wish to do. The expected running time of this algorithm is $L(n)^{1+o(1)}$, which is the same as for the simpler continued fraction method discussed above. In contrast though, the class group method is rigorous. It is surely not practical, being majorized in practice by other methods that will be discussed below.

4 Smoothness tests

When presented with an integer $n \leq x$, how long does it take to determine if n is y-smooth? If one uses trial division with the primes up to y, it takes about $\pi(y)$ steps to determine if most numbers are y-smooth. In factoring algorithms such as the ones above, the overwhelming majority of the auxiliary numbers Q presented are not y-smooth. If it takes us $\pi(y)$ steps per candidate to discover if a number is y-smooth, then this step of the algorithm dominates all others. It is greatly in our interest to find a smoothness test faster than trial division.

In [P1] an "early abort strategy" is described, which suggests that one give up on the trial division of a particular candidate Q if at various strategic points below y not enough of Q has been partially factored. This method loses some y-smooth numbers, but not many. The average time per number is only about $\sqrt{y}$. In addition, trial division may be replaced with a fast Fourier transform method of Pollard and Strassen (see [P4]), which further reduces the amortized time per candidate to about $y^{1/4}$.

The elliptic curve factoring method of Lenstra [L1], [L2] (see Section 6 below) has the feature that its expected running time to completely factor a number is a small function of the second largest prime factor of the number. In particular, it recognizes y-smooth numbers below x in $O(y^\varepsilon \ln x)$ steps, for any $\varepsilon > 0$. It is with this subroutine as a smoothness test that the complexity estimates of the last section are achieved.

A few comments are in order. The elliptic curve method is not completely rigorous. However, it is possible to show that *most* smooth numbers will be factored quickly with the method. Thus, as with the early abort strategy above, it is not crucial that a few y-smooths may pass unrecognized. Nevertheless, it is of theoretical interest if a smoothness test can be devised that recognizes y-smooth numbers in time about y^ε and that has no exceptions. This is provided in a recent method that is similar to the elliptic curve method, but uses Jacobian varieties of hyperelliptic curves of genus 2 (see [LPP]).

Is it ever practical to use the elliptic curve method as a subroutine to recognize smooth auxiliary numbers? We know of no case where it is. This is largely due to the existence of a far better smoothness test that is applicable when the stream of auxiliary numbers presented happens to be the consecutive values of a polynomial with integer coefficients.

Everyone knows the sieve of Eratosthenes as an efficient method of finding all of the primes up to some point. However, this sieve can also be used to find y-smooth numbers. One sieves with the primes up to y (and possibly their powers), and instead of crossing off the multiples of each prime, one keeps a tally at each number of how frequently it has been "hit". This tally may be done by adding the (single precision) logarithms of the primes that hit, and if the sum exceeds a threshold, the number is reported as being y-smooth.

What makes this sieve work is that the multiples of a given prime p occur in a regular pattern, namely an arithmetic progression of difference p. If instead of the consecutive integers to some point, one has the image of this interval under a polynomial with integer coefficients, one still has regularity for the multiples of p. They now form the union of several arithmetic progressions of difference p, and we may sieve just as efficiently as before. For example, no value of $t^2 + 1$ is divisible by 3, and the multiples of 5 are found in the progressions $t \equiv \pm 2 \bmod 5$.

However, the streams of auxiliary numbers Q described in the previous section are not the consecutive values of a polynomial, and there is no discernible regularity to where the multiples of a given prime p appear. In the next section we shall describe two algorithms that can make use of sieving as a smoothness test.

5 The quadratic sieve and the number field sieve

Say we wish to factor the odd number n, which has been already verified to be composite and not a power. Consider the quadratic polynomial $Q(t) = t^2 - n$. For ε small and $|t - \sqrt{n}| < n^\varepsilon$, we have $|Q(t)| < 3n^{1/2+\varepsilon}$. Thus, the values of $Q(t)$ for t close to $\sqrt{n}$ are relatively small. If it could be assumed that the values of $Q(t)$ for t in this range are about as likely to be smooth as random integers of the same size, then Lemma 2.1 suggests that with $x = 3n^{1/2+\varepsilon}$, before $L(x)^{\sqrt{2}+\varepsilon} < L(n)^{1+2\varepsilon}$ values of t are examined, there will be a nonempty subset such that the product of the corresponding values of $Q(t)$ is a square, say u^2. If v is the product of these values of t, then because $Q(t) \equiv t^2 \bmod n$, we have $u^2 \equiv v^2 \bmod n$. So, as before, the greatest common factor of $u - v$ and n may provide a nontrivial factor of n. This algorithm then is exactly the same as the random squares method and the continued fraction method discussed in Section 3, but now the stream of auxiliary numbers $Q(t)$ are the consecutive values of a polynomial with integer coefficients, so that we may use a sieve as a smoothness test.

This is the basic quadratic sieve method (see [P1]). Though the values $Q(t)$ are slightly larger than the auxiliary numbers Q in the continued fraction algorithm, sieving is so good a smoothness test that this small defect is not important. When the quadratic sieve method is used today, we do not use only one polynomial, but a family of many polynomials of the form $at^2 + 2bt + c$, where a, b, c are chosen in a certain range and with $b^2 - ac = n$. This idea of Montgomery (see [P2]) mitigates somewhat the growth of the size of the polynomial values, for when the values of one polynomial become large, we switch to another. The multiple polynomial quadratic sieve currently enjoys the record for the factorization of the largest number of no special form and without small prime factors that has ever been factored. This number is the 129-digit composite announced as a cryptographic challenge in Martin Gardner's *Scientific American* column in 1977. It was factored in 1994 by D. Atkins, M. Graff, A. Lenstra, P. Leyland, and a host of others who volunteered time on their workstations.

A word must be said about the linear algebra subroutines used to assemble the congruences into congruent squares at the final stage of the algorithm. To achieve the complexity estimate $L(n)^{1+o(1)}$ for the quadratic sieve (and the earlier algorithms mentioned), one cannot use Gaussian elimination as the linear algebra subroutine. Instead, there are methods due to Wiedemann, Lanczos, and others that are used. These methods exploit the facts that the matrix of exponent vectors is sparse, and that the algebra is done over the field $\mathbf{F}_2$ of two elements. In practice so far, we have largely been able to get by with Gaussian elimination and variations of it. Although it is easy to distribute the sieving stage of the algorithm to many unextraordinary computers, so far it is awkward to do this with the linear algebra stage, and for the record numbers factored these days, supercomputers are used for the matrix. Clearly more work is needed for this step of combination of congruences algorithms. For more on this subject, see [C], [LO], [M], [PS].

The reader may have noticed that many factoring algorithms seem to end up with either rigorous or heuristic complexity of $L(n)^{1+o(1)}$. This is due to the auxiliary numbers that we examine for smoothness, which in the algorithms we have described so far (except for the random squares method) are all bounded by the common expression $n^{1/2+o(1)}$. If we could reduce the size of these numbers that we hope to find smooth, we could reduce the complexity of the algorithm. The *number field sieve* allows us to do just this. In this algorithm the auxiliary numbers are about $\exp(c(\ln n)^{2/3}(\ln\ln n)^{1/3})$. Putting this bound in for x in Lemma 2.1 suggests that the complexity of the number field sieve is about $\exp((4c/3)^{1/2}(\ln n)^{1/3}(\ln\ln n)^{2/3})$. In the original version of the number field sieve, invented by Pollard in the late 1980's, only numbers near a high power are factored, and in this case, the number c turns out to be $(16/3)^{1/3}$. This method was later generalized to arbitrary numbers n by Buhler, H. Lenstra, and the author, but at the cost of increasing the number c to $(64/3)^{1/3}$.

In sum, the number field sieve is asymptotically fast because it achieves a dramatic reduction in the number of digits of the auxiliary numbers: they have about the 2/3 *power* of the number of digits of n, as opposed to about half the number of digits of n in the quadratic sieve. How then does the number field sieve work?

Suppose f, g are irreducible, monic polynomials over $\mathbf{Z}$ for which there is an integer m with $f(m) \equiv g(m) \equiv 0 \bmod n$. Say α, β are complex numbers with $f(\alpha) = g(\beta) = 0$. Consider the substitution homomorphisms $\phi : \mathbf{Z}[\alpha] \to \mathbf{Z}/(n)$, $\psi : \mathbf{Z}[\beta] \to \mathbf{Z}/(n)$, where $\phi(\alpha) = \psi(\beta) = m + (n)$. Thus, for any integers a, b we have the congruence $\phi(a + b\alpha) \equiv \psi(a + b\beta) \bmod n$. It is via this family of congruences as a, b vary over small (coprime) integers that we hope to assemble our congruent squares. But we actually will construct the squares in the rings $\mathbf{Z}[\alpha]$, $\mathbf{Z}[\beta]$, which is no loss because the homomorphic image of a square is a square. Not only is there no loss, there can be a substantial gain.

We define a member of $\mathbf{Z}[\alpha]$ to be smooth if its norm to $\mathbf{Z}$ is smooth. However, the norm function masks the proliferation of prime ideals that may lie over a rational prime. Taking this into account, and adding some extra information afforded by a few random quadratic characters (to get over the obstructions of possibly complicated class groups, unit groups, and quotient groups of $\mathbf{Z}[\alpha]$, $\mathbf{Z}[\beta]$ in their maximal orders), our above method using exponent vectors allows one to construct squares. Finding the square roots of these squares is not as simple as before, but it is a tractable problem. The auxiliary numbers we wish to find smooth are the products of the norms to $\mathbf{Z}$ of $a + b\alpha$ and $a + b\beta$, where a, b run over small coprime integers. This is a polynomial in a and b, and we may use a sieve as a smoothness test. The size of these auxiliary numbers depends on the largest coefficients of f and g and their degrees.

One way to construct the polynomials f and g is to first pick d, the degree of f, next pick $m = [n^{1/d}]$, and write n in the base m, so that $n = m^d + c_{d-1}m^{d-1} + \cdots + c_0$. Then we let $f(t) = t^d + c_{d-1}t^{d-1} + \cdots + c_0$ and $g(t) = t - m$. There are other strategies too, and in particular it is not essential that the polynomials be monic. For more on the number field sieve see [LL] and [P4].

The largest number of no special form that the number field sieve has factored has 119 digits, a recent accomplishment of Contini, Dodson, A. Lenstra, and Montgomery. It is likely though that this will change soon. The very favorable heuristic complexity estimate has concentrated much attention on the number field sieve, people are beginning to find the improvements necessary to make it a practical algorithm, and it is thought that before long it will replace the quadratic sieve as the champion method for numbers of no special form.

6 The elliptic curve method

The elliptic curve method of H. Lenstra uses smooth numbers in an intrinsically different way than the previous factorization methods discussed. Based on a beautiful method of Pollard to discover those prime factors p in a number for which $p - 1$ is smooth, it makes use of the following observation. If G is a finite group (written additively), then there is a simple algorithm to test if the order of an element $g \in G$ is a y-smooth number below x. Indeed, let M be the least common multiple of the y-smooth numbers below x and form Mg. This calculation can be done in $O(\pi(y)\ln x)$ group operations by the repeated doubling method. Then the order of g is a y-smooth number below x if and only if Mg is the identity.

This observation is used as follows. Suppose $p < q$ are prime factors of the number n we wish to split. Let a, b be integers with $4a^3 + 27b^2$ coprime to n and let $P = (x_0, y_0)$ be an integer point on the elliptic curve $E: \ y^2 = x^3 + ax + b$. Let $E(p)$, $E(q)$ be the elliptic curve groups mod p and mod q, respectively. If P mod p has y-smooth order in $E(p)$, but P mod q does not have y-smooth order in $E(q)$, then we can use this to split n. Indeed, let M be the least common multiple of the y-smooth numbers below $(n^{1/4}+1)^2$. We cannot directly work in the groups $E(p)$ and $E(q)$ because we do not know p and q. However, we can try to add points on E modulo n. If the addition law breaks down it is because we are trying to invert a nonzero, noninvertible residue modulo n. But Euclid's algorithm, which is used for inversion, would in this case split n. The addition law breaks down when we try to add two points R, S such that $R+S$ is the identity modulo some factor of n, but not the identity modulo some other factor of n. This is exactly what happens when we try to compute MP modulo n, because this is the identity in $E(p)$ and it is not the identity in $E(q)$.

We can attempt to do this calculation even if we do not know beforehand that P has y-smooth order in $E(p)$, but not in $E(q)$. If it works we have split n. If it does not work, we have the option of trying again with a larger value of y, or more interestingly, trying again with another triple a, b, P. We can easily generate many such triples by choosing a, x_0, y_0 at random, and solving for b.

This then is the elliptic curve method. If the prime p has sufficiently many smooth numbers near it in the "Hasse interval" $((\sqrt{p}-1)^2, (\sqrt{p}+1)^2)$, then it can be shown rigorously that the method is expected to find p as a prime factor of numbers n divisible by p. It is conjectured that this interval does contain enough smooth numbers, but it has not been proved. It is interesting that in the longer interval $((\sqrt{p}-1)^4, (\sqrt{p}+1)^4)$, we can prove that there are many smooth numbers, which is why the *hyperelliptic* curve method can be rigorously analyzed — see [LPP].

An important contrast between the elliptic curve method and combination of congruences methods, is that in the latter we need to be able to find many smooth numbers for success, but each auxiliary number is quickly dealt with. In the elliptic curve method we are successful if just one auxiliary number (which is hidden from us) is smooth, but it takes a fair amount of time for each trial. The two opposite effects balance out. In the worst case the number n has its least prime factor near $\sqrt{n}$, and so the numbers we hope to find smooth are also near $\sqrt{n}$. So in the worst case, the elliptic curve method takes $L(n)^{1+o(1)}$ steps. However, most numbers are not in the worst case, so that the elliptic curve method can be considerably faster. Thus, when presented with a number to factor, one usually tries the elliptic curve method before attempting the quadratic sieve or the number field sieve.

7 Discrete logarithms and the search for smoothness

Given a cyclic group $G = \langle g \rangle$ (written multiplicatively), and an element h in G, the discrete logarithm problem is to find an integer n with $g^n = h$. In this problem the representation of the group G is of paramount importance. For example, suppose p is a prime. Then $(\mathbf{Z}/(p))^*$ is a cyclic group of order $p-1$, as is the additive group $\mathbf{Z}/(p-1)$. However, solving the discrete logarithm problem in the latter group is a triviality — one uses Euclid's algorithm to solve a linear congruence. But the discrete logarithm problem for the former group is hard, or at least apparently so.

One can find discrete logarithms in the group $(\mathbf{Z}/(p))^*$ by an algorithm similar to the random squares method discussed in Section 3. With g the cyclic generator on which logs are based, consider random powers g^m. Elements of the group are of course residue classes; say we represent these residue classes by their least positive member. That is, we represent group elements by positive integers less than p. If g^m is represented by a smooth integer, we keep it, and otherwise, we discard it. If we can find sufficiently many independent "relations", where a power of g is congruent mod p to a y-smooth number, we can use linear algebra (over the ring $\mathbf{Z}/(p-1)$) to solve for the logs of the primes q up to y. Once this pre-calculation is done, it is now fairly simple to find the log of an element h. Namely, consider $g^m h$, where again m is a random integer. If this is represented by a y-smooth number, say $\prod q_i^{a_i}$, where the q_i's run over the primes up to y, then $\log_g h$ is $-m + \sum a_i \log_g q_i$. To minimize the expected running time we take $y = L(p)^{\sqrt{2}/2}$. Then the running time of the first phase of this algorithm (to compute the logs of all of the primes up to y) is about $L(p)^{\sqrt{2}}$ and the running time to compute an individual log is about $L(p)^{\sqrt{2}/2}$. See [P3] for more details.

Can these ideas be generalized to the multiplicative group of a finite field $\mathbf{F}_q$? In particular, what would it mean to call a member of $\mathbf{F}_q^*$ "smooth"? If $q = p^k$, where k is large, then the usual representation of $\mathbf{F}_q$ is $\mathbf{F}_p[x]/(f)$, where f is an irreducible polynomial in $\mathbf{F}_p[x]$ of degree k. We may represent a group element as the member of the residue class of least degree. Because $\mathbf{F}_p[x]$, like $\mathbf{Z}$, is a Euclidean domain, we may give a definition of a smooth element. Say a polynomial is smooth if it factors completely into low degree irreducibles. There is a theory of the distribution of smooth polynomials in $\mathbf{F}_p[x]$ that is analogous to the distribution of smooth integers — see [Lo1], [O], [So]. We thus obtain a rigorous discrete logarithm algorithm analogous to the one above.

When $q = p^k$ with $k > 1$ and k small, the above representation of $\mathbf{F}_q^*$ is not particularly useful for computing discrete logarithms. Indeed, say $k = 2$. Then every residue class representative has degree 0 or degree 1, and so everything is smooth. Instead, we represent the field as $\mathcal{O}_K/(p)$, where K is an algebraic number field of degree k over the rationals and for which the prime p remains inert. In this case we call a field element smooth, if a canonical representative of the residue class has smooth norm. A problem is how to define a canonical representative. This is solved in the case $k = 2$ in [Lo2] where a rigorous algorithm is described.

Although we have not found a way to use something resembling the elliptic curve method or the quadratic sieve to compute discrete logarithms in $\mathbf{F}_q^*$, we have found a way to use analogs of the number field sieve — see [A], [G], [S]. As with factoring, the analysis is heuristic. Whether these algorithms are practical is unclear.

There are of course many other groups around. For example, one may consider a prime p for which the elliptic curve group $E(p)$ (see Section 6) is cyclic. Does it make sense to say an element of $E(p)$ is smooth? No one has thought of a way to make sense of this (except for some very special cases), and for this reason, we know of no fast ways to compute discrete logs in elliptic curve groups. These groups have been proposed as vehicles for public key cryptography precisely because we have no notion of smoothness for them.

8 Smooth numbers and primality testing

The central problem in primality testing is to decide if a given input is prime or composite. This problem is generally considered much easier than factoring composites. One of the simpler ideas in the subject involves Fermat's "little theorem": $a^p \equiv a \bmod p$ for all integers a. It is computationally easy to compute the residue of $a^p \bmod p$, and if this is not a, then p has been proved composite.

This simple test done with $a = 2$ is enough to recognize most composite numbers. However, for any fixed base a there are infinitely many *pseudoprimes to the base* a, namely composite integers n for which $a^n \equiv a \bmod n$. In fact, there are infinitely many *Carmichael numbers*. These are composite integers n for which $a^n \equiv a \bmod n$ for *every* integer a. It had been conjectured that there are infinitely many Carmichael numbers essentially by Carmichael himself when he introduced them in 1910. The proof that there are infinitely many was accomplished in 1992 by Alford, Granville, and the author [AGP], and is based on a 1956 heuristic argument of Erdős. This heuristic method begins by assuming that there are many primes p for which $p-1$ is y-smooth. In fact, there should be a positive proportion of all primes below y^c with this property, where c is an arbitrary but fixed number. Erdős himself had proved such a result in 1935 for some particular $c > 1$, and recently Friedlander proved it for any $c < 2\sqrt{e}$. With this and other tools, we were able to prove that there are more than $x^{2/7}$ Carmichael numbers up to x, when x is sufficiently large. It is interesting that the Erdős heuristic method in fact suggests that there are more than $x^{1-\varepsilon}$ Carmichael numbers up to x.

There are stronger tests than Fermat's little theorem for which there is no analog of a Carmichael number, and such that on input of a composite number, the test is expected to prove the number composite in only $O(1)$ iterations. One of these is using Selfridge's strong pseudoprime test to random bases, a result of Rabin. From the work of Miller, Bach, and others we know that every odd composite n will fail a strong pseudoprime test to some base less than $2\ln^2 n$, provided that the Riemann hypothesis for Dirichlet L-functions holds. Thus, if this hypothesis holds, we have a deterministic polynomial time primality test. In a sequel to [AGP], the authors show that there are infinitely many odd composites n that pass the strong pseudoprime test for each base up to $(\ln n)^{c/\ln\ln\ln n}$.

Do we have unconditional tests that end up proving a prime input is prime? Surely we should not be satisfied with a probabilistic composite recognition test that fails to recognize our input as composite after several tries.

There are in fact very fast primality proving algorithms. The fastest known deterministic test has complexity $(\ln n)^{c\ln\ln\ln n}$, and so is "almost" polynomial, see [APR]. As with the discussion above on Carmichael numbers, this test uses auxiliary primes p for which $p - 1$ is y-smooth. Certain versions of this test are quite practical, see [BH], [CL]. There is a probabilistic test that expects to find a rigorous proof of primality in expected polynomial time. Though not very practical, simpler, but heuristic versions of it have been used on very large primes, see [AH], [AM], [GK], [L2].

The central unsolved problem in primality testing is to see if there is a deterministic, polynomial time algorithm to distinguish between primes and composites. Towards this end, one may ask for a deterministic, polynomial time algorithm that succeeds in proving prime most or many primes up to a bound x. Recently, Konyagin and the author [KP] have described such an algorithm that proves prime more than $x^{1-\varepsilon}$ primes up to x. It is no mystery on which primes the algorithm works. It works on precisely those primes p for which $p - 1$ has a large smooth divisor.

The author gratefully acknowledges W. Alford, A. Granville, and H. Lenstra for their helpful critical comments on an earlier draft of this paper.

References

[A] L. M. Adleman, *The function field sieve*, in: Algorithmic Number Theory (L. M. Adleman and M.-D. Huang, eds.), Lecture Notes in Comput. Sci. **877**, Springer-Verlag, Berlin, 1994, 108–121.

[AH] L. M. Adleman and M.-D. Huang, Primality testing and abelian varieties over finite fields, Lecture Notes in Math. **1512**, Springer-Verlag, Berlin and New York, 1992.

[APR] L. M. Adleman, C. Pomerance, and R. S. Rumely, *On distinguishing prime numbers from composite numbers*, Ann. of Math. (2) **117** (1983), 173–206.

[AGP] W. R. Alford, A. Granville, and C. Pomerance, *There are infinitely many Carmichael numbers*, Ann. of Math. (2) **140** (1994), 703–722.

[AM] A. O. L. Atkin and F. Morain, *Elliptic curves and primality proving*, Math. Comp. **61** (1993), 29–68.

[BH] W. Bosma and M.-P. van der Hulst, Primality proving with cyclotomy, Ph.D. dissertation at the University of Amsterdam, Amsterdam, The Netherlands, 1990.

[BLP] J. P. Buhler, H. W. Lenstra, Jr., and C. Pomerance, *Factoring integers with the number field sieve*, in [LL], 50–94.

[CEP] E. R. Canfield, P. Erdős, and C. Pomerance, *On a problem of Oppenheim concerning "factorisatio numerorum"*, J. Number Theory **17** (1983), 1–28.

[CL] H. Cohen and H. W. Lenstra, Jr., *Primality testing and Jacobi sums*, Math. Comp. **42** (1984), 297–330.

[C] D. Coppersmith, *Solving homogeneous linear equations over $GF(2)$ via block Wiedemann algorithm*, Math. Comp. **62** (1994), 333–350.

[GK] S. Goldwasser and J. Kilian, *Almost all primes can be quickly certified*, Proc. 18th STOC (Berkeley, May 28–30, 1986), ACM, New York, 1986, 316–329. Also see Chapter 2 in J. Kilian, Uses of Randomness in Algorithms and Protocols, The MIT Press, Cambridge, MA, 1990.

[G] D. Gordon, *Discrete logarithms in $GF(p)$ using the number field sieve*, SIAM J. Discrete Math. **6** (1993), 124–138.

[HT] A. Hildebrand and G. Tenenbaum, *Integers without large prime factors*, J. de Théorie des Nombres de Bordeaux **5** (1993), 411–484.

[KP] S. Konyagin and C. Pomerance, *On primes recognizable in deterministic polynomial time*, in: The Mathematics of Paul Erdős (R. L. Graham and J. Nesetril, eds.), Springer-Verlag, to appear.

[LO] B. A. LaMacchia and A. M. Odlyzko, *Solving large sparse linear systems over finite fields*, in: Advances in Cryptology — CRYPTO 90 (A. J. Menezes and S. A. Vanstone, eds.), Lecture Notes in Comput. Sci. **537**, Springer-Verlag, Berlin and New York, 1991, 109–133.

[LL] A. K. Lenstra and H. W. Lenstra, Jr. (eds.), The development of the number field sieve, Lecture Notes in Math. **1554**, Springer-Verlag, Berlin and New York, 1993.

[L1] H. W. Lenstra, Jr., *Factoring integers with elliptic curves*, Ann. of Math. (2) **126** (1987), 649–673.

[L2] H. W. Lenstra, Jr., *Elliptic curves and number theoretic algorithms*, in: Proc. Int'l Cong. Math., Berkeley, CA, 1986, vol. 1 (A. M. Gleason, ed.), Amer. Math. Soc., Providence, RI, 1987, 99–120.

[LPP] H. W. Lenstra, Jr., J. Pila, and C. Pomerance, *A hyperelliptic smoothness test*, I, in [V], 397–408.

[LP] H. W. Lenstra, Jr., and C. Pomerance, *A rigorous time bound for factoring integers*, J. Amer. Math. Soc. **5** (1992), 483–516.

[Lo1] R. Lovorn, Rigorous, subexponential algorithms for discrete logarithms over finite fields, Ph.D. dissertation at The University of Georgia, Athens, GA, 1992.

[Lo2] R. Lovorn Bender, *Rigorous, subexponential algorithms for discrete logarithms in* $GF(p^2)$, SIAM J. Discrete Math., to appear.

[M] P. L. Montgomery, *A block Lanczos algorithm for finding dependencies over* $GF(2)$, in: Advances in Cryptology — EUROCRYPT 95 (L. C. Guillon and J.-J. Quisquater, eds.), Lecture Notes in Comput. Sci. **921**, Springer-Verlag, Berlin, 1995, 106–120.

[MB] M. Morrison and J. Brillhart, *A method of factoring and the factorization of* F_7, *Math. Comp.* **29** (1975), 183–205.

[O] A. M. Odlyzko, *Discrete logarithms in finite fields and their cryptographic significance*, in: Advances in Cryptology — Proceedings of EUROCRYPT 84 (T. Beth et al., eds.), Lecture Notes in Comput. Sci. **209**, Springer-Verlag, Berlin and New York, 1985, 224–314.

[P1] C. Pomerance, *Analysis and comparison of some integer factoring algorithms*, in: Computational Methods in Number Theory (H. W. Lenstra, Jr. and R. Tijdeman, eds.), Math. Centre Tracts **154/155**, Mathematisch Centrum, Amsterdam, 1982, 89–139.

[P2] C. Pomerance, *The quadratic sieve factoring algorithm*, in: Advances in Cryptology — Proceedings of EUROCRYPT 84 (T. Beth et al., eds.), Lecture Notes in Comput. Sci. **209**, Springer-Verlag, Berlin and New York, 1985, 169–182.

[P3] C. Pomerance, *Fast, rigorous factorization and discrete logarithm algorithms*, in: Discrete Algorithms and Complexity (D. S. Johnson et al., eds.), Academic Press, Orlando, FL, and New York, 1987, 119–143.

[P4] C. Pomerance, *The number field sieve*, in: Mathematics of Computation 1943–1993: A Half-Century of Computational Mathematics (W. Gautschi, ed.), Proc. Sympos. Appl. Math. **48**, Amer. Math. Soc., Providence, 1994, 465–480.

[PS] C. Pomerance and J. W. Smith, *Reduction of huge, sparse matrices over a finite field via created catastrophes*, Experimental Math. **1** (1992), 90–94.

[S] O. Schirokauer, *Discrete logarithms and local units*, in [V], 409–423.

[So] K. Soundararajan, *Asymptotic formulae for the counting function of smooth polynomials*, J. London Math. Soc., to appear.

[V] R. C. Vaughan, ed., *Theory and applications of numbers without large prime factors*, Philos. Trans. Roy. Soc. London **345** (15 November 1993), 327–423.

Non-Archimedean Period Domains

MICHAEL RAPOPORT

Bergische Universität Gesamthochschule Wuppertal
Fachbereich 7: Mathematik
Gaußstraße 20, D-42097 Wuppertal, Germany

The best known example of a non-archimedean period domain is the Drinfeld upper half space Ω_E^d of dimension $d-1$ associated to a finite extension E of $\mathbf{Q}_p$ (complement of all E-rational hyperplanes in the projective space $\mathbf{P}^{d-1}$). Drinfeld [D2] interpreted this rigid-analytic space as the generic fibre of a formal scheme over O_E parametrizing certain p-divisible groups. He used this to p-adically uniformize certain Shimura curves (Cherednik's theorem) and to construct highly nontrivial étale coverings of Ω_E^d. This report gives an account of joint work of Zink and myself [RZ] that generalizes the construction of Drinfeld (Sections 1–3). In the last two sections these results are put in a more general framework (Fontaine conjecture) and the problem of the computation of ℓ-adic cohomology is addressed (Kottwitz conjecture). In this report we return to the subject of Grothendieck's talk at the Nice congress [G, esp. Section 5] where he stressed the relation between the local moduli of p-divisible groups and filtered Dieudonné modules.

Throughout this report we fix a prime number p. Denote by k a fixed algebraically closed field of characteristic p. Let $W(k)$ be its ring of Witt vectors and $K_0 = W(k) \otimes_{\mathbf{Z}} \mathbf{Q}$. Let σ be the Frobenius automorphism of K_0. For most results one must assume $p \neq 2$.

1. Formal moduli of p-divisible groups

If O is a complete discrete valuation ring with uniformizer π, we denote by Nilp_O the category of locally noetherian schemes S over Spec O such that the ideal sheaf $\pi \cdot \mathcal{O}_S$ is locally nilpotent. We denote by $\bar{S}$ the closed subscheme defined by $\pi \cdot \mathcal{O}_S$. A locally noetherian formal scheme over Spf O will be identified with the set-valued functor on Nilp_O it defines. A morphism $\mathcal{X} \to \mathcal{Y}$ of formal schemes is called *locally formally of finite type* if the induced morphism $\mathcal{X}_{\mathrm{red}} \to \mathcal{Y}_{\mathrm{red}}$ between their underlying reduced schemes of definition is locally of finite type.

In what follows, by a *quasi-isogeny between p-divisible groups* X and Y over a scheme $S \in \mathrm{Nilp}_{\mathbf{Z}_p}$ we mean a global section f of the Zariski sheaf $\underline{\mathrm{Hom}}(X, Y) \otimes_{\mathbf{Z}} \mathbf{Q}$ for which there exists locally on S an integer n such that $p^n f$ is an isogeny.

THEOREM 1.1. *Let* $\mathbf{X}$ *be a p-divisible group over* Spec k. *We consider the functor* $\mathcal{M}$ *on* $\mathrm{Nilp}_{W(k)}$, *which associates to* $S \in \mathrm{Nilp}_{W(k)}$ *the set of isomorphism classes of pairs* (X, ϱ) *consisting of a p-divisible group* X *over* S *and a quasi-isogeny* $\varrho : \mathbf{X} \times_{\mathrm{Spec}\, k} \bar{S} \to X \times_S \bar{S}$ *of p-divisible groups over* $\bar{S}$. *Then* $\mathcal{M}$ *is representable by a formal scheme locally formally of finite type over* Spf $W(k)$.

Proceedings of the International Congress
of Mathematicians, Zürich, Switzerland 1994

This representability theorem [RZ, Section 2] allows one to show that certain functors of p-divisible groups endowed with endomorphisms and level structures (case (EL)) resp. with polarizations and endomorphisms and level structures (case (PEL)) are also representable. These functors depend on certain "rational" and "integral" data that we now formulate in both cases.

Case (EL): The *rational data* consists of a 4-tuple (B, V, b, μ), where B is a finite-dimensional semi-simple algebra over $\mathbf{Q}_p$ and V a finite left B-module. Let $G = GL_B(V)$ (algebraic group over $\mathbf{Q}_p$). Then b is an element of $G(K_0)$. The final datum μ is a homomorphism $\mathbf{G}_m \to G_K$ defined over a finite extension K of K_0. Let $V \otimes_{\mathbf{Q}_p} K = \bigoplus V_i$ be the corresponding eigenspace decomposition and $V_K^j = \bigoplus_{i \geq j} V_i$ the associated decreasing filtration. We require that the *filtered isocrystal over* K, $(V \otimes_{\mathbf{Q}_p} K_0, b(\mathrm{id} \otimes \sigma), V_K^\bullet)$, be the filtered isocrystal associated to a p-divisible group over Spec O_K ([G], [Fo1], [Me]). The *integral data* consists of a maximal order O_B in B and an O_B-lattice chain $\mathcal{L}$ in V [RZ, Section 3].

Case (PEL): In this case the *rational data* are given by a 6-tuple $(B, *, V, (\ ,\ ), b, \mu)$. Here B and V are as before. Furthermore, B is endowed with an anti-involution $*$ and V is endowed with a nondegenerate alternating bilinear form $(\ ,\ ) : V \otimes_{\mathbf{Q}_p} V \to \mathbf{Q}_p$ such that

$$(dv, v') = (v, d^* v'), \quad d \in B. \tag{1}$$

The remaining data are as before relative to the algebraic group G over $\mathbf{Q}_p$ whose values in a $\mathbf{Q}_p$-algebra R are

$$G(R) = \{g \in GL_B(V \otimes R);\ (gv, gv') = c(g)(v, v'),\ \ c(g) \in R^\times\}.$$

We require that the rational data define the filtered isocrystal associated to a p-divisible group over Spec O_K endowed with a polarization (= symmetric isogeny to its dual). The *integral data* are as before. We assume that O_B is stable under $*$ and that $\mathcal{L}$ is self-dual w.r.t. $(\ ,\ )$.

In either case let E be the field of definition of the conjugacy class of μ, a finite extension of $\mathbf{Q}_p$ contained in K. Let $\breve{E} = E.K_0$, with ring of integers $O_{\breve{E}}$.

THEOREM 1.2. *We fix data of type (EL) or (PEL). Let* $\mathbf{X}$ *be a p-divisible group with action of* O_B *over* Spec k *with associated isocrystal isomorphic to* $(V \otimes_{\mathbf{Q}_p} K_0, b(\mathrm{id} \otimes \sigma))$. *In the case (PEL) we endow* $\mathbf{X}$ *with an* O_B*-polarization defined by the alternating form on* $V \otimes_{\mathbf{Q}_p} K_0$. *We consider the functor* $\breve{\mathcal{M}}$ *on* $\mathrm{Nilp}_{O_{\breve{E}}}$, *which associates to* S *the isomorphism classes of the following data.*

(1) *A p-divisible group* X_Λ *over* S *with* O_B*-action, for each* $\Lambda \in \mathcal{L}$.
(2) *An* O_B*-quasi-isogeny* $\varrho_\Lambda : \mathbf{X} \times_{\mathrm{Spec}\, k} \bar{S} \to X_\Lambda \times_S \bar{S}$, *for each* $\Lambda \in \mathcal{L}$.

Among the conditions these data are required to satisfy we mention only the following.

(i) *For each* $\Lambda \in \mathcal{L}$ *we have* $\det_{\mathcal{O}_S}(d; \mathrm{Lie}\, X_\Lambda) = \det_K(d;\ V_K^0/V_K^1)$, $d \in B$ *(Kottwitz condition* [Ko3]*).*

(ii) *Let* $M(X_\Lambda)$ *be the Lie algebra of the universal extension of* X_Λ. *Then locally on* S *there is an* O_B*-isomorphism* $M(X_\Lambda) \simeq \Lambda \otimes_{\mathbf{Z}} \mathcal{O}_S$. *If* $\Lambda \subset \Lambda'$, *the quasi-isogeny* $\varrho_{\Lambda'} \circ \varrho_\Lambda^{-1}$ *lifts to an isogeny* $X_\Lambda \to X_{\Lambda'}$ *of height* $\log_p |\Lambda'/\Lambda|$.

The functor $\breve{\mathcal{M}}$ *is representable by a formal scheme locally formally of finite type over* Spf $O_{\breve{E}}$.

REMARKS 1.3. (i) To the pair (G, b) there is associated the algebraic group J over $\mathbf{Q}_p$ with points in a $\mathbf{Q}_p$-algebra R

$$J(R) = \{g \in G(R \otimes_{\mathbf{Q}_p} K_0);\ \sigma(g) = b^{-1}gb\}.$$

The group $J(\mathbf{Q}_p)$ of quasi-isogenies of $\mathbf{X}$ acts on the left of $\breve{\mathcal{M}}$, via

$$g \cdot (X_\Lambda, \varrho_\Lambda) = (X_\Lambda, \varrho_\Lambda \circ g^{-1}).$$

Let Δ be the abelian group dual to the group of $\mathbf{Q}_p$-rational characters of G. The group $J(\mathbf{Q}_p)$ acts on Δ by translations. There is a canonical $J(\mathbf{Q}_p)$-equivariant map $\kappa : \breve{\mathcal{M}} \to \Delta$ [RZ, Section 3].

(ii) We conjecture that $\breve{\mathcal{M}}$ is flat over Spf $O_{\breve{E}}$. This conjecture can be reduced to a similar statement on the *local model*, an explicit closed subscheme of a finite product of Grassmannian varieties over Spec O_E associated to the moduli problem [RZ, Section 3]. In the numerous special cases where this conjecture is proved, the singularities of $\breve{\mathcal{M}}$ have turned out to be roughly of a "determinantal nature", comp. [CN], [dJ]. For a moduli problem of type (EL) the scheme $\breve{\mathcal{M}}_{\mathrm{red}}$ turns out in all known cases to be "elementary". For instance, the zeta function of a model of $\breve{\mathcal{M}}_{\mathrm{red}}$ over a finite field is given by an elementary expression. On the other hand, for type (PEL) there are simple examples where the irreducible components of $\breve{\mathcal{M}}_{\mathrm{red}}$ fibre over nonrational curves [KO]. Laumon has pointed out the similarity with the behaviour of the varieties connected with the local harmonic analysis of G [H].

(iii) The formal scheme $\breve{\mathcal{M}}$ is equipped with a canonical Weil descent datum from Spf $O_{\breve{E}}$ to Spf O_E [RZ, Section 3]. Even though this is not effective, a suitable completion of $\breve{\mathcal{M}}$ can be written in a canonical way as $\mathcal{M} \times_{\mathrm{Spf}\, O_E} \mathrm{Spf}\, O_{\breve{E}}$ for a (pro-)formal scheme $\mathcal{M}$ over Spf O_E [RZ, Section 3].

EXAMPLES 1.4. (i) Let B be a division algebra with invariant $1/d$ over its center F, with maximal order O_B. Set $E = F$ and $\breve{E} = E.K_0$. Drinfeld [D2] considers the moduli problem classifying quasi-isogenies of *special formal* O_B*-modules* (X, ϱ) over schemes $S \in \mathrm{Nilp}_{O_{\breve{E}}}$ (it can be identified with a moduli problem of type (EL)). He exhibits in this case an isomorphism

$$\breve{\mathcal{M}} = \coprod_{n \in \mathbf{Z}} \hat{\Omega}^d_E \times_{\mathrm{Spf}\, O_E} \mathrm{Spf}\, O_{\breve{E}}. \tag{2}$$

Here $\hat{\Omega}^d_E$ is the formal scheme over Spf O_E associated by Deligne to the local field E and the integer d [D2]. The disjoint sum decomposition is induced by the function $(X, \varrho) \mapsto$ height (ϱ).

This example (and trivial variants of it) is the only one we know where the formal scheme $\breve{\mathcal{M}}$ is p-adic, i.e., $p \cdot O_{\breve{\mathcal{M}}}$ is an ideal of definition.

(ii) Let F be a finite extension of $\mathbf{Q}_p$, set $E = F$ and $\breve{E} = E.K_0$. Let $\breve{\mathcal{M}}$ be the moduli problem (of type (EL)) over $\mathrm{Nilp}_{O_{\breve{E}}}$ classifying quasi-isogenies (X, ϱ) of formal O_F-modules of dimension 1 and height d. In this case $\breve{\mathcal{M}}_{\mathrm{red}}$ is a discrete set indexed by height $(\varrho) \in \mathbf{Z}$. The infinitesimal deformation theory of Lubin and Tate yields a (noncanonical) isomorphism

$$\breve{\mathcal{M}} = \coprod_{n\in\mathbf{Z}} \mathrm{Spf}\ O_{\breve{E}}[[T_1,\dots,T_{d-1}]].$$

(iii) Consider the moduli problem $\breve{\mathcal{M}}$ over Spf $W(k)$ (of type (EL)) classifying quasi-isogenies (X,ϱ) of *ordinary* p-divisible groups of height $2n$ and dimension n (i.e. deformations of $\mathbf{X} = \hat{\mathbf{G}}_m^n \times (\mathbf{Q}_p/\mathbf{Z}_p)^n$). There is an isomorphism (Serre-Tate canonical coordinates [Ka])

$$\breve{\mathcal{M}} = \coprod \mathrm{Spf}\ W(k)[[T_{11},T_{12},\dots,T_{nn}]].$$

The index set is equal to $(GL_n(\mathbf{Q}_p)/GL_n(\mathbf{Z}_p))^2$.

2. Non-archimedean uniformization of Shimura varieties

In this section we use slightly different notation. Let B be a finite-dimensional algebra over $\mathbf{Q}$ equipped with a *positive* anti-involution $*$. Let V be a finite B-module with a nondegenerate alternating bilinear form $(\ ,\ )$ with values in $\mathbf{Q}$ satisfying the identity (1). We define the algebraic group G over $\mathbf{Q}$ in complete analogy with the case (PEL) in Section 1. Let $h : \mathrm{Res}_{\mathbf{C}/\mathbf{R}}\,\mathbf{G}_m \to G_{\mathbf{R}}$ be such that (G,h) satisfies the axioms of Deligne defining a Shimura variety over the Shimura field $E \subset \mathbf{C}$. We fix an order O_B of B such that $O_B \otimes \mathbf{Z}_p$ is a maximal order of $B \otimes_{\mathbf{Q}} \mathbf{Q}_p$ stable under $*$, and a self-dual $O_B \otimes_{\mathbf{Z}} \mathbf{Z}_p$-lattice chain $\mathcal{L}$ in $V \otimes_{\mathbf{Q}} \mathbf{Q}_p$. We fix an open compact subgroup $K^p \subset G(\mathbf{A}_f^p)$. Let $\bar{\mathbf{Q}}$ be the field of algebraic numbers and fix an embedding $\bar{\mathbf{Q}} \to \bar{\mathbf{Q}}_p$. Let ν be the corresponding place of E above p and E_ν the completion of E in ν.

These data define a moduli problem of PEL-type parametrizing triples $(A,\bar\lambda,\bar\eta^p)$ consisting of an $\mathcal{L}$-chain of O_B-abelian varieties, a $\mathbf{Q}$-homogeneous O_B-polarization and a K^p-level structure that is representable by a quasi-projective scheme $\mathcal{A}_{K^p}$ over Spec O_{E_ν} (cf. [RZ, Section 6] for details). The generic fibre of $\mathcal{A}_{K^p}$ contains the Shimura variety of (G,h) as an open and closed subscheme.

We take for k the algebraic closure of the residue field of O_{E_ν}. We fix a point $(A_0,\bar\lambda_0,\bar\eta_0^p)$ of $\mathcal{A}_{K^p}(k)$. Let N_0 be the isocrystal associated to A_0. We fix an isomorphism $N_0 \simeq V \otimes_{\mathbf{Q}_p} K_0$ that respects the actions of $B \otimes K_0$ and the alternating bilinear forms on both sides. This allows us to write the Frobenius operator on N_0 as $b(\mathrm{id} \otimes \sigma)$, with $b \in G(K_0)$. Let $\mathcal{M}$ be the (pro-)formal scheme over Spf O_{E_ν} associated to the data of type (PEL), $(B \otimes \mathbf{Q}_p, *, V \otimes \mathbf{Q}_p, (\ ,\ ), b, \mu, O_B \otimes \mathbf{Z}_p, \mathcal{L})$, cf. Remark 1.3, (iii). It is acted on by the group $J(\mathbf{Q}_p)$, Remark 1.3, (i). Here μ denotes a 1-parameter subgroup of G defined over a finite extension K of K_0 in the conjugacy class defined by h.

THEOREM 2.1. *Assume that $(A_0,\bar\lambda_0,\bar\eta_0^p)$ is basic, i.e. the corresponding element $b \in G(K_0)$ is basic* [Ko1]. *Then*

(i) *The set of points $(A,\bar\lambda,\bar\eta^p)$ of $\mathcal{A}_{K^p}(k)$ such that $(A,\bar\lambda)$ is isogenous to $(A_0,\bar\lambda_0)$ is a closed subset Z of $\mathcal{A}_{K^p}$.*

(ii) *Let $\mathcal{A}_{K^p}|Z$ denote the formal completion of $\mathcal{A}_{K^p}$ along Z. There is an isomorphism of formal schemes over* Spf O_{E_ν},

$$I(\mathbf{Q}) \setminus [\mathcal{M} \times G(\mathbf{A}_f^p)/K^p] \xrightarrow{\sim} \mathcal{A}_{K^p}|Z.$$

Here I is an inner form of G such that $I(\mathbf{Q})$ is the group of quasi-isogenies of $(A_0, \bar{\lambda}_0)$ that acts diagonally through suitable natural embeddings of groups,

$$I(\mathbf{Q}) \longrightarrow J(\mathbf{Q}_p), \quad I(\mathbf{Q}) \longrightarrow G(\mathbf{A}_f^p).$$

The source of this isomorphism is a finite disjoint sum of formal schemes of the form $\Gamma \setminus \mathcal{M}$, where $\Gamma \subset J(\mathbf{Q}_p)$ is a discrete subgroup that is cocompact modulo center.

REMARKS 2.2. (i) In the Siegel case (principally polarized abelian varieties with level structure prime to p) the subscheme Z is the *supersingular locus.* In general it may be conjectured that there always exist basic points $(A_0, \bar{\lambda}_0, \bar{\eta}_0^p) \in \mathcal{A}_{K^p}(k)$.

(ii) It is not understood when it happens that the subscheme Z is open in the special fibre (existence of p-adic uniformization), as in Cherednik's theorem [D2]. This is a very subtle property that occurs only rarely (cf. Example 1.4, (i)). However, there are examples [RZ] uniformized by the disjoint sum of unramified forms of formal schemes of the form $\Gamma \setminus \hat{\Omega}_E^d$, where E is any finite extension of $\mathbf{Q}_p$ and d any integer.

(iii) There also exists a version of Theorem 2.1 for nonbasic isogeny classes. Because these do not in general form a closed subset the formulation becomes more technical.

3. The period morphism

In this section we return to the notation of Section 1. Let $\breve{\mathcal{M}}$ be the formal scheme over Spf $O_{\breve{E}}$ given by Theorem 1.2 and let $(X_\Lambda, \varrho_\Lambda;\ \Lambda \in \mathcal{L})$ be the universal object over $\breve{\mathcal{M}}$.

Let $\breve{\mathcal{M}}^{\mathrm{rig}}$ be the rigid-analytic space over $\breve{E}$ associated to $\breve{\mathcal{M}}$ (the generic fibre of $\breve{\mathcal{M}}$ in the sense of Raynaud-Berthelot [RZ, Section 5]). By Grothendieck's rigidity theorem [G] the quasi-isogenies ϱ_Λ yield canonical isomorphisms $V \otimes_{\mathbf{Q}_p} \mathcal{O}_{\breve{\mathcal{M}}^{\mathrm{rig}}} = M(X_\Lambda) \otimes_{\mathcal{O}_{\breve{\mathcal{M}}}} \mathcal{O}_{\breve{\mathcal{M}}^{\mathrm{rig}}}$. The resulting surjective homomorphism of locally free $\mathcal{O}_{\breve{\mathcal{M}}^{\mathrm{rig}}}$-modules

$$V \otimes_{\mathbf{Q}_p} \mathcal{O}_{\breve{\mathcal{M}}^{\mathrm{rig}}} \longrightarrow \mathrm{Lie}(X_\Lambda) \otimes_{\mathcal{O}_{\breve{\mathcal{M}}}} \mathcal{O}_{\breve{\mathcal{M}}^{\mathrm{rig}}} \tag{3}$$

is independent of Λ.

Let $\mathcal{F}$ be the projective algebraic variety over Spec E parametrizing the subspaces of V in the G-conjugacy class of V_K^1. Set $\breve{\mathcal{F}} = \mathcal{F} \times_{\mathrm{Spec}\, E} \mathrm{Spec}\, \breve{E}$. Then (3) corresponds to a rigid-analytic morphism $\breve{\pi}^1 : \breve{\mathcal{M}}^{\mathrm{rig}} \to \breve{\mathcal{F}}^{\mathrm{rig}}$. Let $\breve{\pi}^2 : \breve{\mathcal{M}}^{\mathrm{rig}} \to \Delta$ be the rigid-analytic morphism to the discrete rigid-analytic space associated to the map κ, cf. Remark 1.3, (i). The product morphism

$$\breve{\pi} = \breve{\pi}^1 \times \breve{\pi}^2 : \breve{\mathcal{M}}^{\mathrm{rig}} \longrightarrow \breve{\mathcal{F}}^{\mathrm{rig}} \times \Delta$$

is called the *period morphism associated to the moduli problem* $\breve{\mathcal{M}}$. We list some properties of the period morphism [RZ, Section 5].

Property 3.1. *The morphism $\breve{\pi}$ is étale and $J(\mathbf{Q}_p)$-equivariant.*

Here $J(\mathbf{Q}_p)$ acts diagonally on $\breve{\mathcal{F}} \times \Delta$, via its embedding in $G(\breve{E})$ on $\breve{\mathcal{F}}$ and via translations on Δ. In the rest of this section we shall assume that the algebraic group G is connected. We also make the assumption that $\breve{\mathcal{M}}^{\mathrm{rig}}$ is nonempty (this would certainly follow if the flatness conjecture 1.3, (ii) were true).

Property 3.2. Assume the validity of the conjecture of Fontaine [Fo2] that a weakly admissible filtered isocrystal over a finite extension K of K_0, with filtration steps contained in [0,1] comes from a p-divisible group over Spec O_K. Let $\breve{\mathcal{F}}^{\mathrm{wa}}$ be the admissible open subset of $\breve{\mathcal{F}}^{\mathrm{rig}}$ formed by weakly admissible filtrations, cf. Proposition 4.1 below. *Then the image of $\breve{\pi}$ is of the form $\breve{\mathcal{F}}^{\mathrm{wa}} \times \Delta'$, where Δ' is a union of cosets of a subgroup of finite index in Δ.*

Property 3.3. *For all $\Lambda \in \mathcal{L}$ the rational p-adic Tate module $V_p(X_\Lambda)$ yields a constant $\mathbf{Q}_p$-sheaf on $\breve{\mathcal{M}}^{\mathrm{rig}}$ with typical fibre V.* Let $K_{\mathcal{L}}$ be the subgroup of $G(\mathbf{Q}_p)$ that fixes the lattice chain $\mathcal{L}$. For any subgroup K of finite index in $K_{\mathcal{L}}$ we introduce the finite étale covering $\breve{\mathbf{M}}_K$ of $\breve{\mathcal{M}}^{\mathrm{rig}}$ that classifies the trivializations $\alpha : \{T_p(X_\Lambda)\} \to \mathcal{L} \bmod K$ of the chain of p-adic Tate modules $(T_p(X_\Lambda);\ \Lambda \in \mathcal{L})$. On the tower of étale coverings

$$\{\breve{\mathbf{M}}_K;\ K \subset K_{\mathcal{L}} \text{ of finite index }\}$$

of $\breve{\mathcal{M}}^{\mathrm{rig}} = \breve{\mathbf{M}}_{K_{\mathcal{L}}}$ the group $G(\mathbf{Q}_p)$ acts as a group of Hecke correspondences. Let $\breve{\pi}_K : \breve{\mathbf{M}}_K \to \breve{\mathcal{F}}^{\mathrm{wa}} \times \Delta$ be the resulting morphism. *The fibre of $\breve{\pi}_K$ through a point may be identified with $G(\mathbf{Q}_p)^1/K$.* Here $G(\mathbf{Q}_p)^1$ is the set of points of $G(\mathbf{Q}_p)$ where the values of all $\mathbf{Q}_p$-rational characters of G are units.

Examples 3.4. (i) Historically the first such period morphism was defined by Dwork for Example 1.4, (iii), cf. [Ka].

(ii) The period morphism for Example 1.4, (ii) induces on each connected component of $\breve{\mathcal{M}}^{\mathrm{rig}}$ a *surjective* morphism of the open unit polydisc of dimension $d-1$ to $\mathbf{P}_E^{d-1}$. The period morphism in this case is due to Gross and Hopkins [HG] (their construction is slightly different). Their paper is at the origin of the results of this section. The passage to the rigid category is the essential novelty compared to Grothendieck [G, Section 5].

(iii) In Drinfeld's Example 1.4, (i) the period morphism coincides with the composition of the isomorphism (2) with the natural inclusion $(\hat{\Omega}_E^d)^{\mathrm{rig}} \times \mathbf{Z} \subset (\mathbf{P}_E^{d-1})^{\mathrm{rig}} \times \mathbf{Z}$ induced by the identification $(\hat{\Omega}_E^d)^{\mathrm{rig}} = \Omega_E^d$ (Faltings). This example (and trivial variants of it) is the only known one where the period morphism is quasi-compact. There are examples where the period morphism has finite fibres but is not quasi-compact.

4. Non-archimedean period domains

Let $\mathcal{FI}(K)^{\mathrm{wa}}$ denote the $\mathbf{Q}_p$-tensor category of weakly admissible filtered isocrystals over the finite extension K of K_0 [Fo2], [Fa]. Assuming the validity of his conjecture (weakly admissible $\Rightarrow$ admissible [Fo2]) Fontaine constructs an exact fully faithful functor from this category to the category of p-adic Galois representations of $\mathrm{Gal}(\bar{K}/K)$. Composing this functor with the natural fibre functor on the latter (forgetting the Galois action) we obtain a fibre functor ω of $\mathcal{FI}(K)^{\mathrm{wa}}$ over $\mathbf{Q}_p$.

Let now G be a linear algebraic group over $\mathbf{Q}_p$ and fix $b \in G(K_0)$. Let $\mathcal{REP}(G)$ be the tensor category of $\mathbf{Q}_p$-rational representations of G. Let $\mu : \mathbf{G}_m \to G_K$. Then to any $V \in \mathcal{REP}(G)$ we have an associated filtered isocrystal $\mathcal{I}(V) = \mathcal{I}_\mu(V) = (V \otimes_{\mathbf{Q}_p} K_0, b(\mathrm{id} \otimes \sigma), V_K^\bullet)$ over K, where $V_K^\bullet$ is the filtration associated to μ, cf. Section 1. We call μ *weakly admissible (w.r.t. b)* if $\mathcal{I}(V)$ is weakly admissible for all $V \in \mathcal{REP}(G)$. Let μ be weakly admissible. Then, assuming Fontaine's conjecture and composing the functor $\mathcal{I}$ with ω we obtain a fibre functor of $\mathcal{REP}(G)$ over $\mathbf{Q}_p$. Let ω_0 be the natural fibre functor of $\mathcal{REP}(G)$ over $\mathbf{Q}_p$. Then the right G-torsor $\mathrm{Hom}(\omega_0, \omega \circ \mathcal{I})$ defines a cohomology class

$$\mathrm{cls}(b, \mu) \in H^1(\mathbf{Q}_p, G). \tag{4}$$

When G is a connected reductive group with simply connected derived group there is an explicit expression for this class [RZ, Section 1].

From now on we fix an algebraic closure $\bar{K}_0$ of K_0 and take K to be any finite extension of K_0 inside $\bar{K}_0$. Two 1-parameter subgroups $\mu, \mu' : \mathbf{G}_m \to G_K$ will be called equivalent if they induce the same filtrations on each $V \in \mathcal{REP}(G)$. We fix a conjugacy class $\{\mu\}$ of 1-parameter subgroups of G over $\bar{K}_0$ and denote by $E \subset \bar{K}_0$ its field of definition and $\breve{E} = E.K_0$. Then the equivalence classes of 1-parameter subgroups in $\{\mu\}$ form a projective algebraic variety $\boldsymbol{\mathcal{F}}(G, \{\mu\})$ defined over E that is homogeneous under G_E. We write $\boldsymbol{\mathcal{F}}$ for $\boldsymbol{\mathcal{F}}(G, \{\mu\})$ if this is unambiguous and $\breve{\boldsymbol{\mathcal{F}}} = \boldsymbol{\mathcal{F}} \times_{\mathrm{Spec}\, E} \mathrm{Spec}\, \breve{E}$.

PROPOSITION 4.1. *The weakly admissible points form an admissible open subset* $\breve{\boldsymbol{\mathcal{F}}}^{\mathrm{wa}}$ *of* $\breve{\boldsymbol{\mathcal{F}}}^{\mathrm{rig}}$ *stable under the action of* $J(\mathbf{Q}_p)$.

Here J is the algebraic group associated to (G, b), cf. Remark 1.3, (i). We call $\breve{\boldsymbol{\mathcal{F}}}^{\mathrm{wa}}$ the *non-archimedean period domain associated to* $(G, b, \{\mu\})$. From now on we assume that G and hence J are connected reductive groups over $\mathbf{Q}_p$. We also assume that $\breve{\boldsymbol{\mathcal{F}}}^{\mathrm{wa}} \neq \emptyset$. The fundamental open question in this context is the following. We introduce the free abelian group Δ as in Section 3. Let G' be the inner form of G defined by the image of $cl(b, \mu)$ in $H^1(\mathbf{Q}_p, G_{ad})$, cf. (4) (the class $cl(b, \mu)$ should only depend on $(b, \{\mu\})$).

HOPE 4.2. *There exists a canonical tower of rigid-analytic spaces*

$$\{\breve{\mathbf{M}}_{K'};\ K' \subset G'(\mathbf{Q}_p) \text{ open compact }\} \tag{5}$$

each of which is equipped with an action of $J(\mathbf{Q}_p)$ *and an equivariant étale morphism* $\breve{\pi}_{K'} : \breve{\mathbf{M}}_{K'} \to \breve{\boldsymbol{\mathcal{F}}}^{\mathrm{wa}} \times \Delta$ *with image of the form* $\breve{\boldsymbol{\mathcal{F}}}^{\mathrm{wa}} \times \Delta'$, *where* Δ' *is a union of cosets of a subgroup of finite index in* Δ. *We furthermore want* $G'(\mathbf{Q}_p)$ *to act on* (5) *as a group of Hecke correspondences covering the action on* $\breve{\boldsymbol{\mathcal{F}}}^{\mathrm{wa}} \times \Delta$, *which is trivial on the first factor and by translations on the second factor.*

Heuristically speaking, the tower (5) should be given by the "K'-level structures on the local system on $\breve{\boldsymbol{\mathcal{F}}}^{\mathrm{wa}}$ defined by the fibre functor $\omega \circ \mathcal{I}_x$, as x varies in $\breve{\mathcal{F}}^{\text{wa}}$", but it is not clear how to make sense of this. The tower of Property 3.3 is a typical candidate (in this case we have $G = G'$). In Examples 1.4, (i)–(iii) this tower does indeed exist.

REMARKS 4.3. (i) Assume b is basic. In the few known cases the fibres of $\breve{\pi}^2_{K'}$ over a point in Δ have turned out to be connected. We do not know whether to expect this in general when the derived group is simply connected.

(ii) Assume b basic and that $\text{cls}(b,\mu)$ is trivial. Then the triple (J, b^{-1}, μ^{-1}) satisfies the same assumptions as (G, b, μ) and the group associated to (J, b^{-1}) is G. One might wonder whether there exists a rigid-analytic space X with an action of $G(\mathbf{Q}_p) \times J(\mathbf{Q}_p)$ such that the towers associated to (G, b, μ) resp. (J, b^{-1}, μ^{-1}) are obtained by taking the quotients of X by open compact subgroups of $G(\mathbf{Q}_p)$ resp. $J(\mathbf{Q}_p)$. The pair formed by Example 1.4, (i) and the moduli problem of formal O_F-modules of dimension $d-1$ and height d (dual in some sense to Example 1.4, (ii)) are in this kind of duality and the question was raised in this case by Gross.

One can characterize $\breve{\mathcal{F}}^{\text{wa}}$ by geometric invariant theory as follows. To $\{\mu\}$ there corresponds an essentially unique ample line bundle $\mathcal{L}$ on $\breve{\mathcal{F}}$ that is homogeneous under the derived group $G_{der_{\breve{E}}}$. For any maximal $\mathbf{Q}_p$-split torus $T \subset J$ let $\breve{\mathcal{F}}(T)^{ss}$ be the Zariski-open subset of $\breve{\mathcal{F}}$ formed by the points that are semi-stable w.r.t. the action of $T_{\breve{E}} \cap G_{der_{\breve{E}}}$ on $(\breve{\mathcal{F}}, \mathcal{L})$.

THEOREM 4.4. (Totaro): *We have*

$$\breve{\mathcal{F}}^{\text{wa}} = \bigcap_{T \subset J} \breve{\mathcal{F}}(T)^{ss}.$$

The admissible open subsets of $\breve{\mathcal{F}}^{\text{rig}}$ appearing on the right-hand side have been considered by van der Put and Voskuil [PV].

REMARK 4.5. The period domain $\breve{\Omega}^d_E$ (= complement of all E-rational hyperplanes in $\mathbf{P}^{d-1}_{\breve{E}}$) has the following properties.

(i) It is a Stein space [SS].

(ii) The quotient by any discrete co-compact subgroup of $PGL_d(E)$ exists and is a proper rigid-analytic space over $\breve{E}$. In fact, it is a projective algebraic variety [Mu].

(iii) Let G_x be the stabilizer in $PGL_d(E)$ of a point $x \in \mathbf{P}^{d-1}_{\breve{E}}$. Then $x \in \breve{\Omega}^d_E$ iff G_x is compact. In fact, there is an equivariant map from $\breve{\Omega}^d_E$ to the Bruhat-Tits building of $GL_d(E)$ [D1].

None of these statements continue to hold for general period domains. This raises interesting questions (cohomology of coherent sheaves, stratification by the amount of noncompactness of stabilizers, etc).

5. ℓ-adic cohomology

If X is a rigid-analytic space over the local field E we denote by $H^i_c(X)$ the ith ℓ-adic cohomology group with compact supports, for a fixed prime number $\ell \neq p$

and a fixed algebraic closure $\bar{\mathbf{Q}}_\ell$ of $\mathbf{Q}_\ell$,

$$H^i_c(X) = H^i_c(X \otimes_E \bar{E};\ \bar{\mathbf{Q}}_\ell).$$

We continue with the set-up of the previous section but assume in addition that b is *basic*, i.e. J is an inner form of G. The tower $\{\breve{\mathbf{M}}_{K'};\ K' \subset G'(\mathbf{Q}_p)\}$ (with its Weil descent datum from $\breve{E}$ to E, cf. Remark 1.3, (iii)) — in the cases where it exists, cf. Hope 4.2 — defines the ℓ-adic representation

$$H^i_c((G, b, \{\mu\})) = \varinjlim_{K'} H^i_c(\breve{\mathbf{M}}_{K'})$$

of the product group $G'(\mathbf{Q}_p) \times J(\mathbf{Q}_p) \times W_E$. Here W_E denotes the Weil group of E. Drinfeld [D2] has conjectured in his Example 1.4, (i) that these modules give a geometric realization of the Langlands correspondence on the supercuspidal spectrum of $GL_d(E)$, comp. [C]. Partial results in this direction are due to Carayol, Faltings, Genestier, and Harris. We now describe a conjecture of Kottwitz in the general case, which describes the contribution of the *discrete Langlands parameters* to the Euler-Poincaré characteristic in the appropriate Grothendieck group,

$$H^*_c((G, b, \{\mu\})) = \sum (-1)^i H^i_c((G, b, \{\mu\})). \tag{6}$$

Let $\varphi : W_{\mathbf{Q}_p} \to {}^L G = \hat{G} \rtimes W_{\mathbf{Q}_p}$ be a discrete L-parameter, i.e. the connected component S^0_φ of the centralizer group $S_\varphi = \mathrm{Cent}_{\hat{G}}(\varphi)$ lies in $Z(\hat{G})^\Gamma$. We assume that G is obtained from a quasi-split inner form G^* by twisting with a basic element $b^* \in G^*(K_0)$, i.e. G is the inner form associated by Remark 1.3, (i) to (G^*, b^*). (This is automatic if the center of G is connected.) We use the maps [Ko2]

$$G(K_0) \xrightarrow{\lambda_G} X^*(Z(\hat{G})^\Gamma),\ H^1(\mathbf{Q}_p, G) \longrightarrow X^*(Z(\hat{G})^\Gamma).$$

These maps define elements

$$\lambda_{b^*} = \lambda_{G^*}(b^*),\ \lambda_b = \lambda_G(b), \text{ and } \mathrm{cls}(b, \mu) \in X^*(Z(\hat{G})^\Gamma), \text{ cf. (4)}.$$

According to Kottwitz, generalizing the notion of *strong inner forms* of Vogan, there should be bijections $\pi' \mapsto \tau_{\pi'}$ resp. $\pi \mapsto \tau_\pi$ that yield identifications of the L-packets on G' resp. J corresponding to φ,

$$\begin{aligned} \Pi_\varphi(G') &= \{\text{irreducible repns } \tau \text{ of } S_\varphi;\ \tau | Z(\hat{G})^\Gamma = \lambda_{b^*} + \mathrm{cls}(b, \mu)\}, \\ \Pi_\varphi(J) &= \{\text{irreducible repns } \tau \text{ of } S_\varphi;\ \tau | Z(\hat{G})^\Gamma = \lambda_b + \lambda_{b^*}\}. \end{aligned}$$

Even though these identifications depend on auxiliary choices, the function $(\pi', \pi) \mapsto \check{\tau}_{\pi'} \otimes \tau_\pi$ should be well defined and associate to (π', π) a representation of S_φ. Here and elsewhere $\check{}$ denotes the contragredient representation. Let $r_{\{\mu\}}$ be the finite-dimensional representation of $\hat{G} \rtimes W_E$ defined by $\{\mu\}$ [Ko2]. If φ_E denotes the restriction of φ to W_E, then $r_{\{\mu\}} \circ \varphi_E$ is in a natural way a representation of $S_\varphi \times W_E$, via $r_{\{\mu\}} \circ \varphi_E(s, w) = r_{\{\mu\}}(s \cdot \varphi_E(w))$.

CONJECTURE 5.1. (Kottwitz): *Let $\pi' \otimes \check{\pi} \otimes \varrho$ be an irreducible representation of $G'(\mathbf{Q}_p) \times J(\mathbf{Q}_p) \times W_E$ that contributes in a nontrivial way to* (6). *Then π' lies in an L-packet corresponding to a discrete L-parameter iff π does and then these*

L-packets correspond to the same L-parameter up to equivalence. The total contribution of all (equivalence classes of) discrete L-parameters is given up to sign *by the following expression:*

$$\sum_{\varphi\,\text{discrete}} \quad \sum_{(\pi',\pi)\in\Pi_\varphi(G')\times\Pi_\varphi(J)} \pi'\otimes\check{\pi}\otimes \mathrm{Hom}_{S_\varphi}(\check{\tau}_{\pi'}\otimes\tau_\pi,\ r_{\{\mu\}}\circ\varphi_E).$$

REMARKS 5.2. (i) Kottwitz has the prudence to assume in his conjecture that $\{\mu\}$ is minuscule, as in the examples of Section 3. Based on some heuristical principles he has checked that the above conjecture is compatible (in the sense of Theorem 2.1) with the corresponding global conjecture on Shimura varieties [Ko2].

(ii) Let b be the basic element, unique up to σ-conjugacy such that λ_b coincides with the element of $X^*(Z(\hat{G})^\Gamma)$ defined by $\{\mu\}$. Up to an obvious equivalence relation the representation $H_c^i((G,\{\mu\})) = H_c^i((G,b,\{\mu\}))$ is independent of the choice of b. If the derived group of G is simply connected the above conjectural formula for $H_c^*((G,\{\mu\}))$ simplifies because then $\mathrm{cls}(b,\mu)=0$ and G' is isomorphic to G.

(iii) To extend this conjecture to include certain nondiscrete L-parameters, one might be tempted to replace cohomology with compact supports by some kind of "middle intersection" cohomology (?), as is done in the global case of Shimura varieties.

A problem independent of the determination of (6) is the calculation of the cohomology of the non-archimedean period domains themselves. We describe a result for $H_c^*(\breve{\mathcal{F}}^{\mathrm{wa}}) = H_c^*(\breve{\mathcal{F}}^{\mathrm{wa}}\otimes_{\breve{E}}\bar{E},\bar{\mathbf{Q}}_\ell)$ as a virtual representation of $J(\mathbf{Q}_p)\times \mathrm{Gal}(\bar{E}/\breve{E})$. Let P_0 be a minimal parabolic subgroup of J, M_0 a Levi subgroup of P_0, and A_0 the maximal split torus contained in the center of M_0. Let Δ be the set of simple roots of A_0 in the unipotent radical of P_0. For a parabolic subgroup P containing P_0, with Levi subgroup M containing M_0, let A_P be the maximal split torus contained in the center of M, and set $a_P^J = \dim A_P - \dim A_J$. For $x\in X_*(A_0)\otimes_{\mathbf{Z}}\mathbf{R}$ let

$$\Delta_x = \{\alpha\in\Delta;\ \langle x,\omega_\alpha\rangle > 0\},$$

where ω_α is the fundamental weight corresponding to α, and let P_x be the unique parabolic subgroup containing P_0 such that Δ_x is the set of simple roots occurring in its unipotent radical. Any element $\mu\in\{\mu\}$ factoring through M_0 defines a unique element $\bar{\mu}\in X_*(A_0)\otimes_{\mathbf{Z}}\mathbf{R}$ such that $\langle\bar{\mu},\chi\rangle = \chi\circ\mu$ for all $\mathbf{Q}_p$-rational characters χ of M_0. Let $\{\mu\}_0\subset X_*(A_0)\otimes_{\mathbf{Z}}\mathbf{R}$ be the finite subset of points obtained in this way. Then $\mathrm{Gal}(\bar{E}/\breve{E})$ acts on $\{\mu\}_0$.

THEOREM 5.3. (Kottwitz, Rapoport): *The representation of $J(\mathbf{Q}_p)\times\mathrm{Gal}(\bar{E}/\breve{E})$ on $H_c^i(\breve{\mathcal{F}}^{\mathrm{wa}})$ is admissible for each i. In the Grothendieck group of admissible representations we have*

$$H_c^*(\breve{\mathcal{F}}^{\mathrm{wa}}) = \sum_{\bar{\mu}\in\{\mu\}_0} (-1)^{a^J_{P_{\bar{\mu}}}} v^J_{P_{\bar{\mu}}},$$

where v_P^J denotes the unique irreducible quotient of the representation of $J(\mathbf{Q}_p)$ on $C^\infty(J(\mathbf{Q}_p)/P(\mathbf{Q}_p))$. The action of $\mathrm{Gal}(\bar{E}/\breve{E})$ *is through permutation of the indices.*

REMARKS 5.4. (i) The theorem has been proved with the help of Huber supposing that certain foundational questions concerning the ℓ-adic cohomology of rigid spaces can be resolved (the case of torsion coefficients was developed by Berkovich and by Huber). The proof is modeled on the approach of [AB] to the calculation of the cohomology of the space of semi-stable vector bundles on a Riemann surface.

(ii) All available evidence seems to indicate that the contribution of $\bar{\mu}$ is in degree $a^J_{P_{\bar{\mu}}} + 2\ell(\bar{\mu})$, where $\ell(\bar{\mu})$ is the number of root hyperplanes separating $\bar{\mu}$ from the positive Weyl chamber corresponding to P_0. This is indeed proved by Schneider and Stuhler [SS] in the case of $\breve{\Omega}^d_E$. Their paper is at the origin of the above theorem.

(iii) As we have in general little control over the morphisms $\breve{\pi}_{K'}$ in the tower (5) (cf. Example 3.4, (iii)), the above result gives almost no information on the nature of (6).

In conclusion, I express my strong belief that there exists a theory in the equal characteristic case that closely parallels the one outlined here.

Acknowledgments. *I thank Gross, Kottwitz, Laumon, Messing, and Zink for their help with this report.*

References

[AB] Atiyah, M. F., and Bott, R., *The Yang-Mills equations over Riemann surfaces*, Philos. Trans. Roy. Soc. Lond. Ser. A **308**, 523–615 (1982).

[C] Carayol, H., *Non-abelian Lubin-Tate theory*, in Clozel, L., and Milne, J. S. (eds.), Automorphic forms, Shimura varieties and L-functions, vol. II, Perspect. in Math. **11**, 15–40, Academic Press, Boston, MA, 1990.

[CN] Chai, C. L., and Norman, P., *Singularities of the $\Gamma_0(p)$-level structure*, J. Alg. Geom. **1**, 251–278 (1992).

[Ch] Cherednik, I. V., *Uniformization of algebraic curves by discrete subgroups of $PGL_2(k_w)\ldots$*, Math. USSR-Sb. **29**, 55–78 (1976).

[D1] Drinfeld, V. G., *Elliptic modules*, Math. USSR-Sb. **23**, 561–592 (1974).

[D2] Drinfeld, V. G., *Coverings of p-adic symmetric regions*, Functional Anal. Appl. **10**, 29–40 (1976).

[Fa] Faltings, G., *Mumford-Stabilität in der algebraischen Geometrie*, these proceedings.

[Fo1] Fontaine, J.-M., *Groupes p-divisibles sur les corps locaux*, Astérisque **47–48** (1977).

[Fo2] Fontaine, J.-M., *Modules galoisiens, modules filtrés at anneaux de Barsotti-Tate*, Astérisque **65**, 3–80 (1979).

[H] Hales, Th., *Hyperelliptic curves and harmonic analysis*, Contemp. Math. **177**, 137–169 (1994).

[HG] Hopkins, M., and Gross, B., *Equivariant vector bundles on the Lubin-Tate moduli space*, Contemp. Math. **158**, 23–88 (1994).

[G] Grothendieck, A., *Groupes de Barsotti-Tate et cristaux*, Actes du Congr. Int. Math., Nice (Paris), **1**, Gauthier-Villars, 431–436 (1971).

[dJ] Jong de, A. J., *The moduli spaces of principally polarized abelian varieties with $\Gamma_0(p)$-level structure*, J. Alg. Geom. **2**, 667–688 (1993).

[KO] Katsura, T., and Oort, F., *Supersingular abelian varieties of dimension two or three and class numbers*, Adv. Stud. Pure Math. **10**, 253–281 (1987).

[Ka] Katz, N. M., *Travaux de Dwork*, Sem. Bourbaki 409, Lecture Notes in Math. **317**, 167–200, Springer, Berlin and New York, 1973.

[Ko1] Kottwitz, R. E., *Isocrystals with additional structure*, Comp. Math. **56**, 201–220 (1985).

[Ko2] Kottwitz, R. E., *Shimura varieties and λ-adic representations*, in Clozel, L., and Milne, J. (eds.), Automorphic forms, Shimura varieties and L-functions, vol. I, Perspect. Math. **10**, 161–209, Academic Press, Boston, MA, 1990.

[Ko3] Kottwitz, R. E., *Points on some Shimura varieties over finite fields*, J. Amer. Math. Soc., **5**, 373–444 (1992).

[Me] Messing, W., *The crystals associated to Barsotti-Tate groups ...* , Lecture Notes in Math. **264**, Springer, Berlin and New York, 1972.

[Mu] Mustafin, G. A., *Nonarchimedean uniformization*, Math. USSR-Sb. **34**, 187–214 (1978).

[PV] van der Put, M., and Voskuil, H., *Symmetric spaces associated to split algebraic groups over a local field*, J. Reine Angew. Math. **433**, 69–100 (1992).

[RZ] Rapoport, M., and Zink, Th., *Period spaces for p-divisible groups*, preprint Wuppertal 1994.

[SS] Schneider, P., and Stuhler, U., *The cohomology of p-adic symmetric spaces*, Invent. Math. **105**, 47–122 (1991).

Representations of Galois Groups Associated to Modular Forms

RICHARD TAYLOR

Department of Pure Mathematics and
Mathematical Statistics
Cambridge University
16 Mill Lane, Cambridge, CB2 1SB, United Kingdom

Introduction. We will describe in a special case the conjectural relationship among automorphic forms, l-adic representations, and motives. To make the discussion concrete, we shall restrict ourselves to weight 2 modular forms for GL_2. In this case the modular forms can be thought of either as certain harmonic forms on products of the upper half complex planes and hyperbolic three spaces or as cohomology classes for certain quotients of these products. As such, they are relatively concrete and often computable topological objects. Similarly, we shall restrict attention to irreducible two-dimensional l-adic representations that are de Rham with Hodge-Tate numbers 0 and -1, and to certain abelian varieties.

In the first section, we shall describe in some detail exactly what modular forms we wish to consider. In the second, we shall describe the conjectural relationship to Galois theory and algebraic geometry. Finally, in the third, we shall describe what is currently known about these conjectures.

Of course the situation we consider is very special and the conjectures admit enormous generalization (see for example [Cl]). We concentrate on this very special case because it is more concrete, yet the conjectures are already extraordinary and the difficulties seem immense.

Modular Forms. Let $\mathcal{Z}_2^{\pm}$ denote $\mathbb{C}-\mathbb{R}$ and let $\mathcal{Z}_3$ denote a hyperbolic three space, that is the set of quaternions $z = x + jy \in \mathbb{H} = \mathbb{C} \oplus j\mathbb{C}$ for which $y \in \mathbb{R}_{>0}$. Then $\mathcal{Z}_2^{\pm}$ has an action of $GL_2(\mathbb{R})$ by Möbius transformations

$$\begin{pmatrix} a & b \\ c & d \end{pmatrix} : z \longmapsto (az+b)(cz+d)^{-1}.$$

The same formula defines an action of $SL_2(\mathbb{C})$ on $\mathcal{Z}_3$ and we can extend this to an action of $GL_2(\mathbb{C})$ by letting the center act trivially. We will let c denote complex conjugation.

Now let K be a number field. We decompose the set of embeddings $K \hookrightarrow \mathbb{C}$ as $I_{\mathbb{R}} \cup I_{\mathbb{C}} \cup cI_{\mathbb{C}}$, where $I_{\mathbb{R}}$ denotes those embeddings with image in $\mathbb{R}$, $I_{\mathbb{C}}$ denotes half the remaining embeddings, and $cI_{\mathbb{C}}$ denotes the set of composites $c \circ \tau$ with $\tau \in I_{\mathbb{C}}$. We let $r_{\mathbb{R}}$ and $r_{\mathbb{C}}$ denote the cardinalities of $I_{\mathbb{R}}$ and $I_{\mathbb{C}}$. We let $d = r_{\mathbb{R}} + 2r_{\mathbb{C}}$

Proceedings of the International Congress
of Mathematicians, Zürich, Switzerland 1994

denote the degree of K over $\mathbb{Q}$. We set $\mathcal{Z}_K = (\mathcal{Z}_2^{\pm})^{I_\mathbb{R}} \times \mathcal{Z}_3^{I_\mathbb{C}}$, so that $GL_2(K)$ acts on $\mathcal{Z}_K$ via

$$\gamma((z_\tau)_{\tau \in I_\mathbb{R} \cup I_\mathbb{C}}) = ((^\tau\gamma)(z_\tau))_{\tau \in I_\mathbb{R} \cup I_\mathbb{C}}.$$

We let $\widehat{\mathcal{O}_K}$ denote the direct product of all completions of the ring of integers $\mathcal{O}_K$ of K at finite primes and $\mathbb{A}_K^\infty$ denote the ring of finite adeles of K, i.e. $\widehat{\mathcal{O}_K} \otimes_{\mathcal{O}_K} K$. If $\mathfrak{n}$ is an ideal of $\mathcal{O}_K$ then we define $U_1(\mathfrak{n})$ to be the subgroup of $GL_2(\widehat{\mathcal{O}_K})$ consisting of matrices

$$\begin{pmatrix} a & b \\ c & d \end{pmatrix}$$

with c and $d-1 \in \mathfrak{n}\widehat{\mathcal{O}_K}$. We will write $\Gamma_1(\mathfrak{n})$ for the intersection $U_1(\mathfrak{n}) \cap GL_2(\mathcal{O}_K)$. Our main object of study will be the orbifold

$$Y_1(\mathfrak{n}) = GL_2(K) \backslash ((GL_2(\mathbb{A}_K^\infty)/U_1(\mathfrak{n})) \times \mathcal{Z}_K).$$

It is an orbifold of dimension $2d - r_\mathbb{C}$. If K has strict class number 1, it is nothing other than $\Gamma_1(\mathfrak{n}) \backslash \mathcal{Z}_K$, in general, it is a finite union of quotients of $\mathcal{Z}_K$ by discrete groups. If $K = \mathbb{Q}$, it is the complex points of a modular curve; if K is totally real, it is the complex points of a Hilbert-Blumenthal variety; whereas if K is imaginary quadratic it is an arithmetic hyperbolic 3-orbifold.

We shall be interested in the cohomology groups $H^d(Y_1(\mathfrak{n}), \mathbb{C})$, which are finite-dimensional complex vector spaces. Analytically, $H^d(Y_1(\mathfrak{n}), \mathbb{C})$ can be identified with a space of certain harmonic forms on $(GL_2(\mathbb{A}_K^\infty)/U_1(\mathfrak{n})) \times \mathcal{Z}_K$ that are invariant under the action of $GL_2(K)$ (see [H]). This provides the link with more usual definitions of modular forms. For example if $K = \mathbb{Q}$ then we obtain

$$H^1(Y_1(\mathfrak{n}), \mathbb{C}) \cong M_2(\Gamma_1(\mathfrak{n})) \oplus \overline{S_2(\Gamma_1(\mathfrak{n}))},$$

where M_2 (resp. S_2) denotes the classical elliptic modular forms (resp. cusp forms).

It is not simply the finite-dimensional vector space $H^d(Y_1(\mathfrak{n}), \mathbb{C})$ that is of so much interest, but more importantly, it comes equipped with a natural set of linear operators. Suppose that $\wp$ is a prime of $\mathcal{O}_K$ not dividing $\mathfrak{n}$. Let $\varpi_\wp$ denote a uniformizer of $K_\wp$. Then there are two finite maps π_1 and $\pi_2 : Y_1(\mathfrak{n}\wp) \longrightarrow Y_1(\mathfrak{n})$ induced by the maps Id and $\eta_\wp \times \mathrm{Id} : GL_2(\mathbb{A}_K^\infty)/U_1(\mathfrak{n}) \times \mathcal{Z}_K \longrightarrow GL_2(\mathbb{A}_K^\infty)/U_1(\mathfrak{n}\wp) \times \mathcal{Z}_K$, where $\eta_\wp$ denotes right multiplication by the element

$$\begin{pmatrix} 1 & 0 \\ 0 & \varpi_\wp \end{pmatrix} \in GL_2(K_\wp) \subset GL_2(\mathbb{A}_K^\infty).$$

We let $T_\wp$ denote the composite

$$\pi_{1*} \circ \pi_2^* : H^d(Y_1(\mathfrak{n}), \mathbb{C}) \longrightarrow H^d(Y_1(\mathfrak{n}), \mathbb{C}).$$

Similarly, there is a map $\sigma : Y_1(\mathfrak{n}) \longrightarrow Y_1(\mathfrak{n})$, which is induced by right multiplication by $\varpi_\wp 1_2 \in GL_2(K_\wp)$. We let $S_\wp$ denote σ^*. The linear operators $T_\wp$ and $S_\wp$ for $\wp \not| \mathfrak{n}$ commute among themselves; they are called Hecke operators. We let $\mathbb{T}_1(\mathfrak{n})$ denote the $\mathbb{Z}$-algebra generated by these Hecke operators in the endomorphisms of $H^d(Y_1(\mathfrak{n}), \mathbb{C})$. As the Hecke operators preserve $H^d(Y_1(\mathfrak{n}), \mathbb{Z})$, $\mathbb{T}_1(\mathfrak{n})$ is a finitely generated $\mathbb{Z}$-module. Also $\mathbb{T}_1(\mathfrak{n}) \otimes \mathbb{C}$ embeds in $\mathrm{End}_\mathbb{C}(H^d(Y_1(\mathfrak{n}), \mathbb{C}))$.

It is the eigenvectors of $\mathbb{T}_1(N)$ on $H^d(Y_1(\mathfrak{n}),\mathbb{C})$, or more precisely the corresponding sets of eigenvalues, that will be our main object of study. Equivalently we shall be interested in the ring homomorphisms (characters)

$$\theta : \mathbb{T}_1(\mathfrak{n}) \longrightarrow \mathbb{C}.$$

We shall call θ *trivial* if $\theta(T_{(\varpi)}) = 1 + \#\mathcal{O}_K/(\varpi)$ for every totally positive prime element $\varpi \in \mathcal{O}_K$ with $\varpi - 1 \in \mathfrak{n}$. It is known that the generalised eigenspaces of nontrivial θ are in fact eigenspaces. We call θ and θ' of levels $\mathfrak{n}$ and $\mathfrak{n}'$ *equivalent* if for all primes $\wp$ not dividing $\mathfrak{n}\mathfrak{n}'$ we have $\theta(T_\wp) = \theta'(T_\wp)$.

We remark that if $\theta : \mathbb{T}_1(\mathfrak{n}) \to \mathbb{C}$ then its image generates a number field that we shall denote E_θ.

Conjectures. The expectation is that these characters θ encode subtle arithmetic information of a completely different nature. We will give some of the standard conjectures below. Before doing so we need a couple of definitions.

DEFINITION 1. *If L and F are finite extensions of $\mathbb{Q}_l$ we say that a continuous representation*

$$\rho : \mathrm{Gal}(\overline{L}/L) \longrightarrow GL_2(F)$$

has weight two *if either (or both) of the following conditions hold.*

(1) *There is a finite extension L'/L and an l-divisible group $A/\mathcal{O}_{L'}$ such that $\rho|_{\mathrm{Gal}(\overline{L'}/L')}$ is equivalent to the Tate module $((\varprojlim A[l^n](\overline{L'})) \otimes_{\mathbb{Z}_l} \mathbb{Q}_l)$ of A as a $\mathbb{Q}_l[\mathrm{Gal}(\overline{L'}/L')]$-module.*

(2) *There is a finite extension L'/L and an unramified character χ of* $\mathrm{Gal}(\overline{L'}/L')$ *such that $\rho|_{\mathrm{Gal}(\overline{L'}/L')} \otimes \chi$ is of the form*

$$\begin{pmatrix} \chi_l & * \\ 0 & 1 \end{pmatrix}$$

where χ_l denotes the cyclotomic character.

DEFINITION 2. *Let K be a field and E a number field.*

(1) *By a* generalized elliptic curve *over K with multiplication by E we mean an abelian variety A/K of dimension $[E : K]$ together with an embedding $E \hookrightarrow \mathrm{End}_K^0(A)$.*

(2) *By a* false generalized elliptic curve *over K with multiplication by E we mean an abelian variety A/K of dimension $2[E : K]$, a quaternion algebra (possibly split) D with center E and an embedding $D \hookrightarrow \mathrm{End}_K^0(A)$.*

(3) *We shall say that an abelian variety A/K has* CM over K *if there is a number field of degree $2\dim A$ that embeds in $\mathrm{End}_K^0(A)$.*

We note that there is a natural injection from generalized elliptic curves over K with multiplication by E to false generalized elliptic curves over K with multiplication by E, which sends (A, i) to $(A^2, M_2(E), M_2(i))$. We shall use this without mentioning it.

We are now in a position to state the conjectures that will interest us.

CONJECTURE 1 (Generalized Ramanujan-Peterson Conjecture).
If $\theta : \mathbb{T}_1(\mathfrak{n}) \to \mathbb{C}$ *is a nontrivial character then*

$$|\theta(T_\wp)| \leq 2(\#\mathcal{O}_K/\wp)^{1/2}$$

for all $\wp \not| \mathfrak{n}$.

CONJECTURE 2. *If* $\theta : \mathbb{T}_1(\mathfrak{n}) \to \mathbb{C}$ *is a nontrivial character then there is a finite extension* F_θ/E_θ *and for each prime* λ *of* F_θ *there is a continuous irreducible representation*

$$\rho_{\theta,\lambda} : \mathrm{Gal}(\overline{K}/K) \longrightarrow GL_2(F_{\theta,\lambda}),$$

such that if $\wp \not| \mathfrak{n}l$ *(with* l *the residue characteristic of* λ*) then* $\rho_{\theta,\lambda}$ *is unramified at* $\wp$ *and* $\rho_{\theta,\lambda}(\mathrm{Frob}_\wp)$ *(which is a well-defined conjugacy class in* $GL_2(F_{\theta,\lambda})$*) has characteristic polynomial*

$$X^2 - \theta(T_\wp)X + \theta(S_\wp)(\#\mathcal{O}_K/\wp).$$

Moreover if $\wp | l$ *then* $\rho_{\theta,\lambda}|_{\mathrm{Gal}(\overline{K_\wp}/K_\wp)}$ *has weight 2.*

We remark that in fact for any prime $\wp \not| l$, $\rho_{\theta,\lambda}|_{\mathrm{Gal}(\overline{K_\wp}/K_\wp)}$ should be completely describable in terms of θ as is described in [Ca].

CONJECTURE 3. *If* $\theta : \mathbb{T}_1(\mathfrak{n}) \to \mathbb{C}$ *is a nontrivial character then there is a false generalized elliptic curve* A/K *with multiplication by* E_θ *such that for all* $\wp \not| \mathfrak{n}$ *we have*

$$\#A(\mathcal{O}_K/\wp) = \#\mathcal{O}_E/(1 - \theta(T_\wp) + \theta(S_\wp)(\#\mathcal{O}_K/\wp))^2.$$

Moreover A *does not have CM over* K*. The quaternion algebra* D *implicit here can be taken to be split by* $E.K$*. If* K *has a real place the false generalised elliptic curve arises from a true one.*

Note that Conjecture 3 implies Conjectures 1 and 2. Conjecture 1 follows by using the theorems of Hasse and Weil. Conjecture 2 follows on looking at the Tate module.

Further, it is now standardly conjectured that the constructions implicit in Conjectures 2 and 3 give rise to bijections between the following classes of objects:

(1) Equivalence classes of nontrivial characters $\theta : \mathbb{T}_1(\mathfrak{n}) \to \mathbb{C}$, for variable $\mathfrak{n}$, but for fixed K;
(2) Isogeny classes of false generalized elliptic curves A/K which do not have CM over K;
(3) Continuous irreducible representations

$$\rho : \mathrm{Gal}(\overline{K}/K) \longrightarrow GL_2(\overline{\mathbb{Q}_l})$$

that are unramified outside a finite set of primes and that have the property that whenever $\wp$ is a prime of K above l then $\rho|_{\mathrm{Gal}(\overline{K_\wp}/K_\wp)}$ has weight 2.

Results. If $K = \mathbb{Q}$, Conjectures 1, 2, and 3 are all theorems. The seminal step was taken by Eichler in 1954 [E]. He proved Conjecture 1, but for $\Gamma_0(\mathfrak{n})$ not $\Gamma_1(\mathfrak{n})$, and he only checked it outside an unknown finite set of primes. Shimura generalized this to $\Gamma_1(\mathfrak{n})$ [S1] and Igusa determined the possible bad primes [I]. Although in retrospect Eichler's paper contains most of the ingredients to prove Conjecture 3 (and hence also Conjecture 2), it seems that the idea of decomposing under the action of the Hecke algebra was a long time in arising. To the best of our knowledge, Conjecture 3 was proved by Shimura in his book of 1971 [S4]. We note that Ribet [R] proved the irreducibility of the representations. Carayol [Ca], following Deligne and Langlands, determined the restriction of the l-adic representation to the decomposition group at any prime not dividing l.

All this work relied on realizing the modular curve $Y_1(\mathfrak{n})$ over the rational numbers and finding the desired abelian variety as a factor of the Jacobian of its completion. The hard part is to calculate the action of Frobenius on the Tate module of this Jacobian. Eichler pioneered the so-called Eichler-Shimura congruence relation to do this. Langlands later developed a second approach based on the Selberg trace formula (see [La]).

We note that given an elliptic curve $A/\mathbb{Q}$ one can in practice usually check that A arises from some $\theta : \mathbb{T}_1(\mathfrak{n}) \to \mathbb{C}$. Indeed one can take $\mathfrak{n}$ to be the conductor of A, then one can compute the first few values $\theta(T_{\wp})$ for all θ with that value of $\mathfrak{n}$ and this should allow one to rule out all but one θ. To show that A is in fact associated to this θ it suffices to show that $\rho_{\theta,2}$ is isomorphic to the action of $\mathrm{Gal}(\overline{\mathbb{Q}}/\mathbb{Q})$ on the 2-adic Tate module of A. According to the Faltings-Serre method [Li] it suffices to check that the traces agree on an explicit finite set of Frobenius elements. This is often a rather easy calculation. Whenever one makes such a calculation for an elliptic curve $A/\mathbb{Q}$, one proves that its L-function is entire and satisfies a functional equation.

Over a totally real field the situation is nearly as good. Conjectures 1 and 2 are known, except that the l-adic representations are not always known to be of weight 2 at primes dividing l. Carayol [Ca] has computed the restriction of these l-adic representations to the decomposition groups at all primes not dividing l. Conjecture 3 is known in many cases, but not all. Specifically, it is known if K has odd degree or if θ is discrete series at some finite prime, a condition that we will not make explicit here (see [Ca]).

In the cases where K is of odd degree or where θ is a discrete series at some finite place, Conjecture 3 (and hence Conjectures 1 and 2) follow on combining the results of [S3], [JL], and [A]. Indeed Jacquet and Langlands prove that in these cases θ can be realized as a character of a Hecke algebra for a Shimura curve (part of their argument was only sketched and was completed by Arthur), whereas Shimura had proved the analogue of Conjecture 3 for characters of Hecke algebras on Shimura curves by an analogue of Eichler's method. In the remaining cases Conjecture 1 was proved by Brylinski and Labesse [BL] except that the set of bad primes was not known exactly. They used Langlands' method to analyze the intersection cohomology of a certain compactification of $Y_1(\mathfrak{n})$ defined over $\mathbb{Q}$. Wiles and the author [Wi], [T1] settled Conjecture 2 in the remaining cases, except that they did not show that the l-adic representations were of weight 2.

This was done by constructing congruences to characters that were discrete series at some finite place and then piecing together the l-adic representations already constructed. Combined with [BL] this proves Conjecture 1 at all good primes.

Whenever K is not totally real there is no known direct link to algebraic geometry. The orbifolds $Y_1(\mathfrak{n})$ do not have a complex structure. The only positive results in these cases (except for CM forms or base changed forms) are when K is imaginary quadratic. In this case there has been, for some time, considerable computational evidence to support Conjecture 3 (see [GHM], [EGM] and [Cr]). With Harris and Soudry we have recently proved (see [HST], [T4], and [FH]) the following slightly weakened version of Conjecture 2 in this case.

THEOREM 1. *Suppose that K is an imaginary quadratic field. If $\theta : \mathbb{T}_1(\mathfrak{n}) \to \mathbb{C}$ is a nontrivial character such that $\theta(S_\wp) = \theta(S_{\wp^c})$ for all $\wp$, then there is a finite extension F_θ/E_θ, and for each prime λ of F_θ there is a continuous irreducible representation*

$$\rho_{\theta,\lambda} : \mathrm{Gal}(\overline{K}/K) \longrightarrow GL_2(F_{\theta,\lambda}),$$

such that $\rho_{\theta,\lambda}$ is unramified outside $l\mathfrak{n}\mathfrak{n}^c$ and the discriminant of $K/\mathbb{Q}$. Moreover for all primes outside an explicit set of Dirichlet density zero $\rho_{\theta,\lambda}(\mathrm{Frob}_\wp)$ has characteristic polynomial

$$X^2 - \theta(T_\wp)X + \theta(S_\wp)(\#\mathcal{O}_K/\wp).$$

Combining this theorem with the Faltings-Serre method, one can prove for many explicit pairs θ, A/K that

$$\#A(\mathcal{O}_K/\wp) = 1 + \#(\mathcal{O}_K/\wp) - \theta(T_\wp)$$

for all $\wp$ outside a set of Dirichlet density zero. With Cremona, we carried this out in the case $K = \mathbb{Q}(\sqrt{-3})$, θ the unique homomorphism $\mathbb{T}((17 + \sqrt{-3})/2) \to \mathbb{C}$, and A the elliptic curve $y^2 + xy = x^3 + (3 + \sqrt{-3})x^2/2 + (1 + \sqrt{-3})x/2$. From this one can deduce that for this θ and for all $\wp$ outside a set of Dirichlet density zero we have that

$$|\theta(T_\wp)| \leq 2(\#\mathcal{O}_K/\wp)^{1/2}.$$

To prove the above theorem one considers a set of twists $\theta \otimes \eta$ as η runs over quadratic characters of K and where $\theta \otimes \eta(T_\wp) = \theta(T_\wp)\eta(\wp)$ and $\theta \otimes \eta(S_\wp) = \theta(S_\wp)$. The fact that $GL_2(K)$ is closely related to a fourvariable orthogonal group and the theta lift from that orthogonal group to $GSp_4(\mathbb{Q})$ are used to construct from many of these $\theta \otimes \eta$ a character of the Hecke algebra of a space of holomorphic Siegel modular forms of genus 2 and weight 2 (see [HST] and [FH] for a refinement). Essential use is made of the disconnectedness of the orthogonal group to get a holomorphic lift. To ensure that the lift is nonzero for many η we require a nonvanishing theorem for L-functions. We originally used a result of Waldspurger [Wa] and but for the result quoted above we need a stronger result of Friedberg and Hoffstein [FH]. Congruences between these characters and characters of higher weight Siegel modular forms (which also occur on the l-adic cohomology of Siegel threefolds), the Eichler-Shimura congruence relation for Siegel threefolds [S2], [D], [CF], and the method of pseudo-representations [T2] allow one to construct for many η an l-adic representation R_η of $\mathrm{Gal}(\overline{\mathbb{Q}}/\mathbb{Q})$ such that for almost all rational primes p that split $p = \wp\wp^c$ in K, $R_\eta(\mathrm{Frob}_p)$ has eigenvalues contained in the set of roots of $(X^2 - \theta(T_\wp)X + \theta(S_\wp)p)(X^2 - \theta(T_{\wp^c})X + \theta(S_\wp)p)$, whereas for almost all inert primes p, $R_\eta(\mathrm{Frob}_p)$ has eigenvalues contained in the set of roots of $X^4 - \theta(T_p)X^2 + \theta(S_p)p^2$. See [T2] and [T3] for details. One can then show that there is a two-dimensional l-adic representation r of $\mathrm{Gal}(\overline{K}/K)$ such that for many η we have $R_\eta = r \otimes \eta \oplus (r \otimes \eta)^c$ (see [T4]). It is then not hard to see that r is the desired representation.

References

[A] J. Arthur, *The Selberg trace formula for groups of F-rank one*, Ann. of Math. (2) 100, 326–385 (1974).

[BL] J. L. Brylinski and J.-P. Labesse, *Cohomologie d'intersection et fonctions L de certaines variétés de Shimura*, Ann. Sci. École Norm. Sup. (4) 19, 409–468 (1984).

[Ca] H. Carayol, *Sur les représentations p-adiques associées aux formes modulaires de Hilbert*, Ann. Sci. École Norm. Sup. (4) 19, 409–468 (1986).

[CF] C.-L. Chai and G. Faltings, Degeneration of abelian varieties, Ergeb. Math. Grenzgeb. (3) Folge, Bd. 22 (1990).

[Cl] L. Clozel, *Motifs et formes automorphes: applications du principe de fonctorialité*, in L. Clozel and J. S. Milne (eds.) Automorphic Forms, Shimura Varieties and L-functions, vol. 1, Academic Press, San Diego and New York, 1990.

[Cr] J. Cremona, *Modular symbols for quadratic fields*, Compositio Math. 51, 275–323 (1984).

[D] P. Deligne, letter to Serre, dated 24 June 1968.

[E] M. Eichler, *Quaternäre quadratische Formen und die Riemannsche Vermutung für die Kongruenzzetafunction*, Arch. Math. (Basel) 5, 355–366 (1954).

[EGM] J. Elstrodt, F. Grunewald, and J. Menicke, *On the group $PSL_2(\mathbb{Z}[i])$*, in J. V. Armitage (ed.) Journées Arithmétiques (LMS lecture notes 56) Cambridge University Press, London and New York, 1982.

[FH] S. Friedberg and G. Hoffstein, *Non-vanishing theorems for automorphic L-functions on $GL(2)$*, preprint.

[GHM] F. Gruenewald, H. Helling, and J. Mennicke, *$SL_2(\mathcal{O})$ over complex quadratic numberfiels I*, Algebra i Logica 17, 512–580 (1978).

[H] G. Harder, *Eisenstein cohomology of arithmetic groups. The case GL_2*. Invent. Math. 89, 37–118 (1987).

[HST] M. Harris, D. Soudry, and R. Taylor, *l-adic representations associated to modular forms over imaginary quadratic fields I. Lifting to $GSp_4(\mathbb{Q})$*. Invent. Math. 112, 377–411 (1993).

[I] J.-I. Igusa, *Kroneckerian models of fields of elliptic modular functions*, Amer. J. Math. 81, 561–577 (1959).

[JL] H. Jacquet and R. Langlands, Automorphic forms on GL_2, Lecture Notes in Math. 114, Springer-Verlag, Berlin and New York, 1970.

[La] R. Langlands, *On the zeta-functions of simple Shimura varieties*, Canad. J. Math. XXXI no. 6, 1121–1216 (1979).

[Li] R. Livné, *Cubic exponential sums and Galois representations*, Contemp. Math. 67, 247–261 (1987).

[R] K. Ribet, *Galois representations attached to eigenforms with nebentypus*, Modular Forms of One Variable (eds. J.-P. Serre and D. B. Zagier) SLN 601, Springer-Verlag, Berlin and New York, 1977.

[S1] G. Shimura, *Correspondances modulaires et les fonctions ζ de courbes algébraiques*, J. Math. Soc. Japan 10, 1–28 (1958).

[S2] G. Shimura, *On modular correspondences for $Sp(n, \mathbb{Z})$ and their congruence relations*, Proc. Nat. Acad. Sci. U.S.A. 49, 824–828 (1963).

[S3] G. Shimura, *Construction of class fields and zeta functions of algebraic curves*, Ann. of Math. (2) 85, 58–159 (1967).

[S4] G. Shimura, Introduction to the arithmetic theory of automorphic functions, Publ. Math. Soc. Japan 11, Iwanami Shoten, 1971.

[T1] R. Taylor, *On Galois representations associated to Hilbert modular forms*, Invent. Math. 98, 265–280 (1989).

[T2] R. Taylor, *Galois representations associated to Siegel modular forms of low weight*, Duke Math. J. 63, 281–332 (1991).

[T3] R. Taylor, *On the l-adic cohomology of Siegel threefolds*, Invent. Math. 114, 289–310 (1993).

[T4] R. Taylor, *l-adic representations associated to modular forms over imaginary quadratic fields II*, Invent. Math. 116, 619–643 (1994).

[Wa] J.-L. Waldspurger, *Correspondances de Shimura et quaternions*, Forum Math. 3, 219–307 (1991).

[Wi] A. Wiles, *On ordinary λ-adic representations associated to modular forms*, Invent. Math., 529–573 (1988).

Einstein Metrics and Metrics with Bounds on Ricci Curvature

MICHAEL T. ANDERSON*

Department of Mathematics
State University of New York
Stony Brook, NY 11794, USA

There is a well-developed theory of the behavior of Riemannian metrics on smooth manifolds, which have uniform bounds on the sectional curvature K. The compactness theorem of Cheeger-Gromov [Ch], [Gr] implies that the space of metrics satisfying the bounds

$$|K| \leq \Lambda, \text{ vol} \geq v, \text{ diam} \leq D, \tag{0.1}$$

is $C^{1,\alpha}$ precompact. Thus, given any sequence of metrics g_i satisfying the bounds (0.1), there is a subsequence $\{i'\}$ and a sequence of diffeomorphisms $\phi_{i'}$ of M, such that the isometric metrics $g'_i = (\phi_{i'})^* g_{i'}$ converge, in the $C^{1,\alpha'}$ topology on M, to a $C^{1,\alpha}$ metric g_∞ on M, $\forall \alpha' < \alpha < 1$. If the volume or diameter bounds are removed in (0.1), one no longer has such compactness, but the degeneration of the sequence $\{g_i\}$ is well understood, through the works of Cheeger-Gromov [CG$_1$], [CG$_2$] and Fukaya [F], cf. also [CGF]. The manifolds (M, g_i) divide into two regions, the thick part M^ε and the thin part M_ε. Roughly speaking, on M^ε the metrics converge, as above, to a limit $C^{1,\alpha}$ metric, while the complement M_ε is ε-collapsed along a well-defined topological structure, called an F-structure, or more generally an N-structure.

Here, we are basically concerned with the possible extensions of such a theory to spaces of metrics with bounds imposed on the Ricci curvature, in place of the full curvature. There are several (related) reasons why it is important to consider such extensions. First, both the metric and the Ricci curvature are symmetric bilinear forms. Thus, the problem of controlling the behavior of metrics with bounds on Ricci curvature is, roughly speaking, a determined problem. Assuming bounds on the full curvature corresponds to a highly overdetermined problem. To illustrate this, there are very few manifolds, in general dimensions, that admit metrics of constant curvature. It is not unreasonable to expect that most manifolds admit metrics, or possibly metrics with mild singularities, of constant Ricci curvature, i.e. Einstein metrics. Second, one of the main applications of an understanding of convergence and degeneration of sequences of metrics would be to establish an existence theory for canonical or distinguished metrics on compact manifolds, as for

*) Partially supported by an N.S.F. Grant.

Proceedings of the International Congress
of Mathematicians, Zürich, Switzerland 1994

instance has so beautifully been carried out on the space of connections by Taubes and Uhlenbeck. Such canonical metrics would typically be minima or critical points of natural functionals on the space of metrics. For such variational problems, one needs to understand the behavior of spaces of metrics with (various) bounds on the Ricci curvature; assuming bounds on the sectional curvature is too strong a restriction to be directly applicable for existence questions. Similar remarks apply for the study of the moduli space of such canonical metrics.

In this report, we will survey some aspects of recent progress in this area. The discussion will be concerned, by and large, only with Riemannian metrics; for a survey of the substantial progress for Kähler and Kähler-Einstein metrics, see [Ti].

1 Einstein Metrics

The existence and moduli of Einstein metrics on surfaces, i.e. constant curvature metrics, is classical, and answered by the uniformization theorem and Teichmüller theory. Ideally, one would like to develop a similar theory for higher dimensional manifolds. There are well-known obstructions to the existence of Einstein metrics in dimensions 3 and 4, although presently none are known in higher dimensions. This, together with the generally observed phenomenon that geometry appears to be tied closer to topology in low dimensions, makes the existence and moduli question possibly most interesting for low-dimensional manifolds.

For 3-dimensional manifolds, this essentially corresponds to the Thurston Geometrization program [Th]. In fact, Thurston's Haken manifold theorem [Th] remains to date the sole result (for non-Kähler manifolds) that establishes the existence of an Einstein metric in dimension > 2 under topological hypotheses. It would appear that in dimension 3, one has the best chance to carry out a complete existence program; however, this will not be discussed here, and we refer to [A_7] for further details. In dimensions 4 and above, the existence theory for Einstein metrics still seems to be out of reach. However, there has been much progress on the structure of moduli of Einstein metrics on 4-manifolds. One hopes that these results might also shed light on the existence question.

Let M be a closed 4-manifold. Let $\mathcal{E}$ denote the moduli space of Einstein metrics on M, of volume 1. Let $\mathcal{E}^\lambda$, $\mathcal{E}_\lambda$, and $\mathcal{E}^o$ denote the metrics in $\mathcal{E}$ of scalar curvature $\geq \lambda$, $\leq \lambda$, and 0 respectively. Also, let $\overline{\mathcal{E}^\lambda}$, $\overline{\mathcal{E}_\lambda}$, and $\overline{\mathcal{E}^o}$ denote the completions of these spaces with respect to the Gromov-Hausdorff topology [Gr].

An *Einstein orbifold* (V, g) is a 4-dimensional orbifold with a finite number of singular points $\{q_i\}$, such that on the complement, g is a smooth Einstein metric. The singular points are cones on S^3/Γ, for a finite subgroup of $O(4)$ and, when lifted to the universal cover $B^4 \setminus \{0\}$ of $C(S^3/\Gamma) \setminus \{0\}$, the metric g is required to extend smoothly across $\{0\}$. An *orbifold singular Einstein metric* on M is a symmetric bilinear form of the form $\tilde{g} = \pi^*(g)$, where $\pi : M \to V$ is a resolution of the Einstein orbifold (V, g). Thus, π is a continuous map and, off the degeneration set $D = \cup D_i$, $D_i = \pi^{-1}(q_i)$, π is a diffeomorphism onto its image $V_o = V \setminus \cup\{q_i\}$.

The coarse structure of moduli spaces of Einstein metrics on 4-manifolds bears a strong resemblance to the moduli of constant curvature metrics on surfaces. The descriptions given in Theorems 1–3 below should be compared with the description of moduli of constant curvature metrics on surfaces of genus 0, 1, and $g \geq 2$ respectively.

THEOREM 1. [A_1], [BKN]. *For any $\lambda > 0$, the completion $\overline{\mathcal{E}^\lambda}$ is compact and consists of smooth and orbifold singular Einstein metrics on M.*

THEOREM 2. [A_4]. *The completion $\overline{\mathcal{E}^o}$ is locally compact but generally noncompact, and again consists of smooth and orbifold singular Einstein metrics on M. A sequence $\{g_i\} \in \overline{\mathcal{E}^o}$ diverges in the Gromov-Hausdorff topology if and only if $\{g_i\}$ collapses, in the sense of Cheeger-Gromov [CG_2], metrically on the complement of an a priori bounded number of small balls $\{B_{z_k}(\varepsilon_i)\}$, $\varepsilon_i \to 0$ as $i \to \infty$. In particular, $inj_{g_i}(x) \to 0$, $\forall x \in M$.*

THEOREM 3. [A_4]. *For any $\lambda < 0$, the completion $\overline{\mathcal{E}_\lambda}$ is locally compact and generally noncompact, and again consists of smooth and orbifold singular Einstein metrics on M. A divergent sequence $\{g_i\} \in \overline{\mathcal{E}_\lambda}$ in the Gromov-Hausdorff topology either collapses, as in Theorem 2, or gives rise to an a priori bounded number of cusps N_k. Each N_k is a complete, noncompact (possibly orbifold singular) Einstein 4-manifold, with* $\operatorname{diam} N_k = \infty$ *and* $\operatorname{vol} N_k \leq 1$. *Each N_k can be smoothly embedded in M, and the complement is collapsed to a lower dimensional space under $\{g_i\}$.*

Somewhat more general results than Theorems 1–3 can be found in [A_4]. The orbifold-singular Einstein metrics on M arise from degenerations of noncompact Ricci-flat 4-manifolds, which are asymptotically locally Euclidean (ALE), i.e. asymptotic to the flat metric on the cone $C(S^3/\Gamma)$ at infinity. Roughly speaking, if $\{g_i\}$ is a sequence of Einstein metrics, converging to an orbifold singular metric, then regions where the curvature goes to infinity metrically and topologically resemble rescalings of Ricci-flat ALE spaces. Because the curvature may go to infinity at different rates or scales near a given point, in general one has a finite set of such spaces associated to an orbifold singularity. The detailed analysis of the metric and topological degeneration near such a singularity has been carried out by Bando [B], cf. also [AC_1],[A_4]. In particular, the orbifold singularities typically crush (some) essential 2-spheres in M to points.

Ricci-flat ALE spaces were first systematically studied by physicists, cf. [GH]; the half-conformally flat, Ricci-flat ALE spaces have been classified by Kronheimer [Kr].

There are numerous examples, mostly arising from Kähler-Einstein metrics on complex surfaces, that illustrate the results of Theorems 1–3. One particular application of these results is a new proof of the Global Torelli and Surjectivity theorems for K3 surfaces [A_4].

Relatively few moduli spaces $\mathcal{E}$ of (real) Einstein metrics on 4-manifolds are explicitly known. The Gauss-Bonnet theorem implies that the Euler characteristic $\chi(M) \geq 0$ for Einstein 4-manifolds, with equality iff M is flat. Thus, all Einstein metrics on a flat 4-manifold are necessarily flat. Whether the 4-sphere S^4 admits a nonstandard Einstein metric has been a long-standing open problem. Recently, as a corollary of their very interesting work on the Minimal Volume problem of Gromov, Besson-Courtois-Gallot [BCG] have shown that if M^4 admits a hyperbolic metric, then this metric is the unique Einstein metric on M^4, thus generalizing the Mostow rigidity theorem in this context.

The only other moduli space that is explicitly known is the moduli of Einstein metrics on the K3 surface, and its quotients. The Hitchin-Thorpe inequality c.f. [Be] implies that all Einstein metrics are then hyper-Kähler; this rigidity gives the moduli close ties with the cohomology of the K3 surface.

It is unknown to what extent Theorems 1–3 generalize to higher dimensions. These results do remain valid in all dimensions, for those domains in the moduli spaces on which there is a bound on the scale-invariant norm $\int |R|^{n/2} \leq \Lambda$, $n = \dim M$.

We list several conjectures and questions raised by these results.

CONJECTURE 1. If $\{g_i\}$ is a divergent sequence in $\overline{\mathcal{E}_\lambda}$, for $\lambda < 0$, then the sequence cannot collapse, in the sense of Theorem 2. In particular, there are points $x_i \in M$ such that $inj_{x_i}(M, g_i) \geq \delta = \delta(\lambda) > 0$.

This conjecture is motivated by the case of surfaces, for which it is of course true.

CONJECTURE 2. If $\{g_i\}$ is a sequence in $\mathcal{E}$ that collapses, in the sense of Theorem 2, then M is a resolution of a flat 4-manifold with isolated singularities. In fact, we conjecture that in a collapsing sequence, all of the curvature concentrates in the singularities, i.e. the small balls $\{B_{z_k}(\varepsilon)\}$.

Note that if $\{g_i\}$ collapses, then the set of singularities must be nonempty, if the metrics are not flat. Namely, the absence of singularities implies, by Theorem 2, that M admits an F-structure [CG_1] and thus $\chi(M) = 0$.

QUESTION 1. Are there only finitely many components $\mathcal{C}_i$ to the moduli space $\mathcal{E}$ on 4-manifold M? A related question is whether the completions $\overline{\mathcal{C}_i}$ locally have the structure of a real analytic variety.

If g is a smooth Einstein metric on M, then a neighborhood of g in $\mathcal{E}$ does have the structure of a real analytic variety, c.f. [Ko].

QUESTION 2. Does every simply connected 4-manifold admit an orbifold singular Einstein metric?

This is motivated by the fact that orbifold singularities arise from a collapse of essential surfaces in M, and that the structure of H_2, via the Hitchin-Thorpe inequality, sometimes serves as an obstruction to the existence of Einstein metrics.

A natural collection of questions is to what extent the converses of Theorems 1–3 hold. For instance,

QUESTION 3. Let (V, g) be a 4-dimensional Einstein orbifold. Does there exist a smooth 4-manifold M and a smooth family of Einstein metrics g_t on M that converge to (V, g) in the Gromov-Hausdorff topology?

The only known case where this has been answered (affirmatively) is for hyper-Kähler orbifolds. Recently, Joyce [J] has constructed Ricci-flat metrics, in fact metrics of holonomy G_2 or Spin(7), on closed 7 and 8-manifolds, by resolving certain singular flat metrics.

It is natural to ask if analogues of Theorems 1–3 hold for other "canonical" metrics on 4-manifolds, for instance metrics that are critical for the functionals $\mathcal{R}^2 = \int |R|^2$ or $\mathcal{W}^2 = \int |W|^2$, where R (resp. W) is the Riemann (resp. Weyl) curvature tensor. For example, half-conformally flat metrics are critical (in fact, minima) for $\mathcal{W}^2$. Recently, it has been shown that Theorem 1, with slight modifications, holds for critical metrics of $\mathcal{R}^2$ and $\mathcal{W}^2$, as well as other functionals [A8]. Whether Theorems 2 and 3 hold for these metrics remains unknown.

2 Metrics with $|\,\mathrm{Ric}\,| \leq \lambda$

In this section, we will be concerned with the extension of the results of Section 1, i.e. convergence and degenerations of metrics of constant Ricci curvature, to metrics of bounded Ricci curvature. Besides the intrinsic interest of this question regarding relations of geometry to topology of manifolds, perhaps the main motivation is toward the study of the existence problems mentioned above.

First, the following result shows that Theorem 1 generalizes (via the Myers theorem) to metrics with bounds on Ricci curvature.

THEOREM 4. [A3]. *Let M be a closed 4-manifold. Then for arbitrary positive λ, v, D, the closure in the Gromov-Hausdorff topology of the space of metrics on M satisfying the bounds*

$$|\,\mathrm{Ric}\,| \leq \lambda, \ \mathrm{vol} \geq v, \ \mathrm{diam} \leq D, \tag{2.1}$$

is compact and consists of $C^{1,\alpha}$ smooth and orbifold-singular metrics.

The orbifold singular metrics are as described in Section 1. The same result holds in all higher dimensions, provided one assumes in addition a bound

$$\int |R|^{n/2} \leq \Lambda. \tag{2.2}$$

For instance, if n is odd, then one has a smooth compactness theorem under the bounds (2.1) and (2.2). One may compare this with the Cheeger-Gromov compactness theorem mentioned in the introductory section.

The key ingredient in passing from Theorem 1 to Theorem 4 is the notion of *harmonic radius*, introduced and studied in [A3] and [AC2]. For example, the $L^{2,p}$ harmonic radius of a smooth closed n-manifold (M, g) is the radius r_h of the largest geodesic ball $B = B_x(r_h)$ on which one has harmonic coordinates $U = \{u_i\} : B \to \mathbb{R}^n$, such that the metric g_{ij} in these coordinates satisfies the (scale-invariant) bounds

$$\begin{aligned} e^{-C}\delta_{ij} \leq g_{ij} \leq e^{C}\delta_{ij}, \\ r_h^{2-n/p} \parallel \partial^2 g_{ij} \parallel_{L^p(B)} \leq C. \end{aligned} \tag{2.3}$$

There are similar definitions for the $L^{k,p}$ harmonic radius. A useful feature of this radius is that it is continuous in the (strong) $L^{k,p}$ topology, provided $k - \frac{n}{p} > 0$, corresponding to the Sobolev embedding.

Theorem 4 implies that there is almost a smooth compactness theorem for metrics on fixed 4-manifolds, under the bounds (2.1); the possible degenerations of the metric are well understood. However, an associated finiteness result for the possible topological types under the bounds (2.1) remains unknown. On a fixed 4-manifold, the bounds (2.1), via the Gauss-Bonnet theorem, imply a bound on the L^2 norm of the full curvature; this bound is required in the proof of Theorem 4. We conjecture that such a bound is in fact not necessary:

CONJECTURE 3. For a given (λ, v, D), there are only finitely many diffeomorphism types of 4-manifolds admitting metrics satisfying (2.1).

Whether Conjecture 3 holds also in higher dimensions is of course also an open question. There is however less evidence on which to make a definite conjecture.

The analogues of Theorems 2 and 3 also hold for metrics with bounds on the Ricci curvature. Thus, suppose $\{g_i\}$ is a sequence of metrics on a closed 4-manifold M satisfying $|\mathrm{Ric}| \leq \lambda$, i.e. the volume or diameter bounds in (2.1) are removed. Then, modulo diffeomorphisms, a subsequence either collapses, on the complement of an a priori bounded number of (arbitrarily) small balls, or converges, in the pointed Gromov-Hausdorff (and $C^{1,\alpha}$) topology, to a countable collection of complete cusps, possibly with orbifold singularities, cf. [A_6].

Finally, these results, and their proofs, remain valid for sequences of metrics satisying an L^p bound on the Ricci curvature, in place of an L^∞ bound, provided $p > n/2$. The only distinction is that the cusps may not be complete, and that the limit metrics are no longer $C^{1,\alpha}$, but only $L^{2,p}$. The proofs remain identical, cf. again [A_6], with the use of the isoperimetric inequality of Gallot-Yang [Ga],[Y], compare also [Y].

3 Metrics with $\mathrm{Ric} \geq -\lambda$

There has a been a great deal of progress recently on the behavior of metrics with a lower bound on Ricci curvature. In this section, we briefly report on some of these developments.

Much of this activity began with the celebrated examples of Sha-Yang [SY_1] on metrics of positive Ricci curvature on manifolds of dimension ≥ 7, with arbitrarily large Betti numbers b_k, $k \geq 2$. Shortly thereafter, this result was improved to all dimensions ≥ 4 by Sha-Yang [SY_2] and independently, and by a different method, in [A_2]. For instance, $\#_1^k S^2 \times S^2$, for arbitrary k, carries metrics of positive Ricci curvature.

The construction of examples in [SY_1], [SY_2] is by a delicate metric surgery, using two-fold warped products. The construction in [A_2] is based on the gravitational instantons (Ricci-flat ALE spaces, cf. Section 1), constructed explicitly in [GH]. Somewhat later, it was shown in [A_5] how the Sha-Yang construction can be conceptually understood and recaptured by using other Ricci-flat models, namely the n-dimensional Schwarzschild metric. This is a Ricci-flat manifold, topologically $\mathbb{R}^{n-2} \times S^2$, that is asymptotically locally flat (ALF), i.e. asymptotic to the flat metric on $\mathbb{R}^{n-1} \times S^1$.

For instance, suppose (N, g) is a closed $(n-1)$-manifold with $\mathrm{Ric} > 0$, and let $M' = N \times S^1$, with product metric. For an arbitrary disjoint collection of small balls $B_i \subset N$, one may perform surgery on $B_i \times S^1$, replacing these regions in M' by $\mathbb{R}^{n-2} \times S^2$, to obtain a closed n-manifold M. It is easily seen that a suitably scaled Schwarzschild metric on $\mathbb{R}^{n-2} \times S^2$ may be smoothly joined to the product metric on M' on the complement of these regions, preserving nonnegative Ricci curvature.

When these metrics of positive Ricci curvature are normalized to have diameter 1, all have the property that the volume goes to 0, as some Betti number $b_k \to \infty$. This raises the question (cf. [A$_2$]) of whether there are metrics satisfying

$$\mathrm{Ric} \geq 0, \ \mathrm{vol} \geq v > 0, \ \mathrm{diam} = 1, \tag{3.1}$$

on manifolds with arbitrarily large Betti numbers b_k, $k \geq 2$; compare with Conjecture 3. Recently, Perelman [P$_2$] has constructed such examples in dimensions ≥ 4, for instance $\overset{k}{\#}\,\mathbb{CP}^2$ carries such metrics. The idea is to construct certain complete metrics on $\mathbb{CP}^2\backslash$ball that are asymptotically conical, and rescaling these spaces, glue them metrically onto S^4, with a singular metric of $K > 0$.

All these examples indicate that the degeneration of sequences of metrics satisfying the bounds

$$\mathrm{Ric} \geq -\lambda, \ \mathrm{vol} \geq v, \ \mathrm{diam} \leq D, \tag{3.2}$$

or

$$\mathrm{Ric} \geq -\lambda, \ \mathrm{diam} \leq D, \tag{3.3}$$

are likely to be rather complicated. Namely, the Ricci-flat ALE spaces discussed in Section 1 lead to orbifold singularities in limits of sequences of Einstein metrics, or metrics with two-sided bounds on Ricci curvature, on a fixed 4-manifold. These (topological) singularities of the limit correspond to a collapse of topology in (M, g_i) to a point. On the other hand, in regions where the limit is smooth, one has good convergence to the limit, cf. [A$_3$].

Now, rescalings (blow downs) of the examples above show that the degenerations, both topological and metric, within the classes (3.2), (3.3) are likely to be very complicated locally, and lead to rather complicated singularities of limit spaces.

Thus, consider first the case of sequences satisfying (3.2) or (3.3), whose Gromov-Hausdorff limits are smooth. To what extent are the geometry and topology of the limit related to those of the sequence? The first result along these lines is that the space of metrics satisfying bounds stronger than (3.2), namely the bounds

$$\mathrm{Ric} \geq -\lambda, \ inj \geq i_o, \ \mathrm{diam} \leq D, \tag{3.4}$$

is compact w.r.t. the C^α topology, cf. [AC$_2$]. Thus, both the topology and the metric are well preserved under limits.

A significant advance was achieved by Perelman with the following sphere theorem, answering a question in [A$_2$]:

THEOREM 5. [P$_1$]. *There is an $\varepsilon = \varepsilon(n)$ such that if M^n satisfies $\mathrm{Ric}_M \geq (n-1)$ and $\mathrm{vol}\, M \geq (1-\varepsilon) \cdot \mathrm{vol}\, S^n(1)$, then M^n is a homotopy sphere.*

This improved a previous result [A_3] where in addition an upper bound on the Ricci curvature was assumed. This assumption yields however the stronger conclusion that M is diffeomorphic to the sphere S^n, and the metric $C^{1,\alpha}$ close to $S^n(1)$.

Until recently, it was not known to what extent the metric or topology was preserved for smooth Gromov-Hausdorff limits. Thus, suppose (M_i, g_i) is a sequence of Riemannian n-manifolds satisfying (3.3), which converges in the Gromov-Hausdorff topology to a limit space (X, g), which is a smooth (or C^o) Riemannian n-manifold. The author and Cheeger conjectured that $\mathrm{vol}_{g_i}(M_i) \to \mathrm{vol}_g(X)$, i.e. the volume is continuous for the space of smooth manifolds satisfying (3.3) under the Gromov-Hausdorff topology. If this were true, when combined with Theorem 5, it would follow that the manifolds M_i are homotopy equivalent to the limit manifold X, for i sufficiently large. This conjecture has recently been solved by Colding:

THEOREM 6. [Co]. *Let (M_i, g_i) be a sequence of closed n-manifolds, satisfying the bounds (3.3), that converges to a smooth Riemannian n-manifold (X, g) in the Gromov-Hausdorff topology. Then*

$$\lim_{i\to\infty} \mathop{\mathrm{vol}}_{g_i} M = \mathop{\mathrm{vol}}_{g} X.$$

In fact, Colding has shown in addition that M_i is homeomorphic to X, using the controlled h-cobordism theorem and related results from controlled topology. Thus, when the limit space has no singularities, and is of the same dimension, one has quite good convergence properties for the topology.

We note that the dimension assumption here is sharp. If (M_i, g_i) is a sequence of n-manifolds satisfying (3.3), and converges in the Gromov-Hausdorff topology to a smooth Riemannian manifold (X, g) of dimension $\leq n-1$, then M_i may not be topologically related to X in any simple way, e.g. as a bundle or fibration over X, as shown in [A_5]. In fact, examples constructed there may be chosen to have $b_k(M_i) \geq i$, for any given $k \geq 2$.

Colding's theorem, and its method of proof, have opened up the possibility of studying in more detail the local structure of spaces satisfying $\mathrm{Ric} \geq -\lambda$, $\mathrm{diam} \leq D$, and their limits in the Gromov-Hausdorff topology. For instance, Cheeger-Colding [CC_1] have very recently proved the Splitting conjecture, a generalization of the Cheeger-Gromov splitting theorem. Thus, if (M_i, g_i) is a sequence of Riemannian n-manifolds, with $\mathrm{Ric} \geq -\lambda$, which converges in the Gromov-Hausdorff topology, to a limit space X containing a line L, then L splits off isometrically in X, $X = X' \times L$. In addition, they have begun a study [CC_2] of the regularity of limits X of spaces with a lower bound on Ricci curvature, as has been carried out with such success for Alexandrov spaces.

References

[A_1] M. Anderson, *Ricci curvature bounds and Einstein metrics on compact manifolds*, J. Amer. Math. Soc., vol. 2, (1989), 455–490.

[A_2] ——, *Short geodesics and gravitational instantons*, J. Differential Geom., vol. 31, (1990), 265–275.

[A_3] ——, *Convergence and rigidity of manifolds under Ricci curvature bounds*, Invent. Math., vol. 102, (1990), 429–445.

[A_4] ——, *The L^2 structure of moduli spaces of Einstein metrics on 4-manifolds*, Geom. and Functional Analysis, vol. 2, (1992), 29–89.

[A_5] ——, *Hausdorff perturbations of Ricci-flat manifolds and the splitting theorem*, Duke Math. J., vol. 68, (1992), 67–82.

[A_6] ——, *Degenerations of metrics with bounded curvature and applications to critical metrics of Riemannian functionals*, Proc. Sympos. Pure Math., vol. 54:3, (1993), 53–79.

[A_7] ——, Scalar curvature and geometrization conjectures for 3-manifolds, MSRI Volume on Comparison Geometry, Springer Verlag, to appear.

[A_8] ——, to appear.

[AC_1] M. Anderson and J. Cheeger, *Diffeomorphism finiteness for manifolds with Ricci curvature and $L^{n/2}$ norm of curvature bounded*, Geom. Functional Anal., vol. 1, (1991), 231–251.

[AC_2] ——, *C^α compactness for manifolds with Ricci curvature and injectivity radius bounded below*, J. Differential Geom., vol. 35, (1992), 265–281.

[B] S. Bando, *Bubbling out of Einstein manifolds*, Tôhoku Math. J. (2), vol. 42, (1990), 205–216 and 587–588.

[BKN] S. Bando, A. Kasue, and H. Nakajima, *On a construction of coordinates at infinity on manifolds with fast curvature decay and maximal volume growth*, Invent. Math., vol. 97, (1989), 313–349.

[Be] A. Besse, Einstein Manifolds, Ergebnisse Series 3. Folge Band 10, Springer Verlag, (1987).

[BCG] G. Besson, G. Courtois, and S. Gallot, *Les variétés hyperboliques sont des minima d'entropie topologique*, Invent. Math., vol. 118 (1994).

[Ch] J. Cheeger, *Finiteness theorems for Riemannian manifolds*, Amer. J. Math., vol. 92, (1970), 61–74.

[CC_1] J. Cheeger and T. Colding, *Lower bounds on Ricci curvature and the almost rigidity of warped products*, to appear.

[CC_2] ——, *On the local structure of manifolds with Ricci curvature bounded below*, to appear.

[CGF] J. Cheeger, K. Fukaya, and M. Gromov, *Nilpotent structures and invariant metrics on collapsed manifolds*, J. Amer. Math. Soc., vol. 5, (1992), 327–372.

[CG_1] J. Cheeger and M. Gromov, *Collapsing Riemannian manifolds while keeping their curvature bounded I*, J. Differential Geom., vol. 23, (1986), 309–346.

[CG_2] ——, *Collapsing Riemannian manifolds while keeping their curvature bounded II*, J. Differential Geom., vol. 32, (1990), 269–298.

[Co] T. Colding, *Ricci curvature and volume convergence*, preprint.

[F] K. Fukaya, *Collapsing Riemannian manifolds to ones of lower dimension*, J. Differential Geom., vol. 25, (1987), 139–156.

[Ga] S. Gallot, *Isoperimetric inequalities based on integral norms of Ricci curvature*, Astérisque, vol. 157–158, (1988), 191–217.

[GH] G. Gibbons and S. Hawking, *Gravitational multi-instantons*, Phys. Lett. B, vol. 78, (1978), 430–432.

[Gr] M. Gromov, Structures Metriques pour les Varieties Riemanniennes, Cedic-Fernand/Nathan, Paris, 1981.

[J] D. Joyce, *Compact Riemannian 7-manifolds with holonomy G_2: I, II*, preprints.

[Ko] N. Koiso, *Rigidity and stability of Einstein metrics. The case of compact symmetrics spaces*, Osaka J. Math., vol. 17, (1980), 51–73.

[Kr] P. Kronheimer, *A Torelli-type theorem for gravitational instantons*, J. Differential Geom., vol. 29, (1989), 685–697.

[P_1] G. Perelman, *Manifolds of positive Ricci curvature with almost maximal volume*, J. Amer. Math. Soc., vol. 7, (1994), 299–305.

[P_2] ——, *Construction of manifolds of positive Ricci curvature with big volume and large Betti numbers*, preprint.

[SY_1] J. P. Sha and D. G. Yang, *Examples of metrics of positive Ricci curvature*, J. Differential Geom,, vol. 29, (1989), 95–104.

[SY_2] ——, *Positive Ricci curvature on the connected sums of $S^n \times S^m$*, J. Differential Geom., vol. 33, (1991), 127–137.

[Th] W. Thurston, *Three-dimensional manifolds, Kleinian groups and hyperbolic geometry*, Bull. Amer. Math. Soc., vol. 6, (1982), 357–381.

[Ti] G. Tian, *Kähler-Einstein metrics on algebraic manifolds*, Proc. I.C.M. 1990, Math. Soc. Japan, (1991), 587–598.

[Y] D. Yang, *Convergence of Riemannian manifolds with integral bounds on curvature I, II*, Ann. Sci. École Norm. Sup. (4), vol. 25, (1992), 77–105, 179–199.

Minimal Foliations and Laminations

VICTOR BANGERT

Mathematisches Institut der Universität Freiburg,
Hebelstr. 29, 79104 Freiburg, Germany

1. Introduction

Usually one looks for a compact minimizer to a variational problem, as for example in the classical Plateau problem. Here we concentrate on problems in which the minimizers are noncompact and complete while the surrounding manifold is usually compact. This gives the area a dynamical flavor because notions like limit sets, recurrence, etc., naturally appear. Consequently the theory of dynamical systems and foliations will play an important role. Because of its great interest in Hamiltonian systems the theory of one-dimensional minimizers is best developed, cf. the lectures by Bolotin and Mañé at this congress, the lecture by Mather at the ICM 1986, and Section 3.

Section 2. treats the "Plateau problem at infinity" where one considers the universal cover $\widetilde{M}$ of a compact Riemannian manifold M with infinite fundamental group and assumes that it has some "ideal boundary" $\partial_\infty \widetilde{M}$ at infinity. Then one prescribes the boundary of the sought volume-minimizing submanifold in $\partial_\infty \widetilde{M}$.

The following phenomenon is fundamental for the results surveyed in Section 4. If we have a noncompact complete minimizing hypersurface without self-intersections in M then by the maximum principle its limit hypersurfaces form a lamination of M, i.e., they are pairwise disjoint injectively immersed hypersurfaces whose union is a closed subset K of M. In many cases K will look like a Cantor set in directions transverse to the hypersurfaces. But we may also have $K = M$ so that we obtain a foliation of M by minimal hypersurfaces. This situation is pretty well understood if M is an n-torus: for small perturbations of flat metrics KAM-theory yields foliations by minimal hypersurfaces, whereas in general one can only expect laminations.

In many cases, however, the minimizers may have self-intersections, as, e.g., the geodesics on compact surfaces of negative curvature. In [G] Gromov sketches a framework — worked out in particular instances — that allows for dynamical considerations also in such cases.

Section 5. is devoted to the stable norm on the kth real homology $H_k(M^n, \mathbb{R})$ of M^n — an interesting invariant of a Riemannian manifold that is closely related to the existence of minimizers with certain homological properties. The results presented concern primarily the cases $k = 1$ and $k = n - 1$.

If one tries to characterize standard Riemannian structures by the properties of their minimizers one obtains geometrically interesting rigidity problems — a

Proceedings of the International Congress
of Mathematicians, Zürich, Switzerland 1994

prototype being the Hopf conjecture recently solved by Burago and Ivanov [BI1], see Section 6.

There are interesting relations to geometric measure theory and to the theory of foliations that are not yet worked out: a measured lamination by oriented homological minimizers can be considered as a minimizing real current in the sense of geometric measure theory. On the other hand, there is a rich literature on foliations that admit a Riemannian metric such that the leaves are minimal, see, e.g., the short survey by Sergiescu [S].

2. The asymptotic Plateau problem in hyperbolic manifolds

In this section we consider a complete simply connected manifold (M^n, g_0) of negative curvature $-b^2 \leq K \leq -a^2 < 0$ together with a second Riemannian metric g that is Lipschitz equivalent to g_0, i.e., there exists $c \geq 1$ such that $c^{-2}g_0 \leq g \leq c^2 g_0$. This situation arises if one lifts an arbitrary metric on a compact manifold of negative curvature to the universal cover. In this setting the existence part of the "k-dimensional asymptotic Plateau problem for (M^n, g)" has a very satisfactory solution for every $1 \leq k \leq n-1$ while there will not be uniqueness in general. The fundamental idea goes back to Morse [M].

If we have a g-volume minimizing compact k-dimensional submanifold $D \subseteq M^n$ with boundary, i.e., $\mathrm{vol}_k^g(D) \leq \mathrm{vol}_k^g(D')$ for every compact k-dimensional $D' \subseteq M^n$ with $\partial D = \partial D'$ then obviously D satisfies $\mathrm{vol}_k^{g_0}(D) \leq c^{2k}\,\mathrm{vol}_k^{g_0}(D')$, i.e., D is c^{2k}-quasiminimizing with respect to g_0. The fundamental result in the existence problem is a uniform estimate for quasiminimizing submanifolds in (M^n, g_0). Frequently, the Plateau problem can not be solved in the class of smooth submanifolds; thus, we use notions from geometric measure theory: locally rectifiable $\mathbb{Z}_2$-currents (if $1 < k < n-1$) and locally rectifiable ($\mathbb{Z}$-)currents (if $k = n-1$).

THEOREM 2.1 [BL] *Let (M^n, g_0) be a simply connected complete Riemannian manifold with sectional curvature $-a^2 \leq K \leq -1$. Let S be a k-dimensional λ-quasiminimizing $\mathbb{Z}_2$-current ($\mathbb{Z}$-current if $k = n-1$) with compact support in (M^n, g_0) and let C denote the convex hull of* $\mathrm{spt}(\partial S)$. *Then*

$$\mathrm{dist}(x, C) \leq d$$

for all $x \in \mathrm{spt}(S)$ where d is a constant depending only on a, λ, k and n.

In conjunction with results from [An3] and [Mor] Theorem 2.1 implies that for $k > 1$ the asymptotic Plateau problem is solvable in (M^n, g_0) if the prescribed boundary $W \subseteq \partial_\infty \widetilde{M}^n$ is a compact topological submanifold.

If $k = n-1$ and (M^n, g_0) is hyperbolic space then Lang [La2] determines the optimal constant d in Theorem 2.1, $d = \frac{\pi}{2}\sqrt{\lambda^2 - 1}$. Moreover, he shows that in this case one can solve the asymptotic Plateau problem for all boundaries $W \subseteq \partial_\infty H^n$ that satisfy $W = \partial A$ for some $A \subseteq \partial_\infty H^n$ with $\mathrm{clos}(\mathrm{int}\, A) = A$. Finally he gives an example showing that the asymptotic Plateau problem cannot be solved for slightly more general W. In the case $n = 3$ Lang [La1] also obtains the same optimal estimate for area-quasiminimizing (or "homotopically quasiminimizing") planes in H^3. The Plateau problem at infinity for (M^n, g_0) itself has been treated by Anderson [An1, An2].

In general the solution to the Plateau problem at infinity is not unique. Uniqueness fails even in hyperbolic 3-space for $k = 2$ as follows from [An2]. Under additional assumptions, however, one has uniqueness results [An1], [HL], see also [P]. In general the nonuniqueness prevents one from filling a given foliation of $\partial_\infty M$ by a foliation of M with g-minimizing leaves. However, in this context Gromov [G] proved the following interesting perturbation result.

THEOREM 2.2 *Let (M, g_0) be a compact manifold with sectional curvature $K \equiv -1$ and let g_t be a deformation of g_0 that is continuous in the C^2-topology. Let $\mathrm{Gr}_k(M, g_0)$ denote the bundle of k-planes over M that is foliated into k-dimensional leaves consisting of the tangent planes to totally geodesic immersed submanifolds of (M, g_0). Then there exists $\varepsilon > 0$ and a family of continuous maps λ_t: $\mathrm{Gr}_k(M, g_0) \rightarrow M$, $t \in (-\varepsilon, \varepsilon)$, sending the leaves of $\mathrm{Gr}_k(M, g_0)$ to immersed g_t-minimal submanifolds of M.*

3. The one-dimensional case: Geodesics

We let (M, g) be a compact Riemannian manifold. Geodesics will always be assumed to be parameterized by arclength.

DEFINITION 3.1 *A geodesic c: $\mathbb{R} \rightarrow M$ is minimal if for every pair $s < t$ in $\mathbb{R}$ and every curve γ: $[s, t] \rightarrow M$ homotopic to $c|_{[s,t]}$ with fixed endpoints we have*

$$L^g(c|_{[s,t]}) \leq L^g(\gamma). \tag{1}$$

We say that c is homologically minimal if (1) *even holds for all γ homologous (say over $\mathbb{Z}$) to $c|_{[s,t]}$.*

Observation: A minimal geodesic exists on (M, g) iff M has infinite fundamental group. A homologically minimal geodesic exists on (M, g) iff $H_1(M, \mathbb{R}) \neq 0$.

Minimal geodesics have first been introduced by Morse [M] under the name "geodesics of class A". Their lifts to the universal cover are sometimes called "lines" because they are the shortest connections between any two of their points. A large class of Hamiltonian systems on cotangent bundles can be described by a variational principle that leads to an analogous definition of "action-minimizing orbit". This more general case is intensively studied, cf. [Ma3] and [Her]. Here we only describe some of the results that shed light on the case $k > 1$.

The concepts of "rotation vector" and "stable norm on $H_1(M, \mathbb{R})$" have proved useful in the study of homologically minimal geodesics.

Rotation set of a curve. Suppose γ: $\mathbb{R} \rightarrow M$ is parameterized by arclength. An element $v \in H_1(M, \mathbb{R})$ belongs to the rotation set $R(\gamma) \subseteq H_1(M, \mathbb{R})$ of γ if there exist sequences $s_i < t_i$ in $\mathbb{R}$ with $\lim_{i \to \infty}(t_i - s_i) = \infty$ such that for all closed 1-forms θ on M

$$\langle [\theta], v \rangle = \lim_{i \to \infty} (t_i - s_i)^{-1} \int_{s_i}^{t_i} \theta(\dot{\gamma}(t))\, dt. \tag{2}$$

Here $\langle\ ,\ \rangle$ denotes the bilinear pairing between $H^1(M,\mathbb{R})$ and $H_1(M,\mathbb{R})$. If (2) holds for a set of θ whose cohomology classes generate $H^1(M,\mathbb{R})$ then (2) holds for all closed 1-forms. If we set $\|\theta\|^* = \max_{p\in M}|\theta_p|^*$ where $|\theta_p|^*$ denotes the Riemannian norm of $\theta_p \in TM_p^*$ then $|\theta(\dot\gamma(t))| \le \|\theta\|^*$. This implies $R(\gamma)\neq\emptyset$.

Stable norm on $H_1(M,\mathbb{R})$ **([F],[GLP]).** If $v_{\mathbb{R}} \in H_1(M,\mathbb{R})$ is a real class corresponding to some $v \in H_1(M,\mathbb{Z})$ we define $\|v_{\mathbb{R}}\| = \lim_{m\to\infty}\frac{1}{m}\mathbf{M}(m\cdot v)$ where

$$\mathbf{M}(v) = \inf\Big\{\sum_{i=1}^{l}|n_i|\,L^g(\gamma_i)\ \Big|\ n_i\in\mathbb{Z},\ \gamma_i \text{ are closed curves and } \Big[\sum_{i=1}^{l} n_i\gamma_i\Big] = v\Big\}.$$

It is not difficult to show that $\|\ \|$ has a unique extension to a norm on $H_1(M,\mathbb{R})$.

Note that $R(\gamma)$ is contained in the unit ball B of $\|\ \|$ if $|\dot\gamma|\equiv 1$. This follows from the fact that $\|\ \|$ is dual to the quotient norm $\|\ \|^*$ on $H^1(M,\mathbb{R})$ induced from the norm $\|\ \|^*$ on 1-forms defined above, cf., e.g., [GLP, 4.35].

Now we can state the main results from [Ba5].

Theorem 3.1 *For every homologically minimal geodesic $c\colon \mathbb{R}\to M$ there exists a supporting hyperplane H of B such that the rotation set $R(c)$ is contained in $H\cap B$.*

Theorem 3.2 *For every supporting hyperplane H of B there exists a nonempty closed set of homologically minimal geodesics whose rotation sets are contained in $H\cap B$.*

As there exists a basis of $H_1(M,\mathbb{R})$ consisting of exposed points of ∂B, i.e., points $v\in\partial B$ with a supporting hyperplane H satisfying $H\cap B=\{v\}$, we obtain at least $b_1=\dim H_1(M,\mathbb{R})$ geometrically distinct homologically minimal geodesics. To some extent this result is optimal: [Ba5] determines quite explicitly the minimal geodesics in an example of a Riemannian 3-torus due to Hedlund [He]. Here B is an octahedron and there exist three minimal periodic geodesics c_i such that $R(c_i)=\{v_i\}$ where v_1, v_2, v_3 are linearly independent vertices of B. Moreover, every minimal geodesic c is asymptotic in each of its senses to a reparameterization of one of the c_i.

Elegant proofs for Theorems 3.1 and 3.2 can be given using Mather's theory of minimal measures [Ma1] which in [Ma3] is applied to the more general setting of Lagrangian systems on $TM\times S^1$. In our case one would define minimal measures as follows. Let μ be a Borel probability measure on the unit tangent bundle UM that is invariant under the geodesic flow. We can associate to μ a rotation vector $R(\mu)\in H_1(M,\mathbb{R})$ by requiring

$$\langle[\theta],R(\mu)\rangle = \int_{UM}\theta\,d\mu$$

for all closed 1-forms θ. Note that $R(\mu)\in B$. We call μ minimal iff $R(\mu)\in\partial B$. Now the results from [Ma3] imply:

Theorem 3.3 *Let H be a supporting hyperplane of B, let $\mathcal{M}_H$ be the set of minimal measures with $R(\mu)\in H$, and set $M_H = \mathrm{clos}(\bigcup\{\,\mathrm{spt}(\mu)\mid \mu\in\mathcal{M}_H\,\})\subseteq UM$.*

Then the geodesics with tangent vectors in M_H are homologically minimal and they form a Lipschitz lamination of M.

In the Hedlund example these laminations only consist of the unions of the three periodic geodesics $c_i(\mathbb{R})$.

Now we come to two special situations in which much stronger results are true. We first consider small perturbations of flat metrics. In this case KAM-theory applies, cf. [SZ]:

THEOREM 3.4 *Let (T^n, g_0) be a flat torus. For every $l > 4(n-1)+2$ there exists a neighborhood U of g_0 in the C^l-topology on the space of Riemannian metrics on T^n such that the following holds for all $v \in \mathbb{R}^n$ satisfying diophantine conditions*

$$|\, j \cdot v \,| \geq \gamma\, |j|^{-\tau} \qquad (\textit{for all } 0 \neq j \in \mathbb{Z}^n)$$

with $\gamma > 0$ and $n-1 \leq \tau < \frac{l-2}{4}$: If $g \in U$ there exists a foliation of T^n by g-minimal geodesics that is conjugate to the foliation of T^n by straight lines with direction v. Moreover the conjugating diffeomorphism is of class $C^{l-2\tau-2}$ if $l - 2\tau \notin \mathbb{N}$.

Here g-minimality of the geodesics follows from the fact that KAM-tori are always Lagrangian, cf., e.g., [Her]. The stability (under perturbations) of the (sufficiently) irrational foliations that is guaranteed by this deep theorem is in striking contrast to the instability of foliations with rational direction: these obviously disintegrate into finitely many periodic geodesics for an arbitrarily small generic perturbation of g_0.

Finally we treat two-dimensional tori (T^2, g). This case is closely related to Aubry-Mather theory for monotone twist maps of the annulus, cf. [BP] and [Ba3]. Suppose v_1 and v_2 are linearly independent integer classes in $H_1(T^2, \mathbb{R})$. Then it is an important fact that essentially goes back to the work of Morse [M] and Hedlund [He], see, e.g., [Ba3], that there exist closed geodesics c_i in v_i such that $L^g(c_i) = \|v_i\|$, $i = 1, 2$. Because c_1 and c_2 intersect transversely their conjunction $c_1 * c_2$ is not a geodesic. Hence

$$\|v_1 + v_2\| < L^g(c_1 * c_2) = \|v_1\| + \|v_2\| \;.$$

As a consequence the unit ball B of $\|\;\|$ is strictly convex. Now Theorem 3.2 implies that for every $v \in \partial B$ the set $\mathcal{M}_v$ of minimal geodesics c with $R(c) = \{v\}$ is nonempty. If v is not an $\mathbb{R}$-multiple of some nonzero integer class then the geodesics in $\mathcal{M}_v$ form a lamination (possibly a foliation) of T^2. If v is an integer class then generically $\mathcal{M}_v$ consists of one periodic minimal geodesic c and two intersecting geodesics that are forward and backward asymptotic to reparameterizations of c. In the next section we shall see that much of this situation generalizes to codimension one problems in T^n.

4. Minimal hypersurfaces in tori

It was Moser [Mo1] who noticed that many of the results of Aubry-Mather theory (resp. on minimal geodesics on a two-torus) generalize to variational problems for hypersurfaces in an n-torus. First we mention his stability result [Mo4] for minimal foliations. This is a surprising application of KAM-theory to a partial differential equation problem.

THEOREM 4.1 *Let (T^n, g_0) be a flat torus and for some $\gamma > 0$, $\tau > 0$ let $\alpha \in \mathbb{R}^n$ satisfy the diophantine condition*

$$\sum_{i,j=1}^{n} (\alpha_i k_j - \alpha_j k_i)^2 \geq \gamma |k|^{-\tau}$$

for all $k \in \mathbb{Z}^n \backslash \{0\}$. Then there exists a neighborhood U of g_0 in the C^∞-topology on the set of Riemannian metrics on T^n such that for every $g \in U$ there exists a unique foliation of T^n by g-minimal hypersurfaces that is C^∞-conjugate to the linear foliation of T^n by hyperplanes $\alpha \cdot x = \text{const}$.

Actually this theorem is proved in the following more general framework. We consider a "variational integrand" F: $\mathbb{R}^n \times \mathbb{R} \times \mathbb{R}^n \to \mathbb{R}$ that is $\mathbb{Z}^{n+1}$-periodic in the first $n+1$ variables $\bar{x} = (x, u) \in \mathbb{R}^{n+1}$ and assume that it satisfies

$$\delta |\xi|^2 \leq \sum_{i,j=1}^{n} F_{p_i p_j}(\bar{x}, p) \xi_i \xi_j \leq \delta^{-1} |\xi|^2 \tag{3}$$

for some $\delta \in (0,1)$ and all $(\bar{x}, p) \in \mathbb{R}^{n+1} \times \mathbb{R}^n$, $\xi \in \mathbb{R}^n$. Moreover we assume

$$\left| F_{\bar{x}}(\bar{x}, p) \right| \leq c\,(|p|^2 + 1) \tag{4}$$

for some $c > 0$ and all $(\bar{x}, p) \in \mathbb{R}^{n+1} \times \mathbb{R}^n$. We look for "(F-)minimal solutions" u: $\mathbb{R}^n \to \mathbb{R}$ that minimize

$$\int F(x, u(x), u_x(x))\, dx$$

with respect to arbitrary compactly supported variations of u. The standard example is the Dirichlet integral $F = \frac{1}{2} |p|^2$ whose minimal solutions are precisely the harmonic functions on $\mathbb{R}^n$. If g is a $\mathbb{Z}^{n+1}$-periodic metric on $\mathbb{R}^{n+1}$ then the n-dimensional g-volume of graphs of functions u: $\mathbb{R}^n \to \mathbb{R}$ is obtained by integrating an F as above that does not satisfy (3) uniformly in p. This is irrelevant in the perturbation Theorem 4.1 as it uses $F(\bar{x}, p)$ only for p in a compact neighborhood of some p_0. For "global" problems, however, one cannot reduce the parametric problem to the nonparametric one so that results for an integrand F as above are in this sense weaker than analogous results for minimizing hypersurfaces of a Riemannian metric on T^{n+1}. Note that in both cases the foliation by extremals provided by Theorem 4.1 consists automatically of minimizers, cf. [Mo3].

Prior to his perturbation result Moser [Mo1], see also [Mo3], had studied existence and properties of minimal solutions for an arbitrary $F \in C^{2,\varepsilon}(T^{n+1} \times \mathbb{R}^n, \mathbb{R})$ satisfying (3) and (4). His results will be described subsequently. Because of the periodicity of F we have a $\mathbb{Z}^{n+1}$-action T on the set of F-minimal solutions u by

$$(T_{\bar{k}} u)(x) = u(x - k) + k_{n+1}$$

where $\bar{k} = (k, k_{n+1}) \in \mathbb{Z}^{n+1}$. This action corresponds to translating graph(u) by $\bar{k}$. A special class of minimal solutions consists of those "without self-intersections", i.e., those satisfying for all $\bar{k} \in \mathbb{Z}^{n+1}$:

$$\text{if } \big(T_{\bar{k}}u\big)(x_0) = u(x_0) \text{ for some } x_0 \in \mathbb{R}^n \text{ then } T_{\bar{k}}u = u.$$

If $n = 1$ no minimal solution has self-intersections, see, e.g., [Mo2], whereas for $n \geq 2$ every nonaffine harmonic function minimizes the Dirichlet integral and has self-intersections. Obviously the condition "no self-intersections" is necessary if one wants the graph to belong to a lamination of T^{n+1}.

If a continuous function $u\colon \mathbb{R}^n \to \mathbb{R}$ has no self-intersections in the above sense then in analogy to the rotation number of a circle homeomorphism there exists a "rotation vector" $\alpha \in \mathbb{R}^n$ and $C > 0$ such that

$$\big| u(x+y) - u(x) - \alpha \cdot y \big| < C$$

for all $x, y \in \mathbb{R}^n$. For an integrand F as above we let $\mathcal{M} = \mathcal{M}(F)$ denote the set of minimal solutions without self-intersections. Then $\mathcal{M} = \bigcup_{\alpha \in \mathbb{R}^n} \mathcal{M}_\alpha$ where $\mathcal{M}_\alpha$ consists of the $u \in \mathcal{M}$ with rotation vector α. Moser proved that for every $\alpha \in \mathbb{Q}^n$ there exists a "periodic" $u \in \mathcal{M}_\alpha$, i.e., a u satisfying $T_{\bar{k}}u = u$ for all $\bar{k} \in \mathbb{Z}^{n+1}$ with $\bar{k} \cdot \bar{\alpha} = 0$ where $\bar{\alpha} = (\alpha, -1)$. Moreover, he proved a compactness result for minimal solutions so that one can take limits of the periodic solutions to obtain:

THEOREM 4.2 *For every $\alpha \in \mathbb{R}^n \backslash \mathbb{Q}^n$ there exists a nonempty lamination of T^{n+1} by graphs of minimal solutions in $\mathcal{M}_\alpha$.*

In [Ba1] it is shown that the action of T on $\mathcal{M}_\alpha$ has a unique minimal set if $(-\alpha, 1)$ is rationally independent. This implies that the above laminations are essentially unique. In [Ba4] the possibilities for the dynamics of $T|_{\mathcal{M}_\alpha}$ are completely determined for all $\alpha \in \mathbb{R}^n$. By a sufficiently large perturbation of the Dirichlet integral one can achieve that the laminations in Theorem 4.2 indeed have gaps (i.e., they are not foliations) for all α in a bounded set, cf. [Ba2].

For minimal hypersurfaces in a Riemannian torus (T^n, g) one has results analogous to Theorem 4.2 only in the case $n = 3$ for area-minimizing planes [Ba6], [Ba7].

One of the main open problems in this area is if there is a Liouville type result for minimal solutions. Suppose u is a minimal solution and u grows at most linearly. Is then u without self-intersections, i.e., $u \in \mathcal{M}_\alpha$ for some $\alpha \in \mathbb{R}^n$? Special cases of this question are treated in [Ba4] and [MS].

If the dimension k of the minimizing objects lies between 2 and $n-2$, no general statements — as, e.g., Theorems 3.1, 3.2, 4.2 — are known. However there is a perturbation result by Moser [Mo5] for pseudoholomorphic curves in almost complex $2n$-tori that is related to the case $k = 2$. According to Wirtinger's inequality the pseudoholomorphic curves in Theorem 4.3 below are homologically area-minimizing for a Riemannian metric associated with the almost complex structure.

THEOREM 4.3 *Let J_0 be a constant complex structure on T^{2n} and for some $\gamma > 0$, $\tau > 0$ let $\rho \in \mathbb{R}^{2n}$ satisfy the diophantine condition*

$$|\rho \cdot k| \geq \gamma |k|^{-\tau}$$

for all $k \in \mathbb{Z} \backslash \{0\}$. Then there exists a neighborhood U of J_0 in the C^∞-topology on the set of almost complex structures on T^{2n} such that for every $J \in U$ there exists a unique foliation of T^{2n} by J-holomorphic curves that is C^∞-conjugate to a linear foliation of T^{2n} containing the direction $\mathbb{R}\rho$.

5. The stable norm on $H_k(M, \mathbb{R})$

Existence and properties of homologically volume-minimizing submanifolds are closely related to the stable norm, which is a global invariant of a compact Riemannian manifold (M, g), interesting in itself.

DEFINITION 5.1 [F],[GLP] *The mass $\mathbf{M}(v)$ of $v \in H_k(M, \mathbb{Z})$ is defined by*

$$\mathbf{M}(v) = \inf\Big\{ \sum_i |n_i| \operatorname{vol}_k^g(\sigma_i) \mid \sum n_i \sigma_i \textit{ an integer Lipschitz cycle representing } v \Big\}.$$

The stable norm of $v \in H_k(M, \mathbb{R})$ is defined by

$$\|v\| = \inf\Big\{ \sum_i |r_i| \operatorname{vol}_k^g(\sigma_i) \mid \sum r_i \sigma_i \textit{ a real Lipschitz cycle representing } v \Big\}.$$

Federer [F] proved that for every $v \in H_k(M, \mathbb{Z})$ one has

$$\|v_{\mathbb{R}}\| = \lim_{j \to \infty} \tfrac{1}{j} \mathbf{M}(jv)$$

where $v_{\mathbb{R}}$ denotes the real class corresponding to v. There are simple examples showing that $\frac{1}{j}\mathbf{M}(jv) > \|v_{\mathbb{R}}\|$ for all $j > 0$ is possible. However, if M^n is orientable and $k = n-1$ then $\mathbf{M}(v) = \|v_{\mathbb{R}}\|$ for all $v \in H_{n-1}(M^n, \mathbb{Z})$, cf. [F]. This generalizes the "important fact" mentioned at the end of Section 3..

In general only little is known about the properties of the stable norm. Using the results on minimal geodesics on two-tori one can show, cf. [Au1], [Ba8], [Ma2]:

THEOREM 5.1 *For a Riemannian two-torus the unit ball $B \subseteq H_1(T^2, \mathbb{R})$ of the stable norm is strictly convex, i.e., ∂B does not contain nontrivial line segments. It has a unique supporting line at all points $v \in \partial B$ such that $\lambda v \notin H_1(T^2, \mathbb{Z})$ for all $\lambda \neq 0$. If $v \in \partial B$ and $\lambda v \in H_1(T^2, \mathbb{Z})$ for some $\lambda \neq 0$ then ∂B has a corner at v unless the periodic geodesics in the class λv foliate T^2.*

In a different setting Aubry [Au2] obtained a result that should translate into the following statement for the stable norm: if all minimal geodesics of (T^2, g) are contained in a region of T^2 with negative curvature (it is easy to construct such g) then the sum of the angles at the corners of ∂B is 2π, i.e., all the curvature of ∂B is concentrated in the corners.

There is intensive research on the stable norm on $H_1(T^n, \mathbb{R})$, cf. [BI1]. Katok [K] proved — in a different setting — that ∂B is twice differentiable with positive second fundamental form at a point $v \in \partial B$ if there exists a KAM-torus whose orbits have rotation vector v.

The following surprising result by Burago [Bu] has an interesting proof. For $v \in H_1(M, \mathbb{Z})$ define

$$\mathbf{L}(v) = \inf\Big\{ L^g(\gamma) \mid \gamma \colon S^1 \to M \text{ represents } v \Big\}.$$

Obviously, we have $\mathbf{L}(v) \geq \mathbf{M}(v) \geq \|v_{\mathbb{R}}\|$.

THEOREM 5.2 *For every compact Riemannian manifold (M, g) there exists a constant C such that $\mathbf{L}(v) \leq \|v_{\mathbb{R}}\| + C$ for all $v \in H_1(M, \mathbb{Z})$.*

A complete generalization of Theorem 5.1 and some more results in this direction were obtained by Senn [Se1], [Se2], [Se3], [Se4] for Moser's nonparametric variational problem for hypersurfaces in T^{n+1}, cf. Section 4.. Here the analogue of the stable norm is the "minimal average action" $A \colon \mathbb{R}^n \to \mathbb{R}$ defined by

$$A(\alpha) = \lim_{r \to \infty} |B_r|^{-1} \int_{B_r} F(x, u(x), u_x(x))\, dx$$

where $u \in \mathcal{M}_\alpha(F)$ is any minimal solution with rotation vector $\alpha \in \mathbb{R}^n$ and where $|B_r|$ denotes the volume of the ball of radius r in $\mathbb{R}^n$. To state Senn's results we say that $\mathcal{M}_\alpha$ gives rise to a foliation if the graphs of the functions $u \in \mathcal{M}_\alpha$ form a foliation of $\mathbb{R}^{n+1}$. We let $V_\alpha^\perp$ denote the orthogonal complement in $\mathbb{R}^n$ of $V_\alpha = \operatorname{span}_{\mathbb{R}} \{ k \in \mathbb{Z}^n \mid \alpha \cdot k \in \mathbb{Z} \}$.

THEOREM 5.3 *The minimal average action A is strictly convex. If $\mathcal{M}_\alpha$ gives rise to a foliation then A is differentiable at α. If $\mathcal{M}_\alpha$ does not give rise to a foliation then the directional derivative of A at α in the direction $\beta \in \mathbb{R}^n \backslash \{0\}$ exists iff $\beta \in V_\alpha^\perp$.*

These results are closely related to questions in crystallography, cf. [Se5]. Prior to Senn, Vallet [V] treated a more special two-dimensional problem of this type.

The stable norm in the intermediate dimensions $2 \leq k \leq n-2$ has only been computed in specific examples. Lawson [Law] has shown that the situation can already be surprisingly complicated for flat tori. Furthermore, there is work by Gluck, Morgan et al., see, e.g., [GMM], who — using the method of calibrations — compute minimal representatives and the stable norm for natural metrics on some Grassmann manifolds. Remarkably, the unit balls of the stable norms are polyhedra in all these cases.

6. Rigidity problems

The archetype of the rigidity results in this area is E. Hopf's theorem [Ho] that Riemannian two-tori without conjugate points are flat. The longstanding question of whether this is true also for tori of arbitrary dimensions has recently been answered affirmatively by Burago and Ivanov [BI1]. In a torus without conjugate points every geodesic is minimal and the periodic geodesics in every homology class foliate the torus. The proof in [BI1] is short and elegant, and combines the Birkhoff ergodic theorem, integral geometry, and an inequality on convex sets. Hopf's proof is more dynamical and uses the Riemannian character of the metric only in the final step in the form of the Gauss-Bonnet theorem. Using Hopf's method Dazord [D] was able to extend Hopf's theorem to Finsler two-tori of Landsberg type because for these a Gauss-Bonnet theorem holds. In the general Finsler case the rigidity problem is rather a regularity problem that is unsolved even on T^2: the unit tangent bundle of a Finsler two-torus without conjugate points is foliated into invariant two-tori consisting of the tangent vectors to geodesics with the same rotation vector, cf. the end of Section 3.. How regular is this foliation? The same question can be posed in higher dimensions where it is a result due to Heber [Heb] that, taking limits of the invariant tori formed by the tangent vectors to the periodic geodesics in integer homology classes, one obtains a foliation of the unit tangent bundle.

An interesting application of Hopf's method to a different area is Bialy's [Bi] characterization of the circular billiard by its dynamical properties.

In view of the results in Section 4. it seems reasonable to ask for a rigidity result in the codimension one case.

Problem. Suppose (T^n, g) is a Riemannian torus such that for every prime class $v \in H_{n-1}(T^n, \mathbb{Z})$ there exists a foliation of T^n by g-minimal $(n-1)$-tori in the class v. Is g flat?

Other rigidity results concern the marked length-spectrum [CFF], [Ba8] and the asymptotic volume growth [Bab], [BI2].

References

[An1] M. T. Anderson, *Complete minimal varieties in hyperbolic space*, Invent. Math. **69** (1982), 477–494.

[An2] ——, *Complete minimal hypersurfaces in hyperbolic n-manifolds*, Comment. Math. Helv. **58** (1983), 264–290.

[An3] ——, *The Dirichlet problem at infinity for manifolds of negative curvature*, J. Differential Geom. **18** (1983), 701–721.

[Au1] S. Aubry, *The devil's staircase transformation in incommensurate lattices*, Lecture Notes in Math. **925** (1982), 221–245, Springer-Verlag, Berlin and New York.

[Au2] ——, *The concept of anti-integrability: Definitions, theorems and applications to the standard map*, Twist Mappings and Their Applications (R. McGehee and K.R. Meyer, eds.), Springer-Verlag, Berlin and New York, 1992.

[Bab] I. K. Babenko, *Volume rigidity of two-dimensional manifolds*, Mat. Zametki **48** (1990), 10–14; Engl. transl.: Math. Notes **48** (1990), 629–632.

[Ba1] V. Bangert, *A uniqueness theorem for $\mathbb{Z}^n$-periodic variational problems*, Comment. Math. Helv. **62** (1987), 511–531.

[Ba2] ——, *The existence of gaps in minimal foliations*, Aequationes Math. **34** (1987), 153–166.

[Ba3] ——, *Mather sets for twist maps and geodesics on tori*, Dynamics Reported, vol. 1 (U. Kirchgraber, H.O. Walther, eds.), 1–56, B.G. Teubner, Leipzig - John Wiley, New York, 1988.

[Ba4] ———, *On minimal laminations of the torus*, Ann. Inst. H. Poincaré Anal. Non Linéaire **6** (1989), 95–138.

[Ba5] ———, *Minimal geodesics*, Ergodic Theory Dynamical Systems **10** (1990), 263–286.

[Ba6] ———, *Laminations of 3-tori by least area surfaces*, Analysis, et cetera (P.H. Rabinowitz and E. Zehnder, eds.), Academic Press, 1990.

[Ba7] ———, *Hypersurfaces without selfintersections in the torus*, Twist Mappings and Their Applications (R. McGehee and K.R. Meyer, eds.), Springer-Verlag, Berlin and New York, 1992.

[Ba8] ———, *Geodesic rays, Busemann functions and monotone twist maps*, Calc. Var. **2** (1994), 49–63.

[BL] V. Bangert and U. Lang, *Trapping quasiminimizing submanifolds in manifolds of negative curvature*, in preparation.

[Bi] M. Bialy, *Convex billiards and a theorem by E. Hopf*, Math. Z. **214** (1993), 147–154.

[BP] M. L. Bialy and L. V. Polterovich, *Geodesic flows on the two-dimensional torus and phase transitions "commensurability - noncommensurability"*, Funktional Anal. i. Prilozhen. **20** (1986), 9–16; Engl. transl.: Functional Anal. Appl. **20** (1986), 260–266.

[Bu] D. Burago, *Periodic metrics*, Adv. Soviet Math. **9** (1992), 205–210.

[BI1] D. Burago and S. Ivanov, *Riemannian tori without conjugate points are flat*, Geom. Functional Anal. (GAFA) **4** (1994), 259–269.

[BI2] ———, *An asymptotic volume of tori*, preprint, Univ. Pennsylvania, 1994.

[CFF] C. Croke, A. Fathi, and J. Feldman, *The marked length-spectrum of a surface of nonpositive curvature*, Topology **31** (1992), 847–855.

[D] P. Dazord, *Tores Finslériens sans points conjugués*, Bull. Soc. Math. France **99** (1971), 171–192.

[F] H. Federer, *Real flat chains, cochains and variational problems*, Indiana Univ. Math. J. **24** (1974), 351–407.

[GMM] H. Gluck, D. Mackenzie, and F. Morgan, *Volume-minimizing cycles in Grassmann manifolds*, preprint, Univ. Pennsylvania, 1993.

[G] M. Gromov, *Foliated Plateau problem, part 1: Minimal varieties*, Geom. Functional Anal. (GAFA) **1** (1991), 14–79.

[GLP] M. Gromov, J. Lafontaine, and P. Pansu, *Structures métriques pour les variétés riemanniennes*, CEDIC, Paris 1981.

[HL] R. Hardt and F. H. Lin, *Regularity at infinity for area-minimizing hypersurfaces in hyperbolic space*, Invent. Math. **88** (1987), 217–224.

[Heb] J. Heber, *On the geodesic flow on tori without conjugate points*, Math. Z. **216** (1994), 209–216.

[He] G. A. Hedlund, *Geodesics on a two-dimensional Riemannian manifold with periodic coefficients*, Ann. of Math. (2) **33** (1932), 719–739.

[Her] M. R. Herman, *Inégalités a priori pour des tores Lagrangiens invariants par des difféomorphismes symplectiques*, Publ. Math. IHES **70** (1989), 47–101.

[Ho] E. Hopf, *Closed surfaces without conjugate points*, Proc. Nat. Acad. Sci. **34** (1948), 47–51.

[K] A. Katok, *Minimal orbits for small perturbations of completely integrable Hamiltonian systems*, Twist Mappings and Their Applications (R. McGehee and K.R. Meyers, eds.), Springer-Verlag, Berlin and New York, 1992.

[La1] U. Lang, *Quasi-minimizing surfaces in hyperbolic space*, Math. Z. **210** (1992), 581–592.

[La2] ——, *The existence of complete minimizing hypersurfaces in hyperbolic manifolds*, Internat. J. Math., to appear.

[Law] H. B. Lawson, *The stable homology of a flat torus*, Math. Scand. **36** (1975), 49–73.

[Ma1] J. N. Mather, *Minimal measures*, Comment. Math. Helv. **64** (1989), 375–394.

[Ma2] ——, *Differentiability of the minimal average action as a function of the rotation number*, Bol. Soc. Brasil. Mat. **21** (1990), 59–70.

[Ma3] ——, *Action minimizing invariant measures for positive definite Lagrangian systems*, Math. Z. **207** (1991), 169–207.

[Mor] F. Morgan, *Harnack type mass bounds and Bernstein theorems for area-minimizing flat chains modulo* ν, Comm. Partial Differential Equations **11** (1986), 1257–1283.

[M] M. Morse, *A fundamental class of geodesics on any closed surface of genus greater than one*, Trans. Am. Math. Soc. **26** (1924), 25–60.

[Mo1] J. Moser, *Minimal solutions of variational problems on a torus*, Ann. Inst. H. Poincaré Anal. Non Linéaire **3** (1986), 229–272.

[Mo2] ——, *Recent developments in the theory of hamiltonian systems*, SIAM Rev. **28** (1986), 459–485.

[Mo3] ——, *Minimal foliations on a torus*, Topics in Calculus of Variations (M. Giaquinta, ed.), Lecture Notes in Math. **1365** (1988), 62–99, Springer-Verlag, Berlin and New York.

[Mo4] ——, *A stability theorem for minimal foliations on a torus*, Ergodic Theory Dynamical Systems **8*** (1988), 251–281.

[Mo5] ——, *On the persistence of pseudo-holomorphic curves on an almost complex torus* (with an appendix by Jürgen Pöschel), Invent. Math. **119** (1995), 401–442.

[MS] J. Moser and M. Struwe, *On a Liouville-type theorem for linear and nonlinear elliptic differential equations on a torus*, Bol. Soc. Brasil. Mat. **23** (1992), 1–20.

[P] K. Polthier, *Geometric a priori estimates for hyperbolic minimal surfaces*, preprint, Bonn, 1993.

[S] V. Sergiescu, *Basic cohomology and tautness of Riemannian foliations*, Appendix B in: P. Molino, Riemannian Foliations, Progr. in Math. **73**, Birkhäuser Boston, Boston, MA, 1988.

[SZ] D. Salamon and E. Zehnder, *KAM theory in configuration space*, Comment. Math. Helv. **64** (1989), 84–132.

[Se1] W. Senn, *Strikte Konvexität für* $\mathbb{Z}^n$*-periodische Variationsprobleme auf dem* n*-dimensionalen Torus*, Manuscripta Math. **71** (1991), 45–65.

[Se2] ——, *Über Mosers regularisiertes Variationsproblem auf dem* n*-dimensionalen Torus*, J. Appl. Math. Phys. (ZAMP) **42** (1991), 527–546.

[Se3] ——, *Differentiability properties of the minimal average action*, preprint, Bern, 1993.

[Se4] ——, *Phase-locking in the multidimensional Frenkel-Kontorova model*, preprint, Bern, 1994.

[Se5] ——, *On equilibrium forms of crystals*, preprint, Bern, 1994.

[V] F. Vallet, *Thermodynamique unidimensionelle et structures bidimensionelles de quelques modèles pour des systèmes incommensurables*, Thèse de doctorat, Université Paris 6, 1986.

Harmonic Maps, Rigidity, and Hodge Theory

KEVIN CORLETTE

Mathematics Department, University of Chicago
Chicago, IL 60637, USA

Harmonic maps are nonlinear analogues of harmonic functions or, if one considers their differentials, harmonic 1-forms. As such, one can expect analogues of Hodge-theoretic results about harmonic 1-forms. Harmonic maps arise as critical points for the energy functional on maps between two Riemannian manifolds. If M, N are Riemannian manifolds and $f : M \to N$ is a smooth map between them, then the energy is defined by

$$E(f) = \int_M |df|^2,$$

where df is the differential of f. If f has finite energy, then we can ask whether f is a critical point for E; the corresponding Euler-Lagrange equation is $D^* df = 0$, where D is the exterior derivative operator associated to the natural connection on f^*TN and df is regarded as a 1-form on M with values in f^*TN. The latter is the harmonic map equation. It is a nonlinear analogue of Laplace's equation.

Existence theory for harmonic maps is well behaved when the target manifold N is nonpositively curved. The first important result is due to Eells and Sampson [ES].

THEOREM 1. *If M, N are compact Riemannian manifolds, and N has nonpositive sectional curvature, then any homotopy class of maps from M to N has a harmonic representative.*

In some situations, it is necessary to consider more general classes of maps. Many of the applications to be discussed here are related to representations of the fundamental group of a manifold M in a semisimple Lie group G. In that case, it is natural to consider equivariant maps from the universal cover of M to the symmetric space $X = G/K$ associated to G; here, K is a maximal compact subgroup of G. We shall refer to such a map as a twisted map from M to X. In this setting, the appropriate existence result in this case is proved in [C2]; related results were proved by Diederich-Ohsawa [DO], Donaldson [D], Labourie [L], and Jost-Yau [JY1].

Proceedings of the International Congress
of Mathematicians, Zürich, Switzerland 1994

THEOREM 2. *Suppose M is a compact Riemannian manifold, G a linear semisimple Lie group, and $\rho : \pi_1(M) \to G$ a homomorphism. Then there exists a ρ-equivariant harmonic map from $\tilde{M}$ to X if and only if the Zariski closure of the image of ρ is a reductive group.*

Harmonic maps are a natural tool to apply in trying to prove rigidity theorems for nonpositively curved manifolds. For example, Mostow's strong rigidity theorem for a locally symmetric space would follow if one had techniques for proving that harmonic maps are isometries. However, the first progress in this direction did not occur until the work of Siu [Si], 15 years after that of Eells and Sampson. His basic observation, later extended by Sampson [S], was that there was an analogue for harmonic maps whose domain is a Kähler manifold of the decomposition of a harmonic 1-form into a holomorphic (1,0)-form and an antiholomorphic (0,1)-form. The Siu-Sampson result depends on the notion of complex sectional curvature. N has nonpositive complex sectional curvature if, for any pair X, Y in the complexified tangent space of N at n, we have

$$\langle R(X,Y)\overline{Y}, \overline{X}\rangle \leq 0.$$

The Siu-Sampson result is the following.

THEOREM 3. *Suppose f is a (possibly twisted) harmonic map from a compact Kähler manifold to a manifold of nonpositive complex sectional curvature. Then f satisfies the equation $\overline{\partial}\partial f = 0$, i.e. f is harmonic when restricted to any complex subvariety.*

Siu [Su] used this result to give a strengthening of the Mostow strong rigidity theorem in the case of Hermitian locally symmetric spaces. It has since been applied in a number of different directions, including the development of a nonabelian Hodge theory, as described by Simpson at the 1990 ICM. A recent development is the use of this circle of ideas by Reznikov [R] to prove a conjecture of Bloch [B] concerning the Chern classes of flat vector bundles over smooth projective varieties. Suppose M is a smooth complex projective variety, and $\rho : \pi_1(M) \to SL(n, \mathbb{C})$ is the monodromy of a flat vector bundle E. The flat structure on E induces in particular a holomorphic structure, so E has Chern classes $c_i(E)$ in the Chow group that are torsion in homology. Under the Abel-Jacobi map, these induce classes in $H^{2i-1}(M, \mathbb{R}/\mathbb{Z})$. The result conjectured by Bloch, and proved by Reznikov, is the following.

THEOREM 4. *The images under the Abel-Jacobi map of the Chern classes of a flat bundle with trivial highest exterior power are torsion.*

To prove this, Reznikov first uses the fact that these classes are rigid under deformations, and reduces to the case where the monodromy is contained in $SL(n, \mathcal{O}_S)$, where $\mathcal{O}_S$ is a ring of S-integers in a number field. The classes under consideration can be obtained by pullback of classes in $H^{2i-1}(BSL(n, \mathcal{O}_S), \mathbb{R}/\mathbb{Z})$. To prove they are torsion, it suffices to show that the corresponding map from $H^{2i-1}(BSL(n, \mathcal{O}_S), \mathbb{R})$ to $H^{2i-1}(M, \mathbb{R})$ is zero. By a result of Borel, the first group is generated by the so-called Borel regulators, essentially pullbacks of invariant forms on symmetric spaces. The Siu-Sampson result is used to show that the pullbacks of these classes to a Kähler manifold vanish.

It is natural to ask whether the Siu-Sampson result has an analogue for holonomy groups other than the unitary group. The author [C1] found such analogues by considering Riemannian manifolds with parallel differential forms, generalizing the Kähler case, which corresponds to the existence of a parallel symplectic form. The main result is the following.

THEOREM 5. *Suppose M is a compact Riemannian manifold with a parallel differential form ω. If N is a Riemannian manifold with nonpositive curvature operator and f is a (possibly twisted) harmonic map from M to N, then f satisfies*

$$D^*(df \wedge \omega) = 0.$$

Here the curvature condition on N can often be weakened, as in the Siu-Sampson result, but the precise condition under which the result holds is somewhat complicated to state. The resulting equation is typically highly overdetermined, and thus one can hope that it imposes strong constraints on a harmonic map. As an example, we may consider the case of quaternionic Kähler manifolds, which are Riemannian manifolds of dimension $4n$ whose holonomy group is contained in $\mathrm{Sp}(n)\,\mathrm{Sp}(1)$. These carry a parallel 4-form ω, called the quaternionic Kähler form. A typical example is the quaternionic hyperbolic space $H^n_{\mathbb{H}} = \mathrm{Sp}(n,1)/\,\mathrm{Sp}(n)\,\mathrm{Sp}(1)$, which is a negatively curved Riemannian symmetric space with sectional curvature pinched between -1 and -4. (We shall always be considering the case $n > 1$.) Another holonomy group that shares many features with the quaternionic Kähler case is that of the holonomy group Spin(9) in dimension 16. Here, the only examples are locally isometric to the elliptic Cayley plane $F_4/\,\mathrm{Spin}(9)$ and the hyperbolic Cayley plane $H^2_{\mathbb{O}} = F_4^{-20}/\,\mathrm{Spin}(9)$. In this case, there is a parallel 8-form with which one can work. For these examples, the result above implies the following.

THEOREM 6. *Let M be a compact Riemannian manifold with holonomy* Spin(9) *or* $\mathrm{Sp}(n)\,\mathrm{Sp}(1)$. *If N and f are as in the previous result, then f is necessarily totally geodesic.*

A fairly direct consequence of these ideas (extended slightly so as to allow M to be merely of finite volume) leads to an extension of Margulis' superrigidity results [M] to certain locally symmetric spaces of rank one.

THEOREM 7. *Suppose Γ is a lattice in* $\mathrm{Sp}(n,1)$ *or F_4^{-20}, and $\rho : \Gamma \to G$ is a homomorphism into a semisimple real algebraic group with Zariski dense image. Either G is compact or ρ extends to a homomorphism from the ambient group into G.*

This result has a geometric generalization that leads in particular to a metric rigidity result for manifolds that are locally quaternionic or Cayley hyperbolic.

THEOREM 8. *If M is a finite volume quotient of $H^n_{\mathbb{H}}$ or $H^2_{\mathbb{O}}$, then any complete Riemannian metric on M with nonpositive curvature operator is locally symmetric.*

Subsequently, Mok-Siu-Yeung [MSY] and Jost-Yau [JY2] (in less generality) found formulas that apply in greater generality and with weaker assumptions on curvature. The result of Mok-Siu-Yeung is the following.

THEOREM 9. *Suppose M is a compact locally irreducible symmetric space either of rank at least two or locally isometric to the quaternionic or Cayley hyperbolic space. If N is a Riemannian manifold with, in the former case, nonpositive sectional curvature or, in the latter case, nonpositive complex sectional curvature, and f is a (possibly twisted) harmonic map from M to N, then f is totally geodesic.*

The idea of the proof is related to Matsushima's technique for proving the vanishing of the first cohomology of certain locally symmetric spaces, although it is necessary to make somewhat more careful choices in the nonlinear setting. In the higher rank case, this result allows one to recover Margulis' superrigidity results over the reals for cocompact lattices in simple groups, as well as Gromov's metric rigidity theorem. Hernandez [H] and Yau-Zheng [YZ] have shown that any manifold with negative pointwise $\frac{1}{4}$-pinched sectional curvature has nonpositive complex sectional curvature. Using this and the Siu-Sampson result, they proved that a Riemannian metric on a finite-volume complex hyperbolic manifold with pointwise $\frac{1}{4}$-pinched sectional curvature is necessarily locally symmetric. The result of Mok-Siu-Yeung allows one to extend this to the quaternionic and Cayley hyperbolic cases. Gromov [G] indicated a different method for obtaining this extension based on a theory of harmonic maps from manifolds with foliations.

Margulis' results apply to homomorphisms into p-adic Lie groups as well, and it is natural to ask whether there is an approach to this by means of harmonic maps. This requires one to study harmonic maps from Riemannian manifolds into the Bruhat-Tits building Λ associated to a semisimple p-adic group. The Bruhat-Tits building is a metric simplicial complex whose every simplex is isometric to a simplex in Euclidean space. Furthermore, for each point of Λ, there is at least one subspace containing it that is isometric to a Euclidean space of the same dimension as Λ; these subspaces are called apartments. Gromov and Schoen developed a theory of harmonic maps into such spaces. As in the classical case, it is based on a notion of energy. Suppose f is a (possibly twisted) Lipschitz map from M to Λ. If we consider an isometric embedding of Λ in a Euclidean space $\mathbb{R}^N$ (meaning that the lengths of curves in Λ are the same whether measured in Λ or $\mathbb{R}^N$), then we can define the energy density $e(f)$ to be the pointwise squared norm of the differential of the resulting Lipschitz map from M to $\mathbb{R}^N$. This function is independent of the choice of embedding. The energy of f is then the integral of $e(f)$ over M, and f is said to be harmonic if it minimizes the energy among all nearby maps. In this situation, Λ is to be regarded as an analogue of a nonpositively curved manifold, so one does not expect to have to deal with the more general notion of a critical point for the energy.

Gromov and Schoen [GS] have proved an analogue of Theorem 2 in this setting. To apply this, one needs to be able to apply the vanishing theorems of Siu-Sampson, the author, Jost-Yau, and Mok-Siu-Yeung. The first step toward this goal is to observe that there is a large subset of M on which f can be regarded as a map into a manifold. Define $m \in M$ to be a singular point for f if m has no neighborhood whose image under f is contained in an apartment. An informal examination suggests that the set of singular points has codimension two. Gromov and Schoen show the following.

THEOREM 10. *Suppose f is a (possibly twisted) harmonic map from M to Λ. The singular set of f has Hausdorff codimension two.*

On the complement of the singular set, one can proceed as before: the differential of the map f is well defined away from the singular set, and can be interpreted as a harmonic 1-form with values in a flat orthogonal vector bundle. To prove the analogous vanishing theorems, one needs to perform an integration by parts in order to show that the harmonic form $df \wedge \omega$ is coclosed. This requires the fact above about the size of the singular set and information on the way in which the derivatives of f decay on approach to the singular set. Gromov and Schoen prove such a result, leading in particular to the following consequence.

THEOREM 11. *Suppose Γ is a lattice in $\mathrm{Sp}(n,1)$ or F_4^{-20}. If Γ acts on the Bruhat-Tits building Λ by means of a homomorphism into the corresponding p-adic Lie group, then there is a fixed point for the action, either in Λ itself or at infinity. (In fact, a more refined analysis shows that there must be a fixed point in Λ itself.)*

This is an analogue for these lattices of Margulis' p-adic superrigidity results for higher rank lattices. It implies the following long-conjectured result.

THEOREM 12. *Any lattice in $\mathrm{Sp}(n,1)$ or F_4^{-20} is arithmetic.*

Thus, the question of whether irreducible lattices in a semisimple Lie group are necessarily arithmetic is now open only for the group $SU(n,1)$, $n > 3$.

Twisted harmonic maps from compact Kähler manifolds to trees have been studied by Gromov-Schoen and Simpson. Their basic observation is that such a map factors through a holomorphic map into a holomorphic curve with orbifold singularities (i.e. an orbicurve). Simpson exploited this fact to prove the following.

THEOREM 13. *Suppose M is a smooth complex projective variety and $\rho : \pi_1(M) \to SL(2,\mathbb{C})$ is a homomorphism with Zariski dense image. If ρ is not locally rigid, then there is an orbicurve C and a holomorphic map $f : M \to C$ such that ρ is induced by a homomorphism $\pi_1(M) \to SL(2,\mathbb{C})$. If ρ is locally rigid, then there is a Hilbert modular orbivariety V and a holomorphic map $f : M \to V$ such that ρ is the pullback of one of the standard representations of $\pi_1(V)$ in $SL(2,\mathbb{R})$.*

Simpson and the author have worked on extending this to quasiprojective varieties.

Zimmer has been developing a program of using superrigidity and the ideas behind it to study questions about actions of lattices and semisimple groups on manifolds. One of the principal tools he has used is an extension of superrigidity to cocycles. In joint work, the author and Zimmer [CZ] have extended some of these results to the rank one case. The main technical tool is the theory of foliated harmonic maps first developed by Gromov [G]. As an example of the geometric consequences of these ideas, we mention the following.

THEOREM 14. *Suppose* $m > 2n$. *Then there is no discrete subgroup of* $\mathrm{Sp}(m,2)$ *that acts freely, properly discontinuously, and cocompactly on* $\mathrm{Sp}(m,2)/\mathrm{Sp}(n,1)$.

Korevaar and Schoen have also obtained a superrigidity result for cocycles in the rank one case, based on a generalization of Schoen's work with Gromov. They have developed a theory of harmonic maps from Riemannian manifolds into length spaces of nonpositive curvature. This is a very general class of metric spaces, not requiring, for example, that the target space be locally compact. A particular example would be the space of Riemannian metrics on a compact manifold compatible with a fixed volume form and endowed with an appropriate L^2 metric. Application of the general theory to this example leads to the result on cocycles.

Acknowledgment: To reflect on the way in which my understanding of this subject has developed is to be reminded very forcefully of what is, in some circles, referred to as the dependent arising of phenomena. Many mathematicians have contributed to that understanding in many ways; it would be a hopeless task to try to list them all. All I can do is offer my gratitude.

References

[B] S. Bloch, *Applications of the dilogarithm function in algebraic K-theory and algebraic geometry*, Proc. of the Int. Symp. in Algebraic Geometry, Kyoto 1977, M. Nagata, ed., Kinokuniya, Tokyo, 103–114.

[C1] K. Corlette, *Archimedean superrigidity and hyperbolic geometry*, Ann. of Math. (1) **135** (1992), 165–182.

[C2] K. Corlette, *Flat G-bundles with canonical metrics*, J. Differential Geom. **28** (1988), 361–382.

[CZ] K. Corlette and R. Zimmer, *Superrigidity for cocycles and hyperbolic geometry*, Internat. J. Math. **5** (1994), 273–290.

[DO] K. Diederich and T. Ohsawa, *Harmonic mappings and disc bundles over compact Kähler manifolds*, Publ. Res. Inst. Math. Sci. **21** (1985), 819–833.

[D] S. K. Donaldson, *Twisted harmonic maps and the self-duality equations*, Proc. London Math. Soc. **55** (1987), 127–131.

[ES] J. Eells and J. Sampson, *Harmonic mappings of Riemannian manifolds*, Amer. J. Math. **86** (1964), 109–160.

[G] M. Gromov, *Foliated plateau problem II: Harmonic maps of foliations*, GAFA **1** (1991), 253–320.

[GS] M. Gromov and R. Schoen, *Harmonic maps into singular spaces and p-adic superrigidity for lattices in groups of rank one*, Publ. Math. IHES **76** (1992), 165–246.

[H] L. Hernandez, *Kähler manifolds and $\frac{1}{4}$-pinching*, Duke Math. J. **62** (1991), 601–611.

[JY1] J. Jost and S. T. Yau, *Harmonic maps and group representations*, Differential Geometry, H. B. Lawson and K. Tenenblat, eds., Longman, 241–259.

[JY2] J. Jost and S. T. Yau, *Harmonic maps and superrigidity*, Differential Geometry: Partial Differential Equations on Manifolds, Proc. Sympos. Pure Math. **54**, part 1 (1993), 245–280.

[L] F. Labourie, *Existence d'applications harmoniques tordues à valeurs dans les variétés à courbure négative*, Proc. Amer. Math. Soc. **111** (1991), 877–882.

[M] G. Margulis, Discrete Subgroups of Semisimple Lie Groups, Springer-Verlag, Berlin and New York, 1991.

[MSY] N. Mok, Y.-T. Siu, and S. K. Yeung, *Geometric superrigidity*, Invent. Math. **113** (1993), 57–83.

[R] A. Reznikov, All regulators of flat bundles are torsion, Hebrew University preprint (1993).

[S] J. Sampson, *Applications of harmonic maps to Kähler geometry*, Cont. Math. **49** (1986), 125–133.

[Si] C. T. Simpson, *Nonabelian Hodge Theory*, Proc. Internat. Congr. Math., Kyoto 1990, V. I, Math. Soc. of Japan, Springer-Verlag, 1991.

[Su] Y.-T. Siu, *The complex-analyticity of harmonic maps and the strong rigidity of compact Kähler manifolds*, Ann. of Math. **112** (1980), 73–112.

[YZ] S. T. Yau and F. Zheng, *Negatively $\frac{1}{4}$-pinched Riemannian metric on a compact Kähler manifold*, Invent. Math. **103** (1991), 522–535.

Homological Geometry and Mirror Symmmetry

ALEXANDER B. GIVENTAL*

Department of Mathematics
University of California–Berkeley
Berkeley, CA 94720, USA

0. A popular example

A homogeneous polynomial equation in five variables determines a *quintic* 3-fold in $\mathbb{C}P^4$. Hodge numbers of a nonsingular quintic are known to be: $h^{p,p} = 1$, $p = 0, 1, 2, 3$ (Kähler form and its powers), $h^{3,0} = h^{0,3} = 1$ (a quintic happens to bear a holomorphic volume form), $h^{2,1} = h^{1,2} = 101 = 126 - 25$ (it is the dimension of the space of all quintics modulo projective transformations, and $h^{2,1}$ is responsible here for infinitesimal variations of the complex structure) and all the other $h^{p,q} = 0$.

Consider the family of quintics $x_1 \cdots x_5 = \lambda^{1/5}(x_1^5 + \cdots + x_5^5)$ invariant to 5^4 multiplications of the variables by 5th roots of unity. The quotient by these symmetries will generate singularities. Resolve the singularities. The result is known to be a family Y_λ of 3-folds with the table of Hodge numbers *mirror-symmetric* to that of the quintics X: $h^{p,q}(Y) = h^{3-p,q}(X)$.

Manifolds with mirror-symmetric Hodge tables are called *geometrical mirrors*. Discovered accidentally in a computer experiment, such mirror 3-folds very soon took their place in various string models of the 10-dimensional universe. As it is clear now, the so-called Arnold's strange duality of exceptional singularities [1] was probably the first manifestation of mirror phenomena — for $K3$-surfaces.

Current interest to mirror manifolds is due to the so-called *mirror conjecture* and its first applications to enumerative algebraic geometry. The idea is that along with the equality $h^{1,1}(X) = h^{2,1}(Y)$ of moduli numbers of Kähler structures on X and of complex structures on Y, whole symplectic topology on X is equivalent to complex geometry on Y, and vice versa.

This idea has led to a number of beautiful predictions (see for instance [6], [5]) in enumerative algebraic geometry, in particular – for numbers of rational curves of each degree on the quintics.

1. Singularity theory

Given a complex manifold Y^n, a holomorphic volume form ω, and a holomorphic function $f : Y \to \mathbb{C}$, one can study exponential integrals $I_\hbar = \int_\Gamma e^{f(y)/\hbar}\omega$, their asymptotics at $\hbar \to 0$, and their dependence on parameters.

* Supported by the Alfred P. Sloan Foundation and by NSF Grant DMS-9321915.

Proceedings of the International Congress
of Mathematicians, Zürich, Switzerland 1994

EXAMPLE. Let f be a weighted-homogeneous polynomial of $\deg f = 1$ on n complex variables $(y_1, \ldots, y_n)$ of some positive weights $\deg y_i = \alpha_i > 0$ with an isolated critical point $y = 0$ of multiplicity μ, $a_1, \ldots, a_\mu = 1$ — monomials representing a basis in the local algebra $H = \mathbb{C}[y]/(\partial f/\partial y)$, $f_\lambda = f + \lambda_1 a_1 + \cdots + \lambda_\mu a_\mu$ — a miniversal deformation of the critical point. The formal stationary phase approximation gives

$$I_i(\lambda) = \int e^{f_\lambda(y)/\hbar} a_i(y)\, dy_1 \wedge \cdots \wedge dy_n \sim \hbar^{n/2} e^{f_\lambda(y_*)/\hbar} \frac{a_i(y_*)}{\sqrt{J_\lambda(y_*)}}$$

for each of μ critical points $y_*(\lambda)$ of f_λ, where $J_\lambda = \det(\partial^2 f_\lambda/\partial y^2)$. These asymptotics satisfy $\hbar \partial I_j/\partial \lambda_i \sim \sum_k c_{ij}^k(\lambda) I_k$ where c_{ij}^k are structural constants of the algebra $H_\lambda = \mathbb{C}[y]/(\partial f_\lambda/\partial y)$ of functions on the critical set: $a_i a_j = \sum c_{ij}^k a_k$ in H_λ. The cycles of integration can be described as real n-dimensional Morse-theoretic cycles of the function $\mathrm{Re}\, f$ and thus correspond to the critical points and represent classes in the asymptotical homology group $H_n(\mathbb{C}^n, \mathrm{Re}\, f = -\infty)$. Then the residue pairing

$$(a,b) = \sum_{y_*} \frac{a(y_*) b(y_*)}{J(y_*)} = \frac{1}{(2\pi i)^n} \int_{|\partial f/\partial y| = \mathrm{const}} \frac{a(y) b(y) dy_1 \wedge \cdots \wedge dy_n}{\partial f_\lambda/\partial y_1 \cdots \partial f_\lambda/\partial y_n}$$

becomes an asymptotical intersection pairing between the cohomology for f and $-f$ and is known to extend without singularities to $\lambda = 0$.

THEOREM 1 [16]. *These stationary phase asymptotics can be made exact by a suitable choice of the volume forms ω_λ instead of $dy_1 \wedge \cdots \wedge dy_n$ and in special coordinates $\tilde{\lambda}$ on the parameter space instead of $(\lambda_1, \ldots, \lambda_\mu)$.*

In particular this means that the differential equations $\hbar \partial_i \vec{I} = [a_i] \cdot \vec{I}$ with $[a_i]$, $I(\tilde{\lambda}) \in H_{\tilde{\lambda}}$ form a family $\nabla_\hbar = \hbar d - \sum [a_i] d\tilde{\lambda}_i$ of connections *flat for all* $\hbar$. They are identified with the Gauss-Manin connections in the cohomological bundle. The residue pairing therefore literally coincides with the intersection pairing and induces (see [17]) on the parameter space a flat complex metric. The coordinates $\tilde{\lambda}$ are defined as flat coordinates of this metric.

In the contemporary language this theorem means that the integrals define on the parameter space the structure of a Frobenius manifold [7] and thus satisfy axioms of Topological Conformal Field Theory (TCFT) (*Landau-Ginzburg models* of TCFT).

One can at least try to extend this theorem based on deep properties of variations of Hodge structures to arbitrary families $(Y, f, \omega)_\lambda$. Consider a degenerate case where Y is a compact Kähler manifold. For this, Y should bear a holomorphic volume form ($h^{n,0} = 1$). Then f is necessarily constant, and the exponential integrals turn into periods $\int \omega^{n,0}$ of the volume form. The periods depend on the complex structure on Y and satisfy some linear differential *Picard-Fuchs* equations (describing variation of pure Hodge structures). The algebra H_λ of the critical set Y should be replaced by its cohomology $H^n(Y)$. It is a separate problem whether one can derive from these "massless" Landau-Ginzburg data complete models of

TCFT (they are called B-models, after Witten), but in many cases one can construct flat coordinates on moduli spaces of complex structures. The family of quintic-mirrors Y_λ is one of them.

2. Symplectic topology

Let X^n be a compact Kähler manifold. Given m cycles $A_1, \dots, A_m \subset X$, an integral homology class $D \in H_2(X)$ and a configuration $(x_1, \dots, x_m)$ of m points on $\mathbb{C}P^1$, one may ask: *How many holomorphic maps $\varphi : \mathbb{C}P^1 \to X$ with $\varphi_*[\mathbb{C}P^1] = D$ and $\varphi(x_i) \in A_i$ are there?* The answers (let us denote them $F_{m,D}[A]$), being properly understood as intersection indices in certain moduli spaces of holomorphic maps, turn out to depend only on homology classes of A_i and homotopy type of *almost* Kähler structure on X and provide symplectic invariants of X called *Gromov-Witten* invariants. They are not independent, and the universal identities for them can be interpreted as the associativity constraint of the *quantum cohomology* algebra $H_q^*(X)$ and compatibility of some linear PDEs.

Pick an integral basis $p_1, \dots, p_k$ of symplectic classes in $H^2(X)$, denote $(D_1, \dots, D_k)$ coordinates of D in the dual basis and put $(a_1|a_2|\cdots|a_m) = \sum_D q^D F_{m,D}[A_1, \dots, A_m]$ where A_i are Poincaré dual to cohomology classes a_i.

THEOREM 2 [14], [15]. *Gromov-Witten invariants are well defined at least if $c_1(X) \geq 0$, $(a_1|a_2)$ coincides with the Poincaré pairing (a_1, a_2) on $H_q^*(X) = H^*(X, \mathbb{C}[[q]])$, $(a_1|a_2|a_3)$ are structural constants $(a_1 * a_2, a_3)$ of a skew-commutative associative multiplication $*$ on $H_q^*(X)$, which at $q = 0$ coincides with the usual cup-product, and $(a_1|\cdots|a_m) = (a_1 * \cdots * a_m, 1)$. Besides this, the differential equations $\hbar q_i \partial q_i \vec{I} = p_i * \vec{I}, i = 1, \dots, k$, for a vector-function $I(q) \in H^*(X)$ form a compatible system (i.e. a flat connection) for each $\hbar$.*

The $*$-product is graded if one assigns usual degrees $\deg a_j = \operatorname{codim} A_j$ to the cocycles and nontrivial degrees $\deg q_i = 2d_i$ to the parameters where $d_1 p_1 + \cdots + d_k p_k = c_1(X)$.

Actual definitions of Gromov-Witten invariants involve nonintegrable perturbations of the complex structure on $\mathbb{C}P^1 \times X$, and rigorous computation of quantum cohomology is a nontrivial problem. The following examples, except for the first one, are reasonable conjectures rather than theorems.

EXAMPLES. (1) $H_q^*(\mathbb{C}P^{n-1}) \simeq \mathbb{C}[p,q]/(p^n - q)$, and the differential system is equivalent to the scalar equation $\hbar^n d^n I/dt^n = e^t I$ where $t = \log q$. The intersection pairing is given by the residue formula $\oint a(p)b(p)\, dp/(p^n - q)$ and similar formulas hold in all other examples below.

(2) For the space F_n of complete flags in $\mathbb{C}^n$, denote A_n the $(n \times n)$-matrix with $u_1, \dots, u_n$ on the diagonal, $q_1, \dots, q_{n-1}$ right above, -1's right under the diagonal, and zeros otherwise. Set $\Sigma_i = tr A_n^i$. Then (see [11]) $H_q^*(F_n) \simeq \mathbb{C}[u,q]/(\Sigma)$. In fact Σ_i coincide with conservation laws of a Toda lattice on n particles with potentials $q_i = -e^{t_i - t_{i+1}}$ (see [11]). The question of *why* the algebra $H_q^*(F_n)$ is isomorphic to the algebra of functions on the singular invariant variety of the Toda lattice, is open.

(3) Let $X = \mathbb{C}^N//T^k$ be a toric manifold obtained by the Marsden-Weinstein reduction from the standard Hermitian space by a subtorus in the maximal torus T^N. Denote (m_{ij}) the integral $(k \times N)$-matrix of the natural projection $\mathrm{Lie}^* T^N \to \mathrm{Lie}^* T^k$. Then the quantum cohomology algebra of X is given by the generators $(u_1, \ldots, u_N, p_1, \ldots, p_k, q_1, \ldots, q_k)$ and relations $u_j = \sum_i m_{ij} p_i$, $q_i = \prod_j u^{m_{ij}}$ (Batyrev, see also [9] where a discrete version of quantum cohomology of toric manifolds had been computed as a byproduct of a symplectic fixed point theorem).

(4) Let X^3 be a nonsingular quintic in $\mathbb{C}P^4$. Its hyperplane section p generates in $H^*(X)$ a subalgebra $H^{\mathrm{even}} = \mathbb{C}[p]/(p^4)$ with the intersection form $(p^i, 1) = 0$ for $i \neq 3$ and $(p^3, 1) = 5$. Its quantum deformation is almost the same except for $p*p = K(q)p^2$ or, equivalently, $(p*p, p) = K(q)$ where $K(q) = 5 + \sum_{d=1}^{\infty} n_d d^3 q^d/(1-q^d)$ (see [2]). Here n_d is the number of degree-d, rational curves in X: on a generic (almost)-Kähler 3-fold with $c_1 = 0$ rational curves are discrete and all contribute to the quantum cup-product (which now respects the usual grading, i.e. $\deg q = 0$). The corresponding differential system is equivalent to $(I''/K(q))'' = 0$ with $' = \hbar q d/dq$. It is degenerate in the sense that it is independent on $\hbar$ and easy to solve, except that the numbers n_d with $d > 3$ are unknown! [1]

In fact, we have described only a few of all Gromov-Witten invariants (see [18]), which form a complete set of "correlation functions" of a sigma-model, or A-model of TCFT and, when computable, provide algebraic geometry with very nontrivial new enumerative information [13].

3. The mirror conjecture

The mirror conjecture predicts equivalence of A and B models of TCFT on an algebraic Calabi-Yau manifold to B and A models on its geometrical mirror. In our "down-to-earth" language: for geometrical mirrors X and Y the differential system of $H_q^{\mathrm{even}}(X)$ should coincide with the Picard-Fuchs equation for Y taken in flat coordinates (and vice versa). Authors [6] of this formulation exploited it in order to predict numbers n_d for quintics.

They start with one of the periods $I_1 = \int \omega_\lambda^{3,0} = \sum (5k)!\lambda^k/(k!)^5$, reconstruct the Picard-Fuchs equation: $D^4 I = 5\lambda(5D+1)(5D+2)(5D+3)(5D+4)I$, where $D = \lambda d/d\lambda$, bring it in a neighborhood of the singular point $\lambda = 0$ to the form $(J''/k(q))'' = 0$, conjecture that $k(q) = K(q)$, and find from this $n_1 = 2875$, $n_2 = 609250$, $n_3 = 317206375$ (eventually in coherence with available data) and predict n_4 to be 242467530000.

Bringing the equation to the simple form involves: (1) finding the solution $I_2 = \log(\lambda)I_1 + \tilde{I}$ with $\tilde{I}$ holomorphic and vanishing at $\lambda = 0$, (2) introducing the new local coordinate $q = \lambda \exp(\tilde{I}/I_1)$, and (3) computing the equation satisfied by $J_i = I_i/I_1$, $i = 1, 2, 3, 4$, as functions of q.

Our previous discussion suggests the following generalization of the mirror problem:

Is there a natural map (functor?) from (a class of) TCFT-models (symplectic sigma-models, or Frobenius manifolds, or quantum cohomology algebras) to generalized Landau-Ginzburg data? Simpler, *how to solve the differential equations*

[1] See however [12].

$\hbar dI = A \wedge I$ *by means of exponential holomorphic integrals?* We will partially answer this question for the class of toric manifolds.

From such a point of view the Picard-Fuchs equation for Y should have an *intrinsic* interpretation in terms of the problem of computing Gromov-Witten invariants for X. We will point out some.

The (open) question of why the above computational procedure (for an equation that had already been intrinsically attributed to X) yields quantum cohomology of X, is probably related to the problem of *in what sense* the above "functor" is an *involution* on its invariant subset that the class of algebraic Calabi-Yau manifolds seems to constitute.

All these problems lead to the same question: What is the intrinsic meaning of *solutions* of the quantum cohomology differential system?

4. A project: Equivariant Floer cohomology

Let LX be the space of contractible loops in a compact Kähler manifold X. It inherits the Kähler structures from X and additionally carries the action of S^1 by isometries (translations in the source). The Hamiltonian of this action is the action functional: to a contractible loop it assigns the symplectic area of a disk, contracting the loop, and can be multiple valued. Applying Morse-Novikov theory for the action functionals $H : \tilde{LX} \to \mathbb{R}$ on the universal covering of LX, one comes to the definition of the Floer homology FH of X (isomorphic to $H^*(X, \mathbb{C}[q^{\pm 1}])$) [8]: gradient trajectories of H in LX are holomorphic cylinders in X. If one introduces multiplication in FH using the "map" $LX \times LX \to LX$ of composition of loops (= holomorphic pants in X), it leads to the construction of quantum multiplication in $H_q^*(X)$ (see [11]).

Project: Construct S^1-equivariant Floer cohomology $FH_{S^1}(\tilde{LX})$.

Let $\omega_1, \ldots, \omega_k$ be Kähler forms on LX corresponding to the integral basis of Kähler forms on X, and let $H_1, \ldots, H_k$ be corresponding Hamiltonians (of the same S^1-action!) on the covering $\tilde{LX} \to LX$ with the group of covering transformations $\mathbb{Z}^k = H_2(X, \mathbb{Z})$. The generators $q_1, \ldots, q_k$ of the group $\mathbb{Z}^k$ commute with the S^1-action, preserve the forms ω_i lifted to $\tilde{LX}$ from LX, but transform H_i : $q_i^*(H_j) = H_j - \delta_{ij}$.

Denote $\mathbb{C}[\hbar]$ the coefficient algebra $H^*(BS^1) = H^*(\mathbb{C}P^\infty)$ of the equivariant theory and introduce *Duistermaat-Heckman* equivariantly closed forms $p_i = \omega_i + \hbar H_i$, see [3]).

PROPOSITION. $p_i q_j - q_j p_i = \hbar q_i \delta_{ij}$ *as operators on an equivariant Floer complex.*

COROLLARY. $FH_{S^1}(\tilde{LX})$ *should carry the module structure over the algebra $\mathcal{D}$ of differential operators generated by $q_i = e^{t_i}\cdot$, and $p_j = \hbar\partial/\partial t_j$.*

A semi-classical limit $\hbar \to 0$ should give rise to the subalgebra in $H_q^*(X)$ generated by the Kähler classes p_i and q_i. In particular relations between them should describe a Lagrangian variety with respect to the Poisson bracket $\{p_i, q_j\} = q_i \delta_{ij}$ — the characteristic variety of the $\mathcal{D}$-module.

5. Realization: Toric manifolds

Holomorphic maps $\mathbb{C}P^1 \to X$ to a toric manifold $X = \mathbb{C}^N//T^k$ can be described as equivariant maps $\mathbb{C}^2 \to \mathbb{C}^N$. This compactifies the map spaces up to toric manifolds $X_d = \mathbb{C}^{N+D}//T^k$. Here the homology class d of maps can be identified with an integral point in LieT^k such that $\forall j = 1, \ldots, N$, $D_j = \sum_i m_{ij} d_i \geq 0$, and $D = \sum D_j$. We can interpret the map spaces as approximations of LX by algebraic loops ($S^1_{\mathbb{C}} \subset \mathbb{C}P^1$), define $FH_{S^1}(X)$ as a certain limit (see [10]) of $H^*_{S^1}(X_d)$, and using the explicit toric description of X_d, compute the $\mathcal{D}$-module.

The algebra $H = H^*(X)$ is generated by the integral Kähler classes $P_1, \ldots, P_k$ (see [3]). Denote $(\cdot, \cdot)$ intersection pairing on H, $\Omega : H \to H$ — the automorphism generated by $P_i \mapsto -P_i$. Introduce notation: $U_j = \Sigma m_{ij} P_i$, $D_j = \Sigma m_{ij} d_i$, $t_i = \log q_i$, $\partial_j = \hbar \Sigma m_{ij} \partial/\partial t_i$, and set

$$\Delta_j^r = \partial_j(\partial_j - \hbar)\cdots(\partial_j - (r-1)\hbar), \quad M[x]! = \hbar^M \frac{\prod_{m=-\infty}^{M}(x+m)}{\prod_{m=-\infty}^{0}(x+m)}$$

for any $M \in \mathbb{Z}$.

THEOREM 3 [10]. *Suppose $c_1(X) > 0$. Then*

(1) *$FH_{S^1} \simeq \mathcal{D}/\mathcal{J}$ where the left $\mathcal{D}$-ideal $\mathcal{J}$ is generated by all operators $\Delta_1^{r_1}\cdots\Delta_N^{r_N} - q^d \Delta_1^{l_1}\cdots\Delta_N^{l_N}$ with $r_j \geq 0$, $l_j \geq 0$, and $r_j - l_j = D_j$.*

(2) *The kernel of this linear differential system is generated by the components of the following vector-function of t with values in the cohomology algebra H:*

$$\vec{f}_\hbar(t) = \hbar^{k-N} e^{Pt} \Sigma_{d\in\mathbb{Z}^k} \frac{e^{dt}}{D_1[U_1]!\ldots D_N[U_N]!}.$$

(3) *$\sum_d e^{d\tau} \int_{X_d} e^{p(t-\tau)/\hbar} = \hbar^{N-k}(\vec{f}_\hbar(t), \Omega\vec{f}_{-\hbar}(\tau))$, where p_i are Duistermaat-Heckman forms $\omega_i + \hbar H_i$ on each X_d corresponding to our basis in Lie$^*T^k \simeq H^2(X_d)$ and the S^1-action.*

(4) *Suitable limits to $\hbar = 0$ give rise to the algebra $H_q^*(X)$ and a generating series for symplectic volumes of X_d.*

EXAMPLE. For $k = 1$, $N = n$, and $(m_{ij}) = (1, \ldots, 1)$, we get $X = \mathbb{C}P^{n-1}$. Then $P^n = 0$ and components of $\vec{f} = e^{Pt}\sum_{d=0}^{\infty} e^{dt}/[(P+1)\cdots(P+d)]^n \hbar^{nd}$ give all n solutions of $(\hbar d/dt)^n I = e^t I$, and the first one ($P = 0$) is $\sum q^d/(d!)^n \hbar^{nd}$.

6. Toric complete intersections

Given T^k-invariant homogeneous polynomials on $\mathbb{C}^N$, one can plug components of a rational curve $\mathbb{C}^2 \to \mathbb{C}^N$ into them and equate to zero identically. In the spaces X_d the solutions form the zero locus of a $PSL_2(\mathbb{C})$-invariant holomorphic section of a suitable equivariant vector bundle. If such sections were transverse to the zero section the loci would represent equivariant Euler classes of these bundles. One may hope to reconstruct the $\mathcal{D}$-modules and quantum cohomology algebras of complete intersections $X' \subset X$ from such classes, substituting them for fundamental cycles of the map spaces X'_d.

For simplicity let us consider the case of Calabi-Yau complete intersections in $X = \mathbb{C}P^{n-1}$. Let $l_1, \ldots, l_r > 0$ be Chern numbers of r line bundles with $l_1 +$

$\cdots + l_r = n$. Introduce the algebra $H = \mathbb{C}[P]/(P^{n-r})$ with the intersection pairing $(P^{n-r-1}, 1) = l_1 \dots l_r$ (the image of $H^*(X) \to H^*(X'))$ and denote $E_d^l(p,\hbar)$ the S^1-equivariant Euler class of that "suitable" vector bundle over X_d.

THEOREM 4 [10].

$$\Sigma_{d=0}^{\infty} e^{d\tau} \int_{X_d} e^{p(t-\tau)/\hbar} E_d^l(p,\hbar) = (-1)^{n-1}\hbar^{1+r-n}(\vec{g}_l(t), \Omega\vec{g}_l(\tau))$$

where

$$\vec{g}_l = e^{Pt} \sum_{d=0}^{\infty} e^{dt} \frac{(l_1 d)[P]! \dots (l_r d)[P]!}{(d[P]!)^n}.$$

The $n-r$ components of $\vec{g}_l$ in H provide a complete solution to the differential equation ($D = d/dt$)

$$D^{n-r} I = l_1 \dots l_r e^t (l_1 D + 1) \dots (l_1 D + l_1 - 1) \dots (l_r D + 1) \dots (l_r D + l_r - 1) I.$$

This is exactly the equation that was found in [5] as the *Picard-Fuchs* equation for mirrors of projective Calabi-Yau complete intersections, satisfied by the "hypergeometric" series $\sum_d l_1! \dots l_r! q^d/(d!)^n$.

In particular we have obtained the equation $D^4 I = 5e^t(5D+1)\dots(5D+4)I$ entirely in topological terms of map spaces and not as a Picard-Fuchs equation of a Landau-Ginzburg model. Furthermore, its solution

$$e^{Pt} \sum_{d=0}^{\infty} q^d \frac{(5P+1)\dots(5P+5d)}{(P+1)^5 \dots (P+d)^5},$$

rewritten as

$$e^{Pt}(G_1 + G_2 P + G_3 P^2 + G_4 P^3) = G_1(q) + P(G_1(q)\log q + G_2(q)) + \cdots,$$

yields I_1 as G_1 and $\tilde{I}$ as G_2.

Thus, each coefficient of $G_1, \dots, G_4$ should hide some enumerative information about rational curves in $\mathbb{C}P^4$ relative to (one or many) quintics: what appeared meaningless in the Picard-Fuchs equation because of accidental choice of the coordinate λ in the family of quintic-mirrors, turns out to be related directly to the *exterior* geometry of quintics in $\mathbb{C}P^4$.

PROBLEM. Recover this information.

7. Integral representations

It is not surprising that toric geometry provides integral formulas for some hypergeometric series. However they will illuminate the possible nature of mirror manifolds.

DEFINITION. A function $F : E \to \mathbb{C}$ on the fibered space $\pi : E \to B$ *generates* the Lagrangian variety $L = \{(p,t) \in T^*B \| \exists x \in \pi^{-1}(t) : dF|_x = \pi^*(p)\}$.

LEMMA. *Theorem 1 with $k = N$ and $m_{ij} = \delta_{ij}$ formally gives $P = 0$ and $\vec{f} = \sum_{d \in \mathbb{Z}_+^N} e^{dt}/d_1! \dots d_N! \hbar^{N|d|} = \exp(\sum e^{t_j}/\hbar)$.*

THEOREM 5 [10]. *Let* $X = \mathbb{C}^N//T^k$ *be a compact toric manifold with* $c_1 > 0$. *Then*

(1) *The quantum cohomology algebra* $H_q^*(X)$ *is the algebra of functions on the Lagrangian variety generated by* $F = u_1 + \cdots + u_N : \mathbb{C}^N \to \mathbb{C}$ *with* $\pi : u \mapsto q$ *given by* $q_i = \prod u_j^{m_{ij}}$.

(2) *Introduce the holomorphic volume form on* $\pi^{-1}(q)$ *as*

$$\omega_q = \frac{q_1 \dots q_k}{u_1 \dots u_N} \frac{du_1 \wedge \dots \wedge du_N}{dq_1 \wedge \dots \wedge dq_k}.$$

Then integrals

$$I(\log q) = \int_{\Gamma \subset \pi^{-1}(q)} \omega_q \, e^{(u_1 + \cdots + u_N)/\hbar}$$

over cycles Γ *corresponding to* $\dim H_q^*(X)$ *critical points of* $F|\pi^{-1}(q)$ *provide a complete set of solutions to the differential system of Theorem* 1 (1).

Notice that $\dim \pi^{-1}(q) = N - k = \dim_{\mathbb{C}} X$. According to our formulation of the mirror problem we should call the Landau–Ginzburg data $(E \to B, F, \omega)$ *a family mirror-symmetric to the toric manifold* X.

Furthermore, set $X_q' = F^{-1}(1) \cap \pi^{-1}(q)$, $\omega_q' = \omega_q/d(F|\pi^{-1}(q))$.

THEOREM 6 [5], [10]. *All solutions of the differential equation of Theorem 2 with* $r = 1$ *are integrals* $\int \omega_q'$ *over compact cycles* $\Gamma' \in X_q'$.

In order to obtain the same result for $r > 1$ one should split $u_1 + \cdots + u_N$ into r sums of length $l_1, \dots, l_r$ and consider the sums as equations of a complete intersection X' in the fibers of π.

Theorem 4 matches well to the remarkable Batyrev's construction [4] of geometrical mirror pairs of toric hypersurfaces: fibers $\pi^{-1}(q)$ are the complex tori that, when suitably compactified into a toric variety, meet $F^{-1}(1)$ along Batyrev's Calabi-Yau hypersurface, and ω' extends to its holomorphic volume form. Thus, Batyrev's mirrors of toric hypersurfaces are hypersurfaces in the mirrors of their ambient toric manifolds.

EXAMPLE. Replace the homogeneous equation $q^{1/5}(x_1^5 + \cdots + x_5^5) = x_1 \dots x_5$ with the affine equation in $\mathbb{Z}_5^4$-invariant coordinates $u_i = q^{1/5} x_i^5 / x_1 \dots x_5$. Then $u_1 \dots u_5 = q$, and the equation is $u_1 + \cdots + u_5 = 1$. This corresponds to our matrix $(m_{ij}) = (1, 1, 1, 1, 1)$.

8. Homological geometry

Along with the differential equations of Theorems 1 and 2 the above integral formulas have intrinsic cohomological meaning in toric geometry. In fact the description of $H_q^*(X)$ by $N+k$ generators and relations is a q-deformation of the following similar description of $H^*(X)$. A symplectic quotient $\mathbb{C}^N//T^k$ can be identified with the free quotient $\check{\mathbb{C}}^N/T_{\mathbb{C}}^k$ of some domain in $\mathbb{C}^N$, and $H^*(\mathbb{C}^N//T^k)$ — with the equivariant cohomology $H_{T^k}^*(\check{\mathbb{C}}^N)$. One begins with $H_{T^N}^*(\mathbb{C}^N) = H^*(BT^N) = \mathbb{C}[u]$, then computes $H_{T^N}^*(\check{\mathbb{C}}^N)$, which causes factorization by some "multiplicative" ideal, and finally reduces the group T^N to T^k, which imposes the additive relations.

Therefore our $(u_1, \ldots, u_N)$ are in fact characteristic classes of T^N, and the function $F = \sum u_j$ is the *universal* 1st Chern class c_1 (of all toric manifolds — quotients of $\mathbb{C}^N$). Our solutions to the quantum cohomology differential systems are given by integrals in $\mathrm{Spec}H^*(BT^N)$ over Morse theoretic cycles of the function $\mathrm{Re}(c_1)$. It suggests that in general mirror manifolds should live in some cohomologies of each other.

The integral formulas can be recovered from our $\mathcal{D}$-module approach: one should first compute an "equivariant Floer cohomology" that is equivariant also with respect to the maximal torus T^N/T^k of $\mathrm{Aut}(X)$, and then get rid of this extra structure. The first step adds variables, the second expresses $\vec{f}$ as a De Rham class in excessive variables.

References

[1] V. I. Arnold, *Critical points of smooth functions*, in Proceedings of ICM 74, vol. 1, Vancouver, BC, 1974, 19–40.

[2] P. Aspinwall and D. Morrison, *Topological field theory and rational curves*, preprint, Oxford, 1991.

[3] M. Audin, The Topology of Torus Actions on Symplectic Manifolds, Birkhäuser, Basel and Boston, 1991.

[4] V. V. Batyrev, *Dual polyhedra and mirror symmetry for Calabi-Yau hypersurfaces in toric varieties*, preprint, Essen Univ., 1992.

[5] V. Batyrev and D. van Straten, *Generalized hypergeometric functions and rational curves on Calabi-Yau complete intersections in toric varieties*, preprint, Essen Univ., 1993.

[6] P. Candelas, X.C. de la Ossa, P.S. Green, and L. Parkes, *A pair of Calabi-Yau manifolds as an exactly soluble superconformal field theory*, Nuclear Phys. B, **359** (1991), 21–74.

[7] B. Dubrovin, *Integrable systems and classification of 2-dimensional topological field theories*, preprint, hep-th 9209040.

[8] A. Floer, *Symplectic fixed points and holomorphic spheres*, Comm. Math. Phys. **120** (1989), 575–611.

[9] A. Givental, *A symplectic fixed point theorem for toric manifolds*, preprint, Berkeley, 1992 (to appear in Progr. Math., Floer's memorial volume).

[10] A. Givental, *Homological geometry, I: Projective hypersurfaces*, preprint, 1994. *II: Integral representations*, in preparation.

[11] A. Givental and B. Kim, *Quantum cohomology of flag manifolds and Toda lattices*, preprint, hep-th 9312096 (to appear in Comm. Math. Phys.).

[12] M. Kontsevich, *Enumeration of rational curves via torus actions*, preprint, 1994.

[13] M. Kontsevich and Yu. Manin, *Gromov-Witten invariants, quantum cohomology, and enumerative geometry*, preprint, Max-Plank-Institut, 1994.

[14] D. McDuff and D. Salamon, J-holomorphic Curves and Quantum Cohomology, Amer. Math. Soc., Providence, RI, 1994.

[15] Y. Ruan and G. Tian, *A mathematical theory of quantum cohomology*, preprint, 1994.

[16] K. Saito, *On the periods of primitive integrals I*, RIMS (1982), 1–235.

[17] A. N. Varchenko and A. B. Givental, *Period mappings and intersection forms*, Functional Anal. Appl. **16:2** (1982), 11–25.

[18] E. Witten, *Two-dimensional gravity and intersection theory on moduli space*, Surveys Differential Geom. **1** (1991), 243–310.

Constant Mean Curvature Surfaces in Euclidean Spaces

NIKOLAOS KAPOULEAS*

Department of Mathematics, Brown University
151 Thayer Street, Box 1917
Providence, RI 02912, USA

A variant of the isoperimetric problem is to classify and study the hypersurfaces in the Euclidean space $\mathbb{E}^{n+1}$ that have critical area subject to the requirement that they enclose a fixed volume. In physical terms this is equivalent to having a soap film in equilibrium under its surface tension and a uniform gas pressure applied to one of its sides; hence, such surfaces are often called soap bubbles. The geometric condition for such a surface is that its mean curvature H is a nonzero constant. The precise value of the constant is not important because it can be changed to any desired value by a homothetic expansion. We will be using the abbreviation "CMC surface" to mean "complete smooth hypersurface properly immersed in $\mathbb{E}^{n+1}$ with $H \equiv 1$". Notice that the above definitions do not require embeddedness.

Although it has been known and proven for a long time that the round spheres are the unique answer to the isoperimetric problem, only recently has the corresponding question been settled for soap bubbles. Actually up to 1980 the only known examples of CMC surfaces of finite topological type were the rotationally invariant ones in $\mathbb{E}^3$ studied by Delaunay in 1841 [5]. This, combined with various nonexistence results we will now review, led to the suspicion — sometimes called the Hopf conjecture although it is not clear that Hopf ever took sides on this question — that the round sphere is the only closed bubble in $\mathbb{E}^3$.

Jellet [11] had already proved in 1853 that starshaped closed CMC surfaces are round spheres. A century later Hopf [9] proved that a CMC surface homeomorphic to S^2 is a round sphere. His proof uses the so-called Hopf differential, which is a quadratic holomorphic differential for the underlying Riemann surface structure of the surface M defined by

$$\Phi = \langle X_{zz}, \nu \rangle \, dz^2 = \tfrac{1}{4}(A_{11} - A_{22} - 2iA_{12}) dz^2 ,$$

where $z = u + iv = x_1 + ix_2$ is a local isothermal coordinate, $X : M \to \mathbb{E}^3$ is the immersion in consideration, $\nu : M \to \mathbb{E}^3$ its Gauss map, $\langle \, . \, , \, . \, \rangle$ the standard inner product in $\mathbb{E}^3$, and A_{ij} the second fundamental form. The Cauchy-Riemann equations establishing the holomorphicity amount to the Codazzi equations once $H \equiv 1$ is used. Clearly (by Liouville's theorem for example) holomorphic quadratic

*) Partially supported by NSF grants DMS-9404657 and NYI DMS-9357616 and a Sloan Research Fellowship.

Proceedings of the International Congress
of Mathematicians, Zürich, Switzerland 1994

differentials on S^2 vanish. Hence, $A = Hg$ (g is the first fundamental form), g has constant curvature, and X immerses to a round sphere.

In 1956 Alexandrov [2] proved that all embedded closed CMC surfaces (any n) are round spheres. His method uses moving planes to establish by the use of the maximum principle that there is a plane of symmetry parallel to any given plane; hence, the surface is a round sphere. Both Hopf's and Alexandrov's methods have found numerous applications to other problems; actually there is currently no other method of the wide applicability of Alexandrov's in dealing with questions of uniqueness and symmetry in nonlinear problems. More recently, Barbosa and doCarmo [3] showed that local minimizers of the variational problem are round spheres.

In the 1980s the general picture changed. In 1982 Hsiang [10] demonstrated that Hopf's theorem is not valid in higher dimensions by constructing nonround CMC spheres. His method was to reduce the problem to an ordinary differential equation (ODE) by imposing nonstandard rotational symmetry. In 1984 Wente [27] in a surprising development disproved the so-called Hopf conjecture by producing toroidal soap bubbles in $\mathbb{E}^3$. Because a torus can be covered conformally by $\mathbb{C}$ we can arrange that the Hopf differential lifts to $\Phi = dz^2$ where $z = u + iv$ is the standard coordinate on $\mathbb{C}$. By writing then the fundamental forms as

$$g = \tfrac{1}{4}e^{2w}|dz|^2, \qquad A = \frac{e^{2w}-1}{4}du^2 + \frac{e^{2w}+1}{4}dv^2,$$

the Gauss equation reduces to the sinh-Gordon equation

$$\Delta w + \tfrac{1}{2}\sinh w = 0,$$

hence solutions of this equation integrate on $\mathbb{C}$ to conformal CMC immersions. Wente using partial differential equation (PDE) methods found a 2-parameter family of highly symmetric doubly periodic solutions w and he demonstrated that the parameters can be arranged so that the corresponding CMC immersion X is also doubly periodic. Abresch [1] subsequently realized that one can find the solutions w Wente used by separation of variables and he demonstrated that there are Wente tori with only 3 positively curved regions. Walter [26] expressed the immersion X in closed form using theta functions and gave a very detailed description of the geometry of these surfaces.

Soon afterwards Pinkall and Sterling [22] classified all the doubly periodic solutions w and hence the CMC tori; this result could be thought of as the analogue of the Enneper-Weierstrass representation for minimal surfaces. Bobenko [4] noticed the analogy with the classical soliton theory where one has the sine-Gordon equation instead, and he improved and generalized this classification. Many other people have been working in this direction. An interesting result, for example, of Ercolani, Knörrer, and Trubowitz [7] has been the proof that there are continuous families of CMC tori of arbitrarily large number of parameters. One should mention at this point the Hsiang-Lawson conjecture that the only minimal embedded torus in $S^3(1)$ is the Clifford torus. In spite of the fact that all the attempts have failed up to now, there is still hope that the classification of the minimal tori in $S^3(1)$ [4] will help in settling this conjecture.

All the CMC surfaces of finite topological type known by the above methods are topological spheres, cylinders, planes, tori, or tori with two ends. Attempts to extend to other topological types have been unsuccessful up to now for two reasons. First, umbilics always exist then. Second, the fundamental group is not commutative as in the case of a torus and hence the induced representation into the Euclidean group does not have to consist of Euclidean motions sharing a common axis. Fortunately, there is another general construction [12]–[16] that gives a plethora of examples for almost any finite topological type, including examples of closed surfaces of any genus besides genus 0 and 1, and also embedded (complete) examples of any genus and enough ends. Most of the remaining discussion will concentrate on this construction.

The most ambitious formulation of the construction would be to give general conditions under which the following is possible: start with a collection of unit spheres in $\mathbb{E}^{n+1}$ from each of which a number of small discs is removed, and a collection of complete minimal surfaces rescaled to small size from which a neighborhood of infinity is removed. Span then the existing boundaries with surfaces so that a complete surface is obtained that is subsequently perturbed to a CMC surface. Such a construction seems plausible because all known CMC surfaces come in families containing surfaces that can be decomposed as above: small perturbations of unit spheres minus small discs, rescaled complete minimal surfaces minus a neighborhood of ∞, and regions connecting the two. The main difficulty in having such a general construction is the construction of the connecting surfaces. At the moment we can only borrow the connecting parts from the Delaunay or the Wente cylinders, whose geometry we proceed to describe.

The Delaunay family of surfaces can be parametrized by a single parameter τ to be defined later. Each surface is obtained by rotating a periodic curve around the axis. τ can take both positive and negative values; for positive τ the curve is the graph of a function and the surface is embedded, whereas for negative τ it is not. Actually, Delaunay produced these curves as the loci of a focus of an ellipse ($\tau > 0$) or a hyperbola ($\tau < 0$) rolling on the axis. $\{K = 0\}$ on the surface (K denotes the Gauss curvature) is a union of circles whose removal disconnects it into components that we call almost spherical regions (asr's for short) for reasons that will become clear later. We call an asr positive or negative according to the sign of K on it. If $|\tau|$ is small a positive asr minus a small neighborhood of its boundary approaches a round sphere of radius 1 — recall H has been normalized to be 1 and we assume the conventions that give $H = 1$ on $S^n(1)$ — minus two small antipodal discs. Similarly, an enlarged negative asr by a factor of $|\tau|^{-1}$ suitably placed approaches a catenoid $\{x_2^2 + x_3^2 = \cosh^2 x_1\}$ on any large ball centered at the origin of the coordinate system $Ox_1x_2x_3$ in consideration. In the limit as $\tau \to 0$ the surface becomes a string of unit spheres touching externally and with centers on the axis while the negative asr's shrink to points. The rescaled as above negative asr's tend to catenoids.

There is a translation along the axis of the Delaunay surface that carries a positive asr to one that is separated from the previous one by a single negative asr. The length of this translation is an increasing function of τ, for $\tau = 0$ being 2. Finally, to define τ precisely, consider the corresponding physical system of soap film and enclosed gas in pressure. If one separates it into two components by cutting with a surface, then the part on the left exerts a force on the part on the right. τ is so defined that this force is $\pi\tau\vec{e}$, where $\vec{e}$ is the unit vector in the direction of the axis of the Delaunay surface pointing from right to left. Notice that this force is attractive or repulsive according to the sign of τ and does not depend on the cutting surface because the (finite) part of the system isolated between two different cuts has to be in equilibrium.

We turn our attention now to the Wente surfaces. By suitably choosing one of the free parameters of the Wente construction it is possible to arrange for the immersion $X : \mathbb{C} \to \mathbb{E}^3$ to have a period. This way we obtain a 1-parameter family of Wente cylinders. We call the parameter τ again although we do not have a nice definition for τ as before. In this case τ is restricted to positive values. The Wente cylinders like the Delaunay ones are periodic, the period now is however a rotation around an axis instead of a translation. The angle of this rotation θ varies continuously with τ and $\theta \to 0$ as $\tau \to 0$. The Wente cylinders close up to tori whenever θ/π is rational. There is a plane of symmetry P of the whole surface that is perpendicular to the axis.

We define asr's as before, the only difference being that $\{K = 0\}$ is a highly connected union of curves now. For small τ each positive asr — excluding as usual a small neighborhood of its boundary — approaches a unit sphere minus one small disc. Each negative asr enlarged by a factor of $|\tau|^{-1}$ approaches similarly an Enneper's minimal surface. Asr's come in pairs, each pair contains one positive and one negative asr and their common boundary. The boundary of each pair consists of two circles, each of which is a generator of the fundamental group of the cylinder. Each of these circles immerses as a planar "figure 8" of maximum symmetry. Its plane is parallel to the axis at distance $1/2$. Successive planes form an angle of $\theta/2$. The bisector of the angle of two successive planes is a plane of symmetry of the pair in between, and so is P. The self-intersection q of each figure 8 is the projection on the plane of the figure 8 of the intersection p of the axis with P, each line pq is a line of reflectional symmetry of the whole cylinder. Using such reflections the whole cylinder is generated by a single pair of asr's.

An important difference from the Delaunay case, besides the complicated nature of $\{K = 0\}$, is the fact that each asr is attached to the rest of the surface at one place only. Hence, the force exerted on it by the rest of the physical system has to vanish, whereas in the Delaunay case we have two nonzero forces exerted, one at each of the two antipodal attachments, canceling each other. We could consider the two forces exerted on a pair of asr's (we have two attachments then), but these also vanish.

We go back now to discussing the construction of new CMC surfaces. At the present stage of development of this construction we are restricted to the following approach: fuse positive asr's from Delaunay or Wente cylinders or tori and unit spheres, thereby creating a CMC surface M on which $H - 1$ is supported on some of the positive asr's — called central in the rest of the discussion. Find then a function $\phi : M \to \mathbb{R}$ such that $X_\phi := X + \phi\nu$ is a CMC immersion (X and ν are the initial immersion and its Gauss map). Examples of such M's are a sphere to which a number of Delaunay ends has been attached, two spheres connected by a Delaunay neck, or two Wente tori placed symmetrically so that two particular positive asr's, one from each, are fused together so that a genus 2 closed surface is obtained. It is clear from the second example that there must exist an obstruction to the construction because otherwise we would contradict both Hopf's and Alexandrov's theorems.

The idea for finding ϕ is to use perturbation methods, based on the fact that if the τ parameters of the Delaunay and Wente ingredients are small, then $H-1$ is small (actually of order τ). We would like to linearize and correct for $H-1$, and then correct for the higher order terms. This can be phrased in the language of fixed point theorems. Let H_ϕ be the mean curvature of X_ϕ, then

$$H_\phi = H + \tfrac{1}{2}\mathcal{L}\phi + Q_\phi\,,$$

where $\mathcal{L} = \Delta + |A|^2$, $|A|$ is the length of the second fundamental form, and Q_ϕ is quadratic and higher order in ϕ with geometric invariants of X as coefficients. Given ϕ, let u, v be the solutions to the linear equations

$$\mathcal{L}u = 2(1-H), \qquad \mathcal{L}v = 2Q_\phi.$$

Clearly then, a fixed point of the map $\phi \rightarrow u - v$ provides us with the desired ϕ.

Our approach hence requires us to be able to invert $\mathcal{L}$ and to be able to have good enough estimates for the solutions and the higher order terms. The higher order terms in particular create various technical difficulties because the coefficients blow up as $\tau \rightarrow 0$ on the negative asr's. The main difficulty however is that it is not clear that $\mathcal{L}$ is even invertible. Actually on the unit spheres and the complete minimal surfaces one obtains in the limit as $\tau \rightarrow 0$, $\mathcal{L}$ is not invertible because the translations give rise to a 3-dimensional kernel.

In order to resolve these difficulties it is helpful to employ the conformal covariance of the Laplacian in dimension 2 to appropriately change the equations and facilitate their study. By changing the metric to $h = \frac{1}{2}|A|^2 g$ the linearized equation changes to $\mathcal{L}u = 4|A|^{-2}(1-H)$, where $\mathcal{L}_h := \Delta_h + 2$, while the metric on the positive asr's changes very little and on the negative asr's changes to make them isometric to the positive asr's. $\mathcal{L}_h$ has then 3 small eigenvalues for each asr of M corresponding to the 3-dimensional space of translations. We call the space spanned by the corresponding eigenfunctions an approximate kernel. Moreover, the components of the Gauss map on each asr are functions close in L^2 to this approximate kernel. Hence, there is no hope of bringing our construction to a successful conclusion unless $|A|^{-2}(1-H)$ is (approximately) orthogonal to these functions. This is the obstruction we have been expecting and by using the physical model one can see it amounts to the requirement that the forces exerted on each central asr by the rest of the system have vanishing resultant. This is identical — but in the language of physics — to applying the balancing formula in [18], where such balancing arguments were first developed for CMC surfaces.

As we have already seen, each Delaunay piece attached to a central asr exerts a force $\pi\tau\vec{e}$, where $\vec{e}$ is in the direction of the axis of the Delaunay and points away from the asr towards the piece. The Wente pieces do not contribute to the resultant force. We assume from now on that the balancing condition is satisfied by M; that is, for each central asr we have $\sum \tau \vec{e} = 0$, the summation being over all the Delaunay pieces attached to the asr in consideration. We can then solve the linear equations in our scheme modulo small elements of the approximate kernel. Assuming the required weighted estimates for the smallness of Q_ϕ, we can ensure the existence of a ϕ such that $H_\phi - 1$ is a small element of the approximate kernel.

It remains to correct for this approximate kernel. To this purpose we introduce some further perturbation of M that will introduce $H-1$ with approximate kernel content canceling the one we want to remove. If there are only Delaunay surfaces used in the construction, we can prescribe any small element of approximate kernel by perturbing M as follows: prescribing the projection to the approximate kernel of $H-1$ is approximately the same as prescribing $\int_S (H-1)\nu_i$ on each asr S, where the ν_i's are the components of the Gauss map. This is in turn the ith component of the resultant force exerted on S by the rest of the system. In the Delaunay case the rest of the system has two components; thus, we can create a resultant by changing the parameters of one of them (τ or the direction). The force then exerted by this Delaunay piece does not balance anymore the force exerted by the other.

This approach clearly fails when there are Wente surfaces incorporated in the construction. One needs again to be able to prescribe a resultant force to each asr of the Wente surface, but now there is only one component on the rest of the system and this exerts no force for any value of the parameter. Nevertheless, it is still possible to create such a force as follows: excise most of each asr from the Wente surface to leave a small neighborhood N of $\{K=0\}$. The first Dirichlet eigenvalue of $\mathcal{L}_h$ on N is (large) positive and N is stable. Thinking of the boundary components of N as wires we can move them to new positions without destroying the soap film; this repositioning should be resisted by forces trying to bring the wires to their original positions. By reattaching the excised parts of the asr's at the new positions we should have been successful in creating $|A|^{-2}(H-1)$ with approximate kernel content.

This is indeed so and one way to make it precise is as follows: Let $\vec{Y}$ be a Killing vector field of the ambient $\mathbb{E}^3$. Clearly then $\mathcal{L}_h(\nu \cdot \vec{Y}) = 0$. Hence, if f is some function on some domain Ω of M we have using integration by parts that

$$\int_\Omega d\mu_h \, \nu \cdot \vec{Y} \, \mathcal{L}_h f = \int_{\partial\Omega} \vec{V}(f) \, \nu \cdot \vec{Y} - f \, \vec{V}(\nu \cdot \vec{Y}),$$

where $\vec{V}$ is the outward unit normal to $\partial\Omega$ tangent to M. If we ignore higher order terms the left-hand side is the projection to $\nu \cdot \vec{Y}$ of the $|A|^{-2}(H-1)$ produced by changing X to X_f. For example, suppose that we construct f on the Wente cylinder W as follows: solve the Dirichlet problem $\mathcal{L}_h f = 0$ on N (as above) with $f = 0$ on $\partial\Omega \setminus S$ and $f = \nu \cdot \vec{e}$ on $\partial N \cap S$, where S is an asr and $\vec{e}$ is a unit vector parallel to the intersection of the two planes of symmetry of S. Extend f to be $\nu \cdot \vec{e}$ on $S \setminus N$ and 0 on the rest of W. Finally smooth it out by changing it on a neighborhood of ∂N. Clearly then, the left-hand side above with $\Omega = S$ and $\vec{Y} = \vec{e}$ is up to a constant the approximate kernel created corresponding to $\vec{e}$ on S, whereas the right-hand side can be calculated to be of order $|\log \tau|^{-1}$ by arguing as follows: f is approximately harmonic on N. Most of $S \cap N$ is conformally isometric to a cylinder $S^1(1) \times [0, \ell]$, where $\ell \approx \frac{1}{4}|\log \tau|$. Because harmonicity is conformally invariant, $f \approx 1$ on $\partial N \cap S$, and — as it turns out — $f \approx 0$ on ∂S, we see that the flux of f through the generator of $\pi_1(S \cap N)$ is of order $|\log \tau|^{-1}$. In this case the flux is the dominant term on the right-hand side and the argument is complete. This general approach of creating approximate kernels can be summarized by what we call the Geometric Principle:

GEOMETRIC PRINCIPLE. *Creation of $(H-1)|A|^{-2}$ in the approximate kernel direction amounts to repositioning the asr's relative to each other.*

Notice that in the earlier approach in the Delaunay case this is definitely valid because changing the parameter (and hence the period) or the direction of the axis clearly repositions the asr's. It is conceptually illuminating also to notice that for the above f we have essentially proven that $\mathcal{L}_h f$ has a projection to one of the eigenfunctions of the approximate kernel of order $|\log \tau|^{-1}$. Because f has a projection of order 1, the corresponding eigenvalue is also of order $|\log \tau|^{-1}$. It turns out that the small eigenvalues of $\mathcal{L}_h$ in the fusion of Wente tori are of two kinds: those of order $|\log \tau|^{-1}$, and those of order $\sqrt{\tau}$. From all this we can conclude that our approach amounts to effectively inverting the linearized operator in the direction of the approximate kernel. This method and the "geometric principle" should be widely applicable in all similar problems where small but nonzero eigenvalues appear.

Although the above gives the philosophy of how to prescribe the creation of the kernel in the Wente case, much more work is required to actually do so. The difficulty is due to the fact that in a short distance from an asr there are many (their number tends to ∞ as $\tau \to 0$) asr's. One succeeds by using the precise information one has for $\nu \cdot \vec{Y}$ on the figure 8's separating pairs of asr's, the symmetries of the Wente cylinders, the approximate harmonicity of the various f's as above, the conformal invariance of the harmonic condition, and various other ideas for which we refer the reader to [16]. We only remark here that it fits with the rotational character of the Wente cylinders that in the end one has to prescribe not only forces through the figure 8's but also torques. These are created by a relative rotation of the two components of the complement of the figure 8 in consideration around the axis of the cylinder.

We would like to make some final comments concerning the construction of closed surfaces by this method. If one uses Delaunay surfaces only, then one has to have a number of central spheres. Because of the balancing condition, at least 3 Delaunay pieces have to emanate from each central sphere; hence, there are at least 4 central spheres and the genus is at least 3. To satisfy the balancing condition at an outermost central sphere one has to use Delaunay pieces of both the embedded and nonembedded kind so that some of the forces are repulsive and some attractive. In this case there is another difficulty [14] as well: once all but one of the Delaunay pieces are placed, one has to place a final one to connect central spheres whose positions are fixed already. In general this last piece will not fit. This difficulty can be overcome by using a free parameter one has available to adjust the distance of the two spheres to that required by the piece to fit in. This requires the use of very long Delaunay pieces, which magnifies this effect. Now using Wente cylinders one can produce closed surfaces of genus 2 as well, as in the example we have already mentioned. Here again we have a period problem, which comes into the construction in a subtler fashion; for details see [16].

In the case of complete embedded surfaces we can only use Delaunay pieces connecting central spheres and Delaunay ends. The ends of the surfaces we obtain decay exponentially to the Delaunay ends at ∞. Korevaar, Kusner, and Solomon [19], extending partial results of Meeks [21], proved that this is the case in general; that is, every embedded CMC surface in $\mathbb{E}^3$ has ends that decay exponentially at ∞ to Delaunay ends. More recently, Korevaar and Kusner [18] have further restricted the structure of a general embedded CMC surface to resemble more the structure of the ones constructed as above. Concerning the general structure an extra piece of information comes from a construction of Grosse-Brauckmann [8]. Extending methods of Lawson [20] and Karcher [17] he constructed a 1-parameter continuous family of CMC spheres of maximum symmetry with n ends ($n \geq 3$). The family starts with surfaces like the ones already discussed, where n embedded Delaunay ends are attached to a central sphere. The parameter τ then of the ends increases to a maximum value and then decreases again towards 0. These last surfaces have n Delaunay ends of small $\tau > 0$ joined in the middle by an n-noid.

In summary we can say that in the last few years there has been enormous progress in the subject. We now understand that there is a rich variety of CMC surfaces and we have made substantial progress in understanding the subject in its totality. Still, there are many unanswered questions and some which come to mind are the following:

(1) Very little is known in higher dimensions.

(2) Classify the connected components of the moduli. In the embedded case a lot of progress has been made [19]. In the immersed case the statement should be modified somehow so that all Wente tori count as being in the same component, for example.

(3) Understand better the structure of each component in the spirit of [8]. Perhaps a more general construction based on minimax methods would be useful here.

(4) As an introductory step, understand the geometry of the CMC tori; in particular, what kind of minimal surfaces appear in their degenerations.

Finally, we mention that perturbation methods have been used in other geometric problems, for example with great success for instantons [17], [6] and mininal surfaces [24]. Closer in spirit are constant scalar curvature constructions [23]. There are other proposed constructions, for example for minimal surfaces or Einstein 4-manifolds, which in some respects are very similar to the one we discussed and which are still open.

References

[1] U. Abresch, *Constant mean curvature tori in terms of elliptic functions*, J. Reine Angew. Math. **374** (1987), 169–192.

[2] A. D. Alexandrov, *Uniqueness theorems for surfaces in the large I*, Vestnik Leningrad Univ. Math. **11** (1956), 5–17.

[3] L. Barbosa and M. doCarmo, *Stability of hypersurfaces with constant mean curvature*, Math. Z. **185** (1984), 339–353.

[4] A. I. Bobenko, *All constant mean curvature tori in* $\mathbf{R}^3$, S^3, H^3 *in terms of theta-functions*, Math. Ann. **290** (1991), 209–245.

[5] D. Delaunay, *Sur la surface de révolution dont la courbure moyenne est constante*, J. Math. Pures Appl. (4) **6** (1841), 309–320.

[6] S. K. Donaldson and P. B. Kronheimer, The Geometry of Four-manifolds, Clarendon Press, Oxford, 1990.

[7] Nick Ercolani, Horst Knörrer, and Eugene Trubowitz, *Hyperelliptic curves that generate constant mean curvature tori in* $\mathbf{R}^3$, The Verdier Memorial Conference on Integrable Systems: Actes du colloque international de Luminy (1991), ed. by Olivier Babelon, Pierre Cartier, and Yvette Kosmann-Schwarzbach, Birkhäuser, Basel and Boston.

[8] K. Grosse-Brauckmann, *New surfaces of constant mean curvature*, Math. Z. **214** (1993), 527–565.

[9] H. Hopf, Lectures on differential geometry in the large, Notes by John Gray, Stanford University, Stanford, CA.

[10] Wu-Yi Hsiang, *Generalized rotational hypersurfaces of constant mean curvature in the Euclidean spaces, I.*, J. Differential Geom. **17** (1982), 337–356.

[11] J. H. Jellet, *Sur la surface dont la coubure moyenne est constante*, J. Math. Pures Appl. (9) **18** (1853), 163–167.

[12] N. Kapouleas, *Constant mean curvature surfaces in Euclidean three-space*, Bull. Amer. Math. Soc. **17,** (1987), 318–320.

[13] ———, *Complete constant mean curvature surfaces in Euclidean three-space*, Ann. of Math. (2) **131** (1990), 239–330.

[14] ———, *Compact constant mean curvature surfaces in Euclidean three-space*, J. Differential Geom. **33** (1991), 683–715.

[15] ———, *Constant mean curvature surfaces constructed by fusing Wente tori*, Proc. Nat. Acad. Sci. USA **89** (1992), 5695–5698.

[16] ———, *Constant mean curvature surfaces constructed by fusing Wente tori*, Invent. Math., to appear.

[17] H. Karcher, *The triply periodic minimal surfaces of A. Schoen and their constant mean curvature companions*, Man. Math. **64** (1989), 291–357.

[18] N. Korevaar and R. Kusner, *The structure of complete embedded surfaces with constant mean curvature*, J. Differential Geom. **30** (1989).

[19] N. Korevaar, R. Kusner, and B. Solomon, *The structure of complete embedded surfaces with constant mean curvature*, J. Differential Geom. **30** (1989), 465–503.

[20] H. B Lawson, Jr., *Complete minimal surfaces in* S^3, Ann. of Math. (2) **90** (1970), 335–374.

[21] W. H. Meeks, III, *The topology and geometry of embedded surfaces of constant mean curvature*, J. Differential Geom. **27** (1988), 539–552.

[22] U. Pinkall and I. Sterling, *On the classification of constant mean curvature tori*, Ann. of Math. (2) **130** (1989), 407–451.

[23] R. Schoen, *The existence of weak solutions with prescribed singular behavior for a conformally invariant scalar equation*, Comm. Pure Appl. Math. **XLI** (1988), 317–392.

[24] N. Smale, *A bridge principle for minimal and constant mean curvature submanifolds in* $\mathbb{R}^N$, Invent. Math. **90** (1987), 505–549.

[25] C. H. Taubes, *Self-dual Yang-Mills connections on non-self-dual 4-manifolds*, J. Differential Geom. **17** (1982), 139–170.

[26] R. Walter, *Explicit examples to the H-Problem of Heinz Hopf*, Geom. Dedicata **23** (1987), 187–213.

[27] H. C. Wente, *Counterexample to a conjecture of H. Hopf*, Pacific J. Math. **121** (1986), 193–243.

Intersection Pairings on Quotients and Moduli Spaces, and Witten's Nonabelian Localization

FRANCES KIRWAN

Balliol College, Oxford OX1 3BJ, United Kingdom

Many moduli spaces in complex algebraic geometry can be expressed as quotients, in the sense of Mumford's geometric invariant theory [18], of nonsingular complex projective varieties X by actions of complex reductive groups G. Any such quotient can also be identified with a symplectic quotient (or Marsden-Weinstein reduction) of the variety X by a maximal compact subgroup K of the reductive group G [14], [18], [19]. This symplectic quotient is $\mu^{-1}(0)/K$, where $\mu : X \to \mathbf{k}^*$ is a moment map for the action of K on X equipped with a suitable symplectic form.

Suppose now that X is any compact symplectic manifold acted on by a compact connected Lie group K. If the action of K is Hamiltonian (in other words, if there is a moment map $\mu : X \to \mathbf{k}^*$), then we may form the symplectic quotient $\mu^{-1}(0)/K$. The inclusion map $i_0 : \mu^{-1}(0) \to X$ induces a ring homomorphism

$$i_0^* : H_K^*(X) \to H_K^*(\mu^{-1}(0))$$

on equivariant cohomology (we shall consider only cohomology with complex coefficients throughout). Using Morse theory and the gradient flow of the function $\|\mu\|^2 : X \to \mathbb{R}$, it is proved in [14] that the map i_0^* is *surjective.*

Suppose in addition that 0 is a *regular value* of the moment map μ. This assumption is equivalent to the assumption that the stabilizer K_x of x under the action of K on X is finite for every $x \in \mu^{-1}(0)$, and it implies that the quotient $\mu^{-1}(0)/K$ is an orbifold, or V-manifold, which inherits a symplectic form ω_0 from the symplectic form ω on X. It also implies that there is a canonical isomorphism $H^*(\mu^{-1}(0)/K) \to H_K^*(\mu^{-1}(0))$ (because we are only considering cohomology with complex coefficients). Composing the inverse of this isomorphism with i_0^*, we have a surjective ring homomorphism

$$H_K^*(X) \to H^*(\mu^{-1}(0)/K).$$

Henceforth, if $\eta \in H_K^*(X)$ we shall denote its image in $H^*(\mu^{-1}(0)/K)$ by η_0.

The image of a set of generators of $H_K^*(X)$ is therefore a set of generators of $H^*(\mu^{-1}(0)/K)$. (In fact generators of $H_K^*(X)$ are given by generators of $H^*(BK)$ together with extensions to $H_K^*(X)$ of generators of $H^*(X)$, because the spectral sequence of the fibration $X \times_K EK \to BK$ degenerates [14].) It follows that if the kernel of the map $\eta \mapsto \eta_0$ is known then generators and relations for the cohomology ring $H^*(\mu^{-1}(0)/K)$ can be determined from generators and relations in $H_K^*(X)$ (cf. [4], [16]).

Proceedings of the International Congress
of Mathematicians, Zürich, Switzerland 1994

One way to determine this kernel is based on the observation that because $H^*(\mu^{-1}(0)/K)$ satisfies Poincaré duality, a cohomology class $\eta \in H_K^*(X)$ maps to 0 in $H^*(\mu^{-1}(0)/K)$ if and only if for all $\zeta \in H_K^*(X)$ the intersection pairings

$$\eta_0\zeta_0[\mu^{-1}(0)/K] = (\eta\zeta)_0[\mu^{-1}(0)/K],$$

given by evaluating the product of η_0 and ζ_0 on the fundamental class of $\mu^{-1}(0)/K$, are zero. It therefore suffices to determine these intersection pairings.

There is a natural pushforward map $\int_X : H_K^*(X) \to H_K^*$. Here H_K^* is the equivariant cohomology of a point, which can be identified with the space of K-invariant polynomials on the Lie algebra **k**. This pushforward map can be thought of as integration over X. If T is a compact connected *abelian* group (i.e. a torus) and $\zeta \in H_T^*(X)$, there is a formula (the *abelian localization theorem*) for $\int_X \zeta$ in terms of the restriction of ζ to the components of the fixed point set for the action of T. It says that

$$\int_X \zeta = \sum_{F \in \mathcal{F}} \int_F \frac{i_F^* \zeta}{e_F},$$

where $\mathcal{F}$ is the set of components of the fixed point set of T, and if $F \in \mathcal{F}$ then e_F is the equivariant Euler class of the normal bundle to F in X and $i_F : F \to X$ is the inclusion. The right-hand side is to be interpreted as a rational function on the Lie algebra **t**. This formula was first proved by Berline and Vergne [3], and Atiyah and Bott [2] subsequently gave a cohomological proof motivated by the Duistermaat-Heckman theorem [5] on the pushforward of the symplectic or Liouville measure under the moment map. For a general compact Lie group K with maximal torus T there is a canonical map $H_K^*(X) \to H_T^*(X)$, and we may apply the abelian localization theorem to the image in $H_T^*(X)$ of any $\zeta \in H_K^*(X)$.

Witten in Section 2 of [22] gives a *nonabelian localization theorem*, which applies to actions of arbitrary compact connected Lie groups. This interprets the evaluation $\eta_0[\mu^{-1}(0)/K]$ of η_0 on the fundamental class of the quotient in terms of data on X. Let s be the dimension of K, and let $\| \ \|$ be the norm induced by a fixed invariant inner product $\langle \cdot, \cdot \rangle$ on **k**, which we shall use throughout to identify $\mathbf{k}^*$ with **k**. For $\epsilon > 0$ and $\zeta \in H_K^*(X)$, Witten defines an integral

$$\mathcal{I}^\epsilon(\zeta) = \frac{1}{(2\pi i)^s \operatorname{vol} K} \int_{\phi \in \mathbf{k}} [d\phi] e^{-\epsilon\|\phi\|^2/2} \int_X \zeta(\phi)$$

and he expresses it as a sum of local contributions. In fact it reduces to a sum of integrals localized around the critical set of the function $\|\mu\|^2$ on X. This critical set can be expressed as a disjoint union of closed subsets C_β of X indexed by a finite subset $\mathcal{B}$ of the Lie algebra **t** of the maximal torus T of K, which is explicitly known in terms of the moment map μ_T for the action of T on X [14]. If $\beta \in \mathcal{B}$ then the critical subset C_β is of the form $C_\beta = K(Z_\beta \cap \mu^{-1}(\beta))$, where Z_β is a union of connected components of the fixed point set of the subtorus of T generated by β. The subset $\mu^{-1}(0)$ on which $\|\mu\|^2$ takes its minimum value is C_0. Witten's result can then be expressed in the form

THEOREM 1 *If K acts freely on $\mu^{-1}(0)$, then*

$$\mathcal{I}^\epsilon(\zeta) = \zeta_0 e^{\epsilon\Theta}[\mu^{-1}(0)/K] + \sum_{\beta\in\mathcal{B}-\{0\}} \int_{U_\beta} \zeta'_\beta.$$

Here, the class $\Theta \in H^4(\mu^{-1}(0)/K) \cong H^4_K(\mu^{-1}(0))$ is the image under the natural map $H^*_K \to H^*_K(\mu^{-1}(0))$ of the class in H^4_K represented by the polynomial function $-\|\phi\|^2/2$ of $\phi \in \mathbf{k}$. The U_β are open neighbourhoods in X of the nonminimal critical subsets C_β of the function $\|\mu\|^2$ and the ζ'_β are certain differential forms on U_β obtained from ζ.

In the case of the formal equivariant cohomology class $\zeta = \eta \exp i\bar{\omega}$, where $\eta \in H^*_K(X)$ and $\bar{\omega}(\phi) = \omega + \mu(\phi)$ is the standard extension of the symplectic form ω to an element of $H^2_K(X)$, Witten's results imply the following estimate on the growth of the terms $\int_{U_\beta} \zeta'_\beta$ as $\epsilon \to 0$:

THEOREM 2 *Suppose $\zeta = \eta \exp i\bar{\omega}$ for some $\eta \in H^*_K(X)$. If $\beta \in \mathcal{B} - \{0\}$ then $\int_{U_\beta} \zeta'_\beta = e^{-\|\beta\|^2/2\epsilon} h_\beta(\epsilon)$, where $\|\beta\|^2$ is the value of $\|\mu\|^2$ on the critical set C_β and $|h_\beta(\epsilon)|$ is bounded by a polynomial in ϵ^{-1}.*

Thus, one can think of $\epsilon > 0$ as a small parameter, and then use the asymptotics of the integral $\mathcal{I}^\epsilon$ over X to calculate the pairings $\eta_0 e^{\epsilon\Theta} e^{i\omega_0}[\mu^{-1}(0)/K]$, because the terms in Theorem 1 corresponding to the other critical subsets of $\|\mu\|^2$ vanish exponentially fast as $\epsilon \to 0$. Notice that when $\zeta = \eta \exp i\bar{\omega}$, the vanishing of μ on $\mu^{-1}(0)$ means that $\zeta_0 = \eta_0 \exp i\omega_0$, where ω_0 is the symplectic form induced by ω on $\mu^{-1}(0)/K$.

Of course, if $\eta \in H^d_K(X)$, where d is the real dimension of the quotient $\mu^{-1}(0)/K$, then

$$\eta_0 e^{i\omega_0} e^{\epsilon\Theta}[\mu^{-1}(0)/K] = \eta_0[\mu^{-1}(0)/K].$$

Thus, the intersection pairings we wished to determine can be calculated in this way.

Witten's argument runs along the following lines. He introduces a K-invariant 1-form l on X, and shows that $\mathcal{I}^\epsilon(\zeta) = \mathcal{I}^\epsilon(\zeta \exp sDl)$, where D is the differential in equivariant cohomology and $s \in \mathbb{R}^+$. He then does the integral over $\phi \in \mathbf{k}$ and shows that in the limit as $s \to \infty$, this integral vanishes over any region of X where $l(V^j) \neq 0$ for at least one of the vector fields V^j, $j = 1, \ldots, s$, given by the infinitesimal action of a basis of $\mathbf{k}$ on X indexed by j. Thus, after integrating over $\phi \in \mathbf{k}$, the limit as $s \to \infty$ of $\mathcal{I}^\epsilon(\zeta)$ reduces to a sum of contributions from sets where $l(V^j) = 0$ for all the V^j. In the case when X is a symplectic manifold and the action of K is Hamiltonian, Witten chooses $l(Y) = d\|\mu\|^2(JY)$, where J is a K-invariant almost complex structure on X. Thus, $\mathcal{I}^\epsilon(\zeta)$ reduces to a sum of contributions from the critical sets of $\|\mu\|^2$.

In the case when X is a symplectic manifold and $\zeta = \eta \exp i\bar{\omega}$ for any $\eta \in H^*_K(X)$, there is a way to prove a variant of these results that bypasses the analytical difficulties relating to integrals over neighbourhoods of the critical sets, and reduces the result to fairly well-known facts about Hamiltonian group actions on symplectic manifolds [10]. The key steps in the proof are as follows. One assumes that 0 is a regular value of μ, or equivalently that K acts on $\mu^{-1}(0)$ with finite stabilizers. One observes that integrals of K-invariant functions on $\mathbf{k}$ may be replaced by integrals over the Lie algebra $\mathbf{t}$ of the maximal torus. Then, applying properties of the Fourier transform, one rewrites $\mathcal{I}^\epsilon$ as the integral over $\mathbf{t}^*$ of a Gaussian G_ϵ (which is a constant multiple of $y \mapsto \epsilon^{-s/2}e^{-\|y\|^2/(2\epsilon)}$) multiplied by a function $Q = D_\varpi R$, where R is piecewise polynomial and D_ϖ is a differential operator on $\mathbf{t}^*$:

$$\mathcal{I}^\epsilon = \int_{y\in\mathbf{t}^*} G_\epsilon(y)Q(y)[dy].$$

The function Q is obtained by combining the abelian localization theorem above with facts on Fourier transforms of the sorts of functions that arise in the formula for the pushforward.

The function Q is smooth in a neighbourhood of the origin when 0 is a regular value of μ, so there is a polynomial $Q_0 = D_\varpi R_0$ that is equal to Q near 0. It turns out that the cohomological expression $e^{\epsilon\Theta}e^{i\omega_0}[\mu^{-1}(0)/K]$ is obtained as the integral over $\mathbf{t}^*$ of the Gaussian G_ϵ multiplied not by Q but by the polynomial Q_0:

$$e^{\epsilon\Theta}e^{i\omega_0}[\mu^{-1}(0)/K] = \int_{y\in\mathbf{t}^*} G_\epsilon(y)Q_0(y)[dy].$$

This result can be deduced from an equivariant normal form for ω in a neighbourhood of $\mu^{-1}(0)$, given in [8] as a consequence of the coisotropic embedding theorem.

To obtain an analogue of Witten's estimate for the asymptotics of the difference $\mathcal{I}^\epsilon - e^{\epsilon\Theta}e^{i\omega_0}[\mu^{-1}(0)/K]$ as $\epsilon \to 0$, one then writes

$$\mathcal{I}^\epsilon - e^{\epsilon\Theta}e^{i\omega_0}[\mu^{-1}(0)/K] = \int_{y\in\mathbf{t}^*} G_\epsilon(y)D_\varpi(R - R_0)(y)[dy].$$

Here, $R - R_0$ is piecewise polynomial and supported *away* from 0. By studying the minimum distances from 0 in the support of $R - R_0$ an estimate similar to Witten's can be obtained.

The same methods lead to the following explicit formula (the residue formula, Theorem 8.1 of [10])), in terms of the components F of the fixed point set of T on X, for the evaluation of a class $\eta_0 \in H^*(\mu^{-1}(0)/K)$ on the fundamental class $[\mu^{-1}(0)/K]$, when η_0 comes from a class $\eta \in H^*_K(X)$.

THEOREM 3 (RESIDUE FORMULA) *Let $\eta \in H^*_K(X)$ induce $\eta_0 \in H^*(\mu^{-1}(0)/K)$. Then we can equate*

$$\eta_0 e^{i\omega_0}[\mu^{-1}(0)/K]$$

with

$$\frac{(-1)^{n_+}}{(2\pi)^{s-l}|W|\operatorname{vol}(T)}\operatorname{Res}\left(\varpi^2(\psi)\sum_{F\in\mathcal{F}} e^{i\mu_T(F)(\psi)}\int_F \frac{i_F^*(\eta(\psi)e^{i\omega})}{e_F(\psi)}[d\psi]\right).$$

In this formula, s and l are the dimensions of K and its maximal torus T, and μ_T is the moment map for the T-action. The Weyl group is denoted by W and n_+ is the number of positive roots of K, while $\varpi(\psi) = \prod_{\gamma>0} \gamma(\psi)$ is the product of the positive roots. $\mathcal{F}$ is the set of components of the fixed point set of the maximal torus T on X. If $F \in \mathcal{F}$ then i_F is the inclusion of F in X and e_F is the equivariant Euler class of the normal bundle to F in X.

Here, the residue map Res (whose domain is a certain class of meromorphic differential forms on $\mathbf{t} \otimes \mathbb{C}$) is a linear map, but in order to apply it to the individual terms in the statement of the residue theorem some choices must be made. The choices do not affect the residue of the whole sum. When K is $SU(2)$ or $SO(3)$ acting effectively on X, the formula becomes

$$\eta_0 e^{i\omega_0}[\mu^{-1}(0)/K] = -\frac{1}{2}\mathrm{Res}_0\left(\psi^2 \sum_{F\in\mathcal{F}_+} e^{i\mu_T(F)(\psi)} \int_F \frac{i_F^*(\eta(\psi)e^{i\omega})}{e_F(\psi)}\right),$$

where Res_0 denotes the coefficient of $1/\psi$, and $\mathcal{F}_+$ is the subset of $\mathcal{F}$ consisting of those components F of the T fixed point set for which $\mu_T(F) > 0$. Results for the case when $K = S^1$, which are related to the residue theorem of [10], may be found in the papers of Kalkman [13] and Wu [23].

The case when 0 is not a regular value of the moment map μ and $\mu^{-1}(0)/K$ is singular can be treated using intersection homology, at least when X is a nonsingular complex projective variety [12]. There is now a natural surjection [15]

$$H_K^*(X) \to IH^*(\mu^{-1}(0)/K)$$

and a modification of the residue formula applies to the image $\eta_0 \in IH^*(\mu^{-1}(0)/K)$ of any $\eta \in H_K^*(X)$.

When X is a Kähler manifold with a holomorphic line bundle $\mathcal{L}$ on which K acts, and whose first Chern class is ω, the residue formula is related to an old result of Guillemin and Sternberg [7]. When K acts freely on $\mu^{-1}(0)$ the quotient $\mu^{-1}(0)/K$ is a Kähler manifold and the line bundle $\mathcal{L}$ on X induces a holomorphic line bundle $\mathcal{L}_K$ on $\mu^{-1}(0)/K$. Their result identifies the Riemann-Roch numbers $RR^K(\mathcal{L})$ and $RR(\mathcal{L}_K)$ defined by

$$RR^K(\mathcal{L}) = \sum_{j\geq 0}(-1)^j \dim H^j(X, \mathcal{L})^K$$

and

$$RR(\mathcal{L}_K) = \sum_{j\geq 0}(-1)^j \dim H^j(\mu^{-1}(0)/K, \mathcal{L}_K).$$

Recently Guillemin [6], Meinrenken [17], and independently Vergne [21] have used the ideas behind the residue formula to prove that $RR^K(\mathcal{L}) = RR(\mathcal{L}_K)$ in a wider class of situations where X need not have a K-invariant Kähler structure. A special case is discussed in [11].

Witten applies the ideas in [22] described above in an infinite-dimensional setting to give formulas for intersection pairings of cohomology classes in the moduli spaces $\mathcal{M}(n,d)$ of semistable holomorphic bundles of coprime rank n and degree d over a fixed Riemann surface Σ of genus g. These formulas agree with the calculations of Thaddeus [20] in the rank two case. In order to describe them we need a set of generators for the cohomology ring of $\mathcal{M}(n,d)$. Following [1] we take as generators the slant products

$$a_r \in H^{2r}(\mathcal{M}(n,d)),$$

$$b_r^j \in H^{2r-1}(\mathcal{M}(n,d))$$

and

$$f_r^j \in H^{2r-2}(\mathcal{M}(n,d))$$

of the Chern classes $c_r \in H^{2r}(\Sigma \times \mathcal{M}(n,d))$ with standard bases of $H^0(\Sigma)$, $H^1(\Sigma)$, and $H^2(\Sigma)$, respectively.

In [22] Witten obtains formulas for generating functionals from which it is possible to extract all the intersection pairings

$$\prod_{r=2}^{n} a_r^{m_r} f_r^{n_r} \prod_{k_r=1}^{2g} (b_r^{k_r})^{p_{r,k_r}} [\mathcal{M}(n,d)].$$

For example, if the m_r are sufficiently small to ensure convergence of the sum, he obtains

$$\prod_{r=2}^{n} a_r^{m_r} \exp f_2[\mathcal{M}(n,d)] = \sum_{\lambda} \frac{\operatorname{vol}(K)^{2g-2} c^{-\lambda} \prod_{r=2}^{n} \tau_r(\lambda)^{m_r}}{|\Pi_1(K)|(2\pi)^{(2g-2)\dim K} \mathcal{D}(\lambda)^{2g-2}},$$

where $K = SU(n)$ and $\mathcal{D}$ is the product of the positive roots of K, and τ_r is the rth elementary symmetric polynomial. The sum runs over those elements of the weight lattice that are in the interior of the fundamental Weyl chamber of K. The element

$$c = e^{2\pi i d/n} \operatorname{diag}(1, \dots, 1)$$

is a generator of the centre of K, and if λ is a weight then c^λ is defined to be $e^{i\lambda(\tilde{c})}$ for any $\tilde{c}$ in the Lie algebra of the maximal torus of K such that $\exp \tilde{c} = c$. There are similar formulas for more general pairings.

The moduli spaces of semistable holomorphic bundles over a fixed compact Riemann surface can be regarded as symplectic quotients of finite-dimensional group actions on "extended moduli spaces" [9] and algebro-geometric versions of these, as well as the infinite-dimensional actions used by Witten. It seems that this leads to another derivation of Witten's formulas for pairings on the moduli spaces when the rank and degree of the bundles are coprime, and also to formulas for intersection cohomology when the rank and degree are not coprime [12].

References

[1] M. F. Atiyah and R. Bott, *The Yang-Mills equations over Riemann surfaces*, Philos. Trans. Roy. Soc. London Ser. A 308 (1982) 523–615.

[2] M. F. Atiyah and R. Bott, *The moment map and equivariant cohomology*, Topology 23 (1984) 1–28.

[3] N. Berline and M. Vergne, *Classes caractéristiques équivariantes. Formules de localisation en cohomologie équivariante*, C. R. Acad. Sci. Paris 295 (1982) 539–541.

[4] M. Brion, *Cohomologie équivariante des points semi-stables*, J. Reine Angew. Math. 421 (1991) 125–140.

[5] J. J. Duistermaat and G. Heckman, *On the variation in the cohomology of the symplectic form of the reduced phase space*, Invent. Math. 69 (1982) 259–268; Addendum, 72 (1983) 153–158.

[6] V. Guillemin, *Reduced phase spaces and Riemann-Roch*, in: Lie groups and geometry, in honor of B. Kostant, Birkhäuser, Basel and Boston (Progr. Math.), 1995.

[7] V. Guillemin and S. Sternberg, *Geometric quantization and multiplicities of group representations*, Invent. Math. 67 (1982) 515–538.

[8] V. Guillemin and S. Sternberg, Symplectic Techniques in Physics, Cambridge University Press, London and New York, 1984.

[9] L. Jeffrey, *Extended moduli spaces of flat connections on Riemann surfaces*, Math. Ann. 298 (1994) 667–692.

[10] L. Jeffrey and F. Kirwan, *Localisation for nonabelian group actions*, to appear in Topology.

[11] L. Jeffrey and F. Kirwan, *A note on localization and the Riemann-Roch formula*, preprint.

[12] L. Jeffrey and F. Kirwan, work in progress.

[13] J. Kalkman, *Cohomology rings of symplectic quotients*, to appear in J. Reine Angew. Math.

[14] F. Kirwan, *Cohomology of quotients in symplectic and algebraic geometry*, Princeton Univ. Press, Princeton, NJ, 1984.

[15] F. Kirwan, *Rational intersection cohomology of quotient varieties*, Invent. Math. 86 (1986) 471–505.

[16] F. Kirwan, *The cohomology rings of moduli spaces of vector bundles over Riemann surfaces,* J. Amer. Math. Soc. 5 (1992) 853–906.

[17] E. Meinrenken, *On Riemann-Roch formulas for multiplicities*, MIT preprint (1994).

[18] D. Mumford, J. Fogarty and F. Kirwan, Geometric Invariant Theory, 3rd edition, Springer, Berlin and New York, 1993.

[19] L. Ness, *A stratification of the null cone via the moment map*, Amer. J. Math. 106 (1984) 1281.

[20] M. Thaddeus, *Conformal field theory and the cohomology of the moduli space of stable bundles*, J. Differential Geom. 35 (1992) 131–149.

[21] M. Vergne, *Quantification géometrique et multiplicités*, to appear in C.R.A.S. and *Geometric quantization and multiplicities I*, DMI.ENS preprint (1994).

[22] E. Witten, *Two dimensional gauge theories revisited*, J. Geom. Phys. 9 (1992) 303–368.

[23] S. Wu, *An integral formula for the squares of moment maps of circle actions*, preprint hep-th/9212071.

Anti-Self-Dual Metrics and Kähler Geometry

CLAUDE LEBRUN*

State University of New York
Stony Brook, NY 11794, USA

1. Four-Dimensional Geometry

The fact that the Lie group $SO(4)$ is nonsimple gives 4-dimensional geometry an extremely distinctive flavor. Indeed, the choice of a Riemannian metric g on an oriented 4-manifold M splits the bundle of 2-forms

$$\bigwedge^2 = \bigwedge^+ \oplus \bigwedge^- \tag{1}$$

into the rank-3 bundles of *self-dual* and *anti-self-dual* 2-forms, respectively defined as the ± 1-eigenspaces of the Hodge star operator $\star : \bigwedge^2 \to \bigwedge^2$; this just reflects the fact that the adjoint representation of $SO(4)$ on the skew (4×4)-matrices is the sum of two 3-dimensional representations, as indicated by the Lie algebra isomorphism $so(4) \cong so(3) \oplus so(3)$. The decomposition (1) is *conformally invariant*, in the sense that it is unchanged if g is replaced by ug for any positive function u; but reversing the orientation of M interchanges the bundles $\bigwedge^\pm$.

The central importance of (1) stems from the fact that curvature tensors are bundle-valued 2-forms and thus, on a Riemannian 4-manifold, can be broken up into self-dual and anti-self-dual parts. For the Riemannian curvature of our metric g, however, one can go even further; using the metric to reinterpret the curvature tensor as the *curvature operator* endomorphism $\mathcal{R}$ of the bundle of 2-forms, it is apparent that $\mathcal{R} : \bigwedge^+ \oplus \bigwedge^- \to \bigwedge^+ \oplus \bigwedge^-$ may be considered as consisting of more primitive pieces

$$\mathcal{R} = \left(\begin{array}{c|c} W_+ + \frac{s}{12} & B \\ \hline B^* & W_- + \frac{s}{12} \end{array}\right), \tag{2}$$

where trace $W_\pm = 0$. Whereas s is just the scalar curvature, and $2B$ is just the trace-free Ricci curvature, the remaining two tensors $W_\pm$ may seem less familiar. However, their sum $W = W_+ + W_-$ is exactly the classical *Weyl curvature*, which vanishes identically if and only if g is locally conformally flat; and, as you might therefore expect, $W_\pm$ are both conformally invariant.

*Supported in part by NSF grant DMS-9204093.

Proceedings of the International Congress
of Mathematicians, Zürich, Switzerland 1994

2. Anti-Self-Dual Manifolds

Familiar or not, W_+ plays an important role in 4-dimensional geometry, as will now be explained. For any oriented Riemannian 4-manifold (M, g) one can construct an associated almost-complex manifold (Z, J), the underlying real manifold Z of which is the S^2-bundle $S(\bigwedge^+)$ of length-$\sqrt{2}$ self-dual 2-forms. The automorphism J of TZ, which satisfies $J^2 = -1$ and so makes TZ into a complex vector bundle, preserves the decomposition of TZ into horizontal and vertical components with respect to the Levi-Cività connection:

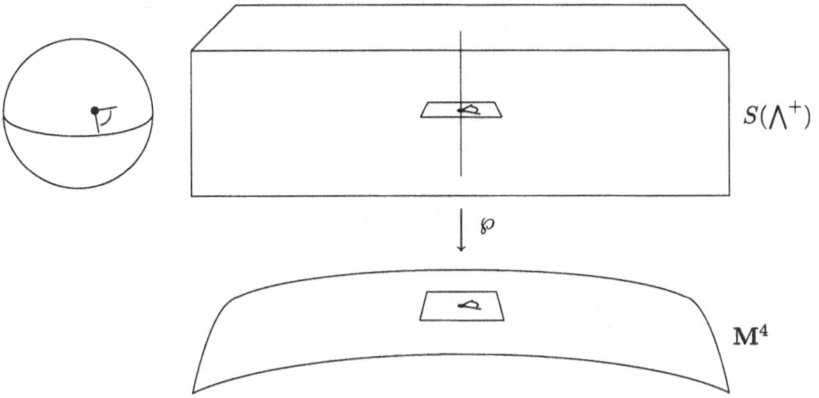

On the tangent spaces of the fiber S^2's, J simply acts by rotation by $-90°$. Meanwhile, in the horizontal subspace at $\phi \in S(\bigwedge^+)$, which we identify with $T_{\wp(\phi)}M$ via the derivative of the canonical projection $\wp$, J acts as $g^{-1} \circ \phi$. Although it is hardly obvious, this almost-complex structure J is actually conformally invariant — despite the fact that replacing g with ug changes the horizontal subspaces on $Z = (\bigwedge^+ - 0)/\mathbb{R}^+$.

Although the action of J on each tangent space is isomorphic to the action of $i = \sqrt{-1}$ on $\mathbb{C}^3$ by multiplication, we may not be able to find $\mathbb{C}^3$-valued coordinate charts on Z that put J in this standard form *simultaneously* at all points of their domains. In general, the obstruction [17] to finding such complex charts on an almost-complex manifold is encoded by a vector-valued 2-form called the Nijenhuis tensor; but in our case [1], [18] the Nijenhuis tensor of (Z, J) is determined by the self-dual part W_+ of the Weyl curvature of (M, g). When this vanishes, the adapted complex charts are then interrelated by biholomorphisms, and Z acquires the structure of a complex manifold:

THEOREM 1 (PENROSE/ATIYAH-HITCHIN-SINGER) *The almost-complex manifold (Z, J) is a complex 3-manifold iff $W_+ = 0$. Moreover, a complex 3-manifold arises by this construction iff it admits a fixed-point-free anti-holomorphic involution $\sigma : Z \to Z$ and a foliation by σ-invariant rational curves $\mathbb{CP}_1$, each of which has normal bundle $\mathcal{O}(1) \oplus \mathcal{O}(1)$. Finally, the complex manifold (Z, J) and the real structure σ suffice to determine the metric g on M up to conformal rescaling.*

This 4-dimensional analogue of the conformal surface/complex curve dictionary clearly merits the introduction of some terminology:

DEFINITION 1 *An oriented Riemannian 4-manifold (M, g) is called* anti-self-dual *if its Weyl curvature satisfies $W_+ = 0$.*

DEFINITION 2 *The complex 3-manifold (Z, J) associated with an anti-self-dual 4-manifold (M, g) by Theorem 1 is called the* twistor space *of M.*

Because reversing the orientation of M interchanges W_+ and W_-, there is but a looking-glass difference between anti-self-dual manifolds and *self-dual* manifolds, which by definition instead satisfy $W_- = 0$. Our preference for anti-self-duality will have certain technical advantages, however, as we now describe the geometry and topology of compact anti-self-dual manifolds.

3. Positive Scalar Curvature

Because the condition $W_+ = 0$ is conformally invariant, classification problems for anti-self-dual metrics have reasonable answers only if one either works modulo conformal rescalings $g \mapsto ug$, or else imposes some extra condition to single out preferred representatives in each conformal class. A particularly natural program of the latter type would have us insist that the metric in question have *constant scalar curvature* in addition to being anti-self-dual. For M compact, this normalization is always possible [21], and the *sign* of the resulting constant s provides an important global invariant of our anti-self-dual conformal class. To determine this sign, it is actually unnecessary to find a representative with $s =$ constant; it is quite enough to produce a representative for which s does not change sign.

This said, let's now examine some anti-self-dual 4-metrics of positive scalar curvature [13] on m-fold connected sums $\overline{\mathbb{CP}}_2 \# \cdots \# \overline{\mathbb{CP}}_2$; here $\overline{\mathbb{CP}}_2$ denotes the oriented 4-manifold obtained from $\mathbb{CP}_2$ by reversing its orientation, while the connected sum operation $\#$ is carried out by deleting balls and identifying the resulting S^3 boundaries.

The metrics in question are determined by a choice of m points $p_1, \ldots, p_m$ in hyperbolic 3-space $\mathcal{H}^3$. Given such a choice, we set $X = \mathcal{H}^3 - \{p_1, \ldots, p_m\}$, and define $V : X \to \mathbb{R}^+$ by

$$V = 1 + \sum_{j=1}^{m} \frac{1}{e^{2r_j} - 1},$$

where r_j denotes the hyperbolic distance from p_j. The latter function is a solution of the Laplace-Beltrami equation $d\star dV = 0$, and the resulting de Rham class $[\frac{1}{2\pi}\star dV]$ is, moreover, an element of the *integral* cohomology $H^2(X, \mathbb{Z}) \subset H^2(X, \mathbb{R})$. Thus, by the Chern-Weil theorem, there exists a circle bundle $P \to X$ with connection 1-form θ whose curvature is $\star dV$. Now let r denote the hyperbolic distance from any reference point. The metric

$$g = (\operatorname{sech}^2 r)\,(Vh + V^{-1}\theta^{\otimes 2}) \tag{3}$$

is then anti-self-dual, and its metric-space completion is a smooth Riemannian 4-manifold, the added points of which consist of a 2-sphere and m points $\hat{p}_1, \ldots, \hat{p}_m$. Indeed, let D^3 denote the closed unit ball in $\mathbb{R}^3$, and identify the interior of D^3 with $\mathcal{H}^3$ via the Klein projective model. Then $M = P \cup S^2 \cup \{\hat{p}_1, \ldots, \hat{p}_m\}$ can be made into a smooth 4-manifold with circle-action in such a manner that $S^2 \cup \{\hat{p}_1, \ldots, \hat{p}_m\}$ is the fixed-point set and D^3 is the orbit space. The projection to D^3 is thus as follows:

$$\begin{array}{ccccccccc} M & = & P & \cup & S^2 & \cup & \{\hat{p}_1, \ldots, \hat{p}_m\} \\ \downarrow & & \downarrow & & \downarrow & & \downarrow \\ D^3 & = & X & \cup & \partial D^3 & \cup & \{p_1, \ldots, p_m\} \end{array}$$

The metric g of equation (3) then extends to M so as to yield a compact anti-self-dual 4-manifold diffeomorphic to $\overline{\mathbb{CP}}_2 \# \cdots \# \overline{\mathbb{CP}}_2$, and the scalar curvature $s = 12V^{-1}$ of this metric is positive almost everywhere, implying that our conformal class has a representative of positive scalar curvature:

THEOREM A *For every $m \geq 0$, there exist anti-self-dual metrics with $s > 0$ on*

$$m\overline{\mathbb{CP}}_2 := \underbrace{\overline{\mathbb{CP}}_2 \# \cdots \# \overline{\mathbb{CP}}_2}_{m}.$$

When $m = 0, 1$, the constructed anti-self-dual metrics are respectively conformal to the standard symmetric-space metrics on $\mathbf{S}^4$ and $\overline{\mathbb{CP}}_2$; when $m = 2$, they coincide with the metrics discovered by Poon [20]. For related results, see [6], [7], [10].

In order to describe the twistor spaces of the above manifolds, let $\mathcal{O}(k, \ell)$ denote the unique holomorphic line-bundle over $\mathbb{CP}_1 \times \mathbb{CP}_1$ with degree k on the first factor and degree ℓ on the second. Because $\mathcal{H}^3$ may be identified with the set of unit future-pointing time-like vectors in Minkowski 4-space $\mathbb{R}^{3,1}$, the data points $p_1, \ldots, p_m \in \mathcal{H}^3 \subset \mathbb{R}^{3,1}$ determine m sections

$$\mathcal{P}_1, \ldots, \mathcal{P}_m \in \Gamma(\mathbb{CP}_1 \times \mathbb{CP}_1, \mathcal{O}(1,1)) \cong \mathbb{C}^4 = \mathbb{R}^{3,1} \otimes \mathbb{C}.$$

Let $\mathcal{B}$ denote the total space of the $\mathbb{CP}_2$-bundle

$$\mathcal{B} := \mathbb{P}(\mathcal{O}(m-1,1) \oplus \mathcal{O}(1, m-1) \oplus \mathcal{O}) \longrightarrow \mathbb{CP}_1 \times \mathbb{CP}_1 ,$$

and define an algebraic variety $\tilde{Z} \subset \mathcal{B}$ by the equation

$$xy = t^2 \prod_{j=1}^{m} \mathcal{P}_j ,$$

where $x \in \mathcal{O}(m-1, 1)$, $y \in \mathcal{O}(1, m-1)$, and $t \in \mathcal{O} = \mathcal{O}(0,0)$. The twistor space Z of the constructed metric is then obtained from $\tilde{Z}$ by making small resolutions of the singular points and blowing down the surfaces $x = t = 0$ and $y = t = 0$ to $\mathbb{CP}_1$'s. Notice that these twistor spaces are all *Moĭshezon*, meaning that they are bimeromorphic to complex projective varieties.

These examples are, in a rough sense, topologically typical:

PROPOSITION 1 *Let (M, g) be a compact anti-self-dual manifold with $s > 0$. Then the intersection form*

$$H^2(M, \mathbb{R}) \times H^2(M, \mathbb{R}) \xrightarrow{\cup} \mathbb{R}$$

is negative definite. If $\pi_1(M)$ is finite, then, as an oriented topological 4-manifold,

$$M \approx \underbrace{\overline{\mathbb{CP}}_2 \# \cdots \# \overline{\mathbb{CP}}_2}_{m}, \quad m \geq 0.$$

Proof: If ω is a self-dual 2-form on any Riemannian 4-manifold, the Weitzenböck formula [2] yields

$$\int_M \|d\omega\|^2 d\mu = \int_M \left[\frac{1}{2}\|\nabla\omega\|^2 - W_+(\omega,\omega) + \frac{s}{6}\|\omega\|^2\right] d\mu \ .$$

If $W_+ = 0$ and $s > 0$, it thus follows that every harmonic 2-form has vanishing self-dual part, and the intersection form is therefore negative definite.

If our manifold has a finite fundamental group, both M and its universal cover have $\chi + \tau = 2(1 - b_1 + b_+) = 2$, and hence $\pi_1(M) = 0$. The conclusion then follows from Donaldson's thesis [5]. □

Geometrically, however, the constructed metrics are *not* so typical; they carry nontrivial Killing fields, in contrast to their generic anti-self-dual deformations for $m \geq 3$. More importantly, the condition $s > 0$ does not impose any *a priori* restriction on the size of the fundamental group of M. Indeed, by replacing hyperbolic space $\mathcal{H}^3$ with a hyperbolic handle-body $\mathcal{H}^3/(\mathbb{Z} * \cdots * \mathbb{Z})$, the above construction yields [11], [14] positive-scalar-curvature anti-self-dual metrics on connected sums $\overline{\mathbb{CP}}_2 \# \cdots \# \overline{\mathbb{CP}}_2 \# (S^1 \times S^3) \# \cdots \# (S^1 \times S^3)$.

4. Kähler Surfaces

When the scalar curvature is positive, we've just seen that an anti-self-dual manifold must have negative-definite intersection form. In the $s = 0$ case, the same argument implies that any harmonic self-dual form ω must satisfy $\nabla\omega = 0$:

PROPOSITION 2 *Let (M, g) be an anti-self-dual 4-manifold with $s = 0$. If M has indefinite intersection form, then (M, g) is a Kähler manifold.*

Recall that a Riemannian $2n$-manifold (M^{2n}, g) is called *Kähler* if it admits an isometric almost-complex structure $J : TM \to TM$, $J^2 = -1$, $J^*g = g$, that is invariant under Riemannian parallel transport. Such an almost-complex structure is automatically integrable, and the 2-form $\omega(\cdot,\cdot) = g(J\cdot,\cdot)$, known as the *Kähler form*, is both closed and of type $(1,1)$, meaning that $J^*\omega = \omega$. Conversely, a closed $(1,1)$-form ω on a complex manifold (M, J) determines a Kähler metric by the formula $g(\cdot,\cdot) = \omega(\cdot, J\cdot)$ provided that this last expression is positive-definite.

Because the complex structure J of a Kähler manifold (M, g) satisfies $\nabla J = 0$, its curvature tensor $\mathcal{R} = \nabla \wedge \nabla$ is J-linear, and so $\mathcal{R} \in \bigwedge^{1,1} \otimes \bigwedge^{1,1}$. But for $2n = 4$ one has $\bigwedge^{1,1} = \mathbb{R}\omega \oplus \bigwedge^{-}$, and (2) this tells us that

$$\left[W_+ + \frac{s}{12}\right] = \frac{s}{16}\omega \otimes \omega \ .$$

As a converse to Proposition 2, one thus observes [8] that *a Kähler manifold of complex dimension 2 is anti-self-dual iff its scalar curvature is zero.* This makes it convenient to introduce the following definition:

DEFINITION 3 *A compact Kähler manifold (M, J, g) of complex dimension 2 and $s = 0$ will be called a* scalar-flat Kähler surface.

Scalar-flat Kähler surfaces have a number of other interesting features, any one of which would justify their study. First, they are solutions of the Einstein-Maxwell equations

$$R_{ab} = 2F_{ac}F_b{}^c - \frac{1}{2}F_{cd}F^{cd}g_{ab} ,$$

where the harmonic 2-form F representing the electromagnetic field is half the sum of the Kähler and Ricci forms. Second, they are L^2-critical; in fact, such metrics are absolute minima of the functional

$$\int_M \|\mathcal{R}\|^2 \, d\mu = -8\pi^2(3\tau + \chi) + \int_M (4\|W_+\|^2 + \frac{s^2}{12}) \, d\mu$$

on the space of all Riemannian metrics on M. Finally, they are solutions of Calabi's extremal-Kähler-metric problem [4].

Scalar-flat Kähler surfaces also exhibit some remarkable properties from the twistor-theoretic perspective, and this will play a key role below. The Kähler form ω obviously gives us a section of $\wp : S(\bigwedge^+) \to M$, and a quick inspection of our construction of the twistor complex structure shows that this map is a holomorphic embedding of M into Z; in particular, the twistor space of a scalar-flat Kähler surface contains a complex submanifold D that meets every fiber of $\wp$ in exactly one point. Conversely [3], [19], if the twistor space Z of a compact anti-self-dual 4-manifold (M, g) with $b_1(M)$ even contains a complex hypersurface D with this property, the metric g is globally conformal to a scalar-flat Kähler metric.

Let us now observe that one can construct many explicit scalar-flat Kähler surfaces by a modification [14] of the "hyperbolic ansatz" of Section 3. Indeed, let Σ be a compact Riemann surface of genus $\mathbf{g} \geq 2$, and observe that $Y = \mathbb{R} \times \Sigma$ can be given a curvature -1 metric by setting $h = dt^2 + (\cosh^2 t)h_\Sigma$, where t is the tautological coordinate on $\mathbb{R}$ and h_Σ is the unique curvature -1 Hermitian metric on Σ. Let $p_1, \ldots, p_m \in Y$ be given, let G_j be the corresponding hyperbolic Green's functions, and let

$$V = 1 + \sum_{j=1}^{m} G_j$$

on $X = Y - \{p_1, \ldots, p_m\}$. If there's a circle-bundle P with connection 1-form θ whose curvature is $\star dV$, then the completion M of (P, g), where

$$g = (\operatorname{sech}^2 t)(Vh + V^{-1}\theta^{\otimes 2}) ,$$

will be a scalar-flat Kähler surface, and the canonical projection $M \to \Sigma$ will be a holomorphic map with generic fiber $\mathbb{CP}_1$; the only catch lies in showing that $[\frac{1}{2\pi}\star dV]$ is an integral cohomology class. This integrality condition is actually non-trivial, but for all $m \neq 1$ there are configurations $\{p_1, \ldots, p_m\}$ that satisfy it, and for these values of m one thus constructs scalar-flat Kähler surfaces diffeomorphic to $(\Sigma \times S^2) \# m\overline{\mathbb{CP}}_2$.

These examples are all *ruled surfaces*, meaning that they are obtained from holomorphic $\mathbb{CP}_1$-bundles over a complex curve Σ by *blowing up* — that is, successively replacing certain points by $\mathbb{CP}_1$'s of self-intersection -1. In particular, deforming these explicit solutions allows one to prove [15]

Proposition 3 *Let Σ be any compact complex curve of genus ≥ 2. If $\Sigma \times \mathbb{CP}_1$ is blown up at $m \geq 2$ points in general position, the resulting complex surface admits scalar-flat Kähler metrics.*

On the other hand, because the total scalar curvature of a Kähler surface is given by $\int_M s\, d\mu = 4\pi c_1 \cup [\omega]$, the Enriques-Kodaira classification of surfaces implies the following [23]:

Proposition 4 (Yau) *Let (M, J) be a compact complex surface that carries a Kähler class $[\omega]$ for which $\int s\, d\mu \geq 0$. Then either M is ruled, or else M is covered by a complex torus or K3 surface.*

As the nonruled cases appearing here admit Ricci-flat Kähler metrics [24], Proposition 3 begins to seem rather satisfying, especially because any ruled surface can be obtained from a product surface $\Sigma \times \mathbb{CP}_1$ by blowing up and down. What we still lack, though, are examples of scalar-flat Kähler metrics on ruled surfaces that fiber over curves of genus ≤ 1. Fortunately, these can also be shown to exist:

Proposition 5 *If $\mathbb{CP}_1 \times \mathbb{CP}_1$ is blown up at 13 suitably chosen points, the resulting complex surface admits scalar-flat Kähler metrics.*

Proposition 6 *Let $\mathbb{E} \approx T^2$ be any curve of genus 1. If $\mathbb{CP}_1 \times \mathbb{E}$ is blown up at 6 suitably chosen points, the resulting complex surface admits scalar-flat Kähler metrics.*

These metrics are produced by the following quotient construction [12]:

Theorem 2 *Let (N, J_N, g_N) be a nonminimal compact complex surface with scalar-flat Kähler metric, and let $\Phi : N \to N$, $\Phi^2 = \mathbf{1}$, be a holomorphic isometry with only isolated fixed points. Let (M, J_M) be obtained from N/Φ by replacing each singular point with a $\mathbb{CP}_1$ of self-intersection -2. Then there exist scalar-flat Kähler metrics g_M on (M, J_M).*

By a variation on the technique of Donaldson and Friedman [6], this result is proved by constructing the desired twistor spaces with divisors as smoothings (Z_t, D_t)

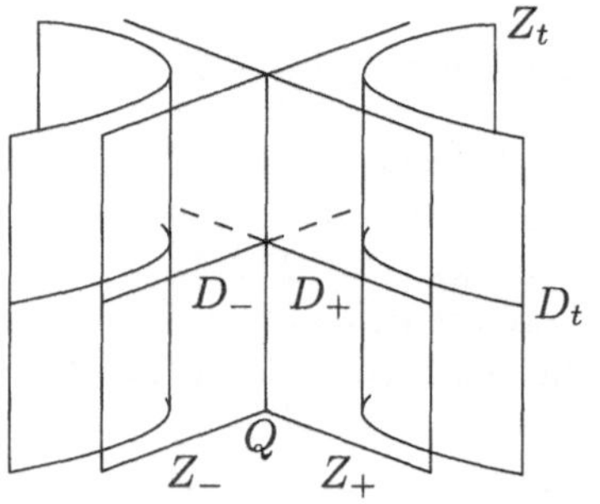

of a singular model $(Z_-, D_-) \cup (Z_+, D_+)$ with normal crossing singularities Q. Here Z_+ is a desingularization of the obvious $\mathbb{Z}_2$-quotient of the twistor space of

N, whereas Z_- is the disjoint union of an appropriate number of copies of Hitchin's Eguchi-Hansen twistor space [9]. The requirement that N be nonminimal — that is, be obtained from another complex surface by blowing up — is used to guarantee that the smoothing problem is unobstructed.

With this tool in hand, one can now construct scalar-flat Kähler metrics on $(\mathbb{CP}_1 \times \mathbb{CP}_1)\#13\overline{\mathbb{CP}}_2 = \mathbb{CP}_2\#14\overline{\mathbb{CP}}_2$ by smoothing a suitable $\mathbb{Z}_2$-quotient

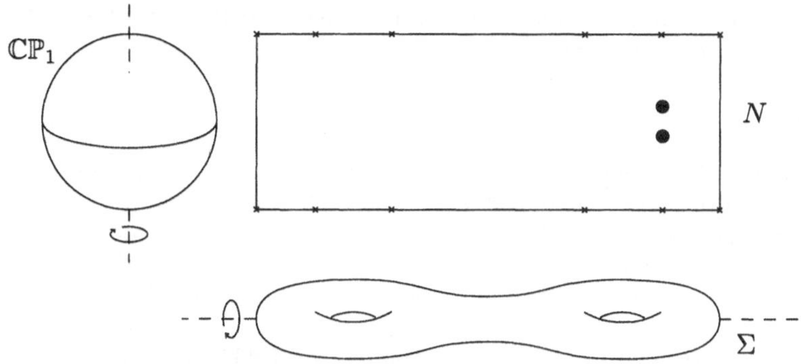

of a 2-point blow-up of $\mathbb{CP}_1 \times \Sigma$, where Σ has genus 2. The proof of Proposition 6 is similar; and in both cases the number of blow-ups can be increased at will. This allows one [12] to prove

THEOREM B *Let (M, J) be a compact complex 2-manifold that admits a Kähler metric for which the integral of the scalar curvature is nonnegative. Then either*

- *(M, J) admits a Ricci-flat Kähler metric; or else*
- *any blow-up of (M, J) has blow-ups $(\tilde{M}, \tilde{J})$ that admit scalar-flat Kähler metrics. Moreover, any blow-up of such an $(\tilde{M}, \tilde{J})$ admits scalar-flat Kähler metrics, too.*

Forming connected sums [6], [7] of the manifolds of Proposition 5 also proves [16]

THEOREM C *Let $m \geq 14n$. Then $n\mathbb{CP}_2\#m\overline{\mathbb{CP}}_2$ admits anti-self-dual metrics.*

COROLLARY 1 *Let M be a smooth simply connected compact oriented 4-manifold. Then the connected sum $M\#m\overline{\mathbb{CP}}_2$ is homeomorphic to an anti-self-dual manifold provided that $m \geq 14b_+(M)$.*

On the other hand, Taubes [22] has proved a much deeper result.

THEOREM 3 (TAUBES) *Let M be a smooth compact oriented 4-manifold. Then the connected sum $M\#m\overline{\mathbb{CP}}_2$ admits anti-self-dual metrics provided that m is sufficiently large.*

Unfortunately, however, the latter theorem gives us no estimate of the necessary size of m. Thus, many further results along the lines of Theorem C would seem to be required to attain a better understanding of the fascinating new diffeomorphism invariant implicit in Taubes' result.

Acknowledgments. Many of the results described here are the fruit of my collaborations with Jongsu Kim, Massimiliano Pontecorvo, and Michael Singer; it is a pleasure to thank them once again for their contributions. I would also like to thank Nigel Hitchin, Dominic Joyce, and Yat-Sun Poon for pointing me in the right direction at some key moments, and the Newton Institute for its hospitality during the preparation of the manuscript.

References

[1] M. Atiyah, N. Hitchin, and I. Singer, *Self-duality in four dimensional Riemannian geometry*, Proc. Roy. Soc. London Ser. A **362** (1978) 425–461.
[2] J.-P. Bourgignon, *Les variétés de dimension 4 à signature non-nulle dont la courbure est harmoniques sont d'Einstein*, Invent. Math. **63** (1981) 263–286.
[3] C. Boyer, *Conformal duality and compact complex surfaces*, Math. Ann. **274** (1986) 517–526.
[4] E. Calabi, *Extremal Kähler metrics*, Seminar on Differential Geometry (ed., S.-T. Yau), Princeton, NJ, 1982.
[5] S. Donaldson, *An application of gauge theory to four-dimensional topology*, J. Differential Geom. **18** (1983) 279–315.
[6] S. Donaldson and R. Friedman, *Connected sums of self-dual manifolds and deformations of singular spaces*, Nonlinearity **2** (1989) 197–239.
[7] A. Floer, *Self-dual conformal structures on $\ell\mathbb{CP}^2$*, J. Differential Geom. **33** (1991) 551–573.
[8] P. Gauduchon, *Surfaces Kählériennes dont la courbure vérifie Certaines conditions de positivité*, in Géometrie Riemannienne en Dimension **4**, (Bérard-Bergery, Berger, and Houzel, eds.), CEDIC/Fernand Nathan, 1981.
[9] N. J. Hitchin, *Polygons and Gravitons*, Math. Proc. Cambridge Philos. Soc. **83** (1979) 465–476.
[10] D. Joyce, *Explicit Construction of Self-Dual 4-Manifolds*, Duke Math. J., to appear.
[11] J.-S. Kim, *On the scalar curvature of self-dual manifolds*, Math. Ann. **297** (1993) 235–251.
[12] J.-S. Kim, C. LeBrun, and M. Pontecorvo, *Scalar-flat Kähler surfaces of all genera*, preprint, 1994.
[13] C. LeBrun, *Explicit self-dual metrics on $\mathbb{CP}_2\#\cdots\#\mathbb{CP}_2$*, J. Differential Geom. **34** (1991) 223–253.
[14] C. LeBrun, *Scalar-flat Kähler metrics on blown-up ruled surfaces*, J. Reine Angew. Math. **420** (1991) 161–177.
[15] C. LeBrun and M. A. Singer, *Existence and deformation theory for scalar-flat Kähler metrics on compact complex surfaces*, Invent. Math. **112** (1993) 273–313.
[16] C. LeBrun and M. A. Singer, *A Kummer-type construction of self-dual 4-manifolds*, Math. Ann. **300** (1994), 165–180.
[17] A. Newlander and L. Nirenberg, *Complex analytic coordinates in almost-complex manifolds*, Ann. Math. **65** (1957) 391–404.
[18] R. Penrose, *Non-linear gravitons and curved twistor theory*, Gen. Relativity Gravitation **7** (1976) 31–52.
[19] M. Pontecorvo, *On twistor spaces of anti-self-dual Hermitian surfaces*, Trans. Amer. Math. Soc. **331** (1992) 653–661.

[20] Y.-S. Poon, *Compact self-dual manifolds with positive scalar curvature*, J. Differential Geom. **24** (1986) 97–132.

[21] R. Schoen, *Conformal deformation of a Riemannian metric to constant scalar curvature*, J. Differential Geom. **20** (1984) 478–495.

[22] C. Taubes, *The existence of anti-self-dual metrics*, J. Differential Geom. **36** (1992) 163–253.

[23] S.-T. Yau, *On the curvature of compact Hermitian manifolds*, Invent. Math. **25** (1974) 213–239.

[24] S.-T. Yau, *On the Ricci-curvature of a complex Kähler manifold and the complex Monge-Ampère equations*, Comm. Pure Appl. Math. **31** (1978) 339–411.

The Geometry of Moduli Spaces of Vector Bundles over Algebraic Surfaces

JUN LI

Mathematics Department
Stanford University
Stanford, CA 94305, USA

The study of moduli problems is one of the central topics in algebraic geometry. After the development of GIT theory, the moduli of vector bundles over curves were constructed and in the 1970's, Gieseker constructed the moduli space of vector bundles over algebraic surfaces. Since then, many mathematicians have studied the geometry of this moduli space. For projective plane, Horrocks discovered the very powerful monad constructions of vector bundles over $\mathbf{CP}^2$. The proof that the moduli space of bundles over $\mathbf{CP}^2$ is either rational or unirational and is irreducible, and the recent development in understanding its cohomology ring rest on this construction. Brosius [Br] gave a simple description of vector bundles over ruled surfaces. In [Mu], Mukai studied the geometry of moduli of vector bundles over K3 surfaces. In particular, he constructed nondegenerate symplectic forms on these moduli spaces. Recently, Friedman has provided us with a description of the global structure of the moduli of bundles over regular elliptic surfaces [Fr1].

As to the geometry of moduli of vector bundles over general surfaces, the picture has emerged only recently. To begin with, Bogomolov's inequality says that the Chern numbers of any stable sheaf E obey

$$\frac{2r}{r-1}c_2(E) - c_1(E)^2 \geq 0, \quad r = \operatorname{rank} E.$$

On the other hand, works of [Gi2], [Ta2] show that when $r = 2$, stable E with $c_1(E) = 0$ does exist if $c_2(E) \geq 2(p_g(X) + 1)$. The major breakthrough in this area comes from Donaldson's generic smoothness theorem, which point out that deformation of general vector bundles of sufficiently large second Chern classes is unobstructed. This discovery contrasts sharply to the counter examples of [Gi2] for small $c_2(E)$.

Speaking of moduli of vector bundles over algebraic surfaces, we have to discuss the influence of gauge theory. In short, based on works of [Do], the understanding of the moduli of vector bundles will provide us valuable information on the topology of the underlining algebraic surfaces.

In this lecture I will report on the recent progress in understanding the geometry of these moduli spaces. I will concentrate on topics that arise from both algebraic geometry and gauge theory and are solved (mainly) by using algebraic geometry means.

Proceedings of the International Congress
of Mathematicians, Zürich, Switzerland 1994

1 Moduli of Stable Vector Bundles

To begin with, let's first recall the notion of stable vector bundles and its relations with gauge theory. Let $X \subset \mathbf{CP}^n$ be a smooth algebraic surface and let H be the pullback hyperplane line bundle on X.

DEFINITION 1.1. (1) (Gieseker [Gi1]) A (coherent) torsion-free sheaf E on X is said to be stable (resp. semistable) if for any proper subsheaf $F \subset E$, we have

$$\frac{1}{\text{rank } F}\chi_F(n) < \frac{1}{\text{rank } E}\chi_E(n) \quad (\text{resp. } \leq) \text{ for } n \gg 0,$$

where $\chi_E(n)$ is the Poincaré polynomial of E as follows:

$$\chi_E(n) = \frac{r}{2}n^2H^2 + nH \cdot (c_1(E) - \frac{r}{2}K_X) + \chi(E), \quad r = \text{rank } E.$$

(2) (Mumford) A vector bundle E on X is said to be μ-stable (resp. μ-semistable) if for any subsheaf $F \subset E$ with $0 < \text{rank } F < \text{rank } E$, we have

$$\frac{1}{\text{rank } F}c_1(F) \cdot H < \frac{1}{\text{rank } E}c_1(E) \cdot H \quad (\text{ resp. } \leq).$$

In 1977, Gieseker first constructed the moduli space of semistable sheaves over X and showed that it is projective [Gi1]. More precisely, for any $r \in \mathbf{N}$ and $(I, d) \in \mathrm{Pic(X)} \times \mathrm{H^4(X, \mathbf{Z})}$, there is a coarse moduli space $\mathbf{M}(r, I, d)$ parameterizing the equivalent classes of rank r semistable sheaves E satisfying $\det E = I$ and $c_2(E) = d$. Because being μ-stable is stronger than being stable, the moduli of rank r μ-stable vector bundles E with $\wedge^r E = I$ and $c_2(E) = d$ is an open subset $\mathbf{M}(r, I, d)^\mu \subset \mathbf{M}(r, I, d)$. Following Kobayashi and Donaldson, $\mathbf{M}(r, I, d)^\mu$ is isomorphic to the space of gauge equivalent classes of irreducible Anti-Self-Dual (ASD) connections on the principal bundle associated to vector bundles in $\mathbf{M}(r, I, d)^\mu$. In particular, $\mathbf{M}(2, 0, d)^\mu$ is the space of gauge equivalent classes of ASD SU(2)-connections.

2 Local Structure of the Moduli Spaces

In the next four sections, we will study the geometry of the moduli of rank 2 vector bundles. Let $r = 2$ and $I \in \mathrm{Pic(X)}$ be fixed throughout and let $\mathbf{M}_{d,I}$ be the moduli space $\mathbf{M}(2, I, d)$. In contrast to the case of vector bundles over a curve, $\mathbf{M}_{d,I}$ in general is singular and there are examples constructed by Gieseker [Gi2] showing that the dimension $\dim_{\mathbf{C}} \mathbf{M}_{d,I}$ can be bigger than the virtue dimension $v_d = 4d - 3\chi(\mathcal{O}_X) - I^2$ predicted by Riemann-Roch. In the 1980's, Donaldson [Do] (later generalized by Friedman [Fr2] and Zuo [Zu]) proved the following generic smoothness result, which opened the gate for studying $\mathbf{M}_{d,I}$ for general surfaces.

THEOREM 2.1. (Donaldson [Do], Friedman [Fr], Zuo [Zu]) *There are constants a_1 and a_2 depending on (X, H, I) so that the singular locus* $\mathrm{Sing}(\mathbf{M}_{d,I})$ *of* $\mathbf{M}_{d,I}$ *has dimension at most* $3d + a_1\sqrt{4d - I^2} + a_2$. *In particular, for sufficiently large d, $\mathbf{M}_{d,I}$ has pure complex dimension* $4d - \chi(\mathcal{O}_X) - I^2$ *and is smooth at general points.*

Based on this generic smoothness result, one can improve the understanding of the singularities of $\mathbf{M}_{d,I}$.

THEOREM 2.2. (Li [Li2]) *There is an N depending on (X, H, I) such that whenever $d \geq N$, then $\mathbf{M}_{d,I}$ is normal and, further, it is a local complete intersection scheme away from the closed points corresponding to strictly semistable sheaves.*

The proof of this theorem is a straightforward application of deformation theory.

3 Compactifications and the Polynomial Invariants

When there is a universal sheaf $\mathcal{E}$ on $X \times \mathbf{M}_{d,I}$, one can use the Künneth component $c_2(\mathcal{E})^{2,2} \in H^2(X) \otimes H^2(\mathbf{M}_{d,I})$ of $c_2(\mathcal{E})$ to define a homomorphism [OG1]

$$\tilde{\mu} = c_2(\mathcal{E})^{2,2} : H_2(X, \mathbf{Z}) \longrightarrow H^2(\mathbf{M}_{d,I}, \mathbf{Z}). \qquad (3.1)_1$$

Otherwise, by applying the descent technique to the moduli space, one obtains a homomorphism $2\mathrm{Pic}(X) \to \mathrm{Pic}(\mathbf{M}_{d,I})$ [Li1] that induces a homomorphism

$$\tilde{\mu} : H^{1,1}(X, \mathbf{R}) \cap H^2(X, \mathbf{Z}) \longrightarrow H^2(\mathbf{M}_{d,I}, \mathbf{Z}[\frac{1}{2}]). \qquad (3.1)_2$$

Assuming $\dim \mathbf{M}_{d,I} = v_d$ is the expected dimension, then the ring structure of $H^*(\mathbf{M}_{d,I}; \mathbf{R})$ and $(3.1)_1$ give rise to a multilinear map

$$\tilde{\varphi}_{v_d} : \mathrm{Symm}^{v_d} H_2(X, \mathbf{Z}) \longrightarrow \mathbf{R}, \qquad (3.2)_1$$

or in the case of $(3.1)_2$,

$$\tilde{\varphi}_{v_d} : \mathrm{Symm}^{v_d} \big(H^{1,1}(X, \mathbf{R}) \cap H^2(X, \mathbf{Z})\big) \to \mathbf{R}. \qquad (3.2)_2$$

These intersection numbers are important in understanding the ring structure of $H^*(\mathbf{M}_{d,I})$.

If we only consider the open set $\mathbf{M}^{\mu}_{d,I} \subset \mathbf{M}_{d,I}$, we recover the μ-map defined by Donaldson in gauge theory. In order to define intersection pairing similar to (3.2), which is the polynomial invariants when $p_g \geq 1$, Donaldson introduced the so-called Uhlenbeck compactification $\mathbf{M}^{\mathrm{Uh}}_{d,I}$ (of $\mathbf{M}^{\mu}_{d,I}$), which is derived by studying the limits of ASD connections.

In order to understand the polynomial invariants in the algebro-geometric setting, one needs to understand the two compactifications of $\mathbf{M}_{d,I}$: the Uhlenbeck compactification $\mathbf{M}^{\mathrm{Uh}}_{d,I}$ and the Gieseker compactification $\mathbf{M}^{\mathrm{Gi}}_{d,I}$ (which is the closure of $\mathbf{M}^{\mu}_{d,I}$ in $\mathbf{M}_{d,I}$ and when $d \gg 0$, $\mathbf{M}^{\mathrm{Gi}}_{d,I} = \mathbf{M}_{d,I}$). Note that $\mathbf{M}^{\mathrm{Uh}}_{d,I}$ is different from $\mathbf{M}^{\mathrm{Gi}}_{d,I}$ as $\mathbf{M}^{\mathrm{Gi}}_{d,I} - \mathbf{M}^{\mu}_{d,I}$ is a real codimension 2 subset whereas $\mathbf{M}^{\mathrm{Uh}}_{d,I} - \mathbf{M}^{\mu}_{d,I}$ is a real codimension 4 subset.

THEOREM 3.1. (Li [Li1]) *The Uhlenbeck compactification $\mathbf{M}^{\mathrm{Uh}}_{d,I}$ is a contraction of the Gieseker compactification $\mathbf{M}^{\mathrm{Gi}}_{d,I}$. More precisely, there is a projective scheme structure on $\mathbf{M}^{\mathrm{Uh}}_{d,I}$ such that under this scheme structure, there is a morphism*

$$F : \mathbf{M}^{\mathrm{Gi}}_{d,I} \longrightarrow \mathbf{M}^{\mathrm{Uh}}_{d,I}$$

that extends the inclusions $\mathbf{M}^{\mu}_{d,I} \subset \mathbf{M}^{\mathrm{Gi}}_{d,I}$ and $\mathbf{M}^{\mu}_{d,I} \subset \mathbf{M}^{\mathrm{Uh}}_{d,I}$.

Although the Uhlenbeck compactification $\mathbf{M}_{d,I}^{\mathrm{Uh}}$ is defined purely by analysis, it is remarkable that it still carries a projective scheme structure and, under this scheme structure, $\mathbf{M}_{d,I}^{\mathrm{Uh}}$ is a contraction of $\mathbf{M}_{d,I}^{\mathrm{Gi}}$. The theorem was proved as follows: By a descent technique, we first construct a (determinant) line bundle L on $\mathbf{M}_{d,I}$, which is an analogy of the theta divisor for moduli of bundles over curves. The difficult part is to construct sufficiently many sections. This is carried out by using a restricting and extending technique: Let $C \in |nH|$ be a smooth divisor and let $\mathbf{M}_C$ be the moduli of vector bundles over C. Then there is a rational map $\mathbf{M}_{d,I} - - \rightarrow \mathbf{M}_C$ so that L is the pullback of an ample line bundle on $\mathbf{M}_C$. So by pulling back sections on $\mathbf{M}_C$, we obtain a lot of meromorphic sections of $L^{\otimes k}$. By using GIT theory, we show that all these sections are regular. In this way, we get a morphism $F_L : \mathbf{M}_{d,I} \rightarrow \mathbf{CP}^N$. It is proved that the image scheme $F_L(\mathbf{M}_{d,I}^{\mathrm{Gi}})$ is homeomorphic to $\mathbf{M}_{d,I}^{\mathrm{Uh}}$. The proof is tricky because the two spaces are defined based on degenerations by different means. This is carried out by using the quotient bundles technique.

The relation of $\tilde{\varphi}_{v_d}$ with the Donaldson polynomial invariants is as follows:

THEOREM 3.2. (Li [Li1], Morgan [Mo]) *Assume $p_g \geq 1$; then there is a constant N depending on (X, I) so that for any $d \geq N$ and d-generic ample divisor H, the polynomials* (3.2) *coincide with the Donaldson polynomial invariants of X.*

I shall point out that Morgan [Mo] has constructed the map $\varphi_d : \mathbf{M}_{d,I}^{\mathrm{Gi}} \rightarrow \mathbf{M}_{d,I}^{\mathrm{Uh}}$ and established its continuity.

Because the polynomial invariants are the top intersection of $H^2(\mathbf{M}_{d,I})$, they are nontrivial because $\tilde{\mu}(H)$ is big. Hence $\varphi_{v_d}(H, \ldots, H) > 0$. In this way, we have recovered the Donaldson nonvanishing theorem of φ_*.

THEOREM 3.3. (Donaldson [Do]) *Let X be any smooth algebraic surface with $p_g \geq 1$ and H an ample divisor. Then for sufficiently large k, $\varphi_k(H, \ldots, H) > 0$.*

There is another nonvanishing theorem that goes as follows. If Θ is a holomorphic symplectic form nondegenerate at general points of $\mathbf{M}_{d,I}$, then when $v_d \equiv 0(2)$,

$$\int_{\mathbf{M}_{d,I}} \wedge^{\frac{v_d}{2}} \Theta \otimes \wedge^{\frac{v_d}{2}} \bar{\Theta} > 0.$$

By using the symplectic form of Mukai, O'Grady proved the following nonvanishing theorem of the polynomial invariants.

THEOREM 3.4. (O'Grady [OG1]) *Assume X is a surface of general type and assume there is a $\theta \in H^0(K_X)$ such that $\theta^{-1}(0) \subset X$ is reduced. Then for sufficiently large k, $\varphi_k(([\theta] + [\bar{\theta}])^k) > 0$ provided $v_d \equiv 0(2)$. Here $[\theta] \in H^2(X; \mathbf{R})$ is the class associated to θ.*

The original assumption (in [OG1]) that $\theta^{-1}(0)$ is smooth can be relaxed by a technical lemma in [Li2].

4 The Kodaira Dimension of the Moduli Spaces

One important question about the geometry of the moduli space $\mathbf{M}_{d,I}$ is to determine its Kodaira dimension. It has been known for a long time that the moduli space of vector bundles over $\mathbf{CP}^2$ is either rational or unirational depending on the choice of I and d. In particular, $\kappa(\mathbf{M}_{d,I}) = -\infty$. The same conclusion holds for all surfaces of $\kappa(X) = -\infty$. For moduli spaces of bundles over K3 surfaces, their Kodaira dimensions are zero following the work of Mukai [Mu]. For elliptic surfaces, the description given by Friedman for bundles over regular elliptic surfaces suggests that $\mathbf{M}_{d,I}$ are birational to fibrations whose generic fibers are Abelian varieties.

For surfaces of general type, we have the following theorem.

THEOREM 4.1. (Li [Li2]) *Let X be a minimal surface of general type. Assume there is a reduced canonical divisor and that $\chi(\mathcal{O}_X) + I^2$ is even; then there is an N depending on (X, H, I) so that for $d \geq N$, $\mathbf{M}_{d,I}$ is of general type.*

The proof consists of two parts. First, one needs a description of the canonical sheaf of $\mathbf{M}_{d,I}$ so that many pluri-canonical forms can be constructed. This was done by using a determinant line bundle similar to the proof of Theorem 3.1. (This was also carried out independently by [Hu].) The second part is to work on a resolution $\tilde{\mathbf{M}}_{d,I}$ of $\mathbf{M}_{d,I}$ (as $\mathbf{M}_{d,I}$ is always singular) and construct pluri-canonical sections of $\tilde{\mathbf{M}}_{d,I}$. The obvious thing one tries to do to obtain pluri-canonical forms of $\tilde{\mathbf{M}}_{d,I}$ is to look at the pullbacks of the pluri-canonical forms of $\mathbf{M}_{d,I}$ and hope that they are regular. We do not know whether $\mathbf{M}_{d,I}$ has only canonical singularities. Thus, the question of whether these pullbacks are regular is a very delicate problem. This problem is solved with the aid of Mukai's symplectic form ω on $\tilde{\mathbf{M}}_{d,I}$. Under the technical conditions on X, one can show that $\det\omega$ is a nontrivial two-canonical form that vanishes along the exceptional divisor of $\pi : \tilde{\mathbf{M}}_{d,I} \to \mathbf{M}_{d,I}$. Hence, for any pluri-canonical form θ on $\mathbf{M}_{d,I}$, $(\det\omega)^{\otimes a} \otimes \pi^*\theta$ is a regular pluri-canonical form of $\tilde{\mathbf{M}}_{d,I}$ for some fixed integer a. Hence, enough pluri-canonical forms of $\tilde{\mathbf{M}}_{d,I}$ can be constructed and the theorem follows.

5 Topology of the Moduli Spaces

The questions we deal with here are how many irreducible components does $\mathbf{M}_{d,I}$ have and what are its Betti numbers. There are examples of surfaces X of which the moduli $\mathbf{M}_{d,I}$ are not irreducible for some d. But this does not occur for large d.

THEOREM 5.1. (Gieseker-Li [GL1]) *There is a constant N depending on (X, H, I) such that whenever $d \geq N$, then the moduli scheme $\mathbf{M}_{d,I}$ is irreducible.*

The crucial part in proving this theorem is to establish the following existence theorem.

THEOREM 5.2. (Gieseker-Li [GL1], O'Grady [OG2]) *There is a constant N depending on (X, H, I) such that whenever $d \geq N$, then every irreducible component $\mathbf{M} \subset \mathbf{M}_{d,I}$ contains nonlocally free sheaves.*

After that, an algebraic geometry analogy of Taubes' work [Ta1] shows that for sufficiently large d, $\mathbf{M}_{d,I}$ is irreducible. As this argument is rather simple, we will sketch it here. First of all, let Λ_d be the set of all irreducible components of $\mathbf{M}_{d,I}$. For large d, we can define a map $f_d : \Lambda_d \to \Lambda_{d+1}$ as follows: Let $\mathbf{M} \in \Lambda_d$ be any component and let $E \in \mathbf{M}$ be a general point. Then a sheaf $F \subset E$ with $E/F \cong \mathbf{C}_x$ is stable and has $c_2 = d+1$. Hence F identifies a unique component in $\mathbf{M}_{d+1}$ that is defined to be $f_d(\mathbf{M})$. According to Theorem 5.2, for sufficiently large d, f_d is surjective. Theorem 5.1 will follow if we can show that for any d, there is an n so that

$$f_{d+n} \circ \cdots \circ f_d(\Lambda_d) = \text{single point}.$$

Indeed, for $\mathbf{M}^1, \mathbf{M}^2 \in \Lambda_d$ and $E_i \in \mathbf{M}^i$, we can find an ample line bundle L so that E_i belong to the exact sequences

$$0 \longrightarrow L^{-1} \longrightarrow E_i \longrightarrow L \otimes I \otimes \mathcal{I}_{z_i} \longrightarrow 0.$$

Then we can find subsheaves $F_i \subset E_i$ so that F_i belongs to the exact sequence

$$0 \longrightarrow L^{-1} \longrightarrow F_i \longrightarrow L \otimes I \otimes \mathcal{I}_{z_1 \cup z_2} \longrightarrow 0.$$

Let $n=$ length(z_1). Clearly, F_i is a sheaf in $f_{d+n} \circ \cdots \circ f_d(\mathbf{M}^i)$. On the other hand, F_1 and F_2 belong to the same irreducible component because they are elements of a family parameterized by $\mathbf{Ext}^1(L \otimes I \otimes \mathcal{I}_{z_1 \cup z_2}, L^{-1})$. Hence F_1 and F_2 are contained in the same irreducible component and thus $\#(\Lambda_{d+n}) < \#(\Lambda_d)$. This completes the proof of Theorem 5.1.

I shall point out that for many surfaces, for instance, projective planes, ruled surfaces [Qi], and K3 surfaces [GH], the moduli $\mathbf{M}_{d,I}$ are irreducible for all choices of I and d.

Next, we shall discuss the higher Betti numbers of $\mathbf{M}_{d,I}$ or $\mathbf{M}^{\mu}_{d,I}$. First of all, works of Taubes show that there is a "weak" limit

$$\lim_d \mathbf{M}^{\mu}_{d,I} = \mathbf{M}_\infty$$

and the stable limit $\lim_d H_i(\mathbf{M}^{\mu}_{d,I})$ is isomorphic to $H_i(\mathbf{M}_\infty)$, which is a homotopy invariant of X. The generalized Atiyah-Jones conjecture states that (in our setting) for any smooth algebraic surface X, there is a sequence $\{q_d\}$, $q_d \to \infty$, such that for $i \le q_d$, $H_i(\mathbf{M}^{\mu}_{d,I})$ is isomorphic to $H_i(\mathbf{M}_\infty)$. Recently, there have been many attempts to solve this conjecture, notably, [BHMM], [Ki], [SE], [Ti], and [Yo1] for projective planes, [HM], [Yo2] for ruled surfaces, [Be] for a large class of rational surfaces, and [GH] for K3 surfaces. For general surfaces, the author recently proved the following.

THEOREM 5.3. (Li [Li3]) *For any algebraic surface (X, H) and $I \in \mathrm{Pic(X)}$, there is a constant N such that for $d \ge N$, $H_1(\mathbf{M}^{\mu}_{d,I}; \mathbf{Q}) \cong H_1(X; \mathbf{Q})$ and $H_2(\mathbf{M}^{\mu}_{d,I}; \mathbf{Q}) \cong \wedge^2 H_1(X; \mathbf{Q}) \oplus H_2(X, \mathbf{Q})$.*

The proof goes as follows: By using the construction in [GL1] and [Ta1], one gets a canonical homomorphism

$$\tau(d)_i : H_i(\mathbf{M}^{\mu}_{d,I};\mathbf{Q}) \longrightarrow H_i(\mathbf{M}^{\mu}_{d+1};\mathbf{Q}), \tag{5.1}$$

which fits into the following commutative diagram

$$\begin{array}{ccc} H_i(\mathbf{M}^{\mu}_{d,I}) & \longrightarrow & H_i(\mathcal{B}^*_d) \\ \downarrow & & \downarrow \\ H_i(\mathbf{M}^{\mu}_{d+1}) & \longrightarrow & H_i(\mathcal{B}^*_{d+1}) \end{array}$$

where $\mathcal{B}^*_d$ is the space of gauge equivalent classes of irreducible connections on an appropriate principal bundle and $\iota(d)_i : H_i(\mathbf{M}^{\mu}_{d,I}) \to H_i(\mathcal{B}^*_d)$ is induced by the inclusion $\mathbf{M}^{\mu}_{d,I} \subset \mathcal{B}^*_d$. Taubes showed that $H_i(\mathcal{B}^*_d) \cong H_i(\mathcal{B}^*_{d+1})$ and for $d \gg i$, $\iota(d)_i$ is surjective and $\mathrm{Ker}\{\iota(d)_i\}$ is contained in

$$\mathrm{Ker}\{\tau(d+l)_i \circ \cdots \circ \tau(d)_i\}$$

for large l (depending on d and i). Hence an easy diagram chasing shows that $H_i(\mathbf{M}^{\mu}_{d,I}) \cong H_i(\mathcal{B}^*_d)$ for large d if we can show that (5.1) is surjective. By using the Lefschetz hyperplane theorem and the deformation constructed in [OG2], the surjectivity of $\tau(d)_1$ and $\tau(d)_2$ for large d is established in [Li3].

6 Moduli of High Rank Vector Bundles

It is interesting to see how many properties of $\mathbf{M}(2,I,d)$ can be generalized to moduli space $\mathbf{M}(r,I,d)$ of high rank vector bundles. As is clear from this exposition, the first challenge is to prove the generic smoothness of $\mathbf{M}(r,I,d)$ for large d. This was first accomplished by Gieseker and the author.

THEOREM 6.1. (Gieseker-Li [GL2]) *For any surface (X,H), $r \in \mathbf{N}$ and $I \in$ Pic(X), and for any constant C, there is an N such that for $d \geq N$,*

$$\dim \mathbf{M}(r,I,d) = 2rd - (r-1)I^2 - (r^2-1)\chi(\mathcal{O}_X),$$

which is the virtue dimension of $\mathbf{M}(r,I,d)$, and further

$$\mathrm{codim}\big(\mathrm{Sing}\mathbf{M}(r,I,d), \mathbf{M}(r,I,d)\big) \geq C.$$

With this generic smoothness result, Theorems 2.2, 3.1, and 5.1 are generalized to $\mathbf{M}(r,I,d)$ [GL2], [OG3].

7 Effective Bounds

One observes that all results mentioned about $\mathbf{M}(r,I,d)$ for general X hold only for sufficiently large d. This is necessary because for small d, we can find counterexamples to each result mentioned. On the other hand, a deeper understanding of $\mathbf{M}(r,I,d)$ inevitably relies on finding an optimal (lower) bound of the second Chern class d for which the previous theorems hold. The recent work of O'Grady has made progress along this direction.

THEOREM 7.1. (O'Grady [OG3]) *For any ample divisor H and $r \in \mathbf{N}$, there are functions $\lambda_2(r)$, $\lambda_1(r,X,H)$ and, $\lambda_0(r,X,H)$ such that $\lambda_2(r) < 2r$ and*

$$\dim \mathrm{Sing}\,\mathbf{M}(r,I,d) \leq \lambda_2\Delta + \lambda_1(r,X,H)\sqrt{\Delta} + \lambda_0(r,X,H),$$

where $\Delta = d - \frac{r-1}{2r}I^2$. Here the functions λ_0, λ_1, and λ_2 are determined effectively depending on the numerical invariants of the arguments involved.

One should compare this bound with the fact that the expected dimension of $\mathbf{M}(r, I, d)$ is $2rd + O(1)$. Based on his technique, O'Grady also derived effective lower bounds that depend only on (r, X, H) so that Theorems 2.2 and 5.2 hold as well.

References

[AJ] M. F. Atiyah, and J. D. Jones, *Topological aspects of Yang-Mills theory*, Comm. Math. Phys. **61** (1978), 97–118.

[Be] A. Beauville, *Sur la cohomologie de certains espaces de modules de fibrés vectoriels*, preprint.

[BHMM] C. P. Boyer, J. C. Hurtubise, B. M. Mann, and R. J. Milgram, *The topology of instanton moduli spaces, I: The Atiyah-Jones conjecture*, Ann. of Math. (2) **137** (1993), 561–609.

[Br] S. Brosius, *Rank-2 vector bundles on a ruled surface I, II*, Math. Ann. **265** (1983), 155–168.

[Do] S. K. Donaldson, *Polynomial invariants for smooth four-manifolds*, Topology **29**, No. 3 (1986), 257–315.

[DN] J.-M. Drezet, and M. S. Narasimhan, *Group de Picard des variétés de modules de fibres semi-stables sur les coubes algébriques*, Invent. Math. **97** (1989), 53–94.

[ES] G. Ellingsrud, and S. A. Strømme, *Toward the Chow ring of the Hilbert scheme of* $\mathbb{P}^2$, J. Reine Angew. Math. **441** (1993), 33–44.

[Fr1] R. Friedman, *Rank two vector bundles over regular elliptic surfaces*, Invent. Math. **96** (1989), 283–332.

[Fr2] R. Friedman, *Vector bundles over surfaces.* To be published.

[Gi1] D. Gieseker, *On the moduli of vector bundles on an algebraic surface*, Ann. of Math. (2), **106** (1977), 45–60.

[Gi2] D. Gieseker, *A construction of stable bundles on an algebraic surface*, J. Differential Geom., **27** (1988), 137–154.

[GL1] D. Gieseker, and J. Li, *Irreducibility of moduli of rank two vector bundles on surfaces*, J. Differential Geom. **40** (1994), 23–104.

[GL2] D. Gieseker, and J. Li, *Moduli of vector bundles over surfaces I*, to appear in J. Amer. Math. Soc.

[GH] L. Göttsche, and D. Huybrechts, *Hodge numbers of moduli spaces of stable bundles on K3 surfaces*, preprint.

[HM] J. C. Hurtubise, and R. J. Milgram, *The Atiyah-Jones conjecture for ruled surfaces*, preprint.

[Hu] D. Huybrechts, *Complete curves in moduli spaces of stable bundles on surfaces*, Math. Ann. **298**, No. 1 (1994), 67–78.

[Ki] F. C. Kirwan, *Geometric invariant theory and the Atiyah-Jones conjecture*, preprint.

[Li1] J. Li, *Algebraic geometric interpretation of Donaldson's polynomial invariants*, J. of Differential Geom. **37** (1993), 417–466.

[Li2] J. Li, *Kodaira dimension of moduli space of vector bundles on surfaces*, Invent. Math. **115** (1994), 1–40.

[Li3] J. Li, *The first two Betti numbers of moduli of vector bundles over surfaces*, preprint.

[Ma] M. Maruyama, *Moduli of stable sheaves I and II*, J. Math. Kyoto Univ. **17** (1977), 91–126; J. Math. Kyoto Univ. **18** (1978), 557–614.

[Mo] J. Morgan, *Comparison of the Donaldson polynomial invariants with their algebro-geometric analogues*, Topology **32** (1993), no. 3, 449–488.

[Mu] S. Mukai, *Symplectic structure on the moduli space of sheaves on an Abelian or $K3$ surfaces*, Invent. Math. **77** (1984), 101–116.

[Mm] D. Mumford, Geometric Invariant Theory, Springer-Verlag, Berlin and New York, 1982.

[OG1] K. G. O'Grady, *Algebro-geometric analogues of Donaldson's polynomials*, Invent. Math. **107** (1992), 351–395.

[OG2] K. G. O'Grady, *The irreducible components of moduli spaces of vector bundles on surfaces*, Invent. Math. **112** (1993), 585–613.

[OG3] K. G. O'Grady, *Moduli of vector bundles on projective surfaces: some basic results*, preprint.

[OSS] C. Okonek, M. Schneider, and H. Spindler, Vector Bundles on Complex Projective Spaces, Progress in Math. **3**, Birkhäuser, Basel, Basel, and Berlin, 1980.

[Qi] Z. Qin, *Moduli spaces of stable rank-2 bundles on ruled surfaces*, Invent. Math. **110** (1992), No. 3, 615–626.

[Ta1] C. Taubes, *The stable topology of self-dual moduli spaces*, J. Differential Geom. **19** (1984), 337–392.

[Ta2] C. Taubes, *Self-dual connections on 4-manifolds with indefinite intersection matrix*, J. Differential Geom. **19** (1984), 517–560.

[Ti1] Y-L. Tian, *The based $SU(n)$-instanton moduli spaces*, Math. Ann. **298** (1994), 117–139.

[Ti2] Y-L. Tian, *The Atiyah-Jones conjecture for classical groups*, preprint.

[Ty] A. N. Tyurin, *Symplectic structures on the moduli variety of vector bundles on an algebraic surface with $p_g \neq 0$*, Izv. Akad. Nauk. SSSR Ser. Mat. **52** (1978), 149–195.

[Yo1] K. Yoshioka, *The Betti numbers of the moduli space of stable sheaves of rank 2 on $\mathbb{P}^2$*, J. Reine Angew. Math. **453** (1994), 193–220.

[Yo2] K. Yoshioka, *The Betti numbers of the moduli space of stable sheaves of rank 2 on a ruled surface*, Math. Ann. **302** (1995) no. 3, 519–540.

[Zu] K. Zuo, *Generic smoothness of the moduli of rank two stable bundles over an algebraic surface*, Math. Z. **207**, No. 4 (1991), 629–643.

Spaces with Curvature Bounded Below

G. PERELMAN

Steklov Institute and Department of Mathematics
Fontanka 27 University of California
St. Petersburg 191011, Russia Berkeley, CA 94720, USA

Introduction

Consider the following problem:

Let M_j be a collapsing sequence (i.e. Vol $M_j \to 0$*) of complete Riemannian manifolds of bounded dimension, with sectional curvatures uniformly bounded below. What can be said about the Gromov-Hausdorff limit M of such a sequence? And what is the relation between the topology and geometry of M and that of manifolds M_j, with large j?*

After the seminal work of Gromov (see [G1],[GLP]), questions of this type, with various assumptions on curvatures, and other geometric characteristics, have been receiving much attention. Cheeger, Gromov, and Fukaya developed a far-reaching theory of collapse under the two-sided curvature bounds (see [CFG]). One of the simplest conclusions of this theory is that the limit space M in this case has a stratification, each stratum being a totally geodesic Riemannian manifold. (This stratification is nontrivial unless M_j admits a structure of locally trivial fibration over M.) This conclusion can be explained by the following argument (see [GLP, 8.30] for details).

Suppose that M_j have sectional curvatures between -1 and 1, and let $p_j \in M_j$ converge to some point $p \in M$. Then the balls B_j of radius $\pi/2$ centered at p_j are covered by convex balls $\tilde{B}_j$ of the same radius in the tangent spaces at p_j, endowed with the lifted metrics. The sequence $\tilde{B}_j$ has the same curvature bounds and, in addition, a uniform lower bound on the injectivity radius; therefore its limit $\tilde{B}$ is a manifold. Now each B_j is a quotient of $\tilde{B}_j$ by an isometric (pseudo)group action, hence a ball B in M centered at p must be a quotient of $\tilde{B}$ by the limit action; this leads to a stratification of B with totally geodesic strata.

No such argument is known for manifolds with curvature bounded only below. Still, it can be proved that M has a natural stratification, with (in some sense) totally geodesic strata; the strata need not be smooth, but each of them is a topological manifold; there is a nontriviality criterion similar to that in the case of the two-sided curvature bound.

The author considers this statement as a kind of experimental result, which needs some (at least heuristic) theoretical explanation. The question can be vaguely formulated as follows: Is there any "algebraic" reason for this stratification to

Proceedings of the International Congress
of Mathematicians, Zürich, Switzerland 1994

appear? The author expects that a solution to this problem may be achieved through an investigation of the spaces of semiconcave functions, but he does not have the slightest idea of how to do it.

The statements about limit spaces are corollaries of the corresponding statements for general Alexandrov spaces with curvature bounded below. In the following sections we review the relevant part of the theory of Alexandrov spaces, and outline the proof of the main result about stratification. The final section contains some speculations.

1 Basic Notions and Results

This section is a summary of [BGP]. Many of the notions discussed here were introduced by Alexandrov [A1], [A2]; see also the surveys [BN], [R].

1.1 Recall that a length space is a metric space where the distance between any two points equals the infimum of the lengths of curves connecting these points. A triangle in a length space M consists of three points, say, p, q, r, and three shortest geodesics $\overline{pq}, \overline{pr}, \overline{qr}$. Given a real number k, a comparison triangle $\tilde{p}\tilde{q}\tilde{r}$ is a triangle on the surface of constant curvature k, with the same side lengths. Its angles are called the comparison angles and denoted by $\tilde{\angle}pqr$, etc. A comparison triangle exists and is unique whenever $k \leq 0$ or $k > 0$ and $|pq| + |pr| + |qr| < 2\pi/\sqrt{k}$.

1.2 Definition. A length space M is called an Alexandrov space of curvature $\geq k$ if any $x \in M$ has a neighborhood U_x, such that for any $a, b, c, d \in U_x$

$$\tilde{\angle}bac + \tilde{\angle}cad + \tilde{\angle}dab \leq 2\pi \; . \tag{1}$$

For locally compact spaces this is equivalent to the more familiar Alexandrov-Toponogov distance comparison.

1.2A Definition. A locally compact length space M is called an Alexandrov space of curvature $\geq k$ if any $x \in M$ has a neighborhood U_x such that for any triangle pqr in U_x and any $q_1 \in \overline{pq}$, $r_1 \in \overline{pr}$, we have $|q_1 r_1| \geq |\tilde{q}_1 \tilde{r}_1|$, where $\tilde{q}_1, \tilde{r}_1$ are the corresponding points on the sides $\tilde{p}\tilde{q}, \tilde{p}\tilde{r}$ of the comparison triangle $\tilde{p}\tilde{q}\tilde{r}$.

1.3 If M is complete then the local conditions in Definitions 1.2 and 1.2A imply the corresponding global conditions. (For Riemannian manifolds this is essentially the content of the celebrated Toponogov comparison theorem.) In particular, if M has curvature $\geq k > 0$, then any triangle in M has perimeter at most $2\pi/\sqrt{k}$, and the diameter of M does not exceed $\pi/\sqrt{k}$.

As (1) is a purely metric condition, we can immediately conclude that a Gromov-Hausdorff limit of complete Alexandrov spaces of curvature $\geq k$ is an Alexandrov space of curvature $\geq k$.

1.4 The Hausdorff dimension of an Alexandrov space is either integer or infinite; it is semicontinuous with respect to Gromov-Hausdorff convergence. In fact, a collection of all compact Alexandrov spaces with diameters $\leq D$, dimensions $\leq n$, and curvatures $\geq k$, for some D, n, k, is compact in Gromov-Hausdorff topology. In this paper we will consider only complete locally compact spaces of finite Hausdorff dimension; general Alexandrov spaces are discussed in [BGP, Sections 2–6], [Pl], [PP2, Appendix].

1.5 EXAMPLES

(a) A complete Riemannian manifold with empty or convex boundary and with sectional curvatures $\geq k$ is an Alexandrov space of curvature $\geq k$.

(b) A convex hypersurface (with its intrinsic metric) in a space of curvature $\geq k$ is expected to have curvature $\geq k$; this is known if the ambient space is Riemannian, see [B].

(c) A cone on a (complete) space of curvature ≥ 1 has a natural metric of nonnegative curvature. A spherical suspension of such a space, or more generally, a join of two such spaces, has a natural metric of curvature ≥ 1. A product of nonnegatively curved spaces is nonnegatively curved.

(d) If a group acts on an Alexandrov space by isometries, and all its orbits are closed, then the quotient space is an Alexandrov space with the same lower curvature bound. (A more general result was obtained by Berestovskii.)

Each of these examples is a guiding one for a particular branch of the theory. Namely, Example (a) provides insight into the global gometry of Alexandrov spaces, Example (b) suggests what kind of analytical results one might or might not expect, Example (c) shows the local topological structure, and Example (d) indicates the properties of the stratification.

At the same time, it is not known whether any Alexandrov space can be obtained from Examples (a), (b) using constructions (c), (d) and taking Gromov-Hausdorff limits.

1.6 The angle between two shortest geodesics $\overline{pq}, \overline{pr}$ in an Alexandrov space M is defined as $\angle qpr = \lim\{\tilde{\angle} q_1 p r_1 : q_1 \in \overline{pq},\ r_1 \in \overline{pr},\ q_1, r_1 \to p\}$; the existence of the limit follows from Definition 1.2A.

The set of equivalence classes of shortest geodesics starting at p, with the angular distance, is a metric space; its completion is called the space of directions at p and denoted by Σ_p. It is an important and nontrivial fact that Σ_p is compact (M-finite dimensional). Moreover, Σ_p is an Alexandrov space of curvature ≥ 1, and $\dim \Sigma_p = \dim M - 1$.

A cone C_p on Σ_p is called the tangent cone of M at p; it is an Alexandrov space of nonnegative curvature, $\dim C_p = \dim M$. Alternatively, the tangent cone can be defined as the pointed Gromov-Hausdorff limit of $(\lambda \cdot M, p)$, $\lambda \to \infty$.

The space of directions Σ_p depends semicontinuously on p; that is if p_i tend to p and Σ_{p_i} converge in the Gromov-Hausdorff sense to some space Σ, then Σ_p admits a noncontracting map into Σ. Moreover, according to a result of Petrunin [Pet2], the space of directions is constant along a shortest geodesic, with endpoints excluded.

1.7 The boundary of an Alexandrov space is defined inductively: if $\dim M = 1$ then M is a 1-dimensional manifold with (possibly empty) boundary, and the definition is clear; if dim $M > 1$ then $\partial M = \{p \in M : \partial\Sigma_p \neq \emptyset\}$. It can eventually be proved that ∂M is a closed codimension 1 subset of M, which can be characterized by local topological properties. Furthermore, the double of an Alexandrov space with nonempty boundary is an Alexandrov space with empty boundary having the same lower curvature bound. Petrunin [Pet1] proved that one can similarly glue two different Alexandrov spaces with (intrinsically) isometric boundaries.

1.8 A point $p \in M$ is called singular if C_p is not isometric to euclidean space. The set of singular points in $M \backslash \partial M$ has Hausdorff codimension at least 2. Moreover, the nonsingular set is totally convex; that is, any shortest geodesic with nonsingular endpoints contains no singular points. On the other hand, the set of singular points may be dense.

A point $p \in M$ can be called weakly singular if the volume (i.e. Hausdorff measure) of Σ_p is close to that of the standard unit sphere, say $\mathrm{Vol}(\Sigma_p) > (1-\delta)\mathrm{Vol}(S^n)$ for some fixed small $\delta > 0$. (Note that nonsingular points are weakly singular according to this definition.) The set of weakly singular points is open and totally convex. Furthermore, each weakly singular point has a neighborhood, almost isometric to a euclidean ball.

1.9 A Lipschitz function f in a domain $U \subset M$ is called λ-concave if for any shortest geodesic γ in U the function $f \circ \gamma(t) + \lambda t^2$ is concave in t. If M has nonempty boundary, it is convenient to modify this definition, requiring that the tautological extension of f to the double of M is λ-concave in the double of U. A function f is semiconcave in U if for any $x \in U$ there exists a neighborhood U_x and a number λ_x, such that f is λ_x-concave in U_x.

The basic examples of semiconcave functions are distance functions: if $P \subset M$ is compact then the distance function from P, dist_P, is semiconcave in $M \backslash P$. Moreover, M has nonnegative curvature iff dist_p^2 is (-1)-concave in M for each $p \in M$; similar characterization exists for other lower curvature bounds.

Semiconcave functions f have (directional) derivatives f'_p at each point p; that is, $f \circ \gamma(t) = f(p) + t f'_p(\xi) + o(t)$ for a shortest geodesic γ starting at p in direction $\xi \in \Sigma_p$. The derivative f'_p is a Lipschitz spherically concave function on Σ_p; the latter condition means

$$\sin|\xi\zeta| f'_p(\eta) \geq \sin|\xi\eta| f'_p(\zeta) + \sin|\eta\zeta| f'_p(\xi) , \quad \text{whenever} \quad \eta \in \overline{\xi\zeta} .$$

The derivative of a distance function $f = \mathrm{dist}_P$ at a point $q \in M \backslash P$ is given by $f'_q(\xi) = -\cos|P'\xi|$, where P' denotes the compact set of directions of all shortest geodesics $\overline{qp}$ such that $p \in P$ and $|pq| = |qP|$.

2 Elementary Morse Theory

In this section we discuss the topology of Alexandrov spaces, according to [P]. Our main tool is the critical points theory for semiconcave maps. Critical points theory for distance functions, in the context of Riemannian manifolds with sectional curvature bounded below, originated in the work of Grove and Shiohama [GS], and was developed further by Gromov [G2], and Grove and Petersen [GP]; similar arguments for concave functions appeared even earlier in the work of Cheeger and Gromoll [CG]. For simplicity we consider only distance maps here.

2.1 Let P be a compact subset of M, $f = \text{dist}_P$. A point $q \in M \backslash P$ is called a regular point for f if $f'_q(\xi) > 0$ for some $\xi \in \Sigma_q$.

More generally, let $P_1, \ldots, P_m$ be compact subsets of M, $f_i = \text{dist}_{P_i}$, and let f denote the map with components f_i.

A point $q \in M \backslash \cup_i P_i$ is called a regular point for f if for some $\xi \in \Sigma_q$ we have $f'_{iq}(\xi) > 0$ for all i, and $\tilde{\angle} p_i q p_j > \pi/2$ for all $p_i \in P_i$, $p_j \in P_j$ such that $|qP_i| = |qp_i|$, $|qP_j| = |qp_j|$, $i \neq j$ (the second condition ensures "independence" of f_i). It is easy to see that the set of regular points is always open.

2.2 PROPOSITION.

(a) *f is a topological submersion near its regular point.*
(b) *if f is proper and has no critical points in some domain $U \subset M$, then $f|_U$ is a locally trivial fibration.*

In the next subsection we indicate how to show that f is open near its regular point.

2.3 LEMMA. *Let Σ be a compact Alexandrov space of curvature ≥ 1, and let $\Pi_1, \ldots, \Pi_{m+1}$ be its closed subsets, such that $|\Pi_i \Pi_j| > \pi/2$ for all $i \neq j$.*

Then for any pair $i \neq j$ there exists a point $\eta_{ij} \in \Sigma$ such that

$$|\Pi_i \eta_{ij}| < \pi/2 , \quad |\Pi_j \eta_{ij}| > \pi/2 , \quad |\Pi_\ell \eta_{ij}| = \pi/2 \quad \textit{for all } \ell \neq i, j .$$

This lemma is proved by induction on m and dimension of Σ.

Now let q be a regular point for f, and let $\xi \in \Sigma_q$ satisfy $f'_{iq}(\xi) > 0$ for all i. Apply our lemma for Σ_q taking $\Pi_{m+1} = \xi$, $\Pi_\ell = P'_\ell$ for $1 \leq \ell \leq m$. We get directions $\eta_i = \eta_{i,m+1} \in \Sigma_q$ such that $f'_{iq}(\eta_i) < 0$, $f'_{iq}(\eta_j) = 0$ for $i \neq j$. This allows us to prove that f is open near its regular point, using consecutive approximations: we can always improve our current approximation by moving in direction η_i if the ith coordinate of the goal is smaller than that of $f(q)$, or in direction ξ, if all the coordinates of the goal are bigger.

2.4 The proof of the proposition is more involved. We use reverse induction on m. The base $m = \dim M$ is relatively easy; in this case f is a local homeomorphism near its regular point. To carry out the step of induction, we assume that f is regular and incomplementable at q (that is, one cannot add to f another component f_{m+1}, so that the map $(f_1, \ldots, f_m, f_{m+1})$ is still regular at q), and construct a function g, strictly concave near q, and such that g has exactly one point of maximum on each "fibre" of f near q, and the map $(f_1, \ldots, f_m, g)$ is regular at all points near q except those maximum points. Thus a neighborhood of q looks like a cylinder, with f_i as the height functions and g as the radial function, and using the inductional assumption (b) for the map $(f_1, \ldots, f_m, g)$ we conclude that such a neighborhood is actually homeomorphic to a "cylinder", thus verifying the assertion (a) for f near q. The implication (a)$\Rightarrow$(b) is automatic for a certain class of stratified spaces, according to a theorem of Siebenmann [S], and our inductive argument allows us to verify the assumptions of this theorem just before it is used.

2.5 Because a distance function from a point p has no critical points in a small punctured neighborhood of p, the proposition implies that a small spherical neighborhood of p is homeomorphic to a cone on its boundary. With some more work, and using another theorem from [S], we can prove that in fact it is homeomorphic to the tangent cone at p. This easily implies by induction that an Alexandrov space is stratified into topological manifolds.

Similiar arguments prove that two compact Alexandrov spaces with the same dimension and lower curvature bound must be homeomorphic, if they are sufficiently close in Gromov-Hausdorff topology.

3 Extremal Subsets

The extremal subsets of Alexandrov spaces were defined and studied in [PP1].

3.1 Loosely speaking, an extremal subset is a subspace whose spaces of normal directions at each point have diameters at most $\pi/2$. More formally, a closed subset $F \subset M$ is called extremal if for each distance function $f = \text{dist}_p, \ p \notin F$, we have $\sup\{f'_q(\xi) : \xi \in \Sigma_q\} \leq 0$ whenever $q \in F$ and $|pF| = |pq|$. (For subsets of Alexandrov spaces of curvature ≥ 1 (usually denoted by Σ) there are additional restrictions in exceptional cases: the empty set is extremal in Σ only if diam $\Sigma \leq \pi/2$, and a single point $\xi \in \Sigma$ is an extremal subset only if diam $\Sigma_\xi \leq \pi/2$ and $|\xi\eta| \leq \pi/2$ for each $\eta \in \Sigma$.)

One can see typical examples of extremal subsets in the quotient spaces; namely, if Δ is a compact group of isometries of an Alexandrov space M and Γ is a closed subgroup of Δ, then the projection of the fixed points set of Γ into M/Δ is extremal there.

3.2 It is not hard to show that $F \subset M$ is extremal if and only if each of its spaces of tangent directions $\Sigma_p F, \ p \in F$, is extremal in the corresponding space Σ_p. Moreover, if $F_1, F_2 \subset M$ are distinct extremal subsets then $F_1 \cup F_2, F_1 \cap F_2$, $\text{clos}(F_1 \backslash F_2)$ are extremal as well. This is proved by induction on dimension.

A compact Alexandrov space has only finitely many extremal subsets.

3.3 An extremal subset is called primitive if it is not covered by its proper extremal subsets. Of course, any extremal subset is a union of primitive ones. The interior of a primitive subset is the complement of its proper extremal subsets. The interiors of all primitive subsets of M form a disjoint covering of M. It turns out that this stratification is finer than the topological one; in particular, the interior of each primitive subset is a topological manifold. This is proved by the Morse-theoretic arguments, outlined in the previous section, the crucial observation being that if under the assumption of Lemma 2.3 Σ contains an extremal subset Φ, then the points η_{ij} can be found in Φ. Here we indicate a simpler argument, which shows that the top stratum is a manifold.

Indeed, we have to check that the set F of all topological singularities of M is an extremal subset. Take $f = \text{dist}_p, \ p \notin F$, and find $q \in F$ such that $|pq| = |pF|$. If the extremality condition is violated then q is a regular point for f. Therefore, according to 2.2(a), there is an isotopy of a small neighborhood of q, which moves q closer to p. This contradicts the definition of F and the assumption $|pq| = |pF|$.

3.4 Our next goal is to relate the extremal subsets to the problem of collapsing, discussed in the introduction. First recall that if M_j collapse to M with lower curvature bound, and M happens to be a Riemannian manifold, then, according to a theorem of Yamaguchi [Y], M_j admits a structure of locally trivial fibration over M. The same is true if M is an Alexandrov space with only weak singularities. In the general case we get a fibration over a "large" open subset of M that consists of weakly singular points, the fibre F_j being a manifold of dimension $\dim M_j - \dim M$.

We expect that the fibration result is true whenever M has no proper extremal subsets. At the moment we can prove a weaker statement, that if M has no proper extremal subsets then the Serre sequence for homotopy groups of M_j, F_j, and M is exact. The proof is based on Morse-theoretic arguments as in the previous section.

3.5 In this subsection we explain in what sense the extremal subsets are totally geodesic. We need the notion of quasigeodesics in Alexandrov spaces, cf. [PP2].

Recall that a geodesic in a Riemannian manifold is both the locally shortest curve and the straightest one. In singular spaces the straightest curves, called quasigeodesics, need not be locally shortest. Alexandrov introduced quasigeodesics on convex surfaces in $\mathbb{R}^3$, see [A1]; an appropriate definition of quasigeodesics in piecewise euclidean spaces of arbitrary dimension was proposed by Milka, and in general Alexandrov spaces by Petrunin. Loosely speaking, a curve, parametrized by the arclength, is called a quasigeodesic if the restriction of each distance function to this curve is at least as concave as the restriction of such a function in a model surface of constant curvature to a geodesic.

It can be shown that this definition is equivalent to a local one, which is independent of the value of the lower curvature bound, and that every quasigeodesic in a Riemannian manifold is a geodesic. Unlike geodesics in Alexandrov spaces, quasigeodesics can be constructed (not always uniquely) for arbitrary initial data, are extendable, and have a natural compactness property.

The totally geodesic property of extremal subsets F means two things:

(a) Every connected component of F has an intrinsic metric, and every shortest geodesic in this intrinsic metric is a quasigeodesic in the ambient space.
(b) Given a point on F and a direction at this point, tangent to F, there is an infinite quasigeodesic (of the ambient space) with prescribed initial data, contained in F.

The property (a) is a generalization of a lemma, proved by Liberman [L] for the intrinsic metric of a convex surface in $\mathbb{R}^3$.

4 Speculations

4.1 The previous discussion makes it natural to expect that the intrinsic metric of the interior of a primitive subset inherits the lower curvature bound of the ambient space. However, Petrunin recently constructed counterexamples to this conjecture; in particular, he has shown that the Veronese embedding of $\mathbb{R}P^2$ into S^5 is an extremal subset of its convex hull there. (The question is still open for extremal subsets of codimension 1 and 2.) Petrunin also proved that the first variation formula is valid for an intrinsic metric of any extremal subset, see [Pet1]. The author believes that extremal subsets should be considered as examples of (yet undefined) spaces with bounded integral curvature.

4.2 Probably the most comprehensive conjecture about Riemannian manifolds of nonnegative sectional curvature is the rational ellipticity conjecture, see [GH]. If one believes in such a conjecture, one should expect that some of its version is valid for nonnegatively curved Alexandrov spaces. Most likely, such a version would take into account not only the topological data, but also the extremal subsets. Perhaps the conjectures about positively curved manifolds can also be included in the same statement because suspensions and joins of such manifolds are positively curved Alexandrov spaces.

4.3 It would be interesting to obtain realistic estimates of the number of primitive subsets in nonnegatively curved Alexandrov spaces of a given dimension. Such an estimate would probably eventually lead to an estimate for Betti numbers; on the other hand, the problem for primitive subsets looks easier.

The author adapted the argument of Danzer and Grunbaum [DG] to prove a sharp estimate $2^{\dim M}$ for the number of zero-dimensional primitive subsets in a nonnegatively curved space M. The proof is very simple:

Let a_i, $1 \leq i \leq N$, be our extremal points; consider "homotheties" of M with respect to a_i with multiple $\frac{1}{2}$, that is, let $M_i = \{x \in M : x$ is a midpoint of some shortest geodesic $a_i\overline{x}_i$, $\overline{x}_i \in M\}$. By the volume comparison, $\mathrm{Vol}(M_i) \geq 2^{-\dim M}\mathrm{Vol}(M)$. On the other hand, the sets M_i are essentially disjoint. (Indeed, if $x \in M_i \cap M_j$ and, say, $|a_i x| > |a_j x|$, then applying the Toponogov angle comparison a couple of times we get $\tilde{\angle} a_i a_j \overline{x}_i > \pi/2$, which is impossible because a_j is extremal.) Therefore, $N \leq 2^{\dim M}$.

References

[A1] A. D. Alexandrov, Intrinsic geometry of convex surfaces, Moscow (1948).

[A2] A. D. Alexandrov, *Über eine Verallgemeinerung der Riemannschen Geometrie*, Schriftenreihe Inst. Math. **1** (1957), 33–84.

[BGP] Yu. Burago, M. Gromov, and G. Perelman, *Alexandrov spaces with curvature bounded below*, Uspekhi Mat. Nauk **47/2** (1992), 3–51.

[B] S. Buyalo, *Shortest lines on convex hypersurfaces in a Riemannian space*, Zap. Nauchn. Sem. LOMI **66** (1976), 114–131.

[BN] V. N. Berestovskii and I. G. Nikolaev, *Multidimensional generalized Riemannian spaces*, Encycl. Math. Sci. **70**, Springer-Verlag, Berlin (1993), 165–243.

[CFG] J. Cheeger, K. Fukaya, and M. Gromov, *Nilpotent structures and invariant metrics on collapsed manifolds*, J. Amer. Math. Soc. **5/2** (1992), 327–372.

[CG] J. Cheeger and D. Gromoll, *On the structure of complete manifolds of nonnegative curvature*, Ann. of Math. (2) **96** (1972), 413–443.

[DG] L. Danzer and B. Grunbaum, *Über zwei Probleme bezüglich konvexer Körper von P. Erdős und von V. L. Klee*, Math. Z. **79** (1962), 95–99.

[G1] M. Gromov, *Synthetic geometry in Riemannian manifolds*, Proc. Internat. Congress Math. 1978, Helsinki, Acad. Sci. Fennica (1980), 415–419.

[G2] M. Gromov, *Curvature, diameter and Betti numbers*, Comment. Math. Helv. **56** (1982), 179–195.

[GLP] M. Gromov, J. Lafontaine, and P. Pansu, *Structures métriques pour les variétés riemanniennes*, Textes Math. no. **1**, Cedic, Paris (1981).

[GH] K. Grove and S. Halperin, *Contributions of rational homotopy theory to global problems in geometry*, Publ. IHES **56** (1983), 379–385.

[GP] K. Grove and P. Petersen, *Bounding homotopy types by geometry*, Ann. of Math. (2) **128** (1988), 195–206.

[GS] K. Grove and K. Shiohama, *A generalized sphere theorem*, Ann. of Math. (2) **106** (1977), 201–211.

[L] I. Liberman, *Geodesic lines on convex surfaces*, Dokl. Akad. Nauk SSSR **32/5** (1941), 310–313.

[P] G. Perelman, *Elementary Morse theory on Alexandrov spaces*, St. Petersburg Math. J. **5/1** (1994), 207–214; see also preprint, G. Perelman, *Alexandrov spaces with curvature bounded below, II.*

[PP1] G. Perelman and A. Petrunin, *Extremal subsets in Alexandrov spaces and a generalized Liberman theorem*, St. Petersburg Math. J. **5/1** (1994), 215–227.

[PP2] G. Perelman and A. Petrunin, *Quasigeodesics and gradient curves in Alexandrov spaces*, preprint.

[Pet1] A. Petrunin, *Applications of quasigeodesics and gradient curves*, preprint.

[Pet2] A. Petrunin, *Parallel transportation and second variation*, preprint.

[Pl] C. Plaut, *Spaces of Wald curvature bounded below*, preprint.

[R] Yu. G. Reshetnyak, *Two-dimensional manifolds of bounded curvature*, Encycl. Math. Sci. **70**, Springer-Verlag, Berlin (1993), 3–163.

[S] L. Siebenmann, *Deformations of homeomorphisms on stratified sets*, Comment. Math. Helv. **47** (1972), 123–163.

[Y] T. Yamaguchi, *Collapsing and pinching under a lower curvature bound*, Ann. of Math. (2) **133** (1991), 317–357.

Lagrangian Intersections, 3-Manifolds with Boundary, and the Atiyah-Floer Conjecture

DIETMAR SALAMON

Mathematics Institute
University of Warwick
Coventry CV4 7AL
United Kingdom

1. Introduction

It has been observed by physicists for a long time that symplectic structures arise naturally from boundary value problems. For example, the **Robbin quotient**

$$V = \operatorname{dom} D^* / \operatorname{dom} D,$$

associated to a symmetric (but not self-adjoint) operator $D : \operatorname{dom} D \to H$ on a Hilbert space H carries a symplectic structure

$$\omega(v, w) = \langle D^* v, w \rangle - \langle v, D^* w \rangle.$$

Self-adjoint extensions of D correspond to Lagrangian subspaces of V and, moreover, the kernel of D^* determines such a Lagrangian subspace whenever D has a closed range. If D is a symmetric differential operator on a manifold with boundary, then, by partial integration, the form ω is given by an integral over the boundary. For example, if D is the Hessian of the symplectic action functional on paths in $\mathbb{R}^{2n}$, then the space $V = \mathbb{R}^{2n} \times \mathbb{R}^{2n}$ corresponds to the two boundary values of the path and the symplectic structure is $(-\omega_0) \oplus \omega_0$ where $\omega_0 = \sum_j dx_j \wedge dy_j$ is the standard symplectic structure on $\mathbb{R}^{2n}$. A more interesting example is given by the Chern-Simons functional on 3-manifolds with boundary, and we shall discuss this in Section 2.

In another closely related direction there is a formal analogy between symplectic manifolds with symplectomorphisms and Lagrangian submanifolds on the one hand and oriented Riemann surfaces with orientation-preserving diffeomorphisms and 3-dimensional bordisms on the other. If Σ is a compact oriented Riemann surface we denote by $\bar{\Sigma}$ the surface with the opposite orientation. Likewise we denote by M a symplectic manifold without mentioning the symplectic form ω explicitly and by $\bar{M}$ the manifold with reversed symplectic form (i.e. ω is replaced by $-\omega$). The following diagram (next page) summarizes the correspondence.

The last correspondence between the gluing operation for 3-manifolds with boundary and symplectic reduction of Lagrangian submanifolds is the most important one. The manifold Y' is obtained from Y by identifying the boundary

Proceedings of the International Congress
of Mathematicians, Zürich, Switzerland 1994

oriented Riemann surface Σ	symplectic manifold M
$\bar{\Sigma}$	$\bar{M}$
or. pres. diffeomorphism	symplectomorphism
$\Sigma = \Sigma_1 \cup \Sigma_2$	$M = M_1 \times M_2$
3-mfld Y with $\partial Y = \Sigma$	Lagrangian submfld $L \subset M$
$Y = Y_1 \cup Y_2$	$L = L_1 \times L_2 \subset M_1 \times M_2$
gluing	symplectic reduction
$\partial Y = \Sigma \cup \bar{\Sigma} \cup \Sigma'$	$L \subset M \times \bar{M} \times M'$,
$\partial Y' = \Sigma'$	$L' = \{x' \in M' \mid \exists x : (x, x, x') \in L\}$.

components Σ and $\bar{\Sigma}$ via the identity map. Similarly, $N = \Delta \times M'$ is a coisotropic submanifold of $M \times \bar{M} \times M'$ with symplectic quotient M', and L' is obtained from L via symplectic reduction. Of course, this analogy can be extended to dimensions other than 2. In the 0-dimensional case where Riemann surfaces are replaced by points and 3-manifolds with boundary by intervals the correspondence is given by the symplectic action. In dimension 2 it is given by the Chern-Simons functional; this will be discussed in Section 2. In Section 3 we shall see that this leads to a natural extension of Floer homology in the form

$$HF^*(Y, L), \qquad L \subset M_\Sigma,$$

where Y is a 3-manifold with boundary $\partial Y = \Sigma$, M_Σ denotes the moduli space of flat SU(2) (or SO(3)) connections over Σ, and $L \subset M_\Sigma$ is a Lagrangian submanifold. Such groups were also considered by Fukaya [11] but his definition differs slightly from the one discussed here. Our goal in this note is to explain the definition of these Floer homology groups and to show how they can be used to prove the Atiyah-Floer conjecture for Heegaard splittings of homology-3-spheres. We shall only outline the main ideas of the proofs. Details will be published elsewhere.

2. Chern-Simons functional

Let Y be a compact 3-manifold with boundary $\partial Y = \Sigma$ and consider the trivial bundle $Y \times \mathrm{G}$ with structure group $\mathrm{G} = \mathrm{SU}(2)$ and Lie algebra $\mathfrak{g} = \mathfrak{su}(2) = \mathrm{Lie}(\mathrm{G})$. The space $\mathcal{A} = \mathcal{A}(Y) = \Omega^1(Y, \mathfrak{g})$ of SU(2)-connections on Y carries a natural 1-form defined by

$$\alpha \mapsto \mathcal{F}_A(\alpha) = \int_Y \langle F_A \wedge \alpha \rangle \tag{1}$$

for $\alpha \in T_A\mathcal{A} = \Omega^1(Y, \mathfrak{g})$. Here $\langle \xi, \eta \rangle = \mathrm{trace}(\xi^* \eta)$ for $\xi, \eta \in \mathfrak{g}$ and $F_A \in \Omega^2(Y, \mathfrak{g})$ denotes the curvature of A. The 1-form (1) is invariant and horizontal with respect to the action of the gauge group $\mathcal{G}(Y) = \mathrm{Map}(Y, \mathrm{G})$. But it is not closed because

$$d\mathcal{F}_A(\alpha, \beta) = \int_Y \langle d_A\alpha \wedge \beta \rangle - \int_Y \langle \alpha \wedge d_A\beta \rangle = \int_{\partial Y} \langle \alpha \wedge \beta \rangle.$$

This is the standard symplectic structure on the space $\mathcal{A}(\Sigma) = \Omega^1(\Sigma, \mathfrak{g})$ of connections on Σ. It reflects the failure of the operator $*d_A : \Omega^1(Y, \mathfrak{g}) \to \Omega^1(Y, \mathfrak{g})$ to be self-adjoint. This operator can be interpreted as the differential of the vector field $A \mapsto *F_A$ on $\mathcal{A}(Y)$ associated to the 1-form $\mathcal{F}$.

In order to obtain a closed 1-form we pick some Lagrangian submanifold $\mathcal{L} \subset \mathcal{A}(\Sigma)$ and consider the subspace $\mathcal{A}(Y, \mathcal{L}) \subset \mathcal{A}(Y)$ of those connections on Y that have boundary values in $\mathcal{L}$. The restriction of $\mathcal{F}$ to this space is closed precisely when $\mathcal{L}$ is Lagrangian. Moreover, in order to preserve the invariance of $\mathcal{F}$ under the gauge group we should also assume that $\mathcal{L}$ is invariant under the action of $\mathcal{G}(\Sigma) = \mathrm{Map}(\Sigma, \mathrm{G})$. But this is equivalent to the condition

$$\mathcal{L} \subset \mathcal{A}_{\mathrm{flat}}(\Sigma) = \{A \in \mathcal{A}(\Sigma) \,|\, F_A = 0\}$$

and thus $\mathcal{L}$ determines a Lagrangian submanifold

$$L = \frac{\mathcal{L}}{\mathcal{G}(\Sigma)} \subset \frac{\mathcal{A}_{\mathrm{flat}}(\Sigma)}{\mathcal{G}(\Sigma)} = M_\Sigma$$

of the moduli space M_Σ of flat SU(2)-connections on Σ. This is a $(6g-6)$-dimensional symplectic manifold (with singularities). We shall assume that L is simply connected and contains the equivalence class of the zero connection. Note that in this case the space $\mathcal{L}$ is not simply connected, but the fundamental group of $\mathcal{L}$ cancels with that of $\mathcal{G}(\Sigma)$. Now the 1-form $\mathcal{F} : T\mathcal{A}(Y, \mathcal{L}) \to \mathbb{R}$ is closed. But because $\mathcal{L}$ is not simply connected $\mathcal{F}$ is not exact. However, it is the differential of the multi-valued **Chern-Simons functional** $\mathcal{CS} : \mathcal{A}(Y, \mathcal{L}) \to \mathbb{R}/4\pi^2\mathbb{Z}$ defined by

$$\mathcal{CS}(A) = \tfrac{1}{2} \int_Y \left(\langle A \wedge dA \rangle + \tfrac{1}{3} \langle [A \wedge A] \wedge A \rangle \right) + \int_0^1 \int_\Sigma \langle A_0(t) \wedge \dot{A}_0(t) \rangle \, dt.$$

Here $A_0(t) \in \mathcal{L}$ is a path with $A_0(0) = 0$ and $A_0(1) = A|_\Sigma$. The homotopy class of this path is not unique and hence the right-hand side is only well defined up to an integer multiple of $4\pi^2$.

Now the 3-manifold Y itself also determines a Lagrangian submanifold

$$L_Y = \frac{\mathcal{L}_Y}{\mathcal{G}(\Sigma)}, \qquad \mathcal{L}_Y = \{A|_\Sigma \,|\, A \in \mathcal{A}_{\mathrm{flat}}(Y)\}.$$

Note that under the correspondence $Y \mapsto L_Y$ (from bordisms to Lagrangian submanifolds) the summing of 3-manifolds along common boundaries translates into symplectic reduction. Note also that the flat connections on Y are in fact the zeros of the 1-form $\mathcal{F} = d\mathcal{CS}$. Hence there is a map

$$\mathrm{Crit}(\mathcal{CS}) \to L_Y \cap L$$

which assigns to every critical point $A \in \mathcal{A}_{\mathrm{flat}}(Y, \mathcal{L})$ of $\mathcal{CS}$ the equivalence class $[A|_\Sigma]$ in $M_\Sigma = \mathcal{A}_{\mathrm{flat}}(\Sigma)/\mathcal{G}(\Sigma)$. In some cases, e.g. when Y is a handlbody, the connection $A \in \mathcal{A}_{\mathrm{flat}}(Y)$ is uniquely determined (up to gauge equivalence) by $A|_\Sigma$ and in this case the above map is a bijection.

3. Floer homology

Let Y be a 3-manifold with boundary $\partial Y = \Sigma$ and $\mathcal{L} \subset \mathcal{A}_{\text{flat}}(\Sigma)$ be a Lagrangian submanifold with simply connected quotient $L = \mathcal{L}/\mathcal{G}(\Sigma)$. Then the gradient flow lines of the Chern-Simons functional $\mathcal{CS} : \mathcal{A}(Y, \mathcal{L}) \to \mathbb{R}/4\pi^2\mathbb{Z}$ are smooth maps $\mathbb{R} \to \Omega^1(Y, \mathfrak{g}) \times \Omega^0(Y, \mathfrak{g}) : t \mapsto (A(t), \Psi(t))$ that satisfy the boundary value problem

$$\dot{A} - d_A\Psi + *F_A = 0, \qquad A|_\Sigma \in \mathcal{L}, \qquad *A|_\Sigma = 0. \tag{2}$$

For any such gradient line the connection $A + \Psi\, dt$ on the 4-manifold $X = Y \times \mathbb{R}$ is a self-dual Yang-Mills instanton with Lagrangian boundary condition on $\partial X = \Sigma \times \mathbb{R}$. Under suitable conditions on Y and $\mathcal{L}$ the Yang-Mills energy of such an instanton is finite if and only if (in a suitable gauge) there are limits

$$\lim_{t\to\pm\infty} A(t) = A^\pm, \qquad \lim_{t\to\pm\infty} \Psi(t) = 0 \tag{3}$$

where $A^\pm \in \mathcal{A}_{\text{flat}}(Y, \mathcal{L})$ are flat connections and hence critical points of $\mathcal{CS}$. If these limits are regular and nondegenerate (i.e. the extended Hessian is bijective) then one can prove that equations (2) and (3) form a well-posed nonlinear elliptic boundary value problem and so, for a generic metric, the space $\mathcal{M}(A^-, A^+)$ of solutions modulo gauge equivalence is a finite-dimensional manifold of dimension

$$\dim \mathcal{M}(A^-, A^+) = \mu(A^-) - \mu(A^+) \pmod 8$$

for some function $\mu : \text{Crit}^*(Y, \mathcal{L}) \to \mathbb{Z}/8\mathbb{Z}$ on the set of irreducible flat connections in $\mathcal{A}(Y, \mathcal{L})$. Here the dimension depends on the component in the space of paths in $\mathcal{A}(Y, \mathcal{L})$ running from A^- to A^+, in contrast to the closed case where any two paths are homotopic and the index ambiguity only comes in after dividing by the gauge group.

REMARK 3.1 The well posedness of (2) and (3) extends to general 4-manifold X with boundary $\partial X = \Sigma \times \mathbb{R}$ and cylindrical ends. The proof involves an estimate for the operator $D = d_A^- \oplus d_A^* : \Omega^1_{\mathcal{L}}(X, \mathfrak{g}) \to \Omega^{2,-}(X, \mathfrak{g}) \oplus \Omega^0(X, \mathfrak{g})$ where $\Omega^1_{\mathcal{L}}(X, \mathfrak{g})$ denotes the subset of all $\alpha \in \Omega^1(X, \mathfrak{g})$ that satisfy

$$\alpha|_{\Sigma\times t} \in \Lambda(t) = T_{A|_{\Sigma\times t}}\mathcal{L}, \qquad \alpha \circ \nu_{\partial X} = 0.$$

There is an inequality

$$\|\alpha\|^2_{W^{1,2}} \leq c\left(\|D\alpha\|^2_{L^2} + \|\alpha\|^2_{L^2}\right) + \int_{\partial X} \langle d_A\alpha \wedge \alpha \rangle$$

and, in view of the Lagrangian boundary conditions, the boundary term can be estimated by

$$\left| \int_{\partial X} \langle d_A\alpha \wedge \alpha \rangle \right| \leq c\,\|\alpha\|^2_{L^2(\partial X)} \leq c'\,\|\alpha\|_{W^{1,2}(X)}\,\|\alpha\|_{L^2(X)}.$$

Now the elliptic estimate $\|\alpha\|_{W^{1,2}} \leq c(\|D\alpha\|_{L^2} + \|\alpha\|_{L^2})$ easily follows. This has to be combined with elliptic regularity at the boundary to obtain the required Fredholm theory.

To obtain finiteness in the case where the index difference is 1 we must employ Uhlenbeck's compactness theorem in the case of bounded curvature and combine this with the usual bubbling argument if there is only an L^2-bound on the curvature. Such a bound is always guaranteed as

$$YM(A+\Psi\,dt) = \int_{-\infty}^{\infty} \left\|F_{A(t)}\right\|_{L^2(Y)}^2 dt = \mathcal{CS}(A^-) - \mathcal{CS}(A^+)$$

for every solution of (2) and (3). However, bubbling near the boundary produces nontrivial finite-energy instantons on a half-space

$$\mathbb{H}^4 = \{x = (x_0, x_1, x_2, x_3) \in \mathbb{R}^4 \,|\, x_0 \geq 0\} \,,$$

which on the boundary $\mathbb{R}^3 = \mathbb{R} \times \mathbb{R}^2$ are flat on each $\mathbb{R}^2$-slice. Such instantons should have Yang-Mills energy equal to an integer multiple of $8\pi^2$.

CONJECTURE 3.2 *Let $A = \sum_{j=0}^3 A_j dx_j \in \Omega^1(\mathbb{H}^4, \mathfrak{g})$ be a connection such that*

$$F_{01} = F_{23}, \quad F_{02} = F_{31}, \quad F_{03} = F_{12}, \qquad A_0|_{\partial\mathbb{H}^4} = 0, \quad F_{23}|_{\partial\mathbb{H}^4} = 0,$$

where $F_{ij} = \partial_i A_j - \partial_j A_i + [A_i, A_j]$. Then either $F_{ij} = 0$ for all i, j or

$$\mathcal{YM}(A) = \frac{1}{2}\int_{\mathbb{H}^4} \sum_{i<j} |F_{ij}|^2 \geq 8\pi^2.$$

The proof will go along the lines of Uhlenbeck's removable singularity theorem. At the time of writing I have not carried out the details. If this holds then in the case of index difference 1 the space $\mathcal{M}(A^-, A^+)$ will consist of only finitely many connecting orbits (moduli time shift) and, as in the case of closed 3-manifolds [9, 6], counting these gives rise to a Floer chain complex

$$CF_*(Y, \mathcal{L}) = \bigoplus_{[A] \in \mathcal{A}^*_{\mathrm{flat}}(Y,\mathcal{L})/\mathcal{G}(Y)} \mathbb{Z}\langle A\rangle$$

generated by the gauge equivalence classes of irreducible flat connections. The boundary operator is defined by

$$\partial\,\langle A^-\rangle = \sum_{A^+} \#\{\mathcal{M}_1(A^-, A^+)/\mathbb{R}\}\,\langle A^+\rangle$$

where the sum runs over all equivalence classes $[A^+] \in \mathcal{A}^*_{\mathrm{flat}}(Y, \mathcal{L})/\mathcal{G}(Y)$ with $\mu(A^+) = \mu(A^-) - 1 (\mathrm{mod}\ 8)$ and $\mathcal{M}_1$ denotes the 1-dimensional components of the moduli space. As in Floer's original work [9] one uses a gluing theorem to prove that $\partial^2 = 0$. The resulting Floer cohomology groups are denoted by $HF^*(Y, L) = H^*(CF, \partial)$. They are graded modulo 8.

REMARK 3.3 **(i)** To make these ideas work we must impose certain conditions on Y and L that guarantee that there are no reducible flat connections in $\mathcal{A}(Y, \mathcal{L})$ other than the equivalence class of the zero connection. Here **reducible** means that the kernel of $d_A : \Omega^0(Y, \mathfrak{g}) \to \Omega^1(Y, \mathfrak{g})$ is zero. For example, we may assume that $L = L_{Y'}$ where Y' is a handlebody with $\partial Y' = \bar{\Sigma}$ and $Y \cup_\Sigma Y'$ is a homology-3-sphere.

(ii) A connection $A \in \mathcal{A}_{\text{flat}}(Y, \mathcal{L})$ is called **nondegenerate** if every $\alpha \in \Omega^1(Y, \mathfrak{g})$ with $d_A\alpha = d_A^*\alpha = 0$ and $*\alpha|_\Sigma = 0$, $\alpha|_\Sigma \in T_A\mathcal{L}$ is equal to zero. If there are degenerate flat connections then we must choose a perturbation of the Chern-Simons functional (as in the case of closed 3-manifolds) to obtain well-defined Floer homology groups.

(iii) The Floer cohomology groups $HF^*(Y, L)$ are independent of the choice of the metric and the perturbation used to define them. They depend on the Lagrangian submanifold L only up to Hamiltonian isotopy. More precisely, for different choices there are natural isomorphisms of Floer homology.

(iv) In [11] Fukaya proposed an alternative construction of Floer homology groups on 3-manifolds with boundary.

CONJECTURE 3.4 *If $Y = Y_0 \cup_\Sigma Y_1$ is a homology-3-sphere, and Y_0 is a handlebody, then there is a natural isomorphism $HF^*(Y) \cong HF^*(Y_1, L_{Y_0})$. If Y_1 is also a handlebody (i.e. in the case of a Heegaard splitting) there is a natural isomorphism*

$$HF^*(Y) \cong HF^*(\Sigma \times [0,1], L_{Y_0} \times L_{Y_1}).$$

At the time of writing the details of the proof have not been carried out. However, here is the main idea. Choose a map $f : \Sigma \times [0,1] \to Y_0$ that identifies $\Sigma \times \{1\}$ with ∂Y_0 and shrinks $\Sigma \times \{0\}$ onto the 1-skeleton of Y_0. Then the pullback of any connection on Y_0 onto $\Sigma \times \{0\}$ is in $\mathcal{L}_{Y_0}$. (A connection on a 1-manifold is just given by the holonomy.) The ASD equation on $\Sigma \times [0,1] \times \mathbb{R}$ with the pullback metric then takes the form

$$\partial_s A - d_A\Phi + *_s(\partial_t A - d_A\Psi) = 0, \quad \partial_s\Psi - \partial_t\Phi + [\Phi, \Psi] + *_s F_A = 0 \qquad (4)$$

where the change of the metric is not in the same conformal class and degenerates at $s = 0$. Note that (4) is just the ASD equation on (half of) the closed manifold $Y \times \mathbb{R}$. If, however, we consider the equation on the interval $s \geq \varepsilon$ for some $\varepsilon > 0$ then we obtain a genuine boundary value problem. The solutions of these should converge to those on the closed manifold as $\varepsilon \to 0$, and this will prove Theorem 3.4. Note that the degeneration of the metric at $s = 0$ is related to the choice of the Lagrangian boundary condition in L_{Y_0}. The case $\Sigma = S^2$ is slightly simpler. In this case $Y_1 = B^3$ and we can take the map $S^2 \times [0,1] \to B^3 : (x, s) \mapsto sx$. Then the change in the metric is conformal and so the Hodge-$*$-operator on 1-forms is independent of s while $*_s F_A = s^{-2} * F_A$. Similar equations were recently studied by Fukaya [12].

4. Atiyah-Floer conjecture

In [8] Floer introduced what is now called Floer homology for Lagrangian intersections. Assume for simplicity that (M, ω) is a compact simply connected symplectic manifold that is **positive** in the sense that the first Chern class $c_1 = c_1(TM)$ (with respect to an ω-compatible almost complex structure) is a positive multiple of the cohomology class $[\omega]$. We also assume that the minimal Chern number N, defined by $\langle c_1, \pi_2(M)\rangle = N\mathbb{Z}$, is at least 2. Then for any two Lagrangian submanifolds $L_0, L_1 \subset M$ with torsion fundamental group there are Floer homology groups

$$HF^*_{\mathrm{symp}}(L_0, L_1)$$

that are graded modulo $2N$. In this theory the critical points are the intersection points $x \in L_0 \cap L_1$ and the connecting orbits are J-holomorphic strips $u : [0,1] \times \mathbb{R} \to M$ that satisfy

$$\partial_s u + J(u)\partial_t u = 0, \quad u(0,t) \in L_0, \quad u(1,t) \in L_1, \quad \lim_{t\to\pm\infty} u(s,t) = x^{\pm}, \tag{5}$$

where $x^{\pm} \in L_0 \cap L_1$ and J is an almost complex structure on M, which is compatible with ω in the sense that $\langle \xi, \eta\rangle = \omega(J\xi, \eta)$ is a Riemannian metric. This construction requires transversal intersections of the Lagrangian submanifolds and surjectivity of the linearized Cauchy-Riemann operators. These conditions can be achieved by a suitable Hamiltonian perturbation and, as before, the resulting Floer homology groups are indepedent of the almost complex structure and the Hamiltonian perturbation used to define them [8], [15].

Now consider the case where $M = M_\Sigma$ is the moduli space of flat SU(2)-connections on a compact oriented Riemann surface Σ and $L_i = L_{Y_i}$ for $i = 0, 1$ where Y_0 and Y_1 are handlebodies with $\partial Y_0 = \Sigma$ and $\partial Y_1 = \bar{\Sigma}$. Then the manifold M_Σ is simply connected and positive in the above sense with minimal Chern number 4. Moreover, the Lagrangian manifolds L_0 and L_1 are diffeomorphic to the quotient of $\mathrm{SU}(2)^g$ by simultaneous conjugacy and hence are obviously simply connected. However, some care must be taken when extending symplectic Floer homology to M_Σ in view of the singularities. To obtain a well-defined theory we must assume that $Y = Y_0 \cup_\Sigma Y_1$ is a homology-3-sphere so that 0 is the only singular intersection point of L_0 and L_1. Another point is to give the right definition of holomorphic curves when they pass through the singular part of M_Σ. The correct definition should be that they can be represented locally by a smooth map $\mathbb{C} \to \mathcal{A}(\Sigma) \oplus \Omega^0(\Sigma) \oplus \Omega^0(\Sigma) : s + it \mapsto (A(s,t), \Phi(s,t), \Psi(s,t))$ such that

$$\partial_s A - d_A\Phi + *(\partial_t A - d_A\Psi) = 0, \qquad F_A = 0. \tag{6}$$

Using the local Coulomb gauge in $\mathcal{A}(\Sigma)$, in a neighborhood of a (possibly singular) connection $A_0 = A(0,0)$, one can show that every $W^{1,p}$-solution of (6) is gauge equivalent to a smooth solution. One should then be able to use a transversality argument in the moduli space $\mathcal{A}_{\mathrm{flat}}(\Sigma)/\mathcal{G}_0(\Sigma)$ of flat connections modulo pointed gauge transformations to prove that generic holomorphic curves avoid the singular set, because it is of codimension larger than 2 if the genus of Σ is at least 3.

As a result there are symplectic Floer cohomology groups for $(M_\Sigma, L_{Y_0}, L_{Y_1})$ whenever $Y = Y_0 \cup_\Sigma Y_1$ is a Heegaard splitting of a homology-3-sphere. It was conjectured by Atiyah and Floer that there should be a natural isomorphism

$$HF^*(Y) = HF^*_{\mathrm{symp}}(M_\Sigma, L_{Y_0}, L_{Y_1}).$$

In view of Conjecture 3.4 this reduces to the following.

CONJECTURE 4.1 *For every Heegaard splitting $Y = Y_0 \cup_\Sigma Y_1$ of a homology-3-sphere there is a natural isomorphism of Floer cohomologies*

$$HF^*(\Sigma \times [0,1], L_{Y_0} \times L_{Y_1}) \cong HF^*_{\mathrm{symp}}(M_\Sigma, L_{Y_0}, L_{Y_1}).$$

The proof of Conjecture 4.1 follows the line of argument in [7] for mapping cylinders. The key idea is to conformally rescale the metric on Σ by a factor $\varepsilon^2 > 0$ and prove that in the limit $\varepsilon \to 0$ the ASD instantons on $\Sigma \times [0,1] \times \mathbb{R}$ degenerate into holomorphic curves. More precisely, the ASD equation on $\Sigma \times [0,1] \times \mathbb{R}$ with respect to the rescaled metric takes the form

$$\partial_s A - d_A \Phi + *(\partial_t A - d_A \Psi) = 0, \quad \partial_s \Psi - \partial_t \Phi + [\Phi, \Psi] + \frac{1}{\varepsilon^2} * F_A = 0, \qquad (7)$$

with boundary conditions

$$A(0,t) \in \mathcal{L}_{Y_0}, \qquad A(1,t) \in \mathcal{L}_{Y_1}, \qquad \Phi(0,t) = \Phi(1,t) = 0. \qquad (8)$$

The proof that for $\varepsilon \to 0$ the solutions of (7) and (8) converge to those of (6) is almost word for word the same as in [7]. An important ingredient in the proof is the observation that the Yang-Mills energy (with respect to the ε-dependent metric) of a connection $\Xi = A + \Phi\, ds + \Psi\, dt$ that satisfies (7) is given by

$$\mathcal{YM}_\varepsilon(\Xi) = \int_{-\infty}^{\infty} \int_0^1 \left(\|\partial_s A - d_A \Phi\|^2_{L^2(\Sigma)} + \frac{1}{\varepsilon^2} \|F_A\|^2_{L^2(\Sigma)} \right) ds\, dt.$$

The main differences in the proof are that, first, the estimates on the curvature in [7], Section 7, must be established near the boundary, second, the bubbling argument requires Conjecture 3.2, and third, care must be taken near the singularities of the moduli space. Details will be carried out elsewhere.

5. Products

There are interesting product structures on Floer cohomology due to Donaldson. Let (M, ω) be a compact simply connected symplectic manifold that is positive in the above sense with minimal Chern number $N \geq 2$. Then there is a **quantum category** $\mathcal{C}_M$ whose objects are the Lagrangian submanifolds $L \subset M$ with torsion fundamental group and whose morphisms are Floer cohomology classes. Thus $\mathrm{Mor}(L_0, L_1) = HF^*_{\mathrm{symp}}(L_0, L_1)$. The composition rule appears as a product structure

$$HF^*_{\mathrm{symp}}(L_0, L_1) \otimes HF^*_{\mathrm{symp}}(L_1, L_2) \to HF^*_{\mathrm{symp}}(L_0, L_2).$$

On the chain level this homomorphism is given by *counting J-holomorphic triangles*. More precisely, one considers J-holomorphic maps $u : \Omega \to M$ defined on a domain $\Omega \subset \mathbb{C}$ with three smooth boundary components and three cylindrical ends that map the boundary components to L_0, L_1, and L_2, respectively, and in the cylindrical ends converge to intersection points. To obtain a well-defined Fredholm theory one can choose Hamiltonian perturbations in the cylindrical ends.

The resulting product is associative in homology but not on the chain level. The proof of associativity involves domains with four cylindrical ends and leads to the A_∞-category of Fukaya [11].

Now there are similar product structures for homology-3-spheres. If Y_0, Y_1, Y_2 are three handlebodies with boundary $\partial Y_i = \Sigma$ such that the manifolds $Y_i \cup \bar{Y}_j$ are homology-3-spheres for $i \neq j$ then there is a product

$$HF^*(Y_0 \cup \bar{Y}_1) \otimes HF^*(Y_1 \cup \bar{Y}_2) \to HF^*(Y_0 \cup \bar{Y}_2).$$

This can be defined in terms of ASD instantons on a cobordism X that is obtained from $\Omega \times \Sigma$ by gluing $Y_0 \times \mathbb{R}$, $Y_1 \times \mathbb{R}$, $Y_2 \times \mathbb{R}$ to the three boundary components (which are all diffeomorphic to $\Sigma \times \mathbb{R}$). The natural extension of the Atiyah-Floer conjecture asserts that these two product structures should correspond under the isomorphisms of Conjectures 3.4 and 4.1 if in the symplectic case we choose $M = M_\Sigma$ and $L_i = L_{Y_i}$. This can be proved with the same techniques as above.

An interesting special case occurs when the symplectic manifold M is replaced by $\bar{M} \times M$ and $L_0 = \Delta$, $L_1 = \mathrm{graph}\,(\phi)$, $L_2 = \mathrm{graph}\,(\psi\phi)$. This gives rise to Floer cohomology groups

$$HF^*_{\mathrm{symp}}(\phi) = HF^*_{\mathrm{symp}}(\bar{M} \times M, \Delta, \mathrm{graph}\,(\phi)).$$

Intuitively, the Floer cohomology of ϕ can be interpreted as the *middle-dimensional* cohomology of the space Ω_ϕ of paths $\gamma : [0,1] \to M$ with $\gamma(1) = \phi(\gamma(0))$

$$HF^*_{\mathrm{symp}}(\phi) = H^{\frac{1}{2}\infty}(\Omega_\phi).$$

These groups are invariant under conjugacy, i.e. $HF^*_{\mathrm{symp}}(\phi) = HF^*_{\mathrm{symp}}(\psi^{-1}\phi\psi)$, and the Donaldson product structure takes the form

$$HF^*_{\mathrm{symp}}(\phi) \otimes HF^*_{\mathrm{symp}}(\psi) \to HF^*_{\mathrm{symp}}(\psi\phi). \tag{9}$$

In the case $\phi = \psi = \mathrm{id}$ there is a natural isomorphism $HF^*(\mathrm{id}) = H^*(M)$ (with the grading made periodic with period $2N$) and the above product reduces to quantum cohomology [17]. (See [14] for an exposition of quantum cohomology.)

Let us now specialize further to the case where $M = M_\Sigma$ is the moduli space of flat connections on the **nontrivial** SO(3)-**bundle** $P \to \Sigma$. The mapping class group of Σ acts on this space by symplectomorphisms $\phi_f : M_\Sigma \to M_\Sigma$ (modulo some finite ambiguity in the choice of a lift). An automorphism $f : P \to P$ also determines a mapping cylinder Y_f and there are corresponding Floer cohomology groups $HF^*(Y_f)$, defined in terms of ASD instantons on $Y_f \times \mathbb{R}$. In [7] it was shown that there are natural isomorphisms

$$HF^*(Y_f) \cong HF^*_{\mathrm{symp}}(\phi_f).$$

Now there is again a product structure

$$HF^*(Y_f) \otimes HF^*(Y_g) \to HF^*(Y_{gf}) \tag{10}$$

defined in terms of ASD instantons on suitable 4-dimensional cobordisms. In [16] it is shown that these agree with the products in (9) under the above isomorphisms.

REMARK 5.1 **(i)** In his thesis [4] Callahan examines these product structures in detail and uses them to find examples of symplectomorphisms $\phi_f : M_\Sigma \to M_\Sigma$ that are homotopic to the identity but not symplectically so. In his examples the automorphism f is generated by a Dehn twist on a loop that divides Σ into two components.

(ii) There is an alternative way to interpret these product structures (in the case $g = \mathrm{id}$) by intersecting the spaces of connecting orbits with suitable submanifolds of finite codimension in either $\mathcal{B}_Y = \mathcal{A}(Y)/\mathcal{G}(Y)$ or Ω_ϕ. In the symplectic context this gives rise to an action

$$H^*(\Omega_\phi) \otimes HF^*_{\mathrm{symp}}(\phi) \to HF^*_{\mathrm{symp}}(\phi).$$

Intuitively, $HF^*_{\mathrm{symp}}(\phi) = H^{\frac{1}{2}\infty}(\Omega_\phi)$ and this is the *cup-product* in Ω_ϕ. The map $\Omega_\phi \to M : \gamma \mapsto \gamma(0)$ induces a homomorphism $H^*(M) \to H^*(\Omega_\phi)$ and the resulting product $H^*(M) \otimes HF^*_{\mathrm{symp}}(\phi) \to HF^*_{\mathrm{symp}}(\phi)$ agrees with (9) in the case $\psi = \mathrm{id}$.

(iii) A loop $\gamma : S^1 \to Y$ determines a submanifold $V_\gamma \subset \mathcal{B}_Y$ via Donaldson's map $\mu : H_1(Y) \to H^3(\mathcal{B}_Y)$, and the induced homomorphism of Floer cohomology appears as the second boundary map in the Fukaya-Floer cohomology groups $HFF^*(Y, \gamma)$ [3]. In the symplectic context these operators correspond to the action of $H^*(\Omega_\phi)$ on $HF^*_{\mathrm{symp}}(\phi)$. If $M = M_\Sigma$ and $\phi = \phi_f$ for some automorphism $f : P \to P$ then a loop $\gamma : S^1 \to Y_f$ determines a codimension-2 submanifold $W_\gamma \subset \Omega_{\phi_f}$ and there is a commuting diagram

$$\begin{array}{ccc} HF^*(Y_f) & \stackrel{V_\gamma}{\to} & HF^*(Y_f) \\ \downarrow & & \downarrow \\ HF^*_{\mathrm{symp}}(\phi_f) & \stackrel{W_\gamma}{\to} & HF^*_{\mathrm{symp}}(\phi_f) \end{array}.$$

If the loop γ lies entirely in $\Sigma \times \{0\}$ then these maps can be interpreted in terms of the product structures (9) and (10) with $g = \mathrm{id}$ and $\psi = \psi_g = \mathrm{id}$. In [5] Donaldson has computed the quantum cohomology of M_Σ for a surface of genus 2.

(iv) In the instanton case the maps in **Floer's exact sequence** can be interpreted in terms of the Donaldson product structures [2]. It was proposed by Donaldson and Callahan [4] that there should be a symplectic analogue of this exact sequence. In special cases this should be related to Floer's original sequence by the Atiyah-Floer conjecture.

(v) There is a related question of what the effect of symplectic reduction is on Floer homology. This should also be related to the quantum product structures discussed here. An interesting example is provided by surgery on a loop $\gamma \subset Y$ in a 3-manifold with boundary $\partial Y = \Sigma$. Cut out a neighborhood U of γ and write $Y = U \cup_T (Y - U)$. Then the disjoint union $U \cup (Y - U)$ has three boundary components $T \cup \bar{T} \cup \partial Y$. Different ways of gluing in U correspond to different symplectic reductions in $M_T \cup \bar{M}_T \cup M_\Sigma$.

(vi) If Y is a 3-manifold with boundary $\partial Y = \Sigma$ then the quantum category $\mathcal{C}_{M_\Sigma}$ acts on the Floer cohomology groups $HF^*(Y, L)$ via natural product type maps

$$HF^*(Y, L_0) \otimes HF^*_{\mathrm{symp}}(L_0, L_1) \to HF^*(Y, L_1).$$

This was already observed by Fukaya [11]. So far these product structures are little understood.

References

[1] M. F. Atiyah, *New invariants of three and four dimensional manifolds*, Proc. Sympos. Pure Math. **48** (1988).
[2] P. Braam and S. Donaldson, *Floer's work on instanton homology, knots and surgery*, in Gauge theory, Symplectic Geometry, and Topology, Essays in Memory of Andreas Floer, edited by H. Hofer, C. Taubes, A. Weinstein, and E. Zehnder, Birkhäuser, Basel and Boston, MA, 1994.
[3] P. Braam and S. Donaldson, *Fukaya-Floer homology and gluing formulae for polynomial invariants*, as in [2].
[4] M. Callahan, Ph.D. thesis, Oxford, in preparation.
[5] S. K. Donaldson, *Floer homology and algebraic geometry*, preprint, 1994.
[6] S. Donaldson, M. Furuta, and D. Kotschick, *Floer homology groups in Yang-Mills theory*, in preparation.
[7] S. Dostoglou and D. A. Salamon, *Self-dual instantons and holomorphic curves*, Ann. of Math. (2), to appear.
[8] A. Floer, *Morse theory for the symplectic action*, J. Differential Geom. **28** (1988), 513–547.
[9] A. Floer, *An instanton invariant for 3-manifolds*, Comm. Math. Phys. **118** (1988), 215–240.
[10] A. Floer, *Symplectic fixed points and holomorphic spheres*, Comm. Math. Phys. **120** (1989), 575–611.
[11] K. Fukaya, *Floer homology for 3-manifolds with boundary*, preprint, University of Tokyo, 1993.
[12] K. Fukaya, *Gauge theory for 4-manifolds with corners*, preprint, Kyoto, 1994.
[13] M. Gromov, *Pseudoholomorphic curves in symplectic manifolds*, Invent. Math. **82** (1985), 307–347.
[14] D. McDuff and D. A. Salamon, J-holomorphic Curves and Quantum Cohomology, Amer. Math. Soc., University Lecture Series **6**, 1994.
[15] Y.-G. Oh, *Floer cohomology of Lagrangian intersections and pseudoholomorphic discs*, Comm. Pure Appl. Math. **46** (1993), 949–994.
[16] D. A. Salamon, *Quantum-cohomology and the Atiyah-Floer conjecture*, in preparation.
[17] M. Schwarz, Ph.D. thesis, ETH-Zürich, in preparation.

Generating Functions, Symplectic Geometry, and Applications

CLAUDE VITERBO

Département de Mathématiques, Bâtiment 425
Université de Paris-Sud and URA 1169 du C.N.R.S.
F-91405 Orsay Cedex, France

Contents

1 Symplectic manifolds, their Lagrange submanifolds and generating functions

A symplectic form on a manifold is a closed two form ω, nondegenerate as a skew-symmetric bilinear form on the tangent space at each point. Integration of the form on a two-dimensional submanifold S with boundary ∂S in M associates to S a real number (positive or negative) the "area of S", which due to Stokes' formula only depends on the curves ∂S, and the homology class of S rel ∂S. If moreover the form is exact, that is $\omega = d\lambda$, the area of S is obtained by integrating λ over ∂S. In this case it is also possible to integrate λ on loops *nonhomologous to zero* and we get the notion of "area enclosed by a loop ". However this area depends on the choice of λ. If this choice is fixed once for all, we shall then talk about an exact manifold. One should be careful about the fact that this notion is slightly different from that of a symplectic manifold with exact symplectic form (because in the latter case we have not chosen the primitive of ω).

It is a theorem of Darboux that the simplest example, $\mathbb{R}^{2n}$ with the constant symplectic form $\sigma = \sum_{j=1}^{n} dx_j \wedge dy^j$, is also the universal local model (i.e. any symplectic manifold is locally symplectomorphic to $(\mathbb{R}^{2n}, \sigma)$.

The main example for us will be the exact manifold T^*N, the phase space of N, where the exact form is the "contraction tensor" given in local coordinates by $\sum_{j=1}^{n} p_j dq^j$, where q_j are local coordinates on N and p_j are the dual coordinates.

Submanifolds of a symplectic manifold inherit naturally the 2 form induced by ω. This is naturally a closed form, but only exceptionally nondegenerate. In this

Proceedings of the International Congress
of Mathematicians, Zürich, Switzerland 1994

case the submanifold is called symplectic. Other remarkable cases occur if the form vanishes on the submanifold, which is then called isotropic (if $2 \cdot \dim(V) < \dim(M)$) or Lagrangian (if $2 \cdot \dim(V) = \dim(M)$). It is then maximal among isotropic submanifolds. In terms of area, any contractible curve on a Lagrange submanifold has zero area. In an exact manifold, any two *homologous curves* on L have the same area. If moreover any two curves have the same area (which is then necessarily zero), the Lagrange submanifold is called exact. This is equivalent to the exactness of the pullback of λ on L.

In T^*N there is a particularly simple family of Lagrange submanifolds. To any closed one form α we may associate $L_\alpha = \{(q, \alpha(q)) \mid q \in N\}$. The form induced by λ on L_α is just α; thus, $\omega = d\lambda$ induces $d\alpha = 0$. In particular, if $\alpha = df$ is exact we get an exact Lagrange submanifold.

A remarkable property of a Lagrange submanifold in T^*N is that it intersects the zero section more often than a differential topologist would expect. For L_f, we see that points in $L_f \cap 0_N$ are in one-to-one correspondence with critical points of f. In any case, as N is compact, there are at least two such points. One of the Arnold conjectures, partially solved by Hofer in 1983 (see [H]), claims that for L exact and obtained from the zero section by a Hamiltonian isotopy, the number of points in $L \cap 0_N$ is bounded from below by the Lusternik-Shnirelman category of N (i.e. the minimal number of critical points for a function on N). For L_f this conjecture is obvious. However, the Lagrange submanifolds that may be written as L_f are exactly those for which the projection $p : T^*N \to N$ restricts to a diffeomorphism.

Our main interest will be on Lagrange manifolds, and we shall represent them through their *generating functions*, an idea first introduced by Hörmander for different purposes (see [Hö]).

A generating function for the Lagrange submanifold L is a function $S : E \to \mathbb{R}$, defined on a vector bundle E over N, and such that

$$L = \left\{ \left(x, \frac{\partial S}{\partial x} \right) \;\middle|\; \frac{\partial S}{\partial \xi} = 0 \right\} \quad \text{where } x \text{ is in } N, \text{ and } \xi \text{ in the fibre}$$

(assuming that 0 is a regular value for $\frac{\partial S}{\partial \xi}$). Generating functions have the advantage of preserving the following interesting property: the points in $L_S \cap 0_N$ are in one-to-one correspondence with the critical points of S. The apparent drawback is that there are of course functions on E with no critical points (because E is noncompact), but this may be restored if we restrict ourselves to *Generating Functions Quadratic at Infinity* (abbreviated as G.F.Q.I.s):

DEFINITION (G.F.Q.I.). A generating function S is a G.F.Q.I. if and only if there exists a fibrewise quadratic nondegenerate form $Q(x, \xi)$ such that $S(x, \xi) - Q(x, \xi)$ has compact support.

The main example is associated to a symplectic diffeomorphism ϕ that we assume to be the time one flow of a compact supported Hamiltonian. Then $\Gamma_\phi = \{(z, \phi(z)) \mid z \in \mathbb{R}^{2n}\}$ is a Lagrange submanifold of $\mathbb{R}^{2n} \times \bar{\mathbb{R}}^{2n}$ ($\bar{\mathbb{R}}^{2n}$ is simply $\mathbb{R}^{2n}$ with the symplectic form $-\omega$). We shall identify $\mathbb{R}^{2n} \times \bar{\mathbb{R}}^{2n}$ with $T^*(\Delta) = T^*(\Gamma_{Id})$

(where Δ is the diagonal). Note that if ϕ_t has compact support, Γ_ϕ coincides with the zero section outside a compact set. We may thus compactify the base of $T^*(\Delta)$ to $T^*(S^{2n})$, and simultaneously compactify Γ_ϕ to $\bar{\Gamma}_\phi$. Then $\bar{\Gamma}_\phi$ coincides with the zero section near the point at infinity on S^{2n}, and is obtained from $0_{S^{2n}}$ by a Hamiltonian isotopy. We shall usually work in this last setting, as it is more pleasant to work with compact bases.

2 Existence and uniqueness theorems for generating functions

From now on, except in Section 4, we shall only consider generating functions quadratic at infinity. Let $\mathcal{L}$ be the space of Lagrange submanifolds Hamiltonianly isotopic to the zero section, $\mathcal{G}$ the set of G.F.Q.I. of elements of $\mathcal{L}$. The obvious projection $\mathcal{G} \to \mathcal{L}$ is denoted by π, and $\mathcal{G}_L$ will be $\pi^{-1}(L)$.

For S, T two G.F.Q.I. on E and F, we denote by $S \oplus T$ the G.F.Q.I. on $E \oplus F$ defined by $(S \oplus T)(x, \xi, \eta) = S(x, \xi) + T(x, \eta)$.

Let us introduce the following two equivalence relations on $\mathcal{G}$:

(a) $S_1 \simeq S_2$ if and only if there are nondegenerate quadratic forms Q_1, Q_2 on F_1, F_2 and a fibre-preserving diffeomorphism $\Phi : E_1 \oplus F_1 \to E_2 \oplus F_2$ such that $(S_1 \oplus Q_1) = (S_2 \oplus Q_2) \circ \Phi$.

(b) $S_1 \simeq S_2$ if and only if there are G.F.Q.I. for the zero section Σ_1, Σ_2 defined on F_1, F_2, and a vector bundle isomorphism Ψ from $E_1 \oplus F_1 \to E_2 \oplus F_2$ such that $(S_1 \oplus \Sigma_1) = (S_2 \oplus \Sigma_2) \circ \Psi$.

Note that in (a) there are more permissible "stabilizations" but fewer isomorphisms than in (b). However the equivalence relation (b) seems weaker than (a). The point of using (b) is that it makes certain proofs easier, without unduly weakening the conclusions. Making the quotient of $\mathcal{G}$ by one of these equivalence relations a topological space and a CW complex is needlessly complicated; it is better to notice that the concept of a continuous map from a cube in $\mathcal{G}/\simeq$ (where $\simeq$ is one of the equivalence relations (a) or (b)) is clear to everyone: it is the projection of a continuous map from the cube to $\mathcal{G}$. It is then clear what a continuous map from a polyhedron to $\mathcal{G}/\simeq$ will be. Smooth maps from a finite-dimensional manifold to $\mathcal{G}/\simeq$ are similarly defined .

The existence and uniqueness results may now be stated as follows.

THEOREM (EXISTENCE AND UNIQUENESS OF G.F.Q.I.) ([LS], [V1], [Th], [V2]). *Let $\tilde{\mathcal{G}}$ be one of the quotient spaces corresponding to* (a) *or* (b). *Then the map $\tilde{\pi} : \tilde{\mathcal{G}} \to \mathcal{L}$ induced by π is a Serre fibration. The fibre is reduced to a line described by the set of $S + C$ with $C \in \mathbb{R}$, and S any element of $\mathcal{G}$.*

As an application, it is easy using the equivalence relation (b) to see that

COROLLARY (GF SYMPLECTIC HOMOLOGY) ([El], [Tr]). *Given any two G.F.Q.I.s for L there is a unique integer m and a unique real number l, such that $H^k(S^b, S^a) = H^{k-m}(S^{b-l}, S^{a-l})$ for any pair of real numbers $a < b$.*

In particular, if we normalize in any reasonable way the constant, we have a well-defined ring $H^*(S^b, S^a)$, for any pair of real numbers (a, b). Note moreover, that there is an action by the cup product of $H^*(E) = H^*(N)$ on this ring (so it is actually an $H^*(N)$-module). This is in particular the case if we consider the Lagrange submanifolds $\bar{\Gamma}_\phi$ associated to a compact supported Hamiltonian isotopy ϕ. Then because $\bar{\Gamma}_\phi$ coincides with the zero section of $T^*(S^{2n})$ outside a compact set, the generating function has a unique critical point over the point at infinity. We shall normalize S so that this critical level is zero.

REMARK. Note that, as was pointed out by Eliashberg and Gromov, G.F.Q.I.s yield more precise information than just their homologies. For instance, if the base space N has a nontrivial fundamental group, the Reidemeister torsion may be recovered from S, a more refined information than just the homology of the space. Also, for families of Lagrange (or Legendre) submanifolds, invariants coming from pseudo-isotopy may be obtained (see [EG]). It also implies a priori more critical points than one may expect from the Morse inequalities, or the ring structure of the cohomology (using Lusternik-Shnirelman theory). In fact, because we are dealing with real topological spaces, we get for free the secondary operations, which yield also more critical points (see [V6]).

As an example of this last fact, we may consider the manifold N, total space of a nontrivial circle bundle over the 2-torus. Even though its cup length is 3, any function on N has four critical points. Moreover, for any Hamiltonian isotopy ϕ_t on T^*N, $\phi_1(0_N) \cap 0_N$ has at least four critical points (and not 3 as one may expect from Floer's theory).

REMARK. There is no reason to limit oneself to the cohomology functor. In fact, any strongly stable homotopy functor is invariant through the equivalence relation (a) (for (b) it is more delicate).[1)] We shall see below how this is connected to the idea that there exists a stable Floer homotopy (see [**CJS**]).

3 Symplectic invariants, solutions of Hamilton-Jacobi equations and applications

We shall briefly sketch a particularly simple construction of a symplectic invariant, related to the "capacity" defined by Ekeland and Hofer in [EH1] and [EH2], and also a new class of solutions for first order Hamilton-Jacobi equations.

Let $S(x, \xi)$ be a G.F.Q.I. for the Lagrange submanifold L. Note that for c large enough, E^c, E^{-c} has the homotopy type of $D(E^-), S(E^-)$, where E^- is the sum of the negative eigenspaces of the quadratic form defined by S at infinity.

Given a cohomology class α in $H^*(N)$, the Thom isomorphism $T : H^*(N) \to H^{*+d}(D(E^-), S(E^-))$ associates the class $T \cup p^*(\alpha)$. Now to S, we may associate the critical level

$$c(\alpha, S) = \inf\{\lambda \mid T \cup \alpha \text{ is nonzero in } H^*(E^\lambda, E^{-c})\}.$$

To simplify notation, we shall again denote by α the class $T \cup p^*(\alpha)$.

---

1) By a strongly stable homotopy functor F we mean that given an orientable vector bundle, E over X, and $D(E), S(E)$ its unit disc and sphere bundle, respectively, then $F(X)$ is isomorphic to $F(D(E)/S(E))$. K-theory, for instance, would be such a functor.

The number $c(\alpha, S)$ is a critical value for S, which implies that the corresponding level of S contains at least one critical point. In turn, this yields an intersection point of L_S and 0_N.

Let μ_L be the generator of $H^n(N)$, and 1 the generator of $H^0(N)$. The number $\gamma(L) = c(\mu, S) - c(1, S)$ is a sort of "norm" for L: $\gamma(L) = 0$ if and only if $L = 0_N$.

REMARK. This holds only for *embedded* Lagrange submanifolds. There are immersed Lagrange submanifolds admitting a G.F.Q.I. , such that $\gamma(L) = 0$ but $L \neq 0_N$ (however it is always true that L contains 0_N).

The most interesting case is when $L = \Gamma_\phi$. Then using the compactification explained at the end of Section 1, we get two critical values $c(1, S)$ and $c(\mu, S)$, which we denote as $c_-(\phi)$ and $c_+(\phi)$, respectively. The main properties of c_+ and c_- are summarized in the following theorem:

THEOREM (PROPERTIES OF $c_-(\phi)$ AND $c_+(\phi)$).
(i) $c_-(\phi) = -c_+(\phi^{-1})$.
(ii) $c_-(\phi) \leq 0 \leq c_+(\phi)$ *and* $c_(\phi) = 0 = c_+(\phi)$ *holds only if* $\phi = Id$.
(iii) $c_+(\phi\psi\phi^{-1}) = c_+(\psi)$ *and the same holds for* c_-.
(iv) $c_+(\phi\psi) \leq c_+(\phi) + c_+(\psi)$, *and the equality is reversed for* c_+ *replaced by* c_-.

The only nontrivial result is *(iv)*.

Most of the basic results of symplectic homology can be proved very simply using this approach. One defines a symplectic invariant, the capacity, by setting $c(U) = \sup\{c_+(\phi) \mid \phi$ is the time one map of a Hamiltonian isotopy generated by a Hamiltonian supported in $U\}$.

Clearly, $c(\psi(U)) = c(U)$ for any symplectic diffeomorphism ψ, and if $U \subset V$, we clearly have $c(U)J \leq c(V)$. As we may also prove that $c(B^{2n}(r)) = c(B^2(r) \times \mathbb{R}^{2n-2}) = \pi r^2$, we immediately get Gromov's theorem

THEOREM (GROMOV). *If there is a symplectic embedding from* $B^{2n}(r)$ *into* $B^2(r') \times \mathbb{R}^{2n-2}$, *then we must have* $r \leq r'$.

This implies for instance that the group of symplectic diffeomorphisms is closed for the C^0 topology in the group for all volume-preserving diffeomorphisms.

A particular feature of the invariant we defined is to be suited for comparing the capacity of a set with the capacity of a symplectic reduction. This follows for instance from the fact that we have the following inequalities. Let L be a Lagrange submanifold in $T^*(N \times P)$, and for each $p \in P$, let $L_p = (L \cap T^*N \times T^*_pP)/T^*_pP$, which we consider as a submanifold of T^*N. Then, if S is a G.F.Q.I. for L, we have that L_p has S_p as a G.F.Q.I. , where $S_p(n, \xi) = S(n, p, \xi)$.

Let us now see how generating functions may be used to find some remarkable solutions of Hamilton-Jacobi equations (this is partially due to Chaperon and Sikorav, [C2]).

The idea is that if L is a Lagrange submanifold in $T^*(N \times J[0, T])$ contained in the hypersurface $\tau + H(t, x, p) = 0$, and if L has a G.F.Q.I. S, then the function defined by

$$u(x) = \inf\{\lambda \mid \text{ the generator of } H^*(E_x^c, E_x^{-c}) \text{ in } H^*(E_x^\lambda, E_x^{-c}) \text{ is nonzero}\}$$

"satisfies the Hamilton-Jacobi equation"

$$\frac{\partial u}{\partial t} + H(t, x, \frac{\partial u}{\partial x}) = 0.$$

Moreover this is a C^0 solution, satisfying the equation almost everywhere, and having a certain number of additional properties. Our main result is that these solutions extend to the case where H and $u_0(x) = u(x, 0)$ are only continuous.

THEOREM. *Let J be the map from $C^k(N) \times C^k([0,T] \times T^*N)$ into $C^{\mathrm{Lip}}([0,T] \times N)$, which associates to (u_0, H) the above constructed solution of the equation*

$$\frac{\partial u}{\partial t} + H(t, x, \frac{\partial u}{\partial x}) = 0 \ ; \ u(0, x) = u_0(x).$$

This map is continuous for the natural topology on each space. Then J extends to a map $\bar{J}$ from $C^0(N) \times C_q^{\mathrm{Lip}}([0,T] \times T^*N)$ to $C^0([0,T] \times N)$ (which also sends $C^{\mathrm{Lip}}(N) \times C^{\mathrm{Lip}}([0,T] \times T^*N)$ to $C^{\mathrm{Lip}}([0,T] \times N)$). Moreover $u = \bar{J}(u_0, H)$ solves almost everywhere the above Hamilton-Jacobi equation.

Here $C_q^{\mathrm{Lip}}([0,T] \times T^*N)$ means continuous in the p variable, and Lipschitz in the q variable.

The remarkable class of solutions thus defined is different from the one defined by Lions under the name of "viscosity solutions", as was shown by Ottolenghi (personal communication).

4 Generalized generating functions and applications to Floer homology computations

The first idea in this section is to give two infinite-dimensional extensions of the notion of a generating function, and to prove a suitably adapted uniqueness theorem.

The first adaptation of the definition of G.F.Q.I. to the infinite-dimensional setting is easy under some natural assumptions. First we may replace the finite-dimensional vector bundle over N, by a Banach bundle over N. Then $S : E \to \mathbb{R}$ is a smooth function satisfying the Palais-Smale condition. The smoothness and transversality assumptions clearly extend to the Banach space setting. And finally, our proof of uniqueness can be also extended without adding any new ingredient to this case. Note however that the G.F.Q.I. should be such that the quadratic function at infinity should have finite index. A typical example is as follows. Let $L(t, q, \dot{q})$ be a Lagrangian on TN such that $\frac{\partial^2}{\partial \dot{q}^2} L$ is invertible.

$$S(q) = \int_0^1 L(t, q, \dot{q}) \, dt$$

defined on the Banach bundle $\mathcal{P} = \{q : [0,1] \to N \mid \dot{q}(0) = 0\}$ and $\pi : \mathcal{P} \to N$ is given by $\pi(q) = q(1)$. We claim that S is a generating function for $\phi_1(0_N)$ where ϕ_t is the flow associated to the Hamiltonian H obtained from L by Legendre duality. The computation is omitted, since a very similar one follows. S will never be quadratic at infinity, but if $L(t, q, \dot{q}) = \|\dot{q}\|^2$ outside a compact set, it is easy to show that $H^*(E^c, E^{-c}) = H^*(N)$ for c large enough (it is easy to prove this by comparison with $L = \|\dot{q}\|^2$).

Now we turn to the more subtle version of infinite-dimensional generating functions, which we shall call *Floer generating functions.*

This relies on Floer's idea to deal with the variational theory of the action functional, as there was not, prior to Floer's work, any reasonable approach to the variational study of this functional on a general manifold (see [R] for a finite-dimensional approach in the case of $\mathbb{R}^{2n}$).

One way to understand the introduction of generating functions is to consider the action functional as such a function. Let $\mathcal{E}$ be the set of paths, $\mathcal{E} = \{\gamma = (q,p) : [0,1] \to T^*N \mid p(0) = 0\}$ (we do not specify the regularity of the path, as it is of no interest for the moment) and $\pi : \mathcal{E} \to N$ be the map $\gamma \to q(1)$. Then consider the function A_H defined as

$$A_H(\gamma) = \int_0^1 [pJ\dot{q} - H(t, q, p)]\, dt.$$

We have that

$$DA_H(\gamma)\delta\gamma = \int_0^1 [(J\dot{q} - \frac{\partial H}{\partial p})\delta p - (J\dot{p} - \frac{\partial H}{\partial q})\delta q](t, q, p)\, dt + p(1)\delta q(1).$$

Thus, the set of γ such that the derivative of A_H in the direction of the fibres of π vanishes corresponds to solutions of

$$\dot{q} - \frac{\partial H}{\partial p} = 0 \ , \ \dot{p} - \frac{\partial H}{\partial q} = 0,$$

that is $\gamma(t) = \phi_t(\gamma(0)) = \phi_t(q(0), 0)$. Now because for such a γ, $\frac{\partial A_H}{\partial q(1)} = p(1)$, so the set of points $(q(1), \frac{\partial A_H}{\partial q(1)})$ is $\phi_1(0_N)$.

The function A_H is particularly difficult to study from a variational point of view, because all its critical points are of infinite index and coindex. As a result, we have for example that $H^*(A_H^b, A_H^a)$ vanishes. In a famous series of papers, Floer explained, using some ideas of Thom, Smale, Conley, and Witten, how to define the groups $FH^*(A_H^b, A_H^a)$ that are a sort of middle-dimensional cohomology of (A_H^b, A_H^a) (see [F1], [F2], and also [FHW]).

In particular these groups only depend on L, so we denote them by $FH^*(L; a, b)$. Now we have:

THEOREM (UNIQUENESS OF SYMPLECTIC HOMOLOGY).

$$FH^*(L; a, b) = GH^*(L; a, b).$$

This circle of ideas may be adapted to the periodic orbit problem. In general one studies the number of fixed points for a Hamiltonian flow on M. The periodic orbits are critical points of the action functional $\int_{D^2} u^*\omega - \int_{S^1} H(t,z)dt$, where u is any map such that $u_{|S^1} = z$. Of course the value of $\int_{D^2} u^*\omega$ may depend on the choice of u, not only on z, but this is not the case, provided ω vanishes on $\pi_2(M)$, that we shall henceforth assume. Then one encounters the same difficulties and cure (through Floer methods) as in the Lagrangian case. With the above assumptions, for compact M, one has the isomorphism

$$FH^*(M) = H^*(M).$$

Of course such an equality is not perfectly honest, unless we say which structures are transported from the right-hand side to the left-hand side. For instance, the original statement of Floer only deals with the additive group structures. In [V3] we proved also that the multiplicative structure extends to $FH^*(M)$ (and the isomorphism between $FH^*(M)$ and $H^*(M)$ preserves this structure). This implies a weak version of the Arnold conjecture for these manifolds, and has been extended to more general manifolds by several authors (Floer, Hofer-Salamon, etc.). Work by Cohen-Jones-Segal and Fukaya seems to indicate that much more structure is defined on $FH^*(M)$ (for example cohomology operations).

On noncompact manifolds, however, the situation is more interesting. For example in T^*N, let us look for periodic orbits of a Hamiltonian H, such that $H(q,p) = \|p\|^2$ at infinity.

We shall denote the Floer homology associated to H as $FH^*(T^*N)$. It is clear as in the compact case, that the Floer homology only depends on the behavior of H at infinity. Now for $H(q,p) = \|p\|^2$, the Hamiltonian flow is the geodesic flow, and we know perfectly well its periodic orbits: they are in one-to-one correspondence with closed geodesics. There is however another functional with the same critical points: the energy functional $E(q) = \int_{S^1} \dot{q}^2$ defined on the loop space of N, ΛN. This is not enough to prove that $FH^*(T^*N) = H^*(\Lambda N)$; for this one would have to show that the connecting trajectories for the gradient flow of E are in one-to-one correspondence with the *Floer trajectories*, solutions of $\partial_J u = -\nabla H$.

As this is not easy to prove, we shall use an approach derived from the ideas of Section 2. Indeed, we introduce the notion of a Floer generating function as a generating function of the form:

$$F_{H,S}(\gamma,\xi) = A_H(\gamma) \oplus S(\gamma(1),\xi)$$

where S is a G.F.Q.I. for L_S. It is not hard to see that, formally, F is a generating function for $\phi^{-1}(L_S)$. But beside this, we want to define $FH^*(F^b, F^a)$. This is done as in Floer homology, by considering critical points and connecting trajectories.

It is easy to see that for $H \equiv 0$, $FH^*(F^b,F^a) = H^*(S^b,S^a)$, whereas for $S = 0$, $FH^*(F^b,F^a) = FH^*(L;a,b)$. These are the main ingredients of the theorem's proof.

Remark. Let us notice that the isomorphism between F and G homology has not been proved to be natural. If this were true, as it seems likely to be, it would have quite interesting consequences, as all cohomology operations that are easily defined for G-homology would automatically be well defined on Floer cohomology, a much less trivial fact. Of course, if one could prove that the spaces S^b/S^a may be *canonically* identified up to suspension, we would have realized the program of Cohen-Jones-Segal of defining the "stable Floer homotopy type" in this setting.

5 Applications to Hamiltonian dynamics and obstructions to Lagrange embeddings

The previous theorem has a number of important applications. The first one will deal with the following question:

Given L, N, is there a Lagrange embedding from L to T^*N?

The answer is of course yes for $L = N$, and no other example is known if we moreover require the embedding to be exact. On the other hand, without the exactness assumption, it is not even known whether there is a Lagrange embedding of the Klein bottle into $\mathbb{R}^4$. In the sequel, all Lagrange embeddings will be assumed to be exact.

A conjecture by Arnold claims that the only such embeddings are obtained by applying a symplectic diffeomorphism to the zero section. At least, since proving this statement seems to be out of reach, we may try to show

CONJECTURE. *Let L be an exact Lagrange submanifold in T^*N. Then the projection of L onto N has nonzero degree.*

Note that this implies Gromov's theorem, claiming, in Sikorav's formulation, that there is no exact embedding into T^*M for open M (remember that L is always compact).

The connection with the former section is as follows. A Lagrange embedding j of L into T^*N extends to an embedding of a neighborhood of the zero section of T^*L into T^*N. This induces a map from the Floer cohomologies, from $\Phi(j) : FH^*(T^*L) = H^*(\Lambda L) \to FH^*(T^*N) = H^*(\Lambda N)$, satisfying the following algebraic property:

$$\Phi(j)(x \cup \Lambda(j)^*(y)) = \Phi(j)(x) \cup y.$$

It is easy to construct, for many pairs of manifolds (L, N), obstructions to the existence of such maps. We refer to [V6] for many examples. Let us just quote the solution to a previously open question (cf. [Lal-S]).

THEOREM. *There is no exact embedding from T^2 to T^*S^2.*

REMARK. The method of proof in [V7] (see also [V5]) is more complicated, because the relation between Floer cohomology and generating functions had not been established yet. Thus, the whole proof is based on generating functions methods.

We also mention another application of the above computation to the Weinstein conjecture in a cotangent bundle. A compact hypersurface in a symplectic manifold is said to be of contact type if there is a conformal vector field (i.e. $L_\xi \omega = \omega$) defined in a neighborhood of the hypersurface, and transverse to it. The characteristic flow on a hypersurface is the flow of a time-independent Hamiltonian having the hypersurface as a regular energy level. The special feature of a contact type hypersurface is that a neighborhood of it is foliated by hypersurfaces having diffeomorphic characteristic flow.

The conjecture of Weinstein claims that such a hypersurface always has a periodic characteristic. This was first proved in $\mathbb{R}^{2n}$ by the author ([V1]), later extended in joint work with Hofer and Floer ([HV1], [FHV], [HV2]). In particular in [HV1], it is proved that the conjecture holds for a contact hypersurface in T^*N surrounding the zero section. But, strangely enough, it was left open for a general contact hypersurface in T^*N, even though, as pointed out by Chaperon, if N has a Lagrange embedding in $\mathbb{R}^{2n}$, then the Weinstein conjecture in T^*N holds as a consequence of it holding in $\mathbb{R}^{2n}$.

The above computation, together with the information on the structure of the ring $H^{\cdot} * (\Lambda N)$ in the simply connected case due to Goodwillie (see [Go]), may be exploited to prove (see [V7]):

THEOREM (WEINSTEIN CONJECTURE IN S.C. COTANGENT BUNDLES). *The Weinstein conjecture holds in T^*N for N a simply connected compact manifold.*

References

[C1] M. Chaperon, *Une idée du type géodésiques brisées pour les systèmes hamiltoniens*, C. R. Acad. Sci. Paris, **298** (1984), 293–296.

[C2] M. Chaperon, *Lois de conservation et géométrie symplectique*, C. R. Acad. Sci. Paris, Série I, **312** (1991), 345–348.

[CJS] R. Cohen, J. Jones, and G. Segal, to appear in Floer Memorial volume, Birkhäuser, Basel and Boston.

[EH2] I. Ekeland and H. Hofer, *Symplectic topology and Hamiltonian dynamics I*, Math. Z. **200** (1989), 355–378.

[EH2] I. Ekeland and H. Hofer, *Symplectic topology and Hamiltonian dynamics II*, Math. Z. **203** (1990), 553–567.

[El] Y. Eliashberg, personal communication.

[EG] Y. Eliashberg and M. Gromov, in preparation.

[F1] A. Floer, *The unregularized gradient flow of the symplectic action*, Comm. Pure Appl. Math. **41** (1988), 775–813.

[F2] A. Floer, *Morse theory for Lagrangian intersections*, J. Differential Geom. **28** (1988), 513–547.

[FHV] A. Floer, H. Hofer, and C. Viterbo, *The Weinstein conjecture in $P \times \mathbb{C}$*, Math. Z. **203** (1989), 355–378.

[Go] T. Goodwillie, *Cyclic homology, derivations and the free loop space*, Topology **24** (1985), 187–215.

[G1] M. Gromov, *Pseudo holomorphic curves in symplectic manifolds*, Invent. Math. **82** (1985), 307–347.

[G2] M. Gromov, *Soft and hard symplectic geometry*, Proc. Internat. Congress Math. 1986 **1** (1987), 81–98.

[H] H. Hofer, *Lagrangian embeddings and critical point theory*, Ann. Inst. H. Poincaré Anal. Non Linéaire **2** (1986), 407–462.

[HV 1] H. Hofer and C. Viterbo, *The Weinstein conjecture in cotangent bundles and related results*, Ann. Scuola Norm. Sup. Pisa Cl. Sci. (4) **20** (1988), 411–445.

[HV 2] H. Hofer and C. Viterbo, *The Weinstein conjecture in the presence of holomorphic curves*, Comm. Pure and Appl. Math. **45** (1992), 583–622.

[Hö] L., Hörmander, *Fourier integral operators I*, Acta Math. **127** (1971), 79–183.

[J] T. Joukovkskaia, Thèse d'université, Université de Paris 7, Denis Diderot, 1993.

[Lal-S] F. Lalonde and J.-C. Sikorav, *Sous-variétés lagrangiennes et lagrangiennes exactes des fibrés cotangents*, Comm. Math. Helv. (1991), 18–33.

[LS] F. Laudenbach and J.-C. Sikorav, *Persistance d'intersection avec la section nulle au cours d'une isotopie hamiltonienne dans un fibré cotangent*, Invent. Math. **82** (1985), 349–357.

[R] P. H. Rabinowitz, *Periodic solutions of Hamiltonian systems*, Comm. Pure Appl. Math. **31** (1978), 157–184.

[S] J.-C. Sikorav, *Problèmes d'intersection et de points fixes en géométrie hamiltonienne*, Comm. Math. Helv. **62** (1987), 61–72.

[Th] D. Théret, *Equivalence globale des fonctions génératrices*, preprint, Université de Paris 7, 2 Place Jussieu, 75230 Paris Cedex 05.

[Tr] L. Traynor, *Symplectic homology via generating functions*, preprint, Math. Department, Bryn Mawr College, Bryn Mawr, PA.

[V1] C. Viterbo, *A proof of Weinstein conjecture in* $\mathbb{R}^{2n}$, Ann. Inst. H. Poincaré Analyse Non Linéaire **4** (1987), 337–356.

[V2] C. Viterbo, *A new obstruction to embedding Lagrangian tori*, Invent. Math. **100** (1990), 301–320.

[V3] C. Viterbo, *The cup product on the Thom-Smale-Witten complex and Floer cohomology*, to appear in Floer Memorial volume, Birkhäuser, Basel and Boston.

[V4] C. Viterbo, *Symplectic topology as the geometry of generating functions*, Math. Ann. **692** (1992), 685–710.

[V5] C. Viterbo, *Properties of embedded Lagrange submanifolds*, Proc. First Europ. Congress Math., Paris 1992; Birkhäuser, Basel and Boston 1994.

[V6] C. Viterbo, *Some remarks on Massey products, tied cohomology classes and Lusternik-Shnirelman category*, preprint.

[V7] C. Viterbo, *Exact Lagrange submanifolds, periodic orbits, and the cohomology of loop spaces*, in preparation.

[V8] C. Viterbo, *Floer generating functions and applications to symplectic topology*, in preparation.

Smooth 4-manifolds and Symplectic Topology

ROBERT E. GOMPF*

Department of Mathematics
The University of Texas at Austin
Austin, TX 78712

One of the famous problems of topology is the classification problem for simply connected 4-manifolds. In the context of topological manifolds (up to homeomorphism), Freedman reduced the problem in 1981 to the classification of $\mathbb{Z}$-quadratic forms ([F]; see also [FQ]). However, for smooth manifolds (up to diffeomorphism) the problem remains wide open, and it is currently the focus of intense research. Henceforth, we only consider smooth (compact, boundaryless) manifolds. Most of our knowledge about such simply connected 4-manifolds has descended from work of Donaldson. In particular, his invariants [D] allow us for the first time to distinguish different diffeomorphism types within a given homotopy type of such manifolds — or equivalently in this context, within a given homeomorphism type [F]. We now know that many homeomorphism types each contain infinitely many diffeomorphism types (a situation that is not possible in any other dimension). This explosion of distinct examples has left topologists struggling to find order amid the confusion.

In this article, we will consider two approaches to the problem of organizing simply connected 4-manifolds. The first approach, which predates Donaldson's work, is easily motivated by considering 2-manifolds. Every orientable 2-manifold admits a complex structure — in fact, diffeomorphism types of orientable 2-manifolds correspond bijectively to deformation types of complex 1-manifolds (i.e., complex manifolds with 1 complex dimension or 2 real dimensions). Similarly, one might try to understand simply connected 4-manifolds by reducing to the theory of complex surfaces (complex manifolds with 2 complex, or 4 real, dimensions). We will see that problems have recently arisen with this approach. An alternative approach will be suggested — that of replacing complex structures in the theory by related structures called *symplectic forms*. Although these forms are less well understood than complex structures, they have the advantage of greater flexibility. It is hoped that a parallel development of 4-manifold theory and symplectic topology can lead to major advances in both fields.

*) Partially supported by NSF grant DMS-9301524.

Proceedings of the International Congress
of Mathematicians, Zürich, Switzerland 1994

1 4-manifolds and complex surfaces

The first question that arises when comparing 4-manifolds and complex surfaces is whether every simply connected 4-manifold admits a complex structure. This is easily seen to be false — in fact, the connected sum of two complex surfaces (with their complex orientations) can never be complex.[1] To bypass this difficulty (and the 4-dimensional Poincaré conjecture) we define a 4-manifold to be *irreducible* if it cannot be split as a connected sum without using a homotopy 4-sphere summand. We can then ask the following question, which was open for many years: Is every irreducible, simply connected 4-manifold (other than the 4-sphere) complex? This question was answered in the negative in 1990 by the author and Mrowka [GM]. Since then, numerous mathematicians have expanded the techniques, producing many other examples of irreducible, simply connected 4-manifolds that do not admit complex structures [FS1], [FS2], [K], [L], [S], [Sz], [Y]. Some of these are not even homotopy equivalent to complex surfaces [FS1]. The evidence now suggests that among irreducible, simply connected 4-manifolds, those that admit complex structures are relatively scarce.

These new manifolds are all created from complex surfaces by cutting and pasting. In practice, it seems that most manifolds produced by such techniques actually split as connected sums of simple pieces (such as complex projective planes with both orientations). Thus, it seems important to examine in detail those constructions that produce irreducible manifolds.

One such construction, called (*generalized*) *logarithmic transformation*, consists of finding a 2-torus T embedded with a trivial normal bundle in a 4-manifold M, cutting out a tubular neighborhood of T, and regluing the neighborhood by any diffeomorphism of the 3-torus boundary. When T is a complex submanifold of M, it is well known that the resulting manifold is frequently complex [Ko]. However, complex surfaces typically contain many essential noncomplex tori, and logarithmic transformations on such tori typically yield infinitely many diffeomorphism types of irreducible manifolds that are homeomorphic to M. In contrast, most homotopy types of 4-manifolds contain at most finitely many diffeomorphism types of irreducible complex surfaces. In this sense, noncomplex 4-manifolds seem to form an overwhelming majority. Fintushel and Stern [FS2] have shown that logarithmic transformation can usually be described in terms of a more general operation, which they call *rationally blowing down*. This consists of removing a neighborhood of a certain transverse collection of embedded spheres, and replacing the neighborhood by a certain plug with trivial rational homology. New irreducible, noncomplex manifolds result from this generalization.

The only other cut-and-paste operation that is currently known to be useful for producing irreducible 4-manifolds is connected-summing along surfaces. Begin with 4-manifolds M_1 and M_2 and embedded orientable surfaces $F_i \subset M_i$ with the same genus g and opposite self-intersection numbers $\pm n$ (so that the normal

[1] This follows from the observation that if the tangent bundle of a 4-manifold reduces to a complex vector bundle, then its first Chern class is a characteristic element of the cup-product pairing [MH], so c_1^2 is congruent to the signature mod 8. It follows that the signature plus the Euler characteristic must be divisible by 4 (because $c_1^2 = 2c_2 + p_1 = 2e + 3\sigma$), a condition that cannot be preserved under connected sums.

bundles are orientation-reversing isomorphic to each other). Remove a tubular neighborhood of F_i from each M_i, and glue the resulting complements together along their boundaries. Without loss of generality, we can assume (unless $g=n=0$) that the gluing diffeomorphism preserves the fibers of the normal circle bundles comprising the boundaries. (For $g > 1$ or $n \neq 0$, no other diffeomorphisms are possible, and for $g = 1$, $n = 0$ we can reduce to this case by first performing a logarithmic transformation.) This technique allows (for example) the construction of irreducible, simply connected 4-manifolds that are not homotopy equivalent to complex surfaces [FS1].

Where does this leave our program for reducing 4-manifold topology to complex surface theory? Most irreducible, simply connected 4-manifolds do not admit complex structures. Of course, there is a canonical procedure for repairing any conjecture by expanding it to include all known counterexamples. In this case, it is already being asked whether all irreducible 4-manifolds are obtained from complex surfaces by the constructions given above. There is actually some justification for this via analogy with 3-manifolds. Thurston's geometrization conjecture asserts that if an arbitrary 3-manifold is decomposed by cutting it into as many nontrivial pieces as possible along spheres and tori, then the resulting pieces should all admit simple geometric structures. One might hope for a similar decomposition of 4-manifolds along circle bundles over surfaces. However, in 3-dimensional topology there is a well-developed theory that shows how to decompose 3-manifolds maximally along spheres and tori. In the absence of any such theory in dimension 4, it seems prudent to examine alternative approaches for organizing simply connected 4-manifolds.

2 Symplectic manifolds

To shift our viewpoint on 4-manifolds, we first consider a different sort of structure. A *symplectic manifold* is a manifold (necessarily of even dimension) endowed with a *symplectic form* — that is, a closed 2-form ω that is nondegenerate as a bilinear form on each tangent space. (In other words, no nonzero vector should be orthogonal to the entire tangent space.) Note that if ω were symmetric instead of skew-symmetric, nondegeneracy would be the condition that we were dealing with a Riemannian or Lorentzian metric. The closure condition, $d\omega = 0$, is analogous to requiring a metric to have vanishing curvature. It guarantees that ω has no local invariants — any point has a neighborhood that is symplectomorphic (diffeomorphic, preserving ω) to an open set in $\mathbb{R}^{2n}$ with the standard form $\sum_{i=1}^{n} dx^{2i-1} \wedge dx^{2i}$. Thus, symplectic manifolds can be thought of as skew-symmetric analogs of flat (or hyperbolic or spherical) Riemannian manifolds.

The relation of symplectic geometry to our previous discussion comes through the notion of a *Kähler manifold*. This is a complex manifold with a symplectic form that is compatible with the complex structure. (Specifically, the form should be the imaginary part of a Hermitian metric.) Any smooth (complex projective) algebraic variety is Kähler. Because any complex surface with b_1 even (hence, any simply connected complex surface) can be deformed into an algebraic surface [Ko], our previous discussion will be nearly unchanged if we shift from the complex to the Kähler viewpoint. Now we are free to reject complex structures as being too rigid, and focus on the underlying symplectic structures as natural tools for understanding 4-manifolds.

We now have an obvious question — have we gained anything? Until recently, few examples of (compact) symplectic manifolds were known, other than Kähler manifolds. Around 1976, Thurston [T] produced some symplectic 4-manifolds with $b_1 = 3$. These could not be homotopy equivalent to Kähler manifolds, because the odd-index Betti numbers of a Kähler manifold are always even. McDuff [Mc] produced simply connected examples in dimensions ≥ 10, with $b_3 = 3$. However, until recently it was unknown whether simply connected symplectic manifolds in dimension 4 (or 6 or 8) could be essentially non-Kähler. In the absence of such examples, our generalization would not be useful. Fortunately, a technique originating in 4-manifold topology has changed the situation radically [G].

The new advance began from the idea that the process of forming connected sums along codimension-2 submanifolds is essentially symplectic in nature. More precisely, suppose that M_1, M_2, and N are symplectic manifolds, with codimension-2 symplectic embeddings $N \hookrightarrow M_i$ $(i = 1,2)$. Suppose that ψ is any orientation-reversing isomorphism of the two normal bundles over N. Then ψ determines a gluing map for a connected sum along N. It can be shown that the resulting manifold always admits a canonical symplectic structure. This is most easily seen (and most useful) in the case where the normal bundles of N are trivial: in this case, Weinstein's technique [W] easily produces symplectic embeddings $N \times D_\varepsilon \hookrightarrow M_i$ (where D_ε is a sufficiently small ε-disk in $\mathbb{R}^2$ with symplectic form $dx^1 \wedge dx^2$). The gluing map is given by $id_N \times \varphi$, where φ symplectically turns $D_\varepsilon - \{0\}$ inside out. (Such maps φ are easy to construct, because a symplectic form on a 2-manifold is just an area form — for example, $\varphi(r,\theta) = (\sqrt{\varepsilon^2 - r^2}, -\theta)$ works.)

The applications of this technique are multitudinous [G]. One can construct many simply connected symplectic 4-manifolds that are not diffeomorphic to complex surfaces. Some of these (including, for example, many of the manifolds constructed in [GM]) are homeomorphic to complex surfaces. Others are not even homotopy equivalent to complex surfaces. A simple construction realizes all possible pairs of integers in a large region of the plane as the Euler characteristic and signature (or equivalently, the Chern numbers) of simply connected, symplectic 4-manifolds. We can also answer the question of which groups are realized as fundamental groups of symplectic 4-manifolds. Although "most" finitely presented groups are *not* realized by Kähler manifolds, one can develop a sort of surgery (by summing along tori) that realizes all finitely presented groups by symplectic 4-manifolds. For any fixed group, the Euler characteristic and signature can be chosen with nearly as much freedom as in the simply connected case. One can also apply the construction profitably in higher dimensions. Combining these constructions, we obtain

THEOREM. [G, Theorem 0.1]. *For any even dimension $n \geq 4$ and any finitely presented group G, there are (compact) symplectic n-manifolds with fundamental group G that are not homotopy equivalent to any Kähler manifold.*

Thus, we have obtained extensive additional freedom in passing from the complex to the symplectic viewpoint on 4-manifolds.

3 4-manifolds and symplectic topology

We have been led to the following general question: Which 4-manifolds admit symplectic structures? In general dimensions, only two obstructions to existence are known. First, nondegeneracy of the symplectic form ω is equivalent to the assertion that the top exterior power of ω be nowhere 0 — i.e., a volume form. Thus, the top exterior power of the cohomology class $[\omega]$ must be nonzero. For example, any symplectic manifold must be oriented and have a nonzero b_2. Second, the form ω determines (up to fiber homotopy) a complex structure on the tangent bundle of the manifold. The existence of such structures is a tractable problem in obstruction theory. For example, the footnote of Section 1 shows that an ordinary connected sum of two symplectic 4-manifolds can never be symplectic.

As in Section 1, we are now led to a more restricted question: Is every irreducible, simply connected 4-manifold (other than the 4-sphere) symplectic? This time, however, our discussion is much different. We have seen that all known irreducible, simply connected 4-manifolds are constructed by means of the operations of connected summing along surfaces and rationally blowing down. We have seen that the first of these operations is essentially symplectic in nature, which shows (for example) that many of the noncomplex examples of [GM] admit symplectic structures. A pivotal question is whether rationally blowing down (and hence, logarithmic transformation) should also turn out to be symplectic under weak hypotheses. If so, it seems reasonable to expect that all known irreducible, simply connected 4-manifolds ($\neq S^4$) will turn out to be symplectic. Perhaps symplectic structures are sufficiently flexible that irreducibility guarantees their existence — if not, then constructing counterexamples may require radically new techniques.

The converse of the question is also interesting: When can we decompose a symplectic 4-manifold as an ordinary connected sum? There is one simple situation in which this is possible. In the context of complex surfaces, we can always *blow up* a point. This has the effect of connected summing any complex surface with a copy of $\mathbb{C}P^2$ with its orientation reversed. A complex surface is called *minimal* if it is not obtained from any other complex surface by blowing up. For symplectic 4-manifolds, an analogous theory exists [Mc]. Thus, we should restrict attention to minimal symplectic 4-manifolds. It is then reasonable to conjecture that every minimal symplectic 4-manifold is irreducible [G].

Why is this conjecture reasonable? After all, the only known obstructions to the existence of a symplectic structure on a 4-manifold depend only on its homotopy type, whereas the conjecture would provide a much more subtle obstruction. For example, the conjecture would imply that the connected sum of 3 projective planes (oriented as complex surfaces) could not admit a symplectic structure, even though this cannot be inferred from the classical obstructions. In addition, a sum of copies of the $K3$ surface and $S^2 \times S^2$ could not be symplectic, even though many such sums are homeomorphic to Kähler surfaces. The reasonableness of the conjecture is based on empirical evidence. Many of the examples of [G] are turning out to be irreducible [FS2], [GM], [S], [Sz], [Y], and none has been shown not to be. In contrast, examples that are constructed by similar but nonsymplectic methods frequently split as connected sums of simple pieces. For example, sums along surfaces tend to split if they mismatch ambient orientations, as do sums along surfaces that cannot be made compatible with a symplectic structure. We have already noted that most cut-and-paste constructions result in such splittings, but the author has been unable to produce any symplectic 4-manifolds that split (other than those that are constructed to be obviously nonminimal). It is tempting to speculate that the remarkable success of logarithmic transformation and connected summation along suitably chosen surfaces results from an underlying symplectic nature of these constructions.

In conclusion, there seems to be an intimate relationship between irreducible 4-manifolds and symplectic topology. The main difficulty with using this approach to understand 4-manifolds is that symplectic topology is in a primitive state compared with the highly developed theory of complex surfaces. However, we have seen that techniques from 4-manifold theory have opened a new avenue through symplectic topology, which has in turn led to new families of irreducible 4-manifolds. The parallel development of 4-manifold theory and symplectic topology should continue to result in new insight into both disciplines.

References

[D] S. Donaldson, *Polynomial invariants for smooth four-manifolds*, Topology **29** (1990), 257–315.

[FS1] R. Fintushel and R. Stern, *Surgery in cusp neighborhoods and the geography of irreducible 4-manifolds*, Invent. Math., **117** (1994), 455–523.

[FS2] ———, *4-manifolds and their rational blow downs*, in preparation.

[F] M. Freedman, *The topology of four-dimensional manifolds*, J. Differential Geom. **17** (1982), 357–453.

[FQ] M. Freedman and F. Quinn, Topology of 4-manifolds, Princeton Math. Ser. no. 39, Princeton Univ. Press, Princeton, NJ, 1990.

[G] R. Gompf, *A new construction of symplectic manifolds*, Ann. of Math., to appear.

[GM] R. Gompf and T. Mrowka, *Irreducible 4-manifolds need not be complex*, Ann. of Math. (2) **138** (1993), 61–111.

[K] Y. Kametani, *Simple invariant and differentiable structures on a Horikawa surface*, preprint.

[Ko] K. Kodaira, *On the structure of compact complex analytic surfaces, I*, Amer. J. Math. **86** (1964), 751–798.

[L] P. Lisca, *On simply connected noncomplex 4-manifolds*, J. Differential Geom. **38** (1993), 217–224.

[Mc] D. McDuff, *Examples of simply-connected symplectic non-Kählerian manifolds*, J. Differential Geom. **20** (1984), 267–277.

[MH] J. Milnor and D. Husemoller, Symmetric Bilinear Forms, Springer-Verlag, Berlin and New York, 1973.

[S] A. Stipsicz, *Donaldson invariants of certain symplectic manifolds*, preprint, 1994.

[Sz] Z. Szabó, *Irreducible four-manifolds with small Euler characteristics*, preprint, 1994.

[T] W. Thurston, *Some simple examples of symplectic manifolds*, Proc. Amer. Math. Soc. **55** (1976), 467–468.

[W] A. Weinstein, *Symplectic manifolds and their lagrangian submanifolds*, Adv. in Math. **6** (1971), 329–346.

[Y] B. Z. Yu, *Some computations of Donaldson's polynomials via flat connections*, dissertation, Univ. California-Berkeley, 1994.

Topological Modular Forms, the Witten Genus, and the Theorem of the Cube

MICHAEL J. HOPKINS

Department of Mathematics, Massachusetts Institute of Technology
Cambridge, MA 02139, USA

1. Introduction

There is a rich mathematical structure attached to the cobordism invariants of manifolds. In the cases described by the index theorem, a generalized cohomology theory is used to express the global properties of locally defined analytic objects. Hirzebruch's theory of multiplicative sequences gives an expression for these invariants in terms of characteristic classes, and brings to focus their remarkable arithmetic properties. Quillen's theory of formal groups and complex oriented cohomology theories illuminates the generalized cohomology theories themselves.

Around eight years ago a new invariant, the *elliptic genus*, was introduced [17]. It is a cobordism invariant of oriented manifolds that takes its values in a certain ring of modular forms. Witten [23], [22] proposed an analytic interpretation of the elliptic genus using analysis on loop spaces. Landweber, Ravenel, and Stong [13] constructed a corresponding cohomology theory (elliptic cohomology), and it is believed that there is an "index" theorem relating analysis on loop space to elliptic cohomology. So far, a satisfying mathematical theory is lacking.

In the same papers [23], [22] Witten introduced a variant of the elliptic genus, now known as the Witten genus. The Witten genus takes its values in modular forms when applied to Spin manifolds with $\frac{p_1}{2} = 0$. The cohomological significance of this invariant has remained unclear.

The point of this note is to describe a generalization of theories of Hirzebruch and Quillen to the cobordism of Spin manifolds with $\frac{p_1}{2} = 0$. It turns out that in the presence of an elliptic curve there is a canonical cobordism invariant. This invariant coincides with the Witten genus in the case where the elliptic curve is the Tate curve, though it is most natural to consider all elliptic curves at once. This leads to a cohomological expression for the modular invariance of the Witten genus (of a family), and to a new generalized cohomology theory. The coefficient ring of this new cohomology theory is the ring of *topological modular forms*. It is related to the ring of modular forms over $\mathbb{Z}$, but is not torsion free. The torsion groups in this ring represent new invariants of Spin manifolds with $\frac{p_1}{2} = 0$, and it would be interesting to describe these invariants in terms of geometry and analysis.

Most of this paper represents joint work with Matthew Ando and Neil Strickland. The construction and computations with the new cohomology theory are

Proceedings of the International Congress
of Mathematicians, Zürich, Switzerland 1994

joint work with Mark Mahowald and Haynes Miller. Some of the results described here represent work in progress.

2. Genera and their characteristic series

Let R be a commutative ring. An R-valued *genus* is a ring homomorphism Φ from some type of cobordism ring to R. Thus a genus is a function Φ that assigns to each manifold M an element $\Phi(M) \in R$, and that satisfies

$$\begin{aligned} \Phi(M_1 \coprod M_2) &= \phi(M_1) + \phi(M_2) \\ \Phi(M_1 \times M_2) &= \phi(M_1)\phi(M_2) \\ \Phi(\partial M) &= 0. \end{aligned}$$

The cobordism rings usually considered are the ring MU_* of cobordism classes of stably almost complex manifolds, and the ring MSO_* of cobordism classes of oriented manifolds. The structure of these rings has been determined [20], [14], [16], [21], and there are isomorphisms

$$\begin{aligned} MU_* \otimes \mathbb{Q} &\approx \mathbb{Q}[\mathbb{C}\mathrm{P}^1, \mathbb{C}\mathrm{P}^2, \dots] \\ MSO_* \otimes \mathbb{Q} &\approx \mathbb{Q}[\mathbb{C}\mathrm{P}^2, \mathbb{C}\mathrm{P}^4, \dots]. \end{aligned}$$

When R is torsion free, a genus is determined by its values on the complex projective spaces. There are two natural generating functions that collect these values

$$\log_\Phi(z) = \sum \Phi(\mathbb{C}\mathrm{P}^n)\frac{z^{n+1}}{n+1} \qquad \text{(logarithm)}$$

$$K_\Phi(z) = \frac{z}{\exp_\Phi(z)}, \qquad \text{(characteristic series)}$$

where $\exp_\Phi(z) = \log_\Phi^{-1}(z)$. A genus Φ with values in a torsion free ring factors through MSO_* if and only if the characteristic series is even

$$K_\Phi(z) = K_\Phi(-z). \tag{2.1}$$

The characteristic series determines a stable exponential characteristic class with values in $H^*(-; R \otimes \mathbb{Q})$ as follows. By the splitting principle, such a class is determined by its value on the complex line bundle L over BS^1 associated to the identity character. Setting $z = c_1(L)$, the characteristic class is then defined by

$$K_\Phi(L) = K_\Phi(z) \in H^*(BS^1; R \otimes \mathbb{Q}).$$

The following formula of Hirzebruch [10] expresses $\Phi(M)$ in terms of characteristic (Pontryagin or Chern) classes:

$$\Phi(M) = \langle K_\Phi(TM), [M] \rangle.$$

Here are some examples.

(1) The genus whose characteristic series is $z/(1 - e^{-z})$ is the Todd genus. The log of the Todd genus is the power series

$$-\log(1 - x) = \sum \frac{x^n}{n},$$

so its value on $\mathbb{C}\mathrm{P}^n$ is 1. It can be shown that the Todd genus of a stably almost complex manifold is an integer.

(2) The genus with characteristic series

$$K(z) = \frac{z/2}{\sinh(z/2)} = \frac{z}{e^{z/2} - e^{-z/2}}$$

is the $\hat{A}$ genus. It is an invariant of oriented manifolds. The $\hat{A}$-genus has the property that it assumes integer values on manifolds that admit a Spin structure.

(3) The genus with logarithm

$$\log_\Phi(z) = \int_0^z (1 - 2\delta t^2 + \varepsilon t^4)^{-\frac{1}{2}}\, dt$$

is the *elliptic genus* of Ochanine [17]. The associated characteristic series is even, so it is an invariant of oriented manifolds.

(4) The *Witten genus* [23], [22] is the genus with characteristic series

$$\frac{z/2}{\sinh(z/2)} \prod_{n\geq 1} \frac{(1-q^n)^2}{(1-q^n e^z)(1-q^n e^{-z})}.$$

This is an even function of z, and so defines a cobordism invariant of oriented manifolds. The Witten genus takes values in $\mathbb{Z}[[q]]$ when applied to manifolds that admit a Spin structure.

There is a dimension 4 characteristic class of Spin bundles, twice which is p_1. Let's denote this class $\frac{p_1}{2}$. If M is a Spin manifold of dimension n, and $\frac{p_1}{2}(TM) = 0$, then the Witten genus of M turns out to be the q-expansion of a modular form for the group $SL_2(\mathbb{Z})$. This means that after setting $q = e^{2\pi i \tau}$, the Witten genus of M can be written as $f(\tau)$, where f is a holomorphic function on the upper half plane $\operatorname{Re}\tau > 0$, and satisfies the functional equation

$$f(-1/\tau) = (-\tau)^n f(\tau). \tag{2.2}$$

3. Genus of a family

The underlying geometry of a genus begins to be revealed when its definition is extended to families. Let M_s be a family of manifolds parameterized by the points of a space S. The manifolds M_s are allowed to transform through cobordisms, but are required to be equidimensional of dimension, say, n. Such a family defines an element of the generalized cohomology group $\mathbf{M}^{-\mathbf{n}}(\mathbf{S})$, where $\mathbf{M}$ is the cohomology theory associated to the type of cobordism being considered.

For a genus Φ the quantities $\Phi(M_s)$ form some kind of structure parameterized by the space S. It is best to think of this structure as representing an element of a generalized cohomology group $E^{-n}(S)$. A *genus for families* of manifolds is then a multiplicative map

$$\mathbf{M} \to \mathbf{E}$$

of generalized cohomology theories.

The process of extending the definition of a genus to families is not at all canonical, and is intimately connected with the expression of the genus in terms of geometry and analysis.

Several kinds of cobordisms will be used in this paper. They are displayed below. The diagram on the left is a diagram of classifying spaces. The map labeled $\frac{p_1}{2}$ is the universal characteristic class of the same name, and the spaces $BU\langle 6\rangle$ and $BO\langle 8\rangle$ are the homotopy fibers of the map $\frac{p_1}{2}$ and its restriction to BSU, respectively. The diagram on the right is the corresponding diagram of cobordism theories. For example, a $BO\langle 8\rangle$-manifold is a manifold equipped with a lift to $BO\langle 8\rangle$ of the map classifying its stable tangent bundle, and $MO\langle 8\rangle$ is the cohomology theory associated to the cobordism of $BO\langle 8\rangle$-manifolds.

$$\begin{array}{ccccc} BU\langle 6\rangle & \longrightarrow & BO\langle 8\rangle & & \\ \downarrow & & \downarrow & & \\ BSU & \longrightarrow & B\mathrm{Spin} & \xrightarrow{\frac{p_1}{2}} & K(\mathbb{Z},4) \\ \downarrow & & \downarrow & & \\ BU & \longrightarrow & BSO & & \end{array} \qquad \begin{array}{ccc} MU\langle 6\rangle & \longrightarrow & MO\langle 8\rangle \\ \downarrow & & \downarrow \\ MSU & \longrightarrow & M\mathrm{Spin} \\ \downarrow & & \downarrow \\ MU & \longrightarrow & MSO. \end{array}$$

The "families" versions of the genera of Section 2 are as follows.

(1) The natural domain for the Todd genus is MU, the theory of complex cobordism. The target of the Todd genus can be taken to be ordinary cohomology with coefficients in the rational numbers. This, however, obscures the fact that the Todd genus of each individual manifold is an integer. If the Todd genus is thought of as a formula for the dimension (Euler characteristic) of certain cohomology sheaves, then the natural target appears as K-theory [4], [1].

(2) The $\hat{A}$ genus is most interesting when applied to Spin manifolds, making the natural domain the cohomology theory MSpin. Atiyah and Singer [2] showed that the $\hat{A}$-genus is the index of the Dirac operator, and portrayed the natural target of the "families $\hat{A}$-genus" as the cohomology theory KO (bundles of vector spaces over $\mathbb{R}$). This refinement represents more than an accounting of the integrality properties of the genus. The groups

$$KO^0\left(S^{8k+1}\right) \approx KO^0\left(S^{8k+2}\right) \approx \mathbb{Z}/2$$

correspond to torsion invariants of families of Spin-manifolds. These invariants can be described in terms of analysis but can not be calculated in terms of Pontryagin classes [3].

(3) In the case of the elliptic genus, it can be shown that the functor

$$\mathrm{Ell}^*(-) = MSO^*(-) \underset{MSO_*}{\otimes} \mathbb{Z}[\tfrac{1}{6}, \delta, \varepsilon, \Delta, \Delta^{-1}]/(2^6\varepsilon(\delta^2-\varepsilon)^2-\Delta)$$

defines a generalized cohomology theory [13], [12], [9] on the category of finite cell complexes. This represents a natural extension of the elliptic genus to families, but, at present, there is no known geometric interpretation of Ell (see, however, the exposé of Segal [19]).

(4) The natural domain for the Witten genus is the cohomology theory $MO\langle 8\rangle$. There is a map

$$MO\langle 8\rangle \to KO[\![q]\!]$$

representing the Witten genus. It accounts for the integrality properties, and has some associated torsion invariants. On the other hand it factors through $M\mathrm{Spin}$, and so cannot possibly express the transformation properties with respect to the modular group. This is related to the fact that behavior with respect to the transformation $\tau \mapsto 1/\tau$ is very difficult to understand from the point of view of power series in q.

4. Cubical Structures

A deeper understanding of the Witten genus of a family requires investigating the genera attached to the cobordism theories $MU\langle 6\rangle$ and $MO\langle 8\rangle$. The result of Hirzebruch, that a genus can be calculated by integrating a stable exponential characteristic class over the manifold, remains valid for these theories. However, a stable characteristic class is not determined by its value on L. In fact it does not even have a value on L, as the structure group of L does not lift to $BO\langle 8\rangle$. On the other hand, the (virtual) bundle

$$V_3 = (L_1 - 1) \otimes (L_2 - 1) \otimes (L_3 - 1), \tag{4.1}$$

over $(BS^1)^3$, admits a canonical lift of its structure group to $BU\langle 6\rangle$. Furthermore, there is a "splitting principle" that allows one to formally express any $BU\langle 6\rangle$ bundle as a sum of trivial line bundles and bundles of this kind. The cohomology of $(BS^1)^3$ is a polynomial algebra in three variables, so one expects the characteristic series of an $MU\langle 6\rangle$ genus to be a function of three variables. This is indeed the case, and the series that arise satisfy a certain functional equation. There is a geometric interpretation of this functional equation that is particularly suited to the study of elliptic spectra. It is known as a *cubical structure*, and was introduced by Breen [5] in order to codify the the rich structure attached to line bundles on abelian varieties coming from the theorem of the cube.

Let G be an abelian group, and $\mathfrak{L}$ a line bundle over G. The group G might be a discrete group, an algebraic group, a topological group, or a group of some other kind. The line bundle $\mathfrak{L}$ consists of a collection of lines $\mathfrak{L}_x$ for $x \in G$, and should be thought of as varying discretely, algebraically, continuously, or in some other manner, depending on the kind of group.

Given G and $\mathfrak{L}$, let $\Theta(\mathfrak{L})$ be the line bundle over G^3 whose fiber at (x, y, z) is

$$\Theta(\mathfrak{L})_{(x,y,z)} = \frac{\mathfrak{L}_{x+y+z}\mathfrak{L}_x\mathfrak{L}_y\mathfrak{L}_z}{\mathfrak{L}_{x+y}\mathfrak{L}_{x+z}\mathfrak{L}_{y+z}\mathfrak{L}_e},$$

where $e \in G$ is the identity element. In this expression, multiplication and division are meant to indicate tensor product of lines and their duals.

The functor Θ is a kind of "second difference" operator. If the terms "line bundle" and "tensor product" are replaced with "function" and "addition," then Θ becomes the operator whose kernel consists of quadratic functions.

A *cubical structure* on $\mathfrak{L}$ is a section s of $\Theta(\mathfrak{L})$ satisfying

$$\begin{array}{lrcl}
\text{(rigid)} & s(e,e,e) & = & 1 \\
\text{(symmetry)} & s(x_{\sigma(1)}, x_{\sigma(2)}, x_{\sigma(3)}) & = & s(x_1, x_2, x_3) \\
\text{(cocycle)} & s(w+x,y,z)s(w,x,z) & = & s(w,x+y,z)s(x,y,z).
\end{array}$$

The two sides of these equations are sections of different bundles, so a comment is in order. In each case a canonical identification can be made. For example, the section $s(e,e,e)$ is an element of $\Theta(\mathfrak{L})_{(e,e,e)}$, which is the tensor product of four copies of $\mathfrak{L}_e$ with its dual. Contracting the lines with their duals gives an identification of this with the trivial line, and it is via this identification that the equation labeled "rigid" takes place. There are similar canonical identifications that need to be made for the other equations.

The set of cubical structures on $\mathfrak{L}$ will be denoted $C^3(G;\mathfrak{L})$.

If the line bundle $\mathfrak{L}$ comes equipped with a symmetry isomorphism

$$t : \mathfrak{L}_x \approx \mathfrak{L}_{-x}$$

then the fiber of $\Theta(\mathfrak{L})$ over the point $(x, y, -x-y)$ admits a canonical trivialization. A Σ-*structure* on $\mathfrak{L}$ is a cubical structure s with the property that

$$s(x, y, -x-y) = 1. \tag{4.2}$$

The set of Σ-structures on $\mathfrak{L}$ will be denoted $C^3_0(G;\mathfrak{L}, t)$.

5. Formal groups and complex orientable spectra

The group that arises in homotopy theory is the formal group attached to a complex orientable cohomology theory [18]. Recall that a cohomology theory E is *complex orientable* if there is a class $x \in E^*BS^1$ whose restriction to $E^*S^2 \approx E^{*-2}(\mathrm{pt})$ is a unit. A choice of such an x gives rise to a very rich structure, and in particular, to a theory of E-valued Chern classes for complex vector bundles.

Suppose that E is a multiplicative, complex orientable cohomology theory with the additional properties that

$$E_*(\mathrm{pt}) \text{ is commutative} \tag{5.1}$$

$$E_2(\mathrm{pt}) \text{ contains a unit.} \tag{5.2}$$

With these assumptions, the ring $E^0(BS^1)^n$ is isomorphic to a formal power series ring in n variables over $E^0(\mathrm{pt})$. The multiplication map

$$BS^1 \times BS^1 \to BS^1$$

gives the formal spectrum $G = \operatorname{spf} E^0BS^1$ the structure of a formal group. In terms of "physical" groups, it provides the abelian group structure on the functor $G = \operatorname{Hom}(E^0BS^1, -)$, from the category of augmented $E^0(\mathrm{pt})$-algebras with nilpotent augmentation ideal, to the category of abelian groups.

The formal group G is the one of interest. The ring of functions on G is isomorphic to a formal power series ring in one variable over $E^0(\mathrm{pt})$. Let $\mathfrak{L}$ be the line bundle $\mathcal{O}(-e)$, whose local sections are functions that vanish at the unit. The module of global sections of $\mathfrak{L}$ is the reduced cohomology group $\widetilde{E}^0BS^1$. This line bundle comes with an obvious symmetry isomorphism $t : \mathfrak{L}_x \approx \mathfrak{L}_{-x}$.

6. A homology calculation

Now let R be a commutative ring, and suppose E is as above. A map

$$E_0 BU\langle 6\rangle \to R$$

can be composed with the map classifying V_3 (4.1), to yield an $E^0(\mathrm{pt})$-module map from $E_0\left(BS^1\right)^3$ to R. This can be thought of as an R-valued function f on G^3. It satisfies the following equations (in which the symbol "+" refers to addition in the group G):

$$\begin{aligned} f(e,e,e) &= 1 \\ f(x_{\sigma(1)},x_{\sigma(2)},x_{\sigma(3)}) &= f(x_1,x_2,x_3) \\ f(w+x,y,z)f(w,x,z) &= f(w,x+y,z)f(x,y,z). \end{aligned}$$

The first two of these equations are obvious. The last arises from the tensor product of (L_4-1) with the equation

$$\begin{aligned} &(L_1L_2-1)(L_3-1)+(L_1-1)(L_2-1) \\ &\qquad = (L_1-1)(L_2L_3-1)+(L_2-1)(L_3-1). \end{aligned}$$

Stated another way, the function f defines a cubical structure on the trivial line bundle $\mathcal{O}_G$.

THEOREM 6.1. *The map described above gives rise to a natural isomorphism*

$$\operatorname{spec} E_0 BU\langle 6\rangle \approx C^3(G;\mathcal{O}_G)$$

of functors on the category of multiplicative complex orientable cohomology theories E *satisfying* (5.1) *and* (5.2).

For multiplicative cohomology theories E and F, let $\mathrm{Mult}(E,F)$ be the set of multiplicative transformations from E to F. In terms of the representing spectra, this is the set of homotopy classes of homotopy multiplicative maps.

The following theorem is proved by applying the Thom isomorphism to Theorem 6.1.

THEOREM 6.2. *The map described above gives rise to a natural isomorphism*

$$\mathrm{Mult}(MU\langle 6\rangle, E) \approx C^3(G;\mathfrak{L})$$

of functors on the category of multiplicative complex orientable cohomology theories E *satisfying* (5.1) *and* (5.2), *with associated formal group* G. *If* $\frac{1}{2}\in E^0(pt)$, *or if* E *is* $K(n)$*-local for some Morava* K*-theory* $K(n)$, *with* $n\le 2$, *then this descends to a natural isomorphism*

$$\mathrm{Mult}(MO\langle 8\rangle, E) \approx C^3_0(G;\mathfrak{L},t).$$

There are even more general criteria guaranteeing the validity of the second assertion, but they involve a lengthy discussion.

Theorem 6.2 is analogous to the result that a genus with values in a torsion free ring is determined by its characteristic series. The role of the characteristic series is played by a cubical structure. The analogue of the symmetry condition (2.1) is condition (4.2).

7. Elliptic spectra

Theorem 6.2 is most interesting when the formal group G is extended to an elliptic curve. For the purposes of this paper, an *elliptic curve* is a generalized elliptic curve in the sense of [8, Définition 1.12], all of whose geometric fibers are irreducible.

DEFINITION 7.1. An *elliptic spectrum* consists of

(1) a complex orientable spectrum E satisfying (5.1) and (5.2), with associated formal group G;

(2) an elliptic curve $\mathbf{E}$ over $E^0(\mathrm{pt})$;

(3) an isomorphism $t : G \to \mathbf{E}^{\mathrm{f}}$ from G to the formal completion of $\mathbf{E}$.

The third condition requires explanation. The elliptic curve $\mathbf{E}$ gives rise to an abelian group-valued functor on $E^0(\mathrm{pt})$-algebras, by associating to an algebra R the abelian group of R-valued points of $\mathbf{E}$. Restricting this functor to the category of augmented $E^0(\mathrm{pt})$-algebras with nilpotent augmentation ideal gives a formal group $\mathbf{E}^{\mathrm{f}}$. This is the *formal completion* of $\mathbf{E}$. The isomorphism $G \to \mathbf{E}^{\mathrm{f}}$ is then an isomorphism of formal groups.

The collection of elliptic spectra forms a category, in which a map consists of a multiplicative map of cohomology theories, and a map of elliptic curves that is compatible with the associated map of formal groups.

THEOREM 7.2. *Attached to each elliptic spectrum E is a multiplicative map*

$$\sigma_E : MU\langle 6\rangle \to E.$$

This map is modular *in the sense that if $f : E \to F$ is a map of elliptic spectra, then $\sigma_F = f \circ \sigma_E$. If $\frac{1}{2} \in E$, or if E has the property that $E^*(pt)$ is torsion free and concentrated in even degrees, then $MU\langle 6\rangle$ can be replaced with $MO\langle 8\rangle$.*

In the case where $E = K[\![q]\!]$, and $\mathbf{E}$ is the Tate curve, the map

$$\pi_*\sigma_E : MO\langle 8\rangle_* \to \mathbb{Z}[\![q]\!]$$

can be shown to be the Witten genus. The *modular* invariance of the genus σ_E is an expression of the modular invariance of the "families" Witten genus. In the next section it will be explained how this reduces to "modular invariance" in the classical sense, when the parameter space S consists of only one point.

The main tool used to deduce Theorem 7.2 from Theorem 6.2 is the theorem of the cube.

THEOREM 7.3 Theorem of the cube. *If $\mathfrak{L}$ is a line bundle over an abelian variety, then $\Theta(\mathfrak{L})$ is trivial.*

Topologically this result follows from the facts that line bundles are classified by $H^2(-;\mathbb{Z})$, and H^2 is a quadratic functor. The theorem of the cube is the analogue of this assertion for algebraic line bundles.

It follows from the theorem of the cube that any line bundle over an abelian variety has a canonical cubical structure. Indeed, the only sections of $\Theta(\mathfrak{L})$ are constants, and any potential cubical structure must assume the value 1 at the unit. The "rigid," "symmetry," and "cocycle" conditions become identities between constant functions that assume a prescribed value at the unit.

Proof of Theorem 7.2, given Theorem 6.2. The unique section s of $\Theta(\mathfrak{L})$ satisfying $s(e,e,e) = 1$, and extending to a section of $\Theta(\mathcal{O}_{\mathbf{E}}(-e))$, is automatically an element of $C_0^3(G;\mathfrak{L},t)$. Take σ_E to be the multiplicative map associated to s by Theorem 6.2. □

8. Modularity

The point of this section is to relate the "modular" invariance of the maps σ_E to modular forms. This leads naturally to two new cohomology theories.

Let $\mathcal{M}_{\mathrm{Ell}}$ be the category whose objects are elliptic curves

$$\mathbf{E} \xrightarrow{\mathbf{p}} \mathbf{S}$$

with identity section e, and in which morphisms are cartesian squares

$$\begin{array}{ccc} \mathbf{E}' & \longrightarrow & \mathbf{E} \\ \downarrow & & \downarrow \\ S' & \longrightarrow & S. \end{array}$$

This is the *elliptic moduli stack* (see [15], [8], [7]), as is denoted $\mathcal{M}_1$ in [8].

For an elliptic curve $\mathbf{E}/\mathbf{S}$ let $\omega_{\mathbf{E}/\mathbf{S}} = e^*\Omega^1_{\mathbf{E}/\mathbf{S}}$ be the line bundle over S consisting of invariant 1-forms along the fibers. For each $k \in \mathbb{Z}$, let ω^k be the functor on $\mathcal{M}_{\mathrm{Ell}}$ whose value on $\mathbf{E}/\mathbf{S}$ is the abelian group of global sections of ${\omega_{\mathbf{E}/\mathbf{S}}}^k$. The collection of functors ω^k forms a functor ω^* on $\mathcal{M}_{\mathrm{Ell}}$ with values in graded rings. The *ring of modular forms over* $\mathbb{Z}$ is the graded ring

$$R_* = \varprojlim{}_{\mathcal{M}_{\mathrm{Ell}}}\omega^*.$$

This ring has been determined [6, Prop. 6.1], and there is an isomorphism

$$R_* \approx \mathbb{Z}[c_4, c_6, \Delta]/(c_4^3 - c_6^2 - 1728\Delta).$$

The grading is such that the class c_n is homogeneous of degree n.

The ring R maps to the classical ring of modular forms by restricting to the inverse limit over the full subcategory of $\mathcal{M}_{\mathrm{Ell}}$ whose only object is the usual family of elliptic curves over the upper half plane. The automorphism group of this object is the group $SL_2(\mathbb{Z})$. This map sends c_4 to $2^4 \cdot 3^2 \cdot 5 \cdot E_2$ and c_6 to $2^5 \cdot 3^3 \cdot 5 \cdot 7 \cdot E_3$, where E_2 and E_3 are the Eisenstein series of weights 4 and 6 respectively. The element Δ maps to the discriminant.

Attached to each elliptic spectrum E is the elliptic curve $\mathbf{E}$ over $S = \operatorname{spec}\pi_0 E$. The isomorphism $G \approx \mathbf{E}^{\mathrm{f}}$ determines an isomorphism

$$E^0(S^{2k}) = E^{-2k}(\mathrm{pt}) \approx \omega^k(\mathbf{E}/\mathbf{S}).$$

It turns out that there are enough elliptic spectra that there are isomorphisms

$$\varprojlim_{\text{E elliptic}} E^{-2k}(\mathrm{pt}) \approx R_k,$$

$$\varprojlim_{\text{E elliptic}} E^{2k+1}(\mathrm{pt}) \approx 0.$$

Moreover, Theorem 7.2 shows that the orientations σ_E give rise to a map

$$MO\langle 8\rangle_* \to \varprojlim_{\mathbf{E}\text{ elliptic}} E_*(\text{pt}).$$

This proves that the Witten genus takes its values in R_*. One can, however, hope for a more refined statement. This is the subject of the next section.

9. Topological modular forms and eo_2

The category of elliptic spectra is closely related to the elliptic moduli stack. There is one important difference. Whereas there is no "good" colimit of the objects in $\mathcal{M}_{\mathrm{Ell}}$, the homotopy inverse limit in spectra, of the category of elliptic spectra, can be formed. The resulting spectrum is no longer elliptic, but it still represents an interesting cohomology theory.

In practice it is necessary to "rigidify" the category of elliptic spectra by working with a certain subcategory of A_∞ etale elliptic spectra. The A_∞ condition has to do with higher homotopy associativity of E, and the etale condition is that the map $\operatorname{spec}\pi_0 E \to \mathcal{M}_{\mathrm{Ell}}$ which classifies $\mathbf{E}$ is etale and open. The other conditions defining this subcategory arise from obstruction theory and will remain unspecified. Though the notation is slightly misleading, the homotopy inverse limits that follow are taken over this subcategory.

Define eo_2 to be the connected cover of

$$\underset{E,\ A_\infty\text{ etale elliptic}}{\underset{\longleftarrow}{\operatorname{holim}}} \quad E$$

and let EO_2 be the spectrum

$$\underset{\substack{E,\ A_\infty\text{ etale elliptic}\\ \mathbf{E}\text{ smooth}}}{\underset{\longleftarrow}{\operatorname{holim}}} \quad E.$$

These spectra are topological models for the moduli space of elliptic curves. There is a spectral sequence

$$\varprojlim{}^s_{\mathcal{M}_{\mathrm{Ell}}}\omega^k \Rightarrow \pi_{2k-s}eo_2, \tag{9.1}$$

so it makes sense to call the ring $eo_{2*}(\text{pt})$ the ring of *topological modular forms.*

The spectrum EO_2 is closely related to a spectrum constructed by the author and Miller, and the spectrum eo_2 is closely related to one constructed by the author and Mahowald [11].

This spectral sequence (9.1) has been computed by the author and Mahowald. It terminates at a finite stage. One interesting feature is that the discriminant Δ is not a permanent cycle, whereas the forms 24Δ and Δ^{24} are. The form Δ^{24} is not a divisor of zero. There is an isomorphism

$$EO_{2*}(-) \approx (\Delta^{24})^{-1}eo_{2*}(-).$$

The cohomology theory EO_2 is periodic with period $24^2 = 576$.

The torsion in eo_{2*} is annihilated by 24. It has a very rich structure. The cohomology theory eo_2 can be used to account for nearly everything that is known about the stable homotopy groups of spheres in dimensions less than 60.

Regarding the Witten genus, the more refined statement for which one can hope is that the maps σ_E assemble to a multiplicative map

$$MO\langle 8\rangle \to eo_2.$$

This is consistent with many calculations, and is the subject of work in progress. It is the most natural target for the "families" Witten genus, and would define new torsion invariants of Spin-manifolds with $\frac{p_1}{2} = 0$. It would be very interesting to have an explanation of these invariants in terms of geometry and analysis.

References

[1] M. F. Atiyah and F. Hirzebruch, *Riemann-Roch theorems for differentiable manifolds*, Bull. Amer. Math. Soc. **65** (1959), 276–281.

[2] M. F. Atiyah and I. M. Singer, *The index of elliptic operators on compact manifolds*, Bull. Amer. Math. Soc. **69** (1963), 422–433.

[3] ———, *The index of elliptic operators: V*, Ann. of Math. **93** (1971), 139–149.

[4] A. Borel and J.-P. Serre, *Le théorème de Riemann-Roch (d'après Grothendieck)*, Bull. Soc. Math. France **86** (1958), 97–136.

[5] L. Breen, Fonctions théta et théorème du cube, Lecture Notes in Math. **980**, Springer-Verlag, Berlin and New York, 1983.

[6] P. Deligne, *Courbes elliptiques: formulaire (d'après J. Tate)*, Modular Functions of One Variable III, Lecture Notes in Math. **476**, Springer-Verlag, Berlin and New York, 1975, pp. 53–73.

[7] P. Deligne and D. Mumford, *Irreducibility of the space of curves of given genus*, Inst. Hautes Études Sci. Publ. Math. **36** (1969), 75–109.

[8] P. Deligne and M. Rapoport, *Les schémas de modules de courbes elliptiques*, Modular Functions of One Variable II, Lecture Notes in Math. **349**, Springer-Verlag, Berlin and New York, 1973, pp. 143–316.

[9] J. Franke, *On the construction of elliptic cohomology*, Max Planck Institut preprint, 1991.

[10] F. Hirzebruch, Topological Methods in Algebraic Geometry, Springer-Verlag, New York, 1956.

[11] M. J. Hopkins and M. Mahowald, *The spectrum* eo_2, to appear.

[12] P. S. Landweber, *Elliptic cohomology and modular forms*, Elliptic Curves and Modular Forms in Algebraic Topology (New York) (P. S. Landweber, ed.), Lecture Notes in Math. **1326**, Springer-Verlag, 1988, pp. 55–68.

[13] P. S. Landweber, D. C. Ravenel, and R. E. Stong, *Periodic cohomology theories defined by elliptic curves*, To appear.

[14] J. W. Milnor, *On the cobordism ring* Ω^* *and a complex analogue, Part I*, Amer. J. Math. **82** (1960), 505–521.

[15] D. Mumford, *Picard groups of moduli problems*, Arithmetic Algebraic Geometry (O. F. G. Schilling, ed.), Harper & Row, New York, 1965.

[16] S. P. Novikov, *Some problems in the topology of manifolds connected with the theory of Thom spaces*, Soviet Mathematics Dokl. **1** (1960), 717–720.

[17] S. Ochanine, *Sur les genres multiplicatifs definis par des integrals elliptiques*, Topology **26** (1987), 143–151.

[18] D. G. Quillen, *On the formal group laws of unoriented and complex cobordism theory*, Bull. Amer. Math. Soc. **75** (1969), 1293–1298.

[19] G. Segal, *Elliptic cohomology*, Séminaire Bourbaki 1987/88, Astérisque, vol. 161-162, Societe Mathematique de France, Fév. 1988, pp. 187–201.

[20] R. Thom, *Quelques propriétés globales des variétés differentiables*, Comm. Math. Helv. **28** (1954), 17–86.

[21] C. T. C. Wall, *Determination of the cobordism ring*, Ann. Math. (2) **72** (1960), 292–311.

[22] E. Witten, *Elliptic genera and quantum field theory*, Comm. Math. Phys. **109** (1987), 525–536.

[23] ———, *The index of the Dirac operator in loop space*, Elliptic Curves and Modular Forms in Algebraic Topology (New York) (P. S. Landweber, ed.), Lecture Notes in Math. **1326**, Springer-Verlag, Berlin and New York, 1988, pp. 161–181.

Applications dont la source est un classifiant

Jean Lannes

Université Paris 7, UFR de Mathématiques, Tour 45-55
2 place Jussieu, F-75251 Paris, France

0 Introduction

Depuis la fin des années 70 un grand nombre de travaux ont été consacrés aux espaces fonctionnels $\mathbf{hom}(BG, -)$ dont la source est le classifiant d'un groupe de Lie compact G (par exemple un groupe fini). Rappelons la terminologie : l'espace fonctionnel $\mathbf{hom}(X, Y)$ est l'espace des applications d'un espace X dans un espace Y. Les motivations originales étaient les conjectures de Segal et Sullivan. La conjecture de Segal concernait l'espace fonctionnel $\mathbf{hom}(BG, QS^0)$, G désignant un groupe fini et QS^0 la limite directe des espaces de lacets $\Omega^n S^n$. Un cas particulier de la conjecture de Sullivan, celui traité en premier par H. R. Miller [Mi], concernait les espaces fonctionnels $\mathbf{hom}(B(\mathbb{Z}/p), Y)$ avec Y un CW-complexe fini.

La théorie des espaces fonctionnels $\mathbf{hom}(BG, -)$ avec G un groupe de Lie compact apparaît comme un jeu de construction dont la pièce élémentaire est le cas $G = \mathbb{Z}/p$, p premier. Nous nous proposons dans cet article de résumer la théorie des espaces fonctionnels $\mathbf{hom}(B(\mathbb{Z}/p), -)$ [La1][La2], et de décrire quelques-unes de ses interventions en théorie de l'homotopie.

Pour alléger la notation, on pose ci-après $\sigma = \mathbb{Z}/p$.

Une généralisation de l'espace fonctionnel $\mathbf{hom}(B\sigma, Y)$ est l'espace des points fixes homotopiques $X^{h\sigma}$ d'une action de σ sur un espace X, c'est-à-dire l'espace fonctionnel $\mathbf{hom}_\sigma(E\sigma, X)$ des applications σ-équivariantes de $E\sigma$ dans X, $E\sigma$ désignant comme d'habitude le revêtement universel de $B\sigma$; $\mathbf{hom}(B\sigma, Y)$ n'est donc rien d'autre que l'espace des points fixes homotopiques de l'action triviale de σ sur Y. En fait commme $\mathbf{hom}_\sigma(E\sigma, X)$ est la fibre en l'identité de l'application $\mathbf{hom}(B\sigma, E\sigma \times_\sigma X) \to \mathbf{hom}(B\sigma, B\sigma)$ l'analyse des points fixes homotopiques d'une action quelconque se ramène à celle d'une action triviale.

Venons-en à la question essentielle : Pourquoi peut-on dire des choses raisonnables sur les espaces fonctionnels $\mathbf{hom}(B\sigma, -)$ alors que l'on sait si peu, par exemple, sur les espaces $\mathbf{hom}(S^n, -)$?

Ceci tient aux propriétés de la cohomologie modulo p de l'espace $B\sigma$, à la fois comme module et algèbre instable sur l'algèbre de Steenrod modulo p, qui en font ce qu'on pourrait appeler un co-espace d'Eilenberg-Mac Lane.

1 Propriétés de $H^*(B\sigma; \mathbf{F}_p)$

Pour exprimer ces propriétés il nous faut introduire les analogues du bifoncteur $\mathbf{hom}(-,-)$ dans les catégories des modules et algèbres instables sur l'algèbre de Steenrod modulo p.

Proceedings of the International Congress
of Mathematicians, Zürich, Switzerland 1994

On note :
- A l'algèbre de Steenrod modulo p ;
- $\mathcal{U}$ la catégorie des A-modules instables (pour $p = 2$, un A-module M est instable si $Sq^i x = 0$ quand i est strictement plus grand que le degré de x) ;
- $\mathcal{K}$ la catégorie des A-algèbres instables (la cohomologie modulo p d'un espace Y, notée H^*Y, est l'exemple type d'une telle algèbre).

Dans la catégorie des espaces (en fait ensembles simpliciaux) on définit le foncteur $Y \mapsto \mathbf{hom}(X, Y)$ comme l'adjoint à droite du foncteur $Z \mapsto X \times Z$:

$$\mathrm{Hom}(X \times Z, Y) = \mathrm{Hom}(Z, \mathbf{hom}(X, Y))\,.$$

Il n'est pas difficile de transposer cette définition dans les catégories $\mathcal{U}$ et $\mathcal{K}$. Soient $\mathcal{C}$ l'une de ces deux catégories et K un objet de $\mathcal{C}$, que l'on suppose de dimension finie en chaque degré, on montre sans peine que le foncteur $\mathcal{C} \to \mathcal{C}$, $N \mapsto K \otimes N$ admet un adjoint à gauche, que l'on note $M \mapsto (M : K)_{\mathcal{C}}$ (et qu'il est raisonnable d'appeler la division par K dans la catégorie $\mathcal{C}$). On a donc par définition

$$\mathrm{Hom}_{\mathcal{C}}(M, K \otimes N) = \mathrm{Hom}_{\mathcal{C}}((M : K)_{\mathcal{C}}, N)\,.$$

Supposons H^*X de dimension finie en chaque degré (rappelons que H^* est une abréviation pour $H^*(\,;\mathbf{F}_p)$). L'application d'évaluation

$$X \times \mathbf{hom}(X, Y) \to Y$$

induit en cohomologie modulo p un $\mathcal{K}$-morphisme naturel

$$H^*Y \to H^*X \otimes H^*\mathbf{hom}(X, Y)$$

qui donne par adjonction un $\mathcal{K}$-morphisme tout aussi naturel (une sorte d'homomorphisme d'Hurewicz)

$$h : (H^*Y : H^*X)_{\mathcal{K}} \to H^*\mathbf{hom}(X, Y)\,.$$

La bonne surprise, comme nous le verrons au §2, est que h est “très souvent” un isomorphisme pour $X = B\sigma$. Mais n'anticipons pas et revenons à $H^*B\sigma$.

Nous notons T le foncteur $\mathcal{U} \to \mathcal{U}$, $M \mapsto (M : H^*B\sigma)_{\mathcal{U}}$; les propriétés de $H^*B\sigma$ qui nous intéressent peuvent s'exprimer de la façon suivante :

THÉORÈME 1.1.
(a) *Le foncteur T est exact.*
(b) *Le foncteur T commute aux produits tensoriels.*
(c) *Si M est une A-algèbre instable, alors TM possède une structure naturelle de A-algèbre instable et TM munie de cette structure coïncide avec $(M : H^*B\sigma)_{\mathcal{K}}$.*

Le point (a) concerne la structure de A-module instable de $H^*B\sigma$. Les points (b) et (c) concernent la structure de A-algèbre instable de $H^*B\sigma$; en particulier on utilise le produit de $H^*B\sigma$ pour définir la transformation naturelle $T(-\otimes -) \to T(-)\otimes T(-)$ dont le point (b), dans sa formulation précise, affirme que c'est un isomorphisme.

Plus généralement, soit V un p-groupe abélien élémentaire (c'est-à-dire un groupe isomorphe à σ^d pour un certain entier d) ; nous notons T_V le foncteur $\mathcal{U} \to \mathcal{U}$, $M \mapsto (M : H^*BV)_{\mathcal{U}}$. Comme H^*BV est isomorphe au produit tensoriel $H^*B\sigma \otimes H^*B\sigma \otimes \ldots \otimes H^*B\sigma$, d fois, le foncteur T_V est équivalent à la composition $T \circ T \circ \ldots \circ T$, d fois, et l'on peut dans le théorème 0.1 remplacer T par T_V :

THÉORÈME 1.2.
(a) *Le foncteur T_V est exact.*
(b) *Le foncteur T_V commute aux produits tensoriels.*
(c) *Si M est une A-algèbre instable, alors $T_V M$ possède une structure naturelle de A-algèbre instable et $T_V M$ munie de cette structure coïncide avec $(M : H^*BV)_{\mathcal{K}}$.*

Les propriétés des A-algèbres instables H^*BV que l'on exprime ainsi sont tout à fait exceptionnelles ; en fait ces algèbres sont "caractérisées" par les "restrictions en degré zéro" de (a) et (b).

Précisons un peu. Considérons une A-algèbre instable de dimension finie en chaque degré K et posons $V = (K^1)^*$, $(K^1)^*$ désignant le $\mathbf{F}_p$-espace vectoriel dual de K^1, alors les deux propriétés suivantes sont équivalentes :
(i) Le $\mathcal{K}$-morphisme canonique $H^*BV \to K$ est un isomorphisme.
(ii) Le foncteur $M \mapsto t(M) = ((M : K)_{\mathcal{U}})^0$ est exact (en d'autres termes, le A-module instable sous-jacent à K est injectif) et commute aux produits tensoriels (i.e. la transformation naturelle $t(-\otimes -) \to t(-)\otimes t(-)$, induite par le produit de K, est un isomorphisme).

2 Conséquences homotopiques

Il existe plusieurs méthodes pour exploiter ces propriétés magiques de H^*BV. La plus facile à esquisser, que l'on doit à E. Dror Farjoun et à J. Smith [DS], procède de la façon suivante.

On considère d'abord la classe des espaces Y, que j'appellerai p-π_*-finis, tels que
- $\pi_0 Y$ est fini ;

et pour tout choix du point base,
- $\pi_n Y$, $n \geq 1$, est un p-groupe fini ;
- $\pi_n Y$ est trivial pour $n >> 0$.

THÉORÈME 2.1. *Pour un espace p-π_*-fini Y l'application naturelle*

$$h_Y : T_V H^*Y \to H^*\mathbf{hom}(BV, Y)$$

est un isomorphisme.

Esquisse de démonstration. On peut supposer Y connexe. Dans ce cas il existe une suite finie de fibrations principales $Y_s \to Y_{s-1}$, $0 \leq s \leq r$, avec groupe structural $K(\mathbb{Z}/p, m_s)$ ($m_s \geq 1$), Y_{-1} étant un point et Y_r ayant le type d'homotopie de Y. On démontre le théorème 2.1 en vérifiant de proche en proche que les h_{Y_s} sont des isomorphismes ; cette escalade s'effectue avec le matériel suivant :

LEMME 2.1.1. *L'application $h_{K(\mathbb{Z}/p,n)}$ est un isomorphisme.*

Démonstration. En fait, pour tout espace X avec H^*X de dimension finie en chaque degré, on a un $\mathcal{K}$-isomorphisme

$$H^*\mathbf{hom}(X, K(\mathbb{Z}/p, n)) \cong (H^*K(\mathbb{Z}/p, n) : H^*X)_{\mathcal{K}}\,.$$

LEMME 2.1.2. *On considère une fibration $Z \to Y \to K(\mathbb{Z}/p, n)$; on suppose H^*Y et $H^*\mathbf{hom}(BV, Y)$ de dimension finie en chaque degré. Alors si h_Y est un isomorphisme, il en est de même pour h_Z.*

Indications sur la démonstration. Soient $E(1)$ et $E(2)$ les suites spectrales d'Eilenberg-Moore des fibrations $Z \to Y \to K(\mathbb{Z}/p, n)$ et $\mathbf{hom}(BV, Z) \to \mathbf{hom}(BV, Y) \to \mathbf{hom}(BV, K(\mathbb{Z}/p, n))$; $E(1)$ et $E(2)$ sont des suites spectrales de A-modules instables. On vérifie que le théorème 1.2 fournit un isomorphisme de suites spectrales $T_V E(1) \to E(2)$ compatible avec h_Z.

Spécialisons le theorème 2.1 en degré zéro. Le fait que h_Y soit en particulier un isomorphisme en degré zéro se traduit par l'énoncé suivant :

THÉORÈME 2.2. *Pour un espace p-π_*-fini Y l'application naturelle*

$$[BV, Y] \to \mathrm{Hom}_{\mathcal{K}}(H^*Y, H^*BV)$$

est une bijection (d'ensembles finis).

(La notation $[-,-]$ désigne l'ensemble des classes d'homotopie libre d'applications).

Voilà pour les espaces p-π_*-finis. On considère ensuite des tours d'espaces p-π_*-finis et on "passe à la limite".

Soit $\widehat{Y}$ le p-complété de Bousfield-Kan de Y. Quand H^*Y est de dimension finie en chaque degré (ce que nous supposerons toujours) $\widehat{Y}$ est la limite inverse d'une tour fonctorielle d'espaces p-π_*-finis (et coïncide à homotopie près avec le p-complété de Sullivan de Y). En passant à la limite respectivement dans les théorèmes 2.2 et 2.1, on obtient :

THÉORÈME 2.3. *Pour tout espace Y dont la cohomologie modulo p est de dimension finie en chaque degré, l'application naturelle*

$$[BV, \widehat{Y}] \to \mathrm{Hom}_{\mathcal{K}}(H^*Y, H^*BV)$$

est une bijection (homéomorphisme d'ensembles profinis).

Théorème 2.4. *Soit Y un espace dont la cohomologie modulo p est de dimension finie en chaque degré et nulle en degré un, alors l'application naturelle*

$$T_V H^*Y \to H^*\mathbf{hom}(BV, \widehat{Y})$$

est un isomorphisme de A-algèbres instables.

(L'existence des applications naturelles ci-dessus résulte par exemple de celle d'une application naturelle $H^*Y \to H^*\widehat{Y}$).

On peut remplacer $\widehat{Y}$ par Y dans 2.3 et 2.4 (et avoir ainsi des énoncés d'apparence moins technique) si l'on est prêt à faire des concessions sur le π_1 de Y, par exemple :

Théorème 2.5. *Pour tout espace simplement connexe Y dont la cohomologie modulo p est de dimension finie en chaque degré, l'application naturelle*

$$[BV, Y] \to \mathrm{Hom}_{\mathcal{K}}(H^*Y, H^*Y, H^*BV)$$

est une bijection.

Théorème 2.6. *Soit Y un espace simplement connexe dont la cohomologie modulo p est de dimension finie en chaque degré. Si $T_V H^*Y$ est de dimension finie en chaque degré et nulle en degré un, alors l'application naturelle*

$$h_Y : T_V H^*Y \to H^*\mathbf{hom}(BV, Y)$$

est un isomorphisme de A-algèbres instables.

Enfin on obtient également le résultat suivant :

Théorème 2.7. *Soient Y et Z deux espaces dont la cohomologie modulo p est de dimension finie en chaque degré ; soit ω une application $BV \times Z \to Y$ (on observera que se donner ω revient à se donner une application $Z \to \mathbf{hom}(BV, Y)$). Alors les deux conditions suivantes sont équivalentes :*

(i) *l'homomorphisme de A-algèbres instables $T_V H^*Y \to H^*Z$, adjoint de ω^* : $H^*Y \to H^*V \otimes H^*Z$, est un isomorphisme ;*

(ii) *l'application $\widehat{Z} \to \mathbf{hom}(BV, \widehat{Y})$ induite par ω est une équivalence d'homotopie.*

L'implication (i) $\Rightarrow$ (ii) fournit une méthode pour déterminer le type d'homotopie de l'espace fonctionnel $\mathbf{hom}(BV, Y)$, à des problèmes de p-complétion près, quand on a un candidat Z pour celui-ci : il suffit de vérifier la condition algébrique (i). En particulier on obtient ainsi une preuve de la conjecture de Sullivan sur les points fixes homotopiques.

3 Quelques utilisations en théorie de l'homotopie

Pour terminer, nous citons quelques travaux dans lequel la théorie précédente intervient plus ou moins directement.

3.1. Soient π un p-groupe fini et X un espace muni d'une action de π. Considérons une suite exacte $1 \to \kappa \to \pi \to \sigma \to 1$; les espaces de points fixes homotopiques satisfont, à homotopie près, le même type de formule que les espaces de points fixes ordinaires :

$$X^{h\pi} \cong (X^{h\kappa})^{h\sigma}.$$

On peut donc obtenir par récurrence des informations sur l'espace des points fixes homotopiques d'une action d'un p-groupe fini π à partir du cas $\pi = \sigma$.

D. Dwyer et A. Zabrodsky ont déterminé ainsi le type d'homotopie de l'espace $\mathbf{hom}(B\pi, BG)$ des applications entre classifiants d'un p-groupe fini et d'un groupe de Lie compact [DZ]. Ils obtiennent en particulier :

THÉORÈME 3.1.1. *Pour tout p-groupe fini π et tout groupe de Lie compact G, l'application naturelle*

$$\mathrm{Rep}(\pi, G) \to [B\pi, BG]$$

est une bijection.

(La notation $\mathrm{Rep}(\pi, G)$ désigne l'ensemble des représentations de π dans G, c'est-à-dire des homomorphismes modulo conjugaison dans G.)

D. Notbohm a montré ensuite que l'on pouvait remplacer dans la théorie de Dwyer-Zabrodsky le p-groupe fini π par un groupe p-toral, c'est-à-dire une extension d'un tore par un p-groupe fini, en passant cette fois à la limite à la source [No1]. En particulier :

THÉORÈME 3.1.2. *Pour tout groupe p-toral P et tout groupe de Lie compact G, l'application naturelle*

$$\mathrm{Rep}(\pi, G) \to [B\pi, BG]$$

est une bijection.

Idée de la démonstration. Expliquons brièvement comment montrer que l'application naturelle $\mathrm{Rep}(S^1, G) \to [BS^1, BG]$ est une bijection ; on suppose, pour simplifier, que G est connexe (ce qui équivaut à supposer que BG est simplement connexe). Comme l'algèbre $H^*(BG; \mathbb{Q})$ est polynômiale, engendrée par des générateurs (de degré pair) en nombre fini, la théorie de Sullivan [Su] permet de ramener sans trop de difficultés la détermination de l'ensemble $[BS^1, BG]$ à celle des ensembles $[BS^1, (BG)^\wedge_p]$, p premier (p étant variable, le p-complété de BG est ici noté $(BG)^\wedge_p$). Soit Π le sous-groupe de S^1 formé des éléments d'ordre une puissance de p ; on a donc $\Pi \cong \mathrm{colim}\{\mathbb{Z}/p^n; n \in \mathbb{N}\}$. Comme l'application canonique $B\Pi \to BS^1$ induit un isomorphisme en homologie modulo p, elle induit également une bijection $[B\Pi, (BG)^\wedge_p] \cong [BS^1, (BG)^\wedge_p]$. On se convainc alors que l'on a la suite de bijections suivante :

$$\begin{aligned}[B\pi, (BG)^\wedge_p] &\cong \lim\{[B(\mathbb{Z}/p^n), (BG)^\wedge_p; n \in \mathbb{N}\} \\ &\cong \lim\{\mathrm{Rep}(\mathbb{Z}/p^n, G)\ ; n \in \mathbb{N}\} \cong \mathrm{Rep}(\Pi, G).\end{aligned}$$

3.2. Le travail de S. Jackowski, J. McClure et R. Oliver [JMO] dont nous allons parler à présent, n'est relié à la théorie des espaces fonctionnels $\mathbf{hom}(B\sigma, -)$ que par l'intermédiaire de celui de Dwyer-Zabrodsky évoqué ci-dessus.

Soit G un groupe de Lie compact. Jackowski, McClure et Oliver montrent d'abord que BG peut être vu, à homotopie et p-complétion près, comme la limite directe homotopique d'un diagramme de classifiants de sous-groupes p-toraux (et donc finalement de p-sous-groupes finis).
Le cas facile est celui où p ne divise pas l'ordre du groupe de Weyl de G. On a alors une équivalence d'homotopie :

$$(BG)^\wedge \cong (EW \times_W BT)^\wedge ,$$

T désignant "le" tore maximal de G et W son groupe de Weyl. Le cas où p divise l'ordre de W est beaucoup plus subtil et son analyse est la contribution principale de [JMO].

Ils sont alors en mesure d'étudier certains espaces fonctionnels dont la source est BG. Par exemple, ils déterminent complètement le monoïde $[BG, BG]$ pour G un groupe de Lie compact connexe simple. Ce monoïde est fait des automorphismes extérieurs de G et des "applications d'Adams instables" ψ^k, k premier à l'ordre de W ; ψ^k est caractérisée, à homotopie près, par la commutativité du diagramme

$$\begin{array}{ccc} BG & \xrightarrow{\psi^k} & BG \\ \uparrow & & \uparrow \\ BT & \xrightarrow{Bk} & BT \, , \end{array}$$

k désignant, dans la notation Bk qui apparaît ci-dessus, l'endomorphisme multiplication par k du groupe abélien T.

3.3. La théorie des espaces $\mathbf{hom}(BV, -)$ est l'un des ingrédients utilisés pour montrer que le type d'homotopie de la p-complétion du classifiant de certains groupes de Lie compact est uniquement déterminé par sa cohomologie modulo p [DMW1][DMW2]. Voir aussi [No2].

3.4. W. Dwyer et C. Wilkerson montrent dans [DW1] que le foncteur T satisfait une "Théorie de Smith algébrique" (voir aussi le travail de S. Zarati et de l'auteur [LZ]). Ils en déduisent dans [DW2] des théorèmes de points fixes homotopiques analogues aux théorèmes de points fixes classiques. Par exemple :

THÉORÈME 3.4. *Soit X un espace muni d'une action de σ. On fait les hypothèses suivantes :*
- *l'homologie $H_*(X; \mathbf{F}_p)$ est finie ;*
- *les espaces X et $X^{h\sigma}$ sont p-complets.*

Alors $H_*(X^{h\sigma}; \mathbf{F}_p)$ est finie et l'on a la congruence

$$\chi(X^{h\sigma}) \equiv \chi(X) \mod p ,$$

$\chi(-)$ désignant la caractéristique d'Euler du $\mathbf{F}_p$-espace vectoriel gradué fini $H_*(-; \mathbf{F}_p)$.

(Un espace X est dit p-complet si l'application canonique $X \to \widehat{X}$ est une équivalence d'homotopie.)

Ils utilisent alors ces théorèmes de points fixes homotopiques pour montrer que les espaces de lacets dont l'homologie modulo p est finie, partagent une grande partie de la riche structure interne des groupes de Lie : tores maximaux, groupes de Weyl ... (la référence est toujours [DW2]).

References

[DMW1] W. G. Dwyer, H. R. Miller and C. W. Wilkerson, *The homotopic uniqueness of* BS^3, Algebraic topology, Barcelona 1986 (proceedings), Springer LNM **1298** (1987), 90–105.

[DMW2] W. G. Dwyer, H. R. Miller and C. W. Wilkerson, *Homotopic uniqueness of classifying spaces*, Topology **31** (1992), 29–45.

[DS] E. Dror Farjoun and J. Smith, *A geometric interpretation of Lannes' functor* T, Théorie de l'homotopie, Astérisque **191** (1990), 87–95.

[DW1] W. G. Dwyer and C. W. Wilkerson, *Smith theory and the functor* T, Comm. Math. Helv. **66** (1991), 1–17.

[DW2] W. G. Dwyer and C. W. Wilkerson, *Homotopy fixed points methods for Lie groups and finite loop spaces*, Annals of Math. **139** (1994), 395–442.

[DZ] W. G. Dwyer and A. Zabrodsky, *Maps between classifying spaces*, Algebraic Topology, Barcelona 1986 (proceedings), Springer LNM **1298** (1987), 106–119.

[JMO] S. Jackowski, J. E. McClure and R. Oliver, *Self-maps of classifying spaces of compact simple Lie groups*, Annals of Math. **135** (1992), 183–270.

[La1] J. Lannes, *Sur la cohomologie modulo p-groupes abéliens élémentaires*, Proc. Durham Symposium on Homotopy Theory 1985, London Math. Soc. LNS **117**, Cambridge Univ. Press 1987, 97–116.

[La2] J. Lannes, *Sur les espaces fonctionnels dont la source est le classifiant d'un p-groupe abélien élémentaire*, Publ. Math. I.H.E.S. **75** (1992), 135–244.

[Mi] H. R. Miller, *The Sullivan conjecture on maps from classifying spaces*, Annals of Math. **120** (1984), and corrigendum, Annals of Math. **121** (1985), 605–609.

[LZ] J. Lannes et S. Zarati, *Théorie de Smith algébrique et classification des* H^*V-$\mathcal{U}$*-injectifs*, Bulletin de la S.M.F. **123** (1995), 189–223.

[No1] D. Notbohm, *Abbildungen zwischen klassifizierenden Räumen*, Dissertation, Göttingen (1988).

[No2] D. Notbohm, *Homotopy uniqueness of classifying spaces of connected Lie groups at primes dividing the order of the Weyl group*, Topology **33** (1994), 271–330.

[Su] D. Sullivan, *Genetics of homotopy theory and the Adams conjecture*, Annals of Math. **100** (1974), 1–79.

Spaces of Algebraic Cycles
Levels of Holomorphic Approximation

H. BLAINE LAWSON, JR.

Mathematics Department
State University of New York
Stony Brook, NY 11794, USA

There is a fascinating phenomenon, observed in recent years in a wide range of contexts, which is roughly that:

As degree increases, holomorphic data approximates topological data. (*)

In certain situations, for example in studying maps from Stein spaces to elliptic spaces, there is a general theory due to Grauert and Gromov [G]. However, in many cases, often ones of particular interest to mathematicians, this phenomenon appears a wonderful surprise, and we have as yet no deep understanding of it. An early example was Segal's Theorem [S_1] that the inclusion $\mathrm{Hol}_d(\mathbb{P}^1, \mathbb{P}^1) \subset \mathrm{Map}_d(\mathbb{P}^1, \mathbb{P}^1) \cong \mathrm{Map}_0(\mathbb{P}^1, \mathbb{P}^1)$ of the holomorphic self-maps of the 2-sphere of degree d into the space of all continuous such maps is d-connected. Subsequently there have been many related results. Another case is the recently established Atiyah-Jones Conjecture for self-dual connections over the 4-sphere [BHMM]. Further examples of current interest come from algebraic cycles and spaces of morphisms.

To understand a phenomenon such as (*) it is useful to study cases where it fails and develop a sensitive measure of this failure. With luck this measure can be encoded in geometric invariants with independent interest and other interpretations.

Hodge Theory provides a good example. Consider the inclusion $i : (\Omega^*_{\mathrm{Hol}}, \partial) \subset (\Omega^*_{C^\infty}, d)$ of the complex of holomorphic differential forms into the smooth forms over $\mathbb{C}$ on a complex manifold X. When X is Stein, this induces an isomorphism in homology. For general X it does not. However there is a filtration $\cdots \subset \mathcal{F}^{p,*} \subset \mathcal{F}^{p-1,*} \subset \cdots$ of $\Omega^*_{C^\infty}$ by subcomplexes, where $\mathcal{F}^{p,*}$ consists of forms $\sum a_{I,J} dz^I \wedge d\bar{z}^J$ with $a_{I,J} = 0$ if $|I| < p$. This induces a filtration $i_* : H^k_{\mathrm{Hol}}(X) \to F^{k,k} \subset F^{k-1,k} \subset \cdots \subset F^{0,k} = H^k_{C^\infty}(X)$. When X is compact and Kaehler (e.g., projective), i_* is an isomorphism. The $F^{p,k}$ represent "levels of holomorphic approximation".

We shall see this again in the more geometric (rather than analytic) cases to be discussed here: spaces of algebraic cycles on an algebraic variety, and spaces of algebraic maps between varieties.

Recall that an algebraic subvariety of a complex projective space is a subset that is defined in homogeneous coordinates by a finite number of homogeneous

Proceedings of the International Congress
of Mathematicians, Zürich, Switzerland 1994

polynomial equations, and that is irreducible in the sense that it cannot be decomposed as a nontrivial union of two such subsets. Connected submanifolds are examples. In fact, on any subvariety $V \subset \mathbb{P}^N$ the set of manifold points is connected and dense. This gives V a well-defined dimension. The complement $\mathrm{Sing}(V)$ of the manifold points in V is a finite union of subvarieties of lower dimension. There is a semi-algebraic triangulation of $\mathbb{P}^N$ such that both V and $\mathrm{Sing}(V)$ are unions of simplices (cf. [H]), and V determines a cycle in the associated complex. In particular V has a homology **degree** in $\mathbb{P}^N$. Furthermore, the set of manifold points of V has finite Hausdorff $2p$-measure, and integration of forms over this set defines an integral current of boundary zero in the sense of Federer [Har].

Fix a complex projective subvariety X of dimension n and an integer p, $0 \leq p \leq n$, and denote by $\mathcal{V}_p(X)$ the set of p-dimensional subvarieties contained in X. These are the p-dimensional points of X in the sense of Grothendieck. It is natural to consider abelianizations of this set, namely

$$\mathcal{C}_p(X) = \mathbb{Z}^+ \cdot \mathcal{V}_p(X) \qquad \text{and} \qquad \mathcal{Z}_p(X) = \mathbb{Z} \cdot \mathcal{V}_p(X),$$

the free abelian monoid and the free abelian group generated by $\mathcal{V}_p(X)$. Points of $\mathcal{Z}_p(X)$ are called **p-cycles** on X, and those in $\mathcal{C}_p(X)$ are called **effective p-cycles**. At this point there enters some magic of fundamental importance: the monoid $\mathcal{C}_p(X)$ can be written as a disjoint union $\mathcal{C}_p(X) = \coprod_d \mathcal{C}_{p,d}(X)$ where each $\mathcal{C}_{p,d}(X)$ has the structure of a *projective algebraic set* for which addition is an algebraic map [ChW], [GKZ]. (The index d corresponds to projective degree.) In particular, each $\mathcal{C}_{p,d}(X)$ is a compact Hausdorff space. This gives $\mathcal{Z}_p(X)$ the structure of an *abelian topological group*. It has been shown by Lima-Filho that the topologies on these spaces have several alternative characterizations that allow their extension to general algebraic varieties [Li_4].

Both $\mathcal{C}_p(X)$ and $\mathcal{Z}_p(X)$ are beautiful and basic objects. The components of $\mathcal{C}_p(X)$, conventionally called **Chow varieties**, are moduli spaces for algebraic p-cycles on X. The associated abelian topological group $\mathcal{Z}_p(X)$ is a CW-complex. In fact it has a filtration (as a group) by finite subcomplexes, and its topology is the one compactly generated by this filtration.

When studying a monoid such as $\mathcal{C}_p(X)$ it is customary to consider its homotopy-theoretic group completion $\Omega B\mathcal{C}_p(X)$ (loops on the classifying space of the monoid) rather than the naïve topological Grothendieck group $\mathcal{Z}_p(X)$. This is because $\Omega B\mathcal{C}_p(X)$ carries the "limiting topology" of the monoid (cf. [McSe]). In our context there is a surprising and very important theorem due to Lima-Filho [Li_2]. (See also [FG].)

THEOREM 1. *The natural map $\Omega B\mathcal{C}_p(X) \to \mathcal{Z}_p(X)$ is a homotopy equivalence.*

We observe as a general principle that in measuring the failure of holomorphic approximation (*) one must let the degree go to infinity. It is a direct consequence of Theorem 1 that, up to homotopy equivalence, $\mathcal{Z}_p(X) \cong \varinjlim \mathcal{C}_{p,d}(X)$ and so $\mathcal{Z}_p(X)$ *is* the limit of the $\mathcal{C}_{p,d}(X)$ as the degree goes to infinity.

Now there is a natural continuous inclusion

$$\mathcal{Z}_p(X) \hookrightarrow \mathfrak{Z}_{2p}(X) \tag{1}$$

into the topological group of all integral $2p$-cycles on X (cf. [Fed]). We want to measure the failure of this map (1) to be a homotopy equivalence.

Note that when $p = 0$ the map (1) is the identity. In positive dimensions this is far from true. Nevertheless, there is a large family of spaces of "Grauert-Gromov" type, for which (1) is a homotopy equivalence. We say that a variety X has a **cell-decomposition** if there is a filtration by subvarieties $X_0 \subset X_1 \subset \cdots \subset X_k = X$ such that for each i, $X_i - X_{i-1}$ is a disjoint union of affine spaces $\coprod_\alpha \mathbb{C}_\alpha^{n_i}$. The following was proved in [L_1] for $X = \mathbb{P}^n$ and in [Li_1] in general.

THEOREM 2. *If X has a cell decomposition, then (1) is a homotopy equivalence.*

Note that Theorem 2 vastly generalizes the well-known fact that if X has a cell decomposition, then every class in $H_*(X;\mathbb{Z})$ is represented by an algebraic cycle. This corresponds to the bijection of connected components in (1).

In general the inclusion (1) is not a homotopy equivalence. For example, it is neither universally surjective nor universally injective on connected components. A basic result of J. Moore states that an abelian topological group is determined up to weak homotopy equivalence by its homotopy groups (up to abstract group isomorphism!). Hence the induced homomorphisms

$$\pi_j \mathcal{Z}_p(X) \longrightarrow \pi_j \mathfrak{Z}_{2p}(X) \cong H_{2p+j}(X;\, \mathbb{Z}) \tag{2}$$

(where the equivalence on the right is a theorem of Almgren-Dold-Thom [A],[DT]) measure the failure of the approximation (1) to be a weak homotopy equivalence, and the groups $\pi_j \mathcal{Z}_p(X)$ are just the invariants we were looking for. They were introduced in [L_1] and in [F_1] in a very general algebraic context, and have been developed by Friedlander, Lima-Filho, Mazur and Gabber.

The invariants have been formulated as "homology groups"

$$L_pH_k(X) \overset{\text{def}}{=} \pi_{k-2p}\mathcal{Z}_p(X) \qquad \text{for} \quad 0 \le 2p \le k$$

where one thinks of k as the *homology degree* and p as the *holomorphic level.* By Dold-Thom [DT] we have a natural equivalence $L_0H_k(X) \cong H_k(X;\mathbb{Z})$. On the other hand, as observed by Friedlander [F_1], there is an isomorphism: $L_pH_{2p}(X) \cong \mathcal{A}_p \equiv$ {algebraic p-cycles on X}/{algebraic equivalence}, a group of interest to algebraists that can sometimes be infinitely generated. This theory is now highly developed and features the following properties.

(i) $L_*H_*(\bullet)$ is a covariant functor on the category of algebraic varieties and proper algebraic maps. There are Gysin "wrong way" maps associated to flat morphisms ([F_1], [Li_1], [Li_4]).
(ii) There is a natural transformation of functors $\Phi : L_pH_k(\bullet) \to H_k(\bullet;\mathbb{Z})$ for all $2p \le k$ ([FM_1], [Li_3]).
(iii) There are localization long exact sequences ([Li_1], [Li_2]).
(iv) There is a ring of functorial operations isomorphic to $\mathbb{Z}[s,h]$ where $s : L_pH_k \to L_{p-1}H_k$ and $h : L_pH_k \to L_{p-1}H_{k-2}$. The natural transformation in (ii) is given by s^p ([FM_1]).
(v) There are filtrations on $H_*(\bullet;\mathbb{Z})$ and on $\mathcal{A}_*(\bullet)$ induced by the s-operation. They are subordinate to filtrations of Grothendieck and Bloch-Ogus and

have geometric characterizations in terms of cycles carried by algebraic correspondences ([FM_1], [FM_2], [F_2]).

(vi) There is a graded ring structure on $L_*H_*(\bullet)$ induced by the intersection of cycles. There is in fact a full intersection theory ([FG]).

(vii) There is a projective bundle theorem leading to Chern class maps from the higher algebraic K-groups of a variety to $L_*H_*(\bullet)$ ([FG]).

(viii) There is a local-to-global spectral sequence for $L_*H_*(\bullet)$ that is analogous to that of Quillen in algebraic K-Theory. If $\mathcal{L}_p\mathcal{H}_k$ denotes the Zariski sheaf on a variety X associated to the presheaf $U \mapsto L_pH_k(U)$, then this spectral sequence has the form ([FG])

$$E^2_{p,q} = H^{n-p}(X, \mathcal{L}_r\mathcal{H}_{n+q}) \Rightarrow L_rH_{p+q}(X).$$

(ix) $L_*H_*(\bullet)$ carry (limits of) mixed Hodge structures (Hain [FM_1]).

(x) There are natural transformations from Bloch's higher Chow groups to $L_*H_*(\bullet)$ ([FG]).

A key result making much of this possible is the following algebraic suspension theorem first proved in [L_1] over $\mathbb{C}$ and algebraicized in [F_1] to work over arbitrary algebraically closed fields. Given a subvariety V in $\mathbb{P}^N$ and a point $\mathbb{P}^0$ disjoint from V, we denote by $\Sigma\!\!\!/ V$ the subvariety consisting of the union of all lines joining $\mathbb{P}^0$ to V. Topologically $\Sigma\!\!\!/ V$ is the Thom space of the dual of the tautological line bundle of $\mathbb{P}^N$ restricted to V. This construction extends by linearity to cycles.

THEOREM 3. *For any projective variety X and any $p \le \dim(X)$, the homomorphism $\Sigma\!\!\!/ : \mathcal{Z}_p(X) \longrightarrow \mathcal{Z}_{p+1}(\Sigma\!\!\!/ X)$ is a homotopy equivalence.*

COROLLARY 4. $\quad \mathcal{Z}_p(\mathbb{P}^n) \cong K(\mathbb{Z},0) \times K(\mathbb{Z},2) \times \cdots \times K(\mathbb{Z},2(n-p))$

This corollary follows from [DT] and the fact that $\Sigma\!\!\!/ \mathbb{P}^n = \mathbb{P}^{n+1}$.

In general for a quasi-projective variety U written as $U = X - X_\infty$ (where X and X_∞ are projective varieties) we define $\mathcal{Z}_p(U) = \mathcal{Z}_p(X)/\mathcal{Z}_p(X_\infty)$ with the quotient topology. This is independent of the choice of compactification X of U [Li_1]. Another version of Corollary 4 is the following intriguing equivalence, which gives an algebraic model for Eilenberg-MacLane spaces:

$$\mathcal{Z}_p(\mathbb{C}^n) \cong K(\mathbb{Z}, 2(n-p)). \tag{3}$$

Algebraic suspension is a special case of an important elementary binary operation on cycles given as follows. Fix $\mathbb{P}^m \coprod \mathbb{P}^n \subset \mathbb{P}^{m+n+1}$ corresponding to a decomposition of homogeneous coordinates $\mathbb{C}^{m+1} \oplus \mathbb{C}^{n+1} = \mathbb{C}^{m+n+2}$. Given irreducible subvarieties $V \subset \mathbb{P}^m$, $W \subset \mathbb{P}^n$, of dimensions p and q, respectively, we define the **algebraic join** $V\#W \subset \mathbb{P}^{m+n+1}$ to be the variety consisting of the union of all lines joining V to W. If $C(V) \subset \mathbb{C}^{m+1}$ denotes the closed cone in homogeneous coordinates sitting above V, then $C(V\#W) = C(V) \times C(W)$, from which it is clear that $V\#W$ is an algebraic subvariety of dimension $p+q+1$. Note that $\Sigma\!\!\!/ V = V\#\mathbb{P}^0$ and $\Sigma\!\!\!/^{n+1} V = V\#\mathbb{P}^n$. For varieties $X \subset \mathbb{P}^m$ and $Y \subset \mathbb{P}^n$ the algebraic join gives a continuous biadditive map $\mathcal{Z}_p(X) \wedge \mathcal{Z}_q(Y) \to \mathcal{Z}_{p+q+1}(X\#Y)$.

In particular, we have the map $\mathcal{Z}_p(X) \wedge \mathcal{Z}_q(\mathbb{P}^n) \to \mathcal{Z}_{p+q+1}(\Sigma^{n+1} X) \cong \mathcal{Z}_{p+q-n}(X)$ where the equivalence comes from Theorem 3. This leads to an induced pairing on homotopy groups. Setting $X = \mathbb{P}^m$ leads to the ring of Friedlander-Mazur operations mentioned in (iv) above, and with general X this pairing gives $L_*H_*(X)$ the structure of a module over this ring.

We now turn our attention from cycles to spaces of holomorphic maps. Let X and Y be complex algebraic varieties and denote by $\mathfrak{Mor}(X, Y)$ the space of **algebraic maps** from X to Y, by which we mean continuous maps whose graphs are subvarieties of $X \times Y$. As above it is natural to abelianize this space by considering multi-valued maps $\mathfrak{Mor}(X, SP^*(Y))$ where $SP^*(Y) = Y \coprod SP^2(Y) \coprod \cdots$ denotes the disjoint union of symmetric products of Y. Group completing this monoid gives a space that we denote $\mathfrak{Mor}(X, \mathcal{Z}_0(Y))$. More generally, we might replace $\mathcal{Z}_0(Y)$ with

$$\mathcal{Z}^s(Y) \stackrel{\text{def}}{=} \mathcal{Z}_{m-s}(Y)$$

where $m = \dim(Y)$. Then letting $Y = \mathbb{C}^m$ we would obtain via (3) an algebraic analogue of the space $\mathrm{Map}(X, K(\mathbb{Z}, 2s))$ whose homotopy groups are $H^{2s-*}(X; \mathbb{Z})$. Taking homotopy groups of $\mathfrak{Mor}(X, \mathcal{Z}^s(\mathbb{C}^m))$ then gives a bigraded theory with a natural transformation to $H^*(X; \mathbb{Z})$, which measures the levels of holomorphic approximation of the inclusion

$$\mathfrak{Mor}(X, \mathcal{Z}^s(\mathbb{C}^m)) \longrightarrow \mathrm{Map}(X, \mathcal{Z}^s(\mathbb{C}^m)).$$

This rough outline is carried through in [FL$_1$]. Here are some details. Let X be a quasi-projective variety and Y a projective variety of dimension m. One defines an **effective algebraic s-cocycle on X** with values in Y to be an algebraic map $\varphi : X \to \mathcal{C}^s(Y) \equiv \mathcal{C}_{m-s}(Y)$, and one denotes by $\mathfrak{Mor}(X, \mathcal{C}^s(Y))$ the space of all such cocycles. There is a *graphing construction* that gives an embedding

$$\mathfrak{Mor}(X, \mathcal{C}^s(Y)) \hookrightarrow \mathcal{C}^s(X \times Y), \tag{4}$$

and we introduce on $\mathfrak{Mor}(X, \mathcal{C}^s(Y))$ the subspace topology. This is the topology of uniform convergence on compacta with bounded degree, and when X is compact it is the usual compact-open topology. It makes $\mathfrak{Mor}(X, \mathcal{C}^s(Y))$ an abelian topological monoid. We denote by $\mathfrak{Mor}(X, \mathcal{Z}^s(Y))$ its associated topological Grothendieck group (naïve topological group completion).

There is an algebraic suspension theorem for these spaces that asserts a homotopy equivalence $\mathfrak{Mor}(X, \mathcal{Z}^s(Y)) \cong \mathfrak{Mor}(X, \mathcal{Z}^s(\Sigma Y))$.

If $Y = \overline{Y} - Y_\infty$ is a difference of projective varieties, then $\mathfrak{Mor}(X, \mathcal{Z}^s(Y))$ is defined via a quotient of $\overline{Y}$-valued cocycles by Y_∞-valued cocycles [FL$_1$], [F$_3$]. One then introduces **morphic cohomology groups**

$$L^s H^k(X) = \pi_{2s-k} \mathfrak{Mor}(X, \mathcal{Z}^s(\mathbb{C}^m))$$

for $2s \geq k$. By the suspension theorem these groups are independent of m. They have the following properties established in [FL$_1$].

(i) $L^*H^*(\bullet)$ is a contravariant functor on the category of quasi-projective varieties and algebraic maps. There are Gysin "wrong way" maps associated to flat morphisms.

(ii) There is a ring structure on $L^*H^*(\bullet)$ whose multiplication is induced by the pointwise join of cocycles.
(iii) There is a natural transformation of ring-valued functors $\Phi : L^sH^k(\bullet) \to H^k(\bullet;\mathbb{Z})$ for all $2s \geq k$.
(iv) There is a ring of functorial operations isomorphic to $\mathbb{Z}[s,h]$ where $s : L^sH^k \to L^{s+1}H^k$ and $h : L^sH^k \to L^{s+1}H^{k+2}$.
(v) There are filtrations on $H^*(\bullet;\mathbb{Z})$ induced by (iii) and (iv) that are subordinate over $\mathbb{Q}$ to the Hodge filtration.
(vi) There is a Kronecker pairing $L^sH^k \otimes L_pH_k \to \mathbb{Z}$ for $2s \geq k \geq 2p$.
(vii) There are Chern classes $c_k \in L^kH^{2k}$ defined for algebraic bundles that transform under Φ to the standard ones.

One of the most intriguing and seductive features of morphic cohomology is that on smooth varieties it satisfies **"Poincaré duality"** with the cycle homology above. The general form of this duality for projective varieties is the following.

THEOREM 5 [FL_3]. *If X and Y are nonsingular, the continuous homomorphism*

$$\mathfrak{Mor}(X, \mathcal{Z}^s(Y)) \hookrightarrow \mathcal{Z}^s(X \times Y), \tag{5}$$

induced by the graphing map (4), is a weak homotopy equivalence.

The proof of this theorem is based, following suggestions of Gabber, on a Chow moving lemma for families. This moving lemma, established in [FL_2], has some independent interest and holds in a broad algebraic context. It is useful for example in rigorizing some of the classical constructions of the Chow ring. An important consequence of the Duality Theorem 5 is the following.

THEOREM 6. *Let X be a smooth projective variety of dimension n. Then the graphing map (5) induces isomorphisms*

$$L^sH^k(X) \xrightarrow{\cong} L_{n-s}H_{2n-k}(X)$$

that carry over under the natural transformations Φ to the Poincaré duality map

$$H^k(X;\mathbb{Z}) \xrightarrow{\cong} H_{2n-k}(X;\mathbb{Z}).$$

These results have many consequences [FL_3]. For example, they yield a **cohomological version of the Dold-Thom theorem**: *For any smooth projective variety X of dimension n there is a natural isomorphism*

$$\pi_*\mathfrak{Mor}(X, \mathcal{Z}_0(\mathbb{C}^n)) \cong H^{2n-*}(X;\mathbb{Z}).$$

This contrasts with the statement $\pi_*\mathcal{Z}_0(X) \cong H_*(X;\mathbb{Z})$ of the classical Dold-Thom theorem. Another consequence is that *if X and Y are smooth projective varieties with cell decompositions, then the inclusion*

$$\mathfrak{Mor}(X, \mathcal{Z}_0(Y)) \longrightarrow \mathrm{Map}(X, \mathcal{Z}_0(Y))$$

is a weak homotopy equivalence. Furthermore, for any smooth projective variety of dimension n and integers p, k with $p+k \leq n$, there is a homotopy equivalence

$$\mathfrak{Mor}(\mathbb{P}^k, \mathcal{Z}_p(X)) \cong \mathcal{Z}_p(X) \times \mathcal{Z}_{p+1}(X) \times \cdots \times \mathcal{Z}_{p+k}(X).$$

In particular, the space of parameterized rational curves on $\mathcal{Z}_p(X)$ is homotopy equivalent to $\mathcal{Z}_p(X) \times \mathcal{Z}_{p+1}(X)$. There are also many consequences within the theory that come by transferring deep results of [FG] to spaces of morphisms.

The Duality Theorem holds also for quasi-projective varieties where even further applications are realized. Details appear in [F_3].

The ideas here concerning algebraic cycles have had some direct applications to homotopy theory. To set the stage we recall the homotopy equivalence

$$\mathcal{Z}^q \stackrel{\text{def}}{=} \mathcal{Z}_{n-q}(\mathbb{P}^n) \cong K(\mathbb{Z},0) \times K(\mathbb{Z},2) \times K(\mathbb{Z},4) \times \cdots \times K(\mathbb{Z},2q) \qquad (6)$$

of Corollary 4. This equivalence is canonically determined by algebraic suspension and the requirement that $\mathcal{Z}_0(\mathbb{P}^q) \to \prod K(\mathbb{Z},2i)$ be generated additively by a map $\mathbb{P}^q \to \prod K(\mathbb{Z},2i)$, each factor of which classifies the canonical generator. The following two theorems were proved with Michelsohn. A rational homotopy version of Theorem 7 was also obtained by Friedlander.

THEOREM 7 [LM_1]. *Under the canonical equivalence (6), the algebraic join pairing*

$$\mathcal{Z}^q \wedge \mathcal{Z}^{q'} \longrightarrow \mathcal{Z}^{q+q'}$$

discussed above classifies the cup product in integral cohomology.

Observe that the Grassmannian $G^q(\mathbb{P}^n)$ of codimension-q planes in $\mathbb{P}^n$ is the component of degree 1 of $\mathcal{C}^q(\mathbb{P}^n)$. This gives a map $G^q(\mathbb{P}^n) \to \mathcal{Z}^q(\mathbb{P}^n)$.

THEOREM 8 [LM_1]. *Under the canonical equivalence (6), the inclusion*

$$c : G^q(\mathbb{P}^n) \hookrightarrow \mathcal{Z}^q(\mathbb{P}^n) \qquad (7)$$

classifies the total Chern class of the tautological q-plane bundle over $G^q(\mathbb{P}^n)$.

Letting $q, n \to \infty$ in (7) we obtain a map

$$c : BU \longrightarrow K(\mathbb{Z}, \mathrm{ev}) \stackrel{\text{def}}{=} \prod_{i \geq 0} K(\mathbb{Z}, 2i) \qquad (8)$$

which corresponds to the universal total Chern class $c = 1 + c_1 + c_2 + \cdots$.

Our next observation is that the join operation, when restricted to linear subspaces, is merely the direct sum. Thus, Theorems 7 and 8 together reprove the classical Chern duality $c(E \oplus E') = c(E)c(E')$ for complex vector bundles. Indeed they lead to much more. The join is a **completely natural extension of the direct sum pairing**. It yields a morphism of topological monoids

$$(G^*(\mathbb{P}^*), \oplus) \to (\mathcal{Z}^*(\mathbb{P}^*), \#). \qquad (9)$$

In 1975 Segal asked whether the map (8) could be extended to a transformation of cohomology theories, with standard K-theory on the left. In dimension zero it would be given by the total Chern class $c : K(X) \to H^{2*}(X;\, \mathbb{Z})$ where the "addition" $\star$ on the right is standard in degree 0 but is given on elements x, y of positive degree by the multiplication of the units $(1+x)\cup(1+y) = 1+x\star y$. Segal's question is equivalent to asking for an infinite loop structure on $K(\mathbb{Z}, \mathrm{ev})$ compatible via c with Bott's delooping of BU. Several such structures were proposed and eventually shown not to work. However, the naturality of the join product yields:

THEOREM 9 [BLLMM]. *The multiplication in $K(\mathbb{Z}, ev)$ induced by the algebraic join enhances to an infinite loop space structure such that the total Chern class (8) is an infinite loop map.*

The proof utilizes the naturality of the join to make the map c compatible with actions of the linear isometries operad of Boardman and Vogt. This task is substantially simplified by invoking May's theory of $\mathcal{I}_*$-functors [May]. This essentially reformulates the morphism (9) in invariant terms. To each hermitian vector space V of dimension v we consider the spaces $G^v(\mathbb{P}(V \oplus V))$ and $\mathcal{C}^v(\mathbb{P}(V \oplus V))$ with distinguished point $\{\mathbb{P}(V \oplus 0)\}$ and with pairings given by $\oplus$ and $\#$. One verifies rapidly that these are $\mathcal{I}_*$- functors and the inclusion $G^v(\mathbb{P}(V \oplus V)) \subset \mathcal{C}^v(\mathbb{P}(V \oplus V))$ is a natural transformation between them. This leads to the theorem and a little more. One can consider the stabilized spaces $\mathcal{D}(d) = \lim_{q,n} \mathcal{C}^q_d(\mathbb{P}^n)$ of cycles of degree d. Note that $\mathcal{D}(1) = BU$ and $\mathcal{D}(\infty) = K(\mathbb{Z}, \mathrm{ev})$. One obtains in [BLLMM] that $\mathcal{D} = \coprod \mathcal{D}(d)$ is an E_∞-ring space in the sense of [May].

All the results of the previous page carry over to algebraic varieties over $\mathbb{R}$ in the modern sense. Here it is necessary to consider a certain "Galois quotient" of cycle spaces. In the analogues of Theorems 7, 8, and 9 one finds the product in $\mathbb{Z}/2\mathbb{Z}$-cohomology and the total Stiefel-Whitney class. See [Lam].

There are also intersection versions of this theory including an intersection Dold-Thom theorem due to Gajer [Ga$_1$], [Ga$_2$].

Much of what has been said above carries over to the category of complex G-varieties where G is a finite group. This world is richly structured and spaces of cycles lead to fascinating invariants. Foundations for the theory on general varieties appear in [LM$_2$]. Concentrating on projective space leads to the construction of new equivariant cohomology theories and an equivariant analogue of Theorem 9. This is done in [LLM$_1$] roughly as follows. Suppose V is a unitary representation space for G of dimension v. Then the inclusion $G^v(\mathbb{P}(V \oplus V)) \subset \mathcal{Z}^v(\mathbb{P}(V \oplus V))$ together with the join pairing becomes a natural transformation of $\mathcal{I}_*$-functors in the category of G-spaces. We let $\mathcal{U} = V_0 \oplus V_0 \oplus \cdots$ where V_0 is the regular representation of G, and we consider the limits $BU_G = \lim_{V\subset\mathcal{U}} G^v(\mathbb{P}(V \oplus V))$ and $\mathcal{Z}_G(\mathcal{U}) = \lim_{V\subset\mathcal{U}} \mathcal{Z}^v(\mathbb{P}(V \oplus V))$.

Using results from [LMS] one shows that $\mathcal{Z}_G(\mathcal{U})$ enhances to become the degree 0 space in an E_∞-ring G-spectrum $\mathcal{Z}_G(\mathcal{U})$. This is a new equivariant cohomology theory with natural geometric origins. It is $R(G)$-graded and admits a natural transformation to Borel cohomology. Coefficients in the theory have been directly computed in many cases. They are also shown to be related to Bredon cohomology by a new equivariant suspension theorem [LLM$_2$] and beautiful equivariant versions of the Dold-Thom theorem due to Lima-Filho [Li$_5$].

There is a natural equivariant Chern class map

$$c : BU_G \longrightarrow \mathcal{Z}_G(\mathcal{U})$$

into the 0-space of this spectrum that transforms to the usual one in Borel cohomology. Applying [LMS] again shows that the map $c : BU_G \longrightarrow \mathcal{Z}_G(\mathcal{U})_1$ into the degree-1 component enhances to a map of G infinite loop spaces, i.e., to a transformation of equivariant cohomology theories (G-spectra)

$$\mathfrak{bu}_G \longrightarrow \mathfrak{z}_G^*$$

where $\mathfrak{bu}_G$ is connective equivariant K-theory. The Borel analogue of $\mathfrak{z}_G^*$ is the group of units $\{1\} \times H_G^{2*}(X;\mathbb{Z})_{\text{Borel}}$ in even Borel cohomology. This gives the equivariant version of Theorem 9.

Rather than enter into more details let me describe the utility of cycles spaces in this setting. For any G-space X there is a natural induced action of G on the topological group $\mathcal{Z}_0(X)$. The assignment $X \mapsto \mathcal{Z}_0(X)$ is a functor (in the world of G-spaces and maps). In fact it is better than this. Assigning to $H < G$ the fixed-point set $\mathcal{Z}_0(X)^H$ has the natural structure of a **topological Mackey functor**, i.e. it satisfies the axioms of Dress [D] in the world of topological groups and continuous homomorphisms. Consequently applying any homotopy functor to this leads to an ordinary Mackey-functor-valued theory on G-spaces. For example, Lima-Filho shows in [Li$_5$] that $\pi_*\mathcal{Z}_0(X)$ is exactly Bredon cohomology with values in the Mackey functor $\mathbb{Z}$. Applying the construction to spectra he obtains the ordinary $RO(G)$-graded cohomology with values in the Burnside ring.

Now if we turn attention to varieties X then for each p, $\mathcal{Z}_p(X)$ gives a topological Mackey functor. Taking π_* (or other homotopy invariants) gives analogous algebraic invariants that measure the levels of G-equivariant holomorphic approximation.

References

[A] Almgren, F. J. Jr., *Homotopy groups of the integral cycle groups*, Topology **1** (1962), 257–299.

[BHMM] Boyer, C. P., J. C. Hurtubise, B. M. Mann, and R. J. Milgram, *The topology of instanton moduli spaces I: The Atiyah-Jones conjecture*, Ann. of Math. (2) **137** (1993),561–609.

[BLLMM] Boyer, C. P., H. B. Lawson, Jr., P. Lima-Filho, B. Mann, and M.-L. Michelsohn, *Algebraic cycles and infinite loop spaces*, Invent. Math. **113** (1993), 373–388.

[ChW] Chow, W.-L., and van der Waerden, *Zur algebraischen Geometrie, IX: Über zugeordnete Formen und algebraische Systeme von algebraischen Mannigfaltigkeiten*, Math. Ann. **113** (1937), 692–704.

[DT] Dold, A., and R. Thom, *Quasifaserungen und unendliche symmetrische Produkte*, Ann. of Math. (2) **67** (1956), 230–281.

[D] Dress, A., *Contributions to the theory of induced representations*, in Algebraic K-Theory, Lecture Notes in Math., Springer-Verlag, Berlin and New York, **342** (1972), 182–240 .

[Fed] Federer, H., Geometric Measure Theory, Springer-Verlag, Berlin and New York, 1969.

[F_1] Friedlander, E., *Algebraic cycles, Chow varieties and Lawson homology*, Compositio Math. **77** (1991), 55–93.

[F_2] —, *Filtrations on algebraic cycles and homology*, to appear in Ann. Ec. Norm. Sup.

[F_3] —, *Algebraic cocycles on normal quasi-projective varieties*, to appear.

[FG] Friedlander, E., and O. Gabber, *Cycle spaces and intersection theory*, Topological Methods in Modern Mathematics, Publish or Perish Press, Austin, TX, 1993, 325–370.

[FL_1] Friedlander, E., and H. B. Lawson, Jr., *A theory of algebraic cocycles*, Ann. of Math. (2) **136** (1992), 361–428.

[FL_2] —, *Moving algebraic cycles of bounded degree*, preprint (1994).

[FL_3] —, *Duality relating spaces of algebraic cocycles and cycles*, preprint (1995).

[FM_1] Friedlander, E., and B. Mazur, *Filtrations on the homology of algebraic varieties*, Memoire of the A. M. S., no. 529 (1994).

[FM_2] —, *Correspondence homomorphisms for singular varieties*, Ann. Inst. Fourier, Grenoble **44**, 3 (1994), 703–727.

[Ga_1] Gajer, P., *The intersection Dold-Thom Theorem*, Ph.D. thesis, SUNY, Stony Brook, 1993.

[Ga_2] —, *Intersection Lawson homology*, M.S.R.I. preprint, 1993.

[GKZ] Gelfand, I. M., M. M. Kapranov, and A. V. Zelevinsky, Discriminants, Resultants and Multidimensional Determinants, Birkhäuser, Basel and Boston, 1994.

[G] Gromov, M., *Oka's principle for holomorphic sections of elliptic bundles*, J. Amer. Math. Soc. **2** (1989), 851–898.

[Har] Harvey, R., *Holomorphic chains and their boundaries*, in Several Complex Variables, vol. 1. Proc. Sympos. Pure Math. **30** (1977), 309–382.

[H] Hironaka, H., *Triangulation of algebraic sets*, in Algebraic Geometry, Proc. Sympos. Pure Math. **29** (1975), 165–185.

[Lam] Lam, T.-K., *Spaces of real algebraic cycles and homotopy theory*, Ph.D. thesis, SUNY, Stony Brook, 1990.

[L_0] Lawson, H. B. Jr., *The topological structure of the space of algebraic varieties*, Bull. Amer. Math. Soc. **17**, no. 2 (1987), 326–330.

[L_1] —, *Algebraic cycles and homotopy theory*, Ann. of Math. (2) **129** (1989), 253–291.

[L_2] —, *Spaces of algebraic cycles*, to appear in Surveys Differential Geom., II, Amer. Math. Soc., Providence, RI, 1995.

[LLM_1] Lawson, H. B., Jr., P. C. Lima-Filho, and M.-L. Michelsohn, *Algebraic cycles and equivariant cohomology theories*, SUNY Stony Brook Preprint, 1995.

[LLM_2] —, *On equivariant algebraic suspension*, SUNY Stony Brook Preprint, 1995.

[LM_1] Lawson, H. B., Jr., and M.-L. Michelsohn, *Algebraic cycles, Bott periodicity, and the Chern characteristic map*, in The Mathematical Heritage of Hermann Weyl, Amer. Math. Soc., Providence, RI, 1988, pp. 241–264.

[LM_2] —, *Algebraic cycles and group actions*, in Differential Geometry, Longman Press, Essex, 1991, pp. 261–278.

[LMS] Lewis, L. G., P. May, and M. Steinberger, *Equivariant stable homotopy theory*, Lecture Notes in Math., Springer-Verlag, Berlin and New York, 1213, 1985.

[Li_1] Lima-Filho, P.C., *Lawson homology for quasiprojective varieties*, Compositio Math. **84** (1992), 1–23.

[Li_2] —, *Completions and fibrations for topological monoids,* Trans. Amer. Math. Soc., **340** no. 1 (1993), 127–146.

[Li_3] —, *On the generalized cycle map*, J. Differential Geom. **38** (1993), 105–130.

[Li_4] —, *The topological group structure of algebraic cycles*, Duke Math. Journal **75** no. 2 (1994), 467–491.

[Li_5] —, *On the equivariant homotopy of free abelian groups on G-spaces and G-spectra*, Texas A. & M. Univ., College Station, TX, preprint, 1994.

[May] May, J. P., E_∞ Ring Spaces and E_∞ Ring Spectra, Lecture Notes in Math., Springer-Verlag, Berlin and New York, 577, 1977.

[McSe] McDuff, D., and G. Segal, *Homology fibrations and the "group completion" theorem*, Invent. Math. **31** (1976), 279–284.

[S_1] Segal, G., *The multiplicative group of classical cohomology*, Quart. J. Math. Oxford Ser. (2) **26** (1975), 289–293.

[S_2] —, *The Topology of Rational Functions*, Acta Math. **143** (1979), 39–72.

Dehn Surgery on Knots in the 3-Sphere

JOHN LUECKE

Department of Mathematics
The University of Texas at Austin
Austin, TX 78712, USA

The Dehn surgery construction is a way of obtaining a closed 3-manifold from a knot in the 3-sphere. The construction depends on two parameters, the knot and the surgery slope, and this article discusses theorems and conjectures describing the way the topology and geometry of the 3-manifold constructed depend on these parameters. We focus on knots in the 3-sphere, but certainly one can consider knots in arbitrary 3-manifolds and the Dehn surgery construction there. Most of the theorems discussed here apply in that context as well. [Go1] is an excellent survey at this level and I recommend it as a companion to this article. My intent here is to update some of the issues in [Go1] as well as to draw attention to some tantalizing aspects of specializing to knots in the 3-sphere.

If M is a manifold we will hereafter use the notation ∂M to mean the boundary of M.

Let K be a knot in the 3-sphere. The *exterior of* K, denoted X_K, is the complement in the 3-sphere of an open tubular neighborhood of K. Thus, X_K is a compact 3-manifold and ∂X_K is a 2-dimensional torus. Let α be a *slope* on ∂X_K, that is, the isotopy class of an essential simple closed curve in ∂X_K. The *Dehn surgery of* K *along* α, denoted $K(\alpha)$, is the closed 3-manifold obtained by attaching a solid torus N to X_K via a homeomorphism from ∂X_K to ∂N whereby a loop of slope α in ∂X_K bounds a disk in N. If α_1 and α_2 are two slopes on ∂X_K denote by $\Delta(\alpha_1, \alpha_2)$ the absolute value of the homological intersection number between α_1 and α_2 on ∂X_K. Finally, note that ∂X_K is parametrized by the meridian and longitude of X_K. Thus, we may write a slope α as p/q, meaning that any loop in class α goes p times meridianally and q times longitudinally around ∂X_K. We will often write $K(p/q)$ for $K(\alpha)$. If $\alpha_1 = p_1/q_1$ and $\alpha_2 = p_2/q_2$ then $\Delta(\alpha_1, \alpha_2) = |p_1 q_2 - p_2 q_1|$.

The focus of this article will be progress on the following:

QUESTION. When does a hyperbolic knot in the 3-sphere admit a nonhyperbolic Dehn surgery?

A 3-manifold will be called *hyperbolic* if it admits a complete Riemannian metric of constant sectional curvature -1. A knot K will be called *hyperbolic* if the interior of X_K is a hyperbolic 3-manifold with finite volume.

Proceedings of the International Congress
of Mathematicians, Zürich, Switzerland 1994

THEOREM 1 [T]. *If K is a hyperbolic knot then $K(p/q)$ will be hyperbolic for all but finitely many p/q.*

This fundamental result of Thurston says that one expects to see a hyperbolic manifold arising from Dehn surgery on a hyperbolic knot.

DEFINITION. If K is a hyperbolic knot and $K(p/q)$ is not a hyperbolic 3-manifold then p/q will be called an *exceptional surgery* on K.

Are those hyperbolic knots that admit exceptional surgeries special? More specifically, if K is a hyperbolic knot that admits an exceptional surgery, p/q, can one say something about p/q and about the topology of K (the topology of X_K)?

The following guiding light of 3-manifold topology (due to Thurston) describes the conjectured topological obstructions to a closed 3-manifold being hyperbolic.

GEOMETRIZATION CONJECTURE. *A closed 3-manifold M is hyperbolic unless one of the following holds:*

(1) *M contains an essential 2-sphere. An embedded 2-sphere in M is* essential *if it does not bound a 3-ball.*
(2) *M contains an essential 2-torus. An embedded 2-torus in M is* essential *if its fundamental group injects under inclusion into the fundamental group of M.*
(3) *M is a Seifert fibered space. A Seifert fibered space is an S^1-bundle over a surface where one allows a finite number of exceptional circle fibers around each of which the fibration is not locally trivial (but a (p,q)-fibration of the solid torus neighborhood off its core).*

Conditions 1–3 are obstructions to M being a hyperbolic 3-manifold. The conjecture part of the above is that these are the only obstructions.

Conditions 2 and 3 overlap. In fact a (closed) Seifert fibered space contains an essential 2-torus unless its orbit surface is a 2-sphere and has at most three exceptional fibers. Furthermore, a Seifert fibered space with orbit surface a 2-sphere has a finite fundamental group if and only if it has at most two exceptional fibers (and is not $S^2 \times S^1$) or has three exceptional fibers whose orders form a Platonic triple.

To get a feel for our problem we describe some examples of knots admitting surgeries that are not hyperbolic manifolds.

Example A. The Trivial Surgery

For any knot K, $K(\text{meridian}) = K(1/0) =$ 3-sphere. One sees this by noting that in the 3-sphere a meridian of ∂X_K bounds a disk in the solid torus neighborhood of K. The 3-sphere is not a hyperbolic manifold (category 3 above). As this surgery does nothing at all to the ambient 3-sphere we call it the *trivial surgery* and will exclude it from consideration. In particular, *it will not be an exceptional surgery.*

Example B. Cable Knots

A cable knot admits a Dehn surgery containing an essential 2-sphere. Let L be any knot in the 3-sphere. Let N_L be a solid torus neighborhood of L and let K be an essential simple closed curve on ∂N_L that winds at least twice around N_L. K is then a knot in the 3-sphere called a *cable knot* or a *cable of* L. The complement of K in ∂N_L restricts to an annulus A properly embedded in X_K. See Figure 1. The boundary of A consists of two curves in ∂X_K.

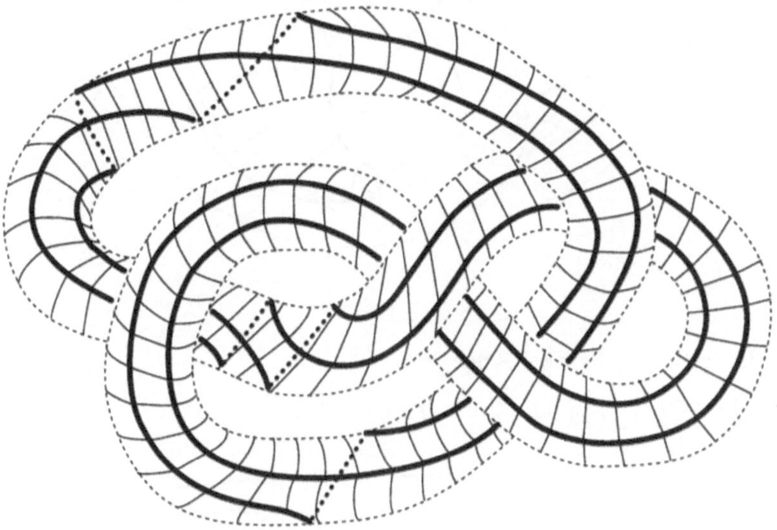

Figure 1

Let α be the slope corresponding to each of these two curves. In $K(\alpha)$ we obtain an embedded 2-sphere by capping off the boundary of A with disks lying in the attached solid torus. This 2-sphere turns out to be essential in $K(\alpha)$.

Example C. The 4/1 Surgery on the Figure-Eight Knot

Let K be the figure-eight knot. Then X_K contains a properly embedded once-punctured Klein bottle F. See Figure 2.
Let α be the slope of the boundary of F in ∂X_K ($\alpha = 4/1$). Then we get an embedded Klein bottle $\widehat{F}$ in $K(\alpha)$ by capping off the boundary of F with a disk lying in the attached solid torus N. The existence of $\widehat{F}$ means that $K(\alpha)$ could not be hyperbolic. For let $\widehat{S}$ be the 2-torus that is the boundary of a neighborhood of $\widehat{F}$ in $K(\alpha)$. One of the following must hold:

(1) $\widehat{S}$ is essential in $K(\alpha)$;
(2) $\widehat{S}$ bounds a solid torus in $K(\alpha)$, hence $K(\alpha)$ is Seifert fibered; or
(3) $K(\alpha)$ contains an essential 2-sphere.

Each of these is an obstruction to $K(\alpha)$ being hyperbolic as described in the Geometrization Conjecture above.

Figure 2

Example D. $(-2, 3, 7)$**-Pretzel Knot**

Let K be the $(-2, 3, 7)$-pretzel knot of Figure 3.

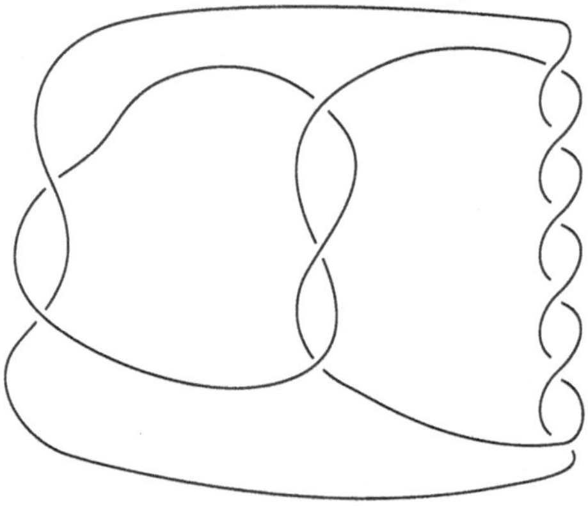

Figure 3

$K(p/q)$ is not hyperbolic when $\frac{p}{q} \in \{\frac{16}{1}, \frac{17}{1}, \frac{18}{1}, \frac{37}{2}, \frac{19}{1}, \frac{20}{1}\}$. In particular, when $\frac{p}{q} \in \{\frac{16}{1}, \frac{37}{2}, \frac{20}{1}\}$, $K(p/q)$ contains an essential 2-torus. When $\frac{p}{q} \in \{\frac{17}{1}, \frac{18}{1}, \frac{19}{1}\}$, $K(p/q)$ is a Seifert fibered space with finite fundamental group. Using Jeff Weeks' computer program Snappea one can see that these are the only exceptional surgeries on this knot [H]. To motivate the theorems of Section 1 we make the following observations about this list of exceptional surgeries. First, note that the denominator of these surgeries has absolute value at most 2. In fact, *usually* the denominator is 1 —

that is, usually the exceptional surgeries are integers. Second, note that there are *six* exceptional surgeries.

Let K be a hyperbolic knot and p/q an exceptional surgery for K. Sections 1 and 2 address restrictions that have been or might be placed on p/q and on the topology of X_K.

Section 1: Restrictions on p/q

Theorem 1 tells us that the number of exceptional surgeries on a hyperbolic knot is finite. But how many can there be and what do these slopes look like? An elementary argument by Thurston and Gromov already goes a long way in this direction. The following is an improved version from [BH].

THEOREM 2. *If K is hyperbolic then $K(p/q)$ has a Riemannian metric of negative sectional curvature for all but at most 24 values of p/q.*

In fact, when K has tunnel number larger than 1, Corollary 2.7 of [A2] shows that all but at most 16 surgeries on K admit metrics of negative sectional curvature.

It is conjectured that a 3-manifold that is negatively curved is hyperbolic. In particular, it is known that a negatively curved manifold cannot contain an essential 2-sphere or 2-torus, nor can it be a Seifert fibered space.

However, it seems as if more can be said. In particular, examples indicate the following conjectures to be good guides.

CONJECTURE 3. *If K is hyperbolic and K is not the figure-eight knot, then K admits at most six exceptional surgeries.*

In fact, the $(-2, 3, 7)$-pretzel knot of Example D seems to be the only knot known to have as many as six exceptional surgeries.

CONJECTURE 4. *If K is a hyperbolic knot and p/q an exceptional surgery on K, then $|q| \leq 2$.*

Over the past ten years, a lot of work has been done in determining accurate restrictions on the slopes of exceptional surgeries by treating separately the different types of obstructions to hyperbolicity (as in conditions 1–3 of the Geometrization Conjecture above).

THEOREM 5 [GLu1], [GLu5]. *If $K(p/q)$ contains an essential 2-sphere then $|q| = 1$. Furthermore, if $K(\alpha_1)$ and $K(\alpha_2)$ each contains an essential 2-sphere then $\Delta(\alpha_1, \alpha_2) \leq 1$. In particular, there are at most two such surgeries on K.*

In [BZ2] the authors give a new proof of this result. Furthermore, they demonstrate that if $K(\alpha_1)$ contains an essential 2-sphere and $K(\alpha_2)$ has a cyclic fundamental group then $\Delta(\alpha_1, \alpha_2) \leq 1$.

THEOREM 6 [GLu3]. *Let K be a hyperbolic knot. If $K(p/q)$ contains an essential 2-torus, then $|q| \leq 2$. Furthermore, [Go2], if $K(\alpha_1)$ and $K(\alpha_2)$ each contains an essential 2-torus, and K is not the figure-eight knot, then $\Delta(\alpha_1, \alpha_2) \leq 5$. In particular, there are at most seven such surgeries.*

The following improves Theorem 6 when $K(p/q)$ is a Seifert fibered space.

THEOREM 7 [BZ2]. *Let K be a hyperbolic knot. Assume $K(\alpha_1)$ contains an essential 2-torus and is also a Seifert fibered space. If $\pi_1(K(\alpha_2))$ is cyclic then $\Delta(\alpha_1, \alpha_2) = 1$. In particular, if $\alpha_1 = p/q$, then $|q| = 1$.*

An important class of Seifert fibered spaces that do not contain essential tori are those with a finite fundamental group.

THEOREM 8 [CGLS]. *Let K be a hyperbolic knot. If $\pi_1(K(\alpha_1))$ and $\pi_1(K(\alpha_2))$ are both cyclic then $\Delta(\alpha_1, \alpha_2) \leq 1$. In particular, if $\pi_1(K(p/q))$ is cyclic then $|q| \leq 1$, and there are at most two (non-trivial) cyclic surgeries on K. If there are two then the slopes are consecutive integers.*

THEOREM 9 [BZ1]. *Let K be a hyperbolic knot.*

(1) *If $\pi_1(K(\alpha_1))$ is cyclic and $\pi_1(K(\alpha_2))$ is finite then $\Delta(\alpha_1, \alpha_2) \leq 2$. In particular if $\pi_1(K(p/q))$ is finite then $|q| \leq 2$.*
(2) *If $\pi_1(K(\alpha_1))$ and $\pi_1(K(\alpha_2))$ are finite then $\Delta(\alpha_1, \alpha_2) \leq 5$. Furthermore, there are at most six finite surgeries on K.*

REMARKS. (1) Even though we are applying Theorems 7, 8, and 9 in the context of obstruction (3) of the Geometrization Conjecture, these theorems assume only information about $\pi_1(K(p/q))$. For example, these theorems would still apply if $K(p/q)$ were a counterexample to the Geometrization Conjecture with finite fundamental group.

(2) Theorems 7–9, and Theorems 11 and 12 below, build on the fundamental work of Culler and Shalen developed in [CS] and [CGLS], which analyzes an algebraic variety of representations of a knot group. A nice exposition of this is given in [S].

(3) [M] categorizes the finite groups that can be fundamental groups of 3-manifolds. When $\pi_1(K(p/q))$ is finite of one of certain special types, [BZ1] gives strong restrictions on p/q beyond those mentioned in Theorem 9.

Theorems 5–9 do not tackle all nonhyperbolic manifolds that could arise by Dehn surgery on a knot. We do not as yet know good restrictions (that is, say, supporting Conjectures 3 and 4) when $K(p/q)$ is a Seifert fibered space over the 2-sphere having exactly three exceptional fibers in which the orders of the exceptional fibers form a parabolic or hyperbolic triple. Theorems 5–9 would also not apply to a $K(p/q)$ that was a counterexample to the Geometrization Conjecture with infinite fundamental group (for example, if $K(p/q)$ were a negatively curved 3-manifold that was not hyperbolic).

More can be said regarding Conjecture 3. Theorems 5–9 show that there are at most seven nontrivial surgeries of a given non-hyperbolic type. These bounds, for the most part, come from restrictions on $\Delta(\alpha_1, \alpha_2)$ when $K(\alpha_1)$ and $K(\alpha_2)$ are nonhyperbolic of the given type. If one could show that $\Delta(\alpha_1, \alpha_2) \leq 5$ for any two nonhyperbolic surgeries α_1, α_2 on a knot K, then we would have that there are at most seven exceptional surgeries on K. In this direction we would like restrictions on $\Delta(\alpha_1, \alpha_2)$ when $K(\alpha_1)$, $K(\alpha_2)$ are nonhyperbolic of different types:

THEOREM 10 [GLi]. *Let K be a hyperbolic knot. If $K(\alpha_1)$ contains an essential 2-sphere and $K(\alpha_2)$ contains an essential 2-torus then $\Delta(\alpha_1, \alpha_2) \leq 5$.*

THEOREM 11 [BZ2]. *Let K be a hyperbolic knot. If $K(\alpha_1)$ contains an essential 2-sphere and $\pi_1(K(\alpha_2))$ is finite, then $\Delta(\alpha_1, \alpha_2) \leq 5$.*

THEOREM 12 [BZ2]. *Let K be a hyperbolic knot. Suppose that $K(\alpha_1)$ contains an essential 2-torus and is also a Seifert fibered space. If $K(\alpha_2)$ has finite fundamental group then $\Delta(\alpha_1, \alpha_2) \leq 5$.*

Theorems 5–9 give strong support to Conjecture 4. Indeed, very often they prove more — that is, that a certain type of exceptional surgery must be integral. The only known examples of exceptional surgeries of the form $p/2$ are where $K(p/2)$ contains an essential 2-torus. This situation is captured by Theorem 13 below. As background we note that the $(-2, 3, 7)$-pretzel knot of Example D is one knot in several infinite families of strongly invertible, hyperbolic knots K such that $K(p/2)$ contains an essential 2-torus [EM].

THEOREM 13 [GLu3], [GLu4]. *Let K be a hyperbolic knot. If $K(p/2)$ contains an essential 2-torus then there is an essential 2-torus that intersects the solid torus N in exactly two meridianal disks (where $K(p/2) = X_K \cup N$). This 2-torus then restricts to an essential, twice-punctured 2-torus in X_K that separates X_K into two genus-2 handlebodies. In particular, K is strongly invertible.*

It seems believable that the only exceptional surgeries on a hyperbolic knot that are nonintegral are those containing an essential 2-torus — hence subject to the description of Theorem 13. If there are other nonintegral, exceptional surgeries, one would expect them to be extremal and consequently subject to a description as in Theorem 13.

The arguments of Theorems 6 and 13 use the beautiful idea of the *thin position* of a knot in the 3-sphere developed by Gabai in his proof of the Property R Conjecture [Ga1]. For a nice exposition of this idea see [Ga2]. These theorems also use the combinatorial techniques developed in [CGLS] and [GLu2].

Section 2: Restrictions on K

If K admits an exceptional surgery, what can you say about K — that is, about the topology of X_K? This is a challenging question, as evidenced by the fact that the answers are conjectural. Theorem 13 is the kind of theorem we are looking for, except that it applies only when the surgery slope is of the form $p/2$. We give a collection of intriguing conjectures that describe potential answers to this question and that suggest directions in which to look for answers.

In Example B we showed how a cable knot admits a Dehn surgery containing an essential 2-sphere. The following Cabling Conjecture says that this is the only way a 2-sphere is created under a Dehn surgery on a knot in the 3-sphere.

CONJECTURE 14 [GS]. *If a Dehn surgery on K contains an essential 2-sphere, then K is a cable knot. In particular, K is not hyperbolic.*

A curve in the boundary of a handlebody is said to be *primitive* if there is a properly embedded disk in the handlebody that intersects it exactly once.

Conjecture 15 [B]. *If a Dehn surgery on K is a lens space then there is a standardly embedded genus-2 handlebody H in the 3-sphere such that K can be isotoped onto ∂H where it is primitive in both H and in the complement of H.*

Let K be as in the conclusion to Conjecture 15, and isotope K onto ∂H. Then ∂H restricts to a twice-punctured 2-torus T in X_K. If α is the slope corresponding to a component of the boundary of T, we can cap off the boundary components of T with disks in $K(\alpha)$ giving a 2-torus $\widehat{T}$. The primitivity condition on K guarantees that each side of $\widehat{T}$ in $K(\alpha)$ is a solid torus. Thus, $K(\alpha)$ is a lens space and $\widehat{T}$ is a Heegaard torus for $K(\alpha)$. We refer to a surface in a closed 3-manifold as a *Heegaard surface* if it divides the manifold into two handlebodies. If K_α denotes the core of N in $K(\alpha) = X_K \cup N$, then K_α intersects $\widehat{T}$ twice. Thus, $\widehat{T}$ separates K_α into two arcs, and the primitivity condition guarantees that each of these arcs is isotopic (rel endpoints) into $\widehat{T}$. An equivalent form of Conjecture 15 is: if $K(\alpha)$ is a lens space then there is a Heegaard torus $\widehat{T}$ of $K(\alpha)$ intersecting the core of the attached solid torus K_α twice; furthermore, each of the two arcs into which $\widehat{T}$ divides K_α is isotopic (rel endpoints) into $\widehat{T}$.

From this point of view, Conjecture 14 may be rephrased as: If $K(\alpha)$ contains an essential 2-sphere then it contains an essential 2-sphere that intersects the core of N, the attached solid torus, exactly two times.

Looking at Conjectures 14 and 15 in this way and recalling Theorem 13 raises:

Question. If K is hyperbolic and $K(\alpha) = X_K \cup N$ contains an essential 2-torus, is there an essential 2-torus in $K(\alpha)$ that intersects the core of N exactly two times?[1]

If the answer is yes then the exterior of any hyperbolic knot K admitting such a surgery would either contain a closed, essential surface of genus 2 (in this case, K sits on a knotted genus-2 handlebody in the 3-sphere) or could be written as the union of two genus-2 handlebodies along a twice-punctured torus (in this case, K is strongly invertible).

The Geometrization Conjecture says that if a closed 3-manifold is not hyperbolic then it contains an embedded surface of small genus (at most two) that is either essential or a Heegaard surface. It seems reasonable to hope that this property in a Dehn surgery on K implies it in X_K.

Conjecture 16. *There is an integer n (hopefully small) such that if K is a hyperbolic knot admitting an exceptional surgery then X_K contains a closed surface of genus at most n that is either essential in X_K or a Heegaard surface in X_K.*

A Heegaard surface for X_K is one that divides X_K into a handlebody and a "compression body" [CG]. This is related to the tunnel number of K.

One way to approach these conjectures is through Theorem 2. The mechanism there shows that if the cusp volume of a hyperbolic knot K is large enough then all nontrivial surgeries on K admit metrics of negative curvature. Thus:

[1] **Added in proof**. Eudare-Munoz has recently shown that the answer to this question is no, by providing a knot K where $K(\alpha)$ contains an essential 2-torus but no such 2-torus may intersect N twice. In this example there is an essential torus intersecting the core of N four times.

QUESTION. How is the cusp volume of a hyperbolic knot expressed in terms of other topological aspects of K or X_K? In particular, is there a good measure of the complexity of a knot so that the more complex the knot the larger the cusp volume?

For work in this direction see [A1],[A2],[ALR],[BH].

Finally, Conjecture 16 is related to the following outstanding conjecture, which describes the topology of $K(p/q)$ in terms of the topology of K.

CONJECTURE 17. *There is a function* $\omega : \mathbb{N} \to \mathbb{N}$ *such that if* $K(p/q)$ *has a Heegaard surface of genus* g *then* X_K *has a Heegaard surface of genus* $\omega(g)$.

In this direction we have the following theorem:

THEOREM 18 [MR]. *Let* K *be a hyperbolic knot and* g *be an integer such that every Dehn surgery on* K *has an irreducible Heegaard splitting of genus at most* g. *Then*

(1) *There is a finite collection of surfaces* $\Sigma_1, \dots, \Sigma_r$ *embedded in* X_K *so that, for all but finitely many slopes* p/q, *every irreducible Heegaard splitting surface of* $K(p/q)$ *of genus at most* g *is isotopic to one of the* Σ_i.

(2) *If* Σ_i *is not a Heegaard surface for* X_K *then there is a unique simple closed curve on* ∂X_K *that is isotopic to a simple closed curve on* Σ_i.

References

[A1] C. Adams, *Dehn filling hyperbolic 3-manifolds*, Pacific J. Math., to appear.

[A2] C. Adams, *Unknotting tunnels in hyperbolic 3-manifolds*, Math. Ann., to appear.

[ALR] I. R. Aitchison, E. Lumsden, and J. H. Rubinstein, *Cusp structure of alternating links*, Invent. Math. **109** (1992), 473–494.

[B] J. Berge, *Some knots with surgeries yielding lens spaces*, preprint.

[BH] S. A. Bleiler and C. G. Hodgson, *Spherical space forms and Dehn surgery*, Topology, to appear.

[BZ1] S. Boyer and X. Zhang, *Finite Dehn surgery on knots*, preprint.

[BZ2] S. Boyer and X. Zhang, *The semi-norm and Dehn filling*, preprint.

[CG] A. Casson and C. McA. Gordon, *Reducing Heegaard splittings*, Topology Appl. **27** (1987), 275–283.

[CGLS] M. Culler, C. McA. Gordon, J. Luecke, and P. B. Shalen, *Dehn surgery on knots*, Ann. of Math. (2) **125** (1987), no. 2, 237–300.

[CS] M. Culler and P. B. Shalen, *Varieties of group representations and splittings of 3-manifolds*, Ann. of Math. (2) **117** (1983), 109–146.

[EM] M. Eudave-Muñoz, *Non-hyperbolic manifolds obtained by Dehn surgery on hyperbolic knots*, Proc. Georgia Int. Top. Conf. (1993), to appear.

[Ga1] D. Gabai, *Foliations and the topology of 3-manifolds III*, J. Differential Geom. **26** (1987), 479–536.

[Ga2] D. Gabai, *Foliations and 3-manifolds*, Proc. ICM Kyoto, 1990, pp. 609–619.

[GS] F. González-Acuña and H. Short, *Knot surgery and primeness*, Math. Proc. Cambridge Philos. Soc. **99** (1986), 89–102.

[Go1] C. McA. Gordon, *Dehn surgery on knots*, Proceedings ICM Kyoto, 1990, pp. 631–642.

[Go2] C. McA. Gordon, *Boundary slopes of punctured tori in 3-manifolds*, preprint.

[GLi] C. McA. Gordon and R. Litherland, *Incompressible planar surfaces in 3-manifolds*, Topology Appl. **18** (1984), 121–144.

[GLu1] C. McA. Gordon and J. Luecke, *Only integral Dehn surgeries can yield reducible manifolds*, Math. Proc. Cambridge Philos. Soc. **102** (1987), 94–101.

[GLu2] C. McA. Gordon and J. Luecke, *Knots are determined by their complements*, J. Amer. Math. Soc. **2** (1989), 371–415.

[GLu3] C. McA. Gordon and J. Luecke, *Dehn surgery on knots creating essential tori, I*, Communications in Analysis and Geometry, to appear.

[GLu4] C. McA. Gordon and J. Luecke, *Dehn surgery on knots creating essential tori, II*, preprint.

[GLu5] C. McA. Gordon and J. Luecke, *Reducible manifolds and Dehn surgery*, Topology, to appear.

[H] Craig Hodgson, private communication.

[M] J. Milnor, *Groups which act on S^n without fixed points*, Amer. J. Math **79** (1957), 623–631.

[MR] Y. Moriah and H. Rubinstein, *Heegaard structures of negatively curved 3-manifolds*, preprint.

[S] P. B. Shalen, *Representations of 3-manifold groups and applications in topology*, Proc. ICM Berkeley, 1986, pp. 607–614.

[T] W. Thurston, *Three dimensional manifolds, Kleinian groups and hyperbolic geometry*, Bull. Amer. Math. Soc. **6** (1982), 357–381.

Cyclic Splittings of Finitely Presented Groups and the Canonical JSJ-Decomposition

E. RIPS

Hebrew University of Jerusalem
Givat Ram
Jerusalem 91904, Israel

The classification of stable actions of finitely presented groups on $\mathbb{R}$-trees has found a number of applications. Perhaps one of the most striking of these applications is the theory of canonical JSJ-decompositions of hyperbolic groups developed by Sela in his seminal paper [Se].

An analysis of the argument of Sela reveals that the assumption of hyperbolicity is not essential for this theory, and, in fact, it can be developed in much greater generality; namely for finitely presented groups, and, possibly, even in a much more general setting.

To simplify the statements, we shall restrict to the case of finitely presented, freely indecomposable, torsion free groups.

The motto of this research is to try to understand groups from the point of view of low dimensional topology. In fact, we develop a little vocabulary that translates topological notions into algebraic ones. This leads to a number of results that are analogs and/or generalizations of the corresponding topological results.

The first item of our vocabulary is:

VOC 1. *A Simple Closed Curve*

(a) For a separating s.c.c., the corresponding algebraic notion is a free product with infinite cyclic amalgamation, $G = A *_C B$, $C \cong \mathbb{Z}$.
(b) For a nonseparating s.c.c., the corresponding notion is an HNN-group over $\mathbb{Z}$; $G = A*_C$, $C \cong \mathbb{Z}$. We call these *elementary $\mathbb{Z}$-splittings.*

To fix notation, we introduce here some general definitions:

DEFINITION. *Graph of Groups*

A graph of groups consists of a graph (V, E) and an assignment of groups: $\mathcal{G}(v)$ for each $v \in V$, $\mathcal{G}(e)$ for each $e \in E$, together with boundary monomorphisms

$$\alpha : \mathcal{G}(e) \longrightarrow \mathcal{G}(\partial_0(e)) \text{ and } \omega : \mathcal{G}(e) \longrightarrow \mathcal{G}(\partial_1(e))$$

where $\partial_0(e)$ and $\partial_1(e)$ denote the initial and terminal vertices of an edge e. We require that $\mathcal{G}(e) = \mathcal{G}(\overline{e})$ so that passing from e to $\overline{e}$ switches α and ω.

Proceedings of the International Congress
of Mathematicians, Zürich, Switzerland 1994

We recall that the fundamental group of a graph of groups $\pi(\mathcal{G}(V,E);T)$ with respect to a maximal subtree T is given by

$$< \mathcal{G}(v) \text{ for all } v \in V, t_e \text{ for all } e \in E \mid t_e t_{\overline{e}} = 1 \text{ for all } e \in E, \\ t_e = 1 \text{ for all } e \in T \text{ and } \alpha(g)t_e = t_e\omega(g) \text{ for all } e \in E, g \in \mathcal{G}(e) > .$$

DEFINITION. *Splitting*

By a *splitting* of a group G, we understand a triple $\mathcal{S} = (\mathcal{G}(V,E),T,\phi)$ where $\mathcal{G}(V,E)$ is a graph of groups, T is a maximal subtree of (V,E), and $\phi : \pi(\mathcal{G}(V,E);T) \longrightarrow G$ is an isomorphism.

With an abuse of language we will omit mention of T and ϕ unless we need to specify them.

DEFINITION. *$\mathbb{Z}$-Splitting*

A splitting is called a *$\mathbb{Z}$-splitting* if $\mathcal{G}(e) \cong \mathbb{Z}$ for all $e \in E$. *Elementary $\mathbb{Z}$-splittings* are $\mathbb{Z}$-splittings for which the graph of groups contains one edge. We shall only consider splittings with V and E finite.

DEFINITION. *Refinements and Collapses of Splittings*

An *elementary refinement* of a graph of groups $\mathcal{G}(V,E)$ consists of an elementary splitting of one of its vertex groups $\mathcal{G}(v) = A *_C B$ or $\mathcal{G}(v) = A*_C$ *compatible* with boundary monomorphisms; that is, for any $e \in E$ with $\partial_0(e) = v$, $\alpha(\mathcal{G}(e)) \subseteq A$ or $\alpha(\mathcal{G}(e)) \subseteq B$. The new graph of groups is obtained from $\mathcal{G}(V,E)$ by replacing the vertex v by a one edge graph; a segment of groups or a loop of groups.

A *refinement* (=blow-up [Ji]) of a graph of groups is the result of a sequence of elementary refinements. A *collapse* is an inverse operation to a refinement.

An *elementary splitting corresponding to an edge* e of a graph of groups is the result of the collapse of all edges other than e. Thus, any elementary splitting of a vertex group compatible with boundary monomorphisms can be extended to an elementary splitting of the whole group: perform the corresponding elementary refinement that introduces a new edge, and collapse all other edges.

This construction can be used for a basic example that will

(a) provide an ample family of elementary $\mathbb{Z}$-splittings of a group, and

(b) demonstrate that $\mathbb{Z}$-splittings need not have common refinements.

EXAMPLE.

Let $F =< x_1,\dots,x_m >$ be a free group on $x_1,\dots,x_m$, and let $\Omega = \{w_1,\dots,w_k\}$ be a *quadratic set of words*; that is, the total number of occurrences of each letter x_i in the words $w_1,\dots,w_k$ is 2 or 0.

Let $\mathcal{G}(V,E)$ be a graph of groups. Let $v \in V$ and let $e_1, e_2,\dots,e_k$ be all edges with $\partial_0(e_i) = v$. We suppose that $\mathcal{G}(v) = F$, that no w_i is conjugate in F to w_j or its inverse. Suppose also that

$$\alpha(\mathcal{G}(e_1)) =< w_1 >, \alpha(\mathcal{G}(e_2)) =< w_2 >, \dots, \alpha(\mathcal{G}(e_k)) =< w_k > .$$

Following [RS1], we call such a vertex a *QH-vertex*.

In the geometric picture, F is the fundamental group of a punctured surface S and $w_1, \dots, w_k$ correspond to cycles around the punctures. So every s.c.c. on S gives an elementary $\mathbb{Z}$-splitting of F compatible with boundary monomorphisms, and hence an elementary $\mathbb{Z}$-splitting of the whole group.

Thus, we obtain a family of elementary splittings in which any two have common refinements if and only if the corresponding s.c.c. are disjoint (up to isotopy).

Let us now consider more closely two distinct elementary $\mathbb{Z}$-splittings of a group: $G = A_1 *_{C_1} B_1$ (or $A_1 *_{C_1}$), and $G = A_2 *_{C_2} B_2$ (or $A_2 *_{C_2}$) where $C_1 =< c_1 >$, $C_2 =< c_2 >$.

The element c_2 is called *elliptic* with respect to the first splitting if it is contained in a conjugate of A_1 or B_1, and *hyperbolic* otherwise, and similarly for c_1 with respect to the second splitting.

LEMMA. *If G is freely indecomposable, then either c_1 and c_2 are simultaneously elliptic (in which case there is a common refinement), or simultaneously hyperbolic.*

Now we are led to the next item of our vocabulary.

VOC 2. *Intersecting Simple Closed Curves*

The algebraic analog is a pair of elementary $\mathbb{Z}$-splittings that are hyperbolic-hyperbolic. We call them *intersecting* elementary $\mathbb{Z}$-splittings.

We now state:

MAIN LEMMA. *Let G be a finitely presented, freely indecomposable, torsion free group, not isomorphic to a surface group. Suppose that G admits two $\mathbb{Z}$-splittings $G = \pi(\mathcal{G}_1(V_1, E_1); T_1)$, $G = \pi(\mathcal{G}_2(V_2, E_2); T_2)$, where $v_1 \in V_1$ and $v_2 \in V_2$ are QH-vertices corresponding to punctured surfaces S_1 and S_2. Assume that for some s.c.c. γ_1 on S_1 and γ_2 on S_2, the corresponding elementary $\mathbb{Z}$-splittings of G are intersecting. Then there exists a $\mathbb{Z}$-splitting $G = \pi(\mathcal{G}(V, E); T)$ and a QH-vertex $v \in V$ such that $\mathcal{G}_1(v_1)$ and $\mathcal{G}_2(v_2)$ are conjugate to subgroups of $\mathcal{G}(v)$, and the elementary $\mathbb{Z}$-splittings corresponding to γ_1 and γ_2 are conjugate to elementary $\mathbb{Z}$-splittings originating from intersecting s.c.c's γ_1' and γ_2' on S.*

To give some indication of the method of proof, consider again two elementary $\mathbb{Z}$-splittings; $G = A_1 *_{C_1} B_1$ (or $A_1 *_{C_1}$), and $G = A_2 *_{C_2} B_2$ (or $A_2 *_{C_2}$) where $C_1 =< c_1 >$, $C_2 =< c_2 >$. The next idea is to introduce:

VOC 3. *Dehn Twists*

If $G = A_1 *_{C_1} B_1$, define $D_1 : G \longrightarrow G$ by

$$D_1(a_1) = a_1 \text{ for all } a_1 \in A_1$$
$$D_1(b_1) = c_1 b_1 c_1^{-1} \text{ for all } b_1 \in B_1.$$

If $G = A_1 *_{C_1} =< A_1, t_1 \mid t_1 c_1 t_1^{-1} = c_1' >$, define $D_1 : G \longrightarrow G$ by

$$D_1(a_1) = a_1 \text{ for all } a_1 \in A_1$$
$$D_1(t) = t c_1.$$

Similarly for $D_2 : G \longrightarrow G$.

Consider now

$$\phi_k = D_2^{n_k} D_1^{m_k} \dots D_2^{n_2} D_1^{m_2} D_2^{n_1} D_1^{m_1}$$

where $1 \ll m_1 \ll n_1 \ll m_2 \ll n_2 \ll \dots \ll m_k \ll n_k$.

Let $\mu : G \times \mathcal{T} \longrightarrow \mathcal{T}$ be the action on a tree $\mathcal{T}$ induced by (say) the first splitting.

Using ϕ_k we define new actions

$$\mu_k : G \times \mathcal{T} \longrightarrow \mathcal{T} \text{ by } (\mu_k(g))(x) = \mu(\phi_k(g))(x).$$

We now apply a method of Paulin [P], [BS] and Bestvina [B] to produce from these actions an action of G on an $\mathbb{R}$-tree using the Gromov-Hausdorff convergence. One shows that this action is nontrivial and small, and has trivial tripod stabilizers.

The next step is to use the classification of stable actions of a finitely presented group on an $\mathbb{R}$-tree [BF2] and [Ri]. As G is freely indecomposable, we are left with one minimal component. For the same reason, the axial, simplicial and Levitt (=thin [BF2]) cases are excluded, and we are left with one minimal surface component that corresponds to a QH-vertex.

With a short additional argument, one shows that both elementary $\mathbb{Z}$-splittings are conjugate to elementary $\mathbb{Z}$-splittings originating from intersecting s.c.c's on this surface.

In order to deal with the general case, one needs more complicated expressions for ϕ_k involving Dehn twists corresponding to various s.c.c's on S_1 and S_2, whereas the rest of the argument is similar.

Definition. *QH-subgroup*

Let H be a subgroup of G. We call H a *QH-subgroup* if it is free, and there exists a $\mathbb{Z}$-splitting $G = \pi(\mathcal{G}(V,E);T)$ and a QH-vertex $v \in V$ with $\mathcal{G}(v) = H$.

Proposition. *Let G be as in the Main Lemma.*

- *(a) Every QH-subgroup is contained in a maximal QH-subgroup.*
- *(b) There are finitely many conjugacy classes of maximal QH-subgroups.*
- *(c) There is a $\mathbb{Z}$-splitting $G = \pi(\mathcal{G}(V,E);T)$ with QH-vertices $v_1, \dots, v_h$ such that $\mathcal{G}(v_1)$, $\mathcal{G}(v_2)$, $\dots$, $\mathcal{G}(v_h)$ are representatives of all conjugacy classes of maximal QH-subgroups.*

This proposition is a consequence of the Main Lemma and a theorem of Bestvina and Feighn [BF1] that bounds the cardinality of vertices and edges for small splittings of G without redundant edges. (Here *redundant* means that $\alpha(\mathcal{G}(e)) = \mathcal{G}(\partial_0(e))$ and $\partial_0(e) \neq \partial_1(e)$.)

Definition. *QH-saturated $\mathbb{Z}$-splittings*

$\mathbb{Z}$-splittings satisfying part (c) of the proposition are called *QH-saturated $\mathbb{Z}$-splittings.*

We introduce a quasi-order on the set of all QH-saturated $\mathbb{Z}$-splittings of G requiring that $\mathcal{S}_1 \leq \mathcal{S}_2$ when $\mathcal{S}_2$ is obtained from $\mathcal{S}_1$ by a refinement of non-QH-vertices, or by a collapse of redundant edges.

DEFINITION AND VOC 4. *JSJ-Splitting*

If G is a closed surface group, we define the trivial (=one vertex, no edges) splitting to be the *JSJ-splitting* of G. If G is *not* a closed surface group, then maximal reduced elements with respect to this quasi-order are called *JSJ* (for Jaco-Shalen-Johannsen)-*splittings* of G.

The existence of JSJ-splittings is a consequence of the above-mentioned theorem of Bestvina and Feighn, and a theorem of Sela stating that there can be no infinite sequence of proper unfoldings of elementary $\mathbb{Z}$-splittings (see below for the definition of folding; unfolding is an inverse operation to folding).

The uniqueness of JSJ-splittings is the subject of the following theorem:

THEOREM. *Any two JSJ-splittings, $\mathcal{S}_1$ and $\mathcal{S}_2$, of G are equivalent in the following sense: $\mathcal{S}_2$ can be obtained from $\mathcal{S}_1$ by a finite sequence of operations from the following list:*
(1) conjugation;
(2) sliding;
(3) modifying boundary monomorphisms by conjugation.

DEFINITION. *Conjugation, Sliding, Modifying Boundary Monomorphisms by Conjugation*

(1) is usual conjugation;
(2) *sliding* is a modification of a graph of groups corresponding to the relation
$$(A *_{C_1} B) *_{C_2} D \cong (A *_{C_1} D) *_{C_2} B$$
in the case when $C_1 \subseteq C_2$;
(3) let $H = A*_C =< A, t \mid t\alpha(c)t^{-1} = \omega(c)$ for all $c \in C >$. For an arbitrary element $a \in A$ we replace α by $ad(a)\cdot\alpha$, (where $ad(a)(g) = aga^{-1}$) and t by ta^{-1}. The corresponding modification of the graph of groups is called *modifying boundary monomorphisms by conjugation* (for edges outside the maximal subtree).

COROLLARY 1. *Let $G = \pi(\mathcal{G}(V,E);T)$ be a JSJ-splitting and let $v_1, \dots, v_h$ be its QH-vertices. All elementary $\mathbb{Z}$-splittings of G are conjugate to ones from the following list:*
(1) elementary $\mathbb{Z}$-splittings corresponding to s.c.c.'s on punctured surfaces corresponding to QH-vertices $v_1, \dots, v_h$;
(2) elementary $\mathbb{Z}$-splittings corresponding to edges of $\mathcal{G}(V,E)$;
(3) foldings of elementary $\mathbb{Z}$-splittings from (2).

DEFINITION. *Folding of a $\mathbb{Z}$-splitting*

If $H = A*_C B$ and $C \subseteq C_1 \subseteq A$, where $C =< c = d^n >$ and $C_1 =< d >$, then the splitting $H = A*_{C_1} B_1$ with $B_1 = C_1 *_C B$ is called a *folding* of $H = A*_C B$ along C_1. The tree associated to the new splitting is obtained from the tree associated to the old splitting by folding.

If $H = A*_C =< A, t \mid tct^{-1} = \gamma(c)$ for all $c \in C >$ and $C \subseteq C_1 \subseteq A$ then $H = A_1 *_{C_1}$ is called a *folding* of the $\mathbb{Z}$-splitting $H = A*_C$ along C_1, where $A_1 = A *_{\gamma(C)} (tC_1t^{-1})$, $\gamma_1 : C_1 \longrightarrow A_1$, and $\gamma_1(g) = tgt^{-1}$ for all $g \in C_1$.

In the following, we assume that G is not a surface group:

COROLLARY 2. *Dehn twists of elementary splittings described in part (1) of Corollary 1 generate in $Out(G)$ a subgroup isomorphic to a direct product of mapping class groups of punctured surfaces.*

REMARKS.

(1) One can remove the restriction that G is torsion free by modifying the definition of a QH-vertex so as to include the case of an orbifold instead of a surface [RS2].
(2) The following are possible directions for further generalization: to replace "finitely presented" by "finitely generated" and "$\mathbb{Z}$-splittings" by "small splittings" (see [BF1]).

ACKNOWLEDGMENTS.

This talk is based on joint work with Z. Sela. While completing this work, the author was on a sabbatical leave from Hebrew University visiting Columbia University. He wishes to express his thanks to the faculty at Columbia for an extremely stimulating atmosphere, especially to Professor Hyman Bass, Professor Mike Shapiro, Lisa Carbone, and Ilya Kapovich.

Bibliography

[B] Bestvina, M., *Degenerations of hyperbolic space*, Duke Math. J. **56** (1988), 143–161.

[BF1] Bestvina, M., and Feighn, M., *Bounding the complexity of simplicial group actions*, Invent. Math **103** (1991), 449–469.

[BF2] Bestvina, M., and Feighn, M., *Stable actions of groups on real trees*, preprint (1992).

[BS] Bridson, M., and Swarup, G., *On Hausdorff-Gromov convergence and a theorem of Paulin*, L'Enseignement Mathématique, 2e Série, t. **40** (1994), Fasc. 3–4, 267–289.

[Ji] Jiang, R., *Collapsing of graphs of groups and length functions*, Michigan Technological University preprint (1994).

[P] Paulin, F., *Outer automorphisms of hyperbolic groups and small actions of groups on real trees*, Arboreal Group Theory, ed. R.C. Alperin, MSRI Series.

[Ri] Rips, E., *Lectures on hyperbolic groups and group actions on $\mathbb{R}$-trees*, Columbia University (1994).

[RS1] Rips, E., and Sela Z., *Structure and rigidity in hyperbolic groups I*, Geom. Functional Anal. **4** (3) (1994), 337–371.

[RS2] Rips, E., and Sela Z., *Cyclic splittings of finitely presented groups and the canonical JSJ-decomposition*, in preparation.

[Se] Sela, Z., *Structure and rigidity in (Gromov) hyperbolic groups and discrete groups in rank 1 Lie groups II*, Princeton University preprint, Princeton, NJ (1992).

[S] Serre, J. P., Trees, Springer-Verlag, Berlin and New York (1980).

An Algorithm to Recognize the 3-Sphere

JOACHIM H. RUBINSTEIN

University of Melbourne
Parkville, Victoria
Australia 3052

1. Introduction

Max Dehn introduced a number of important decision problems in algebra and topology. For groups, the isomorphism problem is to decide if two finite presentations represent isomorphic groups or not. The analogous question for manifolds is to decide if two compact n-dimensional manifolds are homeomorphic or not. For manifolds, the method of presentation could be via triangulations or handle descriptions.

Markov [Ma] (see also [BHP]) showed that both questions are undecidable in the class of finitely presented groups and manifolds of dimension at least 4. On the positive side, recently the isomorphism problem for word hyperbolic groups has been announced to have a positive solution by Sela [Se]. Haken gave a solution to the homeomorphism problem for an important class of 3-manifolds, now called Haken manifolds. This used a solution to the conjugacy problem in the mapping class group of a surface that was found by Hemion [He] and Hatcher and Thurston [HT]. Haken had to assume that both of the manifolds under consideration were irreducible, i.e any embedded 2-sphere bounded a 3-ball. Our solution to the recognition problem for the 3-sphere of course gives an algorithm to recognise the 3-ball. Hence in Haken's work, irreducibility can now be replaced by an assumption of asphericity, i.e the second homotopy group vanishing.

Our method is to use the minimax procedure of Poincaré and Birkhoff. This was employed by Pitts [Pi] to show that any closed Riemannian manifold has an embedded minimal hypersurface that is regular in dimensions up to six. Further refinements were given in the unpublished Ph.D. thesis of Smith [Sm] with Simon, where it is proved that for an arbitrary Riemannian metric on the 3-sphere, there is an embedded minimal 2-sphere. In works of Pitts and the author [PR1]–[PR3], the case of a general class of 3-manifolds that are decomposed into a union of two handlebodies by a strongly irreducible Heegaard surface S is considered. Note that strong irreducibility means that any pair of essential disks for the two handlebodies must have intersecting boundaries. In [PR3] it is established that for a strongly irreducible splitting of a closed orientable irreducible 3-manifold with a bumpy Riemannian metric, the Heegaard surface S can be isotoped to be minimal with index of instability one, or else there is a closed nonorientable embedded minimal surface K and S is isotopic to the boundary of a regular neighborhood of K with an unknotted tube attached running across the regular neighborhood.

Proceedings of the International Congress
of Mathematicians, Zürich, Switzerland 1994

Here a polyhedral version of [PR3] is discussed. In particular, a new theory of unstable polyhedral minimal surfaces is sketched. In [JR1], [JR2], two different approaches to a piecewise linear concept of minimal surface were given. Both these theories were very suitable for discussing *stable* minimal surfaces, which are very useful in many aspects of the topology of 3-manifolds. Heegaard splittings give rise naturally to unstable minimal surfaces. Moreover, deeper applications to the structure of 3-manifolds may be available from a general theory of minimal surfaces with arbitrary index of instability.

In Section 2 of the paper, a discussion is given of polyhedral minimal surfaces of index one. The approach here is to augment Haken's normal surface theory [Ha1]–[Ha3] with the PL minimal surfaces of [JR2] and use minimax to give a constructive version of the main result of [PR3]. Note that such minimal surfaces are called almost normal surfaces as they only differ from normal surfaces in one tetrahedron. We will only treat the case of 2-spheres sweeping out a 3-sphere and refer the reader to [Ru] for more details of the general case. This is the key step in the recognition algorithm for the 3-sphere — the result that any triangulation on the 3-sphere admits a polyhedral minimal 2-sphere of index one, i.e an almost normal 2-sphere.

In the Section 3 various results using almost normal surfaces are given. The two main theorems are that there is an algorithm to find the Heegaard genus of any closed orientable triangulated 3-manifold and a solution of a conjecture of Waldhausen [Wa], namely there is a finite number of inequivalent Heegaard splittings of a fixed genus of any such 3-manifold up to ambient homeomorphisms. A corollary is a solution of the homeomorphism problem for 3-manifolds with Heegaard genus at most 2.

In Section 4 a general theory of higher index polyhedral minimal surfaces is outlined. The index is defined to correspond with critical points for surface complexity under multi-parameter sweepouts. The measure of surface complexity we use is essentially the same as in [JR2], namely the lexicographically ordered pair (w, l), where w is the normal surface weight, i.e the number of intersections of the surface with the edges of the triangulation, and l is the length of the graph where the surface meets the 2-skeleton. When dealing with sweepouts, the surface may compress into several components and the number of such connected components also has to be built into the complexity. For simplicity we assume the surface remains connected here. The 2-skeleton of the triangulation is given a metric by gluing together hyperbolic metrics on triangles for each 2-simplex, so that the angles are all some very small number. This number is chosen so that for a given sweepout, the penalty for a surface going through a vertex is large as compared with an estimate of the sum of the lengths of all the edges of the graph where the surface meets the 2-skeleton. Ideal triangle metrics used in [JR2] are not as convenient for sweepouts as then lengths become infinite.

A remarkable property of index in this context is that locally we see octagonal saddle shaped disks properly embedded in tetrahedra, each contributing index one, or normal disks connected by small unknotted tubes. *In particular, for fixed genus surfaces, there is a bound to their index of instability in this theory.* Note that we are excluding the case of arbitrarily large index arising from many parallel spheres being tubed together, as this case does not occur for critical surfaces in sweepouts. Parallel octahedra in a single tetrahedron also cannot arise for the same reason. Note also it has been conjectured that this index bound is true for analytic minimal surfaces in bumpy Riemannian metrics, but this has only been established for the special case of positive Ricci curvature metrics by the compactness theorem of Choi-Schoen [CS]. A bumpy metric as defined in [Wh] has the property that minimal surfaces have no Jacobi fields, i.e are isolated critical points for area.

In the final section we discuss applications of higher index polyhedral minimal surfaces in 3-dimensional knot and link theory and in the generalized spherical space form problem. For the former, we define higher index thin position pictures of arbitrary knots and links. Thin position was introduced by Gabai [Ga], and has proved an invaluable tool. It can be viewed as an index one minimal 2-sphere arising from a minimax sweepout of the 3-sphere, where the complexity is the number of intersections with the knot or link. By classical Lusternik-Schnirelmann theory, there are canonical n-parameter sweepouts of the 3-sphere by families of 2-spheres, where the number of parameters n varies from 1 to 4. We discuss the interpretation of the corresponding index n thin position pictures. The hope is that these additional thin position descriptions may assist in some of the remaining difficult problems about surgery on knots and links. Notice that the existence of four thin position descriptions of a knot or link would seem to encode in a strong way the fact that the knot or link is actually in the 3-sphere, rather than another 3-manifold.

The generalized 3-dimensional spherical space form problem asks to show that any finite group action on the 3-sphere is topologically conjugate to an orthogonal action. This has been proved for groups where some element has a one-dimensional fixed set, by the orbifold theorem of Thurston (see e.g [Ho]). The latter includes the Smith conjecture as a special case (see [Mo]). Fixed point free actions have only been classified in the case where the group has order $2^k \times 3^m$. (See the discussion in [Ts].) In Hamilton's work on Ricci flow [Hm] the generalized 3-dimensional space form problem is solved if there is an invariant metric of positive Ricci curvature. In particular, any 3-manifold with positive Ricci curvature admits a metric of constant positive curvature. In [PR2] a program was outlined to give an alternative proof of part of Hamilton's theorem using multi-parameter sweepouts, Hatcher's solution of the Smale conjecture [Ht], and the Choi-Schoen compactness theorem referred to above. We finish by outlining a proof of the following result.

THEOREM 1 *Any action of a cyclic group of prime order on the* 3*-sphere is equivalent to an orthogonal action. Any free action of a direct product of a binary polyhedral group, other than the binary dodecahedral group, by a cyclic group of relatively prime order on the* 3*-sphere, is equivalent to an orthogonal action. Any finite group that acts freely on the* 3*-sphere admits a free orthogonal action on the* 3*-sphere.*

Remarks. This theorem is established by combining the program of [PR2] with the theory of higher index polyhedral minimal surfaces.

To complete the classification of free actions of cyclic groups on the 3-sphere it remains to solve the Smale conjecture for all 3-dimensional lens spaces. This should follow from Hatcher's techniques and then only the case of the binary dodecahedral group would remain. This is a nice challenge!

2. Polyhedral minimal surfaces and the recognition algorithm

Recall that a normal surface in a triangulated 3-manifold intersects each tetrahedron in disks that are all either triangles or quadrilaterals. Note that every boundary arc, which is the intersection of such a disk with a 2-simplex, must cross from one edge to a different one. Such an arc is called normal. There are 4 distinct isotopy classes of triangular disks and 3 quadrilaterals. These are called normal disk types. The normal coordinates of such a surface is the vector of nonnegative integers given by the multiplicity or weight of each disk type. The Haken sum of two normal surfaces is defined by adding the corresponding vectors. A normal surface is called fundamental if it is not a nontrivial sum of two normal surfaces. The key result of Haken's theory is that there are finitely many such fundamental surfaces and they can be systematically constructed.

An almost normal surface or polyhedral minimal surface of index one consists of normal disk types in all tetrahedra except one where there is a single exceptional piece, which is either an octagonal disk or two normal disks joined by an unknotted tube. The boundary of the octagon consists of normal arcs, and it is easy to see there are 3 isotopy classes of such octagons in each tetrahedron. The tube being unknotted refers to the fact it is boundary compressible.

THEOREM 2 *Suppose M is a triangulated 3-manifold homeomorphic to the 3-sphere. Then M has an embedded almost normal 2-sphere. Moreover this can be chosen disjoint from a maximal collection of disjoint normal 2-spheres and with an octagon as an exceptional piece.*

Proof. Choose a component of the complement of a maximal collection of embedded normal 2-spheres that has one boundary surface. Moreover this component, which we denote by M^*, can be chosen not to be a cone on a link of a vertex. Let S be the boundary sphere of M^* and consider a sweepout of M^* by a family of embedded 2-spheres denoted by S_t, for t between 0 and 1. Therefore S_0 is a point and S_1 is S. We put a metric on the 2-skeleton of M as described in the introduction, i.e each 2-simplex is identified with an ideal hyperbolic triangle.

Now choose a sweepout of M^* that has the smallest maximum complexity of S_t for some value t_o of t. This is called the *minimax surface.* (In general there are several such surfaces but we assume only one.) The minimax surface has the crucial *2-site property*. This means there cannot be two disjoint pieces of the surface that can be isotoped keeping their boundaries fixed to decrease complexity. The reason is that if the 2-site property fails, then just before the maximum complexity is reached in the sweepout, we can apply one of the local isotopies to decrease complexity in a site disjoint from the last move across the triangulation. Then we do the last move, go on past the minimax, and finally restore the isotopy in the disjoint site. This creates a new sweepout with smaller maximum complexity, a contradiction.

To complete the proof of Theorem 2, the 2-site property can be used to show that the minimax surface has all normal arcs of intersection with the 2-skeleton, which are in fact geodesics in the hyperbolic triangle metrics. This uses the second factor of hyperbolic length in the measure of complexity. There are many possible properly embedded disks and planar surfaces in a tetrahedron with all boundary arcs of this type. However, using the first factor of complexity, i.e the weight or number of intersections with the 1-skeleton, only the octagon or annulus with a single unknotted tube connecting two normal disks can have the 2-site property and also behave like a local maximum for complexity under a sweepout. Finally, by our choice of M^*, the case of an annulus is ruled out because M^* has only its boundary as an embedded normal 2-sphere.[1] □

3. Heegaard splittings and decision problems

In [Ru] a number of results are sketched, using polyhedral minimal surfaces and Heegaard splittings to define sweepouts of general 3-manifolds. We summarize the key points here.

THEOREM 3 *Any strongly irreducible Heegaard splitting of a closed orientable irreducible* 3*-manifold can be isotoped to be an almost normal surface.*

THEOREM 4 *There are finitely many Heegaard splittings of bounded genus of any closed orientable irreducible* 3*-manifold up to ambient homeomorphism.*

THEOREM 5 *There is an algorithm to construct all smallest genus Heegaard splittings of any closed orientable irreducible* 3*-manifold. In particular, there is an algorithm to determine the Heegaard genus of any such* 3*-manifold.*

COROLLARY 1 *There is an algorithm to solve the homeomorphism problem for the class of closed orientable irreducible* 3*-manifolds of Heegaard genus at most two.*

Remarks.

(1) Theorem 3 is proved by a similar technique as that used for Theorem 2 above, with the addition of the iteration minimax technique contained in [PR3]. The problem is that the minimax process may give an almost normal surface of smaller genus than the strongly irreducible splitting if only done once. The trick is to use this almost normal surface as a "barrier" and run the minimax again.

[1]Note added in proof: The minimax surface obtained may not be transverse to the 1-skeleton and a small deformation may be needed to achieve an almost normal surface. I would like to thank W. Haken for painting this out.

(2) Theorem 4 is proved in two steps. First the result is established for strongly irreducible splittings by using Theorem 3 to realize such surfaces as almost normal. Then Haken's technique is adapted to show that such almost normal surfaces can be written as Haken sums of fundamental normal and almost normal surfaces with no summands of positive Euler characteristic and a bounded number with zero Euler characteristic, after possibly Dehn twisting about incompressible tori.

Finally, the general case of irreducible splittings is treated by the method of [PR3], where it is shown that an irreducible but not strongly irreducible Heegaard splitting contains a twice punctured incompressible surface. So such a Heegaard surface can be divided up into strongly irreducible compression body decompositions of the 3-manifold split along some orientable incompressible surfaces.

(3) Theorem 5 and Corollary 1 follow from Theorem 4. The proof of Corollary 1 uses Thurston's orbifold theorem (see e.g [Ho]), plus the well-known observation that 3-manifolds of Heegaard genus 2 are 2-fold branched covers of the 3-sphere over a knot or link.

(4) Hass has recently proved a result similar to Theorem 4 using different methods. Johannson also established Theorem 4 for the important case of Haken manifolds earlier (see [Jo]).

4. Higher index polyhedral minimal surfaces

As in the proof of Theorem 2, we define an n-parameter sweepout of a 3-manifold as a map F from I^n to the space of maps from a surface S to M with the following properties:

- $F(X)$ is an embedding for each point X in intI^n
- $F(X)$ is a map of S to a graph for each point in the boundary of I^n
- F represents a nontrivial cycle in the homology of the space E, which is the union of all embeddings of S and the maps of S to a graph, relative to the subspace G of maps of S to a graph.

A key example arising from Lusternik-Schnirelmann theory is n-parameter sweepouts of the 3-sphere by 2-spheres, for n between 1 and 4. Note that by the Smale conjecture, the space of embeddings of the 2-sphere into the 3-sphere is homotopy equivalent to a real projective 3-space. So the space E above is homotopy equivalent to a real projective 4-space with an open 4-ball removed. There are then obvious nontrivial cycles in $H_q(E, G; Z_2)$ in dimensions $q = 1$ to 4 as asserted, giving the required sweepouts. Jost [Jt] has used these sweepouts to show that for any Riemannian metric on the 3-sphere there are at least four embedded minimal surfaces. Our observation here is that as in Sections 1 and 2, using the complexity of PL minimal surfaces we can show that given such an n-parameter sweepout there is a critical point giving an index n polyhedral minimal surface. Such a surface must meet every 2-simplex in arcs that are hyperbolic geodesics, because otherwise there is a local isotopy decreasing length. We concentrate on describing higher index disks properly embedded in tetrahedra. For the most interesting applications to knots, links, and finite group actions on the 3-sphere, S is a 2-sphere or torus, so there is a bound on the number of tubes connecting disks. (If there are many tubes then it is easy to see that these tubes must connect parallel spheres and this case cannot occur for critical points of sweepouts. Note that a tube must be unknotted if it is to contribute to the index, i.e to allow a boundary compression of the surface across an edge.)

Now a disk is an m-gon for some m. Such m-gons can be conveniently classified by passing to the orbifold universal cover of the 4 punctured sphere, which is 4 equilateral Euclidean triangles glued together. The boundary circle of the disk can be collapsed onto an arc joining two of the vertices in two ways. This arc lifts to an interval in the orbifold universal cover, which can be isotoped to be a geodesic. So taking the equilateral triangle tesselation of the Euclidean plane, the arcs are classified by picking straight lines joining the lifts of two different vertices. It is easy to see then that the corresponding m-gons have $m = 4k$. The case of $m = 8$ gives a disk that contributes index one to the corresponding surface S. For $m = 12$, it can be shown that the disk cannot occur as a minimax in a one parameter sweepout. In fact there is an edge of the tetrahedron met 3 times by the boundary of the disk. There is a pair of 2-gons on either side of the disk giving boundary compressions to this edge. The 2-gons meet at one point and so the disk can be simultaneously boundary compressed using these 2-gons. This "flattens" the maximum complexity occurring at this disk and so the disk cannot be viewed as having index one. Also the disk does not have index two because it cannot be a local maximum in a 2-parameter sweepout.

In the general case of an m-gon, where $m = 4k$ for k at least 3, the disk has k boundary compressions or 2-gons on each side, giving local moves of the disk across edges decreasing the weight by two. However there are always pairs of 2-gons on each side as in the example of the 12-gon with the property that the 2-gons meet at a single point. These 2-gons can be used to perform two boundary compressions of the disk at the same time as before. Using Cerf theory [Ce], a region R in the space E can be found where all the embeddings in R are isotopic to each other without changing the intersection pattern with the 1-skeleton. At the boundary B of R all the embeddings have tangencies with edges of the triangulation. We can assume for simplicity that these tangencies all occur for various boundary compressions of the m-gon under consideration. Note that in general we have to consider births and deaths of critical points, not just saddle tangencies.

As an example, a 16-gon can be used to define a local maximum of a 2-parameter family, using two 2-gons A and A' on one side and J and J' on the other. Here only A and J meet, as do A' and J'. For example, suppose J'' is another 2-gon that meets both A and A' in one point and is disjoint from J and J', so that A, A', and J'' define boundary compressions to an edge intersected by the 16-gon in four points. (See Figure 1.) Then the effect of simultaneously doing the boundary compressions defined by A, A', and J'' is to give a new 2-parameter sweepout where the maximum complexity is decreased. Indeed the boundary compression

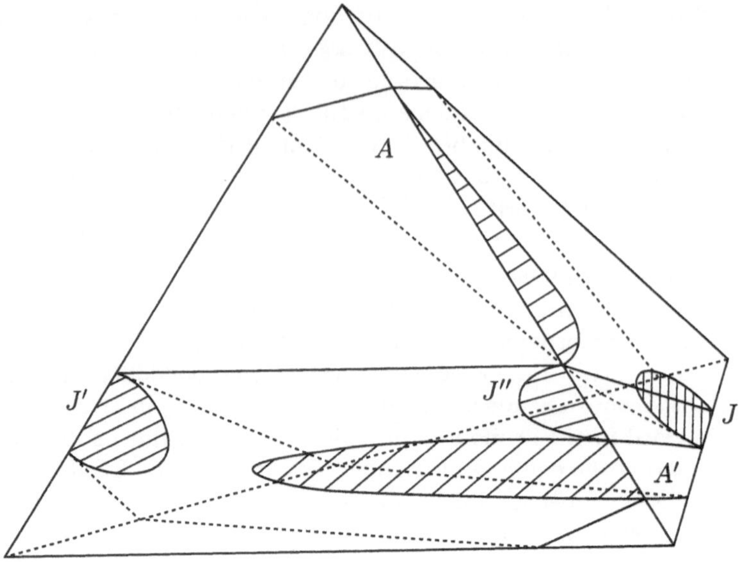

Fig. 1

using J'' is "compatible" with all the other boundary compressions and so can be done throughout the sweepout. Note that in general several different choices of additional boundary compressions may need to be made to achieve this outcome.

Our conclusion is that no m-gon actually occurs as a critical surface of maximum complexity for an n-parameter sweepout for m bigger than 8! I appreciate a very helpful conversation with Thompson on this argument. Note also that if two parallel octagons occur in the same tetrahedron, then the surface cannot occur as a critical point for an index one or two sweepout. Hence no configuration with parallel octagons occurs in an n-parameter sweepout.

To finish we discuss higher order thin position in the sense of Gabai [Ga]. Suppose a knot or link, denoted K, is given in the 3-sphere. As above there are n-parameter sweepouts of the 3-sphere by 2-spheres for n between 1 and 4. We use as a measure of complexity the number of intersections of the moving sphere with K. For a one parameter sweepout, a minimax surface is said to be in thin position. It has the crucial property that if 2-gons are given on either side of the 2-sphere with one boundary arc on K and the other on the sphere then the 2-gons intersect. Note the obvious connection to the 2-site argument. Also, as the thin position surface is a local maximum for complexity there must be such intersecting 2-gons above and below it. (Intersection of 2-gons must be at a point other than a 2-gon vertex — if 2-gons only share such a vertex we consider them to be disjoint.)

Next suppose we choose the canonical 2-parameter sweepout and find a critical surface with regard to this measure of complexity. The analogous property of index 2 thin position is that there are four 2-gons, A and A' above the sphere and J and J' below it, so that only A and J plus A' and J' intersect. Moreover any other 2-gon on the A, A' side must intersect either J or J' and similarly a 2-gon on the other side must meet A or A'. The definitions of index 3 and 4 thin position are similar.

5. Finite group actions on the 3-sphere

We end with a discussion of the proof of Theorem 1. The idea comes from [PR2]. As an illustration consider the case of a prime odd cyclic group acting on the 3-sphere. Let g denote a transformation generating this action. To show that the action is equivalent to a linear action it suffices to show that there is an invariant unknotted torus.

By the Smale conjecture, the space T of such unknotted tori is homotopy equivalent to the product of a 2-sphere with itself divided out by a free Z_2 action. We now get an induced action of the prime cyclic group on T without a fixed point, if the action is not standard. Our aim is to find a contradiction.

There are easily shown to be nontrivial equivariant Z_p cycles in T in all dimensions. We can then realize these by sweepouts, although in general the parameter space will be a chain rather than I^n modulo its boundary, as before. However each such equivariant cycle in dimension n yields a critical surface of index n. Note that this surface may have several pieces, which are one torus and a number of 2-spheres joined by thin tubes. The total index is measured as above, with unknotted tubes contributing as well as octagonal pieces that are "isolated" in tetrahedra.

But then there is a bound on the index of surfaces we can generate in this way and we get a contradiction to the assumption that the action on the torus space T was free. Hence there is an invariant unknotted torus and the action on the 3-sphere is equivalent to a linear action.

For the other assertions in Theorem 1, we need to clarify the argument in [PR2] about how larger finite group actions are built up by successive applications of the same argument. The main problem is to avoid dealing with a space of surfaces for which the homotopy type cannot be computed, as the Smale conjecture has not been shown for all 3-dimensional space forms, only the class of prism manifolds by the work of Ivanov [Iv] and some other examples in [McR].

For example, suppose we want to deal with a dihedral group of order $4pq$ acting freely on the 3-sphere. There is a cyclic normal subgroup of order $2pq$. By our previous discussion, a subgroup of this order p acts linearly. Using a theorem of Myers [My], it then follows that a subgroup of order $2p$ acts linearly, so the quotient space is a standard lens space $L(2p, r)$, for some r. Finally we observe that the dihedral group has a subgroup that is dihedral of order $4p$, which contains this cyclic subgroup of order $2p$ (and is of course not normal). Again by Myers' theorem, the action of this dihedral group of order $4p$ is equivalent to a linear action. Now the surface space we use, which plays the role of T above, is the space of invariant Heegard tori T^* in the lens space $L(2p, r)$, which are flipped over by the involution giving the $D(4p)$ action. By results of [Iv], (see also [McR]), the homotopy type of T^* can be computed because it is the same essentially as the space of Klein bottles embedded in the prism manifold with fundamental group $D(4p)$. There is an induced action of the dihedral group $D(2q)$ on T^* and this can be analyzed by the above technique. The result is the proof of the existence of a $D(2q)$ invariant torus, showing that the quotient manifold under the $D(4pq)$ action on the 3-sphere contains an embedded Klein bottle. This completes a sketch of the main features of the argument for Theorem 1.

Acknowledgement. I would like to express my gratitude to Jon Pitts for sharing with me his insights into the minimax process over many years. I would also like to thank Bus Jaco for teaching me about Haken's normal surface theory and for inspiring me to work on decision problems. Abby Thompson helped me understand the relationship between thin position and minimax and has written an elegant treatment of the algorithm to recognize the 3-sphere using thin position [Tn]. I have had helpful comments from Joel Hass, Peter Shalen, and Marty Scharlemann about aspects of this work. The recognition algorithm for the 3-sphere was announced in lectures at the Technion in 1992. I would like to thank Yoav Moriah and Bronek Wynryb for their hospitality in organizing the program.

References

[BHP] W. Boone, W. Haken, and V. Poenaru, *On recursively unsolvable problems in topology and their classification*, in Contributions to Mathematical Logic, K. Schütte, ed., North-Holland, Amsterdam (1968), 37–74.

[Ce] J. Cerf, Sur les difféomorphismes de la sphére de dimension trois ($\Gamma_4 = 0$), Lecture Notes in Math., **53**, Springer-Verlag, Berlin and New York (1968).

[CS] H. Choi and R. Schoen, *The space of minimal embeddings of a surface into a 3-dimensional manifold of positive Ricci curvature*, Invent. Math., **81** (1985), 387–394.

[Ga] D. Gabai, *Foliations and the topology of 3-manifolds, III*, J. Differential Geom., **26** (1987), 479–536.

[Ha1] W. Haken, *Theorie der Normalflächen*, Acta Math., **105** (1961), 245–375.

[Ha2] W. Haken, *Über das Homöomorphieproblem der 3-Mannigfaltigkeiten I*, Math. Z., **80** (1962), 89–120.

[Ha3] W. Haken, *Some results on surfaces in 3-manifolds*, Studies in Modern Topology, no.5, P. J. Hilton, ed., Math. Assoc. Amer. Washington, D.C. (1968), 39–98.

[Hm] R. Hamilton, *Three-manifolds with positive Ricci curvature*, J. Differential Geom., **17** (1982), 255–306.

[Ht] A. Hatcher, *A proof of the Smale conjecture*, Diff$(S^3) \simeq O(4)$, Ann. of Math. (2), **117** (1983), 553–607.

[HT] A. Hatcher and W. Thurston, *A presentation of the mapping class group of a closed orientable surface*, Topology, **19** (1980), 221–237.

[He] G. Hemion, *On the classification of homeomorphisms of 2-manifolds and the classification of 3-manifolds*, Acta Math., **142** (1979), 123–155.

[Ho] C. Hodgson, *Notes on the orbifold theorem*, preprint.

[Iv] N. Ivanov, *Homotopy of spaces of automorphisms of some 3 dimensional manifolds*, Soviet Math. Dokl., **20** (1979), 47–50.

[JR1] W. Jaco and J. H. Rubinstein, *PL equivariant surgery and invariant decompositions of 3-manifolds*, Adv. in Math., **73** (1989), 149–191.

[JR2] W. Jaco and J. H. Rubinstein, *PL minimal surfaces in 3-manifolds*, J. Differential Geom., **27** (1988), 493–524.

[Jo] K. Johannson, *Heegaard surfaces in Haken 3-manifolds*, Bull. Amer. Math. Soc., **23** (1990), 91–98.

[Jt] J. Jost, *Embedded minimal surfaces in manifolds diffeomorphic to the three-dimensional ball or sphere*, J. Differential Geom., **30** (1989), 555–578.

[Ma] A. Markov, *Unsolvability of the problem of homeomorphy (Russian)*, Proc. Internat. Congress Math. (1958), 300–306.

[McR] D. McCullough and J. H. Rubinstein, *The generalised Smale conjecture for 3-manifolds with genus 2 one-sided Heegaard splittings*, preprint.

[Mo] J. Morgan and H. Bass, eds., The Smith Conjecture, Academic Press (1984), Orlando, Florida.

[My] R. Myers, *Free involutions on lens spaces*, Topology, **20** (1981), 313–318.

[Pi] J. T. Pitts, *Existence and regularity of minimal surfaces on Riemannian manifolds*, Mathematical Notes 27, Princeton University Press, Princeton, NJ (1981).

[PR1] J. T. Pitts and J. H. Rubinstein, *Existence of minimal surfaces of bounded topological type in three-manifolds*, Proc. Centre Math. Anal. Austral. Univ., **10**, Canberra (1985), 163–176.

[PR2] J. T. Pitts and J. H. Rubinstein, *Applications of minimax to minimal surfaces and the topology of 3-manifolds*, Proc. Centre Math. Anal. Austral. Univ., **12**, Canberra (1987), 137–170.

[PR3] J. T. Pitts and J. H. Rubinstein, *Minimal surfaces and finiteness of Heegaard splittings of 3-manifolds*, in preparation.

[Ru] J. H. Rubinstein, *Polyhedral minimal surfaces, Heegaard splittings and decision problems for 3-dimensional manifolds*, preprint.

[Se] Z. Sela, *On the isomorphism problem for hyperbolic groups and the classification of 3-manifolds*, preprint.

[Sm] F. Smith, *On the existence of embedded minimal 2-spheres in the 3-sphere, endowed with an arbitrary metric*, Ph.D thesis with L. Simon, Melbourne (1980).

[Ts] C. Thomas, *Elliptic structures on 3-manifolds*, Cambridge Univ. Press, Cambridge (1986).

[Tn] A. Thompson, *Thin position and an algorithm to recognise the 3-sphere*, preprint.

[Wa] F. Waldhausen, *Some problems on 3-manifolds*, Proc. Sympos. Pure Math., vol. 32., Amer. Math. Soc., Providence, RI (1978), 313–322.

[Wh] B. White, *The space of m-dimensional surfaces that are stationary for parametric elliptic functionals*, Indiana Univ. Math. J., **36** (1987), 567–602.

Characteristic Classes, Lattice Points, and Euler-MacLaurin Formulae

Julius L. Shaneson

Department of Mathematics,
University of Pennsylvania,
Philadelphia, PA, USA

1 Introduction

In this paper I would like to survey and strengthen some of the connections between topology and other areas of mathematics, pure and applied, and areas beyond mathematics as well. One type of problem that appears in many areas of mathematics, applied mathematics, physics, economics, and probably other sciences concerns the summing of the values of a function over a discrete set of points in a prescribed region of space. Slightly more precisely, let L be a lattice in Euclidean space $\mathbb{R}^n$, such as the set of points with integral co-ordinates. Let $S \subset \mathbb{R}^n$ be a region, and let f be a (reasonable) function. Then how can one write (or approximate) the sum $\sum_{X \in L \cap S} f(x)$ in terms of quantities that are continuous or continuously computed from f and the (static) geometry of the region S? For example, if $f \equiv 1$, one is asking for a computation of $\#(L \cap S)$, the number of lattice points in S, in terms of its geometry.

The classical Euler-MacLaurin formula provides a solution to this problem for $S = [a, b] \subset \mathbb{R}^1$, a and b integers:

$$\sum_a^b f(x) = \int_a^b f(x) + \frac{1}{2}[f(a) + f(b)] + \sum_2^\infty (-1)^r \frac{B_r}{r!} \left[f^{(r-1)}(b) - f^{(r-1)}(a)\right], \quad (1.1)$$

for the case when f is a polynomial (or in general when the remainder term goes to zero). Here B_r is the rth Bernoulli number.

In dimension two, for $\Delta \subset \mathbb{R}^2$ a convex polygon with integral vertices (a "lattice polygon"), and $f \equiv 1$, Pick's theorem says

$$\#(\Delta \cap L) = \text{vol}(\Delta) + \text{“Perimeter”}\ \Delta + 1. \quad (1.2)$$

"Perimeter" is the sum of normalized lengths of the edges, where the normalization is to require that adjacent lattice points on each edge have distance one. In dimension three, Mordell gave a formula, involving Dedekind sums, for counting the lattice points of a certain lattice tetrahedron. As one consequence of his work of singular toric varieties, Pommersheim obtained the corresponding result for all tetrahedra.

Proceedings of the International Congress
of Mathematicians, Zürich, Switzerland 1994

For certain special polytopes, Khovanski and Pukhlikov obtained an interesting result with a somewhat different flavor, by considering infinitesimal parallel motions of the codimension one faces. Assume that minimal lattice vectors parallel to the edges meeting any given vertex of the convex lattice polytope Δ form a lattice basis. (This is equivalent to the smoothness of the toric avriety X_Δ; see Section 5.) Let $h = (h_1, \ldots, h_k)$ parametrize parallel motion of the k codimension one faces of Δ, and let Δ_h be the resulting convex polytope. Let

$$\tau(x_1, \ldots, x_k) = \prod_1^k \frac{x_i}{1 - e^{-x_i}}. \tag{1.3}$$

Then
$$\sum_{X \in L \cap \Delta} f(x) = \tau(\partial/\partial h_1, \ldots, \partial/\partial h_k) \int_{\Delta h} f \Big|_{h=0}. \tag{1.4}$$

Cappell and I found a complete and explicit solution for convex lattice polytopes in any dimension. Our results have the following form:

1.5 *There exist infinite order linear partial differential operators P_E, E any face of Δ, with constant rational coefficients, such that for any f,*

$$\sum_{X \in L \cap \Delta} f(x) = \sum_{E \leq \Delta} \int_E P_E f \, d\nu_E.$$

Here ν_E is Lesbesgue measure, normalized so that the fundamental domain in the intersection of L with the plane of E has volume one, and the operators P_E are unaffected by parallel translation of facets ("fan invariance"). The collection $\{P_E \mid E \leq \Delta\}$ will be called an Euler-MacLaurin expansion for Δ.

Cappell and I give an explicit description of the operators P_E in terms of the geometry of Δ ; see Section 6. The coefficients of the operators reflect the geometry of Δ in the same way that Bernoulli numbers reflect the geometry of intervals. The constant terms in the operators P_E yield formulae for the Hilbert-Ehrhart polynomial $P_\Delta(k) = \#(L \cap k\Delta)$ in all dimensions, extending Pick's theorem for dimension two and calculations of Mordell and Pommersheim's calculation dim $\Delta = 3$. Partial expansions with remainder will be discussed elsewhere.

Topologists seek to classify spaces up to homeomorphism or other suitable notions of equivalence (e.g. piecewise linear homeomorphism or diffeomorphism), usually by means of distinguishing invariants. For example, let

$$e(X) = \sum (-1)^i \ \#(\text{cells of dimension}\, i), \tag{1.6}$$

the Euler-Poincaré characteristic. Then $e(X)$ turns out to be a topological invariant, and a compact orientable surface is determined by this invariant.

This classification endeavor, often called "surgery theory", had many successes in the study of manifolds, starting with the work of Kervaire and Milnor, and then of Browder, Novikov, Sullivan, Wall, and many others. The L-classes $L_i(M) \in H_{n-4i}(M^n; \mathbb{Q})$, which will be described in Section 2, play a key role throughout this theory, as illustrated by the celebrated early result of Browder-Novikov.

1.7 *The diffeomorphism type of a closed smooth manifold of dimension at least five is determined up to a finite number of possibilities by its homotopy type and its L-classes.*

Their results also show:

1.8 *In a homotopy type every possible L-class can arise, at least up to a finite index and satisfying the Hirzebruch signature theorem [H].*

Gradually, however, topologists were forced to venture more and more outside the world of manifolds. For example, motivated by questions in transformation groups, Browder and Quinn extended the manifold theory to the category of stratified spaces and isovariant transverse maps. In our work exhibiting topologically conjugate but linearly distinct real representations of finite groups Cappell and I were implicitly working with a stratified situation, and this became explicit in our further work with Steinberger, West, and Weinberger. Most relevant for the present discussion, nontrivial stratifications also arose in my earlier work with Cappell on piecewise linear submanifolds; see Section 3.

The introduction of intersection homology by Goresky and MacPherson gave an enormous impetus to the study of stratified spaces. In particular they and Cheeger defined L-classes for stratified spaces with even codimension strata. For example, Cappell and Weinberger, proved the analogue of 1.7 for stratified spaces with even codimension strata. In this paper I will describe how Cappell and I studied these classes and how we applied our methods and results to invariants and classes of algebraic geometry and to the problems mentioned above.

2 L-classes

Let ξ be a vector bundle over a reasonable space (e.g. a finite complex) X. Factor formally the total Chern class of the complexification of ξ,

$$1 + c_1(\xi \otimes C) + \ldots = (1+\gamma_1)\ldots(1+\gamma_k). \tag{2.1}$$

Then

$$\mathcal{L}(\xi) = \prod_1^k \frac{\frac{1}{2}\gamma_i}{\tanh \frac{1}{2}\gamma_i} = 1 + \mathcal{L}_1(\xi) + \ldots \tag{2.2}$$

actually lies in $H^*(X;\mathbb{Q})$ and defines the Thom-Hirzebruch L-classes of ξ. In fact, these are universal polynomials $\mathcal{L} = \mathcal{L}(p_1, \ldots, p_n, \ldots)$, where $p_i(\xi) = (-1)^i c_{2i}(\xi \otimes \mathbb{C})$. Recall

$$\frac{\frac{1}{2}\gamma}{\tanh \frac{1}{2}\gamma} = 1 + \sum_{k=1}^{\infty} \frac{B_{2k}}{(2k)!}\gamma^{2k}. \tag{2.3}$$

If M is a manifold, we set $\mathcal{L}(M) = \mathcal{L}(TM)$, TM the tangent bundle. For example

$$\mathcal{L}(CP^n) = (1+c^2)^{n+1}, \tag{2.4}$$

where c is a generator of $H^2(CP^n)$.

All this is beautifully explained in $[H]$. For M oriented, the signature $\sigma(M)$ is the index of the intersection pairing on $H_{2k}(M)$ and the famous Hirzebruch signature theorem asserts:

$$\sigma(M) = \int_M \mathcal{L}(M). \tag{2.5}$$

For manifolds let $L(M) = \mathcal{L}(M) \cap [M]$ be the Poincaré dual of $\mathcal{L}(M)$. Just for a moment, let M be a complex n-manifold whose stable tangent bundle splits as a sum of line bundles with Chern classes $a_1, \ldots, a_k$. Let $\mathrm{Td}(M)$ be the Poincaré dual of the Todd class of M. Let $M_{i_1 \ldots i_r}$ be a submanifold of M dual to $a_1^{i_1} \ldots a_k^{i_k}$ (such a submanifold always exists). Let

$$T(M) = [M] + 2\,\mathrm{Td}_{2n-2}(M) + 4\,\mathrm{Td}_{2n-4}(M) + \ldots.$$

Then (omitting inclusion induced maps from the notation)

$$T(M) = \sum_{i_1 < \ldots < i_k} L(M_{a_{i_1} \ldots a_{i_k}}). \tag{2.6}$$

This result is a homological restatement of something Hirzebruch used to derive the Riemann-Roch theorem from the signature theorem.

Let $V^i \subset V \times \mathbb{R}^{n-i} \subset M^n$ be a submanifold with trivial normal bundle, and let $\varphi \cap [M] = [V] \in H_i(M)$ (we omit inclusion induced maps from the notation). Then (2.5) implies

$$\sigma(V) = \int_{L(M)} \varphi. \tag{2.7}$$

Thom showed that (2.7) can be used as a definition of $L(M) \in H_*(M; \mathbb{Q})$ for M a piecewise linear manifold, and for suitable spaces, e.g. algebraic varieties, (2.7) can be viewed as the definition of an invariant $L(X)$ — the Goresky-MacPherson-Cheeger L-class — extending the above Thom-Hirzebruch class to stratified spaces. For more sophisticated definitions, see [Si], [CSW].

One should note that these classes are homeomorphism invariants. For manifolds, this was proven by Novikov. For stratified spaces this essentially is a consequence of the local character of the sheaf-theoretic proof of Gorensky-MacPherson of homeomorphism invariance of intersection homology (cf.[CSW]).

3 Mapping theorems for L-classes

This section will describe some results of Cappell and myself relating $L(X)$ and $L(Y)$, when X and Y are stratified spaces and $f\colon X \longrightarrow Y$ is a map. When X and Y are manifolds and f a bundle map or a smooth embedding, there are many classical results [H], [CHS], [At] but such maps are actually fairly rare.

One could work with various notions of stratified space. Although Quinn's homotopy-theoretic definition is in many ways the most flexible and advantageous, here I will stay closer to geometry. A stratification of a space is a filtration $\phi = X_{-1} \subset X_0 \subset \ldots \subset X_n = X$ such that, first of all, the *open i-stratum* $X_i - X_{i-1}$ is an i-manifold (or empty). Let $\mathcal{V}_X$ be the set of components of open strata. Then, in addition, it is assumed that near $x \in V \in \mathcal{V}_X, X$ looks like $D^i \times \mathrm{cone}(L_V)$, where $i = \dim V$ and L_V is a stratified space of dimension $n-i-1$, with $x = \{0\} \times \{\text{cone point}\}$. It will also be assumed that $n - \dim V$ is even. For example, a complex algebraic variety has such a stratification (even satisfying Whitney's conditions), with the additional property that V is a subvariety; see [GM3] for references.

A (surjective) map $f: X \longrightarrow Y$ is stratified if for each $V \in \mathcal{V}_Y, f^{-1}V$ is a union of components of open strata of X, and $f \mid f^{-1}V$ is a locally trivial map. (In Thom's theory one assumes that the derivative of f is surjective on each stratum in the inverse image and then uses the first isotopy lemma to prove local triviality.) Again, algebraic morphisms are stratified.

Given f stratified, for each V in $\mathcal{V}_Y$, with $\dim V < \dim Y$, we define

$$N_V = N_{V,f} = f^{-1}(\operatorname{cone} L_V) \cup \operatorname{cone}(f^{-1}(L_V)),$$

a stratified space with even codimension strata. For $\dim V = \dim Y$, let $N_V = f^{-1}(v)$, some v in V.

3.1 *Assume each $V \in \mathcal{V}_Y$ is simply connected. Then* $f_*(L(X)) = \sum_{\mathcal{V}} L(\bar{V})\sigma(N_V)$.

The terms on the right in (3.1), for singular strata, can be viewed as describing in terms of the singular structure the difference between $f_*(L(X))$ and what we would expect — $(L(Y)\sigma(f^{-1}(\mathrm{pt}))$ — from the bundle case (see [CHS]).

Without simple connectivity, the terms on the right become L-classes of $\bar{V}$ with coefficients in the nonsingular intersection pairing on $IH_c^{\bar{m}}(N_V), \dim N_V = 2c$, viewed as affording a form-preserving representation of $\pi_1 V$. This situation can sometimes be analyzed as in [At] for bundles, but a completely general description in terms of some kind of product of characteristic classes of the stratum and of the representation remains to be found.

Actually, the interest of Cappell and myself on these questions goes all the way back to our work on piecewise linear embeddings in the early 1970s, in which we discovered an at first somewhat surprising result (see [CS3], also Cappell's ICM talk [C]).

3.2 *Let M^n be a smooth of piecewise submanifold of $W^{n+2}, n \geq 4$. Let $h: N^n \longrightarrow M$ be a homotopy equivalence. Assume M is simply connected. Then the composite $g: N \xrightarrow{h} M \subset W$ is homotopic to a piecewise linear embedding.*

This result was surprising because if g is a smooth or PL locally flat embedding (i.e. locally smoothable), then $g_*L(N) = L(M)$. On the other hand, according to 1.8, there are many N in the homotopy type of M, with different L-classes. Qualitatively, this means that the embedding $N \subset W$ must fail to be locally smoothable in a rather dramatic way: the set of non locally smoothable points must carry a cycle representing $L(N) - g_*^{-1}L(M)$.

In fact, the degree on non-local-smoothability imposes a nontrivial stratification $\mathcal{V}$ on N (with connected strata), such that near each point of $V \in \mathcal{V}$, the pair (W, N) looks like $D^i \times \operatorname{cone}(G_V, F_V)$ with $i = \dim V$. Because W and N are manifolds, $G_V \cong S^{n-i+1}$ and $F_V \cong S^{n-i-1}$.

Using a torsion pairing on intersection homology with local coefficients in the ring $\mathbb{Q}[t, t^{-1}]$ of Laurent polynomials, we were able to recover all the usual invariants, e.g. the signature, of knot cobordism for such very general singular p.l. knots. For any p.l. embedding of a stratified space X^n in a manifold W^{n+2}, with even codimension strata (e.g. a hypersurface in a complex algebraic variety), let $\mathcal{V}'$ be the set of components of singular strata, and assume each Y simply connected. Then the relation between singularities and characteristic classes is made explicit by the formula [CS9]

$$[X] \cap g^* \mathcal{L}\{P(W) \cup (1+\chi^2)^{-1}\} - L(X) = \sum_{\mathcal{V}'} L(\bar{V}) \sigma(G_V; F_V). \tag{3.3}$$

Again, without the simply connected hypothesis one must use L-classes for representations of $\pi_1 V$ preserving the appropriate torsion pairing. Actually, (3.3) is a special case of a result valid for any characteristic class associated to a knot cobordism invariant. The result (3.3) for the term of maximal dimension affected by singularities, i.e. $(G_V; F_V)$ a smooth knot, was proven, but not published, by Cappell and myself in the 1970s. A similar argument proves the top dimensional case of (3.3), at least for X a manifold.

The equivariant class $\Delta^G(X) \in K_*^G(X; \mathbb{Z}_{(\mathrm{odd})})$ defined by Cappell, Weinberger, and myself [CSW], where G is a finite group acting on X, also obeys an equivariant version of the mapping formula.

4 Genera for algebraic varieties

The goal of this section is the study of a morphism of complex varieties. It will be assumed that $f: X \longrightarrow Y$ is a projective morphism, Y is irreducible, and f has been stratified so that the strata are subvarieties; this is always possible (see, e.g. [GM3]). In principle, the results of the previous section are perfectly applicable to this situation. However, (3.2) has the disadvantage that if X and Y are algebraic and f an algebraic morphism, N_V, $V \in \mathcal{V}$, will usually not be a variety at all. To remedy this problem, we use algebraic analogues of the topological notion of neighborhood, and complete by projectivizing instead of coning. More precisely, for Z an irreducible subvariety of a variety Y, defined by a sheaf of ideals ℓ, the normal cone is defined as

$$C_Z W = \operatorname{Spec}\Big\{\bigoplus_{n\geq 0} \ell^n/\ell^{n+1}\Big\}.$$

Let $P(C_Z W \oplus 1)$ be its projective completion. (See [F1, Appendix B]). Let $P_{V,Y}$ be the general fiber of the natural map $P(C_{\bar{V}} \oplus 1) \longrightarrow \bar{V}$, and let $P_{V,f}$ be the general fiber of $P(C_{f^{-1}\bar{V}} \oplus 1) \longrightarrow f^{-1}\bar{V} \xrightarrow{f} \bar{V}$. Define inductively $\hat{L}(\bar{V}) \in H_*(\bar{V}; \mathbb{Q})$ by

$$\hat{L}(\bar{V}) = L(\bar{V}) - \sum_{W<V} \sigma(P_{\bar{V},W}) L(\bar{W}),$$

where the sum is over $W \in \mathcal{V}_Y$ with $W \subset \bar{V}$.

Let F be the general fiber of f. Then assuming either simple connectivity of the strata of Y or some kind of triviality of monodromy, we obtain [CS10]:

$$f_* L(X) = \sigma(F) L(Y) + \sum_{\mathcal{V}'} \{\sigma(P_{V,f}) - \sigma(F)\sigma(P_{V,Y})\} \hat{L}(\bar{V}). \tag{4.1}$$

This property of a characteristic class will be called the *stratified multiplicative property*. When simple-connectivity of strata is not assumed, it must be rephrased in terms of L-classes with coefficients in a representation of π_1, as in Section 3.

If one assumes that the normal cones above actually describe the topological normal structure of strata and their inverse images, then (4.1) can be derived from (3.2), additivity of signatures, and the fact that the signature of a suspension is always zero. However, this assumption is usually not valid, and we prove (4.1) using some much more algebraic geometric techniques.

Moreover, Cappell and I were able to use these methods to prove much more. for X a projective variety over C, Saito [S] proved the existence of a pure Hodge structure. Let $Ih^{p,q}(X)$ be the corresponding Hodge numbers. Let $h_i^{p,q}$ be the Hodge numbers of Deligne's mixed Hodge structure on $H^i(X;C)$. Consider the genera

$$\chi_y(X) = \sum_p \left[\sum_{i,q}(-1)^{i-p}h_i^{p,q}(X)\right] y^p \text{ and } I\chi_y(X) = \sum_p \left[\sum_q(-1)^q h^{p,q}(X)\right] y^p.$$

For example, $\chi_{-1} = e$ is the Euler-Poincaré characteristic, $I\chi_1 = \sigma$ is the signature discussed above, and $I\chi_0$ and χ_0 are two possible extensions of the arithmetic genus to singular varieties. These genera extend to characteristic classes T_y and IT_y whose values in dimension zero are the different genera. For example, T_{-1} is the total MacPherson-Chern class, $IT_1 = L$ is the L-class, and T_0 is the image in homology of the Baum-Fulton-MacPherson Todd class [BFM1,2] that appears in the generalized Grothendieck-Riemann-Roch theorem.

4.2 *The genera $\chi_y, I\chi_y$, and their characteristic classes satisfy the stratified multiplicative property.*

For example, let X be obtained from Y^n by blowing up a point y. Let $D = P(C_{\{y\}}) = f^{-1}(y)$ be the exceptional set. Then

$$e(X) = e(Y) + 2e(D) - e(P(C_{\{y\}} \oplus 1). \tag{4.3}$$

If y is a smooth point, D and $P(C_{\{y\}} \oplus 1)$ are projective spaces, and this is the well-known result

$$e(X) = e(Y) + n - 1 \tag{4.4}$$

for blowing up a smooth point.

The result 4.2 for the Euler characteristic (in the version involving nontrivial monodromy) essentially includes results like the generalization of Riemann-Hurwitz given in [DK], [I], [K, (III, 32)].

5 Toric varieties

Toric varieties provide one way of relating the two problems discussed in the introduction. Let $\Delta^n \subset \mathbb{R}^n$ be a lattice polytope. Let $\mathcal{F}$ be the set of codimension one faces of Δ. For each face $E \leq \Delta$ let $\mathcal{F}_E = \{F \in \mathcal{F} | E \subset F\}$, so that $E = \bigcap\{F | f \in \mathcal{F}_E\}$. Let $n_F \in L$ be a minimal lattice vector orthogonal to L, pointing into Δ. Let $T^n = \mathbb{R}^n/L$ be the n-torus. Let $V_E^\perp = \bigoplus_{\mathcal{F}_E} \mathbb{R}n_F$, E a face of Δ, be the orthogonal subspace to E. Let $L_E^\perp = V_E^\perp \cap L$. Let $T_E \subset T^n$ be the subtorus $T_E = V_E^\perp/L_E^\perp \subset \mathbb{R}^n/L$. (This notation will also be needed in the next section.) Then let X_Δ be defined as the quotient of $\Delta \times T^n$ by the relations $(x,t) \sim (x, st)$ for $x \in E$ and $s \in T_E$.

From this definition it is clear that X_Δ is compact and has an action of T^n without quotient Δ. Let $\pi: X_\Delta \longrightarrow \Delta$. Then $X_E = \pi^{-1}(E)$ is also a toric variety, and X_Δ is stratified with strata $X^\circ{}_E = \pi^{-1}(E^\circ)$. Usually X_Δ is really singular, even for polytopes in the plane. The condition needed in (1.4) is equivalent to smoothness of X_Δ.

Just from topological considerations, it is clear that invariants of X_Δ are related to properties of Δ. For example, the fixed set $(X_\Delta)^{T^n}$ of the T^n-action is finite and

$$e(X_\Delta) = \#\left[(X_\Delta)^{T^n}\right] = \#\,\text{of vertices of}\ \Delta. \tag{5.1}$$

However, the important applications require the algebraic structure which is more readily apparent from the definition of X_Δ in terms of the dual fan Σ_Δ. The cones σ_E of this fan are spanned by rays $\mathbb{R}_{\geq 0} n_F, F \in \mathcal{F}_E$. For any cone σ in a fan Σ, let

$$U_\sigma = \operatorname{Spec}\left\{\mathbb{C}\left[n \in L \mid n \bullet u \geq 0 \text{ for all } u \in \sigma\right]\right\}.$$

These affine pieces fit together to form a variety X_Σ, and $X_{\Sigma_\Delta} = X_\Delta$. The relation between invariants of X_Δ and counting lattice points in Δ is a consequence of the Riemann-Roch theorem:

5.2 *Let* $\operatorname{Td}(X_\Delta) \in H_*(X_\Delta; \mathbb{Q})$ *be the (image of the) Todd class, and suppose that*

$$\operatorname{Td}(X) = \sum_{E \leq \Delta} a_E [X_E].$$

Then for the Ehrhart-Hilbert polynomial $P_\Delta(k) = \#(k\Delta \cap L)$ *one has* $P_\Delta(k) = a_n k^n + \cdots + a_1 k + a_0$ *with* $a_m = \sum_{\dim E = m} a_E \nu(E)$, *where* $\nu(E) = \int_E d\nu_E$ *is the normalized volume as described above.*

All this was well understood before 1980, and is clearly explained in many references, e.g. [Kh], [D], [O], [F2], in chronological order. For ease of exposition we will also assume that Δ is simple (n edges meeting each vertex). In this case, [D], [O], [F2] also include the explicit calculation of $H_*(X_\Sigma; \mathbb{Q}) = A_*(X) \otimes \mathbb{Q} \cong H^*(X_\Delta; \mathbb{Q})$. (For the general case, see [FS]).

Cappell and I realized already in the mid-1980s that sufficiently powerful mapping theorems for characteristic classes, together with what was already known about toric varieties, would lead to a computation of their Todd classes, and hence to the calculation of the Ehrhart-Hilbert polynomial. However, the first significant result for singular toric varieties was obtained by Pommersheim in his 1991 Ph.D. thesis. He calculated Td_{2n-4} (denoted Td^2 in [P]), which is the highest dimensional Todd class that can be affected by singularities. A resolution of X_Σ can be obtained as $X_{\Sigma'}$, where the fan Σ' is obtained by defining Σ by adding more rays to make the cones nonsingular. Pommersheim used the canonical and explicit procedure for doing this through cones of dimension two described, for example, in [O], and the fact, also in the above references, that $\pi_*(\mathrm{Td}(X_{\Sigma'})) = \mathrm{Td}(X_\Sigma)$ to make his calculation. He formulated his results by defining a "mock" class $\mathrm{TD}(X_\Delta)$ as the Todd polynomial in the homology classes $[X_E]$ — which would equal Td in the smooth case — and calculating $\mathrm{Td}_{2n-4}(X_\Delta) - \mathrm{TD}_{2n-4}(X_\Delta)$. In particular, he obtained a formula for a_{n-2}; because $a_n = \mathrm{vol}\,(\Delta), a_{n-1} = \frac{1}{2}\sum_{\mathcal{F}} \nu(F)$, and $a_0 = 1$ were known, the computation of P_Δ for tetrahedra follows. Pommersheim also used his computation in dimension two to reprove and extend the reciprocity formulae [Mo], [R] for classical Dedekind sums. The results of this section and the next imply many reciprocity-type formulae for generalized Dedekind sums. (Cappell and I are grateful to Pommersheim for explaining his work to us prior to publication.)

In contrast to this direct attack, Cappell and I take advantage of the relation of Todd classes and L-classes and use our mapping theorems. From (2.6) and the above mapping theorems, even the topological versions, it follows that

$$T(X_\Delta) = [X_\Delta] + 2\,\mathrm{Td}_{2n-2}(X_\Delta) + 4\,\mathrm{Td}_{2n-4}(X_\Delta) + \dots$$

is a linear combination of the (images under the maps inclusions of the) classes $L(X_E), E \leq \Delta$. But by using the algebraic versions of Section 4 and taking advantage of some known facts about invariants of toric varieties (e.g. $\chi(X_\Sigma) = 1$ from [D] and others), we obtain

$$T(X_\Delta) = \sum_{E \leq \Delta} L(X_E). \tag{5.3}$$

Hence the problem is reduced to computing L-classes, which we do by considering branched coverings corresponding to certain sublattices of L of rank n and applying appropriate versions of methods from transformation groups, such as the G-signature theorem (see [H], [CSW]).

To describe our result, define mock L-classes and T-classes as

$$L^{(m)}(X_\Delta) = \prod_{\mathcal{F}} \frac{\frac{1}{2}[X_F]}{\tanh[X_F]} \tag{5.4}$$

and

$$T^{(m)}(X_\Delta) = \sum_{E \leq \Delta} L^{(m)}(X_E). \tag{5.5}$$

Then let $G_E = L_E^\perp/\mathbb{Z}\{n_F | F \in \mathcal{F}_E\}$, a finite group whose order is by definition the multiplicity $m(E)$. (Note: X_Δ is smooth iff $m(E) = 1$ for $E \leq \Delta$.) Let $m_F^E \in L_E^\perp, F \in \mathcal{F}_E$, be the minimal elements with $m_F^E \bullet n_{F'} = 0$ for $F \neq F'$ and $m_F^E \bullet n_F > 0$. For $g \in G_E$ the coset of $m \in L$, let $\gamma_F^E(g) = (m \bullet m_F^E)/(n_F \bullet m_F^E)^{-1}$, and let

$$G_E^\circ = \{g \in G_E | \gamma_F^E(g) \notin \mathbb{Z} \,\text{for all}\, F \in \mathcal{F}_E\}.$$

Let $\mathcal{A}_\Delta = 1$ and for $E < \Delta$ let

$$\mathcal{A}_E = \frac{1}{m(E)} \sum_{G_E^\circ} \prod_{\mathcal{F}_E} \coth\left\{\pi i \gamma_F^E(g) + \frac{1}{2}[X_F]\right\}. \tag{5.6}$$

Then our general calculation of Todd classes in toric varieties is then given by the formula:

$$T(X_\Delta) = \sum_{E \leq \Delta} T^{(m)}(X_E)\mathcal{A}_E. \tag{5.7}$$

Finally, note that Pommersheim's mock class $\mathrm{TD}(X_\Delta)$ and $T^{(m)}(X_\Delta)$ are not the same, even up to powers of two. In fact

$$TD_i(X_\Delta) = (2)^{i-n} \sum_{E \leq \Delta} m(E)^{-1} L_i^{(m)}(X_E). \tag{5.8}$$

Using this, one recovers the results of [P] on Td_{2n-4} mentioned above. For other valuable perspectives and results on Todd classes and toric varieties, see the work of Brion [B] and Morelli [Mr].

6 Euler-MacLaurin formulae

Finally, here is the promised determination of the operators P_E in (1.7). For f constant; i.e., for counting lattice points, it would be enough to apply (5.7) and the known calculation of $H(X_\Delta; Q)$. In general, one has to consider an infinite family of toric varieties lying over X_Δ, associated to monomials in variables measuring suitably normalized distances from facets, and show that the bits of information obtained from each fit together compatibly to obtain the result that will now be described. In effect one is defining and computing a Todd class in some type of enhanced Chow ring and using it to derive an Euler-MacLaurin formula, but the presentation here will use a language more analogous to the polytope algebra [Mc] (see also [FS]). The role of ground ring will be played by $\mathcal{D} = \mathbb{Q}[[\partial_1, \dots, \partial_n]]$, the ring of formal power series in n variables. For $m = (m_1, \dots, m_n)$ we set $D_m = m_1\partial_1 + \cdots m_n\partial_n$.

Let $\mathcal{H}_E = \{F \in \mathcal{F} - \mathcal{F}_E | F \cap E \neq \phi\}$. Let $\mathcal{P}_0(\Delta)$ be the polynomial algebra over $\mathcal{D}$ on generators $U_F, F \in \mathcal{F}$. Let $\mathcal{Q}(\Delta)$ be the quotient of this algebra by the ideal generated by the elements $U_{F_1}U_{F_2}\dots U_{F_k}$ for $F_1 \cap \dots \cap F_k = \phi$, and the elements

$$D_m + \sum_{\mathcal{F}} (m \bullet n_F) U_F$$

for $m \in L$. (Of course, it suffices to consider any n linearly independent elements of L.) Let ρ be the quotient map, and let $W = \rho_0(U_F)$.

Let $\mathcal{P}(\Delta) \subset \mathcal{P}_0(\Delta)$ be the graded $\mathcal{D}$-submodule generated in dimension k by monomials $U_{F_1} \dots U_{F_k}$ with $F_1, \dots, F_k$ pairwise distinct and $F_1 \cap \dots \cap F_k \neq \phi$. Let $\rho: \mathcal{P}(\Delta) \longrightarrow \mathcal{Q}(\Delta)$ be the $\mathcal{D}$-module map defined by setting

$$\rho(U_{F_1} \dots U_{F_k}) = m(F_1 \cap \dots \cap F_k) W_{F_1} \dots W_{F_k}.$$

6.1 *The map ρ is surjective.*

The proof (see [CS11]) actually provides an inductive procedure for lifting elements. For E a face of Δ let

$$\mathcal{T}(E) = \sum_{K \leq E} m(K) \prod_{\mathcal{F}_K} \frac{1}{2} W_F \prod_{\mathcal{H}_K} \frac{(\frac{1}{2} W_F)}{\tanh(\frac{1}{2} W_F)}.$$

Note from (6.1) that $\mathcal{Q}(\Delta)$ is closed under completion with respect to monomials in the W_F. Let $\mathcal{U}(\Delta) = 1$ and for $E < \Delta$ let

$$\mathcal{U}(E) = \frac{1}{m(E)} \sum_{G_E^{\circ}} \prod_{\mathcal{F}_K} \coth \left\{ \pi i \gamma_F^E(g) + \frac{1}{2} W_F \right\}.$$

Then set
$$T(\Delta) = \sum_{E \leq \Delta} \mathcal{T}(E) \mathcal{U}(E).$$

6.2 *Let ∂ operate on polynomials by partial differentiation $\partial/\partial x_i$. Let*

$$\mathcal{E} = \sum_{E \leq \Delta} P_E U_E \in \mathcal{P}(\Delta)$$

where $P_E \in \mathcal{D}$ and $U_E = \prod_{\mathcal{F}_E} U_F \quad (U_\Delta = 1)$, with $\rho(\mathcal{E}) = T(\Delta)$. Then

$$\sum_{X \in L \cap S} f(x) = \sum_{E \leq \Delta} \int_E P_E f \, d\nu_E.$$

It is not hard to see that for Δ an interval, this result gives the classical Euler-Maclaurin formula. For illustration in detail of the case of a polygon, see [CS11].

References

[At] M. F. Atiyah, *The signature of fibre-bundles*, in Global Analysis, Papers in honour of K. Kodaira (D. C. Spencer and S. Iyanayaga, eds.), Princeton Univ. Press, Princeton, NJ, (1969), 73–84.

[AB] M. F. Atiyah and R. Bott, *A Lefschetz fixed point formula for elliptic complexes II: Applications*, Ann. of Math. (2) **88** (1968), 451–491.

[AS] M. F. Atiyah and I. M. Singer, *The index of elliptic operators, I, III, and IV*, Ann. of Math. (2) **93** (1971), 119–138.

[BFM1] P. Baum, W. Fulton, and R. MacPherson, *Riemann-Roch for singular varieties*, Publ. Math. I.H.E.S. **45** (1975), 101–145.

[BFM2] P. Baum, W. Fulton, and R. MacPherson, *Riemann-Roch and topological K-theory for singular varieties*, Acta Math. **143** (1979), 155–192.

[BBD] A. A. Beilinson, J. Bernstein, and P. Deligne, *Faisceaux pervers, analyse et topologie sur les espaces singuliers*, Astérisque **100** (1982), 1–171.

[Bo] A. Borel et al., Intersection cohomology. Progr. Math **50**, Birkhäuser, Boston, MA, 1984.

[B] M. Brion, *Points entiers dans les polyedres convexes*, Ann. Sci. École Norm. Sup. (4) **21** (1988), 653–663.

[Br] W. Browder, Surgery on Simply Connected Manifolds, Springer Verlag, Berlin and Heidelberg, 1972.

[BQ] W. Browder and F. Quinn, *A surgery theory for G-manifolds and stratified sets*, in Manifolds (Tokyo 1973), University of Tokyo Press (1975), 27–36.

[C] S. E. Cappell, *Submanifolds of small codimension*, Proc. of Internat. Congress Math., Helsinki (1978), 455–462.

[CS1] S. E. Cappell and J. L. Shaneson, *The codimension two placement problem and homology equivalent manifolds*, Ann. of Math. (2) **99** (1974), 277–348.

[CS2] S. E. Cappell and J. L. Shaneson, *An introduction to embeddings, immersions, and singularities in codimension two*, Proc. Sympos. Pure Math. XXXII, part 2 (1978), 129–150.

[CS3] S. E. Cappell and J. L. Shaneson, *Piecewise linear embeddings and their singularities*, Ann. of Math. (2) **103** (1976), 163–228.

[CS4] S. E. Cappell and J. L. Shaneson, *Immersions and Singularities*, Ann. of Math. (2) **105** (1977), 539–552.

[CS5] S. E. Cappell and J. L. Shaneson, *Non-linear similarity*, Ann. of Math. (2) **113** (1981), 315–355.

[CS6] S. E. Cappell and J. L. Shaneson, *Fixed points of periodic differentiable maps*, Invent. Math. **68** (1982), 1–20.

[CS7] S. E. Cappell and J. L. Shaneson, *Characteristic classes, singular embeddings, and intersection homology*, Proc. Nat. Acad. Sci. **84** (1987), 3954–3956.

[CS8] S. E. Cappell and J. L. Shaneson, *Stratifiable maps and topological invariants*, J. Amer. Math. Soc. **4** (1991), 521–551.

[CS9] S. E. Cappell and J. L. Shaneson, *Singular spaces, characteristic classes, and intersection homology*, Ann. of Math. (2) **134** (1991), 325–374.

[CS10] S. E. Cappell and J. L. Shaneson, *Genera of algebraic and counting of lattice points*, Bull. Amer. Math. Soc. **30** (1994), 62–69.

[CS11] S. E. Cappell and J. L. Shaneson, *Euler MacLaurin expansions for lattices above dimension one*, C.R. Acad. Sci. Paris, to appear.

[CSSW] S. E. Cappell, J. L. Shaneson, M. Steinberger, and J. West, *Non-linear similarity begins in dimension six*, Amer. J. Math., to appear.

[CSSWW] S. E. Cappell, J. L. Shaneson, M. Steinberger, J. West, and S. Weinberger, *Classifiying on-linear similarities over* $\mathbb{Z}_2^r$, Bull. Amer. Math. Soc., to appear.

[CSW] S. E. Cappell, J. L. Shaneson, and S. Weinberger, *Classes caractéristiques pour les action des groupes sur les espaces singulières*, C.R. Acad. Sci. Paris **313** (1991), 293–295.

[CW] S. E. Cappell and S. Weinberger, *Classification of certain stratified spaces*, C. R. Acad. Sci. Paris, to appear.

[Ch] J. Cheeger, *On Hodge theory of Riemannian pseudomanifolds*, Proc. Sympos. Pure Math. **36** (1980), Amer. Math. Soc, Providence, RI, 91–146.

[CHS] S. S. Chern, F. Hirzebruch, and J.-P. Serre, *The index of a fibered manifold*, Proc. Amer. Math. Soc. **8** (1957), 587–596.

[D] V. I. Danilov, *The geometry of toric varieties*, Russian Math. Surveys **33** (1978), 97–154.

[DK] P. Deligne and N. Katz, Groupes de Monodromy en Géométrie Algébrique, Springer Lecture Notes in Math. **340** (1973), Exposé XVI, 2, 268–271 (by P. Deligne).

[F1] W. Fulton, Intersection Theory, Springer Verlag, Berlin and New York, 1984.

[F2] W. Fulton, Introduction to toric varieties, Princeton University Press, Princeton, NJ, 1993.

[FS] W. Fulton and M. Sturmfels, Intersection theory on toric varieties, to appear.

[GM1] M. Goresky and R. D. MacPherson, *Intersection homology theory*, Topology **19** (1980), 135–162.

[GM2] M. Goresky and R. D. MacPherson, *Intersection homology II*, Invent. Math. **72** (1983), 77–130.

[GM3] M. Goresky and R. D. MacPherson, Stratified Morse Theory, Ergebnisse der Mathematik, Springer Verlag, Berlin and New York, 1980.

[H] F. Hirzebruch, Topological Methods in Algebraic geometry (3rd ed.), Grundlehren Math. Wiss. 131, Springer Verlag, Berlin and New York, 1978.

[I] B. Iversen, *Critical points of an algebraic function*, Invent. Math. **12** (1971), 210–224.

[KS] M. Kashiwara and P. Schapira, Sheaves on Manifolds, Grundlehren Math. Wiss. **292**, Springer Verlag, Berlin and New York, 1990.

[Kh] G. Khovanski, *Newton polyhedra and toric varieties*, Functional Anal. Appl. **4** (1977), 56–67.

[KP] A. G. Khovanski and A. V. Pukhlikov, *A Riemann-Roch theorem for integrals and sums of quasi polynomials over virtual polytopes*, St. Petersburg Math. Journal **4** (1993), 789–812.

[K] S. Kleiman, *The enumerative theory of singularities*, in Real and Complex Singularities (P. Holm, ed.), Sijthoff and Noordhoff, Alpehn aan den Rijn (1976), 298–384.

[M] R. D. MacPherson, *Global Questions in the Topology of Singular Spaces*, Proc. Internat. Congress Math. (1983), North-Holland, New York, 213–235.

[Mc] P. McMullen, *The polytope algebra*, Adv. in Math. **78** (1989), 76–130.

[Mo] L. J. Mordell, *Lattice points in a tetrahedron and generalized Dedekind sums*, J. Indian Math. **15** (1951), 41–46.

[Mr] R. Morelli, *Pick's Theorem and the Todd Classes of a Toric Variety*, Adv. in Math., to appear.

[O] T. Oda, Convex Bodies and Algebraic Geometry, Springer Verlag, Berlin and New York, 1987.

[Pi] G. Pick, *Geometrisches zur Zahlenlehre*, Sitzungber. Lotos Prag, 2, **19** (1870), 311–319.

[P] J. E. Pommersheim, *Toric varieties, lattice points, and Dedekind sums*, Math. Ann. **296** (1993), 1–24.

[Q] F. Quinn, *Homotopically stratified spaces*, J. Amer. Math. Soc. **1** (1988), 449–499.

[R] H. Rademacher, *On Dedekind sums and lattice points in a tetrahedron*. Stud. Math. Mech., Academic Press, New York (1954), 49–53. (Reprinted in Collected Papers of H. Rademacher, Vol. II, E. Grosswald, editor, M.I.T. Press 1974, 391–398).

[S] M. Saito, *Modules de Hodge polarizables*, Publ. RIMS **24** (1988), 849–995.

[Sh1] J.L. Shaneson, *Wall's surgery obstruction groups for* $\mathbb{Z} \times G$, Ann. of Math. **99** (1969), 296–334.

[Sh2] J. L. Shaneson, *Linear algebra, topology, and number theory*, Proc. Internat. Congr. Math., Warsaw (1983), 685–698.

[Si] P. Siegel, *Witt spaces: A geometric cycle theory for KO-homology theory at odd primes*, Amer. J. Math. **105** (1983), 1067–1105.

[V] J. L. Verdier, *Specialisation de faisceaux et monodromie moderee*, Astérisque **101-102** (1982), 332–364.

[W1] C. T. C. Wall, Surgery on Compact Manifolds, Academic Press, London, 1970.

[We] S. Weinberger, The topological classification of stratified spaces, University of Chicago Press, to appear.

Positive Scalar Curvature Metrics — Existence and Classification Questions

STEPHAN STOLZ*

Department of Mathematics
University of Notre Dame
Notre Dame, Indiana 46556, U.S.A.

1 Introduction

Let M be an n-dimensional manifold (all manifolds considered in this paper are smooth, compact, and, unless otherwise specified, their boundary is empty). In this note we discuss what is known with respect to the following

QUESTIONS.

(1) *Under which (topological) conditions does M admit a positive scalar curvature metric, i.e., a Riemannian metric whose scalar curvature function is positive everywhere?*

(2) *What is the number of connected components of the space of positive scalar curvature metrics on M?*

Concerning the first question, the work of Lichnerowicz [Li], Hitchin [Hi], and Rosenberg [Ro2] shows that the existence of a positive scalar curvature metric on a spin manifold M implies the vanishing of a topological invariant $\alpha(M)$. This invariant is a generalization of the $\hat{A}$-genus $\hat{A}(M)$. It can be interpreted as the "index" of a (generalized) Dirac operator and lives in $KO_n(C_r^*\pi)$, the real K-theory of the (reduced) C^*-algebra of π, the fundamental group of M (cf. 4.3).

According to the Gromov-Lawson-Rosenberg conjecture, the vanishing of $\alpha(M)$ is necessary and sufficient for the existence of a positive scalar curvature metric on spin manifolds M of dimension $n \geq 5$. This conjecture has been verified for simply connected manifolds [St1], and for manifolds with certain fundamental groups (cf. 4.5). A weaker version of the conjecture, the "stable conjecture" (5.2) has been verified for finite fundamental groups [RS2], [RS3], and for large classes of torsion free groups (cf. Corollary 5.5, Theorem 5.6).

The second of the above questions can of course be vastly generalized; one can wonder about the homotopy groups, the homology groups, or the homotopy type of the space $\mathfrak{R}^+(M)$ of positive scalar curvature metrics on M. However, even the classification of the components eludes us so far. To the author's best knowledge, there is not a single manifold M except the 2-sphere or the real projective plane, for which $\pi_0(\mathfrak{R}^+(M))$ is nonempty and has been completely determined!

---

*) Partially supported by NSF Grant DMS–90-02594.

Proceedings of the International Congress
of Mathematicians, Zürich, Switzerland 1994

The situation becomes a little easier if we replace $\pi_0(\mathfrak{R}^+(M))$ by $\widetilde{\pi}_0(\mathfrak{R}^+(M))$, the concordance classes of positive scalar curvature metrics on M. Here, in analogy to the corresponding terminology for diffeomorphisms, we call two positive scalar curvature metrics on M *isotopic* if they are in the same connected component of $\mathfrak{R}^+(M)$ and *concordant* if there is a positive scalar curvature metric on $M \times [0,1]$ that is a product metric near the boundary and restricts to the given metrics on the boundary. It is known that (as for diffeomorphisms) isotopic positive scalar curvature metrics are concordant, but the following question is wide open.

QUESTION. *Are concordant positive scalar curvature metrics isotopic?*

In the case of diffeomorphisms, it is known that the difference between isotopy and concordance is measured by algebraic K-theory, and one might speculate that the same is true here.

The main result concerning $\widetilde{\pi}_0(\mathfrak{R}^+(M))$ is that it depends only on the fundamental group $\pi = \pi_1(M)$ and the first two Stiefel-Whitney classes of M. More precisely, there is an abelian group depending only on the dimension, the fundamental group, and the first two Stiefel-Whitney classes of M, which acts freely and transitively on $\widetilde{\pi}_0(\mathfrak{R}^+(M))$ (cf. Theorem 3.9 and [St3]). This group is defined geometrically as a bordism group. In this paper we restrict ourselves to spin manifolds which simplifies the exposition. The relevant group $R_{n+1}(B\pi)$ is defined in 3.7 (here n is the dimension of M).

These groups are of central importance in the understanding of positive scalar curvature metrics. Besides classifying positive scalar curvature metrics up to concordance, they play the rôle of obstruction groups for the existence of positive scalar curvature metrics (cf. Theorem 3.8), similar to Wall's L_n-groups in surgery theory [Wa]. Unfortunately, none of these groups have been calculated, not even if π is the trivial group! This is basically due to the fact that unlike the L_n-groups, which have an algebraic definition as well as a description as a (relative) bordism group [Wa, Section 9], so far there is no algebraic definition of the R_n-groups.

There are, however, results showing that the groups $R_n(B\pi)$ are frequently nontrivial; e.g., results of Hitchin and Carr (cf. [LM, Chapter IV, Section 7]) concerning the nontriviality of $\pi_0(\mathfrak{R}^+(M))$, or results of Botvinnik-Gilkey [BG] can be reinterpreted in this way. A systematic way to derive all these results is to study the "index homomorphism"

$$\theta\colon R_n(B\pi) \longrightarrow KO_n(C_r^*\pi),$$

which maps the bordism class represented by a manifold M to the index of the Dirac operator on M. It is tempting to conjecture that θ is an isomorphism, as this would imply the Gromov-Lawson-Rosenberg conjecture. In the author's opinion, this is too optimistic, because the K-theory groups are periodic of period eight, whereas there is no obvious periodicity in the groups $R_n(B\pi)$. However, we can enforce periodicity by "inverting the Bott-manifold" B. Here B is any simply connected 8-dimensional spin manifold with $\hat{A}(B) = 1$. Multiplication by B induces a homomorphism $R_n(B\pi) \to R_{n+8}(B\pi)$, and hence we can form $R_n(B\pi)[B^{-1}] \overset{\text{def}}{=} \varinjlim R_{n+8k}(B\pi)$.

CONJECTURE. *θ induces an isomorphism $R_n(B\pi)[B^{-1}] \xrightarrow{\cong} KO_n(C_r^*\pi)$.*

Supporting evidence for this conjecture is that this map is rationally surjective for finite groups π [St3]. Furthermore, injectivity of the map implies the Stable Conjecture, which as mentioned above has been proved for many groups π.

2 Scalar curvature

Let g be a Riemannian metric on a manifold M of dimension n. The scalar curvature is a smooth function $s\colon M \longrightarrow \mathbb{R}$, which is obtained from the curvature tensor by contracting twice. More geometrically, the scalar curvature at a point p is a measure for how fast the volume of the ball of radius r around p is growing with r. More precisely, we compare $\operatorname{vol} B_r(M,p)$, the volume of the ball of radius r around p, with $\operatorname{vol} B_r(\mathbb{R}^n, 0)$, the volume of the ball of radius r in n-dimensional Euclidean space $\mathbb{R}^n$, by expressing their quotient as a power series in r. Up to a dimension dependent normalization constant, the coefficient of r^2 turns out to be the scalar curvature at p [Be, 0.60]:

$$\frac{\operatorname{vol} B_r(M,p)}{\operatorname{vol} B_r(\mathbb{R}^n,0)} = 1 - \frac{s(p)}{6(n+2)} r^2 + \dots . \tag{2.1}$$

In particular, if a Riemannian manifold has positive scalar curvature then the volume of (small) balls grows slower than the volume of Euclidean balls of the same radius. Examples of manifolds with positive scalar curvature are the n-dimensional sphere S^n, with its usual metric, as well as certain quotients of S^n, like projective spaces (real, complex, or quaternionic), equipped with the metric induced by the standard metric on S^n, $n \geq 2$.

3 Bordism results

A method that has been used successfully for studying manifolds is to decompose them into "handles", which is similar to the cell decomposition of a CW-complex. We recall that a "handle" is a product of two discs, say $D^k \times D^{n-k}$. The boundary of this handle consists of the two parts $S^{k-1} \times D^{n-k}$ and $D^k \times S^{n-k-1}$. Given an embedding of $S^{k-1} \times D^{n-k}$ into the boundary ∂W of an n-dimensional manifold W, we can construct a new manifold

$$\widehat{W} = W \cup_{S^{k-1} \times D^{n-k}} D^k \times D^{n-k}$$

by taking the disjoint union of W and the handle $D^k \times D^{n-k}$ and identifying points in $S^{k-1} \times D^{n-k}$ with their image in ∂W. One says that $\widehat{W}$ is obtained by *attaching a k-handle* to W.

It is natural to ask whether a positive scalar curvature metric on W can be extended to a positive scalar curvature metric on $\widehat{W}$. Here the Riemannian metrics we have in mind are product metrics near the boundary; i.e., a neighborhood of ∂W is isometric to the product of ∂W with an interval.

3.1 CONVENTION. All Riemannian metrics considered in this paper are product metrics near the boundary.

3.2 Handle Extension Theorem (Gromov-Lawson [GL1], Schoen-Yau [SY], Gajer [Ga]). *Let W be an n-dimensional manifold with boundary, and let $\widehat{W}$ be obtained from W by attaching a k-handle. If $n-k$, the "codimension" of the handle, is greater than or equal to 3, then any positive scalar curvature metric on W can be extended to $\widehat{W}$.*

We note that $\partial\widehat{W}$ is obtained from ∂W by "surgery", namely by removing $S^{k-1}\times D^{n-k}$ from ∂W, and sewing in $D^k \times S^{n-k-1}$. In particular, considering $W = M\times[0,1]$, the above result implies that if M admits a metric of positive scalar curvature and $\widehat{M}$ is obtained from M by a surgery of codimension $n-k\geq 3$, then also $\widehat{M}$ admits such a metric. In fact, this is what Gromov-Lawson and Schoen-Yau prove (independently); the more general statement (3.2) is due to Gajer.

The idea of the proof of the Handle Extension Theorem is to deform the metric on W near the place where the handle will be attached in such a way that the deformed metric and the product metric on the handle $D^k\times D^{n-k}$ fit together to give a metric on $\widehat{W}$ (here the metric on the discs is given by regarding them as hemispheres of the appropriate spheres). The deformation can be visualized as pulling out a "neck". If the codimension is ≥ 3, this can be done in such a way that the scalar curvature of the deformed metric is still positive.

The Handle Extension Theorem can be generalized as follows.

3.3 Extension Theorem. *Let W be a bordism between n-dimensional closed manifolds M, N; i.e., W is a manifold whose boundary is the disjoint union of M and N. If $n\geq 5$, and the inclusion $M \to W$ is a 2-equivalence (i.e., the induced map on π_i is an isomorphism for $i<2$, and surjective for $i=2$), then any positive scalar curvature metric on N extends to a positive scalar curvature metric on W.*

Proof. Morse theory implies that W can be obtained by successively attaching handles to $N\times[0,1]$. If the handle decomposition can be chosen to contain only handles of codimension $i>2$, then the Extension Theorem implies that every positive scalar curvature metric on N extends to W.

We note that a handle decomposition of W can be interpreted in *two* ways: we either (as above) think of W as being built from $N\times[0,1]$ by attaching handles, or we can think of W as being built from $M\times[0,1]$ by attaching handles. Each handle of codimension i in the first interpretation corresponds to an i-dimensional handle in the second interpretation. Thus, our goal is to build W from $M\times[0,1]$ by attaching handles of *dimension* $i>2$.

Finding such a handle decomposition implies that W is homotopy equivalent to a CW complex obtained by attaching cells of dimension $i>2$ to M. In particular, the inclusion map $M \to W$ is a 2-equivalence. If n, the dimension of M, is greater than or equal to 5, the handle cancellation techniques used in the proof of the s-cobordism Theorem show that the converse is true as well. □

The Extension Theorem shows that under certain conditions if a manifold M is bordant to another manifold N admitting a positive scalar curvature metric, then M admits a positive scalar curvature metric, too. In particular, it suggests that whether a manifold M has a positive scalar curvature metric might depend only on its bordism class in a suitable bordism group. This is in fact true; however, for a general manifold M, the definition of the relevant bordism groups is somewhat technical [RS1, Section 3]. To simplify the exposition, we restrict our discussion to *spin manifolds* and passing remarks on more general cases.

3.4 DEFINITION. Given a space X, let $\Omega_n^{\mathrm{spin}}(X)$ be the set of bordism classes of pairs (N, f), where N is an n-dimensional spin manifold without boundary and $f: N \longrightarrow X$ is a map (two such pairs (N, f), (N', f') are bordant if there is a bordism W between N and N' (with compatible spin structure), and a map $F: W \longrightarrow X$ restricting to f (resp. f') on N (resp. N')). Disjoint union gives $\Omega_n^{\mathrm{spin}}(X)$ the structure of an abelian group known as the *n-dimensional spin bordism of* X.

Let $\Omega_n^{\mathrm{spin},+}(X)$ be the subgroup of $\Omega_n^{\mathrm{spin}}(X)$ represented by pairs (N, f), for which N admits a positive scalar curvature metric.

3.5 BORDISM THEOREM (GROMOV-LAWSON [GL1], ROSENBERG [Ro1]). *Let M be a spin manifold of dimension $n \geq 5$ with fundamental group π, and let $u: M \longrightarrow B\pi$ be the map classifying the universal covering of M. Then M has a positive scalar curvature metric if and only if the bordism class $[M, u]$ is in the subgroup $\Omega_n^{\mathrm{spin},+}(B\pi)$.*

Proof. Assume that $[M, u]$ is in the subgroup $\Omega_n^{\mathrm{spin},+}(B\pi)$. This means that (M, u) is bordant to a pair (N, f) such that N has a positive scalar curvature metric. Let (W, F) be a bordism between (M, u) and (N, f).

After doing surgery on W, we can assume that $F: W \longrightarrow B\pi$ is a 3-equivalence. Here we need the assumption that W is spin; otherwise, it might happen that we cannot kill $\pi_2(W)$ by surgeries because an embedded 2-sphere in W might have a nontrivial normal bundle!

By construction, the inclusion $M \longrightarrow W$ composed with the 3-equivalence $F: W \longrightarrow B\pi$ is the map u, which is a 2-equivalence. This forces the inclusion map to be a 2-equivalence, and hence the Extension Theorem implies that the positive scalar curvature metric on N extends over W. □

There is a relative version of the Bordism Theorem due to Hajduk, which we state after introducing the necessary notation. Let $P_n(X)$ be the bordism group of triples (N, f, g), where N is an n-dimensional closed spin manifold, $f: N \longrightarrow X$ is a map, and g is a positive scalar curvature metric on N (two such triples (N, f, g), (N', f', g') are bordant if there is a bordism (W, F) between (N, f) and (N', f') as in Definition 3.4, and g, g' extend to a positive scalar curvature metric G on W).

Forgetting the metric gives a homomorphism $P_n(X) \longrightarrow \Omega_n^{\mathrm{spin}}(X)$, whose image is the group $\Omega_n^{\mathrm{spin},+}(X)$. As usual for natural transformations between bordism groups, this homomorphism can be embedded in a long exact sequence

$$\longrightarrow P_n(X) \longrightarrow \Omega_n^{\mathrm{spin}}(X) \longrightarrow R_n(X) \xrightarrow{\partial} P_{n-1}(X) \longrightarrow, \tag{3.6}$$

where the 'relative' bordism group $R_n(X)$ is defined as follows.

3.7 DEFINITION OF THE R_n-GROUPS. $R_n(X)$ is the bordism group of triples (N, f, h), where N is an n-dimensional spin manifold (possibly with nonempty boundary), $f: N \longrightarrow X$ is a map, and h is a positive scalar curvature metric *on the boundary* ∂M. Two triples (N, f, h), (N', f', h') are bordant if

(1) There is a bordism (V, F, H) between $(\partial N, f_{|\partial N}, h)$ and $(\partial N', f'_{|\partial N'}, h')$ (considered as triples representing elements of $P_{n-1}(X)$), and
(2) the closed spin manifold $N \cup_{\partial N} V \cup_{\partial N'} N'$, obtained by gluing N, V, and N' along their common boundary components, is the boundary of some spin manifold W. Moreover, there is a map $E: W \longrightarrow X$ restricting to f (resp. f', resp. F) on N (resp. N', resp. V).

Now we can state the following relative version of the Bordism Theorem; its proof is completely analogous.

3.8 EXISTENCE THEOREM (HAJDUK [Ha]). *Let M be a spin manifold of dimension $n \geq 5$ with fundamental group π, and let $u: M \longrightarrow B\pi$ be the map classifying the universal covering of M. A positive scalar curvature metric h on ∂M extends to a positive scalar curvature metric on M if and only if (M, u, h) represents the trivial element of $R_n(B\pi)$.*

In other words, the element $[M, u, h] \in R_n(B\pi)$ is precisely the "obstruction" to extending h over M, and $R_n(B\pi)$ plays the role of an "obstruction group". At the same time, $R_{n+1}(B\pi)$ classifies concordance classes of positive scalar curvature metrics in the following sense.

3.9 CLASSIFICATION THEOREM (STOLZ [St3]). *Let M be a spin manifold of dimension $n \geq 5$ with fundamental group π, and let h be a positive scalar curvature metric on ∂M that extends to a positive scalar curvature metric on M. Then the group $R_{n+1}(B\pi)$ acts freely and transitively on $\widetilde{\pi}_0(\mathfrak{R}^+(M \operatorname{rel} h))$, the concordance classes of positive scalar curvature metrics on M extending h.*

4 The Gromov-Lawson-Rosenberg Conjecture

4.1 THEOREM (LICHNEROWICZ [Li]). *Let M be a spin manifold of dimension $n = 4k$ with a positive scalar curvature metric. Then the Hirzebruch genus $\hat{A}(M)$ vanishes.*

The $\hat{A}$-genus of an orientable manifold M is a rational number, obtained by evaluating a certain polynomial in the Pontryagin classes of M on the fundamental class of M.

Proof. If M is a spin manifold, the "Dirac operator" D on M is defined, and the Atiyah-Singer Index Theorem implies

$$\hat{A}(M) = \operatorname{index}(D) \stackrel{\text{def}}{=} \dim \ker D - \dim \operatorname{coker} D.$$

On the other hand, it follows from the "Weitzenböck formula" that the Dirac operator is invertible if the Riemannian metric used in the construction of D has positive scalar curvature, and in particular the index of D is zero in that case. □

Lichnerowicz' result has been generalized by Hitchin and later by Rosenberg. They consider "Dirac operators" whose index is not just an integer, but rather an element in a K-theory group. More precisely, if M is an n-dimensional spin manifold, and $f: M \longrightarrow B\pi$ is a map to the classifying space of a discrete group π, Rosenberg uses a Dirac operator D whose "index" lives in $KO_n(C_r^*\pi)$ [Ro2]. Here $C_r^*\pi$ is the (reduced) C^*-algebra of π, a suitable completion of the real group ring $\mathbb{R}\pi$, and $KO_n(C_r^*\pi)$ is the real K-homology of $C_r^*\pi$. These K-homology groups are a special case of the KK-groups of Kasparov [Ka]. More topologically, they can be defined by

$$KO_n(C_r^*\pi) = \pi_n(BGL(C_r^*\pi)),$$

where $BGL(C_r^*\pi)$ is the classifying space of the (infinite) general linear group with coefficients in $C_r^*\pi$.

4.2 NOTATION. Given an n-dimensional spin manifold M and a map $f: M \longrightarrow B\pi$ to the classifying space of a discrete group π, we write $\alpha(M, f) \in KO_n(C_r^*\pi)$ for the index of the Dirac operator mentioned above. If M has fundamental group π, and u is the map classifying the universal covering of M, we simply write $\alpha(M)$ instead of $\alpha(M, u)$.

The same argument as in the proof of Lichnerowicz' Theorem proves the following generalization.

4.3 THEOREM (ROSENBERG [Ro2]). *If M is a spin manifold that admits a positive scalar curvature metric, and $f: M \longrightarrow B\pi$ is a map to the classifying space of a discrete group π, then $\alpha(M, f)$ vanishes.*

We remark that for $n \equiv 1, 2 \mod 8$, $n \geq 9$, there are n-dimensional spin manifolds Σ^n, homeomorphic to the n-dimensional sphere S^n with $\alpha(\Sigma^n) \neq 0$ [Hi]. In particular, these manifolds do not admit metrics of positive scalar curvature. This shows that the answer to the question of whether a manifold M admits a positive scalar curvature metric might depend on fairly subtle properties of M; e.g., its differentiable structure.

4.4 CONJECTURE (GROMOV-LAWSON [GL2], ROSENBERG [Ro1]). *Let M be a spin manifold of dimension $n \geq 5$ with fundamental group π. Then M has a positive scalar curvature metric if and only if the element $\alpha(M) \in KO_n(C_r^*\pi)$ vanishes.*

We would like to mention that there is a more general version of the Dirac operator, whose construction doesn't need the assumption that M is spin, but only the weaker assumption that the covering of M classified by some map $f: M \longrightarrow B\pi$ is spin [St3]. The index of this operator lives in $KO_n(C_r^*(\pi, u_1, u_2))$, where $C_r^*(\pi, u_1, u_2)$ is a $\mathbb{Z}/2$-graded real C^*-algebra, which depends on π, as well as cohomology classes $u_i \in H^i(B\pi; \mathbb{Z}/2)$, whose pullback via f are the Stiefel-Whitney classes $w_i(M)$. If the cohomology classes u_i are trivial, $C_r^*(\pi, u_1, u_2)$ agrees with $C_r^*\pi$; in general $C_r^*(\pi, u_1, u_2)$ is constructed from $C_r^*\pi$ by using a 2-cocycle representing u_2 to change the multiplication, while u_1 determines the $\mathbb{Z}/2$-grading.

4.5. The Gromov-Lawson-Rosenberg conjecture has been verified for spin manifolds with the following fundamental groups:

(1) trivial group [St1]
(2) odd order cyclic groups [Ro1], [KwSc]
(3) $\mathbb{Z}/2$ [RS1]
(4) groups whose Sylow subgroups are cyclic or quaternion [BGS]
(5) free groups, free abelian groups, and fundamental groups of orientable surfaces [RS2].

We remark that result (2) is originally stated as saying that a spin manifold with odd order cyclic fundamental group has a positive scalar curvature metric if and only if its universal covering has such a metric. Combining this with result (1) proves the conjecture for these fundamental groups.

Result (2) is proved by showing that for l odd all bordism classes in the reduced bordism group $\widetilde{\Omega}_n^{spin}(B\mathbb{Z}/l) = \ker \Omega_n^{\mathrm{spin}}(B\mathbb{Z}/l) \longrightarrow \Omega_n^{\mathrm{spin}}(pt)$ can be represented by manifolds with positive scalar curvature metrics, and then applying Bordism Theorem 3.5. This strategy works in case (2), and none of the others. The basic reason is that $\Omega_*^{\mathrm{spin}}(pt)$ (and hence $\Omega_n^{\mathrm{spin}}(B\pi)$) contains a lot of 2-torsion that we don't know how to represent by explicit manifolds. All cases except (2) were proved using the following result, which shows that in the Bordism Theorem spin bordism $\Omega_n^{\mathrm{spin}}(X)$ can be substituted by connective real K-theory $ko_n(X)$. We recall that these generalized homology theories are related by means of a natural transformation [ABP]

$$D\colon \Omega_n^{\mathrm{spin}}(X) \longrightarrow ko_n(X).$$

4.6 Theorem (Jung [Ju], Stolz [St2]). *Let M be a spin manifold of dimension $n \geq 5$ with fundamental group π, and let $u\colon M \longrightarrow B\pi$ be the classifying map of the universal covering. Then M has a positive scalar curvature metric if and only if $D[M,u]$ is in $ko_n^+(B\pi)$, the image of D restricted to $\Omega_n^{\mathrm{spin},+}(B\pi)$.*

Sketch of proof. It suffices to show $\ker D \subseteq \Omega_n^{\mathrm{spin},+}(B\pi)$. Away from the prime 2 this is proved by Jung, who gives a Baas-Sullivan description of $ko_*(X)\otimes\mathbb{Z}[\frac{1}{2}]$. In particular, $[M,u] \in \ker D$ implies that the connected sum of 2^r copies of (M,u) for some r bounds a manifold with singularities, and Jung uses this to construct a positive scalar curvature metric on the connected sum.

The result at the prime 2 is due to Stolz, who proves that an odd multiple of a bordism class in the kernel of D can be represented by the total space of an $\mathbb{HP}^2$-bundle. Here an $\mathbb{HP}^2$-bundle is a fiber bundle with fiber the quaternionic projective plane $\mathbb{HP}^2$, whose structure group acts on $\mathbb{HP}^2$ by isometries (with respect to the standard metric). The condition on the structure group implies that these total spaces admit metrics of positive scalar curvature [St3, p. 512], which proves the theorem.

The proof of the statement that (at the prime 2) the kernel of D consists of bordism classes of total spaces of $\mathbb{HP}^2$-bundles is the "technical heart" of Theorem 4.6. After translating the statement into stable homotopy theory, it boils down to showing that a certain map between spectra is a split surjection. In [St1], it is shown that this map induces a surjection on homotopy groups, and this is enough to prove the statement when X is a point. In [St2] the general case is proved by using Adams spectral sequence arguments to show that the spectra in question split into simple pieces, which can be controlled homologically. □

5 The Stable Conjecture

The real K-theory groups of a real C^*-algebra are 8-periodic. This periodicity is given by multiplication by the generator b of $KO_8(\mathbb{R}) \cong \mathbb{Z}$. Geometrically, we can find a simply connected spin manifold B of dimension 8 with $\alpha(B) = b$. There are many possible choices for B, but just pick one, and call it the "Bott manifold". Given an n-dimensional spin manifold M with fundamental group π, Bott periodicity implies that $\alpha(M) \in KO_n(C_r^*\pi)$ vanishes if and only if $\alpha(M \times B) \in KO_{n+8}(C_r^*\pi)$ vanishes. This shows that the Gromov-Lawson-Rosenberg conjecture 4.4 is true if and only if the following two conjectures hold.

5.1 Cancellation Conjecture. *Let M be a spin manifold of dimension $n \geq 5$. Then M has a positive scalar curvature metric if and only if $M \times B$ does.*

5.2 Stable Conjecture. *Let M be a spin manifold. Then the product of M with sufficiently many copies of B has a positive scalar curvature metric if and only if $\alpha(M)$ vanishes.*

We note that Theorem 4.6 has the following corollary (alternatively, this follows from the geometric description of $KO_*(X)$ of Kreck-Stolz (cf. [KrSt], [RS2, Theorem 4.3]).

5.3 Corollary. *Let M be a spin manifold with fundamental group π, and let $u: M \to B\pi$ be the classifying map of the universal covering. Then the product of M with sufficiently many copies of B has a positive scalar curvature metric if and only if $p(D[M,u])$ is in $KO_n^+(B\pi)$. Here $p: ko_n(X) \to KO_n(X)$ is the natural map from connective to periodic K-homology, and $KO_n^+(X) \subseteq KO_n(X)$ is the image of $\bigoplus_{k\geq 0} ko_{n+8k}^+(B\pi)$ under p (we identify $KO_{n+8k}(B\pi)$ with $KO_n(B\pi)$ using periodicity).*

Let M be an n-dimensional spin manifold, and let $f: M \to B\pi$ be a map to the classifying space of a discrete group π. We remark that the invariant $\alpha(M,f) \in KO_n(C_r^*\pi)$ depends only on the class $p(D[M,f]) \in KO_n(B\pi)$. In fact, there is a homomorphism known as the *assembly map*

$$A: KO_n(B\pi) \to KO_n(C_r^*\pi), \tag{5.4}$$

which maps $p(D[M,f])$ to $\alpha(M,f)$ [Ro2]. Combining this with Corollary 5.3, we conclude:

5.5 Corollary. *If the assembly map (5.4) is injective, then the Stable Conjecture 5.2 is true for the group π.*

The Novikov conjecture (or rather, a form of it) claims that A is injective for torsion free groups. It has been proved for many groups, notably for torsion free, discrete subgroups of Lie groups [Ka]. The assembly map is definitely not injective for some groups, e.g., finite groups. Still, there is the following result.

5.6 THEOREM (ROSENBERG-STOLZ [RS2], [RS3]). *The Stable Conjecture 5.2 is true for finite groups.*

To prove this, it suffices by Corollary 5.3 to show that the kernel of the assembly map is contained in $KO_*^+(B\pi)$. This is proved first for cyclic groups by an explicit construction. Using the fact that for a finite group π the KO-homology of $B\pi$ can be expressed in terms of the splitting of $\mathbb{R}\pi$ as a product of matrix rings over $\mathbb{R}$, $\mathbb{C}$, and $\mathbb{H}$, "Artin induction" is used to show that the statement $\ker A = KO_*^+(B\pi)$ for finite cyclic groups implies this statement for a general finite group.

The fact that the Stable Conjecture is true for such a variety of groups, finite groups on one hand, and many torsion free groups on the other hand, gives strong evidence for this conjecture. Concerning the Cancellation Conjecture, the evidence is less convincing, and the author has lingering doubts about it.

At this point the reader might be curious about how the R_n-groups fit into the discussion of the Stable Conjecture. The answer is best expressed in terms of the following commutative diagram, which represents a factorization of the map $\Omega_n^{\text{spin}}(B\pi) \to KO_n(C_r^*\pi)$ given by $[M, f] \mapsto \alpha(M, f)$.

$$\begin{array}{ccccc}
\Omega_n^{\text{spin}}(B\pi) & \xrightarrow{D} & ko_n(B\pi) & \longrightarrow & R_n(B\pi) \\
 & & {\scriptstyle p}\downarrow & & {\scriptstyle \theta}\downarrow \\
 & & KO_n(B\pi) & \xrightarrow{A} & KO_n(C_r^*\pi).
\end{array} \tag{5.7}$$

Here, as mentioned in the introduction, θ maps the bordism class represented by (M, f, h) to the index of the Dirac operator determined by M and f. We note that, unlike the situation for closed manifolds, the index of the Dirac operator on manifolds with boundary in general does *not* depend continuously on the metric. In fact, the ordinary $\mathbb{Z}$-valued index jumps whenever an eigenvalue of the Dirac operator on ∂M crosses the origin. However, by the Weitzenböck formula, this does not happen as long as the metric on ∂M has positive scalar curvature.

Crucial for the proof that θ is well defined is the additivity of the $KO_n(C_r^*\pi)$-valued index, which was proved recently by Bunke [Bu].

The diagram (5.7) and the Existence Theorem 3.8 show that the injectivity of θ implies the Gromov-Lawson-Rosenberg Conjecture. Similarly, the injectivity of the map

$$R_n(B\pi)[B^{-1}] \to KO_n(C_r^*\pi) \tag{5.8}$$

would imply the Stable Conjecture (5.2). Unfortunately, the converse is not true, and presently there are no techniques available to prove injectivity. If π is the trivial group, proving injectivity of the map (5.8) is equivalent to showing that a positive scalar curvature metric h on S^{n-1} can always be extended — after crossing with enough copies of B — to a positive scalar curvature metric on the disc D^n (or rather $D^n \times B \times \cdots \times B$), provided the index of the Dirac operator on D^n (with respect to some metric extending h) vanishes.

However, it is known that the map (5.8) is rationally surjective for finite groups π and this supports the conjecture that it is an isomorphism. Beyond this conjecture, the following might be interesting to know.

5.9 QUESTION. *Is $R_n(B\pi)$ a functor of $C_r^*\pi$ (resp. $\mathbb{Z}\pi$?); i.e., is there an algebraic description of $R_n(B\pi)$ in terms of the C^*-algebra $C_r^*\pi$ or the integral group ring $\mathbb{Z}\pi$?*

References

[ABP] D. W. Anderson, E. H. Brown, Jr., and F. P. Peterson, *Pin cobordism and related topics*, Comment. Math. Helv. **44** (1969), 462–468.

[Be] A. L. Besse, Einstein Manifolds, Springer-Verlag, Berlin and New York, 1986.

[BG] B. Botvinnik and P. B. Gilkey, *The eta invariant and metrics of positive scalar curvature*, preprint.

[BGS] B. Botvinnik, P. B. Gilkey, and S. Stolz, *The Gromov-Lawson-Rosenberg Conjecture for space form groups*, preprint 1994.

[Bu] U. Bunke, *A K-theoretic relative index theorem and Callias-type Dirac operators*, to appear in Math. Ann.

[Ga] P. Gajer, *Riemannian metrics of positive scalar curvature on compact manifolds with boundary*, Ann. Global Anal. Geom. **5** (1987), 179–191.

[GL1] M. Gromov and H. B. Lawson, Jr., *The classification of simply connected manifolds of positive scalar curvature*, Ann. of Math. (2) **111** (1980), 423–434.

[GL2] ———, *Positive scalar curvature and the Dirac operator on complete Riemannian manifolds*, Publ. Math. IHES (1983), no. 58, 83–196.

[Ha] B. Hajduk, *On the obstruction group to the existence of Riemannian metrics of positive scalar curvature*, Proc. International Conf. on Algebraic Topology, Poznań 1989, S. Jackowski, R. Oliver, and K. Pawalowski, eds., Lecture Notes in Math., Springer-Verlag, Berlin and New York, pp. 62–72.

[Hi] N. Hitchin, *Harmonic spinors*, Adv. in Math. **14** (1974), 1–55.

[Ju] R. Jung, Ph.D. thesis, Univ. of Mainz, Germany, in preparation.

[Ka] G. G. Kasparov, *Equivariant KK-theory and the Novikov Conjecture*, Invent. Math. **91** (1988), 147–201.

[KrSt] M. Kreck and S. Stolz, *HP^2-bundles and elliptic homology*, Acta Mathematica **171** (1993), 231–261.

[KwSc] S. Kwasik and R. Schultz, *Positive scalar curvature and periodic fundamental groups*, Comment. Math. Helv. **65** (1990), 271–286.

[LM] H. B. Lawson and M.-L. Michelsohn, Spin geometry, Princeton Mathematical Series, 38, Princeton Univ. Press, Princeton, NJ, 1989.

[Li] A. Lichnerowicz, *Spineurs harmoniques*, C. R. Acad. Sci. Paris, Série A-B **257** (1963), 7–9.

[Ro1] J. Rosenberg, *C*-algebras, positive scalar curvature, and the Novikov Conjecture, II*, Geometric Methods in Operator Algebras, H. Araki and E.G. Effros, eds., Pitman Research Notes in Math., no. 123, Longman/Wiley, Harlow, Essex, England and New York, 1986, pp. 341–374.

[Ro2] ———, *C*-algebras, positive scalar curvature, and the Novikov Conjecture, III*, Topology **25** (1986), 319–336.

[RS1] J. Rosenberg and S. Stolz, *Manifolds of positive scalar curvature*, Algebraic Topology and its Applications (G. Carlsson, R. Cohen, W.-C. Hsiang, and J.D.S. Jones, eds.), M. S. R. I. Publications, vol. 27, Springer-Verlag, Berlin and New York, 1994, pp. 241–267.

[RS2] ———, *A "stable" version of the Gromov-Lawson conjecture*, to appear in the Proc. of the Cech Centennial Homotopy Theory Conference, Cont. Math. **181** (1995), 405–418.

[RS3] ———, *The stable classification of manifolds of positive scalar curvature*, in preparation.

[SY] R. Schoen and S.-T. Yau, *On the structure of manifolds with positive scalar curvature*, Manuscripta Math. **28** (1979), 159–183.

[St1] S. Stolz, *Simply connected manifolds of positive scalar curvature*, Ann. of Math. (2) **136** (1992), 511–540.

[St2] ———, *Splitting certain MSpin-module spectra*, Topology **33** (1994), 159–180.

[St3] ———, *Concordance classes of positive scalar curvature metrics*, in preparation.

[Wa] C.T.C. Wall, Surgery on compact manifolds, Academic Press, London and New York, 1970.

Nonlocally Linear Manifolds and Orbifolds

SHMUEL WEINBERGER[1]

Mathematics Department, University of Pennsylvania
Philadelphia, PA 19104, USA

A *topological manifold* is, by definition, a Hausdorff topological space where each point has a neighborhood homeomorphic to Euclidean space. The geometrical topology of manifolds is a beautiful chapter in mathematics, and a great deal is now known about both the internal structure of manifolds (transversality, isotopy theorems, local contractibility of homeomorphism groups, etc.) and their classification (cobordism theory, surgery theory, etc.). The subject that I would like to explore is the extension of this picture to a larger class of intrinsically interesting spaces (finite-dimensional ANR homology manifolds). Part of our exploration is motivated by an analogy between homology manifolds and orbifolds, that is, spaces that are modeled not on Euclidean space, but rather on the quotients of representation spaces by their finite linear actions.

1 The Topological Characterization of Manifolds

There are several different ways that one can be lead to the nexus of problems considered here. One useful way is to ask: How can one tell whether or not a space that arises in some natural fashion is a manifold?

In low dimensions, there are some classical criteria. A connected space is a circle if no point separates it, but each pair of points separates it. There is a similar characterization of the 2-sphere in terms of nonseparating points and separating circles, due to R. L. Moore. However, in dimension 3 and higher this is not possible because of the existence of homology manifolds: A *homology manifold* will be, until the very last section, a finite-dimensional ANR X with $H_*(X, X - x) = H_*(\mathbb{R}^n, \mathbb{R}^n - 0)$ for every point x in X. Such spaces have many of the properties of manifolds. They satisfy Poincaré duality, and therefore will be separated by exactly the same spaces that would separate a manifold.

The simplest example of a homology manifold that is not a manifold is the cone on a nonsimply connected (manifold) homology sphere. All deleted neighborhoods of the cone point are nonsimply connected, so this space is not a manifold, but it is a trivial calculation to see that the local homology is as required.

Another way to obtain many more and much wilder examples is that of *decomposition spaces*, pioneered by Bing.[2] One starts with a manifold M and

1) Partially supported by the NSF.

2) The earliest striking application was the construction of a nonmanifold whose product with $\mathbb{R}$ is a manifold.

Proceedings of the International Congress
of Mathematicians, Zürich, Switzerland 1994

describes a (suitably semicontinuous) collection of subsets that are in some weak sense contractible (technically, cell-like), and identifies each of these subsets to a point. This identification space X is the image of a natural *CE map*[3] $M \to X$.

Because every homology sphere bounds a contractible manifold (see [K] for high dimensions and [Fr] for dimension 3) the first example is a special case of the second. In fact, for quite some time it seemed as if every homology manifold could be obtained in this fashion:

RESOLUTION CONJECTURE (Cannon): Every homology manifold (of dimension at least five) is the decomposition space of some cellular decomposition of a manifold.

This conjecture was attractive in light of an amazing theorem of Edwards (see [Dav, E]).

THEOREM (Edwards). *A CE map $f : M \to X^n$, $n \geq 5$, can be approximated by homeomorphisms iff X satisfies the disjoint disks property (DDP), that is iff any pair of continuous maps $D^2 \to X$ can be ε-approximated by maps with disjoint images.*

COROLLARY: A resolvable homology manifold is a manifold iff it has the DDP.[4]

COROLLARY: A resolvable homology manifold $\times \mathbb{R}^2$ is a manifold.

With some diffidence, I would like to suggest calling homology manifolds with the DDP *nonlocally linear manifolds*. The conjectures made in [BFMW1] suggest that these will be locally modeled on some new (topologically homogenous) spaces and that they will share many of the geometric properties of manifolds. For instance, in [BFMW2] the resolution conjecture is verified with nonlocally linear manifolds replacing manifolds. However, alone, this conjecture does not give us any insight into what the local geometry of such spaces can be.

An important rigidification of the situation was made by Quinn [Q1]. He showed:

THEOREM (Quinn). *There is a locally defined $i(X) \in H^0(X;\mathbb{Z})$ valued invariant of homology manifolds. Thought of as a function on components, it assumes values in $1 + 8\mathbb{Z}$, and equals 1 (on every component) iff X is resolvable.*

This integer is a signature, and it would be appropriate to think of it as the 0th Pontryagin class of (the "tangent bundle" of) X. We call it the *local index* of X.

Its definition is about the same level of depth as the topological invariance of Pontryagin classes (Novikov's theorem) as it requires defining L-classes in a topological fashion for ANR homology manifolds (and in particular for topological manifolds). L-classes for homology manifolds are constructed in [FP] and [CSW], and we will return to this in Section 3.

---

3) A map is CE if, when restricted to the preimage of any open subset of the range, the map is a homotopy equivalence.

4) For an example of how dramatically the DDP can fail, see [DW].

The locality of Quinn's obstruction implies that for connected X if some open subset of X is resolvable, then X is. In particular, any "manifold with singularities" is resolvable by a manifold. Thus, constructing nonresolvable homology manifolds involves building the whole space simultaneously.

THEOREM ([BFMW]). *For every number $i \in 1 + 8\mathbb{Z}$ there is a homotopy sphere*[5] *that is a DDP homology manifold with local index i.*

We conjecture these spaces to be uniquely determined by i and dimension, at least if the dimension is ≥ 4.

2 Lacunae in the Theory of Topological Manifolds

In this section, I would like to show how the theory of manifolds itself cries out for some missing spaces. The spaces turn out to be supplied by the nonlocally linear manifolds (DDP homology manifolds).

DEFINITION: For M a manifold of dimension ≥ 5, let $S(M)$ denote the set of homotopy equivalences modulo homeomorphisms. That is,

$$\begin{aligned} S(M) \cong &\{h : M' \to M \text{ a simple homotopy equivalence} \\ &\text{with } h : \partial M' \to \partial M \text{ a homeomorphism}\} \\ &/\text{homeomorphism } (\text{rel}\, \partial). \end{aligned}$$

THEOREM (Siebenmann, as corrected by Nicas, [KS, Ni]). *Let M be a manifold of dimension ≥ 5. If ∂M is nonempty, one has*

$$S(M) \cong S(M \times D^4).$$

In general, there is an exact sequence

$$0 \to S(M) \to S(M \times D^4) \to \mathbb{Z}.$$

This means that $S(M \times D^4)$ can be as much as a $\mathbb{Z}$ larger than $S(M)$. The simplest manifold, the sphere, gives an example of this. $S(S^n) \cong 0$, but $S(S^n \times D^4) \cong \mathbb{Z}$. From the point of view of periodicity there should be a $\mathbb{Z}$s worth of homotopy spheres. These are filled in by the homology manifolds. (For manifolds with boundary, the boundary condition forces the domain homology manifold mapping to M to be a homology manifold — essentially because of locality, so that using homology manifolds would not increase the size of S.)

THEOREM ([BFMW1]). *Let X be a homology manifold of dimension ≥ 5, and let $S^H(X)$ denote $\{h : X' \to X$ a simple homotopy equivalence with $h : \partial X' \to \partial X$ a homeomorphism$\}/s$-cobordism $(\text{rel}\, \partial)$. Then if ∂M is nonempty, one has*

$$S^H(X) \cong S^H(X \times D^4) \quad (\cong S(X \times D^4), \text{ if } X \text{ is a manifold}).$$

5) In fact every simply connected manifold has an "evil twin" with given local index, but typically aspherical manifolds do not.

Thus, periodicity is true in the category of homology manifolds. The periodicity map interchanges manifolds and homology manifolds. We have to use s-cobordism rather than homeomorphism as our equivalence relation because we cannot yet prove an s-cobordism theorem for DDP ANR homology manifolds. Nonetheless, this enables us to ignore any DDP conditions in the definitions of our structure sets. (There can be no s-cobordism theorem without assuming DDP because a CE map has as mapping cylinder an s-cobordism, which cannot be a product unless the quotient space is a manifold!)

THEOREM ([BFMW1]). *$S^H(X)$ can be computed as the fiber of the assembly map (see [R]).*[6] *Consequently it is an abelian group, which is functorial for orientation true*[7] *maps between manifolds of dimension that differ by multiples of 4.*[8]

Using $S(M)$ one loses functoriality for maps where the dimension of the target is smaller than that of the domain. So the theory of homology manifolds has better formal properties than the theory of manifolds. In particular, pushing manifolds forward leads to (nonresolvable) homology manifolds.

These theorems imply that the rigidity theory of high-dimensional topology adapts gracefully to include homology manifolds. For instance any DDP homology manifold homotopy equivalent to a nonpositively curved manifold is homeomorphic to it (see [FJ]).[9] Moreover, existence theorems in our larger category work out a little more nicely:

THEOREM ([BFMW]). *Any Poincaré space $\mathbb{Z}$ homology equivalent to a nonpositively curved manifold, is homotopy equivalent to a closed DDP homology manifold, which is unique up to s-cobordism.*

However, such a Poincaré space is not necessarily homotopy equivalent to a closed manifold.[10] Another example where such spaces occur naturally is the following result of Smith theory:

THEOREM ([CW]). *Tame semifree circle actions on a manifold have ANR homology manifolds as fixed sets. If the fixed set has codimension* 2 *mod* 4,[11] *any equivariant homotopy equivalent free circle action on (a manifold homotopy equivalent to) the complement of the fixed set of this action extends to a unique concordance class of circle actions.*

---

6) This is a more modern formulation of surgery theory in the topological category than one finds in Wall's book.

7) That is, a map that preserves the orientation character of curves.

8) There are reindexing tricks that allow one to define functoriality for a related theory for all orientation true maps.

9) A simpler approach would be to show that the local index is necessarily 1, which is of the depth of the Novikov conjecture: it is a kind of tangentiality statement (see [FW] for this point of view, taken in a different direction). Then resolution implies it is a manifold, and the usual rigidity takes over.

10) There is a nonresolvable homology manifold proper homotopy equivalent to a symmetric space of noncompact type iff it has q-rank greater than 2. ([BW])

11) There is a rather different analysis for the case of codimension 0 mod 4.

If one doesn't allow homology manifolds to arise there is a $\mathbb{Z}$ obstruction to the existence of a semifree manifold completion for the free action.

3 Properties of Homology Manifolds

In the previous section we described some systematic features of the class of homology manifolds. In particular, we described the result of surgical classification for these spaces. Next, we describe the cobordism classification (suggested by David Segal).

THEOREM ([BFMW]). *For $n \geq 6$, the formula $\Omega*^H \cong \Omega*^{\mathrm{Top}}[1+8\mathbb{Z}]$ (monoid ring) is correct additively. At least rationally, this formula gives the correct multiplicative structure.*

Note that unlike the classical manifold case, not all bordism classes are represented by connected spaces. Locality implies that there is no analogue of connected sum. I conjecture that the bordism calculation is multiplicatively correct even integrally.

Unlike the usual proofs of bordism results, this is not achieved by direct analyses of transversality, but rather by using the fact that all topological bordism classes have simply connected representatives, and applying the "evil twin" theorem of Section 1.

There are obstructions to transversality, as one can see by the following reasoning. Consider a homotopy equivalence (say) from a manifold to a homology manifold, and assume that it could be made transverse to a point; then the "funny local type" would be present in a manifold giving a contradiction. One can cross this example with a manifold to see that this is not an oddity because of low-dimensional preimages.

Similarly, if one embeds a nonresolvable homology manifold in high-dimensional Euclidean space, one cannot stably hope for any "normal bundle" structure because of multiplicativity properties of the local index.[12]

The following calculation demonstrates the systematic failure of transversality:

THEOREM (Stabilized structure calculation). *One has an isomorphism:*

$$\lim S^H(M \times K \downarrow K) \cong \lim S^H(M \times \mathbb{R}^n \downarrow \mathbb{R}^n)_{(2)}.$$

Here the $\downarrow$ denotes controlled structures, and the K's run over the DDP ANR homology manifolds proper homotopy equivalent to Euclidean spaces ("fake Euclidean spaces"). The right-hand side is a convenient stabilization of S, and only differs from the usual S when there is some algebraic K-theory present (see e.g. [R], [We]) — in particular, for M simply connected it is the structure set $S^H(M)$.

12) There is another kind of structure present: a teardrop neighborhood (see [HTWW]). It is to a bundle what a CE map is to a homeomorphism in the sense that the same homotopical data as is present for the classical notion arises here but with respect to open subsets of the range.

If there were transversality, then the part of $S(M)$ caught by characteristic classes could not die on crossing with any manifold: by projecting down to that manifold, the transverse inverse image of any point would recover this data. However, the theorem asserts that all odd torsion is lost.

Another version of this theorem describes when a map between manifolds is stably homotopic to an s-cobordism: iff it is stably tangential at 2 (it is a classical theorem of Mazur that one has the same condition for stable homeomorphism if stabilization is with Euclidean spaces, except that one does not localize.)

A good analysis of transversality obstructions could lead to the calculation of the multiplicative structure of bordism suggested above. A final ramification of transversality, classically, is Sullivan's $KO[1/2]$ orientation for manifolds (see [Su]). Here we have a variant:

THEOREM (See [CSW]). *For every homology manifold X there is a canonical $KO[1/2]$ class, the signature class, that is an orientation iff X is resolvable.*

Thus, the orientability is closely related to transversality, but independently of transversality, homology characteristic classes can in any case be defined, and used. Sullivan's formulation of surgery in terms of these classes is implicit in the theory adumbrated in Section 2. Away from 2 the homology normal invariant is just the difference of these signature classes.

REMARK: There is a refinement at 2 related to the Morgan-Sullivan class [MS]. The method of proof is to understand the chain complex of X as a self-dual sheaf and recognize the Witt group of such (away from 2) as KO-homology. (This is dependent on the work of Quinn and Yamasaki — or alternatively, Ferry and Pederson — for nonconstructible sheaves, as arise here.) A little thought checks that the local calculation one would do of this invariant agrees with the local index (up to powers of 2).

REMARK: The assignment of characteristic classes to self-dual sheaves has a number of other applications. For instance, one has a close parallel to classical surgery theory for spaces with even codimensional strata and simply connected links by making use of the intersection sheaves (see [GM II]), see [CSW], and also [Sh1] for other applications of cobordism of self-dual sheaves.

4 Conjectured Properties of Homology Manifolds

We have already alluded several times to a conjectured view of homology manifolds.

VAGUE CONJECTURE: High-dimensional homology manifolds with the disjoint disk property (nonlocally linear manifolds) share all the geometrical properties of manifolds.

Obviously one cannot include transversality among the geometrical properties we have in mind! On the other hand, general position holds. We will be guided somewhat by the analogy presented in the next section. Until then, let us be more precise and list the following package of conjectures [BFMW]:

CONJECTURE (Homogeneity): Nonlocally linear manifolds are homogenous.

For manifolds this is a triviality. If the signature class is an orientation then, of course, it is true because of the work of Edwards and Quinn cited above: the space is a manifold. Because the deviation from "manifoldness" is entirely measured by a conventional top locally finite homology class (or 0-dimensional cohomology class), an algebraic sort of homogeneity is guaranteed. Still, the conjecture has eluded us. The next conjectures are somewhat stronger, and we will discuss some progress by analogy in the next section:

CONJECTURE (S-bordism): For nonlocally linear manifolds of dimension ≥ 6, the s-cobordism theorem holds.

In the smooth category this is Smale's theorem (or rather its generalization by Barden-Mazur-Stallings) and is true in the topological category by Kirby-Siebenmann's reinstitution of handlebody theory.

CONJECTURE (CE-approximation): If dim ≥ 5, any CE map from a nonlocally linear manifold to another is approximable by homeomorphisms.

If one assumes the domain and range are conventional manifolds, then this is a theorem of Siebenmann, if just the domain is, then it is Edwards' result, and if just the range is, then one obtains this as a consequence of the work of Edwards and Quinn.

CONJECTURE (Local contractibility of homeomorphism groups): Every homeomorphism sufficiently close to the identity can be canonically isotoped to the identity.

For manifolds, this is due to Cernavski, and Edwards-Kirby.

5 The Analogy to Orbifolds

It is somewhat reassuring that there are other settings in which one can both define objects in terms of explicit models, or alternatively in terms of local homotopy properties, and the latter not only fill in lacunae in the theory of the former, but they themselves possess many nice geometric properties, and homogeneity, in particular.

One such setting is that of orbifolds (although much of what follows is a special case of a general theory of stratified spaces).

DEFINITION: A locally linear orbifold is a space that is locally modeled on the orbit space of an orthogonal representation of a finite group.

DEFINITION: A nonlocally linear orbifold is a space that is locally the quotient of a disjoint disk homology manifold under a finite group, where the fixed sets of all subgroups are (not necessarily locally linear) homology manifolds that are embedded in one another in a "locally homotopically trivial fashion". This local condition (aside from codimension 2) asserts that in the local chart, small 2-disks in one fixed set can be homotoped (in an arbitrarily small way) disjoint from a smaller fixed point set. (See [Q2], [We].)

REMARK: In the second definition, if one is trying to imagine phenomena not stemming directly from the existence of nonresolvable homology manifolds, not much is lost in assuming that one is locally the quotient of an action on a manifold where all of the fixed point sets are submanifolds. The local homotopy condition (by work of Bryant, Chapman, and Quinn) boils down to the assertion that these fixed sets are locally flat.

Now for the analogous theorems to what we discussed above.

(CE APPROXIMATION) There is an equivariant CE approximation theorem[13] (see [StW] for the locally linear case, and [Hu], [We] in general). The analogue of the resolution conjecture would be the coincidence of the two definitions of orbifolds. However, it is quite simple to see that the cone on a fake real projective space (which are produced in profusion by surgery theory, but are not hard to come by explicitly, using linear involutions on Brieskorn spheres, for instance) is never resolvable by a linear orbifold!

(SURGERY) The analogue of the surgery exact sequence was established for odd order locally linear group actions by [MR] by a complex induction depending on transversality methods. Simultaneously with establishing transversality, they proved that there is an equivariant Sullivan orientation $\in KO^G_*[1/2]$ for the actions they considered. Unfortunately for nonlocally linear actions, and for even order groups, equivariant transversality fails, and although subsequently [RtW] (see also [RsW]) a signature class[14] was constructed for more general actions, it was not an orientation. This necessitates a deviation from the Sullivan-Wall exact sequence of classical surgery theory, and is given in [We].

The new sequence boils down to an equivariant extension of the homological form of surgery theory due classically to Quinn and Ranicki (see [R]). Indeed, that theory naturally has Siebenmann's periodicity built into it, and is the one alluded to in Section 2. Moreover, in this formulation, all the theories (including the topological theory for all statified spaces) take a beautiful "local-global" form:

$$\cdots \to L(X \times I) \to S(X) \to H_0(X; L(\text{loc})) \to L(X) \to$$

where L denotes a surgery spectrum (a generalization of the notion of surgery groups, and adapted to stratified spaces), which when applied to open subsets gives a cosheaf of spectra. The difference between the L-cosheaf homology and a global L-spectrum gives rise to the spaces stratified homotopy equivalent but not homeomorphic to X.[15]

As for the conjectures.

(HOMOGENEITY) The version of homogeneity is due to Quinn [Q2]. It asserts that generally a "manifold stratified space" will have all of its connected strata homogenous. Quinn has also established an h-cobordism theorem for these. (Steinberger had independently done the locally linear case.)

13) And even an α-approximation theorem [CF].

14) The original method for doing this was analytic, but the paper [CSW] referred to above sketches a topological approach.

15) There is a deviation at the prime 2 that we are ignoring here.

(LOCAL CONTRACTIBILITY OF THE HOMEOMORPHISM GROUP) Local contractibility is also true, according to [Si2] for locally linear orbifolds, and [Hu] for general ones.

To sum up, the definitions of orbifolds suggested parallel the possibilities of definitions of manifold and DDP homology manifold in the unequivariant setting. (With a little twist: even order locally linear orbifolds are not oriented by their signature classes, and correspond to a "manifold type category in which transversality fails".[16].) In both settings the analogues of simple homotopy theory and surgery are understood.

Regarding the local structure, one is in much better shape in the orbifold setting, assuming the strata are manifolds, ultimately because inductive arguments are possible. Interestingly enough, it is the same basic ideas that are responsible for our advances in both of these directions: the methods of controlled topology. However, as of yet, it does not seem natural to combine the detailed arguments of these two situations. It seems to me that the explicit nature of the stratification in the orbifold setting suggests methods for exploring the problems of homology manifolds at least at the level of conjecture.

On the other hand, in the case of homology manifolds, we have a very good feel of what the aggregate of local structures should be: there are a $\mathbb{Z}$'s worth of them. For the orbifold case, the algebraic problems are much more subtle: see [tD] for the underlying local homotopy theory, and, e.g. [Sh2] for some of the geometry: the part of identifying what role the linear examples play. The signature class can fail, even rationally, to be an orientation, which should mean that the failure of transversality is more striking in the orbifold setting than in the homology manifold case. Consequently, problems in group actions should be addressable by finding their concomitants in the theory of homology manifolds, and working out the easier algebra there. For instance, the stable structure calculation above, in contrast to Mazur's theorem, should lead, by analogy, to interesting phenomena in the cancellation problem for nonlinear similarity.

Finally, and most speculatively, the first method for obtaining signature classes for orbifold was by doing Lipschitz or quasiconformal analysis of signature operators. One would hope that homology manifolds, which have more algebraic signature classes, also support a suitable type of analysis: one that must be based on something other than calculus and linear approximation.

6 Some Remarks on Infinite Dimensions

One can also inquire regarding the nature of infinite-dimensional homology manifolds. These are d-dimensional homology manifolds by the homological definition, but which have infinite covering dimension.

They exist as a consequence of work of Edwards (see [Wl]) and the construction of an infinite-dimensional space of finite homological dimension by Dranishnikov [Dr]. The following begins a study of their geometrical topology

---

16) In the orbifold case, there isn't the same close connection between the signature class being an orientation and local linearity. However, in light of homogeneity, there are locally defined obstructions that tell you when you're locally linear: examine the local structures at a few strategic points!

THEOREM ([DF]). *There are infinite dimensional homology manifolds that do not have any finite-dimensional resolution. When a resolution exists, it need not be unique. However, according to a theorem of Ferry, the number of s-cobordism classes of resolutions is finite.*

This has had applications [DF] to constructing pairs of manifolds that converge to each other in Gromov-Hausdorff space, through metrics with some fixed local contractibility function.

It has also been applied to large-scale geometry in the construction of a uniformly contractible manifold with no degree one Lipschitz map to Euclidean space [DFW] and the failure of a bounded analogue of the rigidity conjecture for aspherical manifolds. Their further study promises to contain many more surprises.

References

[BW] J. Block and S. Weinberger, in preparation.

[BFMW1] J. Bryant, S. Ferry, W. Mio, and S. Weinberger, *Topology of homology manifolds*, Bull. Amer. Math. Soc. and a longer paper, to appear in Ann. of Math.

[BFMW2] J. Bryant, S. Ferry, W. Mio, and S. Weinberger, in preparation.

[Can] J. Cannon, *The characterization of topological manifolds of dimension $n \geq 5$*, Proc Internat. Congress Math. Helsinki 1978, 449–454.

[CSW] S. Cappell, J. Shaneson, and S. Weinberger, *Classes topologiques caracteristiques pour les actions de groupes sur les espaces singuliers*, Comptes Rendus 313 (1991) 293–295.

[CW] S. Cappell and S. Weinberger, *Replacement theorems for fixed sets and normal representations for PL group actions on manifolds*, to appear in the W. Browder Festschrift.

[CF] T. Chapman and S. Ferry, *Approximating homotopy equivalences by homeomorphisms*, Amer. J. Math. 101 (1979) 583–607.

[Dav] R. Daverman, Decompositions of Manifolds, Academic Press, New York and San Diego, CA, 1986.

[DW] R. Daverman and J. Walsh, *A ghastly generalized n-manifold*, Illinois J. Math. 25 (1981) 555–576.

[tD] T. tom Dieck, *Geometric representation theory of compact groups*, Proc. Internat. Congress Math., Berkeley, CA.

[Dr] A. N. Dranishnikov, *Homological dimension theory*, Russian Math. Surveys 43 (1988) 11–55.

[DF] A. N. Dranishnikov and S. Ferry, *Degenerations of manifolds in Gromov-Hausdorff space*, to appear.

[DFW] A. N. Dranishnikov, S. Ferry, and S. Weinberger, *A large flexible manifold*, to appear.

[E] R. Edwards, *The topology of manifolds and cell-like maps*, Proc. Internat. Congress Math., Helsinki, 1978, 111–127.

[FJ] F. T. Farrell and L. Jones, Classical aspherical manifolds, CBMS lecture notes # 75 (1990).

[FP] S. Ferry and E. Pederson, Epsilon Surgery Theory, to appear.

[FW] S. Ferry and S. Weinberger, *Curvature, tangentiality, and controlled topology*, Invent. Math 105 (1991) 401–414.

[Fr] M. Freedman, *The disk theorem in 4-dimensional topology*, Proc. Warsaw Internat. Congress Math., 1983, 647–663.

[GM] M. Goresky and R. MacPherson, *Intersection homology I II*, Topology 19 (1980) 135–162, Invent. Math. 72 (1983) 77–130.

[Hu] B. Hughes, *Geometric topology of stratified spaces*, preprint.

[HTWW] B. Hughes, L. Taylor, S. Weinberger, and B. Williams, *The classification of neighborhoods in a stratified space*, preprint.

[K] M. Kervaire, *Smooth homology spheres and their fundamental groups*, Trans. Amer. Math. Soc. 114 (1969) 67–72.

[KS] R. Kirby and L. Siebenmann, Foundational essays on topological manifolds, smoothings, and triangulations, Princeton University Press, Princeton, NJ, 1977.

[MR] I. Madsen and M. Rothenberg, *On the classification of G-spheres I II III*, Acta Math. 160 (1988) 65–104. Math. Scand. 64 (1989) and several related Aarhus preprints. See in particular Cont. Math. 19 (1983) 193–226 for a summary.

[MS] J. Morgan and D. Sullivan, *The transversality characteristic class and linking cycles in surgery theory*, Ann. of Math. (2) 99 (1974) 384–463.

[Ni] A. Nicas, *Induction theorems for groups of manifold homotopy structure sets*, Memoirs Amer. Math. Soc. 267 (1982).

[Q1] F. Quinn, *Resolution of homology manifolds*, Invent. Math. 72 (1983) 267–284; Corrigendum 85 (1986) 653.

[Q2] F. Quinn, *Homotopically stratified spaces*, J. Amer. Math. Soc. 1 (1988) 441–499.

[R] A. Ranicki, Algebraic L-theory and topological manifolds, Cambridge University Press, 1993.

[RsW] J. Rosenberg and S. Weinberger, *Higher G-indices on Lipschitz manifolds*, K-Theory 7 (1993) 101–132.

[RtW] M. Rothenberg and S. Weinberger, *Group actions and equivariant Lipschitz analysis*, Bull. Amer. Math. Soc. 17 (1987) 109–112.

[Sh1] J. Shaneson, these proceedings.

[Sh2] J. Shaneson, *Linear algebra, topology, and number theory*, Proc. Internat. Congress Math. Warsaw, vol. 1, 685–698.

[Si1] L. Siebenmann, *Topological manifolds*, Proc. Internat. Congress Math., Nice, 1971, vol. 2, 142–163.

[Si2] L. Siebenmann, *Deformations of homeomorphisms on stratified sets*, Comment. Math. Helv. 47 (1972) 123–163.

[StW] M. Steinberger and J. West, *Approximation by equivariant homeomorphisms, I*, Trans. Amer. Math. Soc. 302 (1987) 297–317.

[Su] D. Sullivan, Geometric topology, Part I. Localisation, periodicity, and Galois symmetry, unpublished notes, MIT 1970.

[Wa] C. T. C. Wall, Surgery on compact manifolds, Academic Press, New York and San Diego, CA, 1971.

[Wl] J. Walsh, *Dimension, cohomological dimension, and cell-like mappings*, Lecture Notes in Math. 870 (1981) 105–118.

[We] S. Weinberger, The Topological Classification of Stratified Spaces, University of Chicago Press, Chicago, Il, 1994.

Mumford-Stabilität in der algebraischen Geometrie

GERD FALTINGS

Max-Planck-Institut für Mathematik
Gottfried-Claren-Straße 26
D-53225 Bonn-Beuel, Germany

1. Einleitung

Das Thema des vorliegenden Vortrags ist etwas erklärungsbedürftig. Es bot sich an, weil der Mumfordsche Stabilitätsbegriff ganz überraschend an vielen Stellen in der mathematischen Arbeit des Verfassers auftrat. In der Annahme, daß die Organisatoren des Kongresses einen Bericht über diese erwarteten, und um die verschiedenen Themen des Vortrags einigermaßen zu bündeln, wählte der Verfasser dann die Stabilität als Oberthema. Außerdem ergibt sich damit die Gelegenheit, einmal einige allgemeine Überlegungen zum Thema zusammenzustellen. Der Verfasser entschuldigt sich vorweg dafür, daß der Akzent sehr auf den eigenen Resultaten liegt.

David Mumford hat seinen Stabilitätsbegriff ursprünglich eingeführt, um Modulräume zu konstruieren, siehe sein Buch [M]. Dabei tritt oft das folgende Problem auf:

Man erhält ein Schema X, welches das zugehörige Modulproblem zusammen mit einigen zusätzlichen Eigenschaften darstellt, wie zum Beispiel multikanonisch eingebettete Kurven im projektiven Raum oder Vektorbündel mit einer Basis der globalen Schnitte. Auf diesem Raum operiert eine Gruppe G, und der wirklich gewünschte Modulraum ist der Quotient X/G. Dabei stößt aber die Konstruktion des Quotienten oft auf Schwierigkeiten. Um diese zu behandeln, sind im wesentlichen zwei verschiedene Ansätze entwickelt worden:

Die Theorie der algebraischen Felder ("stacks") kann etwas verkürzend dahingehend beschrieben werden, daß man für viele Betrachtungen den Quotienten gar nicht braucht. Zum Beispiel kann man statt kohärenter Garben auf X/G auch einfach G-äquivariante Garben auf X betrachten, und ob der Quotient eigentlich ist, läßt sich mit Bewertungsringen testen. Dieser Ansatz ist dann sehr flexibel und wird zum Beispiel beim Beweis der Verlinde-Formel angewandt ([Fa3]). Allerdings reicht er, wie zu erwarten, nicht immer aus.

In Mumfords Theorie wird X als quasiprojektiv vorausgesetzt, mit einem G-äquivarianten amplen Geradenbündel. Man findet offene Unterschemata $X^s \subset X^{ss} \subset X$, welche die stabilen (bzw. semistabilen) Punkte darstellen. Der Quotient X^s/G läßt sich sehr gut bilden, der Quotient X^{ss}/G schon etwas schlechter, aber er hat den Vorzug, daß er für projektive X wieder projektiv ist. Dies folgt aus der geometrischen Invariantentheorie. Ein typisches Beipiel für Stabilität erhält

Proceedings of the International Congress
of Mathematicians, Zürich, Switzerland 1994

man wie folgt: Sei V ein endlichdimensionaler Vektorraum über einem Körper k, $\{F_\nu^p | 1 \leq \nu \leq r\}$ eine endliche Familie von (absteigenden) Filtrierungen von V. Das heißt, die F_ν^p sind k-Unterräume, $F_\nu^p \subset F_\nu^{p-1}$, und für genügend großes N ist $F_\nu^N = (0)$ und $F_\nu^{-N} = V$ für alle ν.

Für jeden k-Unterraum $(0) \neq W \subset V$ erhält man induzierte Filtrierungen $F_\nu^p(W) = F_\nu^p \cap W$. Definiere die Invariante $\mu(W)$ als

$$\mu(W) = \dim(W)^{-1} \sum p \cdot \dim(F_\nu^p(W)/F_\nu^{p+1}(W)).$$

Die Summe geht über alle ν und p.

Dann ist der durch die Filtrierungen $\{F_\nu^p\}$ bestimmte Punkt in dem diese klassifizierenden Produkt von Fahnenmannigfaltigkeiten genau dann stabil (bzw. semistabil), falls für alle echten Teilräume $W \subset V$ auch $\mu(W) < \mu(V)$ (bzw. $\mu(W) \leq \mu(V)$) gilt.

Der Hauptzweck dieser Note ist zu zeigen, daß diese Art von Bedingungen auch in vielen anderen Fällen auftritt, in denen man nicht unbedingt an der Konstruktion von Quotienten interessiert ist. Bevor wir dazu kommen, sei noch erwähnt, daß man manchmal die geometrische Invariantentheorie durch die direkte Konstruktion von G-invarianten Funktionen (oder Schnitten von Geradenbündeln) auf X ersetzen kann, wie dies zum Beispiel in [CF], [Fa2] oder [Fa4] geschieht. Die direkte Konstruktion benutzt in diesen Fällen verallgemeinerte Thetafunktionen. Der Begriff der Stabilität bleibt aber auch hier sehr wichtig.

2. Allgemeine Eigenschaften der Stabilität

Es sei wie bisher V ein endlichdimensionaler Vektorraum über einem Körper k, $\{F_\nu^p\}$ eine Familie von Filtierungen. Zur Vereinfachung sprechen wir im folgenden meist nur von V, wenn wir in Wirklichkeit V zusammen mit den Filtrierungen F_ν^p meinen. Diese Räume (mit fester Indexmenge $\{1, \ldots, r\}$ für die ν) bilden eine Kategorie, wobei $\mathrm{Hom}(V_1, V_2)$ aus k-linearen Abbildungen $f{:}V_1 \to V_2$ besteht, für die $f(F_\nu^p(V_1)) \subset F_\nu^p(V_2)$ gilt. Für semistabile V_1, V_2 mit $\mu(V_1) > \mu(V_2)$ verschwindet $\mathrm{Hom}(V_1, V_2)$, während für $\mu(V_1) = \mu(V_2)$ jedes filtrierte $f : V_1 \to V_2$ strikt ist, mit semistabilem Kern, Bild und Kokern.

Falls V nicht schon selbst semistabil ist, so wähle man einen nichttrivialen Unterraum W mit maximaler Invariante $\mu(W)(> \mu(V))$ und mit maximaler Dimension unter den Unterräumen mit dieser μ-Invariante. Dann setze man diese Konstruktion fort mit dem Quotienten V/W und seinen induzierten Quotientenfiltrierungen. Auf diese Weise erhält man eine aufsteigende Filtrierung

$$(0) = V_0 \subset V_1 \subset V_2 \subset \cdots \subset V_n = V,$$

so daß die Quotienten V_i/V_{i-1} semistabil sind mit Invarianten $\mu_i, \mu_1 > \mu_2 > \cdots > \mu_n$. (Kanonischerweise sollte man V_i durch μ_i parametrisieren). Diese Filtrierung heißt die Harder-Narasimhan-Filtrierung von V. Sie ist durch diese Eigenschaften eindeutig bestimmt und wird von jedem filtrierten Homomorphismus respektiert. Somit muß man zum Testen der Semistabilität von V nur Unterräume $W \subset V$ betrachten, die von allen Endomorphismen respektiert werden: Wenn für diese schon stets $\mu(W) \leq \mu(V)$ gilt, so ist die Harder-Narasimhan-Filtrierung trivial, somit V semistabil. Es folgt auch leicht daraus, daß die Harder-Narasimhan-Filtrierung invariant ist unter separablen Körpererweiterungen $k \subset k'$. In der Tat gilt dies für beliebige Körpererweiterungen:

Man reduziert auf den Fall, daß k' durch Adjunktion einer p-ten Wurzel aus k entsteht ($p =$ Charakteristik von k) und daß V schon semistabil ist. Dann erhält man eine Derivation ∂ von k' mit Kern k. Dann ist $\nabla = \partial \otimes_k 1$ ein Endomorphismus von $V' = k' \otimes_k V$, welcher die Filtrierungen respektiert und die Leibnizregel erfüllt:

$$\nabla(\lambda \cdot v) = \partial(\lambda) \cdot v + \lambda \cdot \nabla(v) \qquad (\lambda \in k', v \in V').$$

Falls $W' \subset V'$ die erste Stufe der Harder-Narasimhan-Filtrierung bezeichnet, so induziert ∇ einen filtrierten k'-linearen Homomorphismus $W' \to V'/W'$, welcher notwendigerweise trivial ist. Somit ist W' ∇-stabil, also über k definiert, und $\mu(W') \leq \mu(V) = \mu(V')$.

Trivialerweise ist auch das Duale eines semistabilen Vektorraums V wieder semistabil. Sehr viel interessanter ist das Tensorprodukt.

LEMMA *Seien V_1 und V_2 semistabil. Dann ist auch das Tensorprodukt $V_1 \otimes_k V_2$ (mit den Produktfiltrierungen) semistabil, mit $\mu(V_1 \otimes_k V_2) = \mu(V_1) + \mu(V_2)$.*

Beweis: Wir benutzen die Methoden von G. Lafaille [L]. Wir dürfen annehmen, daß für $V = V_i$ die Invariante $\mu(V)$ ganzzahlig ist. Andernfalls sei N das kleinste gemeinsame Vielfache der Nenner in $\mu(V_i)$, und man wiederholt einfach jede Filtrierung N-mal (so daß man nun insgesamt $N \cdot r$ Filtrierungen hat). Wir erweitern den Körper zu $K = k((t))$ (Laurentreihen über k) und definieren einen Automorphismus $\Phi = t^{\mu(V)} \otimes_k id$ von $K \otimes_k V$.

Sei $M \subset K \otimes_k V$ ein Gitter, d.h. ein endlich erzeugter $k[[t]]$-Untermodul, welcher eine Basis enthält. Für $1 \leq \nu \leq r$ definiert man ein neues Gitter durch

$$M^{(\nu)} = \sum t^{-p} \cdot (M \cap K \otimes_k F_\nu^p).$$

Wendet man diesen Prozeß sukzessive an für alle ν, so erhält man ein Gitter $\widetilde{M} = M^{(1)(2)\dots(r)}$. Man sieht leicht, daß $\Phi(\det(\widetilde{M})) = \det(M)$, und daß für jeden Unterraum W gilt

$$\det(M \cap W) \subseteq t^{\mu(W)\cdot\dim(W)} \cdot \det(\widetilde{M} \cap W).$$

Wir behaupten, daß V genau dann semistabil ist, wenn ein Gitter M existiert mit $\Phi(\widetilde{M}) = M$:

Die eine Richtung folgt aus obiger Betrachtung von Determinanten. Für einen über k definierten Unterraum W gilt nämlich

$$t^{\mu(V)\cdot\dim(W)} \cdot \det(\widetilde{M} \cap W) = \det(\Phi(\widetilde{M} \cap W)) \subseteq t^{\mu(W)\cdot\dim(W)} \cdot \det(\widetilde{M} \cap W),$$

und so weiter.

Für die Umkehrung sei V semistabil. Ausgehend von einem beliebigen Gitter $M = M^0$ definiert man eine absteigende Folge von Gittern M^n durch $M^{n+1} = M^n \cap \Phi(\widetilde{M^n})$. Dann erhält man eine Folge von Injektionen

$$\Phi(\widetilde{M^{n+1}})/M^{n+1} \cap \Phi(\widetilde{M^{n+1}}) \to \Phi(\widetilde{M^n})/M^n \cap \Phi(\widetilde{M^n}).$$

Da alle auftretenden $k[[t]]$-Moduln von endlicher Länge sind, sind diese Injektionen für große n Isomorphismen. Durch Ändern der Indizierung dürfen wir annehmen, daß dies schon für alle $n \geq 0$ gilt. Falls die Moduln dann verschwinden, so ist $M \subseteq \Phi(\widetilde{M})$, also $M = \Phi(\widetilde{M})$, da die Determinanten übereinstimmen. Wenn nicht, so zeigen wir, daß V nicht semistabil ist. Dazu sei $L = M^\infty$ der Durchschnitt aller M^n, $\widetilde{M}^\infty$ der Durchschnitt aller $\widetilde{M}^n$ und W' der von L aufgespannte Unterraum. Wir zeigen, daß $\mu(W') > \mu(V)$ gilt.

Man sieht leicht, daß $\widetilde{L} = \widetilde{M^\infty}$ (für ähnliche Argumente siehe [L], Beweise von Lemma 2.9 und von Lemma 3.10) und daß $\Phi(\widetilde{L})/L = \Phi(\widetilde{M^\infty})/M^\infty \cap \Phi(\widetilde{M^\infty})$ isomorph ist zu $\Phi(\widetilde{M})/(M \cap \Phi(\widetilde{M}))$, also nicht trivial. Eine Betrachtung der Determinanten liefert $\mu(W') > \mu(V)$, und V ist nicht semistabil.

Dies beendet die Konstruktion guter Gitter. Sie impliziert die Behauptung des Lemmas, da das Tensorprodukt guter Gitter wieder gut ist. Es sei bemerkt, daß man in Charakteristik 0 auch die Existenz guter hermitescher Metriken benutzen kann (siehe [FW], vereinfacht von B. Toharo. Ein neuer Beweis findet sich in [T]). Ein wirklich elementarer Beweis des Lemmas ist mir aber nicht bekannt.

Aus dem Lemma folgt die Verträglichkeit der Harder-Narasimhan-Filtrierung mit Tensorprodukten. Man kann dann (in Charakteristik 0) die Überlegung aus [Fa2] auf filtrierte halbeinfache Liealgebren anwenden, und erhält aus der Harder-Narasimhan-Filtrierung eine kanonische instabile parabolische Unteralgebra.

3. Stabilität in der diophantischen Approximation

Diese tritt zuerst in den Arbeiten von W. Schmidt (siehe [S]) auf. Dazu sei k eine Zahlkörper, $\{v_1, \ldots, v_r\}$ eine endliche Menge von Bewertungen von k. Man sucht Endlichkeitsaussagen für die Teilmenge Ω des projektiven Raums $\mathbb{P}^d(k)$, welche durch die folgende Eigenschaft charakterisiert ist:

Sei $x = (x_0, \ldots, x_d) \in \mathbb{P}^d(k)$ ein rationaler Punkt. Ferner sei für jede Stelle v_i eine endliche Anzahl homogener Polynome $\{G_{i,j}\}$ in $k[T_0, \ldots, T_d]$ gegeben (Linearformen bei Schmidt), $d_{i,j} = \operatorname{grad}(G_{i,j})$, und für jedes $G_{i,j}$ eine reelle Zahl $p_{i,j}$. Dann liegt x in Ω, falls für alle (i,j) die folgende Ungleichung gilt, wobei $\| \ \|_i$ die v_i-adische Norm bezeichnet:

$$\|G_{i,j}(x)\|_i / \|x\|^{d_{i,j}} \leq \text{Konstante} \cdot H(x)^{-p_{i,j}}. \qquad (*)$$

Natürlich ist dabei $H(x)$ die Höhe von x.

Das Wesentliche hier sind gewisse Filtrierungen der Polynomalgebra $R = k[T_0, \ldots, T_d]$. Für jede Stelle v_i definiert man nämlich eine Filtrierung F_i^p von R durch die Regel, daß $F_i^p(R) \subset R$ das Ideal ist, welches von allen Produkten von $G_{i,j}$ (mit beliebigen j) erzeugt wird, für die die Summe der $p_{i,j}$ mindestens gleich p ist. Das heißt, G liegt in $F_i^p(R)$, falls aus den Voraussetzungen an x eine Ungleichung

$$\|G(x)\|_i / \|x\|^{d_{i,j}} \leq \text{Konstante} \cdot H(x)^{-p} \qquad (*)$$

folgt.

Seit fast fünf Jahren gibt es eine neue Methode, um solche Endlichkeitsaussagen in höheren Dimensionen anzugehen, nämlich den Produktsatz ([Fa1]). Wenn man diesen hier anwenden will, so benötigt man untere Abschätzungen für die μ-Invarianten von Quotientenfiltrierungen auf Algebren $R(X)$. Dabei ist $X \subset \mathbb{P}^d$ eine über k definierte projektive Untervarietät und $R(X)$ der zugehörige Koordinatenring. Diese Abschätzungen folgen nun leicht aus der Stabilitätstheorie. Wenn zum Beispiel alle $G_{i,j}$ Linearformen sind, so hat man nur für alle i die entsprechenden Filtrierungen F_i^p auf den linearen Polynomen $V = k^{d+1}$ zu betrachten. Falls V semistabil ist, mit μ-Invariante $\mu(V) > 1$ (bei geeigneter Normierung), so sind auch alle symmetrischen Potenzen $S^q(V)$ semistabil, mit μ-Invariante $> q$, und die daraus folgende Abschätzung liefert zusammen mit dem Produktsatz die Endlichkeit von Ω (siehe [FW]).

4. Stabilität und p-adische Galoisdarstellungen

In diesem Abschnitt sei k eine perfekter Körper der Charakteristik $p > 0$, $V_0 = W(k)$ der Ring der Wittvektoren, K_0 sein Quotientenkörper und $K \supseteq K_0$ eine endliche total verzweigte Erweiterung mit ganzen Zahlen V. J.M. Fontaine hat eine Korrespondenz konstruiert zwischen bestimmten p-adischen Darstellungen von $\mathrm{Gal}(\overline{K}/K)$, und bestimmten "kristallinen" Objekten (siehe [Fo]). Diese bestehen aus einem endlichdimensionalen k_0-Vektorraum D_0, einem Frobenius-linearen Automorphismus Φ von D_0 und einer absteigenden Filtrierung $\{F^q\}$ auf $D = K \otimes_{K_0} D_0$. Mit der offensichtlichen Definition von Abbildungen bilden diese eine additive Kategorie, und Fontaines Funktor ist volltreu und respektiert Tensorprodukte und Duale. Es ist aber im allgemeinen nicht bekannt, welche Tupel $(D_0, \Phi, \{F^q\})$ als Bild auftreten können. Eine notwendige Bedingung ("faiblement admissible") ist die folgenden "schwache Zulässigkeit", welche stark an Mumfords Stabilität erinnert:

Sei $E_0 \subset D_0$ ein Φ-stabiler Unterraum. Wir definieren zwei Invarianten $\mu_\Phi(E_0)$ und $\mu_F(E_0)$ wie folgt: $\mu_\Phi(E_0)$ ist die durchschnittliche "Steigung" des Frobenius, das heißt $\dim(E_0) \cdot \mu_\Phi(E_0)$ ist die p-Potenz in der Determinante des Frobenius auf E_0. Die zweite Invariante $\mu_F(E_0)$ benutzt die induzierte Filtrierung auf $E = K \otimes_{K_0} E_0$ und ist gegeben durch

$$\mu_F(E_0) = \dim(E_0)^{-1} \sum_q q \cdot \dim(gr_F^q(E)).$$

Dann lautet die Bedingung, daß für jeden Unterraum $\mu_F(E_0) \leq \mu_\Phi(E_0)$ gilt, mit Gleichheit für $E_0 = D_0$. Falls sie verletzt ist, erhält man wieder eine Harder-Narasimhan-Filtrierung, angefangen mit einem Unterraum E_0 mit maximaler Differenz $\mu_F(E_0) - \mu_\Phi(E_0)$, und so weiter.

Man kann nun mit unseren Methoden zeigen, daß die Menge der dieser Bedingung genügenden $(D_0, \Phi, \{F^q\})$ zumindest stabil unter Tensorprodukten ist. Dies sieht man wie folgt:

Man darf annehmen, daß der Restklassenkörper k algebraisch abgeschlossen ist. Dann kann man D_0 zerlegen in eine direkte Summe

$$D_0 = \bigoplus D_{0,\alpha}\,,$$

wobei der Frobenius auf $D_{0,\alpha}$ reine Steigung $\alpha \in \mathbb{Q}$ hat. Weiter ist jedes $D_{0,\alpha}$ direkte Summe der irreduziblen Objekte U_α mit Steigung α, d.h.

$$D_{0,\alpha} = U_\alpha \otimes_K V_\alpha\,,$$

die V_α sind über $\mathbb{Q}_p$ definiert, und Φ-stabile Unterräume von D_0 sind genau die direkten Summen von $U_\alpha \otimes_K W_\alpha$, $W_\alpha \subset V_\alpha$ beliebige $\mathbb{Q}_p$-rationale Teilräume. Insbesondere ist (D_0, Φ) schon über $\mathbb{Q}_p$ definierbar, und die semistabilen Filtrierungen $\{F^p\}$ sind die K-rationalen Punkte eines rigid-analytischen Raumes $\Omega(D_0)$. $\Omega(D_0)$ ist enthalten in der entsprechenden Fahnenmannigfaltigkeit.

Wähle nun eine ganze Zahl N, so daß $N \cdot \alpha$ ganz ist für alle in $D_{0,1}$ und $D_{0,2}$ vorkommenden Steigungen α und definiere eine Operation ρ von G_m auf $D_{0,1}$ und $D_{0,2}$ durch die Regel, daß $z \in G_m$ auf $D_{0,i,\alpha}$ als $z^{N\cdot\alpha}$ operiert. Dann definieren die Invarianten unter $\rho(p^{-1}) \cdot \Phi^N$ auf $D_{0,i}$ die Struktur eines Vektorraumes über $\mathbb{Q}_{p^N}$, der unverzweigten Erweiterung von $\mathbb{Q}_p$ vom Grad N. Außerdem sind alle Φ-stabilen Unterräume von $D_{0,1} \otimes D_{0,2}$ auch ρ-stabil, und schon über $\mathbb{Q}_{p^N}$ definiert. Falls unsere Behauptung falsch ist, gibt es einen Φ-stabilen Unterraum $W \subset D_{0,1} \otimes D_{0,2}$, und einen K-rationalen Punkt in $\Omega[D_{0,1}] \times \Omega[D_{0,2}]$, so daß die Semistabilitätsbedingung für den Unterraum W und die entsprechenden Filtrierungen verletzt ist. Bei festem W definiert dies eine abgeschlossene analytische Menge in $\Omega[D_{0,1}] \times \Omega[D_{0,2}]$, welche über $\mathbb{Q}_{p^N}$ definiert ist und damit auch einen rationalen Punkt in einer endlichen Erweiterung von $\mathbb{Q}_{p^N}$ besitzt. Wir sind dann in der folgenden Situation: Es gibt eine endliche Galoiserweiterung K von $\mathbb{Q}_p$, und auf $D_i = K \otimes D_{0,i}$ $(i = 1, 2)$ Filtrierungen $\{F^q(D_i)\}$, so daß $(D_{0,i}, \Phi, \{F^q(D_i)\})$ beide schwach zulässig sind, aber nicht ihr Tensorprodukt.

Wir dürfen dabei annehmen, daß K eine N-te Wurzel $p^{1/N}$, die Einheitswurzeln $\mu_{N\cdot N^*}$ der Ordnung $N \cdot N^*$, sowie $\mathbb{Q}_{p^N}$ enthält. Dabei sei N^* eine genügend große natürliche Zahl mit der Eigenschaft, daß die Einschränkung von ρ auf $\mu_{N\cdot N^*}$ schon die verschiedenen Eigenräume von ρ in $D_{0,1}$, $D_{0,2}$ und $D_{0,1} \otimes D_{0,2}$ unterscheidet.

Nunmehr setze man den Frobeniusautomorphismus von $\mathbb{Q}_{p^N}$ zu einem Automorphismus ϕ von K fort und definiere damit semilineare Automorphismen $\phi \otimes \Phi$ auf D_1 und D_2. Sei G die von $\rho(p^{-1/N}) \cdot (\phi \otimes \Phi)$, $\rho(\mu_{N\cdot N^*})$ und Gal $(K/\mathbb{Q}_{p^N}) \otimes id$ erzeugte Untergruppe von Aut $(D_1)\times$ Aut (D_2). G ist endlich (da $\Phi^N = \rho(p)$ auf $\mathbb{Q}_p$-rationalen Punkten in $D_{o,i}$), und Unterräume von D_1, D_2 oder $D_1 \otimes_K D_2$ sind genau dann G-invariant, wenn sie von Φ-stabilen Unterräumen in $D_{0,1} \otimes \mathbb{Q}_{p^N}$, $D_{0,2} \otimes \mathbb{Q}_{P^N}$ bzw. $D_{0,1} \otimes D_{0,2} \otimes \mathbb{Q}_{P^N}$ induziert sind. Die schwache Zulässigkeit ist daher äquivalent zu einer gewichteten Semistabilität bezüglich der endlich vielen

G-Transformierten der Filtrierung $\{F^q\}$, und der Beweis unseres Lemmas liefert Gitter $M_i = \Phi(\widetilde{M_i}) \subset D_i$. Dabei ist wieder $\widetilde{M_i} = M_i^{(1)(2)\dots(r)}$, wobei nun

$$M^{(\nu)} = \sum p^{-q} \cdot (F_\nu^q \cap M_i).$$

Der Beweis der Behauptung ergibt sich nun aus der Betrachtung des Gitters $M_1 \otimes M_2$ in $V_1 \otimes V_2$. Wie bisher folgt daraus die Verträglichkeit der Harder-Narasimhan-Filtrierung (induziert durch $\mu_F - \mu_\Phi$) mit Tensorprodukten mit den üblichen Konsequenzen für halbeinfache Liealgebren.

Man kann die schwach zulässigen Tupel $(D_0, \Phi, \{F^q\})$ durch rigid-analytische Räume parametrisieren, welche ein p-adisches Analogon zu den klassifizierenden Räumen für Hodgestrukturen bilden. Besonders interessant sind darunter die Pendants der symmetrischen Räume. Das wichtigste Beispiel stammt hier von V. Drinfeld (siehe [D]) und parametrisiert gewisse Typen formaler Gruppen. Kürzlich hat H. Voskuil (in [V]) diese für beliebige p-adische reduktive Gruppen verallgemeinert. Ist G eine solche Gruppe, $P \subset G$ eine parabolische Untergruppe, $T \subset G$ ein maximaler Torus, $X = G/P$ der zugehörige homogene Raum und $\mathcal{L}$ ein homogenes amples Geradenbündel auf X, so funktioniert die Konstruktion in den Fällen, wo auf X semistabile und stabile Punkte (bezüglich der T-Operation auf $\mathcal{L}$) übereinstimmen. Insbesondere gilt dies für fast alle $\mathcal{L}$, wenn $P = B$ eine Borelgruppe ist. Der p-adische symmetrische Raum über einem lokalen Grundkörper K ist der Durchschnitt aller T-semistabilen Punkte für alle K-rationalen maximalen Tori T. Voskuil konstruiert dafür ein gutes formales Modell, dessen Singularitäten lokal durch die Toruseinbettung zum Bruhat-Tits-Gebäude von G beschrieben werden. Allerdings parametrisieren diese formalen Modelle im allgemeinen keine formalen Gruppen. Dies wird zum Beispiel in [Fa5] für die Grassmannvarietät $G(2,5)$ gezeigt.

References

[D] V. Drinfeld, *Coverings of p-adic symmetric regions*, Funct. Anal. and Appl. **10** (1976), 107–115.

[CF] C.L. Chai and G. Faltings, Degeneration of abelian varieties, Springer Verlag, Berlin 1990.

[Fa1] G. Faltings, *Diophantine approximation on abelian varieties*, Ann. of Math. **129** (1991), 549–576.

[Fa2] G. Faltings, *Stable G-bundles and projective connections*, J. of Alg. Geometry **2** (1993), 507–568.

[Fa3] G. Faltings, *A proof for the Verlinde formula*, J. of Alg. Geometry **3** (1994), 347–374.

[Fa4] G. Faltings, *Moduli-stacks for bundles on semistable curves*, Manuskript (1993), erscheint in Math. Ann.

[Fa5] G. Faltings, *Integral crystalline cohomology over very ramified valuation rings*, Manuskript (1993), eingereicht bei AMS-Journal.

[FW] G. Faltings and G. Wüstholz, *Diophantine approximations on projective spaces*, Invent.math. **116** (1994), 109–138.

[Fo] J.M. Fontaine, *Modules galoisiens, modules filtrés et anneaux de Barsotti-Tate*, Astérisque **65** (1979), 3–80.

[L] G. Lafaille, *Groupes p-divisibles et modules filtrés: Le cas peu ramifié*, Bull. Soc. math. de France **108** (1980), 187–206.

[M] D. Mumford, Geometric invariant theory, Springer-Verlag, Berlin 1965.

[S] W.M. Schmidt, Diophantine approximation, Springer Verlag LNM **785**, Berlin 1980.

[T] B. Totaro, *Tensor products of weakly admissible filtered isocrystals (as used in p-adic Hodge theory)*, preprint (1994).

[V] H. Voskuil, Ultrametric uniformization and symmetric spaces, Dissertation, Groningen 1990.

Fundamental Groups of Curves in Characteristic p

DAVID HARBATER*

Mathematics Department
University of Pennsylvania,
Philadelphia, PA 19104-6395, USA

1 Introduction

Consider the following general problem: Given a smooth affine curve U over an algebraically closed field k, find the fundamental group $\pi_1(U)$, and its set of (continuous) finite quotients $\pi_A(U)$. When $k = \mathbf{C}$, U is a Riemann surface, and π_1 can be computed using loops. If U is obtained by deleting $S = \{\xi_0, \ldots, \xi_r\}$ from a compact Riemann surface X of genus g, we thus obtain classically that π_1 has generators $a_1, \ldots, a_g, b_1, \ldots, b_g, c_0, \ldots, c_r$ subject to the single relation $\prod_{j=1}^{g}[a_j, b_j]\prod_{i=0}^{r} c_i = 1$. (Here $[a,b] = aba^{-1}b^{-1}$.) This is isomorphic to the free group on $2g+r$ generators, so $\pi_A(U)$ is the set of finite groups with $2g+r$ generators. Thus, these are the Galois groups of finite unramified Galois covers of U, or equivalently of finite branched covers of X with branch locus disjoint from U.

Over other algebraically closed fields k, loops do not make sense. But it does make sense to speak of finite unramified covers of U, and of $\pi_A(U)$. So let $U = X - S$, where X is a smooth projective k-curve of genus $g \geq 0$, and $S = \{\xi_0, \ldots, \xi_r\}$ ($r \geq 0$); we call this an affine curve of *type* (g, r). The result over $\mathbf{C}$ no longer holds if the characteristic of k is $p > 0$, e.g. because of Artin-Schreier covers of the affine line. In 1957, Abhyankar [Ab1] posed:

ABHYANKAR'S CONJECTURE (AC). *In characteristic p, if U is an affine curve of type (g, r), then a finite group G is in $\pi_A(U)$ if and only if every prime-to-p quotient of G has $2g+r$ generators.*

Equivalently, writing $p(G)$ for the subgroup of G generated by the Sylow p-subgroups, AC asserts that $G \in \pi_A(U)$ if and only if $G/p(G)$ is in π_A of a complex curve of type (g, r).

Here Abhyankar allowed $p = 0$. Later, Grothendieck showed [Gr2, XIII, Cor. 2.12] that AC holds for $p = 0$ and that π_1 of a curve of type (g, r) is the same over all algebraically closed fields of characteristic 0. This was proven by specialization techniques, as was a weak form of AC in the $p > 0$ case: that the prime-to-p part of π_1 is the same in characteristic p and in characteristic 0, and that the tame fundamental group $\pi_1^{\mathrm{t}}(U)$ over k is a quotient of π_1 of a complex curve of the same

---

*) Supported in part by NSA grant # MDA 904-92-H-3024.

Proceedings of the International Congress
of Mathematicians, Zürich, Switzerland 1994

type. (If $U = X - S$, $\pi_1^t(U)$ is defined via branched covers of X that are unramified over U and tamely ramified over S.)

Grothendieck's results imply that the forward implication of AC holds; that a prime-to-p group G is in $\pi_A(U)$ if and only if it has $2g + r$ generators; and that not all groups conjectured to be in $\pi_A(U)$ can arise from branched covers of X that are tamely ramified over S and unramified elsewhere. This suggests:

STRONG ABHYANKAR CONJECTURE (SAC). *In characteristic p, if $U = X - \{\xi_0, \ldots, \xi_r\}$ with X of genus g and $r \geq 0$, and if each prime-to-p quotient of G lies in $\pi_A(U)$, then G is the Galois group of a Galois étale cover of U whose smooth completion is tamely ramified over X except possibly at ξ_0.*

As a result of recent work of Raynaud and the author, we now have

THEOREM [Ra2, Ha6]. *SAC (and hence AC) holds for all affine curves.*

Abhyankar's Conjecture was stated in 1957, but evidence began to accumulate only about 1980. The case of $U = \mathbf{A}^1$ was considered first; there AC says that π_A consists of the *quasi-p-groups* (i.e. groups with $p(G) = G$). Nori (cf. [Ka]) and Abhyankar (cf. [Ab2]) showed that various finite groups, especially certain simple groups, lie in $\pi_A(\mathbf{A}^1)$. Later Serre [Se1] proved AC over $\mathbf{A}^1$ for solvable quasi-p-groups. Raynaud [Ra2] then showed the full AC for $\mathbf{A}^1$ using rigid analytic patching and semi-stable reduction. The author's proof of SAC [Ha6] used another form of patching, involving formal schemes, as well as relying on [Ra2].

The structure of the rest of this paper is as follows. Section 2 describes formal and rigid patching, and Section 3 sketches the proof of AC. Finally, Section 4 discusses variants and open problems.

2 Formal and rigid patching

2.1 Formal and rigid geometry. Over the complex numbers, one can construct covers with desired properties by "cutting and pasting." In the proof of AC, analogous (formal or rigid) techniques are used to handle curves in characteristic p. The point is that the Zariski topology is too weak to use in mimicking complex constructions, as there are no "small" open sets. But the formal and rigid approaches provide smaller sets that can be cut and pasted usefully. Here we work over a complete field, e.g. $K = k((t))$, which in some ways is analogous to $\mathbf{C}$.

The formal approach is based on Grothendieck's formal schemes [Gr1, EGA I, Sect. 10]. The rigid setting, due to Tate [Ta] and Kiehl [Ki], is more intuitive, but its foundations have not been worked out as thoroughly. The relationship between these two frameworks has been presented in [Ra1], [Me], [BL], and [BLR].

Consider a curve over $K = k((t))$. One can speak of metric open discs, and can attempt to do analytic geometry, in analogy with complex curves. Unfortunately, it is insufficient to use the naive approach of working with such discs and their rings of holomorphic functions, because the metric topology is totally disconnected, and so the geometry obtained would be "flabby." Instead, the rigid theory introduces a subtler notion of an affinoid set and its ring of functions. (See also [Ra2, Sect. 3].) This enables cutting and pasting that behaves more as desired.

Meanwhile, in the formal context, we begin with a curve over k, and consider "thickenings" to $R = k[[t]]$. If X is a smooth projective k-curve, then such a thickening is $X_R = X \times_k R$, with generic fibre X_K. On the other hand if $U = \mathrm{Spec}(E)$ is an affine curve, then a thickening is $U^* = \mathrm{Spec}(E[[t]])$. This is "smaller" than $U_R = U \times_k R$, which is a Zariski open subset of X_R. For example, if $X = \mathbf{P}^1_k$ and $U = \mathbf{A}^1_k$, then $U^* = \mathrm{Spec}(k[x][[t]])$ and $U_R = \mathrm{Spec}(k[[t]][x])$. As $1 - xt$ is a unit in $k[x][[t]]$ but not in $k[[t]][x]$, the point $(1 - xt)$ in U_R is missing from U^*. Geometrically, we can think of U_R as a "uniformly thick" tubular neighborhood of U, whereas U^* is a neighborhood that "pinches down" near points at infinity. (For projective curves X, there are no points at infinity, and $X^* = X_R$.) We can also consider thickenings of other subschemes of X, e.g. complete local neighborhoods $\mathrm{Spec}(\hat{\mathcal{O}}_{X,\xi})$ of any point ξ of X. In this case we obtain $\mathrm{Spec}(\hat{\mathcal{O}}_{X,\xi}[[t]])$.

For $U \subset X$, the thickening U^* is a surface whose closed fibre is U. Concerning the connection to rigid geometry, consider the generic fibre of U^*, obtained by deleting the closed fibre. This is an affine scheme $\mathrm{Spec}(A)$, where A is the ring of functions of an affinoid subset $\mathcal{U}$ of U_K. For example, if $X = \mathbf{P}^1_k$ and $U = \mathbf{A}^1_k$, then $\mathcal{U}$ is a disc about the origin. And if $X = \mathbf{P}^1_k$ and $U = \mathbf{A}^1_k - (x = 0)$, then $\mathcal{U}$ is a "corona" (annulus) whose complement has two components (one containing the point $(x = 0)$ and the other containing $(x = \infty)$). Under this correspondence, points of $\mathcal{U}$ correspond to curves in U^* not lying in the closed fibre, and two points of $\mathcal{U}$ are "close" if the corresponding curves have a high order of contact.

2.2 Patching. In the proof of Abhyankar's Conjecture, the main idea is to construct G-Galois covers over k by working inductively on the order of G, and to paste together Galois covers having smaller groups. Given an affine k-curve U, if G-Galois covers are constructed over the induced K-curve U_K (where $K = k((t))$), then a specialization argument (the "Lefschetz principle") implies that there is a G-Galois cover of U. Thus, it suffices to work over the complete field K.

Consider the following situation over $\mathbf{C}$, which we wish to mimic in characteristic p. We have a compact Riemann surface $\mathcal{X}$; a subset $\mathcal{U}_1$ obtained by deleting a small disc D; a disc $\mathcal{U}_2$ that is slightly larger than D; and the overlap $\mathcal{U}_0 = \mathcal{U}_1 \cap \mathcal{U}_2$, which is an annulus. Given a structure (a vector bundle, a branched cover, etc.) over $\mathcal{U}_1$ and $\mathcal{U}_2$ together with an agreement over $\mathcal{U}_0$, we wish to patch the data together to obtain such a structure over $\mathcal{X}$.

Analogs of these discs and annuli exist in the rigid setting. Meanwhile, in the formal setting, consider a point ξ on a smooth projective k-curve X. Let $U_1 = X - \{\xi\}$, $U_2 = \mathrm{Spec}(\hat{\mathcal{O}}_{X,\xi})$, and $U_0 = \mathrm{Spec}(\hat{\mathcal{K}}_{X,\xi})$, where $\hat{\mathcal{K}}_{X,\xi}$ is the fraction field of $\hat{\mathcal{O}}_{X,\xi}$. Then the formal analog is given by X^*, U_1^*, U_2^*, and U_0^*. Here, one can patch structures such as vector bundles or Galois covers. This is by a formal patching theorem [Ha5, Theorem 1] which is a variant on Grothendieck's Existence Theorem [Gr1, EGA III, 5.1.6], and can be regarded as a "formal GAGA":

PATCHING THEOREM [Ha5, Theorem 1]. *In the above situation, consider finite projective modules M_1 and M_2 over U_1^* and U_2^*, together with an isomorphism between the induced modules over U_0^*. Then up to isomorphism, there is a unique finite projective $\mathcal{O}_X$-module M inducing M_1 and M_2, compatibly with the identifi-*

cation over U_0^*. Moreover, this association corresponds to an equivalence of categories, and so the result carries over to finite projective algebras, and to covers.

This is proven by reducing to a local analog for modules over discrete valuation rings $\mathcal{O}$ (where projective modules are free). Set $\mathcal{K} = \mathrm{frac}(\mathcal{O})$ and $\hat{\mathcal{K}} = \mathrm{frac}(\hat{\mathcal{O}})$. The problem is to patch together free modules over $\hat{\mathcal{O}}[[t]]$ and $\mathcal{K}[[t]]$ with agreement over $\hat{\mathcal{K}}[[t]]$, and to obtain a free $\mathcal{O}[[t]]$-module inducing the given modules together with the identification. This is done by factoring the transition matrix $M \in \mathrm{GL}_N(\hat{\mathcal{K}}[[t]])$ as a product of change-of-basis matrices in $\mathrm{GL}_N(\hat{\mathcal{O}}[[t]])$ and $\mathrm{GL}_N(\mathcal{K}[[t]])$.

As an application of this patching theorem, we consider the following result, which permits the inductive construction of covers of curves. First we introduce a bit of terminology. Pick roots of unity $\{\zeta_n \mid \mathrm{char}(k) \text{ does not divide } n\} \subset k$ such that $\zeta_{nn'}^{n'} = \zeta_n$. Given a G-Galois cover of curves $Y \to X$, let $\eta \in Y$ be a ramification point lying over $\xi \in X$, with local uniformizers $y \in \hat{\mathcal{O}}_{Y,\eta}$ and $x \in \hat{\mathcal{O}}_{X,\xi}$ satisfying $y^n = x$. We call $g \in G$ the *inertial generator* at η if $g(y) = \zeta_n y$. (If $k = \mathbf{C}$, this can be interpreted via the lifting to Y of counterclockwise loops around ξ.)

COROLLARY. *Let H_1, H_2 be subgroups generating a finite group G; $Y \to X$ a connected H_1-Galois cover of k-curves with branch locus $B \subset X$; and $W \to \mathbf{P}^1$ a connected H_2-Galois cover with m branch points. Let $g \in G$ be the inertial generator at a tame point $\eta \in Y$ over $\xi \in B$, and suppose that g^{-1} is the inertial generator at a tame point $\omega \in W$ over one of the m branch points. Then there is a connected G-Galois cover $Z \to X$ that is branched at B and $m-2$ other points, and whose inertia groups over $B - \{\xi\}$ are the conjugates of those of $Y \to X$.*

To prove this result over $k = \mathbf{C}$, we induce each of the given covers up to G, by taking a disjoint union of copies of the cover, indexed by the cosets of H_i in G. We then cut out small discs around $\xi \in X$ and $\mu \in \mathbf{P}^1$, where ω lies over μ. The two disconnected G-Galois covers agree over the boundaries of the excised discs (because the two boundary orientations are opposite, and the inertial generators are g, g^{-1}); by pasting along the boundaries we obtain the desired cover $Z \to X$. Here, the base is still isomorphic to X, and the pasting can be done so that one of the new branch points coming from $W \to \mathbf{P}^1$ is now positioned at ξ.

For k of characteristic p, using formal geometry, consider the union X' of X and $\mathbf{P}^1$ crossing transversally (identifying $\xi \in X$ with $\mu \in \mathbf{P}^1$). By blowing up the point $(\xi, 0)$ on $X^* = X \times_k k[[t]]$ and pulling back by $t \mapsto t^n$ (where $n = \mathrm{ord}(g)$), we obtain an irreducible $k[[t]]$-thickening X'^* of X' with generic fibre $X_{k((t))}$, and given near the singular point by $xu = t^n$. By choosing a finite morphism $X'^* \to \mathbf{P}^1_{k[[t]]}$ and working over $\mathbf{P}^1$, we can apply the above formal patching theorem [Ha5, Theorem 1]. So there is a G-Galois cover of X'^* consisting of copies of thickenings of Y and W away from the node (first altering W to move a branch point to ∞), and copies of $\mathrm{Spec}\big(k[[y, w, t]]/(y^n - x, w^n - u, yw - t)\big)$ near the node. The generic fibre is a $k((t))$-cover with the desired properties. (Its connectivity follows from that of the closed fibre, which uses $G = H_1 H_2$.) Now apply the Lefschetz principle to obtain such a cover over k.

3 Proof of Abhyankar's Conjecture.

3.1 Outline of the proof of AC. Let k be an algebraically closed field of characteristic p. In 1990, Serre proved the following result:

THEOREM [Se1, Theorem 1]. *Let $1 \to N \to G \to H \to 1$ be an exact sequence of finite groups, with G quasi-p and N solvable. If $H \in \pi_A(\mathbf{A}^1_k)$ then so is G.*

Taking $H = 1$, we obtain Abhyankar's Conjecture for solvable groups over $\mathbf{A}^1$. For the proof, induction reduces to the case of N an elementary abelian l-group on which H acts irreducibly. Because $\mathrm{cd}(\mathbf{A}^1) = 1$ [Se1, Prop. 1], we may replace H by a subgroup of G, and so assume that the exact sequence is split. The proof proceeds cohomologically. The most difficult case is that of $l \neq p$. There, the given H-Galois cover might not be dominated by any G-Galois cover of $\mathbf{A}^1$ (i.e. the corresponding embedding problem over $\mathbf{A}^1$ might have no solution). Instead the H-Galois cover may have to be altered, before obtaining a G-Galois cover dominating it.

Using Serre's result, together with rigid patching and semi-stable reduction, Raynaud [Ra2] proved Abhyankar's Conjecture over $\mathbf{A}^1$ in 1992. That is, he showed that if G is a finite quasi-p-group, then G is a Galois group over $\mathbf{A}^1$. The proof proceeds inductively on the order of G. For P a Sylow p-subgroup of G, let $G(P)$ be the subgroup of G generated by all the proper quasi-p-subgroups $H \subset G$ such that P contains a Sylow p-subgroup of H. There are three cases: (i) G has a nontrivial normal p-subgroup; (ii) $G(P) = G$ for some P; (iii) otherwise.

Case (i) follows from Serre's result and the inductive hypothesis, since p-groups are solvable. Case (ii) uses rigid patching methods; cf. Section 3.2 below. Case (iii) uses semi-stable reduction in mixed characteristic; cf. Section 3.3.

Using Raynaud's result and formal patching, the author proved the general case of AC, including the stronger form SAC. This is discussed in Section 3.4.

3.2 Proof of AC for $\mathbf{A}^1$ in case (ii). As discussed in Section 2 above, it suffices to construct a G-Galois cover of the K-line, where $K = k((t))$. Let $G_1, \dots, G_r$ be the proper quasi-p-subgroups of G having Sylow p-subgroups contained in P. By the inductive hypothesis, each G_i is the Galois group of a cover $X_i \to \mathbf{A}^1$. Pulling back by a cover of the form $y^n = x$ and using Abhyankar's Lemma, we may assume that these G_i-Galois covers have p-groups $Q_i \subset P$ among the inertia groups over infinity. The restriction of X_i to a corona $\mathcal{C}_i$ centered at infinity is a disjoint union of copies of some Q_i-Galois cover $\mathcal{U}_i \to \mathcal{C}_i$.

Choose $r+1$ points $\sigma_1, \dots, \sigma_r, \infty \in \mathbf{P}^1_K$ together with copies of the r coronas $\mathcal{C}_i$ centered at the points σ_i. Also let $C = \mathbf{P}^1_K - \{\sigma_1, \dots, \sigma_r, \infty\}$. These points and coronas can be chosen so that the union $\mathcal{C} = \bigcup_i \mathcal{C}_i$ is disjoint and extends to a disjoint union on the corresponding discs, and so that $(C, \mathcal{C})$ is a *Runge pair* — i.e. so that $\mathbf{P}^1_K - C$ contains a point in each component of the complement of $\mathcal{C}$. Possibly after replacing K by a finite separable extension, there is a P-Galois cover $Y \to C$ whose restriction to each $\mathcal{C}_i$ is a disjoint union of copies of $\mathcal{U}_i \to \mathcal{C}_i$. (This is shown [Ra2, Cor. 4.2.6] using cohomology and induction on the order of the p-group P.) Now induce up to G, pasting each X_i to Y over $\mathcal{C}_i$. This yields a G-Galois cover, which is connected because we are in case (ii).

This case of the proof can also be shown using formal patching. See [Ha8, Application 2.2] for a discussion of this.

3.3 Proof of AC for $\mathbf{A}^1$ in case (iii). Because G is a quasi-p-group, there is a G-Galois cover $Y_K \to \mathbf{P}^1_K$ with p-power inertia groups, over a field K of characteristic 0. Here K can be chosen to be the fraction field of a complete discrete valuation ring R with residue field k. For suitable K and R, there is an R-model $Y \to X$ of this cover with semi-stable reduction and fibre $Y_k \to X_k$, such that X_k is a tree of $\mathbf{P}^1_k$'s; the inertia group I_s at each component s of Y_k is a p-group; and I_s is nontrivial unless s lies over a terminal component of the tree X_k.

Because X_k is a tree of $\mathbf{P}^1_k$'s, there is a natural partial order on the components, with the "base component" o' minimal and terminal components maximal. A partially ordered tree A of components of Y_k is constructed above it, with some o over o' minimal. It is chosen so that $G_o = G$, where for each component s of A, $G_s \subset G$ is the subgroup generated by $\{p(D_t) \mid t \text{ in } A,\ t \geq s\}$ (where D_t is the decomposition group at t and $p(\cdot)$ is as in Section 1).

Let s in A be maximal such that $G_s = G$. If $I_s \neq 1$ then a group theory argument (using the fact that we are not in case (i)) shows $G_s \subset G(P)$ for some P — contradicting $G_s = G$, as we are not in case (ii). So actually $I_s = 1$, and s is a terminal component, with $D_s = G$. Its image s' in X_k is a copy of $\mathbf{P}^1_k$. As s is a terminal component of the tree A, s' meets the rest of the graph at only one point. Deleting this point yields a G-Galois cover of the affine line.

3.4 Proof of SAC for general affine curves. This proof relies on AC for $\mathbf{A}^1$ (which in that case is equivalent to SAC). The key step is to show the result for $\mathbf{A}^1 - \{0\}$. Once that is done, the general case can be shown as follows. Under the hypotheses of SAC, let $Q = p(G)$ and $F = G/Q$. By [Gr2, XIII, Cor. 2.12], there is an F-Galois cover $U \to X$ branched only at $\{\xi_0, \dots, \xi_r\}$. Let C be an inertia group over ξ_0, with inertial generator $g \in G$ (cf. Section 2.2). By group theory, we may assume that the exact sequence $1 \to Q \to G \to F \to 1$ splits and that $E = Q \cdot C$ is a semi-direct product. Using the case of SAC for $\mathbf{A}^1 - \{0\}$, we obtain an E-Galois cover of $\mathbf{A}^1 - \{0\}$ that is tamely ramified over 0, with C an inertia group there and (after pulling back by some $x \mapsto x^j$) inertial generator g^{-1}. Because E and F generate G, the result follows from the Corollary to the Patching Theorem in Section 2.2 above.

To prove SAC for $\mathbf{A}^1 - \{0\}$, let $Q = p(G)$, let $C = G/Q$, and let P be a Sylow p-subgroup of G. Thus, C is cyclic of order n prime to p. By group theory we reduce to the case that $1 \to Q \to G \to C \to 1$ splits and $H = P \cdot C$ is a semi-direct product. By AC for $\mathbf{A}^1$, there is a Q-Galois cover $W \to \mathbf{A}^1 = \operatorname{Spec}(k[x])$. By enlarging inertia (e.g. by [Ha5, Theorem 2]), we may assume that P is an inertia group over $(x = \infty)$. By [Ha1, Cor. 2.4], there is a P-Galois cover $Y \to \mathbf{A}^1$ that agrees locally with $W \to \mathbf{A}^1$ over $\operatorname{Spec}\big(k((x^{-1}))\big)$. Using the moduli space of P-covers of the affine line [Ha1], one may construct a P-Galois cover $Z \to \mathbf{P}^1 \times \mathbf{P}^1$ of the (x,t)-space that is étale over $\mathbf{A}^1 \times \mathbf{A}^1$ and totally ramified elsewhere, whose fibre over $\mathbf{A}^1 \times (t=1)$ agrees with $Y \to \mathbf{A}^1$, and whose composition with $(x,t) \mapsto (x, t^n)$ is H-Galois [Ha6, Prop. 4.1] over the (x,s)-space $\mathbf{P}^1 \times \mathbf{P}^1$ (where $s = t^n$).

For a suitable blow-up T of the (x,s)-space, there is a covering morphism from T to (u,v)-space $\mathbf{P}^1 \times \mathbf{P}^1$ whose fibre over $(v=0)$ consists of two lines X_1 (over $s=1$) and X_2 (over $x=\infty$) crossing at a point τ. The restriction $T^* \to \mathbf{P}^1 \times \mathrm{Spec}(k[[v]])$ has generic fibre isomorphic to the s-line over $K = k((v))$. Pulling back the above H-Galois cover of the (x,s)-space to T^* and normalizing, we obtain an H-Galois cover $B^* \to T^*$. Its fibre over $X_1' = X_1 - \{\tau\}$ is isomorphic to the disconnected H-Galois cover $\mathrm{Ind}_P^H Y \to \mathbf{A}^1$ induced by $Y \to \mathbf{A}^1$. The generic fibre $B^{*o} \to T^{*o}$ is branched precisely at $(s=0)$ and $(s=\infty)$, with inertia groups C and H, respectively [Ha6, Prop. 5.1]. So the cover is unramified over X_1', and the fibre over the thickening $X_1'^*$ (cf. Section 2.1) is $\mathrm{Ind}_P^H Y^*$.

Because the covers $W \to \mathbf{A}^1$ and $Y \to \mathbf{A}^1$ agree locally over $\mathrm{Spec}(k((x^{-1})))$, their thickenings W^* and Y^* agree locally over $\mathrm{Spec}(k((x^{-1})))^*$; hence so do W^* and B^* (over $\mathrm{Spec}(\hat{\mathcal{K}}_{X_1,\tau})^*$). Since the base space T^* is fibred over $\mathbf{P}^1_{k[[v]]}$, we may apply the formal Patching Theorem in Section 2 [Ha5, Theorem 1] to $\mathrm{Ind}_H^G B^*$ and $\mathrm{Ind}_Q^G W^*$, in order to cut out copies of Y^* from B^* and paste in copies of W^*. This yields an irreducible G-Galois cover of T^*. Its general fibre is an irreducible G-Galois cover of the s-line $\mathbf{P}^1_K$ that is branched only at $(s=0)$ and $(s=\infty)$, with inertia groups C and H, respectively. This solves the problem over K, and using the Lefschetz principle we obtain SAC for $\mathbf{A}^1_k - \{0\}$.

The above proof used formal patching, but it is also possible to prove SAC for $\mathbf{A}^1_k - \{0\}$ using rigid methods. Namely, Raynaud has observed that his result on Runge pairs discussed in Section 3.2 above [Ra2, Cor. 4.2.6] can be generalized in a way that can yield the rigid analog of the above construction. See the Remark after [Ha6, Prop. 5.2] for a further discussion of this.

4 Complements and open problems

4.1 Structure of π_1. Abhyankar's Conjecture describes π_A of an affine curve of type (g,r) in characteristic p, and in particular shows that it depends only on the integers (g,r). But the fundamental group π_1 of an affine curve in characteristic p remains unknown, even for the affine line. Moreover, π_1 depends on the cardinality of the field k, as covers in characteristic p can have "moduli" (e.g. consider the family $y^p - y = tx$ of p-cyclic covers of the affine x-line, parametrized by the t-line with $(t=0)$ removed.) And even for a fixed algebraically closed field k, π_1 does not simply depend on the type (g,r). Indeed, even two affine curves of the form $\mathbf{P}^1 - \{0,1,\infty,\lambda\}$ can have nonisomorphic π_1's [Ha7, Theorem 1.8].

Also, for $U = X - \{\xi_0,\ldots,\xi_r\}$ and $G \in \pi_A(U)$, it is unknown which subgroups $G_i \subset G$ can be inertia groups over ξ_i of G-Galois branched covers of X that are étale over U. For $U = \mathbf{A}^1$, it is known that the inertia group can be taken to be a p-group (by Abhyankar's Lemma), and in general it is known that if a p-subgroup can be an inertia group then so can every larger p-subgroup [Ha5, Theorem 2]. Hence, the Sylow p-subgroups can be inertia over infinity for covers of $\mathbf{A}^1$. There is also an obvious necessary condition on a subgroup of a quasi-p-group to arise as inertia over infinity [Ha7, Prop. 1.4]. But it is unclear if this is sufficient.

4.2 Anabelian conjecture. The discussion in Section 4.1 suggests the following problem: For given values of $g, r \geq 0$, consider the moduli space $M_{g,r+1}$ of smooth

k-curves of genus g with $r+1$ points deleted. Is there a dense open subset of $M_{g,r+1}$ on which π_1 of the corresponding affine curves is constant? Or, at the other extreme, does $\pi_1(U)$ essentially determine the curve U? In particular, if $\pi_1(X_1) \approx \pi_1(X_2)$, where X_i is a curve of genus g_i with $r_i > 0$ points deleted, then must $g_1 = g_2$ and $r_1 = r_2$? Also, must X_1 and X_2 be isomorphic over the prime field? If k is the algebraic closure of a finite field, a more precise version of this question is given by Grothendieck's "anabelian conjecture," which here says:

CONJECTURE [Gr3]. *Is an affine curve X over $\mathbf{F}_q$ determined up to $\mathbf{F}_q$-isomorphism by $\pi_1(X)$ together with the surjective homomorphism $\pi_1(X) \to \mathrm{Gal}(\overline{\mathbf{F}}_q/\mathbf{F}_q)$?*

An analogous result of Nakamura [Nm] provides support for this: Two open subsets of $\mathbf{P}^1_{\mathbf{Q}}$ are isomorphic if and only if their fundamental groups are isomorphic as $\mathrm{Gal}(\overline{\mathbf{Q}}/\mathbf{Q})$-modules. Also, birational versions of the conjecture have been proven by Uchida [Uc] and Pop [Po1], and a birational version for number fields (rather than for function fields of curves, as above) is due to Neukirch [Ne].

4.3 Projective case. Although there is no conjecture describing π_A of a projective curve X of genus $g > 1$, Grothendieck [Gr2, XIII, Cor. 2.12] showed that $\pi_1(X)$ is some (unknown) quotient of π_1 of a complex curve of genus g, and he gave an explicit presentation of the maximal prime-to-p quotient of $\pi_1(X)$. Thus, $\pi_1(X)$ is finitely generated, and so is determined by $\pi_A(X)$ [FJ, Proposition 15.4].

For a given genus $g \geq 1$, curves with unequal p-rank (Hasse-Witt invariant) have distinct π_1's; while for $g \geq 2$, even the genus and p-rank do not determine π_1 (or π_A) [Kt], [Nj1]. Also, Nakajima [Nj2] has found a necessary condition for a group G to lie in $\pi_A(g) = \{G \in \pi_A(X) \mid \mathrm{genus}(X) = g\}$, viz. that the ideal $\{\sum_{\gamma \in G} a_\gamma \cdot \gamma \mid \sum a_\gamma = 0\} \subset k[G]$ has g generators.

Recently, formal and rigid patching methods (as discussed in Sections 2 and 3) have been used to obtain more information about $\pi_A(g)$. In their 1994 theses, K. Stevenson [St] and M. Saïdi [Sa] have found quotients of π_1 that are "bigger" than the profinite group on g generators. In particular, $\pi_A(g') \supset \pi_A(g)$ whenever $g' \geq g$; and $\pi_A(g)$ contains all finite groups that have g generators (e.g. all finite simple groups, if $g \geq 2$), among others.

4.4 Embedding problems. Given a finite group G, a quotient map $G \to H$, and an H-Galois unramified cover of k-curves $Y \to X$, we can ask if there is a G-Galois unramified cover $Z \to X$ inducing $Y \to X$. It is necessary that $G \in \pi_A(X)$, but this is not sufficient; cf. the proof of Serre's result on AC for solvable groups (see Section 3.1 above) in the split case with N an elementary abelian l-group, $l \neq p$.

On the other hand, if we instead permit *branched* covers, then this embedding problem can always be solved [Ha8], [Po3] using a patching construction (in fact, with some control on the additional branching). Moreover, for each such embedding problem, the cardinality of the set of solutions is equal to that of the base field k. As a result, the absolute Galois group of the function field of X is a free profinite group. This proves the function field version of a conjecture of Shafarevich: If K is a global field, then the absolute Galois group of its maximal cyclotomic extension is free profinite. This conjecture remains open in the number field case.

4.5 Other base fields. Let Φ be the class of fields K such that every finite group is the Galois group of a (geometrically irreducible) Galois branched cover of $\mathbf{P}^1_K$. It is classical that $\mathbf{C} \in \Phi$. By [Gr2, XIII, Cor. 2.12] and Abhyankar's Conjecture, every algebraically closed field is in Φ. Earlier [Ha2] this was shown (with less control on branching) by formal patching. Similarly [Ha3], the author showed that if R is the completion (or henselization) of a domain at a nonzero maximal ideal, then $K = \mathrm{frac}(R)$ is in Φ. In particular, $\mathbf{Q}_p$ and the algebraic p-adic field lie in Φ, as do $k((t))$ and the algebraic Laurent series field (for any field k).

Many other fields lie in Φ, including the fields $\mathbf{Q}^{\mathrm{tr}}$ of totally real [DF] and $\mathbf{Q}^{\mathrm{t}p}$ of totally p-adic [De] algebraic numbers, as well as PAC fields (see [FV] in the characteristic 0 case). More generally, $k \in \Phi$ (and even a stronger condition holds, concerning embedding problems [Po2, Theorem 1.5]) if k is *existentially closed* in $k((t))$, or equivalently if every geometrically irreducible k-variety with a $k((t))$-point has a k-point. (PAC fields are trivially existentially closed; $\mathbf{Q}^{\mathrm{tr}}$ and $\mathbf{Q}^{\mathrm{t}p}$ are by [GPR,1.4] and [Po2, Lemma 1.8].) The reason is that $k((t)) \in \Phi$, so there is a domain $A \subset k((t))$ of finite type over k and a G-Galois cover $Z \to \mathbf{P}^1_A$ whose k-fibres $Z_0 \to \mathbf{P}^1_k$ are irreducible. Because $A \subset k((t))$, the k-variety $\mathrm{Spec}(A)$ is geometrically irreducible, and taking a k-point yields that $k \in \Phi$.

Combining model theory with the above fact that PAC fields lie in Φ yields the following conclusion (observed by Jarden, Fried-Völklein, and Pop): If G is a finite group, then G is the Galois group of a branched cover of $\mathbf{P}^1_F$ for all but finitely many finite fields F. But it remains unknown whether finite fields lie in Φ.

Similarly, it is unknown if number fields lie in Φ. But by "rigidity," Matzat, Belyi, Thompson, Feit, Fried, Malle, Völklein et al. have realized many finite groups as Galois groups over $\mathbf{P}^1_{\mathbf{Q}}$ and hence over $\mathbf{Q}$. See [Se2, Chap. 8] for more details.

Another approach to the problem over $\mathbf{Q}$ [Ha4] used formal patching to find, for G any finite group, G-Galois (ramified) extensions of domains over $\mathbf{Z}[[t]]$ and $\mathbf{Z}\{t\} := \{f \in \mathbf{Z}[[t]] \mid f \text{ converges on } |t| < 1\}$. (These rings are analogous to $k[x][[t]]$ and $k[[t]][x]$.) Such a G-Galois extension of $\mathbf{Z}\{t\}$ induces G-Galois extensions of $\mathbf{Z}_{r+}[[t]] := \{f \in \mathbf{Z}[[t]] \mid f \text{ holomorphic on } |t| \le r\}$ for all $0 < r < 1$, and these descend to a compatible system of G-Galois extensions of the subrings $\mathbf{Z}_{r+}[[t]]^{\mathrm{h}}$ of algebraic power series. It is tempting to expect that these extensions are induced by a G-Galois extension of $\mathbf{Z}\{t\}^{\mathrm{h}}$, the ring of algebraic power series in $\mathbf{Z}\{t\}$. As $\mathbf{Z}\{t\}^{\mathrm{h}}$ is a subring of $\mathbf{Q}(t)$, this would imply that $\mathbf{Q} \in \Phi$. Unfortunately not all such systems of extensions descend to $\mathbf{Z}\{t\}^{\mathrm{h}}$, but it would suffice to have at least one such system descend for each G (cf. [Ha4]).

Given the fields that are known to be in Φ, and the expectation that number fields and finite fields are in Φ, the following conjecture seems reasonable:

CONJECTURE. *Every field lies in Φ. Hence every finite group is a Galois group over every field of the form $K(x)$, and also over every Hilbertian field.*

References

[Ab1] S. Abhyankar, *Coverings of algebraic curves*, Amer. J. Math. **79** (1957), 825–856.

[Ab2] S. Abhyankar, *Galois theory on the line in nonzero characteristic*, Bull. Amer. Math. Soc. (New Ser.) **27** (1992), 68–133.

[BL] S. Bosch and W. Lütkebohmert, *Formal and rigid geometry*, Part I: Math. Ann. **295** (1993), 291–317. Part II: Math. Ann. **296** (1993) 403–429.

[BLR] S. Bosch, W. Lütkebohmert, and M. Raynaud, *Formal and rigid geometry*, Part III, Math. Ann. **302** (1995), 1–29; Part IV, Invent. Math. **119** (1995), 361–398.

[De] P. Debes, *G-covers of* $\mathbf{P}^1$ *over the p-adics*, in: Recent Developments in the Inverse Galois Problem (S. Abhyankar, M. Fried et al., eds.), AMS Contemporary Math. series **186** (1995), 217–238.

[DF] P. Debes and M. Fried, *Nonrigid constructions in Galois theory*, Pacific J. Math. **163** (1994), 81–122.

[FJ] M. Fried and M. Jarden, Field Arithmetic, Springer-Verlag, Berlin-Heidelberg, 1986.

[FV] M. Fried and H. Völklein, *The embedding problem over a Hilbertian PAC field*, Ann. of Math. **135** (1992), 469–481.

[GPR] B. Green, F. Pop, and P. Roquette, *On Rumely's local-global principle*, manuscript 1993, Heidelberg.

[Gr1] A. Grothendieck, Eléments de géométrie algébrique (EGA), Publ. Math. IHES **4, 11** (1960, 1961).

[Gr2] A. Grothendieck, Revêtements étales et groupe fondamental (SGA 1). Lecture Notes in Math. **224**, Springer-Verlag, Berlin-Heidelberg-New York, 1971.

[Gr3] A. Grothendieck, Esquisse d'un programme, 1984 manuscript.

[Ha1] D. Harbater, *Moduli of p-covers of curves*, Comm. Algebra **8** (1980), 1095–1122.

[Ha2] D. Harbater, *Mock covers and Galois extensions*, J. Algebra **91** (1984), 281–293.

[Ha3] D. Harbater, *Galois covers of the arithmetic line*, in: Number Theory – New York 1984–85, Lecture Notes in Math. **1240** (D.V. and G.V. Chudnovsky, eds.), Springer-Verlag, New York (1987), 165–195.

[Ha4] D. Harbater, *Galois covers of an arithmetic surface*, Amer. J. Math. **110** (1988), 849–885.

[Ha5] D. Harbater, *Formal patching and adding branch points*, Amer. J. Math., **115** (1993), 487–508.

[Ha6] D. Harbater, *Abhyankar's conjecture on Galois groups over curves*, Invent. Math., **117** (1994), 1–25.

[Ha7] D. Harbater, *Galois groups with prescribed ramification*, preprint 1993, in: Arithmetic Geometry (N. Childress and J. Jones, eds.), AMS Contemporary Math. series **174** (1994), 35–60.

[Ha8] D. Harbater, *Fundamental groups and embedding problems in characteristic p*, in: Recent Developments in the Inverse Galois Problem (S. Abhyankar, M. Fried et al., eds.), AMS Contemporary Math. series **186** (1995), 353–369.

[Ka] T. Kambayashi, *Nori's construction of Galois coverings in positive characteristics*, in: Algebraic and Topological Theories (M. Nagata et al., eds.), Kinokuniya Co., Tokyo (1985), 640–647.

[Kt] H. Katsurada, *Generalized Hasse-Witt invariants and unramified Galois extensions of an algebraic function field*, J. Math. Soc. Japan **31** (1979), 101–125.

[Ki] R. Kiehl, *Der Endlichkeitssatz für eigentliche Abbildungen in der nichtarchimedischen Funktionentheorie*, Invent. Math. **2** (1967), 191–214.

[Me] F. Mehlmann, Ein Beweis für einen Satz von Raynaud über flache Homomorphismen affinoïder Algebren, Schriftenreihe Math. Inst. Univ. Münster, Ser. 2 **19** (1981).

[Nj1] S. Nakajima, *On generalized Hasse-Witt invariants of an algebraic curve*, in: Galois groups and their representations, Adv. Stud. Pure Math. **2** (Y. Ihara, ed.), 69–88, North-Holland, 1983.

[Nj2] S. Nakajima, *On Galois module structure of the cohomology groups of an algebraic variety*, Invent. Math. **75** (1984), 1–8.

[Nm] H. Nakamura, *Galois rigidity of the étale fundamental groups of punctured projective lines*, J. Reine Angew. Math. **411** (1990), 205–216.

[Ne] J. Neukirch, *Kennzeichnung der p-adischen und endlichen algebraischen Zahlkörper*, Invent. Math. **6** (1969), 269–314.

[Po1] F. Pop, *On Grothendieck's conjecture of birational anabelian geometry*, Ann. of Math. (2) **138** (1994), 145–182.

[Po2] F. Pop, *Hilbertian fields with a universal local-global principle*, manuscript 1993, Heidelberg.

[Po3] F. Pop, *Étale Galois covers of smooth affine curves*, manuscript 1994, Heidelberg, to appear in Invent. Math.

[Ra1] M. Raynaud, *Géométrie analytique rigide d'après Tate, Kiehl ...*, Mém. Soc. Math. France (New Ser.) **39–40** (1974), 319–327.

[Ra2] M. Raynaud, *Revêtements de la droite affine en caractéristique $p > 0$ et conjecture d'Abhyankar*, Invent. Math. **116** (1994), 425–462.

[Sa] M. Saïdi, *Revêtements étales et groupe fondamental de graphes de groupes*, C. R. Acad. Sci. Paris **318** Série I (1994), 1115–1118.

[Se1] J.-P. Serre, *Construction de revêtements étales de la droite affine en caractéristique p*, Comptes Rendus **311** (1990), 341–346.

[Se2] J.-P. Serre, Topics in Galois Theory, Jones and Bartlett, Boston-London, 1992.

[St] K. Stevenson, Fundamental groups of projective curves in characteristic p, University of Pennsylvania, Ph. D. thesis, 1994.

[Ta] J. Tate, *Rigid analytic spaces*, Invent. Math. **12** (1971), 257–289.

[Uc] K. Uchida, *Isomorphisms of Galois groups of algebraic function fields*, Ann. of Math. (2) **106** (1977), 589–598.

Mixed Motives, Motivic Cohomology, and Ext-Groups

UWE JANNSEN

Universität zu Köln
Mathematisches Institut
Weyertal 86, D-50931 Köln, Germany

The standard cohomology theories for algebraic varieties show some fascinatingly parallel behavior. A well-known instance is the phenomenon of "independence of ℓ" for ℓ-adic cohomology. But there are even more mysterious relations between the latter and Hodge theory. The theory of motives is the suitable setting for describing and studying these issues. Whereas Grothendieck's theory of pure motives is designed as a universal cohomology theory for smooth projective varieties and is linked to algebraic cycles, the theory of mixed motives concerns the cohomology of arbitrary varieties and was related to algebraic K-theory by Beilinson and Deligne. A new phenomenon of the mixed case is the existence of nontrivial extensions. The aim of these notes is to discuss some of these ideas and to report on the problems and a few results.

1 Pure motives and Hom

The following principle underlies the notion of weights for motives.

PRINCIPLE 1.1. *For a smooth projective variety X, $H^i(X)$ is pure of weight i.*

In this paper, we shall consider a base field k of characteristic 0 (unless stated otherwise) and the following cohomology groups:

$H^i_\sigma(X) := H^i(X^{an}_\sigma, \mathbb{Q})$ (singular cohomology), where $\sigma : k \hookrightarrow \mathbb{C}$ is any embedding and $X^{an}_\sigma = (X \times_{k,\sigma} \mathbb{C})(\mathbb{C})$ as a complex manifold,

$H^i_\ell(X) := H^i_{\text{ét}}(\overline{X}, \mathbb{Q}_\ell)$ (ℓ-adic étale cohomology), where ℓ is any prime, and $\overline{X} = X \times_k \overline{k}$ for an algebraic closure $\overline{k}$ of k,

$H^i_{dR}(X) := H^i_{\text{Zar}}(X, \Omega^\cdot_{X/k})$ (de Rham cohomology), the Zariski hypercohomology of the algebraic de Rham complex.

(1.2). The Hodge decomposition $H^i_\sigma(X) \otimes_{\mathbb{Q}} \mathbb{C} = \oplus_{p+q=i} H^{p,q}_\sigma$, $\overline{H^{p,q}_\sigma} = H^{q,p}_\sigma$, defines a pure $\mathbb{Q}$-Hodge structure of weight i on $H^i_\sigma(X)$.

(1.3). By functoriality, $G_k = Gal(\overline{k}/k)$ acts on $H^i_\ell(X)$. If k is finitely generated (as a field), then this finite-dimensional, continuous $\mathbb{Q}_\ell$-representation is pure of weight i, i.e.: there exists a ring R of finite type over $\mathbb{Z}$ having k as its field of fractions such that (1) the action of G_k factors through $Gal(L/k)$, where L is the maximal extension of k which is unramified over R, (2) for every closed point $x \in \text{Spec } R$, the eigenvalues of a corresponding geometric Frobenius

Proceedings of the International Congress
of Mathematicians, Zürich, Switzerland 1994

$Fr_x \in Gal(L/k)$ are algebraic numbers with complex absolute values $(Nx)^{i/2}$, where Nx is the cardinality of the finite residue field of x.

(1.4). By using crystalline cohomology, there is also a structure on $H^i_{dR}(X)$ leading to weights. This will not be discussed in the following.

Weights provide a grading: there are no nontrivial structure preserving morphisms between objects of different weights. Negative weights are introduced by the so-called Tate twists: for $m \in \mathbb{Z}$ and a $\mathbb{Q}_\ell$-G_k-representation V one sets $V(m) = V \otimes_{\mathbb{Q}_\ell} \mathbb{Q}_\ell(1)^{\otimes m}$, where $\mathbb{Q}_\ell(1) = (\lim\limits_{\leftarrow n} \mu_{\ell^n}) \otimes_{\mathbb{Z}_\ell} \mathbb{Q}_\ell$ as a G_k-representation (of weight -2). For a $\mathbb{Q}$-Hodge structure H one defines $H(m) = H \otimes_{\mathbb{Q}} \mathbb{Q}(1)^{\otimes m}$, where $\mathbb{Q}(1) = \mathbb{Q} \cdot (2\pi\sqrt{-1})$ is of type $(-1,-1)$, by definition.

PRINCIPLE 1.5. *The cohomology of X carries a semi-simple structure.*

In fact, $H^i_\sigma(X)$ is a polarizable Hodge structure, and hence completely decomposable. It is conjectured by Grothendieck and Serre that $H^i_\ell(X)$ is a semi-simple representation of G_k, if k is finitely generated, but this is not known in general. The idea of decomposing the cohomology into factors leads to the following

DEFINITION 1.6 *A pure motive (of weight $i - 2m$) over k is a direct factor of $H^i(X)(m)$, for X a smooth, projective variety over k and $i, m \in \mathbb{Z}$.*

To make this precise we have to specify the considered theory and factors, but it is hoped that all natural choices lead to the same answer, and that the factors in different cohomology theories correspond. For example, for the trivial factors one has:

PRINCIPLE 1.7. *Let HS be the category of $\mathbb{Q}$-Hodge structures.*
(i) $r_\sigma = \dim_{\mathbb{Q}} \mathrm{Hom}_{HS}(\mathbb{Q}, H^{2j}_\sigma(X)(j))$ should be independent of σ.
(ii) If k is finitely generated, then $r_\ell = r_\ell(k) = \dim_{\mathbb{Q}_\ell} \mathrm{Hom}_{G_k}(\mathbb{Q}_\ell, H^{2j}_\ell(X)(j))$ should be independent of ℓ, and $r_\sigma = \max\{r_\ell(k') | k'/k \text{ finite}\}$.

To establish links between the different theories, one may use the canonical comparison isomorphisms

$$I_{\infty,\sigma} : H^i_\sigma(X) \otimes_{\mathbb{Q}} \mathbb{C} \xrightarrow{\sim} H^i_{dR}(X) \otimes_{k,\sigma} \mathbb{C} \qquad \text{(G. deRham)}$$
$$I_{\ell,\overline{\sigma}} : H^i_\sigma(X) \otimes_{\mathbb{Q}} \mathbb{Q}_\ell \xrightarrow{\sim} H^i_\ell(X) \qquad \text{(M. Artin),}$$

where $\overline{\sigma} : \overline{k} \hookrightarrow \mathbb{C}$ is any embedding extending $\sigma : k \hookrightarrow \mathbb{C}$, as well as the cycle maps

$$cl_? : CH^j(X) \to H^{2j}_?(X)(j), \qquad ? = \sigma, \ell \text{ or } dR.$$

Here and in the following, $CH^j(X)$ denotes the Chow group of $\mathbb{Q}$-*linear* algebraic cycles of codimension j on X modulo rational equivalence (we shall not need integral coefficients). For $z \in CH^j(X)$, the family $(cl_\sigma(z), cl_\ell(z), cl_{dR}(z))_{\text{all } \sigma,\ell}$ is known to lie in the group $AH^j(X) = \{(x_\sigma, x_\ell, , x_{dR}) | x_\sigma \in H^{2j}_\sigma(X)(j) \cap H^{j,j}_\sigma, x_\ell \in H^{2j}_\ell(X)(\ell)^{G_k}, x_{dR} \in H^{2j}_{dR}(X), I_{\infty,\sigma}(x_\sigma) = x_{dR}, I_{\ell,\overline{\sigma}}(x_\sigma) = x_\ell\}$ of so-called absolute

Hodge cycles. Hence $CH^j(X)_{\text{hom}} = \text{Ker } cl_?$ is independent of $? = \sigma, \ell$ or dR, and via the (injective!) projections π_σ and π_ℓ we obtain a diagram

$$\begin{array}{ccccc} \text{Hom}_{HS}(\mathbb{Q}, H^{2j}_\sigma(X)(j)) & & & & \text{Hom}_{G_k}(\mathbb{Q}_\ell, H^{2j}_\ell(X)(j)) \\ \|\beta_\sigma & & & & \|\beta_\ell \\ H^{2j}(X^{an}_\sigma, \mathbb{Q}(j)) \cap H^{j,j}_\sigma & \overset{\pi_\sigma}{\hookleftarrow} & AH^j(X) & \overset{\pi_\ell}{\hookrightarrow} & H^{2j}(\overline{X}, \mathbb{Q}_\ell(j))^{G_k} \\ \text{Hodge cycles} & & \uparrow\gamma & & \text{Tate cycles} \\ \text{Hom}_{\mathcal{M}_k}(\mathbb{Q}, h(X)(j)) & = & A^j(X) & \overset{\text{def}}{=} & CH^j(X)/CH^j(X)_{\text{hom}} \end{array} \tag{1.8}$$

The identifications $\beta_?$ send morphisms f to $f(1)$, and the bottom equality holds by Grothendieck's definition of an additive category $\mathcal{M}_k$ of (sums of) pure motives over k. This category contains objects $h(X)$ for each smooth, projective variety X such that $\text{Hom}_{\mathcal{M}_k}(h(X), h(Y)) = A^d(X \times Y)$ if X is pure of dimension d, and such that composition is given by composition of correspondences. A general object can be written as a triple (X, p, j), where $p \in A^d(X \times X)$ is a projector ($p^2 = p$) and $j \in \mathbb{Z}$. One has $h(X) = (X, id, 0)$, a natural tensor law with unit $\mathbb{Q} = h(\text{Spec } k)$, and a notion of Tate twists $(X, p, j)(n) := (X, p, j + n)$. The functors $X \mapsto H_?(X) = \oplus_{i \geq 0} H^i_?(X)$ induce covariant functors on $\mathcal{M}_k$ by setting

$$H_?((X, p, j)) := pH_?(X)(j) \quad , \qquad ? = \sigma, \ell \text{ or } dR,$$

the image of p for its natural action on cohomology. In this sense, the motive (X, p, j) corresponds to "taking the factor of cohomology cut out by p and applying the j-fold Tate twist". $\mathcal{M}_k$ should be a semi-simple abelian category. This would follow from Grothendieck's standard conjectures [Kl], but at present is only known for the variant categories $\mathcal{M}_k^{\text{num}}$ and $\mathcal{M}_k^{AH}$ obtained by replacing the groups $A^\cdot(-)$ by their quotients modulo numerical equivalence and by the groups $AH^\cdot(-)$ of absolute Hodge cycles, respectively. In fact, the standard conjectures are trivially true for absolute Hodge cycles [DM], and one has:

THEOREM 1.9. *[J3] A category of motives, defined as above by quotients of Chow groups, is semi-simple abelian if and only if one takes the quotients modulo numerical equivalence.*

A priori the functors $H_?$ might not exist on $\mathcal{M}_k^{\text{num}}$, but the standard conjectures would imply that $\mathcal{M}_k$ and $\mathcal{M}_k^{\text{num}}$ are the same. They would also imply a grading $h(X) = \oplus_{i \geq 0} h^i(X)$ corresponding to the decomposition $H_?(X) = \oplus_{i \geq 0} H^i_?(X)$, giving weights on $\mathcal{M}_k$ such that $h^i(X)(m)$ has weight $i - 2m$. The equality of $\mathcal{M}_k$ and $\mathcal{M}_k^{AH}$ would follow from the Hodge conjecture (stating the bijectivity of $\pi_{id} \circ \gamma$ for $k = \mathbb{C}$) or the Tate conjecture (stating the bijectivity of $(\pi_\ell \circ \gamma) \otimes \mathbb{Q}_\ell : A^j(X) \otimes \mathbb{Q}_\ell \to H^{2j}(\overline{X}, \mathbb{Q}_\ell(j))^{G_k}$ for finitely generated k). Moreover, by (1.8) and the formula $\text{Hom}(M, N) = \text{Hom}(1, M^\vee \otimes N)$ for the natural duals and tensor products on both sides, the Hodge conjecture means that for $k = \mathbb{C}$ the Hodge realization functor $H_\sigma : \mathcal{M}_k \to HS$ is fully faithful and hence that the motivic factors correspond to the factors of $H_{id}(X)$ as a Hodge structure. Something analogous holds for the Tate conjecture, a $\mathbb{Q}_\ell$-linear version of $\mathcal{M}_k$ and the Galois factors of $H_\ell(X)$. Note however, that 1.7 (i) would already follow from the bijectivity of π_{id} for $k = \mathbb{C}$ (this is Deligne's conjecture that every Hodge cycle is absolute Hodge, and was proved by him for abelian varieties) while the first part of 1.7 (ii) would follow from the bijectivity of $\pi_\ell \otimes \mathbb{Q}_\ell$ for finitely generated k.

2. Mixed motives and Ext^1

The cohomology of nonproper or nonsmooth varieties will not in general be pure, but the following holds (cf. [D1]):

PRINCIPLE 2.1 *The cohomology of an arbitrary variety Z over k is mixed. More precisely, there exists an ascending (weight) filtration $\ldots \subseteq W_{m-1} \subseteq W_m \subseteq \ldots$ on $H^i(Z)$ such that $Gr_m^W H^i(Z) := W_m/W_{m-1}$ is pure of weight m.*

First of all, the above data $H^i_{dR}(Z), H^i_\ell(Z), H^i_\sigma(Z), I_{\infty,\sigma}, I_{\ell,\overline{\sigma}}$ also exist for arbitrary varieties (defining $H^i_{dR}(Z)$ suitably for singular Z). But now we have:

(a) By Deligne's fundamental result [D2], $V_\sigma = H^i_\sigma(Z)$ carries a mixed $\mathbb{Q}$-Hodge structure, i.e., an ascending filtration $W.$ and a descending filtration $F^{\cdot}$ on $V_\sigma \otimes_{\mathbb{Q}} \mathbb{C}$ such that $F^{\cdot}$ defines a pure Hodge structure of weight m on $Gr_m^W V_\sigma$.

(b) Using the same techniques one can show that $V_\ell = H^i_\ell(Z)$ carries an ascending G_k-equivariant filtration $W.$ such that $Gr_m^W V_\ell$ is pure of weight m [D1], [J2].

(c) Similarly, $H^i_{dR}(Z)$ carries filtrations $W.$ and $F^{\cdot}$ by k-subspaces, and

(d) the comparison isomorphisms $I_{\infty,\sigma}$ and $I_{\ell,\overline{\sigma}}$ are compatible with the occurring filtrations [J2].

It is useful to formalize these structures by defining an abstract category MR_k of mixed realizations over k consisting of data

$$V = (V_{dR}, V_\ell, V_\sigma, I_{\infty,\sigma}, I_{\ell,\overline{\sigma}})_{\ell \text{ prime}, \sigma: k \hookrightarrow \mathbb{C}, \overline{\sigma}: \overline{k} \hookrightarrow \mathbb{C}}$$

as described in (a)–(d) above, with the obvious notion of morphisms: families $(f_{dR}, f_\ell, f_\sigma)$ of maps $f_{dR} : V_{dR} \to V'_{dR}$ etc. respecting all structures, cf. [D4], [J2]. Then MR_k is an abelian category, it has an obvious tensor product $\otimes$, duals $V^{\vee}$, Tate objects $\mathbb{Q}(n)$ (having the Tate objects of Section 1 as components), and exact functors $W_m : MR_k \to MR_k$ by applying W_m in each component.

THEOREM 2.2. *(i) There are contravariant functors H^i : (varieties $/k$) $\to$ MR_k having the components described in (a)–(d).*
(ii) These functors extend to a twisted Poincaré duality theory in the sense of Bloch and Ogus [BO], with values in MR_k.

The first statement is just a reformulation of (a) - (d) above. In fact, these functors even extend to simplicial varieties $Z.$ over k, and this is an important tool for their construction and the existence of Chern classes. Statement (ii) was proved in [J2] (by using techniques of Deligne and Beilinson) and means that one has also cohomology with supports and homology in the setting of mixed realizations, with the usual functorial properties; so in particular the usual long exact cohomology sequences are compatible with the comparison isomorphisms.

DEFINITION 2.3 *A mixed motive is a successive extension of pure motives, which arises from geometry.*

There is no precise Grothendieck style definition of a category $\mathcal{MM}_k$ of mixed motives over k yet. The only (unconditional) approach so far is to define $\mathcal{MM}_k$ as a certain subcategory of MR_k or some similar category, cf. [D4], [J2]. But in contrast to the pure case there are various, possibly different candidates. Define

$$\mathcal{MM}_k^{\mathrm{var}}, \quad \mathcal{MM}_k^{\mathrm{sv}}, \quad \mathcal{MM}_k^{\mathrm{go}} \tag{2.4}$$

as the full abelian tensor subcategories of MR_k generated by the realizations of all varieties, all simplicial varieties, and all realizations of geometric origin [SM], respectively. Then $\mathcal{MM}_k^{\mathrm{var}}$ is contained in $\mathcal{MM}_k^{\mathrm{sv}}$ and $\mathcal{MM}_k^{\mathrm{go}}$, and it would be interesting to know the relation of the latter two, cf. Theorem 3.9.

The conjectural category $\mathcal{MM}_k$ should possess a realization functor $H : \mathcal{MM}_k \to MR_k$, and the above construction is based on the hope that H is fully faithful. It is further justified by the fact that for each of the three categories above the subcategory of semi-simple objects can be identified with $\mathcal{M}_k^{AH}$, by [J2] for the first two, and private communication of Saito for $\mathcal{MM}_k^{\mathrm{go}}$.

A new phenomenon in the mixed case is the existence of nontrivial extensions: the weight filtration does not split in general. A counterpart of Principle 1.7 is

PRINCIPLE 2.5. *Let $\eta : 0 \to A \to E \to B \to 0$ be exact in $\mathcal{MM}_k$, and denote by η_ℓ and η_σ the associated extensions of ℓ-adic and Hodge realizations, respectively.*
(i) η splits $\Leftrightarrow$ η_σ splits for all σ $\Leftrightarrow$ η_σ splits for one σ.
(ii) For finitely generated k, η splits $\Leftrightarrow$ η_ℓ splits for all ℓ $\Leftrightarrow$ η_ℓ splits for one ℓ.

For example, let X be a smooth projective variety, let z be a cohomologically trivial cycle of codimension j on X, and set $U = X - |z|$, where $|z|$ is the support of z. Then by Theorem 2.2 (ii) and purity one has an exact sequence in $\mathcal{MM}_k^{\mathrm{var}}$

$$0 \to H^{2j-1}(X)(j) \to H^{2j-1}(U)(j) \to H^{2j}_{|z|}(X)(j) \xrightarrow{\delta} H^{2j}(X)(j)$$

and a local cycle class $cl(z) : \mathbb{Q}(0) \to H^{2j}_{|z|}(X)(j)$. The composition of $cl(z)$ and δ gives the cycle class of z in $H^{2j}(X)(j)$, and therefore vanishes. Pulling back with $cl(z)$ we get an exact sequence

$$\eta : 0 \to H^{2j-1}(X)(j) \to E \to \mathbb{Q}(0) \to 0 \tag{2.6}$$

in $\mathcal{MM}_k^{var}$. By passing to the ℓ-adic or Hodge realizations (i.e., by doing the same in ℓ-adic or singular cohomology), we obtain extensions η_ℓ of $\mathbb{Q}_\ell$-G_k-representations and η_σ in the category MHS of mixed $\mathbb{Q}$-Hodge structures, respectively.

PROPOSITION 2.7. *(i)* [Be2], [Ca], [J2] *The class of η_σ is the image of z under the Abel-Jacobi map*

$$cl'_\sigma : CH^j(X)_{\mathrm{hom}} \to \frac{H^{2j-1}(X_\sigma^{an}, \mathbb{C})}{H^{2j-1}(X_\sigma^{an}, \mathbb{Q}(j)) + F^j} \overset{\beta_\sigma}{\cong} \mathrm{Ext}^1_{MHS}(\mathbb{Q}, H^{2j-1}_\sigma(X)(j))$$

(ii) [J2] The class of η_ℓ is the image of z under the ℓ-adic Abel-Jacobi map

$$cl'_\ell : CH^j(X)_{\mathrm{hom}} \to H^1(G_k, H^{2j-1}(\overline{X}, \mathbb{Q}_\ell(j))) \overset{\beta_\ell}{\cong} \mathrm{Ext}^1_{G_k}(\mathbb{Q}_\ell, H^{2j-1}_\ell(X)(j)) .$$

The identification β_σ was constructed independently by Carlson and Beilinson ([Be2], [Ca]) and the ℓ-adic Abel-Jacobi map is defined via (3.4) and (3.5) below. Principle 2.5 predicts that Ker cl'_σ is independent of σ and coincides with Ker cl'_ℓ for all ℓ if k is finitely generated. Because cl'_σ and $cl'_\ell(k$ f.g.) are injective for divisors, and η depends only on the class of z in $CH^j(X)$ [J2,9.8], one gets

COROLLARY 2.8. *Principle 2.5 holds for (2.6) and* $j = 1$, *and* η *splits in this case if and only if* z *is zero in* $CH^1(X)_{\text{hom}}$.

For $j > 1$ the Abel-Jacobi maps are not in general injective as examples of Mumford and Bloch show (cf. [J2]), and one is led to the following considerations.

3. Filtrations on Chow groups and motivic cohomology

Extending ideas of Bloch and Deligne, Beilinson [Be3] stated the following

CONJECTURE 3.1 *There is a descending filtration* $F^\cdot$ *on* $CH^j(X)$ *for each smooth projective variety* X *over* k *and each* $j \geq 0$ *such that*

(1) $F^0 = CH^j(X), F^1 = CH^j(X)_{\text{hom}}$,
(2) $F^\cdot$ *is respected by pull-backs* f^* *and push-forwards* f_* *for morphisms* $f : X \to Y$, *and* $F^r \cdot F^s \subseteq F^{r+s}$ *under intersection,*
(3) $Gr^\nu_F CH^j(X)$ *only depends on the motive* $h^{2j-\nu}(X)$,
(4) $F^\nu = 0$ *for* $\nu \gg 0$.

This assumes the existence of the $h^r(X)$, i.e., that the Künneth components $\pi_0, \dots, \pi_{2d}$ (d = dim X) of the diagonal $\Delta_X = id \in A^d(X \times X)$ are algebraic [Kl]. Note that by (1) and (2), the action of correspondences on $Gr^\cdot_F$ factors through homological equivalence, so $Gr^\cdot_F$ only depends on $h(X)$. The motivic background is as follows. More generally, define motivic cohomology as $H^i_{\mathcal{M}}(X, \mathbb{Q}(j)) = K_{2j-i}(X)^{(j)}_{\mathbb{Q}}$, the subspace of algebraic K-theory where all Adams operators ψ^k act as multiplication by k^j, and recall Grothendieck's isomorphism $H^{2j}_{\mathcal{M}}(X, \mathbb{Q}(j)) = K_0(X)^{(j)}_{\mathbb{Q}} \cong CH^j(X)$. There should be complexes $R\Gamma_{\mathcal{M}}(X)$ in the derived category $D^b(\mathcal{MM}_k)$ with homology objects $h^i(X)$, and isomorphisms

$$H^i_{\mathcal{M}}(X, \mathbb{Q}(j)) = \mathrm{Hom}_{D^b(\mathcal{MM}_k)}(\mathbb{Q}, R\Gamma_{\mathcal{M}}(X)(j)[i]). \tag{3.2}$$

The canonical filtration on $R\Gamma_{\mathcal{M}}(X)$ would induce a descending filtration $F^\cdot$ on $H^i_{\mathcal{M}}(X, \mathbb{Q}(j))$ and isomorphisms

$$Gr^\nu_F H^i_{\mathcal{M}}(X, \mathbb{Q}(j)) = \mathrm{Ext}^\nu_{\mathcal{MM}_k}(\mathbb{Q}, h^{i-\nu}(X)(j), \tag{3.3}$$

and the standard conjectures would imply that $F^\nu H^i_{\mathcal{M}}(X, \mathbb{Q}(j)) = 0$ for $\nu > j$ (in [J4] this was only stated for Chow groups; the general case follows similarly, using Soulé's result [Sou] that $H^i_{\mathcal{M}}(X, \mathbb{Q}(j)) = 0$ for $i > \dim X + j$). The filtration $F^\cdot$ is the motivic analogue of the filtration $F^\cdot_\ell$ induced by the Hochschild-Serre spectral sequence for continuous cohomology

$$E_2^{p,q} = H^p(G_k, H^q(\overline{X}, \mathbb{Q}_\ell(j))) \Rightarrow H^{p+q}(X, \mathbb{Q}_\ell(j)) \tag{3.4}$$

and for fin. generated k should correspond to it via the cycle and regulator maps

$$cl_\ell : CH^j(X) \to H^{2j}(X, \mathbb{Q}_\ell(j)), \quad r_\ell : H^i_{\mathcal{M}}(X, \mathbb{Q}(j)) \to H^i(X, \mathbb{Q}_\ell(j) \tag{3.5}$$

(so that $cl_\ell^{-1}(F_\ell^2) = \text{Ker } cl'_\ell = F^2CH^j(X)$, cf. the previous section).

In [J4] it was shown that 3.1 is equivalent to a conjecture of Murre [Mu] on a "Künneth decomposition" of Chow groups. In particular, there would be idempotents Π_i in $CH^d(X \times X)$ lifting the π_i such that

$$F^\nu CH^j(X) = \text{Ker } \Pi_{2j} \cap \text{Ker } \Pi_{2j-1} \cap \ldots \cap \text{Ker } \Pi_{2j-\nu+1} . \tag{3.6}$$

A related, but unconditional proposal for $F^\cdot$ was given by S. Saito [SS]. Here is yet another description. Set $F^0 = CH^j(X)$, and define F^ν for $\nu > 0$ inductively by setting

$$F^\nu CH^j(X) = \sum \text{Im}(\Gamma_* : F^{\nu-1}CH^r(Y) \to CH^j(X)), \tag{3.7}$$

where Y runs over all smooth projective varieties and Γ over all correspondences in $CH^{\dim Y - r + j}(Y \times X)_{\text{hom}}$, all $r \geq 0$. This filtration satisfies Conjecture 3.1 (1) and (2) (this becomes clear after noticing that (2) is equivalent to the condition that $F^r \circ F^s \subseteq F^{r+s}$ under composition of correspondences), and one can show that it must coincide with Beilinson's filtration given the formalism (3.2) (by using induction starting from $F^\nu = 0 (\nu \gg 0)$ and the fact that a ν-extension is the Yoneda product of a (ν-1)-extension and a 1-extension).

Conjecture 3.1 would have remarkable consequences, cf. the discussion in [J4]. In particular it predicts strong relations between the behavior of Chow groups and properties of cohomology. In recent years several authors have exhibited and studied such relations, in particular between the coniveau or level filtration of cohomology on the one hand, and the injectivity of cycle maps or representability of Chow groups on the other (cf. [Lw], [SS], and [Sch], as well as [J4] and the literature cited therein). All aspects are present in the following extreme case:

THEOREM 3.8. *Let X be a smooth, projective variety over an algebraically closed field k. Then (a) $\Leftrightarrow$ (b) $\Leftrightarrow$ (c) $\Rightarrow$ (d) for the following statements:*

(a) $CH^\cdot(X_\Omega)_{\text{hom}} = 0$ for a universal domain $\Omega \supseteq k$.

(b) The diagonal $\Delta_X \in CH^{\dim X}(X \times X)$ can be written as $\sum \alpha_j \times \beta_j$, where $\alpha_j \times \beta_j$ is the exterior product of cycles α_j and β_j on X.

(c) $CH^\cdot(X) \otimes H^\cdot_{\mathcal{M}}(Y, \mathbb{Q}(\cdot)) \xrightarrow{\sim} H^\cdot_{\mathcal{M}}(X \times Y, \mathbb{Q}(\cdot))$ for any smooth variety Y via exterior product, and $cl_\ell : CH^\cdot(X) \otimes \mathbb{Q}_\ell \xrightarrow{\sim} H^\cdot_\ell(X)$ for $\ell \neq$ char k.

(d) $H^\cdot(X)$ is generated by algebraic cycles for any Weil cohomology [Kl].

The implication (d) $\Rightarrow$ (a) would follow from Conjecture 3.1.

This is a slight extension of [Ja,3.4,3.6], where the last claim and the implications (a) $\Rightarrow$ (b) $\Rightarrow$ (d) are proved. Obviously, (c) implies (b), and from (b) one easily deduces that the motive associated to X in the category $\mathcal{M}_k^{rat}$ of so-called Chow motives (defined by Grothendieck's method, but using the full Chow groups instead of the quotients $A^\cdot(-)$) is isomorphic to $\oplus_{i \geq 0} \mathbb{Q}(-i)^{b_{2i}}$, where

$b_{2i} = \dim_{\mathbb{Q}} CH^i(X) < \infty$. This implies (a), (c) and (d); note that the functor $X \mapsto H^{\cdot}_{\mathcal{M}}(X \times Y, \mathbb{Q}(\cdot))$ only depends on the Chow motive. Property (b) holds for generalized flag manifolds G/P and more generally for the "linear" varieties introduced in [J2]: Using their inductive definition one easily shows the surjectivity of the exterior product $CH.(X) \otimes CH.(Y) \to CH.(X \times Y)$ for (not necessarily smooth or projective) varieties Y and linear varieties X. Cf. [To] for related results.

Hanamura [Ha], Levine [Lv], and Voevodsky [Vo] have proposed different constructions of triangulated categories $\mathcal{D}_k$ of mixed motives such that an analogue of (3.2) holds. The problem is to define a nice t-structure on $\mathcal{D}_k$ such that the associated heart serves as a category $\mathcal{MM}_k$ (and, optimistically, $\mathcal{D}_k = D^b(\mathcal{MM}_k)$). Another line of investigation is to verify parts of the conjectural picture in the setting of mixed realizations. This is also important for several applications, e.g., in connection with the conjectures of Beilinson-Bloch-Kato [Be1], [BK].

Concrete questions are: A. Do elements in K-theory give rise to extension classes in MR_k? B. In which sense are these motivic? C. Can one characterize motivic realizations and extensions intrinsically? As for A and B one has

THEOREM 3.9. *Let X be a smooth variety. (i) [SM] There are natural maps*

$$cl : CH^j(X) \to \mathrm{Hom}_{D^b(\mathcal{MM}_k^{go})}(\mathbb{Q}, R(a_X)_*\mathbb{Q}_X(j)[2j]),$$

where $R(a_X)_*\mathbb{Q}_X$ is a certain complex with homology $H^{\cdot}(X) \in \mathrm{ob}(\mathcal{MM}_k^{var})$. These are surjective if $\mathcal{M}_k = \mathcal{M}_k^{AH}$, e.g., if the Hodge conjecture is true.
(ii) [Hu] There are a full subcategory $\widetilde{\mathcal{MM}}$ of MR_k containing $\mathcal{MM}_k^{sv}$ and $\mathcal{MM}_k^{go}$, a triangulated category $\mathcal{D}_{\widetilde{\mathcal{MM}}}$ with a t-structure and associated heart $\widetilde{\mathcal{MM}}$, objects $R\Gamma(X)$ in $\mathcal{D}_{\widetilde{calMM}}$ with $H_t^i(R\Gamma(X)) = H^{\cdot}(X)$, and natural maps

$$H^i_{\mathcal{M}}(X, \mathbb{Q}(j)) \to \mathrm{Hom}_{\mathcal{D}_{\widetilde{\mathcal{MM}}}}(\mathbb{Q}, R\Gamma(X)(j)[i]).$$

(iii) [Sc1] There are natural maps

$$H^i_{\mathcal{M}}(X, \mathbb{Q}(j))_{\mathrm{hom}} \to \mathrm{Ext}^1_{\widetilde{\mathcal{MM}}_k^{\mathrm{var}}}(\mathbb{Q}, H^{i-1}(X)(j)).$$

Here $H^i_{\mathcal{M}}(X, \mathbb{Q}(j))_{\mathrm{hom}} = H^i_{\mathcal{M}}(X, \mathbb{Q}(j))$ for $i \neq 2j$ and the naturality in particular means that the maps are compatible with the cycle and regulator maps (3.5) and their Hodge theoretic analogues. We refer to the cited papers for more precise and somewhat stronger statements. For (i), Saito uses perverse sheaves, and for (ii), Huber uses an interesting description of K-groups and regulators via hypercoverings. Both extend Theorem 2.2 to a "derived" setting. Scholl obtained (iii) by extending Proposition 2.7 to Bloch's higher Chow groups [Bl].

As for question C, there is only a proposal for the ℓ-adic realization functor:

CONJECTURE 3.10. *(i) If k is a number field, then an ℓ-adic representation V of G_k is motivic (i.e., a subquotient of $H_\ell(M)$ for a mixed motive M) if and only if V is mixed [D3] and potentially semistable [J1], [Fo2] at all places $v \mid \ell$.*

(ii) Let X be a smooth projective variety over a number field k. Then the ℓ-adic regulator maps induce isomorphisms

$$r_\ell : H^i_{\mathcal{M}}(X, \mathbb{Q}(j)) \otimes \mathbb{Q}_\ell \to H^1_{st}(G_k, H^{i-1}(\overline{X}, \mathbb{Q}_\ell(j)))$$

where $H^1_{st}(G_k, V) = \{x \in H^1(G_k, V) | x$ is unramified at almost all places v and restricts to 0 in $H^1(k_v, V \otimes B_{st}(k_v))$ for all $v|\ell\}$.
(iii) Let k be a finitely generated field of characteristic $p > 0$, and let $\ell \neq p$ be a prime. Then an ℓ-adic representation of G_k is motivic if and only if it is mixed and the graded quotients for the weight filtration (cf. [D3]) are semi-simple.

Here $B_{st}(k_v)$ is the ring defined by Fontaine [Fo2] at the completion k_v. The "if" part of (i) is a (weaker) variant of a conjecture of Mazur and Fontaine [Fo1]. The passage from pure to mixed motives is related to (ii), since H^1_{st} classifies potentially semi-stable extensions. A variant of (ii) (with B_{st} replaced by B_{DR}), equivalent in the presence of (i), was formulated by Bloch and Kato [BK]. One may extend (i) to finitely generated fields k of characteristic 0, defining semi-stable representations for these in the spirit of (1.3). Part (iii) is related to the considerations in [J2, Section 9].

4. The category of 1-motives

Deligne defined a category of 1-motives for which many aspects of the conjectural picture can be verified.

DEFINITION 4.1. *[D2] A 1-motive over k is a morphism $u : X \to G$ of group schemes, where X is a finitely generated free $\mathbb{Z}$-module on which G_k acts discretely (regarded as an étale group scheme), and G is a semi-abelian variety over k, i.e., an extension of an abelian variety A by a torus T.*

For example, if $k = \overline{k}$ and $X = \mathbb{Z}^n$, then u corresponds to a tuple $(g_1, \ldots, g_n)$ of elements in $G(k)$. A 1-motive can be regarded as a 2-term complex of group schemes, and morphisms of 1-motives are just morphisms between such complexes.

Each object has a canonical weight filtration

$$\begin{array}{ccccccc} 0 & & 0 & & 0 & & X \\ \downarrow & \subseteq & \downarrow & \subseteq & \downarrow & \subseteq & \downarrow \\ 0 & & T & & G & & G \end{array}$$

with successive quotients $T[-1]$, $A[-1]$, X of weights -2, -1, and 0, respectively. (Thus the category (1-motives $/k$) can only form a small part of the category $\mathcal{MM}_k$.) Note that the morphism sets are finitely generated free abelian groups, and not $\mathbb{Q}$-vector spaces. It is better to compare them to a category $MR^{\mathbb{Z}}_k$ of integral mixed realizations, which is defined by adding $\mathbb{Z}$-lattices $T_\sigma \subseteq V_\sigma$ and G_k-invariant $\mathbb{Z}_\ell$-lattices $T_\ell \subset V_\ell$ to the defining data of MR_k, requiring that $I_{\ell,\overline{\sigma}}(T_\sigma) = T_\ell$.

PROPOSITION 4.2. *(i) There is a canonical, weight preserving, fully faithful realization functor $H : (1\text{-motives}/k) \to MR^{\mathbb{Z}}_k$.*

(ii) For $k = \overline{k}$, each component $H_\sigma : (1\text{-motives}/k) \to$ (mixed $\mathbb{Z}$-Hodge structures) is fully faithful.

This follows from the results in [D2, Section 10]. For the weight graded parts the components can be read off the following table

	H_ℓ	H_σ	H_{dR}
$T[-1]$	$T_\ell T$	$H_1(T_\sigma,\mathbb{Z})$	$\mathrm{Lie}(T)$
$A[-1]$	$T_\ell A$	$H_1(A_\sigma,\mathbb{Z})$	$\mathrm{Lie}(A^\natural)$
X	$\widehat{X}$	X	$X\otimes_{\mathbb{Z}} k$

where $T_\ell G = \lim\limits_{\leftarrow n} G(\overline{k})[\ell^n]$ is the ℓ-Tate module of a group scheme G, $\widehat{X} = \lim\limits_{\leftarrow n} X/\ell^n X$, and $A^\natural$ is the universal vector extension of A.

THEOREM 4.3. *(i)* $H\otimes\mathbb{Q}$: *(1-motives/k)* $\to MR_k$ *has image in* $\mathcal{MM}_k^{var}$.
(ii) For 1-motives M, M' *over a finitely generated field* k *one has isomorphisms*

$$\mathrm{Hom}(M,M')\otimes\mathbb{Z}_\ell \xrightarrow{\sim} \mathrm{Hom}_{G_k}(H_\ell(M),H_\ell(M')). \qquad (\mathit{char}\, k\neq\ell)$$

Indication of proof: As for (i), one reduces to the case that $A = \mathrm{Jac}(C)$, the Jacobian of a smooth, projective, geometrically irreducible curve C over k, and that $X = \mathbb{Z}$ and $T = \mathbb{G}_m$. Then $H(X\to G)_{\mathbb{Q}}$ is a subquotient of $H^1(\tilde{C})(1) \in \mathrm{ob}(\mathcal{MM}_k^{var})$, where the curve $\tilde{C}$ is obtained from C by contracting points $Q_1,\ldots,$ Q_n to one single point and deleting further points $P_1,\ldots,P_m$. The verification of this is largely contained in [D2] and [Br], cf. also [Sc2]. Claim (ii) is trivial on Gr_{-2}^W and Gr_0^W, and on Gr_{-1}^W follows from the Tate conjecture for abelian varieties proved by Faltings [F]. Then by some Ext-sequences it suffices to show the injectivity of the map (for an abelian variety A and a torus T')

$$\mathrm{Ext}^1(A,T')\otimes\mathbb{Z}_\ell \to \mathrm{Ext}^1_{G_k}(H_\ell(A),H_\ell(T'))\,,$$

which holds by the Mordell-Weil theorem, applied to the dual abelian variety $\widehat{A}$. In particular, 4.3 (ii) generalizes Falting's theorem [F] to semi-abelian varieties. This case has been obtained independently by F. Yan [Ya].

Finally, let us mention the following for 1-motives over number fields: the associated ℓ-adic representations are known to be potentially semi-stable [Fo2], and the finiteness of Tate-Shafarevich groups would imply that for 1-motives M_1 and M_2 any extension of $H_\ell(M_1)$ by $H_\ell(M_2)$ comes from a 1-motive.

5. Two arithmetic aspects of Ext^2

In connection with his conjectures on L-functions, Beilinson stated

CONJECTURE 5.1. *If* k *is a number field, then* $\mathrm{Ext}^\nu_{\mathcal{MM}_k} = 0$ *for* $\nu\geq 2$.

More generally it is expected that for a finitely generated field k of Kronecker dimension d (= Krull dimension of the model R as in (1.3)) the motivic cohomological dimension is d. For global fields of positive characteristic, results supporting this conjecture were obtained by Raskind [Ra], and one can show that the category described in 3.10 (iii) has cohomological dimension dim k. It is an open problem to show this for the category in 3.10 (i). For the subcategory of 1-motives one has:

PROPOSITION 5.2. *For any field* k *one has* $\mathrm{Ext}^\nu = 0$ *for* $\nu\geq 2$ *in the abelian category (1-motives/k)*$\otimes\mathbb{Q}$ *of iso-1-motives (obtained from (1-motives/k) by tensoring the morphism sets with* $\mathbb{Q}$*). If* k *is a finite field, then* $\mathrm{Ext}^\nu = 0$ *for* $\nu\geq 1$.

Sketch of proof: An argument of weights shows that the only nontrivial 2-extensions arise from $\mathrm{Ext}^2(\mathbb{Q}, \mathbb{G}_m[-1] \otimes \mathbb{Q})$ and that these extensions are Yoneda products of elements $\chi_1 \in \mathrm{Ext}^1(\mathbb{Q}, A[-1] \otimes \mathbb{Q})$ and $\chi_2 \in \mathrm{Ext}^1(A[-1] \otimes \mathbb{Q}, \mathbb{G}_m[-1] \otimes \mathbb{Q})$ for a suitable abelian variety A. Then it is easy to construct a 1-motive M with successive quotients $\mathbb{G}_m[-1], A[-1]$ and $\mathbb{Z}$ such that $M/W_{-2}M$ realizes $m\chi_1$ and $W_{-1}M$ realizes $n\chi_2$ for suitable $m, n \in \mathbb{N}$. This implies $\chi_2 \cup \chi_1 = 0$. The second claim follows from the fact that semi-abelian varieties have only finitely many points over a finite field.

In view of Theorem 3.9 (iii) the above conjecture suggests that the categories (2.4) coincide for global fields.

So far, we just have discussed motives over fields. The following definition (cf. [DN]) is of arithmetic interest. Let k be a number field and let $\mathcal{O}_k$ be its ring of integers. It is classical and a starting motivation for Arakelov theory that the formal "compactification" $\overline{\mathrm{Spec}\ \mathcal{O}_k} = \mathrm{Spec}\ \mathcal{O}_k \cup \{v|\infty\}$ (obtained by adding the archimedean places v of k) is the analogue of a smooth projective curve.

DEFINITION 5.3 *A motive over* $\overline{\mathrm{Spec}\ \mathcal{O}_k}$ *is a (mixed) motive over* k *such that*
(i) *the weight filtration of* $H_\ell(M)$ *splits over* k_v^{nr} *(the maximal unramified extension of the completion* k_v *of* k *at* v*) for all* $v \not| \ \ell \cdot \infty$,
(ii) *the mixed* $\mathbb{R}$*-Hodge structure* $H_\sigma(M) \otimes \mathbb{R}$ *is split for all* $\sigma : k \hookrightarrow \mathbb{C}$.

There is a canonical isomorphism $k^\times \otimes \mathbb{Q} \xrightarrow{\sim} \mathrm{Ext}^1_{(1-motives/k)\otimes\mathbb{Q}}(\mathbb{Q}, \mathbb{Q}(1))$ sending $x \in k^\times$ to the 1-motive $[\mathbb{Z} \to \mathbb{G}_m,\ 1 \mapsto x]$ (where we have identified $\mathbb{G}_m[-1] \otimes \mathbb{Q}$ with $H(\mathbb{G}_m[-1]) \otimes \mathbb{Q} = \mathbb{Q}(1)$ in MR_k). Let $\mathcal{MM}$ be any full subcategory of MR_k containing (1-motives/k) $\otimes\ \mathbb{Q}$.

THEOREM 5.4. *Assume* $k^\times \otimes \mathbb{Q} \to \mathrm{Ext}^1_{\mathcal{MM}}(\mathbb{Q}, \mathbb{Q}(1))$ *is an isomorphism, and let* $\mathcal{MM}_{\overline{\mathcal{O}}_k}$ *be the full subcategory of* $\mathcal{MM}$ *whose objects satisfy Definition 5.3 (i) and (ii).*
(i) There is a canonical injection $\mathrm{Ext}^2_{\mathcal{MM}_{\overline{\mathcal{O}}_k}}(\mathbb{Q}, \mathbb{Q}(1)) \hookrightarrow CH^1_{\mathrm{Ar}}(\overline{\mathcal{O}}_k)$ *into the Arakelov-Chow group of* $\overline{\mathcal{O}}_k$.
(ii) Let X *be smooth and projective of dimension* d *over* k *and assume that* $\mathcal{MM}$ *contains the extensions (2.6) for* j *and* $d-j$. *Then the associated Abel-Jacobi maps* $CH^r(X)_{\mathrm{hom}} \to \mathrm{Ext}^1_{\mathcal{MM}}(\mathbb{Q}, H^{2r-1}(X)(r))(r = j, d-j)$ *factor through* $\mathrm{Ext}^1_{\mathcal{MM}_{\overline{\mathcal{O}}_k}}$, *and the height pairing* $CH^j(X)_{\mathrm{hom}} \times CH^{d+j-j}(X)_{\mathrm{hom}} \to \mathbb{R}$ *(cf. [Be3], [Sc2]) is induced by the Abel-Jacobi maps and the pairing*

$$\mathrm{Ext}^1_{\mathcal{MM}_{\overline{\mathcal{O}}_k}}(\mathbb{Q}, H^{2j-1}(X)(j)) \times \mathrm{Ext}^1_{\mathcal{MM}_{\overline{\mathcal{O}}_k}}(H^{2j-1}(X)(j), \mathbb{Q}(1))$$
$$\xrightarrow{\text{Yoneda}} \mathrm{Ext}^2_{\mathcal{MM}_{\overline{\mathcal{O}}_k}}(\mathbb{Q}, \mathbb{Q}(1)) \hookrightarrow CH^1_{\mathrm{Ar}}(\overline{\mathcal{O}}_k) \xrightarrow{\deg_{\mathrm{Ar}}} \mathbb{R},$$

where $\deg_{\mathrm{Ar}}$ *is the Arakelov degree map. In particular,* $\mathrm{Ext}^2_{\mathcal{MM}_{\overline{\mathcal{O}}_k}}(\mathbb{Q}, \mathbb{Q}(1)) \neq 0$.

This was proved by Deninger and Nart [DN], building upon results of Scholl [Sc2]. In [DN] $\mathcal{MM}$ is assumed to be a tensor category, but this assumption can be removed. In particular, we can deduce the unconditional result

COROLLARY 5.5. $\mathrm{Ext}^2_{(1\text{-}motives/\overline{\mathcal{O}}_k)\otimes\mathbb{Q}}(\mathbb{Q}, \mathbb{Q}(1)) \neq 0$.

References

[Be1] A. Beilinson, *Higher regulators and values of L-functions*, J. Soviet Math. **30** (1985), 2036–2070.

[Be2] A. Beilinson, *Notes on absolute Hodge cohomology*, in Applications of Algebraic K-theory to Algebraic Geometry and Number Theory, Contemp. Math. **55** (1986), vol. I, Birkhäuser, Boston, MA, pp. 35–68.

[Be3] A. Beilinson, *Height pairing between algebraic cycles*, in K-Theory, Arithmetic and Geometry, Lecture Notes in Math. **1829**, Springer-Verlag, Berlin 1987, pp. 1–26.

[Bl] S. Bloch, *Algebraic Cycles and the Beilinson conjectures*, in The Lefschetz Centennial Conference, Contemp. Math. **58** (1986), vol. I, Birkhäuser, Boston, MA, pp. 65–79

[BK] S. Bloch and K. Kato, *L-functions and Tamagawa numbers of motives*, in The Grothendieck Festschrift, vol. I, Birkhäuser, Boston, MA, 1990, pp. 333–401.

[BO] S. Bloch and A. Ogus, *Gersten's conjecture and the homology of schemes*, Ann. Sci. École Norm. Sup. (4) **7** (1974), 181–202.

[Br] C. Brinkmann, *Die Andersonextension und 1-Motive*, thesis, Bonn, 1991.

[Ca] J. Carlson, *Extensions of mixed Hodge structures*, in Journées de Géométrie Algébrique d'Angers 1979, Sijthoff & Noordhoff, Alphen aan den Rijn, 1980, pp. 107–127.

[D1] P. Deligne, *Poids dans la cohomologie des variétés algébriques*, Actes ICM Vancouver 1974, vol. I, 79–85.

[D2] P. Deligne, *Théorie de Hodge III*, Publ. Math. I.H.E.S. **44** (1974), 5–78.

[D3] P. Deligne, *La conjecture de Weil II*, Publ. Math. I.H.E.S **52** (1981), 313–428.

[D4] P. Deligne, *La groupe fondamental de la droite projective moins trois points*, in Galois Groups over $\mathbb{Q}$, MSRI publications no. 16, Springer-Verlag, Berlin and New York, 1989, pp. 79–297.

[DM] P. Deligne and J. Milne, *Tannakian categories*, in Hodge Cycles, Motives and Shimura Varieties, Lecture Notes in Math. **900**, Springer-Verlag, Berlin and New York, 1982, pp. 101–228.

[DN] C. Deninger and E. Nart, *On* Ext^2 *of motives over arithmetic curves*, preprint 1993.

[F] G. Faltings, *Endlichkeitssätze für abelsche Varietäten über Zahlkörpern*, Invent. Math. **73** (1983), 349–366.

[Fo1] J. M. Fontaine, *Valeurs spéciales des fonctions L des motifs*, Sém. Bourbaki 1991-1992, exp. 751, Astérisque **206** (1992), pp. 205–249.

[Fo2] J. M. Fontaine, *Représentations p-adiques semi-stables*, in Périodes p-Adiques, Séminaire de Bures 1988, Astérisque **223** (1994), pp. 113–184.

[Ha] M. Hanamura, *Mixed motives and algebraic cycles*, preprint 1995.

[Hu] A. Huber, Mixed motives and their realization in derived categories, Lecture Notes in Math. **1604**, Springer-Verlag, Berlin, 1995.

[J1] U. Jannsen, *On the ℓ-adic cohomology of varieties over number fields and its Galois cohomology,* in Galois groups over $\mathbb{Q}$, MSRI Publications, no. 16, Springer-Verlag, Berlin and New York, 1989, pp. 315–360.

[J2] U. Jannsen, Mixed motives and algebraic K-theory, Lecture Notes in Math. **1400**, Springer-Verlag, Berlin 1990.

[J3] U. Jannsen, *Motives, numerical equivalence, and semi-simplicity*, Invent. Math. **107** (1992), 447–452.

[J4] U. Jannsen, *Motivic sheaves and filtrations on Chow groups*, in Motives, Proc. Sympos. Pure Math. **55** (1994), vol. I, 245–302.

[Kl] S. Kleiman, *The standard conjectures*, in Motives, Proc. Sympos. Pure Math. **55** (1994), vol. I, 3–20.

[Lv] M. Levine, *Motivic cohomology and algebraic cycles: A categorical construction I*, preprint 1993.

[Lw] J. Lewis, *Towards a generalization of Mumford's theorem*, J. Math. Kyoto Univ. **29** (1989), 195–204.

[Mu] J. Murre, *On a conjectural filtration on the Chow groups of an algebraic variety*, Indag. Math. **4** (1993), 177–201.

[Ra] W. Raskind, *Higher ℓ-adic Abel-Jacobi maps and filtrations on Chow groups*, Duke Math. J., to appear.

[SM] M. Saito, *On the formalism of mixed sheaves*, preprint RIMS-784, 1991.

[SS] S. Saito, *Motives and filtrations on Chow groups*, preprint 1992.

[Sch] C. Schoen, *On Hodge structures and non-representability of Chow groups*, Compositio Math. **88** (1993), 285–316.

[Sc1] A. Scholl, *Remarks on special values of L-functions*, in L-functions and Arithmetic, Cambridge Univ. Press, Cambridge, 1991, pp. 373–392.

[Sc2] A. Scholl, *Height pairings and special values of L-functions*, in Motives, Proc. Sympos. Pure Math. **55** (1994), vol. I, 571–598.

[Sou] C. Soulé, *Opérations en K-théorie algébrique*, Canad. J. Math. **37** (1985), 488–550.

[To] B. Totaro, *Chow groups, Chow cohomology, and linear varieties*, preprint 1994.

[Vo] V. Voevodsky, *On triangulated categories of mixed motives*, preprint.

[Ya] F. Yan, *Tate property and isogeny estimate for semi-abelian varieties*, thesis, ETH Zürich, 1994.

Rational Curves on Algebraic Varieties

Yoichi Miyaoka

RIMS, Kyoto University
Kyoto 606-01, Japan

Introduction

The aim of this article is to give a brief review on recent developments in the theory of embedded rational curves, which the author believes is a new, useful viewpoint in the study of higher dimensional algebraic varieties. By an *embedded rational curve*, or simply a *rational curve*, on a variety X, we mean the image of the projective line $\mathbb{P}^1$ by a nontrivial morphism to X, hence complete, one-dimensional, but not necessarily smooth.

The topics covered here are summarized into the following two mottos:

- *Find tractable criteria for the existence and/or nonexistence of (sufficiently many) rational curves on a given variety.*
- *Study the structure of varieties with sufficiently many rational curves on them.*

Roughly speaking, the existence of rational curves on a given variety X more or less measures the complexity of X. A very general algebraic variety, which usually has complicated structure, tends to be hyperbolic in the sense of Kobayashi [17] and, in particular, carries no rational (or elliptic) curve. On the contrary, when a variety has simple birational structure, then we can expect to find rational curves on it. If, furthermore, it contains sufficiently many rational curves, then the locus of the rational curves passing through a given general point will extract a simple factor out of the variety, thereby providing us significant information on its geometry.

This principle will be explained in Sections 1 and 2 and two applications will be discussed in Section 3.

Because of the lack of space to elaborate, the reader is referred to the comprehensive treatise [19] by Kollár for details.

For simplicity of the statements, varieties are assumed to be defined over the complex numbers throughout this article. Note, however, that we need reductions modulo p to obtain several results below.

1 Existence of rational curves

Unfortunately, it is categorically impossible to find an ideal characterization of the varieties that carry rational curves, because such a property is not invariant under smooth deformation. For example, a general hypersurface of high degree, say

Proceedings of the International Congress
of Mathematicians, Zürich, Switzerland 1994

$d \geq 50$, in $\mathbb{P}^3$ is known to be hyperbolic (Brody-Green [4]), but it deforms to the Fermat hypersurface $X_0^d + X_1^d + X_2^d + X_3^d = 0$, which clearly contains several lines. Two varieties that are mutually deformations of the other are indistinguishable by means of ordinary invariants, and we have to give up the idea to get a perfect and practical test for the existence of embedded rational curves.

Thus, we will focus on more feasible targets:

(A) *sufficient conditions for varieties to carry rational curves,*

(B) *a criterion for the existence of a sufficiently large family of rational curves*, and

(C) *sufficient conditions for varieties not to carry too many rational curves,* where each condition should fit in with the framework of birational classification theory. Towards these objectives, the most powerful method ever known is the *bend-and-break* technique discovered by Mori [30], the essence of which is the following:

If $f : C \to X$, a morphism from a smooth projective curve to a smooth projective variety, has sufficiently many deformations $\{f_s : C \to X\}$, then we can find a rational curve in some limit of the $\{f_s(C)\}$. If $\deg f^*(-K_X) > 0$, we can produce many deformations of f after taking modulo p reductions and replacing f by the composite of f with geometric Frobenius of high degree, where K_X denotes the canonical divisor.

This shows that a smooth projective variety contains a rational curve whenever its canonical divisor is not nef, thus giving an answer to Problem (A) above:

THEOREM 1.1 (Mori [30], Miyaoka-Mori [29]). *Let X be a smooth projective variety with ample divisor H and let K_X denote the canonical divisor. Assume that there exists an irreducible curve Γ such that $(\Gamma, -K_X) > 0$. Then, for each point $x \in \Gamma$, there is a rational curve $C \subset X$ passing through x such that the degree (C, H) is bounded from above by $\frac{2(\dim X)(\Gamma,H)}{(\Gamma,-K_X)}$ and that $0 < (C, -K_X) \leq \dim X + 1$.*

The bend-and-break technique, and hence the proof of (1.1), hinge on the algebraicity of the variety, because (1) the proof uses reductions to positive characteristics in order to produce sufficiently many deformations, and (2) the compactness of the space of cycles, which guarantees the existence of limits of curves $\{f_s(C)\}$, fails in the category of compact complex manifolds.

An immediate corollary to (1.1) is a characterization of varieties with sufficiently many rational curves, or a solution of Problem (B):

THEOREM 1.2 (Miyaoka-Mori [29]). *A smooth projective variety X is uniruled if and only if there is a family of irreducible curves $\{\Gamma_s\}_{s\in S}$ on X such that*
(1.2.1) $\bigcup \Gamma_s$ *contains an open dense subset of X, and that*
(1.2.2) $(\Gamma_s, -K_X) > 0$.

Recall that a variety X is said to be *uniruled* if there is a family of (possibly singular and reducible) rational curves $\{C_t\}$ on X such that $\bigcup C_t = X$. By definition, a variety consisting of a single point is not uniruled.

Uniruledness is respected by dominant, generically finite rational maps and by smooth, projective (or Kähler) deformation (Fujiki [11], Levine [25]). Uniruled varieties do not have minimal models, and their Kodaira dimension is $-\infty$. The converses are also believed to be true. Namely:

- The minimal model program, completed in dimension three by Mori [32], presumes that a variety without a minimal model will carry a fibre space structure, a general fibre being a $\mathbb{Q}$-Fano variety of positive dimension (called a "Mori fibre space", see Clemens-Kollár-Mori [8]). Here by a $\mathbb{Q}$-*Fano variety* we mean a normal projective variety with only terminal singularities (for the definition, see Reid [38]) and with ample anti-canonical divisor. An arbitrary $\mathbb{Q}$-Fano variety of positive dimension is uniruled, and hence so is a Mori fibre space, see (1.3) below.
- The abundance conjecture, verified up to dimension three (Miyaoka [27], [28], Kawamata [16]), asserts that the class of the varieties without minimal models (or, conjecturally, the uniruled varieties) will be identical with that of the varieties with Kodaira dimension $-\infty$.

As an application of (1.2), we can show that a class of varieties including $\mathbb{Q}$-Fano varieties are uniruled, by looking at suitable smooth models:

COROLLARY 1.3 (Kollár cited in [31] when X is smooth; Miyaoka-Mori [29] for general case). *Let X be a projective variety such that the dualizing sheaf ω_X is invertible in codimension 1. If ω_X^{-1} has positive degree on one-dimensional general complete intersections (on which ω_X is invertible), then X is uniruled. In particular, every irreducible (possibly nonnormal) hypersurface $X \subset \mathbb{P}^{n+1}$ of degree $\leq n+1$ is uniruled.*

The uniruledness results (1.2), (1.3) are stated in terms of the degree of the canonical divisor, which amounts to the positivity of the integration of the Ricci form over curves. There is a refinement, which uses a sort of non-seminegativity of tangent bundles, or of the bisectional curvature:

THEOREM 1.4 (Miyaoka [26], Shepherd-Barron's article in [20]). *Let X be a smooth projective variety, T_X the tangent bundle, and $0 \neq F \subseteq T_X$ a subsheaf. Let $\{\Gamma_s\}$ be a family of irreducible curves on X such that (a) $\bigcup \Gamma_s$ contains an open dense subset $\subseteq X$ and (b) $T_X \otimes \mathcal{O}_{\Gamma_s}/F \otimes \mathcal{O}_{\Gamma_s}$ is locally free on Γ_s. If $(\Gamma_s, \det F) > 0$, then X is uniruled.*

This theorem is usually stated in the following formulation:

COROLLARY 1.5. *If a smooth projective variety X is not uniruled, then the cotangent sheaf Ω_X^1 is semi-positive if restricted to a general complete intersection curve by high multiples of ample divisors.*

Note that the above results (1.1)–(1.5) give no information when a variety X has nef canonical divisor (i.e. X is a minimal variety). The case where X has trivial canonical class is of special interest.

Smooth projective varieties (or compact Kähler manifolds) with trivial canonical divisors are, up to finite étale covering, products of abelian varieties (or complex tori), symplectic manifolds, and **SU**-manifolds (Bogomolov [2], Beauville [1]).

Smooth deformation preserves this decomposition. Though complex tori do not contain rational curves, all known examples of projective symplectic manifolds and **SU**-manifolds (which are automatically projective) contain (perhaps infinitely many) rational curves, thus suggesting the following

QUESTION 1.6. *Does every simply connected, smooth, projective variety with trivial canonical class carry (infinitely many) rational curves? Is the union of the rational curves Zariski dense on it?*

The answer to the first half of this question is affirmative in dimension 2: an arbitrary algebraic K3 surface contains a rational curve as well as a one-dimensional family of elliptic curves (Bogomolov and Mumford, see Mori-Mukai [34]). It is plausible that the second half also holds true in this case. In dimension 3, the question is closely related to the so-called *mirror symmetry* for Calabi-Yau 3-folds. A standard conjecture (Candelas, Green, Morrison, etc.) predicts that deformation data of the mirror manifold compute the number of rational curves of given degree (with respect to a fixed polarization) on the original Calabi-Yau 3-fold. If true, the conjecture would imply that the number of the rational curves on a Calabi-Yau manifold would be infinite and it is rather likely that the union of such curves is Zariski dense on it.

In case (1.6) is affirmatively solved, it would follow that there are always lots of nonconstant holomorphic mappings $\mathbb{C} \to X$ for a projective manifold X with trivial canonical class (the image of such a mapping might not necessarily be an algebraic subvariety in X). Then, by virtue of the Iitaka fibration, the same will then hold true for algebraic varieties of Kodaira dimension 0 through $\dim X - 1$. The varieties with Kodaira dimension $-\infty$ are, conjecturally, uniruled varieties, hence containing a large family of rational curves. In short, if the Kodaira dimension of a variety is not maximal, many holomorphic mappings from $\mathbb{C}$ to X are expected to exist, another potential answer to (a variant of) Problem (A).

When the Kodaira dimension is maximal, the situation is supposed to be quite different:

CONJECTURE 1.7 (Lang). *Let X be a smooth projective variety of general type, i.e. the Kodaira dimension is equal to the complex dimension. Then there exists a proper closed algebraic subset $Y \subset X$ such that every nonconstant holomorphic mapping $\mathbb{C} \to X$ factors through $\mathbb{C} \to Y$. In particular, the locus of the rational curves on X is contained in Y and is not Zariski dense.*

This conjecture, which is not yet proved even for surfaces without an extra condition (Bogomolov [2]), would settle Problem (C) above if verified.

2 Structure of uniruled varieties

We want now to study the structure of uniruled varieties, whose characterization was given by (1.2).

Let us first review the theory of surfaces of Kodaira dimension $-\infty$. The Enriques classification tells us that an algebraic surface X with Kodaira dimension $-\infty$ is either rational or a fibre space over a curve of genus ≥ 1, a generic fibre being $\mathbb{P}^1$. These two cases are distinguished from each other by the dimension (0 or 1) of the image of the Albanese map. In the former case, every two points on X can be connected by a chain of rational curves on X, while in the latter, two points can be joined by rational curves if and only if they sit in the same fibre. Thus, the structure of the locus of embedded rational curves separates rational surfaces from irrational ruled surfaces.

In order to generalize this observation to higher dimensions, we introduce the notion of *rationally connected varieties.* A smooth projective variety X is said to be *rationally connected* if the following four equivalent conditions are satisfied (cf. Kollár-Miyaoka-Mori [23]):

(2.0.1) *Two general points $x, y \in X$ are connected by a chain of rational curves on X.*

(2.0.2) *Two general points $x, y \in X$ are connected by a single rational curve on X.*

(2.0.3) *Every two points $x, y \in X$ are connected by a chain of rational curves on X.*

(2.0.4) *There is an irreducible rational curve $C \subset X$ such that the restriction of the tangent bundle T_X to C is ample.*

In general, a variety is said to be rationally connected if a nonsingular projective model is rationally connected. As convention, a single point is understood to be rationally connected. A rationally connected variety of positive dimension is uniruled.

In view of (2.0.1) or (2.0.2), rational connectedness is respected by dominant rational maps (and, in particular, by birational equivalence). Furthermore, it turns out to be invariant under smooth deformation by (2.0.3) and (2.0.4).

A typical example of rationally connected varieties are *unirational varieties* or, equivalently, the images of projective spaces by rational maps.

Despite its elementary definition, unirationality is hard to deal with in the framework of the classification theory. To begin with, it is extremely hard to check, unless by chance you find an explicit dominant rational map from $\mathbb{P}^N$ to the variety in question, or unless the variety has a very special property (compare, for instance, Morin [35], Ramero [37], Campana-Flenner [6], Ebihara [10] for several sufficient conditions for unirationality). Second, we do not know if smooth deformation preserves unirationality.

Compared with unirationality, rational connectedness is far easier to handle. Moreover, rationally connected varieties enjoy almost every geometric property the unirational varieties are known to share in common. For instance, they have trivial fundamental group and trivial CH_0^0 (Chow group of 0-cycles of degree zero), and all the global holomorphic forms vanish.

A principal structure theorem for uniruled variety is the existence of a canonical rational fibring with rationally connected fibres:

THEOREM 2.1 (Campana [5], Kollár-Miyaoka-Mori [23]). *Let X be a variety. Then there is a dominant rational map $\pi : X \cdots \to Z$, unique up to birational equivalence, that has the following two properties:*

(2.1.1) *A general fibre of π is irreducible and rationally connected;*

(2.1.2) *If a rational curve $C \subset X$ passes through a general point of a general fibre of π, then C is contained in the fibre.*

X is nonuniruled [resp. rationally connected] if and only if Z is birational to X [resp. Z is a single point].

Note that, as rational connectedness is a birational invariant, a "general fibre" makes sense even if π is not a morphism in general. We call the unique dominant rational map π the *MRC (= maximal rationally connected) fibration* of X. When $\dim X = 2$, the MRC fibration of a uniruled (ruled, in this case) surface is identical with the Albanese map. Theorem 2.1 is thus regarded as a natural generalization of the Enriques classification of surfaces of Kodaira dimension $-\infty$.

A dominant rational map $\phi : X \cdots \to Y$ is called a *rationally connected fibration* if it satisfies the condition (2.1.1) above. Given a rationally connected fibration $\phi : X \cdots \to Y$, one can find a dominant rational map $\psi : Y \cdots \to Z$ such that the composite $\psi \circ \phi$ is the MRC fibration π. In other words, the MRC fibration has a universal property. Because rational connectedness behaves nicely under deformation, so does the MRC fibration.

The MRC fibration of X is constructed via a certain equivalence relation between points on X. Let $C \subset X$ be a rational curve and $f : \mathbb{P}^1 \to X$ the generically one-to-one morphism induced by the normalization. C is said to be a *free* rational curve if f^*T_X is semi-positive (i.e. isomorphic to a direct sum of line bundles of nonnegative degrees). Two points on X are defined to be equivalent if they are connected by a chain of free rational curves. If X_x stands for the points on X that are equivalent to x, then the correspondence $x \mapsto [\overline{X_x}]$ defines the MRC fibration $\pi : X \cdots \to Z \subset \mathrm{Chow}(X)$, with generic fibre $\overline{X_x}$ over $[\overline{X_x}]$. Here $\overline{*}$ denotes the Zariski closure of $*$.

If the target Z of the MRC fibration of X is again uniruled, we can take the MRC fibration of Z. Reiterating this process, we can describe uniruled varieties as a tower of fibre spaces, where the bottom is a nonuniruled variety (possibly a point) and general fibres in each step are rationally connected. However, we suspect that a single step might suffice:

QUESTION 2.2. *Is the target Z of the MRC fibration not uniruled?*

One can easily reduce this to the following:

QUESTION 2.3. *Is the total space of a family of rationally connected varieties parametrized by (an open subset of) $\mathbb{P}^1$ rationally connected?*

When the dimension of the fibre is 1 or 2, the answer is affirmative (classical if the dimension is one, and by a theorem of Colliot-Thélène [9] when the dimension is 2), yielding a canonical decomposition of uniruled 3-folds into rationally connected varieties and nonuniruled varieties:

THEOREM 2.4 (Kollár-Miyaoka-Mori [23]). *Let X be a smooth algebraic 3-fold. Then precisely one of the following four cases occurs according to the dimension of the image Z of the MRC fibration:*
(2.4.0) *X is rationally connected;*

(2.4.1) *X is a fibre space over a curve of genus ≥ 1, with rational surfaces as general fibres;*
(2.4.2) *X is birational to a conic bundle over a surface of Kodaira dimension ≥ 0;*
(2.4.3) *X is not uniruled.*

In terms of global differential forms, the four cases above respectively correspond to the clearly distinguished cases below:
(2.4.0′) $H^0(X, (\Omega^1_X)^{\otimes k}) = 0$ *for all* $k > 0$*;*
(2.4.1′) $H^0(X, (\Omega^2_X)^{\otimes k}) = 0$ *for all* $k > 0$ *and* $H^0(X, \Omega^1_X) \neq 0$*;*
(2.4.2′) $H^0(X, (\Omega^3_X)^{\otimes k}) = 0$ *for all* $k > 0$ *and* $H^0(X, (\Omega^2_X)^{\otimes l}) \neq 0$*, for some* $l > 0$*;*
(2.4.3′) $H^0(X, (\Omega^3_X)^{\otimes k}) \neq 0$ *for some* $k > 0$.

The equivalence of (2.4.3) and (2.4.3′) is due to Mori [32] and Miyaoka [27] and the other equivalences are classical.

Theorem 2.4 gives a nice picture of the interplay between the spaces of holomorphic forms and the birational structure of 3-folds.

If one could affirmatively solve (2.2) (or equivalently, (2.3)) and prove the abundance conjecture, one would expect to naturally extend Theorem (2.4) to varieties of arbitrary dimension.

3 Fano manifolds and $\mathbb{Q}$-Fano varieties

From the viewpoint of the minimal model program, the $\mathbb{Q}$-Fano varieties form a specifically important class of varieties because, as mentioned above, the minimal model conjecture asserts that a variety has either a minimal model or the structure of a Mori fibre space with $\mathbb{Q}$-Fano fibres. In this sense, $\mathbb{Q}$-Fano varieties, together with minimal varieties, constitute the fundamental components that build up general algebraic varieties, and we need to understand their structure to further conduct the study of uniruled varieties.

The best way to do so would be to give a complete list of such varieties. The biregular classification of smooth Fano surfaces (known as *Del Pezzo surfaces*) into 10 deformation classes is well known and well understood. Iskovskih [13], [14] and Mori-Mukai [33] extended the classification of Fano manifolds to dimension 3. But the complexity of such lists, as well as the technical difficulty needed, increases very fast as the dimension grows. The number of deformation classes of smooth Fano 3-folds exceeds 100, and it is almost hopeless to completely enumerate Fano 4-folds. In dimension 4 or higher, we should instead be content with the theoretical possibility of the classification, or the *boundedness*.

A set of polarized varieties $\{(X_i, H_i)\}_{i \in I}$ is said to be *bounded* if there is a family of polarized varieties $\{(Y_s, L_s)\}_{s \in S}$ parametrized by a scheme of finite type S such that every (X_i, H_i), up to isomorphisms, appears as a member (Y_s, L_s) in the family. If $\{X_i\}$ is bounded, then discrete invariants of the X_i such as topological invariants take only finitely many values. In particular, there are only finitely many possibilities for their Hilbert polynomials, so that general theory tells us that we can in principle enumerate all X_i's up to isomorphisms.

It is still an open question whether the n-dimensional $\mathbb{Q}$-Fano varieties are bounded or not. However, if we restrict ourselves to the smooth ones (i.e. Fano manifolds), we can affirmatively answer the question.

THEOREM 3.1 (Nadel [36], Kollár-Miyaoka-Mori [22] for manifolds with Picard number one; Campana [5], Kollár-Miyaoka-Mori [24] for the general case). *There is a function $\nu(n)$ in $n \in \mathbb{N}$ such that, for an arbitrary n-dimensional Fano manifold X, the degree K_X^n is bounded by $\nu(n)$ from above. In particular, the n-dimensional Fano manifolds (with anti-canonical polarization) are bounded by a theorem of Kollár-Matsusaka [21].*

This theorem is proved, after an idea of Fano, through the rational connectedness of Fano manifolds and the estimate of the minimum of the degrees of rational curves joining two general points:

THEOREM 3.2 (Campana [5], Kollár-Miyaoka-Mori [24]). *A Fano manifold X is rationally connected. Two given general points on X are connected by an irreducible rational curve $C \subset X$ such that $(C, -K_X)$ is bounded by a function in* $\dim X$.

Our upper bound for $(C, -K_X)$ so far has exponential growth as a function in dimension. There will hopefully be a bound of polynomial order, which is known to be the case when the Picard number is one [36], [22].

The key to the proof of (3.2) is to show that a rational connected fibration of a Fano manifold contains many rational curves that are not contained in the fibres. A modified version of the bend-and-break technique indeed gives the following

PROPOSITION 3.3 (Campana [5], Kollár-Miyaoka-Mori [24]). *Let $\pi : X \cdots \to Y$ be a rational map from a Fano manifold onto a variety of dimension ≥ 1. Then, for a given general point $x \in X$, there is a rational curve $C \subset X$ that passes through x and does not lie in a fibre of π.*

QUESTION 3.4. *Is an arbitrary $\mathbb{Q}$-Fano variety rationally connected? Are the $\mathbb{Q}$-Fano varieties of given dimension bounded?*

This has been checked up to dimension 3 (Kawamata [15]).

In this question, it is essential that only terminal singularities are allowed. A normal Gorenstein variety with ample anti-canonical divisor, though always uniruled by (1.3), is not necessarily rationally connected. A typical example is a cone over a curve of degree 3.

Now that we have the boundedness of Fano manifolds of given dimension, a second problem would be how to characterize "interesting" Fano manifolds in terms of reasonable data.

Embedded rational curves play an important role in overcoming this kind of challenge. A good example is a stronger version of the Hartshorne conjecture formulated as follows: if the restriction of the tangent bundle T_X to C is ample for every rational curve C on a Fano manifold X, then X must be a projective space, the simplest variety of all (Mori [30]).

Mori's theorem is one of the most famous and strongest results in this direction, yet it would be better if one could replace the condition on the tangent bundle with a numerical one, because the former, involving richer information, is far harder to check than the latter.

More explicitly, the following would be an interesting problem:

QUESTION 3.5. *Is a smooth Fano n-fold X isomorphic to* $\mathbb{P}^n$ *if* $(C, -K_X) \geq n+1$ *for every rational curve C on X?*

If true, it would imply several preceding characterizations of projective spaces like the ones (a) of Hirzebruch-Kodaira [12] via the topological type, (b) of Siu-Yau [40] via the positivity of the holomorphic bisectional curvature (Frankel conjecture), (c) of Mori [30] via the ampleness of the tangent bundle (Hartshorne's conjecture), and (d) of Kobayashi-Ochiai [18] via the maximal divisibility of the canonical class.

A possible counterpart for hyperquadrics will be:

QUESTION 3.6. *Is a Fano n-manifold X isomorphic to a quadric hypersurface* $\subset \mathbb{P}^{n+1}$ *if the minimum of the degrees* $(C, -K_X)$ *of the rational curves on X is equal to n?*

Prior works related to (3.6) are: (a) Brieskorn [3], (b) Siu [39], (c) Cho-Sato [7] and (d) Kobayashi-Ochiai [18], which are natural analogs of the corresponding results in the case of projective spaces.

References

[1] A. Beauville, *Variétés kähleriennes dont la première classe de Chern est nulle*, J. Differential Geom. **18** (1983), 755–782.

[2] F. Bogomolov, *Holomorphic tensors and vector bundles on projective varieties*, Math. USSR-Izv. **13** (1979), 499–555.

[3] E. Brieskorn, *Ein Satz über die komplexen Quadriken*, Math. Ann. **155** (1964), 184 –193.

[4] R. Brody and M. Green, *A family of smooth hyperbolic hypersurfaces in* P_3, Duke Math. J. **44** (1977), 873–874.

[5] F. Campana, *Connexité rationnelle des variétés de Fano*, Ann. Sci. École Norm. Sup. (4) **25** (1992), 539–545.

[6] F. Campana and H. Flenner, *Projective threefolds containing a smooth rational surface with ample normal bundle*, J. Reine Angew. Math. **440** (1993), 77–98.

[7] K. Cho and E. Sato, *Manifold with the ample vector bundle* $\wedge^2 T_X$, preprint, Kyushu University.

[8] H. Clemens, J. Kollár, and S. Mori, *Higher dimensional complex geometry*, Astérisque **166** (1988).

[9] J.-L. Colliot-Thélène, *Arithmétique des variétés rationelles et problèmes birationelles*, Proc. Internat. Congress Math. 1986, Berkeley, 641–653.

[10] M. Ebihara, *Formal neighborhoods of a toric variety and unirationality of algebraic varieties,* to appear in J. Math. Soc. Japan.

[11] A. Fujiki, *Deformation of uni-ruled manifolds*, Publ. Res. Inst. Math. Sci. Kyoto Univ. **17** (1981), 687–702.

[12] F. Hirzebruch and K. Kodaira, *On the complex projective spaces*, J. Math. Pures Appl. (9) **36** (1956), 201–216.

[13] V. A. Iskovskih, *Fano 3-folds, I*, Math. USSR-Izv. **11** (1977), 485–527.

[14] ———, *Fano 3-folds, II*, Math. USSR-Izv. **12** (1978), 469–506.

[15] Y. Kawamata, *Boundedness of* $\mathbb{Q}$*-Fano threefolds*, Contemp. Math. **131** (1992), 439–445.

[16] ———, *Abundance theorem for minimal threefolds*, Invent. Math. **108** (1992), 229–246.
[17] S. Kobayashi, Hyperbolic Manifolds and Holomorphic Mappings, Marcel Dekker, New York, 1970.
[18] S. Kobayashi and T. Ochiai, *Characterizations of complex projective spaces and hyperquadrics*, J. Math. Kyoto Univ. **13** (1973), 31–47.
[19] J. Kollár, *Rational curves on algebraic varieties*, preprint, Univ. Utah, Salt Lake City, UT, 1994.
[20] J. Kollár *et al.*, *Flips and abundance for algebraic threefolds*, Astérisque **211** (1993).
[21] J. Kollár and T. Matsusaka, *Riemann-Roch type inequalities*, Amer. J. Math. **105** (1983), 229–252.
[22] J. Kollár, Y. Miyaoka, and S. Mori, *Rational curves on Fano varieties*, Lecture Notes in Math. **1515** (1992), 100–105.
[23] ———, *Rationally connected varieties*, J. Alg. Geom. **1** (1992), 429–448.
[24] ———, *Rational connectedness and boundedness of Fano manifolds*, J. Differential Geom. **36** (1992), 765–779.
[25] M. Levine, *Deformations of uni-ruled varieties*, Duke Math. J. **48** (1981), 467–473.
[26] Y. Miyaoka, *The Chern classes and Kodaira dimension of a minimal variety*, Adv. Stud. Pure Math. **10** (1987), 449–476.
[27] ———, *On the Kodaira dimension of minimal threefolds*, Math. Ann. **281** (1988), 325–332.
[28] ———, *Abundance conjecture for 3-folds — case $\nu = 1$*, Compositio Math. **68** (1988), 203–220.
[29] Y. Miyaoka and S. Mori, *Numerical criterion for uniruledness*, Ann. of Math. (2) **124** (1986), 65–69.
[30] S. Mori, *Projective manifolds with ample tangent bundles*, Ann. of Math. (2) **110** (1979), 593–606.
[31] ———, *Cone of curves and Fano 3-folds*, Proc. Internat. Congress Math., Warszawa, 1983, pp. 747–752.
[32] ———, *Flip theorem and the existence of minimal models for 3-folds*, J. Amer. Math. Soc. **1** (1988), 117–253.
[33] S. Mori and S. Mukai, *Classification of Fano threefolds with $B_2 \geq 2$*, Manuscripta Math. **36** (1981), 147–162.
[34] ———, *Mumford's theorem on curves on K3 surfaces* (Appendix to *The uniruledness of the moduli space of curves of genus 11*), Springer-Verlag, Berlin and New York, Lecture Notes in Math. **1016** (1983), 334–353.
[35] U. Morin, *Sull' unirationalità dell' ipersuperficie algebrica di qualunque ordine e dimensione sufficentemente alta*, Atti dell II. Congresso Unione Math. Italiana, 1940, 298–302.
[36] A. Nadel, *The boundedness of degree of Fano varieties with Picard number one*, J. Amer. Math. Soc. **4** (1991), 681–692.
[37] L. Ramero, *Effective estimates for unirationality*, Manuscripta Math. **68** (1990), 435–445.
[38] M. Reid, *Young person's guide to canonical singularities*, Proc. Sympos. Pure Math. **46** (1987), 345–414.
[39] Y.-T. Siu, *Curvature characterization of hyperquadrics*, Duke Math. J. **47** (1980), 641–654.
[40] Y.-T. Siu and S.-T. Yau, *Compact Kähler manifolds of positive bisectional curvature*, Invent. Math. **59** (1980), 189–204.

Some Eisenstein Cohomology Classes for the Integral Unimodular Group

MADHAV V. NORI

Department of Mathematics, University of Chicago
Chicago, IL 60637, USA

Vector-valued differential forms are frequently used to give cohomology classes on manifolds with values in local systems. In particular, certain automorphic forms serve this purpose on certain locally symmetric spaces. Such a representation does not help in deciding whether of not the cohomology class is rational. For example the normalized holomorphic Eisenstein series of weight $2k$ on the upper half plane gives an elt. of $H^1(SL_2, \mathrm{Sym}^{2k-2}\,\mathbb{C}^2)$, thanks to Eichler. That this comology class is rational was proved by Deligne using weight considerations. A direct proof of this follows from the work of Beilinson and Levin on the elliptic polylogarithm.

It turns out that the Eisenstein cohomology classes alluded to in the previous paragraph can be generalized to obtain elements of $H^{n-1}(\Gamma, \mathrm{Sym}^k V)$, where V is a rational vector space of dimension n and Γ is an arithmetic subgroup of $GL(V)$. These classes result from superimposing the singular chain complex of the complement of S in $V(\mathbb{R})$ on a truncated Koszul complex, where S is a finite union of cosets of a lattice in V, done equivariantly for the action of an arithmetic subgroup of $Aff(V)$. The construction is so simple that we hope it generalizes to other groups and representations as well.

There is no unique way of representing these classes by differential forms. The method we have chosen is connected to the Sullivan minimal model.

Integrating these differential forms then yields the rationality of special values of zeta functions of ideals in totally real fields. These rationality statements were obtained by Klingen and Siegel by studying $SL_2(F)$, where F is a totally real field. Our approach is different — we embed $F^\times$ in $GL_n(\mathbb{Q})$ and then pull back rational cohomology classes (of arithmetic subgroups) of $GL_n(\mathbb{Q})$ to $F^\times$ to obtain the rationality of values of zeta functions. Sczech has defined the notion of an Eisenstein cocycle on $GL_n(\mathbb{Q})$ and used it to calculate values of zeta functions. It is therefore quite likely that this work overlaps with his.

1. Two Standard Chain Complexes

The construction is as follows. $Aff(V)$ is the semi-direct product of $GL(V)$ and V. Set $C = \Delta(V(\mathbb{R}) - S) =$ the singular chain complex of the complement of S in $V(\mathbb{R})$, where S is a finite union of cosets of a lattice L in V, as said earlier. The stabilizer of S in V is an arithmetic subgroup of $Aff(V)$, and the complex C has a natural action of this arithmetic subgroup. Let $\tilde{C}$ denote the kernel of the

Proceedings of the International Congress
of Mathematicians, Zürich, Switzerland 1994

augmentation $\epsilon : C \to \mathbb{Z}$. Let $\mathbb{Z}[S]$ denote the free Abelian group with S as basis. We then have

$$H_i(\tilde{C}) = \mathbb{Z}[S] \otimes H_n(V(\mathbb{R}), V(\mathbb{R}) - 0) \text{ if } i = n - 1$$

and is zero otherwise.

Next set $R = \operatorname{Sym} \widehat{V}$. Thus, R is a power series ring in n variables. Let $\mathbb{Q}[V]$ denote the group ring of V and let $\rho : \mathbb{Q}[V] \to R$ be the ring homomorphism given by $\rho(v) = \exp(v)$ for all v in V. Let K be the standard Koszul complex:

$$\Lambda^n V \otimes R \to \Lambda^{n-1} V \otimes R \to \cdots \to V \otimes R \to R.$$

Let D be the subcomplex of K given by $D_i = K_i$ if $i < n$ and $D_i = 0$ otherwise. Once again, let $\tilde{D}$ denote the kernel of the augmentation $\epsilon : D \to \mathbb{Q}$. All the homologies of $\tilde{D}$ vanish except the $(n-1)$th and this coincides with $K_n = \Lambda^n V \otimes R$. Now, K and D are complexes of R-modules, and R is an algebra over $\mathbb{Q}[V]$, and this gives an action of the group V on both of these complexes. Furthermore, there is a natural action of $GL(V)$ on both. Combining these actions we see that $Aff(V)$ acts on K and on D.

We wish to get a homomorphism from C to D commuting with the augmentations up to a scalar. That would induce an arrow $g : H_{n-1}(\tilde{C}) \to H_{n-1}(\tilde{D})$. Any such g is uniquely given by $g(s) = \exp(s) \cdot f(s)$, where $f : S \to R \otimes \mu(V)$ is any function.

Here $\mu(V) = \Lambda^n V \otimes H^n(V(\mathbb{R}), V(\mathbb{R}) - 0)$.

If $C \to D$ is to commute with the action of some arithmetic subgroup of $Aff(V)$, we need g to have the same property. The conditions (a) and (b) on f below ensure this. The only f we consider satisfy the following:

(a) the image of f is contained in $\mu(V)$, and

(b) f is invariant under translation by some lattice in V.

Now, every lattice L in V defines a canonical element $\delta(L)$ in $\mu(V)$. In fact, a basis v_i of L induces an orientation $\lambda \in H^n(V(\mathbb{R}), V(\mathbb{R}) - 0)$ and $\delta(L) = v_1 \wedge v_2 \wedge \cdots \wedge v_n \otimes \lambda$ defines $\delta(L)$ independent of the choice of basis. Furthermore, if M is a sublattice of L, then $\delta(M) = [L : M]\delta(L)$. So, for f as above, if L is a lattice such that translation by L leaves f invariant, $(\int f)\delta(L) = \Sigma_{s \in S/L} f(s)$ defines $\int f$ independent of the choice of the particular lattice L. This definition of integral also remains unaltered if f is extended by zero to a larger S'.

The exact sequence $0 \to \tilde{C} \to C \to \mathbb{Z} \to 0$ gives rise to a distinguished triangle, denoted by $TR(C)$, and in a like manner, we obtain a distinguished triangle $TR(D)$ from the complex D and its augmentation.

PROPOSITION: Let π be an arithmetic subgroup of $Aff(V)$ leaving the above S and f invariant. Then there is a unique homomorphism of triangles $F(f) : TR(C) \to TR(D)$ in the derived category of complexes of π-modules, so that

(1) $F(f)$ induces multiplication by $\int f$ on $\mathbb{Z} \to \mathbb{Q}$, and

(2) $F(f)$ induces $g : \tilde{C} \to \tilde{D}$.

The proposition is easily deduced from the following observations. The obstructions to the existence and uniqueness of $F(f)$ are elements of $H^i(\pi, K_n)$ for $i = n$ and $i = n-1$ respectively (for the existence, this is a consequence of the axiom (TR 3) for triangulated categories in Verdier's article on Derived Categories, to be found in SGA $4\frac{1}{2}$). The intersection of π with V is a lattice L in V and $H^i(L, K_n)$ vanishes if i is different from n. The Hochschild-Serre spectral sequence now shows that $H^i(\pi, K_n)$ vanishes for $i < n$ and injects into $H^i(L, K_n)$ for $i = n$. The uniqueness of $F(f)$ follows. Furthermore, the existence of $F(f)$ is a consequence of

(A) there is an L-equivariant $TR(C) \to TR(D)$ with the desired properties.

Now, let K_W denote the Koszul complex with coefficients in a field W of characteristic zero, and let D_W denote again the truncated complex. Then $H^n(L, (K_W)_n) = W$ and the other cohomologies vanish. We deduce that $F(f)$ exists if

(B) there is $F(f)_W : TR(C) \to TR(D_W)$ satisfying properties (1) and (2) of the proposition.

In the section on differential forms and Eisenstein series, condition (B) is verified for the field of real numbers. That provides a second proof of the existence statement of the proposition. A quick first proof is obtained by verifying condition (A) directly. For this, consider the simplest case when S is a translate of a lattice L and f is the constant function $\delta(L)$ so that $\int f = 1$. Choose a basis for L and let x_i be the dual basis. Let B be the subset of $V(\mathbb{R})$ given by $x_1 x_2 \dots x_n = 0$ and choose $v \in V$ so that $Y = B + L + v$ is contained in $V(\mathbb{R}) - S$. This inclusion is then a homotopy equivalence. Now Y has a natural L-equivariant cell decomposition and the associated chain complex is once again obtained from the Koszul complex for the group ring $\mathbb{Z}[L]$ and the system $\{v_1 - 1, v_2 - 1, \dots, v_n - 1\}$ by deleting the nth term. Thus, $\Delta(V(\mathbb{R}) - S)$ is quasi-isomorphic to this truncated Koszul complex and producing an L-equivariant homomorphism is now easy. One only needs to observe that the constant coefficient of $(\exp(t) - 1)/t$ is 1. The general case in (A) is obtained by taking linear combinations of this special case. This completes the proof of the proposition.

2. The Eisenstein Operator

We shall apply the proposition as follows. With π as in the proposition, let Γ denote the intersection of π with $GL(V)$. We have homomorphisms

$$i : \mathbb{Z} \to C \text{ and } j : \mathbb{Z} \to D$$

in the derived category of complexes of Γ-modules given by $j(1) = 1$ and $i(1) = [0]$, where $[0]$ denotes the zero-simplex of $V(\mathbb{R}) - S$ at the origin of $V(\mathbb{R})$. Now $E(f) = F(f) \circ i - (\int f) \circ j$, by the remarks following the proposition above, is identified with an element of $H^{n-1}(\Gamma, K_n)$. Composing with the projection $R \to \mathrm{Sym}^k(V)$ we get $E_k(f) \in H^{n-1}(\Gamma, \Lambda^n V \otimes \mathrm{Sym}^k V)$. The $E_k(f)$ depend linearly on the f and thus we get an Eisenstein operator, for each nonnegative integer k:

$$E_k : \mathcal{S}_\circ(\mathbb{A}_f \otimes V) \otimes \mu(V) \to \mathcal{H}^{n-1}(\Lambda^n V \otimes \mathrm{Sym}^k(V)).$$

In the above, $\mathcal{H}^i(X)$ denotes the direct limit of $H^i(\Gamma, X)$ taken over all arithmetic subgroups Γ of $GL(V)$, for X any $GL(V)$-module.

$\mathbb{A}_f$ denotes the finite adeles, $\mathcal{S}(\mathbb{A}_f \otimes V)$ denotes the space of compactly supported locally constant functions with values in $\mathbb{Q}$, and finally, $\mathcal{S}_\circ(A_f \otimes V)$ is the subspace of such functions vanishing at the origin.

The E_k is equivariant for the action of $GL(V)$. In particular, if α is a positive rational number, we see that $E_k(f \circ \alpha^{-1}) = \alpha^k E_k(f)$. If k is positive and if α is different from 1, we see that

$$E_k(f) = (1-\alpha^k)^{-1} E_k(f - f\circ\alpha^{-1}) \text{ for all } f \in \mathcal{S}_\circ(A_f \otimes V).$$

The above formula can be used to define $E_k(f)$ for all $f \in \mathcal{S}(A_f \otimes V)$, because $(f - f\circ\alpha^{-1})$ vanishes at the origin. Thus, for k different from 0, we have:

$$E_k : \mathcal{S}(A_f \otimes V) \to \mathcal{H}^{n-1}(\Lambda^n V \otimes \operatorname{Sym}^k V)$$

3. Differential Forms and Eisenstein Series

We shall next represent the $E_k(f)$ by differential forms. It suffices to consider the simple case considered before where S is the translate of a lattice L by $a \in V$ and f is the constant function $\delta(L)$ on S. The π in the proposition is then the stabilizer of L in $Aff(V)$.

Let P denote the positive definite matrices in $\operatorname{Hom}(V^*, V)\otimes\mathbb{R}$. Let Ω denote the de Rham complex of $V(\mathbb{R})$ with polynomial coefficients, and let Ω_t denote the complex obtained from Ω by deleting the nth term. Thus, $H^{n-1}(\Omega_t) = \Omega^n$. Fix an orientation of $V(\mathbb{R})$ and let θ be the unique constant coefficient n-form on $V(\mathbb{R})$ so that $\int_{V(\mathbb{R})/L}\theta = 1$. Let $\mathcal{A}$ denote the de Rham complex of smooth forms on the product $P \times (V(\mathbb{R}) - S)$. Then $H^{n-1}(\mathcal{A})$ is simply the space of all functions on S, because an orientation has been fixed. We want a π-equivariant homomorphism $U : \Omega_t \to \mathcal{A}$ that takes $h\cdot\theta \in \Omega^n = H^n(\Omega_t)$ to the function $h \mid S : S \to \mathbb{R}$ in $H^{n-1}(\mathcal{A})$, for all polynomials h on $V(\mathbb{R})$. Now, $\operatorname{Hom}(\Omega\ ,\mathbb{R}) = K_{\mathbb{R}}$ and $\operatorname{Hom}(\Omega_t\ ,\mathbb{R}) = D_{\mathbb{R}}$ and thus such a U would induce a homomorphism from the singular chain complex (the real vector space with the *smooth* simplices as basis) of $P\times(V(\mathbb{R})-S)$ to $D_{\mathbb{R}}$, the truncated Koszul complex with real coefficients. This singular chain complex is quasi-isomorphic to $C\otimes\mathbb{R}$, and we thus have a candidate for $F(f)_{\mathbb{R}}$. We find U of the following shape:

$$U(\omega) = \omega \text{ if } \deg.\omega < n-1$$

(the pullback of ω to $P \times (V(\mathbb{R}) - S)$ being denoted by ω again)

$$U(\omega) = \omega + \psi(d\omega) \text{ if } \deg.\omega = n-1.$$

To ensure that U is a homomorphism of complexes that has the right value on H^{n-1}, we need the $\psi : \Omega^n \to \mathcal{A}^{n-1}$ to be a π-equivariant homorphism that satisfies

I: $\quad d\psi(h\cdot\theta) = h(\delta_S - 1)\theta.$

Actually the U and ψ have their images contained in $\oplus_{p+q}\mathcal{A}^{p,q}$, where $\mathcal{A}^{p,q}$ denotes the space of smooth p- forms on P with values in q-currents on $V(\mathbb{R})$ that are smooth on $V(\mathbb{R}) - S$. And $\delta_S = \Sigma_{s\in S}\delta_s$, where δ_s is the Dirac delta distribution on $V(\mathbb{R})$ at s. Choose a basis v_i of L and let x_i be the dual basis so that $\theta = dx_1 \wedge dx_2 \wedge \ldots dx_n$. We define ϕ, a formal power series in the v_i with coefficients in $\oplus_{p+q=n-1}\mathcal{A}^{p,q}$ by the formula

$$\phi = \exp(\Sigma_i - v_i\cdot x_i)\cdot\psi(\exp(\Sigma_i v_i\cdot x_i)\theta).$$

The L-equivariance of ψ is equivalent to the assertion that ϕ is a power series in the v_i with coefficients in forms on $P \times (V(\mathbb{R}) - S)$ that are invariant under translation by the lattice L. Denoting the monomials in the v_i by v^μ and putting $\phi = \Sigma_\mu \phi_\mu \cdot v^\mu$, I above translates into eq. I′ for ϕ and eqs. I″ for ϕ_μ given below:

I′: $d\phi + (\Sigma v_i dx_i) \wedge \phi = (\delta_S - 1)\theta$

I″: $d\phi_0 = (\delta_S - 1)\theta$

$d\phi_i + dx_i \wedge \phi_0 = 0$

$d\phi_{i,j} + dx_i \wedge \phi_j + dx_j \wedge \phi_i = 0$

...

The equations I″ say that the ϕ_μ are the forms required to construct a part of the Sullivan minimal model of the complement of a point in the torus $V(\mathbb{R})/L$. Precisely, if $n > 2$, the Sullivan minimal model has generators the dx_i in degree 1, and the ϕ_μ in degree $(n-1)$, and more generators in higher degrees that need to be studied. (An aside — it is clear that the Sullivan minimal model of our space has a natural $GL(V)$-action. However, there does not exist a Γ-equivariant homomorphism from all of this minimal model to the de Rham complex, at least for $n = 2$; indeed, some of the $E_k(f)$ arise as obstructions to this problem.) Given the ϕ_μ, we can construct the desired ψ. A canonical method of constructing the ϕ will ensure the π-equivariance of ψ. One such method is given below.

Each point of P induces a metric on $V(\mathbb{R})$ and we get d^* from m-currents on $V(\mathbb{R})$ to $(m-1)$-currents on $V(\mathbb{R})$, and this extends to $d^* : \mathcal{A}^{l,m} \to \mathcal{A}^{l,m-1}$. Any system of ϕ_μ contained in d^*(the L-invariant elements of $\oplus_{p+q=n} \mathcal{A}^{p,q}$) and satisfying the equations I″ is unique. The solutions of these equations would then be currents, but the standard regularity theorems (for the Laplacian on $V(\mathbb{R})$ with a constant coefficient metric) would ensure their smoothness outside S. To solve for the ϕ we write its Fourier series:

$$\phi = \Sigma_{l^* \in \mathrm{Hom}(L,\mathbb{Z})} \exp(2\pi\sqrt{-1}(l^*, x)) \cdot \phi_{l^*}.$$

In the above, x denotes a point of $(V\mathbb{R})$, and the ϕ_{l^*} is a formal power series in the v_i with coefficients in the direct sum, taken over $p + q = (n-1)$ of $X^{p,q}$ = smooth p-forms on P tensored with constant coefficient q-forms on $V(\mathbb{R})$. Any $l^* \in \mathrm{Hom}(L, \mathbb{Z})$ extends to a linear form on $V(\mathbb{R})$, and its differential is denoted by $d(l^*)$, and we set $A_{l^*} = (2\pi\sqrt{-1})d(l^*) + \Sigma_i v_i dx_i$. Eq. I′ now reads as

II: $d(\phi_{l^*}) + A_{l^*}\phi_{l^*} = \exp(2\pi\sqrt{-1}(l^*, -s))\theta$ if l^* is nonzero, and the left-hand side vanishes if $l^* = 0$. The right-hand side of this equation is clearly independent of the choice of $s \in S$.

For $T \in P$ and $l^* \in \mathrm{Hom}(L, \mathbb{Z})$ as above, we get $T(l^*) \in V(\mathbb{R})$. We may regard $T(l^*)$ as a vector field on $P \times V(\mathbb{R})$ and interior multiplication with $T(l^*)$ gives an operator $i_{l^*} : X^{p,q} \to X^{p,q-1}$. That ϕ is in the image of d^* is now equivalent to the assertion

III: $\phi_{l^*} \in i_{l^*}(\oplus_{p+q=n} X^{p,q})$.

The operators $C_{l^*} = d + A_{l^*}$ and i_{l^*} satisfy:

IV: $C_{l^*} \circ C_{l^*} = i_{l^*} \circ i_{l^*} = 0$ and $C_{l^*} \circ i_{l^*} + i_{l^*} \circ C_{l^*}$ is an isomorphism.

In fact, $d \circ i_{l^*} + i_{l^*} \circ d = N$ is the Lie derivation on $P \times V(\mathbb{R})$ given by the vector field $T(l^*)$ and because $N(X^{p,q})$ is contained in$X^{p+1,q-1}$ we see that it is nilpotent on $\oplus X^{p,q}$. And $A_{l^*} \circ i_{l^*} + i_{l^*} \circ A_{l^*} = J$ is multiplication by a function $j = t + \Sigma_i t_i v_i$ where t and t_i are complex-valued functions on P, and therefore J and N commute. Explicitly, $t(T) = 2\pi\sqrt{-1}(l^*, Tl^*)$ and $Tl^* = \Sigma_i t_i v_i$. The "constant term" $t(T)$ being nonzero, we see that J and $J + N$ are isomorphisms. Now, C_{l^*} annihilates the R.H.S. of eq. II, and thus the unique soln. for ϕ_{l^*} satisfying II and III is given by evaluating the operator $i_{l^*} \circ (C_{l^*} \circ i_{l^*} + i_{l^*} \circ C_{l^*})^{-1}$ on the R.H.S. of eq. II. We have explicitly

V: $\phi_{l^*} = \Sigma_m(-1)^m(2\pi\sqrt{-1}(l^*, Tl^*) + Tl^*)^{-(m+1)} i_{l^*}(d \circ i_{l^*})^m(\exp 2\pi\sqrt{-1}(-s, l^*))\theta.$

The coefficient of the monomial v^μ in the mth term of the above summation is a function on P bounded above by $\| l^* \|^{-(m+k+1)}$ when T lies in a compact subset of P, where $k = deg.(\mu)$, and this shows that ϕ given by (V) above is a power series in the v_i with coefficients in $\oplus_{p+q=n-1} \mathcal{A}^{p,q}$. Sending T to $(T, 0)$ gives an inclusion of P in $P \times (V(\mathbb{R}) - S)$ and the $E(f)$ are simply obtained by restricting the ϕ to P via this inclusion. The mth term in (V) above belongs to $X^{m,n-m+1}$ and thus vanishes when restricted to P, when m is different from $n - 1$. Denoting the coefficient of the monomial v^μ in the mth term in (V) above by $\phi_{l^*,\mu,m}$, we see that $\Sigma_{l^* \in \mathrm{Hom}(L,\mathbb{Z})-0}\, \phi_{l^*,\mu,(n-1)} \cdot \exp(2\pi\sqrt{-1}(l^*, x))$ actually converges if $\deg .\mu$ is positive. Thus, the coefficient of v^μ in $E_k(f)$ for k positive is obtained by setting $x = 0$ in the above formula. If $\theta^{(n-1)}$ denotes the $(n-1)$-form on $V(\mathbb{R})$ obtained by contracting the volume form θ with the Euler vector field on $V(\mathbb{R})$, and if $e_{l^*} : P \to V(\mathbb{R})$ is given by $e_{l^*}(T) = T(l^*)$, we have explicitly for k positive:

VI: $E_k(f)(T)(-1)^{n+k-1}k!((n+k-1)!^{-1}) =$

$\Sigma_{l^* \in \mathrm{Hom}(L,\mathbb{Z})-0} \exp(2\pi\sqrt{-1}(l^*, -s)) \cdot (2\pi\sqrt{-1}(Tl^*, l^*))^{-(n+k)}(Tl^*)^{\otimes k} \cdot e_{l^*}{}^*\theta^{(n-1)}.$

The above formula is valid when S is a translate of L and f is the constant function $\delta(L)$. The preceding remarks apply when $S \neq L$. The case $S = L$ follows simply from the definition of $E_k(f)$ when f does not vanish at the origin.

Finally, if we set $V = F$, a totally real field of degree n, we have the inclusion of $F - 0 = F^\times$ in $GL(V)$ and we can now restrict the $E_k(f)$ to arithmetic groups of $F^\times$. These cohomology groups vanish if n does not divide k and is isomorphic to $\mathbb{Q}$ otherwise. Restricting the E_{kn} therefore, we get a rationality statement (due to Klingen and Siegel):

Let F be a totally real field of degree n, and let R and D be the regulator and discriminant of F respectively. Let L be a lattice in F and set $M = L^\perp$ and let $a \in F$. Let U be the subgroup of totally positive units of F that stabilize both $L + \mathbb{Z}a$ and L and act trivially on $L + \mathbb{Z}a/L$.

$$(R\sqrt{D})^{-1}(2\pi\sqrt{-1})^{-n(k+1)}\Sigma_{m \in M-0/U}(\exp(2\pi\sqrt{-1})trace_{F/\mathbb{Q}}(am)) \cdot N_{F/\mathbb{Q}}(m)^{k+1}$$

is rational.

The case $k = 0$ has been omitted, and therefore the values of the L-series of F at $s = 1$ have yet to be obtained by this method.

References

[1] A. Beilinson and A. Levin, *The elliptic polylogarithm*, in Motives, Proc. Sympos. Pure Math. 55, part 2, pp. 123–190, Amer. Math. Soc., Providence, RI, 1994.
[2] Robert Sczech, *Eisenstein group cocycles for GL_n and values of L-functions*, Invent. Math. 113, 581–616 (1993), Springer-Verlag, New York.
[3] SGA $4\frac{1}{2}$, Springer Verlag Lecture Notes 569, 1977.

Rigid and Exceptional Vector Bundles and Sheaves on a Fano Variety

ALEXEI N. RUDAKOV*

Independent University of Moscow and
Institute for System Analysis
of the Russian Academy of Science
Moscow, Russia

1. Introduction

Enormous progress was made in the last twenty years in studying the moduli of algebraic vector bundles and sheaves. In the last ten years a considerable understanding was gained for a particular case of the moduli problem — for vector bundles or sheaves with "zero dimensional moduli" or so-called rigid sheaves on a Fano variety and especially on a Del Pezzo surface. Our aim is to descibe results on the construction and properties of rigid sheaves on Fano varieties.

Many of these results were discussed in the author seminar "Vector bundles" and were developed by the people who were for some time participants of the seminar. I am indebted to all of them and especially to A. Bondal, A. Gorodentsev, M. Kapranov, B. Karpov, A. Kuleshov, D. Orlov, A. Tyurin, and S. Zube.

1.1. Rigid sheaves on a Del Pezzo surface. By Fano variety we mean a smooth projective variety X over a field $\boldsymbol{k}$ with an ample anticanonical bundle K_X^{-1}. Traditionally two-dimensional Fano varieties are called Del Pezzo surfaces. Projective spaces $\boldsymbol{P}^n$ are the most common Fano varieties.

All the sheaves in the following are meant to be algebraic coherent sheaves on a variety.

DEFINITION 1.1 *A sheaf F is called rigid if* $\mathrm{Ext}^1(F,F)=0$.

DEFINITION 1.2 *A sheaf E is called exceptional if* $\mathrm{Ext}^i(E,E)=0$ *for* $i\neq 0$ *and* $\mathrm{Ext}^0(E,E)=\boldsymbol{k}$.

It happens to be that on a Del Pezzo surface indecomposable (into a direct sum) rigid sheaves are closely related to exceptional ones.

THEOREM 1.1 (KULESHOV [18]) *Let $\boldsymbol{k}$ be an algebraically closed field. Then any rigid sheaf on a Del Pezzo surface is a direct sum of exceptional ones.*

(For a projective plane this was proven in [5].)

*While writing the paper the author was partially supported by ISF grant MKU000.

Proceedings of the International Congress
of Mathematicians, Zürich, Switzerland 1994

1.2. Exceptional sheaves on a Del Pezzo surface. Let X be a Del Pezzo surface.

THEOREM 1.2 *[18] An exceptional sheaf E on a Del Pezzo surface is either locally free or a torsion sheaf, which is a direct image of an invertible sheaf $\mathcal{O}(k)_C$ from an irreducible exceptional curve C.*

There is a natural anticanonical polarization on a Del Pezzo surface, and in the following, referring to stability, we will mean Giesecker stability relative to this polarization.

THEOREM 1.3 *If a sheaf F on a Del Pezzo surface has no torsion then the following are equivalent:*
(i) F is exceptional,
(ii) F is rigid and stable,
(iii) F is stable and the discriminant $\Delta(F) < 1/2$.

THEOREM 1.4 *If an exceptional sheaf E on a Del Pezzo surface has no torsion then it is uniquely determined among exceptional sheaves by its rank and the first Chern class and among stable sheaves by its rank and the Chern classes.*

These results were first proven for $X = \boldsymbol{P}^2$, then for Del Pezzo surfaces with moving anticanonical class and at last for all Del Pezzo surfaces [5], [22], [9], [18].

To study further exceptional sheaves it is practical to use a braid group action, which is defined in the following, but permit me to tell about properties of the action before describing its definition.

1.3. Properties of the braid group action.

DEFINITION 1.3 *A system $E_0, \ldots, E_m$ of sheaves is called an exceptional system (of a size m) if the sheaves $E_0, \ldots, E_m$ are exceptional and $\operatorname{Ext}^k(E_j, E_i) = 0$ for $i < j$.*

So there is an $m+1$ element in a system of size m. Let a braid group with $m+1$ threads be denoted by Bd_m so that Bd_1 is isomorphic to the additive group $\boldsymbol{Z}$.

THEOREM 1.5 *There exists a braid group Bd_m action on a set of exceptional systems of size m of sheaves on a Del Pezzo surface.*

PROPOSITION 1.4 *Let $\boldsymbol{k}$ be algebraically closed, X be a Del Pezzo surface, and $PicX = \boldsymbol{Z}^t$, where $t = 1, \ldots, 9$. Then there exist exceptional systems of any size up to $l_X = t+1$, which is the maximal possible size.*

THEOREM 1.6 *Let $\boldsymbol{k}$ be algebraically closed. Any exceptional system on a Del Pezzo surface could be enlarged to a system of size l_X. The braid group action on systems of maximal size l_X is transitive.*

In particular this means that any exceptional sheaf is an element in an exceptional system of size l_X.

These results were proven first for $\boldsymbol{P}^2$ also [5], [22], then for a quadric surface [23],[27], and at last in general [18].

1.4. The braid group action for sheaves on a projective plane. The braid group action can be described fairly explicitly for $X = \boldsymbol{P}^2$.

LEMMA 1.5 *Let E_1, E_2 be an exceptional pair on $\boldsymbol{P}^2$.*
Then $\mathrm{Ext}^i(E_1, E_2) = 0$ *for $i > 0$ and the morphism*

$$can : \mathrm{Hom}(E_1, E_2) \otimes E_1 \longrightarrow E_2$$

is an epimorphism and the morphism

$$can^* : E_1 \longrightarrow \mathrm{Hom}(E_1, E_2)^* \otimes E_2$$

is a monomorphism.

DEFINITION 1.6 *The action of Bd_1 of exceptional pairs on $\boldsymbol{P}^2$ is defined by the condition that the generator λ of Bd_1 acts as follows:*

$$(E_1, E_2) \longrightarrow (E_2^+, E_1)$$

where E_2^+ is defined by an exact sequence

$$0 \longrightarrow E_2^+ \longrightarrow \mathrm{Hom}(E_1, E_2) \otimes E_1 \longrightarrow E_2 \longrightarrow 0.$$

It easy to see that λ^{-1} acts as follows:

$$(E_1, E_2) \longrightarrow (E_2, E_1^-)$$

where E_1^- is defined by an exact sequence

$$0 \longrightarrow E_1 \longrightarrow \mathrm{Hom}(E_1, E_2)^* \otimes E_2 \longrightarrow E_1^- \longrightarrow 0\,.$$

PROPOSITION 1.7 *There is an action of Bd_2 on exceptional systems of size 2 that is defined so that its generators λ_1, λ_2 change as above the first or the second pair of a system respectively and do not change the sheaf that does not belong to the pair.*

Similar action exists on $\boldsymbol{P}^n$ for systems of size n and it was practically defined in [11].

THEOREM 1.7 *Any exceptional sheaf on $\boldsymbol{P}^2$ belongs to an exceptional system of size 2. The action of Bd_2 on the set of exceptional systems of size 2 is transitive.*

The field $\boldsymbol{k}$ is not supposed to be algebraically closed here.

THEOREM 1.8 *For an exceptional system E_0, E_1, E_2 on $\boldsymbol{P}^2$ let $r_i = rkE_i$. Then*

$$r_0^2 + r_1^2 + r_2^2 = 3\, r_0\, r_1\, r_2$$

(Markov equation).

The most simple exceptional systems on $\boldsymbol{P}^2$ are line bundle systems. They are of a type:

$$(\mathcal{O}(i), \mathcal{O}(i+1), \mathcal{O}(i+2))$$

and they correspond to the minimal solution of the Markov equation. Thus, all others are made from those by the braid group operators and all exceptional sheaves appear this way.

2. Braid group action in a derived category

Let $\mathcal{D}$ be a derived category $D^b(AlgSh(X))$. It is possible to define a braid group action on exceptional systems in $\mathcal{D}$ for any projective X [10].

DEFINITION 2.1 *Let us call $E \in Obj\, \mathcal{D}$ an exceptional object if*

$$\operatorname{Hom}(E, E) = \boldsymbol{k} \text{ and}$$

$$\operatorname{Ext}^i(E, E) = 0 \text{ for } i \neq 0.$$

DEFINITION 2.2 *A system $E_0, \ldots, E_m$ of elements of $Obj\, \mathcal{D}$ is called an exceptional system (of a size m) if $E_0, \ldots, E_m$ are exceptional and*

$$\operatorname{Ext}^k(E_j, E_i) = 0 \text{ for } i < j.$$

Systems of a type $E_i, \ldots, E_{i+k}$ of size k are called subsystems in $E_0, \ldots, E_m$.

DEFINITION 2.3 *The action of Bd_1 on exceptional pairs in $\mathcal{D}$ is defined by the condition that the generator λ of Bd_1 acts as follows:*

$$(E, F) \longrightarrow (F^+, E)$$

where F^+ is defined by an exact (distinguished) triangle

$$F^+ \longrightarrow \bigoplus_p \operatorname{Hom}(E[p], F) \otimes E[p] \longrightarrow F \longrightarrow F^+[1].$$

There are standard inclusions $Bd_k \longrightarrow Bd_m$ where an image of Bd_k consists of braids having threads outside of k consecutive ones just straight. Bd_m is generated by images of m such inclusions for Bd_1.

PROPOSITION 2.4 *There exist braid group actions of Br_m on a set of exceptional systems of size m where an image of a standard inclusion $Br_1 \longrightarrow Br_m$ acts as was described above on a corresponding pair and does not change elements outside of the pair.*

DEFINITION 2.5 *The minimal full triangular subcategory containing $E_0, \dots, E_m$ is called the subcategory generated by $E_0, \dots, E_m$.*

DEFINITION 2.6 *We will say that an exceptional dimension of a category $\mathcal{D}$ is m, $\operatorname{exdim} \mathcal{D} = m$, if $\mathcal{D}$ coincides with a subcategory generated by an exceptional system of size m.*

It is clear that the Grothendieck group of a subcategory generated by an exceptional system of size m is isomorphic to $\boldsymbol{Z}^{m+1}$ and that the subcategory does not contain such a system of bigger size. All exceptional systems in one Bd_m-orbit generate the same subcategory. Let us remember that the size is less by one than the number of elements in the system, so one could conclude from the Beilinson theorem that

$$\operatorname{exdim} D^b(AlgSh\, \boldsymbol{P}^n) = n$$

and generalizations of Beilinson theorem one could interpret as similar computations of "exdim" for other varieties. It is known that

$$\operatorname{exdim} D^b(AlgSh\, X) = \dim X$$

for projective spaces, odd dimensional quadrics [14] and for some three-dimensional Fano varieties [21].

THEOREM 2.1 *If there is a dualizing sheaf K_X for X and*

$$\operatorname{exdim} D^b(AlgSh\, X) = \dim X$$

and there exists an exceptional system $E_0, \dots, E_m$ of sheaves on X of size m, then all the systems in the Bd_m-orbit of $E_0, \dots, E_m$ are systems of sheaves also.

This is proved by Bondal in [2] as a way to generalize the helix approach, which was used to prove a similar result for $\boldsymbol{P}^n$ in [11].

THEOREM 2.2 *Let $E_0, \dots, E_m$ be an exceptional system that generates $\mathcal{D}$ and such that $\operatorname{Ext}^k(E_i, E_j) = 0$ for $k \neq 0$ and $i \neq j$. Then for $E = \bigoplus E_i$ we have $\operatorname{Ext}^k(E, E) = 0$ for $k \neq 0$ and $\mathcal{D}$ is equivalent to $D^b(\mathrm{mod} - A)$ where $A = \operatorname{Hom}(E, E)$.*

One could interpret this theorem as making a "tilting functor" for a sheaf category [2], [12]. The object E in the theorem is a tilting object in $\mathcal{D}$ (see history remarks in [13, pp. 93–94]).

3. Braid group action for Del Pezzo surfaces

Construction of the braid group action for exceptional systems of sheaves on a Del Pezzo surface is based on the following theorem.

THEOREM 3.1 (GORODENTSEV [9], [18]) *Let X be a Del Pezzo surface and E be an exceptional object from a derived category $\mathcal{D} = D^b(AlgSh\,X)$. Then E is isomorphic to some translation of an exceptional sheaf (provided that as usual a sheaf category $AlgSh\,X$ is considered a full subcategory in $\mathcal{D}$ and translation (or shift) comes for changing numeration in a complex).*

So there is "forgetting of the translation" map from exceptional objects in $\mathcal{D}$ onto exceptional sheaves. It could be applied to exceptional systems also.

PROPOSITION 3.1 *There exists a braid group Bd_m action on a set of exceptional systems of size m of sheaves on a Del Pezzo surface where the result of applying $\beta \in Bd_m$ to $E_0, \dots, E_m$ coincides with "forgetting of translations" in the system*

$$\beta \cdot (E_0, \dots, E_m),$$

which is made by the braid group action in $\mathcal{D}$.

One could compute more explicitly what this means for an exceptional pair.

PROPOSITION 3.2 *If E, F is an exceptional pair of sheaves on a Del Pezzo surface then the generator λ of Bd_1 acts as follows:*

$$(E, F) \longrightarrow (F^+, E)$$

where either

$$0 \longrightarrow F^+ \longrightarrow \mathrm{Hom}(E, F) \otimes E \longrightarrow F \longrightarrow 0$$

is an exact sequence, or

$$0 \longrightarrow \mathrm{Hom}(E, F) \otimes E \longrightarrow F \longrightarrow F^+ \longrightarrow 0$$

is exact, or

$$0 \longrightarrow F \longrightarrow F^+ \longrightarrow \mathrm{Ext}^1(E, F) \otimes E \longrightarrow 0$$

is exact and $\mathrm{Ext}^2(E, F) = 0$ *in any case.*

So one could calculate the results of the braid group action quite explicitly. This permits us to construct all exceptional sheaves on a Del Pezzo surface provided that the field $\boldsymbol{k}$ is algebraically closed because of Theorem 1.6. Those calculations especially for a quadric surface were an inspiration for the results. Some of them are described in [25], [23].

4. Applications and related topics

4.1. Monads and Beilinson type spectral sequences. Relationship between exceptional systems and Beilinson type spectral sequences were first noticed for projective planes and projective spaces of higher dimension [5], [11]. Then a way was found to generalize this to Del Pezzo surfaces where spectral sequences get more complicated [9].

Along with this were found ways to use exceptional sheaves to construct monads in a form

$$M : \bigoplus V_i \otimes E_i \longrightarrow \bigoplus W_j \otimes E_j$$

and thus to get information about moduli varieties of sheaves. The details can be found in [6], [7], [16].

4.2. Chern classes for stable sheaves. The very first paper where exceptional sheaves were considered [8] was devoted to describe ranks and Chern classes possible for stable sheaves on $\boldsymbol{P}^2$. The authors J.-M. Drezet and J. Le Potier found a nice "fractal type" border for points of the form (μ, Δ) where μ is the slope and Δ the discriminant of a nonexceptional stable sheave, and described separately those of exceptional sheaves. The same description is also given for sheaves on a quadric surface [26].

4.3. Quiver representations. It was proven recently in [4] that there could be defined a braid group action for exceptional systems of quiver representations and that the action is transitive on a set of systems of "maximal" size (equal to "exdim" of the category).

5. Open questions and conjectures

There are several things that look promising to do. The first is to look at sheaves on a Del Pezzo surface over a field which is not algebraically closed. Let us say that a sheaf E is quasi-exceptional if $\mathrm{Ext}^i(E, E) = 0$ for $i \neq 0$ and $\mathrm{Hom}(E, E)$ is a division algebra over $\boldsymbol{k}$.

CONJECTURE 5.1 *A rigid sheaf on a Del Pezzo surface is a direct sum of quasi-exceptional ones.*

Probably it is possible to define a braid group action on quasi-exceptional systems and prove that there is a finite number of orbits. To find this number is likely to be more difficult.

The second task is to study the braid group action for $\boldsymbol{P}^n$ where $n > 2$.

CONJECTURE 5.2 *The braid group Bd_n acts transitively on exceptional systems of size n on $\boldsymbol{P}^n$.*

This is known for $n = 1, 2, 3$ (the last is proved in [19]).

CONJECTURE 5.3 *Any exceptional sheaf on $\boldsymbol{P}^n$ can be included into an exceptional system of size n.*

The same questions are open for quadrics, grassmannians, and other varieties of dimension more than two where the existence of exceptional sheaves and exceptional systems is known.

CONJECTURE 5.4 *There is a finite number of orbits for the braid group action on exceptional systems of maximal size in a derived category $D^b(AlgSh\,X)$ of algebraic coherent sheaves on a Fano variety X.*

References

[1] Beilinson, A. A., *Coherent sheaves on P^n and problems of linear algebra*, Funktsional. Anal. i Prilozhen. 12, N3, 68–69 (1978).

[2] Bondal, A. I., *Representation of associative algebras and coherent sheaves*, Izv. Akad. Nauk SSSR, Ser. Mat. 53, N1, 25–44 (1989).

[3] Bondal, A. I., and Kapranov, M. M., *Representable functors, Serre functors and reconstructions*, Izv. Akad. Nauk SSSR, Ser. Mat. 53, N6, 1183–1205 (1989).

[4] Crawley-Boevey, W., *Exceptional sequences of representations of quivers*, Representations of Algebras, Dlab and Lenzing, ed., vol. 14, 117–124, CMS, Amer. Math. Soc., Providence, RI (1994).

[5] Drezet, J.-M., *Fibres exceptionnels et suite spectrale de Beilinson généralisée sur* $\mathbf{P}_2(C)$, Math. Ann. 275, 25–48 (1986).

[6] Drezet, J.-M., *Fibres exceptionnels et varietes de faisceaux semi-stable sur* $\mathbf{P}_2(C)$, J. Reine Angew. Math. 380, 14–58 (1987).

[7] Drezet, J.-M., *Variétés de modules extremales de faisceaux semi-stable sur* $\mathbf{P}_2(C)$, Math. Ann. 290, 727–770 (1991).

[8] Drezet, J. M., and Le Potier, J., *Fibres stables et fibres exceptionnels sur* $\mathbf{P}_2$, Ann. Sci. École Norm. Sup. (4) 18, 193–243 (1985).

[9] Gorodentsev, A. L., *Exceptional vector bundles on a surface with moving anticanonical class*, Izv. Akad. Nauk SSSR, Ser. Mat. 52, N4, 740–757 (1988).

[10] Gorodentsev, A. L., *Exceptional objects and mutations in derived categories* In: [25], 57–74.

[11] Gorodentsev, A. L., and Rudakov, A. N., *Exceptional vector bundles on projective space*, Duke Math. J., 54 (1987), 115–130.

[12] Happel, D., *On the derived category of a finite dimensional algebra*, Comm. Math. Helv. 62, 479–508 (1987).

[13] Happel, D., Triangulated categories in the representation theory of finite dimensional algebras, London Math. Soc. Lecture Note Ser. 119, Cambridge Univ. Press, Cambridge and New York (1988).

[14] Kapranov, M. M., *On the derived categories of coherent sheaves on a quadric*, Funktsional. Anal. i Prilozhen. 20, N2, 67 (1986).

[15] Kapranov, M. M., *On the derived categories of coherent sheaves on some homogeneous spaces*, Invent. Math. 92, 479–508 (1988).

[16] Karpov, B. V., *Semistable sheaves on a two-dimensional quadric and Kronecker modules*, Izv. Ross. Akad. Nauk Ser. Mat. 56, N1, 38–74 (1992); In English: Russian Acad. Sci. Izv. Math. 40, N1, 33–66 (1993).

[17] Kuleshov, S. A., Rigid sheaves on generalized Del Pezzo surfaces, preprints of Moscow Independent University (1994).

[18] Kuleshov, S. A., and Orlov, D. O., *Exceptional sheaves on Del Pezzo surfaces*, Izv. Akad. Nauk SSSR, Ser. Mat. 58, N1, 59–93 (1994).

[19] Nogin, D. Ju., *Helices on a period four and equations of Markov type*, Izv. Akad. Nauk SSSR, Ser. Mat. 54, N4, 862–878 (1990).

[20] Orlov, D. O., *Projective bundles, monoidal transformations and derived categories of coherent sheaves*, Izv. Akad. Nauk SSSR, Ser. Mat. 56, N3, (1992).

[21] Orlov, D. O., *Exceptional systems of vector bundles on a variety* V_5, Vestnik Moskov. Univ., Mat. Mekh. N5, 69–71 (1992).

[22] Rudakov, A. N., *Exceptional vector bundles on* P^2 *and Markov numbers*, Izv. Akad. Nauk SSSR, Ser. Mat. 52, N1, 100–112 (1988); In English: Mathematics of the USSR, IZVESTIYA, 32 (1989), 99–112.

[23] Rudakov, A. N., *Exceptional vector bundles on a quadric*, Izv. Akad. Nauk SSSR, Ser. Mat. 52, N4, 788–812 (1988); In English: Mathematics of the USSR, IZVESTIYA, 33 (1989), 115–138.

[24] Rudakov, A. N., *Integer-valued bilinear forms and vector bundles*, Mat. Sb. (New Ser.) 180, No. 2, 187–194 (1989); In English: Mathematics of The USSR, SBORNIK, 66 (1990), 189–197.

[25] Rudakov, A. N., Helices and Vector Bundles: Seminaire Rudakov, London Math. Soc. Lecture Note Ser. 148, Cambridge Univ. Press, Cambridge and New York (1990).

[26] Rudakov, A. N., *A description of Chern classes of semistable sheaves on a quadric surface*, J. Reine Angew. Math. 453 (1994),113–135.

[27] Zyuzina, S. Yu., *Constructibility of exceptional pairs of vector bundles on a quadric*, Izv. Ross. Akad. Nauk Ser. Mat. 57, N1, 183–191 (1993); In English: Russian Acad. Sci. Izv. Math. 42, N1, 149–161 (1994).

Variations of Hodge Structure and Algebraic Cycles

CLAIRE VOISIN

Université d'Orsay, CNRS, Bât. 425,
F-41405 Orsay, France

0 Introduction

0.1. Let X be a complex projective variety. Then each cohomology group of X admits a Hodge structure, that is a decomposition of $H^k(X,\mathbb{C}) = H^k(X,\mathbb{Z}) \otimes \mathbb{C}$ into a direct sum $\bigoplus_{p+q=k} H^{p,q}(X)$, where $H^{p,q}(X) \simeq H^q(\Omega^p_X) \subset H^k(X,\mathbb{C})$ is the set of classes that can be represented by a closed k-form everywhere of type (p,q). We will be concerned in this paper with the relations between Hodge structures and Chow groups $CH_{.}(X)$, where $CH_\ell(X)$ is the group of ℓ-cycles (= arbitrary integral combinations of ℓ-dimensional subvarieties) modulo rational equivalence [5].

0.2. The simplest way to go from Chow groups to Hodge structures is to use the cycle class map $c : CH_k(X) \to H^{2n-2k}(X)$, which to a cycle $\Gamma = \Sigma n_i W_i$ associates $c(\Gamma) = \Sigma n_i c(W_i)$, where $c(W_i)$ is the Poincaré dual of the current of integration over W_i. The cycle class $c(\Gamma)$ is easily seen to be a Hodge class; that is, to belong to $H^{2k}(X,\mathbb{Z}) \cap H^{k,k}(X)$. The famous Hodge conjecture asserts that $H^{2k}(X,\mathbb{Q}) \cap H^{k,k}(X)$ is equal to $\operatorname{Im} c \otimes \mathbb{Q}$. Not much is known except for the case $k = 1$ (due to Lefschetz) and particular cases for $k > 1$ (see e.g. [35], [36], [37]). But recently an important theoretical evidence for it was given by Cattani, Deligne, and Kaplan, who proved:

0.3. THEOREM [10]. *Let $\mathcal{X} \to B$ be an algebraic family of smooth algebraic varieties $\mathcal{X}_b$ parametrized by a quasi-projective variety B. Then the set $\{(b,\lambda),\ b \in B,\ \lambda \in H^{2k}(\mathcal{X}_b,\mathbb{Z}) \cap H^{k,k}(\mathcal{X}_b)\}$ is a countable union of finite covers of algebraic subvarieties of B.*

These sets are called Hodge loci or Noether-Lefschetz loci and were studied in [4], [34], [IVHS,II].

The class of a cycle is sometimes a very poor invariant: for example the class of a zero-cycle $\Sigma n_i p_i$ is just its degree $\Sigma n_i \in \mathbb{Z}$. Of course a much deeper relation between $CH(X)$ and Hodge structures on X is expected (see [5], [28], [30]); however, for Z a cycle in a family of varieties $(\mathcal{X}_b)_{b\in B}$, the Hodge class of Z carries very much information on the family of cycles $Z_{|\mathcal{X}_b} \in CH(\mathcal{X}_b)$, and this will be the main topic of Section 1.

Proceedings of the International Congress
of Mathematicians, Zürich, Switzerland 1994

0.4. One way to refine the cycle class map is to consider the Deligne cycle class $c_D : CH^p(X) \to H_D^{2p}(X, \mathbb{Z}(p))$ (Deligne cohomology) where $H_D^{2p}(X, \mathbb{Z}(p)) = \mathbb{H}^{2p}(0 \to \mathbb{Z} \to \mathcal{O}_X \to \Omega_X \to \cdots \to \Omega_X^{p-1} \to 0)$ (see [20], [17]). Its restriction to the set of codimension p-cycles homologous to zero was first defined by Griffiths [26] and called the Abel-Jacobi map. It takes values into the pth intermediate jacobian:

$$J^{2p-1}(X) = H^{2p-1}(X, \mathbb{C})/F^p H^{2p-1} \oplus H^{2p-1}(X, \mathbb{Z}).$$

(Here and in the sequel we use the notation $F^k H^\ell(X) := \bigoplus_{p \geq k} H^{p,\ell-p}(X)$.)

0.5. Deligne cohomology groups also appear as the targets of regulator maps, which are defined on higher Chow groups ([6], [27]). Regulators have the same formal properties, from the point of view of infinitesimal variations of Hodge structure, as Abel-Jacobi maps, and we will see in the next section that the result of [23] holds as well for them. To give an idea of what they are, consider for simplicity the case of $K_1(X)^{(p+1)} \simeq CH^{p+1}(X, 1) \simeq H^p_{\mathrm{Zar}}(X, \mathcal{K}_{p+1})$. Using Bloch's definition or using the Gersten resolution of the sheaf $\mathcal{K}_{p+1}$ [5], this group is generated by sums $\alpha = \Sigma(Z_i, \varphi_i)$, where $Z_i \subset X$ is irreducible of codimension p and φ_i is a non-zero rational function on Z_i, subject to the condition: $\Sigma \mathrm{div}(\varphi_i) = 0$ as a cycle of codimension $p+1$ on X. The regulator map R will send it to an element of the partial torus

$$H_D^{2p+1}(X, \mathbb{Z}(p+1)) \simeq H^{2p}(X, \mathbb{C})/F^{p+1} H^{2p}(X, \mathbb{C}) \oplus H^{2p}(X, \mathbb{Z}).$$

Modulo the image of $\langle [Z_i] \rangle \otimes \mathbb{C}$ in this torus, $R(\alpha)$ is constructed as follows: let $Z = \bigcup_i Z_i$, $U = X \backslash Z$. Because $\Sigma \mathrm{div}(\varphi_i) = 0$ it follows that the one forms $w_i = \frac{1}{2i\pi} \frac{d\varphi_i}{\varphi_i}$ on Z_i satisfy: $\mathrm{Res}_{Z_i \cap Z_j} w_i + \mathrm{Res}_{Z_j \cap Z_i} w_j = 0$, hence determine an element w_α of

$$H_Z^{2p+1}(X) \subset \bigoplus_i H^1(Z_i \backslash \bigcup_{j \neq i} Z_j \cap Z_i).$$

$H_Z^{2p+1}(X)$ carries a mixed Hodge structure [14], induced by the mixed Hodge structure on $\bigoplus_i H^1(Z_i \backslash \bigcup_{j \neq i} Z_i \cap Z_j)$, and because w_i have a class in $H^1(Z_i \backslash \bigcup_{j \neq i} Z_i \cap Z_j, \mathbb{Z}) \cap F^1 H^1(Z_i \backslash \bigcup_{j \neq i} Z_i \cap Z_j)$, it follows that

$$w_\alpha \in F^{p+1} H_Z^{2p+1}(X) \cap H_Z^{2p+1}(X, \mathbb{Z}).$$

Consider the exact sequence:

$$0 \longrightarrow H^{2p}(X)/ < Z_i > \longrightarrow H^{2p}(U) \longrightarrow H_Z^{2p+1}(X) \longrightarrow H^{2p+1}(X).$$

Clearly w_α vanishes in $H^{2p+1}(X)$ because $F^{p+1} H^{2p+1}(X) \cap H^{2p+1}(X, \mathbb{Z}) = 0$. So w_α admits liftings in $F^{p+1} H^{2p}(U)$ and in $H^{2p}(U, \mathbb{Z})$, whose difference will give $R(\alpha) \in H^{2p}(X, \mathbb{C})/F^{p+1} H^{2p}(X) \oplus H^{2p}(X, \mathbb{Z}) \oplus \mathbb{C}[Z_i]$. (We have made abstraction here of singularities but the construction works in general [27].)

0.6. One way to study the objects described above is to look at their variation when X varies in a family: suppose $\mathcal{X} \xrightarrow{\pi} B$ is a smooth family of complex projective varieties parametrized by a smooth complex variety B ; then the inclusions $F^pH^k(\mathcal{X}_b) \subset H^k(\mathcal{X}_b, \mathbb{C})$ determine a $\mathcal{C}^\infty$-subbundle $F^p\mathcal{H}^k_\infty \subset \mathcal{H}^k_\infty$ of the bundle $\mathcal{H}^k_\infty$ with fiber $H^k(\mathcal{X}_b, \mathbb{C})$. $\mathcal{H}^k_\infty$ is a flat bundle w.r.t. the Gauss-Manin connection ∇, so in particular it has a natural holomorphic structure, and we will denote by $\mathcal{H}^k$ the sheaf of its holomorphic sections. We have $\mathcal{H}^k = R^k\pi_*\mathbb{C} \otimes \mathcal{O}_B$. The most important results of Griffiths are the following [25]:

0.7. Theorem.

(i) *$F^p\mathcal{H}^k_\infty$ is a holomorphic subbundle of $\mathcal{H}^k_\infty$; we will denote by $F^p\mathcal{H}^k \subset \mathcal{H}^k$ its sheaf of holomorphic sections.*

(ii) *(Transversality) The Gauss-Manin connection $\nabla : \mathcal{H}^k \to \mathcal{H}^k \otimes \Omega_B$ satisfies:*

$$\nabla F^p\mathcal{H}^k \subset F^{p-1}\mathcal{H}^k \otimes \Omega_B \,.$$

(iii) *(Description of the differential of the period map): The $\mathcal{O}_B$-linear map*

$$\begin{array}{lccc} \overline{\nabla}: & F^p/F^{p+1}\mathcal{H}^k & \longrightarrow & F^{p-1}/F^p\mathcal{H}^k \otimes \Omega_B \\ & \| & & \| \\ & \mathcal{H}^{p,k-p} & \longrightarrow & \mathcal{H}^{p-1,k-p+1} \otimes \Omega_B \end{array},$$

obtained from ∇ by passing to the quotient, gives for any $b \in B$ a map: $TB_b \to \mathrm{Hom}(H^{k-p}(\Omega^p_{\mathcal{X}_b}), H^{k-p+1}(\Omega^{p-1}_{\mathcal{X}_b}))$, which identifies to the composite:

$$TB_{(b)} \xrightarrow{\text{Kodaira-Spencer}} H^1(T\mathcal{X}_b) \longrightarrow \mathrm{Hom}(H^{k-p}(\Omega^p_{\mathcal{X}_b}),\ H^{k-p+1}(\Omega^{p-1}_{\mathcal{X}_b}))\,,$$

where the last map is given by the interior product.

0.8. To deduce consequences of this theorem, one needs to know much about the structure of the couplings $H^1(T_{\mathcal{X}_b}) \otimes H^{k-p}(\Omega^p_{\mathcal{X}_b}) \to H^{k-p+1}(\Omega^{p-1}_{\mathcal{X}_b})$. Their description is especially beautiful in the case of hypersurfaces $\{F = 0\}$ in projective space $\mathbb{P}^n$ (and more generally sufficiently ample hypersurfaces in any variety [22]). In this case, the spaces considered (modulo the cohomology of $\mathbb{P}^n$) are homogeneous pieces of the jacobian ring $R(F) = \mathbb{C}[X_0, \ldots, X_n]/ < \partial F/\partial X_i >_{i=0,\ldots,n}$, of F and the coupling is just multiplication [9]. [16], [21] provide a thorough study of the algebraic properties of these rings.

0.9. The Transversality Theorem 0.7 (ii) has its analog for the Abel-Jacobi maps or regulators, known as "quasi-horizontality of normal functions" [44], [IVHS,III], which follows in fact from 0.7 (ii) for variations of mixed Hodge structures, if one constructs the Abel-Jacobi invariants as extension classes [8], [17] (see also 0.5). Concretely it says the following: let $\mathcal{X} \underset{\pi}{\longrightarrow} B$ be a smooth family and let

$\mathcal{Z} \subset \mathcal{X}$ be a codimension p cycle, whose support is flat over B, and such that $\mathcal{Z}_b \subset \mathcal{X}_b$ is homologous to zero, $\forall b \in B$. The family of intermediate jacobians $(J^{2p-1}(\mathcal{X}_b))_{b\in B}$ has a natural complex structure, for which the sheaf of holomorphic sections is $\mathcal{J}^{2p-1} = \mathcal{H}^{2p-1}/F^p\mathcal{H}^{2p-1} \oplus H_{\mathbb{Z}}^{2p-1}$. The cycle $\mathcal{Z}$ gives a normal function $\nu_{\mathcal{Z}} \in \mathcal{J}^{2p-1}$ defined by $\nu_{\mathcal{Z}}(b) = \Phi_{\mathcal{X}_b}(\mathcal{Z}_b)$. (The analog of 0.7 (i) is that $\nu_{\mathcal{Z}}$ is holomorphic.)

0.9.1. The horizontality property is the following: *let $\tilde{\nu}_{\mathcal{Z}} \in \mathcal{H}^{2p-1}$ be a local lifting of $\nu_{\mathcal{Z}}$. Then $\nabla\tilde{\nu}_{\mathcal{Z}} \in F^{p-1}\mathcal{H}^{2p-1} \otimes \Omega_B$.* (Note that this is independent of the choice of the lifting by 0.7 (ii).) A similar statement holds for the regulator.

0.10. In Section 1 we will explain how to exploit this property to study the Abel-Jacobi map in families.

In Section 2, we will state a criterion due to Green for the density of the Noether-Lefschetz locus (0.3), and describe its consequences on the Abel-Jacobi map of certain threefolds. In Section 3, we describe briefly Nori's work, which is the most important recent contribution in the field.

1 Infinitesimal invariants

1.1. Let $\mathcal{X} \xrightarrow{\pi} B$ be a family of smooth complex projective varieties. Let $\mathcal{H}^{p,q} = F^p\mathcal{H}^{p+q}/F^{p+1}\mathcal{H}^{p+q}$ be the Hodge bundles and let $\overline{\nabla} : \mathcal{H}^{p,q} \to \mathcal{H}^{p-1,q+1} \otimes \Omega_B$ be the map of 0.7 (iii). Define $\overline{\nabla}_{(s)} : \mathcal{H}^{p,q} \otimes \Omega_B^{s-1} \to \mathcal{H}^{p-1,q+1} \otimes \Omega_B^s$, by $\overline{\nabla}_{(s)}(\sigma \otimes \alpha) = \overline{\nabla}(\sigma) \wedge \alpha$. Using the fact that $\overline{\nabla}$ is obtained from ∇ by passing to the quotient, and the integrability of ∇, one finds that $\overline{\nabla}_{(s+1)} \circ \overline{\nabla}_{(s)} = 0$. So for fixed (p, q) we get a complex on B:

$$K^{p,q} : O \to \mathcal{H}^{p,q} \xrightarrow{\overline{\nabla}} \mathcal{H}^{p-1,q+1} \otimes \Omega_B \xrightarrow{\overline{\nabla}_{(2)}} \mathcal{H}^{p-2,q+2} \otimes \Omega_B^2 \to \cdots \to \mathcal{H}^{0,p+q} \otimes \Omega_B^p \to 0 .$$

This complex is in fact the pth graded piece of the De Rham complex of $(\mathcal{H}^{p+q}, \nabla)$ for the decreasing filtration (introduced by Deligne and Zucker [44]):

$$K^p(DR\mathcal{H}^{p+q}) :=$$

$$0 \to F^p\mathcal{H}^{p+q} \xrightarrow{\nabla} F^{p-1}\mathcal{H}^{p+q} \otimes \Omega_B \xrightarrow{\nabla_{(2)}} F^{p-2}\mathcal{H}^{p+q} \otimes \Omega_B^2 \to \cdots \to F^0\mathcal{H}^{p+q} \otimes \Omega_B^p \to 0.$$

Now, by the degeneracy of the Leray spectral sequence of π [14], one has (non canonically): $H^n(\mathcal{X}, \mathbb{C}) = \bigoplus_{r+s=n} H^r(B, R^s\pi_*\mathbb{C})$ and the Hodge filtration [15] on $H^n(\mathcal{X}, \mathbb{C})$ induces on $H^r(B, R^s\pi_*\mathbb{C}) = \mathbb{H}^r(B, DR(\mathcal{H}^s))$ a filtration that is the one induced by K^p, if one imposes "logarithmic growth at infinity", that is if one works with the subcomplex $DR(\mathcal{H}^s)(\log \partial B)$.

1.2. The first infinitesimal invariant associated to a Hodge class on $\mathcal{X}$ is a holomorphic section of one of the cohomology sheaves of the complexes $\mathcal{K}^{p,q}$. Precisely let $\alpha \in F^n\mathcal{H}^{2n}(\mathcal{X})$; (integrality of α does not play any rule here). Assume

$\alpha \in H^k(B, R^{2n-k}\pi_*\mathbb{C})$; so $\alpha \in F^n H^k(B, R^{2n-k}\pi_*\mathbb{C}) = \mathbb{H}^k(K^n DR\mathcal{H}^{2n-k}(\log\partial B))$. Then the infinitesimal invariant $\delta\alpha \in H^0(B, \mathcal{H}^k(Gr^n_K(DR\mathcal{H}^{2n-k})))$ is just the image of α under the composite map:

$$\mathbb{H}^k(K^n DR\mathcal{H}^{2n-k}(\log\partial B)) \to \mathbb{H}^k(Gr^n_K DR\mathcal{H}^{2n-k}(\log\partial B)) \to H^0(\mathcal{H}^k(Gr^n_K DR\mathcal{H}^{2n-k})).$$

This is a local invariant of α, which can be as well obtained by looking at the image of α in $H^0(R^n\pi_*\Omega^n_{\mathcal{X}})$, and by studying the spectral sequence associated to the filtration of $\Omega^n_{\mathcal{X}}$ by the subbundles $\pi^*\Omega^p_B \wedge \Omega^{n-p}_{\mathcal{X}}$.

Now we want to describe more concretely these invariants and explain how to use them:

(A) Infinitesimal invariants of normal functions ([23], [44], [IVHS,III]):

1.3. Let $\mathcal{X} \xrightarrow{\pi} B$ be as before and let $\alpha \in H^{2p}(\mathcal{X}, \mathbb{Z}) \cap F^p H^{2p}(\mathcal{X})$; assuming $H^{2p-1}(\mathcal{X}) = 0$, α determines $\alpha_D \in H^{2p}_D(\mathcal{X}, \mathbb{Z}(p))$, and if $\alpha_{|\mathcal{X}_b} = 0$ in $H^{2p}(\mathcal{X}_b, \mathbb{Z})$, $\alpha_{D|\mathcal{X}_b} \in J^{2p-1}(\mathcal{X}_b) \subset H^{2p}_D(\mathcal{X}_b, \mathbb{Z}(p))$, and we get a section ν_α of $\mathcal{J}^{2p-1}$, (cf. 0.9), defined by $\nu_\alpha(b) = \alpha_{D|\mathcal{X}_b}$.

When α is the class of a cycle $\mathcal{Z}$, one has $\nu_\alpha = \nu_{\mathcal{Z}}$. The infinitesimal invariant of α is in the cohomology at the middle of the sequence:

$$\mathcal{H}^{p,p-1} \xrightarrow{\overline{\nabla}} \mathcal{H}^{p-1,p} \otimes \Omega_B \xrightarrow{\overline{\nabla}_{(2)}} \mathcal{H}^{p-2,p+1} \otimes \Omega^2_B,$$

and we construct now the infinitesimal invariant $\delta\nu_\alpha$ of ν_α, which lies in the same sheaf, as follows. Let $\tilde{\nu}_\alpha$ be a local lifting of ν_α in $\mathcal{H}^{2p-1}$; then by 0.9.1 $\nabla\tilde{\nu}_\alpha \in F^{p-1}\mathcal{H}^{2p-1} \otimes \Omega_B$. It is then easily seen that the projection of $\nabla\tilde{\nu}_\alpha$ in $\mathcal{H}^{p-1,p} \otimes \Omega_B$ is in $\operatorname{Ker}\overline{\nabla}_{(2)}$ and well-defined modulo $\operatorname{Im}\overline{\nabla}$. It is shown in [38] that $\delta\nu_\alpha = \delta\alpha$.

1.4. Clearly the vanishing of $\delta\nu_\alpha$ is equivalent to the fact that ν_α has a local lifting $\tilde{\nu}_\alpha \in \mathcal{H}^{2p-1}$ satisfying the stronger horizontality condition: $\nabla\tilde{\nu}_\alpha \in F^p\mathcal{H}^{2p-1} \otimes \Omega_B$. One can then construct a second infinitesimal invariant [23] living in

$$\frac{\operatorname{Ker}\overline{\nabla}_{(2)} : \mathcal{H}^{p,p-1} \otimes \Omega_B \to \mathcal{H}^{p-1,p} \otimes \Omega^2_B}{\operatorname{Im}\overline{\nabla} : \mathcal{H}^{p+1,p-2} \to \mathcal{H}^{p,p-1} \otimes \Omega_B},$$

which measures the obstruction to the existence of a lifting that satisfies: $\nabla\tilde{\nu}_\alpha \in F^{p+1}\mathcal{H}^{2p-1} \otimes \Omega_B$. Finally, if all the cohomology sheaves involved vanish, one can continue this process to get a flat lifting of ν_α in $\mathcal{H}^{2p-1}$. Under mild assumptions on the IVHS, this flat lifting will be unique up to a section of $H^{2p-1}_{\mathbb{Z}}$. Now the necessary vanishing assumptions are true for the universal family of hypersurfaces of degree ≥ 6 in $\mathbb{P}^4$, modulo isomorphisms (one uses there 0.8 and the symmetrizer lemma [16]), and a standard monodromy argument shows that flatness of normal functions implies their triviality mod. torsion, hence we get:

1.5. THEOREM (Green [23], Voisin, unpublished). *Let $X \subset \mathbb{P}^4$ be a general hypersurface of degree ≥ 6. Then the Abel-Jacobi map of X is of torsion.*

Green proved in fact the analogous result for all dimensions.

Green and Müller-Stach have generalized this result to any sufficiently ample linear system in any even dimensional variety [24]. To be precise, they show that for $X \subset Y$, $\dim Y = 2n$, X a general member of a sufficiently ample linear system on Y, the image of the Deligne-Abel-Jacobi map $c_D : CH^n(X) \to H^{2n}_D(X, \mathbb{Z}(n))$ is equal, up to the torsion, to the image of the composite map $CH^n(Y) \to H^{2n}_D(Y, \mathbb{Z}(n)) \to H^{2n}_D(X, \mathbb{Z}(n))$ — and that the last restriction map is injective.

As Bloch and Nori mentioned to me, the same argument applies as well to the regulator map (0.5). This gives the following:

1.6. THEOREM. *Let S be a general surface of degree at least five in $\mathbb{P}^3$; then the image of $R : H^1(\mathcal{K}_2(S)) \to H^2(S, \mathbb{C})/F^2H^2(S) \oplus H^2(S, \mathbb{Z})$ is of torsion modulo* $\operatorname{Pic} S \otimes \mathbb{C}^* = \langle c_1(\mathcal{O}_S(1))\rangle \otimes \mathbb{C}^*$.

As in the previous theorem, the assumption $d \geq 5$ is necessary. In the case $d = 4$ ($K3$-surfaces), Oliva (work in progress) shows the nontriviality of $R(S)$ mod. torsion, using the method of [39].

Theorem 1.6 disproves a conjecture of Beilinson [27], stating that the real Deligne cohomology is generated by the regulator.

As for the geometric content of the infinitesimal invariant δ_ν, we mention the following result of Collino and Pirola:

1.7. Let $\mathcal{M}_3$ be the moduli space of curves of genus three and let $J \xrightarrow{\pi} \mathcal{M}_3$ be the associated jacobian fibration. For $C \in \mathcal{M}_3$, one can choose an Abel-Jacobi embedding $C \subset J_C$, and the Abel-Jacobi image of the one-cycle $C - (-C)$ in the primitive part of the intermediate jacobian of J_C does not depend on the embedding. The normal function so obtained on $\mathcal{M}_3$ has an infinitesimal invariant defined as in 1.3, and one has:

1.8. THEOREM [13]. *This infinitesimal invariant at C lives in a space naturally isomorphic to $S^4H^0(K_C)$, and for C non-hyperelliptic, it is non-zero and gives the equation of C in its canonical embedding.*

(B) Infinitesimal invariants for families of zero-cycles on surfaces:

1.9. Let $\mathcal{S} \xrightarrow{\pi} B$ be a family of smooth regular projective surfaces, and let $Z \subset \mathcal{S}$ be a codimension two cycle, $Z = \Sigma n_i Z_i$, with $Z_i \to B$ flat and $\Sigma n_i d^0 Z_{i/B} = 0$. The class $[Z]$ of Z has then an infinitesimal invariant $\delta[Z]$ in $\mathcal{H}^{0,2} \otimes \Omega^2_B / \overline{\nabla}_{(2)}(\mathcal{H}^{1,1} \otimes \Omega_B)$. If Z satisfies the assumption: $\forall b \in B$, Z_b is rationally equivalent to zero in $\mathcal{S}_b$, a multiple of Z is homologous to a cycle supported over a proper Zariski closed subset of B, and we conclude that $\delta[Z]$ vanishes on a Zariski open set of B.

1.10. Now, using Serre's duality one finds an isomorphism:

$$(\mathcal{H}^{0,2} \otimes \Omega^2_B / \operatorname{Im} \overline{\nabla}_{(2)})_{(b)} \simeq (H^0(\Omega^N_{\mathcal{S}|S_b} \otimes \pi^* K_B^{-1} / \mathcal{O}_{S_b}))^* ,$$

where $N = \dim B$. The geometric content of $\delta[Z]$ is then the following. Suppose $Z = \Sigma n_i \sigma_i(B)$, where $\sigma_i : B \to S$ are sections, and $\Sigma n_i = 0$. At $b \in B$ one has $\sigma_i^* : (\Omega^N_{S|S_b})_{\sigma_i(b)} \to \Omega^N_{B(b)} = K_{B(b)}$, and $\delta[Z]$, as an element of $H^0(\Omega^N_{S|S_b} \otimes \pi^* K_B^{-1})^*$, is given by $\delta[Z] = \Sigma n_i \sigma_i^*$, which factors through the quotient $H^0(\Omega^N_{S|S_b} \otimes \pi^* K_B^{-1}/\mathcal{O}_{S_b})$ by the assumption $\Sigma n_i = 0$.

1.11. In [43] it is shown that if $\mathcal{S}_\pi \to B$ is the family of smooth hypersurfaces of $\mathbb{P}^3$ of degree ≥ 7, modulo isomorphism, the bundle $\Omega^N_{S|S_b}/\pi^* K_{B(b)}$ is very ample on S_b, $\forall b \in B$. From 1.9, 1.10, 1.11 one deduces:

1.12. THEOREM [43]. *Let $S \subset \mathbb{P}^3$ be general of degree ≥ 7. Then two distinct points of S are not rationally equivalent.*

2 Green's infinitesimal criterion and the nontriviality of the Abel-Jacobi map

2.1. Consider a family of surfaces $\mathcal{S} \to B$. Inside B, we have the Noether-Lefschetz loci, characterized by the existence of a certain Hodge class in H^2 of the fiber; that is, by the Lefschetz theorem, by the presence of an "extra" line bundle on the fiber. It is better to consider as in 0.3, the NL loci as contained in the $\mathcal{C}^\infty$ vector bundle $\mathcal{H}^{1,1}_{\mathbb{R}}$, with fiber $H^{1,1}(S_b) \cap H^2(S_b, \mathbb{R})$ at $b \in B$. The NL locus will be then defined as the set $\{(\lambda, b)/\lambda \in H^{1,1}(S_b) \cap H^2(S_b, \mathbb{Q})\}$. Green's lemma gives the following purely algebraic criterion for the density of this locus:

2.2. LEMMA (Green, [29]). *Suppose that for some $b \in B$, $\lambda \in H^1(\Omega_{S_b})$, the map $\overline{\nabla}(\lambda) : TB_{(b)} \to H^2(\mathcal{O}_{S_b})$ is surjective. Then the Noether-Lefschetz locus is dense in $\mathcal{H}^{1,1}_{\mathbb{R}}$.*

2.3. In [40], the criterion was checked for sufficiently ample hypersurfaces in Calabi-Yau threefolds.

2.4. Now this lemma gives a way to produce interesting cycles in threefolds: if $S \underset{j}{\hookrightarrow} X$ and $\lambda \in H^{1,1}(S) \cap \mathrm{Ker}(H^2(S, \mathbb{Z}) \xrightarrow{j_*} H^4(X, \mathbb{Z}))$, λ determines an element of $\mathrm{Pic}\, S$ (assuming S regular), hence a one-cycle on S, which will be homologous to zero in X. The next question is to decide whether the cycles Z_λ so obtained have non-trivial Abel-Jacobi invariants. If the expected dimension of the components $\mathcal{S}_\lambda$ of the NL locus is strictly positive, it is possible to study formally the differential of the Abel-Jacobi map $\Phi : \mathcal{S}_\lambda \to JX$, $\Phi(S, \lambda) = \Phi_X(Z_\lambda)$, and to show that it is nonzero. This method was used in [41] to solve the generalized Hodge-Grothendieck conjecture for certain sub-Hodge structures on certain threefolds. (See [2] for a more geometric solution of a similar example.)

2.5. In the case of a Calabi-Yau threefold, the expected dimension of the NL locus is zero, but one can deform X together with the zero-dimensional components of this NL locus. Using the same construction as above, this will now give normal functions on the family of deformations of X, and the nonvanishing of their infinitesimal invariants gives:

2.6. THEOREM [40]. *Let X be a Calabi-Yau threefold that is nonrigid; then a general deformation of X has a non-torsion Abel-Jacobi map.*

This theorem was known previously for the quintic threefold (see [26] and [12] for a much stronger statement) but the cycles in [26] were easy to get. They are the lines on X.

3 Nori's theorem

3.1. The essential point in 1.4, 1.5 was the vanishing of some cohomology sheaves of the complexes $K^{p,q}$, on the family of all hypersurfaces of sufficiently large degree. Nori realized that these vanishing statements and their generalizations to the case of complete intersections of large degree in any variety are partial aspects of a deep vanishing theorem for the cohomology of the universal hypersurfaces or complete intersections, which is the following:

3.2. CONNECTIVITY THEOREM [32]. *Let X be projective of dimension $n+k$. Let $L_1, \dots, L_k$ be ample line bundles, and for $n_1, \dots, n_k \in \mathbb{N}$, let $S := \prod_1^k H^0(X, L_i^{n_i})$. Let $Y_S \subset X \times S$ be the universal complete intersection. Then for n_i large enough, and for any submersive map $T \to S$, one has $H^k(X \times T, Y_T) = 0$, $k = 0, \dots, 2n$.*

The most striking application of this theorem is the proof of the existence of cycles homologous and Abel-Jacobi equivalent to zero but not algebraically equivalent to zero:

3.3. THEOREM [32]. *Using notation as above, let Z be a cycle on X of codimension $d < n$: suppose that $[Z] \neq 0$ in $H^{2d}(X, \mathbb{Q})$, or that the Abel-Jacobi image of Z is not contained in the algebraic part of JX. Then for n_i such that the conclusion of 3.2 holds, $Z_{|Y_S}$ is not algebraically equivalent to zero, for general s.*

3.4. Griffiths in [26] proved the existence of cycles homologous to zero but not algebraically equivalent to zero, but he used the Abel-Jacobi invariant, which vanishes on cycles algebraically equivalent to zero when the intermediate jacobians do not contain a nontrivial algebraic part.

Albano and Collino [1] have even shown that the kernel of the Abel-Jacobi map can be nonfinitely generated modulo algebraic equivalence. This was obtained as a consequence of 3.3, and of the following result (an analog of Clemens' theorem [12]):

3.5. THEOREM [1]. *Let $X \subset \mathbb{P}^8$ be a general cubic sevenfold; then $J^7(X)$ has no algebraic part and the image of the Abel-Jacobi map $\Phi_X : CH_3(X)_{\text{hom}} \to J^7(X)$ is a countable infinitely generated group.*

References

[1] A. Albano and A. Collino, *On the Griffiths group of the cubic sevenfold*, to appear in Math. Ann.

[2] F. Bardelli, *On Grothendieck generalized Hodge conjecture for a family of threefolds with geometric genus one*, in: Proceedings of the Conference in Algebraic Geometry (Berlin 1985), Teubner-Texte Math. **92** (1986), 14–23.

[3] F. Bardelli and S. Müller-Stach, *Algebraic cycles on certain Calabi-Yau threefolds*, preprint, Pisa 1992.

[4] S. Bloch, *Semi-regularity and De Rham cohomology*, Invent. Math. **17** (1972), 51–66.

[5] S. Bloch, Lectures on algebraic cycles, Duke Univ. Math. Series **IV**, 1980.

[6] S. Bloch, *Algebraic cycles and higher K-theory*, Adv. in Math. **61** (1986), 267–304.

[7] S. Bloch and V. Srinivas, *Remarks on correspondences and algebraic cycles*, Amer. J. Math. **105** (1983), 1235–1253.

[8] J. Carlson, *The geometry of the extension class of a mixed Hodge structure*, Proc. Sympos. Pure Math. **46** (1987), 199–222.

[9] J. Carlson and P. Griffiths, *Infinitesimal variations of Hodge structure and the global Torelli problem*, in: Géométrie Algébrique (A. Beauville, ed.), Sijthoff-Noordhoff, Angers (1980), 51–76.

[10] E. Cattani, P. Deligne, and A. Kaplan, *On the locus of Hodge classes*, preprint.

[11] H. Clemens, *Degeneration technics in the study of threefolds*, in: Algebraic threefolds, Lecture Notes in Math. **947**, Springer, Berlin and New York, 1981.

[12] H. Clemens, *Homological equivalence, modulo algebraic equivalence is not finitely generated*, Publ. Math. IHES **58** (1983), 19–38.

[13] A. Collino and G.P. Pirola, *Griffiths' infinitesimal invariant for a curve in its jacobian*, preprint, 1993.

[14] P. Deligne, *Théorème de Lefschetz et critères de dégénérescence de suites spectrales*, Publ. Math. IHES **35** (1968), 107–126.

[15] P. Deligne, *Théorie de Hodge II*, Publ. Math. IHES **40** (1971), 5–57.

[16] R. Donagi and M. Green, *A new proof of the symmetrizer lemma and a stronger weak Torelli theorem for projective hypersurfaces*, J. Differential Geom. **20** (1984), 459–461.

[17] F. Elzein and S. Zucker, *Extendability of normal functions associated to algebraic cycles*, in: Topics in Transcendental Algebraic geometry (P. Griffiths, ed.), Ann. of Math. Stud. **106**, Princeton Univ. Press, Princeton NJ (1984), 269–288.

[18] H. Esnault, *Hodge type of subvarieties of $\mathbb{P}^n$ of small degrees*, Math. Ann. **288** (1990), 549–551.

[19] H. Esnault, M.V. Nori, and V. Srinivas, *Hodge type of projective varieties of low degree*, Math. Ann. **293** (1992), 1–6.

[20] H. Esnault and E. Viehweg, *Deligne-Beilinson cohomology*, in: Beilinson's conjectures on special values of L-functions (Rapoport, Schappacher, and Schneider, eds.), Perspect. Math. **4**, Academic Press, Boston, MA and New York, (1988), 43–91.

[21] M. Green, Koszul cohomology and the geometry of projective varieties I, II, J. Differential Geom. **20** (1984).

[22] M. Green, *The period map for hypersurfaces sections of high degree of an arbitrary variety*, Compositio Math. (2) **55** (1985), 135–156.

[23] M. Green, *Griffiths infinitesimal invariant and the Abel-Jacobi map*, J. Differential Geom. **29** (1989), 545–555.

[24] M. Green and S. Müller-Stach, to appear.

[25] P. Griffiths, *Periods of integrals on algebraic manifolds I, II*, Amer. J. Math. **90** (1968), 568–626, 805–865.

[26] P. Griffiths, *On the periods of certain rational integrals I, II*, Ann. of Math. **90** (1969), 460–541.
[27] U. Jannsen, *Deligne homology, Hodge D-conjecture and motives*, in: Beilinson's conjectures on special values of L-functions (Rapoport, Schappacher, Schneider, eds.), Perspect. Math. **4**, Academic Press, Boston, MA and New York (1988), 305–372.
[28] U. Jannsen, *Motivic sheaves and filtration on Chow groups*, to appear in the Proceedings of the Seattle conference on Motives.
[29] S. O. Kim, *Noether-Lefschetz locus for surfaces*, Trans. Amer. Math. Soc. **1**, 1991.
[30] D. Mumford, *Rational equivalence of* 0*-cycles on surfaces*, J. Math. Kyoto Univ. **9** (1968), 194–204.
[31] J.-P. Murre, *On a conjectural filtration on the Chow groups of an algebraic variety I, II*, Indag. Math. N.S. **4** (2) (1993), 177–201.
[32] M. V. Nori, *Algebraic cycles and Hodge theoretic connectivity*, Invent. Math. **111** (2) (1993), 349–373.
[33] K.-H. Paranjape, *Cohomological and cycle theoretic connectivity*, to appear in Ann. of Math. (2).
[34] Z. Ran, *Hodge theory and the Hilbert scheme*, J. Differential Geom. **37** (1993), 191–198.
[35] C. Schoen, *Cyclic covers of* $\mathbb{P}^\nu$ *branched along* $\nu + 2$ *hyperplanes and the generalized Hodge conjecture*, in: Lecture Notes in Math. **1399** (Burth and Lange, eds.), Springer-Verlag (1988), 137–154.
[36] T. Shioda, *The Hodge conjecture for Fermat varieties*, Math. Ann. **245** (1979), 175–184.
[37] B. van Geemen, *Theta functions and cycles on some abelian fourfolds*, to appear in Math. Z.
[38] C. Voisin, *Une remarque sur l'invariant infinitésimal des fonctions normales*, C.R. Acad. Sci. Paris, Série I **307**, (1988), 157–160.
[39] C. Voisin, *Une approche infinitésimale du théorème de H. Clemens sur les cycles d'une quintique générale de* $\mathbb{P}^4$, J. Alg. Geom. **1** (1992), 157–174.
[40] C. Voisin, *Densité du lieu de Noether-Lefschetz pour les sections hyperplanes des variétés de Calabi-Yau de dimension trois*, International J. Math., **3**, no. 5 (1992), 699–715.
[41] C. Voisin, *Sur les zéro-cycles de certaines hypersurfaces munies d'un automorphisme*, Ann. Scuola Norm. Sup. Pisa Cl. Sci. (4) **29** (1993), 473–492.
[42] C. Voisin, *Sur l'application d'Abel-Jacobi des variétés de Calabi-Yau de dimension trois*, Ann. Sci. École Norm. Sup. (4) **27** (1994), 129–172.
[43] C. Voisin, *Variations de structure de Hodge et zéro-cycles sur les surfaces générales*, Math. Ann. **299** (1994), 77–103.
[44] S. Zucker, *Hodge theory with degenerating coefficients*, Ann. of Math. (2), **109** (1979), 415–476.

Furthermore:

[IVHS] Infinitesimal variations of Hodge structure, Comp. Math. **50** (1983).
[–,I] J. Carlson, M. Green, P. Griffiths, and J. Harris, *Infinitesimal variations of Hodge structure*, in [IVHS], 109–205.
[–,II] P. Griffiths and J. Harris, *Infinitesimal invariant of Hodge classes*, in [IVHS], 207–265.
[–,III] P. Griffiths, *Determinantal varieties and the infinitesimal invariant of normal functions*, in [IVHS], 267–324.

Author Index

TEX-macros and films by *mathScreen online* CH-4123 Allschwil